Progress in Mathematics

Volume 270

For other titles published in this series, go to
http://www.springer.com/series/4848

Algebra, Arithmetic, and Geometry

In Honor of Yu. I. Manin

Volume II

Yuri Tschinkel
Yuri Zarhin
Editors

Birkhäuser
Boston • Basel • Berlin

Editors
Yuri Tschinkel
New York University
Department of Mathematics
New York, NY 10012-1185
tschinkel@cims.nyu.edu

Yuri Zarhin
Pennsylvania State University
Eberly College of Science
Department of Mathematics
University Park, PA 16802
zarhin@math.psu.edu

ISBN 978-0-8176-4746-9   e-ISBN 978-0-8176-4747-6
DOI 10.1007/978-0-8176-4747-6
Springer New York Dordrecht Heidelberg London

Library of Congress Control Number: 2009939500

Printed on acid-free paper

Birkhäuser is part of Springer Science+Business Media (www.birkhauser.com)

Preface

Yuri Ivanovich Manin has made outstanding contributions to algebra, algebraic geometry, number theory, algorithmic complexity, noncommutative geometry and mathematical physics. His numerous achievements include the proof of the functional analogue of the Mordell Conjecture, the theory of the Gauss–Manin connection, proof with V. Iskovskikh of the nonrationality of smooth quartic threefolds, the theory of p-adic automorphic functions, construction of instantons (jointly with V. Drinfeld, M. Atiyah and N. Hitchin), and the theory of quantum computations.

We hope that the papers in this Festschrift, written in honor of Yu. I. Manin's seventieth birthday, will indicate the great respect and admiration that his students, friends and colleagues throughout the world all have for him.

June 2009
Courant Institute
Penn State University

Yuri Tschinkel
Yuri Zarhin

Contents

Contents of Volume I

Potential Automorphy of Odd-Dimensional Symmetric Powers of Elliptic Curves and Applications

Michael Harris

UFR de Mathématiques, Université Paris 7,
2 Pl. Jussieu, 75251 Paris cedex 05, France
harris@math.jussieu.fr

To Yuri Ivanovich Manin

Summary. I explain how to prove potential automorphy for odd-dimensional symmetric power L-functions.

Key words: elliptic curves, Sato–Tate conjecture, potential automorphy

2000 Mathematics Subject Classifications: 11F80, 11F70, 11G05, 11R37, 22E55

Introduction

The present article was motivated by a question raised independently by Barry Mazur and Nick Katz. The articles [CHT, HST, T] contain a proof of the Sato-Tate conjecture for an elliptic curve E over a totally real field whose j-invariant $j(E)$ is not an algebraic integer. The Sato-Tate conjecture for E is an assertion about the equidistribution of Frobenius angles of E, or equivalently about the number of points $|E(\mathbb{F}_p)|$ on E modulo p as p varies. The precise statement of the conjecture, which is supposed to hold for any elliptic curve without complex multiplication, is recalled in Section 5. Now suppose E and E' are two elliptic curves without complex multiplication, and suppose E and E' are not isogenous. The question posed by Mazur and Katz is roughly the following: are the distributions of the Frobenius angles of E and E', or equivalently of the numbers $p + 1 - |E(\mathbb{F}_p)|$ and $p + 1 - |E'(\mathbb{F}_p)|$, independent?

The Sato-Tate conjecture, in the cases considered in [CHT, HST, T], is a consequence of facts proved there about L-functions of symmetric powers of the Galois representation on the Tate module $T_\ell(E)$ of E, following a strategy

Y. Tschinkel and Y. Zarhin (eds.), *Algebra, Arithmetic, and Geometry*,
Progress in Mathematics 270, DOI 10.1007/978-0-8176-4747-6_1,
© Springer Science+Business Media, LLC 2009

elaborated by Serre in [S]. These facts in turn follow from one of the main theorems of [CHT,HST,T], namely that, if n is *even*, the $(n-1)$st symmetric power of $T_\ell(E)$ is potentially automorphic, in that it is associated to a cuspidal automorphic representation of $\mathrm{GL}(n)$ over some totally real Galois extension of the original base field. The restriction to even n is inherent in the approach to potential modularity developed in [HST], which applies only to even-dimensional representations. The necessary properties of all symmetric power L-functions follow from this result for even-dimensional symmetric powers, together with basic facts about Rankin-Selberg L-functions proved by Jacquet-Shalika-Piatetski-Shapiro and Shahidi. In a similar way, the Mazur-Katz question can be resolved affirmatively if we can prove potential automorphy for *all* symmetric power L-functions of E and E' over the same field.

The main purpose of the present article is to explain how to prove potential automorphy for odd-dimensional symmetric power L-functions, thus providing a response to the question of Mazur and Katz. The principal innovation is a tensor product trick that converts an odd-dimensional representation to an even-dimensional representation. Briefly, in Sections 2 and 3 one tensors with a two-dimensional representation. One has to choose a two-dimensional representation of the right kind, which is not difficult. The challenge is then to recover the odd-dimensional symmetric power unencumbered by the extraneous two-dimensional factor; this is the subject of Section 4. I say "explain how to prove" rather than "prove" because the proofs of the main results of this article make use of stronger modularity theorems than those proved in [CHT] and [T], and are thus conditional. I explain in Section 1 how I expect these modularity results to result from a strengthening of known theorems associating compatible families of ℓ-adic Galois representations to certain kinds of automorphic representations. These stronger theorems, stated as Expected Theorems 1.2 and 1.4, are the subject of work in progress by participants in the Paris automorphic forms seminar, and described in [H]. This work has progressed to a point where it seems legitimate to admit these Expected Theorems. Nevertheless, the present article should be viewed as a promissory note which will not be negotiable until the project outlined in [H] has been completed[1].

One can of course generalize the question and ask whether the distributions of the Frobenius angles of n pairwise non-isogenous elliptic curves without complex multiplication are independent. For $n \geq 3$ this seems completely inaccessible by current techniques in automorphic forms.

The article concludes with some speculations regarding additional applications of the tensor product trick.

---

[1] Note added in proof. All the Expected Theorems have now been proved in articles by Chenevier, Clozel, Guerberoff, Labesse, Shin, and the author, in various combinations. Most of these articles can be consulted at http://fa.institut.math.fr/node/29. An article in preparation by the six authors will make the connections clear.

I met Yuri Ivanovich Manin briefly near the beginning of my career. Later, as a National Academy of Sciences exchange fellow I had the remarkable good fortune of spending a year as his guest in Moscow, and as a (mostly passive but deeply appreciative) participant in his seminar at Moscow State University during what may well have been its final year, and saw first-hand what the Moscow mathematical community owed to his insight and personality. The influence of Yuri Ivanovich on my own work is pervasive, and the present work is no exception: though it is not apparent in what follows, the article [HST], on which all the results presented here are based, can be read as an extended meditation on the Gauss-Manin connection as applied to a particular family of Calabi-Yau varieties. It is an honor to dedicate this article to Yuri Ivanovich Manin.

I thank Barry Mazur and Nick Katz for raising the question that led to this paper. Apart from the tensor product trick, practically all the ideas in this paper are contained in [CHT], [HST], and [T]; I thank my coauthors – Laurent Clozel, Nick Shepherd-Barron, and Richard Taylor – for their collaboration over many years. I thank Richard Taylor specifically for his help with the proof of the crucial Lemma 4.2. Finally, I thank the referee for a careful reading, and for helping me to clarify a number of important points.

1 Reciprocity for n-dimensional Galois representations

All finite-dimensional representations of Galois groups are assumed to be continuous. When E is a number field, contained in a fixed algebraic closure $\overline{\mathbb{Q}}$ of $\mathbb{Q}$, we let Γ_E denote $Gal(\overline{\mathbb{Q}}/E)$. Let ρ be a (finite-dimensional) ℓ-adic representation of Γ_E. Say ρ is *pure* of weight w if for all but finitely many primes v of E the restriction ρ_v of ρ to the decomposition group Γ_v is unramified and if the eigenvalues of $\rho_v(Frob_v)$ are all algebraic numbers whose absolute values equal $q_v^{\frac{w}{2}}$; here q_v is the order of the residue field k_v at v, and $Frob_v$ is geometric Frobenius. If ρ is pure of weight w, the *normalized L-function* of ρ is

$$L^{\mathrm{norm}}(s,\rho) = L(s + \frac{w}{2}, \rho);$$

here we assume we have a way to define the local Euler factors at primes dividing ℓ (for example, ρ belongs to a compatible system of λ-adic representations). Then $L^{\mathrm{norm}}(s,\rho)$ converges absolutely for $Re(s) > 1$.

Let F be a CM field, $F^+ \subset F$ its maximal totally real subfield, so that $[F : F^+] \leq 2$. Let $c \in \mathrm{Gal}(F/F^+)$ be complex conjugation; by transport of structure it acts on automorphic representations of $\mathrm{GL}(n, F)$. The following theorem is the basis for many of the recent results on reciprocity for Galois representations of dimension >2. For the purposes of the following theorem, a unitary Harish-Chandra module σ for $\mathrm{GL}(n, \mathbb{C})$ will be called "cohomological" if $\sigma \otimes || \det ||^{\frac{n-1}{2}}$ is cohomological in the usual sense, i.e., if there is a finite-dimensional irreducible representation W of $\mathrm{GL}(n, \mathbb{C})$ such that

$$H^\bullet(\mathrm{Lie}(\mathrm{GL}(n, \mathbb{C})), U(n); \sigma \otimes || \det ||^{\frac{n-1}{2}} \otimes W) \neq 0.$$

The half-integral twist is required by the unitarity.

Theorem 1.1 ([C2, Ko, HT, TY]). *In what follows, Π denotes a cuspidal automorphic representation of $\mathrm{GL}(n, F)$, and $\{\rho_{\bullet,\lambda}\}$ denotes a compatible family of n-dimensional λ-adic representations of Γ_F.*

There is an arrow $\Pi \mapsto \{\rho_{\Pi,\lambda}\}$, where λ runs through non-archimedean completions of a certain number field $E(\Pi)$, under the following hypotheses:

$$\left\{\begin{array}{c} (1)\ The\ factor\ \Pi_\infty \\ is\ cohomological \\ (2)\ \Pi \circ c \cong \Pi^\vee \\ (3)\ \exists v_0, \Pi_{v_0} \\ discrete\ series \end{array}\right\} \Rightarrow \left\{\begin{array}{c} (a)\ \rho = \rho_{\Pi,\lambda}\ geometric, \\ HT\ regular \\ (b)\ \rho \otimes \rho \circ c \to \mathbb{Q}_\ell(1-n) \\ (c)\ local\ condition \\ at\ v_0 \end{array}\right\}$$

This correspondence has the following properties:

(i) For any finite place v prime to the residue characteristic ℓ of λ,

$$[\rho_{\Pi,\lambda} \mid_{WD_v}]^{Frob-ss} = \mathcal{L}(\Pi_v \otimes |\bullet|_v^{\frac{1-n}{2}}).$$

*Here WD_v is the local Weil-Deligne group at v, $\mathcal{L}$ is the **normalized** local Langlands correspondence, and $Frob - ss$ denotes Frobenius semisimplification;*

(ii) The representation $\rho_{\Pi,\lambda} \mid_{G_v}$ is potentially semistable, in Fontaine's sense, for any v dividing ℓ, and the Hodge-Tate weights at v are explicitly determined by the infinitesimal character of the Harish-Chandra module Π_∞.

The local Langlands correspondence is given the unitary normalization. This means that the correspondence identifies $L(s, \Pi)$ and $L^{norm}(s, \rho_{\Pi,\lambda})$, so that the functional equations always exchange values at s and $1 - s$.

The term "geometric" is used in the sense of Fontaine-Mazur: each $\rho_{\Pi,\lambda}$ is unramified outside a finite set of places of F, in addition to the potential semistability mentioned in the statement of the theorem. The condition "HT regular" (Hodge-Tate) means that the Hodge-Tate weights at v have multiplicity at most one.

For the local condition (c), we can take the condition that the representation of the decomposition group at v_0 is indecomposable as long as v_0 is prime to the residue characteristic of λ, or equivalently that this representation of the decomposition group at v_0 corresponds to a discrete series representation of $\mathrm{GL}(n, F_{v_0})$. The conditions on both sides of the diagram match: $(1) \leftrightarrow (a)$, $(2) \leftrightarrow (b)$, $(3) \leftrightarrow (c)$.

When Π is a base change of a representation Π^+ of $\mathrm{GL}(n, F^+)$, condition (2) just means that Π is self-dual.

In what follows, we will admit the following extension of Theorem 1.1:

Expected Theorem 1.2. *The assertions of Theorem 1.1 remain true provided Π satisfies conditions (1) and (2); then $\rho_{\Pi,\lambda}$ satisfies conditions (a) and (b), as well as (i) and (ii).*

Here "Expected Theorem" means something more than conjecture. The claim of Expected Theorem 1.2 is a very special case of the general Langlands conjectures, in the version for Galois representations developed in Clozel's article [C1]. This specific case is the subject of work in progress on the part of participants in the Paris automorphic forms seminar and others, and an outline of the various steps in the proof can be found in [H]. There are quite a lot of intermediate steps, the most difficult of which involve analysis of the stable trace formula, twisted or not, but I would "expect" that they will all have been verified, and the theorem completely proved, by 2010 at the latest[2].

I single out one of the intermediate steps. By a *unitary group* over F^+ I will mean the group of automorphisms of a vector space V/F preserving a nondegenerate hermitian form. The group is denoted $U(V)$, the hermitian form being understood. We will consider a hermitian vector space V of dimension n with the following properties:

(1.3.1) For every real place σ of F^+, the local group $U(V_\sigma)$ is compact (the hermitian form is totally definite);

(1.3.2) For every finite place v of F^+, the local group $U(V_v)$ is quasi-split and split over an unramified extension.

We write $G_0 = U(V)$. Such a unitary group always exists when n is odd, provided F/F^+ is everywhere unramified. When n is even, there is a sign obstruction that can be removed by replacing F^+ by a totally real quadratic extension. Such restrictions are harmless for applications (see [H]).

We let $K = \prod_v K_v \subset G_0(\mathbf{A}^f)$ be an open compact level subgroup. Hypothesis (1.3.2) guarantees that $G_0(F_v^+)$ contains a hyperspecial maximal compact subgroup for all finite v. If v is split then any maximal compact subgroup is hyperspecial and conjugate to $\mathrm{GL}(n, \mathcal{O}_v)$. We assume

Hypothesis 1.3.3. *K_v is hyperspecial maximal compact for all v that remain inert in F.*

For any ring R, let

$$M_K(G_0, R) = C(G_0(F^+)\backslash G_0(\mathbf{A})/G_0(\mathbb{R}) \cdot K, R),$$

where for any topological space X, $C(X, R)$ means the R-module of continuous functions from X to R, the latter endowed with the discrete topology. The Hecke algebra $\mathcal{H}_K(R)$ of double cosets of K in $G_0(\mathbf{A}^f)$ with coefficients in R acts on $M_K(G_0, R)$. It contains a subring $\mathcal{H}_K^{\mathrm{hyp}}(R)$ generated by the double cosets of K_v in $G_0(F_v^+)$ where v runs over primes that split in F at which K_v is hyperspecial maximal compact. We denote by $\mathbb{T}_K(R)$ the image of $\mathcal{H}_K^{\mathrm{hyp}}(R)$

---

[2]See footnote to introduction.

in $End_R(M_K(G_0, R))$. The algebra $\mathbb{T}_K(R)$ is reduced if R is a semisimple algebra flat over $\mathbb{Z}$ (cf. [CHT], Corollary 2.3.3).

We can also consider $M(G_0, R)$, the direct limit of $M_K(G_0, R)$ over all K, including those not satisfying (1.3.3). This is a representation of $G_0(\mathbf{A}^f)$ and decomposes as a sum of irreducible representations when R is an algebraically closed field of characteristic zero. Let $\pi \subset M(G_0, \mathbb{C})$ be an irreducible summand. Write $\pi = \pi_\infty \otimes \pi_f$, $\pi_f = \otimes'_v \pi_v$, the restricted tensor product over finite primes v of F^+ of representations of $G_0(F_v^+)$. With our hypotheses π_∞ is the trivial representation of $G_{0,\mathbb{R}} = \prod_\sigma U(V_\sigma)$, where σ is as in (1.3.1). Suppose $\pi^K \neq \{0\}$ for some K satisfying (1.3.3). Then for every finite v one can define the local base change $\Pi_v = BC_{F_v/F_v^+}\pi_v$, a representation of $G_0(F_v^+) \overset{\sim}{\longrightarrow} \prod_{w|v} GL(n, F_w)$. If v is inert, then by (1.3.3) we know that π_v is an unramified representation, and so is Π_v. If not, then π_v is a representation of $GL(n, F_v^+)$ and $\Pi_v \overset{\sim}{\longrightarrow} \pi_v \otimes \pi_v^\vee$, with the appropriate normalization. Thus we have the following:

Lemma 1.3.4. *The local factor π_v is uniquely determined by Π_v.*

Expected Theorem 1.4.[3] *There is a cohomological representation Π_∞ of $GL(n, F_\infty) = GL(n, F \otimes_\mathbb{Q} \mathbb{R})$ such that the formal base change*

$$\Pi = BC_{F/F^+}\pi = \Pi_\infty \otimes \bigotimes\nolimits'_v \Pi_v$$

is an automorphic representation of $GL(n, F)$. Moreover, there is a partition $n = a_1 + a_2 + \cdots + a_r$ and an automorphic representation $\bigotimes_j \Pi_j$ of the group $\prod_j GL(a_j, \mathbf{A}_F)$ such that each Π_j is in the discrete automorphic spectrum of $GL(a_j, F)$ and Π, as a representation of $GL(n, \mathbf{A}_F)$, is parabolically induced from the inflation of $\bigotimes_j \Pi_j$ to the standard parabolic subgroup $P(\mathbf{A}) \subset GL(n, \mathbf{A}_F)$ associated to the partition. Moreover, each Π_j satisfies conditions (1) and (2) of Theorem 1.1, where "cohomological" is understood as in the discussion preceding that theorem.

We say π is F/F^+-*cuspidal* if Π is cuspidal, in which case it follows from the classification of generic cohomological representations that Π_∞ is necessarily tempered and is uniquely determined by the condition that π_∞ is trivial. This representation is denoted $\Pi_{\infty,0}$, or $\Pi_{\infty,0}(n, F)$ when this is necessary.

Fix a prime ℓ and let $\mathcal{O}$ be the ring of integers in a finite extension of $\mathbb{Q}_\ell$. The ring $\mathbb{T}_K(\mathcal{O})$ is semilocal and $\mathbb{T}_K(\mathcal{O} \otimes \bar{\mathbb{Q}}_\ell)$ is a product of fields. Let $\mathfrak{m} \subset \mathbb{T}_K(\mathcal{O})$ be a maximal ideal and let $I \subset \mathbb{T}_K(\mathcal{O} \otimes \bar{\mathbb{Q}}_\ell)$ be any prime ideal whose intersection with $\mathbb{T}_K(\mathcal{O})$ is contained in $\mathfrak{m}$. Then I determines an irreducible $G_0(\mathbf{A}^f)$-summand π_0 of $M(G_0, \bar{\mathbb{Q}}_\ell)$, or equivalently of $M(G_0, \mathbb{C})$, if one identifies the algebraic closures of $\mathbb{Q}$ in $\mathbb{C}$ and in $\bar{\mathbb{Q}}_\ell$, as in [HT, p. 20]. More precisely, I determines $\pi_{0,v}$ locally only at v for which K_v is hyperspecial, but this includes all inert primes. Let S be the set at which K_v is not hyperspecial. By Expected Theorem 1.4, I determines a collection of cohomological

---

[3]Note added in proof. This theorem has now been proved by Labesse.

automorphic representations Π of $\mathrm{GL}(n, F)$ which are isomorphic outside the finite set S. By strong multiplicity one, Π is in fact unique; we denote it Π_I. Then π is unique by Lemma 1.3.4.

Combining Expected Theorems 1.4 and 1.2, we thus obtain an n-dimensional representation $\rho_{\Pi,\ell} = \rho_{I,\ell}$ of Γ_F. We say I is *Eisenstein* at ℓ if the reduction mod ℓ, denoted $\bar{\rho}_{\Pi,\ell}$, is not absolutely irreducible. It follows in the usual way from property (i) of the correspondence between $\rho_{\Pi,\ell}$ and Π that if I is Eisenstein at ℓ then every prime ideal of $\mathbb{T}_K(\mathcal{O} \otimes \bar{\mathbb{Q}}_\ell)$ lying above $\mathfrak{m}$ is Eisenstein. In that case we say $\mathfrak{m}$ is Eisenstein at ℓ.

Lemma 1.5. *Admit Expected Theorems 1.4 and 1.2. Fix a prime ℓ, and suppose $\mathfrak{m} \subset \mathbb{T}_K(\mathcal{O})$ is not an Eisenstein ideal at ℓ. Then any prime ideal $I \subset \mathbb{T}_K(\mathcal{O} \otimes \bar{\mathbb{Q}}_\ell)$ lying above $\mathfrak{m}$ has the property that Π_I is cuspidal (in that case we say $\mathfrak{m}$ and I are $F/F^+ - $ cuspidal).*

Sketch of proof. This follows from the classification of automorphic representations of $\mathrm{GL}(n)$ and from properties of base change. Suppose I corresponds to $\pi \subset M(G_0, \mathbb{C})$. If the base change Π of π belongs to the discrete spectrum of $\mathrm{GL}(n, F)$ – that is, if the partition in Expected Theorem 1.4 is a singleton – then it is either cuspidal or in the non-tempered discrete spectrum. In the latter case, the Moeglin-Waldspurger classification implies that n factors as ab, with $a > 1, b > 1$, and Π is the Speh representation attached to a cuspidal automorphic representation Π_1 of $\mathrm{GL}(b, F)$. Clozel has checked in [C3] that Π_1 satisfies properties (1) and (2) of Expected Theorem 1.2 for $\mathrm{GL}(b)$, hence is associated to a b-dimensional ℓ-adic representation ρ_1 of Γ_F. It follows from condition (i) of Expected Theorem 1.2 that the semisimple representation $\rho_{\Pi,\ell}$ decomposes as a sum of a constituents, each of which is an abelian twist of ρ_1.

Suppose Π does not belong to the discrete spectrum of $\mathrm{GL}(n, F)$. Then π is endoscopic, hence is associated to a partition $n = \sum a_j$, with each $a_j > 0$, and an automorphic representation $\otimes_j \Pi_{a_j}$ in the discrete spectrum of $\prod_j \mathrm{GL}(a_j, F)$, such that each Π_{a_j}, satisfies properties (1) and (2) of Expected Theorem 1.2. Then $\rho_{\Pi,\ell}$ decomposes as a sum of $r > 1$ pieces of dimensions a_j. $\square$

I recall the main Modularity Lifting Theorem of [T], whose proof builds on and completes the main results of [CHT].

1.6. Modularity Lifting Theorem. *Let $\ell > n$ be a prime unramified in F^+ (resp. and such that every divisor of ℓ in F^+ splits in F) and let*

$$r : \Gamma_{F^+} \to \mathrm{GL}(n, \bar{\mathbb{Q}}_\ell) \ (\text{resp. } r : \Gamma_F \to \mathrm{GL}(n, \bar{\mathbb{Q}}_\ell))$$

be a continuous irreducible representation satisfying the following properties:

(a) *r ramifies at only finitely many primes, is crystalline at all primes dividing ℓ, and is Hodge-Tate regular;*

(b) *$r \simeq r^\vee(1 - n) \cdot \chi$ (resp. $r^c \simeq r^\vee(1 - n)$) where $(1 - n)$ is the Tate twist and χ is a character whose value is constant on all complex conjugations (resp. c denotes complex conjugation);*

(c) *At some finite place v not dividing ℓ, r_v corresponds to a square-integrable representation of $\mathrm{GL}\,(n, F_v^+)$ under the local Langlands correspondence, and satisfies the final "minimality" hypotheses of [CHT, 4.3.4 (5)] or [T, 5.2 (5)]*

In addition, we assume that $\bar{r}$

(d) *has "big" image in the sense of Definition 3.1 below;*
(e) *is absolutely irreducible;*
(f) *is of the form $\bar{\rho}_{\Pi,\ell}$ for some cuspidal automorphic representation Π of $\mathrm{GL}(n, F^+)$ satisfying conditions (1)-(3) of Theorem 1.1.*

Then r is of the form $\rho_{\Pi',\ell}$ for some cuspidal automorphic representation Π' of $\mathrm{GL}(n, F^+)$ satisfying conditions (i)-(iii) of Theorem 1.1.

I have not written out the last part of hypothesis (c) in detail, because it will be dropped in the remainder of the article. More precisely, if we admit Expected Theorems 1.2 and 1.4, then we obtain the following theorem:

1.7. Expected Modularity Lifting Theorem.[4] *Let ℓ and r be as in Theorem 1.6, but we no longer assume condition (c), and in (f) we drop condition (3). Then r is of the form $\rho_{\Pi',\ell}$ for some cuspidal automorphic representation Π' of $\mathrm{GL}(n, F^+)$ satisfying properties (i)-(ii) of Expected Theorem 1.2.*

In the remainder of the paper, I draw consequences from Theorem 1.7.

2 Potential modularity of a Galois representation

Let F and F^+ be as in Section 1. The article [HST] develops a method for proving that certain n-dimensional ℓ-adic representations ρ of Γ_{F^+} (resp. Γ_F) that look like they arise from automorphic representations via the correspondence of Theorem 1.1, are *potentially* automorphic in the following sense: there exists a totally real Galois extension F'/F^+ such that $\rho \mid_{\Gamma_{F'}}$ (resp. $\rho \mid_{\Gamma_{F \cdot F'}}$) does indeed correspond to a cuspidal automorphic representation of $\mathrm{GL}(n, F')$ (resp. $\mathrm{GL}(n, F \cdot F')$). The relevant result is Theorem 3.1 of [HST], which is in turn based on Theorem 1.6. Although the latter theorem is valid for representations of arbitrary dimension n, Theorem 3.1 only applies to an even-dimensional representation of Γ_{F^+} endowed with an alternating form that is preserved by Γ_{F^+} up to a multiplier.

If we admit the expected results of Section 1, then Theorem 3.1 of [HST] admits the following simplification. The constant $C(n_i)$ is a positive number introduced in [HST], Corollary 1.11; its precise definition is irrelevant to the applications. For a finite prime w, G_w denotes a decomposition group, $I_w \subset G_w$ the inertia group.

---

[4]Note added in proof. This theorem has now been proved by Guerberoff.

Theorem 2.1 [HST]. *Assume the Expected Theorems of Section 1. Let $F^+/F^{0,+}$ be a Galois extension of totally real fields and let $n_1, \dots, n_r$ be even positive integers. Suppose that $\ell > \max\{C(n_i), n_i\}$ is a prime which is unramified in F and which splits in $\mathbb{Q}(\zeta_{n_i+1})$, $i = 1, \dots, r$. Let $\mathcal{L}$ be a finite set of primes of F^+ not containing primes above ℓ and let M be a finite extension of F.*

Suppose that for $i = 1, \dots, r$

$$r_i : \Gamma_{F^+} \to GSp(n_i, \mathbb{Z}_\ell)$$

is a continuous representation with the following properties.

(1) r_i has multiplier $\omega_\ell^{1-n_i}$, where ω_ℓ is the ℓ-adic cyclotomic character.
(2) r_i ramifies at only finitely many primes.
(3) The image $\bar{r}_i(\Gamma_{F^+(\zeta_\ell)})$ is big, in the sense of Definition 3.1 below, where ζ_ℓ is a primitive ℓ-th root of 1.
(4) The fixed field F_i of $\ker \mathrm{ad}(\bar{r}_i) \subset \Gamma_{F^+}$ does not contain $F(\zeta_\ell)$.
(5) r_i is unramified at all primes in $\mathcal{L}$.
(6) If $w \mid \ell$ is a prime of F then $r_i \mid_{G_w}$ is crystalline with Hodge-Tate weights $0, 1, \dots, n_i - 1$, with the conventions of [HST]. Moreover,

$$\bar{r}_i \mid_{I_w} \simeq \bigoplus_{j=0}^{n_i-1} \omega_\ell^{-j}.$$

Then there is a totally real field $F'^{,+}/F^+$, Galois over F_0^+ and linearly disjoint from the compositum of the F_i with M over F^+, with the property that each $r_{i,F'^{,+}} = r_i \mid_{\Gamma_{F'}^+}$ corresponds to an automorphic representation Π_i of $GL(n_i, F'^{,+})$. If $F'/F'^{,+}$ is a CM quadratic extension, then the base change $\Pi_{i,F'}$ has archimedean constituent isomorphic to $\Pi_{\infty,0}(n_i, F')$ (cf. the remarks after Expected Theorem 1.4). Finally, all primes of $\mathcal{L}$ and all primes of F dividing ℓ are unramified in F'.

Apart from a few slight changes in notation, this theorem is practically identical to Theorem 3.1 of [HST]. There is no field M in [HST] but the proof yields an $F'^{,+}$ linearly disjoint over F^+ from any fixed extension. Only condition (7) of Theorem 3.1 of [HST], corresponding to condition (3) of Theorem 1.1, has been eliminated. The proof is identical but simpler: references to Theorem 1.6 are replaced by references to Expected Theorem 1.7, and all arguments involving the primes q and q' in [HST] are no longer necessary.

Let E be an elliptic curve over F^+, and let $\rho_{E,\ell} : \Gamma_{F^+} \to GL(2, \mathbb{Q}_\ell)$ denote the representation on $H^1(E_{\overline{\mathbb{Q}}}, \mathbb{Q}_\ell)$, i.e. the dual of the ℓ-adic Tate module. For $n \geq 1$ let

$$\rho_{E,\ell}^n = Sym^{n-1}\rho_{E,\ell} : \Gamma_{F^+} \to GL(n, \mathbb{Q}_\ell).$$

We will always assume E has no complex multiplication. Then $\rho_{E,\ell}^n$ is irreducible by a theorem of Serre, for all n, and for almost all $\ell > n$, $Im(\bar{\rho})$

contains the image of $SL(2, \mathbb{F}_\ell)$ under the symmetric power representation, and hence is absolutely irreducible. When $F^+ = \mathbb{Q}$ it was proved in the series of papers initiated by Wiles and Taylor-Wiles and completed by Breuil, Conrad, Diamond, and Taylor that $L(s, \rho_{E,\ell})$ is automorphic, which in this case means is attached to a classical new form of weight 2. The prototype for Theorem 2.1 is the theorem proved by Taylor in [T02], which shows that $L(s, \rho_{E,\ell})$ is potentially automorphic for any F^+.

In [HST] and [T], Theorem 2.1 is notably applied to show that $L\left(s, \rho_{E,\ell}^n\right)$ is potentially automorphic for any *even* n, provided E has non-integral j-invariant. One can hardly hope to apply Theorem 2.1 as such when n is odd, given that symplectic groups are only attached to even integers. Moreover, when n is odd $\rho_{E,\ell}^n$ has an orthogonal polarization rather than a symplectic polarization. These two related flaws – the oddness of n and the orthogonality of ρ^n – can be cured simultaneously by tensoring $\rho_{E,\ell}^n$ by a two-dimensional representation $\tau : \Gamma_{F^+} \to GL(2, \mathbb{Z}_\ell)$. Such a representation is necessarily symplectic, with multiplier $\det \tau$. We suppose τ has determinant ω_ℓ^{-n}. In order to preserve the hypotheses of Theorem 2.1, specifically 2.1 (6), we need to assume the following:

Hypothesis 2.2. *If $w \mid \ell$ is a prime of F then $\tau \mid_{G_w}$ is crystalline with Hodge-Tate weights $0, n$, with the conventions of [HST]. Moreover,*

$$\bar{\tau} \mid_{I_w} \simeq 1 \oplus \omega_\ell^{-n}.$$

For example, let f be a classical new form of weight $n + 1$ for $\Gamma_0(N)$, for some integer N. Associated to f is a number field $\mathbb{Q}(f)$, generated by the Fourier coefficients of f, and a compatible system of 2-dimensional λ-adic representations $\tau_{f,\lambda}$ of $\Gamma_{\mathbb{Q}}$ as λ varies over the primes of $\mathbb{Q}(f)$. We choose a prime ℓ that splits completely in $\mathbb{Q}(f)$ and such that $\ell \nmid N$. Fix λ dividing ℓ, and write $\tau = \tau_{f,\lambda}$. Then τ takes values in $GL(2, \mathbb{Z}_\ell)$; since f has trivial nebentypus, the determinant of τ is indeed ω_ℓ^{-n}. The hypothesis regarding $\bar{\tau} \mid_{I_w}$ is in general a serious restriction, but we will find explicit examples.

(2.3)

We say the residual representation $\bar{\tau}$ is "big enough" if its image contains a non-commutative subgroup of the normalizer $N(T)$ of a maximal torus $T \subset SL(2, \mathbb{F}_\ell)$, and more specifically that $\text{Im}(\bar{\tau})$ contains an element h of T with distinct eigenvalues that acts trivially on the cyclotomic field $\mathbb{Q}(\zeta_\ell)$, and an element w of order 2 that does not commute with h.

Here are the representations we will use. Let L be an imaginary quadratic field not contained in F, and let η_L be the corresponding quadratic Dirichlet character, viewed as an idèle class character of $\mathbb{Q}$. Let χ be a Hecke character of (the idèles of) L whose restriction to the idèles of $\mathbb{Q}$ is the product $|\bullet|_{\mathbf{A}}^{-n} \cdot \eta_L$, where $|\bullet|_{\mathbf{A}}$ is the idèle norm. Choose an isomorphism $L_\infty = L \otimes_{\mathbb{Q}} \mathbb{R} \xrightarrow{\sim} \mathbb{C}$, let z be the corresponding coordinate function on L_∞, and assume $\chi_\infty(z) = z^{-n}$.

Then χ is an algebraic Hecke character and is associated to a compatible system of ℓ-adic characters $\chi_\lambda : \Gamma_L \to \mathbb{Q}(\chi)_\lambda^+$, where $\mathbb{Q}(\chi)$ is the field of coefficients of χ, a finite extension of $\mathbb{Q}$, and λ runs through the places of $\mathbb{Q}(\chi)$. Choose a prime $\ell > 2n + 1$ that splits in L and in $\mathbb{Q}(\chi)$; then for any λ dividing ℓ we can view χ_λ as a $\mathbb{Q}_\ell^\times$-valued character of Γ_L; we write $\chi_\ell = \chi_\lambda$.

For such an ℓ, we define the monomial induced representation

$$\tau_\ell = \operatorname{Ind}_{\Gamma_L}^{\Gamma_\mathbb{Q}} \chi_\ell : \Gamma_\mathbb{Q} \to \mathrm{GL}(2, \mathbb{Q}_\ell).$$

Lemma 2.4. *Under the above hypotheses, $\bar\tau_\ell$ is "big enough" in the sense of (2.3).*

Proof. Let v, v' be the two primes of L dividing ℓ. Let $k(v), k(v')$ denote the corresponding residue fields; then the product of the respective Teichmüller characters defines an inclusion

$$k(v)^\times \times k(v')^\times \hookrightarrow \mathcal{O}_v^\times \times \mathcal{O}_{v'}^\times \subset \mathbf{A}_L^\times.$$

Let b and b' denote generators of the images of the cyclic groups $k(v)^\times$ and $k(v')^\times$ in $\mathcal{O}_v^\times$ and $\mathcal{O}_{v'}^\times$, respectively, and let $s(b), s(b')$ denote their images in $\mathrm{Gal}(L^{ab}/L)$ under the reciprocity map. Our hypotheses imply that $s(b)$ acts on $\bar\tau$ with eigenvalues $b^n, 1$, and likewise for $s(b')$. Moreover, the trivial eigenspace for $s(b)$ is the non-trivial eigenspace for $s(b')$, and vice versa. We identify b and b' with roots of unity in $\mathbb{Q}_\ell^\times$. The element $h = \bar\tau(s(b) \cdot s(b')^{-1})$ then belongs to $\mathrm{SL}(2, \mathbb{F}_\ell)$ and has eigenvalues $b^{\pm n}$. Since $\ell - 1 > 2n$, these eigenvalues are distinct; in particular, h does not commute with the image w of complex conjugation in $\mathrm{Im}(\bar\tau)$. Since the action of $\mathcal{O}_v^\times \times \mathcal{O}_{v'}^\times$ on $\mathbb{Q}(\zeta_\ell)$ factors through the norm to $\mathbb{Z}_\ell^\times$, we see that h acts trivially on $\mathbb{Q}(\zeta_\ell)$. This completes the proof of Lemma 2.4.   □

Corollary 2.5. *Let $\tau = \tau_{f,\lambda}$ be as in the preceding paragraph and satisfy Hypothesis 2.2. We continue to admit the Expected Theorems of Section 1. We use the same notation for $\tau\,|_{\Gamma_{F^+}}$. Let E be an elliptic curve over F. Let n be an odd positive integer, and suppose there is a prime $\ell > 2n + 1$, unramified in F^+, such that*

(i) E has good ordinary reduction at all primes w dividing ℓ.
(ii) ℓ does not divide the conductor N of f, and

$$\bar\tau|_{I_w} \simeq 1 \oplus \omega_\ell^{-n}.$$

(iii) ℓ splits in $\mathbb{Q}(\zeta_{2i+1})$, $i = 1, \ldots, n - 1$; in particular, $\ell \equiv 1 \pmod{n}$ (take $i = \frac{n-1}{2}$).
(iv) $\frac{\ell - 1}{n} > 2$.

Suppose $\bar\tau$ is "big enough" in the sense of (2.3). Let M be an arbitrary extension of F^+. For every integer $i \le n$, let $r_i = \rho_{E,\ell}^{2i}$; let $r_\tau = \rho_{E,\ell}^n \otimes \tau$. Then there is a totally real Galois extension $F'^{,+}/F^+$, linearly disjoint from M over

F^+, with the property that for $i = 1, \ldots, n$, $r_{i,F',+} = r_i \mid_{\Gamma_F'^+}$ corresponds to an automorphic representation Π_i of $GL(2i, F'^{,+})$, and such that r_τ corresponds to an automorphic representation Π_τ of $GL(2n, F'^{,+})$. If $F'/F'^{,+}$ is a CM quadratic extension, then the base change $\Pi_{i,F'}$ (resp. $\Pi_{\tau,F'}$) has archimedean constituent isomorphic to $\Pi_{\infty,0}(2i, F')$ (resp. $\Pi_{\infty,0}(2n, F')$).

Proof. Under our hypotheses on ℓ and the image of $\bar\tau$, $\bar r_\tau$ is absolutely irreducible. We first prove the corollary under the hypothesis that for each w dividing ℓ,

$$\bar\rho_{E,\ell}\mid_{I_w} \simeq 1 \oplus \omega_\ell^{-1}. \tag{2.5.1}$$

Conditions (1), (2), and (6) of Theorem 2.1 are clearly satisfied. Since ℓ is unramified in F^+, F^+ and $\mathbb{Q}(\zeta_\ell)$ are linearly disjoint over $\mathbb{Q}$. By condition (1), the intersection of $\mathrm{Im}(\bar r_i)$ with the center of $GSp(n_i, \mathbb{F}_\ell)$ maps onto the subgroup of $\mathrm{Gal}(\mathbb{Q}(\zeta_\ell)/\mathbb{Q})$ generated by $2(1 - n_i)$-th powers, which implies condition (4) for all r_i, and for r_τ as well.

Condition (5) is irrelevant. It remains to verify condition (3). For $\bar r_i$, $i = 1, \ldots, n$, this is Corollary 2.5.4 of [CHT]; the case of $\bar r_\tau$ is Lemma 3.2 below.

This completes the proof under hypothesis (2.5.1). We reduce to this case as in the proof of Theorem 3.3 of [HST], replacing ℓ by a second prime $\ell' > 2n+1$ also split in L and in $\mathbb{Q}(\chi)$, $\bar\tau_\ell$ by $\bar\tau_{\ell'}$, and E by a curve E' (unfortunately denoted E in [HST]) such that $\rho_{E',\ell} \xrightarrow{\sim} \rho_{E,\ell}$ but $\rho_{E',\ell'}$ satisfies hypothesis (2.5.1) at ℓ'. $\qquad\square$

Remark 2.6. Recall that the results of this section are all conditional on the Expected Theorems of Section 1. In particular, we are not assuming that E has potentially multiplicative reduction at some place. If we do assume $j(E)$ is not integral, then the automorphic representations Π_i are constructed unconditionally in [CHT, HST, T]. However, the local condition on $j(E)$ does not suffice to impose a strong enough local condition on r_τ (corresponding to a discrete series representation on the automorphic side).

3. A lemma about certain residual representations

Definition 3.1. Let $V/\bar{\mathbb{F}}_\ell$ be a finite dimensional vector space. Let $ad^0(V) \subset ad(V) = Hom(V,V)$ be the subspace of trace 0 endomorphisms. A subgroup $\Delta \subset GL(V)$ is big if the following hold:

(a) $H^i(\Delta, ad^0 V) = (0)$ for $i = 0, 1$.
(b) For all irreducible $\bar{\mathbb{F}}_\ell[\Delta]$-submodules $W \subset Hom(V,V)$ we can find $h \in \Delta$ and $\alpha \in \bar{\mathbb{F}}_\ell$ with the following properties. The α-generalized eigenspace $V_{h,\alpha}$ of h on V is one-dimensional. Let

$$\pi_{h,\alpha} : V \to V_{h,\alpha}; \quad i_{h,\alpha} : V_{h,\alpha} \hookrightarrow V$$

denote, respectively, the h-equivariant projection and the h-equivariant inclusions of the indicated spaces. Then $\pi_{h,\alpha} \circ W \circ i_{h,\alpha} \neq (0)$.

Remark. It is not the case that if Δ contains a subgroup Δ' which is big in the above sense, then Δ itself is necessarily big (bigger than big is not necessarily big). This is because condition (a) is not preserved under passage to a bigger group. To check condition (b), on the other hand, it clearly suffices to show that it holds for some subgroup $\Delta' \subset \Delta$.

Lemma 3.2. *Let F^+, τ, $\rho^n = \rho^n_{E,\ell}$, and r_τ be as in the statement of Corollary 2.5, with τ as in the proof of Lemma 2.4. Suppose $\ell > 4n - 1$. Then $\bar{r}_\tau(\Gamma_{F^+(\zeta_\ell)})$ is big in the sense of Definition 3.1.*

Proof. We begin by establishing notation. Write $\bar{\rho}^n = \bar{\rho}^n_{E,\ell}$. We define ad^0 as in Definition 3.1, and write

$$ad\ \bar{\rho}^n = ad^0\ \bar{\rho}^n \oplus 1, \quad ad\ \bar{\tau} = ad^0\ \bar{\tau} \oplus 1,$$

where 1 denotes the trivial representation. Then

$$ad^0\ \bar{r}_\tau = ad^0\ \bar{\rho}^n \otimes ad^0\ \bar{\tau} \oplus ad^0\ \bar{\rho}^n \oplus ad^0\ \bar{\tau}. \tag{3.2.1}$$

Let $\Delta = \bar{r}_\tau(\Gamma_{F^+(\zeta_\ell)})$, and define

$$\tilde{\Delta} = \bar{\rho}^n(\Gamma_{F^+(\zeta_\ell)}) \times \bar{\tau}((\Gamma_{F^+(\zeta_\ell)}).$$

The tensor product defines an exact sequence

$$1 \to C \to \tilde{\Delta} \to \Delta \to 1$$

where the kernel C maps injectively to the center of $\mathrm{GL}(2, \mathbb{F}_\ell)$, viewed as the group of linear transformations of the space of $\bar{\tau}$. In particular, the order of C is prime to ℓ. (In fact, as the referee pointed out, under our hypotheses one checks easily that C is trivial, but this makes no difference in the sequel.) Finally, let Δ_ρ denote the image of $\bar{\rho}^n(\Gamma_{F^+(\zeta_\ell)}) \subset \tilde{\Delta}$ in Δ. This is a normal subgroup of Δ isomorphic to the simple finite group $\mathrm{PSL}(2, \mathbb{F}_\ell)$, since n is odd and $\ell > 2n + 1$. Moreover, $\Delta_\tau := \Delta/\Delta_\rho$ is of order prime to ℓ, since the image of $\bar{\tau}$ is contained in the normalizer of a maximal torus. It follows from the inflation-restriction sequence that

$$H^1(\Delta, W) \xrightarrow{\sim} H^0(\Delta_\tau, H^1(\Delta_\rho, W)) \tag{3.2.2}$$

for any summand W of (3.2.1).

Proof of (a): We first note that Δ acts irreducibly on $\bar{r}_\tau$; hence $H^0(\Delta, ad^0\ \bar{r}_\tau) = (0)$. We apply (3.2.2) to show that $H^1(\Delta, W) = 0$ for each summand W of (3.2.1). Indeed, it suffices to show that $H^1(\mathrm{PSL}(2, \mathbb{F}_\ell), W) = 0$ for each W. But as a representation of $\mathrm{PSL}(2, \mathbb{F}_\ell)$, each W is a direct sum of copies of i-dimensional symmetric powers Sym^{i-1} of the standard representation $\mathrm{PSL}(2, \mathbb{F}_\ell)$, where i runs through (odd) integers at most equal to $2n - 1$. Since $\ell > 2n + 1$, it is well known that $H^1(\mathrm{PSL}(2, \mathbb{F}_\ell), Sym^{i-1}) = 0$ for $i \leq 2n - 1$.

Proof of (b): Let b denote a generator of the cyclic group $k(v)^\times \simeq \mu_{\ell-1}$, as in the proof of Lemma 2.4. As remarked above, if $\Delta' \subset \Delta$ is a subgroup that satisfies 3.1(b) for a given summand W of (3.2.1), then Δ also satistifes this property for the given W. We may thus assume $\mathrm{Im}(\bar\tau)$ is contained in the normalizer of a maximal torus T in $\mathrm{SL}(2, \mathbb{F}_\ell)$ and contains an element $t_0 \in T$ with the two distinct eigenvalues b^n, b^{-n} on $\bar\tau$, with corresponding eigenvectors v_1 and v_2, as well as the element $w \notin T$ with eigenvalues 1 and -1. With appropriate normalizations, we can assume the corresponding eigenvectors are $v_1 + v_2$ and $v_1 - v_2$, respectively. Write $k = \mathbb{F}_\ell$. We can write

$$\mathrm{ad}\,\bar\tau = k_+ \oplus k_- \oplus U,$$

where $U = \mathrm{Ind}_{\Gamma_L \cdot F^+}^{\Gamma_{F^+}} \bar\chi_\ell / \bar\chi_\ell^c$ and k_+ and k_- are representations of Γ_{F^+} that factor through $\mathrm{Gal}(L \cdot F^+/F^+)$, with the non-trivial element acting by the indicated sign. We thus have

$$\mathrm{Hom}(\bar\tau_\tau, \bar\tau_\tau) = \mathrm{ad}\,\bar\rho^n \otimes [k_+ \oplus k_- \oplus U]. \qquad (3.2.3)$$

For $i, j \in \{1, 2\})$, let $p_{i,j} \in \mathrm{End}(\bar\tau)$ be the endomorphism that takes v_i to v_j and vanishes on v_k if $k \neq i$. Then k_+ (resp. k_-) is spanned by $p_{1,1} + p_{2,2}$ (resp. $p_{1,1} - p_{2,2}$, whereas U is spanned by $p_{1,2}$ and $p_{2,1}$.

Let $t \in \mathrm{Im}(\bar\rho^n)$ be the image of an element of a split maximal torus of $\mathrm{SL}(2, \mathbb{F}_\ell)$, with n distinct eigenvalues under $\bar\rho^n$, as in the proof of Lemma 3.2 of [HST]. More precisely, we can take t to be the diagonal element $diag(b, b^{-1})$, so that $\bar\rho^n(t)$ has eigenvalues $b^{n-1}, b^{n-3}, \ldots, b^{1-n}$. The formula (3.2.3) expresses $\mathrm{Hom}(\bar\tau_\tau, \bar\tau_\tau)$ as a sum of four copies of $\mathrm{ad}\,\bar\rho^n$, as representation of t. Let h_0, resp h_w denote the image in Δ of $(t, t_0) \in \tilde\Delta$, resp. the image of (t, w). Since $\ell > 4n - 1$, $b^i \neq 1$ for any $i < 4n - 1$, hence no ratio of eigenvalues of $\bar\rho^n(t)$ equals a ratio of eigenvalues of $\bar\tau(t_0)$, nor of $\bar\tau(w)$. It follows that all the generalized eigenspaces of h_0 and h_w in $\bar\tau_\tau$ are of dimension 1. It was shown in the proof of Lemma 3.2 of [HST] that $\mathrm{ad}\,\bar\rho^n$ satisfies 3.1(b), with Δ replaced by $\mathrm{Im}(\bar\rho^n)$ and with t playing the role of h; here we use the hypothesis that $\ell > 2n + 1$. It thus follows that if W is an irreducible summand of $\mathrm{ad}\,\bar\rho^n \otimes [k_+ \oplus k_-]$, then 3.1(b) is satisfied for this W with $h = h_0$. On the other hand, if W is an irreducible summand of $\mathrm{ad}\,\bar\rho^n \otimes U$, then 3.1(b) is satisfied for this W with $h = h_w$. Indeed, it suffices to observe that the element $(p_{1,2} + p_{2,1}) \in U$ takes the eigenvector $v_1 + v_2$ to itself. $\qquad\square$

4. Removing τ

We fix an odd number n as above. The hypotheses of the earlier sections remain in force; in particular, we admit the Expected Theorems of Section 1.

Corollary 4.1. *Let $F^+(\bar\rho_{E,\ell})$ denote the splitting field of $\bar\rho_{E,\ell}$ and $M = L \cdot F^+(\bar\rho_{E,\ell})$. Then there is a totally real Galois extension $F'^{,+}/F^+$, linearly disjoint from M over F^+, with the property that for $i = 0, \ldots, n$, $r_{i, F'^{,+}} = r_i \mid_{\Gamma_F^{,+}}$ corresponds to a cuspidal automorphic representation Π_i*

of $\mathrm{GL}(2i, F'^{,+})$, and such that $r_{\tau, F'^{,+}} = r_\tau \mid_{\Gamma'^{,+}_F}$ corresponds to a cuspidal automorphic representation Π_τ of $\mathrm{GL}(2n, F'^{,+})$.

Let E be a number field, and let ρ be an ℓ-adic representation of Γ_E. We assume ρ to be pure of some weight w, as in Section 1; thus we can define $L^{\mathrm{norm}}(s, \rho)$. We say $L(s, \rho)$ (or $L^{\mathrm{norm}}(s, \rho)$) is *invertible* if it extends to a meromorphic function on $\mathbb{C}$ and if $L^{\mathrm{norm}}(s, \rho)$ has no zeroes for $Re(s) \geq 1$ and no poles for $Re(s) \geq 1$ except for a possible pole at $s = 1$.

Let $L' = L \cdot F'^{,+}$, and let $c \in \mathrm{Gal}(L'/F'^{,+})$ denote complex conjugation. The proof of the following lemma was devised with a great deal of help from Richard Taylor.

Lemma 4.2. *The representation Π_τ of $\mathrm{GL}(2n, F'^{,+})$ is isomorphic to the automorphic induction from L' of some cuspidal automorphic representation $\Pi_1(\tau)$ of $\mathrm{GL}(n, L')$. After possibly replacing $\Pi_1(\tau)$ by its Galois conjugate $\Pi_1(\tau)^c$, the tensor product $\Pi_1(\tau) \otimes \chi$ is isomorphic to its c-conjugate, hence descends to a cuspidal automorphic representation π of $\mathrm{GL}(n, F'^{,+})$.*

Proof. Let $\eta_{L'}$ be the quadratic character of $F'^{,+}$ associated to the extension L'. By construction, $\tau_\ell \mid_{\Gamma_{F'^{,+}}} \otimes \eta_{L'} \xrightarrow{\sim} \tau_\ell \mid_{\Gamma_{F'^{,+}}}$; hence

$$\Pi_\tau \otimes \eta_{L'} \xrightarrow{\sim} \Pi_\tau. \tag{4.2.1}$$

It follows from [AC, Chapter 3, Theorem 4.2 (b)] that there exists a cuspidal automorphic representation $\Pi_1(\tau)$ of $\mathrm{GL}(n, L')$ such that Π_τ is isomorphic to the automorphic induction of $\Pi_1(\tau)$ from L' to $F'^{,+}$. This means in particular that

$$\Pi_1(\tau)^c \not\simeq \Pi_1(\tau),$$

and

$$L(s, \rho_{\tau, L'}) = L(s, \Pi_{\tau, L'}) = L(s, \Pi_1(\tau)) L(s, \Pi_1(\tau)^c) \tag{4.2.2}$$

It follows from Corollary 2.5 that $L\left(s, \rho^m_{E, F}\right)$, and more generally that $L\left(s, \rho^m_{E, F} \otimes \xi_\ell\right)$, is entire for all even $m \leq 2n$, $F = F'^{,+}$ or $F = L'$, when ξ_ℓ is the ℓ-adic Galois avatar of an algebraic Hecke character ξ of $\mathbf{A}_F^\times$. The proof of Theorem 4.2 of [HST] shows that $L\left(s, \rho^m_{E, F} \otimes \xi_\ell\right)$ is invertible for *all* $m \leq 2n$ and for all algebraic Hecke characters ξ, for $F = F'^{,+}$ or $F = L'$. Of course $L(s, r_{\tau, F'^{,+}})$ is also entire and $L(s, r_{\tau, L'})$ is entire unless $n = 1$, which gives rise to the only possible pole at $s = 1$.

Consider the automorphic L-function

$$L(s) = L\left(s, \Pi_{\tau, L'} \times \Pi^\vee_{\tau, L'} \otimes (\chi/\chi^c)\right) \tag{4.2.3}$$

Comparing this to (4.2.2) we find that

$$L(s) = L(s, [\rho^n_{E, L'} \otimes (\chi \oplus \chi^c)] \otimes [\rho^{n, \vee}_{E, L'} \otimes (\chi^{-1} \oplus \chi^{c, -1})] \otimes [\chi/\chi^c])$$

$$= L(s, \left(\rho^n_{E, L'} \otimes \rho^{n, \vee}_{E, L'}\right) \otimes [(\chi/\chi^c) \oplus (\chi/\chi^c) \oplus (\chi/\chi^c)^2]) \cdot L\left(s, \rho^n_{E, L'} \otimes \rho^{n, \vee}_{E, L'}\right)$$

$$\tag{4.2.4}$$

Writing

$$\rho^n_{E,L'} \otimes \rho^{n,\vee}_{E,L'} = \bigoplus_{i=0}^{n-1} \rho^{2i+1}_{E,L'} \otimes \omega^{-i}_\ell$$

we see that the first factor of the last line of (4.2.4) is a product of invertible L-functions without poles at $s = 1$; the final factor of the last line has a simple pole at $s = 1$. Thus $L(s)$ has a simple pole at $s = 1$. But $L(s)$ is an automorphic L-function for $\mathrm{GL}(n) \times \mathrm{GL}(n)$. We rewrite $L(s)$ using (4.2.2):

$$L(s) = L(s, \Pi_1(\tau) \times \Pi_1(\tau)^\vee \otimes (\chi/\chi^c)) \cdot L(s, \Pi_1(\tau)^c \times \Pi_1(\tau)^{c,\vee} \otimes (\chi/\chi^c))$$
$$\cdot L(s, \Pi_1(\tau) \times \Pi_1(\tau)^{c,\vee} \otimes (\chi/\chi^c)) \cdot L(s, \Pi_1(\tau)^c \times \Pi_1(\tau)^\vee \otimes (\chi/\chi^c)).$$

Applying the Jacquet-Shalika classification theorem we see that exactly one of the factors has a simple pole, and in that case the factor is necessarily of the form $L(s, \Pi \times \Pi^\vee)$. Since $(\chi/\chi^c)_\infty$ is a character of infinite order, neither of the first two factors can have a pole; we must therefore either have

$$\Pi_1(\tau)^\vee \xrightarrow{\sim} \Pi_1(\tau)^{c,\vee} \otimes (\chi/\chi^c)$$

or

$$\Pi_1(\tau)^{c,\vee} \xrightarrow{\sim} \Pi_1(\tau)^\vee \otimes (\chi/\chi^c).$$

In other words, up to exchanging $\Pi_1(\tau)$ with $\Pi_1(\tau)^c$, we have

$$\Pi_1(\tau) \otimes \chi \xrightarrow{\sim} (\Pi_1(\tau) \otimes \chi)^c;$$

hence $\Pi_1(\tau) \otimes \chi$ descends to a cuspidal automorphic representation of $\mathrm{GL}(n, F'^{,+})$. This completes the proof.   □

Now $(\Pi_\tau)_\infty$ is cohomological, hence $(\Pi_\tau)_{L',\infty}$ is also cohomological. But $(\Pi_\tau)_{L',\infty}$ is represented as a subquotient of the representation of $\mathrm{GL}(2n, L'_\infty)$ induced from the representation $(\Pi_1(\tau))_\infty \otimes (\Pi_1(\tau)^c)_\infty$ of the Levi factor $\mathrm{GL}(n, L'_\infty) \times \mathrm{GL}(n, L'_\infty)$ of the relevant maximal parabolic, and it follows that $\Pi_1(\tau)_\infty$ is also cohomological. Thus $\Pi_1(\tau)$ satisfies condition (1) of Expected Theorem 1.2.

On the other hand, ρ_τ has a symplectic polarization with multiplier ω^{1-2n}_ℓ, by construction. It follows that the associated automorphic representation Π_τ is self dual. This property is preserved under base change to L'. It thus follows from (4.2.2) and the Jacquet-Shalika classification theorem that

$$\{\Pi_1(\tau), \Pi_1(\tau)^c\} = \{\Pi_1(\tau)^\vee, \Pi_1(\tau)^{c,\vee}\}$$

as sets. Thus either (a) $\Pi_1(\tau)$ satisfies condition (2) of Expected Theorem 1.2, or (b) $\Pi_1(\tau) \xrightarrow{\sim} \Pi_1(\tau)^\vee$.

Assume (a) holds. Then $\Pi_1(\tau)$ satisfies both conditions of Expected Theorem 1.2, hence is associated to an n-dimensional Galois representation $\rho_1(\tau)$ of $\Gamma_{L'}$. It follows from (4.2.2) that

$$\rho_1(\tau) \oplus \rho_1(\tau)^c \xrightarrow{\sim} \rho^n_{E,\ell} \otimes \chi_\ell \oplus \rho^n_{E,\ell} \otimes \chi^c_\ell \qquad (4.3)$$

Since $\rho_{E,\ell}^n$ is irreducible, it follows that it must be equal to either $\rho_1(\tau) \otimes \chi_\ell$ or $\rho_1(\tau) \otimes \chi_\ell^c$. In either case, $\rho_{E,\ell}^n$ is associated to a cuspidal automorphic representation $\Pi_{n,L'}$ of $\mathrm{GL}(n, L')$. Since $\rho_{E,\ell}^n$ descends to a representation of $\Gamma_{F',+}$, $\Pi_{n,L'}$ descends to a cuspidal automorphic representation Π_n of $\mathrm{GL}(n, F'^{,+})$. Thus $\rho_{E,\ell}^n$ is automorphic over $F'^{,+}$.

Assume (b) holds. Then the central character ξ of $\Pi_1(\tau)$ is also self-dual, i.e., $\xi = \xi^{-1}$. It follows from Lemma 4.2 that

$$\Pi_1(\tau) \otimes \chi \xrightarrow{\sim} \Pi_1(\tau)^c \otimes \chi^c.$$

Combining this with (b), we have

$$\xi \cdot \chi \circ \det = \xi^c \cdot \chi^c \circ \det.$$

This implies that χ/χ^c is self-dual, ie.,

$$(\chi^c)^2 = \chi^2.$$

But this is already false for the archimedean components. Thus (b) is impossible.

We have thus proved that $\rho_{E,\ell}^n$ is automorphic over $F'^{,+}$. More generally, Theorem 2.1 allows us to add new r_τ's of different dimensions $2n_i$, with n_i odd, to the list in Corollary 2.5. By adding $\rho_{E,\ell}^{2i+1} \otimes \tau_i$ to the list (we are free to vary the 2-dimensional τ_i if we like), we thus obtain our main theorem:

Theorem 4.4. *Assume the Expected Theorems of Section 1. Let F^+ be a totally real field, and let E be an elliptic curve over F^+. Let n be a positive integer. Then there is a finite totally real Galois extension $F'^{,+}/F$ and, for each positive integer $i \leq n$, a cuspidal automorphic representation Π_i of $\mathrm{GL}(i, F'^{,+})$, satisfying conditions (a) and (b) of Expected Theorem 1.2, such that*

$$\rho_{E,\ell}^i \big|_{\Gamma_{F',+}} = \rho_{\Pi_i,\ell}.$$

In particular, if $i > 1$, $L^{\mathrm{norm}}\left(s, \rho_{E,\ell,F'^{,+}}^i\right) = L(s, \Pi_i)$ is an entire function.

I repeat that we are not assuming that E have non-integral j-invariant; however, all statements are conditional on the Expected Theorems of Section 1.

5. Applications and generalizations

We continue to admit the Expected Theorems of Section 1. Let E and E' be two elliptic curves over F^+ without complex multiplication. Assume E and E' are not isogenous. It then follows from Faltings' isogeny theorem that $\rho_{E,\ell}$ and $\rho_{E',\ell}$ are not isomorphic as representations of Γ_{F^+} for all ℓ. Since the traces of Frob_v, for primes v of good reduction for E and E', are integers that determine $\rho_{E,\ell}$ and $\rho_{E',\ell}$ up to isomorphism, it follows that $\bar\rho_{E,\ell}$ and $\bar\rho_{E',\ell}$ are not isomorphic for sufficiently large ℓ. By Serre's theorem, if ℓ is sufficiently large, $\mathrm{Im}(\bar\rho_{E,\ell}) = \mathrm{Im}(\bar\rho_{E',\ell}) = \mathrm{GL}(2, \mathbb{F}_\ell)$.

Let m, m' be two positive integers. Applying Theorem 2.1, we obtain the analogue of Corollary 2.5 for the collection of representations $r_i = \rho_{E,\ell}^{2i}$, $r'_j = \rho_{E',\ell}^{2j}$, $1 \le i \le m$, $1 \le j \le m'$, together with $r_\tau = \rho_{E,\ell}^m \otimes \tau$ and $r'_{\tau'} = \rho^{m'} \otimes \tau'$ if m or m' is odd, provided $\bar\tau$ and $\bar\tau'$ are "big enough". Bearing in mind the results of Section 3, we have the following statement:

Proposition 5.1. *Let L be as in Section 3, and define*

$$\tau_\ell = \mathrm{Ind}_{\Gamma_L}^{\Gamma_\mathbb{Q}} \chi_\ell; \quad \tau'_\ell = \mathrm{Ind}_{\Gamma_L}^{\Gamma_\mathbb{Q}} \chi'_\ell$$

where χ (resp. χ') is a Hecke character with $\chi_\infty(z) = z^{-m}$, (resp. $\chi'_\infty(z) = z^{-m'}$) if m (resp. m') is odd. We assume

$$\chi\,|_{\mathbf{A}_\mathbb{Q}^\times} = |\bullet|_\mathbf{A}^{-m} \cdot \eta_L; \quad \chi'\,|_{\mathbf{A}_\mathbb{Q}^\times} = |\bullet|_\mathbf{A}^{-m'} \cdot \eta_L$$

Suppose there is a prime $\ell > \sup(2m + 1, 2m' + 1)$, unramified in F^+, such that

(i) E has good ordinary reduction at all primes w dividing ℓ
(ii) ℓ splits in L;
(iii) ℓ splits in $\mathbb{Q}(\zeta_{2i+1})$, $i = 1, \ldots, \sup(m - 1, m' - 1)$; in particular, $\ell \equiv 1 \pmod{mm'}$;
(iv) $\frac{\ell-1}{m} > 2$, $\frac{\ell-1}{m'} > 2$.

Let M be an arbitrary extension of F^+. Define r_i, r'_j, r_τ and $r_{\tau'}$ as above. Then there is a totally real Galois extension $F'^{,+}/F^+$, linearly disjoint from M over F^+, with the property that for $i = 1, \ldots, m$, (resp. $j = 1, \ldots, m'$) $r_{i,F'^{,+}} = r_i\,|_{\Gamma_F^{'+}}$ (resp. $r'_{j,F',+}$) corresponds to an automorphic representation Π_i of $\mathrm{GL}(2i, F'^{,+})$ ((resp. Π'_j of $\mathrm{GL}(2j, F'^{,+})$) and such that r_τ (resp. $r'_{\tau'}$) corresponds to an automorphic representation Π_τ of $\mathrm{GL}(2m, F'^{,+})$ (resp. $\Pi'_{\tau'}$ of $\mathrm{GL}(2m', F'^{,+})$. If $F'/F'^{,+}$ is a CM quadratic extension, then the base change $\Pi_{i,F'}$ (resp. $\Pi_{\tau,F'}$) has archimedean constituent isomorphic to $\Pi_{\infty,0}(2i, F')$ (resp. $\Pi_{\infty,0}(2n, F')$), and likewise for $\Pi'_{j,F'}$, $\Pi'_{\tau',F'}$.

The discussion of Section 4 applies to both Π_τ and Π'_τ, and we obtain the following strengthening of Theorem 4.4.

Theorem 5.2. *Assume the Expected Theorems of Section 1. Let F^+ be a totally real field, let E and E' be elliptic curves over F^+, and assume E and E' do not become isogenous over an abelian extension of F^+. Let m and m' be positive integers. Then there is a finite totally real Galois extension $F'^{,+}/F$ and, for each positive integer $i \le m$, (resp. $j \le m'$) a cuspidal automorphic representation Π_i of $\mathrm{GL}(i, F'^{,+})$ (resp. Π'_j of $\mathrm{GL}(j, F'^{,+})$) satisfying conditions (a) and (b) of Expected Theorem 1.2, such that*

$$\rho_{E,\ell}^i\,|_{\Gamma_{F'^{,+}}} = \rho_{\Pi_i,\ell}; \quad \rho_{E',\ell}^j\,|_{\Gamma_{F',+}} = \rho_{\Pi'_j,\ell}$$

In particular, if $m \cdot m' > 1$,

$$L^{\mathrm{norm}}\left(s, \rho^m_{E,\ell,F',+} \otimes \rho^{m'}_{E',\ell,F',+}\right) = L\left(s, \Pi_m \times \Pi'_{m'}\right)$$

is an entire function.

The Rankin-Selberg L-function has no poles if $m \neq m'$; if $m = m'$ it has a pole if and only if $\Pi'_{m'} \xrightarrow{\sim} \Pi^\vee_m$. which implies that the corresponding Galois representations $\rho^m_{E,\ell}$ and $\rho^m_{E',\ell}$ are isomorphic. The kernel of the map of the standard 2-dimensional representation of $\mathrm{GL}(2)$ to its mth symmetric power is finite and contained in the center. If $\rho^m_{E,\ell} \xrightarrow{\sim} \rho^m_{E',\ell}$, it thus follows that the corresponding adjoint representations $\mathrm{ad}\, \rho_{E,\ell}$ and $\mathrm{ad}\, \rho_{E',\ell}$ are isomorphic, hence that there exists an abelian character η, necessarily finite, such that $\rho_{E',\ell} \xrightarrow{\sim} \rho_{E,\ell} \otimes \eta$. Thus $\rho_{E',\ell}$ and $\rho_{E,\ell}$ become isomorphic over a finite extension of F^+, hence E and E' are isogenous by Faltings' theorem.

Using Brauer's theorem, as in the proof of Theorem 4.2 of [HST], we then obtain:

Theorem 5.3. *Assume the Expected Theorems of Section 1. Let F^+ be a totally real field, let E and E' be elliptic curves over F^+, and assume E and E' do not become isogenous over an abelian extension of F^+. Let m and m' be positive integers. Then the L-function $L\left(s, \rho^m_{E,\ell} \otimes \rho^{m'}_{E',\ell}\right)$ is invertible and satisfies the expected functional equation.*

Proof. This is obtained from Theorem 5.2 by applying Brauer's theorem, as in the proof of Theorem 4.2 of [HST]. It suffices to mention that the non-vanishing of the Rankin-Selberg L-function along the line $\mathrm{Re}(s) = 1$ of a pair of cuspidal automorphic representations (with unitary central characters) is due in general to Shahidi [Shi63]. $\qquad\square$

Finally, here is the precise statement of the question of Mazur and Katz mentioned in the introduction, together with the affirmative response. Recall the notation k_v and q_v of Section 1.

Theorem 5.4. *Assume the Expected Theorems of Section 1. Let F^+ be a totally real field, let E and E' be elliptic curves over F^+, and assume E and E' do not become isogenous over an abelian extension of F^+. For any prime v of F^+ where E and E' both have good reduction, we let*

$$|E(k_v)| = \left(1 - q_v^{\frac{1}{2}} e^{i\phi_v}\right)\left(1 - q_v^{\frac{1}{2}} e^{-i\phi_v}\right)$$

$$|E'(k_v)| = \left(1 - q_v^{\frac{1}{2}} e^{i\psi_v}\right)\left(1 - q_v^{\frac{1}{2}} e^{-i\psi_v}\right)$$

where $\phi_v, \psi_v \in [0, \pi]$.

Then the pairs $(\phi_v, \psi_v) \in [0, \pi] \times [0, \pi]$ are uniformly distributed with respect to the measure

$$\frac{4}{\pi^2} \sin^2\phi \, \sin^2\psi \, d\phi \, d\psi.$$

Proof. Theorem 5.4 follows directly from Theorem 5.3 by the argument in [S, Appendix to Section I]. □

6. Concluding remarks

The author and Richard Taylor have independently noticed that tensoring with an induced representation from a Hecke character may be useful in other situations. For example, let f be an elliptic modular form of weight k, $\rho_{f,\ell} : \Gamma_{\mathbb{Q}} \to \mathrm{GL}(2, \bar{\mathbb{Q}}_\ell)$ the associated two-dimensional Galois representation, and let $\rho_{f,\ell}^n = \mathrm{Sym}^{n-1}\rho_{f,\ell}$. There is no hope of applying the potential modularity technique of [HST] to $\rho_{f,\ell}^n$ if $k > 2$: the series of Hodge-Tate weights at ℓ has gaps for all n, and Griffiths transversality implies it is impossible to obtain families of positive dimension of motives with such Hodge-Tate numbers. However, if k is odd, one can choose a Hecke character χ of an abelian CM extension $L/\mathbb{Q}$ of degree $k-1$ with infinity type so chosen that $\rho_{f,\ell}^n \otimes \mathrm{Ind}_{\Gamma_L}^{\Gamma_{\mathbb{Q}}} \chi_\ell$ has an unbroken series of Hodge-Tate weights. (If k is even one takes L of degree $2(k-1)$.)

Two serious obstacles remain. In the first place, the constructions in Section 4 require that we know in advance that the symmetric power L-functions are invertible, and this information is not available *a priori* for $k > 2$. In the second place, the arguments for finding rational points on moduli spaces over number fields unramified at ℓ break down in higher weights.

References

[AC] J. ARTHUR, L. CLOZEL, *Simple Algebras, Base Change, and the Advanced Theory of the Trace Formula*, Annals of Mathematics Studies **120** (1989).

[C1] L. CLOZEL, *Motifs et formes automorphes: applications du principe de fonctorialité*, in L. Clozel and J. S. Milne, eds., Automorphic Forms, Shimura Varieties, and *L*-functions, New York: Academic Press (1990), Vol I, 77–160.

[C2] L. CLOZEL, *Représentations Galoisiennes associées aux représentations automorphes autoduales de* GL*(n)*, Publ. Math. I.H.E.S., **73**, 97–145 (1991).

[C3] L. CLOZEL, *On the cohomology of Kottwitz's arithmetic varieties*, Duke Math. J., **72** (1993) 757–795.

[CHT] L. CLOZEL, M. HARRIS, R. TAYLOR, *Automorphy for some ℓ-adic lifts of automorphic mod ℓ Galois representations*, Publ. Math. I.H.E.S., **108** (2008) 1–181.

[CL] L. CLOZEL, J.-P. LABESSE, *Changement de base pour les représentations cohomologiques de certains groupes unitaires*, in [L], 119–133.

[H] M. HARRIS, Construction of automorphic Galois representations, manuscript (2006); draft of article for Stabilisation de la formule des traces, variétés de Shimura, et applications arithmétiques, Book 3.

[HST] M. HARRIS, N. SHEPHERD-BARRON, R. TAYLOR, *A family of Calabi-Yau varieties and potential automorphy*, Annals of Math. (in press).

[HT] M. HARRIS, R. TAYLOR, *The geometry and cohomology of some simple Shimura varieties*, Annals of Math. Studies, **151** (2001).

[Ko] R. KOTTWITZ, *On the λ-adic representations associated to some simple Shimura varieties*, Invent. Math., **108**, 653–665 (1992).

[L] J.-P. LABESSE, Cohomologie, stabilisation et changement de base, Astérisque, **257** (1999).

[S] J.-P. SERRE, *Abelian ℓ-adic representations and elliptic curves*, New York: Benjamin (1968).

[Shi63] F. SHAHIDI, *On certain L-functions*, Am. J. Math., **103** (1981) 297–355.

[T] R. TAYLOR, *Automorphy for some ℓ-adic lifts of automorphic mod ℓ Galois representations*, II, Publ. Math. I.H.E.S., **108** (2008) 183–239.

[T02] R. TAYLOR, *Remarks on a conjecture of Fontaine and Mazur*, J. Inst. Math. Jussieu, **1** (2002) 1–19.

[TY] R. TAYLOR, T. YOSHIDA, *Compatibility of local and global Langlands correspondences*, J.A.M.S., **20** (2007), 467–493.

Cyclic Homology with Coefficients

D. Kaledin*

Steklov Math Institute, Moscow, Russia
kaledin@mccme.ru

To Yu. I. Manin, the founder, on the occasion of his 70th birthday

Summary. We propose a category which can serve as the category of coefficients for the cyclic homology $HC_*(A)$ of an associative algebra A over a field k. The construction is categorical in nature, and essentially uses only the tensor category A-bimod of A–*bimodules*; objects of our category are A–bimodules with an additional structure. We also generalize the construction to a more general tensor k-linear tensor category C instead of A-bimod. We add some remarks about Getzler's version of the Gauss–Manin connection for the periodic cyclic homology $HP_*(A)$.

Key words: cyclic homology, Gauss–Manin connection, tensor categories

2000 Mathematics Subject Classifications: 19D55, 18E25, 18G10

Introduction

Ever since it was discovered in 1982 by A. Connes [C1] and B. Tsygan [Ts], cyclic homology has occupied a strange place in the realm of homological algebra. Normally in homological algebra problems, one expects to start from some data, for instance a topological space X, then construct some abelian category, such as the category of sheaves on X, and then define the cohomology of X by computing the derived functors of some natural functor, such as the global sections functor $\Gamma(X, -)$. Admittedly, this is a modern formulation, but it was certainly current already in 1982. Cyclic homology starts with an associative algebra A, and defines its homology groups $HC.(A)$, but there are absolutely no derived functors in sight. Originally, groups $HC.(A)$ were defined as the homology of an explicit complex, which anyone trained to use triangulated categories cannot help but take as an insult. Later, A.

---

*Partially supported by CRDF grant RUM1-2694-MO05.

Y. Tschinkel and Y. Zarhin (eds.), *Algebra, Arithmetic, and Geometry*,
Progress in Mathematics 270, DOI 10.1007/978-0-8176-4747-6_2,
© Springer Science+Business Media, LLC 2009

Connes [C2] improved on the definition by introducing the abelian category of so-called cyclic vector spaces. However, the passage from A to its associated cyclic vector space $A_\#$ is still done by an explicit *ad hoc* formula. It is as if we were to know the bar-complex that computes the homology of a group, without knowing the definition of the homology of a group.

This situation undoubtedly irked many people over the years, but to the best of my knowledge, no satisfactory solution has been proposed, and it may not exist; indeed, many relations to the de Rham homology notwithstanding, it is not clear whether cyclic homology properly forms a part of homological algebra at all (to the point that for instance in [FT], the word "homology" is not used at all for $HC_\bullet(A)$, and it is called instead *additive K-theory* of A). In the great codification of homological algebra done in [GM1], cyclic homology appears only in the exercises. This is not surprising, since the main unifying idea of [GM1] is the ideology of "linearization": homological algebra linearizes geometry, just as functional analysis used to do 50 years ago; triangulated categories and adjoint functors are modern-day versions of Banach spaces and adjoint linear operators. This has been an immensely successful and clarifying point of view in general, but $HC_\bullet(A)$ sticks out on a complete tangent: there is simply no natural place for it in this framework.

This paper arose as one more attempt to propose a solution to the difficulty: to find a natural triangulated category where $HC_\bullet(-)$ would be able to live with a certain level of comfort (and with all the standard corollaries such as the notion of cyclic homology with coefficients, the ability to compute cyclic homology by whatever resolution is convenient, not just the bar resolution, and so on).

In a sense, our attempt has been successful: we define a triangulated category that can serve as the natural "category of coefficients" for cyclic homology of an algebra A, and we prove the comparison theorem that shows that when the coefficients are trivial, the new definition of cyclic homology is equivalent to the old one. In fact, the algebra A enters into the construction only through the category A-bimod of A-bimodules; we also show how to generalize the construction so that A-bimod is replaced with a more general tensor abelian category $\mathcal{C}$.

From a different point of view, though, out attempt failed miserably: the correspondence $A \mapsto A_\#$, being thrown out of the window, immediately returns through the door in a new and "higher-level" disguise: it is now applied not to the algebra A, but to the tensor category $\mathcal{C} = A$-bimod. Then in practice, the freedom to choose an arbitrary resolution to compute the derived functors leads, in our approach to $HC_\bullet(-)$, to complexes that are even larger than the original complex, and at some point the whole exercise starts to look pointless.

Still, we believe that all said and done, some point can be found, and some things are clarified in our approach; one such thing is, for instance, the version of Gauss–Manin connection for cyclic homology discovered by E. Getzler [Ge]. Moreover, we do propose a definition of cyclic homology that – makes sense for a general tensor category; and in some particular questions, even the computations can be simplified. As for the presence of the $A_{\#}$-construction, this might be in the nature of things, after all: not a bug in the theory, but a necessary feature. However, we leave it to the reader to be the judge.

The paper is organized as follows. In Section 1 we recall A. Connes's second definition of cyclic homology, which uses the cyclic category Λ; we also recall some facts about homology of small categories that we will need. We have tried to give only the absolute minimum; the reader not familiar with the material will have to consult the references. In Section 2 we introduce our main object: the notion of a *cyclic bimodule* over an associative algebra A, and the derived category of such bimodules. We also introduce cyclic homology $HC_{\bullet}(A, M)$ with coefficients in a cyclic bimodule M. In Section 3 we give a very short derivation of the Gauss–Manin connection; strictly speaking, the language of cyclic bimodules is not needed for this, but we believe that it shows more clearly what is really going on. In Section 4, we show how to replace the category A-bimod everywhere with a more general tensor abelian category $\mathcal{C}$. Section 5 is a postface, or a "discussion" (as is done in medical journals); we discuss some of the further things one might (and should) do with cyclic bimodules, and how to correct some deficiencies of the theory developed in Sections 2 and 4.

Acknowledgments.

In the course of this work, I have benefited greatly from discussions with A. Beilinson, E. Getzler, V. Ginzburg, A. Kuznetsov, N. Markarian, D. Tamarkin, and B. Tsygan. I am grateful to Northwestern Univeristy, where part of this work was done, and where some of the results were presented in seminars, with great indulgence from the audience toward the unfinished state of the work. And last but not least, it is a great pleasure and a great opportunity to dedicate this paper to Yuri Ivanovich Manin on his birthday. Besides all the usual things, I would like to stress that it is the book [GM1], and [GM2] to a lesser extent, that shaped the way we look at homological algebra today, at least "we" of my generation and of the Moscow school. Without Manin's decisive influence, this paper certainly would not have appeared (as in fact at least half of the papers I have written).

1 Recollection on cyclic homology.

We start by recalling, extremely briefly, A. Connes' approach to cyclic homology, which was originally introduced in [C2] (for detailed overviews, see, e.g., [L, Section 6] or [FT, Appendix]; a brief but complete exposition using the same language and notation as in this paper can be found in [Ka, Section 1]).

Connes' approach relies on the technique of homology of small categories. Fix a base field k. Recall that for every small category Γ, the category $\mathrm{Fun}(\Gamma, k)$ of functors from Γ to the category k-$Vect$ of k-vector spaces is an abelian category with enough projectives and enough injectives, with derived category $\mathcal{D}(\Gamma, k)$. For any object $E \in \mathrm{Fun}(\Gamma, k)$, the homology $H_\bullet(\Gamma, E)$ of the category Γ with coefficients in E is by definition the derived functor of the direct limit functor

$$\varinjlim_{\Gamma} : \mathrm{Fun}(\Gamma, k) \to k\text{-}Vect.$$

Analogously, the cohomology $H^\bullet(\Lambda, E)$ is the derived functor of the inverse limit $\varprojlim_{\Gamma}$. Equivalently,

$$H^\bullet(\Gamma, E) = \mathrm{Ext}^\bullet(k, E),$$

where $k \in \mathrm{Fun}(\Gamma, k)$ is the constant functor (all objects in Γ go to k, all maps go to identity). In particular, $H^\bullet(\Gamma, k)$ is an algebra. For any $E \in \mathrm{Fun}(\Gamma, k)$, the cohomology $H^\bullet(\Gamma, E)$ and the homology $H_\bullet(\Gamma, E)$ are modules over $H^\bullet(\Gamma, k)$.

We also note, although it is not needed for the definition of cyclic homology, that for any functor $\gamma : \Gamma' \to \Gamma$ between two small categories, we have the pullback functor $\gamma^* : \mathrm{Fun}(\Gamma, k) \to \mathrm{Fun}(\Gamma', k)$, and for any $E \in \mathrm{Fun}(\Gamma, k)$, we have natural maps

$$H_\bullet(\Gamma', \gamma^* E) \to H_\bullet(\Gamma, E), \qquad H^\bullet(\Gamma, E) \to H^\bullet(\Gamma', \gamma^* E). \tag{1.1}$$

Moreover, the pullback functor γ^* has a left adjoint $\gamma_! : \mathrm{Fun}(\Gamma', k) \to \mathrm{Fun}(\Gamma, k)$ and a right-adjoint $f_* : \mathrm{Fun}(\Gamma', k) \to \mathrm{Fun}(\Gamma, k)$, known as the left and right Kan extensions. In general, $f_!$ is right-exact, but it need not be left-exact. We will need one particular case in which it is exact. Assume given a covariant functor $V : \Gamma \to Sets$ from a small category Γ to the category of sets, and consider the category Γ' of pairs $\langle [a], v \rangle$ of an object $[a] \in \Gamma$ and an element $v \in V([a])$ (maps in Γ' are those maps $\gamma : [a] \to [a']$ that send $v \in V([a])$ to $v' \in V([a'])$). Such a category is known as a *discrete cofibration* over Γ associated to V; see [Gr]. Then the Kan extension $f_!$ associated to the forgetful functor $f : \Gamma' \to \Gamma$, $\langle [a], v \rangle \mapsto [a]$ is exact, and is easy to compute: for any $E \in \mathrm{Fun}(\Gamma', k)$ and $[a] \in \Gamma$, we have

$$f_! E([a]) = \bigoplus_{v \in V([a])} E(\langle [a], v \rangle). \tag{1.2}$$

Moreover, for any $E \in \mathrm{Fun}(\Gamma, k)$, this imediately gives the projection formula:

$$f_! f^* E \cong E \otimes F_! k, \tag{1.3}$$

where as before, $k \in \mathrm{Fun}(\Gamma', k)$ stands for the constant functor.

For applications to cyclic homology, one starts by introducing the *cyclic category* Λ. This is a small category whose objects $[n]$ are numbered by positive integers $n \geq 1$. One thinks of an object $[n]$ as a circle S^1 with n distinct marked points; we denote the set of these points by $V([n])$. The set of maps $\Lambda([n'], [n])$ from $[n']$ to $[n]$ is then the set of homotopy classes of continuous maps $f : S^1 \rightarrow S^1$ such that

- f has degree 1, sends marked points to marked points, and is nondecreasing with respect to the natural cyclic order on S^1 (that is, if a point $a \in S^1$ lies between points b and c when one counts clockwise, then the same is true for $f(a)$, $f(b)$, and $f(c)$).

In particular, we have $\Lambda([1], [n]) = V([n])$. This topological description of the cyclic category Λ is easy to visualize, but there are also alternative combinatorial descriptions (e.g., [GM1, Exercise II.1.6], [L, Section 6], or [FT, A.2], retold in [Ka, Section 1.4]). All the descriptions are equivalent. Objects in $\text{Fun}(\Lambda, k)$ are usually called *cyclic vector spaces*.

The cyclic category Λ is related to the more familiar *simplicial category* Δ^{opp}, the opposite to the category Δ of finite nonempty linearly ordered sets. To understand the relation, consider the discrete cofibration $\Lambda_{[1]}/\Lambda$ associated to the functor $V : \Lambda \rightarrow \text{Sets}$; equivalently, $\Lambda_{[1]}$ is the category of objects $[n]$ in Λ equipped with a map $[1] \rightarrow [n]$. Then it is easy to check that $\Lambda_{[1]}$ is equivalent to the category Δ^{opp}. From now on, we will abuse notation and identify $\Lambda_{[1]}$ and Δ^{opp}. We then have a natural projection $\Delta^{opp} = \Lambda_{[1]} \rightarrow \Lambda$, $\langle [n], v \rangle \mapsto [n]$, which we denote by $j : \Delta^{opp} \rightarrow \Lambda$.

For any cyclic k-vector space $E \in \text{Fun}(\Lambda, k)$, we have its restriction $j^*E \in \text{Fun}(\Delta^{opp}, E)$, a simplicial vector space. One defines the cyclic homology $HC_\bullet(E)$ and the Hochschild homology $HH_\bullet$ of E by

$$HC_\bullet(E) \stackrel{\text{def}}{=} H_\bullet(\Lambda, E), \qquad HH_\bullet(E) \stackrel{\text{def}}{=} H_\bullet(\Delta^{opp}, j^*E).$$

By (1.1), we have a natural map $HH_\bullet(E) \rightarrow HC_\bullet(E)$ (moreover, since $j : \Delta^{opp} \rightarrow \Lambda$ is a discrete cofibration, the Kan extension $j_!$ is exact, so that we have $HH_\bullet(E) \cong HC_\bullet(j_!j^*E)$, and the natural map is induced by the adjunction map $j_!j^*E \rightarrow E$). It has been shown by A. Connes that this map fits into a long exact sequence

$$HH_\bullet(E) \longrightarrow HC_\bullet(E) \stackrel{u}{\longrightarrow} HC_{\bullet-2}(E) \longrightarrow . \qquad (1.4)$$

Here the map u is the so-called *periodicity map* on $HC_\bullet(E)$: one shows that the algebra $H^\bullet(\Lambda, k)$ is isomorphic to the polynomial algebra $k[u]$ in one generator u of degree 2, and the periodicity map on homology is simply the action of this generator. This allows one to define a third homological invariant, the *periodic cyclic homology* $HP_\bullet(E)$; to do it, one inverts the periodicity map.

Definition 1.1. *For any cyclic k-vector space $E \in \mathrm{Fun}(\Lambda, k)$, the periodic cyclic homology of E is defined by*

$$HP_\bullet(E) = \lim_{\substack{\leftarrow \\ u}}{}^\bullet HC_\bullet(E),$$

where $\lim_{\leftarrow}^\bullet$ *denotes the derived functor of the inverse limit* $\lim_{\leftarrow}$.

Assume now that we are given an associative unital algebra A over k. To define its cyclic homology, we associate to A a canonical cyclic vector space $A_\#$ in the following way. We set $A_\#([n]) = A^{\otimes V([n])}$, the tensor product of n copies of the vector space A numbered by marked points $v \in V([n])$. Then for any map $f \in \Lambda([n'], [n])$, we define

$$A_\#(f) = \bigotimes_{v \in V([n])} m_{f^{-1}(v)} : A^{\otimes V([n'])} = \bigotimes_{v \in V([n])} A^{\otimes f^{-1}(v)} \to A^{\otimes V([n])}, \quad (1.5)$$

where for any linearly ordered finite set S, $m_S : A^{\otimes S} \to A$ is the canonical multiplication map induced by the associative algebra structure on A (and if S is empty, we set $A^{\otimes S} = k$, and m_S is the embedding of the unity). This is obviously compatible with compositions, and it is well defined since for any $v \in V([n])$, its preimage $f^{-1} \subset V([m])$ carries a natural linear order induced by the orientation of the circle S^1.

Definition 1.2. *For any associative unital algebra A over k, its Hochschild, cyclic, and periodic cyclic homologies $HH_\bullet(A)$, $HC_\bullet(A)$, $HP_\bullet(A)$ are defined as the corresponding homologies of the cyclic k-vector space $A_\#$:*

$$HH_\bullet(A) \overset{\mathrm{def}}{=} HH_\bullet(A_\#), \quad HC_\bullet(A) \overset{\mathrm{def}}{=} HC_\bullet(A_\#), \quad HC_\bullet(P) \overset{\mathrm{def}}{=} HP_\bullet(A_\#).$$

2 Cyclic bimodules.

Among all the homology functors introduced in Definition 1.2, Hochschild homology is the most accessible, and this is because it has another definition: for any associative unital algebra A over k, we have

$$HH_\bullet = \mathrm{Tor}^\bullet_{A^{opp} \otimes A}(A, A), \quad (2.1)$$

where $\mathrm{Tor}^\bullet$ is taken over the algebra $A^{opp} \otimes A$ (here A^{opp} denotes A with the multiplication taken in the opposite direction).

This has a version with coefficients: if M is a left module over $A^{opp} \otimes A$, in other words, an A-bimodule, one defines Hochschild homology of A with coefficients in M by

$$HH_\bullet(A, M) = \mathrm{Tor}^\bullet_{A^{opp} \otimes A}(M, A). \quad (2.2)$$

The category A-bimod of A-bimodules is a unital (nonsymmetric) tensor category, with tensor product $- \otimes_A -$ and the unit object A. Hochschild homology is a homological functor from A-bimod to k-Vect.

To obtain a small category interpretation of $HH_\bullet(A, M)$, one notes that for any $n, n' \geq 0$, the A-bimodule structure on M induces a multiplication map

$$A^{\otimes n} \otimes M \otimes A^{\otimes n'} \to M.$$

Therefore, if to any $\langle [n], v \rangle \in \Delta^{opp}$ we associate the k-vector space

$$M_{\#}^{\Delta}([n]) = M \otimes A^{\otimes (V([n]) \setminus \{v\})}, \qquad (2.3)$$

with M filling the place corresponding to $v \in V([n])$, then (1.5) makes perfect sense for those maps $f : [n'] \to [n]$ that preserve the distinguished points. Thus to any $M \in A$-bimod, we can associate a simplicial k-vector space $M_{\#}^{\Delta} \in \mathrm{Fun}(\Delta^{opp}, k)$. In the particular case $M = A$, we have $A_{\#}^{\Delta} = j^* A_{\#}$.

Lemma 2.1. *For any $M \in A$-bimod, we have a canonical isomorphism*

$$HH_\bullet(A, M) \cong H_\bullet(\Delta^{opp}, M_{\#}^{\Delta}). \qquad (2.4)$$

Proof. It is well known that for any simplicial k-vector space E, the homology $H_\bullet(\Delta^{opp}, E)$ can be computed by the standard complex of E (that is, the complex with terms $E([n])$ and the differential $d = \sum_i (-1)^i d_i$, where d_i are the face maps). In particular, $H_0(\Delta^{opp}, M_{\#}^{\Delta})$ is the cokernel of the map $d : A \otimes M \to M$ given by $d(a \otimes m) = am - ma$. The natural projection $M \to M \otimes_{A^{opp} \otimes A} A$ obviously factors through this cokernel, so that we have a natural map

$$\rho_0 : H_0\left(\Delta^{opp}, M_{\#}^{\Delta}\right) \to HH_0(A, M).$$

Both sides of (2.4) are homological functors in M, and $HH_\bullet(A, M)$ is a universal homological functor (= the derived functor of $HH_0(A, M)$); therefore the map ρ_0 extends to a map $\rho_\bullet : H_\bullet(\Delta^{opp}, M_{\#}^{\Delta}) \to HH_\bullet(A, M)$. To prove that $\rho_\bullet$ is an isomorphism for any M, it suffices to prove it when M is free over $A^{opp} \otimes A$, or in fact, when $M = A^{opp} \otimes A$. Then on the one hand, $HH_0(A, M) = A$, and $HH_i(A, M) = 0$ for $i \geq 1$. And on the other hand, the standard complex associated to the simplicial k-vector space $(A^{opp} \otimes A)_{\#}^{\Delta}$ is just the usual bar resolution of the diagonal A-bimodule A.

It is more or less obvious that for an arbitrary $M \in A$-bimod, $M_{\#}^{\Delta}$ does not extend to a cyclic vector space: in order to be able to define $HC_\bullet(A, M)$, we have to equip the bimodule M with some additional structure. To do this, we want to use the tensor structure on A-bimod. The slogan is the following:

- To find a suitable category of coefficients for cyclic homology, we have to repeat the definition of the cyclic vector space $A_{\#} \in \mathrm{Fun}(\Lambda, k)$, but replace the associative algebra A in this definition with the tensor category A-bimod.

Let us explain what this means.

First, consider an arbitrary associative unital monoidal category $\mathcal{C}$ with unit object I (at this point, not necessarily abelian). For any integer n, we have the Cartesian product $\mathcal{C}^n = \mathcal{C} \times \mathcal{C} \times \cdots \times \mathcal{C}$. Moreover, the product on $\mathcal{C}$ induces a product functor

$$m : \mathcal{C}^n \to \mathcal{C},$$

where if $n = 0$, we let $\mathcal{C}^n = \mathsf{pt}$, the category with one object and one morphism, and let $m : \mathsf{pt} \to \mathcal{C}$ be the embedding of the unit object. More generally, for any finite linearly ordered set S with n elements, we have a product functor $m_S : \mathcal{C}^S \to \mathcal{C}$, where $\mathcal{C}^S = \mathcal{C}^n$ with multiples in the product labeled by elements of S. Then for any $[n], [n'] \in \Lambda$, and any $f : [n'] \to [n]$, we can define a functor $f_! : \mathcal{C}^{V([n'])} \to \mathcal{C}^{V([n])}$ by the same formula as in (1.5):

$$f_! = \prod_{v \in V([n])} m_{f^{-1}(v)} : \mathcal{C}^{V([n'])} = \prod_{v \in V([n])} \mathcal{C}^{f^{-1}(v)} \to \mathcal{C}^{V([n])}. \qquad (2.5)$$

The natural associativity isomorphism for the product on $\mathcal{C}$ induces natural isomorphisms $(f \circ f')_! \cong f_! \circ f'_!$, and one checks easily that they satisfy natural compatibility conditions. All in all, setting $[n] \mapsto \mathcal{C}^{V([n])}$, $f \mapsto f_!$ defines a weak functor (a.k.a. Lax functor, a.k.a. 2-functor, a.k.a. pseudofunctor in the original terminology of Grothendieck) from Λ to the category of categories. Informally, we have a "cyclic category."

To work with weak functors, it is convenient to follow Grothendieck's approach in [Gr]. Namely, instead of considering a weak functor directly, we define a *category* $\mathcal{C}_\#$ in the following way: its objects are pairs $\langle [n], M_n \rangle$ of an object $[n]$ of Λ and an object $M_n \in \mathcal{C}^n$, and morphisms from $\langle [n'], M_{n'} \rangle$ to $\langle [n], M_n \rangle$ are pairs $\langle f, \iota_f \rangle$ of a map $f : [n'] \to [n]$ and a bimodule map $\iota_f : f_!(M_{n'}) \to M_n$. A map $\langle f, \iota_f \rangle$ is called *co-Cartesian* if ι_f is an isomorphism. For the details of this construction, in particular, for the definition of the composition of morphisms, we refer the reader to [Gr].

The category $\mathcal{C}_\#$ comes equipped with a natural forgetful projection $\tau : \mathcal{C}_\# \to \Lambda$, and this projection is a *cofibration* in the sense of [Gr]. A *section* of this projection is a functor $\sigma : \Lambda \to \mathcal{C}_\#$ such that $\tau \circ \sigma = \mathrm{id}$ (since Λ is small, there is no harm in requiring that two functors from Λ to itself be equal, not just isomorphic). These sections obviously form a category which we denote by $\mathsf{Sec}(\mathcal{C}_\#)$. Explicitly, an object $M_\# \in \mathsf{Sec}(\mathcal{C}_\#)$ is given by the following:

(i) a collection of objects $M_n = M_\#([n]) \in \mathcal{C}^n$, and
(ii) a collection of transition maps $\iota_f : f_! M_{n'} \to M_n$ for any n, n', and $f \in \Lambda([n'], [n])$,

subject to natural compatibility conditions.

A section $\sigma : \Lambda \to \mathcal{C}_\#$ is called co-Cartesian if $\sigma(f)$ is a cocartesian map for any $[n], [n'] \in \Lambda$ and $f : [n'] \to [n]$, equivalently, a section is cocartesian if all the transition maps ι_f are isomorphisms. co-Cartesian sections form a full subcategory $\mathsf{Sec}_{cart}(\mathcal{C}_\#)$.

Lemma 2.2. *The category* $\mathsf{Sec}_{cart}(\mathcal{C}_{\#})$ *of co-Cartesian objects* $M_{\#} \in \mathsf{Sec}(\mathcal{C}_{\#})$ *is equivalent to the category of the following data:*

(i) an object $M = M_{\#}([1]) \in \mathcal{C}$, *and*
(ii) an isomorphism $\tau : I \times M \to M \times I$ *in the category* $\mathcal{C}^2 = \mathcal{C} \times \mathcal{C}$,

such that if we denote by τ_{ij} *the endomorphism of* $I \times I \times M \in \mathcal{C}^3$ *obtained by applying* τ *to the* i-*th and* j-*th multiples, we have* $\tau_{31} \circ \tau_{12} \circ \tau_{23} = \mathrm{id}$.

Proof. Straightforward and left to the reader.

Thus the natural forgetful functor $\mathsf{Sec}_{\mathrm{cart}}(\mathcal{C}_{\#}) \to \mathcal{C}$, $M_{\#} \mapsto M_{\#}([1])$ is faithful: an object in $\mathsf{Sec}_{cart}(\mathcal{C}_{\#})$ is given by $M_{\#}([1])$ plus some extra structure on it, and all the higher components $M_{\#}([n])$, $n \geq 2$, together with the transition maps ι_f, can be recovered from $M_{\#}([1])$ and this extra structure.

Return now to the abelian situation: we are given an associative unital algebra A over a field k, and our monoidal category is $\mathcal{C} = A$-bimod, with the natural tensor product. Then for every n, the product A-bimod^n has a fully faithful embedding A-bimod$^n \to A^{\otimes n}$-bimod, $M_1 \times M_2 \times \cdots \times M_n \mapsto M_1 \boxtimes M_2 \boxtimes \cdots \boxtimes M_n$, and one checks easily that the multiplication functors m_S actually extend to right-exact functors

$$m_S : A^{\otimes S}\text{-bimod} \to A\text{-bimod};$$

for instance, one can define m_S as

$$m_S(M) = M / \{ a_{v'} m - m a_v \mid v \in S, a \in A, m \in M \},$$

where $a_v = 1 \otimes \cdots \otimes a \otimes \cdots \otimes 1 \in A^{\otimes S}$ with a at the v-th position, and $v' \in S$ is the next element after v. We can therefore define the cofibered category A-bimod$_{\#}/\Lambda$ with fiber $A^{\otimes V([n])}$-bimod over $[n] \in \Lambda$, and transition functors $f_!$ as in (2.5). We also have the category of sections $\mathsf{Sec}(A$-bimod$_{\#})$ and the subcategory of co-Cartesian sections $\mathsf{Sec}_{cart}(A$-bimod$_{\#}) \subset \mathsf{Sec}(A$-bimod$_{\#})$.

Lemma 2.3. *The category* $\mathsf{Sec}(A$-bimod$_{\#})$ *is a* k-*linear abelian category.*

Proof (Sketch of a proof). This is a general fact about cofibered categories; the proof is straightforward. The kernel $\mathsf{Ker}\,\phi$ and cokernel $\mathsf{Coker}\,\phi$ of a map $\phi : M_{\#} \to M'_{\#}$ between objects $M_{\#}, M'_{\#} \in \mathsf{Sec}(A$-bimod$_{\#})$ are taken pointwise: for every n, we have an exact sequence

$$0 \to (\mathsf{Ker}\,\phi)([n]) \to M_{\#}([n]) \overset{\phi}{\to} M'_{\#}([n]) \to (\mathsf{Coker}\,\phi)([n]) \to 0.$$

The transtition maps ι_f for $\mathsf{Ker}\,\phi$ are obtained by restriction from those for $M_{\#}$; for $\mathsf{Coker}\,\phi$, one uses the fact that the functors $f_!$ are right-exact.

Definition 2.4. *A cyclic bimodule* M *over a unital associative algebra* A *is a cocartesian section* $M_{\#} \in \mathsf{Sec}_{cart}(A$-bimod$_{\#})$. *A complex of cyclic bimodules* $M_{\bullet}$ *over* A *is an object in the derived category* $\mathcal{D}(\mathsf{Sec}(A$-bimod$_{\#}))$ *whose homology objects are co-Cartesian.*

Complexes of cyclic bimodules obviously form a full triangulated subcategory in $\mathcal{D}(\mathrm{Sec}(A\text{-bimod}_\#))$; consistent notation for the category would be $\mathcal{D}_{cart}(\mathrm{Sec}(A\text{-bimod}_\#))$, but for simplicity we will denote its $\mathcal{D}\Lambda(A\text{-bimod})$. We have to define complexes separately for the following reasons:

(i) The category $\mathrm{Sec}_{cart}(A\text{-bimod}_\#) \subset \mathrm{Sec}(A\text{-bimod}_\#)$ need not be abelian: since the transition functors $f_!$ are only right-exact, the condition of being cocartesian need not be preserved in passing to kernels.

(ii) Even if $\mathrm{Sec}_{cart}(A\text{-bimod}_\#)$ is abelian, its derived category might be much smaller than $\mathcal{D}\Lambda(A\text{-bimod})$.

Example 2.5. An extreme example of (ii) is the case $A = k$: in this case $\mathrm{Sec}(A\text{-bimod}_\#)$ is just the category of cyclic vector spaces, $\mathrm{Fun}(\Lambda, k)$, and $E \in \mathrm{Fun}(\Lambda, k)$ is co-Cartesian if and only if $E(f)$ is invertible for any map $f : [n'] \to [n]$. One deduces easily that E must be a constant functor, so that $\mathrm{Sec}_{cart}(k\text{-bimod}_\#) = k\text{-}Vect$. Then $\mathcal{D}\Lambda(k\text{-bimod})$ is the full subcategory $\mathcal{D}_{const}(\Lambda, k) \subset \mathcal{D}(\Lambda, k)$ of complexes whose homology is constant. If we were to consider Δ^{opp} instead of Λ, we would have $\mathcal{D}_{const}(\Delta^{opp}, k) \cong \mathcal{D}(k\text{-}Vect)$: since $H^\bullet(\Delta^{opp}, k) = k$, the embedding $\mathcal{D}(k\text{-}Vect) \to \mathcal{D}(\Delta^{opp}, k)$ is fully faithful, and $\mathcal{D}_{const}(\Delta^{opp}, k)$ is its essential image. However, $H^\bullet(\Lambda, k)$ is $k[u]$, not k. Therefore there are maps between constant functors in $\mathcal{D}(\Lambda, k)$ that do not come from maps in $\mathcal{D}(k\text{-}Vect)$, and the cones of these maps give objects in $\mathcal{D}_{const}(\Lambda, k)$ that do not come from $\mathcal{D}(k\text{-}Vect)$.

This phenomenon is quite common in homological algebra; examples are, for instance, the triangulated category of complexes of étale sheaves with constructible homology, the category of complex of $\mathcal{D}$-modules with holonomic homology, and the so-called "equivariant derived category" of sheaves on a topological space X acted upon by a topological group G (which is not in fact the derived category of anything useful). The upshot is that it is the triangulated category $\mathcal{D}\Lambda(A\text{-bimod})$ that should be treated as the basic object, wherever categories are discussed.

Remark 2.6. We note one interesting property of the category $\mathcal{D}_{const}(\Lambda, k)$. Fix an integer $n \geq 1$, and consider the full subcategory $\Lambda_{\leq n} \subset \Lambda$ of objects $[n'] \in \Lambda$ with $n' \leq n$. Then one can show that $H^\bullet(\Lambda_{\leq n}, k) = k[u]/u^n$, so that we have a natural exact triangle

$$H_\bullet(\Lambda_{\leq n}, E^\bullet) \longrightarrow HC_\bullet(E^\bullet) \xrightarrow{u^n} HC_{\bullet+2n}(E) \longrightarrow , \qquad (2.6)$$

for every $E^\bullet \in \mathcal{D}_{const}(\Lambda, k)$. We note that for any $E^\bullet \in \mathcal{D}(\Lambda, k)$, (1.4) extends to a spectral sequence

$$HH_\bullet(E^\bullet)[u^{-1}] \Rightarrow HC_\bullet(E), \qquad (2.7)$$

where the expression on the left-hand side reads as "polynomials in one formal variable u^{-1} of homological degree 2 with coefficients in $HH_\bullet(E^\bullet)$." Then (2.6) shows that for $E^\bullet \in \mathcal{D}_{const}(\Lambda, k)$, the first n differentials in (2.7) depend only on the restriction of $E^\bullet$ to $\Lambda_{\leq(n+1)} \subset \Lambda$. This is useful because in practice, one is often interested only in the first differential in the spectral sequence.

As in Lemma 2.2, a cyclic A-bimodule $M_\#$ essentially consists of an A-bimodule $M = M_\#([1])$ equipped with an extra structure. Explicitly, this structure is a map $\tau : A \otimes_k M \to M \otimes_k A$ that respects the $A^{\otimes 2}$-bimodule structure on both sides and satisfies the condition $\tau_{31} \circ \tau_{12} \circ \tau_{23} = \mathrm{id}$, as in Lemma 2.2.

Another way to view this structure is the following. One checks easily that for any cyclic A-bimodule $M_\#$, the restriction $j^* M_\# \in \mathrm{Fun}(\Delta^{opp}, k)$ is canonically isomorphic to the simplicial k-vector space $M_\#^\Delta$ associated to the underlying A-bimodule M as in (2.3). By adjunction, we have a natural map

$$\tau_\# : j_! M_\#^\Delta \to M_\#.$$

Then $j_! M_\#^\Delta$ in this formula depends only on $M \in A$-bimod, and all the structure maps that turn M into the cyclic bimodule $M_\#$ are collected in the map $\tau_\#$.

We can now define cyclic homology with coefficients. The definition is rather tautological. We note that for any cyclic A-bimodule $M_\#$, or in fact, for any $M_\# \in \mathrm{Sec}(A\text{-bimod}_\#)$, we can treat $M_\#$ as a cyclic vector space by forgetting the bimodule structure on its components M_n.

Definition 2.7. *The cyclic homology $HC_\bullet(A, M_\#)$ with coefficients in a cyclic A-bimodule M is equal to $H_\bullet(\Lambda, M_\#)$.*

Of course, (1.4), being valid for any cyclic k-vector space, also applies to $HC_\bullet(A, M_\#)$, so that we automatically get the whole package: the Connes exact sequence, the periodicity endomorphism, and the periodic cyclic homology $HP_\bullet(A, M)$. By Lemma 2.1, $HH_\bullet(M_\#)$ coincides with $HH_\bullet(A, M)$ as defined in (2.2).

3 Gauss–Manin connection.

To illustate the usefulness of the notion of a cyclic bimodule, let us study the behavior of cyclic homology under deformations.

There are two types of deformation theory objects that one can study for an associative algebra A. The first is the notion of a *square-zero extension* of the algebra A by an A-bimodule M. This is an associative algebra $\widetilde{A}$ that fits into a short exact sequence

$$0 \longrightarrow M \overset{i}{\longrightarrow} \widetilde{A} \overset{p}{\longrightarrow} A \longrightarrow 0,$$

where p is an algebra map, and i is an $\widetilde{A}$-bimodule map, under the $\widetilde{A}$-bimodule structure on M induced from the given A-bimodule structure by means of the map p. In other words, the multiplication on the ideal $\operatorname{Ker} p \subset \widetilde{A}$ is trivial, so that the $\widetilde{A}$-bimodule structure on $\operatorname{Ker} p$ is induced by an A-bimodule structure, and i identifies the A-bimodule $\operatorname{Ker} p$ with M. Square-zero extensions are classified up to an isomorphism by elements in the second Hochschild cohomology group $HH^2(A, M)$, defined as

$$HH^{\bullet}(A, M) = \operatorname{Ext}^{\bullet}_{A^{opp} \otimes A}(A, M).$$

In this setting, we can consider the cyclic homology of the algebra $\widetilde{A}$ and compare it with the cyclic homology of A. T. Goodwillie's theorem [Go] claims that if the base field k has characteristic 0, the natural map

$$HP_{\bullet}(\widetilde{A}) \to HP_{\bullet}(A)$$

is an isomorphism, and there is also some information on the behavior of $HC_{\bullet}(A)$.

A second type of deformation-theory data includes a commutative k-algebra R with a maximal ideal $\mathfrak{m} \subset R$. A *deformation* A_R of the algebra A over R is a flat associative unital algebra A_R over R equipped with an isomorphism $A_R / \mathfrak{m} \cong A$. In this case, one can form the *relative* cyclic R-module $A_{R\#}$ by taking the tensor products over R; thus we have relative homology $HH_{\bullet}(A_R/R)$, $HC_{\bullet}(A_R/R)$, $HP_{\bullet}(A_R/R)$. The fundamental fact discovered by E. Getzler [Ge] is that we have an analogue of the Gauss–Manin connection: if $\operatorname{Spec}(R)$ is smooth, the R-module $HP_i(A_R/R)$ carries a canonical flat connection for every i.

Consider now the case that R is not smooth but, on the contrary, local Artin. Moreover, assume that $\mathfrak{m}^2 = 0$, so that R is itself a (commutative) square-zero extension of k. Then a deformation A_R of A over R is also a square-zero extension of A, by the bimodule $A \otimes \mathfrak{m}$ ($\mathfrak{m}$ here is taken as a k-vector space). But this square-zero extension is special: for a general square-zero extension $\widetilde{A}$ of A by some $M \in A$-bimod, there does not exist any analogue of the relative cyclic R-module $A_{R\#} \in \operatorname{Fun}(\Lambda, R)$.

We observe the following: the data needed to define such an analogue is precisely a cyclic bimodule structure on the bimodule M.

Namely, assume given a square-zero extension $\widetilde{A}$ of the algebra A by some A-bimodule M, and consider the cyclic k-vector space $\widetilde{A}_{\#} \in \operatorname{Fun}(\Lambda, k)$. Let us equip $\widetilde{A}$ with a descreasing two-step filtration $F^{\bullet}$ by setting $F^1\widetilde{A} = M$. Then this induces a decreasing filtration $F^{\bullet}$ on tensor powers $\widetilde{A}^{\otimes n}$. Since $\widetilde{A}$ is square-zero, $F^{\bullet}$ is compatible with the multiplication maps; therefore we also have a filtration $F^{\bullet}$ on $\widetilde{A}_{\#}$. Consider the quotient

$$\overline{A_{\#}} = \widetilde{A}_{\#}/F^2\widetilde{A}_{\#}.$$

One checks easily that $\operatorname{gr}^0_F \widetilde{A}_{\#} \cong A_{\#}$ and $\operatorname{gr}^1_F \widetilde{A}_{\#} \cong j_! M^{\Delta}_{\#}$ in a canonical way, so that $\overline{A_{\#}}$ fits into a canonical short exact sequence

$$0 \longrightarrow j_! M_{\#}^{\Delta} \longrightarrow \overline{A_{\#}} \longrightarrow A_{\#} \longrightarrow 0 \qquad (3.1)$$

of cyclic k-vector spaces.

Now assume in addition that M is equipped with a structure of a cyclic A-bimodule $M_{\#}$, so that $M_{\#}^{\Delta} \cong j^* M_{\#}$, and we have the structure map $\tau_{\#}$: $j_! M_{\#}^{\Delta} \to M_{\#}$. Then we can compose the extension (3.1) with the map $\tau_{\#}$, to obtain a commutative diagram

$$
\begin{array}{ccccccccc}
0 & \longrightarrow & j_! M_{\#}^{\Delta} & \longrightarrow & \overline{A_{\#}} & \longrightarrow & A_{\#} & \longrightarrow & 0 \\
& & \downarrow {\scriptstyle \tau_{\#}} & & \downarrow & & \| & & \\
0 & \longrightarrow & M_{\#} & \longrightarrow & \widehat{A_{\#}} & \longrightarrow & A_{\#} & \longrightarrow & 0
\end{array}
\qquad (3.2)
$$

of short exact sequences in $\mathrm{Fun}(\Lambda, k)$, with Cartesian left square. It is easy to check that when $\widetilde{A} = A_R$ for some square-zero R, so that $M = A \otimes \mathfrak{m}$, and we take the cyclic A-bimodule structure on M induced by the tautological structure on A, then $\widehat{A_{\#}}$ coincides precisely with the relative cyclic object $A_{R\#}$ (which we consider as a k-vector space, forgetting the R-module structure).

We believe that this is the proper generality for the Getzler connection; in this setting, the main result reads as follows.

Proposition 3.1. *Assume given a square-zero extension $\widetilde{A}$ of an associative algebra A by an A-bimodule M, and assume that M is equipped with a structure of a cyclic A-bimodule. Then the long exact sequence*

$$HP_\bullet(A, M) \longrightarrow HP_\bullet(\widehat{A_{\#}}) \longrightarrow HP_\bullet(A) \longrightarrow$$

of periodic cyclic homology induced by the second row in (3.2) admits a canonical splitting $HP_\bullet(A) \to HP_\bullet(\widehat{A_{\#}})$.

Proof. By definition, we have two natural maps

$$
\begin{aligned}
HP_\bullet(\overline{A_{\#}}) &\to HP_\bullet(A_{\#}) = HP_\bullet(A), \\
HP_\bullet(\overline{A_{\#}}) &\to HP_\bullet(\widehat{A_{\#}}),
\end{aligned}
\qquad (3.3)
$$

and the cone of the first map is isomorphic to $HP_\bullet(j_! M_{\#}^{\Delta})$. Since $j_!$ is exact, we have $HC_\bullet(j_! M_{\#}^{\Delta}) \cong HH_\bullet(M_{\#})$, and the periodicity map u : $HC_\bullet(j_! M_{\#}^{\Delta}) \to HC_{\bullet-2}(j_! M_{\#}^{\Delta})$ is equal to 0, so that $HP_\bullet(j_! M_{\#}^{\Delta}) = 0$. Thus the first map in (3.3) is an isomorphism, and the second map is then the required splitting.

Corollary 3.2. *Assume given a commutative k-algebra R with a maximal ideal $\mathfrak{m} \subset R$, and a deformation A_R of the algebra A over R. Then if $\mathrm{Spec}(R)$ is smooth, the R-modules $HP_\bullet(A_R/R)$ carry a natural connection.*

Proof (Sketch of a proof). Consider the $R \otimes R$-algebras $A_R \otimes R$ and $R \otimes A_R$, and their restrictions to the first infinitesimal neighborhood of the diagonal in $\mathrm{Spec}(R \otimes R) = \mathrm{Spec}(R) \times \mathrm{Spec}(R)$. Then Proposition 3.1, suitably generalized, shows that $HP_\bullet(-)$ of these two restrictions are canonically isomorphic. It is well known that giving such an isomorphism is equivalent to giving a connection on $HP_\bullet(A_R/R)$.

We note that we do not claim that the connection is *flat*. It certainly is, at least in characteristic 0; but our present method does not allow one to go beyond square-zero extensions. Thus we cannot analyze the second infinitesemal neighborhood of the diagonal in $\mathrm{Spec}(R \otimes R)$, and we cannot prove flatness.

Unfortunately, at present, we do not understand what is the proper cyclic bimodule context for higher-level infinitesimal extensions. Of course, if one is interested only in an R-deformation $\widetilde{A} = A_R$ over an Artin local base R, not in its cyclic bimodule generalizations, one can use Goodwillie's Theorem: using the full cyclic object $\widetilde{A}_\#$ instead of its quotient $\overline{A}_\#$ in Proposition 3.1 immediately gives a splitting $HP_\bullet(A) \to HP_\bullet(A_R/R)$ of the augmentation map $HP_\bullet(A_R/R) \to HP_\bullet(A)$, and this extends by R-linearity to an isomorphism $HP_\bullet(A_R/R) \cong HP_\bullet(A) \otimes R$. However, this is not quite satisfactory from the conceptual point of view, and it does not work in positive characteristic (where Goodwillie's Theorem is simply not true). If char $k \neq 2$, the latter can be cured by using $\widetilde{A}_\#/F^3\widetilde{A}_\#$, but the former one remains. We plan to return to this elsewhere.

4 Categorical Approach.

Let us now try to define cyclic homology in a more general setting: we will attempt to replace A-bimod with an arbitrary associative unital k-linear tensor category $\mathcal{C}$ with a unit object $I \in \mathcal{C}$. We do not assume that $\mathcal{C}$ is symmetric in any way. However, we will assume that the tensor product $- \otimes -$ is right-exact in each variable, and we will need to impose additional technical assumptions later on.

The first thing to do is to try to define Hochschild homology; so let us look more closely at (2.1). The formula an the right-hand side looks symmetric, but this is an optical illusion; the two copies of A are completely different objects: one is a left module over $A^{opp} \otimes A$, and the other is a right module (A just happens to have both structures at the same time). It is better to separate them and introduce the functor

$$\mathrm{tr} : A\text{-bimod} \to k\text{-}Vect$$

by $\mathrm{tr}(M) = M \otimes_{A^{opp} \otimes A} A$, or equivalently, by

$$\mathrm{tr}(M) = M/\{am - ma \mid a \in A, m \in M\}. \tag{4.1}$$

Then tr is a right-exact functor, and we have $HH_\bullet(A, M) = L^\bullet \mathrm{tr}(M)$.

We want to emphasize that the functor tr cannot be recovered from the tensor structure on A-bimod; this really is an extra piece of data. For a general tensor category $\mathcal{C}$, it does not exist a priori; we have to impose it as an additional structure.

Let us axiomatize the situation. First, forget for the moment about the k-linear and abelian structure on $\mathcal{C}$; let us treat it simply as a monoidal category. Assume given some other category $\mathcal{B}$ and a functor $T : \mathcal{C} \to \mathcal{B}$.

Definition 4.1. *The functor $T : \mathcal{C} \to \mathcal{B}$ is a* trace functor *if it is extended to a functor $\mathcal{C}_\# \to \mathcal{B}$ that sends any cocartesian map $f : M \to M'$ in $\mathcal{C}_\#$ to an invertible map.*

Another way to say the same thing is the following: the categories $\mathrm{Fun}(\mathcal{C}^n, \mathcal{B})$ of functors from $\mathcal{C}^n$ to $\mathcal{B}$ form a fibered category over Λ, and a trace functor is a Cartesian section of this fibration. Explicitly, a trace functor is defined by $T : \mathcal{C} \to \mathcal{B}$ and a collection of isomorphisms

$$T(M \otimes M') \to T(M' \otimes M)$$

for any $M, M' \in \mathcal{C}$ that are functorial in M and M' and satisfy some compatibility conditions analogous to those in Lemma 2.2; we leave it to the reader to write down these conditions precisely. Thus T has a tracelike property with respect to the product in $\mathcal{C}$, and this motivates our terminology.

Recall now that $\mathcal{C}$ is a k-linear abelian category. To define Hochschild homology, we have to assume that it is equipped with a right-exact trace functor $\mathrm{tr} : \mathcal{C} \to k\text{-}Vect$; then for any $M \in \mathcal{C}$, we set

$$HH_\bullet(M) = L^\bullet \, \mathrm{tr}(M). \tag{4.2}$$

Lemma 4.2. *The functor* $\mathrm{tr} : A\text{-bimod} \to k\text{-}Vect$ *canonically extends to a right-exact trace functor in the sense of Definition 4.1.*

Proof. For any object $\langle [n], M_n \rangle \in A\text{-bimod}_\#$, $[n] \in \Lambda$, $M_n \in A^{\otimes n}\text{-bimod}$, let

$$\mathrm{tr}(\langle [n], M_n \rangle) = M_n / \{ a_{v'} m - m a_v \mid v \in V([n]), m \in M_n, a \in A \},$$

where $a_v = 1 \otimes 1 \otimes \cdots \otimes a \otimes \cdots \otimes 1 \in A^{\otimes V([n])}$ has a in the multiple corresponding to $v \in V([n])$, and $v' \in V([n])$ is the next marked point after v counting clockwise. The compatibility with maps in the category $A\text{-bimod}_\#$ is obvious.

We note that here, in the case $\mathcal{C} = A\text{-bimod}$, the category $A\text{-bimod}_\#$ is actually larger than what we would have had purely from the monoidal structure on $\mathcal{C}$: M_n is allowed to be an arbitrary $A^{\otimes n}$-bimodule, not a collection of n A-bimodules. To do the same for general k-linear $\mathcal{C}$, we need to replace $A^{\otimes n}$-bimod with some version of the tensor product $\mathcal{C}^{\otimes n}$. Here we have a difficulty: for various technical reasons, it is not clear how to define tensor products for sufficiently general abelian categories.

One way around it is the following. For any (small) k-linear abelian category $\mathcal{B}$, a k-linear functor $\mathcal{B}^{opp} \to k\text{-}Vect$ is left-exact if and only if it is a sheaf for for the canonical Grothendieck topology on $\mathcal{B}$ ([BD, 5, §10]); the category $\mathrm{Shv}(\mathcal{B})$ of such functors is abelian and k-linear, and $\mathcal{B}$ itself is naturally embedded into $\mathrm{Shv}(\mathcal{B})$ by Yoneda. The embedding is a fully faithful exact functor. Every functor in $\mathrm{Shv}(\mathcal{B})$ is in fact a direct limit of representable functors, so that $\mathrm{Shv}(\mathcal{B})$ is an inductive completion of the abelian category $\mathcal{B}$. Now, if we are given two (small) k-linear abelian categories $\mathcal{B}_1$, $\mathcal{B}_2$, then their product $\mathcal{B}_1 \times \mathcal{B}_2$ is no longer abelian. However, we still have the abelian category $\mathrm{Shv}(\mathcal{B}_1 \times \mathcal{B}_2)$ of bilinear functors $\mathcal{B}_1^{opp} \times \mathcal{B}_2^{opp} \to k\text{-}Vect$, which are left-exact in each variable, and the same goes for polylinear functors.

Moreover, for any right-exact functor $F : \mathcal{B}_1 \to \mathcal{B}_2$ between small abelian categories, we have the restriction functor $F^* : \mathrm{Shv}(\mathcal{B}_2) \to \mathrm{Shv}(\mathcal{B}_1)$, which is left-exact, and its left-adjoint $F_! : \mathrm{Shv}(\mathcal{B}_1) \to \mathrm{Shv}(\mathcal{B}_2)$, which is right-exact. The functor $F_!$ is an extension of the functor F: on Yoneda images $\mathcal{B}_i \subset \mathrm{Shv}(\mathcal{B}_i)$, we have $F_! = F$. And again, the same works for polylinear functors.

In particular, given our k-linear abelian tensor category $\mathcal{C}$, we can form the category $\mathrm{Shv}(\mathcal{C})_\#$ of pairs $\langle E, [n] \rangle$, $[n] \in \Lambda$, $E \in \mathrm{Shv}(\mathcal{C}^n)$, with a map from $\langle E', [n'] \rangle$ to $\langle E, [n] \rangle$ given by a pair of a map $f : [n'] \to [n]$ and either a map $E' \to (f_!)^* E$, or map $(f_!)_! E' \to E$; this is equivalent by adjunction. Then $\mathrm{Shv}(\mathcal{C})_\#$ is a bifibered category over Λ in the sense of [Gr].

The category of sections $\Lambda \to \mathrm{Shv}(\mathcal{C})_\#$ of this bifibration can also be described as the full subcategory $\mathrm{Shv}(\mathcal{C}_\#) \subset \mathrm{Fun}(\mathcal{C}_\#^{opp}, k)$ spanned by those functors $E_\# : \mathcal{C}_\#^{opp} \to k\text{-}Vect$ whose restriction to $(\mathcal{C}^{opp})^n \subset \mathcal{C}_\#^{opp}$ is a sheaf, that is, an object in $\mathrm{Shv}(\mathcal{C}^n) \subset \mathrm{Fun}((\mathcal{C}^{opp})^n, k)$. Since the transition functors $(f_!)_!$ are right-exact, $\mathrm{Shv}(\mathcal{C}_\#)$ is an abelian category (this is proved in exactly the same way as Lemma 2.3).

We denote by $\mathrm{Shv}_{cart}(\mathcal{C}_\#) \subset \mathrm{Shv}(\mathcal{C}_\#)$ the full subcategory of sections $E : \Lambda \to \mathrm{Shv}(\mathcal{C})_\#$ that are co-Cartesian, and moreover, are such that $E([1]) \in \mathrm{Shv}(\mathcal{C})$ actually lies in the Yoneda image $\mathcal{C} \subset \mathrm{Shv}(\mathcal{C})$. We also denote by $\mathcal{D}\Lambda(\mathcal{C}) \subset \mathcal{D}(\mathrm{Shv}(\mathcal{C}_\#))$ the full triangulated subcategory of complexes $E_\#^\cdot \in \mathcal{D}(\mathrm{Shv}(\mathcal{C}_\#))$ with homology in $\mathrm{Shv}_{cart}(\mathcal{C}_\#)$.

If $\mathcal{C}$ is the category of A-bimodules for some algebra A, or better yet, of A-bimodules of cardinality not more than that of $A \times \mathbb{N}$, so that $\mathcal{C}$ is small, then $\mathrm{Shv}(\mathcal{C})$ is equivalent to A-bimod (one shows easily that every sheaf $E \in \mathrm{Shv}(\mathcal{C})$ is completely determined by its value at $A^{opp} \otimes A \in \mathcal{C}$). In this case, $\mathcal{D}\Lambda(\mathcal{C})$ is our old category $\mathcal{D}\Lambda(A\text{-bimod})$.

Now we assume that $\mathcal{C}$ is equipped with a right-exact trace functor $\mathrm{tr} : \mathcal{C} \to k\text{-}Vect$. We would like to define cyclic homology $HC_\cdot(M_\cdot)$ for any $M_\cdot \in \mathcal{D}\Lambda(\mathcal{C})$, and we immediately notice a problem: for a general $\mathcal{C}$, we do not have a forgetful functor to vector spaces. However, it turns out that the forgetful functor *is not needed* for the definition; it can be replaced with the trace functor tr.

We proceed as follows. By definition, tr is extended to a functor $\mathcal{C}_\# \to k\text{-}Vect$; we extend it canonically to a functor $\mathrm{Shv}(\mathcal{C})_\# \to k\text{-}Vect$, and consider the product

$$\text{tr} \times \tau : \text{Shv}(\mathcal{C})_{\#} \to k\text{-}Vect \times \Lambda,$$

where $\tau : \text{Shv}(\mathcal{C})_{\#} \to \Lambda$ is the projection. This is a functor compatible with the projections to Λ, and therefore, it induces a functor of the categories of sections. The category of sections of the projection $k\text{-}Vect \times \Lambda \to \Lambda$ is tautologically the same as $\text{Fun}(\Lambda, k\text{-}Vect)$, so that we have a functor

$$\text{tr}_{\#} : \text{Shv}(\mathcal{C}_{\#}) \to \text{Fun}(\Lambda, k).$$

One checks easily that this functor is right-exact.

Definition 4.3. *For any $M_{\#} \in \text{Sec}(\mathcal{C}_{\#})$, its cyclic homology $HC_{\bullet}(M_{\#})$ is defined by*

$$HC_{\bullet}(M_{\#}) \overset{\text{def}}{=} HC_{\bullet}(L^{\bullet}\, \text{tr}_{\#}(M_{\#})) = H_{\bullet}(\Lambda, L^{\bullet}\, \text{tr}_{\#}(M_{\#})).$$

Definition 4.4. *The pair $\langle \mathcal{C}, \text{tr} \rangle$ is called* homologically clean *if for any n, the category $\text{Shv}(\mathcal{C}^n)$ has enough objects E such that*

(i) E is acyclic both for functors $(f_!)_! : \text{Shv}(\mathcal{C}^n) \to \text{Shv}(\mathcal{C}^{n'})$, for any $f : [n] \to [n']$, and for the trace functor $\text{tr} : \text{Shv}(\mathcal{C}^n) \to k\text{-}Vect$, and

(ii) for any $f : [n] \to [n']$, $(f_!)_! E \in \text{Shv}(\mathcal{C}^{n'})$ is acyclic for $\text{tr} : \text{Shv}(\mathcal{C}^{n'}) \to k\text{-}Vect$.

Example 4.5. Assume that the category $\mathcal{C}$ has enough projectives, and moreover, $P_1 \otimes P_2$ is projective for any projective $P_1, P_2 \in \mathcal{C}$ (this is satisfied, for instance, for $\mathcal{C} = A\text{-bimod}$). Then the pair $\langle \mathcal{C}, \text{tr} \rangle$ is homologically clean, for any trace functor tr. Indeed, $\text{Shv}(\mathcal{C}^n)$ then also has enough projectives, say sums of objects

$$P = P_1 \boxtimes P_2 \boxtimes \cdots \boxtimes P_n \in \text{Shv}(\mathcal{C}^n) \tag{4.3}$$

for projective $P_1, \dots, P_n \in \mathcal{C} \subset \text{Shv}(\mathcal{C})$, and these projectives automatically satisfy the condition (i). To check (ii), one decomposes $f : [n] \to [n']$ into a surjection $p : [n] \to [n'']$ and an injection $i : [n''] \to [n']$. Since the tensor product of projective objects is projective, $(p_!)_!(P) \in \text{Shv}(\mathcal{C}^{n''})$ is also an object of type (4.3), so we may as well assume that f is injective. Then one can find a left-inverse map $f' : [n'] \to [n]$, $f' \circ f = \text{id}$; since $P' = (f_!)_!(P)$ is obviously acyclic for $(f'_!)_!$, and $(f'_!)_! (P') = ((f' \circ f)_!)_!(P) = P$ is acyclic for tr, P' itself is acyclic for $\text{tr} = \text{tr} \circ (f'_!)_!$.

Lemma 4.6. *Assume that $\langle \mathcal{C}, \text{tr} \rangle$ is homologically clean. Then for any object $[n] \in \Lambda$ and any $M_{\#} \in \text{Shv}(\mathcal{C}_{\#})$, we have*

$$L^{\bullet}\, \text{tr}_{\#}(M_{\#})([n]) \cong L^{\bullet}\, \text{tr}(M_{\#}([n])). \tag{4.4}$$

For any $M_{\#}^{\bullet} \in \mathcal{D}\Lambda(\mathcal{C})$, we have $L^{\bullet}\, \text{tr}_{\#}(M_{\#}) \in \mathcal{D}_{const}(\Lambda, k) \subset \mathcal{D}(\Lambda, k)$.

Proof. The natural restriction functor $\mathrm{Shv}(\mathcal{C}_\#) \to \mathrm{Shv}(\mathcal{C}^n)$, $M_\# \mapsto M_\#([m])$ has a left-adjoint functor $I_{n!} : \mathrm{Shv}(\mathcal{C}^n) \to \mathrm{Shv}(\mathcal{C}_\#)$; explicitly, it is given by

$$I_{n!}(E)([n']) = \bigoplus_{f:[n]\to[n']} (f_!)_!(E). \qquad (4.5)$$

Let us say that an object $E \in \mathrm{Shv}(\mathcal{C}^n)$ is admissible if it satisfies the conditions (i), (ii) of Definition 4.4. By assumption, $\mathrm{Shv}(\mathcal{C}^n)$ has enough admissible objects for any n. Then $\mathrm{Shv}(\mathcal{C}_\#)$ has enough objects of the form $I_{n!}E$, $[n] \in \Lambda$, $E \in \mathrm{Shv}(\mathcal{C}^n)$ admissible, and to prove the first claim, it suffices to consider $M_\# = I_{n!}E$ of this form. In degree 0, (4.4) is the definition of the functor $\mathrm{tr}_\#$, and the higher-degree terms on the right-hand side vanish by Definition 4.4 (ii). Therefore it suffices to prove that $M_\# = I_{n!}E$ is acyclic for the functor $\mathrm{tr}_\#$. This is obvious: applying $\mathrm{tr}_\#$ to any short exact sequence

$$0 \longrightarrow M'_\# \longrightarrow M''_\# \longrightarrow M_\# \longrightarrow 0$$

in $\mathrm{Shv}(\mathcal{C}_\#)$, we see that since $M_\#([n'])$ is acyclic for any $[n'] \in \Lambda$, the sequence

$$0 \longrightarrow \mathrm{tr}\, M'_\#([n']) \longrightarrow \mathrm{tr}\, M''_\#([n']) \longrightarrow \mathrm{tr}\, M_\#([n']) \longrightarrow 0$$

is exact; this means that

$$0 \longrightarrow \mathrm{tr}\, M'_\# \longrightarrow \mathrm{tr}\, M''_\# \longrightarrow \mathrm{tr}\, M_\# \longrightarrow 0$$

is an exact sequence in $\mathrm{Fun}(\Lambda, k)$, and this means that $M_\#$ is indeed acyclic for $\mathrm{tr}_\#$.

With the first claim proved, the second amounts to showing that the natural map

$$L^\bullet \mathrm{tr} \circ L^\bullet (f_!)_!(E) \to L^\bullet \mathrm{tr}(E)$$

is a quasi-isomorphism for any $f : [n] \to [n']$ and any $E \in \mathrm{Shv}(\mathcal{C}^n)$. It suffices to prove it for admissible M; then the higher derived functors vanish, and the isomorphism $\mathrm{tr} \circ (f_!)_! \cong \mathrm{tr}$ is Definition 4.1.

Lemma 4.7. *In the assumptions of Lemma 4.6, for any complex $M_\#^\bullet \in D\Lambda(\mathcal{C})$ with the first component $M^\bullet = M_\#^\bullet([1])$ we have*

$$HH_\bullet(M^\bullet) \cong HH_\bullet\left(L^\bullet \mathrm{tr}_\#\left(M_\#^\bullet\right)\right).$$

Proof. By Lemma 4.6, the left-hand side, $HH_\bullet(M^\bullet)$, is canonically isomorphic to the complex $L^\bullet \mathrm{tr}_\#(M_\#^\bullet) \in \mathcal{D}(\Lambda, k)$ evaluated at $[1] \in \Lambda$, and moreover, $L^\bullet \mathrm{tr}_\#(M_\#^\bullet)$ lies in the subcategory $\mathcal{D}_{const}(\Lambda, k) \subset \mathcal{D}(\Lambda, k)$. It remains to apply the following general fact: for any $E^\bullet \in \mathcal{D}_{const}(\Lambda, k)$, we have a natural isomorphism $HH_\bullet(E^\bullet) \cong E^\bullet([1])$. Indeed, by definition we have

$$HH_\bullet(E^\bullet) = H_\bullet(\Delta^{opp}, j^* E^\bullet),$$

and $j^* E^\bullet$ lies in the category $\mathcal{D}_{const}(\Delta^{opp}, k)$, which is equivalent to $\mathcal{D}(k\text{-}Vect)$ (see Example 2.5, and also Remark 2.6: the isomorphism we constructed here is a special case of (2.6) for $n = 1$).

The lemma shows that if the pair $\langle \mathcal{C}, \mathrm{tr} \rangle$ is homologically clean, Definition 4.3 is consistent with (4.2), and we get the whole periodicity package of (1.4): the periodicity map u, the Connes exact sequence

$$HH_{\bullet}(M^{\bullet}) \longrightarrow HC_{\bullet}(M^{\bullet}) \overset{u}{\longrightarrow} HC_{\bullet-2}(M^{\bullet}) \longrightarrow \, ,$$

and the periodic cyclic homology $HP_{\bullet}(M^{\bullet})$.

In general, objects in $\mathcal{D}\Lambda(\mathcal{C})$ may be hard to construct, but we always have at least one: the identity section $\mathsf{I}_{\#} : \Lambda \to \mathrm{Shv}(\mathcal{C})_{\#}$, given by

$$\mathsf{I}_{\#}([n]) = \mathsf{I}^{\boxtimes n} \in \mathcal{C}^{\otimes n},$$

where $\mathsf{I} \in \mathcal{C}$ is the unit object. Thus we can define cyclic homology of a tensor category equipped with a trace functor.

Definition 4.8. *For any k-linear abelian unital tensor category $\mathcal{C}$ equipped with a trace functor* $\mathrm{tr} : \mathcal{C} \to k\text{-}Vect$, *its Hochschild and cyclic homologies are given by*

$$HH_{\bullet}(\mathcal{C}, \mathrm{tr}) \overset{\mathrm{def}}{=} HH_{\bullet}(\mathsf{I}), \qquad HC_{\bullet}(\mathcal{C}, \mathrm{tr}) \overset{\mathrm{def}}{=} HC_{\bullet}(\mathsf{I}_{\#}),$$

where $\mathsf{I} \in \mathcal{C}$ is the unit object, and $\mathsf{I}_{\#} \in \mathcal{D}\Lambda(\mathcal{C})$ is the identity section.

We now have to check that in the case $\mathcal{C} = A\text{-bimod}$, Definition 4.3 is compatible with our earlier Definition 2.7, in other words, that the cyclic homology computed by means of the forgetful functor is the same as the cyclic homology computed by means of the trace. This is not at all trivial. Indeed, if for instance $M_{\#} \in \mathrm{Shv}(\mathcal{C}_{\#})$ is co-Cartesian, then, while $L^{\bullet} \mathrm{tr}^{\#} M_{\#}$ lies in the subcategory $\mathcal{D}_{const}(\Lambda, k) \subset \mathcal{D}(\Lambda, k)$, the same is certainly not true for the object $M_{\#} \in \mathrm{Fun}(\Lambda, k)$ obtained by forgetting the bimodule structure on M_n.

Thus these two objects are different. However, they do become equal after taking cyclic (or Hochschild, or periodic cyclic) homology. Namely, for any $M_{\#} \in \mathrm{Sec}(A\text{-bimod}_{\#})$ we have a natural map

$$M_{\#} \to L^{\bullet} \mathrm{tr}^{\#} M_{\#} \tag{4.6}$$

in the derived category $\mathcal{D}(\Lambda, k)$, and we have the following result.

Proposition 4.9. *For every $M_{\#} \in \mathrm{Sec}(A\text{-bimod}_{\#})$, the natural map (4.6) induces isomorphisms*

$$HH_{\bullet}(M_{\#}) \cong HH_{\bullet}(L^{\bullet} \mathrm{tr}\, M_{\#}),$$
$$HC_{\bullet}(M_{\#}) \cong HC_{\bullet}(L^{\bullet} \mathrm{tr}\, M_{\#}),$$
$$HP_{\bullet}(M_{\#}) \cong HP_{\bullet}(L^{\bullet} \mathrm{tr}\, M_{\#}).$$

Proof. By (1.4), it suffices to consider $HC_{\bullet}(-)$; as in the proof of Lemma 4.6, it suffices to consider $M_{\#} = I_{n!}E$ given in (4.5), with E being the free bimodule

$$E = (A^{opp} \otimes A)^{\otimes n} \in \mathrm{Shv}(\mathcal{C}^n) = A^{\otimes n}\text{-bimod}$$

for some fixed n. Explicitly, we have

$$I_{n!}E([n']) = \bigoplus_{f:[n] \to [n']} \bigotimes_{v' \in V([n'])} A^{opp} \otimes A^{\otimes f^{-1}(v')} \qquad (4.7)$$

for any $[n'] \in \Lambda$. Then $L^p \operatorname{tr}_\# I_{n!}E = 0$ for $p \geq 1$, and one checks easily that

$$\operatorname{tr}_\# I_{n!}E = i_{n!}\operatorname{tr}E = i_{n!}A^{\otimes n} \in \mathrm{Fun}(\Lambda, k),$$

where $i_n : \mathsf{pt} \to \Lambda$ is the embedding of the object $[n] \in \Lambda$ (pt is the category with object and one morphism). Therefore

$$HC_0(L^{\bullet}\operatorname{tr}_\# I_{n!}E) = H_{\bullet}(\Lambda, i_{n!}A^{\otimes n}) = A^{\otimes n},$$

and $HC_p(L^{\bullet}\operatorname{tr}_\# i_{n!}E) = 0$ for $p \geq 1$. We have to compare it with $HC_{\bullet}(i_{n!}E)$.

To do this, consider the category $\Lambda_{[n]}$ of objects $[n'] \in \Lambda$ equipped with a map $[n] \to [n']$, and let $j_n : \Lambda_{[n]} \to \Lambda$ be the forgetful functor. Then j_n is obviously a discrete cofibration. Comparing (1.2) and (4.7), we see that

$$I_{n!}E = j_{n!}E_\#^{[n]}$$

for some $E_\#^{[n]} \in \mathrm{Fun}(\Lambda_{[n]})$. Moreover, fix once and for all a map $[1] \to [n]$. Then we see that the discrete cofibration $j_n : \Lambda_{[n]} \to \Lambda$ factors through the discrete cofibration $j : \Lambda_{[1]} = \Delta^{opp} \to \Lambda$ by means of a discrete cofibration $\gamma_n : \Lambda_{[n]} \to \Lambda_{[1]}$, and we observe that

$$E_\#^{[n]}([n']) = (A^{opp})^{\otimes n'} \otimes A^{\otimes n}$$

depends only on $\gamma_n([n']) \in \Delta^{opp}$. More precisely, we have $E_\#^{[n]} = \gamma_n^* E_n^{\Delta}$, where $E_n^{\Delta} \in \mathrm{Fun}(\Delta^{opp}, k)$ is as in (2.3), and E_n is the free A-bimodule

$$E_n = A^{opp} \otimes A^{\otimes(n-1)} \otimes A.$$

The conclusion: we have

$$HC_{\bullet}(I_{n!}E) = H_{\bullet}\left(\Lambda_{[n]}, E_\#^{[n]}\right) = H_{\bullet}\left(\Delta^{opp}, \gamma_{n!}\gamma_n^* E_n^{\Delta}\right) = H_{\bullet}\left(\Delta^{opp}, E_n^{\Delta} \otimes \gamma_{n!}k\right),$$

where we have used the projection formula (1.3) in the right-hand side. The homology of the category Δ^{opp} can be computed by the standard complex; then by the Künneth formula, the right-hand side is isomorphic to

$$H_{\bullet}\left(\Delta^{opp}, E_n^{\Delta}\right) \otimes H_{\bullet}(\Delta^{opp}, \gamma_{n!}k) \cong H_{\bullet}\left(\Delta^{opp}, E_n^{\Delta}\right) \otimes H_{\bullet}(\Lambda_{[n]}, k).$$

By Lemma 2.1,

$$H_{\bullet}\left(\Delta^{opp}, E_n^{\Delta}\right) \cong HH_{\bullet}(A, E_n) \cong A^{\otimes n}.$$

Since the category $\Lambda_{[n]}$ has an initial object $[n] \in \Lambda_{[n]}$, we have $k = i_{n!}k$, so that the second multiple $H_{\bullet}(\Lambda_{[n]}, k)$ is just k in degree 0.

The essential point of Proposition 4.9 is the following: the cyclic object $A_\#$ associated to an algebra A inconveniently contains two things at the same time: the cyclic structure, which seems to be essential to the problem, and the bar resolution, which is needed only to compute the Hochschild homology $HH_\bullet(A)$. Replacing $A_\#$ with the cyclic complex $L^\bullet \operatorname{tr}_\# A_\# \in \mathcal{D}(\Lambda, k)$ disentangles these two.

We note that while one still has to prove that this does not change the final answer, the construction itself looks pretty straightforward: if one wants to remove the inessential bar resolution from the definition of the cyclic homology, Definition 4.8 seems to be the obvious thing to try. However, it was actually arrived at by a sort of reverse engineering process. To finish the section, perhaps it would be useful to show the reader the first stage of this process.

Assume given an associative algebra A, and fix a projective resolution $P_\bullet$ of the diagonal A-module A. Then $HH_\bullet(A, M)$ can be computed by the complex

$$\operatorname{tr}(P_\bullet) = P_\bullet \otimes_{A^{opp} \otimes A} A.$$

How can one see the cyclic homology in terms of this complex? Or even simpler, what is the first differential in the spectral sequence (2.7), the Connes differential $B : HH_\bullet(A) \to HH_{\bullet+1}(A)$?

There is the following recipe, which gives the answer. Let $\tau : P_\bullet \to A$ be the augmentation map. Consider the tensor product $P_\bullet \otimes_A P_\bullet$. This is also a projective resolution of A, and we actually have *two* natural quasi-isomorphisms

$$\tau_1, \tau_2 : P_\bullet \otimes_A P_\bullet \to P_\bullet,$$

given by $\tau_1 = \tau \otimes \operatorname{id}$, $\tau_2 = \operatorname{id} \otimes \tau$. These quasi-isomorphisms are different. However, since both are maps between projective resolutions of the same object, there should be a chain homotopy between them. Fix such a homotopy $\iota : P_\bullet \otimes_A P_\bullet \to P_{\bullet+1}$.

Now we apply the trace functor tr, and obtain two maps τ_1, τ_2:
$\operatorname{tr}(P_\bullet \otimes P_\bullet) \to \operatorname{tr}(P_\bullet)$, and a homotopy $\iota : \operatorname{tr}(P_\bullet \otimes P_\bullet) \to \operatorname{tr}(P_{\bullet+1})$ between them.

However, by the trace property of τ, we also have an involution $\sigma : \operatorname{tr}(P_\bullet \otimes_A P_\bullet)$ that interchanges the two multiples. This involution obviously also interchages τ_1 and τ_2, but there is no reason why it should fix the homotopy ι – in fact, it sends ι to a second homotopy $\iota' : \operatorname{tr}(P_\bullet \otimes_A P_\bullet) \to \operatorname{tr}(P_{\bullet+1})$ between τ_1 and τ_2.

The difference $\iota' - \iota$ is then a well-defined map of complexes

$$\iota' - \iota : \operatorname{tr}(P_\bullet \otimes_A P_\bullet) \to \operatorname{tr}(P_{\bullet+1}). \tag{4.8}$$

On the level of homology, both sides are $HH_\bullet(A)$; the map $\iota' - \iota$ then induces exactly the Connes differential $B : HH_\bullet(A) \to HH_{\bullet+1}(A)$.

To justify this recipe, we use Proposition 4.9 and identify $HC_\bullet(A)$ with $HC_\bullet(L^\bullet \operatorname{tr}_\#(A_\#))$ rather than $HC_\bullet(A_\#)$. Then $L^\bullet \operatorname{tr}_\#(A_\#)$ is an object in

$\mathcal{D}_{const}(\Lambda, k)$. Therefore, as noted in Remark 2.6, the Connes differential B depends only on the restriction of $L^{\bullet}\,\mathrm{tr}_{\#}(A_{\#})$ to $\Lambda_{\leq 2} \subset \Lambda$. In other words, we do not need to compute the full $L^{\bullet}\,\mathrm{tr}_{\#}(A_{\#})$ and to construct a full resolution $P^{\#}_{\bullet}$ of the cyclic A-bimodule $A_{\#}$; it suffices to construct $P^i_{\bullet} = P^{\#}_{\bullet}([i])$ for $i = 1, 2$ (and then apply the functor tr).

With the choices made above, we set $P^1_{\bullet} = P_{\bullet}$, and we let $P^2_{\bullet}$ be the cone of the map

$$P_{\bullet} \boxtimes P_{\bullet} \xrightarrow{(\tau \boxtimes \mathrm{id}) \oplus (\mathrm{id} \boxtimes \tau)} (A \boxtimes P_{\bullet}) \oplus (P_{\bullet} \boxtimes A).$$

The involution $\sigma : [2] \to [2]$ acts on $P^2_{\bullet}$ in the obvious way. We also need to define the transition maps ι_f for the two injections $d, d' : [1] \to [2]$ and the two surjections $s, s' : [2] \to [1]$. For d_1, the transition map $\iota_d : A \boxtimes P_{\bullet} \to P^2_{\bullet}$ is the obvious embedding, and so is the transition map $\iota_{d'}$. For the surjection s, we need a map ι_s from the cone of the map

$$P_{\bullet} \otimes_A P_{\bullet} \xrightarrow{(\tau \otimes \mathrm{id}) \oplus (\mathrm{id} \otimes \tau)} P_{\bullet} \oplus P_{\bullet}.$$

to $P_{\bullet}$. On $P_{\bullet} \oplus P_{\bullet}$, the map ι_s is just the difference map $a \oplus b \mapsto a - b$; on $P_{\bullet} \otimes_A P_{\bullet}$, ι_s is our fixed homotopy $\iota : P_{\bullet} \otimes_A P_{\bullet} \to P_{\bullet+1}$. And similarly for the other surjection s'.

We leave it to the reader to check that if one computes $L^{\bullet}\,\mathrm{tr}_{\#}(A_{\#})\,|_{\Lambda_{\leq 2}}$ using this resolution $P^{\#}_{\bullet}$, then one obtains exactly (4.8) for the Connes differential B.

5 Discussion

One of the most unpleasant features of the construction presented in Section 4 is the strong assumptions we need to impose on the tensor category $\mathcal{C}$. In fact, the category to which one would really like to apply the construction is the category $\mathrm{End}\,\mathcal{B}$ of endofunctors—whatever that means—of the category $\mathcal{B}$ of coherent sheaves on an algebraic variety X. But if X is not affine, $\mathrm{End}\,\mathcal{B}$ certainly does not have enough projectives, so that Example 4.5 does not apply, and it is unlikely that $\mathrm{End}\,\mathcal{B}$ can be made homologically clean in the sense of Definition 4.4. We note that Definition 4.4 has been arranged so as not to impose anything more than strictly necessary for the proofs; but in practice, we do not know any examples that are not covered by Example 4.5.

As for the category $\mathrm{End}\,\mathcal{B}$, there is an even bigger problem with it: while there are ways to define endofunctors so that $\mathrm{End}\,\mathcal{B}$ is an abelian category with a right-exact tensor product, it cannot be equipped with a right-exact trace functor tr. Indeed, it immediately follows from Definition 4.8 that the Hochschild homology groups $HH_{\bullet}(\mathcal{C})$ of a tensor category $\mathcal{C}$ are trivial in negative homological degrees. If $\mathcal{C} = \mathrm{End}\,\mathcal{B}$, one of course expects $HH_{\bullet}(\mathcal{C}) = HH_{\bullet}(X)$, the Hochschild homology $HH_{\bullet}(X)$ of the variety X, which by now is well understood (see, e.g., [W]). And if X is not affine, $HH_{\bullet}(X)$ typically is nontrivial both in positive and in negative degrees. If X is smooth and proper, $HH_{\bullet}(X)$ in fact carries a nondegenerate pairing, so that it is just as nontrivial in degrees > 0 as in degrees < 0. Thus the case of a nonaffine algebraic variety is far beyond the methods developed in this paper.

The real reason for these difficulties is that we are dealing with abelian categories, while the theory emphatically wants to live in the triangulated world; as we explained in Example 2.5, even our main topic, cyclic bimodules, are best understood as objects of a triangulated category $\mathcal{D}\Lambda(\mathcal{C})$. Unfortunately, we cannot develop the theory from scratch in the triangulated context, since we do not have a strong and natural enough notion of an enhanced triangulated category (and working with the usual triangulated categories is out of the question because, for instance, the category of triangulated functors between triangulated categories is usually not a triangulated category itself). A well-developed theory would probably require a certain compromise between the abelian and the triangulated approaches. We will return to it elsewhere.

Another thing that is very conspicuously not done in the present paper is the combination of Section 4 and Section 3. Indeed, in Section 3, we are dealing with cyclic homology in the straightforward naive way of Section 2, and while we define the cyclic object $\widehat{A_{\#}}$ associated to a square-zero extension $\widetilde{A}$, we make no attempt to find an appropriate category $\mathsf{Sec}(\widehat{A\text{-bimod}_{\#}})$ where it should live. This is essentially the reason why we cannot go further than square-zero extensions. At present, sadly, we do not really understand this hypothetical category $\mathsf{Sec}(\widehat{A\text{-bimod}_{\#}})$.

One suspects that treating this properly would require studying deformations in a much more general context: instead of considering square-zero extensions of an algebra, we should look at the deformations of the abelian category of its modules, or at the deformations of the tensor category of its bimodules. This brings us to another topic completely untouched in the paper: the Hochschild cohomology $HH^{\bullet}(A)$.

Merely *defining* Hochschild cohomology for an arbitrary tensor category $\mathcal{C}$ is in fact much simpler than the definition of $HH_{\bullet}(\mathcal{C})$, and one does not need a trace functor for this: we just set $HH^{\bullet}(\mathcal{C}) = \mathrm{Ext}^{\bullet}(1,1)$, where $1 \in \mathcal{C}$ is the unit object. However, it is well understood by now that just as Hochschild homology always comes equipped with the Connes differential, the spectral sequence (2.7), and the whole cyclic homology package, Hochschild cohomology should be considered not as an algebra but as the so-called *Gerstenhaber* algebra; in fact, the pair $HH_{\bullet}(-), HH^{\bullet}(-)$ should form a version of "noncommutative calculus," as proposed for instance in [TT]. Deformations of the tensor category $\mathcal{C}$ should be controlled by $HH^{\bullet}(\mathcal{C})$, and the behavior of $HH_{\bullet}(\mathcal{C})$ and $HC_{\bullet}(\mathcal{C})$ under these deformations reflects various natural actions of $HH^{\bullet}(-)$ on $HH_{\bullet}(-)$.

We believe that a convenient development of the "noncommutative calculus" for a tensor category $\mathcal{C}$ might be possible along the same lines as our Section 4. Just as our category $\mathcal{D}\Lambda(\mathcal{C})$ is defined as the category of sections of the cofibration $\mathcal{C}_\#/\Lambda$, whose definition imitates the usual cyclic object $A_\#$, one can construct a cofibration $\mathcal{C}^\#/\Delta$ that imitates the standard cosimplicial object computing $HH^\bullet(A)$: for any $[n] \in \Delta$, $\mathcal{C}^\#([n])$ is the category of polylinear right-exact functors from $\mathcal{C}^{n-1}$ to $\mathcal{C}$, and the transition functors between various $\mathcal{C}^\#([n])$ are induced by the tensor product on $\mathcal{C}$. Then one can define a triangulated category $\mathcal{D}\Delta(\mathcal{C})$, the subcategory in $\mathcal{D}(\mathsf{Sec}(\mathcal{C}^\#))$ of complexes with co-Cartesian homology; the higher structures on $HH^\bullet(\mathcal{C})$ should be encoded in the structure of the category $\mathcal{D}\Delta(\mathcal{C})$, and relations between $HH_\bullet(\mathcal{C})$ and $HH^\bullet(\mathcal{C})$ should be reflected in a relation between $\mathcal{D}\Lambda(\mathcal{C})$ and $\mathcal{D}\Delta(\mathcal{C})$. We will proceed in this direction elsewhere. At present, the best we can do is to make the following hopeful observation:

- the category $\mathsf{Sec}_{cart}(\mathcal{C}^\#)$ is naturally a *braided* tensor category over k.

The reason for this is very simple: if one writes out explicitly the definition of $\mathsf{Sec}_{cart}(\mathcal{C}^\#)$ along the lines of Lemma 2.2, one finds that it coincides on the nose with the Drinfeld double of the tensor category $\mathcal{C}$.

References

[BD]   I. BUCUR, A. DELEANU, *Introduction to the theory of categories and functors*, Interscience Publication John Wiley & Sons, Ltd., London-New York-Sydney 1968.

[C1]   A. CONNES, *Non-commutative differential geometry, I, II*, Publ. IHES, **62** (1985), 257–360.

[C2]   A. CONNES, *Cohomologie cyclique et foncteur* Ext^n, Comptes Rendues Ac. Sci. Paris Sér. A-B, **296** (1983), 953–958.

[FT]   B. FEIGIN, B. TSYGAN, *Additive K-Theory*, in Lecture Notes in Math. **1289** (1987), 97–209.

[GM1] S. GELFAND, YU. MANIN, *Methods of homological algebra*, Nauka Publishers, Moscow, 1988 (in Russian).

[GM2] S. GELFAND, YU. MANIN, *Homological algebra*, Itogi vol. **38** (Algebra V), VINITI, Moscow, 1989 (in Russian).

[Ge]   E. GETZLER, *Cartan homotopy formulas and the Gauss-Manin connection in cyclic homology*, in *Quantum deformations of algebras and their representations (Ramat-Gan, 1991/1992; Rehovot, 1991/1992)*, Israel Math. Conf. Proc. **7**, Bar-Ilan Univ., Ramat Gan, 1993, 65–78.

[Go]   T. GOODWILLIE, *Cyclic homology, derivations, and the free loopspace*, Topology **24** (1985), 187–215.

[Gr]   A. GROTHENDIECK, *Expose VI: Catégories fibré et descente*, in *SGAI: Revétements étales et groupe fondamental*, Lecture Notes in Math., **224**, Springer, Berlin; 145–194.

[Ka]   D. KALEDIN, *Non-commutative Hodge-to-de Rham degeneration via the method of Deligne-Illusie*, Pure. Appl. Math. Q., **4** (2008), no. 3, part 2, 785–875.

[L]   J.-L. LODAY, *Cyclic Homology*, second ed., Springer, 1998.

[TT] D. TAMARKIN, B. TSYGAN, *The ring of differential operators on forms in noncommutative calculus*, in *Graphs and patterns in mathematics and theoretical physics.* Proc. Sympos. Pure Math. **73**, AMS, Providence, RI, 2005; 105–131.

[Ts] B. TSYGAN, *Homology of Lie algebras over rings and Hochschild homology*, Uspekhi Mat. Nauk, **38** (1983), 217–218.

[W] C. WEIBEL, *Cyclic homology for schemes*, Proc. AMS **124** (1996), 1655–1662.

Noncommutative Geometry and Path Integrals

Mikhail Kapranov

Department of Mathematics, Yale University, 10 Hillhouse Avenue, New Haven, CT, 06520 USA
mikhail.kapranov@yale.edu

To Yuri Ivanovich Manin on his 70th birthday

Summary. We argue that there should exist a "noncommutative Fourier transform" which should identify functions of noncommutative variables (say, of matrices of indeterminate size) and ordinary functions or measures on the space of paths. Some examples are considered.

Key words: noncommutative geometry, Fourier transform, path integral

2000 Mathematics Subject Classifications: Primary 16S16, Secondary 20E05

Introduction

(0.1) A monomial in noncommutative variables X and Y, say, $X^i Y^j X^k Y^l \ldots$, can be visualized as a lattice path in the plane, starting from 0, going i steps in the horizontal direction, j steps in the vertical one, then again k steps in the horizontal one, and so on. Usual commutative monomials are often visualized as lattice points, for example $x^a y^b$ corresponds to the point (a, b). To lift such a monomial to the noncommutative domain is therefore the same as to choose a "history" for (a, b), i.e., a lattice path originating at 0 and ending at (a, b).

This correspondence between paths and noncommutative monomials can be extended to more general piecewise smooth paths if we deal with exponential functions instead. Let us represent our commutative variables as $x = e^z, y = e^w$; then a monomial will be replaced by the exponential e^{az+bw} and we are free to take a and b to be any real numbers. To lift this exponential to the noncommutative domain, i.e., to a series in Z, W where $X = e^Z, Y = e^W$, one needs to choose a path γ in $\mathbb{R}^2$ joining 0 with (a, b). One can easily see this by approximating γ by lattice paths with step $1/M$, $M \to \infty$, and working with monomials in $X^{1/M} = e^{Z/M}$ and $Y^{1/M} = e^{W/M}$. Denote this exponential series by $E_\gamma(Z, W)$.

Y. Tschinkel and Y. Zarhin (eds.), *Algebra, Arithmetic, and Geometry,*
Progress in Mathematics 270, DOI 10.1007/978-0-8176-4747-6_3,
© Springer Science+Business Media, LLC 2009

This suggests the possibility of a "noncommutative Fourier transform" (NCFT) identifying appropriate spaces of functions of noncommuting variables (say, of matrices of indeterminate size) with spaces of ordinary functions or measures on the space of paths. For example, to a measure μ on the space Π of paths (or some completion of it) we want to associate the function $\mathcal{F}(\mu)$ of Z, W given by

$$(0.1.1) \qquad \mathcal{F}(\mu)(Z, W) = \int_{\gamma \in \Pi} E_\gamma(Z, W) d\mu(\gamma).$$

The basic phenomenon here seems to be that the two types of functional spaces (noncommutative functions of n variables vs. ordinary commutative functions but on the space of paths in $\mathbb{R}^n$) have, on some fundamental level, *the same size.*

The goal of this paper and the ones to follow [K1-2] is to investigate this idea from several points of view.

(0.2) The concept of NCFT seems to implicitly underlie the very foundations of quantum mechanics such as the equivalence of the Lagrangian and Hamiltonian approaches to the theory. Indeed, the Lagrangian point of view deals with path integrals, while the Hamiltonian one works with noncommuting operators. Further, it is very close to the concept of the "Wilson loop" functional (trace of the holonomy) in Yang–Mills theory [Po]. Note that the exponential E_γ, being itself the holonomy of a certain formal connection, is invariant under reparametrization of the path. Quantities invariant under reparametrization are particularly important in string theory, and the reparametrization invariance of the Wilson loop led to conjectural relations between strings and the $N \to \infty$ limit of Yang–Mills theory [Po].

Since the integral transform $\mathcal{F}$ should, intuitively, act between spaces of the same size, it does not lead to any loss of information and can therefore be viewed as "path integration without integration." The actual integration occurs when we restrict the function $\mathcal{F}(\mu)$ to the commutative locus, i.e., make Z and W commute. Alternatively, instead of allowing Z, W to be arbitrary matrices, we take them to be scalars. Then all paths having the same endpoint will contribute to make up a single Fourier mode of the commutativized function. We arrive at the following conclusion: *the natural homomorphism $R \to R_{ab}$ of a noncommutative ring to its maximal commutative quotient is the algebraic analogue of path integration.*

(0.3) The idea that the space of paths is related to the free group and to its various versions was clearly enunciated by K.-T. Chen [C1] in the 1950s and can be traced throughout almost all of his work [C0]. Apparently, much more can be said about this classical subject. Thus, the universal connection with values in the free Lie algebra (known to Chen and appearing in (2.1) below) leads to beautiful nonholonomic geometry on the free nilpotent Lie groups $G_{n,d}$, which is still far from being fully understood; see [G].

Well-known examples of measures on path spaces are provided by probability theory, and we spend some time in §4 below to formulate various results from the probabilistic literature in terms of NCFT. Most importantly, the Fourier transform of the Wiener measure on paths in $\mathbb{R}^n$ is the noncommutative Gaussian series $\exp\left(-\sum_{i=1}^n Z_j^2\right)$, where the Z_i are considered as noncommuting variables. We should mention here the recent book by Baudoin [Ba], who considered the idea of associating a noncommutative series to a stochastic process. It is clearly the same type of construction as our NCFT except in the framework of probability theory: parametrized paths, positive measures, etc.

(0.4) I would like to thank R. Beals, E. Getzler, H. Koch, Y.I. Manin, and M. A. Olshanetsky for useful discussions. I am also grateful to the referee for several remarks that helped improve the exposition. This paper was written during my stay at the Max-Planck-Institut für Mathematik in Bonn, and I am grateful to the institute for support and excellent working conditions. This work was also partially supported by an NSF grant.

1 Noncommutative Monomials and Lattice Paths

(1.1) Noncommutative polynomials and the free semigroup. Consider n noncommuting (free) variables $X_1, \ldots, X_n$ and form the algebra of noncommutative polynomials in these variables. This algebra will be denoted by $\mathbb{C}\langle X_1, \ldots, X_n \rangle$. It is the same as the tensor algebra

$$T(V) = \bigoplus_{d=1}^{\infty} V^{\otimes d}, \quad V = \mathbb{C}^n = \bigoplus_{i=1}^{n} \mathbb{C} \cdot X_i.$$

A noncommutative monomial in $X = (X_1, \ldots, X_n)$ is, as described in the introduction, the same as a monotone lattice path in $\mathbb{R}^n$ starting at 0. We denote by F_n^+ the set of all such paths and write X^γ for the monomial corresponding to a path γ. The set F_n^+ is a semigroup with the following operation. If γ, γ' are two monotone paths as above starting at 0, then $\gamma \circ \gamma'$ is obtained by translating γ so that its beginning meets the end of γ' and then forming the composite path. It is clear that F_n^+ is the free semigroup on n generators. Thus a typical noncommutative polynomial is written as

(1.1.1) $$f(X_1, \ldots, X_n) = f(X) = \sum_{\gamma \in F_n^+} a_\gamma X^\gamma.$$

Along with $\mathbb{C}\langle X_1, \ldots, X_n \rangle$ we will consider the algebra $\mathbb{C}[x_1, \ldots, x_n]$ of usual (commutative) polynomials in the variables $x_1, \ldots, x_n$. A typical such polynomial will be written as

(1.1.2) $$g(x_1, \ldots, x_n) = g(x) = \sum_{\alpha \in \mathbb{Z}_+^n} b_\alpha x^\alpha, \quad x^\alpha = x_1^{\alpha_1} \cdots x_n^{\alpha_n}.$$

The two algebras are related by the *commutativization homomorphism*

$$(1.1.3) \qquad c : \mathbb{C}\langle X_1, \ldots, X_n \rangle \to \mathbb{C}[x_1, \ldots, x_n],$$

which takes X_i to x_i. For a path $\gamma \in \Gamma_n$ let $e(\gamma) \in \mathbb{Z}_+^n$ denote the endpoint of γ. Then we have

$$(1.1.4) \qquad c(X^\gamma) = x^{e(\gamma)}.$$

This means that at the level of coefficients, the commutativization homomorphism is given by the summation over paths with given endpoints: if $g(x) = c(f(X))$, then

$$(1.1.5) \qquad b_\alpha = \sum_{e(\gamma)=\alpha} a_\gamma.$$

(1.2) Noncommutative power series. Let $I \subset \mathbb{C}\langle X_1, \ldots, X_n \rangle$ be the span of monomials of degree ≥ 1. Then clearly I is a 2-sided ideal and I^d is the span of monomials of degree $\geq d$. We define the algebra $\mathbb{C}\langle\langle X_1, \ldots, X_n \rangle\rangle$ as the completion of $\mathbb{C}\langle X_1, \ldots, X_n \rangle$ in the I-adic topology. Explicitly, elements of $\mathbb{C}\langle\langle X_1, \ldots, X_n \rangle\rangle$ can be seen as infinite formal linear combinations of noncommutative monomials, i.e., expressions of the form $\sum_{\gamma \in F_N^+} a_\gamma X^\gamma$. For example,

$$(1.2.1) \qquad e^{X_1} \cdot e^{X_2} = \sum_{i,j=0}^\infty \frac{X_1^i X_2^j}{i!j!}, \qquad \frac{1}{1-(X_1+X_2)} = \sum_{\gamma \in F_2^+} X^\gamma$$

are noncommutative power series. We will also be interested in convergence of noncommutative series. A series $f(X) = \sum_{\gamma \in F_n^+} a_\gamma X^\gamma$ will be called *entire* if

$$(1.2.2) \qquad \lim_{\gamma \to \infty} R^{l(\gamma)} |a_\gamma| = 0, \quad \forall R > 0.$$

Here $l(\gamma)$ is the length of the path γ, and the limit is taken over the countable set F_n^+ (so no ordering of this set is needed). We denote by $\mathbb{C}\langle\langle X_1, \ldots, X_n \rangle\rangle^{\text{ent}}$ the set of entire series. It is clear that this set is a subring.

(1.2.3) Proposition. *The condition (1.2.2) is equivalent to the property that for any N and for any square matrices $X_1^0, \ldots, X_n^0$ of size N the series of matrices $\sum a_\gamma (X^0)^\gamma$ obtained by specializing $X_i \to X_i^0$ converges absolutely.*

(1.3) Noncommutative Laurent polynomials. By a noncommutative Laurent monomial in $X_1, \ldots, X_n$ we will mean a monomial in positive and negative powers of the X_i such as, e.g., $X_1 X_2 X_1^{-1} X_2^5$. In other words, this is an element of F_n, the free noncommutative group on the generators X_i. A noncommutative Laurent polynomial is then a finite formal linear combination of such monomials, i.e., an element of the group algebra of F_n. We will denote this algebra by

$$(1.3.1) \qquad \mathbb{C}\langle X_1^{\pm 1}, \ldots, X_n^{\pm 1}\rangle = \mathbb{C}[F_n].$$

As before, a noncommutative Laurent monomial corresponds to a lattice path in $\mathbb{R}^n$ beginning at 0 but not necessarily monotone. These paths are defined up to cancellation of pieces consisting of a subpath and the same subpath run in the opposite direction immediately afterward.

We retain the notation X^γ for the monomial corresponding to a path γ. We also write $(-\gamma)$ for the path inverse to γ, so $X^{-\gamma} = (X^\gamma)^{-1}$.

(1.4) Noncommutative Fourier transform: discrete case. The usual (commutative) Fourier transform relates the spaces of functions on a locally compact abelian group G and its Pontryagin dual $\widehat{G}$. The "discrete" case $G = \mathbb{Z}^n$, $\widehat{G} = (S^1)^n$ corresponds to the theory of Fourier series.

In the algebraic formulation, the discrete Fourier transform identifies the space of finitely supported functions

$$(1.4.1) \qquad b : \mathbb{Z}^n \to \mathbb{C}, \quad \alpha \mapsto b_\alpha, \quad |\operatorname{Supp}(b)| < \infty,$$

with the space $\mathbb{C}[x_1^{\pm 1}, \ldots, x_n^{\pm 1}]$ of Laurent polynomials. It is given by the well-known formulas

$$(1.4.2) \qquad (b_\alpha) \mapsto f, \quad f(x) = \sum_{\alpha \in \mathbb{Z}^n} b_\alpha x^\alpha,$$

$$(1.4.3) \qquad f \mapsto (b_\alpha), \quad b_\alpha = \int_{|x_1| = \cdots = |x_n| = 1} f(x) x^{-\alpha} d^* x_1 \cdots d^* x_n,$$

where $d^* x$ is the Haar measure on S^1 with volume 1. Our goal in this section is to give a generalization of these formulas for noncommutative Laurent polynomials.

Instead of (1.4.1) we consider the space of finitely supported functions

$$(1.4.4) \qquad a : F_n \to \mathbb{C}, \quad \gamma \mapsto a_\gamma, \quad |\operatorname{Supp}(a)| < \infty.$$

The discrete noncommutative Fourier transform is the identification of this space with $\mathbb{C}\langle X_1^{\pm 1}, \ldots, X_n^{\pm 1}\rangle$ via

$$(1.4.5) \qquad (a_\gamma) \mapsto f, \quad f(X) = \sum_{\gamma \in F_n} a_\gamma X^\gamma.$$

This identification ceases to look like a tautology if we regard a noncommutative Laurent polynomial as a function f that to any n invertible elements $X_1^0, \ldots, X_n^0$ of any associative algebra A associates an element $f(X_1^0, \ldots, X_n^0) \in A$. We want then to recover the coefficients a_γ in terms of the values of f on various elements of various A. Most importantly, we will consider $A = \mathrm{Mat}_N(\mathbb{C})$, the algebra of matrices of size N, and let N be arbitrary. To get a generalization of (1.4.3) we replace the unit circle $|x| = 1$ by the group of unitary matrices $U(N) \subset \mathrm{Mat}_N(\mathbb{C})$. Let d^*X be the Haar measure on $U(N)$ of volume 1.

The following result is a consequence of the so-called asymptotic freedom theorem for unitary matrices due to Voiculescu [V]; see also [HP] for a more elementary exposition.

(1.4.6) Theorem. *If $f(X) = \sum_{\gamma \in F_n} a_\gamma X^\gamma$ is a noncommutative Laurent polynomial, then we have*

$$a_\gamma = \lim_{N \to \infty} \frac{1}{N} \, \mathrm{tr} \int_{X_1,\dots,X_n \in U(N)} f(X_1,\dots,X_n) \, X^{-\gamma} \, d^*X_1 \cdots d^*X_n.$$

As for the commutative case, the theorem is equivalent to the following orthogonality relation. It is this relation that is usually called "asymptotic freedom" in the literature.

(1.4.7) Reformulation. *Let $\gamma \in F_n$ be a nontrivial lattice path. Then*

$$\lim_{N \to \infty} \frac{1}{N} \, \mathrm{tr} \int_{X_1,\dots,X_n \in U(N)} X^\gamma \, d^*X_1 \cdots d^*X_n = 0.$$

Note that for $\gamma = 0$ the integral is equal to 1 for any N.

Passing to the $N \to \infty$ limit is unavoidable here, since for any given N there exist nonzero noncommutative polynomials that vanish identically on $\mathrm{Mat}_N(\mathbb{C})$. An example is provided by the famous Amitsur–Levitsky polynomial

$$f(X_1,\dots,X_{2N}) = \sum_{\sigma \in S_{2N}} \mathrm{sgn}(\sigma) X_{\sigma(1)} \cdots X_{\sigma(2N)}.$$

2 Noncommutative exponential functions.

(2.1) The universal connection and noncommutative exponentials. Let us introduce the "logarithmic" variables $Z_1,\dots,Z_n$, so that we have the embedding

$$(2.1.1) \qquad \mathbb{C}\langle X_1,\dots,X_n \rangle \subset \mathbb{C}\langle\langle Z_1,\dots,Z_n \rangle\rangle, \qquad X_i \mapsto e^{Z_i}.$$

The algebra $\mathbb{C}\langle\langle Z_1,\dots,Z_n \rangle\rangle$ is a projective limit of finite-dimensional algebras, namely

$$(2.1.2) \qquad \mathbb{C}\langle\langle Z_1,\dots,Z_n \rangle\rangle = \varprojlim{}_d \, \mathbb{C}\langle Z_1,\dots,Z_n \rangle / I^d,$$

where the ideal I is as in (1.2).

Consider the space $\mathbb{R}^n$ with coordinates $y_1, \ldots, y_n$. On this space we have the following 1-form with values in $\mathbb{C}\langle\langle Z_1, \ldots, Z_n\rangle\rangle$:

$$(2.1.3) \qquad \Omega \;=\; \sum_i Z_i \cdot dy_i \;\in\; \Omega^1(\mathbb{R}^n) \otimes \mathbb{C}\langle\langle Z_1, \ldots, Z_n\rangle\rangle.$$

We consider the form as a connection on $\mathbb{R}^n$. One can see it as the universal translation-invariant connection on $\mathbb{R}^n$, an algebraic version of the connection of Kobayashi on the path space, see [Si], Section 3.

Let γ be any piecewise smooth path in $\mathbb{R}^n$. We define the noncommutative exponential function corresponding to γ to be the holonomy of the above connection along γ:

$$(2.1.4) \qquad E_\gamma(Z) = E_\gamma(Z_1, \ldots, Z_n) = P\exp \int_\gamma \Omega \;\in\; \mathbb{C}\langle\langle Z_1, \ldots, Z_n\rangle\rangle.$$

The holonomy can be understood by passing to finite-dimensional quotients as in (2.1.2) and solving an ordinary differential equation with values in each such quotient.

It is clear that $E_\gamma(Z)$ becomes unchanged under parallel translations of γ, since the form Ω is translation-invariant. So in the following we will always assume that γ begins at 0.

Further, $E_\gamma(Z)$ is invariant under reparametrizations of γ: this is a general property of the holonomy of any connection. So let us give the following definition.

(2.1.5) Definition. *Let M be a C^∞-manifold. An (oriented) unparametrized path in M is an equivalence class of pairs $(I, \gamma : I \to M)$, where I is a smooth manifold with boundary diffeomorphic to $[0,1]$ and γ is a piecewise smooth map $I \to M$. Two such pairs (I, γ) and (I', γ') are equivalent if there is an orientation-preserving piecewise smooth homeomorphism $\phi : I \to I'$ such that $\gamma = \gamma' \circ \phi$.*

We will denote an unparametrized path simply by γ.

(2.1.6) Example. Let γ be a straight segment in $\mathbb{R}^2$ joining (0,0) and (1,1). Let also δ be the path consisting of the horizontal segment $[(0,0),(0,1)]$ and the vertical segment $[(0,1),(1,1)]$. Let σ be the path consisting of the vertical segment $[(0,0),(1,0)]$ and the horizontal segment $[(1,0),(1.1)]$. Then

$$E_\gamma(Z_1, Z_2) = e^{Z_1 + Z_2}, \quad E_\delta(Z_1, Z_2) = e^{Z_1} e^{Z_2}, \quad E_\sigma(Z) = e^{Z_2} e^{Z_1}.$$

More generally, if γ is a lattice path corresponding to the integer lattice $\mathbb{Z}^n$, then $E_\gamma(Z) = X^\gamma$ is the noncommutative monomial in $X_i = e^{Z_i}$ associated to γ as in Section 1.

Let γ, γ' be two unparametrized paths in $\mathbb{R}^n$ starting at 0. Their product $\gamma \circ \gamma'$ is the path obtained by translating γ so that its beginning meets the end of γ' and then forming the composite path. The set of γ's with this operation forms a semigroup. For a path γ we denote by γ^{-1} the path obtained by translating γ so that its end meets 0 and then taking it with the opposite orientation. Finally, we denote by Π_n the set of paths as above modulo cancellations, i.e., forgetting subpaths of a given path consisting of a segment and then immediately of the same segment run in the opposite direction. Clearly the set Π_n forms a group, which we will call the *group of paths* in $\mathbb{R}^n$.

The standard properties of the holonomy of connections imply the following:

(2.1.7) Proposition. *(a) We have*

$$E_{\gamma \circ \gamma'}(Z) = E_\gamma(Z) \cdot E_{\gamma'}(Z), \quad E_{\gamma^{-1}}(Z) = E_\gamma(Z)^{-1}$$

(equalities in $\mathbb{C}\langle\langle Z_1, \ldots, Z_n \rangle\rangle$).
(b) The series $E_\gamma(Z)$ is entire, i.e., it converges for any given N by N matrices $Z_1^0, \ldots, Z_n^0$.
(c) If $Z_1^0, \ldots, Z_n^0$ are Hermitian, then $E_\gamma\left(iZ_1^0, \ldots, iZ_n^0\right)$ is unitary.

The property (a) implies that $E_\gamma(Z)$ depends only on the image of γ in the group Π_n. Further, let us consider the commutativization homomorphism

$$(2.1.8) \qquad c : \mathbb{C}\langle\langle Z_1, \ldots, Z_n \rangle\rangle \to \mathbb{C}[[z_1, \ldots, z_n]].$$

The following is also obvious.

(2.1.9) Proposition. *If $a = (a_1, \ldots, a_n)$ is the endpoint of γ, then*

$$c(E_\gamma(Z)) = e^{(a,z)}$$

is the usual exponential function.

Thus there are as many ways to lift $e^{(a,z)}$ into the noncommutative domain as there are paths in $\mathbb{R}^n$ joining 0 and a.

(2.2) Idea of a noncommutative Fourier transform. The above observations suggest that there should be a version of Fourier transform that would identify an appropriate space of measures on Π_n with an appropriate space of functions of n noncommutative variables $Z_1, \ldots, Z_n$, via the formula

$$(2.2.1) \qquad \mu \mapsto f(Z_1, \ldots, Z_n) = \int_{\gamma \in \Pi_n} E_\gamma(iZ_1, \ldots, iZ_n) \mathcal{D}\mu(\gamma).$$

The integral in (2.2.1) is thus a path integral. The concept of a "function of noncommuting variables" is of course open to interpretation. Several such interpretations are currently being considered in noncommutative geometry.

In the present paper we adopt a loose point of view that a function of n noncommutative variables is an element of an algebra R equipped with a homomorphism $\mathbb{C}\langle Z_1, \ldots, Z_n \rangle \to R$. We will assume that this homomorphism realizes R as some kind of completion or localization (or both) of $\mathbb{C}\langle Z_1, \ldots, Z_n \rangle$.

In other words, that R does not have "superfluous" elements, independent of the images of the Z_i. See [Ta] for an early attempt to define noncommutative functions in the analytic context.

(2.2.2) Examples. We can take $R = \mathbb{C}\langle\langle Z_1, \ldots, Z_n\rangle\rangle^{\text{ent}}$, the algebra of entire power series. Alternatively we can take R to be the skew field of "noncommutative rational functions" in $Z_1, \ldots, Z_n$ constructed by P. Cohn [Coh]. Thus expressions such as

$$\exp\left(Z_1^2 + Z_2^2\right), \quad \left(Z_1^2 + Z_2^2\right)^{-1}, \quad \left(Z_1 Z_2 - Z_2 Z_1\right)^{-1} + Z_3^{-2} Z_1$$

are considered noncommutative functions.

It will be important for us to be able to view a "function" $f(Z_1, \ldots, Z_n)$ as above as an actual function defined on appropriate subsets of n-tuples of N by N matrices for each N and taking values in matrices of the same size.

Similarly, the group Π_n can also possibly be replaced by various related objects (completions). In this paper we will consider several approaches such as completion by a pro-algebraic group or completion by continuous paths.

Alternatively, functions on Π_n should correspond to "noncommutative measures" or distributions on the space of noncommutative functions. Examples of such "measures" are being studied in free probability theory [HP], [NS], [VDN]. See Section 6 below.

Note that we have a surjective homomorphism of groups

$$(2.2.3) \qquad\qquad e : \Pi_n \to \mathbb{R}^n, \quad \gamma \mapsto e(\gamma).$$

Here $e(\gamma)$ is the endpoint of γ. One important property of the Noncommutative Fourier transform (NCFT) is the following principle, which is just a consequence of Proposition 2.1.9: under the Fourier transform, the integration over paths with given beginning and end, i.e., the pushdown of measures on Π_n to measures on $\mathbb{R}^n$, corresponds to a simple algebraic operation: the commutativization homomorphism

$$(2.2.4) \qquad\qquad c : R \to R/([R, R]),$$

where R is a noncommutative algebra and the right-hand side is the maximal commutative quotient of R.

(2.3) Relation to Chen's iterated integrals. Let us recall the main points of Chen's theory. Let M be a smooth manifold, γ an unparametrized path, and ω a smooth 1-form on M.

Along with the "definite integral" $\int_\gamma \omega$, we can consider the "indefinite integral," which is a function "on γ," or, more precisely, on the abstract interval I such that γ is a map $I \to M$. For any $t \in I$ we have the subpath $\gamma_{\leq t}$ going from the beginning of I until t, and we have the function

$$\int_{(\gamma)} \omega : I \to \mathbb{C}, \quad t \mapsto \int_{\gamma_{\leq t}} \omega.$$

If now ω_1 and ω_2 are two smooth 1-forms on M, we can form a new 1-form on γ by multiplying (the restriction of) ω_2 and the function $\int_{(\gamma)} \omega_1$. Then this form can be integrated along γ. The result is called the *iterated integral*

$$\int_{\gamma}^{\rightarrow} \omega_1 \cdot \omega_2 = \int_{\gamma} \left(\omega_2 \cdot \int_{(\gamma)} \omega_1 \right).$$

Note that if we think of γ as a map $\gamma : I \to M$, then the iterated integral is equal to

$$\int_{t_1 \leq t_2 \in I} \gamma^*(\omega_1)(t_1) \gamma^*(\omega_2)(t_2).$$

Note that integration over all $t_1, t_2 \in I$ would give the product $\left(\int_{\gamma} \omega_1 \right) \cdot \left(\int_{\gamma} \omega_2 \right)$.

Similarly, one defines the d-fold iterated integral of d smooth 1-forms $\omega_1, \ldots, \omega_d$ on M by induction:

$$\int_{\gamma}^{\rightarrow} \omega_1 \cdots \omega_d = \int_{\gamma} \left(\omega_d \cdot \int_{(\gamma)}^{\rightarrow} \omega_1 \cdots \omega_{d-1} \right),$$

where the $(d-1)$-fold indefinite iterated integral is defined as the function on I of the form

$$t \to \int_{\gamma \leq t}^{\rightarrow} \omega_1 \cdots \omega_{d-1}.$$

As before, the iterated integral is equal to the integral over the d-simplex:

$$\int_{\gamma}^{\rightarrow} \omega_1 \cdots \omega_d = \int_{t_1 \leq \cdots \leq t_d \in I} \gamma^* \omega_1(t_1) \cdots \gamma^* \omega_d(t_d).$$

The concept of iterated integrals extends in an obvious way to 1-forms with values in any associative (pro-)finite-dimensional $\mathbb{C}$-algebra R. The well-known Picard series for the holonomy of a connection consists exactly of such iterated integrals. We state this as follows.

(2.3.1) Proposition. *Let R be any (pro-)finite-dimensional associative $\mathbb{C}$-algebra, and A be a smooth 1-form on M with values in R considered as a connection form. Then the parallel transport along an unparametrized path γ has the form*

$$P \exp \int_{\gamma} A = \sum_{d=0}^{\infty} \int_{\gamma}^{\rightarrow} A \cdots A.$$

Here the term corresponding to $d = 0$ is set equal to 1.

Let us specialize this to $M = \mathbb{R}^n$, $R = \mathbb{C}\langle\langle Z_1, \ldots, Z_n \rangle\rangle$, and $\Omega = \sum Z_i dy_i$.

(2.3.2) Corollary. *The coefficient of the series $E_\gamma(Z_1, \ldots, Z_n)$ at any noncommutative monomial $Z_{i_1} \cdots Z_{i_d}$ is equal to the iterated integral*

$$\int_{\gamma}^{\rightarrow} dy_{i_1} \cdots dy_{i_d}.$$

Thus E_γ is the generating function for all the iterated integrals involving constant 1-forms on $\mathbb{R}^n$.

(2.3.3) Example. By the above,

$$E_\gamma(Z) = 1 + \sum a_i Z_i + \sum b_{ij} Z_i Z_j + \cdots,$$

where $a_i = \int_\gamma dy_i$ is the ith coordinate of the endpoint of γ and

$$b_{ij} = \int_\gamma \left(dy_i \cdot \int_{(\gamma)} dy_j \right) = \int_\gamma y_j dy_i.$$

Suppose that γ is closed, so $a_i = 0$. Then $b_{ii} = 0$, and for $i \neq j$ we have that b_{ij} is the oriented area encirlced by γ after the projection to the (i, j)-plane.

The following was proved by Chen [C2].

(2.3.4) Theorem. *The homomorphism $\Pi_n \to \mathbb{C}\langle\langle Z_1, \ldots, Z_n\rangle\rangle^*$ sending γ to E_γ is injective. In other words, if a path γ has all iterated integrals as above equal to 0, then γ is (equivalent modulo cancellations to) a constant path (situated at 0).*

(2.4) Grouplike and primitive elements. Let $FL(Z_1, \ldots, Z_n)$ be the free Lie algebra generated by $Z_1, \ldots, Z_n$. It is characterized by the obvious universal property; see [R] for background. This property implies that we have a Lie algebra homomorphism

(2.4.1) $h : FL(Z_1, \ldots, Z_n) \to \mathbb{C}\langle Z_1, \ldots, Z_n\rangle,$

and this homomorphism identifies $\mathbb{C}\langle Z_1, \ldots, Z_n\rangle$ with the universal enveloping algebra of $FL(Z_1, \ldots, Z_n)$. Further, let us consider the Hopf algebra structure on $\mathbb{C}\langle Z_1, \ldots, Z_n\rangle$ given on the generators by

(2.4.2) $\Delta(Z_i) = Z_i \otimes 1 + 1 \otimes Z_i.$

The following result, originally due to K. Friedrichs, is a particular case of a general property of enveloping algebras.

(2.4.3) Theorem. *The image of h consists precisely of all primitive elements, i.e., of elements f such that $\Delta(f) = f \otimes 1 + 1 \otimes f$.*

We will also use the term *Lie elements* for primitive elements of $\mathbb{C}\langle Z_1, \ldots, Z_n\rangle$.

Further, consider the noncommutative power series algebra $\mathbb{C}\langle\langle Z_1, \ldots, Z_n\rangle\rangle$. It is naturally a topological Hopf algebra with respect to the comultiplication given by (2.4.2) on generators and extended by additivity, multiplicativity, and continuity.

The free Lie algebra is graded:

$$(2.4.3) \qquad \mathrm{FL}(Z_1, \ldots, Z_n) = \bigoplus_{d \geq 1} \mathrm{FL}(Z_1, \ldots, Z_n)_d,$$

where $\mathrm{FL}(Z_1, \ldots, Z_n)_d$ is the span of Lie monomials containing exactly d letters. We denote by

$$(2.4.4) \qquad \mathfrak{g}_n = \prod_{d \geq 1} \mathrm{FL}(Z_1, \ldots, Z_n)_d$$

its completion, i.e., the set of formal *Lie series*. This is a complete topological Lie algebra. We clearly have an embedding of $\mathfrak{g}_n$ into $\mathbb{C}\langle\langle Z_1, \ldots, Z_n \rangle\rangle$ induced by the embedding of the graded components as above. Further, degree-by-degree considerations and Theorem 2.4.3 imply the following:

(2.4.5) Corollary. *A noncommutative power series $f \in \mathbb{C}\langle\langle Z_1, \ldots, Z_n \rangle\rangle$ lies in $\mathfrak{g}_n$ if and only if it is primitive, i.e., $\Delta(f) = f \otimes 1 + 1 \otimes f$ with respect to the topological Hopf algebra structure defined above.*

Along with primitive (or Lie) series in $Z_1, \ldots, Z_n$ we will consider grouplike elements of $\mathbb{C}\langle\langle Z_1, \ldots, Z_n \rangle\rangle$, i.e., series Φ satisfying

$$(2.4.6) \qquad \Delta(\Phi) = \Phi \otimes \Phi.$$

The completed tensor product $\mathbb{C}\langle\langle Z_1, \ldots, Z_n \rangle\rangle \widehat{\otimes} \mathbb{C}\langle\langle Z_1, \ldots, Z_n \rangle\rangle$ consists of series in $2n$ variables $Z_i' = Z_i \otimes 1$ and $Z_i'' = 1 \otimes Z_i$ that satisfy $[Z_i', Z_j''] = 0$ and no other relations. Thus a series $\Phi(Z_1, \ldots, Z_n)$ is grouplike if it satisfies the *exponential property*:

$$(2.4.7) \qquad F(Z_1' + Z_1'', \ldots, Z_n' + Z_n'') = F(Z_1', \ldots, Z_n') \cdot F(Z_1'', \ldots, Z_n''),$$

provided $[Z_i', Z_j''] = 0$, $\forall i, j$. We denote by G_n the set of grouplike elements in $\mathbb{C}\langle\langle Z_1, \ldots, Z_n \rangle\rangle$. Elementary properties of cocommutative Hopf algebras and elementary convergence arguments in the adic topology imply the following:

(2.4.8) Proposition. *(a) G_n is a group with respect to the multiplication. (b) The exponential series defines a bijection*

$$\exp : \mathfrak{g}_n \to G_n,$$

with the inverse given by the logarithmic series.
(c) The image of any series $\Phi \in G_n$ under the commutativization homomorphism (2.1.8) is a formal series of the form $e^{(a,z)}$ for some $a \in \mathbb{C}^n$.
(d) If $\Phi \in G_n$, then

$$\Phi(-Z_1, \ldots, -Z_n) = \Phi(Z_1, \ldots, Z_n)^{-1}$$

(equality of power series).

(2.4.9) Example. The above proposition implies that the series

$$\log(e^{Z_1} \cdot e^{Z_2}) \in \mathbb{C}\langle\langle Z_1, Z_2 \rangle\rangle$$

is in fact a Lie series. It is known as the Campbell–Hausdorff series, and its initial part has the form

$$\log(e^{Z_1} \cdot e^{Z_2}) = Z_1 + Z_2 + \frac{1}{2}[Z_1, Z_2] + \cdots.$$

Let $G_n(\mathbb{R}) \subset G_n$ be the set of grouplike series with real coefficients. Further, the Lie algebra $\mathrm{FL}(Z_1, \ldots, Z_n)$ is in fact defined over rational numbers. In particular, it makes sense to speak about its real part. By taking the completion as above, we define the real part of the completed free algebra $\mathfrak{g}_n(\mathbb{R})$. It is clear that the exponential series establishes a bijection between $\mathfrak{g}_n(\mathbb{R})$ and $G_n(\mathbb{R})$.

The following fact was also pointed out by Chen [C2].

(2.4.9) Theorem. *If $\gamma \in \Pi_n$ is a path in $\mathbb{R}^n$ as above, then $E_\gamma(Z)$ is grouplike. Moreover, it lies in the real part $G_n(\mathbb{R})$.*

Note that a typical element $\Phi = \Phi(Z_1, \ldots, Z_n) \in G_n$ is a priori just a formal power series and does not have to converge for any given matrix values of the Z_i (unless they are all 0). At the same time, series of the form $\Phi = E_\gamma$, $\gamma \in \Pi_n$, converge for all values of the Z_i. This leads to the proposal, formulated by Chen [C3], to view series from G_n with good covergence properties as corresponding to "generalized paths," i.e., paths perhaps more general than piecewise C^∞ ones. The theory of stochastic integrals, see below, provides a step in a similar direction.

(2.5) Finite-dimensional approximations to G_n and $\mathfrak{g}_n$. Let us recall a version of the Malcev theory for nilpotent Lie algebras. Let k be a field of characteristic 0. A Lie algebra $\mathfrak{g}$ over k is called nilpotent of degree d if all d-fold iterated commutators in $\mathfrak{g}$ vanish. Let $U(\mathfrak{g})$ be the universal enveloping algebra of $\mathfrak{g}$. It is a Hopf algebra with the comultiplication given by $\Delta(x) = x \otimes 1 + 1 \otimes x$ for $x \in \mathfrak{g}$. The subspace I in $U(\mathfrak{g})$ generated by all nontrivial Lie monomials in elements of $\mathfrak{g}$ is an ideal, with $U(\mathfrak{g})/I = k$.

(2.5.1) Lemma. *If $\mathfrak{g}$ is nilpotent of some degree, then $\bigcap I^n = 0$.*

Thus the I-adic completion

(2.5.2) $$\widehat{U}(\mathfrak{g}) = \varprojlim U(\mathfrak{g})/I^n$$

is a complete topological algebra containing $U(\mathfrak{g})$. As before, the standard Hopf algebra structure on $U(\mathfrak{g})$ gives rise to a topological Hopf algebra structure on $\widehat{U}(\mathfrak{g})$. We then have the following fact.

(2.5.3) Theorem. *(a)* $\mathfrak{g}$ *is the set of primitive elements of* $\widehat{U}(\mathfrak{g})$.
(b) The set G of grouplike elements in $\widehat{U}(\mathfrak{g})$ is the nilpotent group associated, via the Malcev theory, to the Lie algebra $\mathfrak{g}$.
(c) If $k = \mathbb{R}$ or $\mathbb{C}$, then G is the simply connected real or complex Lie group with Lie algebra $\mathfrak{g}$.
(d) The exponential map establishes a bijection between $\mathfrak{g}$ and G.

Let now $k = \mathbb{C}$ and

$$(2.5.4) \qquad \mathfrak{g}_{n,d} = \mathrm{FL}(X_1, \ldots, X_n)/\mathrm{FL}(X_1, \ldots, X_n)_{\geq d+1}.$$

This is a finite-dimensional Lie algebra known as the free nilpotent Lie algebra of degree d generated by n elements. It satisfies the obvious universal property. Then

$$\mathfrak{g}_n = \varprojlim{}_n \mathfrak{g}_{n,d}.$$

So $\mathfrak{g}_n$ is the free pronilpotent Lie algebra on n generators.

Let $R_{n,d}$ be the quotient of $R_n = \mathbb{C}\langle\langle Z_1, \ldots, Z_n \rangle\rangle$ by the closed ideal generated by all the $(d+1)$-fold commutators of the Z_i. For example, $R_{n,1} = \mathbb{C}[[Z_1, \ldots, Z_n]]$ is the usual (commutative) power series algebra.

The topological Hopf algebra structure on R_n descends to $R_{n,d}$, and we easily see the following:

(2.5.5) Proposition. *$R_{n,d}$ is isomorphic to $\widehat{U}(\mathfrak{g}_{n,d})$ as a topological Hopf algebra.*

We denote by $G_{n,d} \subset R_{n,d}^*$ the group of grouplike elements of $R_{n,d}$. Then the above facts imply:

(2.5.6) Theorem. *(a) $G_{n,d}$ is the simply connected complex Lie group with Lie algebra $\mathfrak{g}_{n,d}$.*
(b) G_n is the projective limit of $G_{n,d}$.

Thus $G_{n,d}$ is the "free unipotent complex algebraic group of degree d with n generators," while G_n is the free prounipotent group with n generators.

As above, taking $k = \mathbb{R}$, we get the real parts $G_{n,d}(\mathbb{R})$ and $\mathfrak{g}_{n,d}(\mathbb{R})$. The homomorphism $E : \Pi_n \to G_n(\mathbb{R})$ gives rise, for any $d \geq 1$, to the homomorphism

$$(2.5.7) \qquad \epsilon_{n,d} : \Pi_n \to G_{n,d}(\mathbb{R}),$$

whose target is a finite-dimensional Lie group.

(2.5.8) Proposition. *For any $d \geq 1$ the homomorphism $\epsilon_{n,d}$ is surjective.*

In other words, the group G_n can be seen as a (pro-)algebraic completion of the path group Π_n.

Proof: Let $\Pi_n^{\text{rect}} \subset \Pi_n$ be the subgroup of rectangular paths, i.e., paths consisting of segments each going in the direction of some particular coordinate. As a group, Π_n^{rect} is the free product of n copies of $\mathbb{R}$. Let $Z_{i,d} \in \mathfrak{g}_{n,d}$ be the image of Z_i. Then the image of Π_n^{rect} in $G_{n,d}(\mathbb{R})$ is the subgroup generated by the 1-parameter subgroups $\exp(t \cdot Z_{i,d})$, $t \in \mathbb{R}$, $i = 1, \ldots, n$. Since the $Z_{i,d}$ generate $\mathfrak{g}_{n,d}$ as a Lie algebra, the corresponding 1-parameter subgroups generate $G_{n,d}(\mathbb{R})$ as a group. Therefore $\epsilon_{n,d}\left(\Pi_n^{\text{rect}}\right) = G_{n,d}(\mathbb{R})$.

(2.6) Complex exponentials. Consider the complexification $\mathbb{C}^n$ of the space $\mathbb{R}^n$ from (2.1). The form Ω from (2.1.3) is then a holomorphic form on $\mathbb{C}^n$ with values in $\mathbb{C}\langle\langle Z_1, \ldots, Z_n\rangle\rangle$. In particular, we have the noncommutative exponential function

$$E_\gamma(Z) \in G_n \subset \mathbb{C}\langle\langle Z_1, \ldots, Z_n\rangle\rangle$$

for any unparametrized path γ in $\mathbb{C}^n$ starting at 0. Because Ω is holomorphic, $E_\gamma(Z)$ is, in addition to invariance under cancellations, also invariant under deformations of subpaths of γ inside holomorphic curves. Let $\Pi_n^{\mathbb{C}}$ be the quotient of Π_{2n}, the group of paths in $\mathbb{C}^n = \mathbb{R}^{2n}$ by the equivalence relation generated by such deformations. Obviously, $\Pi_n^{\mathbb{C}}$ is a group, and the correspondence $\gamma \mapsto E_\gamma$ gives rise to a homomorphism

$$(2.6.1) \qquad\qquad E : \Pi_n^{\mathbb{C}} \to G_n.$$

In contrast to the real case, it seems to be unknown whether (2.6.1) is injective. As before, we see that the composite homomorphism

$$(2.6.2) \qquad\qquad \epsilon_{n,d}^{\mathbb{C}} : \Pi_n^{\mathbb{C}} \to G_{n,d}$$

is surjective.

(2.6.3) Example. Let C be a complex analytic curve, $c_0 \in C$ a point, and $\phi : C \to \mathbb{C}^n$ a holomorphic map such that $\phi(c_0) = 0$. Denote by $p : \widetilde{C} \to C$ the universal covering of C corresponding to the base point c_0. In other words, $\widetilde{C}$ is the space of pairs (c, γ), where $c \in C$ and γ is a homotopy class of paths joinig c_0 and c. Then, by the above, ϕ induces a map $\widetilde{\phi} : \widetilde{C} \to \Pi_n^{\mathbb{C}}$. The composition

$$\varpi_d = \epsilon_{n,d}^{\mathbb{C}} \circ \widetilde{\phi} : \widetilde{C} \to G_{n,d}$$

can be called the period map of degree d. The restriction of ϖ_d to $p^{-1}(c_0) = \pi_1(C, c_0)$ is a homomorphism

$$m_d : \pi_1(C, c_0) \to G_{n,d},$$

called the monodromy homomorphism of degree d. We get then the "Albanese map"

$$\alpha_d : C \to G_{n,d}/\mathrm{Im}(m_d).$$

The particular case that C is the maximal abelian covering of a smooth projective curve of genus n, and ϕ is the Abel–Jacobi map, corresponds to the setting of Parshin [Pa]. Iterated integrals of modular forms were studied by Manin [Ma].

In the subsequent paper [K1] we will use complex noncommutative exponentials to construct invariants of degenerations of families of curves in an algebraic variety.

3 Generalities on the Noncommutative Fourier Transform

(3.0) Formal Fourier Transform on nilpotent groups. Let us start with the general situation of (2.5) with $k = \mathbb{R}$. Thus $\mathfrak{g}$ is a finite-dimensional nilpotent real Lie algebra and G is the corresponding simply connected Lie group. Then G is realized inside $\widehat{U}(\mathfrak{g})$ as the set of grouplike elements. In general, we can think of elements of $\widehat{U}(\mathfrak{g})$ as some kind of formal series (infinite formal linear combinations of elements of a Poincare–Birkhoff–Witt basis of $U(\mathfrak{g})$).

To keep the notation straight, we denote by $E_g \in \widehat{U}(\mathfrak{g})$ the element corresponding to $g \in G$.

(3.0.1) Example. Let $G = \mathbb{R}^n$ with coordinates $y_1, \ldots, y_n$. Then $\widehat{U}(\mathfrak{g})$ is the ring $\mathbb{C}[[z_1, \ldots, z_n]]$ of formal Taylor series. If $g = (y_1, \ldots, y_n) \in G$, then $E_g = E_g(z) = \exp\left(\sum_i y_i z_i\right)$ is the exponential series with the vector of exponents $(y_1, \ldots, y_n)$.

The above example motivates the following definition. Let μ be a measure on G, or, more generally, a distribution (understood as a generalized measure, i.e., as a functional on the space of C^∞-functions). Its formal Fourier transform is the element (formal series) given by

$$(3.0.2) \qquad \widehat{\mathcal{F}}(\mu) = \int_{g \in G} E_g \, d\mu \quad \in \quad \widehat{U}(\mathfrak{g}),$$

whenever the integral is defined.

Recall that for two distributions μ, ν on a Lie group G their convolution is defined by

$$(3.0.3) \qquad \mu * \nu = m_*(\mu \boxtimes \nu),$$

where $m : G \times G \to G$ is the multiplication and $\mu \boxtimes \nu$ is the Cartesian product of μ and ν. Here we assume that the pushdown under m is defined. The following is then straightforward.

(3.0.4) Proposition. *For two (generalized) measures μ, ν on G we have*

$$\widehat{\mathcal{F}}(\mu * \nu) = \widehat{\mathcal{F}}(\mu) \cdot \widehat{\mathcal{F}}(\nu)$$

(product in $\widehat{U}(\mathfrak{g})$).

(3.1) Promeasures and formal NCFT. We now specialize the above to the case $G = G_{n,d}(\mathbb{R})$. In other words, we consider the projective system of Lie groups

(3.1.1) $$\cdots \to G_{n,3}(\mathbb{R}) \to G_{n,2}(\mathbb{R}) \to G_{n,1}(\mathbb{R}) = \mathbb{R}^n$$

with projective limit $G_n(\mathbb{R})$. For $d \geq d'$ let

(3.1.2) $$p_{dd'} : G_{n,d}(\mathbb{R}) \to G_{n,d'}(\mathbb{R})$$

be the projection. By a *promeasure* on $G_n(\mathbb{R})$ we will mean a compatible system of measures on the $G_{n,d}(\mathbb{R})$. In other words, a promeasure is a system $\mu_\bullet = (\mu_d)$ such that each μ_d is a measure on $G_{n,d}(\mathbb{R})$ such that for any $d \geq d'$ the pushdown $(p_{dd'})_*(\mu_d)$ is defined as a measure on $G_{n,d'}(\mathbb{R})$ and is equal to $\mu_{d'}$. Equivalently, this means that for any continuous function f on $G_{n,d'}(\mathbb{R})$ we have

(3.1.3) $$\int_{G_{n,d'}(\mathbb{R})} f \cdot d\mu_{d'} = \int_{G_{n,d}(\mathbb{R})} (f \circ p_{dd'}) \cdot d\mu_d,$$

whenever the left-hand side is defined.

More generally, by a *prodistribution* we mean a system of distributions on the $G_{n,d}(\mathbb{R})$ (understood as generalized measures, i.e., as functionals on C^∞-functions) compatible in a similar sense, i.e., satisfying (3.1.3) for C^∞-functions f.

For $\Phi = \Phi(Z_1, \ldots, Z_n) \in G_n$ we denote by $\Phi_{i_1,\ldots,i_p}$ the coefficient of Φ at $Z_{i_1} \cdots Z_{i_p}$. It is clear that $\Phi_{i_1,\ldots,i_p}$ depends only on the image of Φ in $G_{n,p}$, so it makes sense to speak about $\Psi_{i_1,\ldots,i_p}$ for $\Psi \in G_{n,d}$, $d \geq p$.

Let $\mu_\bullet$ be a prodistribution on $G_n(\mathbb{R})$. Its formal Fourier transform is the formal series $\widehat{\mathcal{F}}(\mu_\bullet) \in \mathbb{C}\langle\langle Z_1, \ldots, Z_n \rangle\rangle$ defined as follows:

(3.1.4) $$\widehat{\mathcal{F}}(\mu_\bullet) = \sum_{p=0}^{\infty} \sum_{i_1,\ldots,i_p} \left(\int_{\Psi \in G_{n,d}(\mathbb{R})} \Psi_{i_1,\ldots,i_p} \cdot d\mu_d \right) Z_{i_1} \cdots Z_{i_p}.$$

Here for each p, the number d is any integer greater than or equal to p, and we assume that all the integrals converge.

The convolution operation extends, in an obvious way, to prodistributions on $G_m(\mathbb{R})$:

(3.1.6) Proposition 1. *If $\mu_\bullet, \nu_\bullet$ are two prodistributions, then*

$$\widehat{\mathcal{F}}(\mu_\bullet * \nu_\bullet) = \widehat{\mathcal{F}}(\mu_\bullet) \cdot \widehat{\mathcal{F}}(\nu_\bullet)$$

(product in $\mathbb{C}\langle\langle Z_1, \ldots, Z_n\rangle\rangle$).

(3.2) Delta functions. In classical analysis, the Fourier transform of $\delta^{(m)}$, the mth derivative of the delta function, is the monomial z^m. We now give a noncommutative analogue of this fact.

First of all, let δ_d be the delta function on $G_{n,d}(\mathbb{R})$ supported at 1. Then $\delta_\bullet = (\delta_d)$ is a prodistribution, and

(3.2.1) $\widehat{\mathcal{F}}(\delta_\bullet) = 1 \in \mathbb{C}\langle\langle Z_1, \ldots, Z_n\rangle\rangle.$

Next, first derivatives of the delta function at a point on a C^∞ manifold correspond to elements of the complexified tangent space to the manifold at this point. This, if $\xi \in \mathrm{FL}(Z_1, \ldots, Z_n)$, and ξ_d is the image of ξ in $\mathfrak{g}_{n,d} = T_1 G_{n,d}(\mathbb{R}) \otimes \mathbb{C}$, then we have the distribution $\partial_{\xi_d}(\delta_d)$ on $G_{n,d}(\mathbb{R})$, and these distributions form a prodistribution $\partial_\xi(\delta_\bullet)$.

Further, for any Lie group G with Lie algebra $\mathfrak{g}$, the iterated derivatives of the delta function at 1 correspond to elements of $U(\mathfrak{g} \otimes \mathbb{C})$, the universal enveloping algebra. Thus for any $\psi \in U(\mathfrak{g}_{n,d})$ we have a punctual distribution $D_\psi(\delta_d)$ on $G_{n,d}(\mathbb{R})$.

Let now $f \in \mathbb{C}\langle Z_1, \ldots, Z_n\rangle$ be a noncommutative polynomial. Recall that $\mathbb{C}\langle Z_1, \ldots, Z_n\rangle$ is the enveloping algebra of $\mathrm{FL}(Z_1, \ldots, Z_d)$. Thus for any d we have the image of f in $U(\mathfrak{g}_{n,d})$, which we denote by f_d. As before, the distributions $D_{f_d}(\delta_d)$ form a prodistribution, which we denote by $D_f(\delta_\bullet)$.

(3.2.2) Theorem. *We have $\widehat{\mathcal{F}}(D_f(\delta_\bullet)) = f$. In other words, $\widehat{\mathcal{F}}$ takes iterated derivatives of the delta function into (noncommutative) polynomials.*

Let $\mathcal{L}_{f,d}$ be the left-invariant differential operator on $G_{n,d}(\mathbb{R})$ corresponding to $f_d \in U(\mathfrak{g}_{n,d})$). Similarly, let $\mathcal{R}_{f,d}$ be the right-invariant differential operator corresponding to f_d. Recall that distributions (volume forms) form a right module over the ring of differential operators. In other words, if P is a differential operator acting on functions by $\phi \mapsto P\phi$, then we write the action of the adjoint operator on volume forms by $\omega \mapsto \omega P$. Thus, if $\mu_\bullet = (\mu_d)$ is a prodistribution, and $f \in \mathbb{C}\langle Z_1, \ldots, Z_n\rangle$, then we have prodistributions $\mu_\bullet \mathcal{L}_f = (\mu_d \mathcal{L}_{f,d})$ and $\mu_\bullet \mathcal{R}_f = (\mu_d \mathcal{R}_{f,d})$. Since applying $\mathcal{R}_{f,d}$ or $\mathcal{L}_{f,d}$ to a distribution is the same as the right or left convolution with $D_{f_d}(\delta_d)$, Proposition 3.1.6 implies the following.

(3.2.3) Proposition. *If $\phi \in \mathbb{C}\langle\langle Z_1, \ldots, Z_n\rangle\rangle$ is the Fourier transform of $\mu_\bullet$, then for any $f \in \mathbb{C}\langle Z_1, \ldots, Z_n\rangle$ the product $f \cdot \phi$ is the Fourier transform of $\mu_\bullet \mathcal{L}_f$, and $\phi \cdot f$ is the Fourier transform of $\mu_\bullet \mathcal{R}_f$.*

(3.3) Measures and convergent NCFT. Let $p_d : G_n(\mathbb{R}) \to G_{n,d}(\mathbb{R})$ be the projection. By a cylindric open set in $G_n(\mathbb{R})$ we mean a set of the form $p_d^{-1}(U)$, where $d \geq 1$ and $U \subset G_{n,d}(\mathbb{R})$ is an open set. These sets thus form a basis of the projective limit topology on $G_n(\mathbb{R})$. We denote by $\mathfrak{S}$ the σ-algebra of sets in $G_n(\mathbb{R})$ generated by cylindric open sets. Its elements will be simply called Borel subsets in $G_n(\mathbb{R})$.

(3.3.1) Example. Let

$$G_n(\mathbb{R})^{\text{ent}} = G_n(\mathbb{R}) \cap \mathbb{C}\langle\langle Z_1, \dots, Z_n \rangle\rangle^{\text{ent}}$$

be the subgroup formed by entire series; see (1.2.2). Since for $\Phi \in G_n(\mathbb{R})$ each given coefficient of f depends on the image of Φ in some $G_{n,d}(\mathbb{R})$, the condition (1.2.2) implies that $G_n(\mathbb{R})^{\text{ent}}$ is a Borel subset. Note further that for $\Phi \in G_n(\mathbb{R})^{\text{ent}}$ and any Hermitian matrices $Z_1^0, \dots, Z_n^0$ (of any size N), the matrix $\Phi(iZ_1^0, \dots, iZ_n^0)$ is unitary. This follows from the reality of the coefficients in Φ and from Proposition 2.4.8(d).

By a *measure* on $G_n(\mathbb{R})$ we mean a complex-valued, countably additive measure on the σ-algebra $\mathfrak{S}$. If μ is such a measure, we define its Fourier transform to be the function of indeterminate Hermitian N by N matrices $Z_1, \dots, Z_n$ (with indeterminate N) given by

$$(3.3.2) \qquad \mathcal{F}(\mu)(Z_1, \dots, Z_n) = \int_{\Phi \in G_n(\mathbb{R})^{\text{ent}}} \Phi(iZ_1, \dots, iZ_n) d\mu(\Phi).$$

As usual, by a *probability measure* on $G_n(\mathbb{R})$ we mean a real, nonnegative-valued measure on $\mathfrak{S}$ of total volume 1.

Given a promeasure $\mu_\bullet = (\mu_d)$ on $G_n(\mathbb{R})$, the correspondence

$$(3.3.3) \qquad p_d^{-1}(U) \mapsto \mu_d(U), \quad U \in G_{n,d}(\mathbb{R}),$$

defines a finite-additive function on cylindric open sets in $G_n(\mathbb{R})$. The following fact is a version of the basic theorem of Kolmogorov ([SW], Theorem 1.1.10) that a stochastic process is uniquely determined by its finite-dimensional distributions.

(3.3.4) Theorem. *If $\mu_\bullet$ is a probability promeasure (i.e., each μ_d is a probability measure), then the correspondence (3.3.3) extends to a unique probability measure $\mu = \lim_{\leftarrow} \mu_d$ on $G_n(\mathbb{R})$, so that $\mu_d = p_{d*}(\mu)$.*

Thus, probability measures and probability promeasures are in bijection.

Proof: The original theorem of Kolmogorov is about probability measures on an infinite product of measure spaces. Now, the projective limit $G_n(\mathbb{R})$ is a closed subset in the infinite product $\prod_d G_{n,d}(\mathbb{R})$. We can then apply Kolmogorov's theorem to this product and get a probability measure supported on this subset.

4 Noncommutative Gaussian and the Wiener Measure

(4.1) Informal overview. By the noncommutative Gaussian we mean the following noncommutative power series:

$$(4.1.1) \qquad \Xi(Z) = \exp\left(-\frac{1}{2}\sum_{i=1}^{n} Z_i^2\right) \quad \in \quad \mathbb{C}\langle\langle Z_1, \ldots, Z_n\rangle\rangle^{\text{ent}}.$$

Since the series is entire, we will denote by the same symbol $\Xi(Z_1, \ldots, Z_n)$ its value on any given square matrices $Z_1, \ldots, Z_n$. In classical (commutative) analysis, the Fourier transform of a Gaussian is another Gaussian. In this section we present a noncommutative extension of this fact. Informally, the answer can be formulated as follows.

(4.1.2) Informal theorem. *"The" measure on the space of paths whose Fourier transform gives $\Xi(Z)$ is the Wiener measure.*

We write "the" in quotes because so far, there is no uniqueness result for NCFT, so (4.1.2) can be read in one direction: that the NCFT of the Wiener measure is $\Xi(Z)$. Still, there are two more issues one has to address in order to make (4.1.2) into a theorem. First, the Wiener measure (see below for a summary) is defined on the space of parametrized paths, while NCFT is defined for measures on the space of unparametrized paths. This can be addressed by considering the pushdown of the Wiener measure (i.e., by performing the integration over the space of parametrized paths).

Second, and more importantly, the Wiener measure is defined on the space of continuous paths, and piecewise smooth paths form a subset of measure 0. On the other hand, the series $E_\gamma(Z)$ is a solution of a differential equation involving the time derivatives of γ and so is a priori not defined if γ is just a continuous path. This difficulty is resolved by using the theory of stochastic integrals and stochastic differential equations, which indeed provides a way of associating $E_\gamma(Z)$ to all continuous γ except those forming a set of Wiener measure 0.

Once these two modifications are implemented, (4.1.2) becomes an instance of the familiar principle in the theory of stochastic differential equations: that the direct image of the Wiener measure under the map given by the solution of a stochastic differential equation is the heat measure for the corresponding (hypo)elliptic operator; see [Bel], [Ok], [Bi1].

(4.2) The hypo-Laplacians and their heat kernels. Let $Z_{i,d}$ be the image of Z_i in $\mathfrak{g}_{n,d}$, and $L_{i,d}$ the left-invariant vector field on $G_{n,d}(\mathbb{R})$ corresponding to $Z_{i,d}$. We consider $L_{i,d}$ as a first-order differential operator on functions. The dth hypo-Laplacian is the operator

$$(4.2.1) \qquad \Delta_d = \sum_{i=1}^{n} L_{i,d}^2$$

in functions on $G_{n,d}(\mathbb{R})$. For $d \geq d'$ the operators Δ_d and $\Delta_{d'}$ are compatible:

$$(4.2.2) \qquad \Delta_d \left(p_{dd'}^* f \right) = p_{dd'}^* (\Delta_{d'} f), \quad \forall f \in C^2(G_{d'}(\mathbb{R})),$$

where the projection $p_{dd'}$ is as in (3.1.2). This follows because a similar compatibility holds for each $L_{i,d}$ and $L_{i,d'}$.

For $d > 1$ the number of summands in (4.2.1) is less than the dimension of $G_{n,d}(\mathbb{R})$, so Δ_d is not elliptic. However, Δ_d is hypoelliptic [Ho], i.e., every distribution solution of $\Delta_d u = 0$ is real analytic. This follows from Theorem 1.1 of Hörmander [Ho], since the $Z_{i,d}$ generate $\mathfrak{g}_{n,d}$ as a Lie algebra. Further, it is obvious that Δ_d is positive:

$$(4.2.3) \qquad (\Delta_d u, u) \geq 0, \quad u \in C_0^\infty(G_{n,d}(\mathbb{R})).$$

General properties of positive hypoelliptic operators [Ho] imply that the heat operator $\exp(-t\Delta_d)$, $t > 0$, is given by a positive C^∞ kernel. Because this operator is left-invariant, we get part (a) of the following theorem:

(4.2.4) Theorem. (a) The operator $\exp(\Delta_d/2)$ is given by convolution with a uniquely defined probability measure θ_d on $G_{n,d}(\mathbb{R})$. This measure is infinitely differentiable with respect to the Haar measure.
(b) For $d \geq d'$ the measures θ_d and $\theta_{d'}$ are compatible: $(p_{dd'})_*(\theta_d) = \theta_{d'}$.

Part (b) above follows from (4.2.2).

Thus we obtain a probability promeasure $\theta_\bullet = (\theta_d)$ on $G_n(\mathbb{R})$ and hence a probability measure $\theta = \varprojlim \theta_d$.

(4.2.5) Examples. (a) the group $G_{n,1}$ is identified with the space $\mathbb{R}^n$ from (2.1) with coordinates $y_1, \ldots, y_n$, and $Z_{i,1} = \partial/\partial y_i$. Therefore Δ_1 is the standard Laplacian on $\mathbb{R}^n$, and

$$\theta_1 = \frac{dy_1 \cdots dy_n}{(2\pi)^{n/2}} \exp\left(-\frac{1}{2} \sum_{i=1}^n y_i^2 \right)$$

is the usual Gaussian measure on $\mathbb{R}^n$. Each θ_d, $d > 1$, is thus a lift of this measure to $G_{n,d}$.

(b) For $d = 2$ an explicit formula for θ_2 was obtained by Gaveau in [G]. Here we consider the case $n = 2$, where the formula was also obtained by Hulanicki [Hu]. In this case $\mathfrak{g}_{2,2}$ is the Heisenberg Lie algebra with basis consisting of $Z_{1,2}, Z_{2,2}$ and the central element $h = [Z_{1,2}, Z_{2,2}]$. Denoting by y_1, y_2, v the corresponding exponential coordinates on $G_{2,2}$, we have

$$\theta_2 = \frac{dy_1 dy_2 dv}{(2\pi)^2} \int_{\tau=-\infty}^\infty \frac{2\tau}{\sinh(2\tau)} \cdot \exp\left(i\tau v - (y_1^2 + y_2^2) \frac{2\tau}{\tanh(2\tau)} \right) d\tau.$$

In fact, all known formulas in the literature (see [BGG] for a survey) involve integration over auxiliary parameters.

(4.2.6) Theorem. *The formal Fourier transform of the promeasure $\theta_\bullet$ is equal to the noncommutative Gaussian $\Xi(Z)$.*

Proof: This follows from the fact that the delta-prodistribution $D_{Z_i}(\delta_\bullet)$ corresponding to the generator $Z_i \in \mathfrak{g}_n$ is taken by $\mathcal{F}$ into the monomial Z_i. For each d the corresponding distribution takes a function f on $G_{n,d}(\mathbb{R})$ into the value of $L_i(f)$ at the unit element of $G_{n,d}(\mathbb{R})$. Further, convolution of such prodistributions corresponds to composition of left-invariant differential operators in the spaces of functions of the $G_{n,d}(\mathbb{R})$. So the system of the heat kernel operators on the $G_{n,d}(\mathbb{R})$, $d \geq 1$, given by $\exp\left(-\frac{1}{2}\sum L_i^2\right)$ has, as a prodistribution, the Fourier transform equal to $\exp\left(-\frac{1}{2}\sum Z_i^2\right)$. $\qquad\square$

(4.3) The Wiener measure. Let P_n be the space of continuous parametrized paths $\gamma : [0,1] \to \mathbb{R}^n$ such that $\gamma(0) = 0$. The Wiener measure w on P_n is first defined on cylindrical open sets $C(t_1,\ldots,t_m,U_1,\ldots,U_m)$, where $0 < t_1 < \cdots < t_m < 1$ and $U_i \subset \mathbb{R}^n$ is open. By definition,

$$C(t_1,\ldots,t_m,U_1,\ldots,U_m) = \{\gamma : \gamma(t_i) \in U_i, \; i = 1,\ldots,m\},$$

and

(4.3.1) $\qquad w\big(C(t_1,\ldots,t_m,U_1,\ldots,U_m)\big)$
$$= \int_{(y^{(1)},\ldots,y^{(m)})\in U_1\times\cdots\times U_m} \prod_{i=0}^{m} \frac{\exp\big(-\|y^{(i+1)} - y^{(i)}\|^2/2(t_{i+1} - t_i)\big)}{\big(2\pi(t_{i+1} - t_i)\big)^{1/2}} dy^{(1)}\cdots dy^{(m)}.$$

Here we put $t_0 = 0, t_{m+1} = 1$ and $y^{(0)} = 0$. Further, it is proved that w extends to a probability measure on the σ-algebra generated by the above subsets.

The Brownian motion is the family of $\mathbb{R}^n$-valued functions (random variables) on P_n parametrized by $t \in [0,1]$:

(4.3.2) $\qquad b(t) = (b_1(t),\ldots,b_n(t)), \quad b(t) : P_n \to \mathbb{R}^n, \; b(t)(\gamma) = \gamma(t).$

let $P_n^{\mathrm{sm}} \subset P_n$ be the subset of piecewuse smooth paths. Then it is well known that $w\left(P_n^{\mathrm{sm}}\right) = 0$.

As is also well known, the Wiener measure has the following intuitive interpretation:

(4.3.3) $\qquad dw(\gamma) = \exp\left(-\int_0^1 \|\gamma'(t)\|^2 dt\right) \mathcal{D}\gamma, \quad \mathcal{D}\gamma = \prod_{t=0}^{1} d\gamma(t).$

In other words, $\mathcal{D}\gamma$ is the (nonexistent) Lebesgue measure on the infinite-dimensional vector space of all paths, while the integral in the exponential is the action of a free particle.

(4.5) Reminder on stochastic integrals. Let $\omega = \sum_{i=1}^{n} \phi_i(y)dy_i$ be a 1-form on $\mathbb{R}^n$ with (complex-valued) C^∞ coefficients. If $\gamma : [0,1] \to \mathbb{R}^n$ is a piecewise smooth path, then we can integrate ω along γ, getting a number

$$(4.5.1) \qquad \int_\gamma \omega = \int_0^1 \gamma^*(\omega) = \int_0^1 \sum_i \phi_i(\gamma(t))\gamma_i'(t)dt.$$

This gives a map (function)

$$(4.5.2) \qquad \int (\omega) : P_n^{\mathrm{sm}} \longrightarrow \mathbb{C}.$$

If $\gamma(t)$ is just a continuous path without any differentiability assumptions, then (4.5.1) is not defined, so there is no immediate extension of the map (4.5.2) to the space P_n. The theory of stochastic integrals provides several (a priori different) ways to construct such an extension. The two best-known approaches are the Ito and Stratonovich integrals over the Brownian motion; see [SW] [KW]. They are the functions

$$(4.5.3) \qquad \int^{\mathrm{Ito}} (\omega), \quad \int^{\mathrm{Str}} (\omega) : \quad P_n \to \mathbb{C},$$

defined everywhere outside some subset of Wiener measure 0, and measurable with respect to this measure.

To construct them, see [Ok], pp. 14–16, one has to consider Riemann sum approximations to the integral but restrict to Riemann sums of some particular type. For a piecewise smooth path γ, the integral is the limit of sums

$$(4.5.4) \qquad \sum_{i=1}^{n} \sum_{\nu=1}^{m} \phi_i(\gamma(\xi_\nu))\big((\gamma_i(t_\nu) - \gamma_i(t_{\nu-1})\big),$$

where $0 = t_0 < t_1 < \cdots < t_m = 1$ is a decomposition of $[0,1]$ into intervals, and $\xi_\nu \in [t_{\nu-1}, t_\nu]$ are some chosen points. In the smooth case the limit exists, provided $\max(t_\nu - t_{\nu-1})$ goes to 0 (in particular, the choice of ξ_ν is inessential).

Now, to obtain $\int^{\mathrm{Ito}}(\omega)$, one chooses the class of Riemann sums with

$$(4.5.5) \qquad t_\nu = \nu/m, \quad \xi_\nu = t_{\nu-1}, \quad m = 2^q, \ q \to \infty.$$

In other words, for each q the above sum defines a function $S_q^{\mathrm{Ito}}(\omega) : P_n \to \mathbb{C}$, and

$$(4.5.6) \qquad \int^{\mathrm{Ito}} (\omega) = \lim_{q \to \infty} S_q^{\mathrm{Ito}}(\omega).$$

To obtain $\int^{\mathrm{Str}}(\omega)$, one chooses the class of Riemann sums with

$$(4.5.7) \qquad t_\nu = \nu/m, \quad \xi_\nu = (t_{\nu-1} + t_\nu)/2, \quad m = 2^q, \ q \to \infty.$$

Each such sum gives a function $\mathcal{S}_q^{\mathrm{Str}}(\omega) : P_n \to \mathbb{C}$, and

$$(4.5.8) \qquad \int^{\mathrm{Str}} (\omega) = \lim_{q \to \infty} \mathcal{S}_q^{\mathrm{Str}}(\omega).$$

It is known that $\int^{\mathrm{Str}}(\omega)$ is invariant under smooth reparametrizations of the path considered as transformations acting on P_n and also satisfies a transparent change of variables formula.

The more common notation for the stochastic integrals (considered as random variables on P_n) is

$$(4.5.9) \qquad \int^{\mathrm{Ito}} (\omega) = \int_0^1 \omega(b(t)) db(t)), \qquad \int^{\mathrm{Str}} (\omega) = \int_0^1 \omega(b(t)) \circ db(t)),$$

where $b(t)$ is the Brownian motion (4.3.2). Thus $db(t)$ and $\circ db(t)$ stand for the two ways (due to Ito and Stratonovich) of regularizing the (a priori divergent) differential of the Brownian path $b(t)$. See [Ok] for the relation between the two regularization schemes. By restricting to the truncated path $[0, s]$, $s \leq 1$, one defines the stochastic integrals $\int_0^s$ in each of the above settings.

(4.6) Stochastic holonomy. Let G be a Lie group, which we suppose to be embedded as a closed subgroup of $\mathrm{GL}_N(\mathbb{C})$ for some N, and let $\mathfrak{g} \subset \mathrm{Mat}_N(\mathbb{C})$ be the Lie algebra of G. Let $A = \sum A_i(y) dy_i$ be a smooth $\mathfrak{g}$-valued 1-form on $\mathbb{R}^n$, which we consider as a connection in the trivial G-bundle over $\mathbb{R}^n$. If $\gamma : [0, 1] \to \mathbb{R}^n$ is a piecewise smooth path, then we have the holonomy of A along γ:

$$(4.6.1) \qquad \mathrm{Hol}_\gamma(A) = P \exp \int_\gamma A \quad \in \quad G.$$

It is the value at $t = 1$ of the solution $U(t) \in \mathrm{GL}_N(\mathbb{C})$ of the differential equation

$$(4.6.2) \qquad \frac{dU}{dt} = U(t)\left(\sum_i A_i(\gamma(t)) \cdot \gamma_i'(t) \right), \quad U(0) = 1.$$

The holonomy defines thus a map

$$(4.6.3) \qquad \mathrm{Hol}(A) : P_n^{sm} \to G.$$

As before, (4.6.2) and thus $\mathrm{Hol}_\gamma(A)$ have no immediate sense without some differentiability assumptions on A.

The theory of stochastic differential equations [Ok], [KW] resolves this difficulty by replacing the above differential equation by an integral equation and understanding the integral in a regularized sense as in (4.5). Thus, one defines the Ito and Stratonovich stochastic holonomies, which are measurable maps

(4.6.4) $\text{Hol}^{\text{Ito}}(A),\ \text{Hol}^{\text{Str}}(A) : P_n \to G,$

defined outside a subset of Wiener measure 0. For example, $\text{Hol}^{\text{Str}}(A)$ is defined as the value at $t = 1$ of the G-valued stochastic process $U(t)$ satisfying the Stratonovich integral equation

(4.6.5) $$U(t) = 1 + \int_0^t U(s)\left(\sum_i A_i(b(s)) \circ db_i(s)\right).$$

We will be particularly interested in the case in which the A_i are constant, i.e., our connection is translation-invariant. In this case, $B(t) = \sum A_i b_i(t)$ is a (possibly degenerate) Brownian motion on $\mathfrak{g}$ and $U(t)$ is the corresponding left-invariant Brownian motion on G as studied by McKean, see [McK, Section 4.7], and also [HL]. In particular, the Stratonovich holonomy can be represented as a "product integral" in the sense of McKean:
(4.6.6)

$$\text{Hol}^{\text{Str}}(A) = \prod_{t\in[0,1]} \exp(dB(t)) \quad := \quad \lim_{q\to\infty} \prod_{\nu=1}^{2^q} \exp\left(B\left(\frac{\nu}{2^q}\right) - B\left(\frac{\nu-1}{2^q}\right)\right);$$

see [HL], Thm. 2. Here the product is taken in the order of increasing ν. In the sequel we will work with $\text{Hol}^{\text{Str}}(A)$.

(4.7) The Malliavin calculus and the Feynman–Kac–Bismut formula. We now specialize (4.6) to the case in which $G = G_{n,d}(\mathbb{R})$, $\mathfrak{g} = \mathfrak{g}_{n,d}(\mathbb{R})$, and $A = \Omega^{(d)}$ is the constant 1-form $\Omega^{(d)} = \sum_{i=1}^n Z_{i,d} dy_i$. We get the stochastic holonomy map

(4.7.1) $\text{Hol}^{\text{Str}}(\Omega^{(d)}) : P_n \to G_{n,d}(\mathbb{R}).$

(4.7.2) Theorem. *The probability measure θ_d on $G_{n,d}(\mathbb{R})$ is equal to*

$$\text{Hol}^{\text{Str}}(\Omega^{(d)})_*(w),$$

the pushdown of the Wiener measure under the holonomy map.

Proof: This is a fundamental property of (hypo)elliptic diffusions holding for any vector fields $\xi_1, \ldots, \xi_n$ on a manifold M such that iterated commutators of the ξ_i span the tangent space at every point. In this case the operator $\Delta = \sum \text{Lie}_{\xi_i}^2$ is hypoelliptic and has a uniquely defined, smooth heat kernel $\Theta(x,y), x,y \in M$, which is a function in x and a volume form in y and represents the operator $\exp(-\Delta/2)$. Further, the heat equation

(4.7.3) $\partial u/\partial t = -\Delta(u)/2$

is the "Kolmogorov backward equation" for the M-valued stochastic process $U(t)$ satisfying the Stratonovich differential equation

$$(4.7.4) \qquad dU = \sum L_{\xi_i}(U) \circ db_i$$

with the $b_i(t)$ being as before. This means that the fundamental solution of (4.7.3) is the pushforward of the Wiener measure under the process $U(t)$. See [Ok], Th. 8.1. Our case is obtained by specializing to $M = G_{n,d}(\mathbb{R})$, $\xi_i = Z_{i,d}$.
$\square$

Further, let θ be the probability measure on $G_n(\mathbb{R}) = \varprojlim_d G_{n,d}(\mathbb{R})$ corresponding to the promeasure (θ_d) by Theorem 3.3.5. Note that the maps $\mathrm{Hol}^{\mathrm{Str}}(\Omega^{(d)})$ for various d unite into a map

$$(4.7.4) \qquad \mathrm{Hol}^{\mathrm{Str}}(\Omega) : P_n \to G_n(\mathbb{R}), \quad \Omega = \sum Z_i dy_i.$$

We get the following corollary.

(4.7.5) Corollary. *The measure θ is the pushdown of the Wiener measure under $\mathrm{Hol}^{\mathrm{Str}}(\Omega)$.*

(4.7.6) Theorem. *(a) The support of the measure θ is contained in $G_n(\mathbb{R})^{\mathrm{ent}}$, the set of entire grouplike power series.*
(b) The convergent Fourier transform of θ is equal to $\Xi(Z)$. In other words (taking into account part (a) and (4.7.5)), for any given Hermitian matrices $Z_1^0, \ldots, Z_n^0$ of any given size N, we have

$$\exp\left(-\frac{1}{2}\sum_{j=1}^n (Z_j^0)^2\right) = \int_{\gamma \in P_n} \mathrm{Hol}^{\mathrm{Str}}_\gamma(A)\left(iZ_1^0, \ldots, iZ_n^0\right) dw(\gamma).$$

(4.8) Stochastic iterated integrals and the proof of Theorem 4.7.6.
In the situation of (4.5), assume that we are given d smooth 1-forms $\omega_1, \ldots, \omega_d$ on $\mathbb{R}^n$. We then define, following Fliess and Normand-Cyrot [FN], the iterated Stratonovich integral

$$(4.8.1) \qquad \int^{\to \mathrm{Str}} (\omega_1 \cdots \omega_d) : P_n \to \mathbb{C}$$

by the same iterative procedure as in (2.3). Like the ordinary Stratonovich integral, it is reparametrization-invariant. This definition extends to the case that each ω_i takes values in a (pro)finite-dimensional associative $\mathbb{C}$-algebra R (with unity). As before, we define the empty iterated integral (corresponding to $d = 0$) to be equal to 1. We will need some extensions of Proposition 2.3.1 to the stochastic case. The first statement deals with the nilpotent case.

(4.8.2) Proposition. *Let $I \subset R$ be a nilpotent ideal, i.e., $I^m = 0$ for some m. Let A be a smooth 1-form on $\mathbb{R}^n$ with values in I. Consider A as a connection form with coefficients in R. Then*

$$\text{Hol}^{\text{Str}}(A) = \sum_{d=0}^{\infty} \int^{\rightarrow \text{Str}} (A \cdots A),$$

the series on the right being terminating.

Proof: This is a consequence of Theorem 2 of [FN].

(4.8.3) Corollary. *The random variable $\text{Hol}^{\text{Str}}(\Omega)$ from (4.7.4), considered as an $\mathbb{R}\langle\langle Z_1, \ldots, Z_n \rangle\rangle$-valued random variable on P_n, has the form*

$$\text{Hol}^{\text{Str}}(\Omega) = \sum_{m=0}^{\infty} \sum_{J=(j_1, \ldots, j_m)} Z_{j_1} \cdots Z_{j_m} \int^{\rightarrow \text{Str}} (dy_{j_1} \cdots dy_{j_m}).$$

Next, we look at convergence of the series in (4.8.3). Questions of this nature ("convergence of stochastic Taylor series") were studied by Ben Arous [Be], and we recall some of his results. Denote by

$$(4.8.4) \qquad B_J = \int^{\rightarrow \text{Str}} (dy_{j_1} \cdots dy_{j_m}) = \int db_{j_1} \circ \cdots \circ db_{j_m}$$

the coefficient in the series (4.8.3) corresponding to the multi-index J. Here the right-hand side is the notation of [Be]. Let $|J| = \sum j_\nu$ be the degree of the monomial corresponding to J.

(4.8.5) Theorem. *Let (x_J) be a collection of real numbers given for each $J = (j_1, \ldots, j_m)$, $m \geq 0$, and satisfying the condition*

$$|x_J| \leq K^J, \quad \text{for some} \quad K > 0.$$

Then the series

$$\sum_m \sum_{|J|=m} |x_J B_J|$$

of random variables on P_n converges almost surely.

This is Corollary 1 of [Be] (with the parameter α from *loc. cit.* taken to be 0).

We now deduce Theorem 4.7.6 from the above results. Let $Z_1^0, \ldots, Z_n^0$ be fixed matrices of any given size N. For a matrix B denote by

$$\|B\| = \max_{v \neq 0} \frac{\|B(v)\|}{\|v\|}$$

the matrix norm of B. Let us apply Theorem 4.8.5 to

$$x_J = \|(Z^0)^J\| = \|Z^0_{j_1} \cdots Z^0_{j_m}\|.$$

Take $K = \max\left(\|Z^0_i\|\right)$. Then $|x_J| \leq K^J$, so (4.8.5) gives that the series $\sum_J B_J(Z^0)^J$ converges absolutely almost surely. This establishes part (a) of Theorem 4.7.6. Part (b) follows from (a) and from Theorem 4.2.6 about the formal Fourier transform.

5 Futher Examples of NCFT

(5.1) Near-Gaussians. In classical analysis, a near-Gaussian is a function of the form $f(z) \cdot e^{-\|z\|^2/2}$, where $f(z)$, $z = (z_1, \ldots, z_n)$, is a polynomial. In that setting, the Fourier transform of a near-Gaussian is another near-Gaussian.

A natural noncommutative analogue of a near-Gaussian is a function of the form

$$(5.1.1) \qquad F(Z) \cdot \Xi(Z) \cdot G(Z), \quad F, G \in \mathbb{C}\langle Z_1, \ldots, Z_n \rangle.$$

It can be represented as a (formal) Fourier transform using Proposition 3.2.3:

$$(5.1.2) \qquad F(Z) \cdot \Xi(Z) \cdot G(Z) = \widehat{\mathcal{F}}\left(\theta_\bullet \mathcal{L}_F \mathcal{R}_G\right),$$

where $\mathcal{L}_F$ is the system of left-invariant differential operators on the $G_{n,d}(\mathbb{R})$, $d \geq 1$, corresponding to F, while $\mathcal{R}_G$ is the system of right-invariant differential operators corresponding to G.

It seems difficult to realize the measures $\theta_d \mathcal{L}_F \mathcal{R}_G$, $d \geq 1$, in terms of some transparent measures on the space P_n, since it requires using group translations on Π_n^{cont}, the group of continuous paths obtained by quotienting P_n by reparametrizations and cancellations.

(5.2) The Green promeasure. Let g_d be the fundamental solution of the dth hypo-Laplacian on $G_{n,d}(\mathbb{R})$ centered at 1, the unit element, i.e.,

$$(5.2.1) \qquad \Delta_d(g_d) = \delta_1.$$

By the general properties of hypoelliptic operators, g_d is a measure (volume form) on $G_{n,d}(\mathbb{R})$ smooth away from 1. In fact, if we denote by $\theta_{d,t}$ the kernel of $\exp(-t\Delta_d/2)$, $t > 0$, i.e., the heat kernel measure at time t, then

$$(5.2.2) \qquad g_d = \int_{t=0}^{\infty} \theta_{d,t} dt.$$

This expresses the fact that the Green measure of a domain is equal to the amount of time a diffusion path spends in the domain. It is clear therefore that $g_\bullet = (g_d)$ is a promeasure on $G_n(\mathbb{R})$.

(5.2.3) Examples. (a) For $d = 1$ we have the Green function of the usual Euclidean Laplacian in $\mathbb{R}^n$, which has the form

$$g_1(y) = \frac{1}{4\pi} \ln\left(y_1^2 + y_2^2\right) dy_1 dy_2, \quad n = 2,$$

$$g_1(y) = -\frac{((n/2) - 2)!}{4\pi^{n/2}} \left(\sum y_i^2\right)^{1-n/2} dy_1 \cdots dy_n, n \geq 3.$$

(b) Consider the case $n = 2, d = 2$ corresponding to the Heisenberg group, and let us use the exponential coordinates y_1, y_2, v as in Example 4.2.5(b). Then

$$g_2(y_1, y_2, v) = \frac{1}{\pi} \frac{1}{\sqrt{(y_1^2 + y_2^2) + v^2}} dy_1 dy_2 dv,$$

as was found by Folland [Fo], see also [G], p. 101.

(5.3) The method of kernels. More generally, if $F(Z_1, \ldots, Z_n)$ is a "noncommutative function" such that the operator $F(L_{1,d}, \ldots, L_{n,d})$ in functions on $G_{n,d}(\mathbb{R})$ makes sense and possesses a distribution kernel $K_d(x, y)dy$, then the distribution $\mu_d = K_d(1, y)dy$ is precisely the dth component of the prodistribution whose Fourier transform is F.

For example, hypoelliptic calculus allows us to consider $F(Z) = \phi\left(\sum_{i=1}^n Z_i^2\right)$, where $\phi : \mathbb{R} \to \mathbb{R}$ is any C^∞ function decaying at infinity such as $\phi(u) = e^{-u^2/2}$, $\phi(u) = 1/u$, or $1/(u^2+1)$. This leads to a considerable supply of prodistributions.

(5.4) Probabilistic meaning. An idea in probability theory very similar to our NCFT, namely the idea of associating a noncommutative power series to a stochastic process, was proposed by Baudoin [Ba], who called this series "expectation of the signature" and emphasized its importance. From the general viewpoint of probability theory one can look at this series (the Fourier transform of a probability measure on the space of paths) as being rather an analogue of the characteristic function of n random variables. Indeed, if $x_1, \ldots, x_n$ are random variables, then their joint distribution is a probability measure on $\mathbb{R}^n$, and the characteristic function is the (usual) Fourier transform of this measure:

(5.4.1) $$f(z_1, \ldots, z_n) = \mathbb{E}\left[e^{i(z,x)}\right],$$

which is an entire function of n variables. Each time we have a natural lifting of the characteristic function to the noncommutative domain, we can therefore expect some n-dimensional stochastic process lurking in the background.

6 Fourier Transform of Noncommutative Measures

(6.1) Nomcommutative measures. Following the general approach of noncommutative geometry [Con], we consider a possibly noncommutative $\mathbb{C}$-algebra R (with unit) as a replacement of a "space" (Spec(A)). A measure on

R is then simply a linear functional ("integration map") $\tau : I \to \mathbb{C}$ defined on an appropriate subspace $I \subset R$ whose elements have the meaning of integrable functions. We will call a measure τ *finite* if $I = R$, and *normalized* if it is finite and $\tau(1) = 1$. If R has a structure of a $*$-algebra, then a finite measure τ is called positive if $\tau(aa^*) \geq 0$ for any $a \in R$. A (noncommutative) *probability measure* on a $*$-algebra A is a normalized positive measure.

(6.1.1) Examples. (a) Let $R = \mathrm{Mat}_N(\mathbb{C})$ with the $*$-algebra structure given by Hermitian conjugation. Then $\tau(a) = \frac{1}{N} \mathrm{Tr}(a)$ is a probability measure.

(b) Let $R = \mathbb{C}\langle Z_1, \ldots, Z_n \rangle$ with the $*$-algebra structure given by $Z_i^* = Z_i$. Let Herm_N be the space of Hermitian N by N matrices. We denote by $dZ = \prod_{i,j=1}^{N} dZ_{ij}$ the standard volume form on Herm_N. Let $\mu = \mu_N$ be a volume form on $(\mathrm{Herm}_N)^n$ of exponential decay at infinity. Then we have a finite measure on R given by

$$\tau(f) = \frac{1}{N} \mathrm{Tr} \int_{Z_1,\ldots,Z_n \in \mathrm{Herm}_N} f(Z_1, \ldots, Z_n) d\mu(Z_1, \ldots, Z_n).$$

If μ_N is a normalized (resp. probability) measure in the usual sense, then τ is a normalized (resp. probability) measure in the noncommutative sense. An important example is

$$\mu_N = \exp\left(-S(Z_1, \ldots, Z_n)\right) dZ_1 \cdots dZ_n,$$

where the "action" $S(Z_1, \ldots, Z_n) \in \mathbb{C}\langle Z_1, \ldots, Z_n \rangle$ is a noncommutative polynomial with appropriate growth conditions at the infinity of Herm_N^n.

(c) Let $R = \mathbb{C}\left\langle X_1^{\pm 1}, \ldots, X_n^{\pm 1} \right\rangle$ with the $*$-algebra structure given by $X_i^* = X_i^{-1}$. If $\mu = \mu_N$ is a finite measure on $U(N)^n$, then we have a finite measure τ on R given by

$$\tau(f) = \frac{1}{N} \mathrm{Tr} \int_{X_1,\ldots,X_n \in U(N)} f(X_1, \ldots, X_n) d\mu(X_1, \ldots, X_n),$$

which is normalized (resp. probability) if μ_N is so in the usual sense.

(6.2) Free products. Let $R_1, \ldots, R_n$ be algebras with unit. Then we have their free product $R_1 \star \cdots \star R_n$. This is an algebra containing all the R_i and characterized by the following universal property: for any algebra B and any homomorphisms $f_i : R_i \to B$ there is a unique homomorphism $f : R_1 \star \cdots \star R_n \to B$ restricting to f_i on R_i for each i. Explicitly, $R_1 \star \cdots \star R_n$ is obtained as the quotient of the free (tensor) algebra generated by the vector space $R_1 \oplus \cdots \oplus R_n$ by the relations saying that the products of elements from each R_i are given by the existing multiplication in R_i. We will also use the notation $\bigstar_{i=1}^{n} R_i$.

(6.2.1) Example. If each $R_i = \mathbb{C}[Z_i]$ is the polynomial algebra in one variable, then $R_1 \star \cdots \star R_n = \mathbb{C}\langle Z_1, \ldots, Z_n \rangle$ is the algebra of noncommutative polynomials. If each $R_i = \mathbb{C}\left[X_i, X_i^{-1}\right]$ is the algebra of Laurent polynomials, then $R_1 \star \cdots \star R_n = \mathbb{C}\left\langle X_1^{\pm 1}, \ldots, X_n^{\pm 1}\right\rangle$ is the algebra of noncommutative Laurent polynomials.

The following description of the free product follows easily from the definition (see [VDN]).

(6.2.2) Proposition. *Suppose that for each i we choose a subspace $R_i^\circ \subset R_i$ that is a complement to $\mathbb{C} \cdot 1$. Then as a vector space,*

$$R_1 \star \cdots \star R_n = \mathbb{C} \cdot 1 \oplus \bigoplus_{k>0} \bigoplus_{i_1 \neq i_2 \neq \cdots \neq i_k} R_{i_1}^\circ \otimes \cdots \otimes R_{i_k}^\circ.$$

The following definition of the free product of (noncommutative) measures is due to Voiculescu; see [VDN].

(6.2.3) Proposition–Definition 1. *Let R_i, $i = 1, \ldots, n$, be associative algebras with 1, and $\tau_i : R_i \to \mathbb{C}$ finite normalized measures. Then there exists a unique finite normalized measure $\tau = \bigstar \tau_i$ on $\bigstar_{i=1}^n R_i$ with the following properties:*
(1) $\tau|_{R_i} = \tau_i$.
(2) If $i_1 \neq \cdots \neq i_k$ and $a_\nu \in R_\nu$ are such that $\tau_{i_\nu}(a_\nu) = 0$, then $\tau(a_{i_1} \cdots a_{i_k}) = 0$.
If the R_i are $*$-algebras and each τ_i is a probability measure, then so is τ.

Both the existence and the uniqueness of τ follow at once from (6.2.2) if we take $R_i^\circ = \mathrm{Ker}(\tau_i)$. The problem of finding $\tau(a_1 \cdots a_k)$ for arbitrary elements $a_\nu \in R_{i_\nu}$ is clearly equivalent to that of writing $a_1 \cdots a_k$ in the normal form (6.2.2). To do this, one writes

(6.2.4) $$a_\nu = \tau_{i_\nu}(a_\nu) \cdot 1 + a_\nu^\circ,$$

with a_ν° defined so as to satisfy (6.2.4), and we have $\phi_{i_\nu}(a_\nu^\circ) = 0$. Then one uses the conditions (1) and (2) to distribute.

(6.2.5) Examples. Suppose we have two algebras A and B and normalized measures $\phi : A \to \mathbb{C}$ and $\psi : B \to \mathbb{C}$. Let $\chi : A \star B \to \mathbb{C}$ be the free product of ϕ and ψ. Then for $a, a' \in A$ and $b, b' \in B$ we have, after some calculations,

$$\chi(ab) = \phi(a)\psi(b), \quad \chi(aba') = \phi(aa')\psi(b),$$
$$\chi(aba'b') = \phi(aa')\psi(b)\psi(b') + \phi(a)\phi(a')\psi(bb') - \phi(a)\phi(a')\psi(b)\psi(b').$$

See [NS], Thm. 14.4, for a general formula for $\chi(a_1 b_1 \cdots a_m b_m)$, $a_i \in A$, $b_i \in B$.

(6.2.6) Examples. (a) Let $R_i = \mathbb{C}[x^{\pm 1}]$, $i = 1, \ldots, n$, and let τ_i be given by the integration over the normalized Haar measure d^*x on the unit circle. Thus

$$\tau_i\left(f(x) = \sum_m a_m x^m\right) = \int_{|x|=1} f(x)d^*x = a_0.$$

The free product of these measures is the functional on $\mathbb{C}\langle X_1^{\pm 1}, \ldots, X_n^{\pm 1}\rangle$ given by

$$\tau\left(f(X) = \sum_{\gamma \in F_n} a_\gamma X^\gamma\right) = a_0,$$

the constant term of a noncommutative Laurent polynomial. The asymptotic freedom theorem for unitary matrices (1.4.7) says that this functional is the limit, as $N \to \infty$, of the functionals from Example 6.1.1(c) with μ_N, for each N, being the normalized Haar measure on $U(N)^n$.

(b) Let $R_i = \mathbb{C}[z]$, $i = 1, \ldots, n$, and let $\tau_i = \delta(z - a_i)$ be the Dirac delta function situated at a point $a_i \in \mathbb{C}$, i.e., $\tau_i(f) = f(a_i)$. Then the free product $\tau = \tau_1 \star \cdots \star \tau_n$ is given by

$$\tau(f(Z_1, \ldots, Z_n)) = f(a_1 \cdot 1, \ldots, a_n \cdot 1);$$

in other words, it depends only on the image of f in the ring of commutative polynomials $\mathbb{C}[z_1, \ldots, z_n]$. This can be seen from the procedure (6.2.4) using the fact that each $\tau_i : \mathbb{C}[z] \to \mathbb{C}$ is a ring homomorphism.

(c) Let $R_i = \mathbb{C}[z]$, $i = 1, \ldots, n$, and let τ_i be integration over the standard Gaussian probability measure

$$\tau_i(f) = \frac{1}{\sqrt{2\pi}} \int_{-\infty}^{\infty} f(z) e^{-z^2/2} dz.$$

Their free product is a probability measure on $\mathbb{C}\langle Z_1, \ldots, Z_n\rangle$ denoted by ξ_n and called the *free Gaussian measure*. The asymptotic freedom for Hermitian Gaussian ensembles [V] can be formulated as follows.

(6.2.7) Theorem. *The measure ξ_n is the limit, as $N \to \infty$, of the measures from Example 6.1.1(b), where for each N, we take for μ_N the Gaussian probability measure on the vector space $(\mathrm{Herm}_N)^n$ corresponding to the scalar product $\sum \mathrm{Tr}(A_i B_i)$ on this vector space:*

$$\mu = \mu_N = \frac{1}{(2\pi)^{nN^2/2}} \exp\left(-\frac{1}{2}\sum_{i=1}^{n} Z_i^2\right) dZ_1 \cdots dZ_n.$$

(6.3) The Fourier transform of noncommutative measures. In classical analysis, the Fourier transform is defined for measures on $\mathbb{R}^n$, not on an arbitrary curved manifold. We will call a measure on $\mathbb{R}_{\mathrm{NC}}^n$ ("noncommutative $\mathbb{R}^n$") a datum consisting of a $*$-algebra R, a $*$-homomorphism $\mathbb{C}\langle Z_1, \ldots, Z_n\rangle \to R$ (i.e., a choice of self-adjoint elements in R, which we will still denote by Z_i), and a measure τ on R. Elements of R for which τ is defined will be thought of as functions integrable with respect to the measure. This concept is thus very similar to that of n noncommutative random variables in noncommutative probability theory, except that we do not require any positivity or normalization.

Let τ be a measure on $\mathbb{R}^n_{NC}$. Its Fourier transform is the complex-valued function $\mathfrak{F}(\tau)$ on the group Π_n of piecewise smooth paths in $\mathbb{R}^n$ defined as follows:

$$(6.3.1) \qquad \mathfrak{F}(\tau)(\gamma) = \tau(E_\gamma(iZ_1, \ldots, iZ_n)), \quad \gamma \in \Pi_n.$$

Here we assume that the "entire function" $E_\gamma(iZ_1, \ldots, iZ_n)$ lies in the domain of definition of τ. In physical terminology, $\mathfrak{F}(\tau)(\gamma)$ is the "Wilson loop functional" (defined here for nonclosed paths as well).

(6.3.2) Example: delta functions. (a) For every $J = (j_1, \ldots, j_m)$ we have the measure $\delta^{(J)}$ on $\mathbb{C}\langle\langle Z_1, \ldots, Z_n \rangle\rangle$ given by

$$\delta^{(J)}\left(\sum_m \sum_{I=(i_1,\ldots,i_m)} a_I Z_{i_1} \cdots Z_{i_m} \right) = a_J.$$

The Fourier transform of $\delta^{(J)}$ is the function $W_J : \Pi_n \to \mathbb{C}$ that associates to a path γ the iterated integral along γ labeled by J:

$$W_J(\gamma) = \int_\gamma^{\longrightarrow} dy_{j_1} \cdots dy_{j_m}.$$

We will call these functions *monomial* functions on Π_n.

(b) If we take for τ the free product of (underived) delta functions $\delta_{a_1} \star \cdots \star \delta_{a_n}$, as in Example 6.2.6(b), then

$$\mathfrak{F}(\tau)(\gamma) = \exp\big(i(e(\gamma), a)\big),$$

where $e(\gamma) \in \mathbb{R}^n$ is the endpoint of γ. This follows from the fact that τ is supported on the commutative locus, i.e., $\tau(E_\gamma(iZ))$ depends only on the image of $E_\gamma(iZ)$ in the commutative power series ring, which is $\exp\big(ie(\gamma), z)\big)$.

(6.4) Convolution and product. Let τ, σ be two measures on $\mathbb{R}^n_{NC}$, so we have homomorphisms

$$\alpha : \mathbb{C}\langle Z_1, \ldots, Z_n \rangle \to R, \quad \beta : \mathbb{C}\langle Z_1, \ldots, Z_n \rangle \to S,$$

and τ is a linear functional on R, while σ is a linear functional on S. Their (tensor) convolution is the measure $\tau * \sigma$, which corresponds to the homomorphism

$$(6.4.1) \qquad \mathbb{C}\langle Z_1, \ldots, Z_n \rangle \to R \otimes S, \quad Z_i \mapsto \alpha(Z_i) \otimes 1 + 1 \otimes \beta(Z_i),$$

and the linear functional

$$\tau * \sigma : R \otimes S \to \mathbb{C}, \quad r \otimes s \mapsto \tau(r) \otimes \sigma(s).$$

For commutative algebras this corresponds to the usual convolution of measures with respect to the group structure on $\mathbb{R}^n$.

(6.4.2) Proposition. *The Fourier transform of the convolution of measures is the product of their Fourier transforms:*

$$\mathfrak{F}(\tau * \sigma) = \mathfrak{F}(\tau) \cdot \mathfrak{F}(\sigma).$$

Proof: This is a consequence of the fact that the elements $E_\gamma(iZ_1, \ldots, iZ_n)$ of $\mathbb{C}\langle\langle Z_1, \ldots, Z_n\rangle\rangle$ are grouplike; see the exponential property (2.4.7). $\quad\square$

(6.5) Formal Fourier transform of noncommutative measures. The product of two monomial functions on Π_n is a linear combination of monomial functions. This expresses Chen's shuffle relations among iterated integrals:

$$(6.5.1) \qquad W_{j_1, \ldots, j_m} W_{j_{m+1}, \ldots, j_{m+p}} = \sum_s W_{j_{s(1)}, \ldots, j_{s(m+p)}},$$

the sum being over the set of (m, p)-shuffles. An identical formula holds for the convolution of the measures $\delta^{(j_1, \ldots, j_m)}$ and $\delta^{(j_{m+1}, \ldots, j_{m+p})}$, since both formulas describe the Hopf algebra structure on $\mathbb{C}\langle\langle Z_1, \ldots, Z_n\rangle\rangle$.

The $\mathbb{C}$-algebra with basis $W_J = W_{j_1, \ldots, j_m}$ and multiplication law (6.5.1) is nothing but the algebra

$$(6.5.2) \qquad \mathbb{C}[G_n] = \varinjlim \mathbb{C}[G_{n,d}]$$

of regular functions on the group scheme $G_n = \varprojlim G_{n,d}$. The multiplication in G_n corresponds to the Hopf algebra structure given by

$$(6.5.3) \qquad \Delta(W_{j_1, \ldots, j_m}) = \sum_{\nu=0}^{m+1} W_{j_1, \cdots, j_\nu} \otimes W_{j_{\nu+1}, \ldots, j_m}.$$

Elements of $\mathbb{C}[G_n]$ can be called polynomial functions on Π_n.

Note that formal infinite linear combinations (series) $\sum_J c_J W_J$ still form a well-defined algebra via (6.5.1), which we denote by $\mathbb{C}[[G_n]]$. This is the algebra of functions on the formal completion of G_n at 1. The rule (6.5.3) makes $\mathbb{C}[[G_n]]$ into a topological Hopf algebra.

Let τ be a measure on $\mathbb{R}^n_{\mathrm{NC}}$. We will call the series

$$(6.5.4) \qquad \widehat{\mathfrak{F}}(\tau) = \sum_{J=(j_1, \ldots, j_m)} \tau(Z_{i_1} \cdots Z_{j_m}) \cdot W_J \quad \in \quad \mathbb{C}[[G_n]]$$

the formal Fourier transform of τ. As before, we see that convolution of measures is taken into the product in $\mathbb{C}[[G_n]]$.

7 Toward the Inverse Noncommutative Fourier Transform

(7.0) In this section we sketch a possible approach to the problem of finding the inverse to the NCFT $\mathcal{F}$ from (2.2) In other words, given a "noncommutative function" $f = f(Z_1, \ldots, Z_n)$, how do we find a measure μ on (possibly some completion of) Π_n such that $\mathcal{F}(\mu) = f$? Note that in contrast to classical analysis, the dual Fourier transform $\mathfrak{F}$ (from noncommutative measures to functions on Π_n) does not provide even a conjectural answer, since there is no natural identification of functions and measures.

So we take as our starting point the case of discrete NCFT (1.4.5), where Theorem 1.4.6 provides a neat inversion formula.

(7.1) Fourier series and Fourier integrals. We recall the classical procedure expressing Fourier integrals as scaling limits of Fourier series; see [W], §5. Let $f(x)$ be a piecewise continuous $\mathbb{C}$-valued function on $\mathbb{R}$ of sufficiently rapid decay. We can restrict f to the interval $[-\pi, \pi]$, which is a fundamental domain for the exponential map $z \mapsto \exp(iz), \mathbb{R} \to S^1$, and then represent f on this interval as a Fourier series in e^{inz}, $n \in \mathbb{Z}$.

Next, let us scale the interval to $[-A, A]$ instead. Then the orthonormal basis of functions is formed by

$$(7.1.1) \qquad \frac{1}{\sqrt{2A}} \exp\left(\frac{n\pi iz}{A}\right), \quad n \in \mathbb{Z},$$

so on the new interval we have

$$(7.1.2) \qquad f(z) = \frac{1}{2A} \sum_{n \in \mathbb{Z}} \exp\left(\frac{n\pi iz}{A}\right) \int_{-A}^{A} f(w) \exp\left(\frac{-n\pi iw}{A}\right) dw.$$

If we associate the Fourier coefficients to the scaled lattice points, putting

$$(7.1.3) \qquad g\left(\frac{n\pi}{A}\right) = \frac{1}{\sqrt{2\pi}} \int_{-A}^{A} f(z) \exp\left(\frac{-n\pi iz}{A}\right) dz,$$

then

$$(7.1.4) \qquad f(z) = \frac{1}{\sqrt{2\pi}} \sum_{n \in \mathbb{Z}} g\left(\frac{n\pi}{A}\right) \exp\left(\frac{n\pi iz}{A}\right) \Delta\left(\frac{n\pi}{A}\right), \quad z \in [-A, A],$$

where $\Delta\left(\frac{n\pi}{A}\right) = \frac{\pi}{A}$ is the step of the dual lattice. So when $A \to \infty$, the formulas (7.1.3) and (7.1.4) "tend to" the formulas for two mutually inverse Fourier transforms for functions on $\mathbb{R}$. In other words, the measures on $\mathbb{R}$ (with coordinate y) given by infinite combinations of shifted Dirac delta functions,

$$(7.1.5) \qquad \frac{1}{\sqrt{2\pi}} \frac{\pi}{A} \sum_{n \in \mathbb{Z}} g\left(\frac{n\pi}{A}\right) \delta\left(y - \frac{n\pi}{A}\right),$$

converge, as $A \to \infty$, to a measure whose Fourier transform is f.

(7.2) Matrix fundamental domains. We now consider the analogue of the above formalism for Hermitian matrices instead of elements of $\mathbb{R}$, and unitary matrices instead of those of S^1. Let $\mathrm{Herm}_N^{\leq A}$ be the set of Hermitian N by N

matrices whose eigenvalues all lie in $[-A, A]$. Then $\mathrm{Herm}_N^{\leq \pi}$ is a fundamental domain for the exponential map

$$(7.2.1) \qquad Z \mapsto X = \exp(iZ), \quad \mathrm{Herm}_N \to U(N).$$

Note that the Jacobian of the map (7.2.1) is given by
$$(7.2.2)$$
$$J(Z) = \det{}_{N^2 \times N^2} \frac{e^{\mathrm{ad}(Z)} - 1}{Z} = \prod_{j,k} \frac{e^{i(\lambda_j - \lambda_k)} - 1}{\lambda_j - \lambda_k} = \prod_{j<k} 2 \frac{1 - \cos(\lambda_j - \lambda_k)}{(\lambda_k - \lambda_k)^2}.$$

Here $\lambda_1, \ldots, \lambda_N$ are the eigenvalues of Z; see [Hel], p. 255. Using the formula for the volume of $U(N)$, see, e.g., [Mac], we can write the normalized Haar measure on $U(N)$ transferred into $\mathrm{Herm}_N^{\leq \pi}$ as

$$(7.2.3) \qquad d^*X = \frac{J(Z)dZ}{V_N}, \quad V_N = \prod_{m=0}^{N-1} \frac{2\pi^{m+1}}{m!}.$$

Let $f(Z_1, \ldots, Z_n)$ be a "good" noncommutative function (for example an entire function or a rational function defined for all Hermitian $Z_1, \ldots, Z_n$ and having good decay at infinity). Then we can restrict f to $(\mathrm{Herm}_N^{\leq \pi})^n$ and transfer it, via the map (7.2.1), to a matrix function on $U(N)^n$. This matrix function is clearly nothing but

$$(7.2.4) \qquad f(-i \log(X_1), \ldots, -i \log(X_n)),$$

where $-i \log : U(N) \to \mathrm{Herm}_N^{\leq \pi}$ is the branch of the logarithm defined using our choice of the fundamental domain. Although (7.2.4) is far from being a noncommutative Laurent polynomial (indeed, it is typically discontinuous), one can hope to use the procedure of Theorem 1.4.6 to expand it into a noncommutative *Fourier series*. In other words, assuming that for each $\gamma \in F_n$ the limit

$$(7.2.5)$$

$$a_\gamma = \lim_{N \to \infty} \frac{1}{N} \mathrm{Tr} \int_{X_1, \ldots, X_n \in U(N)} f(-i \log(X_1), \ldots, -i \log(X_n)) X^{-\gamma} \prod_{j=1}^n d^*X_j$$

$$= \lim_{N \to \infty} \frac{1}{N} \mathrm{Tr} \int_{Z_1, \ldots, Z_n \in \mathrm{Herm}_N^{\leq \pi}} f(Z_1, \ldots, Z_n) E_{\gamma^{-1}}(iZ_1, \ldots, iZ_n)) \prod_{j=1}^n \frac{J(Z_j)dZ_j}{V_N}$$

exists, we can form the series

$$(7.2.6) \qquad \sum_\gamma a_\gamma X^\gamma = \sum_\gamma a_\gamma E_\gamma(iZ_1, \ldots, iZ_n), \quad Z_j \in \mathrm{Herm}_N^{\leq \pi}.$$

By analogy with the classical case one can expect that this series converges to $f|_{(\mathrm{Herm}_N^{\leq \pi})^n}$ away from the boundary.

(7.3) Scaling the period. In the situation of (7.2) let us choose $A > 0$ and restrict f to $(\text{Herm}_N^{\leq A})^n$. The same procedure would then expand the restriction into a series in

$$(7.3.1) \qquad\qquad X_j^{\pi/A} = \exp\left(i\frac{\pi}{A}Z_j\right), \quad j = 1, \ldots, n.$$

Let $F_n^{\pi/A} \subset G_n(\mathbb{R})$ be the group generated by the $X_j^{\pi/A}$. We can think of elements of $F_n^{\pi/A}$ as rectangular paths in $\mathbb{R}^n$ with increments being integer multiples of π/A. The coefficients of the series for the restriction give then a function

$$g_A : F_n^{\pi/A} \to \mathbb{C},$$

so the series will have the form

$$f(Z) = \sum_{\gamma \in F_n^{\pi/A}} g_A(\gamma) E_\gamma(iZ), \quad Z = (Z_1, \ldots, Z_n), \ Z_j \in \text{Herm}_N^{\leq A}.$$

Now, as $A \to \infty$, we would like to say that the g_A, considered as linear combinations of Dirac measures on Π_n (or some completion), tend to a limit measure. Although Π_n is not a manifold, we can pass to finite-dimensional approximations

$$\Pi_n \subset G_n(\mathbb{R}) \xrightarrow{p_d} G_{n,d}(\mathbb{R}).$$

Let $F_{n,d}^{\pi/A} = p_d(F_n^{\pi/A})$. This is a free nilpotent group of degree d on generators $p_d(X_j^{\pi/A})$, and is a discrete subgroup ("lattice") in $G_{n,d}(\mathbb{R})$. As $A \to \infty$, these lattices are getting dense in $G_{n,d}(\mathbb{R})$. Supposing that the direct image (summation over the fibers) $p_{d*}(g_A)$ exists as a function on $F_{n,d}^{\pi/A}$ or, what is the same, a measure on $G_{n,d}(\mathbb{R})$ supported on the discrete subgroup $F_{n,d}^{\pi/A}$, we can then ask for the existence of the limit

$$\mu_d = \lim_{A \to \infty} p_{d*}(g_A) \quad \in \quad \text{Meas}(G_{n,d}(\mathbb{R})).$$

These measures, if they exist, would then form a promeasure $\mu_\bullet$ that is the natural candidate for the inverse Fourier transform of f. The author hopes to address these issues in a future paper.

References

[Ba] F. BAUDOIN, *An Introduction to the Geometry of Stochastic Flows*, World Scientific, Singapore, 2004.

[Bel] D. R. BELL, *Degenerate Stochastic Differential Equations and Hypoellipticity* (Pitman Monographs and Surveys in Pure and Applied Mathematics, 79) Longman, Harlow, 1995.

[Be] G. BEN AROUS, *Flots et séries de Taylor stochastiques*, J. of Prob. Theory and Related Fields, **81** (1989), 29–77.

[BGG] R. Beals, B. Gaveau, P.C. Greiner, *Hamilton-Jacobi theory and the heat kernel on Heisenberg groups*, J. Math. Pures Appl. **79** (2000), 633–689.

[Bi1] J.-M. Bismut, *Large Deviations and the Malliavin Calculus*, Birkhäuser, Boston, 1984.

[Bi2] J.-M. Bismut, *The Atiyah-Singer theorem: a probabilistic approach. I. The index theorem*, J. Funct. Anal. **57** (1984), 56–99.

[C0] K.-T.-Chen, *Collected Papers*, Birkhäuser, Boston, 2000.

[C1] K.-T. Chen, *Integration in free groups*, Ann. Math. **54** (1951), 147–162.

[C2] K.-T. Chen, *Integration of paths—a faithful representation of paths by noncommutative formal power series*, Trans. AMS, **89** (1958), 395–407.

[C3] K.-T. Chen, *Algebraic paths*, J. of Algebra, **10** (1968), 8–36.

[Coh] P. M. Cohn, *Skew fields. Theory of general division rings* (Encyclopedia of Mathematics and Its Applications, 57), Cambridge University Press, Cambridge, 1995.

[Con] A. Connes, *Noncommutative Geometry*, Academic Press, Inc. San Diego, CA, 1994.

[FN] M. Fliess, D. Normand-Cyrot, *Algèbres de Lie nilpotentes, formule de Baker-Campbell-Hausdorff et intégrales iterées de K.-T. Chen*, Sém. de Probabilités XVI, Springer Lecture Notes in Math. **920**, 257–267.

[Fo] G. B. Folland, *A fundamental solution for a subelliptic operator*, Bull. AMS, **79** (1973), 367–372.

[G] B. Gaveau, *Principe de moindre action, propagation de chaleur et estimées sous-elliptiques pour certains groupes nilpotents*, Acta Math. **139** (1977), 95–153.

[HL] M. Hakim-Dowek, D. Lépingle, *L'exponentielle stochastique des groupes de Lie*, Sém. de Probabilites XX, Springer Lecture Notes in Math. **1203**, 352–374.

[Hel] S. Helgason, *Differential Geometry, Lie Groups and Symmetric Spaces*, Academic Press, 1962.

[HP] F. Hiai, D. Petz, *The Semicircle Law, Free Random Variables and Entropy*, Amer. Math. Soc. 2000.

[Ho] L. Hörmander, *Hypoelliptic second order differential equations*, Acta Math. **119** (1967), 141–171.

[Hu] A. Hulanicki, *The distribution of energy in the Brownian motion in the Gaussian field and analytic hypoellipticity of certain subelliptic operators on the Heisenberg group*, Studia Math. **56** (1976), 165–173.

[K1] M. Kapranov, *Free Lie algebroids and the space of paths*, Selecta Math. N.S. **13** (2007), 277–319.

[K2] M. Kapranov, *Membranes and higher groupoids*, in preparation.

[KW] N. Kunita, S. Watanabe, *Stochastic Differential Equations and Diffusion processes*, North-Holland, Amsterdam, 1989.

[Mac] I. G. Macdonald, *The volume of a compact Lie group*, Invent. Math. 56 (1980), 93–95.

[McK] H. P. McKean, *Stochastic Integrals*, Chelsea Publ. Co. 2005.

[Ma] Y. I. Manin, *Iterated Shimura integrals*, preprint math.NT/0507438.

[NS] A. Nica, R. Speicher, *Lectures on the Combinatorics of Free Probability* (London Math. Soc. Lecture Notes vol. 335), Cambridge Univ. Press, 2006.

[Ok] B. Oksendal, *Stochastic Differential Equations*, Springer, Berlin, 1989.

[Pa] A. N. Parshin, *On a certain generalization of the Jacobian manifold*, Izv. AN SSSR, 30(1966), 175–182.

[Po] A.M. Polyakov, *Gauge Fields and Strings*, Harwood Academic Publ. 1987.

[R] C. Reutenauer, *Free Lie Algebras*, Oxford Univ. Press, 1993.

[Si] I. M. Singer, *On the master field in two dimensions,* in: Functional Analysis on the Eve of the 21st Century (In honor of I. M. Gelfand), S. Gindikin et al., eds, Vol. 1, 263–281.

[SW] D. W. Stroock, S. R. S. Varadhan, *Multidimensional Diffusion Processes*, Springer, Berlin, 1979.

[Ta] J. L. Taylor, *Functions of several noncommuting variables,* Bull. AMS, **79** (1973), 1–34.

[V] D. Voiculescu, *Limit laws for random matrices and free products*, Invent. Math. **104** (1991), 201–220.

[VDN] D. Voiculescu, K.J. Dykema, A. Nica, *Free Random Variables*, Amer. Math. Soc. 1992.

[W] N. Wiener, *The Fourier Integral and Certain of Its Applications,* Dover Publ, 1958.

Another Look at the Dwork Family

Nicholas M. Katz

Princeton University, Mathematics, Fine Hall, NJ 08544-1000, USA
nmk@math.princeton.edu

Dedicated to Yuri Manin on his seventieth birthday

Summary. We give a new approach to the cohomology of the Dwork family, and more generally of single-monomial deformations of Fermat hypersurfaces. This approach is based on the surprising connection between these families and Kloosterman sums, and makes use of the Fourier Transform and the theory of Kloosterman sheaves and of hypergeometric sheaves.

Key words: Dwork family, monodromy, hypergeometric, Kloosterman sheaf, Fourier Transform

2000 Mathematics Subject Classifications: 14D10, 14D05, 14C30, 34A20

1 Introduction and a bit of history

After proving [Dw-Rat] the rationality of zeta functions of all algebraic varieties over finite fields nearly fifty years ago, Dwork studied in detail the zeta function of a nonsingular hypersurface in projective space, cf. [Dw-Hyp1] and [Dw-HypII]. He then developed his "deformation theory", cf. [Dw-Def], [Dw-NPI] and [Dw-NPII], in which he analyzed the way in which his theory varied in a family. One of his favorite examples of such a family, now called the Dwork family, was the one parameter (λ) family, for each degree $n \geq 2$, of degree-n hypersurfaces in $\mathbb{P}^{n-1}$ given by the equation

$$\sum_{i=1}^{n} X_i^n - n\lambda \prod_{i=1}^{n} X_i = 0,$$

a family he wrote about explicitly in [Dw-Def, page 249, (i), (ii), (iv), the cases $n = 2, 3, 4$], [Dw-HypII, section 8, pp. 286-288, the case $n = 3$] and [Dw-PC, 6.25, the case $n = 3$, and 6.30, the case $n = 4$]. Dwork of course also considered the generalization of the above Dwork family consisting of

Y. Tschinkel and Y. Zarhin (eds.), *Algebra, Arithmetic, and Geometry*,
Progress in Mathematics 270, DOI 10.1007/978-0-8176-4747-6_4,
© Springer Science+Business Media, LLC 2009

single-monomial deformations of Fermat hypersurfaces of any degree and dimension. He mentioned one such example in [Dw-Def, page 249, (iii)]. In [Dw-PAA, pp. 153–154], he discussed the general single-monomial deformation of a Fermat hypersurface, and explained how such families led to generalized hypergeometric functions.

My own involvement with the Dwork family started (in all senses!) at the Woods Hole conference in the summer of 1964 with the case $n = 3$, when I managed to show in that special case that the algebraic aspects of Dwork's deformation theory amounted to what would later be called the Gauss–Manin connection on relative de Rham cohomology, but which at the time went by the more mundane name of "differentiating cohomology classes with respect to parameters".

That this article is dedicated to Manin on his seventieth birthday is particularly appropriate, because in that summer of 1964 my reference for the notion of differentiating cohomology classes with respect to parameters was his 1958 paper [Ma-ACFD]. I would also like to take this opportunity to thank, albeit belatedly, Arthur Mattuck for many helpful conversations that summer.

I discussed the Dwork family in [Ka-ASDE, 2.3.7.17–23, 2.3.8] as a "particularly beautiful family", and computed explicitly the differential equation satisfied by the cohomology class of the holomorphic $n - 2$ form. It later showed up in [Ka-SE, 5.5, esp. pp. 188–190], about which more below. Ogus [Ogus-GTCC, 3.5, 3.6] used the Dwork family to show the failure in general of "strong divisibility". Stevenson, in her thesis [St-th],[St, end of Section 5, page 211], discussed single-monomial deformations of Fermat hypersurfaces of any degree and dimension. Koblitz [Kob] later wrote on these same families. With mirror symmetry and the stunning work of Candelas et al. [C-dlO-G-P] on the case $n = 5$, the Dwork family became widely known, especially in the physics community, though its occurence in Dwork's work was almost (not entirely, cf. [Ber], [Mus-CDPMQ]) forgotten. Recently the Dwork family turned out to play a key role in the proof of the Sato–Tate conjecture (for elliptic curves over $\mathbb{Q}$ with non-integral j-invariant), cf. [H-SB-T, Section 1, pp. 5–15].

The present paper gives a new approach to computing the local system given by the cohomology of the Dwork family, and more generally of families of single-monomial deformations of Fermat hypersurfaces. This approach is based upon the surprising connection, noted in [Ka-SE, 5.5, esp. pp. 188–190], between such families and Kloosterman sums. It uses also the theory, developed later, of Kloosterman sheaves and of hypergeometric sheaves, and of their behavior under Kummer pullback followed by Fourier Transform, cf. [Ka-GKM] and [Ka-ESDE, esp. 9.2 and 9.3]. In a recent preprint, Rojas-Leon and Wan [RL-Wan] have independently implemented the same approach.

2 The situation to be studied: generalities

We fix an integer $n \geq 2$, a degree $d \geq n$, and an n-tuple $W = (w_1, ..., w_n)$ of strictly positive integers with $\sum_i w_i = d$, and with $gcd(w_1, ..., w_n) = 1$. This data (n, d, W) is now fixed. Let R be a ring in which d is invertible.

Over R we have the affine line $\mathbb{A}_R^1 := Spec(R[\lambda])$. Over $\mathbb{A}_R^1$, we consider certain one-parameter (namely λ) families of degree-d hypersurfaces in $\mathbb{P}^{n-1}$. Given an $n + 1$-tuple $(a, b) := (a_1, ..., a_n, b)$ of invertible elements in R, we consider the one-parameter (namely λ) family of degree-d hypersurfaces in $\mathbb{P}^{n-1}$,

$$X_\lambda(a, b) : \sum_{i=1}^n a_i X_i^d - b\lambda X^W = 0,$$

where we have written

$$X^W := \prod_{i=1}^n X_i^{w_i}.$$

More precisely, we consider the closed subscheme $\mathbb{X}(a, b)_R$ of $\mathbb{P}_R^{n-1} \times_R \mathbb{A}_R^1$ defined by the equation

$$\sum_{i=1}^n a_i X_i^d - b\lambda X^W = 0,$$

and denote by

$$\pi(a, b)_R : \mathbb{X}(a, b)_R \to \mathbb{A}_R^1$$

the restriction to $\mathbb{X}(a, b)_R$ of the projection of $\mathbb{P}_R^{n-1} \times_R \mathbb{A}_R^1$ onto its second factor.

Lemma 2.1. *The morphism*

$$\pi(a, b)_R : \mathbb{X}(a, b)_R \to \mathbb{A}_R^1$$

is lisse over the open set of $\mathbb{A}_R^1$ where the function

$$(b\lambda/d)^d \prod_i (w_i/a_i)^{w_i} - 1$$

is invertible.

Proof. Because d and the a_i are invertible in R, a Fermat hypersurface of the form

$$\sum_{i=1}^n a_i X_i^d = 0$$

is lisse over R. When we intersect our family with any coordinate hyperplane $X_i = 0$, we obtain a constant Fermat family in one lower dimension (because each $w_i \geq 1$). Hence any geometric point $(x, \lambda) \in \mathbb{X}$ at which π is not smooth has all coordinates X_i invertible. So the locus of nonsmoothness of π is defined by the simultaneous vanishing of all the $X_i d/dX_i$, i.e., by the simultaneous equations

$$da_i X_i^d = b\lambda w_i X^W, \text{ for } i = 1, ..., n.$$

Divide through by the invertible factor da_i. Then raise both sides of the i'th equation to the w_i power and multiply together right and left sides separately over i. We find that at a point of nonsmoothness we have

$$X^{dW} = (b\lambda/d)^d \prod_i (w_i/a_i)^{w_i} X^{dW}.$$

As already noted, all the X_i are invertible at any such point, and hence

$$1 = (b\lambda/d)^d \prod_i (w_i/a_i)^{w_i}$$

at any geometric point of nonsmoothness. $\qquad\qquad\Box$

In the Dwork family *per se*, all $w_i = 1$. But in a situation where there is a prime p not dividing $d\ell$ but dividing one of the w_i, then taking for R an $\mathbb{F}_p$-algebra (or more generally a ring in which p is nilpotent), we find a rather remarkable family.

Corollary 2.2. *Let p be a prime which is prime to d but which divides one of the w_i, and R a ring in which p is nilpotent. Then the morphism*

$$\pi(a, b)_R : \mathbb{X}(a, b)_R \to \mathbb{A}^1_R$$

is lisse over all of $\mathbb{A}^1_R$.

Remark 2.3. Already the simplest possible example of the above situation, the family in $\mathbb{P}^1/\mathbb{F}_q$ given by

$$X^{q+1} + Y^{q+1} = \lambda XY^q,$$

is quite interesting. In dehomogenized form, we are looking at

$$x^{q+1} - \lambda x + 1$$

as polynomial over $\mathbb{F}_q(\lambda)$; its Galois group is known to be $PSL(2, \mathbb{F}_q)$, cf. [Abh-PP, bottom of p. 1643], [Car], and [Abh-GTL, Serre's Appendix]. The general consideration of "$p|w_i$ for some i" families in higher dimension would lead us too far afield, since our principal interest here is with families that "start life" over $\mathbb{C}$. We discuss briefly such "$p|w_i$ for some i" families in Appendix II. We would like to call the attention of computational number theorists to these families, with no degeneration at finite distance, as a good test case for proposed methods of computing efficiently zeta functions in entire families.

3 The particular situation to be Studied: details

Recall that the data (n, d, W) is fixed. Over any ring R in which $d \prod_i w_i$ is invertible, we have the family $\pi : \mathbb{X} \to \mathbb{A}_R^1$ given by

$$X_\lambda := X_\lambda(W, d) : \sum_{i=1}^n w_i X_i^d - d\lambda X^W = 0;$$

it is proper and smooth over the open set $U := \mathbb{A}_R^1[1/(\lambda^d - 1)] \subset \mathbb{A}_R^1$ where $\lambda^d - 1$ is invertible.

The most natural choice of R, then, is $\mathbb{Z}[1/(d \prod_i w_i)]$. However, it will be more convenient to work over a somewhat larger cyclotomic ring, which contains, for each i, all the roots of unity of order dw_i. Denote by $lcm(W)$ the least common multiple of the w_i, and define $d_W := lcm(W)d$. In what follows, we will work over the ring

$$R_0 := \mathbb{Z}[1/d_W][\zeta_{d_W}] := \mathbb{Z}[1/d_W][T]/(\Phi_{d_W}(T)),$$

where $\Phi_{d_W}(T)$ denotes the d_W'th cyclotomic polynomial.

We now introduce the relevant automorphism group of our family. We denote by $\mu_d(R_0)$ the group of d'th roots of unity in R_0, by $\Gamma = \Gamma_{d,n}$ the n-fold product group $(\mu_d(R_0))^n$, by $\Gamma_W \subset \Gamma$ the subgroup consisting of all elements $(\zeta_1, ..., \zeta_n)$ with $\prod_{i=1}^n \zeta_i^{w_i} = 1$, and by $\Delta \subset \Gamma_W$ the diagonal subgroup, consisting of all elements of the form $(\zeta, ..., \zeta)$. The group Γ_W acts as automorphisms of $\mathbb{X}/\mathbb{A}_{R_0}^1$, an element $(\zeta_1, ..., \zeta_n)$ acting as

$$((X_1, ..., X_n), \lambda) \mapsto ((\zeta_1 X_1, ..., \zeta_n X_n), \lambda).$$

The diagonal subgroup Δ acts trivially.

The natural pairing

$$(\mathbb{Z}/d\mathbb{Z})^n \times \Gamma \to \mu_d(R_0) \subset R_0^\times,$$

$$(v_1, ..., v_n) \times (\zeta_1, ..., \zeta_n) \to \prod_i \zeta_i^{v_i},$$

identifies $(\mathbb{Z}/d\mathbb{Z})^n$ as the R_0-valued character group $D\Gamma := Hom_{group}(\Gamma, R_0^\times)$. The subgroup

$$(\mathbb{Z}/d\mathbb{Z})_0^n \subset (\mathbb{Z}/d\mathbb{Z})^n$$

consisting of elements $V = (v_1, ..., v_n)$ with $\sum_i v_i = 0$ in $\mathbb{Z}/d\mathbb{Z}$ is then the R_0-valued character group $D(\Gamma/\Delta)$ of Γ/Δ. The quotient group $(\mathbb{Z}/d\mathbb{Z})_0^n/<W>$ of $(\mathbb{Z}/d\mathbb{Z})_0^n$ by the subgroup generated by (the image, by reduction mod d, of) W is then the R_0-valued character group $D(\Gamma_W/\Delta)$ of Γ_W/Δ.

For G either of the groups Γ/Δ, Γ_W/Δ, an R_0-linear action of G on a sheaf of R_0-modules M gives an eigendecomposition

$$M = \bigoplus_{\rho \in D(G)} M(\rho).$$

If the action is by the larger group $G = \Gamma/\Delta$, then $DG = (\mathbb{Z}/d\mathbb{Z})_0^n$, and for $V \in (\mathbb{Z}/d\mathbb{Z})_0^n$ we denote by $M(V)$ the corresponding eigenspace. If the action is by the smaller group Γ_W/Δ, then DG is the quotient group $(\mathbb{Z}/d\mathbb{Z})_0^n/<W>$; given an element $V \in (\mathbb{Z}/d\mathbb{Z})_0^n$, we denote by $V \ mod \ W$ its image in the quotient group, and we denote by $M(V \ mod \ W)$ the corresponding eigenspace.

If M is given with an action of the larger group Γ/Δ, we can decompose it for that action:

$$M = \bigoplus_{V \in (\mathbb{Z}/d\mathbb{Z})_0^n} M(V).$$

If we view this same M only as a representation of the sugroup Γ_W/Δ, we can decompose it for that action:

$$M = \bigoplus_{V \in (\mathbb{Z}/d\mathbb{Z})_0^n/<W>} M(V \ mod \ W).$$

The relation between these decompositions is this: for any $V \in (\mathbb{Z}/d\mathbb{Z})_0^n$,

$$M(V \ mod \ W) = \bigoplus_{r \ mod \ d} M(V + rW).$$

We return now to our family $\pi : \mathbb{X} \to \mathbb{A}_{R_0}^1$, which we have seen is (projective and) smooth over the open set

$$U = \mathbb{A}_{R_0}^1[1/(\lambda^d - 1)].$$

We choose a prime number ℓ, and an embedding of R_0 into $\overline{\mathbb{Q}}_\ell$. [We will now need to invert ℓ, so arguably the most efficient choice is to take for ℓ a divisor of d_W.] We We form the sheaves

$$\mathcal{F}^i := R^i\pi_*\overline{\mathbb{Q}}_\ell$$

on $\mathbb{A}_{R_0[1/\ell]}^1$. They vanish unless $0 \le i \le 2(n-2)$, and they are all lisse on $U[1/\ell]$. By the weak Lefschetz Theorem and Poincaré duality, the sheaves $\mathcal{F}^i|U[1/\ell]$ for $i \ne n-2$ are completely understood. They vanish for odd i; for even $i = 2j \le 2(n-2), i \ne n-2$, they are the Tate twists

$$\mathcal{F}^{2j}|U[1/\ell] \cong \overline{\mathbb{Q}}_\ell(-j).$$

We now turn to the lisse sheaf $\mathcal{F}^{n-2}|U[1/\ell]$. It is endowed with an autoduality pairing (cup product) toward $\overline{\mathbb{Q}}_\ell(-(n-2))$ which is symplectic if $n-2$ is odd, and orthogonal if $n-2$ is even. If $n-2$ is even, say $n-2 = 2m$, then $\mathcal{F}^{n-2}|U[1/\ell]$ contains $\overline{\mathbb{Q}}_\ell(-m)$ as a direct summand (m'th power of the hyperplane class from the ambient $\mathbb{P}$) with nonzero self-intersection. We define

$Prim^{n-2}$ (as a sheaf on $U[1/\ell]$ only) to be the annihilator in $\mathcal{F}^{n-2}|U[1/\ell]$ of this $\overline{\mathbb{Q}}_\ell(-m)$ summand under the cup product pairing. So we have

$$\mathcal{F}^{n-2}|U[1/\ell] = Prim^{n-2} \bigoplus \overline{\mathbb{Q}}_\ell(-m),$$

when $n - 2 = 2m$. When $n - 2$ is odd, we define $Prim^{n-2} := \mathcal{F}^{n-2}|U[1/\ell]$, again as a sheaf on $U[1/\ell]$ only.

The group Γ_W/Δ acts on our family, so on all the sheaves above. For $i \neq n - 2$, it acts trivially on $\mathcal{F}^i|U[1/\ell]$. For $i = n - 2 = 2m$ even, it respects the decomposition

$$\mathcal{F}^{n-2}|U[1/\ell] = Prim \bigoplus \overline{\mathbb{Q}}_\ell(-m),$$

and acts trivially on the second factor. We thus decompose $Prim^{n-2}$ into eigensheaves $Prim^{n-2}(V \bmod W)$. The basic information on the eigensheaves $Prim^{n-2}(V \bmod W)$ is encoded in elementary combinatorics of the coset $V \bmod W$. An element $V = (v_1, \ldots, v_n) \in (\mathbb{Z}/d\mathbb{Z})_0^n$ is said to be totally nonzero if $v_i \neq 0$ for all i. Given a totally nonzero element $V \in (\mathbb{Z}/d\mathbb{Z})_0^n$, we define its degree $deg(V)$ as follows. For each i, denote by $\tilde{v}_i$ the unique integer $1 \leq \tilde{v}_i \leq d - 1$ that mod d gives v_i. Then $\sum_i \tilde{v}_i$ is 0 mod d, and we define

$$deg(V) := (1/d) \sum_i \tilde{v}_i.$$

Thus $deg(V)$ lies in the interval $1 \leq deg(V) \leq n - 1$. The Hodge type of a totally nonzero $V \in (\mathbb{Z}/d\mathbb{Z})_0^n$ is defined to be

$$HdgType(V) := (n - 1 - deg(V), deg(V) - 1).$$

We now compute the rank and the Hodge numbers of eigensheaves $Prim^{n-2}(V \bmod W)$. We have already chosen an embedding of R_0 into $\overline{\mathbb{Q}}_\ell$. We now choose an embedding of $\overline{\mathbb{Q}}_\ell$ into $\mathbb{C}$. The composite embedding $R_0 \subset \mathbb{C}$ allows us to extend scalars in our family $\pi : \mathbb{X} \to \mathbb{A}_{R_0}^1$, which is projective and smooth over the open set $U_{R_0} = \mathbb{A}_{R_0}^1[1/(\lambda^d - 1)]$, to get a complex family $\pi_{\mathbb{C}} : \mathbb{X}_{\mathbb{C}} \to \mathbb{A}_{\mathbb{C}}^1$, which is projective and smooth over the open set $U_{\mathbb{C}} = \mathbb{A}_{\mathbb{C}}^1[1/(\lambda^d - 1)]$. Working in the classical complex topology with the corresponding analytic spaces, we can form the higher direct image sheaves $R^i\pi_{\mathbb{C}}^{an}\mathbb{Q}$ on $\mathbb{A}_{\mathbb{C}}^{1,an}$, whose restrictions to $U_{\mathbb{C}}^{an}$ are locally constant sheaves. We can also form the locally constant sheaf $Prim^{n-2,an}(\mathbb{Q})$ on $U_{\mathbb{C}}^{an}$. Extending scalars in the coefficients from $\mathbb{Q}$ to $\overline{\mathbb{Q}}_\ell$, we get the sheaf $Prim^{n-2,an}(\overline{\mathbb{Q}}_\ell)$. On the other hand, we have the lisse $\overline{\mathbb{Q}}_\ell$-sheaf $Prim^{n-2}$ on $U_{R_0[1/\ell]}$, which we can pull back, first to $U_{\mathbb{C}}$, and then to $U_{\mathbb{C}}^{an}$. By the fundamental comparison theorem, we have

$$Prim^{n-2,an}(\overline{\mathbb{Q}}_\ell) \cong Prim^{n-2}|U_{\mathbb{C}}^{an}.$$

Extending scalars from $\overline{\mathbb{Q}}_\ell$ to $\mathbb{C}$, we obtain

$$Prim^{n-2,an}(\mathbb{C}) \cong \left(Prim^{n-2}|U_{\mathbb{C}}^{an}\right) \otimes_{\overline{\mathbb{Q}}_\ell} \mathbb{C}.$$

This is all Γ_W/Δ-equivariant, so we have the same relation for individual eigensheaves:

$$Prim^{n-2,an}(\mathbb{C})(V \bmod W) \cong \left(Prim^{n-2}(V \bmod W)|U_{\mathbb{C}}^{an}\right) \otimes_{\overline{\mathbb{Q}}_\ell} \mathbb{C}.$$

If we extend scalars on $U_{\mathbb{C}}^{an}$ from the constant sheaf $\mathbb{C}$ to the sheaf $\mathcal{O}_{\mathbb{C}^\infty}$, then the resulting C^∞ vector bundle $Prim^{n-2,an}(\mathbb{C}) \otimes_{\mathbb{C}} \mathcal{O}_{\mathbb{C}^\infty}$ has a Hodge decomposition,

$$Prim^{n-2,an}(\mathbb{C}) \otimes_{\mathbb{C}} \mathcal{O}_{\mathbb{C}^\infty} = \bigoplus_{a \geq 0, b \geq 0, a+b=n-2} Prim^{a,b}.$$

This decomposition is respected by the action of Γ_W/Δ, so we get a Hodge decomposition of each eigensheaf:

$$Prim^{n-2,an}(\mathbb{C})(V \bmod W) \otimes_{\mathbb{C}} \mathcal{O}_{\mathbb{C}^\infty} = \bigoplus_{a \geq 0, b \geq 0, a+b=n-2} Prim^{a,b}(V \bmod W).$$

Lemma 3.1. *We have the following results.*

(1) The rank of the lisse sheaf $Prim^{n-2}(V \bmod W)$ on $U_{R_0[1/\ell]}$ is given by

$$rk(Prim^{n-2}(V \bmod W)) = \#\left\{r \in \mathbb{Z}/d\mathbb{Z} \mid V + rW \text{ is totally nonzero}\right\}.$$

In particular, the eigensheaf $Prim^{n-2}(V \bmod W)$ vanishes if none of the W-translates $V + rW$ is totally nonzero.

(2) For each (a,b) with $a \geq 0, b \geq 0, a+b = n-2$, the rank of the C^∞ vector bundle $Prim^{a,b}(V \bmod W)$ on $U_{\mathbb{C}}^{an}$ is given by

$$rk(Prim^{a,b}(V \bmod W))$$

$$= \#\left\{r \in \mathbb{Z}/d\mathbb{Z} \mid V + rW \text{ is totally nonzero and } deg(V + rW) = b+1\right\}.$$

Proof. To compute the rank of a lisse sheaf on $U_{R_0[1/\ell]}$, or the rank of a C^∞ vector bundle on $U_{\mathbb{C}}^{an}$, it suffices to compute its rank at a single geometric point of the base. We take the $\mathbb{C}$-point $\lambda = 0$, where we have the Fermat hypersurface. Here the larger group $(\mathbb{Z}/d\mathbb{Z})_0^n$ operates. It is well known that under the action of this larger group, the eigenspace $Prim(V)$ vanishes unless V is totally nonzero, e.g., cf. [Ka-IMH, Section 6]. One knows further that if V is totally nonzero, this eigenspace is one-dimensional, and of Hodge type $HdgType(V) := (n-1-deg(V), deg(V)-1)$, cf. [Grif-PCRI, 5.1 and 10.8]. $\square$

The main result of this paper is to describe the eigensheaves

$$Prim^{n-2}(V \bmod W)$$

as lisse sheaves on $U[1/\ell]$, i.e., as representations of $\pi_1(U[1/\ell])$, and to describe the direct image sheaves $j_{U*}(Prim^{n-2}(V \bmod W))$ on $\mathbb{A}_{R_0[1/\ell]}^1$, for $j_U : U[1/\ell] \subset \mathbb{A}_{R_0[1/\ell]}^1$ the inclusion. The description will be in terms of hypergeometric sheaves in the sense of [Ka-ESDE, 8.7.11].

4 Interlude: Hypergeometric sheaves

We first recall the theory in its original context of finite fields, cf. [Ka-ESDE, Chapter 8]. Let k be an $R_0[1/\ell]$-algebra which is a finite field, and

$$\psi : (k, +) \to \overline{\mathbb{Q}}_\ell^\times$$

a nontrivial additive character. Because k is an $R_0[1/\ell]$- algebra, it contains d_W distinct d_W'th roots of unity, and the structural map gives a group isomorphism $\mu_{d_W}(R_0) \cong \mu_{d_W}(k)$. So raising to the $\#k^\times/d_W$'th power is a surjective group homomorphism

$$k^\times \to \mu_{d_W}(k) \cong \mu_{d_W}(R_0).$$

So for any character $\chi : \mu_{d_W}(R_0) \to \mu_{d_W}(R_0)$, we can and will view the composition of χ with the above surjection as defining a multiplicative character of $k^\times$, still denoted χ. Every multiplicative character of $k^\times$ of order dividing d_W is of this form. Fix two non-negative integers a and b, at least one of which is nonzero. Let $\chi_1, ..., \chi_a$ be an unordered list of a multiplicative characters of $k^\times$ of order dividing d_W, some possibly trivial, and not necessarily distinct. Let $\rho_1, ..., \rho_b$ be another such list, but of length b. Assume that these two lists are disjoint, i.e., no χ_i is a ρ_j. Attached to this data is a geometrically irreducible middle extension $\overline{\mathbb{Q}}_\ell$-sheaf

$$\mathcal{H}(\psi; \chi_i{}'s; \rho_j{}'s)$$

on $\mathbb{G}_m/k$, which is pure of weight $a + b - 1$. We call it a hypergeometric sheaf of type (a, b). If $a \neq b$, this sheaf is lisse on $\mathbb{G}_m/k$; if $a = b$ it is lisse on $\mathbb{G}_m - \{1\}$, with local monodromy around 1 a tame pseudoreflection of determinant $(\prod_j \rho_j)/(\prod_i \chi_i)$.

The trace function of $\mathcal{H}(\psi; \chi_i{}'s; \rho_j{}'s)$ is given as follows. For E/k a finite extension field, denote by ψ_E the nontrivial additive character of E obtained from ψ by composition with the trace map $Trace_{E/k}$, and denote by $\chi_{i,E}$ (resp. $\rho_{j,E}$) the multiplicative character of E obtained from χ_i (resp. ρ_j) by composition with the norm map $Norm_{E/k}$. For $t \in \mathbb{G}_m(E) = E^\times$, denote by $V(a, b, t)$ the hypersurface in $(\mathbb{G}_m)^a \times (\mathbb{G}_m)^b/E$, with coordinates $x_1, ..., x_a, y_1, ..., y_b$, defined by the equation

$$\prod_i x_i = t \prod_j y_j.$$

Then

$$Trace(Frob_{t,E} | \mathcal{H}(\psi; \chi_i{}'s; \rho_j{}'s))$$

$$= (-1)^{a+b-1} \sum_{V(n,m,t)(E)} \psi_E \left(\sum_i x_i - \sum_j y_j \right) \prod_i \chi_{i,E}(x_i) \prod_j \overline{\rho}_{j,E}(y_j).$$

In studying these sheaves, we can always reduce to the case $a \geq b$, because under multiplicative inversion we have

$$\text{inv}^* \mathcal{H}(\psi; \chi_i \text{ 's}; \rho_j \text{ 's})) \cong \mathcal{H}(\overline{\psi}; \overline{\rho}_j \text{ 's}; \overline{\chi}_i \text{ 's})).$$

If $a \geq b$, the local monodromy around 0 is tame, specified by the list of χ_i's: the action of a generator γ_0 of I_0^{tame} is the action of T on the $\overline{\mathbb{Q}}_\ell[T]$-module $\overline{\mathbb{Q}}_\ell[T]/(P(T))$, for $P(T)$ the polynomial

$$P(T) := \prod_i (T - \chi_i(\gamma_0)).$$

In other words, for each of the distinct characters χ on the list of the χ_i's, there is a single Jordan block, whose size is the multiplicity with which χ appears on the list. The local monodromy around ∞ is the direct sum of a tame part of dimension b, and, if $a > b$, a totally wild part of dimension $a - b$, all of whose upper numbering breaks are $1/(a-b)$. The b-dimensional tame part of the local monodromy around ∞ is analogously specified by the list of ρ's: the action of a generator γ_∞ of I_∞^{tame} is the action of T on the $\overline{\mathbb{Q}}_\ell[T]$-module $\overline{\mathbb{Q}}_\ell[T]/(Q(T))$, for $Q(T)$ the polynomial

$$Q(T) := \prod_j (T - \rho_j(\gamma_0)).$$

When $a = b$, there is a canonical constant field twist of the hypergeometric sheaf $\mathcal{H} = \mathcal{H}(\psi; \chi_i \text{ 's}; \rho_j \text{ 's})$ which is independent of the auxiliary choice of ψ, which we will call $\mathcal{H}^{can}$. We take for $A \in \overline{\mathbb{Q}}_\ell^\times$ the nonzero constant

$$A = \left(\prod_i (-g(\psi, \chi_i)) \right) \left(\prod_j (-g(\overline{\psi}, \overline{\rho}_j)) \right),$$

and define

$$\mathcal{H}^{can} := \mathcal{H} \otimes (1/A)^{deg}.$$

[That $\mathcal{H}^{can}$ is independent of the choice of ψ can be seen in two ways. By elementary inspection, its trace function is independent of the choice of ψ, and we appeal to Chebotarev. Or we can appeal to the rigidity of hypergeometric sheaves with given local monodromy, cf. [Ka-ESDE, 8.5.6], to infer that with given χ's and ρ's, the hypergeometric sheaves $\mathcal{H}_\psi^{can}$ with different choices of ψ are all geometrically isomorphic. Being geometrically irreducible as well, they must all be constant field twists of each other. We then use the fact that $H^1\left(\mathbb{G}_m \otimes_k \overline{k}, \mathcal{H}_\psi^{can}\right)$ is one dimensional, and that $Frob_k$ acts on it by the scalar 1, to see that the constant field twist is trivial.]

Here is the simplest example. Take $\chi \neq \rho$, and form the hypergeometric sheaf $\mathcal{H}^{can}(\psi; \chi; \rho)$. Then using the rigidity approach, we see that

$$\mathcal{H}^{can}(\psi; \chi; \rho) \cong \mathcal{L}_{\chi(x)} \otimes \mathcal{L}_{(\rho/\chi)(1-x)} \otimes (1/A)^{deg},$$

with A (minus) the Jacobi sum over k,

$$A = -J(k; \chi, \rho/\chi) := -\sum_{x \in k^\times} \chi(x)(\rho/\chi)(1 - x).$$

The object

$$\mathcal{H}(\chi, \rho) := \mathcal{L}_{\chi(x)} \otimes \mathcal{L}_{(\rho/\chi)(1-x)}$$

makes perfect sense on $\mathbb{G}_m/R_0[1/\ell]$, cf. [Ka-ESDE, 8.17.6]. By [We-JS], attaching to each maximal ideal $\mathcal{P}$ of R_0 the Jacobi sum $-J(R_0/\mathcal{P}; \chi, \rho/\chi)$ over its residue field is a grossencharacter, and so by [Se-ALR, Chapter 2] a $\overline{\mathbb{Q}}_\ell$-valued character, call it $\Lambda_{\chi, \rho/\chi}$, of $\pi_1(Spec(R_0[1/\ell]))$. So we can form

$$\mathcal{H}^{can}(\chi, \rho) := \mathcal{H}(\chi, \rho) \otimes (1/\Lambda_{\chi, \rho/\chi})$$

on $\mathbb{G}_m/R_0[1/\ell]$. For any $R_0[1/\ell]$-algebra k which is a finite field, its pullback to $\mathbb{G}_m/k$ is $\mathcal{H}^{can}(\psi; \chi; \rho)$.

This in turn allows us to perform the following global construction. Suppose we are given an integer $a > 0$, and two unordered disjoint lists of characters, $\chi_1, ..., \chi_a$ and $\rho_1, ..., \rho_a$, of the group $\mu_{d_W}(R_0)$ with values in that same group. For a fixed choice of orderings of the lists, we can form the sheaves $\mathcal{H}^{can}(\chi_i, \rho_i), i = 1, ..., a$ on $\mathbb{G}_m/R_0[1/\ell]$. We can then define, as in [Ka-ESDE, 8.17.11], the ! multiplicative convolution

$$\mathcal{H}^{can}(\chi_1, \rho_1)[1] \star_! \mathcal{H}^{can}(\chi_2, \rho_2)[1] \star_! ... \star_! \mathcal{H}^{can}(\chi_a, \rho_a)[1],$$

which will be of the form $\mathcal{F}[1]$ for some sheaf $\mathcal{F}$ on $\mathbb{G}_m/R_0[1/\ell]$ which is "tame and adapted to the unit section". This sheaf $\mathcal{F}$ we call $\mathcal{H}^{can}(\chi_i\text{ 's}; \rho_j\text{ 's})$. For any $R_0[1/\ell]$-algebra k which is a finite field, its pullback to $\mathbb{G}_m/k$ is $\mathcal{H}^{can}(\psi; \chi_i\text{ 's}; \rho_j\text{ 's})$. By Chebotarev, the sheaf $\mathcal{H}^{can}(\psi; \chi_i\text{ 's}; \rho_j\text{ 's})$ is, up to isomorphism, independent of the orderings that went into its definition as an interated convolution. This canonical choice (as opposed to, say, the ad hoc construction given in [Ka-ESDE, 8.17.11], which *did depend* on the orderings) has the property that, denoting by

$$f : \mathbb{G}_m/R_0[1/\ell] \to Spec(R_0[1/\ell])$$

the structural map, the sheaf $R^1 f_! \mathcal{H}^{can}(\chi_i\text{ 's}; \rho_j\text{ 's})$ on $Spec(R_0[1/\ell])$ is the constant sheaf, i.e., it is the trivial one-dimensional representation of $\pi_1(Spec(R_0[1/\ell]))$.

If the unordered lists $\chi_1, ..., \chi_a$ and $\rho_1, ..., \rho_b$ are not disjoint, but not identical, then we can "cancel" the terms in common, getting shorter disjoint lists. The hypergeometric sheaf we form with these shorter, disjoint "cancelled" lists we denote $\mathcal{H}(\psi; \mathbf{Cancel}(\chi_i\text{ 's}; \rho_j\text{ 's}))$, cf. [Ka-ESDE, 9.3.1], where this was denoted $\mathbf{Cancel}\mathcal{H}(\psi; \chi_i\text{ 's}; \rho_j\text{ 's})$. If $a = b$, then after cancellation the shorter disjoint lists still have the same common length, and so we can form the constant field twist $\mathcal{H}^{can}(\psi; \mathbf{Cancel}(\chi_i\text{ 's}; \rho_j\text{ 's}))$. And in the global setting, we can form the object $\mathcal{H}^{can}(\mathbf{Cancel}(\chi_i\text{ 's}; \rho_j\text{ 's}))$ on $\mathbb{G}_m/R_0[1/\ell]$.

5 Statement of the main theorem

We continue to work with the fixed data (n, d, W). Given an element $V = (v_1, ..., v_n) \in (\mathbb{Z}/d\mathbb{Z})_0^n$, we attach to it an unordered list $List(V, W)$ of $d = \sum_i w_i$ multiplicative characters of $\mu_{dw}(R_0)$, by the following procedure. For each index i, denote by χ_{v_i} the character of $\mu_{dw}(R_0)$ given by

$$\zeta \mapsto \zeta^{(v_i/d)dw}.$$

Because w_i divides d_W/d, this character χ_{v_i} has w_i distinct w_i'th roots. We then define

$$List(V, W) = \{\text{all } w_1'\text{th roots of } \chi_{v_1}, ..., \text{all } w_n'\text{th roots of } \chi_{v_n}\}.$$

We will also need the same list, but for $-V$, and the list

$$List(all\ d) := \{all\ characters\ of\ order\ dividing\ d\}.$$

So long as the two lists $List(-V, W)$ and $List(all\ d)$ are not identical, we can apply the **Cancel** operation, and form the hypergeometric sheaf

$$\mathcal{H}_{V,W} := \mathcal{H}^{can}(\mathbf{Cancel}(List(all\ d); List(-V, W)))$$

on $\mathbb{G}_m/R_0[1/\ell]$.

Lemma 5.1. *If $Prim^{n-2}(V \bmod W)$ is nonzero, then the unordered lists $List(-V, W)$ and $List(all\ d)$ are not identical.*

Proof. If $Prim^{n-2}(V \bmod W)$ is nontrivial, then at least one choice of V in the coset $V \bmod W$ is totally nonzero. For such a totally nonzero V, the trivial character is absent from $List(-V, W)$. If we choose another representative of the same coset, say $V - rW$, then denoting by χ_r the character of order dividing d of $\mu_{dw}(R_0)$ given by $\zeta \mapsto \zeta^{(r/d)dw}$, we see easily that

$$List(-(V - rW), W) = \chi_r List(-V, W).$$

Hence the character χ_r is absent from $List(-V + rW, W)$.   □

Lemma 5.2. *If $Prim^{n-2}(V \bmod W)$ is nonzero, then $Prim^{n-2}(V \bmod W)$ and $[d]^\star \mathcal{H}_{V,W}$ have the same rank on $U_{R_0[1/\ell]}$.*

Proof. Choose V in the coset $V \bmod W$. The rank of $Prim^{n-2}(V \bmod W)$ is the number of $r \in \mathbb{Z}/d\mathbb{Z}$ such that $V + rW$ is totally nonzero. Equivalently, this rank is $d - \delta$, for δ the number of $r \in \mathbb{Z}/d\mathbb{Z}$ such that $V + rW$ fails to be totally nonzero. On the other hand, the rank of $\mathcal{H}_{V,W}$ is $d - \epsilon$, for ϵ the number of elements in $List(all\ d)$ which also appear in $List(-V, W)$. Now a given character χ_r in $List(all\ d)$ appears in $List(-V, W)$ if and only if there exists an index i such that χ_r is a w_i'th root of χ_{-v_i}, i.e., such that $\chi_r^{w_i} = \chi_{-v_i}$, i.e., such that $rw_i \equiv -v_i \bmod d$.   □

Theorem 5.3. *Suppose that $Prim^{n-2}(V \mod W)$ is nonzero. Denote by $j_1 : U_{R_0[1/\ell]} \subset \mathbb{A}^1_{R_0[1/\ell]}$ and $j_2 : \mathbb{G}_{m,R_0[1/\ell]} \subset \mathbb{A}^1_{R_0[1/\ell]}$ the inclusions, and by $[d] : \mathbb{G}_{m,R_0[1/\ell]} \to \mathbb{G}_{m,R_0[1/\ell]}$ the d'th power map. Then for any choice of V in the coset $V \mod W$, there exists a continuous character $\Lambda_{V,W} : \pi_1(Spec(R_0[1/\ell])) \to \overline{\mathbb{Q}}_\ell^\times$ and an isomorphism of sheaves on $\mathbb{A}^1_{R_0[1/\ell]}$,*

$$j_{1\star}Prim^{n-2}(V \mod W) \cong j_{2\star}[d]^\star \mathcal{H}_{V,W} \otimes \Lambda_{V,W}.$$

Remark 5.4. What happens if we change the choice of V in the coset $V \mod W$, say to $V - rW$? As noted above,

$$List(-(V - rW), W) = \chi_r List(-V, W).$$

Since $List(all\ d) = \chi_r List(all\ d)$ is stable by multiplication by any character of order dividing d, we find [Ka-ESDE, 8.2.5] that $\mathcal{H}_{V-rW,W} \cong \mathcal{L}_{\chi_r} \otimes \mathcal{H}_{V,W} \otimes \Lambda$, for some continuous character $\Lambda : \pi_1(Spec(R_0[1/\ell])) \to \overline{\mathbb{Q}}_\ell^\times$. Therefore the pullback $[d]^\star \mathcal{H}_{V,W}$ is, up to tensoring with a character Λ of $\pi_1(Spec(R_0[1/\ell]))$, independent of the particular choice of V in the coset $V \mod W$. Thus the truth of the theorem is independent of the particular choice of V.

Question 5.5. There should be a universal recipe for the character $\Lambda_{V,W}$ which occurs in Theorem 5.3. For example, if we look at the Γ_W/Δ-invariant part, both $Prim^{n-2}(0 \mod W)$ and $\mathcal{H}_{0,W}$ are pure of the same weight $n - 2$, and both have traces (on Frobenii) in $\mathbb{Q}$. So the character $\Lambda_{0,W}$ must take $\mathbb{Q}$-values of weight zero on Frobenii in large characteristic. [To make this argument legitimate, we need to be sure that over every sufficiently large finite field k which is an $R_0[1/\ell]$-algebra, the sheaf $Prim^{n-2}(0 \mod W)$ has nonzero trace at some k-point. This is in fact true, in virtue of Corollary 8.7 and a standard equidistribution argument.] But the only rational numbers of weight zero are ± 1. So $\Lambda^2_{0,W}$ is trivial. Is $\Lambda_{0,W}$ itself trivial?

6 Proof of the main theorem: the strategy

Let us admit for a moment the truth of the following characteristic p theorem, which will be proven in the next section.

Theorem 6.1. *Let k be an $R_0[1/\ell]$-algebra which is a finite field, and let $\psi : (k, +) \to \overline{\mathbb{Q}}_\ell^\times$ be a nontrivial additive character of k. Suppose that $Prim^{n-2}(V \mod W)$ is nonzero. Let $j_{1,k} : U_k \subset \mathbb{A}^1_k$ and $j_{2,k} : \mathbb{G}_{m,k} \subset \mathbb{A}^1_k$ be the inclusions. Choose V in the coset $V \mod W$, and put*

$$\mathcal{H}_{V,W,k} := \mathcal{H}^{can}(\psi; \mathbf{Cancel}(; List(all\ d); List(-V, W))).$$

Then on $\mathbb{A}^1_k$ the sheaves $j_{1,k\star}Prim^{n-2}(V \mod W)$ and $j_{2,k\star}[d]^\star \mathcal{H}_{V,W,k}$ are geometrically isomorphic, i.e., they become isomorphic on $\mathbb{A}^1_{\bar{k}}$.

We now explain how to deduce the main theorem. The restriction to

$$U_{R_0} - \{0\} = \mathbb{G}_{m,R_0} - \mu_d$$

of our family

$$X_\lambda : \sum_{i=1}^n w_i X_i^d = d\lambda X^W$$

is the pullback, through the d'th power map, of a projective smooth family over $\mathbb{G}_m - \{1\}$, in a number of ways. Here is one way to write down such a descent $\pi_{desc} : \mathbb{Y} \to \mathbb{G}_m - \{1\}$. Use the fact that $gcd(w_1, ..., w_n) = 1$ to choose integers $(b_1, ..., b_n)$ with $\sum_i b_i w_i = 1$. Then in the new variables

$$Y_i := \lambda^{b_i} X_i$$

the equation of X_λ becomes

$$\sum_{i=1}^n w_i \lambda^{-db_i} Y_i^d = dY^W.$$

Then the family

$$Y_\lambda : \sum_{i=1}^n w_i \lambda^{-b_i} Y_i^d = dY^W$$

is such a descent. The same group Γ_W/Δ acts on this family. On the base $\mathbb{G}_m - \{1\}$, we have the lisse sheaf $Prim_{desc}^{n-2}$ for this family, and its eigensheaves $Prim_{desc}^{n-2}(V \bmod W)$, whose pullbacks $[d]^\star Prim_{desc}^{n-2}(V \bmod W)$ are the sheaves $Prim^{n-2}(V \bmod W)|(\mathbb{G}_{m,R_0} - \mu_d)$.

Lemma 6.2. *Let k be an $R_0[1/\ell]$-algebra which is a finite field. Suppose $Prim_{desc}^{n-2}(V \bmod W)$ is nonzero. Then there exists a choice of V in the coset $V \bmod W$ such that the lisse sheaves $Prim_{desc}^{n-2}(V \bmod W)$ and $\mathcal{H}_{V,W,k}$ on $\mathbb{G}_{m,k} - \{1\}$ are geometrically isomorphic, i.e., isomorphic on $\mathbb{G}_{m,\overline{k}} - \{1\}$.*

Proof. Choose a V in the coset $V \bmod W$. By Theorem 6.1, the lisse sheaves $[d]^\star Prim_{desc}^{n-2}(V \bmod W)$ and $[d]^\star \mathcal{H}_{V,W,k}$ are isomorphic on $\mathbb{G}_{m,\overline{k}} - \mu_d$. Taking direct image by $[d]$ and using the projection formula, we find an isomorphism

$$\bigoplus_{\chi \text{ with } \chi^d \text{ trivial}} \mathcal{L}_\chi \otimes Prim_{desc}^{n-2}(V \bmod W) \cong \bigoplus_{\chi \text{ with } \chi^d \text{ trivial}} \mathcal{L}_\chi \otimes \mathcal{H}_{V,W,k}$$

of lisse sheaves $\mathbb{G}_{m,\overline{k}} - \{1\}$. The right hand side is completely reducible, being the sum of d irreducibles. Therefore the left hand side is completely reducible, and each of its d nonzero summands $\mathcal{L}_\chi \otimes Prim_{desc}^{n-2}(V \bmod W)$ must be irreducible (otherwise the left hand side is the sum of more than d irreducibles). By Jordan-Hölder, the summand $Prim_{desc}^{n-2}(V \bmod W)$ on the left is isomorphic to one of the summands $\mathcal{L}_\chi \otimes \mathcal{H}_{V,W,k}$ on the right, say to the summand $\mathcal{L}_{\chi_r} \otimes \mathcal{H}_{V,W,k}$. As explained in Remark 5.3, this summand is geometrically isomorphic to $\mathcal{H}_{V-rW,W,k}$. □

Lemma 6.3. *Suppose that the sheaf $Prim_{desc}^{n-2}(V \bmod W)$ is nonzero. Choose an $R_0[1/\ell]$-algebra k which is a finite field, and choose V in the coset $V \bmod W$ such that the lisse sheaves $Prim_{desc}^{n-2}(V \bmod W)$ and $\mathcal{H}_{V,W,k}$ on $\mathbb{G}_{m,k} - \{1\}$ are geometrically isomorphic. Then there exists a continuous character*

$$\Lambda_{V,W} : \pi_1(Spec(R_0[1/\ell])) \to \overline{\mathbb{Q}}_\ell^\times$$

and an isomorphism of lisse sheaves on $\mathbb{G}_{m,R_0[1/\ell]} - \{1\}$,

$$Prim_{desc}^{n-2}(V \bmod W) \cong \mathcal{H}_{V,W} \otimes \Lambda_{V,W}.$$

This is an instance of the following general phenomenon, which is well known to the specialists. In our application, the S below is $Spec(R_0[1/\ell])$, C is $\mathbb{P}^1$, and D is the union of the three everywhere disjoint sections $0, 1, \infty$. We will also use it a bit later when D is the union of the $d+2$ everywhere disjoint sections $0, \mu_d, \infty$.

Theorem 6.4. *Let S be a reduced and irreducible normal noetherian $\mathbb{Z}[1/\ell]$-scheme whose generic point has characteristic zero. Let $\overline{s}$ be a chosen geometric point of S. Let C/S be a proper smooth curve with geometrically connected fibres, and let $D \subset C$ be a Cartier divisor which is finite étale over S. Let $\mathcal{F}$ and $\mathcal{G}$ be lisse $\overline{\mathbb{Q}}_\ell$-sheaves on $C - D$. Then we have the following results.*

(1) Denote by $j : C - D \subset C$ and $i : D \subset C$ the inclusions. Then the formation of $j_\star \mathcal{F}$ on C commutes with arbitrary change of base $T \to S$, and $i^\star j_\star \mathcal{F}$ is a lisse sheaf on D.

(2) Denoting by $f : C - D \to S$ the structural map, the sheaves $R^i f_! \mathcal{F}$ on S are lisse.

(3) The sheaves $R^i f_\star \mathcal{F}$ on S are lisse, and their formation commutes with arbitrary change of base $T \to S$.

(4) Consider the pullbacks $\mathcal{F}_{\overline{s}}$ and $\mathcal{G}_{\overline{s}}$ of $\mathcal{F}$ and of $\mathcal{G}$ to $C_{\overline{s}} - D_{\overline{s}}$. Suppose that $\mathcal{F}_{\overline{s}} \cong \mathcal{G}_{\overline{s}}$, and that $\mathcal{G}_{\overline{s}}$ (and hence also $\mathcal{F}_{\overline{s}}$) are irreducible. Then there exists a continuous character $\Lambda : \pi_1(S) \to \overline{\mathbb{Q}}_\ell^\times$ an isomorphism of lisse sheaves on $C - D$,

$$\mathcal{G} \otimes \Lambda \cong \mathcal{F}.$$

Proof. The key point is that because the base S has generic characteristic zero, any lisse sheaf on $C - D$ is automatically tamely ramified along the divisor D; this results from Abhyankar's Lemma. See [Ka-SE, 4.7] for assertions (1) and (2). Assertion (3) results from (2) by Poincaré duality, cf. [De-CEPD, Corollaire, p. 72].

To prove assertion (4), we argue as follows. By the Tame Specialization Theorem [Ka-ESDE, 8.17.13], the geometric monodromy group attached to the sheaf $\mathcal{F}_{\overline{s}}$ is, up to conjugacy in the ambient $GL(rk(\mathcal{F}), \overline{\mathbb{Q}}_\ell)$, independent of the choice of geometric point $\overline{s}$ of S. Since $\mathcal{F}_{\overline{s}}$ is irreducible, it follows that $\mathcal{F}_{\overline{s_1}}$ is irreducible, for every geometric point $\overline{s_1}$ of S. Similarly, $\mathcal{G}_{\overline{s_1}}$ is irreducible, for every geometric point $\overline{s_1}$ of S. Now consider the lisse sheaf $\underline{Hom}(\mathcal{G}, \mathcal{F}) \cong \mathcal{F} \otimes \mathcal{G}^\vee$ on $C - D$. By assertion (3), the sheaf $f_\star \underline{Hom}(\mathcal{G}, \mathcal{F})$ is lisse on S, and its stalk at a geometric point $\overline{s_1}$ of S is the group $\mathrm{Hom}(\mathcal{G}_{\overline{s_1}}, \mathcal{F}_{\overline{s_1}})$. At the chosen geometric point $\overline{s}$, this Hom group is one-dimensional, by hypothesis. Therefore the lisse sheaf $f_\star \underline{Hom}(\mathcal{G}, \mathcal{F})$ on S has rank one. So at every geometric point $\overline{s_1}$, $\mathrm{Hom}(\mathcal{G}_{\overline{s_1}}, \mathcal{F}_{\overline{s_1}})$ is one-dimensional. As source and target are irreducible, any nonzero element of this Hom group is an isomorphism, and the canonical map

$$\mathcal{G}_{\overline{s_1}} \otimes \mathrm{Hom}(\mathcal{G}_{\overline{s_1}}, \mathcal{F}_{\overline{s_1}}) \to \mathcal{F}_{\overline{s_1}}$$

is an isomorphism. Therefore the canonical map of lisse sheaves on $C - D$

$$\mathcal{G} \otimes f^\star f_\star \underline{Hom}(\mathcal{G}, \mathcal{F}) \to \mathcal{F}$$

is an isomorphism, as we see looking stalkwise. Interpreting the lisse sheaf $f_\star \underline{Hom}(\mathcal{G}, \mathcal{F})$ on S as a character Λ of $\pi_1(S)$, we get the asserted isomorphism.

□

Applying this result, we get Lemma 6.3. Now pull back the isomorphism of that lemma by the d'th power map, to get an isomorphism

$$Prim^{n-2}(V \ mod \ W) \cong [d]^\star \mathcal{H}_{V,W} \otimes \Lambda_{V,W}$$

of lisse sheaves on $\mathbb{G}_{m, R_0[1/\ell]} - \mu_d$. Then extend by direct image to $\mathbb{A}^1_{R_0[1/\ell]}$ to get the isomorphism asserted in Theorem 5.3.

7 Proof of Theorem 6.1

Let us recall the situation. Over the ground ring $R_0[1/\ell]$, we have the family $\pi : \mathbb{X} \to \mathbb{A}^1$ given by

$$X_\lambda := X_\lambda(W, d) : \sum_{i=1}^n w_i X_i^d - d\lambda X^W = 0,$$

which is projective and smooth over $U = \mathbb{A}^1 - \mu_d$. We denote by $V \subset \mathbb{X}$ the open set where X^W is invertible, and by $Z \subset \mathbb{X}$ the complementary reduced closed set, defined by the vanishing of X^W. As scheme over $\mathbb{A}^1$, $Z/\mathbb{A}^1$ is the constant scheme with fibre

$$(X^W = 0) \cap \left(\sum_i w_i X_i^d = 0 \right).$$

The group Γ_W/Δ, acting as $\mathbb{A}^1$-automorphisms of $\mathbb{X}$, preserves both the open set V and its closed complement Z. In the following discussion, we will repeatedly invoke the following general principle, which we state here before proceeding with the analysis of our particular situation.

Lemma 7.1. *Let S be a noetherian $\mathbb{Z}[1/\ell]$-scheme, and $f : X \to S$ a separated morphism of finite type. Suppose that a finite group G acts admissibly $(:=$ every point lies in a G-stable affine open set) as S-automorphisms of X. Then in $D_c^b(S, \overline{\mathbb{Q}}_\ell)$, we have a direct sum decomposition of $Rf_!\overline{\mathbb{Q}}_\ell$ into G-isotypical components*

$$Rf_!\overline{\mathbb{Q}}_\ell = \bigoplus_{\text{irred. } \overline{\mathbb{Q}}_\ell \text{ rep.'s } \rho \text{ of } G} Rf_!\overline{\mathbb{Q}}_\ell(\rho).$$

Proof. Denote by $h : X \to Y := X/G$ the projection onto the quotient, and denote by $m : Y \to S$ the structural morphism of Y/S. Then $Rh_!\overline{\mathbb{Q}}_\ell = h_\star\overline{\mathbb{Q}}_\ell$ is a constructible sheaf of $\overline{\mathbb{Q}}_\ell[G]$ modules on Y, so has a G-isotypical decomposition

$$Rh_!\overline{\mathbb{Q}}_\ell = h_\star\overline{\mathbb{Q}}_\ell = \bigoplus_{\text{irred. } \overline{\mathbb{Q}}_\ell \text{ rep.'s } \rho \text{ of } G} h_\star\overline{\mathbb{Q}}_\ell(\rho).$$

Applying $Rm_!$ to this decomposition gives the asserted decomposition of $Rf_!\overline{\mathbb{Q}}_\ell$. □

We now return to our particular situation. We are given a $R_0[1/\ell]$-algebra k which is a finite field, and a nontrivial additive character $\psi : (k, +) \to \overline{\mathbb{Q}}_\ell^\times$. We denote by

$$\pi_k : \mathbb{X}_k \to \mathbb{A}_k^1$$

the base change to k of our family. Recall that the Fourier Transform FT_ψ is the endomorphism of the derived category $D_c^b(\mathbb{A}_k^1, \overline{\mathbb{Q}}_\ell)$ defined by looking at the two projections $\mathrm{pr}_1, \mathrm{pr}_2$ of $\mathbb{A}_k^2$ onto $\mathbb{A}_k^1$, and at the "kernel" $\mathcal{L}_{\psi(xy)}$ on $\mathbb{A}_k^2$, and putting

$$\mathrm{FT}_\psi(K) := R(\mathrm{pr}_2)_!\left(\mathcal{L}_{\psi(xy)} \otimes \mathrm{pr}_1^\star K[1]\right);$$

cf. [Lau-TFCEF, 1.2]. One knows that FT_ψ is essentially involutive,

$$\mathrm{FT}_\psi(\mathrm{FT}_\psi(K)) \cong [x \mapsto -x]^\star K(-1),$$

or equivalently

$$\mathrm{FT}_{\overline{\psi}}(\mathrm{FT}_\psi(K)) \cong K(-1),$$

that FT_ψ maps perverse sheaves to perverse sheaves and induces an exact autoequivalence of the category of perverse sheaves with itself.

We denote by $K\left(\mathbb{A}_k^1, \overline{\mathbb{Q}}_\ell\right)$ the Grothendieck group of $D_c^b\left(\mathbb{A}_k^1, \overline{\mathbb{Q}}_\ell\right)$. One knows that K is the free abelian group on the isomorphism classes of irreducible perverse sheaves, cf. [Lau-TFCEF, 0.7, 0.8]. We also denote by FT_ψ the endomorphism of $K\left(\mathbb{A}_k^1, \overline{\mathbb{Q}}_\ell\right)$ induced by FT_ψ on $D_c^b\left(\mathbb{A}_k^1, \overline{\mathbb{Q}}_\ell\right)$.

The key fact for us is the following, proven in [Ka-ESDE, 9.3.2], cf. also [Ka-ESDE, 8.7.2 and line -4, p. 327].

Theorem 7.2. *Denote by $\psi_{-1/d}$ the additive character $x \mapsto \psi(-x/d)$, and denote by $j : \mathbb{G}_{m,k} \subset \mathbb{A}_k^1$ the inclusion. Denote by $\Lambda_1, ..., \Lambda_d$ the list $List(all\ d)$ of all the multiplicative characters of $k^\times$ of order dividing d. For any unordered list of d multiplicative characters $\rho_1, ... \rho_d$ of $k^\times$ which is different from $List(all\ d)$, the perverse sheaf*

$$\mathrm{FT}_\psi j_*[d]^* \mathcal{H}(\psi_{-1/d}; \rho_1, ... \rho_d; \emptyset)[1]$$

on $\mathbb{A}_k^1$ is geometrically isomorphic to the perverse sheaf

$$j_*[d]^* \mathcal{H}(\psi; \mathbf{Cancel}(List(all\ d); \overline{\rho_1}, ..., \overline{\rho_d}))[1].$$

Before we can apply this result, we need some preliminaries. We first calculate the Fourier Transform of $R\pi_{k,!}\overline{\mathbb{Q}}_\ell$, or more precisely its restriction to $\mathbb{G}_{m,k}$, in a Γ_W/Δ-equivariant way. Recall that $V_k \subset \mathbb{X}_k$ is the open set where X^W is invertible, and $Z_k \subset \mathbb{X}_k$ is its closed complement. We denote by

$$f := \pi_k|V_k : V_k \to \mathbb{A}_k^1$$

the restriction to V_k of π_k. Concretely, V_k is the open set $\mathbb{P}_k^{n-1}[1/X^W]$ of $\mathbb{P}_k^{n-1}$ (with homogeneous coordinates $(X_1, ..., X_n)$) where X^W is invertible, and f is the map

$$(X_1, ..., X_n) \mapsto \sum_i (w_i/d) X_i^d / X^W.$$

Lemma 7.3. *For any character $V \bmod W$ of Γ_W/Δ, the canonical map of ρ-isotypical components $Rf_!\overline{\mathbb{Q}}_\ell(V \bmod W) \to R\pi_{k,!}\overline{\mathbb{Q}}_\ell(V \bmod W)$ induced by the $\mathbb{A}_k^1$-linear open immersion $V_k \subset \mathbb{X}_k$ induces an isomorphism in $D_c^b(\mathbb{G}_{m,k}, \overline{\mathbb{Q}}_\ell)$,*

$$(\mathrm{FT}_\psi Rf_!\overline{\mathbb{Q}}_\ell)(V \bmod W)|\mathbb{G}_{m,k} \cong (\mathrm{FT}_\psi R\pi_{k,!}\overline{\mathbb{Q}}_\ell)(V \bmod W)|\mathbb{G}_{m,k}.$$

Proof. We have an "excision sequence" distinguished triangle

$$Rf_!\overline{\mathbb{Q}}_\ell(V \bmod W) \to R\pi_{k,!}\overline{\mathbb{Q}}_\ell(V \bmod W) \to R(\pi|Z)_{k,!}\overline{\mathbb{Q}}_\ell(V \bmod W) \to$$

The third term is constant, i.e., the pullback to $\mathbb{A}_k^1$ of a an object on $Spec(k)$, so its FT_ψ is supported at the origin. Applying FT_ψ to this distinguished triangle gives a distinguished triangle

$$\mathrm{FT}_\psi Rf_!\overline{\mathbb{Q}}_\ell(V \bmod W) \to \mathrm{FT}_\psi R\pi_{k,!}\overline{\mathbb{Q}}_\ell(V \bmod W)$$

$$\to \mathrm{FT}_\psi R(\pi|Z)_{k,!}\overline{\mathbb{Q}}_\ell(V \bmod W) \to$$

Restricting to $\mathbb{G}_{m,k}$, the third term vanishes.   □

We next compute $(\mathrm{FT}_\psi Rf_!\overline{\mathbb{Q}}_\ell)|\mathbb{G}_{m,k}$ in a Γ_W/Δ-equivariant way. We do this by working upstairs, on V_k with its Γ_W/Δ-action.

Denote by $T_W \subset \mathbb{G}_{m,k}^n$ the connected (because $gcd(w_1, ... w_n) = 1$) torus of dimension $n - 1$ in $\mathbb{G}_{m,k}^n$, with coordinates $x_i, i = 1, ..., n$, defined by the equation $x^W = 1$. Denote by $\mathbb{P}_k^{n-1}[1/X^W] \subset \mathbb{P}_k^{n-1}$ the open set of $\mathbb{P}_k^{n-1}$ (with homogeneous coordinates $(X_1, ..., X_n)$) where X^W is invertible. Our group Γ_W is precisely the group $T_W[d]$ of points of order dividing d in T_W. And the subgroup $\Delta \subset \Gamma_W$ is just the intersection of T_W with the diagonal in the ambient $\mathbb{G}_{m,k}^n$. We have a surjective map

$$g : T_W \to \mathbb{P}_k^{n-1}[1/X^W], (x_1, ..., x_n) \mapsto (x_1, ..., x_n).$$

This map g makes T_W a finite étale galois covering of $\mathbb{P}_k^{n-1}[1/X^W]$ with group Δ. The d'th power map $[d] : T_W \to T_W$ makes T_W into a finite étale galois covering of itself, with group Γ_W. We have a beautiful factorization of $[d]$ as $h \circ g$, for

$$h : \mathbb{P}_k^{n-1}[1/X^W] \to T_W, (X_1, ..., X_n) \mapsto \left(X_1^d/X^W, ..., X_n^d/X^W\right).$$

This map h makes $\mathbb{P}_k^{n-1}[1/X^W]$ a finite étale galois covering of T_W with group Γ_W/Δ. Denote by m the map

$$m : T_W \to \mathbb{A}_k^1, (x_1, ..., x_n) \mapsto \sum_i (w_i/d)x_i.$$

Let us state explicitly the tautology which underlies our computation.

Lemma 7.4. *The map $f : V_k = \mathbb{P}_k^{n-1}[1/X^W] \to \mathbb{A}_k^1$ is the composition*

$$f = m \circ h : \mathbb{P}_k^{n-1}[1/X^W] \xrightarrow{h} T_W \xrightarrow{m} \mathbb{A}_k^1.$$

Because h is a a finite étale galois covering of T_W with group Γ_W/Δ, we have a direct sum decomposition on T_W,

$$Rh_!\overline{\mathbb{Q}}_\ell = h_\star\overline{\mathbb{Q}}_\ell = \bigoplus_{char's\ V\ mod\ W\ of\ \Gamma_W/\Delta} \mathcal{L}_{V\ mod\ W}.$$

More precisely, any V in the coset $V\ mod\ W$ is a character of Γ/Δ, hence of Γ, so we have the Kummer sheaf $\mathcal{L}_V$ on the ambient torus $\mathbb{G}_{m,k}^n$. In the standard coordinates $(x_1, ..., x_n)$ on $\mathbb{G}_{m,k}^n$, this Kummer sheaf $\mathcal{L}_V$ is $\mathcal{L}_{\prod_i \chi_{v_i}(x_i)}$. The restriction of $\mathcal{L}_V$ to the subtorus T_W is independent of the choice of V in the coset $V\ mod\ W$; it is the sheaf denoted $\mathcal{L}_{V\ mod\ W}$ in the above decomposition.

Now apply $Rm_!$ to the above decomposition. We get a direct sum decomposition

$$Rf_!\overline{\mathbb{Q}}_\ell = Rm_!h_\star\overline{\mathbb{Q}}_\ell = \bigoplus_{char's\ V\ mod\ W\ of\ \Gamma_W/\Delta} Rm_!\mathcal{L}_{V\ mod\ W}$$

into eigenobjects for the action of Γ_W/Δ.

Apply now $\mathrm{FT}_{\overline{\psi}}$. We get a direct sum decomposition

$$\mathrm{FT}_{\overline{\psi}}Rf_!\overline{\mathbb{Q}}_\ell = \bigoplus_{\substack{char's\ V\ mod\ W\ of\ \Gamma_W/\Delta}} \mathrm{FT}_{\overline{\psi}}Rm_!\mathcal{L}_{V\ mod\ W}$$

into eigenobjects for the action of Γ_W/Δ; we have

$$(\mathrm{FT}_{\overline{\psi}}Rf_!\overline{\mathbb{Q}}_\ell)(V\ mod\ W) = \mathrm{FT}_{\overline{\psi}}Rm_!\mathcal{L}_{V\ mod\ W}$$

for each character $V\ mod\ W$ of Γ_W/Δ.

Theorem 7.5. *Given a character $V\ mod\ W$ of Γ_W/Δ, pick V in the coset $V\ mod\ W$. We have a geometric isomorphism*

$$(\mathrm{FT}_{\overline{\psi}}Rf_!\overline{\mathbb{Q}}_\ell)(V\ mod\ W)|\mathbb{G}_{m,k} \cong [d]^\star\mathcal{H}(\psi_{-1/d}; List(V,W); \emptyset)[2-n].$$

Proof. By the definition of $\mathrm{FT}_{\overline{\psi}}$, and proper base change for $Rm_!$, we see that $\mathrm{FT}_{\overline{\psi}}Rm_!\mathcal{L}_{V\ mod\ W}$ is obtained as follows. Choose V in the coset $V\ mod\ W$. Endow the product $T_W \times \mathbb{A}_k^1$, with coordinates $(x = (x_1, ..., x_n); t)$ from the ambient $\mathbb{G}_{m,k}^n \times \mathbb{A}_k^1$. The product has projections $\mathrm{pr}_1, \mathrm{pr}_2$ onto T_W and $\mathbb{A}_k^1$ respectively. On the product we have the lisse sheaf $\mathcal{L}_{\overline{\psi}(t\sum_i(w_i/d)x_i)} \otimes \mathrm{pr}_1^\star \mathcal{L}_V$. By definition, we have

$$\mathrm{FT}_{\overline{\psi}}Rm_!\mathcal{L}_{V\ mod\ W} = R\mathrm{pr}_{2,!}(\mathcal{L}_{\overline{\psi}(t\sum_i(w_i/d)x_i)} \otimes \mathrm{pr}_1^\star\mathcal{L}_{\prod_i \chi_{v_i}(x_i)})[1].$$

If we pull back to $\mathbb{G}_{m,k} \subset \mathbb{A}_k^1$, then the source becomes $T_W \times \mathbb{G}_{m,k}$. This source is isomorphic to the subtorus Z of $\mathbb{G}_{m,k}^{n+1}$, with coordinates $(x = (x_1, ..., x_n); t)$, defined by

$$x^W = t^d,$$

by the map

$$(x = (x_1, ..., x_n); t) \mapsto (tx = (tx_1, ..., tx_n); t).$$

On this subtorus Z, our sheaf becomes $\mathcal{L}_{\overline{\psi}(\sum_i(w_i/d)x_i)} \otimes \mathrm{pr}_1^\star\mathcal{L}_{\prod_i \chi_{v_i}(x_i)}[1]$. [Remember that V has $\sum_i v_i = 0$, so $\mathcal{L}_{\prod_i \chi_{v_i}(x_i)}$ is invariant by $x \mapsto tx$.] Thus we have

$$\mathrm{FT}_{\overline{\psi}}Rm_!\mathcal{L}_{V\ mod\ W}|\mathbb{G}_{m,k} = R\mathrm{pr}_{n+1,!}(\mathcal{L}_{\overline{\psi}(\sum_i(w_i/d)x_i)} \otimes \mathrm{pr}_1^\star\mathcal{L}_{\prod_i \chi_{v_i}(x_i)}[1]).$$

This situation,

$$\mathcal{L}_{\overline{\psi}(\sum_i(w_i/d)x_i)} \otimes \mathrm{pr}_1^\star\mathcal{L}_{\prod_i \chi_{v_i}(x_i)}[1]\ on\ Z := (x^W = t^d) \overset{\mathrm{pr}_{n+1}}{\to} \mathbb{G}_{m,k},$$

is the pullback by the d'th power map on the base of the situation

$$\mathcal{L}_{\overline{\psi}(\sum_i(w_i/d)x_i)} \otimes \mathrm{pr}_1^\star\mathcal{L}_{\prod_i \chi_{v_i}(x_i)}[1]\ on\ \mathbb{G}_{m,k}^n \overset{x^W}{\to} \mathbb{G}_{m,k}.$$

Therefore we have

$$\mathrm{FT}_{\overline{\psi}}Rm_!\mathcal{L}_V \ mod \ w \,|\mathbb{G}_{m,k} \cong [d]^\star R(x^W)_! \left(\mathcal{L}_{\overline{\psi}(\sum_i (w_i/d)x_i)} \otimes \mathrm{pr}_1^\star \mathcal{L}_{\prod_i \chi_{v_i}(x_i)}[1] \right).$$

According to [Ka-GKM, 4.0,4.1, 5.5],

$$R^a(x^W)_! \left(\mathcal{L}_{\overline{\psi}(\sum_i (w_i/d)x_i)} \otimes \mathrm{pr}_1^\star \mathcal{L}_{\prod_i \chi_{v_i}(x_i)} \right)$$

vanishes for $a \neq n - 1$, and for $a = n - 1$ is the multiple multiplicative !
convolution

$$Kl(\psi_{-w_1/d}; \chi_{v_1}, w_1) \star_! Kl(\psi_{-w_2/d}; \chi_{v_2}, w_2) \star_! \cdots \star_! Kl(\psi_{-w_n/d}; \chi_{v_n}, w_n).$$

By [Ka-GKM, 4.3, 5.6.2], for each convolvee we have geometric isomorphisms

$$Kl(\psi_{-w_i/d}; \chi_{v_i}, w_i) = [w_i]_\star Kl(\psi_{-w_i/d}; \chi_{v_i}) \cong Kl(\psi_{-1/d}; \text{all } w_i'\text{th roots of } \chi_{v_i}).$$

So the above multiple convolution is the Kloosterman sheaf

$$Kl(\psi_{-1/d}; \text{all } w_1'th \text{ roots of } \chi_{v_1}, ..., \text{all } w_n'\text{th roots of } \chi_{v_n})$$

$$:= \mathcal{H}(\psi_{-1/d}; \text{all } w_1'\text{th roots of } \chi_{v_1}, ..., \text{all } w_n'\text{th roots of } \chi_{v_n}; \emptyset).$$

Recall that by definition

$$List(V, W) := (\text{all } w_1'\text{th roots of } \chi_{v_1}, ..., \text{all } w_n'\text{th roots of } \chi_{v_n}).$$

Putting this all together, we find the asserted geometric isomorphism

$$(\mathrm{FT}_{\overline{\psi}}Rf_!\overline{\mathbb{Q}}_\ell)(V \ mod \ W)|\mathbb{G}_{m,k} \cong [d]^\star \mathcal{H}(\psi_{-1/d}; List(V,W); \emptyset)[2 - n].$$

$\square$

We are now ready for the final step in the proof of Theorem 6.1. Recall that $j_{1,k} : U_k := \mathbb{A}^1_k - \mu_d \subset \mathbb{A}^1_k$, and $j_{2,k} : \mathbb{G}_{m,k} \subset \mathbb{A}^1_k$ are the inclusions. We must prove.

Theorem 7.6. *(Restatement of 6.1) Let $V \ mod \ W$ be a character of Γ_W/Δ for which $Prim^{n-2}(V \ mod \ W)$ is nonzero. Pick V in the coset $V \ mod \ W$. Then we have a geometric isomorphism of perverse sheaves on $\mathbb{A}^1_k$,*

$$j_{1,k,\star}Prim^{n-2}(V \ mod \ W)[1] \cong j_{2,k,\star}[d]^\star \mathcal{H}_{V,W,k}[1].$$

Proof. Over the open set U_k, we have seen that sheaves $R^i\pi_{k,\star}\overline{\mathbb{Q}}_\ell|U_k$ are geometrically constant for $i \neq n - 2$, and that $R^{n-2}\pi_{k,\star}\overline{\mathbb{Q}}_\ell|U_k$ is the direct sum of $Prim^{n-2}$ and a geometrically constant sheaf. The same is true for the Γ_W/Δ-isotypical components. Thus in $K(U_k, \overline{\mathbb{Q}}_\ell)$, we have

$$R\pi_{k,\star}\overline{\mathbb{Q}}_\ell(V \bmod W)|U_k := \sum_i (-1)^i R^i\pi_{k,\star}\overline{\mathbb{Q}}_\ell(V \bmod W)|U_k$$

$$= (-1)^{n-2}Prim^{n-2}(V \bmod W) + (geom.\ const.).$$

Comparing this with the situation on all of $\mathbb{A}^1_k$, we don't know what happens at the d missing points of μ_d, but in any case we will have

$$R\pi_{k,\star}\overline{\mathbb{Q}}_\ell(V \bmod W) = (-1)^{n-2}j_{1,k,\star}Prim^{n-2}(V \bmod W)$$

$$+(geom.\ const.) + (punctual,\ supported\ in\ \mu_d)$$

in $K\left(\mathbb{A}^1_k, \overline{\mathbb{Q}}_\ell\right)$.

Taking Fourier Transform, we get

$$\mathrm{FT}_{\overline{\psi}}j_{1,k,\star}Prim^{n-2}(V \bmod W) =$$

$$(-1)^{n-2}\mathrm{FT}_{\overline{\psi}}R\pi_{k,\star}\overline{\mathbb{Q}}_\ell(V \bmod W)+(punctual,\ supported\ at\ 0)+(sum\ of\ \mathcal{L}_{\psi'_\zeta}\mathrm{s})$$

in $K\left(\mathbb{A}^1_k, \overline{\mathbb{Q}}_\ell\right)$.

By Lemma 7.3 , we have

$$(\mathrm{FT}_\psi R\pi_{k,!}\overline{\mathbb{Q}}_\ell)(V \bmod W)|\mathbb{G}_{m,k} \cong \mathrm{FT}_\psi Rf_!\overline{\mathbb{Q}}_\ell(V \bmod W)|\mathbb{G}_{m,k},$$

so we have

$$\mathrm{FT}_{\overline{\psi}}j_{1,k,\star}Prim^{n-2}(V \bmod W) =$$

$$(-1)^{n-2}\mathrm{FT}_\psi Rf_!\overline{\mathbb{Q}}_\ell(V \bmod W) + (punctual,\ supported\ at\ 0) + (sum\ of\ \mathcal{L}'_{\psi_\zeta}\mathrm{s})$$

in $K\left(\mathbb{A}^1_k, \overline{\mathbb{Q}}_\ell\right)$.

By the previous theorem, we have

$$(\mathrm{FT}_{\overline{\psi}}Rf_!\overline{\mathbb{Q}}_\ell)(V \bmod W)|\mathbb{G}_{m,k} = (-1)^{n-2}[d]^\star\mathcal{H}(\psi_{-1/d}; List(V,W); \emptyset)$$

in $K(\mathbb{G}_{m,\overline{k}}, \overline{\mathbb{Q}}_\ell)$. We don't know what happens at the origin, but in any case we have

$$(\mathrm{FT}_{\overline{\psi}}Rf_!\overline{\mathbb{Q}}_\ell)(V \bmod W)$$

$$= (-1)^{n-2}j_{2,k,\star}[d]^\star\mathcal{H}(\psi_{-1/d}; List(V,W); \emptyset) + (punctual,\ supported\ at\ 0)$$

in $K\left(\mathbb{A}^1_k, \overline{\mathbb{Q}}_\ell\right)$. So we find

$$\mathrm{FT}_{\overline{\psi}}j_{1,k,\star}Prim^{n-2}(V \bmod W)$$

$$= j_{2,k,\star}[d]^\star\mathcal{H}(\psi_{-1/d}; List(V,W); \emptyset)$$

$$+(punctual,\ supported\ at\ 0) + (sum\ of\ \mathcal{L}_{\psi_\zeta}\ \mathrm{s})$$

in $K\left(\mathbb{A}^1_k, \overline{\mathbb{Q}}_\ell\right)$. Now apply the inverse Fourier Transform FT_ψ. By Theorem 7.2, we obtain an equality

$j_{1,k,\star} Prim^{n-2}(V \ mod \ W)[1]$

$$= j_{2,k,\star}[d]^\star \mathcal{H}_{V,W,k}[1] + (geom. \ constant) + (punctual)$$

in the group $K\left(\mathbb{A}_{\overline{k}}^{1}, \overline{\mathbb{Q}}_\ell\right)$. This is the free abelian group on isomorphism classes of irreducible perverse sheaves on $\mathbb{A}_{\overline{k}}^{1}$. So in any equality of elements in this group, we can delete all occurrences of any particular isomorphism class, and still have an equality.

On the open set U_k, the lisse sheaves $Prim^{n-2}(V \ mod \ W)$ and $[d]^\star \mathcal{H}_{V,W,k}$ are both pure, hence completely reducible on $U_{\overline{k}}$ by [De-Weil II, 3.4.1 (iii)]. So both perverse sheaves $j_{1,k,\star} Prim^{n-2}(V \ mod \ W)[1]$ and $j_{2,k,\star}[d]^\star \mathcal{H}_{V,W,k}[1]$ on $\mathbb{A}_{\overline{k}}^{1}$ are direct sums of perverse irreducibles which are middle extensions from $U_{\overline{k}}$, and hence have no punctual constituents. So we may cancel the punctual terms, and conclude that we have

$$j_{1,k,\star} Prim^{n-2}(V \ mod \ W)[1] - j_{2,k,\star}[d]^\star \mathcal{H}_{V,W,k}[1] = (geom. \ constant)$$

in the group $K\left(\mathbb{A}_{\overline{k}}^{1}, \overline{\mathbb{Q}}_\ell\right)$. By Lemma 5.2, the left hand side has generic rank zero, so there can be no geometrically constant virtual summand. Thus we have an equality of perverse sheaves

$$j_{1,k,\star} Prim^{n-2}(V \ mod \ W)[1] = j_{2,k,\star}[d]^\star \mathcal{H}_{V,W,k}[1]$$

in the group $K\left(\mathbb{A}_{\overline{k}}^{1}, \overline{\mathbb{Q}}_\ell\right)$. Therefore the two perverse sheaves have geometrically isomorphic semisimplifications. But by purity, both are geometrically semisimple. This concludes the proof of Theorem 6.1, and so also the proof of Theorem 5.3. □

8 Appendix I: The transcendental approach

In this appendix, we continue to work with the fixed data (n, d, W), but now over the groundring $\mathbb{C}$. We give a transcendental proof of Theorem 5.3, but only for the Γ_W/Δ-invariant part $Prim^{n-2}(0 \ mod \ W)$. Our proof is essentially a slight simplification of an argument that Shepherd-Barron gave in a November, 2006 lecture at MSRI, where he presented a variant of [H-SB-T, pages 5–22]. We do not know how to treat the other eigensheaves $Prim^{n-2}(V \ mod \ W)$, with $V \ mod \ W$ a nontrivial character of Γ_W/Δ, in an analogous fashion.

First, let us recall the bare definition of hypergeometric D-modules. We work on $\mathbb{G}_m$ (always over $\mathbb{C}$), with coordinate λ. We write $D := \lambda d/d\lambda$. We denote by $\mathcal{D} := \mathbb{C}[\lambda, 1/\lambda][D]$ the ring of differential operators on $\mathbb{G}_m$. Fix nonnegative integers a and b, not both 0. Suppose we are given an unordered list of a complex numbers $\alpha_1, ..., \alpha_a$,not necessarily distinct. Let $\beta_1, ..., \beta_b$ be a second such list, but of length b. We denote by $Hyp\left(\alpha_i's; \beta_j's\right)$ the differential operator

$$\text{Hyp}\,(\alpha_i{}'\text{s}; \beta_j{}'\text{s}) := \prod_i (D - \alpha_i) - \lambda \prod_j (D - \beta_j)$$

and by $\mathcal{H}(\alpha_i'\text{s}; \beta_j'\text{s})$ the holonomic left D-module

$$\mathcal{H}(\alpha_i'\text{s}; \beta_j'\text{s}) := \mathcal{D}/\mathcal{D}\text{Hyp}(\alpha_i'\text{s}; \beta_j'\text{s}).$$

We say that $\mathcal{H}(\alpha_i'\text{s}; \beta_j'\text{s})$ is a hypergeometric of type (a, b).

One knows [Ka-ESDE, 3.2.1] that this $\mathcal{H}$ is an irreducible D-module on $\mathbb{G}_m$, and remains irreducible when restricted to any dense open set $U \subset \mathbb{G}_m$, if and only if the two lists are disjoint "mod $\mathbb{Z}$", i.e., for all i, j, $\alpha_i - \beta_j$ is not an integer. [If we are given two lists $List_1$ and $List_2$ which are not identical mod $\mathbb{Z}$, but possibly not disjoint mod $\mathbb{Z}$, we can "cancel" the common (mod $\mathbb{Z}$) entries, and get an irreducible hypergeometric $\mathcal{H}(\mathbf{Cancel}(List_1, List_2))$.]

We will assume henceforth that this disjointness mod $\mathbb{Z}$ condition is satisfied, and that $a = b$. Then $\mathcal{H}(\alpha_i'\text{s}; \beta_j'\text{s})$ has regular singular points at $0, 1, \infty$. If the α_i and β_j lie in $\mathbb{Q}$, pick a common denominator N, and denote by χ_{α_i} the character of $\mu_N(\mathbb{C})$ given by

$$\chi_{\alpha_i}(\zeta) := \zeta^{\alpha_i N}.$$

Similarly for χ_{β_j}. For any prime number ℓ, the Riemann-Hilbert partner of $\mathcal{H}(\alpha_i'\text{s}; \beta_j'\text{s})$ is the $\overline{\mathbb{Q}}_\ell$ perverse sheaf $\mathcal{H}^{can}(\chi_{\alpha_i}{}'\text{s}; \chi_{\beta_j}{}'\text{s})[1]$ on $\mathbb{G}_m$, cf. [Ka-ESDE, 8.17.11].

We denote by $\mathcal{D}_\eta := \mathbb{C}(\lambda)[D]$ the ring of differential operators at the generic point. Although this ring is not quite commutative, it is near enough to being a one-variable polynomial ring over a field that it is left (and right) Euclidean, for the obvious notion of long division. So every nonzero left ideal in $\mathcal{D}_\eta$ is principal, generated by the monic (in $\mathcal{D}_\eta$) operator in it of lowest order. Given a left $\mathcal{D}_\eta$-module M and an element $m \in M$, we denote by $Ann(m, M)$ the left ideal in $\mathcal{D}_\eta$ defined as

$$Ann(m, M) := \{\,operators\ L \in \mathcal{D}_\eta | L(m) = 0\ in\ M\,\}.$$

If $Ann(m, M) \neq 0$, we define $L_{m,M} \in \mathcal{D}_\eta$ to be the lowest order monic operator in $Ann(m, M)$.

We have the following elementary lemma, whose proof is left to the reader.

Lemma 8.1. *Let N and M be left $\mathcal{D}_\eta$-modules, $f : M \to N$ a horizontal (:= $\mathcal{D}_\eta$-linear) map, and $m \in M$. Suppose that $Ann(m, M) \neq 0$. Then $Ann(m, M) \subset Ann(f(m), N)$, and $L_{m,M}$ is right-divisible by $L_{f(m),N}$.*

We now turn to our complex family $\pi : \mathbb{X} \to \mathbb{A}^1$, given by

$$X_\lambda := X_\lambda(W, d) : \sum_{i=1}^{n} w_i X_i^d - d\lambda X^W = 0.$$

We pull it back to $U := \mathbb{G}_m - \mu_d \subset \mathbb{A}^1$, over which it is proper and smooth, and form the de Rham incarnation of $Prim^{n-2}$, which we denote $Prim_{\mathrm{dR}}^{n-2}$.

We also have the relative de Rham cohomolgy of $(\mathbb{P}^{n-1} \times U - \mathbb{X}_U)/U$ over the base U in degree $n-1$, which we denote simply $H_{dR}^{n-1}((\mathbb{P} - \mathbb{X})/U)$. Both are $\mathcal{O}$-locally free D-modules (Gauss–Manin connection) on U, endowed with a horizontal action of Γ_W/Δ. The Poincaré residue map gives a horizontal, Γ_W/Δ-equivariant isomorphism

$$Res : H_{dR}^{n-1}((\mathbb{P} - \mathbb{X})/U) \cong Prim_{dR}^{n-2}.$$

As in the discussion beginning Section 6, we write $1 = \sum_i b_i w_i$ to obtain a descent of our family through the d power map: the family $\pi_{desc} : \mathbb{Y} \to \mathbb{G}_m$ given by

$$Y_\lambda : \sum_{i=1}^{n} w_i \lambda^{-b_i} Y_i^d = dX^W.$$

The same group Γ_W/Δ acts on this family, which is projective and smooth over $\mathbb{G}_m - \{1\}$. So on $\mathbb{G}_m - \{1\}$, we have $Prim_{dR,desc}^{n-2}$ for this family, and its fixed part $Prim_{dR,desc}^{n-2}(0 \bmod W)$, whose pullback $[d]^* Prim_{dR,desc}^{n-2}(0 \bmod W)$ is the sheaf $Prim_{dR}^{n-2}(0 \bmod W)|(\mathbb{G}_m - \mu_d)$.

Our next step is to pull back further, to a small analytic disk. Choose a real constant $C > 4$. Pull back the descended family to a small disk $\mathcal{U}_{an,C}$ around C. We take the disk small enough that for $\lambda \in \mathcal{U}_{an,C}$, we have $|C/\lambda|^{b_i} < 2$ for all i. The extension of scalars map

$$H_{dR}^{n-1}((\mathbb{P} - \mathbb{Y})/(\mathbb{G}_m - \{1\})) \to H_{dR}^{n-1}((\mathbb{P} - \mathbb{Y})/(\mathbb{G}_m - \{1\})) \otimes_{\mathcal{O}_{\mathbb{G}_m - \{1\}}} \mathcal{O}_{\mathcal{U}_{an,C}}$$

is a horizontal map; we view both source and target as D-modules.

Over this disk, the C^∞ closed immersion

$$\gamma : (S^1)^n/Diagonal \to \mathbb{P}^{n-1},$$
$$(z_1, \ldots, z_n) \mapsto (C^{b_1/d} z_1, \ldots, C^{b_{n-1}/d} z_{n-1}, C^{b_n/d} z_n)$$

lands entirely in $\mathbb{P} - \mathbb{Y}$: its image is an $(n-1)$-torus $Z \subset \mathbb{P}^{n-1}$ that is disjoint from Y_λ for $\lambda \in \mathcal{U}_{an,C}$. Restricting to the Γ_W/Δ-invariant part

$$H_{dR}^{n-1}((\mathbb{P} - \mathbb{Y})/(\mathbb{G}_m - \{1\}))(0 \bmod W),$$

we get a horizontal map

$$H_{dR}^{n-1}((\mathbb{P} - \mathbb{Y})/(\mathbb{G}_m - \{1\}))(0 \bmod W) \to H^0(\mathcal{U}_{an,C}, \mathcal{O}_{\mathcal{U}_{an,C}}), \quad \omega \mapsto \int_Z \omega.$$

Write $y_i := Y_i/Y_n$ for $i = 1, \ldots, n-1$. Denote by

$$\omega \in H_{dR}^{n-1}((\mathbb{P} - \mathbb{Y})/(\mathbb{G}_m - \{1\}))(0 \bmod W)$$

the (cohomology class of the) holomorphic $(n-1)$-form

$$\omega := (1/2\pi i)^{n-1} \left(\frac{dY^W}{dY^W - \sum_{i=1}^{n} w_i \lambda^{-b_i} Y_i^d} \right) \prod_{i=1}^{n-1} dy_i/y_i.$$

Our next task is to compute the integral

$$\int_Z \omega.$$

The computation will involve the Pochammer symbol. For $\alpha \in \mathbb{C}$, and $k \geq 1$ a positive integer, the Pochammer symbol $(\alpha)_k$ is defined by

$$(\alpha)_k := \Gamma(\alpha + k)/\Gamma(\alpha) = \prod_{i=0}^{k-1} (\alpha + i).$$

We state for ease of later reference the following elementary identity.

Lemma 8.2. *For integers $k \geq 1$ and $r \geq 1$, we have*

$$(kr)!/r^{kr} = \prod_{i=1}^{r} (i/r)_k.$$

Lemma 8.3. *We have the formula*

$$\int_Z \omega = 1 + \sum_{k \geq 1} \left(\frac{\prod_{i=1}^{d} (i/d)_k}{\prod_{i=1}^{n} \prod_{j=1}^{w_i} (j/w_i)_k} \right) (1/\lambda)^k.$$

Proof. Divide top and bottom by dY^W, expand the geometric series, and integrate term by term. This is legitimate because at a point $z \in Z$, the function $\sum_{i=1}^{n} (w_i/d) \lambda^{-b_i} Y_i^d / Y^W$ has the value

$$\sum_{i=1}^{n} (w_i/d) \lambda^{-b_i} C^{b_i} z_i^d / C z^W = \sum_{i=1}^{n} (w_i/d)(C/\lambda)^{b_i} z_i^d / C z^W,$$

which has absolute value $\leq 2 \left(\sum_{i=1}^{n} (w_i/d) \right)/C = 2/C \leq 1/2$. Because each term in the geometric series is homogeneous of degree zero, the integral of the k'th term in the geometric series is the coefficient of z^{kW} in $\left(\sum_{i=1}^{n} (w_i/d)(\lambda)^{-b_i} z_i^d \right)^k$. This coefficient vanishes unless k is a multiple of d (because $\gcd(w_1, ..., w_n) = 1$). The integral of the dk'th term is the coefficient of z^{kdW} in $\left(\sum_{i=1}^{n} (w_i/d)(\lambda)^{-b_i} z_i^d \right)^{dk}$, i.e., the coefficient of z^{kW} in $\left(\sum_{i=1}^{n} (w_i/d)(\lambda)^{-b_i} z_i \right)^{dk}$. Expanding by the multinomial theorem, this coefficient is

$$(dk)! \prod_{i=1}^{n} \left(((w_i/d)\lambda^{-b_i})^{kw_i} / (kw_i)! \right) = (\lambda)^{-k}((dk)!/d^{dk}) \prod_{i=1}^{n} \left((kw_i)!/w_i^{kw_i} \right),$$

which, by the previous lemma, is as asserted. $\square$

This function

$$F(\lambda) : \int_Z \omega = 1 + \sum_{k \geq 1} \left(\frac{\prod_{i=1}^d (i/d)_k}{\prod_{i=1}^n \prod_{j=1}^{w_i} (j/w_i)_k} \right) (1/\lambda)^k$$

is annihilated by the following differential operator. Consider the two lists of length d.

$$List(all\ d) := \{1/d, 2/d, ..., d/d\},$$

$$List(0, W) := \{1/w_1, 2/w_1, ..., w_1/w_1, ..., 1/w_n, 2/w_n, ..., w_n/w_n\}.$$

These lists are certainly not identical mod $\mathbb{Z}$; the second one contains 0 with multiplicity n, while the first contains only a single integer. Let us denote the cancelled lists, whose common length we call a,

$$\mathbf{Cancel}(List(all\ d); List(0, W)) = (\alpha_1, ..., \alpha_a); (\beta_1, ..., \beta_a).$$

So we have

$$F(\lambda) : \int_Z \omega = 1 + \sum_{k \geq 1} \left(\frac{\prod_{i=1}^a (\alpha_i)_k}{\prod_{i=1}^a (\beta_i)_k} \right) (1/\lambda)^k,$$

which one readily checks is annihilated by the differential operator

$$\mathrm{Hyp}_{0,W} := \mathrm{Hyp}(\alpha_i's; \beta_i - 1's) := \prod_{i=1}^a (D - \alpha_i) - \lambda \prod_{i=1}^a (D - (\beta_i - 1)).$$

Theorem 8.4. *We have an isomorphism of D-modules on $\mathbb{G}_m - \{1\}$,*

$$H_{\mathrm{dR}}^{n-1}((\mathbb{P} - \mathbb{Y})/(\mathbb{G}_m - \{1\}))(0\ mod\ W) \cong \mathcal{H}_{0,W}|(\mathbb{G}_m - \{1\})$$
$$:= \mathcal{H}(\alpha_i's; \beta_i - 1's)|(\mathbb{G}_m - \{1\}).$$

Proof. Both sides of the alleged isomorphism are $\mathcal{O}$-coherent D-modules on $\mathbb{G}_m - \{1\}$, so each is the "middle extension" of its restriction to any Zariski dense open set in $\mathbb{G}_m - \{1\}$. So it suffices to show that both sides become isomorphic over the function field of $\mathbb{G}_m - \{1\}$, i.e., that they give rise to isomorphic $\mathcal{D}_\eta$-modules. For this, we argue as follows. Denote by $\mathcal{A}$ the ring

$$\mathcal{A} := H^0(\mathcal{U}_{an,C}, \mathcal{O}_{\mathcal{U}_{an,C}}) \otimes_{\mathcal{O}_{\mathbb{G}_m - \{1\}}} \mathbb{C}(\lambda),$$

which we view as a $\mathcal{D}_\eta$-module. We have the horizontal map

$$H_{dR}^{n-1}((\mathbb{P} - \mathbb{Y})/(\mathbb{G}_m - \{1\}))(0\ \mod\ W) \xrightarrow{\int_Z} H^0(\mathcal{U}_{an,C}, \mathcal{O}_{\mathcal{U}_{an,C}}).$$

Tensoring over $\mathcal{O}_{\mathbb{G}_m - \{1\}}$ with $\mathbb{C}(\lambda)$, we obtain a horizontal map

$$H_{\mathrm{dR}}^{n-1}((\mathbb{P} - \mathbb{Y})/\mathbb{C}(\lambda))(0\ \mod\ W) \xrightarrow{\int_Z} \mathcal{A}.$$

By (the *Hyp* analogue of) Lemma 5.2, we know that the source has $\mathbb{C}(\lambda)$-dimension $a :=$ the order of $Hyp(\alpha_i's; \beta_i - 1's)$. So the element ω in the source is annihilated by some operator in $\mathcal{D}_\eta$ of order at most a, simply because ω and its first a derivatives must be linearly dependent over $\mathbb{C}(\lambda)$. So the lowest order operator annihilating ω in $H_{\mathrm{dR}}^{n-1}((\mathbb{P} - \mathbb{Y})/\mathbb{C}(\lambda))(0 \mod W)$, call it $L_{\omega, H_{\mathrm{dR}}}$, has order at most a. On the other hand, the irreducible operator $Hyp(\alpha_i's; \beta_i - 1's)$ annihilates $\int_Z \omega \in \mathcal{A}$. But $\int_Z \omega \neq 0$, so $Ann(\int_Z \omega, \mathcal{A})$ is a proper left ideal in $\mathcal{D}_\eta$, and hence is generated by the irreducible monic operator $(1/(1-\lambda))Hyp(\alpha_i's; \beta_i - 1's)$. By Lemma 8.2, we know that $L_{\omega, H_{\mathrm{dR}}}$ is divisible by $(1/(1-\lambda))Hyp(\alpha_i's; \beta_i - 1's)$. But $L_{\omega, H_{\mathrm{dR}}}$ has order at most a, the order of $Hyp(\alpha_i's; \beta_i - 1's)$, so we conclude that $L_{\omega, H_{\mathrm{dR}}} = (1/(1-\lambda))Hyp(\alpha_i's; \beta_i - 1's)$. Thus the $\mathcal{D}_\eta$-span of ω in the group $H_{\mathrm{dR}}^{n-1}((\mathbb{P} - \mathbb{Y})/\mathbb{C}(\lambda))(0 \mod W)$ is $\mathcal{D}_\eta/\mathcal{D}_\eta Hyp(\alpha_i's; \beta_i - 1's)$. Comparing dimensions we see that this $\mathcal{D}_\eta$-span is all of $H_{\mathrm{dR}}^{n-1}((\mathbb{P} - \mathbb{Y})/\mathbb{C}(\lambda))(0 \mod W)$.   □

Corollary 8.5. *For the family*

$$X_\lambda := X_\lambda(W, d) : \sum_{i=1}^{n} w_i X_i^d - d\lambda X^W = 0,$$

its $Prim_{\mathrm{dR}}^{n-2}(0 \mod W)$ as D-module on $\mathbb{A}^1 - \mu_d$ is related to the D-module $[d]^\star(\mathcal{H}_{0,W}|(\mathbb{G}_m - \{1\}))$ on $\mathbb{G}_m - \mu_d$ as follows.

(1) We have an isomorphism of D-modules on $\mathbb{G}_m - \mu_d$,

$$Prim_{\mathrm{dR}}^{n-2}(0 \mod W)|(\mathbb{G}_m - \mu_d) \cong [d]^\star(\mathcal{H}_{0,W}|(\mathbb{G}_m - \{1\})).$$

(2) Denote by $j_1 : \mathbb{A}^1 - \mu_d \subset \mathbb{A}^1$ and $j_2 : \mathbb{G}_m - \mu_d \subset \mathbb{A}^1$ the inclusions. Then we have an isomorphism of D-modules on $\mathbb{A}^1$ of the middle extensions

$$j_{1,!,\star}(Prim_{\mathrm{dR}}^{n-2}(0 \mod W)) \cong j_{2,!,\star}([d]^\star(\mathcal{H}_{0,W}|(\mathbb{G}_m - \{1\}))).$$

Proof. The first isomorphism is the pullback by d'th power of the isomorphism of the theorem above. We obtain the second isomorphism as follows. Denote by $j_3 : \mathbb{G}_m - \mu_d \subset \mathbb{A}^1 - \mu_d$ the inclusion. Because $Prim_{\mathrm{dR}}^{n-2}(0 \mod W)$ is an $\mathcal{O}$-coherent D-module on $\mathbb{A}^1 - \mu_d$, it is the middle extension

$$j_{3,!,\star}\left(Prim_{\mathrm{dR}}^{n-2}(0 \mod W)|(\mathbb{G}_m - \mu_d)\right).$$

Because $j_2 = j_1 \circ j_3$, we obtain the second isomorphism by applying $j_{2,!,\star}$ to the first isomorphism.   □

Theorem 8.6. *Suppose $n \geq 3$. For either the family*

$$X_\lambda := X_\lambda(W, d) : \sum_{i=1}^{n} w_i X_i^d - d\lambda X^W = 0$$

over $\mathbb{A}^1 - \mu_d$ or the descended family

$$Y_\lambda : \sum_{i=1}^{n} w_i \lambda^{-b_i} Y_i^d = dX^W$$

over $\mathbb{G}_m - \{1\}$ consider its $Prim_{\mathrm{dR}}^{n-2}(0 \bmod W)$ (resp. $Prim_{dR,desc}^{n-2}(0 \bmod W)$)
as a D-module, and denote by a its rank. For either family, its differential
galois group G_{gal} (which here is the Zariski closure of its monodromy group)
is the symplectic group $Sp(a)$ if $n-2$ is odd, and the orthogonal group $O(a)$
if $n-2$ is even.

Proof. Poincaré duality induces on $Prim_{\mathrm{dR}}^{n-2}(0 \bmod W)$ (resp. on the module
$Prim_{dR,desc}^{n-2}(0 \bmod W)$) an autoduality which is symplectic if $n-2$ is odd,
and orthogonal if $n-2$ is even. So we have a priori inclusions $G_{gal} \subset Sp(a)$
if $n-2$ is odd, $G_{gal} \subset O(a)$ if $n-2$ is even. It suffices to prove the theorem
for the descended family. This is obvious in the Sp case, since the identity
component of G_{gal} is invariant under finite pullback. In the O case, we must
rule out the possibility that the pullback has group $SO(a)$ rather than $O(a)$.
For this, we observe that an orthogonally autodual hypergeometric of type
(a, a) has a true reflection as local monodromy around 1 (since in any case
an irreducible hypergeometric of type (a, a) has as local monodromy around
1 a pseudoreflection, and the only pseudoreflection in an orthogonal group is
a true reflection). As the d'th power map is finite étale over 1, the pullback
has a true reflection as local monodromy around each $\zeta \in \mu_d$. So the group
for the pullback contains true reflections, so must be $O(a)$.

We now consider the descended family. So we are dealing with $\mathcal{H}_{0,W} :=$
$\mathcal{H}(\alpha_i's; \beta_i - 1's)$. From the definition of $\mathcal{H}_{0,W}$, we see that $\beta = 1 \bmod \mathbb{Z}$ occurs
among the β_i precisely $n-1$ times ($n-1$ times and not n times because of
a single cancellation with $List(all\ d)$). Because $n-1 \geq 2$ by hypothesis,
local monodromy around ∞ is not semisimple [Ka-ESDE, 3.2.2] and hence
$\mathcal{H}(\alpha_i's; \beta_j's)$ is not Belyi induced or inverse Belyi induced, cf. [Ka-ESDE, 3.5],
nor is its $G^{0,der}$ trivial.

We next show that $\mathcal{H}_{0,W}$ is not Kummer induced of any degree $r \geq 2$.
Suppose it is not. As the α_i all have order dividing d in $\mathbb{C}/\mathbb{Z}$, r must divide
d, since $1/r \bmod \mathbb{Z}$ is a difference of two α_i's, cf. [Ka-ESDE, 3.5.6]. But the
$\beta_j \bmod \mathbb{Z}$ are also stable by $x \mapsto x + 1/r$, so we would find that $1/r \bmod \mathbb{Z}$
occurs with the same multiplicity $n-1$ as $0 \bmod \mathbb{Z}$ among the $\beta_j \bmod \mathbb{Z}$.
So r must divide at least $n-1$ of the w_i; it cannot divide all the w_i because
$gcd(w_1, ..., w_n) = 1$. But this $1/r$ cannot cancel with $List(all\ d)$, otherwise its
multiplicity would be at most $n-2$. This lack of cancellation means that r
does not divide d, contradiction.

Now we appeal to [Ka-ESDE, 3.5.8]: let $\mathcal{H}(\alpha_i's; \beta_j's)$ be an irreducible hypergeometric of type (a, a) which is neither Belyi induced nor inverse Belyi
induced not Kummer induced. Denote by G its differential galois group G_{gal},
G^0 its identity component, and $G^{0,der}$ the derived group ($:=$ commutator subgroup) of G^0. Then $G^{0,der}$ is either trivial or it is one of $SL(a)$ or $SO(a)$ or,
if a is even, possibly $Sp(a)$.

In the case of $\mathcal{H}_{0,W}$, we have already seen that $G_{gal}^{0,der}$ is not trivial. Given
that G_{gal} lies in either $Sp(a)$ or $O(a)$, depending on the parity of $n-2$, the only

possibility is that $G_{gal} = Sp(a)$ for $n - 2$ odd, and that $G_{gal} = O(a)$ or $SO(a)$ if $n - 2$ is even. In the even case, the presence of a true reflection in G_{gal} rules out the SO case.   □

Corollary 8.7. *In the context of Theorem 5.3, on each geometric fibre of $U_{R_0[1/\ell]}/Spec(R_0[1/\ell])$, the geometric monodromy group G_{geom} of the sheaf $Prim^{n-2}(0 \bmod W)$ is the full symplectic group $Sp(a)$ if $n - 2$ is odd, and is the full orthogonal group $O(a)$ if $n - 2$ is even.*

Proof. On a $\mathbb{C}$-fibre, this is just the translation through Riemann–Hilbert of the theorem above. The passage to other geometric fibres is done by the Tame Specialization Theorem [Ka-ESDE, 8.17.3].   □

When does it happen that $Prim_{dR}^{n-2}(0 \bmod W)$ has rank $n - 1$ and all Hodge numbers 1?

Lemma 8.8. *The following are equivalent.*

(1) $Prim_{dR}^{n-2}(0 \bmod W)$ has rank $n - 1$.
(2) Every w_i divides d, and for all $i \neq j$, $gcd(w_i, w_j) = 1$.
(3) Local monodromy at ∞ is a single unipotent Jordan block.
(4) Local monodromy at ∞ is a single Jordan block.
(5) All the Hodge numbers $Prim_{dR}^{a,b}(0 \bmod W)_{a+b=n-2}$ are 1.

Proof. (1)⇒(2) The rank is at least $n - 1$, since this is the multiplicity of 0 mod $\mathbb{Z}$ as a β in $\mathcal{H}_{0,W}$. If the rank is no higher, then each w_i must divide d, so that the elements $1/w_i, ..., (w_i - 1)/w_i$ mod $\mathbb{Z}$ can cancel with $List(all\ d)$. And the w_i must be pairwise relatively prime, for if a fraction $1/r$ mod $\mathbb{Z}$ with $r \geq 2$ appeared among both $1/w_i, ..., (w_i - 1)/w_i$ and $1/w_j, ..., (w_j - 1)/w_j$, only one of its occurrences at most can cancel with $List(all\ d)$.

(2)⇒(1) If all w_i divide d, and if the w_i are pairwise relatively prime, then after cancellation we find that $\mathcal{H}_{0,W}$ has rank $n - 1$.

(1)⇒(3) If (1) holds, then the β_i's are all 0 mod $\mathbb{Z}$, and there are $n - 1$ of them. This forces $\mathcal{H}_{0,W}$ and also $[d]^\star \mathcal{H}_{0,W}$ to have its local monodromy around ∞, call it T, unipotent, with a single Jordan block, cf. [Ka-ESDE, 3.2.2].

(3)⇒(4) is obvious.

(4)⇒(3) Although d'th power pullback may change the eigenvalues of local monodromy at ∞, it does not change the number of distinct Jordan blocks. But there is always one unipotent Jordan block of size $n - 1$, cf. the proof of (1)⇒(2).

(3)⇒(5) If not all the $n-1$ Hodge numbers are 1, then some Hodge number vanishes, and at most $n - 2$ Hodge numbers are nonzero. But by [Ka-NCMT, 14.1] [strictly speaking, by projecting its proof onto Γ_W/Δ-isotypical components] any local monodromy is quasiunipotent of exponent of nilpotence $\leq h :=$ the number of nonzero Hodge numbers. So our local monodromy T around ∞, already unipotent, would satisfy $(T - 1)^{n-2} = 0$. But as we have already remarked, $\mathcal{H}_{0,W}$ always has unipotent Jordan block of size $n-1$. Therefore all the Hodge numbers are nonzero, and hence each is 1.

(5)$\Rightarrow$(1) is obvious.   □

Remark 8.9. Four particular $n = 5$ cases where condition (2) is satisfied, namely $W = (1,1,1,1,1)$, $W = (1,1,1,1,2)$, $W = (1,1,1,1,4)$, and $W = (1,1,1,2,5)$, were looked at in detain in the early days of mirror symmetry, cf. [Mor, Section 4, Table 1].

Whatever the rank of $Prim_{dR}^{n-2}(0 \ mod \ W)$, we have:

Lemma 8.10. *All the Hodge numbers* $Prim_{dR}^{a,b}(0 \ mod \ W)_{a+b=n-2}$ *are nonzero.*

Proof. Repeat the proof of (3)$\Rightarrow$(5).   □

9 Appendix II: The situation in characteristic p, when p divides some w_i

We continue to work with the fixed data (n, d, W). In this appendix, we indicate briefly what happens in a prime-to-d characteristic p which divides one of the w_i. For each i, we denote by w_i° the prime-to-p part of w_i, i.e.,

$$w_i = w_i^{\circ} \times (\text{a power of } p),$$

and we define

$$W^{\circ} := (w_1^{\circ}, ..., w_n^{\circ}).$$

We denote by $d_{W^{\circ}}$ the integer

$$d_{W^{\circ}} := lcm(w_1^{\circ}, ..., w_n^{\circ})d,$$

and define

$$d' := \sum_i w_i^{\circ}.$$

For each i, we have $w_i \equiv w_i^{\circ}$ mod $p-1$, so we have the congruence, which will be used later,

$$d \equiv d' \ \text{mod} \ p-1.$$

We work over a finite field k of characteristic p prime to d that contains the $d_{W^{\circ}}$'th roots of unity. We take for ψ a nontrivial additive character of k that is of the form $\psi_{\mathbb{F}_p} \circ Trace_{k/\mathbb{F}_p}$, for some nontrivial additive character $\psi_{\mathbb{F}_p}$ of $\mathbb{F}_p$. The signifigance of this choice of ψ is that for $q = p^e, e \geq 1$, any power of p, under the q'th power map we have

$$[q]_{\star}\mathcal{L}_{\psi} = \mathcal{L}_{\psi}, [q]^{\star}\mathcal{L}_{\psi} = \mathcal{L}_{\psi}$$

on $\mathbb{A}_k^1$.

The family we study in this situation is $\pi : \mathbb{X} \to \mathbb{A}^1$,

$$X_\lambda := X_\lambda(W, d) : \sum_{i=1}^{n} w_i^\circ X_i^d - d\lambda X^W = 0.$$

The novelty is that, because p divides some w_i, this family is projective and smooth over all of $\mathbb{A}^1$.

The group Γ_W/Δ operates on this family. Given a character V *mod* W of this group, the rank of the eigensheaf $Prim^{n-2}(V \; mod \; W)$ is still given by the same recipe as in Lemma 3.1(1), because at $\lambda = 0$ we have a smooth Fermat hypersurface of degree d.

Given an element $V = (v_1, ..., v_n) \in (\mathbb{Z}/d\mathbb{Z})_0^n$, we attach to it an unordered list $List(V, W)$ of $d' = \sum_i w_i^\circ$ multiplicative characters of $k^\times$, by the following procedure. For each index i, we denoted by χ_{v_i} the character of $k^\times$ given by

$$\zeta \mapsto \zeta^{(v_i/d)\#k^\times}.$$

This character χ_{v_i} has w_i° (as opposed to w_i) distinct w_i'th roots. We then define

$$List(V, W) = \{\text{all } w_1'\text{th roots of } \chi_{v_1}, ..., \text{all } w_n'\text{th roots of } \chi_{v_n}\}.$$

We will also need the same list, but for $-V$, and the list

$$List(all \; d) := \{all \; characters \; of \; order \; dividing \; d\}.$$

The two lists $List(-V, W)$ and $List(all \; d)$ are not identical, as they have different lengths d' and d respectively, so we can apply the **Cancel** operation, and form the hypergeometric sheaf

$$\mathcal{H}_{V,W} := \mathcal{H}^{can}(\textbf{Cancel}(List(all \; d); List(-V, W)))$$

on $\mathbb{G}_{m,k}$. Exactly as in Lemma 5.2, if $Prim^{n-2}(V \; mod \; W)$ is nonzero, its rank is the rank of $\mathcal{H}_{V,W}$.

An important technical fact in this situation is the following variant of Theorem 7.2, cf. [Ka-ESDE, 9.3.2], which "works" because $\mathbb{F}_p^\times$ has order $p - 1$.

Theorem 9.1. *Denote by $\psi_{-1/d}$ the additive character $x \mapsto \psi(-x/d)$, and denote by $j : \mathbb{G}_{m,k} \subset \mathbb{A}_k^1$ the inclusion. Denote by $\Lambda_1, ..., \Lambda_d$ the list $List(all \; d)$ of all the multiplicative characters of $k^\times$ of order dividing d. Let d' be a strictly positive integer with $d' \equiv d \mod p - 1$. For any unordered list of d' multiplicative characters $\rho_1, ... \rho_{d'}$ of $k^\times$ which is not identical to $List(all \; d)$, the perverse sheaf*

$$FT_\psi j_\star[d]^\star \mathcal{H}(\psi_{-1/d}; \rho_1, ... \rho_{d'}; \emptyset)[1]$$

on $\mathbb{A}_k^1$ is geometrically isomorphic to the perverse sheaf

$$j_\star[d]^\star \mathcal{H}(\psi; \textbf{Cancel}(List(all \; d); \overline{\rho_1}, ..., \overline{\rho_{d'}}))[1].$$

The main result is the following.

Theorem 9.2. *Suppose that $Prim^{n-2}(V \bmod W)$ is nonzero and denote by $j : \mathbb{G}_m \subset \mathbb{A}^1$ the inclusion. Choose V in the coset $V \bmod W$. There exists a constant $A_{V,W} \in \overline{\mathbb{Q}}_\ell^\times$ and an isomorphism of lisse sheaves on $\mathbb{A}^1_k$,*

$$Prim^{n-2}(V \bmod W) \cong j_\star [d]^\star \mathcal{H}_{V,W} \otimes (A_{V,W})^{\deg}.$$

Proof. Because our family is projective and smooth over all of $\mathbb{A}^1$, Deligne's degeneration theorem [De-TLCD, 2.4] gives a decomposition

$$R\pi_\star \overline{\mathbb{Q}}_\ell \cong Prim^{n-2}[2-n] \oplus (geom.\ constant).$$

So applying Fourier Transform, we get

$$\mathrm{FT}_{\overline{\psi}} R\pi_\star \overline{\mathbb{Q}}_\ell(V \bmod W)|\mathbb{G}_m \cong \mathrm{FT}_{\overline{\psi}} Prim^{n-2}(V \bmod W)[2-n]|\mathbb{G}_m.$$

On the open set $V \subset \mathbb{X}$ where X^W is invertible, the restriction of π becomes the map f, now given by

$$(X_1, ..., X_n) \mapsto \sum_i (w_i^\circ/d) X_i^d / X^W.$$

Then the argument of Lemma 7.3 gives

$$\mathrm{FT}_{\overline{\psi}} Prim^{n-2}(V \bmod W)[2-n]|\mathbb{G}_m \cong \mathrm{FT}_{\overline{\psi}} Rf_! \overline{\mathbb{Q}}_\ell(V \bmod W)|\mathbb{G}_m.$$

Theorem 7.5 remains correct as stated. [In its proof, the only modification needed is the analysis now of the sheaves $Kl(\psi_{-w_i^\circ/d}; \chi_{v_i}, w_i)$. Pick for each i a w_i'th root ρ_i of χ_{v_i}. We have geometric isomorphisms

$$Kl(\psi_{-w_i^\circ/d}; \chi_{v_i}, w_i) = [w_i]_\star Kl(\psi_{-w_i^\circ/d}; \chi_{v_i}) = \mathcal{L}_{\rho_i} \otimes [w_i]_\star \mathcal{L}_{\psi_{-w_i^\circ/d}}$$

$$= \mathcal{L}_{\rho_i} \otimes [w_i^\circ]_\star \mathcal{L}_{\psi_{-w_i^\circ/d}}$$

$$\cong \mathcal{L}_{\rho_i} \otimes Kl(\psi_{-1/d}; all\ the\ w_i^\circ\ char's\ of\ order\ dividing\ w_i)$$

$$\cong Kl(\psi_{-1/d}; all\ the\ w_i^\circ\ w_i'th\ roots\ of\ \chi_{v_i}).]$$

At this point, we have a geometric isomorphism

$$\mathrm{FT}_{\overline{\psi}} Prim^{n-2}(V \bmod W)[2-n]|\mathbb{G}_m \cong [d]^\star \mathcal{H}(\psi_{-1/d}; List(V,W); \emptyset)[2-n].$$

So in the Grothendieck group $K(\mathbb{A}^1_{\overline{k}}, \overline{\mathbb{Q}}_\ell)$, we have

$$\mathrm{FT}_{\overline{\psi}} Prim^{n-2}(V \bmod W)$$

$$= j_\star [d]^\star \mathcal{H}(\psi_{-1/d}; List(V,W); \emptyset) + (punctual,\ supported\ at\ 0).$$

Applying the inverse Fourier Transform, we find that in $K(\mathbb{A}^1_{\bar{k}}, \overline{\mathbb{Q}}_\ell)$ we have

$$Prim^{n-2}(V \bmod W) = j_\star[d]^\star \mathcal{H}_{V,W} + (geom.\ constant).$$

As before, the fact that $Prim^{n-2}(V \bmod W)$ and $j_\star[d]^\star \mathcal{H}_{V,W}$ have the same generic rank shows that there is no geometically constant term, so we have an equality of perverse sheaves in $K(\mathbb{A}^1_{\bar{k}}, \overline{\mathbb{Q}}_\ell)$,

$$Prim^{n-2}(V \bmod W) = j_\star[d]^\star \mathcal{H}_{V,W}.$$

So these two perverse sheaves have isomorphic semisimplifications. Again by purity, both are geometrically semisimple. So the two sides are geometrically isomorphic. To produce the constant field twist, we repeat the descent argument of Lemma 6.2 to reduce to the case when both descended sides are geometrically irreducible and geometrically isomorphic, hence constant field twists of each other.   □

10 Appendix III: Interesting pieces in the original Dwork family

In this appendix, we consider the case $n = d, W = (1, 1, ..., 1)$. We are interested in those eigensheaves $Prim^{n-2}(V \bmod W)$ that have unipotent local monodromy at ∞ with a single Jordan block. In view of the explicit description of $Prim^{n-2}(V \bmod W)|(\mathbb{G}_m - \mu_d)$ as $[d]^\star \mathcal{H}_{V,W}$, and the known local monodromy of hypergeometric sheaves, as recalled in Section 4, we have the following characterization.

Lemma 10.1. *In the case $n = d, W = (1, 1, ..., 1)$, let $V \bmod W$ be a character of Γ_W/Δ such that $Prim^{n-2}(V \bmod W)$ is nonzero. The following are equivalent.*

(1) *Local monodromy at ∞ on $Prim^{n-2}(V \bmod W)$ has a single Jordan block.*
(2) *Local monodromy at ∞ on $Prim^{n-2}(V \bmod W)$ is unipotent with a single Jordan block.*
(3) *Every $V = (v_1, ..., v_n)$ in the coset $V \bmod W$ has the following property: there is at most one v_i which occurs more than once, i.e., there is at most one $a \in \mathbb{Z}/d\mathbb{Z}$ for which the number of indices i with $v_i = a$ exceeds 1.*
(4) *A unique $V = (v_1, ..., v_n)$ in the coset $V \bmod W$ has the following property: the value $0 \in \mathbb{Z}/d\mathbb{Z}$ occurs more than once among the v_i, and no other value $a \in \mathbb{Z}/d\mathbb{Z}$ does.*

Proof. In order for $Prim^{n-2}(V \bmod W)$ to be nonzero, the list $List(-V, W)$ must differ from $List(all\ d)$. In this $n = d$ case, that means precisely that $List(-V, W)$ must have at least one value repeated. Adding a suitable multiple of $W = (1, 1, ..., 1)$, we may assume that the value 0 occurs at least twice among the v_i. So (3) $\Leftrightarrow$ (4).

For a hypergeometric $\mathcal{H}^{can}(\chi_i's; \rho_j's)$ of type (a, a), local monodromy at ∞ has a single Jordan block if and only if all the ρ_j's coincide, in which case the common value of all the ρ_j's is the eigenvalue in that Jordan block. And $[d]^\star \mathcal{H}^{can}(\chi_i's; \rho_j's)$'s local monodromy at ∞ has the same number of Jordan blocks (possibly with different eigenvalues) as that of $\mathcal{H}^{can}(\chi_i's; \rho_j's)$. In our situation, if we denote by $(\chi_1, ..., \chi_d)$ all the characters of order dividing d, and by $(\chi_{-v_1}, ..., \chi_{-v_a})$ the list $List(-V, W)$, then

$$\mathcal{H}_{V,W} = \mathcal{H}^{can}(\mathbf{Cancel}((\chi_1, ..., \chi_d); (\chi_{-v_1}, ..., \chi_{-v_a}))).$$

So in order for local monodromy at ∞ to have a single Jordan block, we need all but one of the characters that occur among the χ_{v_i} to cancel into $List(all\ d)$. But those that cancel are precisely those which occur with multiplicity 1. So (1) $\Leftrightarrow$ (3). Now (2) $\Rightarrow$ (1) is trivial, and (2) $\Rightarrow$ (4) by the explicit description of local monodromy at ∞ in terms of the ρ_j's. $\qquad\square$

Lemma 10.2. *Suppose the equivalent conditions of Lemma 10.1 hold. Denote by a the rank of $Prim^{n-2}(V\ mod\ W)$. Then on any geometric fibre of $(\mathbb{A}^1 - \mu_d)/Spec(\mathbb{Z}[\zeta_d][1/d\ell])$, the geometric monodromy group G_{geom} attached to $Prim^{n-2}(V\ mod\ W)$ has identity component either $SL(a)$ or $SO(a)$ or, if a is even, possibly $Sp(a)$.*

Proof. By the Tame Specialization Theorem [Ka-ESDE, 8.17.13], the group is the same on all geometric fibres. So it suffices to look in some characteristic $p > a$. Because on our geometric fibre $\mathcal{H}_{V,W}$ began life over a finite field, and is geometrically irreducible, G_{geom}^0 is semisimple. The case $a = 1$ is trivial. Suppose $a \geq 2$. Because its local monodromy at ∞ is a single unipotent block, the hypergeometric $\mathcal{H}_{V,W}$ is not Belyi induced, or inverse Belyi induced, or Kummer induced, and $G_{geom}^{0,der}$ is nontrivial. The result now follows from [Ka-ESDE, 8.11.2]. $\qquad\square$

Lemma 10.3. *Suppose the equivalent conditions of Lemma 10.1 hold. Denote by a the rank of $Prim^{n-2}(V\ mod\ W)$. Suppose $a \geq 2$. Denote by V the unique element in the coset $V\ mod\ W$ in which $0 \in \mathbb{Z}/d\mathbb{Z}$ occurs with multiplicity $a+1$, while no other value occurs more than once. Then we have the following results.*

(1) Suppose that $-V$ is not a permutation of V. Then

$$G_{geom} = SL(a)$$

if $n - 2$ is odd, and

$$G_{geom} = \{A \in GL(a) | det(A) = \pm 1\}$$

if $n - 2$ is even.
(2) If $-V$ is a permutation of V and $n - 2$ is odd, then a is even and

$$G_{geom} = Sp(a).$$

(3) If $-V$ is a permutation of V and $n-2$ is even, then a is odd and

$$G_{geom} = O(a).$$

Proof. That these results hold for $\mathcal{H}_{V,W}$ results from [Ka-ESDE, 8.11.5, 8.8.1, 8.8.2]. In applying those results, one must remember that $\sum_i v_i = 0 \in \mathbb{Z}/d\mathbb{Z}$, which implies that ("even after cancellation") local monodromy at ∞ has determinant one. Thus in turn implies that when d, or equivalently $n-2$, is even, then ("even after cancellation") local monodromy at 0 has determinant the quadratic character, and hence local monodromy at 1 also has determinant the quadratic character. So in the cases where the group does not have determinant one, it is because local monodromy at 1 is a true reflection. After $[d]^\star$, which is finite étale over 1, we get a true reflection at each point in μ_d.   □

Lemma 10.4. *If the equivalent conditions of the previous lemma hold, then over $\mathbb{C}$ the Hodge numbers of $Prim^{n-2}(V \bmod W)$ form an unbroken string of 1's, i.e., the nonzero among the $Prim^{b,n-2-b}(V \bmod W)$ are all 1, and the b for which $Prim^{b,n-2-b}(V \bmod W)$ is nonzero form (the integers in) an interval $[A, A-1+a]$ for some A.*

Proof. From the explicit determination of G_{geom}, we see in particular that $Prim^{n-2}(V \bmod W)$ is an irreducible local system. Looking in a $\mathbb{C}$-fibre of $(\mathbb{A}^1 - \mu_d)/Spec(\mathbb{Z}[\zeta_d][1/d\ell])$ and applying Riemann-Hilbert, we get that the D-module $Prim_{dR}^{n-2}(V \bmod W)$ is irreducible. By Griffiths transversality, this irreducibility implies that the b for which $Prim^{b,n-2-b}(V \bmod W)$ is nonzero form (the integers in) an interval. The fact that local monodromy at ∞ is unipotent with a single Jordan block implies that the number of nonzero Hodge groups $Prim^{b,n-2-b}(V \bmod W)$ is at least a, cf. the proof of Lemma 8.8, (3) ⇔ (5).   □

References

[Abh-GTL] ABHYANKAR, S., "Galois theory on the line in nonzero characteristic", *Bull. Amer. Math. Soc. (N.S.)* **27** (1992), no. 1, 68–133.

[Abh-PP] ABHYANKAR, S., "Projective polynomials", *Proc. Amer. Math. Soc.* **125** (1997), no. 6, 1643–1650.

[Ber] BERNARDARA, M., "Calabi-Yau complete intersections with infinitely many lines", preprint, math.AG/0402454.

[Car] CARLITZ, L., "Resolvents of certain linear groups in a finite field", *Canad. J. Math.* **8** (1956), 568–579.

[C-dlO-RV] CANDELAS, P., DE LA OSSA, X., RODRIGUEZ-VILLEGAS, F., "Calabi–Yau manifolds over finite fields, II", Calabi–Yau varieties and mirror symmetry (Toronto, ON, 2001), 121–157, Fields Inst. Commun., 38, Amer. Math. Soc., Providence, RI, 2003.

[C-dlO-G-P] CANDELAS, P., DE LA OSSA, X., GREEN, P., PARKES, L., "A pair of Calabi-Yau manifolds as an exactly soluble superconformal theory", *Nuclear Phys. B* **359** (1991), no. 1, 21–74.

[De-CEPD]   DELIGNE, P., "Cohomologie étale: les points de départ", redigé par J.F. Boutot, pp. 6–75 in SGA 4 1/2, cited below.

[De-ST]   DELIGNE, P., "Applications de la formule des traces aux sommes trigonométriques", pp. 168–232 in SGA 4 1/2, cited below.

[De-TLCD]   DELIGNE, P., "Théorème de Lefschetz et critres de dégénérescence de suites spectrales", *Publ. Math. IHES* **35** (1968) 259–278.

[De-Weil I]   DELIGNE, P., "La conjecture de Weil", *Publ. Math. IHES* **43** (1974), 273–307.

[De-Weil II]   DELIGNE, P., "La conjecture de Weil II", *Publ. Math. IHES* **52** (1981), 313–428.

[Dw-Def]   DWORK, B., "A deformation theory for the zeta function of a hypersurface" *Proc. Internat. Congr. Mathematicians (Stockholm, 1962)*, 247–259.

[Dw-Rat]   DWORK, B., "On the rationality of the zeta function of an algebraic variety", *Amer. J. Math* **82** (1960), 631–648.

[Dw-Hyp1]   DWORK, B., "On the Zeta Function of a Hypersurface", *Publ. Math. IHES* **12** (1962), 5–68.

[Dw-HypII]   DWORK, B., "On the zeta function of a hypersurface, II", *Ann. of Math.* (2) **80** (1964), 227–299.

[Dw-HypIII]   DWORK, B., "On the zeta function of a hypersurface, III", *Ann. of Math.* (2) **83** (1966), 457–519.

[Dw-PAA]   DWORK, B., "On p-adic analysis", *Some Recent Advances in the Basic Sciences, Vol. 2 (Proc. Annual Sci. Conf., Belfer Grad. School Sci., Yeshiva Univ., New York, 1965–1966)*, 129–154.

[Dw-PC]   DWORK, B., "p-adic cycles", *Publ. Math. IHES* **37** (1969), 27–115.

[Dw-NPI]   DWORK, B., "Normalized period matrices, I, Plane curves", *Ann. of Math.* (2) **94** (1971), 337–388.

[Dw-NPII]   DWORK, B., "Normalized period matrices, II", *Ann. of Math.* (2) **98** (1973), 1–57.

[Grif-PCRI]   GRIFFITHS, P., "On the periods of certain rational integrals, I, II", *Ann. of Math.* (2) **90** (1969), 460–495; ibid. (2) **90** (1969), 496–541.

[Gr-Rat]   GROTHENDIECK, A., "Formule de Lefschetz et rationalité des fonctions L", *Séminaire Bourbaki*, Vol. 9, Exp. No. 279, 41–55, Soc. Math. France, 1995.

[H-SB-T]   HARRRIS, M., SHEPHERD-BARRON, N., TAYLOR, R., "A family of Calabi-Yau varieties and potential automorphy", preprint, June 19, 2006.

[Ka-ASDE]   KATZ, N., "Algebraic solutions of differential equations (p-curvature and the Hodge filtration)", *Invent. Math.* **18** (1972), 1–118.

[Ka-ESES]   KATZ, N., "Estimates for "singular" exponential sums", *IMRN* **16** (1999), 875–899.

[Ka-ESDE]   KATZ, N., "Exponential sums and differential equations", *Annals of Math. Study* **124**, Princeton Univ. Press, 1990.

[Ka-GKM]   KATZ, N., "Gauss sums, Kloosterman sums, and monodromy groups", *Annals of Math. Study* **116**, Princeton Univ. Press, 1988.

[Ka-IMH]   KATZ, N., "On the intersection matrix of a hypersurface", *Ann. Sci. École Norm. Sup.* (4) **2** (1969), 583–598.

[Ka-NCMT] KATZ, N., "Nilpotent connections and the monodromy theorem: Applications of a result of Turrittin", *Publ. Math. IHES* **39** (1970), 175–232.

[Ka-SE] KATZ, N., "Sommes Exponentielles", *Astérisque* **79**, Soc. Math. Fr., 1980.

[Kob] KOBLITZ, N., "The number of points on certain families of hypersurfaces over finite fields", *Compositio Math.* **48** (1983), no. 1, 3–23.

[Ma-ACFD] MANIN, YU. I., "Algebraic curves over fields with differentiation", (Russian) *Izv. Akad. Nauk SSSR. Ser. Mat.* **22** (1958), 737–756.

[Lau-TFCEF] LAUMON, G., "Transformation de Fourier, constantes d'équations fonctionnelles et conjecture de Weil", *Publ. Math. IHES* **65** (1987), 131–210.

[Mor] MORRISON, D. R., "Picard-Fuchs equations and mirror maps for hypersurfaces", *Essays on mirror manifolds*, 241–264, Int. Press, Hong Kong, 1992. Also available at http://arxiv.org/pdf/hep-th/9111025.

[Mus-CDPMQ] MUSTATA, A., "Degree 1 Curves in the Dwork Pencil and the Mirror Quintic" preprint, math.AG/0311252.

[Ogus-GTCC] OGUS, A., "Griffiths transversality in crystalline cohomology", *Ann. of Math.* (2) **108** (1978), no. 2, 395–419.

[RL-Wan] ROJAS-LEON, A., AND WAN, D., "Moment zeta functions for toric calabi-yau hypersurfaces", preprint, 2007.

[Se-ALR] SERRE, J.-P., "Abelian *l*-adic representations and elliptic curves", W. A. Benjamin, Inc., New York-Amsterdam 1968.

[SGA 4 1/2] Cohomologie Etale. Séminaire de Géométrie Algébrique du Bois Marie SGA 4 1/2. par P. Deligne, avec la collaboration de J. F. Boutot, A. Grothendieck, L. Illusie, et J. L. Verdier. Lecture Notes in Mathematics, Vol. 569, Springer-Verlag, 1977.

[SGA 1] Revêtements étales et groupe fondamental. Séminaire de Géométrie Algébrique du Bois Marie 1960–1961 (SGA 1). Dirigé par Alexandre Grothendieck. Augmenté de deux exposés de M. Raynaud. Lecture Notes in Mathematics, Vol. 224, Springer-Verlag, 1971.

[SGA 4 Tome 3] Théorie des Topos et Cohomologie Etale des Schémas, Tome 3. Séminaire de Géométrie Algébrique du Bois Marie 1963–1964 (SGA 4). Dirigé par M. Artin, A. Grothendieck, J. L. Verdier. Lecture Notes in Mathematics, Vol. 305, Springer-Verlag, 1973.

[SGA 7 II] Groupes de monodromie en géométrie algébrique. II. Séminaire de Géométrie Algébrique du Bois Marie 1967–1969 (SGA 7 II). Dirigé par P. Deligne et N. Katz. Lecture Notes in Mathematics, Vol. 340. Springer-Verlag, 1973.

[St] STEVENSON, E., "Integral representations of algebraic cohomology classes on hypersurfaces" *Pacific J. Math.* **71** (1977), no. 1, 197–212.

[St-th] STEVENSON, E., "Integral representations of algebraic cohomology classes on hypersurfaces", Princeton thesis, 1975.

[We-JS] WEIL, A., "Jacobi sums as Grössencharaktere", *Trans. Amer. Math. Soc* **73**, (1952), 487–495.

Graphs, Strings, and Actions

Ralph M. Kaufmann

Purdue University, Department of Mathematics, 150 University St.,
West Lafayette, IN 47907, USA
rkaufman@math.purdue.edu

To my teacher Yuri Ivanovich Manin on the occasion of his 70th birthday

Summary. In this paper, we revisit the formalism of graphs, trees, and surfaces which allows one to build cell models for operads of algebraic interest and represent them in terms of a dynamical picture of moving strings—hence relating string dynamics to algebra and geometry. In particular, we give a common framework for solving the original version of Deligne's conjecture, its cyclic, A_∞, and cyclic–A_∞ versions. We furthermore study a question raised by Kontsevich and Soibelman about models of the little discs operad. On one hand, we give a new smooth model and on the other hand, a minimally small cell model for the A_∞ case. Further geometric results these models provide are novel decompositions and realizations of cyclohedra as well as explicit simple cell representatives for Dyer–Lashof–Cohen operations. We also briefly discuss the generalizations to moduli space actions and applications to string topology as well as further directions.

Key words: moduli spaces, operads, string theory, Hochschild cohomology, cohomology operations, manifolds, cell models, foliations, non-commutative geometry

2000 Mathematics Subject Classifications: 18D50, 55P48, 55S12

Introduction

As often happens in pure mathematics, a dynamical physical point of view can be very helpful in solving complex problems. One instance of these dynamics which has been particularly useful is string theory. There are many incarnations of this theory given by highly developed mathematical tools, such as Gromov–Witten theory or singularity theory. We will take a less algebraic and more geometric point of view in the following. Surprisingly, this approach turns out to have far-reaching algebraic and topological implications. The basic idea is to treat a string as an interval or a circle with a measure. As these

Y. Tschinkel and Y. Zarhin (eds.), *Algebra, Arithmetic, and Geometry*,
Progress in Mathematics 270, DOI 10.1007/978-0-8176-4747-6_5,
© Springer Science+Business Media, LLC 2009

types of strings move, split, and recombine, they give rise to a surface *with a partially measured foliation*. These ideas are completely described in [KP], where actually we are considering strings that move on an oriented surface with boundary. We will consider only the closed case here and furthermore restrict ourselves to surfaces with no internal punctures.

The first step in obtaining applications to algebra and topology is to represent these surfaces by certain types of ribbon graphs. The measure of the foliation translates to weights on the edges of these graphs. To be precise, there are two types of graphs. One is called the *arc graph*, which is obtained by replacing each band of parallel leaves of the foliation with one edge called an arc. There is a dual picture, provided the foliation sweeps out the surface. The condition for this to occur is that the complementary regions of the arcs be polygonal. This condition is called quasi–filling. In this case there is the natural notion of a dual graph. This dual graph is again a ribbon graph with weights on its edges, and furthermore the surface it defines is precisely of the same topological type as the underlying surface. We stress that in general this need not be the case. Usually we call this dual graph if it exists the associated ribbon graph or simply the ribbon graph. In these considerations we take the closed strings to be pointed, which induces marked points on the boundary and marked points on the cycles of the ribbon graph.

Now it is striking that with this picture one obtains several well-known algebraic and topological objects in one fell swoop. The first object is an operad [KLP] which is defined when all the boundaries are hit by arcs. This contains the moduli space of genus g curves with n marked points and a tangent vector at each of its points as a rational suboperad. Here rational means densely defined. Furthermore, taking a different route and using $\mathbb{R}_{\geq 0}$ graded operads instead [KP], one can even induce a modular operad structure on cohomology. We will forego this option and concentrate on the cell level instead. This cell level is described by graphs, one ribbon graph of the above type for each cell. Focusing our attention on different types of graphs, we obtain natural operads, cyclic operads, PROPs, and other algebraic structures.[1]

Moreover, we are naturally led to cell models for various known and important operads such as *the little discs, and the framed little discs*. Extending the graphs, we are led to the definition of a ribbon graph operad for a cell model of moduli space and a model for a PROP which can be called the Sullivan PROP.[2]

In the current note, we wish to present the results as a reverse engineering of sorts, starting with the combinatorics and building spaces out of them. This is contrary to the historical genesis and the dynamic approach mentioned above, but it is a purely algebraic/combinatorial formulation which matches up beautifully with natural operations on the Hochschild complex of various

---

[1]See, e.g., [MSS] for a review of these notions and the operads mentioned below.

[2]There are actually several versions of this PROP on the topological and chain level see, e.g., [CS, S1, S2, CG, TZ]; our version is that of [K4, K5].

algebras. We will treat the associative, A_∞, Frobenius, and Frobenius A_∞ algebra cases. The latter has sometimes been called [Ko2] a cyclic A_∞ algebra. The classical case has been solved in [Ko3, T, MS1, Vo1, KS, MS2, BF, K2], the cyclic case was first established in [K3] and then extended in [TZ] (see also [MS4] for an announcement of a different proof), the A_∞ case has been treated in [KS]. The plethora of proofs goes back to the possibility of choosing suitable chain models. In our approach the chain models are all CW models which are minimal in a sense we explain below. Moreover, they all appear naturally in a geometric picture dictated by string dynamics. The desire to have such operations has three main sources: string topology [CS, Vo2, CJ, CG, Ch, Me, S1, S2], Deligne's conjecture, and D-brane considerations [KR, KLi1, KLi2]; see [K4, K5] for details.

Taking this approach, there are algebraic questions and obstacles, but it turns out that each time the geometry tells us how to overcome them. Along the way, we introduce new cell models for the little discs and framed little discs, some of which are smooth. This partially answers a question of Kontsevich and Soibelman on this subject. Finally, our cell models also cast light on the Dyer–Lashof Araki–Kudo [AK, Co, DL] operations on the Hochschild cohomology, which thanks to the affirmative answer to Deligne's conjecture formally has the structure of a double loop space. Here we give the explicit cells that are responsible for the operations, naturally reproducing the results of [We, Tou].

Finally, we comment on a new natural geometric stabilization for our surface operad. This lends itself to exploring all of the above constructions in a stable limit.

The paper is organized as follows:

§1 contains all the necessary details about graphs. §2 contains the construction of various cell models of the little discs and framed little discs using trees and graphs. In this paragraph, in particular, we also give a new smooth cellular model for the little discs and the framed little discs and a cell model for the minimal operad of [KS]. We furthermore identify the cells responsible for the Dyer–Lashof operations. To illustrate our approach to operations using trees, §3 contains a full self-contained proof of the cyclic version of Deligne's conjecture for a Frobenius algebra. In §3 we also go on to treat the A_∞ and cyclic A_∞ versions. §4 contains the extensions to moduli space and the Sullivan PROP, hence string topology. It also contains the important new notion of operadic correlation functions. We close the discussion in §5 with an outlook and complementary results on the higher loop spaces and stabilization.

Acknowledgments

This paper is dedicated to my teacher Yuri Ivanovich Manin, who inspired me by sharing his deep insight into the inner structure of mathematics, the mathematical structure of physics and the beautiful results which one can obtain by combining them.

Conventions

We fix k to be a field of arbitrary characteristic. We let $\bar{n}$ be the set $\{0, \ldots, n\}$. I will denote the interval $[0, 1]$, and Δ^n the standard n-simplex. Furthermore, K_n is the n-th Stasheff polytope or associahedron, and W_n is the nth cyclohedron or Bott–Taubes polytope, see, e.g., [MSS] for the definitions of these polytopes.

1 Graphs, Spaces of Graphs, and Cell Models

1.1 Classes of Graphs

In this section, we formally introduce the graphs and the operations on graphs which we will use in our analysis.

We will use several types of trees and ribbon graphs.

Graphs

A graph Γ is a tuple $(V_\Gamma, F_\Gamma, \imath_\Gamma : F_\Gamma \to F_\Gamma, \partial_\Gamma : F_\Gamma \to V_\Gamma)$, where $\imath_\Gamma$ is an involution $\imath_\Gamma^2 = \mathrm{id}$ without fixed points. We call V_Γ the vertices of Γ and F_Γ the flags of Γ. The edges E_Γ of Γ are the orbits of the flags under the involution $\imath_\Gamma$. A directed edge is an edge together with an order of the two flags which define it. In case there is no risk of confusion, we will drop the subscripts Γ. Notice that $f \mapsto (f, \imath(f))$ gives a bijection between flags and directed edges.

We also call $F_{\Gamma,v} := \partial^{-1}(v) \subset F_\Gamma$ the set of flags of the vertex v. If Γ is clear from the context, we will just write F_v, and we also call $|F_v|$ the valence of v and denote it by $\mathrm{val}(v)$. We also let $E(v) = \{\{f, \imath(f)\} | f \in F(v)\}$ and call these edges the edges incident to v.

The geometric realization of a graph is given by considering each flag as a half-edge and glueing the half-edges together using the involution $\imath$. This yields a one-dimensional CW complex whose realization we call the realization of the graph.

As usual, a tree is a graph whose image is contractible. A black and white graph, b/w for short, is a graph with a map $V_\Gamma \to \{0, 1\}$. The inverse image of 1 is called the set of white vertices and denoted by V_w, while the inverse image of 0 is called the black vertices, and denoted by V_b.

Ribbon Graphs

A ribbon graph with tails is a connected graph together with a cyclic order of the set of flags $F_\Gamma(v)$ of the vertex v for every vertex v. A ribbon graph with tails that satisfies $\mathrm{val}(v) \geq 2$ for all vertices v will simply be called a ribbon graph. Notice that we do *not* fix $\mathrm{val}(v) \geq 3$. We will call a ribbon graph stable if it does satisfy this condition.

For a ribbon graph with tails, the tail vertices are $V_{\text{tail}} = \{v \in V_\Gamma | \text{val(v)} = 1\}$, the tail edges $E_{\text{tail}}(\Gamma)$ are the edges incident to the tail vertices, and the tail flags $F_{\text{tail}}(\Gamma)$ are those flags of the tail edges which are *not* incident to the tail vertices.

A tree that is a ribbon graph with tails is called a planar tree.

A graph with a cyclic order of the flags at each vertex gives rise to bijections $\text{Cyc}_v : F_v \to F_v$, where $\text{Cyc}_v(f)$ is the next flag in the cyclic order. Since $F = \amalg F_v$, one obtains a map $\text{Cyc} : F \to F$. The orbits of the map $N := \text{Cyc} \circ \imath$ are called the cycles or the boundaries of the graph. These sets have the induced cyclic order.

Notice that each boundary can be seen as a cyclic sequence of directed edges. The directions are as follows. Start with any flag f in the orbit. In the geometric realization go along this half-edge starting from the vertex $\partial(f)$, continue along the second half-edge $\imath(f)$ until you reach the vertex $\partial(\imath(f))$, then continue starting along the flag $\text{Cyc}(\imath(f))$ and repeat.

An angle is a pair of flags $(f, \text{Cyc}(f))$; we denote the set of angles by $\angle_\Gamma$. It is clear that $f \mapsto (f, \text{Cyc}(f))$ yields a bijection between F_Γ and $\angle_\Gamma$. It is, however, convenient to keep both notions.

By an angle marking we mean a map $\text{mk}^\angle : \angle_\Gamma \to \mathbb{Z}/2\mathbb{Z}$.

The genus of a ribbon graph and its surface

The genus $g(\Gamma)$ of a ribbon graph Γ is given by $2 - 2g(\Gamma) = |V_\Gamma| - |E_\Gamma| + \text{Cyc}(\Gamma) = \chi(\Gamma) + \text{Cyc}(\Gamma)$, where $\text{Cyc}(\Gamma) = \#\text{cycles}$.

The surface $\Sigma(\Gamma)$ of a ribbon graph Γ is the surface obtained from the realization of Γ by thickening the edges to ribbons. That is, replace each 0-simplex v by a closed oriented disc $D(v)$ and each 1-simplex e by $e \times I$ oriented in the standard fashion. Now glue the boundaries of $e \times I$ to the appropriate discs in their cyclic order according to the orientations. This is a surface whose boundary components are given by the cycles of Γ. The graph Γ is naturally embedded as the spine of this surface $\Gamma \subset \Sigma(\Gamma)$. Let $\bar{\Sigma}(\Gamma)$ be the surface obtained from $\Sigma(\Gamma)$ by filling in the boundaries with discs. Notice that the genus of the $\bar{\Sigma}(\Gamma)$ is $g(\Gamma)$ and $\chi(\Gamma) = 2 - 2g(\Sigma(\Gamma))$.

Treelike, normalized Marked ribbon graphs

Definition 1.1. A ribbon graph together with a distinguished cycle c_0 is called *treelike* if

(i) the graph is of genus 0 and
(ii) for all flags either $f \in c_0$ or $\imath(f) \in c_0$ (and not both).

In other words, each edge is traversed exactly once by the cycle c_0. Therefore there is a cyclic order on all (undirected) edges, namely the cyclic order of c_0.

The data above are called *almost treelike* if the condition (i) holds and in condition (ii) the exclusive "or" is replaced by the logical "or". This means that there might be edges both of whose flags belong to c_0. We call these edges the black edges of the graph.

Definition 1.2. A *marked ribbon graph* is a ribbon graph together with a map mk : {cycles} → F_Γ satisfying the conditions

(i) For every cycle c the directed edge mk(c) belongs to the cycle.
(ii) All vertices of valence two are in the image of mk, that is $\forall v, \text{val}(v) = 2$ implies $v \in \text{Im}(\partial \circ \text{mk})$.

Notice that on a marked treelike ribbon graph there is a linear order on each of the cycles c_i. This order is defined by upgrading the cyclic order to the linear order $\prec_i$ in which mk(c_i) is the smallest element.

The intersection tree of an almost treelike ribbon graph

Notice that an almost treelike ribbon graph need not be a tree. Indeed, if it has more than two cycles it won't be. But the following construction yields a black and white tree. The following definition of a dual tree is indeed a duality, since one can recover the ribbon graph from its dual tree. For the gory combinatorial details, see the appendix of [K2].

Dual b/w tree of a Marked ribbon graph

Given a marked almost treelike ribbon graph Γ, we define its dual tree to be the colored graph whose black vertices are given by V_Γ and whose set of white vertices is the set of cycles c_i of Γ. The set of flags at c_i consists of the flags f with $f \in c_i$, and the set of flags at v consists of the flags $\{f : f \in c_0, \partial(f) = v\}$. The involution is given by $\imath_\tau(f) = N(f)$ if $f \in c_0$ and $\imath_\tau(f) = N^{-1}(f)$ otherwise.

This graph is a tree and is b/w and bipartite by construction. It is also planar, since the c_i and the sets $F(v)$ have a cyclic order and therefore also so does $F_v \cap c_0$. It is furthermore rooted by declaring $\partial(\text{mk}(c_0))$ to be the root vertex. Declaring mk(c_0) to be the smallest element makes it into a planted tree.

An equivalent definition is given by defining that there be an edge between a pair of a black and a white vertex if and only if the vertex corresponding to b is on the boundary of the cycle c_i, i.e., $v \in \partial(c_i) := \{\partial(f) : f \in c_i\}$ and two black vertices are connected if there was a black edge between them.

Spineless marked ribbon graphs

A marked almost treelike ribbon graph is called *spineless* if

(i) There is at most one vertex of valence 2. If there is such a vertex v_0 then $\partial(\text{mk}(c_0)) = v_0$.
(ii) The induced linear orders on the c_i are (anti)compatible with that of c_0, i.e., $f \prec_i f'$ if and only if $\imath(f') \prec_0 \imath(f)$.

1.2 Operations on graphs

In this section, we will give the basic definitions of the operations on graphs that we will need.

Contracting Edges

The contraction $\Gamma/e = (\bar{V}_\Gamma, \bar{F}_\Gamma, \bar{\imath}, \bar{\partial})$ of a graph $\Gamma = (V_\Gamma, F_\Gamma, \imath, \partial)$ with respect to an edge $e = \{f, \imath(f)\}$ is defined as follows. Let $\sim$ be the equivalence relation induced by $\partial(f) \sim \partial(\imath(f))$. Then let $\bar{V}_\Gamma := V_\Gamma/\sim$, $\bar{F}_\Gamma = F_\Gamma \setminus \{f, \imath(f)\}$ and $\bar{\imath} : \bar{F}_\Gamma \to \bar{F}_\Gamma, \bar{\partial} : \bar{F}_\Gamma \to \bar{V}_\Gamma$ be the induced maps.

For a ribbon graph, the cyclic order is the one which descends naturally.

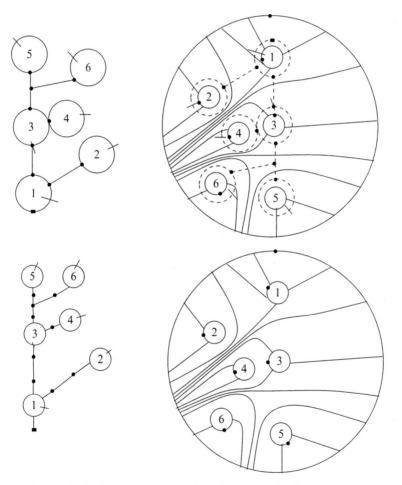

Fig. 1. Counterclockwise: An example of an element in $\mathcal{C}acti_\infty$, the construction of its dual arc graph and its dual black and white tree.

For a marked ribbon graph, we define the marking of $(\bar{V}_\Gamma, \bar{F}_\Gamma, \bar{\imath}, \bar{\partial})$ to be $\overline{\mathrm{mk}}(\bar{c}) = \mathrm{mk}(c)$ if $\mathrm{mk}(c) \notin \{f, \imath(f)\}$ and $\overline{\mathrm{mk}}(\bar{c}) = \overline{N \circ \imath}(\mathrm{mk}(c))$ if $\mathrm{mk}(c) \in \{f, \imath(f)\}$, viz. the image of the next flag in the cycle.

If there is an angle marking, set $f' = N^{-1}(f), f'' = \mathrm{Cyc}(f), g' = N^{-1}(\imath(f))$ and $g'' = \mathrm{Cyc}(\imath(f))$, let $\mathrm{mk}^{\angle}(f', f) = a, \mathrm{mk}^{\angle}(f, f'') = b$, $\mathrm{mk}^{\angle}(g', \imath(f)) = c$ and $\mathrm{mk}^{\angle}(\imath(f), g'') = d$, after the contraction we set $\mathrm{mk}^{\angle}(f', g'') = \bar{\bar{a}}\bar{d}$ and $\mathrm{mk}^{\angle}(g', f'') = \bar{\bar{b}}\bar{c}$, where we use the notation $\bar{a} = 1 - a \in \mathbb{Z}/2\mathbb{Z}$.

1.3 Spaces of Graphs with Metrics

Notation 1.3. We will write $\mathcal{R}ib_{n,g}$ for the set of marked ribbon graphs of genus g with n cycles and, by abuse of notation, also for the free Abelian group generated by this set.

We set $\mathcal{R}ib := \amalg_{n,g}\mathcal{R}ib_{n,g}$, and we will again not distinguish in notation between the set $\mathcal{R}ib$, the free Abelian group generated by it, and the set $\{\amalg_g \mathcal{R}ib_{n,g} : n \in \mathbb{N}\}$ to avoid unnecessary clutter. We also write $\mathcal{R}ib(n)$ for the set of marked ribbon graphs with $n + 1$ cycles together with a labelling by $\{0, \ldots, n\}$ of these cycles. Again we also denote the free Abelian group generated by this set as $\mathcal{R}ib(n)$. Finally, to streamline the notation, we will denote the collection $\{\mathcal{R}ib(n)|n \in \mathbb{N}\}$ simply by $\mathcal{R}ib$.

The meaning of the symbols will always be clear from the context.

Graphs with a Metric

A metric w_Γ for a graph is a map $E_\Gamma \to \mathbb{R}_{>0}$. The (global) rescaling of a metric w by λ is the metric $\lambda w : \lambda w(e) = \lambda w(e)$. The length of a cycle c is the sum of the lengths of its edges $\mathrm{length}(c) = \sum_{f \in c} w(\{f, \imath(f)\})$. A metric for a treelike ribbon graph is called normalized if the length of each undistinguished cycle is 1. We will write $\mathcal{MR}ib_{n,g}$ for the set of metric marked ribbon graphs of genus g with n boundary cycles.

Projective Metrics

Notice that there is an $\mathbb{R}_{>0}$-action on $\mathcal{MR}ib$ which scales the metric μ by an overall factor. This action of course preserves the genus and number of boundaries. We set $\mathbb{PR}ib := \mathcal{MR}ib/\mathbb{R} > 0$. The elements of $\mathbb{PR}ib$ are called graphs with a projective metric. Notice that one can always choose a normalized representative for any projective metric. We set $\mathbb{PR}ib_{n,g} = \mathcal{MR}ib_{n,g}/\mathbb{R}_{>0}$.

Remark 1.4. Now and in the following, we do not wish to dwell on distinguishing projective and non-projective metrics.

Definition 1.5. By a local scaling at a cycle i, we mean that the metric is scaled only on the edges belonging to the cycle i.

The Space of Metric Ribbon Graphs

We endow these above sets with a topology by constructing $\mathbb{P}Rib_{n,g}$ in the standard fashion. That is, we realize them as a subspace of the quotient of the disjoint union of simplices by an equivalence relation. For each graph $\Gamma \in Rib_{n,g}$ with $|E(\Gamma)| = k + 1$ we fix a k-simplex Δ_Γ. Using barycentric coordinates for this simplex, a point of this simplex can be identified with a choice of projective weights on the edges. The points of $\mathbb{P}Rib_{n,g}$ can thus be identified with the interior of the disjoint union over all $\Delta_\Gamma : \Gamma \in Rib_{n,g}$. Furthermore, the faces of Δ_Γ correspond to the edges of Γ. We use the following identifications: A face of Δ_Γ is identified with $\Delta_{\Gamma/e}$ if $\Gamma/e \in Rib_{n,g}$. We give the resulting space the quotient topology (this is actually a CW complex) and identify $\mathbb{P}Rib$ with the image of the interiors of the Δ_Γ. Then we give $\mathcal{M}Rib := \mathbb{P}Rib \times \mathbb{R}_{>0}$ the product topology.

Cacti and Spineless Cacti and Thickened Cacti

Definition 1.6. We let $Cacti(n)$ denote the subspace of the treelike ribbon graphs with n labeled cycles (that is, excluding the distinguished cycle c_0). Furthermore, we let $Cact(n) \subset Cacti(n)$ be the subset of spineless cacti.

We let $Cacti_\infty$ be the almost treelike ribbon graphs and $Cact_\infty$ be the almost treelike spineless ribbon graphs.

Marked Ribbon Graphs with Metric and Maps of Circles

For a marked ribbon graph with a metric, let c_i be its cycles, let $|c_i|$ be their image in the realization, and let r_i be the length of c_i. Then there are natural maps $\phi_i : S^1 \to |c_i|$ which map S^1 onto the cycle by starting at the vertex $v_i := \partial(mk(c_i))$ and going around the cycle mapping each point $\theta \in S^1$ to the point at distance $\frac{\theta}{2\pi} r_i$ from v_i along the cycle c_i. This observation connects the current constructions to those involving a more geometric definition of $Cacti$ in terms of configurations of circles [Vo2, K1] and other geometric constructions involving such configurations such as the map $\mathcal{L}oop$ used for the $\mathcal{A}rc$ operad [KLP]. In particular, the treelike ribbon graphs correspond to $Cacti$, and the spineless treelike ribbon graphs correspond to $Cact$.

This observation is also useful in order to describe the glueing operations

$$\circ_i : Cacti_\infty(n) \times Cacti_\infty(m) \to Cacti_\infty(n + m - 1), \tag{1}$$
$$(\Gamma_1, \Gamma_2) \mapsto \Gamma_1 \circ_i \Gamma_2, \tag{2}$$

which are given by scaling Γ_2 to the size of the length of the ith cycle of Γ_1 and then glueing together the graphs using the identification given by the corresponding maps of S^1s parameterizing the scaled cycle c_0 of Γ_2 and the cycle c_i of Γ_1.

For a purely combinatorial version of this construction we refer to the appendix of [K2]. The version presented there which pertains to $Cacti$ can easily be adapted to the current case of $Cacti_\infty$.

Proposition 1.7. *The spaces $Cacti_\infty(n)$ together with the $\mathbb{S}_n$ action permuting the labels and the glueing operations $\circ_i$ of (1) form a topological operad, and the subspaces $Cact_\infty(n), Cacti(n), Cact(n)$ form suboperads.*

Proof. Straightforward.

Recall that two operads are equivalent as operads if there is a chain of quasi-isomorphisms connecting them.

Theorem 1.8. $Cacti_\infty(n)$ *and* $Cacti(n)$ *are equivalent to the framed little discs operad, and* $Cact_\infty(n)$ *and* $Cact(n)$ *are equivalent to the little discs operad.*

Proof. The statements about $Cacti$ and $Cact$ are contained in [K1]. The corresponding statements about $Cacti_\infty$ and $Cact_\infty$ follow from the observation that these spaces are homotopic to $Cacti$ and $Cact$ by the homotopy which contracts all the edges both of whose flags are elements of the distinguished cycle c_0.

Cactus Terminology

The edges of a cactus are traditionally called arcs or segments, and the cycles of a cactus are traditionally called lobes. The vertices are sometimes called the marked or special points. Furthermore, the distinguished cycle c_0 is called the outside circle or the perimeter, and the vertex $\partial(\mathrm{mk}(c_0))$ is called the global zero. And the vertices $\partial(\mathrm{mk}(c_i)), i \neq 0$, are called the local zeros. In pictures these are represented by lines rather than fat dots.

Normalized Treelike and Almost Treelike Ribbon Graphs and Their Cell Complexes

Definition 1.9. An element of $Cacti_\infty$ is called normalized if the length of all the cycles except for possibly the distinguished cycle are 1 and the lengths of all of the black edges are less than or equal to 1. We use the superscript 1 on the spaces above to indicate the subset of normalized elements, e.g., $Cacti_\infty^1$.

Notation 1.10. We will call an element of the set $\{Cacti, Cact, Cact_\infty, Cacti_\infty\}$ simply a species of cactus.

Lemma 1.11. *Every species of cactus is homotopy equivalent to its subspace of normalized elements.*

Proof. The homotopy is given by locally scaling each lobe to size 1. Notice that this is possible, because the graphs are almost treelike.

The normalized versions have their good side and their bad side. On the bad side, we see that they are not stable under glueing, but we can modify the glueing as follows to obtain a topological quasioperad, that is, an operad which is associative only up to homotopy:

$$\circ_i : Cacti^1_\infty(n) \times Cacti^1_\infty(m) \to Cacti_\infty(n+m-1), \tag{3}$$

$$(\Gamma_1, \Gamma_2) \mapsto \Gamma_1 \circ_i \Gamma_2. \tag{4}$$

Here the composition is given by first locally scaling the lobe i of Γ_1 to the length of the distinguished cycle of Γ_2 and then glueing.

Proposition 1.12. *The normalized elements of any species of cactus together with the $\mathbb{S}_n$ action of relabelling and the glueings above form a topological quasioperad.*

Proof. Tedious but straightforward. See [K1] for $Cact$ and $Cacti$, the more general version is covered under the Sullivan PROP in [K4], see also Section 4.

The relations between the species are as follows:

Theorem 1.13. *[K1] The operad of cacti is the bicrossed product of the operad $Cact$ of spineless cacti with the operad S^1 based on the monoid S^1. Furthermore, this bicrossed product is homotopic to the semidirect product of the operad of cacti without spines with the circle group S^1:*

$$Cacti \cong Cact \bowtie S^1 \simeq Cact \rtimes S^1. \tag{5}$$

The same holds true for the thickened versions

$$Cacti_\infty \cong Cact_\infty \bowtie S^1 \simeq Cact_\infty \rtimes S^1. \tag{6}$$

The details of the semidirect products and bicrossed products are given below.

Proof. The proof of the first statement is given by verifying that the two operad structures coincide. For the second statement, one notices that the homotopy diagonal is homotopy equivalent to the usual one and that one can find homotopies to the diagonal which continuously depend on the cactus. The third statement follows from contracting the factors $\mathbb{R}^n_{>0}$ and using Theorem 1.15. Full details are given in [K1] for the non–thickened species. They go over *mutatis mutandis* for the thickened species.

Corollary 1.14. *The homology operad of $Cacti$ is the semidirect product of $Cact$ and the homology of the operad S^1 built on the monoid S^1. The same holds true for $Cacti_\infty$.*

Theorem 1.15. *Every species of cactus is homotopy equivalent through quasioperads to its normalized version.*

Proof. The statement for regular cacti is contained in [K1], and the argument carries over *mutatis mutandis* to the thickened versions.

Corollary 1.16. *Every species of cactus is quasi-isomorphic as quasi-operads to its normalized version, and in particular, the induced homology quasi-operads are operads and are isomorphic as operads.*

Details of the Bicrossed Product Structure for Cacti

In this section we recall the construction of the bicrossed product as it was given in [K1], to which we refer the reader for more details.

First notice that there is an action of S^1 on $Cact(n)$ given by rotating the base point *clockwise* (i.e., in the orientation opposite the usual one of c_0) around the perimeter. We denote this action by

$$\rho^{S^1} : S^1 \times Cact(n) \to Cact(n).$$

With this action we can define the twisted glueing

$$\circ_i^{S^1} : Cact(n) \times S^1(n) \times Cact(m) \to Cact(n+m-1),$$
$$(C, \theta, C') \mapsto C \circ \rho^{S^1}(\theta_i, C') =: C \circ_i^{\theta_i} C'. \quad (7)$$

Given a cactus without spines $C \in Cact(n)$, the orientation-reversed perimeter (i.e., going around the outer circle *clockwise*, i.e., reversing the orientation of the source of ϕ_0) gives a map $\Delta_C : S^1 \to (S^1)^n$.

As one goes around the perimeter the map goes around each circle once, and thus the map Δ_C is homotopic to the diagonal $\Delta_C(S^1) \sim \Delta(S^1)$.

We can use the map Δ_C to give an action of S^1 and $(S^1)^{\times n}$:

$$\rho^C : S^1 \times (S^1)^{\times n} \overset{\Delta_C}{\to} (S^1)^{\times n} \times (S^1)^{\times n} \overset{\mu^n}{\to} (S^1)^{\times n}; \quad (8)$$

here μ_n is the diagonal multiplication in $(S^1)^{\times n}$ and $\bar{o}_i$ is the operation that forgets the ith factor and shuffles the last m factors to the ith, ..., $(i+m-1)$st places. Set

$$\circ_i^C : (S^1)^{\times n} \times (S^1)^{\times m} \overset{(id \times \pi_i)(\Delta) \times id}{\longrightarrow} (S^1)^{\times n} \times S^1 \times (S^1)^{\times m}$$
$$\overset{id \times \rho^C}{\longrightarrow} (S^1)^{\times n} \times (S^1)^{\times m} \overset{\bar{o}_i}{\longrightarrow} (S^1)^{\times n+m-1}. \quad (9)$$

These maps are to be understood as perturbations of the usual maps

$$\circ_i : (S^1)^{\times n} \times (S^1)^{\times m} \overset{(id \times \pi_i)(\Delta) \times id}{\longrightarrow} (S^1)^{\times n} \times S^1 \times (S^1)^{\times m}$$
$$\overset{id \times \rho}{\longrightarrow} (S^1)^{\times n} \times (S^1)^{\times m} \overset{\bar{o}_i}{\longrightarrow} (S^1)^{\times n+m-1}. \quad (10)$$

where now ρ is the diagonal action of S^1 on $(S^1)^{\times n}$. The maps $\circ_i$ and the permutation action on the factors give the collection $\{\mathcal{S}^1(n)\} = (S^1)^{\rangle\, n}$ the structure of an operad. In fact, this is exactly the usual construction of an operad built on a monoid.

The multiplication in the bicrossed product is given by

$$(C, \theta) \circ_i (C', \theta') = \left(C \circ_i^{\theta_i} C', \theta \circ_i^{C'} \theta' \right). \tag{11}$$

The multiplication in the semidirect product is given by

$$(C, \theta) \circ_i (C', \theta') = \left(C \circ_i^{\theta_i} C', \theta \circ_i \theta' \right). \tag{12}$$

Also, normalized cacti are homotopy equivalent to cacti that are homotopy equivalent to the bicrossed product of normalized cacti with $\mathcal{S}^1$ and the semidirect product with $\mathcal{S}^1$, where all equivalences are as quasioperads:

$$\mathcal{C}acti^1 \sim \mathcal{C}acti \cong \mathcal{C}act \bowtie \mathcal{S}^1 \sim \mathcal{C}act^1 \bowtie \mathcal{S}^1 \sim \mathcal{C}act^1 \rtimes \mathcal{S}^1. \tag{13}$$

2 The Tree Level: Cell Models for (Framed) Little Discs and Their Operations

The virtue of the normalized species is that they provide cellular models. In order to give the cell model, we will use the dualized trees.

2.1 A First Cell Model for the Little Discs: $\mathcal{C}act^1$

In this section we will give a cell model for $\mathcal{C}act^1$. It will be indexed by the dual trees of the ribbon graphs. The specific type of trees we need are given by the sets $\mathcal{T}^{bp}(n)$, that is, planar planted bipartite black and white trees with only white leaves. Here as usual a leaf is a vertex of valence one that is not the root. Since the tree is rooted, the edges have a natural direction toward the root, and we call the edges that are incoming to white vertices the white edges and denote the set they form by E_w.

Notice that the differential on the ribbon graphs induces a differential on the dual trees.

Definition 2.1. We define $\mathcal{T}^{bp}(n)^k$ to be the elements of $\mathcal{T}^{bp}(n)$ with $|E_w| = k$.

Definition 2.2. For $\tau \in \mathcal{T}^{bp}$ we define $\Delta(\tau) := \times_{v \in V_w(\tau)} \Delta^{|v|}$. We define $C(\tau) = |\Delta(\tau)|$. Notice that $\dim(C(\tau)) = |E_w(\tau)|$.

Given $\Delta(\tau)$ and a vertex x of any of the constituting simplices of $\Delta(\tau)$, we define the xth face of $C(\tau)$ to be the subset of $|\Delta(\tau)|$ whose points have the xth coordinate equal to zero.

Definition 2.3. We let $K(n)$ be the CW complex whose k-cells are indexed by $\tau \in T^{bp}(n)^k$ with the cell $C(\tau) = |\Delta(\tau)|$ and the attaching maps e_τ defined as follows. We identify the xth face of $C(\tau)$ with $C(\tau')$, where $\tau' = \partial_x(\tau)$ is the local contribution of the differential contracting the corresponding white edge. This corresponds to contracting an edge of the cactus if its weight goes to zero so that $\Delta(\partial\tau)$ is identified with $\partial(\Delta(\tau))$.

Lemma 2.4. $K(n)$ is a CW composition for $Cact$.

Proof. It is straightforward to see that the differential on the graphs which contracts an edge on the tree side collapses an angle.

Proposition 2.5. $K(n)$ is a cellular chain model for the little discs.

Proof. The claim is that already on the cell level the induced quasioperad is an operad. This is indeed the case, since in a cell all possible positions of the lobes are possible and the composition again gives all possible positions; see [K1] for details.

2.2 A CW Decomposition for $Cacti^1$ and a Cellular Chain Model for the Framed Little Discs

Definition 2.6. A $\mathbb{Z}/2\mathbb{Z}$ decoration for a black and white bipartite tree is a map $dec^\pm : V_w \to \mathbb{Z}/2\mathbb{Z}$.

Proposition 2.7. *The quasi–operad of normalized cacti $Cacti^1$ has a CW decomposition which is given by cells indexed by planar planted bipartite trees with a $\mathbb{Z}/2\mathbb{Z}$ decoration. The k-cells are indexed by trees with $k - i$ white edges and i vertices marked by 1.*

Moreover, cellular chains are a chain model for the framed little discs operad and form an operad. This operad is isomorphic to the semidirect product of the chain model of the little discs operad given by $CC_*(Cact)$ of [K2] and the cellular chains of the operad built on the monoid S^1.

Proof. For the CW decomposition we note that as spaces, $Cacti^1(n) = Cact^1(n) \times (S^1)^{\times n}$. Now viewing $S^1 = [0, 1]/0 \sim 1$ as a 1-cell together with the 0-cell given by $0 \in S^1$, the first part of the proposition follows immediately, by viewing the decoration by 1 as indicating the presence of the 1-cell of S^1 for that labeled component in the product of cells.

To show that the cellular chains indeed form an operad, we use the fact that the bicrossed product is homotopy equivalent to the semidirect product in such a way that the action of a cell S^1 in the bicrossed product is homotopic to the diagonal action. This is just the observation that the diagonal and the diagonal defined by a cactus are homotopic. Since a semidirect product of a monoid with an operad is an operad, the statement follows. Alternatively, one could just remark that there is also an obvious functorial map induced by the diagonal for these cells.

The chains are a chain model for the framed little discs operad since $\mathcal{C}acti^1(n)$ and $\mathcal{C}acti(n)$ are homotopy equivalent and the latter is equivalent to the framed little discs operad.

Although the above chain model is the one one would expect to use for framed little discs, it does not have enough cells for our purposes. In order to translate the proofs in the arc complex given in [KLP] into statements about the Hochschild complex, we will need a slightly finer cell structure than the one above. After having used the larger structure one can reduce to the cell model with few cells, since they are obviously equivalent.

Definition 2.8. A spine decoration dec′ for a planted planar bipartite tree is a $\mathbb{Z}/2\mathbb{Z}$ decoration together with the marking of one angle at each vertex labeled by one and a flag at each vertex labeled by zero. We denote the set of such trees which are n-labeled by $T^{\mathrm{bp},\mathrm{dec}'}(n)$ and again use this notation as well for the free Abelian group and the k vector space generated by these sets. We let $T^{\mathrm{bp},\mathrm{dec}'}$ be their union respectively direct sum. In pictures we show the angle marking as a line emanating from the vertex that lies between the marked edges and an edge marking by a line through the respective edge. For an example see Figure 2 (VI). We sometimes omit the edge marking if the marked edge is the outgoing edge, e.g., in Figure 3.

The realization $\hat{\tau}$ of a planar planted bipartite tree τ with a spine decoration is the realization of τ as a planar planted tree (the root is fixed to be black) together with one additional edge inserted into each marked angle connecting to a new vertex. We call the set of these edges spine edges and denote them by E_{spine}. Likewise, set V_{spine} to be the set of new vertices called the spine vertices, which are defined to be black. The spine edges are then white edges. As for tails, we will consider only the flags of E_{spine}, which are not incident to the spine vertices. We call the set of these flags F_{spine}.

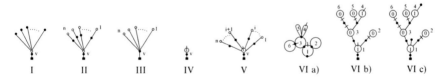

Fig. 2. I. the tree l_n. II. the tree τ_n. III. the tree τ_n^b. IV. the tree O'. V. the tree $\tau'_{n,i}$. VI.(a). a marked treelike ribbon graph. (b) the corresponding decorated tree. (c) its realization.

Fig. 3. The decomposition of the BV operator.

Notice that this tree is the dual tree of a cactus with an explicit marking of the flags $\mathrm{mk}(c_i)$. Given a cactus, we call its dual tree with explicit markings its topological type. If τ had tails, we will split the set of tails of the realization into spines and free tails, which are the images of the original tails: $E_{\mathrm{tails}}(\hat{\tau}) = E_{\mathrm{ftails}}(\hat{\tau}) \amalg E_{\mathrm{spine}}(\hat{\tau})$; and we proceed likewise for the respective flags.

A spine decoration induces a new linear order on the flags incident to the white vertices of its realization. This order $\prec'_v$ is given by the cyclic order at v and declaring the smallest element to be the spine flag in case $\mathrm{dec}^{\pm}(v) = 1$ and the marked flag in case $\mathrm{dec}^{\pm}(v) = 0$. This gives a canonical identification of $F_{\prec'_v} : F_v \to \{0, \ldots, |v|\}$.

Proposition 2.9. *The spaces $Cacti^1(n)$ of the quasi–operad of normalized cacti $Cacti^1$ have CW decompositions $K'(n)$ whose cells are indexed by spine decorated planar planted bipartite trees $(\tau, \mathrm{dec}') \in \mathcal{T}^{\mathrm{bp},\mathrm{dec}'}$ corresponding to the topological type of the cacti. The k-cells are indexed by n-labeled trees with $k - i$ white edges and i markings by 1.*

Moreover, cellular chains of the complex above are a chain model for the framed little discs operad and form an operad.

Proof. The decomposition is almost as in the preceding proposition except that in the product $Cact^1(n) \times (S^1)^{\times n}$ we decompose each factor S^1 as indicated by the lobe it presents. That is, for the S^1 associated to the nth lobe we chose the 0-cells to correspond to the marked points and 1-cells to correspond to the arcs with glueing given by attaching the 1-cells to the 0-cells representing the endpoints of the arcs (e.g., four 0-cells and four 1-cells for the lobe 1 in Figure 2 (VIa)). In terms of trees, the arcs correspond to the angles, and thus we take a marking of an arc to be the inclusion of the corresponding 1-cell in the tensor product of the cell complexes. Likewise, the edges correspond to the marked points, and we take a marking of an edge to be the inclusion of the corresponding 0-cell in the tensor product of the cell complexes.

For the operadic properties, we remark that moving the spine along an arc and then glueing, which is what is parameterized by marking an angle on the lobe i of c when calculating $c \circ_i c'$, has the effect of moving the base point of c' along a complete sequence of arcs until it coincides with a marked point in the composition of the two cacti. This is one side of the bicrossed product. The effect on the local zeros of c' of the movement of the base point is to move them corresponding to structure maps of the bicrossed product above. The local zeros thus move through a full arc if the global zero passes through the arc on which they lie. Therefore the $\circ_i$ product of two cells results in sums of cells. Marking an arc of c' obviously gives rise to a sum of cells. Alternatively, one can again just remark that there is a functorial map for the diagonal for this cell model, since there is such a map on the first factor by [K2] and its existence is obvious on the second factor.

The associativity follows from the associativity of cacti. Let $C(\tau)$, $\tau \in \mathcal{T}^{\mathrm{bp},\mathrm{dec}'}(n)$ be the cells in the CW-complex and $\dot{C}(\tau)$ their interior.

Then $P(\tau) = \dot{C}(\tau) \times \mathbb{R}^n_{>0}, \tau \in \mathcal{T}^{\text{bp,dec}'}$ give a pseudocell decomposition $Cacti(n) = \amalg_\tau P(\tau)$. It is easy to see that $Im(P(\tau) \circ_i P(\tau')) = \amalg_k P(\tau_k)$ for some τ_k and $\circ_i$ is a bijection onto its image. Let $\circ_i^{\text{comb}}$ be the quasioperad structure pulled back from K' to $\mathcal{T}^{\text{bp,dec}'}$ and let $\circ_i^+$ be the operad structure pulled back from the pseudocell decomposition of $Cacti$ to $\mathcal{T}^{\text{bp,dec}'}$. Then these two operad structures coincide over $\mathbb{Z}/2\mathbb{Z}$, thus yielding associativity up to signs. The signs are just given by shuffles, cf. Section 3.1, and are associative as well.

Remark 2.10. Pulling back the operadic compositions, the differential, and the grading yields a dg-operad structure on $\mathcal{T}^{\text{bp,dec}'}$ which is isomorphic to that of $CC_*(Cacti^1) := \bigoplus_n CC_*(K'(n))$, where $CC_*(K'(n))$ are the cellular chains of the CW model $K'(n)$ of $Cacti^1(n)$.

The operation is briefly as follows: given two trees $\tau, \tau' \in \mathcal{T}^{\text{bp,dec}'}$ the product is $\tau \circ_i^{\text{comb}} \tau' = \sum \pm \tau_k$, where the τ_k are the trees obtained by the following procedure. Delete v_i to obtain an ordered collection of trees $(\tau_l^c, \prec_v')$, then graft these trees to τ', keeping their order by first identifying the spine edge or marked edge of v_i with the root edge of τ' and then grafting the rest of the branches to τ' so that their original order is compatible with that of τ'. Lastly, contract the image of the root edge of τ' and declare the image of the root of τ to be the new root. The sign is as explained in Section 3.1. Due to the isomorphism between $CC_*(Cacti^1)$ and $\mathcal{T}^{\text{bp,dec}'}$ we will drop the superscript comb.

2.3 The GBV Structure

The picture for the GBV structure is essentially that of [KLP] and goes back to [CS]. It appears here in another guise, however, since we are now dealing with cells in $CC_*(Cacti^1)$.

First notice that there is a product on the chain level induced by the spineless cactus given by the rooted tree τ_n depicted in Figure 2. Explicitly: $a \cdot b \mapsto \gamma\left(\tau_2^b; a, b\right)$, where γ is the usual operadic composition. This product gives $CC_*(Cacti^1)$ the structure of an associative algebra with unit. Moreover, the product is commutative up to homotopy. The homotopy is given by the usual operation, which is induced by $\gamma(\tau_1; a, b)$. This also induces a bracket which is Gerstenhaber up to homotopy. This can be seen by translating the statements from [KLP, K2], but it also follows from the BV description of the bracket below (Figure 5).

To give the BV structure, let O' be the tree with one white vertex, no additional black edges, no free tails, and a spine. Notice that the operation δ induced by $a \mapsto \gamma(O', a)$ on $CC_*(Cacti^1)$ breaks up on products of chains as follows, see Figure 3:

$$\delta(ab) \sim \delta(a, b) + (-1)^{|a||b|}\delta(b, a),$$
$$\delta(abc) \sim \delta(a, b, c) + (-1)^{|a|(|b|+|c|)}\delta(b, c, a) + (-1)^{|c|(|a|+|b|)}\delta(c, a, b), \quad (14)$$

$$\delta(a_1 a_2 \cdots a_n) \sim \sum_{i=0}^{n-1} (-1)^{\sigma(c^i, a)} \delta(a_{c^i(1)}, \ldots, a_{c^i(n)}), \tag{15}$$

where c is the cyclic permutation and $\sigma(c^i, a)$ is the sign of the cyclic permutations of the graded elements a_i.

Lemma 2.11.

$$\delta(a, b, c) \sim (-1)^{(|a|+1)|b|} b \delta(a, c) + \delta(a, b)c - \delta(a)bc. \tag{16}$$

Proof. The proof is contained in Figure 4.

Proposition 2.12. *The chains* $CC_*(Cacti^1)$ *are a GBV algebra up to homotopy. That is, there are a bracket and a BV operator that satisfy the usual equations up to homotopy. Taking coefficients in k when k is of characteristic zero, the homology of* $Cacti$ *hence becomes a GBV algebra.*

Proof. The BV structure follows from Lemma 2.11 via the calculation

$$
\begin{aligned}
\delta(abc) \sim{} & \delta(a, b, c) + (-1)^{|a|(|b|+|c|)} \delta(b, c, a) + (-1)^{|c|(|a|+|b|)} \delta(c, b, a) \\
\sim{} & (-1)^{(|a|+1)|b|} b \delta(a, c) + \delta(a, b)c - \delta(a)bc + (-1)^{|a|} a \delta(b, c) \\
& + (-1)^{|a||b|} \delta(b, a)c - (-1)^{|a|} a \delta(b)c + (-1)^{(|a|+|b|)|c|} a \delta(b, c) \\
& + (-1)^{|b|(|a|+1)+|a||c|} b \delta(c, a) - (-1)^{|a|+|b|} ab \delta(c) \\
\sim{} & \delta(ab)c + (-1)^{|a|} a \delta(bc) + (-1)^{|a|+1|b|} b \delta(ac) - \delta(a)bc \\
& - (-1)^{|a|} a \delta(b)c - (-1)^{|a|+|b|} ab \delta(c) \tag{17}
\end{aligned}
$$

Figure 5 contains the homotopy relating the BV operator to the bracket.

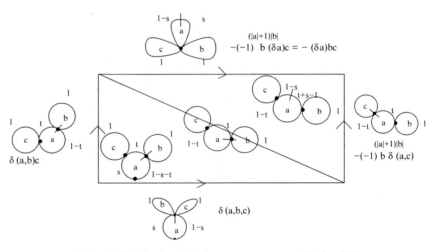

Fig. 4. The basic chain homotopy responsible for BV.

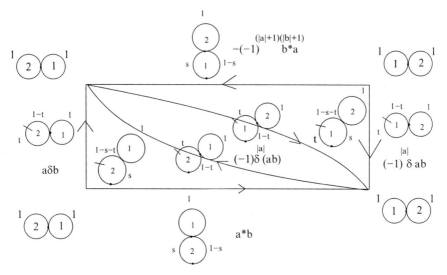

Fig. 5. The compatibility of the BV operator and the bracket.

2.4 Cells for the Araki–Kudo–Cohen, Dyer–Lashof Operations

By the general theory, see, e.g., [Tou], we need to find elements

$$\xi_1 \in H_{p-1}(\mathcal{C}act^1(p)/\mathbb{S}_p, \pm\mathbb{Z}/p\mathbb{Z}),$$

that is, homology classes with values in the sign representation.

Now taking coinvariants on $\mathcal{C}act^1$, we see that the iteration of the product $*$, that is, the operation given by $^n* := \gamma(\gamma(\ldots(\gamma(\tau_1),\tau_1),\ldots,\tau_1),\tau_1)$, gives a class that is the sum over all trees of the highest dimension, where the partial order on the labeled vertices when considered in the usual tree partial order is compatible with the linear order on $\bar{n}$.

Proposition 2.13. $^n*$ *is the cohomology class* ξ_1 *in* $H_{p-1}(\mathcal{C}act^1(p)/\mathbb{S}_p, \pm\mathbb{Z}/p\mathbb{Z})$.

Proof. First we could reengineer the result from the proof of Tourtchine [Tou], but it also follows from a straightforward calculation of the boundary of said cell.

The first example for $p = 2$ is given by the operation of τ_1, which has boundaries in the multiplication and its opposite, cf. Figure 5, and the example for $p = 3$ is the hexagon of Figure 6 with $i = 1$.

Remark 2.14. We wish to point out two interesting facts. First, the class is solely induced by an operation for $p = 2$, and second, the resulting cell description is just the left iteration of $*$, whereas the right iteration of $*$ is the simple class given by a cube.

146   Ralph M. Kaufmann

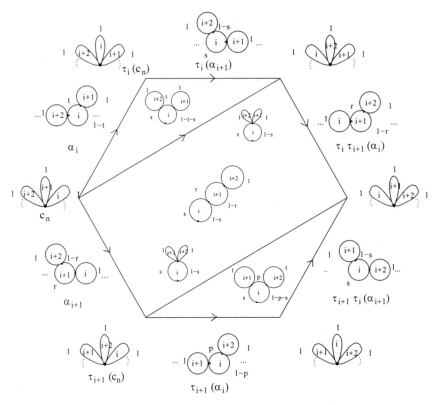

Fig. 6. The hexagon that gives the Dyer–Lashof operation, establishes that *Cacti* is a braid operad, and shows the associativity up to homotopy of the multiplication.

2.5 A Smooth Cellular Model for the Framed Little Discs: $\mathcal{Cacti}_\infty$

The above CW model for the little discs is actually the smallest model which solves Deligne's conjecture and has enough cells to support the brace and the multiplication operations. However, the model is easily seen to be nonsmooth starting at $n = 3$. We can read this off Figure 6, since two of these hexagons glue to give a "cylinder with wings."

There is, however, a surprisingly small CW model that is smooth. It is given by considering $\mathcal{Cact}_\infty$. This is not the minimal model that yields a solution for the A_∞-Deligne conjecture, which we will discuss later.

The Relevant Trees

Again the cells of this model will be indexed by certain types of trees that are the dual trees of the ribbon graphs that are elements of $\mathcal{Cact}^1_\infty$. These are planar planted b/w trees with heights.

It will be convenient to use the convention that these trees have a black root with valence one and call the unique incident edge the root edge. We will call the edges which are *incident to a white vertex* the *white edges*, E_w, the other edges are considered to be black, E_b. The exception is the root edge, which is not considered to be black in case it is not white. We will also fix the stability condition that there are no black vertices of valence 2 with two black edges.[3] We will call such trees $\mathcal{T}_{b/w}$.

Definition 2.15. We let $\mathcal{HT}(n)$ be pairs (τ, ht) of a planar planted b/w tree with white leaves only and a black root (τ) and a function $\mathrm{ht} : E_b \to \{1, \mathrm{var}\}$. We will let E_{var} be the inverse image of var, and call them variable edges. Likewise, let $\mathcal{HT}^{\mathrm{top}}(n)$ be pairs $(\tau, \mathrm{ht}^{\mathrm{top}})$ with τ as above and $\mathrm{ht}^{\mathrm{top}} : E \to [0, 1]$ such that the sum of the weights of the edges adjacent to a white vertex is 1.

Remark 2.16. Notice that there is a natural differential on the underlying ribbon graphs, which can also be considered to have white and black edges. The latter are labeled by 1, var. The differential is given by summing (with the appropriate sign) over contractions of the white edges, contractions of the black edges labeled by var, and relabellings of these edges by 1.

Definition 2.17. We define $\mathcal{HT}(n)^k$ to be the elements of $\mathcal{HT}(n)$ with $|E_{\mathrm{var}}| + |E_w| - |V_w| = k$.

Definition 2.18. For $\tau \in \mathcal{HT}$ we define $\Delta(\tau) := \times_{v \in V_w(\tau)} \Delta^{|v|} \times \times_{e \in E_{\mathrm{var}}(\tau)} I$. We define $C(\tau) = |\Delta(\tau)|$. Notice that $\dim(C(\tau)) = |E_w(\tau)| + |E_b(\tau)|$.

Given $\Delta(\tau)$ and a vertex x of any of the constituting simplices of $\Delta(\tau)$, we define the xth face of $C(\tau)$ to be the subset of $|\Delta(\tau)|$ whose points have the xth coordinate equal to zero. The boundaries of the intervals are taken to be 0 and 1.

Definition 2.19. We let $K_\infty(n)$ be the CW complex whose k-cells are indexed by $\tau \in \mathcal{HT}(n)^k$ with the cell $C(\tau) = |\Delta(\tau)|$ and the attaching maps e_τ defined as follows. We identify the xth face of $C(\tau)$ with $C(\tau')$, where $\tau' = \partial_x(\tau)$. This corresponds to contracting a white edge of the cactus as its weight goes to zero so that $\Delta(\partial \tau)$ is identified with $\partial(\Delta(\tau))$ for these edges. For the black edges, passing to the boundaries of the intervals corresponds to letting the weight of the edge go to 1 or 0, and the latter is taken to mean that the relevant edge is contracted.

Lemma 2.20. $K_\infty(n)$ *is a CW composition for* $\mathcal{Cact}^1_\infty$.

Proof. For this it suffices to remark that the dual ribbon graph of the tree indexing a cell and an element of this cell has a natural metric on the corresponding graph given by the barycentric coordinates on the simplices for the white edges and the natural coordinates on the intervals taking values

---

[3]This means no parallel arcs in the dual picture.

between 0 and 1 on the black edges. Conversely, using the dual tree construction turns any element of $Cacti_\infty$ into a tree of the given type, and the metric determines a unique point in the open cell.

Proposition 2.21. $K_\infty(n)$ *is a cellular chain model for the little discs.*

Proof. The proof is analogous to the one for normalized spineless cacti. $\blacksquare$

Theorem 2.22. *The space $Cact^1_\infty$ is smooth, that is, it is a manifold with corners.*

Proof. The easiest way to see this is to use the dual description in terms of arc graph. The arc graph is the dual graph on the surface $\Sigma(\Gamma)$ to Γ, where Γ is embedded as the spine of this surface (more details are contained in Section 4.1). Now fix an element $p \in Cacti^1_\infty(n)$. If it has the maximal number of edges, that is, the complementary regions of the arc graph are triangles,[4] then we can vary the weights of the white edges freely and the weights of the black variable edges as well, while the ones for the black edges with weight 1 can only decrease. So for the interior of the maximal cells we are done. If we are in the interior of a cell of lower dimension, some of the complementary regions are not triangles, but other polygons. Now, not all the diagonals are allowed, since we have to take care that the resulting ribbon graph is still treelike. To be precise, the vertices of the polygons are labeled by $i \neq 0$ or by 0 and the diagonals are not allowed to connect two vertices with non–zero labels. But the vertices adjacent to a vertex with a non–zero label have to be labeled by zero. See Figure 8 for an example. The relevant space is a subspace of the product of the spaces of the diagonals of all of the polygons. Now the space of diagonals of a polygon near the point without diagonals is homeomorphic to a neighborhood of zero in the corresponding Stasheff polytope. There is a subpolygon given by connecting the nonzero labeled vertices. Removing these points corresponds to collapsing cubes in the cubical decomposition of the Stasheff polytope in such a fashion that the result is again a polytope. See Figure 7 for an example. The image of 0 can, however, now lie on a face of the polytope. Nevertheless, we again have found a neighborhood that is homeomorphic to a neighborhood of 0 in $\mathbb{R}^n \times \mathbb{R}^k_{\geq 0}$. $\blacksquare$

Remark 2.23. This cell model almost answers a question of Kontsevich and Soibelman in [KS]. Namely, the existence of a certain smooth CW model for the Fulton–MacPherson configuration spaces. In fact, this is a minimal thickening of a minimal cell model of the little discs, which is minimal in the sense that it contains all the cells for the A_∞ multiplications and the brace operations, that is, a cell incarnation of the minimal operad $\mathcal{M}$ of [KS], which we construct in the next section. See also Remark 2.32.

---

[4]We contract the edges of the polygon which lie on the boundary and label them by the corresponding boundary component.

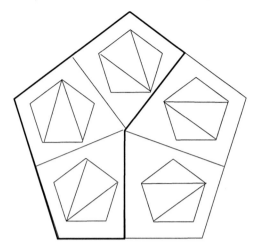

Fig. 7. The Stasheff polytope K_4, its cubical decomposition, and the polytope of the cells avoiding one diagonal of the underlying polygon.

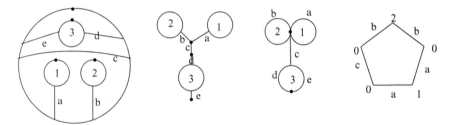

Fig. 8. An arc graph, its tree, cactus representation, and one of its polygons.

2.6 The KW Cell Model for the Little Discs

In this section, we construct two more cell models for the little discs operad. The first will be a cell incarnation of the minimal operad of [KS], and the second will be a cacti-based model that is a contraction of the model $Cact^1_\infty$ above. We need this second model only as a mediator, to establish the equivalence to the little discs operad.

Trees

The relevant trees are the stable b/w planar planted trees of [KS] with white leaves $\mathcal{T}_\infty$. Here stability means that there are no black vertices of valence 2.

The Minimal A_∞ Complex

Let K_n denote the nth Stasheff polytope (associahedron) of dimension $n - 2$ and let W_n denote the nth Bott–Taubes polytope (cyclohedron) of dimension $n - 1$.

We will now construct the following CW complex $\mathcal{KW}$. The cells are indexed by $\tau \in \mathcal{T}_\infty$, and the cell for τ is given by

$$\Delta(\tau) := \times_{v \in V_{\text{white}}} W_{|v|} \times \times_{v \in V_{\text{black}}} K_{|v|-1}. \tag{18}$$

The boundary of this cell is given by

$$\partial(\Delta(\tau)) = \sum_{v' \in V_{\text{white}}} \pm \partial W_{|v|} \times \times_{v \neq v' \in V_{\text{white}}} W_{|v|} \times \times_{v \in V_{\text{black}}} K_{|v|}$$

$$+ \sum_{v' \in V_{\text{black}}} \pm \times_{v \in V_{\text{white}}} W_{|v|} \times \partial K_{|v'|} \times \times_{v \neq v' \in V_{\text{black}}} K_{|v|}$$

$$= \Delta(\partial(\tau)) \tag{19}$$

Fixing n, we inductively glue the cells corresponding to $\tau \in \mathcal{T}_\infty(n)$ to the existing skeleton by identifying the boundary pieces with the cells of lower dimension. For this we have to remark that indeed the cell differential given above agrees with the differential on $\mathcal{T}_\infty$, which is straightforward.

We call the resulting CW complex $\mathcal{KW}(n)$.

Lemma 2.24. *The collection $CC_*(\mathcal{KW}(n))$ forms an operad isomorphic to the minimal operad of [KS].*

Proof. Since $\mathcal{T}_\infty$ is an operad, we just pull back the operad structure, since as Abelian groups, $CC_*(\mathcal{KW}(n)) \simeq \mathcal{T}_\infty(n)$.

2.7 A Finer Cell Model, the Generalized Boardman–Vogt Decomposition

In order to connect the above cell model with the little discs, we need to transform it slightly by subdividing the cells. We will call the corresponding model $\mathcal{KS}$. First, we identify the spaces of the two CW models and then afterward, we can contract to the model $\mathcal{C}act^1$. The full details are in [KSch].

Decomposing the Stasheff Polytope

For this we need two basic decompositions. First we decompose the associ-ahedron into its Boardman–Vogt decomposition (see, e.g., [MSS]). We will actually need a topological realization, which is given by trees with heights. In this case, we consider a planar planted tree as used in this construction as a b/w tree with white leaves and topologically realize the cubical cells by using a height function on the black edges. This means that a point in this cubical model of K_n is an element of $\mathcal{HT}^{\text{top}}$.

Decomposing the Cyclohedra

We actually decompose the cyclohedra as a blowup of the simplex. For this we again use b/w trees in $\mathcal{HT}^{\text{top}}$ as above. The basic simplex is given by taking a tree with one internal vertex. Now we glue to this simplex the cells that allow black edges to appear. This is again easiest to describe in the arc graph. We consider all arc graphs corresponding to at most one internal white vertex, but we allow diagonals, that is, edges between 0 and 0, that do not form a triangle two of whose sides are identified. An example of such a complementary region is given in Figure 8.

Trees and Their Cell Complex

In other words, we consider trees of $\mathcal{T}_{b/w}$ with the following restrictions. There are no black vertices of valence two such that one edge is a leaf edge and the other is black.[5]

We call this subset $\mathcal{T}_{b/w}^{\text{rig}}$. For the height functions, we have one more restriction. A height function for $\mathcal{T}_{b/w}^{\text{rig}}$ is compatible if the height of a black edge, both of whose vertices are of valence 2, has to be 1. This restriction is needed, but in a sense is ungeometric. Omitting it, one is led to the thickened model above. It is necessary to make the incidence relations of the cells match.

It is clear that we can again glue a cell complex from these trees. This time

$$\Delta(\tau) := \times_{v \in V_w(\tau)} \Delta^{|v|} \times \times_{e \in E_{\text{var}}} I. \tag{20}$$

In particular, there is a new subdivision of cyclohedra, that is, not the Boardman–Vogt subdivision. The cells are products of cubes *and* simplices. This also allows for a partially linear realization in terms of trees with heights. Here the restriction for the cyclohedron is that there is only one nonleaf white vertex. See Figure 9 for an example in the language of arc graphs; for further details we refer to [KSch].

Proposition 2.25. *Each element of* $\mathcal{KW}$ *corresponds to a pair* (τ, ht) *with* $\tau \in \mathcal{T}_{b/w}^{\text{rig}}$ *and* ht *a compatible height function. That is, the* $\mathcal{KS}$ *and* $\mathcal{KW}$ *are cell models for the same space. In the description in terms of* $\mathcal{KS}$, *an element is given by the tree of its cell and a compatible topological height function.*

Proof. Any element of $\mathcal{KW}$ lies in a unique maximal cell. This corresponds to a tree $\tilde{\tau} \in \mathcal{T}_\infty$. Now each cyclohedron and associahedron of the product making up $|D(\tau)|$ has a decomposition as above, and our element inside the cell $\Delta(\tilde{\tau})$ lies inside one of these finer cells. Inside this product the element is given by a tree with height satisfying the given conditions. Moreover, given a pair (τ, ht) satisfying the above conditions, it is easy to see that this element in the above description belongs to the cell $\Delta(\tilde{\tau})$, where $\tilde{\tau}$ is the tree in which all the black edges with ht < 1 are contracted.

---

[5]This means that there is no triangle with two sides given by the same arc in the polygon picture.

Fig. 9. The subdivision of W_2 into a simplex and cubes.

The Homotopy from $\mathcal{KS}$ to $\mathcal{C}act^1$

Definition 2.26. We define the flow $\Psi : I \times \mathcal{KS} \to \mathcal{C}act^1$ by $1 \geq t > 0$:
$\Psi(t)((\tau, \mathrm{ht})) = (\tau, \psi(t)(\mathrm{ht}))$, where

$$\psi(t)(\mathrm{ht})(e) = \begin{cases} \mathrm{ht}(e) & \text{if } e \notin E_{\mathrm{black}}, \\ t\,\mathrm{ht}(e) & \text{if } e \in E_{\mathrm{black}}, \end{cases}$$

and $\Psi(0)(\tau, \mathrm{ht}) = (\tilde{\tau}, \tilde{\mathrm{ht}})$, where $\tilde{\tau}$ is the tree τ with all black edges contracted and $\tilde{\mathrm{ht}}$ is ht descended to $\tilde{\tau}$.

Definition 2.27. We define $i_\infty^{\mathrm{top}} : \mathcal{C}act^1(n) \to \mathcal{KW}(n)$ by mapping (τ, ht) to itself.

Proposition 2.28. *The spaces $\mathcal{C}act^1(n)$ and $\mathcal{KS}(n)$ are homotopy equivalent, and hence $\mathcal{KW}$ is too.*

Proof. Using the flow Ψ and the maps i_∞^{top}, the statement is straightforward.

The Cell Level: Maps π_∞ and i_∞

On the cell level this induces the following maps. There are maps $\pi_\infty : \mathcal{T}_\infty \to \mathcal{T}^{\mathrm{bp}}$ and $i_\infty : \mathcal{T}^{\mathrm{bp}} \to \mathcal{T}_\infty$.

The first π_∞ is given as follows. If there is a black vertex of valence > 3, then the image is set to be 0. If all black vertices are of valence 3, then contract all black edges and then insert a black vertex into each white edge. It is clear that the leaves will stay white. The global marking is defined to be the image under the contraction.

The second map i_∞ is given as follows. Remove all black vertices with valence $= 2$ and replace each black vertex of valence > 2 by the binary tree, with all branches to the left.

It is clear that π_∞ is surjective and $\pi_\infty \circ i_\infty = id$.

Lemma 2.29. *These maps behave well with respect to the differential. We have* $\pi_\infty(\partial(\tau)) = \partial\pi_\infty(\tau)$ *and the same for* i_∞. *And* π_∞ *is an operadic map.*

Proof. It is straightforward to check.

We now come to the main statement of this section:

Theorem 2.30. *The topological spaces* $\mathcal{KW}(n)$ *and* $\mathcal{C}act^1(n)$ *are homotopy equivalent. Moreover, the homotopy is given by an explicit contraction* Ψ, *which descends to the chain level operadic map* $\pi_\infty : CC_*(\mathcal{KW}) \to CC_*(\mathcal{C}act^1)$, *where we used the isomorphisms of operads* $CC_*(\mathcal{KW}) \simeq \mathcal{T}_\infty$ *and* $CC_*(\mathcal{C}act^1) \simeq \mathcal{T}^{\mathrm{bp}}$ *to pull back the map* π_∞.

Proof. First it is clear that Φ contracts onto the image of i_∞^{top}, which gives the desired statement about homotopies. We see that any cell of $\mathcal{T}_\infty$ is contracted to a cell of lower dimension as soon as there is a black vertex whose valence is greater than 3, so that these cells are sent to zero. If the vertices only have valence 3, then the black subtrees are contracted onto the image of i_∞, which yields a cell of the same dimension indexed by the tree $\pi_\infty(\tau)$. Finally, since π_∞ is an operadic map and $CC_*(\mathcal{C}act^1)$ is an operadic chain model for the little discs, we deduce that $CC_*(\mathcal{KW})$ also has this property.

Corollary 2.31. $\mathcal{KW}$ *is a cell model for the little discs operad whose cells are indexed by* $\mathcal{T}_\infty$.

Remark 2.32. This remark should be seen in conjunction with Remark 2.23. We have identified a natural cell model for the minimal operad of [KS]. This is, however, not smooth. We can thicken it by the procedure above to obtain the smooth model $\mathcal{C}act_\infty$. Its dimension is, however, too large, and some cells will have to operate as 0. We do wish to point out that there is an inclusion of the cells of $\mathcal{KW}$ into $\mathcal{C}act_\infty$, and indeed there are cells which correspond to $\mathcal{T}_\infty$. So it seems that finding a smooth *and* minimal cell model for the A_∞-Deligne conjecture is not possible.

The Versions for the Framed Little Discs

We do not wish to go through all of the details again. Going over to the framed versions means taking a bicrossed product on the topological level, which on the cell level can be realized by inducing $\mathbb{Z}/2\mathbb{Z}$ decorations as in Section 2.2.

3 Operations of the Cell Models on Hochschild Complexes

In this section we use the tree language in order to naturally obtain operations on the Hochschild complex.

3.1 The Cyclic Deligne Conjecture

In this subsection we give the full details of an action of the model $\mathcal{C}acti$ of the framed little discs on the Hochschild complex.

Assumption

Now we fix A to be a finite–dimensional associative algebra with unit 1 together with an inner product $\eta : A \otimes A \to k$ which is non-degenerate and both (i) invariant: $\eta(ab, c) = \eta(a, bc)$ and (ii) symmetric: $\eta(a, b) = \eta(b, a)$. Such an algebra is called a Frobenius algebra.

We will use CH to stand for Hochschild cochains $CH^n(A, A) := \mathrm{Hom}(A^{\otimes n}, A)$.

Actually, it would be enough to have a non-degenerate inner product η on $A \simeq CH^0(A, A)$ for which (i) holds on $HH^0(A, A)$, that is, up to homotopy for A. The condition (ii) will then hold automatically up to homotopy, since $CH^0(A, A)$ is commutative up to homotopy [G].

If one wishes to relax furthermore the other conditions "up to homotopy," one can fix that η needs to be non-degenerate only on $HH^0(A, A)$ and require only that $HH^0(A, A)$ be finite–dimensional. In this case, the operadic operations defined below will give operations $f : A^{\otimes n} \to HH^0(A, A)$ and will thus give actions only up to homotopy. This is enough to get the BV structure on $CH^*(A, A)$, but not quite enough to lift the action to the chain level.

Notation

Let (e_i) be a basis for A and let $C := e_i \eta^{ij} \otimes e_j$ be the Casimir element, i.e., η^{ij} is the inverse to $\eta_{ij} = \eta(e_i, e_j)$.

With the help of the non–degenerate bilinear form, we identify

$$CH^n(A, A) = \mathrm{Hom}(A^{\otimes n}, A) \cong A \otimes A^{*\otimes n} \cong A^{*\otimes n+1}. \qquad (21)$$

We would like to stress the order of the tensor products we choose. This is the order from right to left, which works in such a way that one does not need to permute tensor factors in order to contract.

If $f \in \mathrm{Hom}(A^{\otimes n}, A)$, we denote by $\tilde{f}$ its image in $A^{*\otimes n+1}$, explicitly $\tilde{f}(a_0, \ldots, a_n) = \eta(a_0, f(a_1, \ldots, a_n))$.

With the help of (21) we can pull back the Connes operators b and B (see, e.g., [L]) on the spaces $A^{\otimes n}$ to their duals and to $\mathrm{Hom}(A^{\otimes n}, A)$.

Also let $t : A^{\otimes n} \to A^{\otimes n}$ be the operator given by performing a cyclic permutation $(a_1, \ldots, a_n) \mapsto (-1)^{n-1}(a_n, a_1, \ldots, a_{n-1})$ and $N := 1 + t + \cdots + t^{n-1} : A^{\otimes n} \to A^{\otimes n}$.

It is easy to check that the operator induced by b is exactly the Hochschild differential; we will denote this operator by ∂. We write Δ for the operator induced by B. It follows that $\Delta^2 = 0$ and $\Delta\partial + \partial\Delta = 0$.

Assumption

To make the formulas simpler we will restrict to normalized Hochschild cochains $\overline{\mathrm{CH}}^n(A, A)$, which are the $f \in \mathrm{CH}^n(A, A)$ that vanish when evaluated on any tensor containing $1 \in A$ as a tensor factor (see, e.g., [L]). On the normalized chains the operator Δ is explicitly defined as follows: for $f \in \overline{\mathrm{CH}}^n(A, A)$,

$$\eta(a_0, (\Delta f)(a_1, \ldots, a_{n-1})) := \eta(1, f \circ N(a_0, \ldots, a_n)). \tag{22}$$

Correlators from Decorated Trees

We will use the notation of tensor products indexed by arbitrary sets; see, e.g., [D]. For a linearly ordered set I denote by $\bigcup_I a_i$ the product of the a_i in the order dictated by I.

Definition 3.1. Let τ be the realization of a spine-decorated planted planar b/w tree, $v \in V_w$, and $f \in \overline{\mathrm{CH}}^{|v|}(A, A)$. We define $Y(v, f) : A^{F_v(\tau)} \to k$ by

$$Y(v, f)\left(\bigotimes_{i \in F_v(\tau)} a_i \right) := \eta\left(a_{F_{\prec'_v}^{-1}(0)}, f\left(a_{F_{\prec'_v}^{-1}(1)} \otimes \cdots \otimes a_{F_{\prec'_v}^{-1}(|v|)} \right) \right).$$

Set $V_{b\text{-}int} := V_b(\tau) \setminus (V_{\text{tail}} \cup \{v_{\text{root}}\} \cup V_{\text{spine}})$. For $v \in V_{b\text{-}int}$ we define $Y(v) := A^{F_v(\tau)} \to k$ by

$$Y(v)\left(\bigoplus_{i \in F_v(\tau)} a_i \right) = \eta\left(1, \bigcup_{i \in F_v} a_i \right).$$

Definition 3.2. Let τ be the realization of a planar planted b/w tree with n free tails and k labels and $f_i \in \overline{\mathrm{CH}}^{n_i}(A, A)$. For such a tree there is a canonical identification $\{v_{\text{root}}\} \cup V_{\text{ftail}} \to \{0, 1, \ldots, |V_{\text{ftail}}|\}$ which is given by sending v_{root} to 0 and enumerating the tails in the linear order induced by the planted planar tree. Set $E_{\text{int}}(\tau) := E(\tau) \setminus (E_{\text{tail}} \cup E_{\text{root}} \cup E_{\text{spine}})$ and for $(a_0, \ldots, a_n) \in A^{\otimes(\{v_{\text{root}}\} \cup V_{\text{ftail}})}$ set

$$Y(\tau)(f_1, \ldots, f_k)(a_0, \ldots, a_n)$$

$$:= \left(\bigotimes_{v \in V_w(\tau)} Y(v, f_{\mathrm{Lab}(v)}) \bigotimes_{v \in V_{b\text{-int}}} Y_v \right)$$

$$\times \left(\left(\bigotimes_{i \in F_{\mathrm{ftail}}(\tau) \cup \{F_{\mathrm{root}}\}} a_i \right) \left(\bigotimes_{j \in F_{\mathrm{spine}}} 1 \right) \otimes C^{\otimes E_{\mathrm{int}}(\tau)} \right).$$

$$(23)$$

In other words, decorate the root flag by a_0, the free tail flags by $a_1, \ldots, a_n$, the spines by 1, and the edges by C and then contract tensors according to the decoration at the white vertices while using the product at the black vertices.

Definition 3.3. We extend the definition above by

$$Y(\tau)(f_1, \ldots, f_k)(a_0, \ldots, a_n) = 0 \quad \text{if } |v_{\mathrm{Lab}^{-1}(i)}| \neq n_i =: |f_i|. \quad (24)$$

The Foliage Operator

Let F be the foliage operator of [K2] applied to trees. This means that $F(\tau)$ is the formal sum over all trees obtained from τ by glueing an arbitrary number of free tails to the white vertices. The extra edges are called free tail edges E_{ftail}, and the extra vertices V_{ftail} are defined to be black and are called free tail vertices.

Using the trees defined in Figure 2, this corresponds to the formal sum $F(\tau) := \sum_n l_n \circ_v \tau$, where the operadic composition is the one for b/w trees that are not necessarily bipartite (see [K2]). In our current setup we should first form $\tilde{F}(\tau) := \sum_n \tau_n \circ_v \tau$ and then delete the images of all leaf edges together with their white vertices of the τ_n to obtain $F(\tau)$.

Signs

The best way to fix signs of course is to work with tensors indexed by edges as in [K2,KS]. For this, one fixes a free object L (free $\mathbb{Z}$-module or k-vector space) generated by one element of degree ± 1 and calculates signs using $L^{\otimes E_w(\tau)}$ before applying the foliage operator while using $L^{\otimes E_{\mathrm{weight}}}$ after applying the foliage operator, where $E_{\mathrm{weight}} = E_w \cup E_{\mathrm{root}} \cup E_{\mathrm{ftail}} \cup E_{\mathrm{spine}}$.

Explicitly, we fix the signs to be given as follows. For any tree τ' in the linear combination above, we take the sign of τ' to be the sign of the permutation which permutes the set E_{weight} in the order induced by $\prec$ to the order where at each vertex one first has the root if applicable, then all non–tail edges, then all the free tails, and if there is a spine edge, the spine.

The explicit signs above coincide with usual signs [L] for the operations and the operators b and B and also coincide with the signs of [G] for the $\circ_i$

and hence for the brace operations [Ge,Kad,GV]. The signs for the operations corresponding to operations on the Hochschild side are fixed by declaring the symbols "," and "{" to have degree one.

Definition 3.4. For $\tau \in \mathcal{T}^{\mathrm{bp,dec}'}$ let $\hat{\tau}$ be its realization. We define the operation of τ on $\overline{\mathrm{CH}}(A, A)$ by

$$\eta(a_0, \tau(f_1, \ldots, f_n)(a_1, \ldots, a_N)) := Y(F(\hat{\tau}))(f_1, \ldots, f_n)(a_0, \ldots, a_N). \quad (25)$$

Notice that due to Definition 3.3, the right-hand side is finite.

Examples

We will first regard the tree O' with one white vertex, no additional black edges, no free tails, and a spine; see Figure 2. For a function $f \in \overline{\mathrm{CH}}^n$ we obtain

$$
\begin{aligned}
Y(F(O'))(f)(a_0, \ldots, a_{n-1}) &= \eta \left(1, f(a_0, \ldots a_{n-1}) \right. \\
&\quad \left. + (-1)^{n-1} f(a_{n-1}, a_0, \ldots, a_{n-2}) + \cdots \right) \\
&= \eta(a_0, \Delta(f)(a_1, \ldots, a_{n-1}))
\end{aligned}
$$

Let $\tau'_{n,i}$ be the tree of Figure 2. Then the operation corresponds to

$$
\begin{aligned}
Y(F(\tau'_{n,i}))(f; g_1, \ldots, g_n)(a_0, \ldots, a_N) \\
= \eta(1, f\{'g_{i+1}, \ldots, g_n, g_1, \ldots, g_i\}(a_{(2)}, a_0, a_{(1)})),
\end{aligned}
$$

where $N = |f| + \sum |g_i| - n - 1$ and we used the shorthand notation

$$
\begin{aligned}
f\{'g_{j+1}, \ldots, g_n, g_1, \ldots, g_j\}(a_{(2)}, a_0, a_{(1)}) = \sum \pm f(a_{k+1}, \ldots, a_{i_{j+1}-1}, \\
g_{j+1}(a_{i_{j+1}}, \ldots, a_{i_{j+1}+|g_{j+1}|}), \ldots, a_{i_n-1}, g_n(a_{i_n}, \ldots, a_{i_n+|g_n|}), \ldots, a_N, a_0, \\
a_1, \ldots, a_{i_1-1}, g_1(a_{i_1}, \ldots, a_{i_1+|g_1|}), \ldots, a_{i_j-1}, g_j(a_{i_j}, \ldots, a_{i_j+|g_j|}), \ldots, a_k),
\end{aligned}
$$

where the sum runs over $1 \leq i_1 \leq \cdots \leq i_j \leq \cdots \leq k \leq \cdots \leq i_{j+1} \leq \cdots \leq i_n \leq N : i_l + |g_l| \leq i_{l+1}, i_j + |g_j| \leq k$ and the signs are as explained above.

Theorem 3.5. *[K3]* (The cyclic Deligne conjecture) *The Hochschild cochains of a finite-dimensional associative algebra with a non–degenerate, symmetric, invariant bilinear form are an algebra over the chains of the framed little discs operad. This operation is compatible with the differentials.*

Proof. We will use the cellular chains $CC_*(Cacti^1)$ as a model for the chains of the framed little discs operad. It is clear that Definition 3.4 defines an action. On the Hochschild side, the $\circ_i$ operations are substitutions of the type $f_i = \psi(g_1, \ldots, g_n)$. For $CC_*(Cacti^1)$ the $\tau \circ_i \tau'$ operations are the pullback via the foliage operator of all possible substitutions of elements of $F(\tau), \tau \in$

$CC_*(Cacti^1)$ into the position i of $F(\tau')$. The action Y then projects onto the substitution $f_i = \psi(g_1, \ldots, g_n)$, so that the action is operadic. Explicitly, the substitution $t \circ_i^s t'$ for planted planar bipartite trees with a decoration dec' and additional free tails is given as follows: Say the number of tails of t' coincides with $|F(v_i)|$. In this case replace the vertex v_i of t, its edges, and the black vertices corresponding to the edges with the tree t' matching the flags of v_i with the tails of t' by first matching the root edge with the marked flag of v_i and then using the linear order. Lastly, contract the image of the root flag. Otherwise, set $t \circ_i^s t' = 0$. With this definition it is easy to see that $F(\tau \circ \tau') = F(\tau) \circ_i^s F(\tau')$.

The compatibility of the Hochschild differential with the differential of the cell complex follows from the relevant statements for τ_n and τ_n^b, which are straightforward but lengthy calculations (see, e.g., [11,K2]), together with the calculations above Section 3.1, which are easily modified to show that $(\partial O')(f) = \Delta(\partial(f))$ and that $(\partial\tau'_{n,i})(f, g_1, \ldots, g_n) = (\partial\tau'_{n,i})(f, g_1, \ldots, g_n) \pm (\tau'_{n,i})(\partial f, g_1, \ldots, g_n) + \sum_i \pm (\tau'_{n,i})(f, g_1, \ldots, \partial(g_i), \ldots, g_n)$ via an even more lengthy but still straightforward calculation. This then verifies the claim in view of the compatibility of the differentials and the respective operad structures.

Alternatively, in view of the operation of the foliage operator, the compatibilities follow from a straightforward translation of trees with tails into operations on the Hochschild complex. The compatibility of the differential then follows from the almost identical definition of the differential for trees with tails of [K2] and that in the Hochschild complex as $\partial(f) = f \circ \cup - (-1)^{|f|} \cup \circ f$.

Corollary 3.6. *The normalized Hochschild cochains of an algebra as above are a GBV algebra up to homotopy in the sense of Proposition 2.12.*

This could of course have been checked directly without recourse to the operation of a chain model, but we do not know of any source for this result. It also seems to be difficult to guess the right homotopies as Gerstenhaber did in the non-cyclic case [G].

Corollary 3.7. *Over a field of characteristic zero, the Hochschild cohomology of an algebra as above is a BV algebra such that the induced bracket is the Gerstenhaber bracket.*

Lastly, since our second version of cellular chains of Proposition 2.9 is a subdivision of the cell decomposition of Proposition 2.7, we can also use the latter cell decomposition.

Corollary 3.8. *The normalized Hochschild cochains of an algebra as above are an algebra over the semidirect product over a chain model of the little discs operad and a chain model for the operad S built on the monoid S^1.*

Remark 3.9. The operation of the little discs operad by braces, viz. the original Deligne conjecture as discussed in [K2] for Frobenius algebras, corresponds to the decorations in which $\mathrm{dec}^{\pm} \equiv 0$ and the decorated edge is always the outgoing edge.

Remark 3.10. In Theorem 3.5 we can relax the conditions and implications as explained in Section 3.1.

3.2 The Araki–Kudo–Cohen, Dyer–Lashof Operations on the Hochschild Complex

By the positive answer to Deligne's conjecture, the Hochschild complex behaves as if it were a double loop space. So we should expect operations ξ_1 and ζ_1 on it. Indeed, they were found by Westerland [We] for $p = 2$ and by Tourtchine [Tou] for general p. We wish to point out that the cells of Section 2.4 naturally induce these operations. It is easy to see that ξ_1 is just the iterated $\circ$ product and ζ_1 is the product of such iterations. That is,

$$\xi_1(x) = x \circ (x \circ (\cdots \circ x) \cdots). \tag{26}$$

Note that the result is not novel, only the cells of Section 2.4 are. This description, however, simplifies matters very much.

3.3 The A_∞-Deligne Conjecture

Theorem 3.11. *There is an action of the cellular chains model* $\mathrm{CC}_*(\mathcal{KW})$ *on the Hochschild cochain complex of an* A_∞*-algebra.*

Proof. This follows from the theorem above in conjunction with the theorem of [KS] that the operad $\mathcal{T}_\infty$ acts in a dg fashion on $C^*(A, A)$.

Remark 3.12. We recall that the action is given by viewing the tree as a flow chart. Given functions $f_1, \ldots, f_n$, the action of $\tau \in \mathcal{T}_\infty(n)$ is defined as follows. First "insert" the functions f_i into the vertex labeled by i and then view the tree as a flow chart using the operations μ_n of the A_∞-algebra at each black vertex of arity n and the brace operation $h\{g_1, \ldots, g_k\}$ at each white vertex marked by h of arity k to concatenate the function. Here the brace operation [Ge, Kad, GV] is given by

$$
\begin{aligned}
&h\{g_1, \ldots, g_n\}(x_1, \ldots, x_N) \\
&:= \sum_{\substack{1 \le i_1 \le \cdots \le i_n \le |h| : \\ i_j + |g_j| \le i_{j+1}}} \pm h(x_1, \ldots, x_{i_1-1}, g_1(x_{i_1}, \ldots, x_{i_1+|g_1|}), \\
&\qquad \ldots, x_{i_n-1}, g_n(x_{i_n}, \ldots, x_{i_n+|g_n|}), \ldots, x_N). \tag{27}
\end{aligned}
$$

3.4 The Cyclic A_∞ Case

We assume that we have an A_∞-algebra A which is Frobenius in the sense that there is a nondegenerate symmetric inner product such that the higher multiplications μ_n are all cyclic with respect to the inner product. These are sometimes called cyclic A_∞-algebras, see [Ko2].

Theorem 3.13. *The cyclic A_∞ conjecture holds.*

Sketch of proof. For the proof of this statement use spine-decorated stable trees, that is, trees in $\mathcal{T}_\infty$ together with a spine decoration. First they give compatible operations, and second, they index a cell model of the framed little discs. Both these claims follow from constructions completely analogous to the ones above. □

4 The Moduli Space vs. the Sullivan PROP

There are two generalizations of interest for the construction of the previous paragraph. The first is given by generalizing the restriction In to Out to the case of several "Out"s, and the second is given by going to the full moduli space. Surprisingly, these lead to slightly different results. The first route leads one into the realm of Penner's combinatorial compactification, and it fits perfectly with the algebra of the Hochschild complex. However, it does not exhaust moduli space. Alternatively, one can expand to moduli space and even omit invoking the compactification, but the price one pays is in terms of further construction on the Hochschild side to make things match.

4.1 Ribbon Graphs and Arc Graphs

A Short Introduction to the Arc Operad

In this section, we start by giving a brief review of the salient features of the $\mathcal{A}rc$ operad of [KLP] which is reasonably self-contained. The presentation of the material closely follows Appendix B of [K1]. For full details, we refer to [KLP]. In addition to this review, we furthermore introduce an equivalent combinatorial language which will be key for the following, in particular for [K5]. Simultaneously, we introduce new cell-level structures and then go on to define new cell-level operads and extensions of the $\mathcal{A}rc$ operad structure.

4.2 Spaces of Graphs on Surfaces

Fix an oriented surface $F_{g,r}^s$ of genus g with s punctures and r boundary components which are labeled from 0 to $r-1$, together with marked points on the boundary, one for each boundary component. We call this data F for short if no confusion can arise.

The piece of the $\mathcal{A}rc$ operad supported on F will be an open subspace of a space $\mathcal{A}^s_{g,r}$. The latter space is a CW complex whose cells are indexed by graphs on the surface $F^s_{g,r}$ up to the action of the pure mapping class group PMC, which is the group of orientation-preserving homeomorphisms of $F^s_{g,r}$ modulo homotopies that pointwise fix the set which is the union of the set of the marked points on the boundary and the set of punctures. A quick review in terms of graphs follows.

Embedded Graphs

By an embedding of a graph Γ into a surface F, we mean an embedding $i : |\Gamma| \to F$ with the conditions

(i) Γ has at least one edge.
(ii) The vertices map bijectively to the marked points on the boundaries.
(iii) No images of two edges are homotopic to each other by homotopies fixing the endpoints.
(iv) No image of an edge is homotopic to a part of the boundary, again by homotopies fixing the endpoints.

Two embeddings are equivalent if there is a homotopy of embeddings of the above type from one to the other. Note that such a homotopy is necessarily constant on the vertices.

The images of the edges are called arcs. And the connected components of $F \setminus i(\Gamma)$ are called complementary regions.

Changing representatives in a class yields natural bijections of the sets of arcs and connected components of $F \setminus i(\Gamma)$ corresponding to the different representatives. We can therefore associate to each equivalence class of embeddings its sets of arcs together with their incidence conditions and connected components—strictly speaking, of course, the equivalence classes of these objects.

Definition 4.1. By a graph γ on a surface we mean a triple $(F, \Gamma, [i])$, where $[i]$ is an equivalence class of embeddings of Γ into that surface. We will denote the isomorphism class of complementary regions by $\mathrm{Comp}(\gamma)$. We will also set $|\gamma| = |E_\Gamma|$. Fixing the surface F, we will call the set of graphs on a surface $\mathcal{G}(F)$.

A Linear Order on Arcs

Notice that due to the orientation of the surface, the graph inherits an induced linear order of all the flags at every vertex $F(v)$ from the embedding. Furthermore, there is even a linear order on all flags by enumerating the flags first according to the boundary components on which their vertex lies and then according to the linear order at that vertex. This induces a linear order on all edges by enumerating the edges by the first appearance of a flag of that edge.

The Poset Structure

The set of such graphs on a fixed surface F is a poset. The partial order is given by writing $(F, \Gamma', [i']) \prec (F, \Gamma, [i])$ if Γ' is a subgraph of Γ with the same vertices and $[i']$ is the restriction of $[i]$ to Γ'. In other words, the first graph is obtained from the second by deleting some arcs.

We associate a simplex $\Delta(F, \Gamma, [i])$ to each such graph, where Δ is the simplex whose vertices are given by the set of arcs/edges enumerated in their linear order. The face maps are then given by deleting the respective arcs. This allows us to construct a CW complex out of this poset.

Definition 4.2. Fix $F = F_{g,n}^s$. The space $\mathcal{A}_{g,n}^{\prime s}$ is the space obtained by gluing the simplices $\Delta(F, \Gamma', [i'])$ for all graphs on the surface according to the face maps.

The pure mapping class group naturally acts on $\mathcal{A}_{g,n}^{\prime s}$ and has finite isotropy [KLP].

Definition 4.3. The space $\mathcal{A}_{g,r}^s$ is defined to be $\mathcal{A}_{g,r}^{\prime s}/\text{PMC}$.

CW Structure of $\mathcal{A}_{g,r}^s$

Definition 4.4. Given a graph on a surface, we call its PMC orbit its arc graph. If γ is a graph on a surface, we denote by $\bar{\gamma}$ its arc graph or PMC orbit. We denote the set of all arc graphs of a fixed surface F by $\overline{\mathcal{G}}(F)$. A graph is called exhaustive if there are no vertices v with $\text{val}(v) = 0$. This condition is invariant under PMC, and hence we can speak about exhaustive arc graphs. The set of all exhaustive arc graphs on F is denoted by $\overline{\mathcal{G}}^e(F)$.

Notice that since the incidence conditions are preserved, we can set $|\bar{\gamma}| = |\gamma|$, where γ is any representative, and likewise define $\text{Comp}(\bar{\gamma})$. We call an arc graph exhaustive if it contains no isolated vertices, that is, vertices with $\text{val}(v) = 0$.

Now by construction it is clear that $\mathcal{A}_{g,r}^s$ is realized as a CW complex that has one cell for each arc graph $\bar{\gamma}$ of dimension $|\gamma| - 1$. Moreover, the cell for a given class of graphs is actually a map of a simplex whose vertices correspond to the arcs in the order discussed above. The attaching maps are given by deleting edges and identifying the resulting face with its image. Due to the action of PMC, some of the faces might become identified by these maps, so that the image will not necessarily be a simplex. The open part of the cell will, however, be an open simplex. Let $C(\bar{\alpha})$ be the image of the cell, and $\dot{C}(\bar{\alpha})$ its interior. Then

$$\mathcal{A}_{g,r}^s = \cup_{\bar{\alpha} \in \overline{\mathcal{G}}(F_{g,r}^s)} C(\bar{\alpha}), \quad \mathcal{A}_{g,r}^s = \amalg_{\bar{\alpha} \in \overline{\mathcal{G}}(F_{g,r}^s)} \dot{C}(\bar{\alpha}). \tag{28}$$

Let Δ^n denote the standard n-simplex and $\dot{\Delta}$ its interior. Then $\dot{C}(\gamma) = \mathbb{R}_{>0}^{|E_\Gamma|}/\mathbb{R}_{>0} = \dot{\Delta}^{|E_\Gamma|-1} =: C(\Gamma)$, which depends only on the underlying graph Γ of γ.

This also means that the space $\mathcal{A}_{g,r}^s$ is filtered by the cells of dimension less than or equal to k. We will use the notation $(\mathcal{A}_{g,r}^s)^{\leq k}$ for the pieces of this filtration.

Open-Cell Cell Complex

It is clear by construction that the $\mathcal{A}rc$ operad again has a decomposition into open cells:

$$\mathcal{A}rc_g^s(n) = \amalg_{\gamma \in \overline{\mathcal{G}}^e}(F_{g,n+1}^s)\dot{C}(\gamma). \tag{29}$$

Again $\dot{C}(\gamma) = \mathbb{R}_{>0}^{|E_\Gamma|}/\mathbb{R}_{>0} = \dot{\Delta}^{|E_\Gamma|-1} := \dot{C}(\Gamma)$ depends only on the underlying graph Γ of γ.

We will denote the free Abelian group generated by the $C(\alpha)$ as above by $\mathcal{C}_o^*(\mathcal{A}rc)_g^s(n)$. We will write $\mathcal{C}_o^*(\mathcal{A}rc)(n) = \amalg_{g,s}\mathcal{C}_o^*(\mathcal{A}rc)_g^s(n)$ and $\mathcal{C}_o^*(\mathcal{A}rc) = \amalg_n\mathcal{C}_o^*(\mathcal{A}rc)(n)$. We choose the notation to reflect the fact that we are strictly speaking not dealing with cellular chains; however, see [K4].

The group $\mathcal{C}_o^*(\mathcal{A}rc)(n)$ is also graded by the dimension of the cells; we will write $\mathcal{C}_o^*(\mathcal{A}rc)(n)^k$ for the subgroup generated by cells of dimension k, and we will also write $\mathcal{C}_o^*(\mathcal{A}rc)(n)^{\leq k}$ for the subgroup of cells of dimension $\leq k$. It is clear that $\mathcal{C}_o^*(\mathcal{A}rc)(n)^{\leq k}$ induces a filtration on $\mathcal{C}_o^*(\mathcal{A}rc)(n)$ and that the associated graded is isomorphic to the direct sum of the $\mathcal{C}_o^*(\mathcal{A}rc)(n)^k$:

$$\mathrm{Gr}(\mathcal{C}_o^*(\mathcal{A}rc)) := Gr\left(\mathcal{C}_o^*(\mathcal{A}rc)(n), \leq\right) \simeq \bigoplus_k \mathcal{C}_o^*(\mathcal{A}rc)^k(n). \tag{30}$$

The differential ∂ of $\mathcal{A}_{g,r}^s$ also descends to $\mathcal{C}_o^*(\mathcal{A}rc)$ and $\mathrm{Gr}(\mathcal{C}_o^*(\mathcal{A}rc))$ by simply omitting the cells which are not in $\mathcal{A}rc$. Applying the differential twice will kill two arcs, and each original summand will either be twice treated as zero or appear with opposite sign as in $\mathcal{A}_{g,r}^s$. Hence the differential squares to zero.

Relative Cells

The complex $\mathcal{C}_o^*(\mathcal{A}rc)_g^s(n)$ and the isomorphic complex $\mathrm{Gr}(\mathcal{C}_o^*(\mathcal{A}rc))_G^s(n)$ can be identified with the complex of relative cells $CC_*(A, A \setminus \mathcal{A}rc)$.

Elements of the $\mathcal{A}_{g,r}^s$ as Projectively Weighted Graphs

Using barycentric coordinates for the open part of the cells, the elements of $\mathcal{A}_{g,r}^s$ are given by specifying an arc graph together with a map w from the edges of the graph E_Γ to $\mathbb{R}_{>0}$ assigning a weight to each edge such that the sum of all weights is 1.

Alternatively, we can regard the map $w : E_\Gamma \to \mathbb{R}_{>0}$ as an equivalence class under the equivalence relation $w \sim w'$ if $\exists \lambda \in \mathbb{R}_{>0} \forall e \in E_\Gamma \ w(e) = \lambda w'(e)$. That is, w is a projective metric. We call the set of $w(e)$ the projective weights of the edges. In the limit, when the projective weight of an edge goes to zero, the edge/arc is deleted; see [KLP] for more details. For an example see Figure 10, which is discussed below.

An element $\alpha \in \mathcal{A}_{g,r}^s$ can be described by a tuple $\alpha = (F, \Gamma, \overline{[i]}, w)$, where F and Γ are as above, $\overline{[i]}$ is a PMC orbit of an equivalence class of embeddings, and w is a projective metric for Γ. Alternatively, it can be described by a tuple $(\bar{\gamma}, w)$, where $\bar{\gamma} \in \overline{\mathcal{G}}(F)$ and w is a projective metric for the underlying abstract graph Γ.

Example 4.5. $\mathcal{A}_{0,2}^0 = S^1$. Up to PMC there is a unique graph with one edge and a unique graph with two edges. The former gives a zero-cell and the latter gives a one-cell whose source is a 1-simplex. Its two subgraphs with one edge that correspond to the boundary lie in the same orbit of the action of PMC and thus are identified to yield S^1. The fundamental cycle is given by Δ of Figure 10.

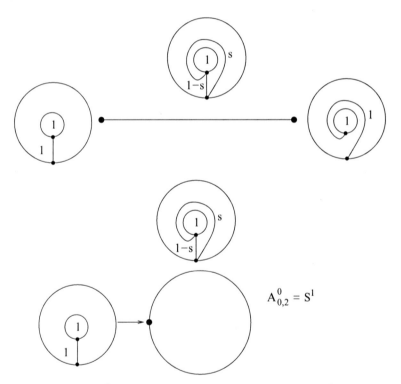

Fig. 10. The space $\mathcal{A}_{0,2}^0$ is given as the CW decomposition of S^1 with one 0-cell and one 1-cell. It can be thought of as the quotient of the interval in which the endpoints are identified by the action of the pure mapping class group. The generator of $CC_*(S^1)$ is called Δ.

4.3 Topological Operad Structure

The Spaces $\mathcal{A}rc(n)$

We begin by reviewing the construction of [KLP]. We then recast it in a purely combinatorial way. This will allow us to define the actions of [K5] more simply, but also allow us to show that although $\mathcal{A}rc_\#$ is not an operad on the topological level, it is a rational operad and gives rise to a cellular operad.

Definition 4.6. We define $\mathcal{A}rc_g^s(n) \subset \mathcal{A}_{g,n+1}^s$ to be the subset of those weighted arc graphs whose arc graph is exhaustive. We define $\mathcal{A}rc(n) := \coprod_{s,g \in \mathbb{N}} \mathcal{A}rc_g^s(n)$.[6]

Notice that the space $\mathcal{A}rc(n)$ carries a natural operation of $\mathbb{S}_n$ which permutes the labels $\{1, \ldots, n\}$ and one of $\mathbb{S}_{n+1}$ which permutes the labels $\{0, \ldots, n\}$. Also notice that the spaces $\mathcal{A}rc_g^s(n)$ inherit the grading and filtration from $\mathcal{A}_g^s(n)$. This is also true for their unions $\mathcal{A}rc(n)$, and we will write $\mathcal{A}rc(n)^{\leq k}$ for these pieces. That is, if $\alpha \in \mathcal{A}rc(n)^{\leq k}$ then $|E(\Gamma(\alpha))| \leq k + 1$.

Topological Description of the Glueing [KLP]

To give the composite $\alpha \circ_i \alpha'$ for two arc families $\alpha = (F, \Gamma, \overline{[i]}, w) \in \mathcal{A}rc(m)$ and $\alpha' = (F', \Gamma', \overline{[i']}, w') \in \mathcal{A}rc(n)$ one most conveniently chooses metrics on F and F'. The construction does not depend on the choice. With this metric, one produces a partially measured foliation in which the arcs are replaced by bands of parallel leaves (parallel to the original arc) of width given by the weight of the arc. For this we choose the window representation and also make the window tight in the sense that there is no space between the bands and between the endpoints of the window and the bands. Finally, we put in the separatrices. The normalization we choose is that the sum of the weights at boundary i of α coincides with the sum of the weights at the boundary 0; we can also fix them both to be one. Now when glueing the boundaries, we match up the windows, which have the same width, and then just glue the foliations. This basically means that we glue two leaves of the two foliations if they end on the same point. We then delete the separatrices. Afterward, we collect together all parallel leaves into one band. In this procedure, some of the original bands might be split or "cut" by the separatrices. We assign to each band one arc with weight given by the width of the consolidated band. If arcs occur, which do not hit the boundaries, then we simply delete these arcs. We call these arcs or bands "closed loops" and say that "closed loops appear in the glueing."

---

[6]Unfortunately there is a typo in the definition of $\mathcal{A}rc(n)$ in [KLP], where $\coprod$ was inadvertently replaced by the direct limit.

Theorem 4.7. *[KLP] Together with the glueing operations above, the spaces* $\mathcal{A}rc$ *form a cyclic operad.*

In [KLP] we furthermore obtained the following theorem.

Theorem 4.8. *[KLP] The chains of the $\mathcal{A}rc$ operad carry the structure of a GBV algebra up to homotopy. That is, it has a natural Gerstenhaber algebra structure up to homotopy and a BV operator up to homotopy, and they are compatible.*

The Dual Graph

Informally the dual graph of an element in $\mathcal{A}rc_\#$ is given as follows. The vertices are the complementary regions. Two vertices are joined by an edge if the complementary regions border the same arc. Due to the orientation of the surface, this graph is actually a ribbon graph via the induced cyclic order. Moreover, the marked points on the boundary make this graph into a marked ribbon graph. A more precise formal definition is given in [K4].

4.4 $\mathcal{D}\mathcal{A}rc$

The whole theory of arc graphs can be looked at in two flavors, either with projective metrics as we did, or with metrics proper, that is, without modding out by the overall scaling. This results in a completely equivalent theory. Here the operad $\mathcal{A}rc$ is replaced by the operad $\mathcal{D}\mathcal{A}rc$, where the "D" stands for "deprojectivized."

The Relation to Moduli Space

An interesting subspace of $\mathcal{D}\mathcal{A}rc$ is the space $\mathcal{D}\mathcal{A}rc_\#$, which consists of the arc graphs whose complementary regions are all polygons.

Theorem 4.9. *[K4] The space $\mathcal{D}\mathcal{A}rc_\#$ is equivalent to $M_{g,n+1}^{1^{n+1}}$, that is, the moduli space of curves of genus g with n marked points and a tangent vector at each of these points. The glueing operations on $\mathcal{D}\mathcal{A}rc$ induce the structure of a rational operad on $M_{g,n+1}^{1^{n+1}}$.*

4.5 Cells

There are several cell models hidden in this construction. First $\mathcal{A}$ is a cell complex from the start.

Second, we wish to point out that the arc graphs actually index cells of a relative cell complex. This is in complete analogy to the graph complex that describes the moduli space $M_{g,n}$ [Ko1, CV], with the addition that we are not dealing with a projectivized version, since the tangent vectors have real lengths.

4.6 Digraphs and Sullivan Chord Diagrams

Ribbon Digraphs

A ribbon graph is a digraph Γ together with a $\mathbb{Z}/2\mathbb{Z}$ labelling of the cycles of Γ: i/o : {cycles of Γ} $\to \mathbb{Z}/2\mathbb{Z}$. We call the cycles $i/o^{-1}(0) =:$ Out$_\Gamma$ the outgoing ones and $i/o^{-1}(1) =:$ In$_\Gamma$ the incoming ones. A digraph is said to be of type (n, m) if $|\text{In}_\Gamma| = n$ and $|\text{Out}_\Gamma| = m$. We will denote the set of these graphs by $\mathcal{R}ib^{i/o}$.

A ribbon digraph is called perfectly partitioned if $i/o(\imath(f)) = 1 - i/o(f)$ for every flag f. That is, each edge is part of one input and one output cycle. We will call the set of these graphs $\mathcal{R}ib^{i\leftrightarrow o}$.

An (S_1, S_2)-labeled ribbon digraph is a ribbon digraph together with bijective maps In $\to S_1$ and Out $\to S_2$. We denote the induced map on In $\amalg$ Out by Lab. If (S_1, S_2) is not mentioned, we will use $S_1 = \bar{n}$ and $S_2 = \bar{m}$ as the default indexing sets for a graph of type (n, m).

Sullivan Chord and Ribbon Diagrams

There are many definitions of Sullivan chord diagrams in the literature; we will use the following conventions.

Definition 4.10. A Sullivan chord diagram is a marked labeled ribbon digraph which satisfies the following condition:

(i) after deleting the edges of the incoming cycles one is left with a forest, i.e., a possibly disconnected set of contractible graphs.

Remark 4.11. In terms of the dual arc picture, this means that there is a partition of the boundary components of the surface into In and Out and arcs only run from In to Out and Out to Out. A complete list of all versions of Sullivan chord diagrams and their dual $\mathcal{A}rc$ pictures can be found in [K4].

The most important candidate for us will be a homotopically equivalent version of contracted diagrams.

Definition 4.12. We let $\overline{\mathcal{A}rc}_1^{i\leftrightarrow o}$ be weighted arc graphs on surfaces with marked inputs and outputs such that

1. All arcs run from In to Out.
2. The sum of the weights on each In boundary is 1.

The importance of this space is that it is the analogue of the normalized cacti, that is, it gives a cell model for the Sullivan PROP.

Theorem 4.13. [K4] The subspaces $\overline{\mathcal{A}rc}_1^{i\leftrightarrow o}$ when bigraded by the number of In and Out boundaries and endowed with the symmetric group actions permuting the labels form a topological quasi-PROP, i.e., a PROP up to homotopy. It is naturally a CW complex whose cells are indexed by the corresponding graphs, and the induced quasi-PROP structure on the cell level is already a PROP structure.

4.7 Graph Actions, Feynman Rules, and Correlation Functions

Operadic Correlation Functions

In this section, we introduce operadic correlation functions, which can be thought of as the generalization of an algebra over a cyclic operad to the dg setting. In order to get to the main definition, we first set up some notation.

Given a pair (A, C) where A is a vector space and $C = \sum c^{(1)} \otimes c^{(2)} \in A \otimes A$, we define the following operations:

$$\circ_i : \mathrm{Hom}(A^{\otimes n+1}, k) \otimes \mathrm{Hom}(A^{\otimes m+1}, k) \to \mathrm{Hom}(A^{\otimes n+m}, k), \qquad (31)$$

where for $\phi \in \mathrm{Hom}(A^{\otimes n+1}, k)$ and $\psi \in \mathrm{Hom}(A^{\otimes m+1}, k)$,

$$\phi \circ_i \psi(a_1 \otimes \cdots \otimes a_{n+m})$$
$$= \sum \phi(a_1 \otimes \cdots \otimes a_{i-1} \otimes c^{(1)} \otimes a_{i+m} \otimes \cdots \otimes a_{m+n}) \psi(c^{(2)} \otimes a_i \otimes \cdots \otimes a_{i+m-1}). \qquad (32)$$

Definition 4.14. A set of operadic correlation functions for a cyclic linear operad $\mathcal{O}$ is a tuple $(A, C, \{Y_n\})$, where A is a vector space, $C = \sum c^{(1)} \otimes c^{(2)} \in A \otimes A$ is a fixed element, and $Y_{n+1} : \mathcal{O}(n) \to \mathrm{Hom}(A^{\otimes n+1}, k)$ is a set of multilinear maps. The maps $\{Y_n\}$ should be $\mathbb{S}_{n+1}$ equivariant, and for $op_n \in \mathcal{O}(n), op_m \in \mathcal{O}(m)$,

$$Y_{n+m}(op_n \circ_i op_m) = Y_{n+1}(op_n) \circ_i Y_{m+1}(op_m), \qquad (33)$$

where the $\circ_i$ on the left is the multiplication of equation (31) for the pair (A, C).

We call the data $(A, \{Y_n\})$ of an algebra and the $\mathbb{S}_{n+1}$ equivariant maps correlation functions or simply correlators for $\mathcal{O}$.

Example 4.15. *Correlators for algebras over cyclic operads.* An example is given by an algebra over a cyclic operad. Recall that this a triple $(A, \langle\ ,\ \rangle, \{\rho_n\})$, where A is a vector space, $\langle\ ,\ \rangle$ is a non–degenerate bilinear pairing, and $\rho_n : \mathcal{O}(n) \to \mathrm{Hom}(A^{\otimes n}, A)$ are multilinear maps, also called correlators, that satisfy

(i) $\rho(op_n \circ_i op_m) = \rho(op_n) \circ_i \rho(op_m)$, where $\circ_i$ is the substitution in the ith variable.
(ii) The induced maps $Y_{n+1} : \mathcal{O}(n) \to \mathrm{Hom}(A^{\otimes n+1}, k)$ given by

$$Y_{n+1}(op_n)(a_0 \otimes \cdots \otimes a_n) := \langle a_0, \rho(op_n)(a_1 \otimes \cdots \otimes a_n)\rangle \qquad (34)$$

are $\mathbb{S}_{n+1}$ equivariant.

Notation 4.16. Given a finite dimensional vector space A with a non-degenerate pairing $\langle\ ,\ \rangle = \eta \in \check{A} \otimes \check{A}$, let $C \in A \otimes A$ be dual to η under the isomorphism induced by the pairing and call it the Casimir element. It has the following explicit expression: Let e_i be a basis of V, let $\eta_{ij} := \langle e_i, e_j\rangle$ be the matrix of the metric, and let η^{ij} be the inverse matrix. Then $C = \sum_{ij} e_i \eta^{ij} \otimes e_j$.

4.8 Operadic Correlation Functions with Values in a Twisted $\mathcal{H}$om Operad

Definition 4.17. Let $(A, \langle\ ,\ \rangle, \{Y_n\})$ be as above. And let $\mathcal{H} = \{\mathcal{H}(n)\}$ with $\mathcal{H}(n) \subset \mathrm{Hom}(A^{\otimes n}, A)$ as k-modules be an operad where the $\mathbb{S}_n$ action is the usual action, but the operad structure is *not necessarily* the induced operad structure. Furthermore, assume that $\rho_{Y_{n+1}} \in \mathcal{H}(n)$. We say that the $\{Y_n\}$ are operadic correlation functions for $\mathcal{O}$ with values in $\mathcal{H}$ if the maps ρ are operadic maps from $\mathcal{O}$ to $\mathcal{H}$. We will also say that we get an action of $\mathcal{O}$ with values in $\mathcal{H}$.

Signs

As in the case of the Deligne conjecture, one twist which we have to use is dictated by picking sign rules. In the case of Deligne's conjecture this could be done by mapping to the brace operad $\mathcal{B}race$ (see, e.g., [K2]) or by twisting the operad $\mathcal{H}$om by lines of degree 1 (see, e.g., [KS]). In what follows, our actions will take values on operads that are naturally graded, *and moreover*, we will identify the grading with the geometric grading by, e.g., the number of edges or the number of angles. The signs will then automatically match up if we use the procedure at the same time for *both* the graph side and the $\mathcal{H}$om side, i.e., for the operad $\mathcal{H}$. In fact, this approach unifies the two sign conventions mentioned above on the subspace of operations corresponding to $\mathcal{L}Tree_{cp}$.

Definition 4.18. A quasi-Frobenius algebra is a triple $(A, d, \langle\ ,\ \rangle)$, where (A, d) is a unital *dg*-algebra whose homology algebra $H := H(A, d)$ is finite-dimensional and has a non–degenerate pairing $\langle\ ,\ \rangle$ and is a Frobenius algebra for this pairing. A quasi-Frobenius algebra with an integral is a triple $(A, d, \int)$, where $\int : A \to k$ is a linear map such that

(i) $\forall a \in A : \int da = 0$;

(ii) $(A, d, \langle\ ,\ \rangle)$ is a quasi-Frobenius algebra, where $\langle a, b \rangle := \int ab$. The cocycles of a quasi-Frobenius algebra with an integral are the subalgebra $Z = \ker(d) \subset A$ of the algebra above.

4.9 $\mathcal{A}rc^{\angle}$ Correlation Functions

In order to present the correlation functions, we need to partition the arc graphs and endow them with angle markings. Given an arc graph α, it gives rise to a formal sum of arc graphs $\mathcal{P}(\alpha)$, where each summand is obtained from α by inserting finitely many parallel edges. See Figure 11 for one such summand. This operation is the analogue of the foliage operator. An angle marking is an angle marking of the arc graph. The corresponding space is called $\mathcal{A}^{\angle}$. In keeping with the notation already in place, $\mathcal{A}rc^{\angle}$ is the subspace of graphs that hit all boundaries, and elements of $\mathcal{A}rc^{\angle}_{\#}$ are also quasifilling.

Given an arc graph, there are two standard angle markings. The first marks all angles by 0 except the angles spanned by the smallest and biggest element at each boundary. The second marking marks all angles by 1. When partitioning an angle-marked graph, we mark all new angles by 1.

The idea of how to obtain the correlation functions for the tensor algebra is very nice in the $\mathcal{A}rc$ picture, where it is based on the polygon picture. This polygon picture can be thought of as an IRF (interaction 'round a face) picture for a grid on a surface which is dual to the ribbon picture. For this we would modify the arc graph by moving the arcs a little bit apart as described. Then the complementary regions of partitioned quasifilling arc graphs, denoted by $\mathcal{P}\overline{\mathcal{G}}_\#$, are $2k$-gons whose sides alternately correspond to arcs and pieces of the boundary. The pieces of the boundary correspond to the angles of the graph, and of course any polygonal region corresponds to a cycle of the arc graph. If the graph α^p has an angle marking, then the sides of the polygons corresponding to the boundaries will also be marked. We fix the following notation. For an angle-marked partitioned arc graph α^p that is quasifilling, let Poly(α^p) be the set of polygons given by the complementary regions of α^p when treated as above. See Figure 8 for an example. For $\pi \in$ Poly(α^p), let Sides$'$ be the sides corresponding to the angles which are marked by 1, and Sides$'(\alpha^p)$ be the union of all of these sides. If we set $\angle^+(\Gamma) = (\mathrm{mk}^\angle)^{-1}(1)$, let there is a natural bijection between $\angle^+(\alpha^p)$ and Sides$'(\alpha^p)$.

For some purposes it is convenient to contract the edges of the $2k$-gon that belong to pieces of the boundary and label the resulting vertex by the corresponding boundary label.

Correlation Functions on the Tensor Algebra of an Algebra

Fix an algebra A with a cyclic trace, i.e., a map $\int : A \to k$ which satisfies $\int a_1 \cdots a_n = \pm \int a_n a_1 \cdots a_{n-1}$, where $\pm$ is the standard sign.

Now for $\pi \in$ Poly(α^p), set

$$Y(\pi)\left(\bigotimes_{s\in\mathrm{Sides}'(\pi)} a_s\right) = \int \prod_{s\in\mathrm{Sides}'(\pi)} a_s. \tag{35}$$

Notice that we only have a cyclic order for the sides of the polygon, but $\int$ is (super)-invariant under cyclic permutations, so that if we think of the tensor product and the product as indexed by sets (35), it is well defined.

For an angle marked partitioned arc family α^p set

$$Y(\alpha^p)\left(\bigotimes_{s\in(\mathrm{mk}^\angle)^{-1}(1)} a_s\right) = \bigotimes_{\pi\in\mathrm{Poly}(\alpha^p)} Y(\pi)\left(\bigotimes_{s\in\mathrm{Sides}'(\pi)} a_s\right), \tag{36}$$

where we used the identification of the set Sides$'(\alpha^p) = \amalg_{\pi\in\mathrm{Poly}(\alpha^p)}\mathrm{Sides}'(\pi)$ with $\angle^+(\alpha^p)$. Since for each $\alpha^p \in \mathcal{P}\angle\overline{\mathcal{G}}^e(n)$ the set of all flags has a linear order, we can think of $Y(\alpha^p)$ as a map $A^{\otimes|F(\alpha^p)|} = \bigotimes_{i=1}^{n} A^{\otimes|F(v_i)|} \to k$ and furthermore as a map to $TA^{\otimes n} \to k$ by letting it be equal to equation (36) as a map from $\bigotimes_{i=1}^{n} A^{\otimes|F(v_i)|} \subset TA^{\otimes n}$ and setting it to zero outside of this subspace.

Extending linearly, for an angle-marked arc family $\alpha \in \mathcal{A}rc^{\angle}$, we finally define

$$Y(\alpha) := Y(\mathcal{P}(\alpha)). \tag{37}$$

Correlators for the Hochschild Cochains of a Frobenius Algebra

Let A be an algebra and let $C^n(A, A) = \mathrm{Hom}(A^{\otimes n}, A)$ be the Hochschild cochain complex of A. We denote the cyclic cochain complex by $CC^n(A, k) = \mathrm{Hom}(A^{\otimes n+1}, k)$. Then one has a canonical isomorphism of $CC^*(A) \cong C^*(A, \check{A})$ as complexes and hence also $HC^*(A) \cong H^*(A, A)$, where HC is Connes' cyclic cohomology and H is the Hochschild cohomology.

Lemma 4.19. *For any Frobenius algebra $(A, \langle \, , \, \rangle)$, we have canonical isomorphisms $CC^*(A) \cong C^*(A, \check{A}) \cong C^*(A, A)$ and $HC^*(A) \cong H^*(A, A) \cong H^*(A, \check{A})$ induced by the isomorphism of A and $\check{A}$ which is defined by the non-degenerate pairing of A.*

Proof. The only statement to prove is the last isomorphism. As mentioned, the map on the chain level is induced by the isomorphism of A and $\check{A}$ defined by the nondegenerate pairing of A. The fact that the complexes are isomorphic follows from the well-known fact that the invariance of the pairing $\langle ab, c \rangle = \langle a, bc \rangle$ implies that the isomorphism between A and $\check{A}$ is an isomorphism of A bimodules, where the bimodule structure of functions $f \in \check{A}$ is given by $a'fa''(c) = f(a''ca')$; see, e.g., [L]. $\qquad \square$

For any $f \in C^n(A, A)$ let $\tilde{f} \in \check{A}^{\otimes n}$ be its image under the isomorphism of $\check{A}$ with A defined by the Frobenius structure of A.

Given pure tensors $f_i = f^{0i} \otimes f_{1i} \otimes \cdots \otimes f_{in_i} \in C^{n_i}(A, A), i \in \{0, \ldots, n\}$, we write $\tilde{f}_i = f_{0i} \otimes \cdots \otimes f_{in_i}$ for their image in $CC^{n_i}(A)$. Fix $\alpha \in \mathcal{A}rc^{\angle}(n)$. Now decorate the sides $s \in \mathrm{Sides}'(\alpha) := (\mathrm{mk}^{\angle})^{-1}(1)$ of the complementary regions, which correspond to pieces of the boundary, by elements of A as follows: for a side $s \in \mathrm{Sides}'$ let j, be its position in its cycle c_i counting only the sides of c_i in Sides' starting at the side corresponding to the unique outside angle at the boundary given by the cycle. If the number of such sides at the boundary i is $n_i + 1$, then set $f_s := f_{ij}$.

Now we set

$$Y(\alpha)(f_1, \ldots, f_n) := Y(\mathcal{P}(\alpha)) \left(\bigotimes_{s \in \angle^+(\alpha^p)} f_s \right). \tag{38}$$

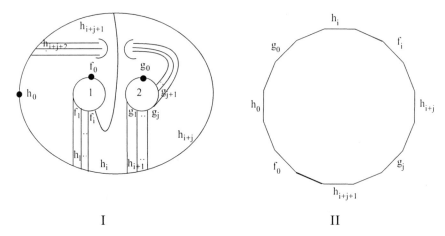

$$\text{I} \qquad\qquad\qquad\qquad \text{II}$$

Fig. 11. A partitioned arc graph with decorations by elements of A and one of its decorated polygons. The bold line corresponds to the bold edges.

We extend this definition by linearity if $f_i \in C^{n_i}(A, A), i \in \bar{n}$. If the condition that $n_i + 1$ equals the number of Sides' at the boundary i is not met, we set $Y(\alpha)(f_0, \ldots, f_n) = 0$. An example of a decorated partitioned surface and its polygons is given in Figure 11.

Theorem 4.20. *[K5] Let A be a Frobenius algebra and let $C(A, A)$ be the Hochschild complex of the Frobenius algebra. Then the cyclic chain operad of the open cells of $Arc^{\angle}$ acts on $C(A, A)$ via correlation functions. Hence so do all the suboperads, subdioperads, and PROPs of [K4] mentioned in the introduction. In particular, the graph complex of $M_{g,n+1}^{1^{n+1}}$, the Moduli space of pointed curves with fixed tangent vectors at each point, act on $\mathrm{CH}(A, A)$ by its two embeddings into $Arc_\#^{\angle}$. Furthermore, there is a natural operad structure on the corresponding partitioned graphs $\mathcal{P}^{\angle}Arc_\#$, and for this operad structure the correlation functions are operadic correlation functions with values in $Gr\mathcal{CM}$. Moreover, the operations of the suboperad $Tree_{cp}$ correspond to the operations $\sqcup$ and $\square_i$ induced by Ξ_2 as defined in [MS3].*

The same formalism also yields operadic correlation functions for the tensor algebra of the cocycles of a differential algebra (A, d) over k with a cyclically invariant trace $\int : A \to k$ that satisfies $\int da = 0$ and whose induced pairing on $H = H(A, d)$ turns H into a Frobenius algebra, i.e., they are chain-level operadic correlation functions with values in $Gr\mathcal{CM}$.

Here $Gr\mathcal{CM}$ is the associated graded operad of a filtered suboperad of Hom, which is essentially generated by products, co–products, and shuffles.

Remark 4.21. We wish to point out that strictly speaking, Deligne's original conjecture also only yields correlation functions with values in the $Brace$ suboperad. This is due to the necessary fixing of signs.

The Sullivan–Chord Diagram Case

ASSUMPTION: For the rest of the discussion of this subsection let A be a commutative Frobenius algebra.

4.10 Correlators for $\mathcal{A}^{\measuredangle}$

In general, we extend the action as follows. Notice that given an arc graph α, each complementary region $S \in \mathrm{Comp}(G)$ has the following structure: it is a surface of some genus g with $r \geq 1$ boundary components whose boundaries are identified with a $2k$-gons. Alternating sides belong to arcs and boundaries as above, and the sides come marked with 1 or 0 by identifying them with the angles of the underling arc graph. Now let $\mathrm{Sides}'(S)$ be the sides which have an angle marking by 1 and let χ be the Euler characteristic of S. We set

$$Y(S)\left(\bigotimes_{s \in \mathrm{Sides}'(S)} a\right) := \int \left(\prod_{s \in \mathrm{Sides}'(S)} a_s\right) e^{-\chi+1}, \tag{39}$$

where $e := \mu(\Delta(1))$ is the Euler element. For an angle marked partitioned arc graph α^p we set

$$Y(\alpha^p)\left(\bigotimes_{S \in \mathrm{Comp}(\alpha_i)}\left(\bigotimes_{s \in \mathrm{Sides}'(S)} a_s\right)\right) = \bigotimes_{S \in \mathrm{Comp}(\alpha_i)} Y(S)\left(\bigotimes_{s \in \mathrm{Sides}'(S)} a_s\right) \tag{40}$$

Again, for $\alpha \in \mathrm{CC}_*(\mathcal{A}^{\measuredangle})$ we simply set

$$Y(\alpha) = Y(\mathcal{P}(\alpha)). \tag{41}$$

Theorem 4.22. *The $Y(\alpha)$ defined in equation (41) give operadic correlation functions for $\mathrm{CC}_*(\overline{\mathcal{A}rc}_1^{i \leftrightarrow o})$ and induces a dg-action of the dg-PROP $\mathrm{CC}_*(\overline{\mathcal{A}rc}_1^{i \leftrightarrow o})$ on the dg-algebra $\overline{\mathrm{CH}}^*(A, A)$ of reduced Hochschild cochains for a commutative Frobenius algebra A.*

The $Y(\alpha)$ also yield correlation functions on the tensor algebra of the cocycles of a differential algebra (A, d) over k with a cyclically invariant trace $\int : A \to k$ that satisfies $\int da = 0$ and whose induced pairing on $H = H(A, d)$ turns H into a Frobenius algebra. These correlation functions are operadic chain-level correlation functions.

Corollary 4.23. *The operadic correlation functions descend to give a PROP action of $H_*(\overline{\mathcal{A}rc}_1^{i \leftrightarrow o})$ on $H^*(A)$ for a commutative Frobenius algebra A.*

4.11 Application to String Topology

Let M be a simply connected compact manifold M and denote the free loop space by $\mathcal{L}M$ and let $C_*(M)$ and $C^*(M)$ be the singular chains and (co)-chains of M. We know from [J, CJ] that $C_*(\mathcal{L}M) = C^*(C^*(M), C_*(M))$ and $H_*(\mathcal{L}M) \simeq H^*(C^*(M), C_*(M))$. Moreover, $C^*(M)$ is an associative dg algebra with unit, differential d, and an integral (M was taken to be a compact manifold) $\int : C^*(M) \to k$ such that $\int d\omega = 0$. By using the spectral sequence and taking field coefficients, we first obtain operadic correlation functions Y for $\mathcal{T}ree$ on $E^1 \simeq C^*(H, H)$, where $H = H^*(M)$. The spectral sequence converges to $H_*(\mathcal{L}M)$ and the operadic correlation functions Y descend to induce an operadic action on the homology of the loop space. Except for the last remark, this was established in [K3].

Theorem 4.24. *When taking field coefficients, the above action gives a dg action of a dg-PROP of Sullivan chord diagrams on the E^1-term of a spectral sequence converging to $H_*(LM)$, that is, the homology of the loop space if a simply connected compact manifold and hence induces operations on this loop space.*

Proof. Recall from [CJ] that the isomorphism $C_*(\mathcal{L}M) = C^*(C^*(M), C_*(M))$ comes from dualizing the isomorphism $C_*(\mathcal{L}M) = C_*(C^*(M))[J]$. Calculating the latter with the usual bicomplex [L], we see that the E^1-term is given by $CH_*(H^*(M))$, and dualizing the corresponding E^1 spectral sequence, we get $CH^*(H^*(M), H_*(M))$, so we get an operation of the E^1 level. Since the operation of $\mathcal{T}ree$ was dg, it is compatible with the E^1 differential and hence gives an action on the convergent spectral sequence computing $H_*(\mathcal{L}M)$ and hence on its abutment.

5 Stabilization and Outlook

We have shown that the above methods are well suited to treat the double loop space nature of the Hochschild complex, string topology, and a moduli space generalization. The $\mathcal{A}rc$ operad is manifestly BV, and since it describes string topology, it should not go beyond the double loop space. To go to higher loop spaces we need a stabilization of the arc operad. In the following, we will give an outlook of the results we aim to prove in the higher loop case.

In this section $s = 0$.

Definition 5.1. The elements in the complement of $\mathcal{A}rc_\#$ are called non-effective. Let $\mathcal{A}rc^{\text{ctd}}$ be the suboperad of connected arc families.

Definition 5.2. We define $St\mathcal{A}rc_0(n) := \varinjlim \mathcal{A}rc^{\text{ctd}}$, where the limit is taken with respect to the system $\alpha \to \alpha \circ_i Op_g, \alpha \to Op_g \circ_i \alpha$, where $Op_g \in \mathcal{A}rc_1^{\text{ctd}}(2)$ is non-effective.

Claim. The spaces $St\mathcal{A}rc_0(n)$ form an operad.

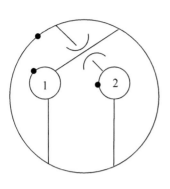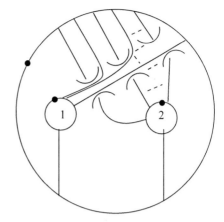

Fig. 12. The $\cup_2$ and the $\cup_i$ operations.

Claim. The operad *StArc0(n)* detects infinite loop spaces, i.e., if X admits an operadic action of *StArc0(n)*, then it has the homotopy type of an infinite loop space.

Sketch of proof. We can give a hemispherical construction a la Fiedorowicz by using the arc graphs for the $\cup_i$ products as given in Figure 12.

Corollary 5.3. *StArc$_0$(n) has the homotopy type of an infinite loop space.*

This can be compared to the theorems of Tillmann and Madsen on infinite loop spaces and Segal's approach to CFT.

Notice that the construction above uses only the tree part and indeed, we make the following claim.

Claim. The suboperad of stabilized linear Chinese trees (cf. [KLP]) has an operadic filtration *StGTree$_g$* in terms of effective genus. The operad linear *StGTree$_g$* is isomorphic to the little $2g$ cubes operad. That is, we get cells for the $\cup_i$-operations. A finer filtration gives all k-cubes.

This fits well with the slogan that strings yield all higher-dimensional objects. It also gives tools to describe the cells for the higher Dyer–Lashof–Cohen operations.

References

[AK] S. Araki, T. Kudo. *Topology of H_n-spaces and H-squaring operations.* Mem. Fac. Sci. Kyūsyū Univ. Ser. A, 1956, pp. 85–120.

[BF] C. Berger and B. Fresse. *Une décomposition prismatique de l'opérade de Barratt-Eccles.* C. R. Math. Acad. Sci. Paris 335 (2002), no. 4, 365–370.

[Co] F. R. Cohen. *The homology of* CC_{n+1}*-spaces,* $n \geq 0$*, The homology of iter-ated loop spaces*, Lecture Notes in Mathematics 533, Springer-Verlag, Berlin, 1976, pp.207–351.

[Ch] D. Chataur. *A bordism approach to string topology*. Int. Math. Res. Not. 2005, no. 46, 2829–2875.

[CG] R. L. Cohen and V. Godin. *A polarized view of string topology*. Topology, geometry and quantum field theory, 127–154, London Math. Soc. Lecture Note Ser., 308, Cambridge Univ. Press, Cambridge, 2004.

[CJ] R. L. Cohen and J. D. S. Jones. *A homotopy theoretic realization of string topology*. Math. Ann. 324 (2002), no. 4, 773–798.

[CS] M. Chas and D. Sullivan. *String Topology*. Preprint math.GT/9911159. To appear in Ann. of Math.

[CV] J. Conant and K. Vogtmann. *On a theorem of Kontsevich*. Algebr. Geom. Topol. 3 (2003), 1167–1224.

[D] P. Deligne. *Catégories tannakiennes*. The Grothendieck Festschrift, Vol. II, 111–195.

[DL] E. Dyer, R. K. Lashof. *Homology of iterated loop spaces*. Amer. J. Math., 1962, **84**, 35–88.

[G] M. Gerstenhaber. *The cohomology structure of an associative ring*, Ann. of Math. 78 (1963), 267–288.

[Ge] E. Getzler. *Cartan homotopy formulas and the Gauss-Manin connection in cyclic homology*. Quantum deformations of algebras and their representations (Ramat-Gan, 1991/1992; Rehovot, 1991/1992), 65–78, Israel Math. Conf. Proc., 7, Bar-Ilan Univ., Ramat Gan, 1993.

[GV] A. A. Voronov and M. Gerstenkhaber. *Higher-order operations on the Hochschild complex*. (Russian) Funktsional. Anal. i Prilozhen. 29 (1995), no. 1, 1–6, 96; translation in Funct. Anal. Appl. 29 (1995).

[J] J. D. S. Jones. *Cyclic homology and equivariant homology*. Inv. Math. 87 (1987), 403–423.

[Kad] T. Kadeishvili. The structure of the A(∞)-algebra, and the Hochschild and Harrison cohomologies. Trudy Tbilis. Mat. Inst. Razmadze Akad. Nauk Gruzin. SSR 91 (1988).

[K1] R. M. Kaufmann. *On several varieties of cacti and their relations*. Algebraic & Geometric Topology 5 (2005), 2–300.

[K2] R. M. Kaufmann. *On Spineless Cacti, Deligne's Conjecture and Connes-Kreimer's Hopf Algebra*. Topology 46, 1 (2007), 39–88.

[K3] R. M. Kaufmann. *A proof of a cyclic version of Deligne's conjecture via Cacti*. Math. Res. Letters 15, 5(2008), 901–921.

[K4] R. M. Kaufmann. *Moduli space actions on the Hochschild CoChains of a Frobenius algebra I: Cell Operads*. Journal of Noncommutative Geometry 1, 3(2007) 333–384.

[K5] R. M. Kaufmann. *Moduli space actions on the Hochschild cochains of a Frobenius algebra II: Correlators* 2, 3(2008), 283–332.

[KP] R. M. Kaufmann and R. C. Penner. *Closed/open string diagrammatics*. Nucl. Phys. B748 (2006), 3, 335–379.

[KLP] R. M. Kaufmann, M. Livernet and R. C. Penner. *Arc Operads and Arc Algebras*. Geometry and Topology 7 (2003), 511–568.

[KSch] R. M. Kaufmann and R. Schwell. *Associahedra, Cyclohedra, and a Topological Solution to the A_∞-Deligne Conjecture*. Advances in Math. to appear.

[Ko1] M. Kontsevich. *Formal (non)commutative symplectic geometry*. The Gel'fand Mathematical Seminars, 1990–1992, 173–187, Birkhäuser Boston, Boston, MA, 1993.

[Ko2] M. Kontsevich. *Feynman diagrams and low-dimensional topology*. First European Congress of Mathematics, Vol. II (Paris, 1992), 97–121, Progr. Math., 120, Birkhäuser, Basel, 1994.

[Ko3] M. Kontsevich. *Operads and Motives in Deformation Quantization*. Lett. Math. Phys. 48 (1999) 35–72.

[KS] M. Kontsevich and Y. Soibelman. *Deformations of algebras over operads and Deligne's conjecture*. Conférence Moshé Flato 1999, Vol. I (Dijon), 255–307, Math. Phys. Stud., 21, Kluwer Acad. Publ., Dordrecht, 2000.

[KLi1] A. Kapustin and Y. Li. *D-branes in Landau-Ginzburg models and algebraic geometry*. JHEP **0312**, 005 (2003).

[KLi2] A. Kapustin and Y. Li. *Topological correlators in Landau-Ginzburg models with boundaries*. Preprint hep-th/0305136.

[KR] A. Kapustin and L. Rozansky. *On the relation between open and closed topological strings*. Commun. Math. Phys. 252 (2004) 393–414.

[L] J–L. Loday. *Cyclic Homology*. Appendix E by María O. Ronco. Second edition. Chapter 13 by the author in collaboration with Teimuraz Pirashvili. Grundlehren der Mathematischen Wissenschaften, 301. Springer-Verlag, Berlin, 1998.

[Me] S. A. Merkulov. *De Rham model for string topology*. Int. Math. Res. Not. 2004, no. 55, 2955–2981.

[MS1] J. E. McClure and J. H. Smith. *A solution of Deligne's Hochschild cohomology conjecture*. Recent progress in homotopy theory (Baltimore, MD, 2000), 153–193, Contemp. Math., 293, Amer. Math. Soc., Providence, RI, 2002.

[MS2] J. E. McClure and J. H. Smith. *Multivariable cochain operations and little n-cubes*. J. Amer. Math. Soc. 16 (2003), no. 3, 681–704.

[MS3] James E. McClure and Jeffrey H. Smith. *Cosimplicial objects and little n-cubes. I.* Amer. J. Math. 126 (2004), no. 5, 1109–1153.

[MS4] James E. McClure and Jeffrey H. Smith. *Operads and cosimplicial objects: an introduction*. Axiomatic, Enriched and Motivic Homotopy Theory. Proceedings of a NATO Advanced Study Institute. Edited by J.P.C. Greenlees. Kluwer 2004.

[MSS] M. Markl, S. Shnider and J. Stasheff. *Operads in algebra, topology and physics*. Mathematical Surveys and Monographs, 96. American Mathematical Society, Providence, RI, 2002.

[S1] D. Sullivan. *Sigma models and string topology*. Graphs and patterns in mathematics and theoretical physics, 1–11, Proc. Sympos. Pure Math., 73, Amer. Math. Soc., Providence, RI, 2005.

[S2] D. Sullivan. *Open and closed string field theory interpreted in classical algebraic topology*. Topology, geometry and quantum field theory, 344–357, London Math. Soc. Lecture Note Ser., 308, Cambridge Univ. Press, Cambridge, 2004.

[T] D. Tamarkin. *Another proof of M. Kontsevich formality theorem*. Preprint math/9803025.
Formality of Chain Operad of Small Squares. Lett. Math. Phys. 66 (2003), no. 1-2, 65–72.

[Tou] V. Tourtchine. *Dyer-Lashof-Cohen operations in Hochschild cohomology*. Algebr. Geom. Topol. 6 (2006), 875–894.

[TZ] T. Tradler and M. Zeinalian. *On the cyclic Deligne conjecture.* J. Pure Appl. Algebra 204 (2006), no. 2, 280–299.

[Vo1] A. A. Voronov. *Homotopy Gerstenhaber algebras.* Conférence Moshé Flato 1999, Vol. II (Dijon), 307–331, Math. Phys. Stud., 22, Kluwer Acad. Publ., Dordrecht, 2000.

[Vo2] A. A. Voronov. *Notes on universal algebra.* Graphs and Patterns in Mathematics and Theoretical Physics (M. Lyubich and L. Takhtajan, eds.), Proc. Sympos. Pure Math., vol. 73. AMS, Providence, RI, 2005, pp. 81–103.

[We] C. Westerlan. *Dyer-Lashof operations in the string topology of spheres and projective spaces.* Math. Z. 250 (2005), no. 3, 711–727.

Quotients of Calabi–Yau Varieties

János Kollár[1] and Michael Larsen[2]

[1] Department of Mathematics, Princeton University, Fine Hall, Washington Road
Princeton, NJ 08544-1000, U.S.A.
kollar@math.princeton.edu

[2] Department of Mathematics, Indiana University, Bloomington, IN 47405, U.S.A.
larsen@math.indiana.edu

Summary. Let X be a complex Calabi–Yau variety, that is, a complex projective variety with canonical singularities whose canonical class is numerically trivial. Let G be a finite group acting on X and consider the quotient variety X/G. The aim of this paper is to determine the place of X/G in the birational classification of varieties. That is, we determine the Kodaira dimension of X/G and decide when it is uniruled or rationally connected. If G acts without fixed points, then $\kappa(X/G) = \kappa(X) = 0$; thus the interesting case is when G has fixed points. We answer the above questions in terms of the action of the stabilizer subgroups near the fixed points. We give a rough classification of possible stabilizer groups which cause X/G to have Kodaira dimension $-\infty$ or equivalently (as we show) to be uniruled. These stabilizers are closely related to unitary reflection groups.

Key words: Calabi–Yau, uniruled, rationally connected, reflection group

2000 Mathematics Subject Classifications: 14J32, 14K05, 20E99 (Primary) 14M20, 14E05, 20F55 (Secondary)

Let X be a Calabi–Yau variety over $\mathbb{C}$, that is, a projective variety with canonical singularities whose canonical class is numericaly trivial. Let G be a finite group acting on X and consider the quotient variety X/G. The aim of this paper is to determine the place of X/G in the birational classification of varieties. That is, we determine the Kodaira dimension of X/G and decide when it is uniruled or rationally connected.

If G acts without fixed points, then $\kappa(X/G) = \kappa(X) = 0$; thus the interesting case is that in which G has fixed points. We answer the above questions in terms of the action of the stabilizer subgroups near the fixed points.

The answer is especially nice if X is smooth. In the introduction we concentrate on this case. The precise general results are formulated later.

Definition 1. *Let V be a complex vector space and $g \in \mathrm{GL}(V)$ an element of finite order. Its eigenvalues (with multiplicity) can be written as*

Y. Tschinkel and Y. Zarhin (eds.), *Algebra, Arithmetic, and Geometry*,
Progress in Mathematics 270, DOI 10.1007/978-0-8176-4747-6_6,
© Springer Science+Business Media, LLC 2009

$e(r_1), \ldots, e(r_n)$, where $e(x) := e^{2\pi i x}$ and $0 \le r_i < 1$. Following [IR96, Rei02], we define the age of g as

$$\mathrm{age}(g) := r_1 + \cdots + r_n.$$

Let G be a finite group and (ρ, V) a finite-dimensional complex representation of G. We say that $\rho : G \to \mathrm{GL}(V)$ satisfies the (local) Reid–Tai condition if $\mathrm{age}(\rho(g)) \ge 1$ for every $g \in G$ for which $\rho(g)$ is not the identity (cf. [Rei80, 3.1]).

Let G be a finite group acting on a smooth projective variety X. We say that the G-action satisfies the (global) Reid–Tai condition if for every $x \in X$, the stabilizer representation $\mathrm{Stab}_x(G) \to \mathrm{GL}(T_x X)$ satisfies the (local) Reid–Tai condition.

Our first result relates the uniruledness of X/G to the Reid–Tai condition.

Theorem 2. Let X be a smooth projective Calabi–Yau variety and G a finite group acting on X. The following are equivalent:

1. $\kappa(X/G) = 0$.
2. X/G is not uniruled.
3. The G-action satisfies the global Reid–Tai condition.

The equivalence of (1) and (3) is essentially in [Rei80, Sec. 3]. It is conjectured in general that being uniruled is equivalent to having Kodaira dimension $-\infty$. The main point of Theorem 2 is to establish this equivalence for varieties of the form X/G.

It can happen that X/G is uniruled but not rationally connected. The simplest example is that in which $X = X_1 \times X_2$ is a product, G acts trivially on X_1, and X_2/G is rationally connected. Then $X/G \cong X_1 \times (X_2/G)$ is a product of the Calabi–Yau variety X_1 and of the rationally connected variety X_2/G. We show that this is essentially the only way that X/G can be uniruled but not rationally connected. The key step is a description of rational maps from Calabi–Yau varieties to lower-dimensional nonuniruled varieties.

Theorem 3. Let X be a smooth, simply connected projective Calabi–Yau variety and $f : X \dashrightarrow Y$ a dominant map such that Y is not uniruled. Then one can write

$$f : X \xrightarrow{\pi} X_1 \dashxrightarrow{g} Y,$$

where π is a projection to a direct factor of $X \cong X_1 \times X_2$ and $g : X_1 \dashrightarrow Y$ is generically finite.

Note: C. Voisin pointed out that the smooth case discussed above also follows from the Beauville–Bogomolov–Yau structure theorem; see [Bea83b, Bea83a]. The structure theorem is conjectured to hold also for singular Calabi–Yau varieties. In any case, we prove the singular version Theorem 14 in Section 2 by other methods.

Applying this to the MRC-fibration of X/G, we obtain the following.

Corollary 4. *Let* X *be a smooth, simply connected projective Calabi–Yau variety that is not a nontrivial product of two Calabi–Yau varieties. Let* G *be a finite group acting on* X. *The following are equivalent:*

1. X/G *is uniruled.*
2. X/G *is rationally connected.*
3. *The* G-*action does not satisfy the global Reid–Tai condition.*

Next we turn our attention to a study of the local Reid–Tai condition. For any given representation it is relatively easy to decide whether the Reid–Tai condition is satisfied. It is, however, quite difficult to get a good understanding of all representations that satisfy it. For instance, it is quite tricky to determine all ≤ 4-dimensional representations of cyclic groups that satisfy the Reid–Tai condition; cf. [MS84, Mor85, MMM88, Rei87]. These turn out to be rather special.

By contrast, we claim that every representation of a "typical" nonabelian group satisfies the Reid–Tai condition. The groups that have some representation violating the Reid–Tai condition are closely related to complex reflection groups. In the second part of the paper we provide a kit for building all of them, using basic building blocks, all but finitely many of which are (up to projective equivalence) reflection groups.

Let G be a finite group and (ρ, V) a finite-dimensional complex representation of G such that (ρ, V) does not satisfy the (local) Reid–Tai condition. That is, G has an element g such that $0 < \mathrm{age}(\rho(g)) < 1$. We say that such a pair (G, V) is a *non-RT pair* and g is an *exceptional element*. There is no essential gain in generality in allowing $\rho\colon G \to \mathrm{GL}(V)$ not to be faithful. We therefore assume that ρ is faithful, and remove it from the notation, regarding G as a subgroup of $\mathrm{GL}(V)$ (which is to be classified up to conjugation). If the conjugacy class of g does not generate the full group G, it must generate a normal subgroup H of G such that (H, V) is again a non-RT pair. After classifying the cases for which the conjugacy class of g generates G, we can take the normalization of each such G in $\mathrm{GL}(V)$; all finite subgroups intermediate between G and this normalizer give further examples. If V is reducible, then for every irreducible factor V_i of V on which g acts nontrivially, (G, V_i) is again a non-RT pair with exceptional element g. Moreover, if the conjugacy class of g generates G, then g must be an exceptional element for every nontrivial factor V_i of V. These reduction steps motivate the following definition:

Definition 5. *A basic non-RT pair is an ordered pair* (G, V) *consisting of a finite group* G *and a faithful irreducible representation* V *such that the conjugacy class of any exceptional element* $g \in G$ *generates* G.

Given a basic non-RT pair (G, V) and a positive integer n, we define $G_n = G \times \mathbb{Z}/n\mathbb{Z}$ and let V_n denote the tensor product of V with the character of $\mathbb{Z}/n\mathbb{Z}$ sending 1 to $e(1/n)$. Then V_n is always an irreducible representation of G_n and is faithful if n is prime to the order of G. Also if the conjugacy class of $g \in G$ generates the whole group, $a \in \mathbb{Z}$ is relatively prime to n, and n is prime to the order of G, then the conjugacy class of $(g, \bar{a}) \in G_n$ generates G_n. Finally, if g is an exceptional element, $(a, n) = 1$, and $a/n > 0$ is sufficiently small, then $(g, \bar{a})$ is exceptional. Thus for each basic non-RT pair (G, V), there are infinitely many other basic non-RT pairs that are projectively equivalent to it. To avoid this complication, we seek to classify basic non-RT pairs only up to projective equivalence:

Definition 6. *Pairs (G_1, V_1) and (G_2, V_2) are* projectively equivalent *if there exists an isomorphism* $\mathrm{PGL}(V_1) \to \mathrm{PGL}(V_2)$ *mapping the image of G_1 in* $\mathrm{PGL}(V_1)$ *isomorphically to the image of G_2 in* $\mathrm{PGL}(V_2)$.

We recall that a *pseudoreflection* $g \in \mathrm{GL}(V)$ is an element of finite order that fixes a codimension-1 subspace of V pointwise. A (complex) reflection group is a finite subgroup of $\mathrm{GL}(V)$ that is generated by pseudoreflections. We say that (G, V) is a reflection group if G is a reflection group in $\mathrm{GL}(V)$.

Definition 7. *We say that a basic non-RT pair (G, V) is* of reflection type *if (G, V) is projectively equivalent to some reflection group (G', V').*

The reflection groups are classified in [ST54]. Note that every pseudoreflection is of exceptional type, so every reflection group is a non-RT pair. On the other hand, there may be elements of exceptional type in a reflection group that are *not* pseudoreflections. Moreover, not every irreducible reflection group gives rise to a basic non-RT pair; it may happen that a particular conjugacy class of pseudoreflections fails to generate the whole group, since some irreducible reflection groups have multiple conjugacy classes of pseudoreflections.

We are interested in the other direction, however, and here we have the following theorem.

Theorem 8. *Up to projective equivalence, there are only finitely many basic non-RT pairs that are* not *of reflection type.*

We give an example in Section 5 below. Guralnick and Tiep [GT07] have proved that all exceptions are 4-dimensional and given two additional examples. In principle our proof provides an effective way (via the classification of finite simple groups) to determine all examples.

Finally, we study in detail the case in which $X = A$ is an Abelian variety. In fact, this case was the starting point of our investigations. For Abelian varieties, the induced representations $\mathrm{Stab}_x G \to V = T_x A$ have the property that $V + V^*$ is isomorphic to the rational representation of $\mathrm{Stab}_x G$ on $H^1(A, \mathbb{Q})$. This property substantially reduces the number of cases that we need to consider and allows us to show that a basic non-RT pair arising from an Abelian variety that is not of reflection type is of the unique known type. A more precise statement is given in Theorem 36 below. Unlike the proof of

Theorem 8, the proof of this theorem does not make use of the classification of finite simple groups.

A consequence of the analysis that ultimately gives Theorem 36 is the following easier statement.

Theorem 9. *Let A be a simple Abelian variety of dimension ≥ 4 and G a finite group acting on A. Then A/G has canonical singularities and $\kappa(A/G) = 0$.*

Acknowledgments. We thank Ch. Hacon, S. Kovács, and C. Voisin for helpful comments and references and De-Qi Zhang for correcting an error in an earlier version of Theorem 3. Partial financial support to JK and ML was provided by the NSF under grant numbers DMS-0500198 and DMS-0354772.

1 Uniruled quotients

Let X be a projective Calabi–Yau variety and $G \subset \mathrm{Aut}(X)$ a finite group of automorphisms. Our aim is to decide when the quotient variety X/G is uniruled or rationally connected. Our primary interest is in the case that X is smooth, but the proof works without change whenever X has canonical singularities and K_X is numerically trivial. (Note that by [Kaw85, 8.2], K_X numerically trivial implies that K_X is torsion.)

Theorem 10. *Let X be a projective Calabi–Yau variety and G a finite group acting on X. The following are equivalent:*

1. *G acts freely outside a set of codimension ≥ 2 and X/G has canonical singularities.*
2. *$\kappa(X/G) = 0$.*
3. *X/G is not uniruled.*

Proof. Set $Z := X/G$ and let $D_i \subset Z$ be the branch divisors of the quotient map $\pi : X \to Z$ with branching index e_i. Set $\Delta := \sum \left(1 - \frac{1}{e_i}\right) D_i$. By the Hurwitz formula,

$$K_X \sim_{\mathbb{Q}} \pi^* \left(K_Z + \Delta\right),$$

where $\sim_{\mathbb{Q}}$ means that some nonzero integral multiples of the two sides are linearly equivalent. Thus $K_Z + \Delta \sim_{\mathbb{Q}} 0$, and hence Theorem 10 is a special case of Theorem 11.   □

Theorem 11. *Let Z be a projective variety and Δ an effective divisor on Z such that $K_Z + \Delta \sim_{\mathbb{Q}} 0$. The following are equivalent:*

1. *$\Delta = 0$ and Z has canonical singularities.*
2. *$\kappa(Z) = 0$.*
3. *Z is not uniruled.*

Proof. If $\Delta = 0$ then K_Z is numerically trivial. Let $g : Y \to Z$ be a resolution of singularities and write

$$K_Y \sim_{\mathbb{Q}} g^* K_Z + \sum a_i E_i \sim_{\mathbb{Q}} \sum a_i E_i,$$

where the E_i are g-exceptional and $a_i \geq 0$ for every i iff Z has canonical singularities.

Thus if (1) holds then $K_Y \sim_{\mathbb{Q}} \sum a_i E_i$ is effective and so $\kappa(Y) \geq 0$. Since $\sum a_i E_i$ is exceptional, no multiple of it moves; hence $\kappa(Y) = \kappa(Z) = 0$.

The implication (2) $\Rightarrow$ (3) always holds (cf. [Kol96, IV.1.11]). It is conjectured that in fact (2) is equivalent to (3), but this is known only in dimensions ≤ 3 (cf. [KM98, 3.12–13]).

Thus it remains to prove that if (1) fails then Z is uniruled. We want to use the Miyaoka–Mori criterion [MM86] to get uniruledness. That is, a projective variety Y is uniruled if an open subset of it is covered by projective curves $C \subset Y$ such that $K_Y \cdot C < 0$ and $C \cap \operatorname{Sing} Y = \emptyset$.

If $\Delta \neq 0$ then we can take $C \subset Z$ to be any smooth complete intersection curve that does not intersect the singular locus of Z.

Thus we can assume from now on that $\Delta = 0$ and, as noted before, $K_Y \sim_{\mathbb{Q}} \sum a_i E_i$, where the E_i are g-exceptional and $a_i < 0$ for some i, since Z does not have canonical singularities by assumption. For notational convenience assume that $a_1 < 0$.

Ideally, we would like to find curves $C \subset Y$ such that C intersects E_1 but no other E_i. If such a C exists then

$$(K_Y \cdot C) = \left(\sum a_i E_i \cdot C \right) = a_1 (E_1 \cdot C) < 0.$$

We are not sure that such curves exist. (The condition $K_Z \equiv 0$ puts strong restrictions on the singularities of Z and creates a rather special situation.)

Fortunately, it is sufficient to find curves C such that $(C \cdot E_1)$ is big and the other $(C \cdot E_i)$ are small. This is enough to give $K_Y \cdot C < 0$.

Lemma 12. *Let Z be a normal projective variety over a field of arbitrary characteristic, $g : Y \to Z$ a birational morphism, and $E = \sum a_i E_i$ a noneffective g-exceptional Cartier divisor. Then Y is covered by curves C such that $(E \cdot C) < 0$.*

Proof. Our aim is to reduce the problem to a carefully chosen surface $S \to Y$ and then construct such curves directly on S. At each step we make sure that S can be chosen to pass through any general point of Y, so if we can cover these surfaces S with curves C, then the resulting curves also cover Y.

Assume that $a_1 < 0$. If $\dim g(E_1) > 0$ then we can cut Y by pullbacks of hypersurface sections of Z and use induction on the dimension.

Thus assume that $g(E_1)$ is a point. The next step would be to cut with hypersurface sections of Y. The problem is that in this process some of the divisors E_i may become nonexceptional, even ample. Thus first we need to kill all the other E_i such that $\dim g(E_i) > 0$.

To this end, construct a series of varieties $\sigma_i : Z_i \to Z$ starting with $\sigma_0 : Z_0 = Z$ as follows. Let $W_i \subset Z_i$ be the closure of $\left(\sigma_i^{-1} \circ g \right)_* (E_1)$, $\pi_i : Z_{i+1} \to Z_i$ the blowup of $W_i \subset Z_i$, and $\sigma_{i+1} = \sigma_i \circ \pi_i$.

By Abhyankar's lemma (in the form given in [KM98, 2.45]), there is an index j such that if $Y' \subset Z_j$ denotes the main component and $g' := \sigma_j : Y' \to Z$ the induced birational morphism then the following hold:

1. g' is an isomorphism over $Z \setminus g(E_1)$, and
2. $h := g^{-1} \circ g' : Y' \dashrightarrow Y$ is a local isomorphism over the generic point of E_1.

Thus $h^*(\sum a_i E_i)$ is g'-exceptional and not effective (though it is guaranteed to be Cartier only outside the indeterminacy locus of h).

Now we can cut by hypersurface sections of Y' to get a surface $S' \subset Y'$. Let $\pi : S \to S'$ be a resolution such that $h \circ \pi : S \to Y$ is a morphism and $f := (g' \circ \pi) : S \to T := g'(S') \subset Z$ the induced morphism. Then $(h \circ \pi)^* E$ is exceptional over T and not effective. Thus it is sufficient to prove Lemma 12 in case $S = Y$ is a smooth surface.

Fix an ample divisor H on S such that

$$H^1(S, \mathcal{O}_S(K_S + H + L)) = 0$$

for every nef divisor L. (In characteristic 0 any ample divisor works by the Kodaira vanishing theorem. In positive characteristic, one can use for instance [Kol96, Sec. II.6] to show that any H such that $(p-1)H - K_S - 4(\text{some ample divisor})$ is nef has this property.)

Assume next that H is also very ample and pick $B \in |H|$. Using the exact sequence

$$0 \to \mathcal{O}_S(K_S + 2H + L) \to \mathcal{O}_S(K_S + 3H + L) \to \mathcal{O}_B(K_B + 2H|_B + L|_B) \to 0,$$

we conclude that $\mathcal{O}_S(K_S + 3H + L)$ is generated by global sections.

By the Hodge index theorem, the intersection product on the curves E_i is negative definite, hence nondegenerate. Thus we can find a linear combination $F = \sum b_i E_i$ such that $F \cdot E_1 > 0$ and $F \cdot E_i = 0$ for every $i \neq 1$. Choose H_Z ample on Z such that $F + g^* H_Z$ is nef.

Thus the linear system $|K_S + 3H + m(F + g^* H_Z)|$ is base-point-free for every $m \geq 0$. Let $C_m \in |K_S + 3H + m(F + g^* H_Z)|$ be a general irreducible curve. Then

$$(C_m \cdot E_1) = m(F \cdot E_1) + (\text{constant}), \quad \text{and}$$
$$(C_m \cdot E_i) = (\text{constant}) \quad \text{for } i > 1.$$

Thus $(C_m \cdot E) \to -\infty$ as $m \to \infty$. $\qquad\square$

The following consequence of Lemma 12 is of independent interest. In characteristic zero, Corollary 13 is equivalent to Lemma 12 by [BDPP]; thus one can use Corollary 13 to give an alternative proof of Lemma 12. The equivalence should also hold in any characteristic.

Corollary 13 (Lazarsfeld, unpublished). *Let Z be a normal projective variety, $g : Y \to Z$ a birational morphism, and $E = \sum a_i E_i$ a g-exceptional Cartier divisor. Then E is pseudoeffective iff it is effective.*   □

2 Maps of Calabi–Yau Varieties

Every variety has many different dominant rational maps to projective spaces, but usually very few dominant rational maps whose targets are not unirational. The main result of this section proves a version of this for Calabi–Yau varieties.

Theorem 14. *Let X be a projective Calabi-Yau variety and $g : X \dashrightarrow Z$ a dominant rational map such that Z is not uniruled.*
Then there are

1. *a finite Calabi-Yau cover $h_X : \tilde{X} \to X$,*
2. *an isomorphism $\tilde{X} \cong \tilde{F} \times \tilde{Z}$ where $\tilde{F}, \tilde{Z}$ are Calabi-Yau varieties and π_Z denotes the projection onto $\tilde{Z}$, and*
3. *a generically finite map $g_Z : \tilde{Z} \dashrightarrow Z$*

such that $g \circ h_X = g_Z \circ \pi_Z$.

Remark 15. 1. De-Qi Zhang pointed out to us that in general g_Z cannot be chosen to be Galois (contrary to our original claim). A simple example is given as follows. Let A be an Abelian surface, $K = A/\{\pm 1\}$ the corresponding smooth Kummer surface, and $m > 1$. Then multiplication by m on A descends to a rational map $K \dashrightarrow K$, but it is not Galois.

2. Standard methods of the Iitaka conjecture (see especially [Kaw85]) imply that for any dominant rational map $X \dashrightarrow Z$, either $\kappa(Z) = -\infty$ or π is an étale locally trivial fiber bundle with Calabi–Yau fiber over an open subset of Z. Furthermore, [Kaw85, Sec. 8] proves Theorem 14 for the Albanese morphism. The papers [Zha96, Zha05] also contain related results and techniques.

First we explain how to modify the standard approach to the Iitaka conjecture to replace $\kappa(Z) = -\infty$ with Z uniruled.

The remaining steps are more subtle, since we have to construct a suitable birational model of Z and then to extend the fiber bundle structure from the open set to everywhere, at least after a finite cover.

Proof. As we noted before, there is a finite Calabi–Yau cover $X' \to X$ such that $K_{X'} \sim 0$. In particular, X' has canonical singularities. We can further replace Z

by its normalization in $\mathbb{C}(X)$. Thus we can assume to start with that $K_X \sim 0$, X has canonical singularities, and g has irreducible general fibers.

If g is not a morphism along the closure of the general fiber of g then Z is uniruled. If X is smooth, this is proved in [Kol96, VI.1.9]; the general case follows from [HM05]. Thus there are open subsets $X^* \subset X$ and $Z^* \subset Z$ such that $g : X^* \to Z^*$ is proper. We are free to shrink Z^* in the sequel if necessary.

Let us look at a general fiber $F \subset X$ of g. It is a local complete intersection subvariety whose normal bundle is trivial. So, by the adjunction formula, the canonical class K_F is also trivial. F has canonical singularities by [Rei80, 1.13].

Choose smooth birational models $\sigma : X' \to X$ and $Z' \to Z$ such that the corresponding $g' : X' \to Z'$ is a morphism that is smooth over the complement of a simple normal crossing divisor $B' \subset Z'$. We can also assume that the image of every divisor in $X \setminus X^*$ is a divisor in Z'. Thus we can choose smooth open subvarieties $X^0 \subset X$ and $Z^* \subset Z^0 \subset Z'$ such that

1. $X \setminus X^0$ has codimension ≥ 2 in X, and
2. $g_0 := g'|_{X^0} : X^0 \to Z^0$ is flat and surjective (but not proper).

The proof proceeds in three steps.

First, we show that $\omega_{X'/Z'}|_{X^0}$ is the pullback of a line bundle L from Z^0 that is $\mathbb{Q}$-linearly equivalent to 0.

Second, we prove that there are an étale cover $Z^1 \to Z^0$ and a birational map $Z^1 \times F \dashrightarrow X^0 \times_{Z^0} Z^1$ that is an isomorphism in codimension 1.

Third, we show that if $\tilde{X} \to X$ is the corresponding cover, then $\tilde{X}$ is a product of two Calabi–Yau varieties, as expected.

In order to start with Step 1, we need the following result about algebraic fiber spaces for which we could not find a convenient simple reference.

Proposition 16. *Notation as above. Then $g'_* \omega_{X'/Z'}$ is a line bundle and one can write the corresponding Cartier divisor*

$$\text{divisor class of } \left(g'_* \omega_{X'/Z'} \right) \sim_{\mathbb{Q}} J_g + B_g,$$

where

1. *B_g is an effective $\mathbb{Q}$-divisor supported on B', and*
2. *J_g is a nef $\mathbb{Q}$-divisor such that*
   (a) *either $J_g \sim_{\mathbb{Q}} 0$ and g is an étale locally trivial fiber bundle over some open set of Z^*,*
   (b) *or $(J_g \cdot C) > 0$ for every irreducible curve $C \subset Z'$ that is not contained in B' and is not tangent to a certain foliation of Z^*.*

Proof. Over the open set where g' is smooth, the results of [Gri70, Thm. 5.2] endow $g'_* \omega_{X'/Z'}$ with a Hermitian metric whose curvature is semipositive. This metric degenerates along B', but this degeneration is understood [Fuj78, Kaw81], giving the decomposition $J_g + B_g$, where J_g is the curvature term and B_g comes from the singularities of the metric along B'.

Set $d = \dim F$, where F is a general fiber of g. If F is smooth, we can assume that F is also a fiber of g'. In this case the curvature is flat in the directions corresponding to the (left) kernel of the Kodaira–Spencer map

$$H^1(F, T_F) \times H^0\left(F, \Omega_F^d\right) \to H^1\left(F, \Omega_F^{d-1}\right).$$

If $\Omega_F^d \cong \mathcal{O}_F$ then this is identified with the Serre duality isomorphism

$$\left(H^{d-1}\left(F, \Omega_F^1\right)\right)^* \cong H^1\left(F, \Omega_F^{d-1}\right);$$

hence the above (left) kernel is zero. Thus $(J_g \cdot C) = 0$ iff the deformation of the fibers $g^{-1}(C) \to C$ is trivial to first order over every point of $C \setminus B'$. This holds iff the fibers of g are all isomorphic to each other over $C \setminus B'$.

The corresponding result for the case that F has canonical singularities is worked out in [Kaw85, Sec. 6].   □

Let us now look at the natural map

$$g'^*\left(\omega_{Z'} \otimes g'_* \omega_{X'/Z'}\right) \to \omega_{X'},$$

which is an isomorphism generically along F, thus nonzero. Hence there is an effective divisor D_1 such that

$$g'^*\left(\omega_{Z'} \otimes g'_* \omega_{X'/Z'}\right) \cong \omega_{X'}(-D_1).$$

Write $\omega_{X'} \cong \sigma^* \omega_X(D_2) \cong \mathcal{O}_{X'}(D_2)$, where D_2 is σ-exceptional.

Let $C \subset X^0$ be a general complete intersection curve. Then σ^{-1} is defined along C, and setting $C' := \sigma^{-1}(C)$ we get that

$$\deg_{C'} g'^*\left(\omega_{Z'} \otimes g'_* \omega_{X'/Z'}\right) = \deg_{C'} \omega_{X'}(-D_1) = \left(C' \cdot (D_2 - D_1)\right) = -\left(C' \cdot D_1\right).$$

By the projection formula this implies that

$$\left(g_0(C) \cdot K_{Z'}\right) + \left(g_0(C) \cdot J_g\right) + \left(g_0(C) \cdot B_g\right) + \left(C \cdot \sigma_*(D_1)\right) = 0. \qquad (*)$$

If $\left(g_0(C) \cdot K_{Z'}\right) < 0$ then Z' is uniruled by the Miyaoka–Mori criterion [MM86], contrary to our assumptions. Thus all the terms on the left-hand side are nonnegative; hence they are all zero.

Since C is a general curve, it can be chosen to be not tangent to any given foliation. Therefore J_g is torsion in $\mathrm{Pic}(Z')$ and $X^* \to Z^*$ is an étale locally trivial fiber bundle for a suitable Z^*. A general complete intersection curve intersects every divisor in X; thus $(C \cdot \sigma_*(D_1)) = 0$ implies that $\sigma_*(D_1) = 0$, that is, D_1 is σ-exceptional.

Similarly, $g_0(C)$ intersects every irreducible component of $Z^0 \setminus Z^*$. Thus $(g_0(C) \cdot B_g) = 0$ implies that $B_g|_{Z^0} = 0$. These together imply that $L := \left(g'_* \omega_{X'/Z'}\right)|_{Z^0}$ is $\mathbb{Q}$-linearly equivalent to 0 and $\omega_{X'/Z'}|_{X^0} \cong g_0^* L$. This completes the first step.

Now to Step 2. Apply Lemma 17 to $X^* \to Z^*$. We get a finite cover $\pi : Z^1 \to Z^0$ such that $X' \times_{Z^0} Z^1$ is birational to $F \times Z^1$. By shrinking Z^0, we may assume that Z^1 is also smooth. Eventually we prove that π is étale over Z^0, but for now we allow ramification over $Z^0 \setminus Z^*$.

Let $n : X^1 \to X' \times_{Z^0} Z^1$ be the normalization. We compare the relative dualizing sheaves

$$\omega_{F \times Z^1 / Z^1} \cong \mathcal{O}_{F \times Z^1} \quad \text{and} \quad \omega_{X^1 / Z^1}.$$

Let $X^1 \xleftarrow{u} Y \xrightarrow{v} F \times Z^1$ be a common resolution. We can then write

$$\omega_{Y/Z^1} \cong v^* \omega_{F \times Z^1 / Z^1}(E_v) \cong \mathcal{O}_Y(E_v)$$

for some divisor E_v supported on $\mathrm{Ex}(v)$ and also

$$\omega_{Y/Z^1} \cong u^* \omega_{X^1/Z^1}(E_u') \cong (g' \circ n \circ u)^* L(E_u)$$

for some divisors E_u', E_u. Since $g^* L|_{X^0}$ is $\mathbb{Q}$-linearly equivalent to zero, we conclude from these that

$$u_*(E_u - E_v)|_{X^0} \sim_{\mathbb{Q}} 0.$$

Next we get some information about E_u and E_v. Since $F \times Z^1$ has canonical singularities, every irreducible component of E_v is effective. Furthermore, an irreducible component of $\mathrm{Ex}(v)$ appears with positive coefficient in E_v unless it dominates Z^0. Thus we see that $u_*(E_v)|_{X^0}$ is supported in $X^0 \setminus X^*$ and an irreducible component of $X^0 \setminus X^*$ appears with positive coefficient in $u_*(E_v)$, unless $v \circ u^{-1}$ is a local isomorphism over its generic point.

On the other hand, since $\omega_{X'/Z'}|_{X^0}$ is the pullback of L, $\omega_{X' \times_{Z^0} Z^1 / Z^1}|_{X^0}$ is the pullback of $\pi^* L$. As we normalize, we subtract divisors corresponding to the nonnormal locus. Every other irreducible component of E_u is u-exceptional, hence gets killed by u_*. Thus we obtain that $u_*(E_u)|_{X^0}$ is also contained in $X^0 \setminus X^*$, and either an irreducible component of $X^0 \setminus X^*$ appears with negative coefficient or $X' \times_{Z'} Z^1$ is normal over that component and the coefficient is 0.

Thus $u_*(E_u - E_v)|_{X^0}$ is a nonpositive linear combination of the irreducible components of $X^0 \setminus X^*$ and it is also $\mathbb{Q}$-linerally equivalent to 0. Since $X \setminus X^0$ has codimension ≥ 2, we conclude that $u_*(E_u - E_v)|_{X^0} = 0$. That is, $X^0 \times_{Z^0} Z^1$ is normal in codimension 1 and isomorphic to $F \times Z^1$, again only in codimension 1.

We may as well assume that $Z^1 \to Z^0$ is Galois with group G. We then have a corresponding G-action on $F \times Z^1$ for a suitable G-action on F. By taking the quotient, we obtain a birational map

$$\phi : W := (F \times Z^1)/G \to Z^0 \dashrightarrow X^0,$$

which is an isomorphism in codimension 1. In particular,

$$\omega_{W/Z^0} \cong \phi^* \omega_{X^0/Z^0} \cong g_W^* L,$$

where $g_W : W \to Z^0$ is the quotient of the projection map to Z^1. Using (18.1) we conclude that $g_W : W \to Z^0$ is in fact an étale locally trivial fiber bundle with fiber F, at least outside a codimension 2 set.

Then Lemma 17 shows that $Z^1 \to Z^0$ is also étale at every generic point of $Z^0 \setminus Z^*$; thus it is a finite étale cover.

Let now $\tilde{X} \to X$ be the normalization of X in the function field of X^1. Since $Z^1 \to Z^0$ is étale, we see that $\tilde{X} \to X$ is étale over X^0. Thus $\tilde{X} \to X$ is étale outside a set of codimension ≥ 2. In particular, $\tilde{X}$ is a Calabi–Yau variety.

Furthermore, the birational map $\phi : \tilde{X} \to F \times Z^1$ is an open embedding outside a set of codimension ≥ 2. That is, ϕ does not contract any divisor. This completes Step 2.

By Proposition 18, $\tilde{X}$ is itself a product $\tilde{F} \times \tilde{Z}$. Note that $F = \tilde{F}$, since ϕ is an isomorphism along $F \cong \pi^{-1}(z)$ for $z \in Z^*$. Thus $\tilde{X} \cong F \times \tilde{Z}$ and there is a generically finite map $\tilde{Z} \dashrightarrow Z$.   □

Lemma 17. *Let $f : U \to V$ be a projective morphism between normal varieties, V smooth. Assume that f is an étale locally trivial fiber bundle with typical fiber F that is a Calabi–Yau variety. Then there is a finite étale cover $V' \to V$ such that the pullback $U \times_V V' \to V'$ is globally trivial. Moreover, we can choose $V' \to V$ such that its generic fiber depends only on the generic fiber of f.*

Proof. Let H be an ample divisor on U. Let $\pi : \mathrm{Isom}(F \times V, U, H) \to V$ denote the V-scheme parametrizing V-isomorphisms $\phi : F \times V \to U$ such that $\phi^* H$ is numerically equivalent to H_F. The fiber of $\mathrm{Isom}(F \times V, U, H) \to V$ over $v \in V$ is the set of isomorphisms $\phi : F \to U_v$ such that $\phi^*(H|_{U_v})$ is numerically equivalent to H_F.

Note that $\mathrm{Isom}(F \times V, U, H) \to V$ is an étale locally trivial fiber bundle with typical fiber $\mathrm{Aut}_H(F)$. Any étale multisection of π gives a required étale cover $V' \to V$.

Thus we need to find an étale multisection of a projective group scheme (in characteristic 0). The Stein factorization of π gives an étale cover $V_1 \to V$, and if we pull back everything to V_1, then there is a well-defined identity component. Thus we are reduced to the case that $\pi : I \to V$ is a torsor over an Abelian scheme $A \to V$.

Let I_g be the generic fiber and let $P \in I_g$ be a point of degree d. Let $S_d \subset I_g$ be the set of geometric points p such that $dp - P = 0 \in A_g$. Then S_d is defined over $k(V)$ and it is a principal homogeneous space over the subgroup of d-torsion points of A_g. We claim that the closure of S_d in I is finite and étale over V. Indeed, it is finite over codimension-1 points and also étale over codimension-1 points since the limit of nonzero d-torsion points cannot be zero. Thus it is also étale over all points by the purity of branch loci.   □

The $K_X = 0$ of the following lemma is proved in [Pet94, Thm. 2].

Proposition 18. *Let X, U, V be normal projective varieties. Assume that X has rational singularities. Let $\phi : X \dashrightarrow U \times V$ be a birational map that does not contract any divisor. Then there are normal projective varieties U' birational to U and V' birational to V such that $X \cong U' \times V'$.*

Proof. We can replace U, V by resolutions; thus we may assume that they are smooth.

Let $X \xleftarrow{p} Y \xrightarrow{q} U \times V$ be a factorization of g. By assumption, $\mathrm{Ex}(q) \subset \mathrm{Ex}(p)$. Let H be a very ample divisor on X and $\phi_* H = q_* p^* H$ its birational transform. Then $|q_* p^* H| = |p^* H + m \,\mathrm{Ex}(q)|$ for $m \gg 1$. On the other hand,

$$|H| = |p^* H| \subset |p^* H + m\,\mathrm{Ex}(q)| \subset |p^* H + m\,\mathrm{Ex}(p)| = |H|.$$

Thus $|H| = |\phi_* H|$.

Assume that there are divisors H_U on U and H_V on V such that $\phi_* H \sim \pi_U^* H_U + \pi_V^* H_V$. Then

$$H^0(U \times V, \mathcal{O}_{U \times V}(\phi_* H)) = H^0(U, \mathcal{O}_U(H_U)) \otimes H^0(V, \mathcal{O}_V(H_V)).$$

Since X is the closure of the image of $U \times V$ under the linear system $|\phi_* H|$, we see that $X \cong U' \times V'$, where U' is the image of U under the linear system $|H_U|$ and V' is the image of V under the linear system $|H_V|$.

If $H^1(U, \mathcal{O}_U) = 0$, then $\mathrm{Pic}(U \times V) = \pi_U^* \,\mathrm{Pic}(U) + \pi_V^* \,\mathrm{Pic}(V)$, and we are done. In general, however, $\mathrm{Pic}(U \times V) \supsetneq \pi_U^* \,\mathrm{Pic}(U) + \pi_V^* \,\mathrm{Pic}(V)$, and we have to change H.

Fix points $u \in U$ and $v \in V$ and let $D_U := \phi_* H|_{U \times \{v\}}$ and $D_V := \phi_* H|_{\{u\} \times V}$. Set $D' := \phi_* H - \pi_U^* D_U - \pi_V^* D_V$. Then D' restricted to any $U \times \{v'\}$ is in $\mathrm{Pic}^0(U)$ and D' restricted to any $\{u'\} \times V$ is in $\mathrm{Pic}^0(V)$. Thus there is a divisor B on $\mathrm{Alb}(U \times V)$ such that $D' = \mathrm{alb}_{U \times V}^* B$, where, for a variety Z, $\mathrm{alb}_Z : Z \to \mathrm{Alb}(Z)$ denotes the Albanese map.

Choose divisors B_U on $\mathrm{Alb}(U)$ and B_V on $\mathrm{Alb}(V)$ such that $\pi_U^* B_U + \pi_V^* B_V - B$ is very ample, where, somewhat sloppily, π_U, π_V also denote the coordinate projections of $\mathrm{Alb}(U \times V)$.

Since X has rational singularities, $\mathrm{Alb}(X) = \mathrm{Alb}(U \times V)$. Replace H by

$$H^* := H + \mathrm{alb}_X^* \left(\pi_U^* B_U + \pi_V^* B_V - B \right).$$

Then

$$\phi_* H^* = \phi_* H + \mathrm{alb}_{U \times V}^* \left(\pi_U^* B_U + \pi_V^* B_V - B \right)$$
$$= \pi_U^* \left(H_U + \mathrm{alb}_U^* B_U \right) + \pi_V^* \left(H_V + \mathrm{alb}_V^* B_V \right).$$

Since H^* is again very ample, we are done. $\qquad\square$

18.1 (Quotients of trivial families). We consider families $X \to C$ over a smooth pointed curve germ $0 \in C$ such that after a finite base change and normalization we get a trivial family. This means that we start with a trivial family $F \times D$ over a disk D, an automorphism τ of F of order dividing m, and take the quotient $X := (F \times D)/(\tau, e(1/m))$.

If the order of τ is less than m then there is a subgroup that acts trivially on F and the quotient is again a trivial family. Thus we may assume that the order of τ is precisely m.

Fix a top form ω on F. Pulling back by τ gives an isomorphism $\omega = \eta\tau^*\omega$ for some mth root of unity η. If $\eta \neq 1$ then on the quotient family the monodromy around $0 \in C$ has finite order $\neq 1$, and the boundary term B in Proposition 16 is nonzero.

Finally, if $\omega = \tau^*\omega$ then $F_0 := F/(\tau)$ also has trivial canonical class. Thus by the adjunction formula we see that $\omega_{X/C} \cong \mathcal{O}_X((m-1)F_0)$ is not trivial.

Next we apply Theorem 14 to study those quotients of Calabi–Yau varieties that are uniruled but not rationally connected. Let us see first some examples of how this can happen. Then we will see that these trivial examples exhaust all possibilities.

Example 19. Let $\Pi : X' \to Z'$ be an étale locally trivial fiber bundle whose base Z' and typical fiber F' are both projective Calabi–Yau varieties. Then X' is also a projective Calabi–Yau variety. Let G' be a finite group acting on X' and assume that Π is G'-equivariant. Assume that

1. $\kappa(Z'/G') = 0$, and
2. for general $z \in Z'$, the quotient $\Pi^{-1}(z)/\operatorname{Stab}_z G'$ is rationally connected.

Then $\Pi/G' : X'/G' \to Z'/G'$ is the MRC fibration of X'/G'.

More generally, let $H \subset G'$ be a normal subgroup such that $X := X'/H$ is a Calabi–Yau variety and set $G := G'/H$. Then $X/G \cong X'/G'$, and its MRC fibration is given by $\Pi/G' : X'/G' \to Z'/G'$.

Theorem 20. Let X be a projective Calabi–Yau variety and G a finite group acting on X. Assume that X/G is uniruled but not rationally connected. Let $\pi : X/G \dashrightarrow Z$ be the MRC fibration. Then there are

1. a finite, Calabi–Yau, Galois cover $X' \to X$,
2. a proper morphism $\Pi : X' \to Z'$ that is an étale locally trivial fiber bundle whose base Z' and typical fiber F' are both projective Calabi–Yau varieties, and
3. a group G' acting on X', where $G \supset \operatorname{Gal}(X'/X)$ and $G'/\operatorname{Gal}(X'/X) = G$,

such that $\Pi/G' : X'/G' \to Z'/G'$ is birational to the MRC fibration $\pi : X/G \dashrightarrow Z$.

Proof. Let $X/G \dashrightarrow Z$ be the MRC fibration and let $\pi : X \to X/G \dashrightarrow Z$ be the composite. Since Z is not uniruled by [GHS03]; Theorem 14 applies and we get a direct product $F \times Z$ mapping to X. Since both X and $F \times Z$ have trivial canonical class, $F \times Z \to X$ is étale in codimension 1.

In order to lift the G-action from X to a cover, we need to take the Galois closure of $F \times Z \to X/G$. Let G' be its Galois group. This replaces $F \times Z$ with a finite cover that is étale in codimension 1. The latter need not be globally a product, only étale locally a product. $\qquad\qquad \square$

Corollary 21. *Let A be an Abelian variety and G a finite group acting on A. There is a unique maximal G-equivariant quotient $A \to B$ such that $A/G \to B/G$ is the MRC quotient.*

Proof. Let $\Pi : A^0 \to Z^0$ be the quotient constructed in Theorem 20. Its fibers F_z are smooth subvarieties of A with trivial canonical class. Thus each F_z is a translation of a fixed Abelian subvariety $C \subset A$ (cf. [GH79, 4.14]). Set $B = A/C$. $\qquad\qquad \square$

Definition 22. *Let G be a finite group acting on a vector space V. Let $G^{RT} < G$ be the subgroup generated by all elements of age < 1 and V^{RT} the complement of the fixed space of G^{RT}.*

Definition 23. *Let A be an Abelian variety and G a finite group acting on A. For every $x \in A$, let $G_x := \mathrm{Stab}(x) < G$ denote the stabilizer and $i_x : A \to A$ the translation by x. Consider the action of G_x on $T_x A$, the tangent space of A at x. Let G_x^{RT} and $(T_x A)^{RT}$ be as above. Note that $(T_x A)^{RT}$ is the tangent space of a translate of an Abelian subvariety $A_x \subset A$, since it is the intersection of the kernels of the endomorphisms $g - 1_A$ for $g \in G_x^{RT}$. Set*

$$G^{RT} := \left\langle G_x^{RT} : x \in A \right\rangle \quad and \quad (TA)^{RT} := \left\langle i_x^*(T_x A)^{RT} : x \in A \right\rangle.$$

Then $(TA)^{RT}$ is the tangent space of the Abelian subvariety generated by the A_x. Denote it by A_1^{RT}.

The group G/G^{RT} acts on the quotient Abelian variety $q_1 : A \to A/A_1^{RT}$. If $q_i : A \to A/A_i^{RT}$ is already defined, set

$$A_{i+1}^{RT} := q_i^{-1} \left(\left(A/A_i^{RT} \right)_1^{RT} \right)$$

and let $q_{i+1} : A \to A/A_{i+1}^{RT}$ be the quotient map. The increasing sequence of Abelian subvarieties $A_1^{RT} \subset A_2^{RT} \subset \cdots$ eventually stabilizes to $A_{\mathrm{stab}}^{RT} \subset A$.

Corollary 24. *Let A be an Abelian variety and G a finite group acting on A. Then*

1. *$\kappa(A/G) = 0$ iff $G^{RT} = \{1\}$, and*
2. *A/G is rationally connected iff $A_{\mathrm{stab}}^{RT} = A$.*

3 Basic Non-Reid–Tai Pairs

Our goal in this section is to classify basic non-RT pairs.

There is a basic dichotomy:

Proposition 25. *If (G, V) is a basic non-RT pair, then either (G, V) is primitive or G respects a decomposition of V as a direct sum of lines:*

$$V = L_1 \oplus \cdots \oplus L_n.$$

In the latter case, the homomorphism $\phi \colon G \to S_n$ given by the permutation action of G on $\{L_1, \ldots, L_n\}$ is surjective, and every exceptional element in G maps to a transposition.

Proof. Suppose that G respects the decomposition $V \cong V_1 \oplus \cdots \oplus V_m$ for some $m \geq 2$. If there is more than one such decomposition, we choose one such that $m \geq 2$ is minimal. By irreducibility, G acts transitively on the set of V_i. Since the conjugacy class of any exceptional element g generates G, it follows that g permutes the V_i nontrivially. Suppose that for $2 \leq k \leq m$, we have

$$g(V_1) = V_2, \; g(V_2) = V_3, \; g(V_k) = V_1.$$

Then g and $e(1/k)g$ are isospectral on $V_1 \oplus \cdots \oplus V_k$. Thus the eigenvalues of g constitute a union of $\dim V_1$ cosets of the cyclic group $\langle e(1/k) \rangle$. Such a union of cosets can satisfy the Reid–Tai condition only if $k = 2$ and $\dim V_1 = 1$, and then g must stabilize V_i for every $i \geq 3$. Thus g induces a transposition on the V_i, each of which must be of dimension 1. A transitive subgroup of S_m that contains a transposition must be of the form $S_a^b \rtimes T$, where $ab = m$, T is a transitive subgroup of S_b, and $a \geq 2$. It corresponds to a decomposition of the set of factors V_i into b sets of cardinality a. If W_j denote the direct sums of the V_i within each of the a-element sets of this partition, it follows that G respects this coarser decomposition, contrary to the assumption that m is minimal.

To analyze the primitive case, it is useful to quantify how far a unitary operator is from the identity.

Definition 26. *Let H be a Hilbert space, T a unitary operator on H, and B an orthonormal basis of H. The* deviation *of T with respect to B is given by*

$$d(T, B) := \sum_{b \in B} \|T(b) - b\|.$$

The deviation *of T is*

$$d(T) := \inf_B d(T, B),$$

as B ranges over all orthonormal bases. If $d(T) < \infty$, we say that T has finite *deviation.*

Since the arc of a circle cut off by a chord is always longer than the chord, if H is a finite-dimensional Hilbert space and $g\colon H \to H$ a unitary operator of finite order not satisfying the Reid–Tai condition, we have $d(g) < 2\pi$. This is the primary motivation for our definition of deviation.

For any space H, unitary operator T, basis B, and real number $x > 0$, we define

$$S(T, B, x) = \{b \in B \mid \|T(b) - b\| \geq x\}.$$

If 1_I denotes the characteristic function of the interval I, then

$$\int_0^\infty |S(T, B, x)| dx = \int_0^\infty \sum_{b \in B} 1_{[0, \|T(b) - b\|]} dx$$

$$= \sum_{b \in B} \int_0^\infty 1_{[0, \|T(b) - b\|]} dx = \sum_{b \in B} \|T(b) - b\| = d(T, B).$$

It is obvious that deviation is symmetric in the sense that $d(T) = d(T^{-1})$. Next we prove a lemma relating d to multiplication in the unitary group.

Proposition 27. *If $T_1, T_2, \ldots, T_n$ are unitary operators of finite deviation on a Hilbert space H, then*

$$d(T_1 T_2 \cdots T_n) \leq n(d(T_1) + d(T_2) + \cdots + d(T_n)).$$

Proof. Let $B_1, B_2, \ldots, B_n$ denote orthonormal bases of H. We claim that there exists an orthonormal basis B such that for all $x > 0$,

$$|S(T_1 T_2 \cdots T_n, B, nx)| \leq |S(T_1, B_1, x)| + |S(T_2, B_2, x)| + \cdots + |S(T_n, B_n, x)|. \tag{1}$$

This claim implies the proposition, by integrating over x.

Given T_i, B_i and $x > 0$, we define

$$V_x = \operatorname{Span} \bigcup_{i=1}^n S(T_i, B_i, x).$$

Since all T_i are of finite deviation, the set of "jumps" (x such that V_x is not constant in a neighborhood of x) is discrete in $(0, \infty)$. Arranging them in reverse order, we see that there exists a (possibly infinite) increasing chain of finite-dimensional subspaces W_i of H such that each V_x is equal to one of the W_i. We choose B to be any orthonormal basis adapted to $W_1 \subset W_2 \subset \cdots$ in the sense that $B \cap W_i$ is an orthonormal basis of W_i for all i.

For all $b \in B$, by the triangle inequality,

$$\|T_1 T_2 \cdots T_n(b) - b\| \leq \sum_{i=1}^n \|T_1 T_2 \cdots T_{i-1}(T_i(b) - b)\| = \sum_{i=1}^n \|T_i(b) - b\|.$$

If $b \notin V_x$, then b is orthogonal to every element of $S(T_i, B_i, x)$ for $i = 1, \ldots, n$, and therefore $\|T_i(b) - b\| \leq x$ for all i. It follows that

$$\|T_1 T_2 \cdots T_n(b) - b\| \leq nx,$$

or $b \notin S(T_1 T_2 \cdots T_n, B, nx)$. This implies (1).

Proposition 28. *If T_1 and T_2 are operators on a Hilbert space H such that T_1 is of bounded deviation, then*

$$d\left(T_1^{-1} T_2^{-1} T_1 T_2\right) \leq 4d(T_1).$$

Proof. Since $d\left(T_1^{-1}\right) = d(T_1) = d\left(T_2^{-1} T_1 T_2\right)$, the proposition follows from Proposition 27.

Lemma 29. *Let G be a compact group and (ρ, V) a nontrivial representation of G such that $V^G = (0)$. Then there exists $g \in G$ with $d(g) \geq \dim V$.*

Proof. Since V has no G-invariants,

$$\int_G \mathrm{tr}(\rho(g)) \, dg = 0,$$

there exists $g \in G$ with $\Re(\mathrm{tr}(\rho(g))) \leq 0$. If $d(g) < \dim V$, there exists an orthonormal basis B of V such that $\sum_{b \in B} |g(b) - b| < \dim V$. If a_{ij} is the matrix of $\rho(g)$ with respect to such a basis then $\Re(\sum_i a_{ii}) < 0$, so

$$\sum_{b \in B} \|b - g(b)\| > \sum_i |1 - a_{ii}| \geq \sum_i (1 - \Re(a_{ii})) > \dim V,$$

which gives a contradiction.

Lemma 30. *Let (G, V) be a finite group and a representation such that $V^G = (0)$. Suppose that for some integer k, every element of G can be written as a product of at most k elements conjugate to g or g^{-1}. Then $d(g) \geq \frac{\dim V}{k^2}$.*

Proof. This is an immediate consequence of Proposition 27 and Lemma 29.

We recall that a *characteristically simple* group G is isomorphic to a group of the form K^r, where K is a (possibly abelian) finite simple group.

Proposition 31. *There exists a constant C such that if H is a perfect central extension of a characteristically simple group K^r and the conjugacy class of $h \in H$ generates the whole group, then every element in H is the product of no more than $C \log |H|$ elements conjugate to h or h^{-1}.*

Proof. If H is perfect, then K^r is perfect, so K is a nonabelian finite simple group and therefore perfect. Let $\tilde{K}$ denote the universal central extension of K, so $\tilde{K}^r$ is the universal central extension of K^r as well as of H.

Since the only subgroup of $\tilde{K}^r$ mapping onto H is $\tilde{K}^r$, it suffices to prove that if $\tilde{h} = (x_1, \ldots, x_r) \in \tilde{K}^r$ has the property that its conjugacy class generates $\tilde{K}^r$, then every element of $\tilde{K}^r$ is the product of at most $C \log |H|$ conjugates of $\tilde{h}$ or $\tilde{h}^{-1}$. For any t from 1 to r, we can choose $(1, \ldots, 1, y, 1, \ldots, 1) \in \tilde{K}^r$ (with y in the tth coordinate) whose commutator with $\tilde{h}$ is an element $(1, \ldots, 1, z, 1, \ldots, 1)$ not in the center of $\tilde{K}^r$. This element is the product of a conjugate of $\tilde{h}$ and a conjugate of $\tilde{h}^{-1}$. If we can find an absolute constant A such that for every finite simple group K and every noncentral element $z \in \tilde{K}$, every element of $\tilde{K}$ is the product of at most $A \log |K|$ elements conjugate to z or z^{-1}, the proposition holds with $C = 2A$.

To prove the existence of A, we note that we may assume that K has order greater than any specified constant. In particular, we may assume that K is either an alternating group A_m, $m \geq 8$, or a group of Lie type. It is known that the *covering number* of A_m is $\lfloor m/2 \rfloor$ (see, e.g., [ASH85]), so every element of the group can be written as a product of $\leq m/2$ elements belonging to any given nontrivial conjugacy class X. The universal central extension of A_m is of order $m!$, and at least half of those elements are products of $\leq m/2$ elements in any fixed conjugacy class X, so all of the elements are products of $\leq m < \log m!/2$ elements of X. For the groups of Lie type and their perfect central extensions, we have an upper bound linear in the absolute rank ([EGH99], [LL98]) and therefore sublogarithmic in order.

Lemma 32. *For every integer $n > 0$, there are only finitely many classes of primitive finite subgroups $G \subset \mathrm{GL}_n(\mathbb{C})$ up to projective equivalence.*

Proof. Since G is primitive, a normal abelian subgroup of G lies in the center of $\mathrm{GL}_n(\mathbb{C})$. By Jordan's theorem, G has a normal abelian subgroup whose index can be bounded in terms of n. Thus the image of G in $\mathrm{PGL}_n(\mathbb{C})$ is bounded in terms of n. For each isomorphism class of finite groups, there are only finitely many projective n-dimensional representations.

32.1 (Proof of Theorem 8).

First we assume that G stabilizes a set $\{L_1, \ldots, L_n\}$ of lines that give a direct sum decomposition of V. We have already seen that the resulting homomorphism $\phi \colon G \to S_n$ is surjective. Let $t \in G$ lie in $\ker \phi$, so $t(v_i) = \lambda_i v_i$ for all i and all $v_i \in L_i$. The commutator of t with any preimage of the transposition $(i\, j) \in S_n$ gives an element of $\ker \phi$ that has eigenvalues λ_i/λ_j, λ_j/λ_i, and 1 (of multiplicity $n - 2$). The G-conjugacy class of this element consists of all diagonal matrices with this multiset of eigenvalues. Thus $\ker \phi$ contains $\ker \det_C \colon C^n \to C$, where C is the group generated by all ratios of eigenvalues of all elements of $\ker \phi$. It follows that $\ker \phi$ is the product of $\ker \det_C$ and a group of scalar matrices. If we pass to $\mathrm{PGL}(V)$, therefore, the image of G is an extension of C^{n-1} by S_n.

We claim that this extension is split if n is sufficiently large. To prove this, it suffices to prove $H^2(S_n, C^{n-1}) = 0$. This follows if we can show that $H^2(S_n, \mathbb{Z}^{n-1}) = H^3(S_n, \mathbb{Z}^{n-1}) = 0$, or that the sum-of-coordinate maps

$$H^i(S_n, \mathbb{Z}^n) \to H^i(S_n, \mathbb{Z})$$

are isomorphisms for $i = 1, 2, 3$, where S_n acts on $\mathbb{Z}^n$ by permutations. By Shapiro's lemma, the composition of restriction and sum-of-coordinates gives an isomorphism $H^i(S_n, \mathbb{Z}^n) \tilde{\to} H^i(S_{n-1}, \mathbb{Z})$, so we need to know that the restriction homomorphisms $H^i(S_n, \mathbb{Z}) \to H^i(S_{n-1}, \mathbb{Z})$ are isomorphisms when n is large compared to i, which follows from [Nak60].

Thus the image of G in $\mathrm{PGL}(V)$ is $C^{n-1} \rtimes S_n$, which is the same as the image in $\mathrm{PGL}(V)$ of the imprimitive unitary reflection group $G(|C|, k, n)$, where k is any divisor of G.

It remains to consider the primitive case. Let Z denote the center of G. Since Z is abelian and has a faithful isotypic representation, it must be cyclic. If G is abelian, then $G = Z$, and we are done. (This can be regarded as a subcase of the case that G stabilizes a decomposition of V into lines.) Otherwise, let $\overline{H} \cong K^r$ denote a characteristically simple normal subgroup of $\overline{G} := G/Z$, where K is a (possibly abelian) finite simple group and $r \geq 1$. If K is abelian, we let $H \subset G$ denote the inverse image of $\overline{H}$ in G. In the nonabelian case, we let H denote the derived group of the inverse image of $\overline{H}$ in G, which is perfect and again maps onto $\overline{H}$. We know that H is not contained in the center of G, so some inner automorphism of G acts nontrivially on H. It follows that conjugation by g acts nontrivially on H. By Proposition 28, there exists a nontrivial element $h \in H$ with $d(h) < 8\pi$.

We consider five cases:

1. H is abelian.
2. $\overline{H}$ is abelian but H is not.
3. K is a group of Lie type.
4. K is an alternating group A_m, where m is greater than a sufficiently large constant.
5. K is nonabelian but not of type (3) or (4).

We prove that case (4) leads to reflection groups, and all of the other cases contribute only finitely many solutions.

If H is abelian, then the restriction of V to H is isotypical and H is central, contrary to the definition of H.

If $\overline{H}$ is abelian and H is not, then H is a central extension of a vector group and is therefore the product of its center Z by an extraspecial p-group H_p for some prime p. The kernel $Z[p]$ of multiplication by p on Z is isomorphic to $\mathbb{Z}/p\mathbb{Z}$, and the commutator map $\overline{H} \times \overline{H} \to Z[p]$ gives a nondegenerate pairing. Therefore, conjugation by any element in $G \backslash H$ induces a nontrivial map on $\overline{H}$. Thus we can take h with $d(h) < 8\pi$ to lie outside the center of H.

The image of every element of H in $\mathrm{Aut}(V)$ is the product of a scalar matrix and the image of an element of H_p in a direct sum of $m \geq 1$ copies of one of its faithful irreducible representations. By the Stone–von Neumann theorem, a faithful representation of an extraspecial p-group is determined by a central character; its dimension is p^n, where $|H_p| = p^{2n+1}$, and every noncentral element has eigenvalues $\omega, \omega e(1/p), \omega e(2/p), \ldots, \omega e(-1/p)$, each occurring with multiplicity p^{n-1}, where $\omega^{p^2} = 1$. Since h is a scalar multiple of an element with these eigenvalues, we have

$$8\pi > d(h) \geq 2\pi m p^{n-1} \frac{p(p-1)}{2p} \geq \frac{2\pi m p^n}{4},$$

so $\dim V = mp^n < 16$. By Lemma 32, there are only finitely many possibilities for (G, V) up to projective equivalence.

In cases (3)–(5), H is perfect. If the conjugacy class of h in G does not generate H, it generates a proper normal subgroup of H, which is a central extension of a subgroup of K^r that is again normal in $\overline{G}$. Such a subgroup is of the form K^s for $s < r$. Replacing H if necessary by a smaller group, we may assume that the G-conjugacy class of h generates H. By Proposition 31, every element of H is the product of at most $C \log |H|$ elements conjugate to h or h^{-1}, and by Lemma 30, this implies

$$\dim V < 8\pi C^2 \log^2 |H|. \tag{2}$$

For case (3), we note that by [SZ93, Table 1], a faithful irreducible projective representation of a finite simple group K that is not an alternating group always has dimension at least $e^{c_1 \sqrt{\log |K|}}$ for some positive absolute constant c_1. A faithful irreducible representation of K^r is the tensor power of r faithful irreducible representations of K, so its dimension is at least $e^{c_1 \sqrt{\log |\overline{H}|}} > e^{c_1 \sqrt{\log |H|}/2}$. For $|H| \gg 0$, this is in contradiction with (2). Thus there are only finitely many possibilities for H up to isomorphism, and this gives an upper bound for $\dim V$. By Lemma 32, there are only finitely many possibilities for (G, V) up to projective equivalence.

For case (4), we need to consider both ordinary representations of A_m and spin representations (i.e., projective representations that do not lift to linear representations). By [KT04], the minimal degree of a spin representation of A_m grows faster than any polynomial, in particular, faster than $m^{5/2}$. To every irreducible linear representation of A_m one can associate a partition λ of m for which the first part λ_1 is greater than or equal to the number of parts. There may be one or two representations associated to λ, and in the latter case their degrees are equal and their direct sum is irreducible as an S_m-representation. By [LS04, 2.1, 2.4], if $\lambda_1 \leq m - 3$, the degree of any S_m-representation associated to λ is at least $\binom{m-3}{3}$, so the degree of any A_m-representation associated to λ is at least half of that. Thus, for $m \gg 0$, the only faithful representations of A_m that have degree less than $m^{5/2}$ are $V_{m-1,1}$, $V_{m-2,2}$, and $V_{m-2,1,1}$ of degrees $m - 1$, $\frac{(m-1)(m-2)}{2}$, and $\frac{m(m-3)}{2}$ respectively. Now, $\log |H| \leq \log(m!)^r < rm^{1.1}$ for $m \gg 0$, and the minimal degree of any faithful representation of H is at least $(m - 1)^r \gg r^2 m^{2.2}$ for $r \geq 3$.

By (2), there are only four possibilities that need be considered. If $r = 2$, then $H = \overline{H} = A_m^2$, and V must be the tensor product of two copies of $V_{m-1,1}$. Otherwise, $r = 1$, $H = \overline{H} = A_m$, and V is $V_{m-1,1}$, $V_{m-2,2}$, or $V_{m-2,1,1}$. In the first case, the normalizer of H in $\mathrm{GL}(V)$ is S_n^2; in the remaining cases, it is S_n. Since representations of S_n^2 or S_n respectively are all self-dual, if $g \in G$ or any scalar multiple thereof satisfies the Reid–Tai condition, all eigenvalues of g must be 1 except for a single -1. By the classification of reflection groups, the only one of these possibilities that can actually occur is the case $V = V_{m-1,1}$, which corresponds to the Weyl group of type A_{m-1}.

For case (5), there are only finitely many possibilities for K, and for each K, we have $\log |H| \leq r \log |\tilde{K}|$, while the minimal dimension of a faithful irreducible representation of H grows exponentially. Thus, (2) gives an upper bound on $\dim V$. The theorem follows from Lemma 32.

4 Quotients of Abelian Varieties

Let us now specialize to the case in which $X = A$ is an abelian variety and G a finite group acting on A. For any $x \in A$, the dual of the tangent space $T_x A$ can be canonically identified with $H^0(A, \Omega_A)$. By Hodge theory the representation of $\mathrm{Aut}_x(A)$ on $H^1(A, \mathbb{Q}) \otimes \mathbb{C}$ is isomorphic to the direct sum of the dual representations on $H^0(A, \Omega_A)$ and on $H^1(A, \mathcal{O}_A)$.

Definition 33. *A pair (G, V) is of* AV-type *if $V \oplus V^*$ is isomorphic to the complexification of a rational representation of G.*

We have the following elementary proposition.

Proposition 34. *Let (G, V) denote a non-RT pair of AV-type. Let $G_1 \subset G$ be a subgroup and $V_1 \subset V$ a G_1-subrepresentation such that (G_1, V_1) is a basic non-RT pair. Let $g_1 \in G_1$ be exceptional for V, and let S_1 denote the set of eigenvalues of g_1 acting on V_1, excluding 1. Then every element of S_1 is a root of unity whose order lies in*

$$\{2, 3, 4, 5, 6, 7, 8, 10, 12, 14, 18\}. \tag{3}$$

If $|S_1| > 1$, then S_1 is one of the following:

1. $\{e(1/6), e(1/3)\}$.
2. $\{e(1/6), e(1/2)\}$.
3. $\{e(1/6), e(2/3)\}$.
4. $\{e(1/3), e(1/2)\}$.
5. $\{e(1/8), e(3/8)\}$.
6. $\{e(1/8), e(5/8)\}$.
7. *A subset of $\{e(1/12), e(1/4), e(5/12)\}$.*

Proof. Let $\Sigma \subset \mathrm{Gal}(\overline{\mathbb{Q}}/\mathbb{Q})$ be the set of automorphisms σ such that V_1^σ is a G_1-subrepresentation of V. Since $V \oplus V^*$ is $\mathrm{Gal}(\overline{\mathbb{Q}}/\mathbb{Q})$-stable, $\Sigma \cup c\Sigma = \mathrm{Gal}(\overline{\mathbb{Q}}/\mathbb{Q})$, where c denotes complex conjugation. Let $e(r_1)$ be an element of S_1, set

$$S_0 = \{\sigma(e(r_1)) \mid \sigma \in \Sigma\},$$

and let $r_1, \ldots, r_k$ denote distinct rational numbers in $(0,1)$ such that $S_0 = \{e(r_1), \ldots, e(r_k)\}$. The r_i have a common denominator d, and we write $a_i = dr_i$. Since $\mathrm{age}(g_1) < 1$,

$$d > a_1 + \cdots + a_k \geq 1 + 2 + \cdots + k \geq \binom{k+1}{2} \geq \left(\frac{\phi(d)/2 + 1}{2}\right) \geq \frac{\phi(d)^2}{8}.$$

On the other hand,

$$\phi(d) = d \prod_{p \mid d} \frac{p-1}{p} \geq \frac{d}{3} \prod_{p \mid d, p \geq 5} p^{\frac{\log 4}{\log 5} - 1} \geq \frac{d^{\frac{\log 4}{\log 5}}}{3}.$$

Thus, $d < 372$, and an examination of cases by machine leads to the conclusion that d belongs to the set (3).

If α and β are two distinct elements of S_1, then there exists Σ for which $\Sigma(\alpha) \cup \Sigma(\beta)$ satisfies the Reid–Tai condition. On the other hand, $\alpha^f = \beta^f = 1$ for some $f \leq 126$. We seek to classify triples of integers (a, b, f), $0 < a < b < f \leq 126$, for which there exists a subset $\Sigma \subset \mathrm{Gal}(\overline{\mathbb{Q}}/\mathbb{Q})$ with $\Sigma \cup c\Sigma = \mathrm{Gal}(\overline{\mathbb{Q}}/\mathbb{Q})$ for which $\Sigma(e(a/f)) \cup \Sigma(e(b/f))$ satisfies the Reid–Tai condition. A machine search for such triples is not difficult and reveals that the only possibilities are given by the first six cases of the proposition together with the three pairs obtained by omitting a single element from $\{e(1/12), e(1/4), e(5/12)\}$. The proposition follows.

34.1 (Proof of Theorem 9). If A is a simple Abelian variety of dimension ≥ 4, $V = T_0 A$, and G is a finite automorphism group of A that constitutes an exception to the statement of the theorem, then (G, V) is a non-RT pair of AV-type. For every $g \in G$ and every integer k, the identity component of the kernel of $g^k - 1$, regarded as an endomorphism of A, is an Abelian subvariety of A and therefore either trivial or equal to the whole of A. It follows that all eigenvalues of g are roots of unity of the same order.

Let g denote an exceptional element. Let S_g be the multiset of eigenvalues of g on V and $S_g \cup \overline{S}_g$ the multiset of eigenvalues of g on $V \oplus V^*$. Then $S_g \cup \overline{S}_g$ can be partitioned into a union (in the sense of multisets) of $\mathrm{Gal}(\overline{\mathbb{Q}}/\mathbb{Q})$-orbits of roots of unity. Thus S_g can be partitioned into subsets $S_{g,i}$ such that for each $X_{g,i}$ either $S_{g,i} \cup \overline{S}_{g,i}$ or $S_{g,i}$ itself is a single $\mathrm{Gal}(\overline{\mathbb{Q}}/\mathbb{Q})$-orbit. If each $S_{g,i}$ is written as a set $\{e(r_{i,1}), \ldots, e(r_{i,j_i})\}$, where the $r_{i,j}$ lie in $(0,1)$, then for some i, the mean of the values $r_{i,j}$ is less than $\frac{1}{4}$. Since every root of unity is Galois-conjugate to its inverse and $\mathrm{age}(g) < 1$, if $S_{g,i}$ consists of a single $\mathrm{Gal}(\overline{\mathbb{Q}}/\mathbb{Q})$-orbit, then $S_{g,i}$ is $\{1\}$ or $\{-1\}$, in which case all eigenvalues of S_g

Table 1. Means of fractions in $[0, 1/2]$

n	$\phi(n)/2$	Values of r_j	Mean of r_j
3	1	$\dfrac{1}{3}$	$\dfrac{1}{3}$
4	1	$\dfrac{1}{4}$	$\dfrac{1}{4}$
5	2	$\dfrac{1}{5}, \dfrac{2}{5}$	$\dfrac{3}{10}$
6	1	$\dfrac{1}{6}$	$\dfrac{1}{6}$
7	3	$\dfrac{1}{7}, \dfrac{2}{7}, \dfrac{3}{7}$	$\dfrac{2}{7}$
8	2	$\dfrac{1}{8}, \dfrac{3}{8}$	$\dfrac{1}{4}$
9	3	$\dfrac{1}{9}, \dfrac{2}{9}, \dfrac{4}{9}$	$\dfrac{7}{27}$
10	2	$\dfrac{1}{10}, \dfrac{3}{10}$	$\dfrac{1}{5}$
12	2	$\dfrac{1}{12}, \dfrac{5}{12}$	$\dfrac{1}{4}$
14	3	$\dfrac{1}{14}, \dfrac{3}{14}, \dfrac{5}{14}$	$\dfrac{3}{14}$
18	3	$\dfrac{1}{18}, \dfrac{5}{18}, \dfrac{7}{18}$	$\dfrac{13}{54}$

are equal, so the mean of the values $r_{i,j}$ is always $\geq \frac{1}{2}$. For each $n > 2$ in the set (3), we present in Table 1 the set $S = \{e(r_1), e(r_2), \ldots, e(r_{\phi(n)/2})\}$ such that $S \cup \overline{S}$ contains all primitive nth roots of unity and $\sum_j r_j$ is minimal, $r_j \geq 0$:

Inspection of this table reveals that the mean of the values r_j is less than $\frac{1}{4}$ only if $n = 18$, $n = 14$, $n = 10$, or $n = 6$. In the first two cases, the condition $\dim A \geq 4$ implies that there must be at least two subsets $S_{g,i}$ in the partition, which implies $\dim A \geq 6$. Since the mean of the r_j exceeds $\frac{1}{6}$ for $n = 14$ and $n = 18$, this is impossible. If $n = 6$, all the eigenvalues of g must be $\frac{1}{6}$, and there could be as many as five. However, an abelian variety A with an automorphism that acts as the scalar $e(1/6)$ on $T_0 A$ is of the form $\mathbb{C}^g / \Lambda$, where Λ is a torsion-free $\mathbb{Z}[e(1/6)] = \mathbb{Z}[e(1/3)]$-module with the inclusion $\Lambda \to \mathbb{C}^g$ equivariant with respect to $\mathbb{Z}[e(1/3)]$. Every finitely generated torsion-free module over $\mathbb{Z}[e(1/3)]$ is free (since $\mathbb{Z}[e(1/3)]$ is a PID), so A decomposes as a product of elliptic curves with CM by $\mathbb{Z}[e(1/3)]$, contrary to hypothesis. If $n = 10$, the only possibility is that $\dim A = 4$, and the eigenvalues of g are $e(1/10), e(1/10), e(3/10), e(3/10)$. Again, $\mathbb{Z}[e(1/5)]$ is a PID, so $A = \mathbb{C}^4 / \Lambda$, where $\Lambda \cong \mathbb{Z}[e(1/5)] \oplus \mathbb{Z}[e(1/5)]$. Let $\Lambda_1 \subset \Lambda$ denote the first summand. Since A is simple, the $\mathbb{C}$-span of Λ_1 must have dimension > 2. However, letting $\lambda \in \Lambda_1$ be a generator, we can write λ as a sum of two eigenvectors for

$e(1/10) \in \mathbb{Z}[e(1/5)]$. Every element of $\mathbb{Z}[e(1/5)]\lambda$ is then a complex linear combination of these two eigenvectors.

Lemma 35. *Let (G, V) denote a non-RT pair of AV-type. Let $G_1 \subset G$ be a subgroup and $V_1 \subset V$ a G_1-subrepresentation such that (G_1, V_1) is an imprimitive basic non-RT pair. Then (G_1, V_1) is a complex reflection group.*

Proof. By Proposition 25, V_1 decomposes as a direct sum of lines that G_1 permutes. Let g_1 be an element of G_1 that is exceptional for V_1. By Proposition 25, after renumbering the L_i, g_1 interchanges L_1 and L_2 and stabilizes all the other L_i. The eigenvalues of g_1 acting on $L_1 \oplus L_2$ are therefore of the form $e(r)$ and $e(r + 1/2)$ for some $r \in [0, 1/2)$. By Proposition 34, this means that $r = 0$, $r = 1/6$, or $r = 1/8$. In the first case, g might have an additional eigenvalue $e(1/6)$ or $e(1/3)$ on one of the lines L_i, $i \geq 3$, and fix all the remaining lines pointwise. In all other cases, g_1 must fix L_i pointwise for $i \geq 3$. If g_1 has eigenvalues $-1, 1, e(1/6), 1, \ldots, 1$, eigenvalues $-1, 1, e(1/3), 1 \ldots, 1$, eigenvalues $e(1/6), e(2/3), 1, \ldots, 1$, or eigenvalues $e(1/8), e(5/8), 1, \ldots, 1$, then g_1^2 is again exceptional but stabilizes all of the lines L_i, which is impossible by Proposition 25. In the remaining case, g_1 has eigenvalues $-1, 1, \ldots, 1$, so g_1 is a reflection, and G_1 is a complex reflection group.

Theorem 36. *Let (G, V) be a non-RT pair of AV-type, $G_1 \subset G$ a subgroup, and $V_1 \subset V$ a G_1-subrepresentation such that (G_1, V_1) is a basic non-RT pair. If (G_1, V_1) is not of reflection type, then $\dim V_1 = 4$, and G_1 is contained in the reflection group G_{31} in the Shephard–Todd classification.*

Proof. Let $g_1 \in G_1$ be an exceptional element for V. If g_1 or any of its powers is a pseudoreflection on V_1, then G_1 is generated by pseudoreflections and is therefore a reflection group on V_1. We may therefore assume that every power of g_1 that is nontrivial on V_1 has at least two nontrivial eigenvalues in its action on V_1. Also, by Lemma 35, (G_1, V_1) may be assumed primitive.

We consider first the case that S_1 consists of a single element of order n. By a well-known theorem of Blichfeldt (see, e.g., [Coh76, 5.1]), a nonscalar element in a primitive group cannot have all of its eigenvalues contained in an arc of length $\pi/3$. If follows that $n \leq 5$. If $n = 5$, the spectrum of g, on V contains at least two different fifth roots of unity, and since g_1 is exceptional on V, it follows that the multiplicity of the nontrivial eigenvalue of g_1 on V_1 is 1, contrary to hypothesis. If $n = 4$, then by [Wal01], the eigenvalues 1 and i have the same multiplicity (which must be at least 2), and by [Kor86], $\dim V_1$ is a power of 2. Since the multiplicity of i is at most 3, the only possibility is that $\dim V_1 = 4$ and the eigenvalues of g_1 are $1, 1, i, i$. The classification of primitive 4-dimensional groups [Bli17] shows that the only such groups containing such an element are contained in the group G_{31} in the Shephard–Todd classification. If $n = 3$, the multiplicity of the eigenvalue $e(1/3)$ must be 2, and [Wal01] shows that there are only two possible examples, one in dimension 3 (which is projectively equivalent to the Hessian reflection group G_{25}) and one in dimension 5 (which is projectively equivalent to the reflection group G_{33}). The case $n = 2$ does not arise, since the nontrivial eigenvalue multiplicity is ≥ 2.

Thus, we need only consider the cases that $|S_1| \geq 2$. The possibilities for the multiset of nontrivial eigenvalues of g_1 acting on V_1 that are consistent with g_1 being exceptional for an AV-pair (G_1, V) are as follows:

a. $e(1/6), e(1/3)$
b. $e(1/6), e(1/6), e(1/3)$
c. $e(1/6), e(1/6), e(1/6), e(1/3)$
d. $e(1/6), e(1/3), e(1/3)$
e. $e(1/6), e(1/2)$
f. $e(1/6), e(1/6), e(1/2)$
g. $e(1/6), e(2/3)$
h. $e(1/3), e(1/2)$
i. $e(1/8), e(3/8)$
j. $e(1/8), e(5/8)$
k. $e(1/12), e(1/4)$
l. $e(1/12), e(5/12)$
m. $e(1/4), e(5/12)$
n. $e(1/12), e(1/4), e(5/12)$

Cases (a), (d), (e), (g), (h), (k), and (m) are ruled out because no power of g_1 may be a pseudoreflection. In case (f), g_1^2 has two nontrivial eigenvalues, both equal to $e(1/3)$, and we have already treated this case. Likewise, cases (j) and (l) are subsumed in our analysis of the case that there are two nontrivial eigenvalues, both equal to i.

For the four remaining cases, we observe that the conjugacy class of g_1 generates the nonabelian group G_1, so g_1 fails to commute with some conjugate h_1. The group generated by g_1 and h_1 fixes a subspace W_1 of V_1 of codimension at most 6, 8, 4, and 6 in cases (b), (c), (i), and (n) respectively that g_1 and h_1 fix pointwise. Let $U_1 \subset V_1/W_1$ denote a space on which $\langle g_1, h_1 \rangle$ acts irreducibly and on which g_1 and h_1 do not commute. The nontrivial eigenvalues of g_1 and h_1 on U_1 form subsets of the nontrivial eigenvalues of g_1 and h_1 on V_1, and the action of $\langle g_1, h_1 \rangle$ on U_1 is primitive because the eigenvalues of g_1 do not include a coset of any nontrivial subgroup of $\mathbb{C}^\times$.

We claim that if $\dim U_1 > 1$, all the nontrivial eigenvalues of g_1 on V_1 occur already in U_1. In cases (i) and (n), we have already seen that no proper subset of indicated sets of eigenvalues can appear, together with the eigenvalue 1 with some multiplicity, in any primitive irreducible representation. In cases (b) and (c), Blichfeldt's $\pi/3$ theorem implies that if the eigenvalues of some element in a primitive representation of a finite group are 1 with some multiplicity, $e(1/6)$ with some multiplicity, and possibly $e(1/3)$, then $e(1/3)$ must actually appear. Therefore, the factor U_1 must have $e(1/3)$ as eigenvalue, and every other irreducible factor of V_1 must be 1-dimensional. If no eigenvalue $e(1/6)$ appears in g_1 acting on U_1, then g_1^3 and h_1^3 commute. If all conjugates of g_1^3 commute, then G has a normal abelian subgroup. Such a subgroup must consist of scalar elements of $\mathrm{End}(V_1)$, but this is not possible given that at least one eigenvalue of g_1^3 on V_1 is 1 and at least one eigenvalue is -1. Without loss of generality, therefore, we may assume that g_1^3 and h_1^3 fail to commute. It follows that both $e(1/3)$ and $e(1/6)$ are eigenvalues of g_1 on U_1. Since case (a) has already been disposed of, the multiplicity of $e(1/6)$ as an eigenvalue of g_1 on U_1 is at least 2. In the case (b), this proves the claim. Once it has been shown that there are no solutions of type (b), it will follow that the eigenvalue $e(1/6)$ must appear with multiplicity 3, which proves the claim for (c).

Finally, we show that for each of the cases (b), (c), (i), and (n) there is no finite group G_1 with a primitive representation U_1 and an element g_1 whose multiset of nontrivial eigenvalues is as specified. First we consider whether G_1 can stabilize a nontrivial tensor decomposition of U_1. The only possibilities for g_1 respecting such a decomposition are case (b) with eigenvalues $1, e(1/6), e(1/6), e(1/3)$ decomposing as a tensor product of two representations with eigenvalues $1, e(1/6)$ and case (c) with eigenvalues $1, 1, e(1/6), e(1/6), e(1/6), e(1/3)$ decomposing as a tensor product of representations with eigenvalues $1, e(1/6)$ and $1, 1, e(1/6)$. Since U_1 is a primitive representation of G_1, Blichfeldt's theorem rules out both possibilities.

Next we rule out the possibility that G_1 normalizes a tensor decomposition with g_1 permuting tensor factors nontrivially. Given that $\dim U_1 \leq 8$, this can happen only if there are two or three tensor factors, each of dimension 2. It is easy to see that if $T_1, \ldots, T_n$ are linear transformations on a vector space V, the transformation on $V^{\otimes n}$ defined by $v_1 \otimes \cdots \otimes v_n \mapsto T_n(v_n) \otimes T_1(v_1) \otimes \cdots \otimes T_{n-1}(v_{n-1})$ has the same trace as $T_1 T_2 \cdots T_n$. It follows that any unitary transformation T on $V^{\otimes n}$ that normalizes the tensor decomposition but permutes the factors nontrivially satisfies

$$|\mathrm{tr}(T)| \leq (\dim V)^{n-1},$$

with equality only if the permutation is a transposition (ij), $T_i T_j$ is scalar, and all other factors T_i are scalar; in particular, equality implies that T^2 is scalar. Table 2 gives for each case the absolute value of the trace of g_1 acting on U_1 in terms of the dimension of U_1.

In each case, except (b) and $\dim U_1 = 4$, $\mathrm{tr}(T)$ violates the inequality, and in this case, T^2 is not scalar.

Table 2. Absolute traces of g_1 on u_1.

Case	3	4	5	6	7	8
(b)		$\sqrt{7}$	3	$\sqrt{13}$	$\sqrt{19}$	
(c)			$\sqrt{13}$	4	$\sqrt{21}$	$2\sqrt{7}$ $\sqrt{37}$
(i)		$\sqrt{3}$	$\sqrt{6}$			
(n)		2	$\sqrt{5}$	$2\sqrt{2}$	$\sqrt{13}$	

Let H_1 denote a characteristically simple normal subgroup of G_1. Since G_1 does not normalize a tensor decomposition, U_1 is an irreducible representation of H_1. Either H_1 is the product of an extraspecial p-group H_p and a group Z of scalars or H_1 is a central extension of a product of mutually isomorphic finite simple groups by a scalar group Z. Since $\dim U_1 \leq 8$, in the former case, $|H_p| \in \{2^3, 2^5, 2^7, 3^3, 5^3, 7^3\}$. In the latter case, $\overline{H}_1 = H_1/Z$ is isomorphic to K^r for some finite simple group K, and $r = 1$, since G_1 does not normalize a tensor decomposition. For a list of possibilities for H_1, we use the tables of Hiss and Malle [HM01], which are based on the classification of finite simple groups. Note that primitive groups were classified up through dimension 10 before the classification of finite simple groups was available (see, e.g., [Fei71], [Fei76] and the references therein). Table 3 enumerates the possibilities for $\overline{H}_1$, where representation numbering is that of [CCN+85] and asterisks indicate a Stone–von Neumann representation:

For the finite simple groups $\overline{H}_1$, we consult character tables [CCN+85]. This is easy to do so, since only a few of the characters in Table 3 take values whose absolute values are large enough to appear in Table 2. There are no cases in which an element of order 6 has a character absolute value as given in

Table 3. Degree ≤ 8 projective representations of $\overline{H}_1$.

Group	2	3	4	5	6	7	8
$(\mathbb{Z}/2\mathbb{Z})^2$	*						
$(\mathbb{Z}/3\mathbb{Z})^2$		*					
$(\mathbb{Z}/2\mathbb{Z})^4$			*				
$(\mathbb{Z}/5\mathbb{Z})^2$				*			
$(\mathbb{Z}/7\mathbb{Z})^2$						*	
A_5	6	3	4,8	5	9		
$(\mathbb{Z}/2\mathbb{Z})^6$							*
$L_2(7)$		2	7		4,9	5	6,11
A_6		14	8	2	16,19		4,10
$L_2(8)$						2,3	6
$L_2(11)$				2	9		
$L_2(13)$					10	2	
$L_2(17)$							12
A_7			10		2,17,24		
$U_3(3)$					2	3,4	
A_8						2	15
$L_3(4)$					41		19
$U_4(2)$			21	2	4		
A_9							2,19
J_2					22		
$S_6(2)$						2	31
$U_4(3)$					72		
$O_8^+(2)$							54

row (b) or (c) of Table 2, an element of order 8 has an absolute value as given by row (i), or an element of order 12 has an absolute value as given by row (n).

For the case that H_1 is an extraspecial p-group, every nonzero character value is an integral power of $\sqrt{p}$. This was proved for $p > 2$ by Howe [How73, Prop. 2(ii)]. For lack of a reference for $p = 2$, we sketch a proof that works in general. The embedding $H_1 \to \mathrm{GL}(U_1)$ is a Stone–von Neumann representation with central character χ. Let GG_1 denote the group of pairs $(g_1, g_2) \in G_1^2$ such that $g_1 H_1 = g_2 H_1$. There is a natural action of GG_1 on $\mathbb{C}[H_1]$ given by

$$(g_1, g_2)([h_1]) = \left[g_1 h_1 g_2^{-1} \right].$$

The restriction of this representation to $H_1^2 \subset GG_1$ is of the form $\bigoplus V_i \boxtimes V_i^*$, where the sum is taken over all irreducible representations V_i of H_1. The factor $U_1 \boxtimes U_1^*$ is the $\chi \boxtimes \chi^*$ eigenspace of the center Z^2 of H_1^2 acting on $\mathbb{C}[H_1]$, where χ is the central character of Z on U_1. Since the action of GG_1 on this eigenspace of $\mathbb{C}[H_1]$ extends the irreducible representation of H_1^2 on $U_1 \boxtimes U_1^*$, any other extension of $\left(H_1^2, U_1 \boxtimes U_1^* \right)$ to GG_1 is projectively equivalent to this one. The particular extension we have in mind is obtained by letting $(g_1, g_2) \in GG_1$ act on $U_1 \boxtimes U_1^*$ according to the action of g_1 on U_1 and the action of g_2 on U_1^* coming from the inclusion $G_1 \subset \mathrm{GL}(U_1)$. From this it is easy to see that the character value of (g_1, g_1) on each Z^2-eigenspace of $\mathbb{C}[H_1]$ is either 0 or $\left| \overline{H}_1^{g_1} \right|$. Since $\overline{H}_1^{g_1}$ is a vector space over $\mathbb{Z}/p\mathbb{Z}$, $\mathrm{tr}(g_1|U_1)\mathrm{tr}\,(g_1|U_1^*)$ is either 0 or an integer power of p. Consulting Table 2, we see that this rules out every possibility except a character value 2 and $\dim U_1 = 4$. This can actually occur, but not with the eigenvalues of case (b).

5 Examples

We conclude with some examples to illustrate various aspects of the classification given above. We begin with some examples from group theory.

In principle, all non-RT pairs can be built up from basic pairs, by reversing the operations that led to constructing basic pairs in the first place, i.e., by replacing G by an extension $\tilde{G}$ of G whose image in $\mathrm{Aut}(V)$ is the same as that of G; by combining (G, V_1) and (G, V_2) to give the pair $(G, V_1 \oplus V_2)$ (which may or may not be non-RT); and by replacing (G, V) by (G', V), where G' lies between G and its normalizer in $\mathrm{Aut}(V)$. To illustrate this, we observe that all non-RT pairs of the form $((\mathbb{Z}/2\mathbb{Z})^n \rtimes H, \mathbb{C}^n)$, where $H \subset S_n$ is a transitive group, arise from the basic non-RT pair $(\mathbb{Z}/2\mathbb{Z}, \mathbb{C})$. This accounts for the series of Weyl groups of type B_n/C_n, but not for the Weyl groups of type D_n, which are primitive. This construction can be used more generally to build non-RT pairs of the form $(G^n \rtimes H, V^n)$ starting with a non-RT pair (G, V) and a transitive permutation group $H \subset S_n$.

It may happen that a basic non-RT pair (G, V) of reflection type nevertheless fails to have an exceptional element that is a scalar multiple of a pseudoreflection. Consider the case $G = U_4(2) \times \mathbb{Z}/3\mathbb{Z}$ and V is a faithful irreducible 5-dimensional representation of G. Then G has an exceptional element g whose eigenvalues are $1, 1, 1, e(1/3), e(1/3)$ and whose conjugacy class generates G. It has another element h with eigenvalues $1, -1, -1, -1, -1$ that is not exceptional. Since $-h$ is a reflection, it is easy to see that $U_4(2) \times \mathbb{Z}/2\mathbb{Z}$ is a 5-dimensional reflection group (in fact, it is G_{33} in the Shephard–Todd classification), and of course this reflection group is projectively equivalent to (G, V).

There really does exist a primitive 4-dimensional non-RT pair (G, V) that is not of reflection type. By [Bli17], there is a short exact sequence

$$0 \to I_4 \to G_{31} \to S_6 \to 0,$$

where I_4 is the central product of $\mathbb{Z}/4\mathbb{Z}$ and any extraspecial 2-group of order 32. The group S_6 contains two nonconjugate subgroups isomorphic to S_5, whose inverse images in G_{31} are primitive. One is the reflection group G_{29}, and one contains elements with eigenvalues $1, 1, i, i$. The question arises whether these two groups are conjugate. The character table of G_{29}, provided by the software package CHEVIE [GHL$^+$96], reveals that this group has two faithful 4-dimensional representations. One has reflections and the other has elements with spectrum $1, 1, i, i$. It follows that G_{29} with respect to this nonreflection representation, or equivalently, the nonreflection index-6 subgroup of G_{31}, gives the desired example. This example (in fact all of G_{31}) can actually be realized inside $\mathrm{GL}_4(\mathbb{Z}[i])$, as shown in [Bli17].

The set of projective equivalence classes of basic non-RT pairs that are of AV-type is infinite, as is the set of basic non-RT pairs that are not. We have already mentioned the Weyl groups of type D_n as examples of the first kind; the reflection groups $(\mathbb{Z}/k\mathbb{Z})^{n-1} \times S_n$ are never of AV-type if $k > 4$.

We conclude with some geometric examples.

If (G, V) is a non-RT pair and $V = V_0 \otimes_{\mathbb{Q}} \mathbb{C}$ for some rational representation V_0 of G, then there exist an Abelian variety A and a homomorphism $G \to \mathrm{Aut}(A)$ such that A/G is uniruled and the Lie algebra of A is isomorphic to V as a G-module. Indeed, we may choose any integral lattice $\Lambda_0 \subset V_0$ that is G-stable and define $A = \mathrm{Hom}(\Lambda_0, E)$ for any elliptic curve E. If V is irreducible, then A/G is rationally connected. This includes all examples in which G is the Weyl group of a root system and V_0 is the $\mathbb{Q}$-span of the root system. When Λ_0 is taken to be the root system, the quotients A/G are in fact weighted projective spaces by a theorem of E. Looijenga [Loo77], which was one of the motivations for this paper.

Let Λ_0 denote the (12-dimensional) Coxeter–Todd lattice, which we regard as a free module of rank 6 over $R := \mathbb{Z}[e(1/3)]$. Let $G = G_{34}$ denote the group of R-linear isometries of this lattice. If E denotes the elliptic curve over $\mathbb{C}$ with complex multiplication by R, then $\mathrm{Hom}_R(\Lambda_0, E)/G$ is rationally connected. The group G is a reflection group but not a Weyl group, and we do not know whether this variety is rational or even unirational.

We have already observed that there is a 4-dimensional basic non-RT pair (G, V) that is not of reflection type and such that $G \subset \mathrm{GL}_4(\mathbb{Z}[i]) \subset \mathrm{GL}(V)$. If E denotes the elliptic curve with CM by $\mathbb{Z}[i]$, then E^4/G is rationally connected. Again, we do not know about rationality or unirationality.

Let A be an abelian variety with complex multiplication by $\mathbb{Z}[e(1/7)]$ with CM type chosen so that some automorphism g of order 7 has eigenvalues $e(1/7), e(2/7), e(3/7)$ acting on $T_0 A$. Then $A/\langle g \rangle$ is rationally connected, but once again we do not know whether it is rational or unirational.

Let E be any elliptic curve, $A = E^3$, and $G = S_3 \times \{\pm 1\}$. Let the factors S_3 and $\{\pm 1\}$ of G act on A by permuting factors and by multiplication respectively. If $V = T_0 A = \mathbb{C}^3$ and W denotes the plane in which coordinates sum to zero, then the images of G in W and in V/W are reflection groups. Thus $A_{\mathrm{stab}}^{RT} = A$, so A/G is rationally connected by Corollary 24. Yet again, we do not know about the rationality or unirationality of the quotient.

There are imprimitive basic non-RT triples of arbitrarily large degree that can be realized by automorphism groups of Calabi–Yau varieties but not as automorphism groups of Abelian varieties. For example,

$$G_{n+2,1,n} = (\mathbb{Z}/(n+2)\mathbb{Z})^n \rtimes S_n$$

acts on the n-dimensional Fermat hypersurface $x_0^{n+2} + \cdots + x_{n+1}^{n+2} = 0$ fixing the coordinates x_0 and x_1 and therefore the point

$$P = \left(1 : e^{\frac{\pi i}{n+2}} : 0 : \cdots : 0 \right).$$

The action of G on the tangent space to P gives the reflection representation of $G_{n+2,1,n}$.

References

[ASH85] Z. ARAD, J. STAVI, and M. HERZOG – "Powers and products of conjugacy classes in groups," Products of conjugacy classes in groups, Lecture Notes in Math., vol. 1112, Springer, Berlin, 1985, pp. 6–51.

[BDPP] S. BOUCKSOM, J.-P. DEMAILLY, M. PAUN, and T. PETERNELL – "The pseudo-effective cone of a compact Kähler manifold and varieties of negative Kodaira dimension," arXiv:math.AG/0405285.

[Bea83a] A. BEAUVILLE – "Some remarks on Kähler manifolds with $c_1 = 0$," Classification of algebraic and analytic manifolds (Katata, 1982), Progr. Math., vol. 39, Birkhäuser Boston, Boston, MA, 1983, pp. 1–26.

[Bea83b] — , "Variétés Kähleriennes dont la première classe de Chern est nulle," J. Differential Geom. 18 (1983), no. 4, pp. 755–782 (1984).

[Bli17] H. F. BLICHFELDT – Finite collineation groups, with an introduction to the theory of groups of operators and substitution groups, University of Chicago Press, Chicago, Ill., 1917.

[CCN+85] J. H. CONWAY, R. T. CURTIS, S. P. NORTON, R. A. PARKER, and R. A.
WILSON – *Atlas of finite groups*, Oxford University Press, Eynsham,
1985, Maximal subgroups and ordinary characters for simple groups,
with computational assistance from J. G. Thackray.

[Coh76] A. M. COHEN – "Finite complex reflection groups," *Ann. Sci. École
Norm. Sup. (4)* **9** (1976), no. 3, p. 379–436.

[EGH99] E. W. ELLERS, N. GORDEEV, and M. HERZOG – "Covering numbers for
Chevalley groups," *Israel J. Math.* **111** (1999), pp. 339–372.

[Fei71] W. FEIT – "The current situation in the theory of finite simple
groups," Actes du Congrès International des Mathématiciens (Nice,
1970), Tome 1, Gauthier-Villars, Paris, 1971, pp. 55–93.

[Fei76] — , "On finite linear groups in dimension at most 10", *Proceedings of
the Conference on Finite Groups (Univ. Utah, Park City, Utah, 1975)*
(New York), Academic Press, 1976, pp. 397–407.

[Fuj78] T. FUJITA – "On Kähler fiber spaces over curves," *J. Math. Soc. Japan*
30 (1978), no. 4, pp. 779–794.

[GH79] P. GRIFFITHS and J. HARRIS – "Algebraic geometry and local differential
geometry," *Ann. Sci. École Norm. Sup. (4)* **12** (1979), no. 3, pp. 355–452.

[GHL+96] M. GECK, G. HISS, F. LÜBECK, G. MALLE, and G. PFEIFFER –
"CHEVIE—a system for computing and processing generic character
tables," *Appl. Algebra Engrg. Comm. Comput.* **7** (1996), no. 3,
pp. 175–210, Computational methods in Lie theory (Essen, 1994).

[GHS03] T. GRABER, J. HARRIS, and J. STARR – "Families of rationally connected varieties," *J. Amer. Math. Soc.* **16** (2003), no. 1, pp. 57–67
(electronic).

[Gri70] P. A. GRIFFITHS – "Periods of integrals on algebraic manifolds. III.
Some global differential-geometric properties of the period mapping,"
Inst. Hautes Études Sci. Publ. Math. (1970), no. 38, pp. 125–180.

[GT07] R. M. GURALNICK and P. H. TIEP – "A Problem of Kollár and Larsen
on Finite Linear Groups and Crepant Resolutions," May 2009.

[HM01] G. HISS and G. MALLE – "Low-dimensional representations of quasi-
simple groups," *LMS J. Comput. Math.* **4** (2001), pp. 22–63 (electronic).

[HM05] C. D. HACON and J. MCKERNAN – "Shokurov's Rational Connectedness
Conjecture," Duke Math., 138 (2007), vol, 119–136.

[How73] R. E. HOWE – "On the character of Weil's representation," *Trans. Amer.
Math. Soc.* **177** (1973), pp. 287–298.

[IR96] Y. ITO and M. REID – "The McKay correspondence for finite subgroups
of SL(3, **C**)," Higher-dimensional complex varieties (Trento, 1994), de
Gruyter, Berlin, 1996, pp. 221–240.

[Kaw81] Y. KAWAMATA – "Characterization of abelian varieties," *Compositio
Math.* **43** (1981), no. 2, pp. 253–276.

[Kaw85] — , "Minimal models and the Kodaira dimension of algebraic fiber
spaces," *J. Reine Angew. Math.* **363** (1985), pp. 1–46.

[KM98] J. KOLLÁR and S. MORI – *Birational geometry of algebraic varieties*,
Cambridge Tracts in Mathematics, vol. 134, Cambridge University Press,
Cambridge, 1998.

[Kol96] J. KOLLÁR – *Rational curves on algebraic varieties*, Ergebnisse der Mathematik und ihrer Grenzgebiete. 3. Folge. A Series of Modern Surveys in
Mathematics [Results in Mathematics and Related Areas. 3rd Series.

A Series of Modern Surveys in Mathematics], vol. 32, Springer-Verlag, Berlin, 1996.

[Kor86] A. V. KORLYUKOV – "Finite linear groups generated by quadratic elements of order 4," *Vestsī Akad. Navuk BSSR Ser. Fīz.-Mat. Navuk* (1986), no. 4, pp. 38–40, 124.

[KT04] A. S. KLESHCHEV and P. H. TIEP – "On restrictions of modular spin representations of symmetric and alternating groups," *Trans. Amer. Math. Soc.* **356** (2004), no. 5, pp. 1971–1999 (electronic).

[LL98] R. LAWTHER and M. W. LIEBECK – "On the diameter of a Cayley graph of a simple group of Lie type based on a conjugacy class," *J. Combin. Theory Ser. A* **83** (1998), no. 1, pp. 118–137.

[Loo77] E. LOOIJENGA – "Root systems and elliptic curves," *Invent. Math.* **38** (1976/77), no. 1, pp. 17–32.

[LS04] M. W. LIEBECK and A. SHALEV – "Fuchsian groups, coverings of Riemann surfaces, subgroup growth, random quotients and random walks," *J. Algebra* **276** (2004), no. 2, pp. 552–601.

[MM86] Y. MIYAOKA and S. MORI – "A numerical criterion for uniruledness," *Ann. of Math. (2)* **124** (1986), no. 1, pp. 65–69.

[MMM88] S. MORI, D. R. MORRISON, and I. MORRISON – "On four-dimensional terminal quotient singularities," *Math. Comp.* **51** (1988), no. 184, pp. 769–786.

[Mor85] D. R. MORRISON – "Canonical quotient singularities in dimension three," *Proc. Amer. Math. Soc.* **93** (1985), no. 3, pp. 393–396.

[MS84] D. R. MORRISON and G. STEVENS – "Terminal quotient singularities in dimensions three and four," *Proc. Amer. Math. Soc.* **90** (1984), no. 1, pp. 15–20.

[Nak60] M. NAKAOKA – "Decomposition theorem for homology groups of symmetric groups," *Ann. of Math. (2)* **71** (1960), pp. 16–42.

[Pet94] T. PETERNELL – "Minimal varieties with trivial canonical classes. I," *Math. Z.* **217** (1994), no. 3, pp. 377–405.

[Rei80] M. REID – "Canonical 3-folds," Journées de Géometrie Algébrique d'Angers, Juillet 1979/Algebraic Geometry, Angers, 1979, Sijthoff & Noordhoff, Alphen aan den Rijn, 1980, pp. 273–310.

[Rei87] — , "Young person's guide to canonical singularities," Algebraic geometry, Bowdoin, 1985 (Brunswick, Maine, 1985), Proc. Sympos. Pure Math., vol. 46, Amer. Math. Soc., Providence, RI, 1987, pp. 345–414.

[Rei02] — , "La correspondance de McKay," *Astérisque* (2002), no. 276, pp. 53–72, Séminaire Bourbaki, Vol. 1999/2000.

[ST54] G. C. SHEPHARD and J. A. TODD – "Finite unitary reflection groups," *Canadian J. Math.* **6** (1954), pp. 274–304.

[SZ93] G. M. SEITZ and A. E. ZALESSKII – "On the minimal degrees of projective representations of the finite Chevalley groups. II," *J. Algebra* **158** (1993), no. 1, pp. 233–243.

[Wal01] D. WALES – "Quasiprimitive linear groups with quadratic elements," *J. Algebra* **245** (2001), no. 2, pp. 584–606.

[Zha96] Q. ZHANG – "On projective manifolds with nef anticanonical bundles," *J. Reine Angew. Math.* **478** (1996), pp. 57–60.

[Zha05] — , "On projective varieties with nef anticanonical divisors," *Math. Ann.* **332** (2005), no. 3, pp. 697–703.

Notes on Motives in Finite Characteristic

Maxim Kontsevich

IHES, 35 route de Chartres, Bures-sur-Yvette 91440, France
maxim@ihes.fr

To Yuri Manin on the occasion of his 70th birthday, with admiration

Summary. Motivic local systems over a curve in finite characteristic form a countable set endowed with an action of the absolute Galois group of rational numbers commuting with the Frobenius map. I will discuss three series of conjectures about such sets, based on an analogy with algebraic dynamics, on a formalism of commutative algebras of motivic integral operators, and on an analogy with 2-dimensional lattice models in statistical physics.

Key words: finite fields, Langlands correspondence, algebraic dynamics, motivic sheaves, lattice models, 2-categories

2000 Mathematics Subject Classifications: 11G25, 37K20, 37C30, 11R39, 82B20, 18D05

Introduction and an example

These notes grew from an attempt to interpret a formula of Drinfeld (see [3]) enumerating the absolutely irreducible local systems of rank 2 on algebraic curves over finite fields, obtained as a corollary of the Langlands correspondence for GL(2) in the function field case, together with the trace formula.

Let C be a smooth projective geometrically connected curve defined over a finite field $\mathbb{F}_q$, with a base point $v \in C(\mathbb{F}_q)$. The geometric fundamental group $\pi_1^{\text{geom}}(C, v) := \pi_1(C \times_{\text{Spec}\,\mathbb{F}_q} \text{Spec}\,\overline{\mathbb{F}}_q, v)$ is a profinite group on which the Galois group $\widehat{\mathbb{Z}} = \text{Gal}(\overline{\mathbb{F}}_q/\mathbb{F}_q)$ (with the canonical generator $\text{Fr} := \text{Fr}_q$) acts. In what follows we will omit the base point from the notation.

Theorem 1. (Drinfeld) *Under the above assumptions, for any integer $n \geq 1$ and any prime $l \neq \text{char}(\mathbb{F}_q)$, the set of fixed points*

Y. Tschinkel and Y. Zarhin (eds.), *Algebra, Arithmetic, and Geometry*,
Progress in Mathematics 270, DOI 10.1007/978-0-8176-4747-6_7,
© Springer Science+Business Media, LLC 2009

$$X_n^{(l)} := \left(\mathrm{IrrRep}\left(\pi_1(C \times_{\mathrm{Spec}\,\mathbb{F}_q} \mathrm{Spec}\,\overline{\mathbb{F}}_q) \to \mathrm{GL}(2, \overline{\mathbb{Q}}_l)\right)/\mathrm{conjugation}\right)^{\mathrm{Fr}^n}$$

is finite. Here $\mathrm{IrrRep}(\ldots)$ *denotes the set of conjugacy classes of irreducible continuous 2-dimensional representations of* $\pi_1^{\mathrm{geom}}(C)$ *defined over finite extensions of* $\mathbb{Q}_l$. *Moreover, there exist a finite collection* $(\lambda_i) \in \overline{\mathbb{Q}}^{\times}$ *of algebraic integers and signs* $(\epsilon_i) \in \{-1, +1\}$ *depending only on* C *such that for any* n, l *one has the equality*

$$\#X_n^{(l)} = \sum_i \epsilon_i \lambda_i^n \ .$$

From the explicit formula that one can extract from [3], one can see that numbers λ_i are q-Weil algebraic integers whose norm for any embedding $\overline{\mathbb{Q}} \hookrightarrow \mathbb{C}$ belongs to $q^{\frac{1}{2}\mathbb{Z}_{\geq 0}}$. Therefore, the number of elements of $X_n^{(l)}$, $n = 1, 2, \ldots$, looks like the number of $\mathbb{F}_{q^n}$-points on some variety over $\mathbb{F}_q$. The largest exponent is q^{4g-3}, which indicates that this variety has dimension $4g - 3$. A natural guess is that it is closely related to the moduli space of stable bundles of rank 2 over C. At least the dimensions coincide, and Weil numbers that appear are essentially the same; they are products of the eigenvalues of the Frobenius acting on the motive defined by the first cohomology of C.

The Langlands correspondence identifies $X_n^{(l)}$ with the set of $\overline{\mathbb{Q}}_l$-valued unramified cuspidal automorphic forms for the adelic group $\mathrm{GL}(2, \mathbb{A}_{\mathbb{F}_q(C)})$. These forms are eigenvectors of a collection of commuting matrices (Hecke operators) with integer coefficients. Therefore, for a given $n \geq 1$ one can identify[1] all sets $X_n^{(l)}$ for various primes l with one set X_n endowed with an action of the absolute Galois group $\mathrm{Gal}(\overline{\mathbb{Q}}/\mathbb{Q})$, extending the obvious actions of $\mathrm{Gal}(\overline{\mathbb{Q}}_l/\mathbb{Q}_l)$ on $X_n^{(l)}$.

Today, the Langlands correspondence in the function field case has been established for all the groups $\mathrm{GL}(N)$ by L. Lafforgue. To my knowledge, almost no attempts have been made to extend Drinfeld's calculation to the case of higher rank, or even to the $\mathrm{GL}(2)$ case with nontrivial ramification.

It is convenient to take the inductive limit $X_\infty := \varinjlim X_n$, $X_{n_1} \hookrightarrow X_{n_1 n_2}$, which is an infinite countable set endowed with an action of the product[2]

$$\mathrm{Gal}(\overline{\mathbb{Q}}/\mathbb{Q}) \times \mathrm{Gal}(\overline{\mathbb{F}}_q/\mathbb{F}_q) \ .$$

The individual set X_n can be reconstructed from this datum as the set of fixed points of $\mathrm{Fr}^n \in \mathrm{Gal}(\overline{\mathbb{F}}_q/\mathbb{F}_q)$.

In spite of the numerical evidence, it would be too naive to expect a natural identification of X_∞ with the set of $\overline{\mathbb{F}}_q$-points of an algebraic variety defined

---

[1]It is expected that all representations from $X_n^{(l)}$ are motivic, i.e., they arise from projectors with coefficients in $\overline{\mathbb{Q}}$ acting on l-adic cohomology of certain projective varieties defined over the field of rational functions $\mathbb{F}_{q^n}(C)$.

[2]One can replace $\mathrm{Gal}(\overline{\mathbb{Q}}/\mathbb{Q})$ by its quotient $\mathrm{Gal}(\mathbb{Q}^{q-\mathrm{Weil}}/\mathbb{Q})$ where $\mathbb{Q}^{q-\mathrm{Weil}} \subset \overline{\mathbb{Q}}$ is CM-field generated by all q-Weil numbers.

over $\mathbb{F}_q$, since there is no obvious mechanism producing a nontrivial $\mathrm{Gal}(\overline{\mathbb{Q}}/\mathbb{Q})$-action on the latter.

The main question addressed here is the following:

Question 2. Does there exist some alternative way to construct the set X_∞ with the commuting action of two Galois groups?

In the present notes I will offer three different hypothetical constructions. The first construction comes from the analogy between the Frobenius acting on $\pi_1^{\mathrm{geom}}(C)$ and an element of the mapping class group acting on the fundamental group of a closed oriented surface; the second one is almost tautological and arises from the contemplation on the shape of explicit formulas for Hecke operators (see an example in Section 0.1); the third one is based on an analogy with lattice models in statistical physics.

I propose several conjectures, which should be better considered as guesses in the first and in the third part, as there is almost no experimental evidence in their favor. In a sense, the first and the third part should be regarded as science fiction, but even if the appropriate conjectures are wrong (as I strongly suspect), there should be some grains of truth in them.

On the contrary, I feel quite confident that the conjectures made in the second part are essentially true, the output is a higher-dimensional generalization of the Langlands correspondence in the functional field case. At the end of the second part I will show how to make a step in the arithmetic direction, extending the formulas to the case of an arbitrary local field.

In the fourth part I will describe briefly a similarity between a modification of the category of motives based on non-commutative geometry, and two other categories introduced in the second and the third part. Also I will make a link between the proposal based on polynomial dynamics and the one based on lattice models.

Finally, I apologize to the reader that the formulas in Sections 0.1 and 1.3 are given without explanations, this is the result of my poor knowledge of the representation theory. The formulas were polished with the help of computer.

Acknowledgements: I am grateful to many people for useful discussions, especially to Vladimir Drinfeld (on the second part of this paper), to Laurent Lafforgue who told me about [3] and explained some basic stuff about automorphic forms, to Vincent Lafforgue who proposed an argument for the Conjecture 5 based on lifting, to James Milne for consultations about the category of motives over a finite field, to Misha Gromov who suggested the idea to imitate Dwork's methods in the third part, also to Mitya Lebedev, Sasha Goncharov, Dima Grigoriev, Dima Kazhdan, Yan Soibelman and Don Zagier. Also I am grateful to the referee for useful remarks and corrections.

0.1 An explicit example

Here we will show explicit formulas for the tower $(X_n)_{n\geq 1}$ in the simplest truly non-trivial case. Consider the affine curve $C = \mathbb{P}^1_{\mathbb{F}_q} \setminus \{0, 1, t, \infty\}$ for a given element $t \in \mathbb{F}_q \setminus \{0,1\}$. We are interested in motivic local $SL(2)$-systems on C with tame non-trivial unipotent monodromies around all punctures $\{0,1,t,\infty\}$.

A lengthy calculation lead to the following explicit formulas[3] for the Hecke operators for cuspidal representations. In what follows we assume char $\mathbb{F}_q \neq 2$. The Hecke operators act on the spaces of functions on certain double coset space for the adelic group, which can be identified with the set of equivalence classes of vector bundles of rank 2 over $\mathbb{P}^1_{\mathbb{F}_q}$ together with a choice of one-dimensional subspaces of fibers at $\{0,1,t,\infty\}$. This double coset space is infinite, but the eigenfunctions of Hecke operators corresponding to cuspidal representations have finite support, which one can bound a priori.

For any $x \in \mathbb{F}_q$, the Hecke operator T_x (on cuspidal forms) can be written as an integral $q \times q$ matrix whose rows and columns are labeled by elements of $\mathbb{F}_q$ (i.e., $T_x \in \mathrm{Mat}(\mathbb{F}_q \times \mathbb{F}_q; \mathbb{Z})$). Coefficients of T_x are given by the formula

$$(T_x)_{yz} := 2 - \#\{w \in \mathbb{F}_q \,|\, w^2 = f_t(x,y,z)\} + (\text{correction term}),$$

where $f_t(x,y,z)$ is the following universal polynomial with integral coefficients:

$$f_t(x,y,z) := (xy + yz + zx - t)^2 + 4xyz(1 + t - (x + y + z)) .$$

The correction term is equal to

$$-\begin{cases} q+1 & x = y \in \{0,1,t\} \\ 1 & x = y \notin \{0,1,t\} \\ 0 & x \neq y \end{cases} + \begin{cases} q & \text{if } x \notin \{0,1,t\} \text{ and } \begin{cases} y = \dfrac{t}{x}, & z = 0, \\ y = \dfrac{t-x}{1-x}, & z = 1, \\ y = \dfrac{t(1-x)}{t-x}, & z = t, \end{cases} \\ 0 & \text{otherwise.} \end{cases}$$

Operators $(T_x)_{x \in \mathbb{F}_q}$ satisfy the following properties:

1. $[T_{x_1}, T_{x_2}] = 0$;
2. $\sum_{x \in \mathbb{F}_q} T_x = \mathbf{1} = id_{\mathbb{Z}^{\mathbb{F}_q}}$;
3. $T_x^2 = \mathbf{1}$ for $x \in \{0,1,t\}$; moreover, $\{\mathbf{1}, T_0, T_1, T_t\}$ form a group under multiplication, isomorphic to $\mathbb{Z}/2\mathbb{Z} \oplus \mathbb{Z}/2\mathbb{Z}$,
4. for any $x \notin \{0,1,t\}$ the spectrum of T_x is real and belongs to $[-2\sqrt{q}, +2\sqrt{q}]$; any element of $\mathrm{Spec}(T_x)$ can be written as $\lambda + \overline{\lambda}$, where $|\lambda| = \sqrt{q}$ is a q-Weil number;
5. for any $\xi = \lambda + \overline{\lambda} \in \mathrm{Spec}(T_x)$ and any integer $n \geq 1$ the spectrum of the matrix $T_x^{(n)}$ corresponding to $x \in \mathbb{F}_q \subset \mathbb{F}_{q^n}$ (if we pass to the extension $\mathbb{F}_{q^n} \supset \mathbb{F}_q$) contains the element $\xi^{(n)} := \lambda^n + \overline{\lambda}^n$;

---

[3]I was informed by V. Drinfeld that a similar calculation for the case of $SL(2)$ local systems on $\mathbb{P}^1_{\mathbb{F}_q} \setminus \{4 \text{ points}\}$, with tame non-trivial *semisimple* monodromy around punctures was performed few years ago by Teruji Thomas.

6. the vector space generated by $\{T_x\}_{x \in \mathbb{F}_q}$ is closed under the product; the multiplication table is

$$T_x \cdot T_y = \sum_{z \in \mathbb{F}_q} c_{xyz} T_z \text{ where } c_{xyz} = (T_x)_{yz}.$$

Typically (for "generic" t, x) the characteristic polynomial of T_x splits into the product of four irreducible polynomials of almost the same degree. The splitting is not surprising, since we have a group[4] of order 4 commuting with all operators T_x (see property 3). Computer experiments indicate that the Galois groups of these polynomials (considered as permutation groups) tend to be rather large, typically the full symmetric groups if q is prime, and the corresponding number fields have huge factors in the prime decomposition of the discriminant.

Notice that in the theory of automorphic forms one usually deals with infinitely many commuting Hecke operators corresponding to all places of the global field, i.e. to closed points of C (in other words, to orbits of $\mathrm{Gal}(\overline{\mathbb{F}}_q/\mathbb{F}_q)$ acting on $C(\overline{\mathbb{F}}_q)$). Here we are writing formulas only for the points defined over $\mathbb{F}_q$. The advantage of our example is that the number these operators coincides with the size of Hecke matrices, hence one can try to write formulas for structure constants, which by luck turn out to coincide with the matrix coefficients of matrices T_x (property 6).

1 First proposal: algebraic dynamics

As was mentioned before, it is hard to imagine a mechanism for a non-trivial action of the absolute Galois group of $\mathbb{Q}$ on the set of points of a variety over a finite field. One can try to exchange the roles of fields $\mathbb{Q}$ and $\mathbb{F}_q$. The first proposal is the following one:

Conjecture 3. *For a tower* $(X_n)_{n \geq 1}$ *arising from automorphic forms (or from motivic local systems on curves), as defined in the Introduction, there exists a variety X defined over $\mathbb{Q}$ and a map $F : X \to X$ such that there is a family of bijections*

$$X_n \simeq (X(\overline{\mathbb{Q}}))^{F^n}, \quad n \geq 1$$

covariant with respect to $\mathrm{Gal}(\overline{\mathbb{Q}}/\mathbb{Q}) \times \mathbb{Z}/n\mathbb{Z}$ actions, and with respect to inclusions $X_{n_1} \subset X_{n_1 n_2}$ for integers $n_1, n_2 \geq 1$.

1.1 The case of $GL(1)$

Geometric class field theory gives a description of the sets $(X_n)_{n \geq 1}$ in terms of the Jacobian of C:

$$X_n = (\mathrm{Jac}_C(\mathbb{F}_{q^n}))^{\vee}(\overline{\mathbb{Q}}) = \mathrm{Hom}(\mathrm{Jac}_C(\mathbb{F}_{q^n}), \overline{\mathbb{Q}}^{\times}).$$

---

[4]This is the group of automorphisms of $\mathbb{P}^1 \setminus \{4 \text{ points}\}$ for the generic cross-ratio.

The number of elements of this set is equal to

$$\# \operatorname{Jac}_C(\mathbb{F}_{q^n}) = \det(\operatorname{Fr}^n_{H^1(C)} - 1)$$

where $\operatorname{Fr}_{H^1(C)}$ is the Frobenius operator acting on, say, l-adic first cohomology group of C.

One can propose a blatantly non-canonical candidate for the corresponding dynamical system (X, F). Namely, let us choose a semisimple $(2g \times 2g)$ matrix $A = (A_{i,j})_{1 \le i,j \le 2g}$ (where g is the genus of C) with coefficients in $\mathbb{Z}$, whose characteristic polynomial is equal to the characteristic polynomial of $\operatorname{Fr}_{H^1(C)}$. Define $X/\mathbb{Q}$ to be the standard $2g$-dimensional torus $\mathbb{G}_m^{2g} = \operatorname{Hom}(\mathbb{Z}^{2g}, \mathbb{G}_m)$, and the map F to be the dual to the map $A : \mathbb{Z}^{2g} \to \mathbb{Z}^{2g}$:

$$F(z_1, \dots, z_{2g}) = \left(\prod_i z_i^{A_{i,1}}, \dots, \prod_i z_i^{A_{i,2g}} \right).$$

Moreover, one can choose A in such a way that

$$F^*\omega = q\omega, \text{ where } \omega = \sum_{i=1}^{g} \frac{dz_i}{z_i} \wedge \frac{dz_{g+i}}{z_{g+i}}.$$

On the set of fixed points of F^n, the groups $\operatorname{Gal}(\overline{\mathbb{Q}}/\mathbb{Q})$ (via the cyclotomic quotient) and $\mathbb{Z}/n\mathbb{Z}$ (by powers of F) act simultaneously. Nothing contradicts the existence of an equivariant isomorphism between two towers of finite sets.

1.2 Moduli of local systems on surfaces

One can interpret the scheme $\mathbb{G}_m^{2g}$ as the moduli space of rank-1 local systems on an oriented closed topological surface S of genus g; the form ω is the natural symplectic form on this moduli space.

In general, for any $N \ge 1$, one can make an analogy between the action of the Frobenius Fr on the the set of l-adic irreducible representations of $\pi_1^{\operatorname{geom}}(C)$ of rank N and the action of the isotopy class of a homeomorphism $\varphi : S \to S$ on the set of irreducible complex representations of $\pi_1(S)$ of the same rank. Sets of representations are similar to each other, for it is known that the maximal quotient of $\pi_1^{\operatorname{geom}}(C)$ coprime to q is isomorphic to the analogous quotient of the profinite completion $\widehat{\pi}_1(S)$ of $\pi_1(S)$. Also, if we assume that there are only finitely many fixed points of φ acting on

$$\operatorname{IrrRep}(\pi_1(S) \to GL(2, \mathbb{C}))/\text{conjugation},$$

then the sets

$$X^{(l)} := \left(\operatorname{IrrRep}(\widehat{\pi}_1(C \times_{\operatorname{Spec}\mathbb{F}_q} \operatorname{Spec}\overline{\mathbb{F}}_q) \to GL(2, \overline{\mathbb{Q}}_l))/\text{conjugation} \right)^{\varphi}$$

do not depend on the prime l for l large enough.

All this leads to the following conjecture (which is formulated a bit sloppily), a strengthening of Conjecture 1:

Conjecture 4. *For any smooth compact geometrically connected curve $C/\mathbb{F}_q$ of genus $g \geq 2$ there exists an endomorphism Φ_C of the tensor category of finite-dimensional complex local systems on S such that*

- *Φ_C is algebraic and defined over $\mathbb{Q}$, in the sense that it acts on the moduli stack of irreducible local systems of any given rank $N \geq 1$ by a rational map defined over $\mathbb{Q}$;*
- *Φ_C multiplies the natural symplectic form on the moduli space of irreducible local systems of rank N by the constant q;*
- *for every $n, N \geq 1$ there exists an identification of the set of isomorphism classes of irreducible motivic local systems of rank N on $C \times_{\mathrm{Spec}\mathbb{F}_q} \mathrm{Spec}\,\overline{\mathbb{F}}_q$ invariant under Fr^n with the set of isomorphism classes of $\overline{\mathbb{Q}}$-local systems of rank N on S invariant under Φ_C^n, compatible with the relevant Galois symmetries and tensor constructions.*

One cannot expect that Φ_C comes from an actual endomorphism φ of the fundamental group $\pi_1(S)$, since it is known that for $g \geq 2$, any such φ is necessarily an automorphism. That is a rationale for replacing a putative endomorphism of $\pi_1(S)$ by a more esoteric endomorphism of the tensor category of its finite-dimensional representations.

Example: SL(2)-local systems on the sphere with three punctures

A generic $\mathrm{SL}(2, \mathbb{C})$-local system on $\mathbb{C}P^1 \setminus \{0, 1, \infty\}$ is uniquely determined by three traces of monodromies around punctures. A similar statement holds for l-adic local systems with tame monodromy in the case of finite characteristic. Motivic local systems correspond to the case that all the eigenvalues of the monodromies around punctures are roots of unity, i.e., that the traces of monodromies are twice cosines of rational angles. This leads to the following prediction:

$$X = \mathbb{A}^3, \quad F(x_1, x_2, x_3) = (T_q(x_1), T_q(x_2), T_q(x_3)),$$

where $T_q \in \mathbb{Z}[x]$ is the qth Chebyshev polynomial,

$$T_q(\lambda + \lambda^{-1}) = \lambda^q + \lambda^{-q} .$$

In this case the identifications between the fixed points of F and motivic local systems on $\mathbb{P}^1_{\mathbb{F}_q} \setminus \{0, 1, \infty\}$ exist, and can be extracted form the construction of these local systems (called hypergeometric) as summands in certain direct images of abelian local systems (analogous to the classical integral formulas for hypergeometric functions). The identification is ambiguous; it depends on a choice of a group embedding $\overline{\mathbb{F}}_q^{\times} \hookrightarrow \mathbb{Q}/\mathbb{Z}$.

1.3 Equivariant bundles and Ruelle-type zeta functions

The analogy with an element of the mapping class group acting on a surface S suggests the following addition to Conjecture 1. Let us fix the curve $C/\mathbb{F}_q$ and the rank $N \geq 1$ of the local systems under consideration. For a given point $x \in C(\mathbb{F}_q)$ we have a sequence of Hecke operators $T_x^{(n)}$ associated with curves $C \times_{\mathrm{Spec}\,\mathbb{F}_q} \mathrm{Spec}\,\mathbb{F}_{q^n}$. The spectrum of $T_x^{(n)}$ is a $\overline{\mathbb{Q}}$-valued function on X_n, i.e., according to Conjecture 1, a function on the set of fixed points of F^n. We expect that the collection of these functions for $n = 1, 2 \ldots$ comes from an F-equivariant vector bundle on X.

Conjecture 5. *Using the notation of Conjecture 1, for given $x \in C(\mathbb{F}_q)$ there exists a pair $(\mathcal{E}, g)$, where $\mathcal{E}$ is a vector bundle on X of rank N together with an isomorphism $g : F^*\mathcal{E} \to \mathcal{E}$ (defined over $\mathbb{Q}$) such that the eigenvalue of $T_x^{(n)}$ at the point of its spectrum corresponding to $z \in X_n$ coincides with*

$$\mathrm{Trace}(\mathcal{E}_z = \mathcal{E}_{F^n(z)} \to \cdots \to \mathcal{E}_{F(z)} \to \mathcal{E}_z),$$

where arrows are isomorphisms of fibers of $\mathcal{E}$ coming from g.

In particular, one can ask for an explicit formula for the F-equivariant bundle $\mathcal{E}$ in the case of SL(2)-local systems on the sphere with three punctures where we have an explicit candidate for (X, F).

In the limiting simplest nonabelian case when the monodromy is unipotent around two punctures, and arbitrary semisimple around the third puncture, one can make the above question completely explicit:

Question 6. For a given $x \in \mathbb{F}_q \setminus \{0, 1\}$, does there exist a rational function $R = R_x$ on $\mathbb{C}P^1$ with values in $q^{1/2}\mathrm{SL}(2, \mathbb{C}) \subset \mathrm{Mat}(2 \times 2, \mathbb{C})$ that has no singularities on the set

$$\left(\cup_{n \geq 1} \left\{ z \in \mathbb{C} | z^{q^n - 1} = 1 \right\} \right) \setminus \{1\}$$

such that for any $n \geq 1$ two sets of complex numbers (with multiplicities),

$$X_n := \left\{ \sum_{y \in \mathbb{F}_{q^n} \setminus \{0, 1, x\}} \chi \left(\frac{y(1 - xy)}{1 - y} \right) \mid \chi : \mathbb{F}_{q^n}^\times \to \mathbb{C}^\times, \chi \neq 1 \right\},$$

where χ runs through all nontrivial multiplicative characters of $\mathbb{F}_{q^n}$, and

$$X'_n := \left\{ \mathrm{Trace}\left(R(z)R(z^q) \cdots R\left(z^{q^{n-1}}\right) \right) \mid z^{q^n - 1} = 1, z \neq 1 \right\},$$

coincide?

Elements of the set X_n are real numbers of the form $\lambda + \overline{\lambda}$, where $\lambda \in \overline{\mathbb{Q}}$ is a q-Weil number with $|\lambda| = q^{1/2}$. Therefore it is natural to expect that $R(z)$ belongs to $q^{1/2}\mathrm{SU}(2)$ if $|z| = 1$.

The Galois symmetry does not forbid the function R (as a rational function with values in 2×2 matrices) to be defined over $\mathbb{Q}$ after conjugation by a constant matrix. Moreover, the existence of such a function over $\mathbb{Q}$ leads to certain choice of generators of the multiplicative groups $\left(\mathbb{F}_{q^n}^{\times}\right)$ for all $n \geq 1$, well defined modulo the action of the Frobenius Fr_q (the Galois group $\mathrm{Gal}(\mathbb{F}_{q^n}/\mathbb{F}_q) = \mathbb{Z}/n\mathbb{Z}$), as in a sense we identify roots of unity in $\mathbb{C}$ and multiplicative characters of $\mathbb{F}_{q^n}$. In particular, there will be a *canonical* irreducible polynomial over $\mathbb{F}_q$ of degree n for every $n \geq 1$. This is something almost too good to be true.

Reminder: Trace formula and Ruelle-type zeta function

Let X now be a smooth *proper* variety (say, over $\mathbb{C}$), endowed with a map $F : X \to X$, and $\mathcal{E}$ a vector bundle on X together with a morphism (not necessarily invertible) $g : F^*\mathcal{E} \to \mathcal{E}$. Let us assume that for any $n \geq 1$ all fixed points z of $F^n : X \to X$ are isolated and *nondegenerate*, i.e., the tangent map

$$(F^n)'_z : T_z X \to T_z X$$

has no nonzero invariant vectors (in other words, all eigenvalues of $(F^n)'_z$ are not equal to 1). Then one has the following identity (Atiyah–Bott fixed-point formula):

$$\sum_{v \in X : F^n(z)=z} \frac{\mathrm{Trace}(\mathcal{E}_z = \mathcal{E}_{F^n(z)} \to \cdots \to \mathcal{E}_z)}{\det\left(1 - (F^n)'_z\right)}$$
$$= \mathrm{Trace}((g_* \circ F^*)^n : H^\bullet(X, \mathcal{E}) \to H^\bullet(X, \mathcal{E})).$$

The trace on the right-hand side is understood in the super sense, as the alternating sum of the ordinary traces in individual cohomology spaces.

If one wants to eliminate the determinant factor in the denominator on the left-hand side, one should replace $\mathcal{E}$ by the superbundle $\mathcal{E} \otimes \wedge^\bullet (T_X^*)$.

The trace formula implies that the series in t

$$\exp\left(-\sum_{n \geq 1} \frac{t^n}{n} \sum_{z \in X : F^n(z)=z} \frac{\mathrm{Trace}(\mathcal{E}_z = \mathcal{E}_{F^n(z)} \to \cdots \to \mathcal{E}_z)}{\det\left(1 - (F^n)'_z\right)}\right)$$

is the Taylor expansion of a rational function in t. It seems that in many cases, for *noncompact* varieties X a weaker form of rationality holds as well when no equivariant compactification can be found. Namely, the above series (called the Ruelle zeta function in general, not necessarily algebraic, case) admits a meromorphic continuation to $\mathbb{C}$; also the zeta function in the version without the denominator is often rational in the noncompact case.

Rationality conjecture for motivic local systems

In the case hypothetically corresponding to motivic local systems on curves (in the setting of Conjecture 3), one can make a natural a priori guess about the denominator in the left-hand side of the trace formula. Namely, for a fixed point z of the map F^n corresponding to a fixed point $[\rho]$ in the space of representations of $\pi_1(C \times_{\mathrm{Spec}\,\mathbb{F}_q} \mathrm{Spec}\,\overline{\mathbb{F}}_q)$ in $\mathrm{GL}(N, \overline{\mathbb{Q}}_l)$, we expect that the vector space $T_z X$ together with the automorphism $(F^n)'_z$ should be isomorphic (after the change of scalars) to

$$H^1(C \times_{\mathrm{Spec}\,\mathbb{F}_q} \mathrm{Spec}\,\overline{\mathbb{F}}_q, \mathrm{End}(\rho)) = \mathrm{Ext}^1(\rho, \rho)$$

endowed with the Frobenius operator.

Eigenvalues of Fr^n in this case have norm $q^{n/2}$ by the Weil conjecture, hence not equal to 1, and the denominator in the Ruelle zeta function does not vanish (meaning that the fixed points are nondegenerate).

In our basic example from Section 0.1 one can propose an explicit formula for the denominator term. Define (in notation from Section 0.1) for given $t \in \mathbb{F}_q \setminus \{0, 1\}$ a matrix $T_{\mathrm{tan}} \in \mathrm{Mat}(\mathbb{F}_q \times \mathbb{F}_q, \mathbb{Q})$ by the formula

$$T_{\mathrm{tan}} := -\frac{1}{q} \sum_{x \in \mathbb{F}_q} (T_x)^2 + (q - 3 - 1/q) \cdot \mathrm{id}_{\mathbb{Q}^{\mathbb{F}_q}} \ .$$

This matrix satisfies the same properties as Hecke operators.[5] Namely, all eigenvalues of T_{tan} belong to $[-2\sqrt{q}, +2\sqrt{q}]$, any element of $\mathrm{Spec}(T_{\mathrm{tan}})$ can be written as $\lambda + \overline{\lambda}$ where $|\lambda| = \sqrt{q}$ is a q-Weil number, and for any $\xi = \lambda + \overline{\lambda} \in \mathrm{Spec}(T_x)$ and any integer $n \geq 1$ the spectrum of the matrix $T_{\mathrm{tan}}^{(n)}$ corresponding to $x \in \mathbb{F}_q \subset \mathbb{F}_{q^n}$ (if we pass to the extension $\mathbb{F}_{q^n} \supset \mathbb{F}_q$) contains the element $\xi^{(n)} := \lambda^n + \overline{\lambda}^n$.

We expect that the eigenvalue of T_{tan} at the point of the spectrum corresponding to the motivic local system ρ is equal to the trace of the Frobenius in a two-dimensional submotive of the motive $H^1(C, \mathrm{End}(\rho))$, corresponding to the deformations of ρ preserving the unipotency of the monodromy around punctures.

Notice that any motivic local system ρ on C can be endowed with a nondegenerate skew-symmetric pairing with values in the Tate motive. This explains the main term of the formula:

- the sum of squares of Hecke operators means that we are using the trace formula for the Frobenius in the cohomology of C with coefficients in the tensor square of ρ;
- the factor $1/q$ comes from the Tate twist;
- the minus sign comes from the odd (first) cohomology.

---

[5]The only difference is that eigenvalues of operators T_x are algebraic integers, while eigenvalues of T_{tan} are algebraic integers divided by q.

The candidate for the denominator term in the putative Ruelle zeta function is the following operator commuting with the Hecke operators (we write the formula only for the first iteration, $n = 1$), considered as a function on the spectrum:

$$D := (q + 1 - T_{\tan})^{-1} \ .$$

The reason is that the eigenvalue of D at the eigenvector corresponding to the motivic local system ρ is equal to

$$\frac{1}{(1 - \lambda)(1 - \overline{\lambda})} = \frac{1}{1 + q - \xi},$$

where $\lambda, \overline{\lambda}$ are Weil numbers, eigenvalues of the Frobenius in $H^1(C, \mathrm{End}(\rho))$ satisfying the equations

$$\lambda + \overline{\lambda} = \xi, \ \lambda + \overline{\lambda} = q \ .$$

The l.h.s. of the putative trace formula for the equivariant vector bundle $\mathcal{E}_{x_1} \otimes \cdots \otimes \mathcal{E}_{x_k}$ (here $\mathcal{E}_x$ is the F-equivariant vector bundle corresponding to point $x \in C(\mathbb{F}_q)$, see Conjecture 3), is given (for the n-th iteration) by the formula

$$\mathrm{Trace}\left(T_{x_1}^{(n)} \dots T_{x_k}^{(n)} D^{(n)}\right) \ .$$

It looks that in order to achieve the rationality of the putative Ruelle zeta-function one has to add by hand certain contributions corresponding to "missing fixed points". For example, for any $x \in \mathbb{F}_q \setminus \{0, 1, t\}$ one has

$$\mathrm{Trace}(T_x D) = \frac{q}{(q - 1)^2}$$

and the corresponding zeta-function

$$\exp\left(-\sum_{n \geq 1} \frac{t^n}{n} \frac{q^n}{(q^n - 1)^2}\right) = \prod_{m \geq 1}(1 - q^{-m}t)^m \in \mathbb{Q}[[t]]$$

is meromorphic but not rational. The above zeta-function looks like the contribution of just one[6] fixed point z_0 on an algebraic dynamical system $z \mapsto F(z)$ on a two-dimensional variety, with the spectrum of $(F')_{z_0}$ equal to (q, q), and the spectrum of the map on the fiber $\mathcal{E}_{z_0} \to \mathcal{E}_{z_0}$ equal to $(q, 0)$. Here is the precise conjecture coming from computer experiments:

Conjecture 7. *For any* $x_1, \dots, x_k \in \mathbb{F}_q \setminus \{0, 1, t\}$, $k \geq 1$ *the series*

$$\exp\left(-\sum_{n \geq 1} \frac{t^n}{n}\left\{\mathrm{Trace}\left(T_{x_1}^{(n)} \dots T_{x_k}^{(n)} D^{(n)}\right) + \mathrm{Corr}(n, k)\right\}\right)$$

---

[6]Maybe the complete interpretation should be a bit more complicated as one can check numerically that $\mathrm{Trace}(D) = \frac{q^2(q-2)}{(q-1)^2(q+1)}$.

where

$$\text{Corr}(n, k) := -\frac{(-1 - q^n)^k}{(1 - q^{-n})(1 - q^{2n})},$$

is a rational function.

The rational function in the above conjecture should be an L-function of a motive over $\mathbb{F}_q$, all its zeroes and poles should be q-Weil numbers.

Finally, if one considers Ruelle zeta-functions *without* the denominator term, then rationality is elementary, as will become clear in the next section.

2 Second proposal: formalism of motivic function spaces and higher-dimensional Langlands correspondence

The origin of this section is property 6 (the multiplication table) of Hecke operators in our example from Section 0.1.

2.1 Motivic functions and the tensor category $\mathcal{C}_\mathbf{k}$

Let S be a Noetherian scheme.

Definition 8. The commutative ring $\text{Fun}^{\text{poor}}(S)$ of poor man's motivic functions[7] on S is the quotient of the free abelian group generated by equivalence classes of schemes of finite type over S, modulo relations

$$[X \to S] = [Z \to S] + [(X \setminus Z) \to S],$$

where Z is a closed subscheme of X over S. The multiplication on $\text{Fun}^{\text{poor}}(S)$ is given by the fibered product over S.

In the case that S is the spectrum of a field $\mathbf{k}$, we obtain the standard definition[8] of the Grothendieck ring of varieties over $\mathbf{k}$. Any motivic function on S gives a function on the set of points of S with values in the Grothendieck rings corresponding to the residue fields.

For a given field $\mathbf{k}$ let us consider the following additive category $\mathcal{C}_\mathbf{k}$. Its objects are schemes of finite type over $\mathbf{k}$; the abelian groups of homomorphisms are defined by

$$\text{Hom}_{\mathcal{C}_\mathbf{k}}(X, Y) := \text{Fun}^{\text{poor}}(X \times Y).$$

The composition of two morphisms represented by schemes is given by the fibered product as below,

$$[B \to Y \times Z] \circ [A \to X \times Y] := [A \times_Y B \to X \times Z],$$

---

[7]The name was suggested by V. Drinfeld.

[8]Usually one extends the Grothendieck ring of varieties by inverting the class $[\mathbb{A}^1_\mathbf{k}]$ of the affine line, which is the geometric counterpart of the inversion of the Lefschetz motive $L = H_2(\mathbb{P}^1_k)$ in the construction of Grothendieck pure motives. Here also we can do the same thing.

and extended by additivity to all motivic functions. The identity morphism id_X is given by the diagonal embedding $X \hookrightarrow X \times X$.

One can start from the beginning from constructible sets over $\mathbf{k}$ instead of schemes. The category of constructible sets over $\mathbf{k}$ is a full subcategory of $\mathcal{C}_{\mathbf{k}}$; the morphism in $\mathcal{C}_{\mathbf{k}}$ corresponding to a constructible map $f : X \to Y$ is given by $[X \overset{(\mathrm{id}_X, f)}{\longrightarrow} X \times Y]$, the graph of f.

Finite sums (and products) in $\mathcal{C}_{\mathbf{k}}$ are given by the disjoint union.

We endow the category $\mathcal{C}_{\mathbf{k}}$ with the following tensor structure on objects:

$$X \otimes Y := X \times Y,$$

and by a similar formula on morphisms. The unit object $\mathbf{1}_{\mathcal{C}_{\mathbf{k}}}$ is the point $\mathrm{Spec}(\mathbf{k})$. The category $\mathcal{C}_{\mathbf{k}}$ is rigid, i.e., for every object X there exists another object $X^{\vee}$ together with morphisms $\delta_X : X \otimes X^{\vee} \to 1$, $\epsilon_X : 1 \to X^{\vee} \otimes X$ such that both compositions:

$$X \overset{\mathrm{id}_X \otimes \epsilon_X}{\longrightarrow} X \otimes X^{\vee} \otimes X \overset{\delta_X \otimes \mathrm{id}_X}{\longrightarrow} X, \quad X^{\vee} \overset{\epsilon_X \otimes \mathrm{id}_{X^{\vee}}}{\longrightarrow} X^{\vee} \otimes X \otimes X^{\vee} \overset{\mathrm{id}_{X^{\vee}} \otimes \delta_X}{\longrightarrow} X^{\vee}$$

are identity morphisms. The dual object $X^{\vee}$ in $\mathcal{C}_{\mathbf{k}}$ coincides with X; the duality morphisms δ_X, ϵ_X are given by the diagonal embedding $X \hookrightarrow X^2$.

As in any tensor category, the ring $\mathrm{End}_{\mathcal{C}_{\mathbf{k}}}(\mathbf{1}_{\mathcal{C}_{\mathbf{k}}})$ is commutative, and the whole category is linear over this ring, which is nothing but the Grothendieck ring of varieties over $\mathbf{k}$.

Fiber functors for finite fields

If $\mathbf{k} = \mathbb{F}_q$ is a finite field then there is an infinite chain $(\phi_n)_{n \geq 1}$ of tensor functors from $\mathcal{C}_{\mathbf{k}}$ to the category of finite-dimensional vector spaces over $\mathbb{Q}$. It is defined on objects by the formula

$$\phi_n(X) := \mathbb{Q}^{X(\mathbb{F}_{q^n})} \ .$$

The operator corresponding by ϕ_n to a morphism $[A \to X \times Y]$ has the following matrix coefficient with indices $(x, y) \in X(\mathbb{F}_{q^n}) \times Y(\mathbb{F}_{q^n})$:

$$\#\{a \in A(\mathbb{F}_{q^n}) \mid a \mapsto (x, y)\} \in \mathbb{Z}_{\geq 0} \subset \mathbb{Q}.$$

The functor ϕ_n is not canonically defined for $n \geq 2$; the ambiguity is the cyclic group $\mathbb{Z}/n\mathbb{Z} = \mathrm{Gal}(\mathbb{F}_{q^n}/\mathbb{F}_q) \subset \mathrm{Aut}(\phi_n)$.

Extensions and variants

The abelian group $\mathrm{Fun}^{\mathrm{poor}}(S)$ of poor man's motivic functions can (and probably should) be replaced by the K_0 group of the triangulated category $\mathrm{Mot}_{S, \mathbb{Q}}$ of "constructible motivic sheaves" (with coefficients in $\mathbb{Q}$) on S. Although the latter category is not yet rigorously defined, one can envision a reasonable candidate for the elementary description of $K_0(\mathrm{Mot}_{S,\mathbb{Q}})$. This group should be generated by equivalence classes of families of Grothendieck motives (with coefficients in $\mathbb{Q}$) over closed subschemes of S, modulo a suitable equivalence relation. Moreover, the group $K_0(\mathrm{Mot}_{S,\mathbb{Q}})$ should be filtered by the dimension of support, and the associated graded group should be canonically isomorphic to the direct sum over all points $x \in S$ of K_0 groups of categories of pure motives (with coefficients in $\mathbb{Q}$) over the residue fields.[9]

Similarly, one can extend the coefficients of the motives from $\mathbb{Q}$ to any field of zero characteristic. This change will affect the group K_0 and give a different algebra of motivic functions.

Finally, one can add formally images of projectors to the category $\mathcal{C}_{\mathbf{k}}$.

Question 9. Are there interesting nontrivial projectors in $\mathcal{C}_{\mathbf{k}}$?

I do not know at the moment any example of an object in the Karoubi closure of $\mathcal{C}_{\mathbf{k}}$ that is not isomorphic to a scheme. Still, there are interesting nontrivial isomorphisms between objects of $\mathcal{C}_{\mathbf{k}}$, for example the following version of the Radon transform.

Example: motivic Radon transform

Let $X = \mathbb{P}(V)$ and $Y = \mathbb{P}(V^\vee)$ be two dual projective spaces over $\mathbf{k}$. We assume that $n := \dim V$ is at least 3.

The incidence relation gives a subvariety $Z \subset X \times Y$, which can be interpreted as a morphism in $\mathcal{C}_{\mathbf{k}}$ in two ways:

$$f_1 := [Z \hookrightarrow X \times Y] \in \mathrm{Hom}_{\mathcal{C}_{\mathbf{k}}}(X, Y),$$

$$f_2 := [Z \hookrightarrow Y \times X] \in \mathrm{Hom}_{\mathcal{C}_{\mathbf{k}}}(Y, X).$$

The composition $f_2 \circ f_1$ is equal to

$$[\mathbb{A}^{n-2}] \cdot \mathrm{id}_X + [\mathbb{P}^{n-3}] \cdot [X \to \mathrm{pt} \to X] \ .$$

The reason is that the scheme of hyperplanes passing through points $x_1, x_2 \in X$ is either $\mathbb{P}^{n-3}$ if $x_1 \neq x_2$, or $\mathbb{P}^{n-2}$ if $x_1 = x_2$. On the level of constructible sets one has $\mathbb{P}^{n-2} = \mathbb{P}^{n-3} \sqcup \mathbb{A}^{n-2}$.

The first term is the identity morphism multiplied by the $(n-2)$nd power of the Tate motive, while the second term is in a sense a rank-1 operator. It can be killed after passing to the quotient of X by pt, which is in fact a direct summand in $\mathcal{C}_{\mathbf{k}}$:

$$X \simeq \mathrm{pt} \oplus (X \setminus \mathrm{pt}) \ .$$

Here we have to choose a point $\mathrm{pt} \in X$. Similar arguments work for Y, and as a result we obtain an isomorphism (inverting the Tate motive)

$$X \setminus \mathrm{pt} \simeq Y \setminus \mathrm{pt}$$

in the category $\mathcal{C}_{\mathbf{k}}$ that is not a geometric isomorphism of constructible sets.

---

[9] I do not know how to fill in all the details in the above sketch.

2.2 Commutative algebras in $\mathcal{C}_{\mathbf{k}}$

By definition, a unital commutative associative algebra A in the tensor category $\mathcal{C}_{\mathbf{k}}$ is given by a scheme of finite type $X/\mathbf{k}$, and two elements

$$1_A \in \mathrm{Fun}^{\mathrm{poor}}(X), \ m_A \in \mathrm{Fun}^{\mathrm{poor}}(X^3)$$

(the unit and the product in A) satisfying the usual constraints of unitality, commutativity, and associativity.

The formula for the structure constants $c_{xyz} = (T_x)_{yz}$ of the algebra of Hecke operators in our basic example (see Section 0.1) is given explicitly by counting points on constructible sets depending constructibly on a point $(x, y, z) \in X^3$, where $X = \mathbb{A}^1$, for any $t \in \mathbf{k} \setminus \{0, 1\}$ (one should replace factors q by bundles with fiber $\mathbb{A}^1$). Hence we obtain a motivic function on X^3 that gives the structure of a commutative algebra on X for any $t \in \mathbf{k} \setminus \{0, 1\}$, for arbitrary field $\mathbf{k}$. A straightforward check (see Proposition 1 in Section 2.4 below for a closely related statement) shows that this algebra is associative.

Elementary examples of algebras

The first example of a commutative algebra is given by an arbitrary scheme X (or a constructible set) of finite type over $\mathbf{k}$. The multiplication tensor is given by the diagonal embedding $X \hookrightarrow X^3$; the unit is given by the identity map $X \to X$. If $\mathbf{k} = \mathbb{F}_q$ is finite, then for any $n \geq 1$ the algebra $\phi_n(X)$ is the algebra of $\mathbb{Q}$-valued functions on the finite set $X(\mathbb{F}_{q^n})$, with pointwise multiplication.

The next example corresponds to the case that X is an abelian group scheme (e.g., $\mathbb{G}_a$, $\mathbb{G}_m$, or an abelian variety). We define the multiplication tensor $m_A \in \mathrm{Fun}^{\mathrm{poor}}(X^3)$ as the graph of the multiplication morphism $X \times X \to X$. Again, if $\mathbf{k}$ is finite then the algebra $\phi_n(X)$ is the group algebra with coefficients in $\mathbb{Q}$ of the finite abelian group $X(\mathbb{F}_{q^n})$. Its points in $\overline{\mathbb{Q}}$ are additive (resp. multiplicative) characters of $\mathbf{k}$ if $X = \mathbb{G}_a$ (resp. $X = \mathbb{G}_m$).

Also, one can see that the algebra in $\mathcal{C}_{\mathbf{k}}$ corresponding to the group scheme $\mathbb{G}_a$ is isomorphic to the direct sum of $1_{\mathcal{C}_{\mathbf{k}}}$ (corresponding to the trivial additive character of $\mathbf{k}$) and another algebra A, which can be thought of as parameterizing *nontrivial* additive characters of the field, with the underlying scheme $\mathbb{A}^1 \setminus \{0\}$.

Finally, one can make "quotients" of abelian group schemes by finite groups of automorphisms. For example, for $\mathbb{G}_a$ endowed with the action of the antipodal involution $x \to -x$, the formula for the product for the corresponding algebra is the sum of the "main term"

$$[Z \hookrightarrow (\mathbb{A}^1)^3], \quad Z = \{(x, y, z) \mid x^2 + y^2 + z^2 - 2(xy + yz + zx) = 0\}$$

(the latter equation means that $\sqrt{x} + \sqrt{y} = \sqrt{z}$) and certain correction terms. Similarly, for the antipodal involution $(x, w) \to (x, -w)$ on the elliptic curve

$E \subset \mathbb{P}^1 \times \mathbb{P}^1$ given by $w^2 = x(x-1)(x-t)$ (with (∞, ∞) serving as zero for the group law), the quotient is $\mathbb{P}^1$ endowed with a multiplication law similar to that from Section 0.1. The main term is given by the hypersurface $f_t(x, y, z) = 0$ in the notation of Section 0.1. The spectrum of the corresponding algebra is rather trivial, in comparison to our example. The difference is that in Section 0.1 we consider the twofold cover of $(\mathbb{A}^1)^3$ ramified at the hypersurface $f_t(x, y, z) = 0$.

Categorification

One may wonder whether a commutative associative algebra A in $\mathcal{C}_{\mathbf{k}}$ (for general field $\mathbf{k}$, not necessarily finite) is in fact a materialization of the structure of a symmetric (or only braided) monoidal category on a triangulated category, i.e., whether the multiplication morphism is the class in K_0 of a bifunctor defining the monoidal structure. The category under consideration should be either the category of constructible mixed motivic sheaves on the underlying scheme of A or some small modification of it not affecting the group K_0 (e.g., both categories could have semiorthogonal decompositions with the same factors).

2.3 Algebras parameterizing motivic local systems

As we have already observed, the example of Section 0.1 can be interpreted as a commutative associative algebra in $\mathcal{C}_{\mathbf{k}}$ parameterizing in a certain sense (via the chain of functors $(\phi_n)_{n \geq 1}$) motivic local systems on a curve over $\mathbf{k} = \mathbb{F}_q$. Here we will formulate a general conjecture, which goes beyond the case of curves.

Preparations on ramification and motivic local systems

Let Y be a smooth geometrically connected projective variety over a finitely generated field $\mathbf{k}$. Let us denote by K the field of rational functions on X and by K' the field of rational functions on $Y' := Y \times_{\operatorname{Spec} \mathbf{k}} \operatorname{Spec} \overline{\mathbf{k}}$. We have an exact sequence

$$1 \to \operatorname{Gal}(\overline{K}/K') \to \operatorname{Gal}(\overline{K}/K) \to \operatorname{Gal}(\overline{\mathbf{k}}/\mathbf{k}) \to 1.$$

For a continuous homomorphism

$$\rho : \operatorname{Gal}(\overline{K}/K') \to \operatorname{GL}(N, \overline{\mathbb{Q}}_l),$$

where $l \neq \operatorname{char}(\mathbf{k})$, which factorizes through the quotient $\pi_1^{\operatorname{geom}}(U)$ for some open subscheme $U \subset Y'$, one can envision some notion of ramification divisor (similar to the notion of the conductor in the one-dimensional case) that should be an effective divisor on Y'.

One expects that for a pure motive of rank N over K with coefficients in $\overline{\mathbb{Q}}$, the ramification divisor of the corresponding l-adic local system does not depend on the prime $l \neq \mathrm{char}(\mathbf{k})$, at least for large l.

Denote by $\mathrm{IrrRep}_{Y',N,l}$ the set of conjugacy classes of irreducible representations $\rho : \mathrm{Gal}(\overline{K}/K) \to \mathrm{GL}(N, \overline{\mathbb{Q}}_l)$ factoring through $\pi_1^{\mathrm{geom}}(U)$ for some open subscheme $U \subset Y'$ as above. The Galois group $\mathrm{Gal}(\overline{\mathbf{k}}/\mathbf{k})$ acts on this set.

Denote by $\mathrm{IrrRep}_{Y,N}^{\mathrm{mot,geom}}$ the set of equivalence classes of pure motives in the sense of Grothendieck (defined using the numerical equivalence) of rank N over K, with coefficients in $\overline{\mathbb{Q}}$, that are absolutely simple (i.e., remain simple after the pullback to K'), modulo the action of the Picard group of rank-1 motives over $\mathbf{k}$ with coefficients in $\overline{\mathbb{Q}}$. This set is endowed with a natural action of $\mathrm{Gal}(\overline{\mathbb{Q}}/\mathbb{Q})$. The superscript *geom* indicates that we are interested only in representations of the geometric fundamental group.

One expects that the natural map from $\mathrm{IrrRep}_{Y,N}^{\mathrm{mot,geom}}$ to the set of fixed points $(\mathrm{IrrRep}_{Y',N,l})^{\mathrm{Gal}(\overline{\mathbf{k}}/\mathbf{k})}$ is a bijection. In particular, it implies that one can define the ramification divisor for an element of $\mathrm{IrrRep}_{Y,N}^{\mathrm{mot,geom}}$. Presumably, one can give a purely geometric definition of it, without referring to l-adic representations.

Conjecture on algebras parameterizing motivic local systems

Conjecture 10. *For a smooth projective geometrically connected variety Y over a finite field $\mathbf{k} = \mathbb{F}_q$, an effective divisor D on Y, and a positive integer N, there exists a commutative associative unital algebra $A = A_{Y,D,N}$ in the category $\mathcal{C}_{\mathbf{k}}$ satisfying the following property:*
For any $n \geq 1$ the algebra $\phi_n(A)$ over $\mathbb{Q}$ is semisimple (i.e., it is a finite direct sum of number fields) and for any prime l, $(l, q) = 1$, there exists a bijection between $\mathrm{Hom}_{\mathbb{Q}-\mathrm{alg}}(\phi_n(A), \overline{\mathbb{Q}})$ and the set of elements of $\mathrm{IrrRep}_{Y \times_{\mathrm{Spec}\,\mathbb{F}_q} \mathrm{Spec}\,\mathbb{F}_{q^n},N}^{\mathrm{mot,geom}}$ for which the ramification divisor is D. Moreover, the above bijection is equivariant with respect to the natural $\mathrm{Gal}(\overline{\mathbb{Q}}/\mathbb{Q}) \times \mathbb{Z}/n\mathbb{Z}$-action.

One can also try to formulate a generalization of the above conjecture, allowing not an individual variety Y but a family, i.e., a smooth projective morphism $\mathcal{Y} \to B$ to a scheme of finite type over k, with geometrically connected fibers, together with a flat family of ramification divisors. The corresponding algebra should parameterize choices of a point $b \in B(\mathbb{F}_{q^n})$, an irreducible motivic system of given rank, and a given ramification on the fiber $\mathcal{Y}_b$. This algebra should map to the algebra of functions with the pointwise product (see Section 2.2) associated with the base B.

In the above conjecture we did not describe how to associate a *tower* of finite sets to the algebra A, since a priori we have just a *sequence* of finite sets

$X_n := \mathrm{Hom}_{\mathbb{Q}-\mathrm{alg}}(\phi_n(A), \overline{\mathbb{Q}})$ with no obvious maps between them. This leads to the following question:

Question 11. Which property of an associative commutative algebra A in $\mathcal{C}_{\mathbb{F}_q}$ gives naturally a chain of embeddings

$$\mathrm{Hom}_{\mathbb{Q}-\mathrm{alg}}(\phi_{n_1}(A), \overline{\mathbb{Q}}) \hookrightarrow \mathrm{Hom}_{\mathbb{Q}-\mathrm{alg}}(\phi_{n_1 n_2}(A), \overline{\mathbb{Q}})$$

for all integers $n_1, n_2 \geq 1$?

It appears that this holds automatically, by a kind of trace morphism.

Arguments in favor, and extensions

First of all, there is good reason to believe that Conjecture 5 holds for curves. Also, it would be reasonable to consider local systems with an arbitrary structure group G instead of $\mathrm{GL}(N)$. The algebra parameterizing motivic local systems on a curve $Y = C$ with structure group G should be (roughly) equal to some finite open part of the moduli stack Bun_{G^L} of G^L-bundles on C, where G^L is the Langlands dual group. The multiplication should be given by the class of a motivic constructible sheaf on

$$(\mathrm{Bun}_{G^L})^3 = \mathrm{Bun}_{G^L} \times \mathrm{Bun}_{G^L \times G^L},$$

which should be a geometric counterpart to the lifting of automorphic forms corresponding to the diagonal embedding

$$G^L \to G^L \times G^L \ .$$

Presumably, the multiplication law from Example 0.1 corresponds to the lifting.

If we believe in Conjecture 5 in the case of curves, then it is very natural to believe in it in general. The reason is that for a higher-dimensional variety Y (not necessarily compact) there exists a curve $C \subset Y$ such that $\pi_1^{\mathrm{geom}}(Y)$ is a *quotient* of $\pi_1^{\mathrm{geom}}(C)$. Such a curve can be, e.g., a complete intersection of ample divisors; the surjectivity is a particular case of the Lefschetz theorem on hyperplane sections. Therefore, the set of equivalence classes of absolutely irreducible motivic local systems on $Y \times_{\mathrm{Spec}\,\mathbb{F}_q} \mathrm{Spec}\,\mathbb{F}_{q^n}$ should be a *subset* of the corresponding set for C for any $n \geq 1$, and invariant under the $\mathrm{Gal}(\overline{\mathbb{Q}}/\mathbb{Q})$-action as well. It seems very plausible that such a collection of subsets should arise from a quotient algebra in $\mathcal{C}_{\mathbb{F}_q}$.

From the previous discussion it appears that the motivic local systems in the higher-dimensional case are "less interesting"; the 1-dimensional case is the richest one. Nevertheless, there is definitely nontrivial higher-dimensional information about local systems that cannot be reduced to 1-dimensional data. Namely, for any motivic local system ρ^{arith} on Y and an integer $i \geq 0$, the cohomology space

$$H^i(Y', \rho),$$

where ρ is the pullback of ρ^{arith} to Y', is a motive over the finite field $\mathbf{k} = \mathbb{F}_q$. We can calculate the trace of the Nth power of the Frobenius on it for a given $N \geq 1$, and get a $\overline{\mathbb{Q}}$-valued function[10] on the set

$$X_n := \mathrm{Hom}_{\mathbb{Q}-\mathrm{alg}}(\phi_n(A), \overline{\mathbb{Q}})$$

for each $n \geq 1$. This leads to a natural addition to Conjecture 5. Namely, we expect that systems of $\overline{\mathbb{Q}}$-valued functions on X_n associated with higher cohomology spaces arise from elements in $\mathrm{Hom}_{\mathcal{C}_{\mathbb{F}_q}}(1, A)$ (i.e., from motivic functions on the constructible set underlying the algebra A).

More generally, one can expect that the motivic constructible sheaves with some kind of boundedness will be parameterized by commutative algebras.

Formulas from the example from Section 0.1 make sense and give an algebra in $\mathcal{C}_{\mathbf{k}}$ for an arbitrary field $\mathbf{k}$. This leads to the following question:

Question 12. Can one construct algebras $A_{Y,D,N}$ for an arbitrary ground field $\mathbf{k}$, not necessarily finite? In what sense will these algebras "parameterize" motivic local systems?

In general, it seems that the natural source of commutative algebras in $\mathcal{C}_{\mathbf{k}}$ is not the representation theory, but (quantum) algebraic integrable systems.

2.4 Toward integrable systems over local fields

Here we will describe briefly an analogue of commutative algebras of integral operators as above for arbitrary local fields, i.e., $\mathbb{R}$, $\mathbb{C}$, or finite extensions of $\mathbb{Q}_p$ or $\mathbb{F}_p((x))$. Let us return to our basic example. The check of the associativity of the multiplication law given by the formula from Section 0.1 in the case of finite fields is reduced to an identification of certain varieties. The most nontrivial part is the following:

Proposition 13. *For generic parameters* t, x_1, x_2, x_3, x_4, *the two elliptic curves*

$$E : f_t(x_1, x_2, y) = w_{12}^2, \quad f_t(y, x_3, x_4) = w_{34}^2,$$
$$\tilde{E} : f_t(x_1, x_3, \tilde{y}) = \tilde{w}_{13}^2, \quad f_t(\tilde{y}, x_2, x_4) = \tilde{w}_{24}^2,$$

given by equations in variables (y, w_{12}, w_{34}) *and* $(\tilde{y}, \tilde{w}_{13}, \tilde{w}_{24})$ *respectively, are canonically isomorphic over the ground field. Moreover, one can choose such an isomorphism that identifies the abelian differentials*

$$\frac{dy}{w_{12}w_{34}} \quad and \quad \frac{d\tilde{y}}{\tilde{w}_{13}, \tilde{w}_{24}} .$$

---

[10]Here there is a small ambiguity that should be resolved somehow, since one can multiply ρ^{arith} by a one-dimensional motive over $\mathbf{k}$ with coefficients in $\overline{\mathbb{Q}}$.

In fact, it is enough to check the proposition over an algebraically closed field and observe that the curves $E, \tilde{E}$ have points over the ground field.[11]

Let now $\mathbf{k}$ be a local field. For a given $t \in \mathbf{k} \setminus \{0, 1\}$ we define a (nonnegative) half-density c_t on $\mathbf{k}^3$ by the formula

$$c_t := \pi_* \left(\frac{|dx_1|^{1/2}|dx_2|^{1/2}|dx_3|^{1/2}}{|w|} \right),$$

where

$$\pi : Z(\mathbf{k}) \to \mathbb{A}^3(\mathbf{k}), \quad \pi(x_1, x_2, x_3, w) = (x_1, x_2, x_3)$$

is the projection of the hypersurface

$$Z \subset \mathbb{A}^4_{\mathbf{k}} : \quad f_t(x_1, x_2, x_3) = w^2 \ .$$

We will interpret c_t as a half-density on $(\mathbb{P}^1(\mathbf{k}))^3$ as well.

One can deduce from the above proposition the following:

Theorem 14. *The operators T_x, $x \in \mathbf{k} \setminus \{0, 1, t\}$, on the Hilbert space of half-densities on $\mathbb{P}^1(\mathbf{k})$, given by*

$$T_x(\phi)(y) = \int_{z \in \mathbb{P}^1(\mathbf{k})} c_t(x, y, z) \ \phi(z),$$

are commuting compact self-adjoint operators.

Moreover, in the nonarchimedean case one can show that the joint spectrum of commuting operators as above is discrete and consists of densities locally constant on $\mathbb{P}^1(\mathbf{k}) \setminus \{0, 1, t, \infty\}$. In particular, all eigenvalues of operators T_x are algebraic complex numbers. Passing to the limit over finite extensions of $\mathbf{k}$, we obtain a countable set on which acts

$$\mathrm{Gal}(\overline{\mathbb{Q}}/\mathbb{Q}) \times \mathrm{Gal}(\overline{\mathbf{k}}/\mathbf{k}) \ .$$

Also notice that in the case of local fields the formula is much simpler than the motivic one: there is no correction term. On the other hand, one has a new ingredient, the local density of an integral operator. In general, one can imagine a new formalism[12] in which the structure of an algebra is given by data (X, Z, π, ν), where X is a (birational type of) variety over a given field $\mathbf{k}$, Z is another variety, $\pi : Z \to X^3$ is a map (defined only at generic points of Z), and ν is a rational section of the line bundle $K_Z^{\otimes 2} \otimes \pi^* \left(K_{X^3}^{\otimes -1} \right)$. If $\mathbf{k}$ is a local field, then the pushdown by π of $|\nu|^{1/2}$ is a half-density on X^3. The condition of the associativity would follow from a property of certain data formulated purely in terms of birational algebraic geometry.

---

[11]The curve E has 16 rational points with coordinate $y \in \{0, 1, t, \infty\}$, and the same for $\tilde{E}$.

[12]A somewhat similar formalism was proposed by Braverman and Kazhdan (see [1], who had in mind orbital integrals in the usual local Langlands correspondence.

Presumably, the spectrum for the case of the finite field is just a "low-frequency" part of a much larger spectrum for p-adic fields, corresponding to some mysterious objects.[13]

The commuting integral operators in the archimedean case $\mathbf{k} = \mathbb{R}, \mathbb{C}$ are similar to ones found recently in the usual algebraic quantum integrable systems; see [5].

3 Third proposal: lattice models

3.1 Traces depending on two indices

Let X be a constructible set over $\mathbb{F}_q$ and let M be an endomorphism of X in the category $\mathcal{C}_{\mathbb{F}_q}$ (e.g., a Hecke operator). What kind of object can be called the "spectrum" of M?

Applying the functors ϕ_n for $n \geq 1$, we obtain an infinite sequence of finite matrices, of exponentially growing size. We would like to understand the behavior of spectra of operators $\phi_n(M)$ as $n \to +\infty$. A similar question arises in some models in quantum physics where one is interested in the spectrum of a system with finitely many states, with the dimension of the Hilbert space depending exponentially on the "number of particles."

The spectrum of an operator acting on a finite-dimensional space can be reconstructed from the traces of all positive powers. This leads us to the consideration of the following collection of numbers:

$$Z_M(n, m) := \operatorname{Trace}((\phi_n(M))^m),$$

where $n \geq 1$ and $m \geq 0$ are integers. It will be important later to restrict attention only to strictly positive values of m, which means that we are interested only in nonzero eigenvalues of matrices $\phi_n(M)$, and want to ignore the multiplicity of the zero eigenvalue.

Observation 1. For a given $n \geq 1$ there exists a finite collection of nonzero complex numbers (λ_i) such that for any $m \geq 1$ one has

$$Z_M(n, m) = \sum_i \lambda_i^m \ .$$

Observation 2. For a given $m \geq 1$ there exists a finite collection of nonzero complex numbers (μ_j) and signs $(\epsilon_j \in \{-1, +1\})$ such that for any $n \geq 1$ one has

$$Z_M(n, m) = \sum_j \epsilon_j \mu_j^n \ .$$

---

[13]It appears that all this goes beyond motives, and on the automorphic side is related to some kind of Langlands correspondence for two-(or more) dimensional mixed local–global fields.

The symmetry between the parameters n and m (modulo a minor difference with signs) is quite striking.

The first observation is completely trivial. For a given n the numbers (λ_i) are all nonzero eigenvalues of the matrix $\phi_n(M)$.

Let us explain the second observation. By functoriality we have

$$Z_M(n, m) = \mathrm{Trace}(\phi_n(M^m)) \ .$$

Let us assume first that M is given by a constructible set Y that maps to $X \times X$:

$$Y \to X \times X, \ \ y \mapsto (\pi_1(y), \pi_2(y)) \ .$$

Then M^m is given by the consecutive fibered product

$$Y^{(m)} = Y \times_X Y \times_X \cdots \times_X Y \subset Y \times \cdots \times Y$$

of m copies of Y:

$$\begin{aligned} Y^{(m)}(\overline{\mathbb{F}}_q) &= \{(y_1, \ldots, y_m) \in (Y(\overline{\mathbb{F}}_q))^m \,|\, \pi_2(y_1) \\ &= \pi_1(y_2), \ldots, \pi_2(y_{m-1}) = \pi_1(y_m)\} \end{aligned}$$

The projection to $X \times X$ is given by $(y_1, \ldots, y_m) \mapsto (\pi_1(y_1), \pi_2(y_m))$. To take the trace we should intersect $Y^{(m)}$ with the diagonal. The conclusion is that $Z_M(n, m)$ is equal to the number of $\mathbb{F}_{q^n}$-points of the constructible set

$$\widetilde{Y}^{(m)} := Y^{(m)} \times_{X \times X} X \ ,$$
$$\widetilde{Y}^{(m)}(\overline{\mathbb{F}}_q) = \{(y_1, \ldots, y_m) \in Y^{(m)}(\overline{\mathbb{F}}_q) \,|\, \pi_1(y_1) = \pi_2(y_m)\} \ .$$

The second observation is now an immediate corollary of the Weil conjecture on numbers of points of varieties over finite fields[14]. The general case when M is given by a formal *integral* linear combination $\sum_\alpha n_\alpha [Y_\alpha \to X \times X]$ can be treated in a similar way.

3.2 Two-dimensional translation invariant lattice models

There is another source of numbers depending on two indices with a similar behavior with respect to each of indices when another one is fixed. It comes from the so-called lattice models in statistical physics. A typical example is the Ising model. There is a convenient way to encode Boltzmann weights of a general lattice model on $\mathbb{Z}^2$ in terms of linear algebra.

Definition 15. Boltzmann weights of a 2-dimensional translation invariant lattice model are given by a pair V_1, V_2 of finite-dimensional vector spaces over $\mathbb{C}$ and a linear operator

$$R : V_1 \otimes V_2 \to V_1 \otimes V_2 \ .$$

---

[14]Here we mean only the fact that the zeta-function of a variety over is rational in q^s, and not the more deep statement about the norms of Weil numbers.

Such data give a function (called the partition function) on a certain set of graphs. Namely, let Γ be a finite oriented graph whose edges are colored by $\{1,2\}$ in such a way that for every vertex v there are exactly two edges colored by 1 and 2 with head v, and also there are exactly two edges colored by 1 and 2 with tail v. Consider the tensor product of copies of R labelled by the set $Vert(\Gamma)$ of vertices of Γ. It is an element $v_{R,\Gamma}$ of the vector space

$$(V_1^\vee \otimes V_2^\vee \otimes V_1 \otimes V_2)^{\otimes Vert(\Gamma)} .$$

The structure of an oriented colored graph gives an identification of the above space with

$$(V_1 \otimes V_1^\vee)^{\otimes Edge_1(\Gamma)} \otimes (V_2 \otimes V_2^\vee)^{\otimes Edge_2(\Gamma)}$$

where $Edge_1(\Gamma), Edge_2(\Gamma)$ are the sets of edges of Γ colored by 1 and by 2. The tensor product of copies of the standard pairing gives a linear functional u_Γ on the above space. We define the *partition function* of the lattice model on Γ as

$$Z_R(\Gamma) = u_\Gamma(v_{R,\Gamma}) \in \mathbb{C} .$$

An oriented colored graph Γ as above is the same as a finite set with two permutations τ_1, τ_2. The set here is $\mathrm{Vert}(\Gamma)$, and permutations τ_1, τ_2 correspond to edges colored by 1 and 2 respectively.

In the setting of *translation-invariant* 2-dimensional lattice models we are interested in the values of the partition function only on graphs corresponding to pairs of commuting permutations. Such a graph (if it is nonempty and connected) corresponds to a subgroup $\Lambda \subset \mathbb{Z}^2$ of finite index. We will denote the partition function[15] of the graph corresponding to Λ by $Z_R^{\mathrm{lat}}(\Lambda)$.

Finally, Boltzmann data make sense in an arbitrary rigid tensor category $\mathcal{C}$. The partition function of a graph takes values in the commutative ring $\mathrm{End}_{\mathcal{C}}(1)$. In particular, one can speak about *super* Boltzmann data for the category $\mathrm{Super}_{\mathbb{C}}$ of finite-dimensional complex super vector spaces.

Transfer matrices

Let us consider a special class of lattices $\Lambda \subset \mathbb{Z}^2$ depending on two parameters. Namely, we set

$$\Lambda_{n,m} := \mathbb{Z} \cdot (n,0) \oplus \mathbb{Z} \cdot (0,m) \subset \mathbb{Z}^2 .$$

Proposition 16. *For any Boltzmann data (V_1, V_2, R) and a given $n \geq 1$ there exists a finite collection of nonzero complex numbers (λ_i) such that for any $m \geq 1$ one has*

$$Z_R^{\mathrm{lat}}(\Lambda_{n,m}) = \sum_i \lambda_i^m .$$

---

[15]In the physical literature it is called the partition function with periodic boundary conditions.

The proof is the following. Let us introduce a linear operator (called the *transfer matrix*) by the formula

$$T_{(2),n} := \text{Trace}_{V_1^{\otimes n}}((\sigma_n \otimes id_{V_2^{\otimes n}}) \circ R^{\otimes n}) \in \text{End}\left(V_2^{\otimes n}\right),$$

where $\sigma_n \in \text{End}\left(V_1^{\otimes n}\right)$ is the cyclic permutation. Here we interpret $R^{\otimes n}$ as an element of

$$(V_1^\vee)^{\otimes n} \otimes (V_2^\vee)^{\otimes n} \otimes V_1^{\otimes n} \otimes V_2^{\otimes n} = \text{End}\left(V_1^{\otimes n}\right) \otimes \text{End}\left(V_2^{\otimes n}\right) .$$

It follows from the definition of the partition function that

$$Z_R^{\text{lat}}(\Lambda_{n,m}) = \text{Trace}\left(T_{(2),n}\right)^m$$

for all $m \geq 1$. The collection (λ_i) is just the collection of all *nonzero* eigenvalues of $T_{(2),n}$ taken with multiplicities.

Similarly, one can define transfer matrices $T_{(1),m}$ such that $Z_R^{\text{lat}}(\Lambda_{n,m}) = \text{Trace}\left(T_{(1),m}\right)^n$ for all $n, m \geq 1$. We see that the function $(n,m) \mapsto Z_R^{\text{lat}}(\Lambda_{n,m})$ has the same two properties as the function $(n,m) \mapsto Z_M(n,m)$ from Section 3.1. For super Boltzmann data one obtains sums of exponents with signs.

3.3 Two-dimensional Weil conjecture

Let us return to the case of an endomorphism $M \in \text{End}_{C_{\mathbb{F}_q}}(X)$. In Section 3.1 we have defined numbers $Z_M(n,m)$ for $n, m \geq 1$. The results of Section 3.2 indicate that one should interpret pairs (n,m) as parameters for a special class of "rectangular" lattices in $\mathbb{Z}^2$. A general lattice $\Lambda \subset \mathbb{Z}^2$ depends on three integer parameters

$$\Lambda = \Lambda_{n,m,k} = \mathbb{Z} \cdot (n,0) \oplus \mathbb{Z} \cdot (k,m), \quad n, m \geq 1, \ 0 \leq k < n .$$

Here we propose an extension of the function Z_M to all lattices in $\mathbb{Z}^2$:

$$Z_M(\Lambda_{n,m,k}) := \text{Trace}((\phi_n(M))^m (\phi_n(\text{Fr}_X))^k),$$

where $\text{Fr}_X \in \text{End}_{C_{\mathbb{F}_q}}(X)$ is the graph of the Frobenius endomorphism of the scheme X. Notice that $\phi_n(\text{Fr}_X)$ is periodic with period n for any $n \geq 1$.

Proposition 17. *The function Z_M on lattices in $\mathbb{Z}^2$ defined as above satisfies the following property: for any two vectors $\gamma_1, \gamma_2 \in \mathbb{Z}^2$ such that $\gamma_1 \wedge \gamma_2 \neq 0$ there exist a finite collection of nonzero complex numbers (λ_i) and signs (ϵ_i) such that for any $n \geq 1$ one has*

$$Z_M(\mathbb{Z} \cdot \gamma_1 \oplus \mathbb{Z} \cdot n\gamma_2) = \sum_i \epsilon_i \lambda_i^n .$$

In other words, the series

$$\exp\left(-\sum_{n \geq 1} Z_M(\mathbb{Z} \cdot \gamma_1 \oplus \mathbb{Z} \cdot n\gamma_2) \cdot t^n / n\right)$$

in the formal variable t is rational.

The proof is omitted here; we'll just indicate that it follows from a consideration of the action of the Frobenius element and of cyclic permutations on the (étale) cohomology of spaces $\widetilde{Y}^{(m)}$ introduced in Section 3.1.

Also, it is easy to see that the same property holds for the partition function $Z_R^{\text{lat}}(\Lambda_{m,n,k})$ for arbitrary (super) lattice models.[16] The analogy leads to a two-dimensional analogue of the Weil conjecture (the name will be explained in the next section):

Conjecture 18. *For any endomorphism $M \in \text{End}_{\mathcal{C}_{\mathbb{F}_q}}(X)$ there exist super Boltzmann data (V_1, V_2, R) such that for any $\Lambda \subset \mathbb{Z}^2$ of finite index one has*

$$Z_M(\Lambda) = Z_R^{\text{lat}}(\Lambda) \ .$$

Up to now there is no hard evidence for this conjecture; there are just a few cases in which one can construct a corresponding lattice model in an ad hoc manner. For example, it is possible (and not totally trivial) to do so for the case that $X = \mathbb{A}_{\mathbb{F}_q}^1$ and M is the graph of the map $x \to x^c$, where $c \geq 1$ is an integer.

The above conjecture means that one can see matrices $\phi_n(M)$ as analogues of transfer matrices.[17] In the theory of integrable models, people are interested in systems in which the Boltzmann weights R depend nontrivially on a parameter ρ (spaces V_1, V_2 do not vary), and the horizontal transfer matrices commute with each other,

$$[T_{(2),n}(\rho_1), T_{(2),n}(\rho_2)] = 0,$$

because of the Yang–Baxter equation. The theory of automorphic forms seems to produce families of commuting endomorphisms in the category $\mathcal{C}_{\mathbb{F}_q}$, which is quite analogous to the integrability in lattice models. There are still serious differences. First of all, commuting operators in the automorphic forms case depend on discrete parameters, whereas in the integrable model case they depend algebraically on continuous parameters. Secondly, the spectrum of a Hecke operator in its nth incarnation (such as $T_x^{(n)}$ in Section 0.1) has typically n-fold degeneracy, which does not happen in the case of the usual integrable models with period n.

3.4 Higher-dimensional lattice models and a higher-dimensional Weil conjecture

Let $d \geq 0$ be an integer.

---

[16] In general, one can show that for any lattice model given by the operator R, and for any matrix $A \in \text{GL}(2, \mathbb{Z})$, there exists another lattice model with operator R' such that for any lattice $\Lambda \subset \mathbb{Z}^2$ one has $Z_R^{\text{lat}}(\Lambda) = Z_{R'}^{\text{lat}}(A(\Lambda))$.

[17] At least if one is interested in the nonzero part of spectra. In general, the size of the transfer matrix depends on n as an exact exponent, while the size of $\phi_n(M)$ is a finite alternating sum of exponents.

Definition 19. The Boltzmann data of a d-dimensional translation-invariant lattice model are given by a collection $V_1, \ldots, V_d$ of finite-dimensional vector spaces over $\mathbb{C}$ and a linear operator

$$R : V_1 \otimes \cdots \otimes V_d \to V_1 \otimes \cdots \otimes V_d .$$

Similarly, one can define a d-dimensional lattice model in an arbitrary rigid tensor category. The partition function is a function on finite sets endowed with the action of the free group with d generators. In particular, for abelian actions, it gives a function $\Lambda \mapsto Z_R^{\mathrm{lat}}(\Lambda) \in \mathbb{C}$ on the set of subgroups of finite index in $\mathbb{Z}^d$. Also, for any lattice $\Lambda_{d-1} \subset \mathbb{Z}^d$ of rank $(d-1)$ and a vector $\gamma \in \mathbb{Z}^d$ such that $\gamma \notin \mathbb{Q} \otimes \Lambda_{d-1}$, the function

$$n \geq 1 \mapsto Z_R^{\mathrm{lat}}(\Lambda_{d-1} \oplus \mathbb{Z} \cdot n\gamma)$$

is a finite sum of exponents. Analogously, for any d-dimensional lattice model R and any integer $n \geq 1$ there exists its dimensional reduction, periodic with period n in the dth coordinate, which is a $(d-1)$-dimensional lattice model $R_{(n)}$ satisfying the property

$$Z_{R_{(n)}}(\Lambda_{d-1}) = Z_R(\Lambda_{d-1} \oplus \mathbb{Z} \cdot n\, e_d), \quad \forall \Lambda_{d-1} \subset \mathbb{Z}^{d-1},$$

where $e_d = (0, \ldots, 0, 1) \in \mathbb{Z}^d = \mathbb{Z}^{d-1} \oplus \mathbb{Z}$ is the last standard basis vector.

Conjecture 20. *For any $(d-1)$-dimensional lattice model $(X_1, \ldots, X_{d-1}, M)$, $d \geq 1$, in the category $\mathcal{C}_{\mathbb{F}_q}$, there exists a d-dimensional super lattice model $(V_1, \ldots, V_d, R)$ in $\mathrm{Super}_{\mathbb{C}}$ such that for any $n \geq 1$ the numerical $(d-1)$-dimensional model $\phi_n(M)$ gives the same partition function on the set of subgroups of finite index in $\mathbb{Z}^{d-1}$ as the dimensional reduction $R_{(n)}$.*

In the case $d = 1$ this conjecture follows from the usual Weil conjecture. Namely, a 0-dimensional Boltzmann data in $\mathcal{C}_{\mathbb{F}_k}$ is just an element

$$M \in \mathrm{End}_{\mathcal{C}_{\mathbb{F}_q}}(\mathbf{1}) = \mathrm{End}_{\mathcal{C}_{\mathbb{F}_q}}(\otimes_{i \in \emptyset} X_i)$$

of the Grothendieck group of varieties over $\mathbb{F}_k$ (or of K_0 of the category of pure motives over $\mathbb{F}_k$ with rational coefficients). The corresponding numerical lattice models $\phi_n(M)$ are just numbers, counting $\mathbb{F}_{q^n}$-points in M. By the usual Weil conjecture these numbers are traces of powers of an operator in a super vector space, i.e., values of the partition function for 1-dimensional super lattice model.

Similarly, for $d = 2$ one gets the 2-dimensional Weil conjecture from the previous section.

Evidence: p-adic Banach lattice models

Let K be a complete nonarchimedean field (e.g., a finite extension of $\mathbb{Q}_p$). We define a d-dimensional *contracting* Banach lattice model as follows. The Boltzmann data consists of

- $2d$ countable generated K-Banach spaces $V_1^{\text{in}}, \ldots, V_d^{\text{in}}, V_1^{\text{out}}, \ldots, V_d^{\text{out}}$,
- a bounded operator $R^{\text{vertices}} : V_1^{\text{in}} \widehat{\otimes} \cdots \widehat{\otimes} V_d^{\text{in}} \to V_1^{\text{out}} \widehat{\otimes} \cdots \widehat{\otimes} V_d^{\text{out}}$,
- a collection of compact operators $R_i^{\text{edges}} : V_i^{\text{out}} \to V_i^{\text{in}}$, $i = 1, \ldots, d$.

Such data again give a function on oriented graphs with colored edges; in the definition one should insert the operator R_i^{edges} for each edge colored by the index i, $i = 1, \ldots, d$. In the case of *finite-dimensional* spaces $\left(V_i^{\text{in}}, V_i^{\text{out}}\right)_{i=1,\ldots,d}$ we obtain the same partition function as for a usual finite-dimensional lattice model. Namely, one can set

$$R := \left(\otimes_{i=1}^d R_i^{\text{edges}}\right) \circ R^{\text{vertices}}, \quad V_i = V_i^{\text{in}}, \ \forall i = 1, \ldots, d,$$

or alternatively,

$$\tilde{R} := R^{\text{vertices}} \circ \left(\otimes_{i=1}^d R_i^{\text{edges}}\right), \quad \tilde{V}_i := V_i^{\text{out}}, \ \forall i = 1, \ldots, d \ .$$

In particular, for any contracting Banach model one gets a function $\Lambda \mapsto Z_R^{\text{lat}}(\Lambda) \in K$ on the set of sublattices of $\mathbb{Z}^d$. This function satisfies the property that for any lattice $\Lambda_{d-1} \subset \mathbb{Z}^d$ of rank $(d-1)$ and a vector $\gamma \in \mathbb{Z}^d$ such that $\gamma \notin \mathbb{Q} \otimes \Lambda_{d-1}$, one has

$$Z_R^{\text{lat}}(\Lambda_{n-1} \oplus \mathbb{Z} \cdot n\gamma) = \sum_\alpha \lambda_\alpha^n, \quad \forall n \geq 1,$$

where (λ_α) is a (possibly) countable $\text{Gal}(\overline{K}/K)$-invariant collection of nonzero numbers in $\overline{K}$ (eigenvalues of transfer operators) whose norms tend to zero. Similarly, one can define super Banach lattice models.

Here we announce a result supporting Conjecture 7; the proof is a straightforward extension of the Dwork method for the proving the rationality of zeta the function of a variety over a finite field.

Theorem 21. *Conjecture 7 holds if one allows contracting Banach super lattice models over a finite extension of $\mathbb{Q}_p$, where p is the characteristic of the finite field $\mathbb{F}_q$.*

3.5 Tensor category $\mathcal{A}$ and the Master Conjecture

Let us consider the following rigid tensor category $\mathcal{A}$. Objects of $\mathcal{A}$ are finite-dimensional vector spaces over $\mathbb{C}$. The set of morphisms $\text{Hom}_{\mathcal{A}}(V_1, V_2)$ is defined as the group K_0 of the category of finite-dimensional representations of the free (tensor) algebra

$$T(V_1 \otimes V_2^\vee) := \bigoplus_{n \geq 0} (V_1 \otimes V_2^\vee)^{\otimes n} \ .$$

A representation of the free algebra by operators in a vector space U is the same as an action of its generators on U, i.e., a linear map

$$V_1 \otimes V_2^\vee \otimes U \to U .$$

Using duality we interpret it as a map

$$V_1 \otimes U \to V_2 \otimes U .$$

The composition of morphisms is defined by the following formula on generators:

$$[V_1 \otimes U \to V_2 \otimes U] \circ [V_2 \otimes U' \to V_3 \otimes U']$$

is equal to

$$[V_1 \otimes (U \otimes U') \to V_3 \otimes (U \otimes U')],$$

where the expression in the brackets is the obvious composition of linear maps

$$V_1 \otimes U \otimes U' \to V_2 \otimes U \otimes U' \to V_3 \otimes U \otimes U' .$$

The tensor product in $\mathcal{A}$ coincides on objects with the tensor product in $\mathrm{Vect}_{\mathbb{C}}$, the same for the duality. The formula for the tensor product on morphisms is an obvious one; we leave the details to the reader.

As in Section 2.1 (Question 3), we can ask the following question:

Question 22. Are there interesting nontrivial projectors in $\mathcal{A}$?[18]

We denote by $\mathcal{A}^{\mathrm{kar}}$ the Karoubi closure of $\mathcal{A}$.

There exists an infinite chain of tensor functors $\left(\phi_n^{\mathcal{A}}\right)_{n \geq 1}$ from $\mathcal{A}$ to the category of finite-dimensional vector spaces over $\mathbb{C}$ given by

$$\phi_n^{\mathcal{A}}(V) := V^{\otimes n}$$

on objects, and by

$$[f : V_1 \otimes U \to V_2 \otimes U] \xmapsto{\phi_n} \mathrm{Trace}_{U^{\otimes n}}((\sigma_n \otimes \mathrm{id}_{V_2^{\otimes n}} f^{\otimes n}) \in \mathrm{Hom}_{\mathrm{Vect}_{\mathbb{C}}}\left(V_1^{\otimes n}, V_2^{\otimes n}\right)$$

on morphisms, where $\sigma_n : U^{\otimes n} \to U^{\otimes n}$ is the cyclic permutation. The cyclic group $\mathbb{Z}/n\mathbb{Z}$ acts by automorphisms of $\phi_n^{\mathcal{A}}$. Moreover, the generator of the cyclic group acting on $V^{\otimes n} = \phi_n^{\mathcal{A}}(V)$ is the image under $\phi_n^{\mathcal{A}}$ of a certain central element Fr_V in the algebra of endomorphisms $\mathrm{End}_{\mathcal{A}}(V)$. This "Frobenius" element is represented by the linear map $\sigma : V \otimes U \to V \otimes U$, where $U := V$ and $\sigma = \sigma_2$ is the permutation. As in the case of $\mathcal{C}_{\mathbb{F}_q}$, for any V the operator $\phi_n^{\mathcal{A}}(\mathrm{Fr}_V)$ is periodic with period n.

---

[18] A similar question about commuting endomorphisms in $\mathcal{A}$ is almost equivalent to the study of finite-dimensional solutions of the Yang–Baxter equation.

Let us introduce a small modification $\mathcal{A}'$ of the tensor category $\mathcal{A}$. Namely, it will have the same objects (finite-dimensional vector spaces over $\mathbb{C}$), and the morphism groups will be the quotients

$$\mathrm{Hom}_{\mathcal{A}'}(V_1, V_2) := K_0\left(T\left(V_1 \otimes V_2^\vee\right) - \mathrm{mod}\right)/\mathbb{Z} \cdot [\mathrm{triv}]$$

where triv is the trivial one-dimensional representation of $T\left(V_1 \otimes V_2^\vee\right)$ given by the zero map

$$V_1 \otimes \mathbf{1} \xrightarrow{0} V_2 \otimes \mathbf{1}.$$

All the previous considerations extend to the case of $\mathcal{A}'$.

The amazing similarities between the categories $\mathcal{C}_{\mathbb{F}_q}$ and $\mathcal{A}'$ suggest the following conjecture:

Conjecture 23. *For any prime p there exists a tensor functor*

$$\Phi_p : \mathcal{C}_{\mathbb{F}_p} \to \mathcal{A}'^{kar}$$

and a sequence of isomorphisms of tensor functors from $\mathcal{C}_{\mathbb{F}_p}$ to $Vect_{\mathbb{C}}$ for all $n \geq 1$

$$iso_{n,p} : \phi_n^{\mathcal{A}'} \circ \Phi_p \simeq i_{Vect_{\mathbb{Q}} \to Vect_{\mathbb{C}}} \circ \phi_n$$

where $i_{Vect_{\mathbb{Q}} \to Vect_{\mathbb{C}}}$ is the obvious embedding functor from the category of vector spaces over $\mathbb{Q}$ to the one over $\mathbb{C}$. Moreover, for any $X \in \mathcal{C}_{\mathbb{F}_p}$ the functor Φ_p maps the Frobenius element $\mathrm{Fr}_X \in \mathrm{End}_{\mathcal{C}_{\mathbb{F}_p}}(X)$ to $\mathrm{Fr}_{\Phi_p(V)}$.

This conjecture we call the Master Conjecture because it implies simultaneously *all* higher-dimensional versions of the Weil conjecture at once, as one has the bijection (essentially by definition)

$$\{(d-1)\text{-dimensional super lattice models in } \mathcal{A}'\} \simeq$$
$$\simeq \{d\text{-dimensional super lattice models in } Vect_{\mathbb{C}}\} \ .$$

Remark 24. One can consider a larger category $\mathcal{A}^{super}$ adding to objects of $\mathcal{A}$ super vector spaces as well. The group K_0 in the super case should be defined as the naive K_0 modulo the relation

$$[V_1 \otimes U \to V_2 \otimes U] = -[V_1 \otimes \Pi(U) \to V_2 \otimes \Pi(U)]$$

where Π is the parity changing functor.

It suffices to verify the Master Conjecture only on the full symmetric monoidal subcategory of $\mathcal{C}_{\mathbb{F}_p}$ consisting of powers $\left(\mathbb{A}_{\mathbb{F}_p}^n\right)_{n \geq 0}$ of the affine line. The reason is that any scheme of finite type can be embedded (by a constructible map) in an affine space $\mathbb{A}_{\mathbb{F}_p}^n$, and the characteristic function of the image of such an embedding as an idempotent in $\mathrm{End}_{\mathcal{C}_{\mathbb{F}_p}}(\mathbb{A}_{\mathbb{F}_p}^n)$.

Machine modelling finite fields

Let us fix a prime p. The object $A := \mathbb{A}^1_{\mathbb{F}_p}$ of $\mathcal{C}_{\mathbb{F}_p}$ is a commutative algebra (as well as any scheme of finite type, see 2.2.1), with the product given by the diagonal in its cube. The category $\mathrm{Aff}(\mathcal{C}_{\mathbb{F}_p})$ of "affine schemes" in $\mathcal{C}_{\mathbb{F}_p}$ (i.e. the category opposite to the category of commutative associative unital algebras in $\mathcal{C}_{\mathbb{F}_p}$) is closed under finite products. In particular, it makes sense to speak about group-like etc. objects in $\mathrm{Aff}(\mathcal{C}_{\mathbb{F}_p})$. Affine line A is a commutative ring-like object in $\mathrm{Aff}(\mathcal{C}_{\mathbb{F}_p})$, with the operations of addition and multiplication corresponding to the graphs of the usual addition and multiplication on $\mathbb{A}^1_{\mathbb{F}_p}$. In plain terms, this means that besides the commutative algebra structure on A

$$m : A \otimes A \to A$$

we have two coproducts (for the addition and for the multiplication)

$$co - a : A \to A \otimes A, \; co - m : A \to A \otimes A,$$

which are homomorphisms of algebras and satisfy the usual bunch of rules for commutative associative rings, including the distributivity law.

If the master conjecture is true, then it gives an object $V_p := \Phi_p(A) \in \mathcal{A'}^{\mathrm{kar}}$, with one product and two coproducts. One can expect that it is just $\mathbb{C}^p$ as a vector space. For any $n \geq 1$, the $\mathcal{A'}$-product on V_p defines a commutative algebra structure on $V_p^{\otimes n}$. Its spectrum should be a finite set consisting of p^n elements. Two coproducts give operations of addition and multiplication on this set, and we will obtain a *canonical* construction[19] of the finite field $\mathbb{F}_{p^n}$ uniformly for all $n \geq 1$.

Even in the case $p = 2$ the construction of such V_p is a formidable task: one should find three finite-dimensional super representations of the free algebra in eight generators, satisfying nine identities in various K_0 groups.

3.6 Corollaries of the Master Conjecture

Good sign: Bombieri–Dwork bound

One can deduce easily from the master conjecture that for any given p and any system of equations in an arbitrary number of variables (x_i), where each of the equations is of an elementary form like $x_{i_1} + x_{i_2} = x_{i_3}$ or $x_{i_1} x_{i_2} = x_{i_3}$ or $x_i = 1$, the number of solutions of this system over $\mathbb{F}_{p^n}$ is an alternating sum of exponents in n, with the total number of terms bounded by C^N, where $C = C_p$ is a constant depending on p, and N is the number of equations. In fact, it is a well-known Bombieri–Dwork bound (and C is an absolute constant)[20]; see [2].

---

[19] Compare with Question 2 in Section 1.3, and remarks thereafter.

[20] A straightforward application of [2] gives the upper bound $C \leq 17^4$, which is presumably very far from the optimal one.

Bad sign: cohomology theories for motives over finite fields

Any machine-modeling finite field should be defined over a finitely generated commutative ring. In particular, there should be a machine defined over a number field K_p depending only on the characteristic p. A little thinking shows that the enumeration of the number of solutions of any given system of equations in the elementary form as above will be expressed as a super trace of an operator in a finite-dimensional super vector space defined over K_p. On the other hand, it looks very plausible that the category of motives over any finite field $\mathbb{F}_q$ does not have any fiber functor defined over a number field; see [9] for a discussion. I think that this is a strong sign indicating that the master conjecture is just wrong!

4 Categorical afterthoughts

4.1 Decategorifications of 2-categories

The two categories $\mathcal{C}_\mathbf{k}$ and $\mathcal{A}$ introduced in this paper have a common feature that is also shared (almost) by the category of Grothendieck motives. The general framework is the following.

Let $\mathcal{B}$ be a 2-category such that for any two objects $X, Y \in \mathcal{B}$ the category of 1-morphisms $\mathrm{Hom}_\mathcal{B}(X, Y)$ is a small *additive* category, and the composition of 1-morphisms is a biadditive functor. In practice, we may ask for categories $\mathrm{Hom}_\mathcal{B}(X, Y)$ to be triangulated categories (enriched by spectra, or by complexes of vector spaces). Moreover, the composition could be only a weak functor (e.g., A_∞-functor), and the associativity of the composition could hold only up to (fixed) homotopies and higher homotopies. The rough idea is that objects of $\mathcal{B}$ are "spaces" (nonlinear in general), whereas objects of the category $\mathrm{Hom}_\mathcal{B}(X, Y)$ are linear things on the "product" $X \times Y$ interpreted as kernels of some additive functors transforming some kind of sheaves from X to Y, by taking the pullback from X, the tensor product with the kernel on $X \times Y$, and then the direct image with compact support to Y.

In such a situation one can define a new (1-)category $K^{\mathrm{tr}}(\mathcal{B})$, which is in fact a triangulated category. This category will be called the *decategorification* of $\mathcal{B}$.

The first step is to define a new 1-category $K(\mathcal{B})$ enriched by spectra. It has the same objects as $\mathcal{B}$; the morphism spectrum $\mathrm{Hom}_{K(\mathcal{B})}(X, Y)$ is defined as the spectrum of K-theory of the triangulated category $\mathrm{Hom}_\mathcal{B}(X, Y)$.[21]

The second step is to make a formal triangulated envelope of this category. This step needs nothing; it can be performed for an arbitrary category enriched

---

[21]It is well known that in order to define a correct K-theory one needs either an appropriate enrichment on $\mathrm{Hom}_\mathcal{B}(X, Y)$ or a model structure in the sense of Quillen; see, e.g., [10].

by spectra. Objects of the new category are finite extensions of formal shifts of the objects of $K(\mathcal{B})$, such as twisted complexes by Bondal and Kapranov.

At the third step one adds formally direct summands for projectors. The resulting category $K^{\text{tr}}(\mathcal{B})$ is the same as the full category of compact objects in the category of exact functors from $K(\mathcal{B})^{\text{opp}}$ to the triangulated category of spectra (enriched by itself).

Finally, one can define a more elementary preadditive[22] category $K_0(\mathcal{B})$ by defining $\text{Hom}_{K_0(\mathcal{B})}(X, Y)$ to be the K_0 group of the triangulated category $\text{Hom}_{\mathcal{B}}(X, Y)$. Then we add formally to it finite sums and images of projectors. The resulting additive Karoubi-closed category will be denoted by $K_0^{\text{kar}}(\mathcal{B})$ and called the K_0-decategorification of $\mathcal{B}$. In what follows we list several examples of decategorifications.

Noncommutative stable homotopy theory

R. Meyer and R. Nest introduced in [8] a noncommutative analogue of the triangulated category of spectra. Objects of their category are not necessarily unital C^*-algebras; the morphism group from A to B is defined as the bivariant Kasparov theory $\text{KK}(A, B)$. One of the main observations in [8] is that this gives a structure of a triangulated category on C^*-algebras. Obviously this construction has a flavor of the K_0-decategorification.

Elementary algebraic model of bivariant K-theory

One can define a toy algebraic model of the construction by Meyer and Nest. For a given base field $\mathbf{k}$ consider the preadditive category whose objects are unital associative $\mathbf{k}$-algebras, and the group of morphisms from A to B is defined as K_0 of the exact category consisting of bimodules ($A^{\text{op}} \otimes B$-modules) that are projective and finitely generated as B-modules. This is obviously a K_0-decategorification of a 2-category.

Noncommutative pure and mixed motives

Let us consider the quotient of the category of Grothendieck–Chow motives $\text{Mot}_{\mathbf{k}, \mathbb{Q}}$ over a given field $\mathbf{k}$ with rational coefficients by an autoequivalence given by the invertible functor $\mathbb{Q}(1) \otimes \cdot$. The set of morphisms in this category between motives of two smooth projective schemes X, Y is given by

$$\text{Hom}_{\text{Mot}_{\mathbf{k}, \mathbb{Q}} / \mathbb{Z}^{\mathbb{Q}(1) \otimes \cdot}}(X, Y) = \bigoplus_{n \in \mathbb{Z}} \text{Hom}_{\text{Mot}_{\mathbf{k}, \mathbb{Q}}}(X, \mathbb{Q}(n) \otimes Y)$$

$$= \left(\mathbb{Q} \otimes_{\mathbb{Z}} \bigoplus_{n \in \mathbb{Z}} \text{Cycles}_n(X \times Y) \right) / (\text{rational equivalence})$$

$$= \mathbb{Q} \otimes_{\mathbb{Z}} \bigoplus_{n \in \mathbb{Z}} \text{CH}^n(X \times Y) = \mathbb{Q} \otimes_{\mathbb{Z}} K^0(X \times Y)$$

---

[22] Enriched by abelian groups in the plain sense (without higher homotopies).

because the Chern character gives an isomorphism modulo torsion between the sum of all Chow groups and $K^0(X) = K_0(D^b(\mathrm{Coh}\, X))$, the K_0 group of the bounded derived category $D(X) := D^b(\mathrm{Coh}\, X)$ of coherent sheaves on X. Finally, the category $D(X \times Y)$ can be interpreted as the category of functors $D(Y) \to D(X)$.

Triangulated categories of type $D(X)$, where X is a smooth projective variety over $\mathbf{k}$, belong to a larger class of *smooth proper* triangulated $\mathbf{k}$-linear dg-categories (another name is "saturated categories"); see, e.g., [7], [12]. We see that the above quotient category of pure motives is a full subcategory of K_0-decategorification (with $\mathbb{Q}$ coefficients) of the 2-category of smooth proper $\mathbf{k}$-linear dg-categories. This construction was described recently (without mentioning the relation to motives) in [11].

Analogously, if one takes the quotient of the Voevodsky triangulated category of mixed motives by the endofunctor $\mathbb{Q}(1)[2] \otimes \cdot$, the resulting triangulated category seems to be similar to a full subcategory of the full decategorification of the 2-category of smooth proper $\mathbf{k}$-linear dg-categories.

Motivic integral operators

We mentioned already in Section 2.1 that the category $\mathcal{C}_{\mathbf{k}}$ should be considered as a K_0-decategorification of a 2-category of motivic sheaves. A similar 2-category was considered in [6] in the relation to questions in integral geometry and calculus of integral operators with holonomic kernels.

Correspondences for free algebras

The category $\mathcal{A}$ is a K_0-decategorification by definition.

4.2 Trace of an exchange morphism

Let G_1, G_2 be two endofunctors of a triangulated category $\mathcal{C}$, and an exchange morphism (a natural transformation)

$$\alpha : G_1 \circ G_2 \to G_2 \circ G_1$$

is given[23]. Under the appropriate finiteness condition (e.g. when $\mathcal{C}$ is smooth and proper) one can define the *trace* of α, which can be calculated in two ways, as the trace of endomorphism of $\mathrm{Tor}(G_1, id_{\mathcal{C}})$ associated with G_2 and α, and as a similar trace with exchanged G_1 and G_2 (see [4] for a related stuff). Passing to powers and natural exchange morphisms constructed from nm copies of α:

$$\alpha_{(n,m)} : G_1^n \circ G_2^m \to G_2^m \circ G_1^n$$

---

[23] We do not assume that α is an isomorphism.

we obtain a collection of numbers $Z_\alpha(n,m) := \mathrm{Trace}(\alpha_{(n,m)})$ for $n, m \geq 1$. It is easy to see that these numbers come from a 2-dimensional super lattice model.

Let $\mathcal{C} = D(X)$ for smooth projective X, and functors are given by F^* and by $\mathcal{E} \otimes \cdot$ where $F : X \to X$ is a map, and $\mathcal{E}$ is a vector bundle endowed with a morphism $g : F^*\mathcal{E} \to \mathcal{E}$ (as in Section 1.3). In this case $Z_\alpha(n,m)$ is the trace (without the denominator) associated with the map F^n and the bundle $\mathcal{E}^{\otimes m}$. For example, one can construct a 2-dimensional super lattice model with the partition function

$$Z_R^{lat}(\Lambda_{n,m}) = \sum_{x \in \mathbb{C}: F^n(x) = x} \prod_{i=1}^{n} \left(F^i(x)\right)^m$$

where $F : \mathbb{C} \to \mathbb{C}$ is a polynomial map,[24] e.g., $F(x) = x^2 + c$.

The conclusion is that we have two different proposals concerning motivic local systems in positive characteristic: the first (algebraic dynamics) and the third one (lattice models), are ultimately related. It is enough to find the dynamical realization, and then the lattice model will pop out. As was mentioned already, most probably these two proposals will fail, but they still can serve as sources of analogies.

References

1. A. BRAVERMAN, D. KAZHDAN, *Gamma-functions of representations and lifting*, with appendix by V. Vologodsky, GAFA 200 (Tel Aviv, 1999), Geom. Funct. Anal. **2000**, Special Volume, Part I, 237–278, math.AG/9912208.

2. E. BOMBIERI, *On exponential sums in finite fields. II.* Invent. Math. **47** (1978), no.1, 29–39.

3. V. DRINFELD, *The number of two-dimensional irreducible representations of the fundamental group of a curve over a finite field*, Functional Anal. Appl. **15** (1981), no. 4, 294–295 (1982).

4. N. GANTER, M. KAPRANOV, *Representation and character theory in 2-categories*, math.KT/0602510.

5. A. GERASIMOV, S. KHARCHEV, D. LEBEDEV, S. OBLEZIN, *On a Gauss-Givental Representation of Quantum Toda Chain Wave Function*, Int. Math. Res. Not. 2006, Art. ID 96489, 23 pp., math.RT/0505310.

6. A. GONCHAROV, *Differential equations and integral geometry*, Adv. Math. **131** (1997), no.2, 279–343.

7. M. KONTSEVICH, Y. SOIBELMAN, *Notes on A_∞-algebras, A_∞-categories and non-commutative geometry. I*, math.RA/0606241.

8. R. MEYER, R. NEST, *The Baum-Connes Conjecture via Localization of Categories*, Topology, vol. **45** (2006), no. 2, pp. 209–259, math.KT/0312292.

9. J. S. MILNE, *Motives over $\mathbb{F}_p$*, math.AG/0607569.

---

[24]This seems to be a new type of integrability in lattice models, different from the usual Yang–Baxter ansatz.

10. M. SCHLICHTING, *A note on K-theory and triangulated categories*, Invent. Math. **150** (2002), no. 1, 111–116.

11. G. TABUADA, *Invariants additifs de dg-categories*, Int. Math. Res. Not. **2005**, no. 53, 3309-3339, `math.KT/0507227`.

12. B. TOËN, M. VAQUIÉ, *Moduli of objects in dg-categories*, `math.AG/0503269`.

PROPped-Up Graph Cohomology

M. Markl[*1] and A.A. Voronov[†2]

[1] Institute of Mathematics, Academy of Sciences, Žitná 25, 115 67 Praha 1, The Czech Republic
markl@math.cas.cz

[2] School of Mathematics, University of Minnesota, 206 Church St. S.E., Minneapolis, MN 55455, USA
voronov@math.umn.edu

Dedicated to Yuri I. Manin on the occasion of his seventieth birthday

Summary. We consider graph complexes with a flow and compute their cohomology. More specifically, we prove that for a PROP generated by a Koszul dioperad, the corresponding graph complex gives a minimal model of the PROP. We also give another proof of the existence of a minimal model of the bialgebra PROP from [14]. These results are based on the useful notion of a $\frac{1}{2}$PROP introduced by Kontsevich in [9].

Key words: operads, graph complexes, cohomology

2000 Mathematics Subject Classifications: 18D50, 55P48

Introduction

Graph cohomology is a term coined by M. Kontsevich [7,8] for the cohomology of complexes spanned by graphs of a certain type with a differential given by vertex expansions (also known as splittings), i.e., all possible insertions of an edge in place of a vertex. Depending on the type of graphs considered, one gets various "classical" types of graph cohomology. One of them is the graph cohomology implicitly present in the work of M. Culler and K. Vogtmann [2]. It is isomorphic to the rational homology of the "outer space," or equivalently, the rational homology of the outer automorphism group of a free group. Another type is the "fatgraph," also known as "ribbon graph," cohomology of R.C. Penner [17], which is isomorphic to the rational homology of the moduli spaces of algebraic curves.

---

[*]Partially supported by the grant GA ČR 201/02/1390 and by the Academy of Sciences of the Czech Republic, Institutional Research Plan No. AV0Z10190503.

[†]Partially supported by NSF grant DMS-0227974.

Y. Tschinkel and Y. Zarhin (eds.), *Algebra, Arithmetic, and Geometry*,
Progress in Mathematics 270, DOI 10.1007/978-0-8176-4747-6_8,
© Springer Science+Business Media, LLC 2009

These types of graph cohomology appear to be impossible to compute, at least at this ancient stage of development of mathematics. For example, the answer for ribbon graph cohomology is known only in a "stable" limit, as the genus goes to infinity; see a recent "hard" proof of the Mumford conjecture by I. Madsen and M.S. Weiss [11]. No elementary method of computation seems to work: the graph complex becomes very complicated combinatorially in higher degrees. Even the apparently much simpler case of tree cohomology had been quite a tantalizing problem (except for the associative case, when the computation follows from the contractibility of the associahedra) until V. Ginzburg and M.M. Kapranov [6] attacked it by developing Koszul duality for operads.

This paper originated from a project of computing the cohomology of a large class of graph complexes. The graph complexes under consideration are "PROPped up," which means that the graphs are directed, provided with a flow, and decorated by the elements of a certain vector space associated to a given PROP. When this PROP is IB, the one describing infinitesimal bialgebras, see M. Aguiar [1], we get a directed version of the ribbon graph complex, while the PROP LieB describing Lie bialgebras gives a directed commutative version of the graph complex. In both cases, as well as in more general situations of a directed graph complex associated to a PROP coming from a Koszul dioperad in the sense of W.L. Gan [4] and of a similar graph complex with a differential perturbed in a certain way, we prove that the corresponding graph complex is acyclic in all degrees but one, see Corollary 28, answering a question of D. Sullivan in the Lie case. This answer stands in amazing contrast with anything one may expect from the nondirected counterparts of graph cohomology, such as the ones mentioned in the previous paragraphs: just putting a flow on graphs in a graph complex changes the situation so dramatically!

Another important goal of the paper is to construct free resolutions and minimal models of certain PROPs, which might be thought of as Koszul-like, thus generalizing both the papers of Ginzburg–Kapranov [6] and Gan [4], from trees (and operads and dioperads, respectively) to graphs (and PROPs). This is the content of Theorem 37 below. This theory is essential for understanding the notions of strongly homotopy structures described by the cobar construction for Koszul dioperads in [4] and the resolution of the bialgebra PROP in [14].

We also observe that axioms of some important algebraic structures over PROPs can be seen as perturbations of axioms of structures over $\frac{1}{2}$PROPs, objects in a way much smaller than PROPs and even smaller than dioperads, whose definition, suggested by Kontsevich [9], we give in Section 1.1. For example, we know from [14] that the PROP B describing bialgebras is a perturbation of the $\frac{1}{2}$PROP $\frac{1}{2}$b for $\frac{1}{2}$bialgebras (more precisely, B is a perturbation of the PROP generated by the $\frac{1}{2}$PROP $\frac{1}{2}$b). Another important perturbation of $\frac{1}{2}$b is the dioperad IB for infinitesimal bialgebras and, of course, also the PROP IB generated by this dioperad. In the same vein, the dioperad $LieB$

describing Lie bialgebras and the corresponding PROP LieB are perturbations of the $\frac{1}{2}$PROP $\frac{1}{2}$lieb for $\frac{1}{2}$Lie bialgebras introduced in Example 20.

As we argued in [14], every minimal model of a PROP or dioperad that is a perturbation of a $\frac{1}{2}$PROP can be expected to be a perturbation of a minimal model of this $\frac{1}{2}$PROP. However, there might be some unexpected technical difficulties in applying this principle, such as the convergence problem in the case of the bialgebra PROP; see Section 1.6.

The above observation can be employed to give a new proof of Gan's results on Koszulness of the dioperads describing Lie bialgebras and infinitesimal bialgebras. First, one proves that the $\frac{1}{2}$PROPs $\frac{1}{2}$b and $\frac{1}{2}$lieb are Koszul in the sense of Section 1.4, simply repeating Gan's proof in the simpler case of $\frac{1}{2}$PROPs. This means that the $\frac{1}{2}$PROP cobar constructions on the quadratic duals of these $\frac{1}{2}$PROPs are minimal models thereof. Then one treats the dioperadic cobar constructions on the dioperadic quadratic duals of IB and $LieB$ as perturbations of the dg dioperads freely generated by the $\frac{1}{2}$PROP cobar constructions and applies our perturbation theory to show that these dioperadic cobar constructions form minimal models of the corresponding dioperads, which is equivalent to their Koszulness.

This paper is based on ideas of the paper [14] by the first author and an e-mail message [9] from Kontsevich. The crucial notion of a $\frac{1}{2}$PROP (called in [9] a small PROP) and the idea that generating a PROP out of a $\frac{1}{2}$PROP constitutes a polynomial functor belong to him.

Acknowledgment: We are grateful to Wee Liang Gan, Maxim Kontsevich, Sergei Merkulov, Jim Stasheff, and Dennis Sullivan for useful discussions.

1.1 PROPs, Dioperads, and $\frac{1}{2}$PROPs

Let k denote a ground field, which will always be assumed of characteristic zero. This guarantees the complete reducibility of finite group representations. A PROP is a collection $\mathsf{P} = \{\mathsf{P}(m,n)\}$, $m, n \geqslant 1$, of *differential graded* (dg) (Σ_m, Σ_n)-bimodules (left Σ_m- right Σ_n-modules such that the left action commutes with the right one), together with two types of compositions, *horizontal*

$$\otimes : \mathsf{P}(m_1, n_1) \otimes \cdots \otimes \mathsf{P}(m_s, n_s) \to \mathsf{P}(m_1 + \cdots + m_s, n_1 + \cdots + n_s),$$

defined for all $m_1, \ldots, m_s, n_1, \ldots, n_s > 0$, and *vertical*,

$$\circ : \mathsf{P}(m, n) \otimes \mathsf{P}(n, k) \to \mathsf{P}(m, k),$$

defined for all $m, n, k > 0$. These compositions respect the dg structures. One also assumes the existence of a *unit* $1 \in \mathsf{P}(1, 1)$.

PROPs should satisfy axioms that could be read off from the example of the *endomorphism* PROP $\mathcal{E}nd_V$ of a vector space V, with $\mathcal{E}nd_V(m, n)$ the space of linear maps $Hom(V^{\otimes n}, V^{\otimes m})$ with n "inputs" and m "outputs,"

$1 \in \mathcal{E}nd_V(1,1)$ the identity map, horizontal composition given by the tensor product of linear maps, and vertical composition by the ordinary composition of linear maps. For a precise definition see [10, 12].

Let us denote, for later use, by $_j\circ_i : \mathsf{P}(m_1, n_1) \otimes \mathsf{P}(m_2, n_2) \to \mathsf{P}(m_1 + m_2 - 1, n_1 + n_2 - 1)$, $a, b \mapsto a_{\,j}\circ_i b$, $1 \leqslant i \leqslant n_1$, $1 \leqslant j \leqslant m_2$, the operation that composes the jth output of b to the ith input of a. Formally,

$$a_{\,j}\circ_i b := (\mathbf{1} \otimes \cdots \otimes \mathbf{1} \otimes a \otimes \mathbf{1} \otimes \cdots \otimes \mathbf{1})\sigma(\mathbf{1} \otimes \cdots \otimes \mathbf{1} \otimes b \otimes \mathbf{1} \otimes \cdots \otimes \mathbf{1}), \quad (1.1)$$

where a is at the jth place, b is at the ith place, and $\sigma \in \Sigma_{n_1+m_2-1}$ is the block permutation $((12)(45))_{i-1,j-1,m_2-j,n_1-i}$; see [4], where this operation was in fact denoted by $_i\circ_j$, for details.

It will also be convenient to introduce special notation for $_1\circ_i$ and $_j\circ_1$, namely $\circ_i := {_1\circ_i} : \mathsf{P}(m_1, n_1) \otimes \mathsf{P}(1, l) \to \mathsf{P}(m_1, n_1 + l - 1)$, $1 \leqslant i \leqslant n_1$, which can be defined simply by

$$a \circ_i b := a \circ (\mathbf{1} \otimes \cdots \otimes \mathbf{1} \otimes b \otimes \mathbf{1} \otimes \cdots \otimes \mathbf{1}) \quad (b \text{ at the } i\text{th position}), \quad (1.2)$$

and similarly, $_j\circ := {_j\circ_1}\mathsf{P}(k, 1) \otimes \mathsf{P}(m_2, n_2) \to \mathsf{P}(m_2 + k - 1, n_2)$, $1 \leqslant j \leqslant m_2$, which can be expressed as

$$c_{\,j}\circ d := (\mathbf{1} \otimes \cdots \otimes \mathbf{1} \otimes c \otimes \mathbf{1} \otimes \cdots \otimes \mathbf{1}) \circ d \quad (c \text{ at the } j\text{th position}). \quad (1.3)$$

A general iterated composition in a PROP is described by a "flow chart," which is a not necessarily connected graph of arbitrary genus, equipped with a "direction of gravity" or a "flow"; see Section 1.2 for more details. PROPs are in general gigantic objects, with $\mathsf{P}(m, n)$ infinite-dimensional for any m and n. W.L. Gan [4] introduced dioperads, which avoid this combinatorial explosion. Roughly speaking, a dioperad is a PROP in which only compositions along contractible graphs are allowed.

This can be formally expressed by saying that a *dioperad* is a collection $D = \{D(m, n)\}$, $m, n \geqslant 1$, of dg (Σ_m, Σ_n)-bimodules with compositions

$$_j\circ_i : D(m_1, n_1) \otimes D(m_2, n_2) \to D(m_1 + m_2 - 1, n_1 + n_2 - 1),$$

$1 \leqslant i \leqslant n_1$, $1 \leqslant j \leqslant m_2$, that satisfy the axioms satisfied by operations $_j\circ_i$, see (1.1), in a general PROP. Gan [4] observed that some interesting objects, such as Lie bialgebras and infinitesimal bialgebras, can be defined using algebras over dioperads.

M. Kontsevich [9] suggested an even more radical simplification consisting in considering objects for which only $\circ_i$ and $_j\circ$ compositions and their iterations are allowed. More precisely, he suggested the following definition:

Definition 1. *A* $\frac{1}{2}$*PROP is a collection* $\mathsf{s} = \{\mathsf{s}(m, n)\}$ *of dg* (Σ_m, Σ_n)*-bimodules* $\mathsf{s}(m, n)$ *defined for all pairs of natural numbers except* $(m, n) = (1, 1)$, *together with compositions*

$$\circ_i : \mathsf{s}(m_1, n_1) \otimes \mathsf{s}(1, l) \to \mathsf{s}(m_1, n_1 + l - 1), \quad 1 \leqslant i \leqslant n_1, \quad (1.4)$$

and

$$_j\circ : \mathsf{s}(k,1) \otimes \mathsf{s}(m_2,n_2) \to \mathsf{s}(m_2+k-1,n_2), \quad 1 \leqslant j \leqslant m_2, \qquad (1.5)$$

that satisfy the axioms satisfied by operations $\circ_i$ and $_j\circ$, see (1.2) and (1.3), in a general PROP.

We suggest as an exercise to unwrap the above definition, write the axioms explicitly, and compare them to the axioms of a dioperad in [4]. Observe that $\frac{1}{2}$PROPs cannot have units, because $\mathsf{s}(1,1)$ is not there. Later we will also use the notation

$$\circ := \circ_1 = {}_1\circ : \mathsf{s}(k,1) \otimes \mathsf{s}(1,l) \to \mathsf{s}(k,l), \ k,l \geqslant 2. \qquad (1.6)$$

The category of $\frac{1}{2}$PROPs will be denoted by $\frac{1}{2}$PROP.

Example 2. Since $\frac{1}{2}$PROPs do not have units, their nature is close to that of *pseudo-operads* [15, Definition 1.16], which are, roughly, operads without units, with axioms defined in terms of $\circ_i$-operations. More precisely, the category of $\frac{1}{2}$PROPs s with $\mathsf{s}(m,n) = 0$ for $m \geqslant 2$ is isomorphic to the category of pseudo-operads $\mathcal{P}$ with $\mathcal{P}(0) = \mathcal{P}(1) = 0$. This isomorphism defines a faithful embedding $\iota : \mathtt{Oper} \mapsto \frac{1}{2}$PROP from the category $\mathtt{Oper}$ of pseudo-operads $\mathcal{P}$ with $\mathcal{P}(0) = \mathcal{P}(1) = 0$ to the category of $\frac{1}{2}$PROPs. To simplify the terminology, by "operad" we will, in this paper, always understand a pseudo-operad in the above sense.

Example 3. Given a PROP P, there exists the "opposite" PROP $\mathsf{P}^\dagger$ with $\mathsf{P}^\dagger(m,n) := \mathsf{P}(n,m)$, for each $m,n \geqslant 1$. A similar duality exists also for dioperads and $\frac{1}{2}$PROPs. Therefore one may define another faithful embedding, $\iota^\dagger : \mathtt{Oper} \mapsto \frac{1}{2}$PROP, by $\iota^\dagger(\mathcal{P}) := \iota(\mathcal{P})^\dagger$, where ι was defined in Example 2. The image of this embedding consists of all $\frac{1}{2}$PROPs s with $\mathsf{s}(m,n) = 0$ for all $n \geqslant 2$.

Every PROP defines a dioperad by forgetting all compositions that are not allowed in a dioperad. In the same vein, each dioperad defines a $\frac{1}{2}$PROP if we forget all compositions not allowed in Definition 1. These observations can be organized into the following diagram of forgetful functors, in which $\mathtt{diOp}$ denotes the category of dioperads:

$$\mathtt{PROP} \xrightarrow{\square_1} \mathtt{diOp} \xrightarrow{\square_2} \tfrac{1}{2}\mathtt{PROP}. \qquad (1.7)$$

The left adjoints $F_1 : \mathtt{diOp} \to \mathtt{PROP}$ and $F_2 : \frac{1}{2}\mathtt{PROP} \to \mathtt{diOp}$ exist by general nonsense. In fact, we give, in Section 1.3, an explicit description of these functors. Of primary importance for us will be the composition

$$F := F_1 \circ F_2 : \tfrac{1}{2}\mathtt{PROP} \to \mathtt{PROP}, \qquad (1.8)$$

which is clearly the left adjoint to the forgetful functor $\square := \square_2 \circ \square_1 :$ $\mathtt{PROP} \to \frac{1}{2}\mathtt{PROP}$. Given a $\frac{1}{2}$PROP s, $F(\mathsf{s})$ could be interpreted as the *free* PROP *generated by the* $\frac{1}{2}$PROP s.

Recall [10, 12] that an *algebra* over a PROP P is a morphism $P \to \mathcal{E}nd_V$ of PROPs. The adjoints above offer an elegant way to introduce algebras over $\frac{1}{2}$PROPs and dioperads: an algebra over a $\frac{1}{2}$PROP s is simply an algebra over the PROP $F(s)$, and similarly, an algebra over a dioperad D is defined to be an algebra over the PROP $F_1(D)$.

The following important theorem, whose proof we postpone to Section 1.3, follows from the fact, observed by M. Kontsevich in [9], that F and F_2 are, in a certain sense, *polynomial functors*; see (1.10) and (1.11).

Theorem 4. *The functors $F : \frac{1}{2}PROP \to PROP$ and $F_2 : \frac{1}{2}PROP \to diOp$ are exact. This means that they commute with homology, that is, given a differential graded $\frac{1}{2}$PROP s, $H_*(F(s))$ is naturally isomorphic to $F(H_*(s))$. In particular, for any morphism $\alpha : s \to t$ of dg $\frac{1}{2}$PROPs, the diagram of graded PROPs*

$$
\begin{array}{ccc}
H_*(F(s)) & \xrightarrow{\;\;H_*(F(\alpha))\;\;} & H_*(F(t)) \\
\Big\downarrow{\scriptstyle \cong} & & \Big\downarrow{\scriptstyle \cong} \\
F(H_*(s)) & \xrightarrow{\;\;F(H_*(\alpha))\;\;} & F(H_*(s))
\end{array}
$$

is commutative. A similar statement is also true for F_2 in place of F.

Let us emphasize here that we *do not know* whether the functor F_1 is exact. As a consequence of Theorem 4 we immediately obtain the following:

Corollary 5. *A morphism $\alpha : s \to t$ of dg $\frac{1}{2}$PROPs is a homology isomorphism if and only if $F(\alpha) : F(s) \to F(t)$ is a homology isomorphism. A similar statement is also true for F_2.*

Let us finish our catalogue of adjoint functors by the following definitions. By a *bicollection* we mean a sequence $E = \{E(m, n)\}_{m,n \geqslant 1}$ of differential graded (Σ_m, Σ_n)-bimodules such that $E(1, 1) = 0$. Let us denote by bCol the category of bicollections. Display (1.7) then can be completed into the following diagram of obvious forgetful functors:

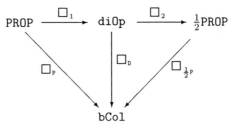

Denote finally by $\Gamma_P : bCol \to PROP$, $\Gamma_D : bCol \to diOp$, and $\Gamma_{\frac{1}{2}P} : bCol \to \frac{1}{2}PROP$ the left adjoints of the functors $\Box_P$, $\Box_D$, and $\Box_{\frac{1}{2}P}$, respectively.

Notation. We will use capital calligraphic letters $\mathcal{P}$, $\mathcal{Q}$, etc. to denote operads, small sans serif fonts s, t, etc. to denote $\frac{1}{2}$PROPs, capital italic fonts S, T, etc. to denote dioperads, and capital sans serif fonts S, T, etc. to denote PROPs.

1.2 Free PROPs

To deal with free PROPs and resolutions, we need to fix a suitable notion of a graph. Thus, in this paper a *graph* or an (m, n)-*graph*, $m, n \geqslant 1$, will mean a directed (i.e., each edge is equipped with direction) finite graph satisfying the following conditions:

1. the valence $n(v)$ of each vertex v is at least three;
2. each vertex has at least one outgoing and at least one incoming edge;
3. there are no directed cycles;
4. there are precisely m outgoing and n incoming *legs*, by which we mean edges incident to a vertex on one side and having a "free end" on the other; these legs are called the *outputs* and the *inputs*, respectively;
5. the legs are labeled, the inputs by $\{1, \ldots, n\}$, the outputs by $\{1, \ldots, m\}$.

Note that the graphs considered are not necessarily connected. Graphs with no vertices are also allowed. Those will be precisely the disjoint unions $\uparrow \uparrow \cdots \uparrow$ of a number of directed edges. We will always assume the flow to go from bottom to top when we sketch graphs.

Let $v(G)$ denote the set of vertices of a graph G, $e(G)$ the set of all edges, and $Out(v)$ (respectively, $In(v)$) the set of outgoing (respectively, incoming) edges of a vertex $v \in v(G)$. With an (m, n)-graph G, we will associate a *geometric realization* $|G|$, a CW complex whose 0-cells are the vertices of the graph G, as well as one extra 0-cell for each leg, and 1-cells are the edges of the graph. The 1-cells of $|G|$ are attached to its 0-cells, as given by the graph. The *genus* gen(G) of a graph G is the first Betti number $b_1(|G|) = \operatorname{rank} H_1(|G|)$ of its geometric realization. This terminology derives from the theory of modular operads, but is not perfect, e.g., our genus is not what one usually means by the genus for ribbon graphs, which are discussed in Section 1.7.

An *isomorphism* between two (m, n)-graphs G_1 and G_2 is a bijection between the vertices of G_1 and G_2 and a bijection between the edges thereof preserving the incidence relation, the edge directions, and fixing each leg. Let $Aut(G)$ denote the group of automorphisms of graph G. It is a finite group, being a subgroup of a finite permutation group.

Let $E = \{E(m, n) \mid m, n \geqslant 1, (m, n) \neq (1, 1)\}$, be a bicollection; see Section 1.1. A standard trick allows us to extend the bicollection E to pairs (A, B) of finite sets:

$$E(A, B) := Bij([m], A) \times_{\Sigma_m} E(m, n) \times_{\Sigma_n} Bij(B, [n]),$$

where *Bij* denotes the set of bijections, $[k] = \{1, 2, \ldots, k\}$, and A and B are any m- and n-element sets, respectively. We will mostly ignore such subtleties as distinguishing finite sets of the same cardinality in the sequel and hope this will cause no confusion. The inquisitive reader may look up an example of careful treatment of such things and what came out of it in [5].

For each graph G, define a vector space

$$E(G) := \bigotimes_{v \in v(G)} E(\mathit{Out}(v), \mathit{In}(v)).$$

Note that this is an unordered tensor product (in other words, a tensor product "ordered" by the elements of an index set), which makes a difference for the sign convention in a graded algebra; see [15, page 64]. By definition, $E(\uparrow) = k$. We will refer to an element of $E(G)$ as a *G-monomial*. One may also think of a G-monomial as a *decorated graph*. Finally, let

$$\Gamma_{\mathsf{P}}(E)(m, n) := \bigoplus_{G \in Gr(m,n)} E(G)_{\mathit{Aut}(G)}$$

be the (m, n)-space of the free PROP on E for $m, n \geqslant 1$, where the summation runs over the set $Gr(m, n)$ of isomorphism classes of all (m, n)-graphs G and

$$E(G)_{\mathit{Aut}(G)} := E(G)/Span(ge - e \mid g \in \mathit{Aut}(G), e \in E(G))$$

is the space of *coinvariants* of the natural action of the automorphism group $\mathit{Aut}(G)$ of the graph G on the vector space $E(G)$. The appearance of the automorphism group is due to the fact that the "right" definition would involve taking the colimit over the diagram of all graphs with respect to isomorphisms, see [5]. The space $\Gamma_{\mathsf{P}}(E)(m, n)$ is a (Σ_m, Σ_n)-bimodule via the action by relabeling the legs. Moreover, the collection $\Gamma_{\mathsf{P}}(E) = \{\Gamma_{\mathsf{P}}(E)(m, n) \mid m, n \geqslant 1\}$ carries a natural PROP structure via disjoint union of decorated graphs as horizontal composition, and grafting the outgoing legs of one decorated graph to the incoming legs of another one as vertical composition. The unit is given by $1 \in k = E(\uparrow)$. The PROP $\Gamma_{\mathsf{P}}(E)$ is precisely the free PROP introduced at the end of Section 1.1.

1.3 From $\frac{1}{2}$PROPs to PROPs

Let us emphasize that in this article a dg free PROP means a dg PROP whose underlying (non-dg) PROP is freely generated (by a $\frac{1}{2}$PROP, dioperad, bicollection, ...) in the category of (non-dg) PROPs. Such PROPs are sometimes also called *quasi-free* or *almost-free* PROPs.

We are going to describe the structure of the functors $F : \frac{1}{2}\mathsf{PROP} \to \mathsf{PROP}$ and $F_2 : \frac{1}{2}\mathsf{PROP} \to \mathtt{diOp}$ and prove that they commute with homology, i.e., prove Theorem 4. It is precisely the sense of equations (1.9), (1.10), and (1.11), in which we say that the functors F and F_2 are *polynomial*.

Let s be a dg $\frac{1}{2}$PROP. Then the dg free PROP $F(\mathsf{s})$ generated by s may be described as follows. We call an (m, n)-graph G, see Section 1.2, *reduced* if it has no internal edge that is either a unique output or unique input edge of a vertex. It is obvious that each graph can be modified to a reduced one by contracting all the edges violating this condition, i.e., the edges like this:

$$\text{⧖} \quad \& \quad \text{⧗},$$

where a triangle denotes a graph with at least one vertex and exactly one leg in the direction pointed by the triangle, and a box denotes a graph with at least one vertex. For each reduced graph G, define a vector space

$$\mathsf{s}(G) := \bigotimes_{v \in v(G)} \mathsf{s}(Out(v), In(v)). \tag{1.9}$$

We claim that the PROP $F(\mathsf{s})$ is given by

$$F(\mathsf{s})(m, n) = \bigoplus_{G \in \overline{Gr}(m,n)} \mathsf{s}(G)_{Aut(G)}, \tag{1.10}$$

where the summation runs over the set $\overline{Gr}(m, n)$ of isomorphism classes of all *reduced (m, n)-graphs* G, and $\mathsf{s}(G)_{Aut(G)}$ is the space of coinvariants of the natural action of the automorphism group $Aut(G)$ of the graph G on the vector space $\mathsf{s}(G)$. The PROP structure on the whole collection $\{F(\mathsf{s})(m, n)\}$ will be given by the action of the permutation groups by relabeling the legs and the horizontal and vertical compositions by disjoint union and grafting, respectively. If grafting creates a nonreduced graph, we will contract the bad edges and use suitable $\frac{1}{2}$PROP compositions to decorate the reduced graph appropriately.

A unit in the PROP $F(\mathsf{s})$ is given by $1 \in \mathsf{s}(\uparrow)$. A differential is defined as follows. Define a differential on $\mathsf{s}(G) = \bigotimes_{v \in v(G)} \mathsf{s}(Out(v), In(v))$ as the standard differential on a tensor product of complexes. The action of $Aut(G)$ on $\mathsf{s}(G)$ respects this differential, and therefore the space $\mathsf{s}(G)_{Aut(G)}$ of coinvariants inherits a differential. Then we take the standard differential on the direct sum (1.10) of complexes.

Proposition 6. *The dg PROP $F(\mathsf{s})$ is the dg free PROP generated by a dg $\frac{1}{2}$PROP s, as defined in Section 1.1.*

Proof. What we need to prove is that this construction delivers a left adjoint functor for the forgetful functor $\square : \text{PROP} \to \frac{1}{2}\text{PROP}$. Let us define two maps

$$Mor_{\frac{1}{2}\text{PROP}}(\mathsf{s}, \square(\mathsf{P})) \underset{\psi}{\overset{\phi}{\rightleftarrows}} Mor_{\text{PROP}}(F(\mathsf{s}), \mathsf{P}),$$

which will be inverses of each other. For a morphism $f : \mathsf{s} \to \square(\mathsf{P})$ of $\frac{1}{2}$PROPs and a reduced graph decorated by elements $s_v \in \mathsf{s}(m, n)$ at each vertex v, we can always compose $f(s_v)$'s in P as prescribed by the graph. The associativity of PROP compositions in P ensures the uniqueness of the result. This way we get a PROP morphism $\phi(f) : F(\mathsf{s}) \to \mathsf{P}$.

Given a PROP morphism $g : F(\mathsf{s}) \to \mathsf{P}$, restrict it to the sub-$\frac{1}{2}$PROP $\mathsf{s}' \subset F(\mathsf{s})$ given by decorated graphs with a unique vertex, such as ✕. We define $\psi(g)$ as the resulting morphism of $\frac{1}{2}$PROPs.   □

Remark 7. The above construction of the dg free PROP $F(\mathsf{s})$ generated by a $\frac{1}{2}$PROP s does not go through for the free PROP $F_1(D)$ generated by a dioperad D. The reason is that there is no unique way to reduce an (m,n)-graph to a graph with all possible dioperadic compositions, i.e., all interior edges, contracted, as the following figure illustrates:

This suggests that the functor F_1 may not be polynomial.

There is a similar description of the dg free dioperad $F_2(\mathsf{s})$ generated by a dg $\frac{1}{2}$PROP s:

$$F_2(\mathsf{s})(m,n) = \bigoplus_{T \in \overline{Tr}(m,n)} \mathsf{s}(T). \qquad (1.11)$$

Here the summation runs over the set $\overline{Tr}(m,n)$ of isomorphism classes of all *reduced* contractible (m,n)-graphs T. The automorphism groups of these graphs are trivial and therefore do not show up in the formula. The following proposition is proven by an obvious modification of the proof of Proposition 6.

Proposition 8. *The dg PROP $F_2(\mathsf{s})$ is the dg free dioperad generated by a dg $\frac{1}{2}$PROP s, as defined in Section 1.1.*

Proof of Theorem 4. Let us prove Theorem 4 for the dg free PROP $F(\mathsf{s})$ generated by a dg $\frac{1}{2}$PROP s. The proof of the statement for $F_2(\mathsf{s})$ will be analogous and even simpler, because of the absence of the automorphism groups of graphs.

Proposition 6 describes $F(\mathsf{s})$ as a direct sum (1.10) of complexes $\mathsf{s}(G)_{Aut(G)}$. Thus the homology $H_*(F(\mathsf{s}))$ is naturally isomorphic to

$$\bigoplus_{G \in \overline{Gr}(m,n)} H_*(\mathsf{s}(G)_{Aut(G)}).$$

The automorphism group $Aut(G)$ is finite, acts on $\mathsf{s}(G)$ respecting the differential, and therefore, by Maschke's theorem (remember, we work over a field of characteristic zero), the coinvariants commute naturally with homology:

$$H_*(\mathsf{s}(G)_{Aut(G)}) \xrightarrow{\sim} H_*(\mathsf{s}(G))_{Aut(G)}.$$

Then, using the Künneth formula, we get a natural isomorphism

$$H_*(\mathsf{s}(G)) \xrightarrow{\sim} \bigotimes_{v \in v(G)} H_*(\mathsf{s}(Out(v), In(v))).$$

Finally, combination of these isomorphisms results in a natural isomorphism

$$H_*(F(\mathsf{s})) \xrightarrow{\sim} \bigoplus_{G \in \overline{Gr}(m,n)} \bigotimes_{v \in v(G)} H_*(\mathsf{s}(Out(v), In(v)))_{Aut(G)} = F(H_*(\mathsf{s})).$$

The diagram in Theorem 4 is commutative, because of the naturality of the above isomorphisms. $\square$

1.4 Quadratic Duality and Koszulness for $\frac{1}{2}$PROPs

W.L. Gan defined in [4], for each dioperad D, a dg dioperad $\Omega_{\mathsf{D}}(D) = (\Omega_{\mathsf{D}}(D), \partial)$, the *cobar dual* of D (**D**D in his notation). He also introduced quadratic dioperads, quadratic duality $D \mapsto D^!$, and showed that for each quadratic dioperad, there exists a natural map of dg dioperads $\alpha_{\mathsf{D}} : \Omega_{\mathsf{D}}(D^!) \to D$. He called D *Koszul* if α_{D} was a homology isomorphism. His theory is a dioperadic analogue of a similar theory for operads developed in 1994 by V. Ginzburg and M.M. Kapranov [6]. The aim of this section is to build an analogous theory for $\frac{1}{2}$PROPs. Since the passage from $\frac{1}{2}$PROPs to PROPs is given by an exact functor, resolutions of $\frac{1}{2}$PROPs constructed with the help of this theory will induce resolutions in the category of PROPs.

Let us pause a little and recall, following [4], some facts about quadratic duality for dioperads in more detail. First, a *quadratic dioperad* is a dioperad D of the form

$$D = \Gamma_{\mathsf{D}}(U, V)/(A, B, C), \tag{1.12}$$

where $U = \{U(m, n)\}$ is a bicollection with $U(m, n) = 0$ for $(m, n) \neq (1, 2)$, $V = \{V(m, n)\}$ is a bicollection with $V(m, n) = 0$ for $(m, n) \neq (2, 1)$, and $(A, B, C) \subset \Gamma_{\mathsf{D}}(U, V)$ denotes the dioperadic ideal generated by (Σ, Σ)-invariant subspaces $A \subset \Gamma_{\mathsf{D}}(U, V)(1, 3)$, $B \subset \Gamma_{\mathsf{D}}(U, V)(2, 2)$, and $C \subset \Gamma_{\mathsf{D}}(U, V)(3, 1)$. Notice that we use the original terminology of [6], where quadraticity refers to arities of generators and relations, rather than just relations. The *dioperadic quadratic dual* $D^!$ is then defined as

$$D^! := \Gamma_{\mathsf{D}}(U^\vee, V^\vee)/(A^\perp, B^\perp, C^\perp), \tag{1.13}$$

where $U^\vee$ and $V^\vee$ are the linear duals with the action twisted by the sign representations (the *Czech duals*, see [15, p. 142]), and $A^\perp$, $B^\perp$, and $C^\perp$ are the annihilators of spaces A, B, and C in

$$\Gamma_{\mathsf{D}}(U^\vee, V^\vee)(i, j) \cong \Gamma_{\mathsf{D}}(U, V)(i, j)^*,$$

where $(i, j) = (1, 3), (2, 2)$, and $(3, 1)$, respectively. See [4, Section 2] for details.

Quadratic $\frac{1}{2}$PROPs and their quadratic duals can then be defined in exactly the same way as sketched above for dioperads, only replacing everywhere Γ_D by $\Gamma_{\frac{1}{2}P}$. We say therefore that a $\frac{1}{2}$PROP s is *quadratic* if it is of the form

$$s = \Gamma_{\frac{1}{2}P}(U, V)/(A, B, C),$$

with U, V, and $(A, B, C) \subset \Gamma_{\frac{1}{2}P}(U, V)$ having a similar obvious meaning as for dioperads. The *quadratic dual* of s is defined by a formula completely analogous to (1.13):

$$s^! := \Gamma_{\frac{1}{2}P}(U^\vee, V^\vee)/(A^\perp, B^\perp, C^\perp).$$

The apparent similarity of the above definitions, however, hides one very important subtlety. While

$$\Gamma_D(U^\vee, V^\vee)(1, 3) \cong \Gamma_{\frac{1}{2}P}(U^\vee, V^\vee)(1, 3)$$

and

$$\Gamma_D(U^\vee, V^\vee)(3, 1) \cong \Gamma_{\frac{1}{2}P}(U^\vee, V^\vee)(3, 1),$$

the (Σ_2, Σ_2)-bimodules $\Gamma_D(U^\vee, V^\vee)(2, 2)$ and $\Gamma_{\frac{1}{2}P}(U^\vee, V^\vee)(2, 2)$ are substantially different, namely

$$\Gamma_D(U^\vee, V^\vee)(2, 2) \cong \Gamma_{\frac{1}{2}P}(U^\vee, V^\vee)(2, 2) \oplus \mathrm{Ind}_{\{1\}}^{\Sigma_2 \times \Sigma_2}(U^\vee \otimes V^\vee),$$

where $\Gamma_{\frac{1}{2}P}(U^\vee, V^\vee)(2, 2) \cong V^\vee \otimes U^\vee$; see [4, section 2.4] for details.

We see that the annihilator of $B \subset \Gamma_{\frac{1}{2}P}(E, F)(2, 2)$ in $\Gamma_{\frac{1}{2}P}(E^\vee, F^\vee)(2, 2)$ is much smaller than the annihilator of the same space taken in $\Gamma_D(E^\vee, F^\vee)(2, 2)$. A consequence of this observation is the rather stunning fact that quadratic duals *do not commute* with the functor $F_2 : \frac{1}{2}$PROP $\rightarrow$ diOp, that is, $F_2(s^!) \neq F_2(s)^!$. The relation between $s^!$ and $F_2(s)^!$ is much finer and can be described as follows.

For a $\frac{1}{2}$PROP t, let $\jmath(t)$ denote the dioperad that coincides with t as a bicollection and whose structure operations are those of t if they are allowed for $\frac{1}{2}$PROPs, and are trivial if they are not allowed for $\frac{1}{2}$PROPs. This clearly defines a functor $\jmath : \frac{1}{2}$PROP $\rightarrow$ diOp.

Lemma 9. *Let* s *be a quadratic* $\frac{1}{2}$PROP. *Then* $F_2(s)$ *is a quadratic dioperad and*

$$F_2(s)^! \cong \jmath(s^!).$$

Proof. The proof immediately follows from the definitions and we may safely leave it to the reader. □

Remark 10. Obviously $\jmath(s) = F_2(s^!)^!$. This means that the restriction of the functor $\jmath : \frac{1}{2}$PROP $\rightarrow$ diOp to the full subcategory of quadratic $\frac{1}{2}$PROPs can in fact be defined using quadratic duality.

The cobar dual $\Omega_{\frac{1}{2}P}(s)$ of a $\frac{1}{2}$PROP s and the canonical map $\alpha_{\frac{1}{2}P}$: $\Omega_{\frac{1}{2}P}(s^!) \to s$ can be defined by mimicking mechanically the analogous definitions for dioperads in [4], and we leave this task to the reader. The following lemma, whose proof is completely straightforward and hides no surprises, may in fact be interpreted as a characterization of these objects.

Lemma 11. *For an arbitrary $\frac{1}{2}$PROP* t, *there exists a functorial isomorphism of dg dioperads*

$$\Omega_D(\jmath(t)) \cong F_2\left(\Omega_{\frac{1}{2}P}(t)\right).$$

If s *is a quadratic $\frac{1}{2}$PROP, then the canonical maps*

$$\alpha_{\frac{1}{2}P} : \Omega_{\frac{1}{2}P}(s^!) \to s$$

and

$$\alpha_D : \Omega_D(F_2(s)^!) \to F_2(s)$$

are related by

$$\alpha_D = F_2\left(\alpha_{\frac{1}{2}P}\right). \tag{1.14}$$

We say that a quadratic $\frac{1}{2}$PROP s is *Koszul* if the canonical map $\alpha_{\frac{1}{2}P}$: $\Omega_{\frac{1}{2}P}(s^!) \to s$ is a homology isomorphism. The following proposition is not unexpected, though it is in fact based on the rather deep Theorem 4.

Proposition 12. *A quadratic $\frac{1}{2}$PROP* s *is Koszul if and only if $F_2(s)$ is a Koszul dioperad.*

Proof. The proposition immediately follows from (1.14) of Lemma 11 and Corollary 5 of Theorem 4. $\qquad\square$

We close this section with a couple of important constructions and examples. Let $\mathcal{P}$ and $\mathcal{Q}$ be two operads. Recall from Examples 2 and 3 that $\mathcal{P}$ and $\mathcal{Q}$ can be considered as $\frac{1}{2}$PROPs, via embeddings $\iota : \mathsf{Oper} \to \frac{1}{2}$PROP and $\iota^\dagger : \mathsf{Oper} \to \frac{1}{2}$PROP, respectively. Let us denote by

$$\mathcal{P} * \mathcal{Q}^\dagger := \iota(\mathcal{P}) \sqcup \iota^\dagger(\mathcal{Q})$$

the coproduct ("free product") of $\frac{1}{2}$PROPs $\iota(\mathcal{P})$ and $\iota^\dagger(\mathcal{Q})$. We will need also the quotient

$$\mathcal{P} \diamond \mathcal{Q}^\dagger := (\iota(\mathcal{P}) \sqcup \iota^\dagger(\mathcal{Q}))/(\iota^\dagger(\mathcal{Q}) \circ \iota(\mathcal{P})),$$

with $(\iota^\dagger(\mathcal{Q}) \circ \iota(\mathcal{P}))$ denoting the ideal generated by all $q^\dagger \circ p$, $p \in \iota(\mathcal{P})$, and $q^\dagger \in \iota^\dagger(\mathcal{Q})$; here $\circ$ is as in (1.6).

Exercise 13. Let $\mathcal{P} = \Gamma_{\mathsf{Op}}(F)/(R)$ and $\mathcal{Q} = \Gamma_{\mathsf{Op}}(G)/(S)$ be quadratic operads [15, Definition 3.31]; here $\Gamma_{\mathsf{Op}}(-)$ denotes the free operad functor. If we interpret F, G, R, and S as bicollections with

$$F(1,2) := F(2), \ G(2,1) := G(2), \ R(1,3) := R(3) \text{ and } S(3,1) := S(3),$$

then we clearly have presentations (see (1.12))

$$\mathcal{P} * \mathcal{Q}^\dagger = \Gamma_{\frac{1}{2}\mathrm{P}}(F,G)/(R,0,S) \text{ and } \mathcal{P} \diamond \mathcal{Q}^\dagger = \Gamma_{\frac{1}{2}\mathrm{P}}(F,G)/(R, G \circ F, S),$$

which show that both $\mathcal{P} * \mathcal{Q}^\dagger$ and $\mathcal{P} \diamond \mathcal{Q}^\dagger$ are quadratic $\frac{1}{2}$PROPs.

Exercise 14. Let $\mathcal{A}ss$ be the operad for associative algebras [15, Definition 1.12]. Verify that algebras over the $\frac{1}{2}$PROP $\mathcal{A}ss * \mathcal{A}ss^\dagger$ are given by a vector space V, an associative multiplication $\bullet : V \otimes V \to V$, and a coassociative comultiplication $\Delta : V \to V \otimes V$, with no relation between these two operations. Verify also that the algebra over $\frac{1}{2}\mathrm{b} := \mathcal{A}ss \diamond \mathcal{A}ss^\dagger$ consists of an associative multiplication $\bullet$ and a coassociative comultiplication Δ as above, with the exchange rule

$$\Delta(a \bullet b) = 0, \text{ for each } a, b \in V.$$

These are exactly the $\frac{1}{2}$*bialgebras* introduced in [14]. The PROP $F\left(\frac{1}{2}\mathrm{b}\right)$ generated by the $\frac{1}{2}$PROP $\frac{1}{2}\mathrm{b}$ is precisely the PROP $\frac{1}{2}\mathrm{B}$ for the $\frac{1}{2}$bialgebras considered in the same paper.

Exercise 15. Let $\mathcal{P}$ and $\mathcal{Q}$ be quadratic operads [15, Definition 3.31], with quadratic duals $\mathcal{P}^!$ and $\mathcal{Q}^!$, respectively. Prove that the quadratic dual of the $\frac{1}{2}$PROP $\mathcal{P} \diamond \mathcal{Q}^\dagger$ is given by

$$(\mathcal{P} \diamond \mathcal{Q}^\dagger)^! = \mathcal{P}^! * (\mathcal{Q}^!)^\dagger.$$

Example 16. The quadratic dual of the $\frac{1}{2}$PROP $\frac{1}{2}\mathrm{b}$ introduced in Exercise 14 is $\mathcal{A}ss * \mathcal{A}ss^\dagger$. Let $\mathcal{L}ie$ denote the operad for Lie algebras [15, Definition 1.28] and $\mathcal{C}om$ the operad for commutative associative algebras [15, Definition 1.12]. The quadratic dual of the $\frac{1}{2}$PROP $\frac{1}{2}\mathrm{lieb} := \mathcal{L}ie \diamond \mathcal{L}ie^\dagger$ is $\mathcal{C}om * \mathcal{C}om^\dagger$.

Gan defined a monoidal structure $(E, F) \mapsto E \square F$ on the category of bicollections such that dioperads were precisely monoids for this monoidal structure. Roughly speaking, $E \square F$ was a sum over all directed contractible graphs G equipped with a level function $\ell : v(G) \to \{1, 2\}$ such that vertices of level one (that is, vertices with $\ell(v) = 1$) were decorated by E and vertices of level two were decorated by F. See [4, Section 4] for precise definitions. Needless to say, this $\square$ should not be mistaken for the forgetful functors of Section 1.1.

Let $D = \Gamma_{\mathrm{D}}(U, V)/(A, B, C)$ be a quadratic dioperad as in (1.12), $\mathcal{P} := \Gamma_{\mathrm{Op}}(U)/(A)$, and $\mathcal{Q} := \Gamma_{\mathrm{Op}}(V)/(C)$. Let us interpret $\mathcal{P}$ as a bicollection with $\mathcal{P}(1, n) = \mathcal{P}(n)$, $n \geq 1$, and let $\mathcal{Q}^{\mathrm{op}}$ be the bicollection with $\mathcal{Q}^{\mathrm{op}}(n, 1) := \mathcal{Q}(n)$, $n \geq 1$, trivial for other values of (m, n). Since dioperads are $\square$-monoids in the category of bicollections, there are canonical maps of bicollections

$$\varphi : \mathcal{P} \square \mathcal{Q}^{\mathrm{op}} \to D \text{ and } \vartheta : \mathcal{Q}^{\mathrm{op}} \square \mathcal{P} \to D.$$

Let us formulate the following useful proposition.

Proposition 17. *The canonical maps*

$$\varphi : \mathcal{P}\square\mathcal{Q}^{\mathrm{op}} \to F_2(\mathcal{P} \diamond \mathcal{Q}^\dagger) \text{ and } \vartheta : (\mathcal{Q}^!)^{\mathrm{op}}\square\mathcal{P}^! \to \mathcal{P}^! * (\mathcal{Q}^!)^\dagger$$

are isomorphisms of bicollections.

Proof. The fact that φ is an isomorphism follows immediately from the definitions. The second isomorphism can be obtained by quadratic duality: according to [4, Proposition 5.9(b)], $F_2(\mathcal{P} \diamond \mathcal{Q}^\dagger)^! \cong (\mathcal{Q}^!)^{\mathrm{op}}\square\mathcal{P}^!$, while $F_2(\mathcal{P} \diamond \mathcal{Q}^\dagger)^! \cong \jmath(\mathcal{P}^! * (\mathcal{Q}^!)^\dagger) \cong \mathcal{P}^! * (\mathcal{Q}^!)^\dagger$ (isomorphisms of bicollections) by Lemma 9 and Exercise 15. $\square$

The following theorem is again not surprising, because $\mathcal{P} \diamond \mathcal{Q}^\dagger$ was constructed from operads $\mathcal{P}$ and $\mathcal{Q}$ using the relation

$$q^\dagger \circ p = 0, \text{ for } p \in \mathcal{P} \text{ and } q \in \mathcal{Q}^\dagger,$$

which is a rather trivial *mixed distributive law* in the sense of [3, Definition 11.1]. As such, it cannot create anything unexpected in the derived category; in particular, it cannot destroy the Koszulness of $\mathcal{P}$ and $\mathcal{Q}$.

Theorem 18. *If $\mathcal{P}$ and $\mathcal{Q}$ are Koszul quadratic operads, then $\mathcal{P} \diamond \mathcal{Q}^\dagger$ is a Koszul $\frac{1}{2}$PROP. This implies that the bar construction $\Omega_{\frac{1}{2}\mathrm{P}}(\mathcal{P}^! * (\mathcal{Q}^!)^\dagger)$ is a minimal model, in the sense of Definition 30, of $\frac{1}{2}$PROP $\mathcal{P} \diamond \mathcal{Q}^\dagger$.*

Proof. We will use the following result of Gan [4]. Given a quadratic dioperad D, suppose that the operads $\mathcal{P}$ and $\mathcal{Q}$ defined by $\mathcal{P}(n) := D(1, n)$ and $\mathcal{Q} := D(n, 1)$, $n \geqslant 2$, are Koszul and that $D \cong \mathcal{P}\square\mathcal{Q}^{\mathrm{op}}$. Proposition 5.9(c) of [4] then states that D is a Koszul dioperad.

Since by Proposition 17, $F_2(\mathcal{P}\diamond\mathcal{Q}^\dagger) \cong \mathcal{P}\square\mathcal{Q}^{\mathrm{op}}$, the above-mentioned result implies that $F_2(\mathcal{P} \diamond \mathcal{Q}^\dagger)$ is a Koszul quadratic dioperad. Theorem 18 now immediately follows from Proposition 12 and Exercise 15. $\square$

Example 19. The following example is taken from [14], with signs altered to match the conventions of the present paper. The minimal model (see Definition 30) of the $\frac{1}{2}$PROP $\frac{1}{2}$b for $\frac{1}{2}$bialgebras, given by the cobar dual $\Omega_{\frac{1}{2}\mathrm{P}}(Ass * Ass^\dagger)$, equals

$$\left(\Gamma_{\frac{1}{2}\mathrm{P}}(\Xi), \partial_0\right) \xrightarrow{\alpha_{\frac{1}{2}\mathrm{P}}} \left(\tfrac{1}{2}\mathrm{b}, \partial = 0\right),$$

where Ξ denotes the bicollection freely (Σ, Σ)-generated by the linear span $Span\left(\{\xi_n^m\}_{m,n \in I}\right)$ with

$$I := \{m, n \geqslant 1, (m, n) \neq (1, 1)\}.$$

The generator ξ_n^m of biarity (m, n) has degree $n + m - 3$. The map $\alpha_{\frac{1}{2}\mathrm{P}}$ is defined by

$$\alpha_{\frac{1}{2}\mathrm{P}}\left(\xi_2^1\right) := \curlywedge, \quad \alpha_{\frac{1}{2}\mathrm{P}}\left(\xi_1^2\right) := \curlyvee,$$

while $\alpha_{\frac{1}{2}P}$ is trivial on all remaining generators. The differential ∂_0 is given by the formula

$$\partial_0\left(\xi_n^m\right) := (-1)^m \xi_1^m \circ \xi_n^1 + \sum_U (-1)^{i(s+1)+m+u-s} \xi_u^m \circ_i \xi_s^1 \qquad (1.15)$$

$$+ \sum_V (-1)^{(v-j+1)(t+1)-1} \xi_1^t {}_j\circ \xi_n^v,$$

where we set $\xi_1^1 := 0$,

$$U := \{u, s \geqslant 1, \ u + s = n + 1, \ 1 \leqslant i \leqslant u\},$$

and

$$V := \{t, v \geqslant 1, \ t + v = m + 1, \ 1 \leqslant j \leqslant v\}.$$

If we define $\xi_2^1 = \curlywedge$ and $\xi_1^2 = \curlyvee$, then $\partial_0(\curlywedge) = \partial_0(\curlyvee) = 0$. If $\xi_2^2 = \mathsf{X}$, then

$$\partial_0(\mathsf{X}) = \mathsf{X}.$$

Under obvious, similar notation,

$$\partial_0(\curlywedge) = \curlywedge - \curlywedge,$$
$$\partial_0(\curlywedge) = -\curlywedge + \curlywedge - \curlywedge + \curlywedge + \curlywedge,$$
$$\partial_0(\curlyvee) = -\curlyvee + \curlyvee,$$
$$\partial_0(\mathsf{X}) = \mathsf{X} - \mathsf{X} + \mathsf{X},$$
$$\partial_0(\mathsf{X}) = -\mathsf{X} - \mathsf{X} + \mathsf{X},$$
$$\partial_0(\mathsf{X}) = -\mathsf{X} + \mathsf{X} - \mathsf{X} - \mathsf{X} + \mathsf{X},$$
$$\partial_0(\mathsf{X}) = \mathsf{X} - \mathsf{X} - \mathsf{X} + \mathsf{X} - \mathsf{X} + \mathsf{X}, \text{ etc.}$$

Example 20. In this example we discuss a minimal model of the $\frac{1}{2}$PROP $\frac{1}{2}$lieb introduced in Example 16. The $\frac{1}{2}$PROP $\frac{1}{2}$lieb describes $\frac{1}{2}$*Lie bialgebras* given by a vector space V with a Lie multiplication $[-, -] : V \otimes V \to V$ and Lie comultiplication (diagonal) $\delta : V \to V \otimes V$ tied together by

$$\delta[a, b] = 0 \quad \text{for all } a, b \in V.$$

A minimal model of $\frac{1}{2}$lieb is given by the cobar dual $\Omega_{\frac{1}{2}P}(\mathcal{C}om * \mathcal{C}om^\dagger)$. It is clearly of the form

$$\left(\Gamma_{\frac{1}{2}P}(\Upsilon), \partial_0\right) \xrightarrow{\alpha_{\frac{1}{2}P}} \left(\tfrac{1}{2}\mathsf{lieb}, \partial = 0\right),$$

where Υ is the bicollection such that $\Upsilon(m, n)$ is the ground field placed in degree $m + n - 3$ with the sign representation of (Σ_m, Σ_n) for $(m, n) \neq 1$, while $\Upsilon(1, 1) := 0$. If we denote by 1_n^m the generator of $\Upsilon(m, n)$, then the map $\alpha_{\frac{1}{2}P}$ is defined by

$$\alpha_{\frac{1}{2}P}\left(1_2^1\right) := \curlywedge, \quad \alpha_{\frac{1}{2}P}\left(1_1^2\right) := \curlyvee,$$

while it is trivial on all remaining generators. There is a formula for the differential ∂_0 that is in fact an antisymmetric version of (1.15). We leave writing this formula, which contains a summation over unshuffles, as an exercise to the reader.

1.5 Perturbation Techniques for Graph Cohomology

Let E be a bicollection. We are going to introduce, for an arbitrary fixed m and n, three very important gradings of the piece $\Gamma_P(E)(m,n)$ of the free PROP $\Gamma_P(E)$. We know, from Section 1.2, that $\Gamma_P(E)(m,n)$ is the direct sum, over the graphs $G \in Gr(m,n)$, of the vector spaces $E(G)_{Aut(G)}$. Recall that we refer to elements of $E(G)_{Aut(G)}$ as G-monomials.

The first two gradings are of a purely topological nature. The *component grading* of a G-monomial f is defined by $\mathrm{cmp}(f) := \mathrm{cmp}(G)$, where $\mathrm{cmp}(G)$ is the number of connected components of G minus one. The *genus grading* is given by the topological genus $\mathrm{gen}(G)$ of graphs (see Section 1.2 for a precise definition), that is, for a G-monomial f we put $\mathrm{gen}(f) := \mathrm{gen}(G)$. Finally, there is another *path grading*, denoted by $\mathrm{pth}(G)$, implicitly present in [9], defined as the total number of directed paths connecting inputs with outputs of G. It induces a grading of $\Gamma_P(E)(m,n)$ by setting $\mathrm{pth}(f) := \mathrm{pth}(G)$ for a G-monomial f.

Exercise 21. Prove that for each G-monomial $f \in \Gamma_P(E)(m,n)$,

$$\mathrm{gen}(f) + \max\{m,n\} \leqslant \mathrm{pth}(f) \leqslant mn(\mathrm{gen}(f)+1)$$

and

$$\mathrm{cmp}(f) \leqslant \min\{m,n\} - 1.$$

Find examples that show that these inequalities cannot be improved and observe that our assumption that $E(m,n)$ is nonzero only for $m,n \geqslant 1$, $(m,n) \neq (1,1)$, is crucial.

Properties of these gradings are summarized in the following proposition.

Proposition 22. *Suppose E is a bicollection of finite-dimensional (Σ, Σ)-bimodules. Then for any fixed d, the subspaces*

$$Span\{f \in \Gamma_P(E)(m,n); \ \mathrm{gen}(f) = d\} \tag{1.16}$$

and

$$Span\{f \in \Gamma_P(E)(m,n); \ \mathrm{pth}(f) = d\}, \tag{1.17}$$

where $Span\{-\}$ is the k-linear span, are finite-dimensional. The subspace $\Gamma_D(E)(m,n) \subset \Gamma_P(E)(m,n)$ can be characterized as

$$\Gamma_D(E)(m,n) = Span\{f \in \Gamma_P(E)(m,n); \ \mathrm{cmp}(f) = \mathrm{gen}(f) = 0\}. \tag{1.18}$$

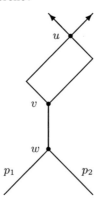

Fig. 1.1. Three branching points u, v, and w of paths p_1 and p_2.

For each $f \in \Gamma_{\mathrm{D}}(E)(m,n)$, $\mathrm{pth}(f) \leqslant mn$, and the subspace $\Gamma_{\frac{1}{2}\mathrm{P}}(E)(m,n) \subset \Gamma_{\mathrm{D}}(E)(m,n)$ can be described as

$$\Gamma_{\frac{1}{2}\mathrm{P}}(E)(m,n) = Span\{f \in \Gamma_{\mathrm{D}}(E)(m,n); \ \mathrm{pth}(f) = mn\}. \qquad (1.19)$$

Proof. Since all vertices of our graphs are at least trivalent, it follows from standard combinatorics that there is only a finite number of (m,n)-graphs with a fixed genus. This proves the finite-dimensionality of the space in (1.16). Description (1.18) follows immediately from the definition of a dioperad. Our proof of the finite-dimensionality of the space in (1.17) is based on the following argument taken from [9].

Let us say that a vertex v is a *branching vertex* for a pair of directed paths p_1, p_2 of a graph $G \in Gr(m,n)$ if v is a vertex of both p_1 and p_2 and if it has the property that either there exist two different input edges f_1, f_2 of v such that $f_s \in p_s$, $s = 1, 2$, or there exist two different output edges e_1, e_2 of v such that $e_s \in p_s$, $s = 1, 2$. See also Figure 1.1. Denote by $\mathrm{br}(p_1, p_2)$ the number of all branching vertices for p_1 and p_2. A moment's reflection convinces us that a pair of paths p_1 and p_2 with b branching points generates at least 2^{b-1} different paths in G; therefore $2^{\mathrm{br}(p_1,p_2)-1} \leqslant d$, where d is the total number of directed paths in G. This implies that

$$\mathrm{br}(p_1, p_2) \leqslant \log_2(d) + 1.$$

Now observe that each vertex is a branching point for at least one pair of paths. We conclude that the number of vertices of G must be less than or equal to $d^2 \cdot (\log_2(d) + 1)$.

The graph G cannot have vertices of valence bigger than d, because each vertex of valence k generates at least $k - 1$ different paths in G. Since there are only finitely many isomorphism classes of graphs with the number of vertices bounded by a constant and with the valences of its vertices bounded by another constant, the finite-dimensionality of the space in (1.17) is proven.

Let us finally demonstrate (1.19). Observe first that for a graph $G \in Gr(m,n)$ of genus 0, mn is actually an upper bound for $\mathrm{pth}(G)$, because for each output–input pair (i,j) there exists at most one path joining i with j (genus 0 assumption). It is also not difficult to see that $\mathrm{pth}(f) = mn$ for a G-monomial $f \in \Gamma_{\frac{1}{2}\mathrm{P}}(E)$. So it remains to prove that $\mathrm{pth}(f) = mn$ implies $f \in \Gamma_{\frac{1}{2}\mathrm{P}}(E)$.

Suppose that f is a G-monomial such that $f \in \Gamma_{\mathrm{D}}(E)(m,n) \backslash \Gamma_{\frac{1}{2}\mathrm{P}}(E)(m,n)$. This happens exactly when G contains a configuration shown in Figure 1.2, forbidden for $\frac{1}{2}$PROPs. Then there certainly exists a path p_1 containing edges e and a, and another path p_2 containing edges b and g. Suppose that p_s connects output i_s with input j_s, $i = 1,2$, as in Figure 1.2. It is then clear that there is no path that connects i_2 with j_1, which means that the total number of paths in G is not maximal. This finishes the proof of the proposition. $\qquad \square$

Remark 23. As we already know, there are various "restricted" versions of PROPs characterized by types of graphs along which the composition is allowed. Thus $\frac{1}{2}$PROPs live on contractible graphs without "bad" edges as in Figure 1.2, and Gan's dioperads live on all contractible graphs. A version of PROPs for which only compositions along *connected* graphs are allowed was studied by Vallette, who called these PROPs *properads* [19]. All this can be summarized by a chain of inclusions of full subcategories

$$\mathtt{Oper} \subset \tfrac{1}{2}\mathtt{PROP} \subset \mathtt{diOp} \subset \mathtt{Proper} \subset \mathtt{PROP}.$$

Let $\Gamma_{\mathrm{pth}}(E) \subset \Gamma_{\mathrm{P}}(E)$ be the subspace spanned by all G-monomials such that G is contractible and contains at least one "bad" edge as in Figure 1.2. By Proposition 22, one might equivalently define $\Gamma_{\mathrm{pth}}(E)$ by

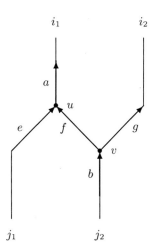

Fig. 1.2. A configuration forbidden for $\frac{1}{2}$PROPs – f is a "bad" edge. Vertices u and v might have more input or output edges, which we did not indicate.

$\Gamma_{\mathrm{pth}}(E)(m,n)$
$= Span\{f \in \Gamma_{\mathrm{D}}(E)(m,n);\ \mathrm{cmp}(f) = \mathrm{gen}(f) = 0,\ \text{and pth}(f) < mn\}.$

If we write

$$\Gamma_{\mathrm{c+g}}(E) := Span\{f \in \Gamma_{\mathrm{P}}(E);\ \mathrm{cmp}(f) + \mathrm{gen}(f) > 0\},$$

then there is a natural decomposition

$$\Gamma_{\mathrm{P}}(E) = \Gamma_{\frac{1}{2}\mathrm{P}}(E) \oplus \Gamma_{\mathrm{pth}}(E) \oplus \Gamma_{\mathrm{c+g}}(E),$$

in which clearly $\Gamma_{\frac{1}{2}\mathrm{P}}(E) \oplus \Gamma_{\mathrm{pth}}(E) = \Gamma_{\mathrm{D}}(E)$. Let $\pi_{\frac{1}{2}\mathrm{P}}$, π_{pth}, and $\pi_{\mathrm{c+g}}$ denote the corresponding projections. For a degree-(-1) differential ∂ on $\Gamma_{\mathrm{P}}(E)$, introduce derivations ∂_0, ∂_{pth}, and $\partial_{\mathrm{c+g}}$ determined by their restrictions to the generators E as follows:

$$\partial_0\big|_E := \pi_{\frac{1}{2}\mathrm{P}} \circ \partial\big|_E, \quad \partial_{\mathrm{pth}}|_E := \pi_{\mathrm{pth}} \circ \partial|_E, \quad \text{and } \partial_{\mathrm{c+g}}|_E := \pi_{\mathrm{c+g}} \circ \partial|_E.$$

Let us define also $\partial_{\mathrm{D}} := \partial_0 + \partial_{\mathrm{pth}}$, the *dioperadic part* of ∂. The decompositions

$$\partial = \partial_{\mathrm{D}} + \partial_{\mathrm{c+g}} = \partial_0 + \partial_{\mathrm{pth}} + \partial_{\mathrm{c+g}} \tag{1.20}$$

are fundamental for our purposes. We will call them the *canonical decompositions* of the differential ∂. The following example shows that in general, ∂_0, ∂_{D}, and $\partial_{\mathrm{c+g}}$ *need not* be differentials, since they may not square to zero.

Example 24. Let us consider the free PROP $\Gamma_{\mathrm{P}}(a,b,c,u,x)$, where the generator a has degree 1 and biarity $(4,2)$, b degree 0 and biarity $(2,1)$, c degree 1 and biarity $(4,1)$, u degree 0 and biarity $(2,1)$, and x degree 2 and biarity $(4,1)$. Define a degree-(-1) differential ∂ by the following formulas, whose meaning is, as we believe, clear:

$$\partial\left(\,\underset{x}{\mathsf{Y}}\,\right) := \underset{b}{\overset{a}{\mathsf{\bigvee}}} + \mathsf{Y}c,\ \partial\left(\overset{a}{\mathsf{X}}\right) := \mathsf{Y}u \otimes \mathsf{Y}u,\ \partial\left(\underset{c}{\mathsf{Y}}\right) := -\ \overset{u}{\underset{b}{\mathsf{V}}}\overset{u}{\mathsf{}},$$

while $\partial(b) = \partial(u) = 0$. One can easily verify that $\partial^2 = 0$. By definition,

$$\partial_0\left(\,\underset{x}{\mathsf{Y}}\,\right) = \mathsf{Y}c,\ \partial_0\left(\overset{a}{\mathsf{X}}\right) = 0\ \partial_0\left(\underset{c}{\mathsf{Y}}\right) = -\ \overset{u}{\underset{b}{\mathsf{V}}}\overset{u}{\mathsf{}},$$

and of course, $\partial_0(b) = \partial_0(u) = 0$. A simple calculation shows that

$$\partial_0^2\left(\,\underset{x}{\mathsf{Y}}\,\right) = -\ \overset{u}{\underset{b}{\mathsf{V}}}\overset{u}{\mathsf{}};$$

therefore $\partial_0^2 \neq 0$. Since $\partial_0 = \partial_{\mathrm{D}}$, we conclude that also $\partial_{\mathrm{D}}^2 \neq 0$.

Let us formulate some conditions that guarantee that the derivations ∂_0 and ∂_D square to zero. We say that a differential ∂ in $\Gamma_P(E)$ is *connected* if $\operatorname{cmp}(\partial(e)) = 0$ for each $e \in E$. Similarly, we say that ∂ *has genus zero* if $\operatorname{gen}(\partial(e)) = 0$ for $e \in E$. Less formally, connectivity of ∂ means that $\partial(e)$ is a sum of G-monomials with all G's connected, and ∂ has genus zero if $\partial(e)$ is a sum of G-monomials with all G's of genus 0 (but not necessarily connected).

Proposition 25. *In the canonical decomposition (1.20) of a differential ∂ in a free PROP $\Gamma_P(E)$, $\partial_D^2 = 0$ always implies that $\partial_0^2 = 0$.*

If, moreover, either (i) the differential ∂ is connected or (ii) ∂ has genus zero, then $\partial_D^2 = 0$; therefore both ∂_0 and ∂_D are differentials on $\Gamma_P(E)$.

Proof. For a G-monomial f, write

$$\partial_D(f) = \sum_{H \in U} g_H, \tag{1.21}$$

the sum of H-monomials g_H over a finite set U of graphs. Since ∂_D is a derivation, each $H \in U$ is obtained by replacing a vertex $v \in v(G)$ of biarity (s,t) by a graph R of the same biarity. It follows from the definition of the dioperadic part ∂_D that each such R is contractible. This implies that all graphs $H \in U$ that nontrivially contribute to the sum (1.21) have the property that $\operatorname{pth}(H) \leqslant \operatorname{pth}(G)$ (∂_D *does not increase the path grading*) and that

$$\partial_0(f) = \sum_{H \in U_0} g_H, \text{ where } U_0 := \{H \in U;\ \operatorname{pth}(H) = \operatorname{pth}(G)\}. \tag{1.22}$$

This can be seen as follows. It is clear that a replacement of a vertex by a contractible graph cannot increase the total number of paths in G. This implies that ∂_D does not increase the path grading. Equation (1.22) follows from the observation that decreasing the path grading locally at a vertex decreases the path grading of the whole graph. By this we mean the following.

Assume that a vertex v of biarity (s,t) is replaced by a contractible graph R such that $\operatorname{pth}(R) < st$. This means that there exists an output–input pair (i,j) of R for which there is no path in R connecting output i with input j. On the other hand, in G there certainly existed a path that ran through output i and input j of vertex v and broke apart when we replaced v by R.

Now we see that ∂_0^2 is precisely the part of ∂_D^2 that preserves the path grading. This makes the implication $(\partial_D^2 = 0) \implies (\partial_0^2 = 0)$ completely obvious and proves the first part of the proposition.

For the proof of the second half, it will be convenient to introduce still another grading by putting

$$\operatorname{grad}(G) := \operatorname{cmp}(G) - \operatorname{gen}(G) = +|v(G)| - |e(G)| - 1, \tag{1.23}$$

where $|v(G)|$ denotes the number of vertices and $|e(G)|$ the number of internal edges of G. Let f be a G-monomial as above. Let us consider a sum similar to (1.21), but this time for the entire differential ∂:

$$\partial(f) = \sum_{H \in S} g_H,$$

where S is a finite set of graphs. We claim that under assumption (i) or (ii),

$$\partial_{\mathrm{D}}(f) = \sum_{H \in S_{\mathrm{D}}} g_H, \text{ where } S_{\mathrm{D}} := \{H \in S;\ \mathrm{grad}(H) = \mathrm{grad}(G)\}. \qquad (1.24)$$

This would clearly imply that ∂_{D}^2 is exactly the part of ∂^2 that preserves the grad-grading; therefore $\partial_{\mathrm{D}}^2 = 0$.

As in the first half of the proof, each $H \in S$ is obtained from G by replacing $v \in v(G)$ by some graph R. In case (i), R is connected, that is, $\mathrm{cmp}(G) = \mathrm{cmp}(H)$ for all $H \in S$. It follows from elementary algebraic topology that $\mathrm{gen}(H) \geqslant \mathrm{gen}(G)$ and that $\mathrm{gen}(G) = \mathrm{gen}(H)$ if and only if $\mathrm{gen}(R) = 0$. This proves (1.24) for connected differentials.

Assume now that ∂ has genus zero, that is, $\mathrm{gen}(R) = 0$. This means that R can be contracted to a disjoint R' union of $\mathrm{cmp}(R)+1$ corollas. Since $\mathrm{grad}(-)$ is a topological invariant, we may replace R inside H by its contraction R'. We obtain a graph H' for which $\mathrm{grad}(H) = \mathrm{grad}(H')$. It is obvious that H' has the same number of internal edges as G and that $|v(H')| = |v(G)| + \mathrm{cmp}(R)$. Therefore $\mathrm{grad}(G) = \mathrm{grad}(H) + \mathrm{cmp}(R)$. This means that $\mathrm{grad}(G) = \mathrm{grad}(H)$ if and only if $\mathrm{cmp}(R) = 0$, i.e., if R is connected. This proves (1.24) in case (ii) and finishes the proof of the proposition. $\qquad \square$

The following theorem will be our basic tool to calculate the homology of free differential graded PROPs in terms of the canonical decomposition of the differential.

Theorem 26. *Let $(\Gamma_{\mathrm{P}}(E), \partial)$ be a dg free PROP and m, n fixed natural numbers.*

(i) Suppose that the differential ∂ is connected. Then the genus grading defines, by

$$F_p^{\mathrm{gen}} := Span\{f \in \Gamma_{\mathrm{P}}(E)(m, n);\ \mathrm{gen}(f) \geqslant -p\}, \qquad (1.25)$$

an increasing ∂-invariant filtration of $\Gamma_{\mathrm{P}}(E)(m, n)$.

(ii) If the differential ∂ has genus zero, then

$$F_p^{\mathrm{grad}} := Span\{f \in \Gamma_{\mathrm{P}}(E)(m, n);\ \mathrm{grad}(f) \geqslant -p\}$$

is also an increasing ∂-invariant filtration of $\Gamma_{\mathrm{P}}(E)(m, n)$.

The spectral sequences induced by these filtrations both have the first term isomorphic to $(\Gamma_{\mathrm{P}}(E)(m, n), \partial_{\mathrm{D}})$ and they both abut to $H_*(\Gamma_{\mathrm{P}}(E)(m, n), \partial)$.

(iii) Suppose that $\partial_{\mathsf{D}}^2 = 0$. Then the path grading defines an increasing ∂_{D}-invariant filtration

$$F_p^{\mathrm{pth}} := Span\{f \in \Gamma_{\mathsf{P}}(E)(m,n); \ \mathrm{pth}(f) \leqslant p\}.$$

This filtration induces a first quadrant-spectral sequence whose first term is isomorphic to $(\Gamma_{\mathsf{P}}(E)(m,n), \partial_0)$ and that converges to $H_*(\Gamma_{\mathsf{P}}(E)(m,n), \partial_{\mathsf{D}})$.

Proof. The proof easily follows from Proposition 25 and the analysis of the canonical decomposition given in the proof of that proposition. □

The following proposition describes an important particular case in which the spectral sequence induced by the filtration (1.25) converges.

Proposition 27. *If ∂ is connected and preserves the path grading, then the filtration (1.25) induces a second-quadrant spectral sequence whose first term is isomorphic to $(\Gamma_{\mathsf{P}}(E)(m,n), \partial_0)$ and that converges to $H_*(\Gamma_{\mathsf{P}}(E)(m,n), \partial)$.*

Proof. Under the assumptions of the proposition, the path grading is a ∂-invariant grading, compatible with the genus filtration (1.25), by finite-dimensional pieces; see Proposition 22. This guarantees that the generally ill-behaved second-quadrant spectral sequence induced by (1.25) converges. The proof is finished by observing that the assumption that ∂ preserves the path grading implies that $\partial_0 = \partial_{\mathsf{D}}$. □

In most applications either ∂ is connected or $\partial = \partial_{\mathsf{D}}$, though there are also natural examples of PROPs with disconnected differentials, such as the *deformation quantization* PROP DefQ introduced by Merkulov in [16]. The following corollary immediately follows from Theorem 26(iii) and Proposition 27.

Corollary 28. *Let P be a graded PROP concentrated in degree 0 and $\alpha : (\Gamma_{\mathsf{P}}(E), \partial) \to (\mathsf{P}, 0)$ a homomorphism of dg PROPs. Suppose that α induces an isomorphism $H_0(\Gamma_{\mathsf{P}}(E), \partial) \cong \mathsf{P}$ and that $\Gamma_{\mathsf{P}}(E)$ is ∂_0-acyclic in positive degrees. Suppose moreover that either*

   (i) ∂ is connected and preserves the path grading, or
   (ii) $\partial(E) \subset \Gamma_{\mathsf{D}}(E)$.

Then α is a free resolution of the PROP P.

Remark 29. In Corollary 28 we assumed that the PROP P was concentrated in degree 0. The case of a general nontrivially graded nondifferential PROP P can be treated by introducing the *Tate–Jozefiak grading*, as was done, for example, for bigraded models of operads in [13, page 1481].

1.6 Minimal Models of PROPs

In this section we show how the methods of this paper can be used to study minimal models of PROPs. Let us first give a precise definition of this object.

Definition 30. *A* minimal model *of a dg PROP* P *is a dg free PROP* $(\Gamma_P(E), \partial)$ *together with a homology isomorphism*

$$P \xleftarrow{\alpha} (\Gamma_P(E), \partial).$$

We also assume that the image of ∂ *consists of decomposable elements of* $\Gamma_P(E)$ *or, equivalently, that* ∂ *has no "linear part" (the* minimality condition*). Minimal models for* $\frac{1}{2}$PROP*s and dioperads are defined in exactly the same way, only replacing* $\Gamma_P(-)$ *by* $\Gamma_{\frac{1}{2}P}(-)$ *or* $\Gamma_D(-)$.

The above definition generalizes minimal models for operads introduced in [13]. While we proved, in [13, Theorem 2.1], that each operad admits, under some very mild conditions, a minimal model, and while the same statement is probably true also for dioperads, a similar statement for a general PROP would require some way to handle a divergence problem (see also the discussion in [14] and below).

Bialgebras. Recall that a *bialgebra* is a vector space V with an associative multiplication $\cdot : V \otimes V \to V$ and a coassociative comultiplication $\Delta : V \to V \otimes V$ that are related by

$$\Delta(a \cdot b) = \Delta(a) \cdot \Delta(b), \text{ for } a, b \in V.$$

The PROP B describing bialgebras has a presentation $B = \Gamma_P(\curlywedge, \curlyvee)/I_B$, where I_B denotes the ideal generated by

$$\curlywedge\!\!\!\curlywedge - \curlywedge\!\!\!\curlywedge, \quad \curlyvee\!\!\!\curlyvee - \curlyvee\!\!\!\curlyvee, \quad \text{and} \quad \times - \langle\!\!\!\rangle.$$

In the above display we have that

$$\curlywedge\!\!\!\curlywedge := \curlywedge(\curlywedge \otimes 1), \quad \curlywedge\!\!\!\curlywedge := \curlywedge(1 \otimes \curlywedge), \quad \curlyvee\!\!\!\curlyvee := (\curlyvee \otimes 1)\curlyvee, \quad \curlyvee\!\!\!\curlyvee := (1 \otimes \curlyvee)\curlyvee,$$

$$\times := \curlyvee \circ \curlywedge, \quad \text{and} \quad \langle\!\!\!\rangle := (\curlywedge \otimes \curlywedge) \circ \sigma(2,2) \circ (\curlyvee \otimes \curlyvee),$$

where $\sigma(2,2) \in \Sigma_4$ is the permutation

$$\sigma(2,2) = \begin{pmatrix} 1\,2\,3\,4 \\ 1\,3\,2\,4 \end{pmatrix}.$$

As we argued in [14], the PROP B can be interpreted as a perturbation of the PROP $\frac{1}{2}B = F\left(\frac{1}{2}b\right)$ for $\frac{1}{2}$bialgebras mentioned in Example 14. More precisely, let ϵ be a formal parameter, I_B^ϵ the ideal generated by

$$\curlywedge\!\!\!\curlywedge - \curlywedge\!\!\!\curlywedge, \quad \curlyvee\!\!\!\curlyvee - \curlyvee\!\!\!\curlyvee, \quad \text{and} \quad \times - \epsilon\,\langle\!\!\!\rangle,$$

and $B_\epsilon := \Gamma_P(\curlywedge, \curlyvee)/I_B^\epsilon$. Then B_ϵ is a one-dimensional family of deformations of $\frac{1}{2}B = B_0$ whose specialization (value) at $\epsilon = 1$ is B. Therefore, every minimal model for B can be expected to be a perturbation of the minimal model for $\frac{1}{2}B$ described in the following theorem:

Theorem 31 ([14]). *The dg free* PROP

$$(\mathsf{M}, \partial_0) = (\Gamma_{\mathsf{P}}(\Xi), \partial_0), \tag{1.26}$$

where the generators $\Xi = Span\left(\{\xi_m^n\}_{m,n \in I}\right)$ *are as in Example 19 and the differential* ∂_0 *is given by formula (1.15), is a minimal model of the* PROP $\frac{1}{2}\mathsf{B}$ *for* $\frac{1}{2}$*bialgebras.*

Proof. Clearly, $(\mathsf{M}, \partial_0) = F\left(\Omega_{\frac{1}{2}\mathsf{P}}(\mathcal{A}ss * \mathcal{A}ss^\dagger)\right)$. The theorem now follows from Theorem 18 (see also Example 19) and from the fact that the functor F preserves homology isomorphisms; see Corollary 5.   □

The methods developed in this paper were used in [14] to prove the following:

Theorem 32. *There exists a minimal model* (M, ∂) *of the* PROP B *for bialgebras that is a perturbation of the minimal model* (M, ∂_0) *of the* PROP $\frac{1}{2}\mathsf{B}$ *for* $\frac{1}{2}$*bialgebras described in Theorem 31,*

$$(\mathsf{M}, \partial) = (\Gamma_{\mathsf{P}}(\Xi), \partial_0 + \partial_{pert}),$$

for some perturbation ∂_{pert} *that raises the genus and preserves the path grading.*

Proof. As shown in [14], a perturbation ∂_{pert} can be constructed using standard methods of the homological perturbation theory because we know, by Theorem 31, that $\Gamma_{\mathsf{P}}(\Xi)$ is ∂_0-acyclic in positive degrees. The main problem was to show that the procedure converges. This was achieved by finding a subspace $X \subset \Gamma_{\mathsf{P}}(\Xi)$ of *special elements* whose pieces $X(m, n)$ satisfy the conditions that:

(i) each $X(m, n)$ is a finite-dimensional space spanned by G-monomials with connected G,
(ii) each $X(m, n)$ is ∂_0-closed and ∂_0-acyclic in positive degrees,
(iii) each $X(m, n)$ is closed under vertex insertion (see below), and
(iv) both $\curlyvee$ and $\curlyvee\!\!\!\curlywedge$ belong to $X(2, 2)$.

Item (iii) means that X is stable under all derivations (not necessarily differentials) ω of $\Gamma_{\mathsf{P}}(\Xi)$ such that $\omega(\Xi) \subset X$. The perturbation problem was then solved in X instead of $\Gamma_{\mathsf{P}}(\Xi)$. It remained to use, in an obvious way, Corollary 28(i) to prove that the object we constructed is really a minimal model of B.   □

Dioperads. In this part we prove that the cobar duals of dioperads with a replacement rule induce, via the functor $F_1 : \mathtt{diOp} \to$ PROP introduced in Section 1.1, minimal models in the category of PROPs. Since we are unable to prove the exactness of F_1, we will need to show first that these models are perturbations of minimal models of quadratic Koszul $\frac{1}{2}$PROPs and then use Corollary 28(ii). This approach applies to the main examples of [4], i.e., Lie bialgebras and infinitesimal bialgebras.

Let $\mathcal{P}$ and $\mathcal{Q}$ be quadratic operads, with presentations $\mathcal{P} = \Gamma_{0p}(F)/(R)$ and $\mathcal{Q} = \Gamma_{0p}(G)/(S)$. We will consider dioperads created from $\mathcal{P}$ and $\mathcal{Q}$ by a *dioperadic replacement rule*. By this we mean the following.

As in Example 13, interpret F, G, R, and S as bicollections. We already observed in Section 1.4 that

$$\Gamma_{\mathsf{D}}(F,G)(2,2) \cong \Gamma_{\frac{1}{2}\mathsf{P}}(F,G)(2,2) \oplus \mathrm{Ind}_{\{1\}}^{\Sigma_2 \times \Sigma_2}(F \otimes G) \cong G \circ F \oplus \mathrm{Ind}_{\{1\}}^{\Sigma_2 \times \Sigma_2}(F \otimes G);$$

see also [4, Section 2.4] for details. The above decomposition is in fact a decomposition of $\Gamma_{\mathsf{D}}(F,G)(2,2)$ into pth-homogeneous components, namely

$$G \circ F = Span\{f \in \Gamma_{\mathsf{D}}(F,G)(2,2);\ \mathrm{pth}(f) = 4\}$$

and

$$\mathrm{Ind}_{\{1\}}^{\Sigma_2 \times \Sigma_2}(F \otimes G) = Span\{f \in \Gamma_{\mathsf{D}}(F,G)(2,2);\ \mathrm{pth}(f) = 3\}.$$

Given a (Σ_2, Σ_2)-equivariant map

$$\lambda : G \circ F \to \mathrm{Ind}_{\{1\}}^{\Sigma_2 \times \Sigma_2}(F \otimes G), \qquad (1.27)$$

one might consider a subspace

$$B = B_\lambda := Span\{f - \lambda(f);\ f \in G \circ F\} \subset \Gamma_{\mathsf{D}}(F,G)(2,2)$$

and a quadratic dioperad

$$D_\lambda := \Gamma_{\mathsf{D}}(F,G)/(R, B_\lambda, S). \qquad (1.28)$$

We say that the map λ in (1.27) is a *replacement rule* [3, Definition 11.3] if it is coherent in the sense that it extends to a *mixed distributive law* between operads $\mathcal{P}$ and $\mathcal{Q}$; see [3, Section 11] for details. An equivalent way to express this coherence is to say that D_λ and $F_2(\mathcal{P} \diamond \mathcal{Q}^\dagger)$ are isomorphic as bicollections or, in the terminology of [4, Proposition 5.9], that $D_\lambda \cong \mathcal{P} \square \mathcal{Q}^{\mathrm{op}}$; see Proposition 17.

Example 33. An important example is given by an *infinitesimal bialgebra* (which we called in [3, Example 11.7] a *mock bialgebra*). It is a vector space V together with an associative multiplication $\cdot : V \otimes V \to V$ and a coassociative comultiplication $\Delta : V \to V \otimes V$ such that

$$\Delta(a \cdot b) = \sum \left(a_{(1)} \otimes a_{(2)} \cdot b + a \cdot b_{(1)} \otimes b_{(2)} \right)$$

for any $a, b \in V$.

The dioperad IB describing infinitesimal bialgebras is given by $IB = \Gamma_{\mathsf{D}}(\curlywedge, \curlyvee)/I_{IB}$, where I_{IB} denotes the dioperadic ideal generated by

$$\curlywedge - \curlywedge, \quad \curlyvee - \curlyvee, \quad \text{and} \quad \times - \curlyvee\!\curlywedge - \curlywedge\!\curlyvee .$$

The dioperad IB is created from two copies of the operad $\mathcal{A}ss$ for associative algebras using a replacement rule given by

$$\lambda(\chi) := \curlyvee\!\curlywedge + \curlywedge\!\curlyvee\ ;$$

see [3, Example 11.7] for details. As before, one may consider a one-parameter family $IB_\epsilon := \Gamma_{\mathsf{D}}(\curlywedge, \curlyvee)/I_{IB}^\epsilon$, where I_{IB}^ϵ is the dioperadic ideal generated by

$$\curlywedge\!\curlywedge - \curlywedge\!\curlywedge,\quad \curlyvee\!\curlyvee - \curlyvee\!\curlyvee,\quad \text{and}\quad \chi - \epsilon\left(\curlyvee\!\curlywedge + \curlywedge\!\curlyvee\right)$$

given by the one-parameter family of replacement rules

$$\lambda_\epsilon(\chi) := \epsilon\left(\curlyvee\!\curlywedge + \curlywedge\!\curlyvee\right).$$

Let $\mathsf{IB} := F_1(IB)$ be the PROP generated by the dioperad IB. It follows from the above remarks that IB is another perturbation of the PROP $\frac{1}{2}\mathsf{B}$ for $\frac{1}{2}$bialgebras.

Example 34. Recall that a *Lie bialgebra* is a vector space V, with a Lie algebra structure $[-,-] : V \otimes V \to V$ and a Lie diagonal $\delta : V \to V \otimes V$. As in Example 20 we assume that the bracket $[-,-]$ is antisymmetric and satisfies the Jacobi equation and that δ satisfies the obvious duals of these conditions, but this time $[-,-]$ and δ are related by

$$\delta[a,b] = \sum \left([a_{(1)},b] \otimes a_{(2)} + [a,b_{(1)}] \otimes b_{(2)} + a_{(1)} \otimes [a_{(2)},b] + b_{(1)} \otimes [a,b_{(2)}]\right)$$

for any $a,b \in V$, where we used, as usual, the Sweedler notation $\delta a = \sum a_{(1)} \otimes a_{(2)}$ and $\delta b = \sum b_{(1)} \otimes b_{(2)}$.

The dioperad $LieB$ for Lie bialgebras is given by $LieB = \Gamma_{\mathsf{D}}(\curlywedge, \curlyvee)/I_{LieB}$, where $\curlywedge$ and $\curlyvee$ are now *antisymmetric* generators and I_{LieB} denotes the ideal generated by

$$\curlywedge_{123} + \curlywedge_{231} + \curlywedge_{312},\quad \curlyvee^{123} + \curlyvee^{231} + \curlyvee^{312},\quad \text{and}\quad \chi - \curlyvee\!\curlywedge - \curlywedge\!\curlyvee + \curlyvee\!\curlywedge + \curlywedge\!\curlyvee,$$

with labels indicating, in the obvious way, the corresponding permutations of the inputs and outputs. The dioperad $LieB$ is a combination of two copies of the operad $\mathcal{L}ie$ for Lie algebras, with the replacement rule

$$\lambda\left(\chi\right) := \curlyvee\!\curlywedge + \curlywedge\!\curlyvee - \curlyvee\!\curlywedge - \curlywedge\!\curlyvee\ ;$$

see [3, Example 11.6]. One may obtain, as in Example 33, a one-parameter family $LieB_\epsilon$ of dioperads generated by a one-parameter family λ_ϵ of replacement rules such that $LieB_1 = LieB$ and $LieB_0 = \frac{1}{2}LieB := F_2\left(\frac{1}{2}\text{lieb}\right)$, where $\frac{1}{2}$lieb is the $\frac{1}{2}$PROP for $\frac{1}{2}$Lie bialgebras introduced in Example 20. Thus, the PROP $\mathsf{LieB} := F_1(LieB)$ is a perturbation of the PROP $\frac{1}{2}\mathsf{LieB}$ governing $\frac{1}{2}$Lie bialgebras.

Examples 33 and 34 can be generalized as follows. Each replacement rule λ as in (1.27) can be extended to a one-parameter family of replacement rules by defining $\lambda_\epsilon := \epsilon \cdot \lambda$. This gives a one-parameter family $D_\epsilon := D_{\lambda_\epsilon}$ of dioperads such that $D_1 = D_\lambda$ and $D_0 = \mathcal{P} \diamond \mathcal{Q}^\dagger$. Therefore D_λ is a perturbation of the dioperad generated by the $\frac{1}{2}$PROP $\mathcal{P} \diamond \mathcal{Q}^\dagger$. This suggests that every minimal model of the PROP $F_1(D_\lambda)$ is a perturbation of a minimal model for $F_2(\mathcal{P} \diamond \mathcal{Q}^\dagger)$, which is, as we already know from Section 1.4, given by $F_2\left(\Omega_{\frac{1}{2}\mathrm{P}}((\mathcal{P} \diamond \mathcal{Q}^\dagger)^!)\right) = F_2\left(\Omega_{\frac{1}{2}\mathrm{P}}(\mathcal{P}^! * (\mathcal{Q}^!)^\dagger)\right)$. The rest of this section makes this idea precise.

For any quadratic dioperad D, there is an obvious candidate for a minimal model of the PROP $F_1(D)$ generated by D, namely the dg PROP $\Omega_\mathrm{P}(D^!) = (\Omega_\mathrm{P}(D^!), \partial) := F_1((\Omega_\mathrm{D}(D^!), \partial))$ generated by the dioperadic cobar dual $\Omega_\mathrm{D}(D^!) = (\Omega_\mathrm{D}(D^!), \partial)$ of $D^!$.

The following proposition, roughly speaking, says that the dioperadic cobar dual of D_λ is a perturbation of the cobar dual of the $\frac{1}{2}$PROP $(\mathcal{P} \diamond \mathcal{Q}^\dagger)^! = \mathcal{P}^! * (\mathcal{Q}^!)^\dagger$.

Proposition 35. *Let $D = D_\lambda$ be a dioperad constructed from Koszul quadratic operads $\mathcal{P}$ and $\mathcal{Q}$ using a replacement rule λ. Consider the canonical decomposition*

$$(\Omega_\mathrm{D}(D^!), \partial_0 + \partial_\mathrm{pth})$$

of the differential in the dioperadic bar construction $(\Omega_\mathrm{D}(D^!), \partial)$. Then

$$(\Omega_\mathrm{D}(D^!), \partial_0) \cong F_2\left(\Omega_{\frac{1}{2}\mathrm{P}}(\mathcal{P}^! * (\mathcal{Q}^!)^\dagger)\right). \tag{1.29}$$

Proof. We already observed that, in the terminology of [4], $D \cong \mathcal{P} \square \mathcal{Q}^\mathrm{op}$. This implies, by [4, Proposition 5.9(b)], that $D^! \cong (\mathcal{Q}^!)^\mathrm{op} \square \mathcal{P}^!$, which clearly coincides, as a bicollection, with our $\mathcal{P}^! * (\mathcal{Q}^!)^\dagger$. The rest of the proposition follows from the description of $D^!$ given in [4], the behavior of the replacement rule λ with respect to the path grading, and definitions. $\square$

Remark 36. Since as a nondifferential dioperad, $\Omega_\mathrm{D}(D) = \Lambda^{-1}\Gamma_\mathrm{D}(\uparrow \bar{D}^*)$, where $\uparrow$ denotes the suspension of a graded bicollection, Λ^{-1} the sheared desuspension of a dioperad, and $\bar{D}^*$ the linear dual of the augmentation ideal of D, see Sections 1.4, 2.3, and 3.1 of [4] for details, the PROP $(\Omega_\mathrm{P}(D), \partial)$ may be constructed from scratch as $\Omega_\mathrm{P}(D^!) = \Lambda^{-1}\Gamma_\mathrm{P}(\uparrow \bar{D}^*)$ with a differential coming from the "vertex expansion" (also called edge insertion). Thus, the PROP $(\Omega_\mathrm{P}(D), \partial)$ may be thought of as a *naive cobar dual* of $F_1(D)$, as opposed to the categorical cobar dual [6, Section 4.1.14].

Perhaps one can successfully develop quadratic and Koszul duality theory for PROPs using this naive cobar dual by analogy with [4,6]. We are reluctant to emphasize $(\Omega_\mathrm{P}(D), \partial)$ as a PROP cobar dual of the PROP $F_1(D)$, because we do not know how this naive cobar dual is related to the categorical one.

The following theorem generalizes a result of Kontsevich [9] for $D = LieB$.

Theorem 37. *Under the assumptions of Proposition 35, $(\Omega_P(D^!), \partial)$ is a minimal model of the PROP $F_1(D)$.*

Proof of Theorem 37. We are going to use Corollary 28(ii). It is straightforward to verify that $H_0((\Omega_P(D^!), \partial) \cong F_1(D)$. Equation (1.29) gives

$$\Omega_P(D^!) \cong F\left(\Omega_{\frac{1}{2}P}(\mathcal{P}^! * (\mathcal{Q}^!)^\dagger)\right),$$

and therefore the ∂_0-acyclicity of $\Omega_P(D^!)$ follows from the exactness of the functor F stated in Theorem 4.   □

Example 38. By Theorem 37, the dg PROP $\Omega_P(IB^!)$, where the quadratic dual $IB^!$ of the dioperad IB for infinitesimal bialgebras is described in [4] as $IB^! = Ass^{\mathrm{op}} \square Ass$, is a minimal model of the PROP $IB = F_1(IB)$ for infinitesimal bialgebras. The dg PROP $\Omega_P(IB^!)$ has a form $(\Gamma_P(\Xi), \partial_0 + \partial_{\mathrm{pth}})$, where Ξ and ∂_0 are the same as in Example 19. The path part ∂_{pth} of the differential is trivial on generators ξ_n^m with $m + n \leqslant 4$; therefore the easiest example of the path part is provided by

$$\partial(\mathsf{X}) = \partial_0(\mathsf{X}) + \mathsf{X} + \mathsf{X} + \mathsf{X} - \mathsf{X},$$

where

$$\partial_0(\mathsf{X}) = \mathsf{X} - \mathsf{X} + \mathsf{X}$$

is the same as in Example 19. We encourage the reader to verify that

$$\mathrm{pth}(\mathsf{X}) = \mathrm{pth}(\mathsf{X}) = \mathrm{pth}(\mathsf{X}) = \mathrm{pth}(\mathsf{X}) = 6,$$
$$\mathrm{pth}(\mathsf{X}) = \mathrm{pth}(\mathsf{X}) = 5, \quad \text{and} \quad \mathrm{pth}(\mathsf{X}) = \mathrm{pth}(\mathsf{X}) = 4.$$

Similarly, the dg PROP $\Omega_P(LieB^!) = \Omega_P(Com^{\mathrm{op}} \square Com)$ is a minimal model of the PROP $\mathsf{LieB} := F_1(LieB)$ for Lie bialgebras.

1.7 Classical Graph Cohomology

Here we will reinterpret minimal models for the Lie bialgebra PROP $\mathsf{LieB} = F_1(LieB)$ and the infinitesimal bialgebra PROP $\mathsf{IB} = F_1(IB)$ given by Theorem 37 and Example 38 as graph complexes.

The commutative case. Consider the set of connected (m, n)-graphs G for $m, n \geqslant 1$ in the sense of Section 1.2. An *orientation* on an (m, n)-graph G is an orientation on $\mathbb{R}^{v(G)} \oplus \mathbb{R}^m \oplus \mathbb{R}^n$, i.e., the choice of an element in $\det \mathbb{R}^{v(G)} \otimes \det \mathbb{R}^m \otimes \det \mathbb{R}^n$ up to multiplication by a positive real number. This is equivalent to an orientation on $\mathbb{R}^{e(G)} \oplus H_1(|G|; \mathbb{R})$, where $e(G)$ is the set of (all) edges of G; to verify this, consider the cellular chain complex of the geometric realization $|G|$; see for example [18, Proposition B.1] and [15, Proposition 5.65].

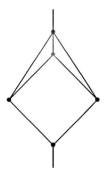

Fig. 1.3. A graph vanishing in the quotient by the automorphism group.

Thus, an orientation on a connected (m, n)-graph G is equivalently given by an ordering of the set $e(G)$ along with the choice of an orientation on $H_1(|G|; \mathbb{R})$ up to permutations and changes of orientation on $H_1(|G|; \mathbb{R})$ of even total parity. Consider the set of isomorphism classes of oriented (m, n)-graphs and take its k-linear span. More precisely, we should rather speak about a colimit with respect to graph isomorphisms, as in Section 1.2. In particular, if a graph G admits an orientation-reversing automorphism, such as the graph in Figure 1.3, then G gets identified with G^-, which will vanish after passing to the following quotient. Let $G(m, n)$ be the quotient of this space by the subspace spanned by

$$G + G^- \qquad \text{for each oriented graph } G,$$

where G^- is the same graph as G, taken with the opposite orientation. Each space $G(m, n)$ is bigraded by the genus and the number of *interior edges* (i.e., edges other than legs) of the graph. Let $G_g^q = G_g^q(m, n)$ denote the subspace spanned by graphs of genus g with q interior edges for $g, q \geqslant 0$. Computing the Euler characteristic of $|G|$ in two ways, we get an identity $|v(G)| - q = 1 - g$. A graph $G \in G_g(m, n)$ has a maximal number of interior edges if each vertex of G is trivalent, in which case we have $3|v(G)| = 2q + m + n$, whence $q = 3g - 3 + m + n$ is the top degree in which $G_g^q(m, n) \neq 0$.

Define a differential

$$\partial : G_g^q \to G_g^{q+1},$$

so that $\partial^2 = 0$, as follows:

$$\partial G := \sum_{\{G' \mid G'/e = G\}} G',$$

where the sum is over the isomorphism classes of connected (m, n)-graphs G' whose contraction along an edge $e \in e(G')$ is isomorphic to G. We will induce an orientation on G' by first choosing an ordering of the set of edges of G and an orientation on $H_1(|G|; \mathbb{R})$ in a way compatible with the orientation of G.

Then we will append the edge e that is being contracted at the end of the list of the edges of G. Since we have a canonical isomorphism $H_1(|G'|; \mathbb{R}) \xrightarrow{\sim} H_1(|G|; \mathbb{R})$, an orientation on the last space induces one on the first. This gives an orientation on G'. An example is given below:

$$\partial\left(\begin{array}{c}1 \quad 2 \\ \times \\ 1 \quad 2\end{array}\right) = \begin{array}{c}1 \quad 2 \\ \curlyvee \\ 1 \quad 2\end{array} - \begin{array}{c}1 \quad 2 \\ \curlyvee\!\curlyvee \\ 1 \quad 2\end{array} - \begin{array}{c}1 \quad 2 \\ \curlyvee\!\curlyvee \\ 1 \quad 2\end{array} + \begin{array}{c}2 \quad 1 \\ \curlyvee\!\curlyvee \\ 1 \quad 2\end{array} + \begin{array}{c}2 \quad 1 \\ \curlyvee\!\curlyvee \\ 1 \quad 2\end{array}$$

In this figure we have oriented graphs, which are provided with a certain canonical orientation that may be read off from the picture. The rule of thumb is as follows. An *orientation on the composition* of two graphs is given by (1) reordering the edges of the first, lower, graph in such a way that the output legs follow the remaining edges, (2) reordering the edges of the second, upper, graph in such a way that the input legs precede the remaining edges, and (3) after grafting, putting the edges of the second graph after the edges of the first graph. The resulting ordering should look like this: the newly grafted edges in the middle, preceded by the remaining edges of the first graph and followed by the remaining edges of the second graph. We remind the reader that we place the inputs at the bottom of a graph and the outputs on the top.

Theorem 39. *The graph complex in the commutative case is acyclic everywhere but at the top term $G_g^{3g-3+m+n}$. The graph cohomology can be computed as follows:*

$$H^q\left(G_g^*(m,n), \partial\right) = \begin{cases} \mathsf{LieB}_g^0(m,n) & \text{for } q = 3g - 3 + m + n, \\ 0 & \text{otherwise,} \end{cases}$$

where $\mathsf{LieB}_g^0(m,n)$ is the subspace of the (m,n)th component of the Lie bialgebra PROP $\mathsf{LieB} = F_1(LieB)$ consisting of linear combinations of connected graphs of genus g; see the presentation of the corresponding dioperad LieB in Example 34.

Remark 40. The acyclicity of the graph complex $G_g^*(m,n)$ has been proven in Kontsevich's message [9], whose method we have essentially used in this paper.

Proof. The dioperad *LieB* may be represented as a $\square$ product of the Lie operad $\mathcal{L}ie$ and the Lie co-operad $\mathcal{L}ie^{\mathrm{op}}$: $LieB = \mathcal{L}ie\square\mathcal{L}ie^{\mathrm{op}}$; see [4, Section 5.2]. The dioperadic quadratic dual $LieB^!$ is then $\mathcal{C}om^{\mathrm{op}}\square\mathcal{C}om$, so that $LieB^!(m,n) \cong k$ with a trivial action of (Σ_n, Σ_n) for each pair (m,n), $m, n \geq 1$. Then the subcomplex $\left(\Omega_\mathsf{P}^0\left(LieB^!\right), \partial\right) \subset (\Omega_\mathsf{P}(LieB^!), \partial)$ spanned by connected graphs is isomorphic to the graph complex $(G^*(m,n), \partial)$. Now the result follows from Theorem 37. $\square$

The associative case. Consider connected, oriented (m,n)-graphs G for $m, n \geq 1$, as above, now with a *ribbon* structure at each vertex, by which we mean orderings of the set $In(v)$ of incoming edges and the set $Out(v)$

of outgoing edges at each vertex $v \in v(G)$. It is convenient to think of an equivalent cyclic ordering (i.e., ordering up to cyclic permutation) of the set $e(v) = In(v) \cup Out(v)$ of all the edges incident to a vertex v in a way that elements of $In(v)$ precede those of $Out(v)$. Let $RG(m,n)$ be the linear span of isomorphism classes of connected oriented ribbon (m,n)-graphs modulo the relation $G + G^- = 0$, with $RG_g^q(m,n)$ denoting the subspace of graphs of genus g with q interior edges. The same formula

$$\partial G := \sum_{\{G' \mid G'/e=G\}} G'$$

defines a differential, except that in the ribbon case, when we contract an edge $e \in e(G')$, we induce a cyclic ordering on the set of edges adjacent to the resulting vertex by an obvious operation of insertion of the ordered set of edges adjacent to the edge e through one of its vertices into the ordered set of edges adjacent to e through its other vertex. An orientation is induced on G' in the same way as in the commutative case. An example is shown in the following display:

$$\partial(\text{✕}) = \text{✕} - \text{✕} - \text{✕} + \text{✕} - \text{✕} + \text{✕} - \text{✕} - \text{✕} - \text{✕} - \text{✕} - \text{✕} - \text{✕}$$

A vanishing theorem, see below, also holds in the ribbon-graph case. The proof is similar to the commutative case: it uses Theorem 37 and the fact that $IB = Ass \square Ass^{op}$ and $IB^! = Ass^{op} \square Ass$; see Example 38.

Theorem 41. *The ribbon graph complex is acyclic everywhere but at the top term $RG_g^{3g-3+m+n}$. The ribbon graph cohomology can be computed as follows:*

$$H^q\left(RG_g^*(m,n), \partial\right) = \begin{cases} \mathsf{IB}_g^0(m,n) \text{ for } q = 3g - 3 + m + n, \\ 0 \qquad\qquad\quad otherwise, \end{cases}$$

where $\mathsf{IB}_g^0(m,n)$ is the subspace of the (m,n)th component of the infinitesimal bialgebra PROP $\mathsf{IB} = F_1(IB)$ consisting of linear combinations of connected ribbon graphs of genus g; see the presentation of the corresponding dioperad IB in Example 33.

Remark 42. Note that our notion of the genus is not the same as the one coming from the genus of an oriented surface associated to the graph, usually used for ribbon graphs. Our genus is just the first Betti number of the surface.

References

1. M. AGUIAR, *Infinitesimal Hopf algebras*, New trends in Hopf algebra theory (La Falda, 1999), Contemporary Math., vol. 267, Amer. Math. Soc., 2000, pp. 1–29.

2. M. CULLER, K. VOGTMANN, *Moduli of graphs and automorphisms of free groups*, Invent. Math. **84** (1986), pp. 91–119.

3. T. FOX, M. MARKL, *Distributive laws, bialgebras, and cohomology*, Operads: Proceedings of Renaissance Conferences (J.-L. Loday, J. Stasheff, A. Voronov, eds.), Contemporary Math., vol. 202, Amer. Math. Soc., 1997, pp. 167–205.

4. W. GAN, *Koszul duality for dioperads*, Math. Res. Lett. **10** (2003), no. 1, pp. 109–124.

5. E. GETZLER, M. KAPRANOV, *Modular operads*, Compositio Math. **110** (1998), no. 1, pp. 65–126.

6. V. GINZBURG, M. KAPRANOV, *Koszul duality for operads*, Duke Math. J. **76** (1994), no. 1, pp. 203–272.

7. M. KONTSEVICH, *Formal (non)commutative symplectic geometry*, The Gel'fand mathematics seminars 1990–1992, Birkhäuser, 1993.

8. M. KONTSEVICH, *Feynman diagrams and low-dimensional topology*, First European Congress of Mathematics II, Progr. Math., vol. 120, Birkhäuser, Basel, 1994.

9. M. KONTSEVICH, *An e-mail message to M. Markl*, November 2002.

10. S. MAC LANE, *Natural associativity and commutativity*, Rice Univ. Stud. **49** (1963), no. 1, pp. 28–46.

11. I. MADSEN, M. S. WEISS, *The stable moduli space of Riemann surfaces: Mumford's conjecture*, Preprint math.AT/0212321, December 2002.

12. M. MARKL, *Cotangent cohomology of a category and deformations*, J. Pure Appl. Algebra **113** (1996), no. 2, pp. 195–218.

13. M. MARKL, *Models for operads*, Comm. Algebra **24** (1996), no. 4, pp. 1471–1500.

14. M. MARKL, *A resolution (minimal model) of the PROP for bialgebras*, J. Pure Appl. Algebra **205** (2006), no. 2, pp. 341–374.

15. M. MARKL, S. SHNIDER, J. D. STASHEFF, Operads in algebra, topology and physics, Mathematical Surveys and Monographs, vol. 96, American Mathematical Society, Providence, Rhode Island, 2002.

16. S. MERKULOV, *PROP profile of deformation quantization*, Preprint math.QA/0412257, December 2004.

17. R. PENNER, *Perturbative series and the moduli space of Riemann surfaces*, J. Differential Geom. **27** (1988), no. 1, pp. 35–53.

18. D. THURSTON, *Integral expressions for the Vassiliev knot invariants*, Preprint math.QA/9901110, January 1999.

19. B. VALLETTE, *Dualité de Koszul des PROPs*, Thesis, Université Louis Pasteur, 2003.

Symboles de Manin et valeurs de fonctions L

Loïc Merel

Institut de Mathématiques de Jussieu, Université Paris-Diderot, UFR de
Mathématiques, case 7012, 2 place Jussieu, 75251 Paris cedex 05, France
merel@math.jussieu.fr

À Yuri Ivanovich Manin à l'occasion de son soixante-dixième anniversaire

Summary. En utilisant les symboles de Manin, nous observons d'abord qu'une
forme modulaire primitive f de niveau N et de poids 2 est caractérisée par un nombre
fini d'invariants associés aux fonctions L obtenues en tordant f par des caractères
de Dirichlet de niveau divisant N ; il s'agit des valeurs en 1 prises par ces fonctions
L et de quelques invariants locaux, qui concernent purement les places divisant N.
Nous établissons ensuite quelques relations numériques suivant le principe suivant.
Considérons la fonction L d'un objet de la catégorie tensorielle engendrée par f ;
la valeur en un nombre entier de cette fonction L s'exprime m écaniquement en
fonction des invariants considérés ci-dessus.

Key words: Dirichlet series, functional equations, modular forms, Special
values of automorphic L-series, periods of modular forms, modular symbols

2000 Mathematics Subject Classifications: 11FXX, 11F66, 11F67

1 Introduction

1.1 Les symboles de Manin

Soit N un entier >0. Soit f une forme modulaire parabolique de poids $k = 2$
pour le groupe de congruence $\Gamma_1(N)$. Dans [10], Yu. Manin lui associe une
fonction $\xi_f : (\mathbf{Z}/N\mathbf{Z})^2 \to \mathbf{C}$ dont les valeurs sont les *symboles de Manin de* f.
Cette fonction est définie ainsi.

Soit $(u, v) \in (\mathbf{Z}/N\mathbf{Z})^2$. On pose $\xi_f(u, v) = 0$ si (u, v) n'est pas d'ordre N
dans le groupe additif $(\mathbf{Z}/N\mathbf{Z})^2$. Sinon on considère une matrice $g = \begin{pmatrix} a & b \\ c & d \end{pmatrix} \in$
$\mathrm{SL}_2(\mathbf{Z})$ telle que $(c, d) \in (u, v)$ et on pose

$$\xi_f(u, v) = -i \int_{g0}^{g\infty} f(z)dz,$$

Y. Tschinkel and Y. Zarhin (eds.), *Algebra, Arithmetic, and Geometry*,
Progress in Mathematics 270, DOI 10.1007/978-0-8176-4747-6_9,
© Springer Science+Business Media, LLC 2009

où l'intégrale est prise le long d'un chemin continu du demi-plan de Poincaré. L'application $f \mapsto \xi_f$ est injective. On peut même être plus précis. Si on pose $\xi_f^+(u,v) = (\xi_f(u,v) + \xi_f(-u,v))/2$ et $\xi_f^-(u,v) = (\xi_f(u,v) - \xi_f(-u,v))/2$, les applications $f \mapsto \xi_f^+$ et $f \mapsto \xi_f^-$ sont injectives [10].

Rappelons en quoi cette construction est utile à l'étude des symboles modulaires : elle a notamment permis a Manin d'établir sa loi de réciprocité [10, 11, 14] et est souvent le fondement des calculs sur ordinateur concernant les formes modulaires (voir [5] et les tables de W. Stein).

Par ailleurs, les expressions (si utiles en vue de construire des fonctions L p-adique, d'établir des théorèmes de non-annulation...) en termes de symboles modulaires des valeurs en $s = 1$ des fonctions L des tordues de f ne font pas intervenir les symboles de Manin. C'est à ce lien manquant que fait référence notre titre. Notre objectif est, en réalité, inverse de ce qui est obtenu par la démarche classique : lorsque f est une forme primitive (*i.e.* propre pour l'algèbre de Hecke, nouvelle et normalisée), nous exprimons les symboles de Manin de f purement en termes des fonctions L des tordues f. Nous verrons même qu'il suffit de tordre f par des caractères de niveaux divisant N.

1.2 Analyse de Fourier multiplicative

Supposons désormais f primitive de niveau N. Nous calculons la transformée de Fourier multiplicative de ξ_f en le sens suivant.

Pour tout entier $m \geq 1$, on note Σ_m le support de m dans l'ensemble des nombres premiers. Supposons $(u,v) \in (\mathbf{Z}/N\mathbf{Z})^2$ d'ordre N. Notons N' l'ordre de uv dans $\mathbf{Z}/N\mathbf{Z}$. Soit S un sous-ensemble de Σ_N contenant le support de u mais disjoint du support de v. Posons $\bar{S} = \Sigma_N - S$. On identifie $(\mathbf{Z}/N\mathbf{Z})$ à $\cup_{d|N}(\mathbf{Z}/d\mathbf{Z})^*$ (par $w \mapsto wN'_w/N \pmod{N'_w}$, où N'_w est l'ordre de w dans $(\mathbf{Z}/N\mathbf{Z})$). Les images de u et v par cette identification sont uN'_S/N_S et $vN'_{\bar{S}}/N_{\bar{S}}$ respectivement ; les entiers N_S/N'_S et $N_{\bar{S}}/N'_{\bar{S}}$ ne dépendent pas du choix de S. Toute fonction $\xi : (\mathbf{Z}/N\mathbf{Z})^2 \to \mathbf{C}$ s'écrit sous la forme

$$\xi(u,v) = \sum_{\alpha,\beta} c_{\alpha,\beta}\, \alpha\left(N'_{\bar{S}}v/N_{\bar{S}}\right) \beta\left(N'_S u/N_S\right),$$

où $c_{\alpha,\beta}$ dépend seulement de ξ, α, β et N' et où α et β parcourent les caractères de Dirichlet primitifs de niveaux divisant N'. Le théorème 1 donne une forme explicite aux coefficients $c_{\alpha,\beta}$ lorsque $\xi = \xi_f$.

Soit χ un caractère de Dirichlet de conducteur à support dans Σ_N. Notons $f \otimes \chi$ la forme primitive dont le p-ième coefficient de Fourier est $a_p(f)\chi(p)$ (p nombre premier ne divisant pas N). Notons N_χ le niveau de $f \otimes \chi$. Notons $L(f \otimes \chi, s)$ la fonction L de $f \otimes \chi$. Elle admet un développement en série de Dirichlet $\sum_{n=1}^{\infty} a_n(f \otimes \chi)/n^s$ et en produit eulerien $\prod_p L_p(f \otimes \chi, p^{-s})$, où $L_p(f \otimes \chi, X) = 1/(1 - a_p(f \otimes \chi)X + a_{p,p}(f \otimes \chi)p^{k-1}X^2)$ (p nombre premier) ; on complète ce produit pour former

$$\Lambda(f \otimes \chi, s) = (2\pi)^{-s}\Gamma(s)N_\chi^{s/2}L(f \otimes \chi, s).$$

On pose $a_p = a_p(f)$ et $a_{p,p} = a_{p,p}(f)$. Notons ψ le caractère de Dirichlet vérifiant $\psi(p) = a_{p,p}(f)$ (p nombre premier ne divisant pas N). On pose $\bar{f} = f \otimes \bar{\psi}$, et on a $a_n(\bar{f}) = \bar{a}_n(f)$ (n entier ≥ 1).

Lorsque T^+ et T^- sont des ensembles finis de nombres premiers, on prive $\Lambda(f \otimes \chi, s)$ de certains facteurs d'Euler en posant

$$\Lambda^{[T^+,T^-]}(f \otimes \chi, s) = \frac{\Lambda(f \otimes \chi, s)}{\prod_{p \in T^+} L_p(f \otimes \chi, p^{-s}) \prod_{p \in T^-} L_p(\bar{f} \otimes \bar{\chi}, p^{s-k})}.$$

Lorsque R^+ et R^- sont des sous-ensembles de T^+ et T^- respectivement, on pose

$$\Lambda^{\left[\frac{T^+}{R^+},\frac{T^-}{R^-}\right]}(f \otimes \chi, s) = \frac{\Lambda^{[T^+-R^+,T^--R^-]}(f \otimes \chi, s)}{\prod_{p \in R^+} L_p(f \otimes \chi, p^{-s-1}) \prod_{p \in R^-} L_p(\bar{f} \otimes \bar{\chi}, p^{s-k+1})}.$$

Nous dirons que les nombres premiers p qui vérifient $v_p(N) = 1$ (où v_p est la valuation p-adique) et ψ non ramifié en p sont *spéciaux* pour f (ils correspondent aux représentations spéciales de $\mathrm{GL}_2(\mathbf{Q}_p)$). Notons Σ_f l'ensemble des nombres premiers spéciaux pour f. Le cas qui nous intéressera est le cas où R^+ et R^- sont composés de nombres premiers spéciaux pour f.

Pour S sous-ensemble de Σ_N et M nombre entier ≥ 1 de support $\Sigma_M \subset \Sigma_N$, posons $M = M_S M_{\bar{S}}$ où M_S et $M_{\bar{S}}$ sont à supports dans S et $\bar{S}$ respectivement. On pose $S(M) = \Sigma_M \cap S$ et $\bar{S}(M) = \Sigma_M \cap \bar{S}$. On note $w_S(\bar{f} \otimes \chi)$ la pseudo-valeur propre de $\bar{f} \otimes \chi$ pour l'opérateur d'Atkin-Lehner associé à S (voir [1] ou la mise au point de la section 2.2). On note de plus $w(f \otimes \chi) = w_{\Sigma_N}(f \otimes \chi)$; on a

$$\Lambda^{[T^+,T^-]}(f \otimes \chi, s) = i^k w(f \otimes \chi) \Lambda^{[T^-,T^+]}(\bar{f} \otimes \bar{\chi}, k - s).$$

Pour α caractère de niveau à support dans Σ_N, on convient de décomposer α sous la forme $\alpha = \alpha_S \alpha_{\bar{S}}$, où α_S et $\alpha_{\bar{S}}$ sont des caractères de Dirichlet de niveaux à supports dans S et $\bar{S}$ respectivement.

Pour $p \in \Sigma_N$ et χ caractère de Dirichlet, notons m_χ le conducteur du caractère primitif associé à χ divisant N et $Q_{p,f,\chi}(X)$ la fraction rationnelle suivante :

$$Q_{p,f,\chi}(X) = (\bar{a}_p p^{1-k/2})^{v_p(N'/m_\chi)}$$

sauf si $a_p = 0$, $v_p(N') = 1$ et $v_p(m_\chi) = 0$, auquel cas on a

$$Q_{p,f,\chi}(X) = -\bar{\chi}(p) X^{-1}.$$

Cet objet désagréable dépend de p, a_p, $\chi(p)$, $v_p(m_\chi)$, k et $v_p(N')$; c'est donc un objet local. De plus on note $\tau'(\chi)$ la somme de Gauss associée au caractère primitif provenant de χ. Notons ϕ la fonction indicatrice d'Euler.

Théorème 1. *On a*

$$\xi_f(u,v) = \frac{w(f)}{\phi(N')} \sum_\chi \chi_{\bar S}(m_{\chi,S})(\bar\psi_S\bar\chi_S)(m_{\bar\psi_{\bar\chi},\bar S})\chi_S(-1)\frac{\tau'(\chi_S)\tau'(\bar\psi_{\bar S}\bar\chi_{\bar S})}{\sqrt{N_\chi}}$$

$$\times \left(\prod_{p\in S(N')} Q_{p,f,\bar\chi}(1)\right)\left(\prod_{p\in \bar S(N')} Q_{p,f,\chi\psi}(1)\right)\left(\bar\psi_{\bar S}\bar\chi_{\bar S}^2\right)(N_{\chi,S})\overline{w_S(f\otimes\chi)}$$

$$\times\Lambda\left[\frac{\Sigma_{N'}-S(m_\chi)}{(S(N')-S(m_\chi))\cap\Sigma_f},\frac{\Sigma_{N'}-\bar S(m_{\bar\psi_{\bar\chi}})}{(\bar S(N')-\bar S(m_{\bar\psi_{\bar\chi}}))\cap\Sigma_f}\right](f\otimes\chi,1)(\bar\psi\bar\chi)\left(\frac{N'_{\bar S}v}{N_{\bar S}}\right)\chi\left(\frac{N'_S u}{N_S}\right),$$

où χ parcourt les caractères de Dirichlet primitifs tels que $m_{\chi,S}m_{\psi_\chi,\bar S}|N'$. La formule analogue pour ξ_f^+ (resp. ξ_f^-) est obtenue en faisant disparaître les termes pour χ impair (resp. pair).

1.3 Interprétation arithmétique

La formule du théorème 1 est si peu aisément manipulable, si inapte à s'insérer dans le langage naturel, que le lecteur séduit par le point de vue exposé par Manin dans son essai "Mathematics as Metaphor" [12] pourrait penser qu'elle présente bien peu d'intérêt. Nous espérons que les conséquences qui suivent peuvent effacer cette impression.

Il procède d'un examen superficiel du théorème 1 et de l'injectivité des applications $f \mapsto \xi_f^+$ et $f \mapsto \xi_f^-$ l'énoncé suivant.

Corollaire 2. *La forme modulaire f primitive de poids 2 pour $\Gamma_1(N)$ est caractérisée par les données suivantes, où on fait parcourir à χ les caractères de Dirichlet pairs (resp. impairs) de conducteurs divisant N :*

(i) le caractère de f,

(ii) les niveaux des formes primitives $f\otimes\chi$,

(iii) les pseudo-valeurs propres $w_S(f\otimes\chi)$,

(iv) les facteurs d'Euler $L_p(f\otimes\chi,p^{-s})$, pour $p\in\Sigma_N$ et

(v) les nombres complexes $\Lambda(f\otimes\chi,1)$.

Nous sommes donc tentés de voir la fonction ξ_f comme une façon commode de comprimer et de manipuler les données (*i*), (*ii*), (*iii*), (*iv*) et (*v*).

On pourrait rendre l'énoncé et la démonstration du théorème 1 plus agréables en employant le langage adélique. Les données (*i*), (*ii*) (*iii*) et (*iv*) sont équivalentes à celles issues des facteurs d'Euler et des facteurs ϵ associés aux représentations irréductibles de $\mathrm{GL}_2(\mathbf{Q}_p)$ provenant de f après torsion par des caractères de $\mathbf{Q}_p^*$ de conducteur $\le v_p(N)$, pour $p\in\Sigma_N$.

On est tenté de rapprocher le corollaire 2 du théorème de Hecke–Weil [17] sur la caractérisation, par les prolongements analytiques et les équations fonctionnelles, des séries de Dirichlet qui proviennent des formes modulaires, voir la section 3.3. Nous avons à l'esprit tout spécialement la version précise due à W. C. W. Li [8] qui, comme le corollaire 2, ne fait intervenir que les tordues par les caractères de conducteur divisant le niveau N. On précisera dans la section 3.2, comment, à partir de ξ_f, on peut retrouver les invariants (i), (ii), (iii), (iv) et (v) notamment lorsque f est de niveau minimal parmi ses tordues par des caractères de Dirichlet, auquel cas nous essaierons d'indiquer en quoi l'information contenue dans ξ_f est optimale.

Notre travail ne semble pas présenter de lien avec le théorème de Luo et Ramakrishnan [9] qui caractérise f par les nombres complexes $L(f \otimes \chi, 1)$ où χ parcourt une infinité de caractères quadratiques.

Il résulte du théorème 1 un énoncé de théorie analytique des nombres.

Corollaire 3. *Il existe des caractères de Dirichlet χ^+ et χ^-, pair et impair respectivement, de conducteurs divisant N et tels que $L(f \otimes \chi^+, 1) \neq 0$ et $L(f \otimes \chi^-, 1) \neq 0$.*

En raison de la modularité des courbes elliptiques [3] et des résultats obtenus par Kato [6] sur la conjecture de Birch et Swinnerton-Dyer et ses variantes, on obtient des conséquences diophantiennes dont voici l'exemple type.

Corollaire 4. *Soit E une courbe elliptique sur $\mathbf{Q}$ de conducteur N. Soit $\mathbf{Q}(\mu_N)$ une extension cyclotomique de $\mathbf{Q}$ engendrée par une racine primitive N-ième de l'unité. Notons $\mathbf{Q}(\mu_N)^+$ le plus grand sous-corps totalement réel de $\mathbf{Q}(\mu_N)$. La représentation régulière du groupe de Galois $\mathrm{Gal}(\mathbf{Q}(\mu_N)^+/\mathbf{Q})$ n'intervient pas dans le $\mathrm{Gal}(\mathbf{Q}(\mu_N)^+/\mathbf{Q})$-module $E(\mathbf{Q}(\mu_N)^+)$.*

Le lecteur pourra trouver des généralisations du corollaire 4 pour les groupes de Selmer p-adiques des motifs associés aux formes modulaires, en s'appuyant sur [6]. Nous nous demandons s'il existe une direction directe (*i.e.* ne faisant pas appel aux formes modulaires) du corollaire 4.

1.4 Perspectives

Comme ξ_f détermine f, on peut, en principe, exprimer tout invariant associé à f en terme de ξ_f, puis en combinant avec le théorème 1, en terme des données (i), (ii), (iii), (iv) et (v). Ce principe appliqués aux valeurs de fonctions L construites à partir de f (via des puissances tensorielles, la torsion par des caractères etc) produirait alors des identités numériques entre valeurs de fonctions L. Nous donnons un exemple de telles identités dans la section 4. Dans sa thèse, F. Brunault exprime $L(f, 2)$ en termes des ξ_f [4], qu'on peut combiner avec le théorème 1. Y a-t-il une théorie systématique ?

2 Formulaire préliminaire

Cette section consiste en des mises au points concernant des questions essentiellement déjà connues. Elles concernent en 2.1 la suppression des facteurs d'Euler des fonctions L, en 2.2 les opérateurs d'Atkin–Lehner, en 2.3 et 2.4 la torsion des formes modulaires par des caractères non nécessairement primitifs, en 2.5 la translation des formes modulaires par des nombres rationnels. Dans les sections 2.6 à 2.8, qui ne seront pas utiles avant la section 3.2, nous rappelons ce qui est connu sur le comportement aux mauvaises places des formes modulaires tordues. Pour tout cela nous avons trouvé une aide précieuse dans un article d'Atkin et Li [2].

2.1 Suppression des facteurs d'Euler

On note $\mathrm{GL}_2(\mathbf{Q})^+$ le sous-groupe de $\mathrm{GL}_2(\mathbf{Q})$ formé par les matrices de déterminant >0. On pose, pour $\begin{pmatrix} A & B \\ C & D \end{pmatrix} \in \mathrm{GL}_2(\mathbf{Q})^+$, et F forme primitive de poids k et de niveau M :

$$F_{|\begin{pmatrix} A & B \\ C & D \end{pmatrix}}(z) = \frac{(AD - BC)^{k/2}}{(Cz + D)^k} F\left(\frac{Az + B}{Cz + D}\right).$$

Cette opération s'étend $\mathbf{C}$-linéairement à $\mathbf{C}[\mathrm{GL}_2(\mathbf{Q})^+]$; elle se factorise par $\mathbf{C}[\mathrm{PGL}_2(\mathbf{Q})^+]$. Gardons à l'esprit la formule suivante

$$(2\pi)^{-s}\Gamma(s)L(F, s) = \int_0^\infty F(iy)y^s \frac{dy}{y}.$$

On a, pour $h = \begin{pmatrix} A & 0 \\ 0 & D \end{pmatrix} \in \mathrm{GL}_2(\mathbf{Q})^+$,

$$\int_0^\infty F_{|h}(iy)y^s \frac{dy}{y} = \left(\frac{A}{D}\right)^{k/2-s} \int_0^\infty F(iy)y^s \frac{dy}{y} = \left(\frac{A}{D}\right)^{k/2-s}(2\pi)^{-s}\Gamma(s)L(F, s).$$

Soient T^+ et T^- deux ensembles de nombres premiers. On pose

$$F^{[T^+, T^-]} = F_{|\prod_{p\in T^+} L_p\left(F, p^{-k/2}\begin{pmatrix} p & 0 \\ 0 & 1 \end{pmatrix}\right)^{-1} \prod_{p\in T^-} L_p\left(\bar{F}, p^{-k/2}\begin{pmatrix} 1 & 0 \\ 0 & p \end{pmatrix}\right)^{-1}},$$

de telle sorte que

$$\int_0^\infty F^{[T^+, T^-]}(iy)y^s \frac{dy}{y} = \frac{(2\pi)^{-s}\Gamma(s)L(F, s)}{\prod_{p\in T^+} L_p(F, p^{-s}) \prod_{p\in T^-} L_p(\bar{F}, p^{s-k})}$$

$$= M^{-s/2}\Lambda^{[T^+, T^-]}(F, s).$$

On pose, lorsque R^+ et R^- sont des sous-ensembles de T^+ et T^- respectivement,

$$F^{\left[\frac{T^+}{R^+},\frac{T^-}{R^-}\right]} = F^{[T^+-R^+,T^--R^-]}$$
$$\mid \prod_{p\in R^+} L_p\left(F,p^{1-k/2}\begin{pmatrix} p & 0 \\ 0 & 1 \end{pmatrix}\right)^{-1} \prod_{p\in R^-} L_p\left(\bar{F},p^{1-k/2}\begin{pmatrix} 1 & 0 \\ 0 & p \end{pmatrix}\right)^{-1}$$

si bien que

$$\int_0^\infty F^{\left[\frac{T^+}{R^+},\frac{T^-}{R^-}\right]}(iy)y^s\frac{dy}{y} = M^{-s/2}\Lambda^{\left[\frac{T^+}{R^+},\frac{T^-}{R^-}\right]}(F,s).$$

2.2 Opérateurs d'Atkin–Lehner

Mettons au point les normalisations pour les opérateurs d'Atkin-Lehner. Notons ψ' le caractère de nebentypus de F. Notons M' le conducteur de ψ'. Supposons que M soit à support dans Σ_N. Soit S un sous-ensemble de Σ_M. Notons $\bar{S} = \Sigma_M - S$. Posons $M = M_S M_{\bar{S}}$ et $M' = M'_S M'_{\bar{S}}$ et $\psi' = \psi'_S \psi'_{\bar{S}}$. Soit $\begin{pmatrix} A & B \\ C & D \end{pmatrix} \in M_2(\mathbf{Z})$ telle que $M_S|A$, $M_S|D$, $M|C$, $M_{\bar{S}}|B$, $AD-BC = M_S$, $A \equiv M_S \pmod{M'}$ et $B \equiv 1 \pmod{M'_S}$; on pose alors, comme Atkin et Li dans [2], $W_S F = F_{\mid \begin{pmatrix} A & B \\ C & D \end{pmatrix}}$ et il existe un nombre complexe $w_S(F)$ de module 1 tel que $W_S F = w_S(F)F \otimes \bar{\psi}'_S$. Lorsque $\begin{pmatrix} A & B \\ C & D \end{pmatrix} \in M_2(\mathbf{Z})$ avec $M_S|A$, $M_S|D$, $M|C$, $M_{\bar{S}}|B$ et $AD - BC = M_S$, on a de plus [2]

$$F_{\mid \begin{pmatrix} A & B \\ C & D \end{pmatrix}} = \psi'_S(B)\psi'_{\bar{S}}(A/M_S)W_S F.$$

Lorsque $M|NN'$, $M'|N$ et lorsque $\begin{pmatrix} A & B \\ C & D \end{pmatrix} \in M_2(\mathbf{Z})$ vérifie les conditions $N_S N'_S|A, N_S N'_S|D, NN'|C, N_{\bar{S}}N'_{\bar{S}}|B$ et $AD - BC = N_S N'_S$, on a

$$F_{\mid \begin{pmatrix} A & B \\ C & D \end{pmatrix}} = w_S(F)\bar{\psi}'_S(B)\bar{\psi}'_{\bar{S}}(A/(N_S N'_S)) F_{\mid \begin{pmatrix} N_S N'_S/M_S & 0 \\ 0 & 1 \end{pmatrix}}.$$

Lorsque S est égal au support de M, on pose $w_S(F) = w(F)$.
On a de plus [2] proposition 1.1,

$$w_S(F)w_S\left(F \otimes \bar{\psi}'_S\right) = \psi'_S(-1)\bar{\psi}'_{\bar{S}}(M_S). \tag{1}$$

Mentionnons enfin la formule, pour S_1 et S_2 deux sous-ensembles disjoints de Σ_M,

$$W_{S_2}(W_{S_1}F) = \psi'_{S_2}(M_{S_1})W_{S_1 \cup S_2}F.$$

Cela permet de ramener le calcul de $w_S(F)$ aux cas où S est un singleton.

Ajoutons la formule suivante. Soit p un nombre premier tel que $a_p(F) \neq 0$ (c'est le cas si et seulement si $v_p(M) = v_p(m_\psi)$ ou si $v_p(M) \leq 1$). On a

$$w_{\{p\}}(F) = \frac{p^{v_p(N)(k/2-1)}\tau\left(\psi'_S\right)}{a_p(F)^{v_p(N)}}$$

où $\tau\left(\psi'_S\right)$ est la somme de Gauss du caractère (non nécessairement primitif) ψ'_S. Si p est spécial pour F, on a $a_p(F)\bar{a}_p(F) = p^{k-2}$.

2.3 Torsion des formes modulaires par des caractères quelconques

Revenons maintenant sur la torsion des formes modulaires par des caractères. Soit α un caractère de Dirichlet de niveau m, de caractère de Dirichlet primitif associé ω, lui-même de conducteur m_ω. Notons $\bar{f}_\alpha$ la forme modulaire (non nécessairement primitive) donnée par le développement

$$\bar{f}_\alpha(z) = \sum_{n=1}^{\infty} \bar{a}_n\alpha(n)q^n.$$

Elle est liée à la forme primitive $\bar{f} \otimes \omega$ par la formule

$$\bar{f}_\alpha = (\bar{f} \otimes \omega)^{[\Sigma_m, \emptyset]}.$$

Posons de plus

$$S_\alpha\bar{f} = \sum_{a \bmod m} \bar{\alpha}(a)\bar{f}_{|\left(\begin{smallmatrix} 1 & a/m \\ 0 & 1 \end{smallmatrix}\right)}.$$

On a, lorsque α est primitif (et donc égal à ω),

$$S_\omega\bar{f} = \tau(\bar{\omega})\bar{f}_\omega.$$

Soit p un nombre premier divisant m/m_ω. Notons β le caractère de Dirichlet de niveau m/p qui coïncide avec α sur les entiers premiers à p. On a

$$S_\alpha\bar{f} = \bar{a}_p p^{1-k/2}(S_\beta\bar{f})_{|\left(\begin{smallmatrix} p & 0 \\ 0 & 1 \end{smallmatrix}\right)} - \bar{\beta}(p)S_\beta\bar{f}.$$

Posons, dans $\mathbf{C}[X]$,

$$R_p(X) = (\bar{a}_p p^{1-k/2}X)^{v_p(m/m_\omega)-1}(\bar{a}_p p^{1-k/2}X - \bar{\omega}(p)). \qquad (2)$$

Par une application répétée de la formule 2, on obtient

$$S_\alpha\bar{f} = \tau(\bar{\omega})(\bar{f}_\omega)_{|\Pi_p R_p\left(\left(\begin{smallmatrix} p & 0 \\ 0 & 1 \end{smallmatrix}\right)\right)},$$

où le produit porte sur les nombres premiers divisant m/m_ω.

Il est nécessaire maintenant de distinguer plusieurs cas. Si $v_p(m/m_\omega) = 0$, on a $R_p = 1$. Si $v_p(m/m_\omega) = 1$ et $a_p = 0$, on a $R_p = 0$. Si $v_p(m/m_\omega) > 1$ et $a_p = 0$, on a $R_p = -\bar\omega(p)$.

Or on a, lorsque $a_p \neq 0$ et $p|N$ non spécial pour $\bar f$, $a_p \bar a_p = p^{k-1}$ et donc, lorsque de plus $p|m$ on a, dans $\mathbf{C}[\mathrm{PGL}_2(\mathbf{Q})^+]$,

$$R_p\left(\begin{pmatrix} p & 0 \\ 0 & 1 \end{pmatrix}\right) = \left(\bar a_p p^{1-k/2}\begin{pmatrix} p & 0 \\ 0 & 1 \end{pmatrix}\right)^{v_p(m/m_\omega)}\left(1 - \bar\omega(p)a_p p^{-k/2}\begin{pmatrix} 1 & 0 \\ 0 & p \end{pmatrix}\right).$$

Cette dernière formule est encore valable lorsque $a_p = 0$ et $v_p(m) > 1$.

Lorsque $p|(m/m_\omega)$ et p est spécial pour $\bar f$, on a $a_p \bar a_p = p^{k-2}$ (et donc $a_p \neq 0$). On a donc

$$R_p\left(\begin{pmatrix} p & 0 \\ 0 & 1 \end{pmatrix}\right) = \left(\bar a_p p^{1-k/2}\begin{pmatrix} p & 0 \\ 0 & 1 \end{pmatrix}\right)^{v_p(m/m_\omega)}\left(1 - \bar\omega(p)a_p p^{1-k/2}\begin{pmatrix} 1 & 0 \\ 0 & p \end{pmatrix}\right).$$

Lorsque $a_p = 0$, $v_p(m) = 1$ et $v_p(m_\omega) = 0$, on a

$$R_p\left(\begin{pmatrix} p & 0 \\ 0 & 1 \end{pmatrix}\right) = -\bar\omega(p).$$

On a donc

$$S_\alpha \bar f = \tau(\bar\omega)(\bar f \otimes \omega)^{\left[\Sigma_m, \frac{\Sigma_{m/m_\omega}}{\Sigma_{m/m_\omega} \cap \Sigma_f}\right]}_{\mid \prod_p P_p\left(\begin{pmatrix} p & 0 \\ 0 & 1 \end{pmatrix}\right)}, \tag{3}$$

où le monôme $P_p(X)$ vaut $(\bar a_p p^{1-k/2} X)^{v_p(m/m_\omega)}$ sauf si $a_p = 0$, $v_p(m) = 1$ et $v_p(m_\chi) = 0$, auquel cas on a $P_p\left(\begin{pmatrix} p & 0 \\ 0 & 1 \end{pmatrix}\right) = -\bar\omega(p)$.

2.4 La torsion des formes modulaires par des caractères de niveaux divisant N

Reprenons la situation laissée en 2.3 en nous plaçant dans le cas où $N' = m$ est un diviseur de N.

Lemme 5. *Soit p un nombre premier tel que $p|m_\omega$ et $p|(N'/m_\omega)$. On a*

$$(\bar f \otimes \omega)^{[0,p]}_{\Big| P_p\left(\begin{pmatrix} p & 0 \\ 0 & 1 \end{pmatrix}\right)} = (\bar f \otimes \omega)_{\Big| P_p\left(\begin{pmatrix} p & 0 \\ 0 & 1 \end{pmatrix}\right)}.$$

Démonstration. Il suffit de montrer que $P_p = 0$ ou que $a_p(\bar f \otimes \omega) = 0 = a_{p,p}(\bar f \otimes \omega)$. Supposons $P_p \neq 0$. Si $a_p = 0$, on a $v_p(m_\omega) = 0$ et $v_p(N') = 1$, ce qui entraîne $a_p(\bar f \otimes \omega) = 0 = a_{p,p}(\bar f \otimes \omega)$. Restreignons maintenant notre attention au cas où $a_p \neq 0$. Rappelons d'abord que cela entraîne que le conducteur de ψ a pour valuation p-adique $v_p(N)$ (ce qui entraîne $a_{p,p}(\bar f \otimes \omega) = 0$) ou que $v_p(N) = 1$. Les hypothèses excluent le cas $v_p(N) = 1$. On a de plus $a_p(\bar f \otimes \omega) \neq 0$ si et seulement si ω est de conducteur premier à p (impossible par hypothèse) ou $\bar\psi/\omega$ est de conducteur premier à p; ce dernier cas est impossible, en effet on a $p|(N'/m_\omega)$, et donc $p|(N/m_\omega)$ et les valuations p-adiques des conducteurs de ψ et $\bar{\psi}/\omega$ sont égales et donc non nulles. On a bien $a_p(\bar{f} \otimes \omega) = 0$.

On a donc, par applications répétées du lemme 5 à la formule 3,

$$S_\alpha \bar{f} = \tau(\bar{\omega})(\bar{f} \otimes \omega) \frac{\left[\Sigma_{N'}, \frac{\Sigma_{N'} - \Sigma_{m_\omega}}{(\Sigma_{N'} - \Sigma_{m_\omega}) \cap \Sigma_f}\right]}{|\prod_p P_p\left(\begin{pmatrix} p & 0 \\ 0 & 1 \end{pmatrix}\right)}, \tag{4}$$

2.5 La torsion des formes modulaires par des caractères additifs

Soit $n \in \mathbf{Z}$. Récrivons la forme modulaire $\bar{f}_{\mid\begin{pmatrix} 1 & n/N \\ 0 & 1 \end{pmatrix}}$ comme combinaison linéaire de $F_{\begin{pmatrix} d & 0 \\ 0 & 1 \end{pmatrix}}$ où d parcourt les diviseurs de N et où F parcourt les formes primitives de niveau divisant N^2/d. Nous ne savons pas si un pareil calcul a déjà été rédigé. Notons n' le nombre entier et N' le diviseur > 0 de N qui vérifient $n'/N' = n/N$. On a, par inversion de Fourier,

$$\bar{f}_{\mid\begin{pmatrix} 1 & n'/N' \\ 0 & 1 \end{pmatrix}} = \sum_\alpha \frac{\alpha(n')}{\phi(N')} S_\alpha \bar{f},$$

où α parcourt les caractères de Dirichlet de niveau N'. En combinant avec la formule 4, on obtient

$$\bar{f}_{\mid\begin{pmatrix} 1 & n/N \\ 0 & 1 \end{pmatrix}} = \sum_\omega \frac{\omega(n')}{\phi(N')} \tau(\bar{\omega})(\bar{f} \otimes \omega) \frac{\left[\Sigma_{N'}, \frac{\Sigma_{N'} - \Sigma_{m_\omega}}{(\Sigma_{N'} - \Sigma_{m_\omega}) \cap \Sigma_f}\right]}{|\prod_p P_p\left(\begin{pmatrix} p & 0 \\ 0 & 1 \end{pmatrix}\right)} \tag{5}$$

où ω parcourt les caractères de Dirichlet primitifs de conducteur m_ω divisant N', le produit portant sur les nombres premiers divisant N'/m_ω.

2.6 Invariants locaux des tordues de formes modulaires, première analyse

On reprend les notations de 2.1 et 2.2. Soient n un entier > 0. Supposons m_χ et N_S premiers entre eux. On a $a_n(f \otimes \chi) = a_n\chi(n)$ et

$$w_S(f \otimes \chi) = \bar{\chi}(m_\chi)w_S(f).$$

De plus si N est premier à m_χ, le caractère de $f \otimes \chi$ est $\psi\chi^2$ et on a $N_\chi = Nm_\chi^2$.

Nous allons maintenant étudier les cas où m_χ, N_S et n ne sont pas premiers entre eux.

Soit p un nombre premier. On dit que f est p-*primitive par torsion* si pour tout caractère de Dirichlet χ de conducteur une puissance de p, le niveau de $f \otimes \chi$ est $\geq N$. C'est évidemment une propriété locale, qui serait peut-être davantage mise en valeur par le langage adélique. On suppose que m_χ, N_S et n sont des puissances de p et que f est p-primitive par torsion.

Notons $(\psi\chi)_0$ le caractère primitif associé à $\psi\chi$. On a

$$L_p(f \otimes \chi, X)^{-1} = 1 - \frac{\bar{a}_p(\psi\chi)_0(p)}{p} X.$$

Comme on a

$$f_\chi = f \otimes \chi_{\mid L_p\left(f\otimes\chi,\left(\begin{smallmatrix} p & 0 \\ 0 & 1 \end{smallmatrix}\right)\right)^{-1}}.$$

On a donc, puisque χ est primitif,

$$f \otimes \chi = f_{\chi \mid \left(1 - \frac{\bar{a}_p(\psi\chi)(p)}{p}\left(\begin{smallmatrix} p & 0 \\ 0 & 1 \end{smallmatrix}\right)\right)^{-1}}.$$

Pour progresser dans l'étude des invariants de $f \otimes \chi$, il faut distinguer deux cas [2],
– on a $a_p(f) \neq 0$ (cas de *série principale*), cela assure que f est primitive par torsion, ou
– on a $a_p(f) = 0$ et le conducteur du caractère de Dirichlet primitif associé à ψ est de valuation p-adique $\leq v_p(N)/2$ (cas *supercuspidal*). (Cette dernière condition n'entraîne pas que f est p-primitive, voir [2].)

2.7 Invariants locaux des tordues de formes modulaires, cas de série principale

Reprenons la situation laissée en 2.6 en supposant $a_p \neq 0$. On suppose χ non trivial. Le niveau N_χ de $f \otimes \chi$ est donné par la recette suivante [2]. On a

$$v_p(N_\chi) = v_p(m_\chi m_{\psi\chi}).$$

On a [2] théorème 4.1,

$$w_S(f \otimes \chi) = \bar{\psi}_{\bar{S}}(m_\chi)\chi(-1)\frac{\tau(\psi_S\chi)}{\tau(\bar{\chi})}$$

si $v_p(m_\chi) \geq v_p(N)$ et $v_p(m_{\chi\psi}) = v_p(m_\chi)$. On a de plus, [2] théorème 4.2,

$$w_S(f \otimes \chi) = \bar{\psi}_{\bar{S}}(m_\chi)\chi(-1)\left(\frac{N_S}{m_\chi}\right)^{k/2-1}\frac{\tau(\psi_S\chi)}{a_{N_S/m_\chi}(f)\tau(\bar{\chi})}$$

si $v_p(m_\chi) < v_p(N)$. On a enfin, [2] corollaire 4.2,

$$w_S(f \otimes \chi) = \bar{\psi}_{\bar{S}}(m_\chi)\chi(-1)\frac{\tau(\psi_S\chi)}{\tau(\bar{\chi})}$$

si $v_p(m_\chi) = v_p(N)$ et $v_p(m_{\chi\psi}) < v_p(m_\chi)$.

2.8 Invariants locaux des tordues de formes modulaires, cas supercuspidal

Reprenons la situation laissée en 2.6 en supposant $a_p = 0$. On a

$$v_p(N_\chi) = \max\left(v_p\left(m_\chi^2\right), v_p(m_\psi m_\chi), v_p(N)\right)$$

Posons au préalable, pour tout caractère de Dirichlet primitif ω $m'_\omega = m_\omega$ si $\omega(p) \neq 0$ et $m'_\omega = pm_\omega$ si $\omega(p) = 0$ On a, [2] théorème 4.1,

$$w_S(f \otimes \chi) = \bar{\psi}_{\bar{S}}(m_\chi)\chi(-1)\frac{\tau(\psi_S\chi)}{\tau(\bar{\chi})}$$

si $v_p(m_\chi) \geq v_p(N)$. On a enfin, [2] théorème 4.5,

$$w_S(f \otimes \chi) = w_S(f)\frac{\psi_{\bar{S}}(N_S/m_\chi)\chi(-1)\psi_S(-1)}{\left(N''_{\chi,S}/N_S\right)\phi(N_S/m_\chi)}\frac{1}{\tau(\bar{\chi})}\sum_\omega \tau(\omega)\tau(\chi\psi_S\omega)w_{f\otimes\bar{\psi}_S\bar{\omega}}$$

si $v_p(m_\chi) < v_p(N)$, où $N''_{\chi,S} = \max\left(N_S, N_Sm_\psi/m_\chi, N_S^2/m_\chi^2\right)$ et où ω parcourt les caractères de Dirichlet tels que $m'_\omega = N_S/m_\chi$ et $m_{\chi\psi_S\omega} = N''_{\chi,S}m_\chi/N_S$. En particulier, lorsque $v_p(m_\chi) > v_p(N)/2$, $w_S(f \otimes \chi)$ se déduit de la collection des $w_S(f \otimes \omega)$, pour ω caractère vérifiant $m_\omega^2|N_S$; en particulier, lorsque $v_p(m_\chi) > v_p(N)/2$, $w_S(f \otimes \chi)$ se déduit de la collection des $w_S(f \otimes \omega)$ pour $N_{\omega,S} = N_S$.

2.9 Invariants locaux des tordues de formes primitives par torsion, conclusion

Supposons f primitive par torsion (c'est-à-dire p-primitive par torsion pour tout nombre premier p). La donnée de $a_p(f)$ et de $w_S(f \otimes \omega)$ ($p \in \Sigma_N$, $S \subset \Sigma_N$, et ω caractère primitif tel que $N_\omega = N$ et $a_r(f) = 0$ ($r \in S$ et $r|m_\chi$)) détermine N_χ, $a_p(f \otimes \chi)$, $w_S(f \otimes \chi)$ ($p \in \Sigma_N$, $S \subset \Sigma_N$).

3 Le théorème 1 et ses corollaires

3.1 La démonstration du théorème 1

Soit $g = \left(\begin{smallmatrix} a & b \\ c & d \end{smallmatrix}\right) \in \mathrm{SL}_2(\mathbf{Z})$ telle que la classe modulo N de (c,d) soit égale à (u,v).

On peut comprendre notre démarche ainsi. La fonction $f_{|g}$ est une forme modulaire pour le groupe de congruence $\Gamma(N)$, si bien que la fonction $f_{|g\left(\begin{smallmatrix} N & 0 \\ 0 & 1 \end{smallmatrix}\right)}$ est modulaire pour le groupe de congruence $\Gamma_1(N)\cap\Gamma_0(N^2)$. Cette dernière forme modulaire s'écrit donc comme combinaison linéaire de fonctions du type $F_{\left|\begin{pmatrix} d & 0 \\ 0 & 1 \end{pmatrix}\right.}$, où F parcourt les formes primitives de niveau M divisant N^2 et d les entiers divisant N^2/M. Nous allons montrer que les formes primitives qui interviennent dans cette écriture sont de la forme $f \otimes \chi$, où χ parcourt les caractères de Dirichlet de niveau divisant N et donner explicitement les coefficients de cette combinaison linéaire.

Lorsque $k = 2$, on a $\xi_f(u,v) = \int_0^\infty f_{|g}(iy)\,dy$. Lorsque, de plus, $s = 1$ et h est une matrice diagonale de $\mathrm{PGL}_2(\mathbf{Q})^+$, et χ est un caractère de Dirichlet de conducteur divisant N, on a

$$\int_0^\infty (f \otimes \chi)_{|h}(iy)\,dy = \frac{1}{2\pi}L(f \otimes \chi, 1) = \frac{1}{\sqrt{N_\chi}}\Lambda(f \otimes \chi, 1).$$

C'est pourquoi le théorème 1 se déduit de la proposition 6 suivante, par intégration de chaque membre de l'égalité ci-dessous le long de la géodésique reliant 0 à ∞ dans le demi-plan de Poincaré.

Remarquons que la proposition 6 permet de démontrer des analogues du théorème 1 pour les formes modulaires de poids $\neq 2$. (Voir par exemple la thèse de F. Martin lorsque $k = 1$ [13].)

Proposition 6. *On a*

$$f_{|g} = \frac{w(f)}{\phi(N')} \sum_\chi \chi_{\bar{S}}(m_{\chi,s})(\bar{\psi}_S \bar{\chi}_S)(m_{\psi\chi,\bar{s}})(\psi_S \chi_S)(-1)\tau'(\chi_S)\tau'(\bar{\psi}_{\bar{S}}\bar{\chi}_{\bar{S}})$$

$$(\bar{\psi}_{\bar{S}}\bar{\chi}_{\bar{S}}^2)\,(N_{\chi,s})\overline{w_S(f \otimes \chi)}(\bar{\psi}\bar{\chi})\left(\frac{N'_{\bar{S}}v}{N_{\bar{S}}}\right)\chi\left(\frac{N'_S u}{N_S}\right)$$

$$(f \otimes \chi)_{\left|\left[\frac{\Sigma_{N'} - S(m_\chi)}{(S(N')-S(m_\chi))\cap\Sigma_f}, \frac{\Sigma_{N'} - S(m_{\bar{\psi}\bar{\chi}})}{(\bar{S}(N')-\bar{S}(m_{\bar{\psi}\bar{\chi}}))\cap\Sigma_f}\right]\right.}\begin{pmatrix} \frac{N'_{\bar{S}}}{N_{\chi,S}N_{\bar{S}}m_{\psi\chi,\bar{S}}} & 0 \\ 0 & \frac{N'_S}{N_S m_{\bar{\chi},S}} \end{pmatrix}(**)\,,$$

où χ parcourt les caractères de Dirichlet primitifs tel que $m_{\chi,s}m_{\psi\chi,\bar{s}}|N'$ et où

$$** = \prod_{p\in S(N')} Q_{p,f,\bar{\chi}}\begin{pmatrix} 1 & 0 \\ 0 & p \end{pmatrix}\prod_{p\in\bar{S}(N')} Q_{p,f,\chi\psi}\begin{pmatrix} p & 0 \\ 0 & 1 \end{pmatrix}.$$

Démonstration. Considérons $\begin{pmatrix} A & B \\ C & D \end{pmatrix} \in \mathrm{M}_2(\mathbf{Z})$ telle que $N_S N'_S|A, N_S N'_S|D$, $NN'|C, N_{\bar{S}}N'_{\bar{S}}|B, AD - BC = N_S N'_S, A \equiv uN'_S \pmod{N_{\bar{S}}}$ et $B \equiv v/N_{\bar{S}} \pmod{N_S}$. Soit $k \in \mathbf{Z}$ tel que $n \equiv uv \pmod{N_{\bar{S}}}$ et $n \equiv -uv \pmod{N_S}$. Notre point de départ réside dans l'identité

$$\Gamma_1(N)g = \Gamma_1(N)\begin{pmatrix} 0 & -1 \\ N & 0 \end{pmatrix}\begin{pmatrix} 1 & n/N \\ 0 & 1 \end{pmatrix}\begin{pmatrix} A & B \\ C & D \end{pmatrix}\begin{pmatrix} NN'_S & 0 \\ 0 & N_S \end{pmatrix}^{-1},$$

que le lecteur vérifiera grâce au lemme chinois. Comme $w(f)\bar{f} = f_{|\begin{pmatrix} 0 & 1 \\ -N & 0 \end{pmatrix}}$, on a la formule

$$f_{|g} = w(f)\bar{f}_{|\begin{pmatrix} 1 & n/N \\ 0 & 1 \end{pmatrix}\begin{pmatrix} A & B \\ C & D \end{pmatrix}\begin{pmatrix} NN'_S & 0 \\ 0 & N_S \end{pmatrix}^{-1}},$$

et donc, d'après la formule 5,

$$f_{|g} = w(f) \sum_{\omega} \frac{\omega(n')}{\phi(N')} \tau(\bar{\omega})(\bar{f} \otimes \omega)^{\left[\Sigma_{N'}, \frac{\Sigma_{N'} - \Sigma_{m_\omega}}{(\Sigma_{N'} - \Sigma_{m_\omega}) \cap \Sigma_f} \right]}_{|\prod_p P_p\left(\begin{pmatrix} p & 0 \\ 0 & 1 \end{pmatrix}\right)\begin{pmatrix} A & B \\ C & D \end{pmatrix}\begin{pmatrix} NN'_S & 0 \\ 0 & N_S \end{pmatrix}^{-1}}$$

$$(6)$$

où ω parcourt les caractères de Dirichlet primitifs de conducteur m_ω divisant N', le produit portant sur les nombres premiers divisant N'.

Appliquons les formules de 2.2 à $F = \bar{f} \otimes \omega$: on a $M = N_\omega$, $\psi' = \bar{\psi}\omega^2$ et $F \otimes \bar{\psi}'_S = \bar{f} \otimes \omega \psi_S \bar{\omega}_S^2 = \bar{f} \otimes \psi_S \bar{\omega}_S \omega_S = f \otimes \bar{\psi}_S \bar{\omega}_S \omega_S$.

Soit $p \in \Sigma_N$. Soit r un entier ≥ 0. On a

$$(\bar{f} \otimes \omega)_{|\begin{pmatrix} p^r & 0 \\ 0 & 1 \end{pmatrix}\begin{pmatrix} A & B \\ C & D \end{pmatrix}} = (\bar{f} \otimes \omega)_{|\begin{pmatrix} p^r A & B \\ C & D/p^r \end{pmatrix}\begin{pmatrix} 1 & 0 \\ 0 & p^r \end{pmatrix}}$$

si $p \in S$ et

$$(\bar{f} \otimes \omega)_{|\begin{pmatrix} p^r & 0 \\ 0 & 1 \end{pmatrix}\begin{pmatrix} A & B \\ C & D \end{pmatrix}} = (\bar{f} \otimes \omega)_{|\begin{pmatrix} A & p^r B \\ C/p^r & D \end{pmatrix}\begin{pmatrix} p^r & 0 \\ 0 & 1 \end{pmatrix}}$$

si $p \in \bar{S}$. On a de plus les formules

$$(\bar{f} \otimes \omega)_{|\begin{pmatrix} p^r A & B \\ C & D/p^r \end{pmatrix}} = (\psi_S \bar{\omega}_S^2)(B)(\psi_{\bar{S}} \bar{\omega}_{\bar{S}}^2)(p^r A/(N_S N'_S))$$

$$w_S(\bar{f} \otimes \omega)(\bar{f} \otimes \omega_{\bar{S}} \bar{\omega}_S \psi_S)_{|\begin{pmatrix} N_S N'_S / N_{\bar{\omega}, S} & 0 \\ 0 & 1 \end{pmatrix}}$$

lorsque $p^r \mid (N_S N'_S / N_{\omega, S})$ et

$$(\bar{f} \otimes \omega)_{|\begin{pmatrix} A & p^r B \\ C/p^r & D \end{pmatrix}} = (\psi_S \bar{\omega}_S^2)(p^r B)(\psi_{\bar{S}} \bar{\omega}_{\bar{S}}^2)(A/(N_S N'_S))$$

$$w_S(\bar{f} \otimes \omega)(\bar{f} \otimes \omega_{\bar{S}} \bar{\omega}_S \psi_S)_{|\begin{pmatrix} N_S N'_S / N_{\bar{\omega}, S} & 0 \\ 0 & 1 \end{pmatrix}}$$

lorsque $p^r \mid (N_{\bar{S}} N'_{\bar{S}} / N_{\omega, \bar{S}})$. Soit $P \in \mathbf{C}[X]$. On a alors

$$(\bar{f} \otimes \omega)_{|P\begin{pmatrix} p & 0 \\ 0 & 1 \end{pmatrix}\begin{pmatrix} A & B \\ C & D \end{pmatrix}} = (\psi_S \bar{\omega}_S^2)(B)(\psi_{\bar{S}} \bar{\omega}_{\bar{S}}^2)(A/(N_S N'_S)) w_S(\bar{f} \otimes \omega)$$

$$(\bar{f} \otimes \omega_{\bar{S}} \bar{\omega}_S \psi_S)_{|P\left((\psi_{\bar{S}} \bar{\omega}_{\bar{S}}^2)(p)\begin{pmatrix} 1 & 0 \\ 0 & p \end{pmatrix}\right)\begin{pmatrix} \frac{N_S N'_S}{N_{\bar{\omega}, S}} & 0 \\ 0 & 1 \end{pmatrix}}.$$

si $p \in S$ et P de degré $\leq v_p\left(N_S N'_S / N_{\omega,S}\right)$ et on a

$$(\bar{f} \otimes \omega)_{\vert P \begin{pmatrix} p & 0 \\ 0 & 1 \end{pmatrix} \begin{pmatrix} A & B \\ C & D \end{pmatrix}} = \left(\psi_S \bar{\omega}_S^2\right)(B) \left(\psi_{\bar{S}} \bar{\omega}_{\bar{S}}^2\right)\left(A/\left(N_S N'_S\right)\right) w_S(\bar{f} \otimes \omega)$$

$$(\bar{f} \otimes \omega_{\bar{S}} \bar{\omega}_S \psi_S)_{\vert P \left(\left(\psi_S \bar{\omega}_S^2\right)(p) \begin{pmatrix} p & 0 \\ 0 & 1 \end{pmatrix}\right) \begin{pmatrix} \frac{N_S N'_S}{N_{\bar{\omega},S}} & 0 \\ 0 & 1 \end{pmatrix}}.$$

si $p \in \bar{S}$ et P de degré $\leq v_p\left(N_{\bar{S}} N'_{\bar{S}} / N_{\omega,\bar{S}}\right)$.

On en déduit que

$$(\bar{f} \otimes \omega)^{[\{p\},\emptyset]}_{\vert \begin{pmatrix} A & B \\ C & D \end{pmatrix}} = \left(\psi_S \bar{\omega}_S^2\right)(B) \left(\psi_{\bar{S}} \bar{\omega}_{\bar{S}}^2\right)\left(A/\left(N_S N'_S\right)\right)$$

$$w_S(\bar{f} \otimes \omega)(\bar{f} \otimes \omega_{\bar{S}} \bar{\omega}_S \psi_S)^{[\emptyset,\{p\}]}_{\vert \begin{pmatrix} N_S N'_S / N_{\bar{\omega},S} & 0 \\ 0 & 1 \end{pmatrix}}$$

si et $p \in S$ et

$$(\bar{f} \otimes \omega)^{[\{p\},\emptyset]}_{\vert} \begin{pmatrix} A & B \\ C & D \end{pmatrix} = \left(\psi_S \bar{\omega}_S^2\right)(B) \left(\psi_{\bar{S}} \bar{\omega}_{\bar{S}}^2\right)\left(A/\left(N_S N'_S\right)\right)$$

$$w_S(\bar{f} \otimes \omega)(\bar{f} \otimes \omega_{\bar{S}} \bar{\omega}_S \psi_S)^{[\{p\},\emptyset]}_{\vert \begin{pmatrix} N_S N'_S / N_{\bar{\omega},S} & 0 \\ 0 & 1 \end{pmatrix}}$$

si $p \in \bar{S}$. Un calcul analogue donne les formules

$$(\bar{f} \otimes \omega)^{[\emptyset,\{p\}]}_{\vert \begin{pmatrix} A & B \\ C & D \end{pmatrix}} = \left(\psi_S \bar{\omega}_S^2\right)(B) \left(\psi_{\bar{S}} \bar{\omega}_{\bar{S}}^2\right)\left(A/\left(N_S N'_S\right)\right)$$

$$w_S(\bar{f} \otimes \omega)(\bar{f} \otimes \omega_{\bar{S}} \bar{\omega}_S \psi_S)^{[\{p\},\emptyset]}_{\vert \begin{pmatrix} N_S N'_S / N_{\bar{\omega},S} & 0 \\ 0 & 1 \end{pmatrix}}$$

si et $p \in S$ et

$$(\bar{f} \otimes \omega)^{[\emptyset,\{p\}]}_{\vert} \begin{pmatrix} A & B \\ C & D \end{pmatrix} = \left(\psi_S \bar{\omega}_S^2\right)(B) \left(\psi_{\bar{S}} \bar{\omega}_{\bar{S}}^2\right)\left(A/\left(N_S N'_S\right)\right)$$

$$w_S(\bar{f} \otimes \omega)(\bar{f} \otimes \omega_{\bar{S}} \bar{\omega}_S \psi_S)^{[\emptyset,\{p\}]}_{\vert \begin{pmatrix} N_S N'_S / N_{\bar{\omega},S} & 0 \\ 0 & 1 \end{pmatrix}}$$

si $p \in \bar{S}$.

Dans les quatre formules qui précèdent, on peut remplacer, partout où il intervient, le symbole $\{p\}$ par $\frac{\{p\}}{\{p\} \cap \Sigma_f}$.

Remarquons qu'on a, dans $\mathbf{C}(X)$, $Q_{p,f,\omega}(X) = X^{-v_p(N'/m_X)} P_p(X)$. On a

$$\prod_{p \in S(N')} P_p\left(\left(\psi_{\bar{S}} \bar{\omega}_{\bar{S}}^2\right)(p) \begin{pmatrix} 1 & 0 \\ 0 & p \end{pmatrix}\right) \prod_{p \in \bar{S}(N')} P_p\left(\left(\psi_{\bar{S}} \bar{\omega}_{\bar{S}}^2\right)(p) \begin{pmatrix} p & 0 \\ 0 & 1 \end{pmatrix}\right)$$

$$= \left(\psi_{\bar{S}} \bar{\omega}_{\bar{S}}^2\right)\left(N'_S / m_{\omega,S}\right) \left(\psi_{\bar{S}} \bar{\omega}_{\bar{S}}^2\right)\left(N'_{\bar{S}} / m_{\omega,\bar{S}}\right) \begin{pmatrix} N'_{\bar{S}} / m_{\omega,S} & 0 \\ 0 & N'_S / m_{\omega,\bar{S}} \end{pmatrix}$$

$$\times \prod_{p \in S(N')} Q_{p,f,\omega}\left(\left(\psi_{\bar{S}} \bar{\omega}_{\bar{S}}^2\right)(p) \begin{pmatrix} 1 & 0 \\ 0 & p \end{pmatrix}\right) \prod_{p \in \bar{S}(N')} Q_{p,f,\omega}\left(\left(\psi_{\bar{S}} \bar{\omega}_{\bar{S}}^2\right)(p) \begin{pmatrix} p & 0 \\ 0 & 1 \end{pmatrix}\right).$$

Revenons à la formule 6. On a

$$(\bar{f} \otimes \omega)^{\left[\Sigma_{N'}, \frac{\Sigma_{N'}-\Sigma_{m_\omega}}{(\Sigma_{N'}-\Sigma_{m_\omega})\cap\Sigma_f}\right]}_{\mid \prod_p P_p(\begin{pmatrix} p & 0 \\ 0 & 1 \end{pmatrix})\begin{pmatrix} A & B \\ C & D \end{pmatrix}\begin{pmatrix} NN'_S & 0 \\ 0 & N_S \end{pmatrix}^{-1}}$$

$$= (\psi_S\bar{\omega}_S^2)\left(\frac{N'_{\bar{S}}B}{m_{\omega,\bar{S}}}\right)(\psi_{\bar{S}}\bar{\omega}_{\bar{S}}^2)\left(\frac{A}{N_S m_{\omega,S}}\right)$$

$$w_S(\bar{f} \otimes \omega)(\bar{f} \otimes \omega_{\bar{S}}\bar{\omega}_S\psi_S)^{\left[\frac{\Sigma_{N'}-S(m_\chi)}{(S(N')-S(m_\chi))\cap\Sigma_f}, \frac{\Sigma_{N'}-\bar{S}(m_{\bar{\psi}_{\bar{\chi}}})}{(\bar{S}(N')-\bar{S}(m_{\bar{\psi}_{\bar{\chi}}}))\cap\Sigma_f}\right]}_{\mid \begin{pmatrix} \frac{N'_{\bar{S}}}{N_{\bar{\omega},S}N_{\bar{S}}m_{\omega,S}} & 0 \\ 0 & \frac{N'_S}{N_S m_{\omega,S}} \end{pmatrix}(**)}, \qquad (7)$$

où

$$** = \prod_{p\in S(N')} Q_{p,f,\omega}\left((\psi_{\bar{S}}\bar{\omega}_{\bar{S}}^2)(p)\begin{pmatrix} 1 & 0 \\ 0 & p \end{pmatrix}\right) \prod_{p\in\bar{S}(N')} Q_{p,f,\omega}\left((\psi_S\bar{\omega}_S^2)(p)\begin{pmatrix} p & 0 \\ 0 & 1 \end{pmatrix}\right).$$

Par ailleurs, on a les formules

$$(\psi_S\bar{\omega}_S^2)\left(N'_{\bar{S}}B\right) = (\psi_S\bar{\omega}_S^2)\left(N'_{\bar{S}}v/N_{\bar{S}}\right), (\psi_{\bar{S}}\bar{\omega}_{\bar{S}}^2)(A/N_S) = (\psi_{\bar{S}}\bar{\omega}_{\bar{S}}^2)(uN'_S/N_S)$$

et

$$\omega(n') = \omega(nN'/N) = \omega_S(nN'/N)\omega_{\bar{S}}(nN'/N) = \omega_S(-uvN'/N)\omega_{\bar{S}}(uvN'/N)$$

et donc

$$\omega(n') = \omega_S(-1)\omega\left(uN'_S/N_S\right)\omega\left(vN'_{\bar{S}}/N_{\bar{S}}\right).$$

On a donc la simplification

$$\omega(n')\left(\psi_S\bar{\omega}_S^2\right)\left(N'_{\bar{S}}B\right)(\psi_{\bar{S}}\bar{\omega}_{\bar{S}}^2)\left(\frac{A}{N_S}\right) = \omega_S(-1)\psi_S\bar{\omega}_S\omega_{\bar{S}}\left(\frac{N'_{\bar{S}}v}{N_{\bar{S}}}\right)\psi_{\bar{S}}\bar{\omega}_{\bar{S}}\omega_S\left(\frac{N'_S u}{N_S}\right).$$

De plus on a

$$Q_{p,f,\omega}\left((\psi_{\bar{S}}\bar{\omega}_{\bar{S}}^2)(p)\begin{pmatrix} 1 & 0 \\ 0 & p \end{pmatrix}\right) = Q_{p,f,\bar{\omega}_{\bar{S}}\omega_S\psi_S}\left(\begin{pmatrix} 1 & 0 \\ 0 & p \end{pmatrix}\right)$$

si $p \in S$ et

$$Q_{p,f,\omega}\left((\psi_S\bar{\omega}_S^2)(p)\begin{pmatrix} p & 0 \\ 0 & 1 \end{pmatrix}\right) = Q_{p,f,\omega_{\bar{S}}\bar{\omega}_S\psi_S}\left(\begin{pmatrix} p & 0 \\ 0 & 1 \end{pmatrix}\right)$$

si $p \in \bar{S}$.

En combinant avec la formule 7, on obtient

$$f_{|g} = \frac{w(f)}{\phi(N')} \sum_{\omega} \left(\bar\psi_{\bar S}\omega_{\bar S}^2\right)(m_{\omega,S})\left(\bar\psi_S\omega_S^2\right)(m_{\omega,\bar S})(\psi_S\bar\omega_S\omega_S)\left(\frac{N'_{\bar S}v}{N_{\bar S}}\right)(\psi_{\bar S}\bar\omega_{\bar S}\omega_{\bar S})\left(\frac{N'_S u}{N_S}\right)$$

$$\times \tau(\bar\omega)\omega_S(-1)w_S(\bar f\otimes\omega)(\bar f\otimes\omega_{\bar S}\bar\omega_S\psi_S)\left[\begin{smallmatrix}\frac{\Sigma_{N'}-S(m_\chi)}{(S(N')-S(m_\chi))\cap\Sigma_f},\frac{\Sigma_{N'}-\bar S(m_{\bar\psi\bar\chi})}{(\bar S(N')-\bar S(m_{\bar\psi\bar\chi}))\cap\Sigma_f}\end{smallmatrix}\right],$$
$$\left(\begin{array}{cc}\frac{N'_{\bar S}}{N_{\bar\omega,S}N_{\bar S}m_{\omega,\bar S}} & 0 \\ 0 & \frac{N'_S}{N_S m_{\omega,S}}\end{array}\right)(**),$$

$$(8)$$

où ω parcourt les caractères de Dirichlet primitifs de conducteur divisant N' et où

$$** = \prod_{p\in S(N')} Q_{p,f,\bar\omega_S\omega_S\psi_S}\begin{pmatrix}1 & 0 \\ 0 & p\end{pmatrix}\prod_{p\in \bar S(N')} Q_{p,f,\omega_{\bar S}\bar\omega_S\psi_S}\begin{pmatrix}p & 0 \\ 0 & 1\end{pmatrix}.$$

Simplifions encore cette formule. On a la relation entre sommes de Gauss

$$\tau(\bar\omega) = \bar\omega_S(m_{\bar\omega,\bar S})\bar\omega_{\bar S}(m_{\bar\omega,S})\tau(\bar\omega_S)\tau(\bar\omega_{\bar S}).$$

Cela donne

$$\tau(\bar\omega)\left(\bar\psi_{\bar S}\omega_{\bar S}^2\right)(m_{\omega,S})\left(\bar\psi_S\omega_S^2\right)(m_{\omega,\bar S}) = \tau(\bar\omega_S)\tau(\bar\omega_{\bar S})(\bar\psi_{\bar S}\omega_{\bar S})(m_{\omega,S})(\bar\psi_S\omega_S)(m_{\omega,\bar S}).$$

Récrivons 8 en notant χ le caractère de Dirichlet primitif associé à $\omega_{\bar S}\bar\omega_S\bar\psi_{\bar S}$. On a donc $\chi_S = \bar\omega_S$ et $\chi_{\bar S} = \omega_{\bar S}\bar\psi_{\bar S}$, $S(m_\omega) = S(m_\chi)$, $\bar S(m_\omega) = \bar S(m_{\psi\chi})$, $N_{\bar\omega,S} = N_{\chi,S}$ et $\omega_S(-1) = \chi_S(-1)$.

On obtient

$$f_{|g} = \frac{w(f)}{\phi(N')} \sum_{\chi} \tau'(\chi_S)\tau'(\bar\psi_{\bar S}\bar\chi_{\bar S})\chi_S(-1)\chi_{\bar S}\chi_{\bar S}(m_{\chi,S})(\bar\psi_S\chi_S)(m_{\psi\chi,\bar S})$$

$$(\bar\psi\bar\chi)\left(\frac{N'_{\bar S}v}{N_{\bar S}}\right)\chi\left(\frac{N'_S u}{N_S}\right)w_S(f\otimes\bar\psi_S\bar\chi_S\chi_S)(f\otimes\chi)\left[\begin{smallmatrix}\frac{\Sigma_{N'}-S(m_\chi)}{(S(N')-S(m_\chi))\cap\Sigma_f},\frac{\Sigma_{N'}-\bar S(m_{\bar\psi\bar\chi})}{(\bar S(N')-\bar S(m_{\bar\psi\bar\chi}))\cap\Sigma_f}\end{smallmatrix}\right],$$
$$\left(\begin{array}{cc}\frac{N'_{\bar S}}{N_{\chi,S}N_{\bar S}m_{\psi\chi,\bar S}} & 0 \\ 0 & \frac{N'_S}{N_S m_{\bar\chi,S}}\end{array}\right)(**)$$

$$(9)$$

où χ parcourt les caractères de Dirichlet primitifs de conducteur divisant N' et où

$$** = \prod_{p\in S(N')} Q_{p,f,\bar\chi}\begin{pmatrix}1 & 0 \\ 0 & p\end{pmatrix}\prod_{p\in \bar S(N')} Q_{p,f,\chi\psi}\begin{pmatrix}p & 0 \\ 0 & 1\end{pmatrix}.$$

Appliquons la relation 1 pour $F = f\otimes\chi$ (et donc $\psi' = \psi\chi^2$). On obtient

$$w_S(f\otimes\chi)w_S(f\otimes\bar\psi_S\bar\chi_S\chi_S) = \psi_S(-1)\left(\bar\psi_{\bar S}\bar\chi_{\bar S}^2\right)(N_{\chi,S}).$$

Cela permet de substituer $w_S(f\otimes\bar\psi_S\bar\chi_S\chi_S)$ dans 9 pour obtenir la proposition 6.

3.2 Réciproque du corollaire 2 et observations algorithmiques sur les aspects locaux

Le lecteur vérifiera sans peine que les termes qui apparaissent dans la formule du théorème 1 se déduisent des invariants dont la liste est donnée dans le corollaire 2.

Nous nous proposons dans cette section d'étudier la réciproque : comment la fonction ξ_f permet de retrouver les données (i), (ii), (iii), (iv), et (v). La procédure à suivre pour cette étude nous paraît plus agréable lorsque ξ_f est primitive par torsion.

a. Les invariants de f en termes de la fonction ξ_f

On obtient le caractère de nebentypus ψ de f grâce à la formule $\xi_f(\lambda u, \lambda v) = \bar{\psi}(\lambda)\xi_f(u,v)$ $(u,\, v \in (\mathbf{Z}/N\mathbf{Z}),\, \lambda \in (\mathbf{Z}/N\mathbf{Z})^*)$. On peut déterminer $a_p\xi_f$ (et donc $L_p(f, p^{-s})$) et $\xi W_S f$ lorsque p est un nombre premier et S un sous-ensemble de Σ_N grâce aux formules données dans [15], théorèmes 2 et 5.

b. Torsion de f par des caractères ω tels que $N = N_\omega$

Supposons f primitive par torsion.

Proposition 7. *Soit ω un caractère de Dirichlet primitif tel que $N = N_\omega$ et tel que pour tout nombre premier p divisant m_ω on a $a_p(f) = 0$. Soit $(u,v) \in (\mathbf{Z}/N\mathbf{Z})^2$ d'ordre N. Choisissons un sous-groupe cyclique C d'ordre m_ω de $(\mathbf{Z}/N\mathbf{Z})^2$ tel que $C \cap \mathbf{Z}(u,v) = \{0\}$. On a alors*

$$w(f)w(\bar{f} \otimes \bar{\omega})\xi_{f \otimes \omega}(u,v) = \frac{1}{\tau(\omega)} \sum_{(x,y) \in (u,v)+C} \omega((yu - xv)m/N)\xi_f(x,y).$$

Démonstration. Puisque f est primitive par torsion, et que $a_p(f) = 0$ pour tout nombre premier p divisant m_ω, on a $f \otimes \bar{\omega} = f_{\bar{\omega}}$. Appliquons la section 2.3,

$$f \otimes \bar{\omega} = f_{\bar{\omega}} = \frac{1}{\tau(\omega)}S_{\bar{\omega}}f = \frac{1}{\tau(\omega)}\sum_{t=0}^{m-1}\omega(t)f_{\left|\begin{pmatrix} 1 & t/m \\ 0 & 1 \end{pmatrix}\right.}.$$

On a donc, puisque $N_\omega = N$,

$$w(f)w(\bar{f} \otimes \bar{\omega})f \otimes \omega = \frac{1}{\tau(\omega)}\sum_{t=0}^{m-1}\omega(t)f_{\left|\begin{pmatrix} 0 & -1 \\ N & 0 \end{pmatrix}\begin{pmatrix} 1 & t/m \\ 0 & 1 \end{pmatrix}\begin{pmatrix} 0 & -1 \\ N & 0 \end{pmatrix}\right.}$$

$$= \frac{1}{\tau(\omega)}\sum_{t=0}^{m-1}\omega(t)f_{\left|\begin{pmatrix} 1 & 0 \\ -tN/m & 1 \end{pmatrix}\right.}.$$

Soit $\begin{pmatrix} a & b \\ c & d \end{pmatrix} x \in \mathrm{SL}_2(\mathbf{Z})$ telle que $(c,d) \in (u,v)$. On a

$$w(f)w(\bar{f} \otimes \bar{\omega})f \otimes \omega_{\left| \begin{pmatrix} a & b \\ c & d \end{pmatrix} \right.} = \frac{1}{\tau(\omega)} \sum_{t=0}^{m-1} \omega(t)f_{\left| \begin{pmatrix} 1 & 0 \\ -tN/m & 1 \end{pmatrix} \begin{pmatrix} a & b \\ c & d \end{pmatrix} \right.}.$$

Cela se traduit par la formule :

$$w(f)w(\bar{f} \otimes \bar{\omega})\xi_{f \otimes \omega}(u,v) = \frac{1}{\tau(\omega)} \sum_{t=0}^{m-1} \omega(t)\xi_f(c - atN/m, d - btN/m).$$

Posons $(x,y) = (c - atN/m, d - btN/m)$ dans $(\mathbf{Z}/N\mathbf{Z})^2$. On a alors la relation $xv - yu = -tN/m$. Lorsque t décrit les entiers de 0 à $m-1$, $(atN/m, btN/m)$ décrit bien un sous-groupe C comme indiqué dans l'énoncé de la proposition.

c. Les invariants locaux des tordues de f par des caractères χ tels que $N_\omega = N$

Soit ω un caractère de Dirichlet tel que $N_\omega = N$ et tel que pour tout nombre premier p divisant m_ω on a $a_p(f) = 0$. Supposons f primitive par torsion. L'étape **b.** permet de connaître la fonction $\xi_{f \otimes \omega}$ à multiplication par un scalaire près. Cela permet de déterminer le facteur $L_p(f \otimes \chi, p^{-s})$ et $w_S(f \otimes \chi)$ pour $p \in \Sigma_N$ et $S \subset \Sigma_N$ grâce encore à [15], théorèmes 2 et 5.

d. Les invariants locaux des tordues de $f \otimes \chi$ pour χ caractère quelconque

Supposons encore f primitive par torsion. Les invariants locaux de $f \otimes \chi$, pour χ caractère quelconque, se déduisent des invariants locaux de $f \otimes \omega$ pour ω parcourant les caractères tels que $N_\omega = N$, d'après la conclusion 2.9 des sections 2.6, 2.7, et 2.8.

e. Les nombres $\Lambda(f \otimes \chi, 1)$ pour χ caractère de niveau divisant N

Supposons encore f primitive par torsion. Après l'étape **d**, tous les termes qui interviennent dans la formule du théorème 1 sont déterminés, excepté les valeurs $\Lambda(f \otimes \chi, 1)$ pour χ caractère de niveau divisant N. Ces derniers nombres sont eux aussi déterminés par le théorème 1, il suffit de s'assurer qu'ils interviennent au moins une fois précédés d'un coefficient non nul. C'est bien le cas, si on choisit u et v tels que $N' = m_\chi$.

f. Que faire lorsque f n'est pas primitive par torsion ?

Indiquons succinctement comment on peut ramener le cas général (*i.e.* f n'est pas primitive par torsion) au cas où f est primitive par torsion. Pour cela il faut déterminer quel caractère χ_0 est tel que N_{χ_0} soit minimal, puis déterminer

N_{χ_0} et la fonction $\xi_{f \otimes \chi_0} : (\mathbf{Z}/N_{\chi_0}\mathbf{Z})^2 \to \mathbf{C}$. Pour cela on peut considérer tous les caractères χ de niveau divisant N et les formes modulaires f_χ qui sont propres pour presque tous les opérateurs de Hecke d'indice premiers. On peut déterminer les symboles de Manin associés à f_χ.

Le problème revient, alors à la question suivante. Soit F une forme modulaire propre pour presque tout opérateur de Hecke d'indice premier de forme primitive associée F_0. Étant donné ξ_F, comment trouver F_0 ? C'est possible si on connaît l'action des morphismes de dégénérescence sur les symboles de Manin. Les formules dans ce sens sont données dans [15], proposition 15, 16 et 17. Cette manipulation est moins agréable que celles que nous avons faites lorsque f est primitive par torsion.

3.3 Équations fonctionnelles et relations de Manin

Dans [10], Manin décrit deux familles d'équations fonctionnelles satisfaites par ξ_f (dites *relations de Manin*) :

$$\xi_f(u,v) + \xi_f(-v,u) = 0 \tag{10}$$

et

$$\xi_f(u,v) + \xi_f(v,-u-v) + \xi(-u-v,u) = 0 \tag{11}$$

$((u,v) \in (\mathbf{Z}/N\mathbf{Z})^2)$. Observons comment se traduit la relation 10 en utilisant la formule du théorème 1. Pour cela remarquons que le théorème 1 a la forme suivante :

$$\xi_f(u,v) = \sum_\chi c_{\chi,S} c_{\bar{\psi}\bar{\chi},\bar{S}} \chi_S(-1) \left(\bar{\psi}_{\bar{S}} \bar{\chi}_{\bar{S}}^2 \right) (N_{\chi,S}) \overline{w_S(f \otimes \chi)}$$

$$\Lambda \left[\frac{\Sigma_{N'} - S(m_\chi)}{(S(N') - S(m_\chi)) \cap \Sigma_f}, \frac{\Sigma_{N'} - \bar{S}(m_{\bar{\psi}\bar{\chi}})}{(\bar{S}(N') - \bar{S}(m_{\bar{\psi}\bar{\chi}})) \cap \Sigma_f} \right] (f \otimes \chi, 1)(\bar{\psi}\bar{\chi}) \left(\frac{N'_{\bar{S}} v}{N_{\bar{S}}} \right) \chi \left(\frac{N'_S u}{N_S} \right),$$

où χ parcourt les caractères de Dirichlet primitifs tels que $m_{\chi,S} m_{\psi\chi,\bar{S}} | N'$ et où $c_{\chi,S}$ et $c_{\bar{\psi}\bar{\chi},\bar{S}}$ dépendent de f, χ, N' et S mais pas de (u,v) et sont échangés par $(\chi, S) \mapsto (\bar{\chi}\bar{\psi}, \bar{S})$.

On peut appliquer le théorème 1 à $\xi(u,v)$ et $\xi(-v,u)$. L'identité 10 se traduit alors par une identification des χ-èmes termes pour chaque caractère χ. Après échanges simultanés de u et $-v$, de S et $\bar{S}$ et des termes correspondant à χ et $\bar{\psi}\bar{\chi}$ on retrouve alors l'équation fonctionnelle de la fonction $s \mapsto \Lambda(f \otimes \chi, s)$ en $s = 1$. Après vérification de la non nullité de suffisamment de coefficients (comme dans la section 3.2 étape **e**), on peut même établir qu'il y a équivalence entre la relation 10 et la totalité des équations fonctionnelle des $\Lambda(f \otimes \chi, s)$ en $s = 1$ pour χ parcourant les caractères de Dirichlet de conducteurs divisant N.

On peut se demander, en ayant à l'esprit les théorèmes inverses, comment interpréter la relation 11 en termes de fonctions L. Nous n'avons de réponse satisfaisante à cette question. Nous pouvons tout juste rappeler que les relations 10 et 11 sont équivalentes à 10 et

$$\xi_f(u,v) - \xi_f(v, u+v) - \xi(u+v, v) = 0 \qquad (12)$$

$((u,v) \in (\mathbf{Z}/N\mathbf{Z})^2)$. La famille de relations 12 a été mise en évidence par Lewis [7].

4 Produit de formes modulaires

4.1 Le produit scalaire de Petersson

Soit Γ un sous-groupe d'indice fini de $\mathrm{SL}_2(\mathbf{Z})$. Soient f_1 et f_2 deux formes modulaires paraboliques de poids 2 pour Γ. Rappelons que le produit scalaire de Petersson de f_1 et f_2 est donné par la formule :

$$\langle f_1, f_2 \rangle = \frac{1}{[\mathrm{SL}_2(\mathbf{Z}) : \Gamma]} \int_{D_\Gamma} f_1(z)\overline{f}_2(z) dx\, dy,$$

où D_Γ est un domaine fondamental pour Γ dans le demi-plan de Poincaré $\mathcal{H}$. Posons $\tau = \begin{pmatrix} 0 & -1 \\ 1 & -1 \end{pmatrix}$ et $\sigma = \begin{pmatrix} 0 & -1 \\ 1 & 0 \end{pmatrix}$. Posons de plus $\rho = e^{2i\pi/3} \in \mathcal{H}$. Soit R un système de représentants de $\Gamma\backslash\mathrm{SL}_2(\mathbf{Z})$.

Théorème 8. *On a*

$$\langle f_1, f_2 \rangle = \frac{1}{2i[\mathrm{SL}_2(\mathbf{Z}) : \Gamma]} \sum_{g \in R} \int_{g0}^{g\infty} f_1(z)\, dz \int_{gi}^{g\rho} \overline{f_2(z)\, dz},$$

et

$$\langle f_1, f_2 \rangle$$
$$= \frac{-i}{12[\mathrm{SL}_2(\mathbf{Z}) : \Gamma]} \sum_{g \in R} \int_{g\tau0}^{g\tau\infty} f_1(z)\, dz \int_{g0}^{g\infty} \overline{f_2(z)\, dz} - \int_{g0}^{g\infty} f_1(z)\, dz \int_{g\tau0}^{g\tau\infty} \overline{f_2(z)\, dz}.$$

Démonstration. Posons $\omega_1 = f_1(z)\, dz$ et $\omega_2 = f_2(z)\, dz$. Pour g dans $\mathrm{SL}_2(\mathbf{Z})$, posons $\omega_{j|g} = f_{j|g}\, dz$ $(j \in \{1, 2\})$. Considérons le domaine fondamental D_0 pour $\mathrm{SL}_2(\mathbf{Z})$ constitué par le triangle hyperbolique de sommets ∞, 0 et ρ. On a

$$\langle f_1, f_2 \rangle = \frac{1}{2i[\mathrm{SL}_2(\mathbf{Z}) : \Gamma]} \int_{D_\Gamma} \omega_1 \wedge \overline{\omega_2} = \frac{1}{2i[\mathrm{SL}_2(\mathbf{Z}) : \Gamma]} \sum_{g \in R} \int_{D_0} \omega_{1|g} \wedge \overline{\omega_{2|g}}.$$

Posons $F_g(z) = \int_\rho^z \overline{f_{2|g}(u)}\,du$. On a $df_{1|g}F_g(z)\,dz = \omega_1 \wedge \overline{\omega_2}$. Cela donne, par la formule de Stokes,

$$2i[\mathrm{SL}_2(\mathbf{Z}) : \Gamma]\,\langle f_1, f_2 \rangle = \sum_{g \in R} \int_{\partial D_0} f_{1|g}F_g(z)\,dz$$

$$= \sum_{g \in R} \int_\infty^0 f_{1|g}F_g(z)\,dz + \int_0^\rho f_{1|g}F_g(z) + \int_0^\rho f_{1|g}F_g(z).$$

Utilisons que σ est d'ordre 2 dans $\mathrm{PSL}_2(\mathbf{Z})$ et qu'on a $\tau\rho = \rho$ et $\tau\infty = 0$. Cela donne

$$2i[\mathrm{SL}_2(\mathbf{Z}) : \Gamma]\,\langle f_1, f_2 \rangle$$

$$= \frac{1}{2}\sum_{g \in R} \int_\infty^0 f_{1|g}F_g(z)\,dz + \int_\infty^0 f_{1|g\sigma}F_{g\sigma}(z)\,dz + \sum_{g \in R} \int_\infty^\rho f_{1|g}F_g(z)\,dz$$

$$+ \int_\rho^\infty f_{1|g\tau}F_{g\tau}(z)\,dz.$$

Utilisons la relation $F_{gh}(hz) = \int_{h^{-1}\rho}^z \overline{f_{2|g}(u)}\,du$. On obtient

$$2i[\mathrm{SL}_2(\mathbf{Z}) : \Gamma]\,\langle f_1, f_2 \rangle = \frac{1}{2}\sum_{g \in R} \int_\infty^0 \omega_{1|g} \int_{\sigma\rho}^\rho \overline{\omega_{2|g}} + \int_\rho^\infty \omega_{1|g\tau} \int_{\tau^2\rho}^\rho \overline{\omega_{2|g}}.$$

Le dernier terme est nul. Décomposons le deuxième facteur du premier terme. On a

$$2i[\mathrm{SL}_2(\mathbf{Z}) : \Gamma]\,\langle f_1, f_2 \rangle = \frac{1}{2}\sum_{g \in R} \int_0^\infty \omega_{1|g} \left(\int_{\sigma\rho}^{\sigma i} \overline{\omega_{2|g}} + \int_{\sigma i}^\rho \overline{\omega_{2|g}} \right).$$

Comme $\sigma i = i$, et comme $\int_0^\infty \omega_{1|g} \left(\int_{\sigma\rho}^{\sigma i} \overline{\omega_{2|g}} \right) = \int_0^\infty \omega_{1|g\sigma} \left(\int_i^\rho \overline{\omega_{2|g\sigma}} \right)$, on a la première formule du théorème.

Démontrons maintenant la deuxième formule. On a

$$2i[\mathrm{SL}_2(\mathbf{Z}) : \Gamma]\,\langle f_1, f_2 \rangle = \sum_{g \in R} \int_0^\infty \omega_{1|g} \int_i^\infty \overline{\omega_{2|g}} - \int_0^\infty \omega_{1|g} \int_\rho^\infty \overline{\omega_{2|g}}.$$

Calculons séparément les deux séries de termes. On a

$$\sum_{g \in R} \int_0^\infty \omega_{1|g} \int_i^\infty \overline{\omega_{2|g}} = \frac{1}{2}\sum_{g \in R} \int_0^\infty \omega_{1|g} \int_i^\infty \overline{\omega_{2|g}} + \int_0^\infty \omega_{1|g\sigma} \int_i^\infty \overline{\omega_{2|g\sigma}},$$

et comme $\sigma\infty = 0$,

$$\sum_{g \in R} \int_0^\infty \omega_{1|g} \int_i^\infty \overline{\omega_{2|g}} = \frac{1}{2}\sum_{g \in R} \int_0^\infty \omega_{1|g} \int_0^\infty \overline{\omega_{2|g}}.$$

Par ailleurs, on a

$$\int_0^\infty \omega_{1|g} \int_\rho^\infty \overline{\omega_{2|g}}$$

$$= \frac{1}{3} \sum_{g \in R} \int_0^\infty \omega_{1|g} \int_\rho^\infty \overline{\omega_{2|g}} + \int_0^\infty \omega_{1|g\tau} \int_\rho^\infty \overline{\omega_{2|g\tau}} + \int_0^\infty \omega_{1|g\tau^2} \int_\rho^\infty \overline{\omega_{2|g\tau^2}}.$$

Remarquons qu'on a $\int_0^\infty \omega_{1|g} + \int_0^\infty \omega_{1|g\tau} + \int_0^\infty \omega_{1|g\tau^2} = 0$. C'est pourquoi on a

$$\int_0^\infty \omega_{1|g} \int_\rho^\infty \overline{\omega_{2|g}}$$

$$= \frac{1}{3} \sum_{g \in R} \int_0^\infty \omega_{1|g\tau} \left(\int_\rho^\infty \overline{\omega_{2|g\tau}} - \int_\rho^\infty \overline{\omega_{2|g}} \right) + \int_0^\infty \omega_{1|g\tau^2} \left(\int_\rho^\infty \overline{\omega_{2|g\tau^2}} - \int_\rho^\infty \overline{\omega_{2|g}} \right).$$

Comme $\left(\int_\rho^\infty \overline{\omega_{2|g\tau}} - \int_\rho^\infty \overline{\omega_{2|g}} \right) = - \int_0^\infty \overline{\omega_{2|g}}$ et comme $\int_\rho^\infty \overline{\omega_{2|g\tau^2}} - \int_\rho^\infty \overline{\omega_{2|g}} = \int_0^\infty \overline{\omega_{2|g\tau^2}}$, on obtient, en posant $\lambda_j(g) = \int_0^\infty \omega_j$ $(j \in \{1,2\})$,

$$\int_0^\infty \omega_{1|g} \int_i^\rho \overline{\omega_{2|g}} = \frac{1}{2} \sum_{g \in R} \lambda_1(g)\overline{\lambda_2(g)} + \frac{1}{3} \sum_{g \in R} \lambda_1(g\tau)\overline{\lambda_2(g)} - \frac{1}{3} \sum_{g \in R} \lambda_1(g\tau)\overline{\lambda_2(g)}$$

$$= \frac{1}{6} \sum_{g \in R} \lambda_1(g)\overline{\lambda_2(g)} + \frac{1}{3} \sum_{g \in R} \lambda_1(g\tau)\overline{\lambda_2(g)}.$$

En utilisant la relation $\lambda_1(g) + \lambda_1(g\tau) + \lambda_1(g\tau^2) = 0$, on obtient finalement

$$\int_0^\infty \omega_{1|g} \int_i^\rho \overline{\omega_{2|g}} = \frac{1}{6} \sum_{g \in R} \lambda_1(g\tau)\overline{\lambda_2(g)} - \frac{1}{6} \sum_{g \in R} \lambda_1(g\tau)\overline{\lambda_2(g\tau^2)}.$$

Cela achève la démonstration.

Corollaire 9. *Supposons qu'on ait $\Gamma = \Gamma_1(N)$, on a*

$$\langle f_1, f_2 \rangle$$

$$= \frac{i}{12[\mathrm{SL}_2(\mathbf{Z}) : \Gamma_1(N)]} \sum_{(u,v) \in (\mathbf{Z}/N\mathbf{Z})^2} \xi_{f_1}(v, -u-v)\overline{\xi_{f_2}(u,v)} - \xi_{f_1}(u,v)\overline{\xi_{f_2}(v, -u-v)}.$$

Démonstration. C'est une application directe de la deuxième formule du théorème 8, en tenant compte de la formule $(j \in \{1,2\})$

$$\xi_{f_j}(u,v) = -i \int_{g0}^{g\infty} f_j(z) \, dz,$$

où $g = \begin{pmatrix} a & b \\ c & d \end{pmatrix} \in \mathrm{SL}_2(\mathbf{Z})$ vérifie $(c,d) \in (u,v)$.

4.2 La fonction L du carré tensoriel

Supposons la forme modulaire f de caractère de nebentypus trivial et de niveau N premier. Considérons la série de Dirichlet

$$L(f \otimes f, s) = \sum_{n=1}^{\infty} \frac{a_n^2}{n^s}.$$

Elle admet un prolongement méromorphe au plan complexe, avec un pôle simple en $s = 2$ et on a [16] (notre normalisation pour le produit scalaire de Petersson est en rapport $\pi/3 = \mathrm{vol}(\mathrm{SL}_2(\mathbf{Z})\backslash\mathcal{H})$ avec celle de [16])

$$\mathrm{Res}_{s=2} L(f \otimes f, s) = 12\pi \langle f, f \rangle.$$

Théorème 10. *On a*

$$\mathrm{Res}_{s=2} L(f \otimes f, s) = \frac{2\pi i}{(N+1)(N-1)^2} \sum_{\chi,\chi',\chi\chi'(-1)=-1} \frac{\Lambda(f \otimes \chi', 1)\Lambda(f \otimes \chi, 1)}{\tau(\chi\chi')},$$

où χ et χ' parcourent les caractères de Dirichlet primitifs de conducteur N tels que $\chi\chi'$ soit impair.

Démonstration. Partons de la formule donnée dans le corollaire 9. On a, puisque N est premier, $[\mathrm{SL}_2(\mathbf{Z}) : \Gamma_1(N)] = (N^2 - 1)$ et donc

$$\langle f, f \rangle = \frac{i}{12(N^2-1)} \sum_{(u,v)\in(\mathbf{Z}/N\mathbf{Z})^2} \xi_f(v,-u-v)\overline{\xi_f(u,v)} - \xi_f(u,v)\overline{\xi_f(v,-u-v)}.$$

Venons-en à la fonction ξ_f. Puisque le nebentypus de f est trivial, la fonction ξ_f est homogène. C'est pourquoi on pose, pour $u/v \in \mathbf{P}_1(\mathbf{Z}/N\mathbf{Z}) = \mathbf{Z}/N\mathbf{Z} \cup \{\infty\}$,

$$\xi_f(u/v) = \xi_f(u,v).$$

Cela permet d'écrire

$$\langle f, f \rangle = \frac{i}{12(N+1)} \sum_{x\in\mathbf{P}_1(\mathbf{Z}/N\mathbf{Z})} \xi_f\left(-\frac{1}{x+1}\right)\overline{\xi_f(x)} - \xi_f(x)\overline{\xi_f\left(-\frac{1}{x+1}\right)}.$$

Utilisons la relation de Manin $\xi_f(x) + \xi_f(-1/x) = 0$ ($x \in \mathbf{P}_1(\mathbf{Z}/N\mathbf{Z})$). On obtient

$$\langle f, f \rangle = \frac{i}{12(N+1)} \sum_{x\in\mathbf{P}_1(\mathbf{Z}/N\mathbf{Z})} \xi_f(x)\overline{\xi_f(x+1)} - \xi_f(x+1)\overline{\xi_f(x)}.$$

Le terme correspondant à $x = \infty$ est nul dans la somme qui précède. Par application des relations de Manin, on a les relations $\xi_f(1/0) + \xi_f(0/1) = 0$ et $\xi_f(1/0) + \xi_f(0/1) + \xi_f(-1/1) = 0$. On en déduit $\xi_f(-1) = 0$ et $\xi_f(1) = 0$.

C'est pourquoi on a également la nullité des termes correspondant à $x = 0$ et $x = -1$. On a donc, si on ne conserve que les termes correspondant à $x \neq 0$, $x \neq 1$ et $x \neq \infty$ et si on change de plus x en $-x$,

$$\langle f, f \rangle = \frac{i}{12(N+1)} \sum_{x \in (\mathbf{Z}/N\mathbf{Z})^* - \{1\}} \xi_f(-x)\overline{\xi_f(1-x)} - \xi_f(1-x)\overline{\xi_f(-x)}.$$

Comme N est premier, rendons plus explicite le théorème 1. On a, pour $x \in (\mathbf{Z}/N\mathbf{Z})^*$, $N' = N$ et on peut choisir $S = \emptyset$ et $\bar{S} = \{N\}$. Le terme associé au caractère χ dans la formule du théorème 1 est nul lorsque $\chi = 1$, car N est spécial pour f et on a que $\Lambda^{\left[\frac{\{N\}}{\emptyset}, \frac{\{N\}}{\{N\}}\right]}(f, 1)$ est multiple de $(1 - a_N)\Lambda(f, 1)$ qui est nul (en effet $a_N = -w(f)$ et $\Lambda(f, 1) = 0$ si $w(f) \neq -1$). Lorsque χ est non trivial, le terme associé à χ se simplifie car on a $\chi_S = 1$, $Q_{p,f,\chi} = 1$, $N_\chi = N^2$, $L_p(f \otimes \chi, X) = 1$. On obtient, pour $x \in \mathbf{Z}/N\mathbf{Z}^*$,

$$\xi_f(x) = \frac{w(f)}{N-1} \sum_{\chi \neq 1} \frac{\tau(\bar{\chi})}{N} \chi(x)\Lambda(f \otimes \chi, 1)$$

où la somme porte sur les caractères de Dirichlet primitifs de conducteur N.

Comme f est de nebentypus trivial, on a

$$\overline{\Lambda(f \otimes \chi, 1)} = \Lambda(f \otimes \bar{\chi}, 1).$$

On en déduit, en utilisant que $|w(f)| = 1$,

$\langle f, f \rangle$

$$= \frac{i}{12(N+1)(N-1)^2 N^2} \sum_{x \in (\mathbf{Z}/N\mathbf{Z})^* - \{1\}} \sum_{\chi, \chi'} \tau(\bar{\chi})\overline{\tau(\bar{\chi}')}\Lambda(f \otimes \chi, 1)\Lambda(f \otimes \bar{\chi}', 1))F(\chi, \chi'),$$

où

$$F(\chi, \chi') = \sum_{x \in (\mathbf{Z}/N\mathbf{Z})^* - \{1\}} (\chi(-x)\bar{\chi}'(1-x) - \chi(1-x)\bar{\chi}'(-x)).$$

Changeons χ' en $\bar{\chi}'$ dans la somme. On obtient, en tenant compte de l'égalité $\overline{\tau(\chi')} = \chi'(-1)\tau(\bar{\chi}')$,

$\langle f, f \rangle$

$$= \frac{i}{12(N+1)(N-1)^2 N^2} \sum_{\chi, \chi'} \tau(\bar{\chi})\tau(\bar{\chi}')\chi'(-1)\Lambda(f \otimes \chi', 1)\Lambda(f \otimes \chi, 1)F(\chi, \bar{\chi}').$$

Cette somme est antisymétrique (resp. symétrique) en χ et χ' lorsque $\chi'(-1) = \chi(-1)$ (resp. $\chi'(-1) = -\chi(-1)$). C'est pourquoi on a

$\langle f, f \rangle$

$$= \frac{i}{6(N+1)(N-1)^2 N^2} \sum_{\chi, \chi', \chi\chi'(-1)=-1} \tau(\bar{\chi})\tau(\bar{\chi}')\chi'(-1)\Lambda(f \otimes \chi', 1))\Lambda(f \otimes \chi, 1))E(\chi, \chi'),$$

où $E(\chi, \chi') = \sum_{x \in (\mathbf{Z}/N\mathbf{Z})^* - \{1\}} \chi(-x)\chi'(1-x)$. On reconnaît dans cette dernière expression une somme de Jacobi donnée par la formule

$$E(\chi, \chi') = \chi(-1)\frac{\tau(\chi)\tau(\chi')}{\tau(\chi\chi')}.$$

On obtient donc, en tenant compte de l'imparité de $\chi\chi'$,

$$\langle f, f \rangle$$

$$= \frac{i}{6(N+1)(N-1)^2 N^2} \sum_{\chi, \chi', \chi\chi'(-1)=-1} \frac{\tau(\chi)\tau(\chi')\tau(\bar{\chi})\tau(\bar{\chi}')}{\tau(\chi\chi')} \Lambda(f \otimes \chi', 1)\Lambda(f \otimes \chi, 1).$$

Utilisons les identités $\tau(\chi)\tau(\bar{\chi}) = \chi(-1)N = -\chi'(-1)N = -\tau(\chi')\tau(\bar{\chi}')$. On obtient

$$\langle f, f \rangle = \frac{i}{6(N+1)(N-1)^2} \sum_{\chi, \chi', \chi\chi'(-1)=-1} \frac{\Lambda(f \otimes \chi', 1)\Lambda(f \otimes \chi, 1)}{\tau(\chi\chi')}.$$

Cela donne la formule annoncée lorsque l'on combine avec l'identité rappelée avant l'énoncé du théorème 10.

Références

1. ATKIN A.O.L., LEHNER J. Hecke operators on $\Gamma_0(m)$. *Math. Ann.*, 185 :134–160, 1970.

2. ATKIN A.O.L., LI W. C. W. Twists of newforms and pseudo-eigenvalues of W-operators. *Invent. Math.*, 48(3) :221–243, 1978.

3. BREUIL C., CONRAD B., DIAMOND F., TAYLOR, R. On the modularity of elliptic curves over $\mathbf{Q}$: wild 3-adic exercises. *J. Amer. Math. Soc.*, 14(2) :489–549, 2000.

4. BRUNAULT F. *Sur la valeur en $s = 2$ de la fonction L d'une courbe elliptique.* Thèse. Université Denis Diderot, 2005.

5. CREMONA J. *Computations of modular elliptic curves and the Birch-Swinnerton-Dyer conjecture.* Cambridge University Press, 1992.

6. KATO K. p-adic Hodge theory and values of zeta functions of modular forms. In *Cohomologies p-adiques et applications arithmétiques. III*, number 295 in Astérisque, pages 117–290. Société Mathématiques de France, 2004.

7. LEWIS J. Spaces of holomorphic functions equivalent to the even Maass cusp forms. *Invent. Math.*, 127(2) :271–306, 1997.

8. LI W. C. W. On converse theorems for GL(2) and GL(1). *Amer. J. of Math.*, 103(5) :851–885, 1981.

9. LUO W., RAMAKRISHNAN D. Determination of modular forms by twists of critical L-values. *Invent. Math.*, 130(2) :371–398, 1997.

10. MANIN Y. Parabolic points and zeta-functions of modular curves. *Math. USSR Izvestija*, 6(1) :19–64, 1972.

11. MANIN Y. Explicit formulas for the eigenvalues of Hecke operators. *Acta arithmetica*, XXIV :239–249, 1973.

12. MANIN Y. Mathematics as metaphor. In *Proceedings of the International Congress of Mathematicians (Kyoto, 1990)*, volume II, pages 1665–1671. Math. Soc. Japan, Tokyo, 1991.

13. MARTIN F. *Périodes des formes modulaires de poids* 1. Thèse. Université Denis Diderot, 2001.

14. MEREL L. Opérateurs de Hecke pour $\Gamma_0(N)$ et fractions continues. *Ann. Inst. Fourier*, 41(3), 1991.

15. MEREL L. Universal Fourier expansions of modular forms. In Gerhard Frey, editor, *On Artin's conjecture for 2-dimensional, odd Galois representations*, number 1585 in Lecture Notes in Mathematics, pages 59–94. Springer Verlag, 1994.

16. SHIMURA G. The special values of the zeta functions associated with cusp forms. *Comm. Pure Appl. Math.*, 29(6) :783–804, 1976.

17. WEIL A. Über die Bestimmung Dirichletscher Reihen durch Funktional gleichungen. *Math. Ann.*, 168 :149–156, 1967.

Graph Complexes with Loops and Wheels

S.A. Merkulov

Department of Mathematics, Stockholm University, 10691 Stockholm, Sweden
sm@math.su.se

To Yuri Ivanovich Manin on his 70th birthday

Summary. Motivated by the problem of deformation quantization we introduce and study directed graph complexes with oriented loops and wheels – differential graded (dg) wheeled props. We develop a new technique for computing cohomology groups of such graph complexes and apply it to several concrete examples such as the wheeled completion of the operad of strongly homotopy Lie algebras and the wheeled completion of the dg prop of Poisson structures. The results lead to a new notion of a wheeled Poisson structure and to a new theorem on deformation quantization of arbitrary wheeled Poisson manifolds.

Key words: operads, props, deformation quantization, Poisson manifolds

2000 Mathematics Subject Classifications: 17B66, 18D50, 53D17, 53D55

1 Introduction

The first instances of graph complexes were introduced in the theory of operads and props, which have found recently many applications in algebra, topology, and geometry. Another set of examples was introduced by Kontsevich [Kon93, Kon02] as a way to expose highly nontrivial interrelations between certain infinite-dimensional Lie algebras and topological objects, including moduli spaces of curves, invariants of odd-dimensional manifolds, and the group of outer automorphisms of a free group.

Motivated by the problem of deformation quantization we introduce and study directed graph complexes with oriented loops and wheels. We show that universal quantizations of Poisson structures can be understood as morphisms of dg props, $Q : \mathsf{P}\langle \mathcal{D} \rangle \to \mathsf{Lie}^1\mathsf{B}_\infty^{\circlearrow}$, where

Y. Tschinkel and Y. Zarhin (eds.), *Algebra, Arithmetic, and Geometry*,
Progress in Mathematics 270, DOI 10.1007/978-0-8176-4747-6_10,
© Springer Science+Business Media, LLC 2009

- $P\langle\mathcal{D}\rangle$ is the dg free prop whose representations in a vector space V describe Maurer–Cartan elements of the Hochschild dg Lie algebra on $\mathcal{O}_V := \widehat{\odot}^\bullet V^*$ (see Section 2.7 for a precise definition), and
- $\mathsf{Lie}^1\mathsf{B}_\infty^{\circlearrowleft}$ is the *wheeled* completion of the minimal resolution, $\mathsf{Lie}^1\mathsf{B}_\infty$, of the prop, $\mathsf{Lie}^1\mathsf{B}$, of Lie 1-bialgebras; it is defined explicitly in Section 2.6 and is proven to have the property that its representations in a *finite-dimensional* vector space V correspond to Maurer–Cartan elements in the Schouten Lie algebra $\wedge^\bullet \mathcal{T}_V$, where $\mathcal{T}_V := \mathrm{Der}\mathcal{O}_V$.

In the theory of props one is most interested in those directed graph complexes that contain *no* loops and wheels. A major advance in understanding the cohomology groups of such complexes was recently accomplished in [Kon02, MV03, V] using key ideas of $\frac{1}{2}$prop and Koszul duality. In particular, these authors were able to compute cohomologies of directed versions (without loops and wheels though) of Kontsevich's ribbon graph complex and the "commutative" graph complex, and show that both are acyclic almost everywhere. This paper is an attempt to extend some of the results of [Kon02, MV03, V] to a more difficult situation in which the directed graphs are allowed to contain loops and wheels (i.e., directed paths that begin and end at the same vertex). In this case the answer differs markedly from the unwheeled case: we prove, for example, that while the cohomology of the wheeled extension, $\mathsf{Lie}_\infty^{\circlearrowleft}$, of the operad of Lie_∞-algebras remains acyclic almost everywhere (see Theorem 4.1.1 for a precise formula for $H^\bullet\left(\mathsf{Lie}_\infty^{\circlearrowleft}\right)$), the cohomology of the wheeled extension of the operad Ass_∞ gets more complicated. Both these complexes describe irreducible summands of directed "commutative" and, respectively, ribbon graph complexes with the restriction on absence of wheels dropped.

The wheeled complex $\mathsf{Lie}_\infty^{\circlearrowleft}$ is a subcomplex of the above-mentioned graph complex $\mathsf{Lie}^1\mathsf{B}_\infty^{\circlearrowleft}$, which plays a central role in deformation quantizations of Poisson structures. Using Theorem 4.1.1 on $H^\bullet\left(\mathsf{Lie}_\infty^{\circlearrowleft}\right)$ we show in Section 4.2 that a subcomplex of $\mathsf{Lie}^1\mathsf{B}_\infty^{\circlearrowleft}$ that is spanned by graphs with at most genus-1 wheels is also acyclic almost everywhere. However, this acyclicity breaks down for graphs with higher-genus wheels: we find an explicit cohomology class with three wheels in Section 4.2.4, which proves that the natural epimorphism $\mathsf{Lie}^1\mathsf{B}_\infty \to \mathsf{Lie}^1\mathsf{B}$ *fails* to stay a quasi-isomorphism when extended to the wheeled completions $\mathsf{Lie}^1\mathsf{B}_\infty^{\circlearrowleft} \to \mathsf{Lie}^1\mathsf{B}^{\circlearrowleft}$.

This paper grew out of the project on props and quantizations that was launched in 2004 with an attempt to create a prop profile of deformation quantization of Poisson structures (and continued on with [MMS] on minimal wheeled resolutions of the main classical operads and with [Mer06a] on a propic proof of a formality theorem associated with quantizations of Lie bialgebras). It is organized as follows. In Section 2 we recall some basic facts about props and graph complexes and describe a universal construction that associates dg props to a class of sheaves of dg Lie algebras on smooth formal manifolds, and apply that construction to the sheaf of polyvector fields and the sheaf of polydifferential operators, creating thereby associated dg props

Lie$^!$B$_\infty$ and, respectively, P$\langle \mathcal{D} \rangle$. In Section 3 we develop new methods for computing cohomology of directed graph complexes with wheels, and prove several theorems on cohomology of wheeled completions of minimal resolutions of dioperads. In Section 4 we apply these methods and results to compute cohomology of several concrete graph complexes. In Section 5 we use ideas of cyclic homology to construct a cyclic multicomplex computing cohomology of wheeled completions of dg operads.

A few words about our notation. The cardinality of a finite set I is denoted by $|I|$. The degree of a homogeneous element a of a graded vector space is denoted by $|a|$ (this should never lead to confusion). $\mathbb{S}_n$ stands for the group of all bijections $[n] \to [n]$, where $[n]$ denotes (here and everywhere) the set $\{1, 2, \ldots, n\}$. The set of positive integers is denoted by $\mathbb{N}^*$. If $V = \oplus_{i \in \mathbb{Z}} V^i$ is a graded vector space, then $V[k]$ is a graded vector space with $V[k]^i := V^{i+k}$.

We work throughout over a field k of characteristic 0.

2 Dg Props Versus Sheaves of dg Lie Algebras

2.1. Props. An $\mathbb{S}$-bimodule E is, by definition, a collection of graded vector spaces $\{E(m, n)\}_{m,n \geq 0}$ equipped with a left action of the group $\mathbb{S}_m$ and with a right action of $\mathbb{S}_n$ that commute with each other. For any graded vector space M, the collection $\mathsf{End}\langle M \rangle = \{\mathsf{End}\langle M \rangle(m, n) := \mathrm{Hom}(M^{\otimes n}, M^{\otimes m})\}_{m,n \geq 0}$ is naturally an $\mathbb{S}$-bimodule. A *morphism* of $\mathbb{S}$-bimodules $\phi : E_1 \to E_2$ is a collection of equivariant linear maps $\{\phi(m, n) : E_1(m, n) \to E_2(m, n)\}_{m,n \geq 0}$. A morphism $\phi : E \to \mathsf{End}\langle M \rangle$ is called a *representation* of an $\mathbb{S}$-bimodule E in a graded vector space M.

There are two natural associative binary operations on the $\mathbb{S}$-bimodule $\mathsf{End}\langle M \rangle$,

$$\otimes : \mathsf{End}\langle M \rangle(m_1, n_1) \otimes \mathsf{End}\langle M \rangle(m_2, n_2) \longrightarrow \mathsf{End}\langle M \rangle(m_1 + m_2, n_1 + n_2),$$
$$\circ : \mathsf{End}\langle M \rangle(p, m) \otimes \mathsf{End}\langle M \rangle(m, n) \longrightarrow \mathsf{End}\langle M \rangle(p, n),$$

and a distinguished element, the identity map $1 \in \mathsf{End}\langle M \rangle(1, 1)$.

Axioms of prop ("*products and permutations*") are modeled on the properties of $(\otimes, \circ, 1)$ in $\mathsf{End}\langle M \rangle$ (see [McL65]).

2.1.1. Definition. A *prop* E is an $\mathbb{S}$-bimodule $E = \{E(m, n)\}_{m,n \geq 0}$ equipped with the following data:

- a linear map called *horizontal composition*,

$$\otimes : E(m_1, n_1) \otimes E(m_2, n_2) \longrightarrow E(m_1 + m_2, n_1 + n_2),$$
$$\mathfrak{e}_1 \otimes \mathfrak{e}_2 \longrightarrow \mathfrak{e}_1 \otimes \mathfrak{e}_2,$$

such that $(\mathfrak{e}_1 \otimes \mathfrak{e}_2) \otimes \mathfrak{e}_3 = \mathfrak{e}_1 \otimes (\mathfrak{e}_2 \otimes \mathfrak{e}_3)$ and $\mathfrak{e}_1 \otimes \mathfrak{e}_2 = (-1)^{|\mathfrak{e}_1||\mathfrak{e}_1|} \sigma_{m_1,m_2}$ $(\mathfrak{e}_2 \otimes \mathfrak{e}_1) \sigma_{n_2,n_1}$ where σ_{m_1,m_2} is the following permutation in $\mathbb{S}_{m_1+m_2}$:

$$\begin{pmatrix} 1 & , \dots, & m_2 & , m_2+1, \dots, m_2+m_1 \\ 1+m_1, & \dots, m_2+m_1, & 1 & , \dots, & m_1 \end{pmatrix};$$

- a linear map called *vertical composition*,

$$\circ : E(p,m) \otimes E(m,n) \longrightarrow E(p,n),$$
$$\mathfrak{e}_1 \otimes \mathfrak{e}_2 \longrightarrow \mathfrak{e}_1 \circ \mathfrak{e}_2,$$

such that $(\mathfrak{e}_1 \circ \mathfrak{e}_2) \circ \mathfrak{e}_3 = \mathfrak{e}_1 \circ (\mathfrak{e}_2 \circ \mathfrak{e}_3)$ whenever both sides are defined;
- an algebra morphism $i_n : k[\mathbb{S}_n] \to (E(n,n), \circ)$ such that (i) for any $\sigma_1 \in \mathbb{S}_{n_1}$, $\sigma_2 \in \mathbb{S}_{n_2}$ one has $i_{n_1+n_2}(\sigma_1 \times \sigma_2) = i_{n_1}(\sigma_1) \otimes i_{n_2}(\sigma_2)$, and (ii) for any $\mathfrak{e} \in E(m,n)$ one has $1^{\otimes m} \circ \mathfrak{e} = \mathfrak{e} \circ 1^{\otimes n} = \mathfrak{e}$, where $1 := i_1(\mathrm{Id})$.

A morphism of props $\phi : E_1 \to E_2$ is a morphism of the associated $\mathbb{S}$-bimodules that respects, in the obvious sense, all the prop data.

A *differential* in a prop E is a collection of degree-1 linear maps $\{\delta : E(m,n) \to E(m,n)\}_{m,n \geq 0}$ such that $\delta^2 = 0$ and

$$\delta(\mathfrak{e}_1 \otimes \mathfrak{e}_2) = (\delta \mathfrak{e}_1) \otimes \mathfrak{e}_2 + (-1)^{|\mathfrak{e}_1|} \mathfrak{e}_1 \otimes \delta \mathfrak{e}_2,$$
$$\delta(\mathfrak{e}_3 \circ \mathfrak{e}_4) = (\delta \mathfrak{e}_3) \circ \mathfrak{e}_4 + (-1)^{|\mathfrak{e}_3|} \mathfrak{e}_3 \circ \delta \mathfrak{e}_4,$$

for any $\mathfrak{e}_1, \mathfrak{e}_2 \in E$ and any $\mathfrak{e}_3, \mathfrak{e}_4 \in E$ such that $\mathfrak{e}_3 \circ \mathfrak{e}_4$ makes sense. Note that $d1 = 0$.

For any dg vector space (M,d) the associated prop $\mathsf{End}\langle M \rangle$ has a canonically induced differential, which we always denote by the same symbol d.

A *representation* of a dg prop (E, δ) in a dg vector space (M, d) is, by definition, a morphism of props $\phi : E \to \mathsf{End}\langle M \rangle$ that commutes with differentials: $\phi \circ \delta = d \circ \phi$. (Here and elsewhere $\circ$ stands for the composition of maps; it will always be clear from the context whether $\circ$ stands for the composition of maps or for the vertical composition in props.)

2.1.2. Remark. If $\psi : (E_1, \delta) \to (E_2, \delta)$ is a morphism of dg props, and $\phi : (E_2, \delta) \to (\mathsf{End}\langle M \rangle, d)$ is a representation of E_2, then the composition $\phi \circ \psi$ is a representation of E_1. Thus representations can be "pulled back."

2.1.3. Free props. Let $\mathfrak{G}^\uparrow(m,n)$, $m, n \geq 0$, be the space of *directed (m,n)-graphs* G, that is, connected 1-dimensional CW complexes satisfying the following conditions:

(i) each edge (that is, 1-dimensional cell) is equipped with a direction;
(ii) if we split the set of all vertices (that is, 0-dimensional cells) that have exactly one adjacent edge into a disjoint union $V_{\mathrm{in}} \sqcup V_{\mathrm{out}}$, with V_{in} being the subset of vertices with the adjacent edge directed from the vertex and V_{out} the subset of vertices with the adjacent edge directed toward the vertex, then $|V_{\mathrm{in}}| \geq n$ and $|V_{\mathrm{out}}| \geq m$;

(iii) precisely n of vertices from V_{in} are labeled by $\{1, \ldots, n\}$ and are called *inputs*;

(iv) precisely m of vertices from V_{out} are labeled by $\{1, \ldots, m\}$ and are called *outputs*;

(v) there are no oriented *wheels*, i.e., directed paths that begin and end at the same vertex; in particular, there are no *loops* (oriented wheels consisting of one internal edge). Put another way, directed edges generate a continuous flow on the graph, which we always assume in our pictures to go from bottom to the top.

Note that $G \in \mathfrak{G}^{\uparrow}(m, n)$ may not be connected. Vertices in the complement,

$$v(G) := \overline{\text{inputs} \sqcup \text{outputs}},$$

are called *internal vertices*. For each internal vertex v we denote by $\text{In}(v)$ (resp., by $\text{Out}(v)$) the set of those adjacent half-edges whose orientation is directed toward (resp., from) the vertex. Input (resp., output) vertices together with adjacent edges are called input (resp., output) *legs*. The graph with one internal vertex, n input legs, and m output legs is called the (m, n)-*corolla*.

We set $\mathfrak{G}^{\uparrow} := \sqcup_{m,n} \mathfrak{G}^{\uparrow}(m, n)$.

The *free* prop $\mathsf{P}\langle E \rangle$ generated by an $\mathbb{S}$-module $E = \{E(m, n)\}_{m,n \geq 0}$ is defined by (see, e.g., [MV03, V])

$$\mathsf{P}\langle E \rangle(m, n) := \bigoplus_{G \in \mathfrak{G}^{\uparrow}(m,n)} \left(\bigotimes_{v \in v(G)} E(Out(v), In(v)) \right)_{AutG},$$

where

- $E(Out(v), In(v)) := \text{Bij}([m], Out(v)) \times_{\mathbb{S}_m} E(m, n) \times_{\mathbb{S}_n} \text{Bij}(In(v), [n])$ with Bij standing for the set of bijections,
- $\text{Aut}(G)$ stands for the automorphism group of the graph G.

An element of the summand above, $G\langle E \rangle := \left(\bigotimes_{v \in v(G)} E(Out(v), In(v)) \right)_{AutG}$, is often called a *graph G with internal vertices decorated by elements of E*, or just a *decorated graph*.

A differential δ in a free prop $\mathsf{P}\langle E \rangle$ is completely determined by its values,

$$\delta : E(Out(v), In(v)) \longrightarrow \mathsf{P}\langle E \rangle(|Out(v)|, |In(v)|),$$

on decorated corollas (whose unique internal vertex is denoted by v).

The prop structure on an $\mathbb{S}$-bimodule $E = \{E(m, n)\}_{m,n \geq 0}$ provides us, for any graph $G \in \mathfrak{G}^{\uparrow}(m, n)$, with a well-defined *evaluation* morphism of $\mathbb{S}$-bimodules,

$$\text{ev} : G\langle E \rangle \longrightarrow E(m, n).$$

In particular, if a decorated graph $C \in \mathsf{P}\langle E \rangle$ is built from two corollas $C_1 \in \mathfrak{G}(m_1, n_1)$ and $C_2 \in \mathfrak{G}(m_2, n_2)$ by gluing the jth output leg of C_2 to ith input leg of C_1, and if the vertices of these corollas are decorated, respectively, by elements $a \in E(m_1, n_1)$ and $b \in E(m_2, n_2)$, then we reserve a special notation,

$$a\ _i \circ_j b := ev(C) \in E(m_1 + m_2 - 1, n_1 + n_2 - 1),$$

for the resulting evaluation map.

2.1.4. Completions. Any free prop $\mathsf{P}\langle E \rangle$ is naturally a direct sum $\mathsf{P}\langle E \rangle = \oplus_{n \geq 0} \mathsf{P}_n \langle E \rangle$ of subspaces spanned by the number of vertices of the underlying graphs. Each summand $\mathsf{P}_n \langle E \rangle$ has a natural filtration by the genus g of the underlying graphs (which is, by definition, equal to the first Betti number of the associated CW complex). Hence each $\mathsf{P}_n \langle E \rangle$ can be completed with respect to this filtration. Similarly, there is a filtration by the number of vertices. We shall always work in this paper with free props completed with respect to these filtrations and hence use the same notation, $\mathsf{P}\langle E \rangle$, and the same name, *free prop*, for the completed version.

2.2. Dioperads and $\frac{1}{2}$props. A *dioperad* is an $\mathbb{S}$-bimodule, $E = \{E(m, n)\}_{\substack{m,n \geq 1 \\ m+n \geq 3}}$, equipped with a set of compositions

$$\left\{ \ _i \circ_j : E(m_1, n_1) \otimes E(m_2, n_2) \longrightarrow E(m_1 + m_2 - 1, n_1 + n_2 - 1) \right\}_{\substack{1 \leq i \leq n_1 \\ 1 \leq j \leq m_2}}$$

that satisfy the axioms imitating the properties of the compositions $\ _i \circ_j$ in a generic prop. We refer to [Gan03], where this notion was introduced, for a detailed list of these axioms. The *free dioperad* generated by an $\mathbb{S}$-bimodule E is given by

$$\mathsf{D}\langle E \rangle(m, n) := \bigoplus_{G \in \mathfrak{T}(m,n)} G\langle E \rangle,$$

where $\mathfrak{T}(m, n)$ is a subset of $\mathfrak{G}(m, n)$ consisting of connected trees (i.e., connected graphs of genus 0).

Another and less obvious reduction of the notion of prop was introduced by Kontsevich in [Kon02] and studied in detail in [MV03]: a $\frac{1}{2}$*prop* is an $\mathbb{S}$-bimodule $E = \{E(m, n)\}_{\substack{m,n \geq 1 \\ m+n \geq 3}}$ equipped with two sets of compositions,

$$\left\{ \ _1 \circ_j : E(m_1, 1) \otimes E(m_2, n_2) \longrightarrow E(m_1 + m_2 - 1, n_2) \right\}_{1 \leq j \leq m_2}$$

and

$$\left\{ \ _i \circ_1 : E(m_1, n_1) \otimes E(1, n_2) \longrightarrow E(m_1 + m_2 - 1, n_2) \right\}_{1 \leq i \leq n_1},$$

satisfying the axioms that imitate the properties of the compositions $\ _1 \circ_j$ and $\ _i \circ_1$ in a generic dioperad. The *free $\frac{1}{2}$prop* generated by an $\mathbb{S}$-bimodule E is given by

$$\frac{1}{2}\mathsf{P}\langle E \rangle(m, n) := \bigoplus_{G \in \frac{1}{2}\mathfrak{T}(m,n)} G\langle E \rangle.$$

where $\frac{1}{2}\mathfrak{T}(m,n)$ is a subset of $\mathfrak{T}(m,n)$ consisting of those directed trees that, for each pair of internal vertices (v_1, v_2) connected by an edge directed from v_1 to v_2 have either $|Out(v_1)| = 1$ or/and $|In(v_2)| = 1$. Such trees have at most one vertex v with $|Out(v)| \geq 2$ and $|In(v)| \geq 2$.

Axioms of the dioperad (resp., $\frac{1}{2}$prop) structure on an $\mathbb{S}$-bimodule E ensure that there is a well-defined evaluation map

$$\mathrm{ev} : G\langle E \rangle \longrightarrow E(m,n)$$

for each $G \in \mathfrak{T}(m,n)$ (resp., $G \in \frac{1}{2}\mathfrak{T}(m,n)$).

2.2.1. Free resolutions. A *free resolution* of a dg prop P is, by definition, a dg free prop $(\mathsf{P}\langle E \rangle, \delta)$ generated by some $\mathbb{S}$-bimodule E together with a morphism of dg props $\alpha : (\mathsf{P}\langle E \rangle, \delta) \to P$, which is a homology isomorphism.

If the differential δ in $\mathsf{P}\langle E \rangle$ is decomposable (with respect to the prop's vertical and /or horizontal compositions), then $\alpha : (\mathsf{P}\langle E \rangle, \delta) \to P$ is called a *minimal model* of P.

Similarly one defines free resolutions and minimal models $(\mathsf{D}\langle E \rangle, \delta) \to P$ and $\left(\frac{1}{2}\mathsf{P}\langle E \rangle, \delta\right) \to P$ of dioperads and $\frac{1}{2}$props.

2.2.2. Forgetful functors and their adjoints. There is an obvious chain of forgetful functors $\mathsf{Prop} \longrightarrow \mathsf{Diop} \longrightarrow \frac{1}{2}\mathsf{Prop}$. Let

$$\Omega_{\frac{1}{2}\mathsf{P}\to\mathsf{D}} : \frac{1}{2}\mathsf{Prop} \longrightarrow \mathsf{Diop}, \quad \Omega_{\mathsf{D}\to\mathsf{P}} : \mathsf{Diop} \longrightarrow \mathsf{Prop}, \quad \Omega_{\frac{1}{2}\mathsf{P}\to\mathsf{P}} : \frac{1}{2}\mathsf{Prop} \longrightarrow \mathsf{Prop},$$

be the associated adjoints. The main motivation behind introducing the notion of $\frac{1}{2}$prop is the very useful fact that the functor $\Omega_{\frac{1}{2}\mathsf{P}\to\mathsf{P}}$ is exact [Kon02,MV03], i.e., it commutes with the cohomology functor, which in turn is due to the fact that for any $\frac{1}{2}$prop P, there exists a kind of PBW lemma that represents $\Omega_{\frac{1}{2}\mathsf{P}\to\mathsf{P}}\langle P \rangle$ as a vector space *freely* generated by a family of decorated graphs

$$\Omega_{\frac{1}{2}\mathsf{P}\to\mathsf{P}}\langle P \rangle(m,n) := \bigoplus_{G \in \overline{\mathfrak{G}}(m,n)} G\langle E \rangle,$$

where $\overline{\mathfrak{G}}(m,n)$ is a subset of $\mathfrak{G}(m,n)$ consisting of so-called *reduced* graphs G that satisfy the following defining property [MV03]: for each pair of internal vertices (v_1, v_2) of G that are connected by a single edge directed from v_1 to v_2 one has $|Out(v_1)| \geq 2$ and $|In(v_2)| \geq 2$. The prop structure on $\Omega_{\frac{1}{2}\mathsf{P}\to\mathsf{P}}\langle P \rangle$ is given by

(i) horizontal compositions := disjoint unions of decorated graphs,
(ii) vertical compositions := graftings followed by $\frac{1}{2}$prop compositions of all those pairs of vertices (v_1, v_2) that are connected by a single edge directed from v_1 to v_2 and have either $|Out(v_1)| = 1$ or/and $|In(v_2)| = 1$ (if there are any).

2.3. Graph complexes with wheels. Let $\mathfrak{G}^{\circlearrowleft}(m,n)$ be the set of all directed (m,n)-graphs that satisfy conditions 2.1.3(i)–(iv), and set $\mathfrak{G}^{\circlearrowleft} := \sqcup_{m,n}\mathfrak{G}(m,n)$. A vertex (resp., edge or half-edge) of a graph $G \in \mathfrak{G}^{\circlearrowleft}$ that belongs to an oriented wheel is called a *cyclic* vertex (resp., edge or half-edge).

Note that for each internal vertex of $G \in \mathfrak{G}^{\circlearrowleft}(m,n)$ there is still a well-defined separation of adjacent half-edges into input and output ones, as well as a well-defined separation of legs into input and output ones.

For any $\mathbb{S}$-bimodule $E = \{E(m,n)\}_{m,n\geq 0}$, we define an $\mathbb{S}$-bimodule

$$\mathsf{P}^{\circlearrowleft}\langle E\rangle(m,n) := \bigoplus_{G\in\mathfrak{G}(m,n)} \left(\bigotimes_{v\in v(G)} E(Out(v), In(v))\right)_{AutG},$$

and notice that $\mathsf{P}^{\circlearrowleft}\langle E\rangle$ has a natural prop structure with respect to disjoint union and grafting of graphs. Clearly, this prop contains the free prop $\mathsf{P}\langle E\rangle$ as a natural subprop.

A *derivation* in $\mathsf{P}^{\circlearrowleft}\langle E\rangle$ is, by definition, a collection of linear maps $\delta : \mathsf{P}^{\circlearrowleft}\langle E\rangle(m,n) \to \mathsf{P}^{\circlearrowleft}\langle E\rangle(m,n)$ such that for any $G \in \mathfrak{G}$ and any element of $\mathsf{P}^{\circlearrowleft}\langle E\rangle(m,n)$ of the form

$$\mathfrak{e} = \operatorname*{coequalizer}_{\text{orderings of } v(G)}\ \left(\mathfrak{e}_1 \otimes \mathfrak{e}_2 \otimes \cdots \otimes \mathfrak{e}_{|v(G)|}\right),$$

with $\mathfrak{e}_k \in E(Out(v_k), In(v_k))$ for $1 \leq k \leq |v(G)|$, one has

$$\delta\mathfrak{e} = \operatorname*{coequalizer}_{\text{orderings of } v(G)} \left(\sum_{k=1}^{|v(G)|} (-1)^{|\mathfrak{e}_1|+\cdots+|\mathfrak{e}_{k-1}|}\mathfrak{e}_1 \otimes \cdots \otimes \delta\mathfrak{e}_k \otimes \cdots \otimes \mathfrak{e}_{|v(G)|}\right).$$

Put another way, a graph derivation is completely determined by its values on decorated corollas

$$a \in E(m,n),$$

that is, by linear maps

$$\delta : E(m,n) \longrightarrow \mathsf{P}^{\circlearrowleft}\langle E\rangle(m,n).$$

A *differential* in $\mathsf{P}^{\circlearrowleft}\langle E\rangle$ is, by definition, a degree-1 derivation δ satisfying the condition $\delta^2 = 0$.

2.3.1. Remark. If $(\mathsf{P}\langle E\rangle, \delta)$ is a free dg prop generated by an $\mathbb{S}$-bimodule E, then δ extends naturally to a differential on $\mathsf{P}^{\circlearrowleft}\langle E\rangle$, which we denote by the same symbol δ. It is worth pointing out that such an induced differential may *not* preserve the number of oriented wheels. For example, if δ applied to an element $a \in E(m,n)$ (which we identify with the a-decorated (m,n)-corolla) contains a summand of the form

$$\delta\left(\!\!\begin{array}{c} {}^{i_1\ i_2}\!\cdots \\ \boxed{a} \\ {}_{j_1\ j_2}\cdots \end{array}\!\!\right) = \cdots + \begin{array}{c} \cdots \\ \boxed{c} \\ {}^{j_1}\!\cdots \\ {}^{i_1\ i_2}\!\cdots\!\cdots \\ \boxed{b} \\ {}_{j_2}\cdots \end{array} + \cdots,$$

then the value of δ on the graph obtained from this corolla by gluing output i_1 to input j_1 into a loop,

$$\delta\left(\!\!\begin{array}{c} {}^{i_2}\cdots \\ \bigcirc\!\boxed{a} \\ {}_{j_2} \end{array}\!\!\right) = \cdots + \begin{array}{c} \cdots \\ \boxed{c} \\ {}^{i_2}\!\cdots\!\cdots \\ \boxed{b} \\ {}_{j_2} \end{array} + \cdots,$$

contains a term with no oriented wheels at all. Thus propic differential can, in general, *decrease* the number of wheels. Notice in this connection that if δ is induced on $\mathsf{P}^{\circlearrowleft}\langle E\rangle$ from the minimal model of a $\frac{1}{2}$prop, then such summands are impossible, and hence the differential preserves the number of wheels.

The vector spaces $\mathsf{P}\langle E\rangle$ and $\mathsf{P}^{\circlearrowleft}\langle E\rangle$ have a natural positive gradation

$$\mathsf{P}\langle E\rangle = \bigoplus_{k\geq 1} \mathsf{P}_k\langle E\rangle, \quad \mathsf{P}^{\circlearrowleft}\langle E\rangle = \bigoplus_{k\geq 1} \mathsf{P}^{\circlearrowleft}\langle E\rangle,$$

by the number k of internal vertices of the underlying graphs. In particular, $\mathsf{P}_1\langle E\rangle(m,n)$ is spanned by decorated (m,n)-corollas and can be identified with $E(m,n)$.

2.3.2. Representations of $\mathsf{P}^{\circlearrowleft}\langle E\rangle$. Any representation $\phi : E \to \mathrm{End}\langle M\rangle$ of an $\mathbb{S}$-bimodule E in a finite-dimensional vector space M can be naturally extended to representations of props $\mathsf{P}\langle E\rangle \to \mathrm{End}\langle M\rangle$ and $\mathsf{P}^{\circlearrowleft}\langle E\rangle \to \mathrm{End}\langle M\rangle$. In the latter case, decorated graphs with oriented wheels are mapped into appropriate traces.

2.3.3. Remark. The prop structure on an $\mathbb{S}$-bimodule $E = \{E(m,n)\}$ can be defined as a family of evaluation linear maps

$$\mu_G : G\langle E\rangle \longrightarrow E(m,n), \quad \forall\, G \in \mathfrak{G}^{\uparrow},$$

satisfying a certain associativity axiom (cf. Section 2.1.3). Analogously, one can define a *wheeled prop structure* on E as a family of linear maps

$$\mu_G : G\langle E\rangle \longrightarrow E(m,n), \quad \forall\, G \in \mathfrak{G}^{\circlearrowleft},$$

such that

(i) $\mu_{(m,n)-\text{corolla}} = \text{Id}$,
(ii) $\mu_G = \mu_{G/H} \circ \mu_H$ for every subgraph $H \in \mathfrak{G}^{\circlearrowleft}$ of G, where G/H is obtained from G by collapsing to a single vertex every connected component of H, and $\mu_H : G\langle E \rangle \to G/H\langle E \rangle$ is the evaluation map on the subgraph H and the identity on its complement.

Claim. For every finite-dimensional vector space M, the associated endomorphism prop $\text{End}\langle M \rangle$ *has a natural structure of a wheeled prop.*

The notion of representation of $\mathsf{P}^{\circlearrowleft}\langle E \rangle$ in a finite-dimensional vector space M introduced above is just a morphism of *wheeled* props $\mathsf{P}^{\circlearrowleft}\langle E \rangle \to \text{End}\langle M \rangle$. We shall discuss these issues in detail elsewhere, since in the present paper we are most interested in computing cohomology of *free* dg wheeled props $(\mathsf{P}^{\circlearrowleft}\langle E \rangle, \delta)$, where the composition maps μ_G are tautological.

2.4. Formal graded manifolds. For a finite-dimensional vector space M, we denote by $\mathcal{M}$ the associated formal graded manifold. The distinguished point of the latter is always denoted by $*$. The structure sheaf $\mathcal{O}_{\mathcal{M}}$ is (non-canonically) isomorphic to the completed graded symmetric tensor algebra $\hat{\odot}M^*$. A choice of a particular isomorphism $\phi : \mathcal{O}_{\mathcal{M}} \to \hat{\odot}M^*$ is called a choice of a local coordinate system on $\mathcal{M}$. If $\{e_\alpha\}_{\alpha \in I}$ is a basis in M and $\{t^\alpha\}_{\alpha \in I}$ the associated dual basis in M^*, then one may identify $\mathcal{O}_{\mathcal{M}}$ with the graded commutative formal power series ring $\mathbb{R}[[t^\alpha]]$.

Free modules over the ring $\mathcal{O}_{\mathcal{M}}$ are called locally free sheaves (= vector bundles) on $\mathcal{M}$. The $\mathcal{O}_{\mathcal{M}}$-module $\mathcal{T}_{\mathcal{M}}$ of derivations of $\mathcal{O}_{\mathcal{M}}$ is called the tangent sheaf on $\mathcal{M}$. Its dual $\Omega^1_{\mathcal{M}}$ is called the cotangent sheaf. One can form their (graded skew-symmetric) tensor products such as the sheaf of polyvector fields $\wedge^\bullet \mathcal{T}_{\mathcal{M}}$ and the sheaf of differential forms $\Omega^\bullet_{\mathcal{M}} = \wedge^\bullet \Omega^1_{\mathcal{M}}$. The first sheaf is naturally a sheaf of Lie algebras on $\mathcal{M}$ with respect to the Schouten bracket.

One can also define a sheaf of polydifferential operators, $\mathcal{D}_{\mathcal{M}} \subset \oplus_{i \geq 0}$ $\text{Hom}_{\mathbb{R}} (\mathcal{O}_{\mathcal{M}}^{\otimes i}, \mathcal{O}_{\mathcal{M}})$. The latter is naturally a sheaf of dg Lie algebras on $\mathcal{M}$ with respect to the Hochschild differential d_H and brackets $[\ ,\ ]_H$.

2.5. Geometry $\Rightarrow$ graph complexes. We shall sketch here a simple trick that associates a dg prop $\mathsf{P}^{\circlearrowleft}\langle E_{\mathcal{G}} \rangle$ generated by a certain $\mathbb{S}$-bimodule $E_{\mathcal{G}}$ to a sheaf of dg Lie algebras $\mathcal{G}_{\mathcal{M}}$ over a smooth graded formal manifold $\mathcal{M}$.

We assume that

(i) $\mathcal{G}_{\mathcal{M}}$ is built from direct sums and tensor products of (any order) jets of the sheaves $\mathcal{T}_{\mathcal{M}}^{\otimes \bullet} \otimes \Omega^1_{\mathcal{M}}{}^{\otimes \bullet}$ and their duals (thus $\mathcal{G}_{\mathcal{M}}$ can be defined for *any* formal smooth manifold $\mathcal{M}$, i.e., its definition does not depend on the dimension of $\mathcal{M}$);
(ii) the differential and the Lie bracket in $\mathcal{G}_M$ can be represented, in a local coordinate system, by polydifferential operators and natural contractions between duals.

The motivating examples are $\wedge^{\bullet}\mathcal{T}_{\mathcal{M}}$, $\mathcal{D}_{\mathcal{M}}$, and the sheaf of Nijenhuis dg Lie algebras on $\mathcal{M}$ (see [Mer05]).

By assumption (i), a choice of a local coordinate system on $\mathcal{M}$ identifies $\mathcal{G}_{\mathcal{M}}$ with a subspace in

$$\bigoplus_{p,m\geq 0} \hat{\odot}^{\bullet} M^* \otimes \mathrm{Hom}(M^{\otimes p}, M^{\otimes m}) = \prod_{p,q,m\geq 0} \mathrm{Hom}(\odot^p M \otimes M^{\otimes q}, M^{\otimes m})$$

$$\subset \prod_{m,n\geq 0} \mathrm{Hom}(M^{\otimes n}\, M^{\otimes m}).$$

Let Γ be a degree-1 element in $\mathcal{G}_{\mathcal{M}}$. Denote by $\Gamma_{p,q}^m$ the bit of Γ that lies in $\mathrm{Hom}(\odot^p M \otimes M^{\otimes q}, M^{\otimes m})$ and set $\Gamma_n^m := \oplus_{p+q=n}\Gamma_{p,q}^m \in \mathrm{Hom}(M^{\otimes n}\, M^{\otimes m})$.

There exists a uniquely defined finite-dimensional S-bimodule $E_{\mathcal{G}} = \{E_{\mathcal{G}}(m,n)\}_{m,n\geq 0}$ whose representations in the vector space M are in one-to-one correspondence with Taylor components $\Gamma_n^m \in \mathrm{Hom}(M^{\otimes n}\, M^{\otimes m})$ of a degree-1 element Γ in $\mathcal{G}_{\mathcal{M}}$. Set $\mathsf{P}^{\circlearrowleft}\langle \mathcal{G}\rangle := \mathsf{P}^{\circlearrowleft}\langle E_{\mathcal{G}}\rangle$ (see Section 2.3).

Next we employ the dg Lie algebra structure in $\mathcal{G}_{\mathcal{M}}$ to introduce a differential δ in $\mathsf{P}^{\circlearrowleft}\langle \mathcal{G}\rangle$. The latter is completely determined by its restriction to the subspace of $\mathsf{P}_1^{\circlearrowleft}\langle \mathcal{G}\rangle$ spanned by decorated corollas (without attached loops).

First we replace the Taylor coefficients Γ_n^m of the section Γ by the decorated (m,n)-corollas

- with the unique internal vertex decorated by a basis element, $\{\mathbf{e}_r\}_{r\in J}$, of $E_{\mathcal{G}}(m,n)$,
- with input legs labeled by basis elements $\{e_\alpha\}$ of the vector space M and output legs labeled by the elements of the dual basis $\{t^\beta\}$.

Next we consider a formal linear combination

$$\overline{\Gamma}_n^m = \sum_r \sum_{\substack{\alpha_1,\dots,\alpha_n \\ \beta_1,\dots,\beta_m}} \raisebox{-1em}{\begin{picture}(0,0)\end{picture}} \quad t^{\alpha_1} \otimes \dots \otimes t^{\alpha_n} \otimes e_{\beta_1} \otimes \dots \otimes e_{\beta_m}$$

$$\in \mathsf{P}_1^{\circlearrowleft}\langle \mathcal{G}\rangle \otimes \mathrm{Hom}(M^{\otimes n}, M^{\otimes m}).$$

This expression is essentially a component of the Taylor decomposition of Γ,

$$\Gamma_n^m = \sum_{\substack{\alpha_1\dots\alpha_n \\ \beta_1\dots\beta_m}} \Gamma_{\alpha_1\dots\alpha_n}^{\beta_1\dots\beta_m} t^{\alpha_1} \otimes \dots \otimes t^{\alpha_n} \otimes e_{\beta_1} \otimes \dots \otimes e_{\beta_m},$$

in which the numerical coefficient $\Gamma_{\alpha_1\dots\alpha_n}^{\beta_1\dots\beta_m}$ is a replacement for the decorated labeled graph. More precisely, the interrelation between $\overline{\Gamma} = \oplus_{m,n\geq 0}\overline{\Gamma}_n^m$ and $\Gamma = \oplus_{m,n\geq 0}\Gamma_n^m \in \mathcal{G}_{\mathcal{M}}$ can be described as follows: a choice of any particular representation of the S-bimodule $E_{\mathcal{G}}$,

$$\phi : \big\{E_{\mathcal{G}}(m,n) \to \mathrm{Hom}(M^{\otimes n}, M^{\otimes m})\big\}_{m,n\geq 0},$$

defines an element $\Gamma = \phi(\overline{\Gamma}) \in \mathcal{G}_\mathcal{M}$ that is obtained from $\overline{\Gamma}$ by replacing each graph

by the value $\Gamma_{\alpha_1...\alpha_n}^{r\,\beta_1...\beta_m} \in \mathbb{R}$, of $\phi(\mathfrak{e}_r\} \in \mathrm{Hom}(M^{\otimes n}, M^{\otimes m})$ on the basis vector $e_{\alpha_1} \otimes \cdots \otimes e_{\alpha_n} \otimes t^{\beta_1} \otimes \cdots \otimes t^{\beta_m}$ (so that $\Gamma_{\alpha_1...\alpha_n}^{\beta_1...\beta_m} = \sum_r \Gamma_{\alpha_1...\alpha_n}^{r\,\beta_1...\beta_m}$) .

In a similar way one can define an element

$$\overline{[\cdots[[d\Gamma, \Gamma], \Gamma]\cdots]} \in \mathsf{P}_n^{\circlearrowleft}\langle\mathcal{G}\rangle \otimes \mathrm{Hom}(M^{\otimes \bullet}, M^{\otimes \bullet})$$

for any Lie word

$$[\cdots[[d\Gamma, \Gamma], \Gamma]\cdots]$$

built from Γ, $d\Gamma$, and $n-1$ Lie brackets. In particular, there are uniquely defined elements

$$\overline{d\Gamma} \in \mathsf{P}_1^{\circlearrowleft}\langle\mathcal{G}\rangle \otimes \mathrm{Hom}(M^{\otimes\bullet}, M^{\otimes\bullet}), \quad \overline{\tfrac{1}{2}[\Gamma, \Gamma]} \in \mathsf{P}_2^{\circlearrowleft}\langle\mathcal{G}\rangle \otimes \mathrm{Hom}(M^{\otimes\bullet}, M^{\otimes\bullet}),$$

whose values $\phi(\overline{d\Gamma})$ and $\phi\left(\overline{\tfrac{1}{2}[\Gamma, \Gamma]}\right)$ for any particular choice of representation ϕ of the $\mathbb{S}$-bimodule $E_\mathcal{G}$ coincide respectively with $d\Gamma$ and $\tfrac{1}{2}[\Gamma, \Gamma]$.

Finally, one defines a differential δ in the graded space $\mathsf{P}^{\circlearrowleft}\langle\mathcal{G}\rangle$ by setting

$$\delta\overline{\Gamma} = \overline{d\Gamma} + \frac{1}{2}\overline{[\Gamma, \Gamma]}, \qquad (\star\star)$$

i.e., by equating the graph coefficients of both sides. That $\delta^2 = 0$ is clear from the following calculation,

$$
\begin{aligned}
\delta^2\overline{\Gamma} &= \delta\left(\overline{d\Gamma} + \frac{1}{2}\overline{[\Gamma, \Gamma]}\right) \\
&= \delta\overline{d\Gamma} + \overline{\left[d\Gamma + \frac{1}{2}[\Gamma, \Gamma], \Gamma\right]} \\
&= -\overline{d\left(d\Gamma + \frac{1}{2}[\Gamma, \Gamma]\right)} + \overline{[d\Gamma, \Gamma]} + \frac{1}{2}\overline{[[\Gamma, \Gamma], \Gamma]} \\
&= -\overline{[d\Gamma, \Gamma]} + \overline{[d\Gamma, \Gamma]} \\
&= 0,
\end{aligned}
$$

where we used both the axioms of dg Lie algebra in $\mathcal{G}_\mathcal{M}$ and the axioms of the differential in $\mathsf{P}^{\circlearrowleft}\langle\mathcal{G}\rangle$. This completes the construction of $(\mathsf{P}^{\circlearrowleft}\langle\mathcal{G}\rangle, \delta)$[1].

---

[1] As a first approximation to the propic translation of *nonflat* geometries (Yang–Mills, Riemann, etc.) one might consider the following version of the "trick": in

2.5.1. Remarks. (i) If the differential and Lie brackets in $\mathcal{G}_\mathcal{M}$ contain no traces, then the expression $\overline{d\Gamma} + \frac{1}{2}\overline{[\Gamma, \Gamma]}$ does not contain graphs with oriented wheels. Hence formula $(\star\star)$ can be used to introduce a differential in the free prop $\mathsf{P}\langle\mathcal{G}\rangle$ generated by the $\mathbb{S}$-bimodule $E_\mathcal{G}$.

(ii) If the differential and Lie brackets in $\mathcal{G}_\mathcal{M}$ contain no traces and are given by first-order differential operators, then the expression $\overline{d\Gamma} + \frac{1}{2}\overline{[\Gamma, \Gamma]}$ is a tree. Therefore formula $(\star\star)$ can be used to introduce a differential in the free dioperad $\mathsf{D}\langle\mathcal{G}\rangle$.

2.5.2. Remark. The above trick also works for sheaves

$$(\mathcal{G}_\mathcal{M}, \mu_n : \wedge^n \mathcal{G}_\mathcal{M} \to \mathcal{G}_\mathcal{M}[2-n], n = 1, 2, \ldots)$$

of L_∞ algebras over $\mathcal{M}$. The differential in $\mathsf{P}^\circlearrowleft\langle\mathcal{G}\rangle$ (or in $\mathsf{P}\langle\mathcal{G}\rangle$, if appropriate) is defined by

$$\delta\overline{\Gamma} = \sum_{n=1}^{\infty} \frac{1}{n!} \overline{\mu_n(\Gamma, \ldots, \Gamma)}.$$

The term $\overline{\mu_n(\Gamma, \ldots, \Gamma)}$ corresponds to decorated graphs with n internal vertices.

2.5.3. Remark. Any sheaf of dg Lie subalgebras $\mathcal{G}'_\mathcal{M} \subset \mathcal{G}_\mathcal{M}$ defines a dg prop $(\mathsf{P}^\circlearrowleft\langle\mathcal{G}'\rangle, \delta)$ that is a quotient of $(\mathsf{P}^\circlearrowleft\langle\mathcal{G}\rangle, \delta)$ by the ideal generated by decorated graphs lying in the complement $\mathsf{P}^\circlearrowleft\langle\mathcal{G}\rangle \setminus \mathsf{P}^\circlearrowleft\langle\mathcal{G}'\rangle$. A similar observation holds true for $\mathsf{P}\langle\mathcal{G}\rangle$ and $\mathsf{P}\langle\mathcal{G}'\rangle$ (if they are defined).

2.6. Example (polyvector fields). Let us consider the sheaf of polyvector fields $\wedge^\bullet \mathcal{T}_\mathcal{M} := \sum_{i \geq 0} \wedge^i \mathcal{T}_\mathcal{M}[1-i]$ equipped with the Schouten Lie bracket $[\,,\,]_S$ and vanishing differential. A degree-one section Γ of $\wedge^\bullet \mathcal{T}_\mathcal{M}$ decomposes into a direct sum $\oplus_{i \geq 0} \Gamma_i$ with $\Gamma_i \in \wedge^i \mathcal{T}_\mathcal{M}$ having degree $2-i$ with respect to the grading of the underlying manifold. In a local coordinate system Γ can be represented as a Taylor series

$$\Gamma = \sum_{m,n \geq 0} \sum_{\substack{\alpha_1 \ldots \alpha_n \\ \beta_1 \ldots \beta_m}} \Gamma^{\beta_1 \ldots \beta_m}_{\alpha_1 \ldots \alpha_n} (e_{\beta_1} \wedge \cdots \wedge e_{\beta_m}) \otimes (t^{\alpha_1} \odot \cdots \odot t^{\alpha_n}).$$

Since $\Gamma^{\beta_1 \ldots \beta_m}_{\alpha_1 \ldots \alpha_n} = \Gamma^{[\beta_1 \ldots \beta_m]}_{(\alpha_1 \ldots \alpha_n)}$ has degree $2-m$, we conclude that the associated $\mathbb{S}$-bimodule $E_{\wedge^\bullet \mathcal{T}}$ is given by

$$E_{\wedge^\bullet \mathcal{T}}(m,n) = \mathrm{sgn}_\mathbf{m} \otimes \mathbf{1}_\mathbf{n}[m-2], \quad m, n \geq 0,$$

---

addition to a generic element $\Gamma \in \mathcal{G}_\mathcal{M}$ of degree 1 take into consideration a (probably *nongeneric*) element $F \in \mathcal{G}_\mathcal{M}$ of degree 2, extend appropriately the $\mathbb{S}$-bimodule $E_\mathcal{G}$ to accommodate the associated "curvature" F-corollas, and then (attempt to) define the differential δ in $\mathsf{P}^\circlearrowleft\langle E_\mathcal{G}\rangle$ by equating graph coefficients in the expressions $\delta\overline{\Gamma} = \overline{F} + \overline{d\Gamma} + \frac{1}{2}\overline{[\Gamma, \Gamma]}$ and $\delta\overline{F} = \overline{dF} + \overline{[\Gamma, F]}$.

where $\mathbf{sgn_m}$ stands for the one-dimensional sign representation of Σ_m, and $\mathbf{1_n}$ stands for the trivial one-dimensional representation of Σ_n. Then a generator of $P\langle \wedge^\bullet \mathcal{T}\rangle$ can be represented by the directed planar (m,n)-corolla

with skew-symmetric outgoing legs and symmetric ingoing legs. The formula $(\star\star)$ in Section 2.5 gives the following explicit expression for the induced differential δ in $P\langle \wedge^\bullet \mathcal{T}\rangle$:

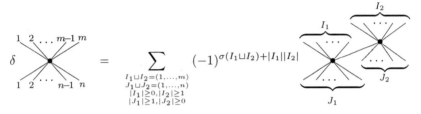

where $\sigma(I_1 \sqcup I_2)$ is the sign of the shuffle $I_1 \sqcup I_2 = (1,\dots,m)$.

2.6.1. Proposition. *There is a one-to-one correspondence between representations*

$$\phi : (P\langle \wedge^\bullet \mathcal{T}\rangle, \delta) \longrightarrow (\mathsf{End}\langle M\rangle, d)$$

of $(P\langle \wedge^\bullet \mathcal{T}\rangle, \delta)$ *in a dg vector space* (M,d) *and Maurer–Cartan elements* γ *in* $\wedge^\bullet \mathcal{T}_\mathcal{M}$, *that is, degree-one elements satisfying the equation* $[\gamma, \gamma]_S = 0$.

Proof. Let ϕ be a representation. Images of the above (m,n)-corollas under ϕ provide us with a collection of linear maps $\Gamma_n^m : \odot^n M \to \wedge^m M[2-m]$ that we assemble, as in Section 2.5, into a section $\Gamma = \sum_{m,n} \Gamma_n^m$, of $\wedge^\bullet \mathcal{T}_\mathcal{M}$.

The differential d in M can be interpreted as a linear (in the coordinates $\{t^\alpha\}$) degree-one section of $\mathcal{T}_\mathcal{M}$, which we denote by the same symbol.

Finally, the commutativity of ϕ with the differentials implies

$$[-d + \Gamma, -d + \Gamma]_S = 0.$$

Thus setting $\gamma = -d + \Gamma$, one gets a Maurer–Cartan element in $\wedge^\bullet \mathcal{T}_\mathcal{M}$.

Conversely, if γ is a Maurer–Cartan element in $\wedge^\bullet \mathcal{T}_\mathcal{M}$, then decomposing the sum $d + \gamma$ into a collection of its Taylor series components as in Section 2.5, one gets a representation ϕ. $\square$

Let $\wedge_0^\bullet \mathcal{T}_\mathcal{M} = \sum_{i \geq 1} \wedge_0^i \mathcal{T}_\mathcal{M}[1-i]$ be a sheaf of Lie subalgebras of $\wedge^\bullet \mathcal{T}_\mathcal{M}$ consisting of those elements that vanish at the distinguished point $* \in \mathcal{M}$, have no $\wedge^0 \mathcal{T}_\mathcal{M}[2]$-component, and whose $\wedge^1 \mathcal{T}_\mathcal{M}[1]$-component is at least quadratic in the coordinates $\{t^\alpha\}$. The associated dg free prop $P\langle \wedge_0^\bullet \mathcal{T}\rangle$ is generated by

(m, n)-corollas with $m, n \geq 1$, $m + n \geq 1$, and has a surprisingly small co-homology, a fact that is of key importance for our proof of the deformation quantization theorem.

2.6.2. Theorem. *The cohomology of* $(\mathsf{P}\langle \wedge_0^\bullet \mathcal{T}\rangle, \delta)$ *is equal to a quadratic prop* $\mathsf{Lie}^1\mathsf{B}$, *which is a quotient*

$$\mathsf{Lie}^1\mathsf{B} = \frac{\mathsf{P}\langle A\rangle}{\mathsf{Ideal}\langle R\rangle}$$

of the free prop generated by the following $\mathbb{S}$-*bimodule* A:

- *all* $A(m, n)$ *vanish except* $A(2, 1)$ *and* $A(1, 2)$,
- $A(2, 1) := \mathbf{sgn_2} \otimes \mathbf{1_1} = \mathrm{span}\left(\begin{array}{c}\end{array}\right)$
- $A(1, 2) := \mathbf{1_1} \otimes \mathbf{1_2}[-1] = \mathrm{span}\left(\begin{array}{c}\end{array}\right)$

modulo the ideal generated by the following relations R:

R_1 : $+$ $+$ $\in \mathsf{P}\langle A\rangle(3, 1)$,

R_2 : $+$ $+$ $\in \mathsf{P}\langle A\rangle(1, 3)$,

R_3 : $-$ $+$ $-$ $+$ $\in \mathsf{P}\langle A\rangle(2, 2)$.

Proof. The cohomology of $(\mathsf{P}\langle \wedge_0^\bullet \mathcal{T}\rangle, \delta)$ cannot be computed directly. At the dioperadic level, the theorem was established in [Mer06b]. That this result extends to the level of props can be shown easily using either ideas of per-turbations of $1/2$ props and path filtrations developed in [Kon02, MV03] or the idea of Koszul duality for props developed in [Val03]. One can argue, for example, as follows: for any $f \in \mathsf{P}\langle \wedge_0^\bullet \mathcal{T}\rangle$, define the natural number

$$|f| := \begin{array}{l}\text{number of directed paths in the graph } f \\ \text{that connect input legs with output ones,}\end{array}$$

and notice that the differential δ preserves the filtration

$$F_p\mathsf{P}\langle \wedge_0^\bullet \mathcal{T}\rangle := \{\mathrm{span}\ f \in \mathsf{P}\langle \wedge_0^\bullet \mathcal{T}\rangle : |f| \leq p\}.$$

The associated spectral sequence $\{E_r\mathsf{P}\langle \wedge_0^\bullet \mathcal{T}\rangle, \delta_r\}_{r\geq 0}$ is exhaustive and bounded below, so that it converges to the cohomology of $(\mathsf{P}\langle \wedge_0^\bullet \mathcal{T}\rangle, \delta)$.

By Koszulness of the operad Lie and exactness of the functor $\Omega_{\frac{1}{2}\mathsf{P}\to\mathsf{P}}$, the zeroth term of the spectral sequence, $(E_0\mathsf{P}\langle \wedge_0^\bullet \mathcal{T}\rangle, \delta_0)$, is precisely the minimal resolution of a quadratic prop $\mathsf{Lie}^1 B'$ generated by the $\mathbb{S}$-bimodule A modulo the ideal generated by relations R_1, R_2, and the following one:

$$R_3' : \quad \Yvertex = 0.$$

Since the differential δ vanishes on the generators of A, this spectral sequence degenerates at the first term, $(E_1 \mathsf{P}\,\langle \wedge_0^\bullet T \rangle, d_1 = 0)$, implying the isomorphism

$$\bigoplus_{p \geq 1} \frac{F_{p+1} H\,(\mathsf{P}\,\langle \wedge_0^\bullet T \rangle, \delta)}{F_p H\,(\mathsf{P}\,\langle \wedge_0^\bullet T \rangle, \delta)} = \mathsf{Lie}^1 B'.$$

There is a natural surjective morphism of dg props $p : (\mathsf{P}\,\langle \wedge_0^\bullet T \rangle, \delta) \to \mathsf{Lie}^1 B$. Define the dg prop (X, δ) via an exact sequence

$$0 \longrightarrow (X, \delta) \xrightarrow{\;i\;} (\mathsf{P}\,\langle \wedge_0^\bullet T \rangle, \delta) \xrightarrow{\;p\;} (\mathsf{Lie}^1 B, 0) \longrightarrow 0.$$

The filtration on $(\mathsf{P}\,\langle \wedge_0^\bullet T \rangle, \delta)$ induces filtrations on sub- and quotient complexes,

$$0 \longrightarrow (F_p X, \delta) \xrightarrow{\;i\;} (F_p \mathsf{P}\,\langle \wedge_0^\bullet T \rangle, \delta) \xrightarrow{\;p\;} (F_p \mathsf{Lie}^1 B, 0) \longrightarrow 0,$$

and hence an exact sequence of 0th terms of the associated spectral sequences,

$$0 \longrightarrow (E_0 X, \delta_0) \xrightarrow{\;i_0\;} (E_0 \mathsf{P}\,\langle \wedge_0^\bullet T \rangle, \delta_0) \xrightarrow{\;p_0\;} \left(\bigoplus_{p \geq 1} \frac{F_{p+1} \mathsf{Lie}^1 B}{F_p \mathsf{Lie}^1 B}, 0 \right) \longrightarrow 0.$$

By the above observation,

$$E_1 \mathsf{P}\,\langle \wedge_0^\bullet T \rangle = \bigoplus_{p \geq 1} \frac{F_{p+1} H\,(\mathsf{P}\,\langle \wedge_0^\bullet T \rangle, \delta)}{F_p H\,(\mathsf{P}\,\langle \wedge_0^\bullet T \rangle, \delta)} = \mathsf{Lie}^1 B'.$$

On the other hand, it is not hard to check that

$$\bigoplus_{p \geq 1} \frac{F_{p+1} \mathsf{Lie}^1 B}{F_p \mathsf{Lie}^1 B} = \mathsf{Lie}^1 B'.$$

Thus the map p_0 is a quasi-isomorphism implying vanishing of $E_1 X$ and hence acyclicity of (X, δ). Thus the projection map p is a quasi-isomorphism. $\qquad \square$

2.6.3. Corollary. The dg prop $(\mathsf{P}\,\langle \wedge_0^\bullet T \rangle, \delta)$ is a minimal model of the prop $\mathsf{Lie}^1 B$: the natural morphism of dg props

$$p : (\mathsf{P}\,\langle \wedge_0^\bullet T \rangle, \delta) \longrightarrow (\mathsf{Lie}^1 B, \text{vanishing differential})$$

that sends to zero all generators of $\mathsf{P}\langle\wedge_0^\bullet\mathcal{T}\rangle$ except those in $A(2,1)$ and $A(1,2)$ is a quasi-isomorphism. Hence we can and shall by denote $\mathsf{P}\langle\wedge_0^\bullet\mathcal{T}\rangle$ $\mathrm{Lie}^!\mathsf{B}_\infty$.

2.7. Example (polydifferential operators). Let us consider the sheaf of dg Lie algebras

$$\mathcal{D}_{\mathcal{M}} \subset \bigoplus_{k\geq 0} \mathrm{Hom}\left(\mathcal{O}_{\mathcal{M}}^{\otimes k}, \mathcal{O}_{\mathcal{M}}\right)[1-k]$$

consisting of polydifferential operators on $\mathcal{O}_{\mathcal{M}}$ that for $k \geq 1$, vanish on every element $f_1 \otimes \cdots \otimes f_k \in \mathcal{O}_{\mathcal{M}}^{\otimes k}$ with at least one function, f_i, $i = 1, \ldots, k$, constant. A degree-one section $\Gamma \in \mathcal{D}_{\mathcal{M}}$ decomposes into a sum $\sum_{k\geq 0}\Gamma_k$ with $\Gamma_k \in \mathrm{Hom}_{2-k}\left(\mathcal{O}_{\mathcal{M}}^{\otimes k}, \mathcal{O}_{\mathcal{M}}\right)$. In a local coordinate system $(t^\alpha, \partial/\partial t^\alpha \simeq e_\alpha)$ on $\mathcal{M}$, Γ can be represented as a Taylor series

$$\Gamma = \sum_{k\geq 0}\sum_{I_1,\ldots,I_k,J} \Gamma_J^{I_1,\ldots,I_k} e_{I_1} \otimes \cdots e_{I_k} \otimes t^J,$$

where for each fixed k and $|J|$ only a finite number of coefficients $\Gamma_J^{I_1,\ldots,I_k}$ are nonzero. The summation runs over multi-indices $I := \alpha_1\alpha_2\ldots\alpha_{|I|}$ and $e_I := e_{\alpha_1} \odot \cdots \odot e_{\alpha_{|I|}}$, $t^I := t^{\alpha_1} \odot \cdots \odot t^{\alpha_{|I|}}$. Hence the associated $\mathbb{S}$-bimodule $E_{\mathcal{D}}$ is given by

$$E_{\mathcal{D}}(m,n) = \mathbf{E}(m) \otimes \mathbf{1_n}, \quad m, n \geq 0,$$

where

$$\mathbf{E}(0) := \mathbb{R}[-2],$$

$$\mathbf{E}(m \geq 1) := \bigoplus_{\substack{k\geq 1 \\ [m]=I_1\sqcup\ldots\sqcup I_k \\ |I_1|,\ldots,|I_k|\geq 1}} \mathrm{Ind}_{\mathbb{S}_{|I_1|}\times\cdots\times\mathbb{S}_{|I_k|}}^{\mathbb{S}_m} \mathbf{1}_{|I_1|} \otimes \cdots \otimes \mathbf{1}_{|I_k|}[k-2].$$

The basis of $\mathsf{P}_1\langle\mathcal{D}\rangle(m,n)$ can be represented by directed planar corollas of the form

$\simeq \Gamma_J^{I_1,\ldots,I_k},$

where

- the input legs are labeled by the set $[n] := \{1, 2, \ldots, n\}$ and are symmetric (so that it does not matter how labels from $[n]$ are distributed over them);
- the output legs (if there are any) are labeled by the set $[m]$ partitioned into k disjoint nonempty subsets

$$[m] = I_1 \sqcup \cdots \sqcup I_i \sqcup I_{i+1} \sqcup \cdots \sqcup I_k,$$

and legs in each I_i-bunch are symmetric (so that it does not matter how labels from the set I_i are distributed over legs in the I_ith bunch).

The $\mathbb{Z}$-grading in $\mathsf{P}\langle\mathcal{D}\rangle$ is defined by associating degree $2-k$ to such a corolla. The formula $(\star\star)$ in Section 2.5 provides us with the following explicit expression for the differential δ in $\mathsf{P}\langle\mathcal{D}\rangle$:

where the first sum comes from the Hochschild differential d_H, and the second sum comes from the Hochschild brackets $[\,,\,]_H$. The s-summation in the latter runs over the number s of edges connecting the two internal vertices. Since can be zero, the right-hand side above contains disconnected graphs (more precisely, disjoint unions of two corollas).

2.7.1. Proposition. *There is a one-to-one correspondence between representations*

$$\phi : (\mathsf{P}\langle\mathcal{D}\rangle, \delta) \longrightarrow (\mathsf{End}\langle M\rangle, d)$$

of $(\mathsf{P}\langle\mathcal{D}\rangle, \delta)$ in a dg vector space (M, d) and Maurer–Cartan elements γ in $\mathcal{D}_{\mathcal{M}}$, that is, degree-one elements satisfying the equation $d_H\gamma + \frac{1}{2}[\gamma, \gamma]_H = 0$.

Proof The proof is similar to the proof of Proposition 2.6.1.

2.8. Remark. Kontsevich's formality map [Kon03] can be interpreted as a morphism of dg props

$$F_\infty : (\mathsf{P}\langle\mathcal{D}\rangle, \delta) \longrightarrow (\mathsf{P}\langle\wedge^\bullet T\rangle^{\circlearrowleft}, \delta).$$

Conversely, any morphism of the above dg props gives rise to a *universal* formality map in the sense of [Kon03]. Note that the dg prop of polyvector fields $\mathsf{P}\langle\wedge^\bullet T\rangle$ appears above in the wheel extended form $\mathsf{P}\langle\wedge^\bullet T\rangle^{\circlearrowleft}$. This is not accidental: it is not hard to show (by quantizing, e.g., a pair consisting of a linear Poisson structure and quadratic homological vector field) that there does *not* exist a morphism between ordinary (i.e., unwheeled) dg props $(\mathsf{P}\langle\mathcal{D}\rangle,\delta) \longrightarrow (\mathsf{P}\langle\wedge^\bullet\mathcal{T}\rangle,\delta)$ satisfying the quasiclassical limit. Thus wheeled completions of classical dg props are absolutely necessary at least from the point of view of applications to geometric problems.

We shall next investigate how wheeled completion of directed graph complexes affects their cohomology.

3 Directed Graph Complexes with Loops and Wheels

3.1. $\mathfrak{G}^{\circlearrowleft}$ versus $\mathfrak{G}^{\uparrow}$. One of the most effective methods for computing cohomology of dg free props (that is, $\mathfrak{G}^{\uparrow}$-graph complexes) is based on the idea of interpreting the differential as a perturbation of its $\frac{1}{2}$propic part, which, in this $\mathfrak{G}^{\uparrow}$-case, can often be singled out by the path filtration [Kon02, MV03]. However, one cannot apply this idea directly to graphs with wheels—it is shown below that a filtration that singles out the $\frac{1}{2}$propic part of the differential does *not* exist in general even for dioperadic differentials. Put another way, if one takes a $\mathfrak{G}^{\uparrow}$-graph complex $(\mathsf{P}\langle E\rangle,\delta)$, enlarges it by adding decorated graphs with wheels while keeping the original differential δ unchanged, then one ends up in a very different situation in which the idea of $\frac{1}{2}$props is no longer directly applicable.

3.2. Graphs with back-in-time edges. Here we suggest the following trick to solve the problem: we further enlarge our set of graphs with wheels, $\mathfrak{G}^{\circlearrowleft} \rightsquigarrow \mathfrak{G}^{+}$, by putting a mark on one (and only one) of the edges in each wheel, and then study the natural "forgetful" surjection $\mathfrak{G}^{+} \to \mathfrak{G}^{\circlearrowleft}$. The point is that $\mathfrak{G}^{+}$-graph complexes again admit a filtration that singles out the $\frac{1}{2}$prop part of the differential, and hence their cohomology is often easily computable.

More precisely, let $\mathfrak{G}^{+}(m,n)$ be the set of all directed (m,n)-graphs G that satisfy conditions 2.1.3(i)–(iv) and the following one:

(v) every oriented wheel in G (if any) has one and only one of its internal edges marked (say, dashed) and called the *back-in-time* edge.

For example,

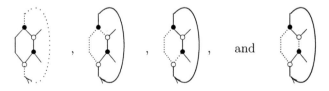

are four different graphs in $\mathfrak{G}^{+}(1,1)$.

Clearly, we have a natural surjection

$$u : \mathfrak{G}^{+}(m,n) \longrightarrow \mathfrak{G}^{\circlearrowleft}(m,n),$$

which forgets the markings. For example, the four graphs above are mapped under u into the same graph

$$\in \mathfrak{G}^{\circlearrowleft}(1,1),$$

and in fact, span its preimage under u.

3.3. Graph complexes. Let $E = \{E(m,n)\}_{m,n\geq 1, m+n\geq 3}$ be an $\mathbb{S}$-bimodule and let $(\mathsf{P}\langle E\rangle, \delta)$ be a dg free prop on E with the differential δ that preserves connectedness and genus, that is, δ applied to any decorated (m,n)-corolla creates a connected (m,n)-tree. Such a differential can be called *dioperadic*, and from now on we restrict ourselves to dioperadic differentials only. This restriction is not that dramatic: every dg free prop with a nondioperadic but connected[2] differential always admits a filtration that singles out its dioperadic part [MV03]. Thus the technique we develop here in Section 3 can, in principle, be applied to a wheeled extension of any dg free prop with a connected differential.

We enlarge the $\mathfrak{G}^{\uparrow}$-graph complex $(\mathsf{P}\langle E\rangle, \delta)$ in two ways:

$$\mathsf{P}^{\circlearrowleft}\langle E\rangle(m,n) := \bigoplus_{G\in\mathfrak{G}^{\circlearrowleft}(m,n)} \left(\bigotimes_{v\in v(G)} E(Out(v), In(v))\right)_{AutG},$$

$$\mathsf{P}^{+}\langle E\rangle(m,n) := \bigoplus_{G\in\mathfrak{G}^{+}(m,n)} \left(\bigotimes_{v\in v(G)} E(Out(v), In(v))\right)_{AutG},$$

and notice that both $\mathsf{P}^{\circlearrowleft}\langle E\rangle := \{\mathsf{P}^{\circlearrowleft}\langle E\rangle(m,n)\}$ and $\mathsf{P}^{+}\langle E\rangle := \{\mathsf{P}^{\circlearrowleft}\langle E\rangle(m,n)\}$ have a natural structure of dg prop with respect to disjoint unions, grafting of graphs, and the original differential δ. Clearly, they contain $(\mathsf{P}\langle E\rangle, \delta)$ as a dg subprop. There is a natural morphism of dg props

$$u : (\mathsf{P}^{+}\langle E\rangle, \delta) \longrightarrow (\mathsf{P}^{\circlearrowleft}\langle E\rangle, \delta)$$

that forgets the markings. Let $(\mathsf{L}^{+}\langle E\rangle, \delta) := \mathrm{Ker}\, u$ and denote the natural inclusion $\mathsf{L}^{+}\langle E\rangle \subset \mathsf{P}^{+}\langle E\rangle$ by i.

3.4. Fact. There is a short exact sequence of graph complexes

$$0 \longrightarrow (\mathsf{L}^{+}\langle E\rangle, \delta) \xrightarrow{i} (\mathsf{P}^{+}\langle E\rangle, \delta) \xrightarrow{u} (\mathsf{P}^{\circlearrowleft}\langle E\rangle, \delta) \longrightarrow 0,$$

---

[2]A differential δ in $\mathsf{P}\langle E\rangle$ is called *connected* if it preserves the filtration of $\mathsf{P}\langle E\rangle$ by the number of connected (in the topological sense) components.

where u is the map that forgets markings. Thus, if the natural inclusion of complexes $i : (\mathsf{L}^+\langle E\rangle, \delta) \to (\mathsf{P}^+\langle E\rangle, \delta)$ induces a monomorphism in cohomology $[i] : H(\mathsf{L}^+\langle E\rangle, \delta) \to H(\mathsf{P}^+\langle E\rangle, \delta)$, then

$$H(\mathsf{P}^{\circlearrowleft}\langle E\rangle, \delta) = \frac{H(\mathsf{P}^+\langle E\rangle, \delta)}{H(\mathsf{L}^+\langle E\rangle, \delta)}.$$

Put another way, if $[i]$ is a monomorphism, then $H(\mathsf{P}^{\circlearrowleft}\langle E\rangle, \delta)$ is obtained from $H(\mathsf{P}^+\langle E\rangle, \delta)$ simply by forgetting the markings.

3.5. Functors that adjoin wheels. We are interested in this paper in dioperads D that are either free, $\mathsf{D}\langle E\rangle$, on an $\mathbb{S}$-bimodule $E = \{E(m,n)\}_{m,n\geq 1, m+n\geq 3}$, or are naturally represented as quotients of free dioperads

$$D = \frac{\mathsf{D}\langle E\rangle}{\langle I\rangle},$$

modulo ideals generated by some relations $I \subset \mathsf{D}\langle E\rangle$. Then the free prop $\Omega_{\mathsf{D}\to\mathsf{P}}\langle E\rangle$ generated by D is simply the quotient of the free prop $\mathsf{P}\langle E\rangle$,

$$\Omega_{\mathsf{D}\to\mathsf{P}}\langle D\rangle := \frac{\mathsf{P}\langle E\rangle}{\langle I\rangle},$$

by the ideal generated by the same relations I. Now we define two other props[3]

$$\Omega_{\mathsf{D}\to\mathsf{P}\circlearrowleft}\langle D\rangle := \frac{\mathsf{P}^{\circlearrowleft}\langle E\rangle}{\langle I\rangle^{\circlearrowleft}}, \qquad \Omega_{\mathsf{D}\to\mathsf{P}+}\langle D\rangle := \frac{\mathsf{P}^+\langle E\rangle}{\langle I\rangle^+},$$

where $\langle I\rangle^{\circlearrowleft}$ (resp., $\langle I\rangle^+$) is the subspace of those graphs G in $\mathsf{P}^{\circlearrowleft}\langle E\rangle$ (resp., in $\mathsf{P}^+\langle E\rangle$) that satisfy the following condition: there exists a (possibly empty) set of cyclic edges whose breaking up into two legs produces a graph lying in the ideal $\langle I\rangle$ that defines the prop $\Omega_{\mathsf{D}\to\mathsf{P}}\langle D\rangle$.

Analogously, one defines functors $\Omega_{\frac{1}{2}\mathsf{P}\to\mathsf{P}\circlearrowleft}$ and $\Omega_{\frac{1}{2}\mathsf{P}\to\mathsf{P}+}$.

From now on we abbreviate notation as

$$D^\uparrow := \Omega_{\mathsf{D}\to\mathsf{P}}\langle D\rangle, \quad D^+ := \Omega_{\mathsf{D}\to\mathsf{P}+}\langle D\rangle, \quad D^{\circlearrowleft} := \Omega_{\mathsf{D}\to\mathsf{P}\circlearrowleft}\langle D\rangle,$$

for values of the above-defined functors on dioperads, and respectively

$$D_0^\uparrow := \Omega_{\frac{1}{2}\mathsf{P}\to\mathsf{P}}\langle D_0\rangle, \quad D_0^+ := \Omega_{\frac{1}{2}\mathsf{P}\to\mathsf{P}\circlearrowleft}\langle D_0\rangle, \quad D_0^{\circlearrowleft} := \Omega_{\frac{1}{2}\mathsf{P}\to\mathsf{P}\circlearrowleft}\langle D_0\rangle$$

for their values on $\frac{1}{2}$props.

---

[3]These props are particular examples of *wheeled props*, which will be discussed in detail elsewhere.

3.5.1. Facts. (i) If D is a dg dioperad, then both $D^{\circlearrowleft}$ and D^+ are naturally dg props. (ii) If D is an operad, then both $D^{\circlearrowleft}$ and D^+ may contain at most one wheel.

3.5.2. Proposition. *Any finite-dimensional representation of the dioperad D lifts to a representation of its wheeled prop extension $D^{\circlearrowleft}$.*

Proof. If $\phi : D \rightarrow \mathrm{End}\langle M \rangle$ is a representation, then we first extend it to a representation $\phi^{\circlearrowleft}$ of $\mathsf{P}^{\circlearrowleft}\langle E \rangle \subset$ as in Section 2.3.2 and then notice that $\phi^{\circlearrowleft}(f) = 0$ for any $f \in \langle I \rangle^{\circlearrowleft}$. $\qquad\qquad\square$

3.5.3. Definition. Let D be a Koszul dioperad with $(D_\infty, \delta) \rightarrow (D, 0)$ being its minimal resolution. The dioperad D is called *stably Koszul* if the associated morphism of the wheeled completions,

$$\left(D_\infty^{\circlearrowleft}, \delta \right) \longrightarrow \left(D^{\circlearrowleft}, 0 \right),$$

remains a quasi-isomorphism.

3.5.4. Example. The notion of stable Koszulness is nontrivial. Just adding oriented wheels to a minimal resolution of a Koszul operad while keeping the differential unchanged may alter the cohomology group of the resulting graph complex, as the following example shows.

Claim. *The operad Ass of associative algebras is not stably Koszul.*

Proof. The operad Ass can be represented as a quotient

$$\mathsf{Ass} = \frac{\mathsf{Oper}\langle E \rangle}{\mathsf{Ideal}\langle R \rangle}$$

of the free operad $\mathsf{Oper}\langle E \rangle$ generated by the following $\mathbb{S}$-module E:

$$E(n) := \begin{cases} k[\mathbb{S}_2] = \mathrm{span} \left({}^1 \underset{}{\curlyvee}{}^2 , \ {}^2 \underset{}{\curlyvee}{}^1 \right) & \text{for } n = 2, \\ 0 & \text{otherwise,} \end{cases}$$

modulo the ideal generated by the relations

$$= 0, \quad \forall \sigma \in \mathbb{S}_3.$$

Hence the minimal resolution $(\mathsf{Ass}_\infty, \delta)$ of Ass contains a degree-1 corolla such that

Therefore, in its wheeled extension $\left(\mathsf{Ass}_\infty^\circlearrowleft, \delta\right)$, one has

$= 0,$

implying the existence of a nontrivial cohomology class in $\mathsf{H}\left(\mathsf{Ass}_\infty^\circlearrowleft, \delta\right)$

that does *not* belong to $\mathsf{Ass}^\circlearrowleft$. Thus Ass is Koszul, but *not* stably Koszul. $\square$

It is instructive to see explicitly how the map $[i] : \mathsf{H}(\mathsf{L}^+\langle\mathsf{Ass}_\infty\rangle, \delta) \rightarrow \mathsf{H}(\mathsf{P}^+\langle\mathsf{Ass}_\infty\rangle, \delta)$ fails to be a monomorphism. As $\mathsf{L}^+\langle\mathsf{Ass}_\infty\rangle$ does not contain loops, the element

defines a non-trivial cohomology class, $[a]$, in $\mathsf{H}(\mathsf{L}^+\langle\mathsf{Ass}_\infty\rangle, \delta)$, whose image, $[i]([a])$, in $\mathsf{H}(\mathsf{P}^+\langle\mathsf{Ass}_\infty\rangle, \delta)$ vanishes.

3.6. Koszul substitution laws. Let $P = \{P(n)\}_{n\geq 1}$ and $Q = \{Q(n)\}_{n\geq 1}$ be two quadratic Koszul operads generated,

$$P := \frac{\mathsf{P}\langle E_P(2)\rangle}{< I_P >}, \qquad Q := \frac{\mathsf{P}\langle E_Q(2)\rangle}{< I_Q >},$$

by $\mathbb{S}_2$-modules $E_P(2)$, and, respectively, $E_Q(2)$.

One can canonically associate [MV03] to such a pair the $\frac{1}{2}$prop, $P \diamond Q^\dagger$, with

$$P \diamond Q^\dagger(m, n) = \begin{cases} P(n) & for\ m = 1, n \geq 2, \\ Q(m) & for\ n = 1, m \geq 2, \\ 0 & otherwise, \end{cases}$$

and the $\frac{1}{2}$prop compositions,

$$\left\{ {}_1\circ_j : P \diamond Q^\dagger(m_1, 1) \otimes P \diamond Q^\dagger(m_2, n_2) \longrightarrow P \diamond Q^\dagger(m_1 + m_2 - 1, n_2)\right\}_{1\leq j\leq m_2}$$

being zero for $n_2 \geq 2$ and coinciding with the operadic composition in Q for $n_2 = 1$, and

$$\left\{ {}_i\circ_1 : P \diamond Q^\dagger(m_1, n_1) \otimes P \diamond Q^\dagger(1, n_2) \longrightarrow P \diamond Q^\dagger(m_1 + m_2 - 1, n_2) \right\}_{1 \leq i \leq n_1}$$

being zero for $m_1 \geq 2$ and otherwise coinciding with the operadic composition in P for $m_1 = 1$.

Let $D_0 = \Omega_{\frac{1}{2}P \to D}\langle P \diamond Q^\dagger \rangle$ be the associated free dioperad, $D_0^!$ its Koszul dual dioperad, and $\left(D_{0\infty} := \mathbf{D}D_0^!, \delta_0 \right)$ the associated cobar construction [Gan03]. As D_0 is Koszul [11, MV03], the latter provides us with the dioperadic minimal model of D_0. By exactness of $\Omega_{\frac{1}{2}P \to P}$, the dg free prop, $\left(D_{0\infty}^\uparrow := \Omega_{D \to P}\langle D_{0\infty} \rangle, \delta_0 \right)$, is the minimal model of the prop $D_0^\uparrow := \Omega_{D \to P}\langle D_0 \rangle \simeq \Omega_{\frac{1}{2}P \to P}\langle P \diamond Q^\dagger \rangle$.

3.6.1. Remark. The prop $D_0^\uparrow$ can equivalently be defined as the quotient,

$$\frac{P * Q^\dagger}{I_0}$$

where $P * Q^\dagger$ is the free product of props associated to operads P and[4] $Q^\dagger$, and the ideal I_0 is generated by graphs of the form

$$I_0 = \mathrm{span}\left\langle \vcenter{\hbox{\includegraphics{}}} \right\rangle \simeq D_0(2,1) \otimes D_0(1,2) = E_Q(2) \otimes E_P(2)$$

with white vertex decorated by elements of $E_Q(2)$ and black vertex decorated by elements of $E_P(2)$.

Let us consider a morphism of $\mathbb{S}_2$-bimodules

$$\lambda : D_0(2,1) \otimes D_0(1,2) \longrightarrow \qquad D_0(2,2)$$

$$\mathrm{span}\left\langle \vcenter{\hbox{\includegraphics{}}} \right\rangle \longrightarrow \mathrm{span}\left\langle \vcenter{\hbox{\includegraphics{}}}, \vcenter{\hbox{\includegraphics{}}} \right\rangle$$

and define [MV03] the dioperad D_λ as the quotient of the free dioperad generated by the two spaces of binary operations $D_0(2,1) = E_Q(2)$ and $D_0(1,2) = E_P(2)$ modulo the ideal generated by relations in P, relations in Q, as well as the followings ones:

$$I_\lambda = \mathrm{span}\{ f - \lambda f : \forall f \in D_0(2,1) \otimes D_0(1,2) \}.$$

Note that in the notation of Section 3.6.1, the associated prop $D_\lambda^\uparrow := \Omega_{D \to P}\langle D_\lambda \rangle$ is just the quotient $P * Q^\dagger / I_\lambda$.

---

[4]the symbol † stands for the functor on props $P = \{P(m,n)\} \to P^\dagger = \{P^\dagger(m,n)\}$ that reverses "time flow," i.e., $P^\dagger(m,n) := P(n,m)$.

The substitution law λ is called *Koszul* if D_λ is isomorphic to D_0 as an S-bimodule, which implies that D_λ is Koszul [Gan03]. Koszul duality technique provides $\mathbf{DD}_0^! \simeq \mathbf{DD}_\lambda^!$ with a perturbed differential δ_λ such that $\left(\mathbf{DD}_0^!, \delta_\lambda\right)$ is the minimal model $(D_{\lambda\infty}, \delta_\lambda)$ of the dioperad D_λ.

3.7. Theorem [MV03, V]. *The dg free prop* $D_{\lambda\infty}^\uparrow := \Omega_{D\to P}\langle D_{\lambda\infty}\rangle$ *is the minimal model of the prop* $D_\lambda^\uparrow$, *i.e., the natural morphism*

$$\left(D_{\lambda\infty}^\uparrow, \delta_\lambda\right) \longrightarrow \left(D_\lambda^\uparrow, 0\right),$$

which sends to zero all vertices of $D_{\lambda\infty}^\uparrow$ *except binary ones decorated by elements of* $E_P(2)$ *and* $E_Q(2)$, *is a quasi-isomorphism.*

Proof. The main point is that

$$F_p := \left\{ \operatorname{span} \left\langle f \in D_{\lambda\infty}^\uparrow \right\rangle : \begin{array}{l} \text{number of directed paths in the graph } f \\ \text{that connect input legs with output ones} \end{array} \leq p \right\}$$

defines a filtration of the complex $D_{\lambda\infty}^\uparrow$. The associated spectral sequence $\{E_r, d_r\}_{r\geq 0}$ is exhaustive and bounded below, so that it converges to the cohomology of $\left(D_{\lambda\infty}^\uparrow, \delta_\lambda\right)$.

The zeroth term of this spectral sequence is isomorphic to $\left(D_{0\infty}^\uparrow, \delta_0\right)$, and hence by Koszulness of the dioperad D_0 and exactness of the functor $\Omega_{\frac{1}{2}P\to P}$ has the cohomology E_1 isomorphic to $D_0^\uparrow$, which, by Koszulness of D_λ, is isomorphic to $D_\lambda^\uparrow$ as an S-bimodule. Since $\{d_r = 0\}_{r\geq 1}$, the result follows along the same lines as in the second part of the proof of Theorem 2.6.2. $\square$

3.8. Cohomology of graph complexes with marked wheels. In this section we analyze the functor $\Omega_{\frac{1}{2}P\to P+}$. The following statement is one of the motivations for its introduction (it does *not* hold true for the "unmarked" version $\Omega_{\frac{1}{2}P\to P\circlearrowright}$).

3.8.1. Theorem. *The functor* $\Omega_{\frac{1}{2}P\to P+}$ *is exact.*

Proof. Let T be an arbitrary dg $\frac{1}{2}$prop. The main point is that we can use $\frac{1}{2}$prop compositions and presence of marks on cyclic edges to represent $\Omega_{\frac{1}{2}P\to P+}\langle T\rangle$ as a vector space *freely* generated by a family of decorated graphs

$$\Omega_{\frac{1}{2}P\to P+}\langle T\rangle(m,n) = \bigoplus_{G\in\overline{\mathfrak{G}^+}(m,n)} G\langle P\rangle,$$

where $\overline{\mathfrak{G}^+}(m, n)$ is a subset of $\mathfrak{G}^+(m, n)$ consisting of so-called *reduced* graphs G that satisfy the following defining property: for each pair of internal vertices (v_1, v_2) of G that are connected by an unmarked edge directed from v_1 to v_2 one has $|Out(v_1)| \geq 2$ and $|In(v_2)| \geq 2$. Put another way, given an arbitrary

T-decorated graph with wheels, one can perform $\frac{1}{2}$prop compositions ("contractions") along all unmarked internal edges (v_1, v_2) that do not satisfy the above conditions. The result is a reduced decorated graph (with wheels) that is uniquely defined by the original one. Notice that marks are vital for this contraction procedure, e.g.,

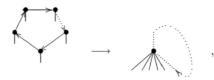

to be well defined.

Then we have

$$H^* \left(\Omega_{\frac{1}{2}\mathsf{P}\to\mathsf{P}^+}\langle T\rangle(m, n) \right) = H^* \left(\bigoplus_{G\in\bar{\mathfrak{G}}^+(m,n)} \left(\bigotimes_{v\in v(G)} T(Out(v), In(v)) \right)_{AutG} \right)$$

$$\text{(by Maschke's theorem)} = \bigoplus_{G\in\bar{\mathfrak{G}}^+(m,n)} H^* \left(\bigotimes_{v\in v(G)} T(Out(v), In(v)) \right)_{AutG}$$

$$\text{(by the Künneth formula)} = \bigoplus_{G\in\bar{\mathfrak{G}}^+(m,n)} \left(\bigotimes_{v\in v(G)} H^*(T)(Out(v), In(v)) \right)_{AutG}$$

$$= \Omega_{\frac{1}{2}\mathsf{P}\to\mathsf{P}^+}\langle H^*(T)\rangle(m, n).$$

In the second line we used the fact that the group $AutG$ is finite. □

Another motivation for introducing graph complexes with *marked* wheels is that they admit a filtration that singles out the $\frac{1}{2}$propic part of the differential, a fact that we use heavily in the proof of the following theorem.

3.8.2. Theorem. *Let D_λ be a Koszul dioperad with Koszul substitution law and let $(D_{\lambda\infty}, \delta)$ be its minimal resolution. The natural morphism of graph complexes*

$$\left(D_{\lambda\infty}^+, \delta_\lambda\right) \longrightarrow \left(D_\lambda^+, 0\right)$$

is a quasi-isomorphism.

Proof. Consider first a filtration of the complex $\left(D_{\lambda\infty}^+, \delta_\lambda\right)$ by the number of marked edges, and let $\left(D_{\lambda\infty}^+, b\right)$ denote the 0th term of the associated spectral sequence (which, as we shall show below, degenerates at the first term). To any decorated graph $f \in D_{\lambda\infty}^+ = \Omega_{\mathsf{D}\to\mathsf{P}^+}\langle D_{\lambda\infty}\rangle$ one can associate a graph without wheels $\bar{f} \in \Omega_{\mathsf{D}\to P}\langle D_{\lambda\infty}\rangle$ by breaking every marked cyclic edge into two legs (one of which is input and the other, output). Let $|\overline{f}|$ be the number of directed paths in the graph $\overline{f}$ that connect input legs with output ones. Then

$$F_p := \{ f \in D_{\lambda\infty}^+ \ : \ |\overline{f}| \le p \}$$

defines a filtration of the complex $\left(D_{\lambda\infty}^+, b \right)$.

The zeroth term of the spectral sequence $\{E_r, d_r\}_{r \ge 0}$ associated to this filtration is isomorphic to $\left(\Omega_{\frac{1}{2}\mathrm{P}\to\mathrm{P}+}\left\langle D_\infty^0 \right\rangle, \delta_0 \right)$ and hence, by Theorem 3.8.1, has the cohomology E_1 equal to $D_0^+ := \Omega_{\frac{1}{2}\mathrm{P}\to\mathrm{P}+}\langle D_0 \rangle$, which by Koszulness of D_λ is isomorphic as a vector space to D_λ^+. Since differentials of all higher terms of both our spectral sequences vanish, the result follows. $\qquad\square$

3.8.3. Remark. In the proof of Theorem 3.8.2, the $\frac{1}{2}$propic part δ_0 of the differential δ_λ was singled out in two steps: first we introduced a filtration by the number of marked edges, and then a filtration by the number of paths $|\overline{f}|$ in the unwheeled graphs $\overline{f}$. As the following lemma shows, one can do it in one step. Let $w(f)$ stand for the number of marked edges in a decorated graph $f \in D_{\lambda\infty}^+$.

3.8.4. Lemma *The sequence of vector spaces spaces $p \in \mathbb{N}$,*

$$\mathsf{F}_p := \left\{ span \left\langle f \in D_{\lambda\infty}^+ \right\rangle \ : \ \|f\| := 3^{w(f)} |\overline{f}| \le p \right\},$$

defines a filtration of the complex $\left(D_{\lambda\infty}^+, \delta_\lambda \right)$ whose spectral sequence has 0th term isomorphic to $\left(D_{0\infty}^+, \delta_0 \right)$.

Proof. It is enough to show that for any graph f in $D_{\lambda\infty}^+$ with $w(f) \ne 0$ one has $\|\delta_\lambda f\| \le \|f\|$.

We can, in general, split $\delta_\lambda f$ into two groups of summands,

$$\delta_\lambda f = \sum_{a \in I_1} g_a + \sum_{b \in I_2} g_b,$$

where $w(g_a) = w(f) \ \forall a \in I_1$, and $w(g_b) = w(f) - p_b$ for some $p_b \ge 1$ and all $b \in I_2$.

For any $a \in I_1$,

$$\|g_a\| = 3^{w(f)} |\overline{g_a}| \le 3^{w(f)} |\overline{f}| = \|f\|.$$

So it remains to check the inequality $\|g_b\| \le \|f\|$, $\forall b \in I_2$.

We can also split $\delta_\lambda \overline{f}$ into two groups of summands

$$\delta_\lambda \overline{f} = \sum_{a \in I_1} h_a + \sum_{b \in I_2} h_b,$$

where $\{h_b\}_{b \in I_2}$ is the set of all those summands that contain two-vertex sub-graphs of the form

that contain half-edges of the type x and y corresponding to broken wheeled paths in f. Every graph g_b is obtained from the corresponding h_b by gluing some number of output legs connected to y with the same number of input legs connected to x into new internal noncyclic edges. This gluing operation creates p_b new paths connecting some internal vertices in $\overline{h_b}$, and hence may increase the total number of paths in $\overline{h_b}$, but by no more than the factor of $p_b + 1$, i.e., $|\overline{g_b}| \leq (p_b + 1)|h_b|$, $\forall b \in I_2$.

Finally, we have

$$\|g_b\| = 3^{w(f)-p_b}|\overline{g_b}| \leq 3^{w(f)-p_b}(p_b + 1)|h_b| < 3^{w(f)}|\overline{f}| = \|f\|, \quad \forall b \in I_2.$$

The part of the differential δ_λ that preserves the filtration must in fact preserve both the number of marked edges $w(f)$ and the number of paths $|\overline{f}|$ for any decorated graph f. Hence this is precisely δ_0.   □

3.9. Graph complexes with unmarked wheels built on $\frac{1}{2}$props. Let $\left(T = \frac{1}{2}P\langle E \rangle / \langle I \rangle, \delta\right)$ be a dg $\frac{1}{2}$prop. In Section 3.5 we defined its wheeled extension

$$\left(T^{\circlearrowleft} := \frac{P^{\circlearrowleft}\langle E \rangle}{\langle I \rangle^{\circlearrowleft}}, \delta\right).$$

Now we specify a dg subprop, $\Omega_{\text{no-oper}}\langle T \rangle \subset T^{\circlearrowleft}$, whose cohomology is easy to compute.

3.9.1. Definition. Let $E = \{E(m, n)\}_{m,n \geq 1, m+n \geq 3}$ be an $\mathbb{S}$-bimodule, and $P^{\circlearrowleft}\langle E \rangle$ the associated prop of decorated graphs with wheels. We say that a wheel W in a graph $G \in P^{\circlearrowleft}\langle E \rangle$ is *operadic* if all its cyclic vertices are decorated either by elements of $\{E(1, n)\}_{n \geq 2}$ only or by elements $E(n, 1)_{n \geq 2}$ only. Vertices of operadic wheels are called *operadic cyclic vertices*. Notice that operadic wheels can be of geometric genus 1 only.

Let $P^{\circlearrowleft}_{\text{no-oper}}\langle E \rangle$ be the subspace of $P^{\circlearrowleft}\langle E \rangle$ consisting of graphs with no operadic wheels, and let

$$\Omega_{\text{no-oper}}\langle T \rangle = \frac{P^{\circlearrowleft}_{\text{no-oper}}\langle E \rangle}{\langle I \rangle^{\circlearrowleft}}$$

be the associated dg subprop of $(T^{\circlearrowleft}, \delta)$.

Clearly, $\Omega_{\text{no-oper}}$ is a functor from the category of dg $\frac{1}{2}$props to the category of dg props. It is worth pointing out that this functor can*not* be extended to dg dioperads, since the differential can, in general, create operadic wheels from nonoperadic ones.

3.9.2. Theorem. *The functor* $\Omega_{no-oper}$ *is exact.*

Proof. Let (T, δ) be an arbitrary dg $\frac{1}{2}$prop. Every wheel in $\Omega_{no-oper}\langle T \rangle$ contains at least one cyclic edge along which $\frac{1}{2}$prop composition in T is not possible. This fact allows one to unambiguously perform such compositions along all those cyclic and noncyclic edges at which such a composition makes sense, and hence represent $\Omega_{no-oper}\langle T \rangle$ as a vector space *freely* generated by a family of decorated graphs,

$$\Omega_{no-oper}\langle T \rangle(m,n) = \bigoplus_{G \in \overline{\mathfrak{G}^{\circlearrowleft}}(m,n)} G\langle T \rangle,$$

where $\overline{\mathfrak{G}^{\circlearrowleft}}(m,n)$ is a subset of $\mathfrak{G}^{\circlearrowleft}(m,n)$ consisting of *reduced* graphs G that satisfy the following defining properties: (i) for each pair of internal vertices (v_1, v_2) of G that are connected by an edge directed from v_1 to v_2 one has $|Out(v_1)| \geq 2$ and $|In(v_2)| \geq 2$; (ii) there are no operadic wheels in G. The rest of the proof is exactly the same as in Section 3.8.1.   □

Let P and Q be Koszul operads and let D_0 be the associated Koszul dioperad (defined in Section 3.6), whose minimal resolution is denoted by $(D_{0\infty}, \delta_0)$.

3.9.3. Corollary. $H(\Omega_{no-oper}\langle D_{0\infty} \rangle, \delta_0) = \Omega_{\frac{1}{2}P \to P}\langle D_0 \rangle$.

Proof. By Theorem 3.9.2,

$$H(\Omega_{no-oper}\langle D_{0\infty} \rangle, \delta_0) = \Omega_{no-oper}\langle H(D_{0\infty}, \delta_0) \rangle = \Omega_{no-oper}\langle D_0 \rangle.$$

But the latter space cannot have graphs with wheels, since any such wheel would contain at least one "nonreduced" internal cyclic edge corresponding to composition,

$$\circ_{1,1} : D_0(m,1) \otimes D_0(1,n) \longrightarrow D_0(m,n),$$

which is zero by the definition of D_0 (see Section 3.6).   □

3.10. Theorem. *For any Koszul operads* P *and* Q,
   (i) *the natural morphism of graph complexes*

$$\left(D_{0\infty}^{\circlearrowleft}, \delta_0 \right) \longrightarrow \left(D_0^{\circlearrowleft}, 0 \right)$$

is a quasi-isomorphism if and only if the operads P *and* Q *are stably Koszul;*
   (ii) *there is, in general, an isomorphism of* $\mathbb{S}$-*bimodules*

$$H\left(D_{0\infty}^{\circlearrowleft}, \delta_0 \right) = \frac{H\left(P_{\infty}^{\circlearrowleft} \right) * H\left(Q_{\infty}^{\circlearrowleft} \right)^{\dagger}}{I_0},$$

where $H\left(P_{\infty}^{\circlearrowleft} \right)$ *and* $H\left(Q_{\infty}^{\circlearrowleft} \right)$ *are cohomologies of the wheeled completions of the minimal resolutions of the operads* P *and* Q, $*$ *stands for the free product of PROPs, and the ideal* I_0 *is defined in Section 3.6.1.*

Proof. (i) The necessity of the condition is obvious. Let us prove its sufficiency.

Let P and Q be stably Koszul operads such that the natural morphisms

$$\left(P_\infty^\circlearrowleft, \delta_P\right) \to P^\circlearrowleft \quad \text{and} \quad \left(Q_\infty^\circlearrowleft, \delta_Q\right) \to Q^\circlearrowleft$$

are quasi-isomorphisms, where (P_∞, δ_P) and (Q_∞, δ_Q) are minimal resolutions of P and Q respectively.

Consider a filtration of the complex $\left(D_{0\infty}^\circlearrowleft, \delta_0\right)$,

$$F_p := \left\{\text{span}\left\langle f \in D_{0\infty}^\circlearrowleft\right\rangle : |f|_2 - |f|_1 \le p\right\},$$

where

- $|f|_1$ is the number of cyclic vertices in f that belong to operadic wheels;
- $|f|_2$ is the number of noncyclic half-edges attached to cyclic vertices in f that belong to operadic wheels.

Note that $|f|_2 - |f|_1 \ge 0$. Let $\{E_r, d_r\}_{r \ge 0}$ be the associated spectral sequence. The differential d_0 in E_0 is given by its values on the vertices as follows:

(a) on every noncyclic vertex and on every cyclic vertex that does *not* belong to an operadic wheel one has $d_0 = \delta_0$;

(b) on every cyclic vertex that belongs to an operadic wheel one has $d_0 = 0$.

Hence modulo the action of finite groups (which we can ignore by Maschke's theorem) the complex (E_0, d_0) is isomorphic to the complex $(\Omega_{no-oper}\langle D_{0\infty}\rangle, \delta_0)$, tensored with a trivial complex (i.e., one with vanishing differential). By Corollary 3.9.3 and the Künneth formula we obtain,

$$E_1 = H(E_0, d_0) = W_1/h(W_2),$$

where

- W_1 is the subspace of $P^\circlearrowleft\left\langle E_P \oplus E_Q^\dagger\right\rangle$ consisting of graphs whose wheels (if any) are operadic; here the $\mathbb{S}$-bimodule $E_P \oplus E_Q^\dagger$ is given by

$$\left(E_P \oplus E_Q^\dagger\right)(m,n) = \begin{cases} E_P(2), \text{ the space of generators of } P, \text{ if } m=1, n=2, \\ E_Q(2), \text{ the space of generators of } Q, \text{ if } m=2, n=1, \\ 0, \qquad\qquad\qquad\qquad\qquad\qquad\qquad \text{otherwise;} \end{cases}$$

- W_2 is the subspace of $P^\circlearrowleft\left\langle E_P \oplus E_Q^\dagger \oplus I_P \oplus I_Q^\dagger\right\rangle$ consisting of graphs G whose wheels (if any) are operadic and satisfy the following condition: the elements of I_P and $I_Q^\dagger$ are used to decorate at least one noncyclic vertex in G. Here I_P and $I_Q^\dagger$ are $\mathbb{S}$-bimodules of relations of the quadratic operads P and $Q^\dagger$ respectively.

- the map $h : W_2 \to W_1$ is defined to be the identity on vertices decorated by elements of $E_P \oplus E_Q^\dagger$, and the tautological (in the obvious sense) morphism on vertices decorated by elements of I_P and $I_Q^\dagger$.

To understand all the remaining terms $\{E_r, d_r\}_{r \geq 1}$ of the spectral sequence we step aside and contemplate for a moment a purely operadic graph complex with wheels, say $(P_\infty^\circlearrowleft, \delta_P)$.

The complex $(P_\infty^\circlearrowleft, \delta_P)$ is naturally a subcomplex of $(D_{0\infty}^\circlearrowleft, \delta_0)$. Let

$$F_p := \left\{ \text{span} \left\langle f \in P_\infty^\circlearrowleft \right\rangle : \ |f|_2 - |f|_1 \ \leq p \right\}$$

be the induced filtration, and let $\left\{ E_r^P, d_r^P \right\}_{r \geq 0}$ be the associated spectral sequence. Then $E_1^P = H\left(E_0^P, d_0^P\right)$ is a subcomplex of E_1.

The main point is that modulo the action of finite groups, the spectral sequence $\{E_r, d_r\}_{r \geq 1}$ is isomorphic to the tensor product of spectral sequences of the form $\left\{E_r^P, d_r^P\right\}_{r \geq 1}$ and $\left\{E_r^Q, d_r^Q\right\}_{r \geq 1}$. By assumption, the latter converge to $P^\circlearrowleft$ and $Q^\circlearrowleft$ respectively, which implies the result.

(ii) The argument is exactly the same as in (i) except for the very last paragraph: the spectral sequences of the form $\left\{E_r^P, d_r^P\right\}_{r \geq 1}$ and $\left\{E_r^Q, d_r^Q\right\}_{r \geq 1}$ converge, respectively, to $H\left(P_\infty^\circlearrowleft\right)$ and to $H\left(Q_\infty^\circlearrowleft\right)$ (rather than to $P^\circlearrowleft$ and $Q^\circlearrowleft$). $\qquad\square$

3.11. Operadic wheeled extension. Let D_λ be a dioperad and $D_{\lambda\infty}$ its minimal resolution. Let $D_{\lambda\infty}^{\looparrowright}$ be a dg subprop of $D_{\lambda\infty}^\circlearrowleft$ spanned by graphs with at most operadic wheels (see Section 3.9.1). Similarly, one defines a subprop $D_\lambda^{\looparrowright}$ of $D_\lambda^\circlearrowleft$.

3.11.1. Theorem. *For any Koszul operads P and Q and any Koszul substitution law λ,*

(i) the natural morphism of graph complexes

$$\left(D_{\lambda\infty}^{\looparrowright}, \delta_\lambda\right) \longrightarrow D_\lambda^{\looparrowright}$$

is a quasi-isomorphism if and only if the operads P and Q are both stably Koszul;

(ii) there is, in general, an isomorphism of $\mathbb{S}$-bimodules

$$H\left(D_{\lambda\infty}^{\looparrowright}, \delta_\lambda\right) = H\left(D_{0\infty}^{\looparrowright}, \delta_0\right) = \frac{H\left(P_\infty^\circlearrowleft\right) * H\left(Q_\infty^\circlearrowleft\right)^\dagger}{I_0},$$

where $H\left(P_\infty^\circlearrowleft\right)$ and $H\left(Q_\infty^\circlearrowleft\right)$ are cohomologies of the wheeled completions of the minimal resolutions of the operads P and Q.

Proof. Use the spectral sequence of a filtration $\{F_p\}$ defined similarly to the one introduced in the proof of Theorem 3.10. We omit full details, since they are analogous to Section 3.10. $\qquad\square$

In the next section we apply some of the above results to compute cohomology of several concrete graph complexes with wheels.

4 Examples

4.1. Wheeled operad of strongly homotopy Lie algebras. Let $(\mathsf{Lie}_\infty, \delta)$ be the minimal resolution of the operad Lie of Lie algebras. It can be identified with the subcomplex of $(\mathsf{Lie^!B}_\infty, \delta)$ spanned by connected trees built on degree-one $(1, n)$-corollas, $n \geq 2$,

with the differential given by

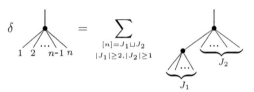

Let $\mathsf{Lie}^\circlearrowleft_\infty$ and $\mathsf{Lie}^\circlearrowleft$ be wheeled extensions of Lie_∞ and, respectively, Lie (see Section 3.5 for precise definitions).

4.1.1. Theorem. *The operad* Lie *of Lie algebras is stably Koszul, i.e.,* $\mathsf{H}\left(\mathsf{Lie}^\circlearrowleft_\infty\right) = \mathsf{Lie}^\circlearrowleft$.

Proof. We shall show that the natural morphism of dg props

$$\left(\mathsf{Lie}^\circlearrowleft_\infty, \delta\right) \longrightarrow \left(\mathsf{Lie}^\circlearrowleft, 0\right)$$

is a quasi-isomorphism. Consider a surjection of graph complexes (cf. Section 3.4)

$$u : \left(\mathsf{Lie}^+_\infty, \delta\right) \longrightarrow \left(\mathsf{Lie}^\circlearrowleft_\infty, \delta\right),$$

where Lie^+_∞ is the marked extension of $\mathsf{Lie}^\circlearrowleft_\infty$, i.e., the one in which one cyclic edge in every wheel is marked. This surjection respects the filtrations

$$F_p\mathsf{Lie}^+_\infty := \left\{\text{span}\left\langle f \in \mathsf{Lie}^+_\infty\right\rangle : \text{total number of cyclic vertices in } f \geq p\right\},$$

$$F_p\mathsf{Lie}^\circlearrowleft_\infty := \left\{\text{span}\left\langle f \in \mathsf{Lie}^\circlearrowleft_\infty\right\rangle : \text{total number of cyclic vertices in } f \geq p\right\},$$

and hence induces a morphism of the associated 0th terms of the spectral sequences

$$u_0 : \left(\mathsf{E}^+_0, \partial_0\right) \longrightarrow \left(\mathsf{E}^\circlearrowleft_0, \partial_0\right).$$

The point is that the (pro-)cyclic group acting on $\left(\mathsf{E}^+_0, \partial_0\right)$ by shifting the marked edge one step further along according to the orientation commutes with the differential ∂_0, so that u_0 is nothing but the projection to the coinvariants with respect to this action. Since we work over a field of characteristic 0, coinvariants can be identified with invariants in $\left(E_0^+, \partial_0\right)$. Hence we get, by Maschke's theorem,

$$H\left(E_0^{\circlearrowleft}, \partial_0\right) = \text{cyclic invariants in } H\left(E_0^+, \partial_0\right).$$

The next step is to compute the cohomology of the complex $\left(E_0^+, \partial_0\right)$. Consider its filtration

$$\mathcal{F}_p := \left\{ \text{span}\left\langle f \in E_0^+ \right\rangle : \begin{array}{l} \text{total number of noncyclic input} \\ \text{edges at cyclic vertices in } f \end{array} \leq p \right\},$$

and let $\{\mathcal{E}_r, \delta_r\}_{r \geq 0}$ be the associated spectral sequence. We shall show below that the latter degenerates at the second term (so that $\mathcal{E}_2 \simeq H\left(E_0^+, \partial_0\right)$). The differential δ_0 in $\mathcal{E}_0$ is given by its values on the vertices as follows:

(i) on every noncyclic vertex one has $\delta_0 = \delta$, the differential in Lie_∞;
(ii) on every cyclic vertex, $\delta_0 = 0$.

Hence the complex $(\mathcal{E}_0, \delta_0)$ is isomorphic to the direct sum of tensor products of complexes $(\text{Lie}_\infty, \delta)$. By the Künneth theorem, we get

$$\mathcal{E}_1 = V_1/h(V_2),$$

where

- V_1 is the subspace of Lie_∞^+ consisting of all those graphs whose every noncyclic vertex is ⋏;
- V_2 is the subspace of Lie_∞^+ whose every noncyclic vertex is either ⋏ or ⋏ with the number of vertices of the latter type ≥ 1;
- the map $h : V_2 \to V_1$ is given on noncyclic vertices by

$$h\left(\includegraphics{}\right) = \includegraphics{}, \qquad h\left(\includegraphics{}\right) = \includegraphics{} + \includegraphics{} + \includegraphics{},$$

and on all cyclic vertices, h is set to be the identity.

The differential δ_1 in $\mathcal{E}_1$ is given by its values on vertices as follows:

(i) on every noncyclic vertex one has $\delta_1 = 0$;
(ii) on every cyclic $(1, n+1)$-vertex with cyclic half-edges denoted by x and y, one has

$$\delta_1 \;\; \includegraphics{} \;\; = \sum_{\substack{[n]=J_1 \sqcup J_2 \\ |J_1|=2, |J_2| \geq 0}} \includegraphics{} .$$

To compute the cohomology of $(\mathcal{E}_1, \delta_1)$ let us step aside and compute the cohomology of the minimal resolution $(\mathsf{Lie}_\infty, \delta)$ (which we of course already know to be equal to Lie) in a slightly unusual way:

$$F_p^{\mathsf{Lie}} := \{\mathrm{span}\, \langle f \in \mathsf{Lie}_\infty \rangle : \text{number of edges attached to the root vertex of } f \leq p\}$$

is clearly a filtration of the complex $(\mathsf{Lie}_\infty, \delta)$. Let $\{E_r^{\mathsf{Lie}}, d_r^{\mathsf{Lie}}\}_{r \geq 0}$ be the associated spectral sequence. The cohomology classes of $E_1^{\mathsf{Lie}} = \mathsf{H}\left(E_0^{\mathsf{Lie}}, d_0^{\mathsf{Lie}}\right)$ resemble elements of $\mathcal{E}_1$: they are trees whose root vertex may have any number of edges while all other vertices are binary, $\bigwedge$. The differential d_1^{Lie} is nontrivial only on the root vertex, on which it is given by

The cohomology of $\left(E_1^{\mathsf{Lie}}, d_1^{\mathsf{Lie}}\right)$ is equal to the operad of Lie algebras. The differential d_1^{Lie} is identical to the differential δ_1 above except for the term corresponding to $|J_2| = 0$. Thus let us define another complex $\left(E_1^{\mathsf{Lie}+}, d_1^{\mathsf{Lie}+}\right)$ by adding to E_1^{Lie} trees whose root vertex is a degree-(-1) corolla $\bullet$ while all other vertices are binary $\bigwedge$. The differential $d_1^{\mathsf{Lie}+}$ is defined on root $(1, n)$-corollas with $n \geq 2$ by formally the same formula as for d_1^{Lie} except that the summation range is extended to include the term with $|I_1| = 0$. We also set $d_1^{\mathsf{Lie}+} \bullet = 0$.

Claim. The cohomology of the complex $\left(E_1^{\mathsf{Lie}+}, d_1^{\mathsf{Lie}+}\right)$ is a one-dimensional vector space spanned by $\bullet$.

Proof of the claim. Consider the 2-step filtration $F_0 \subset F_1$ of the complex $\left(E_1^{\mathsf{Lie}+}, d_1^{\mathsf{Lie}+}\right)$ by the number of $\bullet$'s. The zeroth term of the associated spectral sequence is isomorphic to the direct sum of the complexes,

$$\left(E_1^{\mathsf{Lie}}, d_1^{\mathsf{Lie}}\right) \oplus \left(E_1^{\mathsf{Lie}}[1], d_1^{\mathsf{Lie}}\right) \oplus (\mathrm{span}\langle \bullet \rangle, 0),$$

so that the next term of the spectral sequence is

$$\mathsf{Lie} \oplus \mathsf{Lie}[1] \oplus \langle \bullet \rangle$$

with the differential being zero on $\mathsf{Lie}[1] \oplus \langle \bullet \rangle$ and the the natural isomorphism

$$\mathsf{Lie} \longrightarrow \mathsf{Lie}[1]$$

on the remaining summand. Hence the claim follows.

The point of the above claim is that the graph complex $(\mathcal{E}_1, \delta_1)$ is isomorphic to the tensor product of a trivial complex with complexes of the form

$\left(E_1^{\mathsf{Lie}+}, d_1^{\mathsf{Lie}+}\right)$, which immediately implies that $\mathcal{E}_2 = \mathcal{E}_\infty \simeq H\left(E_0^+, \partial_0\right)$ is the direct sum of Lie and the vector space spanned by marked wheels of the type

whose every vertex is cyclic. Hence the cohomology group $H\left(E_0^{\circlearrowleft}, \partial_0\right) = E_1^{\circlearrowleft}$ we started with is equal to the direct sum of Lie and the space Z spanned by unmarked wheels of the type

whose every vertex is cyclic. Since every vertex is binary, the induced differential ∂_1 on $E_1^{\circlearrowleft}$ vanishes, the spectral sequence by the number of cyclic vertices we began with degenerates, and we conclude that this direct sum $\mathsf{Lie} \oplus Z$ is isomorphic to the required cohomology group $H\left(\mathsf{Lie}_\infty^{\circlearrowleft}, d\right)$.

Finally, one checks using Jacobi identities that every element of $\mathsf{Lie}^{\circlearrowleft}$ containing a wheel can be *uniquely* represented as a linear combination of graphs from Z, implying

$$\mathsf{Lie}^{\circlearrowleft} \simeq \mathsf{Lie} \oplus Z \simeq H\left(\mathsf{Lie}_\infty^{\circlearrowleft}, d\right)$$

and completing the proof. □

4.2. Wheeeled prop of polyvector fields.

Let $\mathsf{Lie}^1\mathsf{B}$ be the prop of Lie 1-bialgebras and $\left(\mathsf{Lie}^1\mathsf{B}_\infty, \delta\right)$ its minimal resolution (see Section 2.6.3). We denote their wheeled extensions by $\mathsf{Lie}^1\mathsf{B}^{\circlearrowleft}$ and $\left(\mathsf{Lie}^1\mathsf{B}_\infty^{\circlearrowleft}, \delta\right)$ respectively (see Section 3.5), and their operadic wheeled extensions by $\mathsf{Lie}^1\mathsf{B}^{\looparrowright}$ and $\left(\mathsf{Lie}^1\mathsf{B}_\infty^{\looparrowright}, \delta\right)$ (see Section 3.11). By Theorems 3.11.1 and 4.1.1, we have the followings result.

4.2.1. Proposition.
The natural epimorphism of dg props

$$\left(\mathsf{Lie}^1\mathsf{B}_\infty^{\looparrowright}, \delta\right) \longrightarrow \left(\mathsf{Lie}^1\mathsf{B}^{\looparrowright}, 0\right)$$

is a quasi-isomorphism.

We shall study next a subcomplex (*not* a subprop!) of the complex $\left(\mathsf{Lie}^1\mathsf{B}_\infty^{\circlearrowleft}, \delta\right)$ that is spanned by directed graphs with at most one wheel, i.e., with at most one closed path that begins and ends at the same vertex. We denote this subcomplex by $\mathsf{Lie}^1\mathsf{B}_\infty^{\circ}$. Similarly we define a subspace $\mathsf{Lie}^1\mathsf{B}^{\circ} \subset \mathsf{Lie}^1\mathsf{B}^{\circlearrowleft}$ spanned by equivalence classes of graphs with at most one wheel.

4.2.2. Theorem. $\mathsf{H}\left(\mathrm{Lie}^1\mathsf{B}^\circ_\infty, \delta\right) = \mathrm{Lie}^1\mathsf{B}^\circ$.

Proof. (a) Consider a two-step filtration $F_0 \subset F_1 := \mathrm{Lie}^1\mathsf{B}^\circ_\infty$ of the complex $(\mathrm{Lie}^1\mathsf{B}^\circ_\infty, \delta)$, with $F_0 := \mathrm{Lie}^1\mathsf{B}_\infty$ being the subspace spanned by graphs with no wheels. We shall show below that the cohomology of the associated direct sum complex

$$F_0 \oplus \frac{F_1}{F_0}$$

is equal to $\mathrm{Lie}^1\mathsf{B} \oplus \frac{\mathrm{Lie}^1\mathsf{B}^\circ}{\mathrm{Lie}^1\mathsf{B}}$. In fact, the equality $\mathsf{H}(F_0) = \mathrm{Lie}^1\mathsf{B}$ is obvious, so that it is enough to show below that the cohomology of the complex $\mathsf{C} := \frac{F_1}{F_0}$ is equal to $\frac{\mathrm{Lie}^1\mathsf{B}^\circ}{\mathrm{Lie}^1\mathsf{B}}$.

(b) Consider a filtration of the complex (C, δ),

$$F_p\mathsf{C} := \{\mathrm{span}\langle f \in \mathsf{C}\rangle : \text{ number of cyclic vertices in } f \geq p\},$$

and a similar filtration

$$F_p\mathsf{C}^+ := \{\mathrm{span}\langle f \in \mathsf{C}^+\rangle : \text{ number of cyclic vertices in } f \geq p\}$$

of the marked version of C. Let $\{\mathsf{E}_r, \partial_r\}_{r \geq 0}$ and $\{\mathsf{E}^+_r, \partial_r\}_{r \geq 0}$ be the associated spectral sequences. There is a natural surjection of *complexes*

$$u_0 : \left(\mathsf{E}^+_0, \partial_0\right) \longrightarrow \left(\mathsf{E}^\circlearrowright_0, \partial_0\right).$$

It is easy to see that the differential ∂_0 in E^+_0 commutes with the action of the (pro-)cyclic group on $\left(\mathsf{E}^+_0, \delta_0\right)$ by shifting the marked edge one step further along according to the orientation, so that u_0 is nothing but the projection to the coinvariants with respect to this action. Since we work over a field of characteristic 0, we get by Maschke's theorem

$$\mathsf{H}\left(\mathsf{E}^\circlearrowright_0, \partial_0\right) = \text{cyclic invariants in } \mathsf{H}\left(\mathsf{E}^+_0, \partial_0\right),$$

so that at this stage we can work with the complex $\left(\mathsf{E}^+_0, \partial_0\right)$. Consider a filtration of the latter,

$$\mathcal{F}_p := \left\{\mathrm{span}\,\langle f \in \mathsf{E}^+_0\rangle : \begin{array}{l} \text{total number of noncyclic input} \\ \text{edges at cyclic vertices in } f \end{array} \leq p\right\},$$

and let $\{\mathcal{E}_r, d_r\}_{r \geq 0}$ be the associated spectral sequence. The differential δ_0 in $\mathcal{E}_0$ is given by its values on the vertices as follows:

(i) on every noncyclic vertex one has $d_0 = \delta_\lambda$, the differential in $\mathrm{Lie}^1\mathsf{B}_\infty$;
(ii) on every cyclic vertex, $d_0 = 0$.

Hence the complex $(\mathcal{E}_0, d_0)$ is isomorphic to the direct sum of tensor products of complexes $(\mathrm{Lie}_\infty, \delta)$ with trivial complexes. By the Künneth theorem, we conclude that $\mathcal{E}_1 = \mathsf{H}(\mathcal{E}_0, d_0)$ can be identified with the quotient of the subspace in C spanned by graphs whose every noncyclic vertex is ternary, e.g., either $\curlyvee$ or $\curlywedge$, with respect to the equivalence relation generated by the following equations among noncyclic vertices:

$$(\star) \quad \raisebox{-1em}{[tree diagram]} + \raisebox{-1em}{[tree diagram]} + \raisebox{-1em}{[tree diagram]} = 0 \;, \quad \raisebox{-1em}{[tree diagram]} + \raisebox{-1em}{[tree diagram]} + \raisebox{-1em}{[tree diagram]} =$$

$$0\;, \quad \raisebox{-1em}{[tree diagram]} = 0.$$

The differential d_1 in $\mathcal{E}_1$ is nonzero only on cyclic vertices,

$$d_1 \;\raisebox{-1em}{[vertex diagram]} = \sum_{\substack{[m]=I_1 \sqcup I_2 \\ |I_1| \geq 0, |I_2| = 2}} (-1)^{\sigma(I_1 \sqcup I_2)+1} \;\raisebox{-1em}{[diagram]} \; + \sum_{\substack{[n]=J_1 \sqcup J_2 \\ |J_1|=2, |J_2|=2}} \;\raisebox{-1em}{[diagram]}$$

$$+ \sum_{\substack{[m]=I_1 \sqcup I_2 \\ [n]=J_1 \sqcup J_2 \\ |I_1| \geq 0, |I_2|=1 \\ |J_1| \geq 0, |J_2|=1}} (-1)^{\sigma(I_1 \sqcup I_2)} \;\raisebox{-1em}{[diagram]}$$

$$+ \sum_{\substack{[m]=I_1 \sqcup I_2 \\ [n]=J_1 \sqcup J_2 \\ |I_1|=1, |I_2| \geq 0 \\ |J_1|=1, |J_2| \geq 0}} (-1)^{\sigma(I_1 \sqcup I_2)+m} \;\raisebox{-1em}{[diagram]}$$

where cyclic half-edges (here and below) are dashed. Then $\mathcal{E}_1$ can be interpreted as a bicomplex $\left(\mathcal{E}_1 = \bigoplus_{m,n} \mathcal{E}_1^{m,n}, d_1 = \partial + \bar{\partial} \right)$ with, say, m counting the number of vertices attached to cyclic vertices in an "operadic" way (as in the first two summands above), and n counting the number of vertices attached to cyclic vertices in a "nonoperadic" way (as in the last two summands in the above formula). Note that the assumption that there is only *one* wheel in C is vital for this splitting of the differential d_1 to have sense. The differential ∂ (respectively, $\bar{\partial}$) is equal to the first (respectively, last) two summands in d_1.

Using the claim in the proof of Theorem 4.1.1, it is not hard to check that $H(\mathcal{E}_1, \partial)$ is isomorphic to the quotient of the subspace of C spanned by graphs whose

— every noncyclic vertex is ternary (i.e., the total number of attached half-edges equals 3), e.g., either $\raisebox{-0.3em}{[diagram]}$ or $\raisebox{-0.3em}{[diagram]}$;

— every cyclic vertex is either $\raisebox{-0.3em}{[diagram]}$ or $\raisebox{-0.3em}{[diagram]}$, or $\raisebox{-0.3em}{[diagram]}$, with respect to the equivalence relation generated by equations $(\star)$ and the following ones:

$(\star\star)$ $= 0$, $= 0$, $= 0$, $= 0$.

The differential $\bar{\partial}$ is nonzero only on cyclic vertices of the type , on which it is given by

$$\bar{\partial} \;\; \text{} \;\; = \; - \; \text{} \; - \; \text{}$$

Hence $\mathsf{A} := \mathsf{H}(\mathsf{H}(\mathcal{E}_1, \partial), \bar{\partial})$ can be identified with the quotient of the subspace of C spanned by graphs whose

— every noncyclic vertex is ternary, e.g., either or ;

— every cyclic vertex is also ternary, e.g., either or .

with respect to the equivalence relation generated by equations $(\star)$, $(\star\star)$, and, say, the following one:

$$\text{} = 0 .$$

Since all vertices are ternary, all higher differentials in our spectral sequences vanish, and we conclude that

$$\mathsf{H}(\mathsf{C}, \delta) \simeq \mathsf{A},$$

which proves the theorem.   □

4.2.3. Remark. As an independent check of the above arguments one can show using relations $R_1 - R_3$ in Section 2.6.2 that every element of $\frac{\mathsf{Lie}^1\mathsf{B}^\circ}{\mathsf{Lie}^1\mathsf{B}}$ can indeed be uniquely represented as a linear combinations of graphs from the space A.

4.2.4. Remark. Proposition 4.2.1 and Theorem 4.2.2 cannot be extended to the full wheeled prop $\mathsf{Lie}^1\mathsf{B}^\circlearrowleft_\infty$, i.e., the natural surjection

$$\pi : \left(\mathsf{Lie}^1\mathsf{B}^\circlearrowleft_\infty, \delta\right) \longrightarrow \left(\mathsf{Lie}^1\mathsf{B}, 0\right)$$

is *not* a quasi-isomorphism. For example, the graph

represents a nontrivial cohomology class in $\mathsf{H}^1\left(\mathsf{Lie}^1\mathsf{B}^\circlearrowleft_\infty, \delta\right)$.

4.3. Wheeled prop of Lie bialgebras. Let LieB be the prop of Lie bialgebras that is generated by the dioperad very similar to $\mathsf{Lie}^1\mathsf{B}$ except that both generating Lie and coLie operations, are in degree zero. This dioperad is again Koszul with Koszul substitution law, so that the analogue of Proposition 4.2.1 holds true for the operadic wheelification $\mathsf{LieB}_\infty^\circlearrowleft$. In fact, the analogue of Theorem 4.2.2 holds true for $\mathsf{LieB}_\infty^\circ$.

4.4. Prop of infinitesimal bialgebras. Let IB be the dioperad of infinitesimal bialgebras [MV03] that can be represented as a quotient

$$\mathsf{IB} = \frac{\mathsf{D}\langle E \rangle}{\mathsf{Ideal}\langle R \rangle}$$

of the free prop generated by the following $\mathbb{S}$-bimodule E:

- all $E(m, n)$ vanish except $E(2, 1)$ and $E(1, 2)$,
- $E(2, 1) := k[\mathbb{S}_2] \otimes \mathbb{1}_1 = \mathrm{span}\left(\begin{array}{c} \end{array} \right)$,
- $E(1, 2) := \mathbb{1}_1 \otimes k[\mathbb{S}_2] = \mathrm{span}\left(\begin{array}{c} \end{array} \right)$,

modulo the ideal generated by the associativity conditions for $\wedge$, coassociativity conditions for $\vee$, and

$$\vee\!\!\wedge - \wedge\!\!\vee - \wedge\!\!\vee = 0.$$

This is a Koszul dioperad with a Koszul substitution law. Its minimal prop resolution $(\mathsf{IB}_\infty, \delta)$ is a dg prop freely generated by the $\mathbb{S}$-bimodule $\mathsf{E} = \{\mathsf{E}(m, n)\}_{m, n \geq 1, m+n \geq 3}$, with

$$\mathsf{E}(m, n) := k[\mathbb{S}_m] \otimes k[\mathbb{S}_n][3 - m - n] = \mathrm{span}\left\langle \begin{array}{c} 1 \ \ 2 \ \ \ldots \ m{-}1 \ m \\ \\ 1 \ \ 2 \ \ \ldots \ n{-}1 \ n \end{array} \right\rangle.$$

By Claim 3.5.4, the analogue of Proposition 4.2.1 does *not* hold true for $\mathsf{IB}_\infty^\leftrightarrow$. Moreover, it is not hard to check that the graph

$$\mathord{\text{—}}$$

represents a nontrivial cohomology class in $\mathsf{H}^{-1}\left(\mathsf{IB}_\infty^\circlearrowleft \right)$. Thus the analogue of Theorem 4.2.2 does not hold true for IB_∞°, nor is the natural surjection $\mathsf{IB}_\infty^\circlearrowleft \to \mathsf{IB}^\circlearrowleft$ a quasi-isomorphism. This example is of interest because the wheeled dg prop $\mathsf{IB}_\infty^\circlearrowleft$ controls the cohomology of a directed version of Kontsevich's ribbon graph complex.

4.5. Wheeled quasiminimal resolutions. Let P be a graded prop with zero differential admitting a minimal resolution

$$\pi : (P_\infty = P\langle E\rangle, \delta) \to (P, 0).$$

We shall use in the following discussion of this pair of props P_∞ and P a *Tate–Jozefak* grading,[5] which, by definition, assigns degree zero to *all* generators of P and hence makes P_∞ into a nonpositively graded differential prop $P_\infty = \bigoplus_{i\leq 0} P_\infty^i$ with cohomology concentrated in degree zero, $H^0(P_\infty, \delta) = P$. Both props P_∞ and P admit canonically wheeled extensions,

$$P_\infty^\circlearrowleft := \bigoplus_{G\in\mathcal{G}^\circlearrowleft} G\langle E\rangle,$$

$$P^\circlearrowleft := H^0\left(P_\infty^\circlearrowleft, \delta\right).$$

However, the natural extension of the epimorphism π,

$$\pi^\circlearrowleft : \left(P_\infty^\circlearrowleft, \delta\right) \to (P, 0),$$

fails in general to stay a quasi-isomorphism.

Note that the dg prop $(P_\infty^\circlearrowleft, \delta)$ defined above is a *free* prop

$$P_\infty^\circlearrowleft := \bigoplus_{G\in\mathcal{G}^\uparrow} G\langle E^\circlearrowleft\rangle$$

on the S-bimodule $E^\circlearrowleft = \{E(m, n)\}_{m,n\geq 0}$,

$$E^\circlearrowleft(m, n) := \bigoplus_{G\in\mathcal{G}^\circlearrowleft_{indec}(m,n)} G\langle E\rangle,$$

generated by indecomposable (in the propic sense) decorated wheeled graphs. Note that the induced differential is *not* quadratic with respect to the generating set $E^\circlearrowleft$.

4.5.1. Theorem–definition. *There exists a dg free prop $([P^\circlearrowleft]_\infty, \delta)$ that fits into a commutative diagram of morphisms of props,*

$$
\begin{array}{ccc}
[P^\circlearrowleft]_\infty & \xrightarrow{\ \alpha\ } & P_\infty^\circlearrowleft \\
& \searrow^{qis} & \Big\downarrow \pi^\circlearrowleft \\
& & P^\circlearrowleft
\end{array}
$$

where α is an epimorphism of (nondifferential) props and q is a quasi-isomorphism of dg props. The prop $[P^\circlearrowleft]_\infty$ is called a quasiminimal resolution *of $P^\circlearrowleft$.*

---

[5]The Tate–Josefak grading of props Lie^1B_∞ and Lie^1B, for example, assigns to generating (m, n) corollas degree $3 - m - n$.

Proof. Let $s_1 : H^{-1}(P_\infty^{\circlearrowleft}) \to P_\infty^{\circlearrowleft}$ be any representation of degree-(-1) cohomology classes (if there are any) as cycles. Set $E_1 := H^{-1}(P_\infty^{\circlearrowleft})[1]$ and define a differential graded prop

$$P_1 := P\langle E^{\circlearrowleft} \oplus E_1 \rangle$$

with the differential δ extended to new generators as $s_1[1]$. By construction, $H^0(P_1) = P^{\circlearrowleft}$ and $H^{-1}(P_1) = 0$.

Let $s_2 : H^{-2}(P_1) \to P_1$ be any representation of degree-(-2) cohomology classes (if there are any) as cycles. Set $E_2 := H^{-2}(P_1)[1]$ and define a differential graded prop

$$P_2 := P\langle E^{\circlearrowleft} \oplus E_1 \oplus E_2 \rangle$$

with the differential δ extended to new generators as $s_2[1]$. By construction, $H^0(P_2) = P^{\circlearrowleft}$ and $H^{-1}(P_2) = H^{-2}(P_2) = 0$.

Continuing by induction, we construct a dg free prop $[P^{\circlearrowleft}]_\infty := P\langle E^{\circlearrowleft} \oplus E_1 \oplus E_2 \oplus E_3 \oplus \cdots \rangle$ with all the cohomology concentrated in Tate–Jozefak degree 0 and equal to $P^{\circlearrowleft}$. □

4.5.2. Example. The prop $[Ass^{\circlearrowleft}]_\infty$ has been explicitly described in [MMS] (it is denoted there by $Ass_\infty^{\circlearrowleft}$).

5 Wheeled Cyclic Complex

5.1. Genus-1 wheels. Let $(P\langle E \rangle, \delta)$ be a dg free prop, and let $(P^{\circlearrowleft}\langle E \rangle, \delta)$ be its wheeled extension. We assume in this section that the differential δ preserves the number of wheels.[6] Then it makes sense to define a subcomplex $(T^{\circlearrowleft}\langle E \rangle \subset P^{\circlearrowleft}\langle E \rangle$ spanned by graphs with precisely one wheel. In this section we use the ideas of cyclic homology to define a new cyclic bicomplex that computes cohomology of $(T^{\circlearrowleft}\langle E \rangle, \delta)$.

All the above assumptions are satisfied automatically if $(P\langle E \rangle, \delta)$ is the free dg prop associated with a free dg operad.

We denote by $T^+\langle E \rangle$ the obvious "marked wheel" extension of $T^{\circlearrowleft}\langle E \rangle$ (see Section 3.2).

5.2. Abbreviated notation for graphs in $T^+\langle E \rangle$. The half-edges attached to any internal vertex of split into, say m, ingoing and, say n, outgoing ones. The differential δ is uniquely determined by its values on such (m, n)-vertices for all possible $m, n \geq 1$. If the vertex is cyclic, then one of its input half-edges is cyclic and one of its output half-edges is also cyclic. In this section we show in pictures only those (half-)edges attached to vertices that are cyclic (unless otherwise explicitly stated), so that

---

[6]This is not that dramatic a loss of generality in the sense that there always exists a filtration of $(P^{\circlearrowleft}\langle E \rangle, \delta)$ by the number of wheels whose spectral sequence has zeroth term satisfying our condition on the differential.

- $\boxed{e}$ stands for a noncyclic (m,n)-vertex decorated by an element $e \in E(m,n)$;

- $\boxed{e}$ is a decorated cyclic (m,n)-vertex with no input or output cyclic half-edges marked;

- $\boxed{e}$ is a decorated cyclic (m,n)-vertex with the output cyclic half-edge marked;

- $\boxed{e}$ is a decorated cyclic (m,n)-vertex with the input cyclic half-edge marked.

The differential δ applied to any vertex of the last three types can be uniquely decomposed into the sum of the following three groups of terms:

$$\delta \boxed{e} = \sum_{\alpha \in I_1} \boxed{e_{a'}} \!\!\!{}^{\boxed{e_{a''}}} + \sum_{a \in I_2} \boxed{e_{a'}} \!\!\!{}^{\boxed{e_{a''}}} + \sum_{b \in I_3} \boxed{e_{b'}} \boxed{e_{b''}}$$

where we have shown also noncyclic *internal* edges in the last two groups of terms. The differential δ applied to $\boxed{e}$ and $\boxed{e}$ is given by exactly the same formula except for the presence/position of dashed markings.

5.3. New differential in $\mathsf{T}^+\langle E \rangle$. Let us define a new derivation b in $\mathsf{T}^+\langle E \rangle$: as follows:

- $b\,\boxed{e} := \delta\,\boxed{e}$,

- $b\,\boxed{e} := \delta\,\boxed{e}$,

- $b\,\boxed{e} = \delta\,\boxed{e}$,

- $b\,\boxed{e} := \delta\,\boxed{e} + \sum_{\alpha \in I_1} \boxed{e_{a'}}\!\!{}^{\boxed{e_{a''}}}$.

5.3.1. Lemma. *The derivation b satisfies $b^2 = 0$, i.e., $(\mathsf{T}^+\langle E \rangle, b)$ is a complex.*

The proof is a straightforward but tedious calculation based solely on the relation $\delta^2 = 0$.

5.4. Action of cyclic groups. The vector space $\mathsf{T}^+\langle E \rangle$ is naturally bigraded,

$$\mathsf{T}^+\langle E \rangle = \sum_{m \geq 0, n \geq 1} \mathsf{T}^+\langle E \rangle_{m,n},$$

where the summand $\mathsf{T}^+\langle E \rangle_{m,n}$ consists of all graphs with m noncyclic and n cyclic vertices. Note that $\mathsf{T}^+\langle E \rangle_{m,n}$ is naturally a representation space of the cyclic group $\mathbb{Z}_n$ whose generator t moves the mark to the next cyclic edge along the orientation. Define also the operator $N := 1 + t + \cdots + t^n$: $\mathsf{T}^+\langle E\rangle_{m,n} \to \mathsf{T}^+\langle E\rangle_{m,n}$, which symmetrizes the marked graphs.

5.4.1. Lemma. $\delta(1 - t) = (1 - t)b$ and $N\delta = bN$.

The proof is a straightforward calculation based on the definition of b.

Following the ideas of the theory of cyclic homology (see, e.g., [Lod98]), we introduce a fourth quadrant bicomplex

$$C_{p,q} := C_q, \quad C_q := \sum_{m+n=q} \mathsf{T}^+\langle E\rangle_{m,n}, \quad p \leq 0, q \geq 1,$$

with the differentials given by the following diagram:

$$
\begin{array}{ccccccccc}
\vdots & & \vdots & & \vdots & & \vdots & \\
\uparrow b & & \uparrow \delta & & \uparrow b & & \uparrow \delta & \\
\cdots \xrightarrow{N} & C_4 & \xrightarrow{1-t} & C_4 & \xrightarrow{N} & C_4 & \xrightarrow{1-t} & C_4 \\
\uparrow b & & \uparrow \delta & & \uparrow b & & \uparrow \delta & \\
\cdots \xrightarrow{N} & C_3 & \xrightarrow{1-t} & C_3 & \xrightarrow{N} & C_3 & \xrightarrow{1-t} & C_3 \\
\uparrow b & & \uparrow \delta & & \uparrow b & & \uparrow \delta & \\
\cdots \xrightarrow{N} & C_2 & \xrightarrow{1-t} & C_2 & \xrightarrow{N} & C_2 & \xrightarrow{1-t} & C_2 \\
\uparrow b & & \uparrow \delta & & \uparrow b & & \uparrow \delta & \\
\cdots \xrightarrow{N} & C_1 & \xrightarrow{1-t} & C_1 & \xrightarrow{N} & C_1 & \xrightarrow{1-t} & C_1 \\
\end{array}
$$

5.4.2. Theorem. *The cohomology group of the unmarked graph complex* $H(\mathsf{T}^{\circlearrowleft}\langle E\rangle, \delta)$ *is equal to the cohomology of the total complex associated with the cyclic bicomplex* $C_{\bullet,\bullet}$.

Proof. The complex $(\mathsf{T}^{\circlearrowleft}\langle E\rangle, \delta)$ can be identified with the cokernel $C_{\bullet}/(1-t)$ of the endomorphism $(1 - t)$ of the total complex $C_{\bullet}$ associated with the bicomplex $C_{\bullet,\bullet}$. Since the rows of $C_{\bullet,\bullet}$ are exact [Lod98], the claim follows. $\square$

Acknowledgment. It is a pleasure to thank Sergei Shadrin and Bruno Vallette for helpful discussions.

References

[Gan03] W. L. GAN, *Koszul duality for dioperads*, Math. Res. Lett. **10** (2003), no. 1, 109–124.

[Kon93] M. KONTSEVICH, *Formal (non)commutative symplectic geometry*, Gel'fand mathematical seminars, 1990–1992, Birkhäuser, 1993, pp. 173–187.

[Kon02] M. KONTSEVICH, Letter to M. Markl, 2002.

[Kon03] M. KONTSEVICH, *Deformation quantization of Poisson manifolds*, Lett. Math. Phys. **66** (2003), no. 3, 157–216.

[Lod98] J.-L. LODAY, *Cyclic homology*, Springer-Verlag, Berlin, 1998.

[McL65] S. MCLANE, *Categorical algebra*, Bull. Amer. Math. Soc. **71** (1965), 40–106.

[Mer05] S.A. MERKULOV, *Nijenhuis infinity and contractible dg manifolds*, *math.ag/0403244*, Compositio Mathematica (2005), no. 141, 1238–1254.

[Mer06a] S.A. MERKULOV, *Deformation quantization of strongly homotopy lie algebras*, 2006.

[Mer06b] S.A. MERKULOV, *Prop profile of Poisson geometry*, *math.dg/0401034*, Commun. Math. Phys. (2006), no. 262, 117–135.

[MMS] M. MARKL, S. MERKULOV, AND S. SHADRIN, *Wheeled PROPs, graph complexes and the master equation*.

[MV03] M. MARKL AND A. VORONOV, *PROPped-up graph cohomology*, arXiv:math.QA/0307081 (2003) and this volume.

[Val03] B. VALLETTE, *A koszul duality for props*, To appear in Trans. of Amer. Math. Soc. (2003).

Yang–Mills Theory and a Superquadric

Mikhail V. Movshev

Mathematics Department, Stony Brook University,
Stony Brook NY, 11794-3651, USA
mmovshev@math.sunysb.edu

Dedicated to Yu. I. Manin on his 70th birthday

Summary. We construct a real-analytic CR supermanifold $\mathcal{R}$, holomorphically embedded into a superquadric $\mathcal{Q} \subset \mathbf{P}^{3|3} \times \mathbf{P}^{*3|3}$. A CR distribution $\mathcal{F}$ on $\mathcal{R}$ enables us to define a tangential CR complex $\left(\Omega_{\mathcal{F}}^{\bullet}, \bar{\partial} \right)$.

We define a $\bar{\partial}$-closed trace functional $\int : \Omega_{\mathcal{F}}^{\bullet} \to \mathbb{C}$ and conjecture that a Chern-Simons theory associated with a triple $\left(\Omega_{\mathcal{F}}^{\bullet} \otimes \mathrm{Mat}_n, \bar{\partial}, \int \mathrm{tr} \right)$ is equivalent to $N = 3$, $D = 4$ Yang–Mills theory with a gauge group $\mathrm{U}(n)$. We give some evidences to this conjecture.

Key words: supersymmetry, Lagrangian, Yang–Mills theory, supermanifold

2000 Mathematics Subject Classifications: 53C80, 53C28, 81R25, 32C11, 58C50

1 Introduction

Twistor methods in gauge theory have a long history (summarized in [P90]). A common feature of these methods is that spacetime is replaced by a twistor (or ambitwistor) analytic manifold **T**. Equations of motion "emerge" (in the terminology of Penrose) from complex geometry of **T**.

The twistor approach turns out to be a very useful technical innovation. For example, difficult questions of classical gauge theory, e.g., those that appear in the theory of instantons, admit a translation into considerably more simple questions of analytic geometry of space **T**. In this way classification theorems in the theory of instantons has been obtained [AHDM].

Quantum theory has not been given a simple reformulation in the language of geometry of the space **T** so far. One of the reasons is that the quantum theory formulated formally in terms of a path integral requires a

Y. Tschinkel and Y. Zarhin (eds.), *Algebra, Arithmetic, and Geometry*,
Progress in Mathematics 270, DOI 10.1007/978-0-8176-4747-6_11,
© Springer Science+Business Media, LLC 2009

Lagrangian. Classical theory, as was mentioned earlier, provides only equations of motion whose definition needs no metric. In contrast, a typical Lagrangian requires a metric in order to be defined. Thus a task of finding the Lagrangian in (ambi)twistor setup is not straightforward. In this paper we present a Lagrangian for $N = 3$ $D = 4$ Yang–Mills (YM) theory formulated in terms of ambitwistors.

Recall that $N = 3$ YM theory coincides in components with $N = 4$ YM theory. The easiest way to obtain $N = 4$ theory is from $N = 1$ $D = 10$ YM theory by dimensional reduction. The Lagrangian of this ten-dimensional theory is equal to

$$(\langle F_{ij}, F_{ij} \rangle + \langle \not{D}\chi, \chi \rangle) dvol. \tag{1}$$

In the last formula, F_{ij} is a curvature of connection ∇ in a principal $U(n)$-bundle over $\mathbb{R}^{10}$. An odd field χ is a section of $S \otimes \mathrm{Ad}$, where S is a complex sixteen-dimensional spinor bundle, Ad is the adjoint bundle, $\not{D}$ is the Dirac operator, $\langle ., . \rangle$ is a Killing pairing on $u(n)$. The measure $dvol$ is associated with a flat Riemannian metric on $\mathbb{R}^{10}$, F_{ij} are coefficients of the curvature in global orthonormal coordinates. The $N = 4$ theory is obtained from this by considering fields invariant with respect to translations in six independent directions. The theory is conformally invariant and can be defined on any conformally flat manifold, e.g., S^4 with a round metric.

In 1978, E. Witten [W78] discovered that it is possible to encode solutions of the $N = 3$ supersymmetric YM-equation by holomorphic structures on a vector bundle defined over an open subset U in a superquadric $\mathcal{Q}$. We shall call the latter a complex ambitwistor superspace. In this description the action of $N = 3$ superconformal symmetry on the space of solutions is manifest. The symbol $n|m$ denotes the dimension of a supermanifold. More precisely, the quadric $\mathcal{Q} \subset \mathbf{P}^{3|3} \times \mathbf{P}^{*3|3}$ is defined by the equation

$$\sum_{i=0}^{3} x_i x^i + \sum_{i=1}^{3} \psi_i \psi^i = 0, \tag{2}$$

in bihomogeneous coordinates

$$x_0, x_1, x_2, x_3, \psi_1, \psi_2, \psi_3; \ x^0, x^1, x^2, x^3, \psi^1, \psi^2, \psi^3 \tag{3}$$

in $\mathbf{P}^{3|3} \times \mathbf{P}^{*3|3}$ (x_i, x^j even, ψ_i, ψ^j odd, the symbol $*$ in the superscript stands for the dual space). The quadric is a complex supermanifold. It makes sense therefore to talk about differential (p, q)-forms $\Omega^{p,q}(\mathcal{Q})$.

Let $\mathcal{G}$ be a holomorphic vector bundle on U. Denote by

$$\Omega^{0\bullet} \mathrm{End}\mathcal{G}, \tag{4}$$

a differential graded algebra of smooth sections of $\mathrm{End}\mathcal{G}$ with coefficients in $(0, p)$-forms.

Let $\bar{\partial}$ and $\bar{\partial}'$ be two operators corresponding to two holomorphic structures in $\mathcal{G}$. It is easy to see that $(\bar{\partial}' - \bar{\partial})b = ab$, where $a \in \Omega^{0,1}\mathrm{End}\mathcal{G}$. The integrability condition $\bar{\partial}'^2 = 0$ in terms of $\bar{\partial}$ and a becomes a Maurer–Cartan (MC) equation:

$$\bar{\partial}a + \frac{1}{2}\{a,a\} = 0. \tag{5}$$

A first guess would be that the space of fields of the ambitwistor version of $N = 3$ YM is $\Omega^{0,1}(U)\mathrm{End}\mathcal{G}$, where $\mathcal{G}$ is a vector bundle on U of some topological type. Witten suggested [W03] that the Lagrangian in question should be similar to a Lagrangian of holomorphic Chern–Simons theory

$$CS(a) = \int \mathrm{tr}\left(\frac{1}{2}a\bar{\partial}a + \frac{1}{6}a^3\right)\mathrm{Vol} \tag{6}$$

where Vol is some integral form. The action (6) reproduces equations of motion (5). The hope is that perturbative analysis of this quantum theory will give some insights on the structure of $N = 3$ YM.

The main result of the present note is that we give a precise meaning to this conjecture.

Introduce a real supermanifold $\mathcal{R} \subset \mathcal{Q}$ of real superdimension 8|12. It is defined by the equation

$$x_1\bar{x}^2 - x_2\bar{x}^1 + x_3\bar{x}^4 - x_4\bar{x}^3 + \sum_{i=1}^{3}\psi_i\bar{\psi}^i = 0. \tag{7}$$

In Section 5.1 we discuss the meaning of reality in superalgebra and geometry.

Definition 1. *Let M be a C^∞ supermanifold, equipped with a subbundle H of the tangent bundle T. We say that M is equipped with a CR structure if H carries a complex structure defined by a fiberwise transformation J.*

The operator J defines a decomposition of the complexification $H^{\mathbb{C}}$ into a direct sum of eigensubbundles $\mathcal{F} + \bar{\mathcal{F}}$.

We say that the CR structure J is integrable if the sections of $\mathcal{F}$ form a Lie subalgebra of $T^{\mathbb{C}}$ under the bracket of vector fields.

The tautological embedding of $\mathcal{R}$ into the complex manifold $\mathcal{Q}$ induces a CR structure specified by the distribution $\mathcal{F}$. Properties of this CR structure are discussed in Section 2.1. A global holomorphic supervolume form *vol* on $\mathcal{Q}$ is constructed in Proposition 12. When restricted on $\mathcal{R}$ it defines a section of $^{\mathrm{int}}\Omega_{\mathcal{F}}^{-3}$, a CR integral form. Functorial properties of this form are discussed in Section 5.3. For any CR holomorphic vector bundle $\mathcal{G}$ we define a differential graded algebra $\Omega_{\mathcal{F}}^{\bullet}\mathrm{End}(\mathcal{G})$. It is the tangential CR complex. We equip it with the trace

$$\int : a \to \int_{\mathcal{R}} \mathrm{tr}(a)\ vol \tag{8}$$

We define a CS-action of the form (6), where we replace an element of $\Omega^{0\bullet}(U)\mathrm{End}(\mathcal{G})$ by an element of $\Omega_{\mathcal{F}}^{\bullet}(\mathcal{R})\mathrm{End}(\mathcal{G})$. The integral is taken with respect to the measure *vol.* We make some assumptions about the topology of $\mathcal{G}$ as it is done in classical twistor theory. The space $\mathcal{S}$ is a superextension of a four dimensional sphere S^4 (see Section 2.2 for details). There is a projection

$$p : \mathcal{R} \to \mathcal{S}. \tag{9}$$

We require that $\mathcal{G}$ be topologically trivial along the fibers of p. It is an easy exercise in algebraic topology to see that topologically, all such bundles are pullbacks from S^4. On S^4, unitary vector bundles are classified by their second Chern classes.

Conjecture 2. Suppose $\mathcal{G}$ is a CR holomorphic vector bundle on $\mathcal{R}$ of rank n. Under the above assumptions, a CS theory defined by the algebra $\Omega^\bullet_{\mathcal{F}}\mathrm{End}(\mathcal{G})$ is equivalent to $N = 3$ YM theory on S^4 in a principal $U(n)$ bundle with the second Chern class equal to $c_2(\mathrm{End}(\mathcal{G}))$.

For perturbative computations in YM theory it is convenient to work in BV formalism. See [Sch00] for mathematical introduction and [MSch06] for applications to YM.

Conjecture 3. In the assumptions of Conjecture 2 we believe that $N = 3$ YM theory in the BV formulation is equivalent to a CS theory defined by the algebra $\Omega^\bullet_{\mathcal{F}}\mathrm{End}(\mathcal{G})$, where the field $a \in \Omega^\bullet_{\mathcal{F}}\mathrm{End}(\mathcal{G})$ has a mixed degree.

The following abstract definition will be useful.

Definition 4. *Suppose we are given a differential graded algebra (dga)(A, d) with a d-closed trace functional $\int$. We can consider A as a space of fields in some field theory with Lagrangian defined by the formula*

$$CS(a) = \int \left(\frac{1}{2} ad(a) + \frac{1}{6} a^3 \right). \tag{10}$$

We call it a Chern–Simons CS theory associated with a triple $(A, d, \int)$.

We say that two theories $(A, d, \int)$ and $(A', d', \int')$ are classically (formally) equivalent if there is a quasi-isomorphism of algebras with trace $f : (A, d, \int) \to (A', d', \int')$.

See the appendix of [MSch05] for an extension of this definition to A_∞ algebras with a trace.

Thus the matrix-valued Dolbeault complex $(\Omega^\bullet_{\mathcal{F}}(\mathcal{R}) \otimes \mathrm{Mat}_n, \bar{\partial})$ with a trace defined by the formula $\int(a) = \int_{\mathcal{R}} \mathrm{tr}(a) vol$ is an example of such, algebra.

We shall indicate existence of classical equivalence of $N = 3$ YM defined over $\Sigma = \mathbb{R}^4 \subset S^4$ and a CS theory defined over $p^{-1}(U)$, where U is an open submanifold of $\mathcal{S}$ with $U_{\mathrm{red}} = \Sigma$.

Here is the idea of the proof.

We produce a supermanifold Z and an integral form Vol on it. We show that a CS theory constructed using differential graded algebra with a trace $A(Z)$, associated with manifold a Z is classically equivalent to $N = 3$ YM theory. We interpret the algebra $A(Z)$ as a tangential CR complex on Z.

We shall construct a manifold Z and algebra A in two steps.

Here is a description of the steps in more details:

Step 1. We define a compact analytic supermanifold $\widetilde{\Pi F}$ and construct an integral form μ on it in the spirit of [MSch05]. Let A_{pt} be the Dolbeault complex of $\widetilde{\Pi F}$. Integration of an element $a \in A_{pt}$ against μ over $\widetilde{\Pi F}$ defines a $\bar{\partial}$-closed trace functional on A_{pt}. Recall that there is a canonical isomorphism of Lie algebras $\mathrm{Lie}(\mathrm{U}(n)) \otimes \mathbb{C} \cong \mathrm{Mat}_n$. We show that the Chern–Simons theory associated with dga $(A_{pt} \otimes \mathrm{Mat}_n, \bar{\partial}, \int \mathrm{tr}_{\mathrm{Mat}_n})$ is classically equivalent to $N = 3$ Yang–Mills theory with gauge group $\mathrm{U}(n)$ reduced to a point.

Step 2. From the algebra A_{pt} we reconstruct a differential algebra A. The algebra $A \otimes \mathrm{Mat}_n$ conjecturally encodes full $N = 3$ Yang–Mills theory with gauge group $\mathrm{U}(n)$ in the sense of Definition 4. If we put aside the differential d, A is equal to $A_{pt} \otimes C^{\infty}(\Sigma)$. The integral form we are looking for is equal to $\mathrm{Vol} = \mu \, dx_1 \, dx_2 \, dx_3 \, dx_4$, where Σ is equipped with global coordinates x_1, x_2, x_3, x_4.

The manifold Z is a CR submanifold. We identify it with an open subset of $\mathcal{R}$.

Finally, we would like to formulate an unresolved question. The restriction of a holomorphic vector bundle $\mathcal{G}$ over U on $U \cap \mathcal{R}$ defines a CR vector bundle over the intersection. Is it true that every CR holomorphic vector bundles can be obtained this way? The answer would be affirmative if we impose some analyticity conditions on the CR structure on $\mathcal{G}$. Presumably, super Levi forms will play a role in a solution of this problem.

It is tempting to speculate that there is a string theory on $\mathcal{Q}$ and $\mathcal{R}$ defines a D-brane in it.

We need to say a few words about the structure of this note. In Section 2 we make some definitions and provide some constructions used in the formulation of conjectures 2 and 3.

In Section 3 we give a geometric twistor-like description of $N = 3$ YM theory reduced to a point (Step 1). In Section 4 we do Step 2.

The appendix contains some useful definitions concerning reality in super-algebra and CR structures.

2 Infinitesimal constructions

In this section we shall show that the space $\mathcal{R}$ is homogeneous with respect to the action of a real form of the $N = 3$ superconformal algebra $\mathfrak{gl}(4|3)$. Here we also collect facts that are needed for a coordinate-free description of the space $\mathcal{R}$ in terms of Lie algebras of the symmetry group and isotropy subgroup.

2.1 Real structure on the Lie algebra $\mathfrak{gl}(4|3)$

In this section we describe a graded real structure on $\mathfrak{gl}(4|3)$. It will be used later in the construction of the CR structure on the real super-ambitwistor space.

The reader might wish to consult Section 5.1 for the definition of a graded real structure. There the reader will find an explanation of some of our notations. By definition, $\mathfrak{gl}(4|3)$ is a super Lie algebra of endomorphisms of $\mathbb{C}^{4|3} = \mathbb{C}^4 + \Pi\mathbb{C}^3$. The symbol Π stands for parity change. This Lie algebra consists of matrices of block form $\begin{pmatrix} A & B \\ C & D \end{pmatrix}$ with $A \in \mathrm{Mat}(4 \times 4, \mathbb{C})$, $D \in$ $\mathrm{Mat}(3 \times 3, \mathbb{C})$, $C \in \mathrm{Mat}(3 \times 4, \mathbb{C})$, $B \in \mathrm{Mat}(4 \times 3, \mathbb{C})$. The elements $\begin{pmatrix} A & 0 \\ 0 & D \end{pmatrix}$ belong to the even part $\mathfrak{gl}_0(4|3)$, the elements $\begin{pmatrix} 0 & B \\ C & 0 \end{pmatrix}$ to the odd $\mathfrak{gl}_1(4|3)$.

In the following, the symbol $\mathfrak{g}(\mathbb{K})$ will stand for a Lie algebra defined over a field $\mathbb{K}$. If the field is not present, it means that the algebra is defined over $\mathbb{C}$. The same applies to Lie groups.

Let $\mathfrak{g}$ be a complex super Lie algebra. By definition, a map ρ defines a graded real structure on a super Lie algebra $\mathfrak{g}$ if ρ is a homomorphism: $\rho[a, b] = [\rho(a), \rho(b)]$. In [Man], Yu. I. Manin suggested several definitions of a real structure on a (Lie) superalgebra. In the notation of [Man], these definitions are parametrized by a triple $(\epsilon_1, \epsilon_2, \epsilon_3), \epsilon_i = \pm$. Our real structure corresponds to the choice $\epsilon_1 = -, \epsilon_2 = \epsilon_3 = +$.

The reader will find a complete classification of graded real structures of simple Lie algebras in [Serg].

Define a matrix J as:

$$J = \begin{pmatrix} 0 & \mathrm{id} \\ -\mathrm{id} & 0 \end{pmatrix} \tag{11}$$

where id is the 2×2 identity matrix. A map ρ is defined as

$$\rho \begin{pmatrix} A & B \\ C & D \end{pmatrix} = \begin{pmatrix} J & 0 \\ 0 & \mathrm{id} \end{pmatrix} \begin{pmatrix} \overline{A} & \overline{B} \\ \overline{C} & \overline{D} \end{pmatrix} \begin{pmatrix} -J & 0 \\ 0 & \mathrm{id} \end{pmatrix} = \begin{pmatrix} -J\overline{A}J & J\overline{B} \\ -\overline{C}J & \overline{D} \end{pmatrix}. \tag{12}$$

The identity[1] $\rho^2 = \mathrm{sid}$ is a corollary of equation $J^2 = -\mathrm{id}$.

It is useful to analyze the Lie subalgebra $\mathfrak{gl}_0(4|3)^\rho$ of real points in $\mathfrak{gl}_0(4|3) = \mathfrak{gl}(4, \mathbb{C}) \times \mathfrak{gl}(3, \mathbb{C})$. Because of (12), we have $\mathfrak{gl}(3)^\rho = \mathfrak{gl}(3, \mathbb{R})$. To identify $\mathfrak{gl}^\rho(4)$ we interpret $\mathbb{C}^4 = \mathbb{C}^2 + \mathbb{C}^2$ (whose algebra of endomorphisms is $\mathfrak{gl}(4)$) as a two-dimensional quaternionic space $\mathbb{H} + \mathbb{H}$. Let $1, i, j, k$ be the standard $\mathbb{R}$-basis in quaternions, $\langle e_1, e_2 \rangle$ an $\mathbb{H}$-basis in $\mathbb{H} + \mathbb{H}$. The space $\mathbb{H} + \mathbb{H} = \mathbb{C}^2 + \mathbb{C}^2$ has a complex structure defined by right multiplication on i. Right multiplication on j defines an i-antilinear map. In a $\mathbb{C}$-basis $e_1, e_2, e_1 j, e_2 j$, a matrix of right multiplication on j is equal to J. From this it is straightforward to deduce that $\mathfrak{gl}^\rho(4) = \mathfrak{gl}(2, \mathbb{H})$.

---

[1]The operator sid is defined in the appendix in Definition 24.

Definition 5. *Let M be a C^∞ supermanifold with the tangent bundle T. Let $H \subset T$ be a subbundle equipped with a complex structure J. This data defines a (nonintegrable) CR structure on M. There is a decomposition[2] $H^\mathbb{C} = \mathcal{F} + \overline{\mathcal{F}}$. A CR structure (H, J) is integrable if a space of sections of $\mathcal{F}$ is closed under the commutator. In this case we also say that $\mathcal{F}$ is integrable.*

Definition 6. *Let M_{red} denote the underlying manifold of supermanifold M.*

If M is a real submanifold of a complex supermanifold N, then at any $x \in M$ the tangent space T_x to M contains a maximal complex subspace H_x. If rank H_x is constant along M, then a family of spaces H defines an integrable CR structure. In our case the manifold $\mathcal{R} \subset Q$ is defined by equation (7).

Denote by GL(4|3) an affine supergroup with Lie algebra Lie(GL(4|3)) equal to $\mathfrak{gl}(4|3)$.[3] We will show later that $\mathcal{R}$ is a homogeneous space of the real form of GL(4|3) described above. The induced CR structure is real-analytic and homogeneous with respect to the group action.

A CR structure on a supermanifold enables us to define an analogue of the Dolbeault complex. Suppose a supermanifold M carries a CR structure $\mathcal{F} \subset T^\mathbb{C}$. A space of complex 1-forms Ω^1_M contains a subspace I of forms pointwise orthogonal to $\mathcal{F}$. It is easy to see that $\mathcal{F}$ is integrable iff the ideal (I) is closed under d. Define the tangential CR complex $(\Omega^\bullet_\mathcal{F}, \bar\partial)$ to be $(\Omega^\bullet/(I), d)$.

A vector bundle $\mathcal{G}$ is CR holomorphic if the gluing cocycle g_{ij} satisfies $\bar\partial g_{ij} = 0$. In such a case we can define a $\mathcal{G}$-twisted CR complex $\Omega^\bullet_\mathcal{F}\mathcal{G}$.

Remark 7. Denote by σ the operation of complex conjugation. Define an antilinear map

$$s = \sigma \circ \begin{pmatrix} J & 0 \\ 0 & \mathrm{id} \end{pmatrix} : \mathbb{C}^{4|3} \to \mathbb{C}^{4|3}. \tag{13}$$

The map $a \to sas^{-1}$, $a \in \mathfrak{gl}(4|3)$, coincides with the real structure ρ. Let us think about the left-hand side of equation (2) as a quadratic function associated with an even bilinear form (a, b). It is easy to see that the left-hand side of equation (7) is equal to $(a, s(a)) = 0$. Naively, it would appear that the centralizer of the operator s would be precisely the real form of $(\mathfrak{gl}(4|3), \rho)$ and that it would preserve equations (2) and (7). The problem is that we cannot work pointwise in supergeometry. Instead, we consider equations (2), (7) as a system of real algebraic equations. We interpret them as a system of sections of some line bundles on the $C_\mathbb{H}$ manifold $M = \mathbf{P}^{3|3} \times \mathbf{P}^{*3|3} \times \overline{\mathbf{P}}^{3|3} \times \overline{\mathbf{P}}^{*3|3}$ (see Section 5.1 for a discussion of reality in supergeometry). The space M carries the canonical graded real structure ρ that leaves the space of equations invariant. The ρ-twisted diagonal action of $\mathfrak{gl}(4|3)$ also leaves the equations invariant.

The graded real structure induces a graded real structure ρ on $\mathfrak{gl}(4|3)^\rho$ and makes a supermanifold $\mathcal{R}$ an algebraic graded real supermanifold.

---

[2]In the following, the letter $\mathbb{C}$ in a superscript denotes complexification.
[3]For a global description of (GL(4|3), ρ) see [Pel].

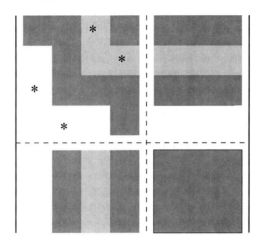

Fig. 1.

2.2 Symmetries of the ambitwistor space

We define the space $\mathcal{R}_{\mathrm{GL}(4|3)}$ as the homogeneous space of a real supergroup $(\mathrm{GL}(4|3), \rho)$. In this section we establish an isomorphism $\mathcal{R}_{\mathrm{GL}(4|3)} \cong \mathcal{R}$.

In Figure 1 the reader can see a graphical presentation of some matrix $\begin{pmatrix} A & B \\ C & D \end{pmatrix} \in \mathfrak{gl}(4|3)$.

The isotropy subalgebra $\mathfrak{a} \subset \mathfrak{gl}(4|3)$ of a base point in the space $\mathcal{R}_{\mathrm{GL}(4|3)}$ is defined as a linear space of matrices whose nonzero entries are in the darkest shaded area of the matrix in Figure 1.

Lemma 8. *The subspace $\mathfrak{a} \subset \mathfrak{gl}(4|3)$ is a ρ-invariant subalgebra.*

Proof. Direct inspection. □

Let A be an algebraic subgroup of $\mathrm{GL}(4|3)$ with the Lie algebra $\mathfrak{a}$.

The space $\mathcal{R}_{\mathrm{GL}(4|3)}$ carries a homogeneous CR structure (see Section 5.2 for a related discussion). Define a subspace $\mathfrak{p} \subset \mathfrak{gl}(4|3)$ as a set of matrices with nonzero entries in the gray and the dark gray areas in Figure 1.

Lemma 9. *The subspace $\mathfrak{p} \subset \mathfrak{gl}(4|3)$ is a subalgebra. It satisfies $\mathfrak{p} \cap \rho(\mathfrak{p}) = \mathfrak{a}$.*

Proof. Direct inspection. □

Let P denote an algebraic subgroup with Lie algebra $\mathfrak{p}$. A complex supermanifold $X = \mathrm{GL}(4|3)/P$ has an explicit description.

Equation (2) is preserved by the action of $\mathrm{GL}(4|3)$.

Proposition 10. *There is a $\mathrm{GL}(4|3)$-equivariant isomorphism $X = \mathcal{Q}$.*

Proof. We can identify the quadric $\mathcal{Q}$ with the space of partial flags $\mathbb{C}^{4|3}$ as it is done in the purely even case (see [GH], for example). A spaces $\mathcal{Q}$ is a connected component of the flag space containing the flag

$$F_1 \subset F_2 \subset \mathbb{C}^{4|3} \tag{14}$$

with $F_1 \cong \mathbb{C}^{1|0}$ and $F_2 \cong \mathbb{C}^{3|3}$. This flag can be interpreted as a pair of points $F_1 \in \mathbf{P}^{3|3}$, $F_2 \in \mathbf{P}^{*3|3}$. The condition (14) is equivalent to (2).

Let us choose a standard basis $e_1, \ldots, e_7$ of $\mathbb{C}^{4|3}$ such that the parities of elements are $\varepsilon(e_1) = \varepsilon(e_2) = \varepsilon(e_3) = \varepsilon(e_7) = 1$, $\varepsilon(e_4) = \varepsilon(e_5) = \varepsilon(e_6) = -1$. In this notations the standard flag F has the following description:

$$\begin{aligned} F_1 &= \operatorname{span}\langle e_7 \rangle \\ F_2 &= \operatorname{span}\langle e_2, \ldots e_7 \rangle. \end{aligned} \tag{15}$$

The flag defines a point in the space $\mathcal{Q}$. It is easy to compute the shape of the matrix of an element from the stabilizer P_F of F. The following picture is useful:

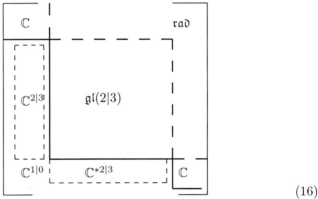

$$\tag{16}$$

The Lie algebra $\mathfrak{p}_F$ of the stabilizer P_F is formed by matrices with zero entries below the thick solid line in picture (16). Conjugating with a suitable permutation of coordinates t we see that $\mathfrak{p}_F^t = \mathfrak{p}$. $\qquad\square$

Remark 11. The transformation s defined in equation (13) acts on the space of flags. By definition, an s-invariant flag belongs to the subvariety $\mathcal{R}$. Direct inspection shows that the nonzero entries of the matrix of an element of the stabilizer are located in the darkest shaded area of Figure 1. The manifold $\mathcal{R}_{\mathrm{red}}$ fibers over $\mathbf{P}^3$ with connected fibers. Thus $\mathcal{R}$ is connected. From this and a simple dimension count we conclude that the subvariety $\mathcal{R}$ coincides with $\mathcal{R}_{\mathrm{GL}(4|3)}$. Identification of the CR structure also follows from this.

The Lie algebra $\mathfrak{gl}(4|3)$ contains a subalgebra $\mathfrak{l}$. The elements of this subalgebra have nonzero entries in the darkest area of Figure 1 and also the points marked by $*$. This algebra is invariant with respect to the real structure ρ. Denote by L an algebraic subgroup of $\mathrm{GL}(4|3)$ with the Lie algebra $\mathfrak{l}$.

The quotient $(\mathcal{S}_{\mathrm{GL}(4|3)}, \rho) = ((\mathrm{GL}(4|3)/L), \rho)$ is a supermanifold with $(\mathcal{S}_{\mathrm{GL}(4|3)})^\rho_{\mathrm{red}} = S^4$. Indeed, the real points L^ρ_{red} of the group L_{red} are conjugated to quaternionic matrices of the form $\begin{pmatrix} a & b \\ 0 & d \end{pmatrix} \in \mathrm{GL}(2, \mathbb{H})$. Thus the quotient space $\mathrm{GL}(2, \mathbb{H})/L^\rho_{\mathrm{red}} = \mathbb{H}\mathbf{P}^1$ is isomorphic to S^4. Denote by p the projection

$$\mathcal{R}_{\mathrm{GL}(4|3)} \to \mathcal{S}_{\mathrm{GL}(4|3)}. \tag{17}$$

An easy local exercise with Lie algebras reveals that the fibers of the projection p are CR holomorphic and are isomorphic to $\mathbf{P}^1 \times \mathbf{P}^1$.

The following direct geometric description of ambitwistor space will be useful. Let M be a C^∞ 4-dimensional Riemannian manifold. A metric g defines a relative quadric (the ambitwistor space) in the projectivisation of a complexified tangent bundle $A(M) \subset \mathbf{P}(T^\mathbb{C})$. By construction there is a projection $p: A(M) \to M$. The space $A(M)$ carries a CR structure (it could be nonintegrable). Indeed, a fiber of the distribution $\mathcal{F}$ at a point $x \in A(M)$ is a direct sum of the holomorphic tangent space to the fiber through x and a complex line in $T^\mathbb{C}(M)$ spanned by x. From the point of view of topology, the space $A(M)$ coincides with a relative Grassmannian of oriented 2-planes in T_M. A constructed complex distribution depends only on the conformal class of the metric. From this we conclude that $A(S^4)$ (S^4 has a round metric induced by the standard embedding into $\mathbb{R}^5$) is a homogeneous space of $\mathrm{Conf}(S^4) = \mathrm{PGL}(2, \mathbb{H})$, $A(S^4) = \mathrm{PGL}(2, \mathbb{H})/A^\rho_{\mathrm{red}}$, and the CR structure is integrable.

An appropriate super generalization of this construction is as follows. We have an isomorphism

$$\mathcal{W}_l \otimes \mathcal{W}_r \overset{\Gamma}{\cong} T^\mathbb{C}_M. \tag{18}$$

In the last formula, $\mathcal{W}_l, \mathcal{W}_r$ are complex two-dimensional spinor bundles on M (we assume that M has a spinor structure). The isomorphism Γ is defined by Clifford multiplication. Let T be a 3-dimensional linear space. This vector space will enable us to implement $N = \dim(T) = 3$ supersymmetry. To simplify notations we keep $\mathcal{W}_l \otimes T + \mathcal{W}_r \otimes T^*$ for the pullback $p^*(\mathcal{W}_l \otimes T + \mathcal{W}_r \otimes T^*)$. Define a split, holomorphic in odd directions[4] supermanifold $\tilde{A}(M)$ associated with a vector bundle $\Pi(\mathcal{W}_l \otimes T + \mathcal{W}_r \otimes T^*)$ over $A(M)$. To complete the construction we define a superextension of the CR structure. Introduce odd local coordinates $\theta^i_\alpha, \tilde{\theta}^{j\beta}$ ($1 \leq i, j \leq 2, 1 \leq \alpha, \beta \leq 3$) on fibers of $\Pi(\mathcal{W}_l \otimes T + \mathcal{W}_r \otimes T^*)$. We decompose local complex vector fields $\Gamma\left(\frac{\partial}{\partial \theta^i_\alpha} \otimes \frac{\partial}{\partial \tilde{\theta}^{j\beta}}\right)$ in a local real basis $\frac{\partial}{\partial x^s}$ as $\delta^\beta_\alpha \Gamma^s_{ij} \frac{\partial}{\partial x^s}$, $1 \leq s \leq 4$. The odd part of the CR distribution $\mathcal{F}$ is locally spanned by vector fields

$$\begin{aligned} &\frac{\partial}{\partial \theta_{i\alpha}} + \tilde{\theta}^j_\alpha \Gamma^s_{ij} \frac{\partial}{\partial x^s}, \\ &\frac{\partial}{\partial \tilde{\theta}^j_\alpha} + \theta^{j\alpha} \Gamma^s_{ij} \frac{\partial}{\partial x^s}. \end{aligned} \tag{19}$$

---

[4]The reader might wish to consult Section 5.1 about this.

This construction of a superextension of the ordinary CR structure depends only on the conformal class of the metric. It is convenient to formally add the complex conjugate odd coordinates. This way we get $\mathcal{A}(M) = \Pi(\mathcal{W}_l \otimes T + \mathcal{W}_r \otimes T^* + \overline{\mathcal{W}_l} \otimes T + \overline{\mathcal{W}_r} \otimes T^*)$, equipped with a graded real structure. As in the even case the symmetry analysis allows us to identify the CR space $\mathcal{A}(S^4)$ with $\mathcal{R}$.

The tangent space $\mathfrak{m}$ to the quadric $\mathcal{Q}$ at a point fixed by $\mathfrak{p}$ is formed by elements with nonzero entries below the thick solid line in the picture (16). It decomposes into a sum $\mathbb{C}^{2|3} + \mathbb{C}^{*2|3} + \mathbb{C}^{1|0}$ of irreducible GL(2|3) representations.

The elements $a_{21}, a_{31}, \alpha_{41}, \alpha_{51}, \alpha_{61}, a_{71}, a_{72}, a_{73}, \alpha_{74}, \alpha_{75}, \alpha_{76}$ stand for matrix coordinate functions on the linear space $\mathfrak{m}$ (coordinates a are even, α are odd).

Proposition 12. *An element*

$$vol = da_{21} \wedge da_{31} \wedge d\alpha_{41} \wedge d\alpha_{51} \wedge d\alpha_{61} \wedge da_{71} \wedge da_{72} \wedge da_{73} \wedge d\alpha_{74} \wedge d\alpha_{75} \wedge d\alpha_{76} \tag{20}$$

belongs to the Berezinian $\mathrm{Ber}(\mathfrak{m}^*)$. *It is invariant with respect to the action of P.*

Proof. Simple weight count. □

We spread a generator of $\mathrm{Ber}(\mathfrak{m}^*)$ by the action of GL(4|3) over $\mathcal{Q}$ and form a GL(4|3)-invariant section *vol* of the Berezinian bundle $\mathrm{Ber}_\mathbb{C}(\mathcal{Q})$.

3 Reduced theory

As a preliminary step in the construction of the superspace Z we introduce a "holomorphic" manifold $\widetilde{\Pi F}$ and an integral form on it (see Section 3.2 for an explanation of the quotation marks). The form defines a functional $\int$ on the Dolbeault complex of this manifold. We prove that the CS theory constructed by the triple $(\Omega^{0\bullet}(\widetilde{\Pi F}) \otimes \mathrm{Mat}_n, \bar{\partial}, \int \mathrm{tr}_{\mathrm{Mat}_n})$ is classically equivalent to $N = 3$ YM theory with the gauge group U(n) reduced to a point.

3.1 The manifold ΠF

A manifold $\widetilde{\Pi F}$ is a deformation of a simple manifold ΠF. In this section we give relevant definitions concerning ΠF.

Denote a product $\mathbf{P}^1 \times \mathbf{P}^1$ by X. It has two projections $p_i : X \to \mathbf{P}^1, i = l, r$. Let $\mathcal{O}(1)$ denote the dual to the Hopf line bundle over $\mathbf{P}^1$. The Picard group of X is $\mathbb{Z} + \mathbb{Z}$. It is generated by the classes of line bundles $\pi_l^* \mathcal{O}(1) = \mathcal{L}_l, \pi_r^* \mathcal{O}(1) = \mathcal{L}_r$ which can serve as coordinates in $Pic(X)$. Let $\mathcal{O}(a,b)$ denote a line bundle $\mathcal{L}_l^{\otimes a} \otimes \mathcal{L}_r^{\otimes b}$.

Convention. We denote by $H^\bullet(Y, \mathcal{G})$ the cohomology of the (super)manifold Y with coefficients in the vector bundle $\mathcal{G}$. It can be computed as cohomology of the Dolbeault complex $\Omega^{0\bullet}(Y)\mathcal{G}$. It is tacitly assumed that in Section 3, the omitted argument Y in $\Omega^{0\bullet}(Y)\mathcal{G}$ implies $Y = X$. If the $\mathcal{G}$-argument is missing, we assume that $\mathcal{G} = \mathcal{O}$.

Denote by $\mathrm{Sym}V, \Lambda V$ symmetric and exterior algebras of a vector space (bundle).

Denote by Θ a line bundle isomorphic to $\mathcal{O}(1,1)$. We construct a vector bundle F over X as a direct sum:

$$F = T \otimes \mathcal{L}_l + T^* \otimes \mathcal{L}_r + \Theta^*,$$
$$H = T \otimes \mathcal{L}_l + T^* \otimes \mathcal{L}_r. \tag{21}$$

As before, T is a three-dimensional vector space.

The reader may have noticed that the manifold X has also appeared as a fiber of projection (9). We shall see that this is not accidental.

3.2 Properties of the manifold ΠF

In this section we devise an infinitesimal deformation of a complex structure on ΠF. This deformation will be promoted to the actual deformation, and denote the corresponding complex main fold by $\widetilde{\Pi F}$. The algebra A_{pt} from the introduction is equal to $\Omega^{0\bullet}(\widetilde{\Pi F})$. We construct on $\widetilde{\Pi F}$ an integral form that will enable us to define a functional $\int_D : \Omega^{0\bullet}(\widetilde{\Pi F}) \to \mathbb{C}$.

The manifold ΠF is a complex split supermanifold. The Dolbeault complex $(\Omega^{0\bullet}(\Pi F), \bar{\partial})$ is defined on the supermanifold ΠF, considered as a graded real supermanifold. The complex $(\Omega^{0\bullet}(\Pi F), \bar{\partial})$ contains as a differential subalgebra the Dolbeault complex $\Omega^{0\bullet}\Lambda F^*$.

Proposition 13. *The differential algebras $\Omega^{0\bullet}(\Pi F)$ and $\Omega^{0\bullet}\Lambda F^*$ are quasi-isomorphic.*

Proof. The same as the proof of Proposition 26. □

The canonical line bundle $\mathcal{K}_X$ is equal to $\mathcal{O}(-2,-2)$. There is a nontrivial cohomology class, the "fundamental" class: $\alpha \in H^2(X, \mathcal{O}(-2,-2)) \overset{\mathrm{id}}{\subset} H^2(X, \mathcal{O}(-2,-2) \otimes T \otimes T^*) \subset H^2(X, \Lambda^2(H^*) \otimes \Theta^*) \subset H^2(X, \Lambda F^* \otimes \Theta^*)$. We interpret $\Lambda(F^*) \otimes \Theta^*$ as a sheaf of local holomorphic differentiations of ΠF in the direction of Θ^*.

A representative $\alpha = f d\bar{z}_l d\bar{z}_r \frac{\partial}{\partial \theta}$ (z_l, z_r are local coordinates on X) of the class $[\alpha]$ can be extended to a differentiation of $\Omega^{0\bullet}(\Pi F)$. The main properties of $D = \bar{\partial} + \alpha$ are:

(1) it is a differentiation of $\Omega^{0\bullet}(\Pi F)$,
(2) the equation $D^2 = 0$ holds.

These are corollaries of the $\bar{\partial}$-cocycle equation for α. The operator D defines a new "holomorphic" structure on ΠF.[5] This new complex manifold will be denoted by $\widetilde{\Pi F}$.

---

[5]This definition is not standard, because usually a deformation cocycle α is an element of $\Omega^{0,1}\mathcal{T}$ ($\mathcal{T}$ is a holomorphic tangent bundle), whereas in our case $\alpha \in$

The manifold $\varPi F$ is Calabi–Yau. By this we mean that $\mathrm{Ber}_{\mathbb{C}}$ is trivial. Indeed, the determinant line bundle of F is equal to $\det(T \otimes \mathcal{L}_l) \otimes \det(T^* \otimes \mathcal{L}_r) \otimes \det(\mathcal{O}(-1,-1)) = \mathcal{O}(3,0) \otimes \mathcal{O}(0,3) \otimes \mathcal{O}(-1,-1) = \mathcal{O}(2,2)$; $\mathrm{Ber}_{\mathbb{C}} \varPi F = \mathcal{K}_X \otimes \det F = \mathcal{O}$ is trivial.

It implies that the bundle $\mathrm{Ber}_{\mathbb{C}} \varPi F$ admits a nonvanishing section $vol_{\varPi F}$.

This section is SO(4)-invariant. The action of $\mathfrak{u} = \mathbb{C} + \mathbb{C}$, the unipotent subgroup of the Borel subgroup $B \subset \mathrm{SO}(4)$ on the large Schubert cell of X, is transitive and free.

Hence the section of $\mathrm{Ber}_{\mathbb{C}} \varPi F$ in $\mathfrak{u}$-coordinates is

$$vol_{\varPi F} = dz_l \wedge dz_r \wedge d\alpha_1 \wedge d\alpha_2 \wedge d\alpha_3 \wedge d\tilde{\alpha}_1 \wedge d\tilde{\alpha}_2 \wedge d\tilde{\alpha}_3 \wedge d\theta,$$

where $\alpha_1, \ldots, \tilde{\alpha}_3, \theta$ are $\mathfrak{u}$-invariant coordinates on the odd fiber. The section $vol_{\varPi F}$ is in the kernel of D by construction.

We can construct on manifold $\varPi F$ a global holomorphic integral -2-form in the way explained in remark (9). In our case it is equal to $c_{\varPi F} = \alpha_1 \ldots \tilde{\alpha}_3 \theta d\alpha_1 \wedge \cdots \wedge d\tilde{\alpha}_3 \wedge d\theta$.

Proposition 14. *The form* $\mu = vol_{\varPi F} \otimes \bar{c}_{\varPi F}$ *is a D-closed nontrivial integral* $(0,-2)$-*form on the underlying real graded* $\varPi F$.

Proof. Direct inspection in local coordinates. □

An integral form μ defines a $(\bar{\partial} + \alpha)$-closed trace on $\Omega^{0\bullet}(\varPi F) \otimes \mathrm{Mat}_n$,

$$\int a = \int_{\varPi F} \mathrm{tr}(a)\mu.$$

Definition 15. *By definition, an* A_∞ *algebra is a graded linear space, equipped with a series of maps* $\mu_n : A^{\otimes n} \to A, n \geq 1$, *of degree* $2 - n$ *that satisfy the quadratic relation*

$$\sum_{i+j=n+1} \sum_{0 \leq l \leq i} \epsilon(l,j)$$
$$\times \mu_i(a_0, \ldots, a_{l-1}, \mu_j(a_l, \ldots, a_{l+j-1}), a_{l+j}, \ldots, a_n) = 0, \quad (22)$$

where $a_m \in A$, *and* $\epsilon(l,j) = (-1)^{j \sum_{0 \leq s \leq l-1} \deg(a_s) + l(j-1) + j(i-1)}$. *In particular,* $\mu_1^2 = 0$.

Remark 16. Suppose we have an A_∞ algebra A equipped with a projector π. A homotopy H such that $\{d, H\} = \mathrm{id} - \pi$ can be used as input data for construction of a new A_∞ structure on $\mathrm{Im}\,\pi$ (see [Kad], [Markl] for details). The homotopy H is not unique. The resulting A_∞ algebras will have different

---

$\Omega^{0,2}\mathcal{T}$. We, however, continue to use traditional wording and call it a deformation of a complex structure, though a more precise term would be deformation of the Dolbeault algebra $(\Omega^{0\bullet}, \bar{\partial})$. This algebra in our approach becomes a substitute for the underlying manifold.

multiplications, depending on H. All of them will be A_∞ equivalent. An additional structure on A helps to prevent ambiguity in the choice of H. In our case, the algebra A is a Dolbeault complex of a manifold with the operator π being an orthogonal projection on cohomology. If the manifold is compact, Kähler, and G-homogeneous, there is a natural choice of $H : H = \bar{\partial}^*/\Delta'$. The operators $\bar{\partial}^*$, Δ' are constructed by a G-invariant metric. The operator Δ' is equal to Δ on $\mathrm{Ker}\Delta^\perp$ and equal to the identity on $\mathrm{Ker}\Delta$.

Remark 17. The construction described in Remark 16 admits a generalization. Suppose an A_∞ algebra A has a differential d that is the sum of two anticommuting differentials d_1 and d_2. Assume that $\{d_1, H\} = \mathrm{id} - \pi$ and the composition $d_2 H$ is a nilpotent operator. Then $\mathrm{Im}\ \pi$ carries the structure of an A_∞ algebra quasi-isomorphic to A. The same statement is true for A_∞ algebras with a trace. The proof goes along the same lines as in [Markl], but we allow two-valent vertices.

Technically, it is more convenient to work not with algebra $\Omega^{0\bullet}(\widetilde{\Pi F})$ but with a quasi-isomorphic subalgebra $(\Omega^{0\bullet}\Lambda F, D)$.

In application of the constructions from the remarks (3), (4) we choose π to be an orthogonal projector from Hodge theory, corresponding to the $SO(4, \mathbb{R})$-invariant metric on X. We also use the decomposition $D = d_1 + d_2 = \bar{\partial} + \alpha$.

The algebra of cohomology of $(\Omega^{0\bullet}\Lambda F, \bar{\partial})$ carries an A_∞-algebra structure. We denote it by $C = H^\bullet(X, \Lambda(H^*) \otimes \Lambda(\Theta))$. Denote by ψ a quasi-isomorphism $(\Omega^{0\bullet}\Lambda F, \bar{\partial}) \to C$. We shall describe some properties of C. Let W_l, W_r be spinor representations of $SO(4)$. The vector representation V is equal to $W_l \otimes W_r$.

The differential α induces a differential $[\alpha]$ on C. The ghost grading of the group $H^i(X, \Lambda^k(H^*) \otimes \Lambda^s(\Theta))$ is equal to $i + s$; the additional grading is equal to $k + 2s$ (preserved by $\bar{\partial}$ and α). We used a nonstandard ghost grading that differs from the one used in physics by a shift by one. In particular, the ghost grading of the gauge (labeled by V) and spinor (labeled by spinors W_l, W_r) and matter ($SO(4)$ action is trivial) fields is equal to one. In the table below you will find the field content (representation theoretic description) of C:

gh	deg		gh	deg	
0	0	$H^0(X, \Lambda^0(H^*)) = \mathbb{C}$	1	2	$H^0(X, \Lambda^0(H^*) \otimes \Theta) = V$
1	2	$H^1(X, \Lambda^2(H^*)) = \Lambda^2(T) + \Lambda^2(T^*)$	1	3	$H^0(X, \Lambda^1(H^*) \otimes \Theta) = W_l \otimes T$
					$+W_r \otimes T^*$
1	3	$H^1(X, \Lambda^3(H^*)) = W_l + W_r$	1	4	$H^0(X, \Lambda^2(H^*) \otimes \Theta) = T \otimes T^*$
2	4	$H^2(X, \Lambda^4(H^*)) = T \otimes T^*$	2	5	$H^1(X, \Lambda^3(H^*) \otimes \Theta) = W_l + W_r$
2	5	$H^2(X, \Lambda^5(H^*)) = W_l \otimes \Lambda^2(T) + W_r \otimes \Lambda^2(T^*)$	2	6	$H^1(X, \Lambda^4(H^*) \otimes \Theta) = T + T^*$
2	6	$H^2(X, \Lambda^6(H^*)) = V$	3	8	$H^2(X, \Lambda^6(H^*) \otimes \Theta) = \mathbb{C}$

$$(23)$$

The groups $H^0(X, \Lambda^2(H^*) \otimes \Theta)$ and $H^2(X, \Lambda^4(H^*))$ are contracting pairs, they are killed by the differential $[\alpha]$ and should be considered as auxiliary fields in the related CS theory.

An A_∞ algebra C has, besides, the differential $[\alpha]$ and multiplication, a higher multiplication on three arguments (corresponding to the cubic nonlinearity of the YM equation). However operations in more than three arguments are not present. This can be deduced from the homogeneity of $\bar{\partial}$ and α with respect to the additional grading. Finally, representation theory fixes structure maps up to a finite number of parameters. The integral $\int$ defines a nonzero map tr : $H^2(X, \Lambda^6(H^*) \otimes \Theta) \to \mathbb{C}$.

Presumably, it is possible to complete this line of arguments to a full description of multiplications in C. We prefer do it indirectly through the relation to the Berkovits construction [Berk].

Remark 18. Let $\mathbb{R}^{10}$ be a linear space, equipped with a positive definite dot product. Denote by S an irreducible complex spinor representation of the orthogonal group $SO(10)$. Denote by $\Gamma^i_{\alpha\beta}$ the coefficients of the nontrivial intertwiner $\mathrm{Sym}^2(S) \to \mathbb{C}^{10}$ in some basis of S and an orthonormal basis of $\mathbb{C}^{10}$. We assume that the $\mathbb{C}^{10}$-basis is real.

On a superspace $(\mathbb{R}^{10} + \Pi S) \otimes \mathfrak{u}(n)$ we define a superfunction (Lagrangian)

$$S(A, \chi) = \sum_{i<j} \mathrm{tr}([A_i, A_j][A_i, A_j]) + \sum_{\alpha\beta i} \mathrm{tr}(\Gamma^i_{\alpha\beta}[A_i, \chi^\alpha]\chi^\beta); \qquad (24)$$

$A_1, \ldots, A_{10}$ is a collection of anti-hermitian matrices labeled by the basis of $\mathbb{C}^{10}$. Similarly, odd matrices $\chi^1, \ldots, \chi^{16}$ are labeled by the basis of S. This can be considered as a field theory, obtained from $D = 10$, $N = 1$ YM theory by reduction to zero dimensions. We call it IKKT after the paper [IKKT], where it has been studied. IKKT theory has a gauge invariance: invariance with respect to conjugation. A BV version of IKKT coincides with a CS theory associated with the A_∞ algebra $\mathcal{A}_{IKKT}$ that we shall introduce presently.

Definition 19. *An A_∞ algebra $\mathcal{A}_{IKKT}$ can be considered as a vector space spanned by symbols $x_k, \xi^\alpha, c, x^{*k}, \xi^*_\alpha, c^*, 1 \le k \le 10, 1 \le \alpha \le 16$, with operations μ_2 (multiplication), μ_3 (Massey product) defined by the following formulas:*

$$\mu_2(\xi^\alpha, \xi^\beta) = \Gamma^{\alpha\beta}_k x^{*k}, \qquad (25)$$

$$\mu_2(\xi^\alpha, x_k) = \mu_2(x_k, \xi^\alpha) = \Gamma^{\alpha\beta}_k \xi^*_\beta, \qquad (26)$$

$$\mu_2(\xi^\alpha, \xi^*_\beta) = \mu_2(\xi^*_\beta, \xi^\alpha) = c^*, \qquad (27)$$

$$\mu_2(x_k, x^{*k}) = \mu_2(x^{*k}, x_k) = c^*, \qquad (28)$$

$$\mu_3(x_k, x_l, x_m) = \delta_{kl} x^{*m} - \delta_{km} x^{*l}, \qquad (29)$$

$$\mu_2(c, \bullet) = \mu_2(\bullet, c) = \bullet. \qquad (30)$$

All other products are equal to zero. An element c is a unit.

All operations μ_k with $k \ne 2, 3$ vanish. The algebra carries a trace functional tr equal to one on c^* and zero on the rest of the generators. It induces a dot product by the formula $(a, b) = \mathrm{tr}(\mu_2(a, b))$, compatible with μ_k.

By definition, an A_∞ algebra has a grading (we call it a ghost grading) such that the operation μ_n has degree $2 - n$. An A_∞ algebra might also have an additional grading such that all operations have degree zero with respect to it. See [MSch05] for details on gradings of A_{IKKT}.

Proposition 20. *A differential graded algebra with a trace* $(\Omega^{0\bullet}(\widetilde{\Pi F}), \bar{\partial} + \alpha, \mathrm{tr}_\mu)$ *is quasi-isomorphic to* A_{IKKT}.

Proof. We shall employ methods developed in [MSch05]. Recall that the manifold of pure spinors in dimension ten is equal to $\mathcal{P} = SO(10, \mathbb{R})/U(5)$. As a complex manifold it is defined as the space of solutions of homogeneous equations $\Gamma^i_{\alpha\beta}\lambda^\alpha\lambda^\beta = 0$, where λ^α are homogeneous coordinates on $\mathbf{P}^{15}$. A space $\mathbf{P}^{15}$ is the projectivisation of the irreducible complex spinor representation of $\mathrm{Spin}(10, \mathbb{R})$. We denote by R the restriction on $\mathcal{P}$ of the twisted tangent bundle $T_{\mathbf{P}^{15}}(-1)$. Denote by A the coordinate algebra $\mathbb{C}[\lambda^1, \dots, \lambda^{16}]/\Gamma^i_{\alpha\beta}\lambda^\alpha\lambda^\beta$ of $\mathcal{P}$. Denote by B the Koszul complex $A \otimes \Lambda[\theta^1, \dots \theta^{16}]$ with differential $\lambda^\alpha \frac{\partial}{\partial\theta^\alpha}$.

The algebra B can be equipped with various gradings. The cohomological grading is defined on generators as $|\lambda^\alpha| = 2$, $|\theta^\alpha| = 1$; the grading by the degree (or homogeneous grading) is $\deg\lambda^\alpha = \deg\theta^\alpha = 1$. The differential has degree one with respect to $\|$ and zero with respect to deg. As a result, the cohomology groups are bigraded: $H(B) = H^{ij}$, where i corresponds to $\|$-grading, j corresponds to deg.

In [MSch05] we proved that $\bigoplus_{i-j=k} H^{ij}(B) = H^k(\Omega^{0\bullet}(\mathcal{P})\Lambda(R^*))$. In fact, we proved that the identification map is a quasi-isomorphism of differential graded algebras with a trace. In the language of supermathematics we may say that cohomology $H^k(\Omega^{0\bullet}(\mathcal{P})\Lambda(R^*))$ is Dolbeault cohomology of a split supermanifold ΠR.

In the course of the proof of quasi-isomorphism we have identified the algebra of functions on the fiber of the projection

$$\Pi R \to \mathcal{P} \tag{31}$$

over a point $pt = (\lambda_0^\alpha) \in \mathcal{P}$ with cohomology of the algebra $B^\bullet_{pt} = (\Lambda[\theta^1, \dots \theta^{16}], d)$, where $d = \lambda_0^\alpha \frac{\partial}{\partial\theta^\alpha}$. An analogue of the complex $B^\bullet_{pt}$ can be defined for any subscheme U of $\mathcal{P}$ as a tensor product $B^\bullet_U = A_U \otimes \Lambda[\theta^1, \dots \theta^{16}]$. The algebra A_U is equal to $\bigoplus_{i\geq 0} H^0(U, \mathcal{O}(i))$. The cohomology of B_U is equal to $H^\bullet B_{pt} \otimes \mathcal{O}(U)$ if $\mathcal{O}(1)$ is trivial on U.

We proved in [MSch06] that the algebra B with Berkovits trace tr is quasi-isomorphic to the algebra A_{IKKT}.

We need to present a useful observation from [MSch05]. Let us decompose the set $\{\lambda^1, \dots, \lambda^{16}\}$ into a union $\{\lambda^{\alpha_1}, \dots, \lambda^{\alpha_s}\} \cup \{\lambda^{\beta_1}, \dots, \lambda^{\beta_k}\}$ such that $\{\lambda^{\alpha_1}, \dots, \lambda^{\alpha_s}\}$ is a regular sequence. Then the algebras (B, d) and $(B', d) = (A/(\lambda^{\alpha_1}, \dots, \lambda^{\alpha_s}) \otimes \Lambda[\theta^{\beta_1}, \dots, \theta^{\beta_k}], d)$ are quasi-isomorphic.

The following construction has been described in [MSch05]. The spin representation S of $\mathfrak{so}(10)$ splits after restriction on $\mathfrak{gl}(3) \times \mathfrak{sl}_l(2) \times \mathfrak{sl}_r(2)$ into

$T \otimes W_l + T^* \otimes W_r + W_l + W_r$. We choose coordinates on $W_l + W_{r-}$ equal to $(\lambda^{\alpha_i}) = (\tilde{w}_l^+, \tilde{w}_l^-, \tilde{w}_r^+, \tilde{w}_r^-)$. They form a regular subsequence of λ^α. The manifold corresponding to $A/(\lambda^{\alpha_i})$ is equal to $Q \cap \mathbf{P}(T \otimes W_l + T^* \otimes W_r)$. The intersection is isomorphic to $F(1,2) \times X$. The algebra of homogeneous functions $A' = A(F(1,2) \times X)$ on $F(1,2) \times X$ is generated by $s^{i\alpha}, t_\alpha^j (1 \leq \alpha\beta \leq 3, 1 \leq ij \leq 2)$. The relations are

$$\sum_\alpha s^{i\alpha} t_\alpha^j = 0, \quad \det(s^{i\alpha}) = 0, \quad \det(t_\alpha^j) = 0. \tag{32}$$

In the formula det stands for a row of 2×2 minors of a 2×3 matrix.

We plan to follow almost the same method of construction of supermanifold as for the Koszul complex of pure spinors. Denote by g a projection

$$g : F(1,2) \times X \to X$$

We fix a point $x \in X$. The algebra $A'_{g^{-1}(x)}$ is isomorphic to $\mathbb{C}[p_1, \ldots, p_3, u^1, \ldots, u^3]/(p_i u^i)$. The algebra $B'_{p^{-1}(x)}$ is isomorphic to

$$A'_{g^{-1}(x)} \otimes \Lambda[\pi_1, \ldots, \pi_3, \nu^1, \ldots, \nu^3, \tilde{\pi}_1, \ldots, \tilde{\pi}_3, \tilde{\nu}^1, \ldots, \tilde{\nu}^3], d \tag{33}$$

with a differential

$$d = p_i \frac{\partial}{\partial \tilde{\pi}_i} + u^i \frac{\partial}{\partial \tilde{\nu}^i} \tag{34}$$

The cohomology of this differential is equal to

$$\Lambda[E_x] = \Lambda[\pi_1, \ldots, \pi_3, \nu^1, \ldots, \nu^3, \theta]$$

The induced A_∞ algebra structure on cohomology has no higher multiplications. The element θ is represented by a cocycle $\tilde{\pi}_i u^i$. The linear space E_x coincides with the fiber F_x of the vector bundle F (21). The main distinction between this computation and a computation with pure spinors is that we encountered a noncanonical A_∞ morphism $\iota : B'_{p^{-1}(x)} \to \Lambda[E_x]$, which may not be a homomorphism of associative algebras.

Recall that we viewed the cohomology of B_{pt} as functions of the fiber of projections $p : \Pi R \to \mathcal{P}$. We have a natural identification of fibers over different patches of $\mathcal{P}$. It gives us a consistent construction of a split manifold ΠR.

In the case of the manifold X, if we ignore the issues related to ambiguities of choice of morphism ι, it is not hard to see that an isomorphism of fibers $F_x \cong H_x$ can be extended to a Spin(4) equivariant isomorphism of the vector bundles $F_x \cong H_x$ (use homogeneity of both vector bundles with respect to $\mathfrak{sl}^l(2) \times \mathfrak{sl}^r(2)$ action).

In this way, we recover the manifold ΠF. In reality, when we try to glue rings of functions on different patches, the structure isomorphisms will be A_∞-morphisms. We may claim on general grounds that we get an A_∞ structure on the space of Čech chains of $\Pi H \cong \Pi F$. This structure can be trivialized by a twist on a local A_∞-morphism (reduced to the standard multiplication in the Grassmann algebra) on every double intersection U_{ij} of patches (if $U_{ij} \subset X$ is sufficiently small). An ambiguity in the choice of such a twist leads to the appearance of a Čech 2-cocycle β_{ijk} with values in infinitesimal (not A_∞) transformations of the fiber ΠF. In the Dolbeault picture this cocycle corresponds to α. Finally, we use Čech–Dolbeault equivalence. This proves the claim. □

Remark 21. The algebra C carries a differential $d = [\alpha]$. The minimal model of C (by definition it is a quasi-isomorphic A_∞ algebra without a differential) constructed for an obvious homotopy of the differential $d = [\alpha]$ is quasi-isomorphic to $\mathcal{A}_{IKKT}$. If we ask for a quasi-isomorphism to be compatible with all gradings that exist on both algebras this quasi-isomorphism is an isomorphism.

From this it is quite easy to recover all multiplications in the algebra C.

4 Nonreduced theory

A manifold Z with an integral form is constructed in this section.

4.1 Construction of the algebra $A(Z)$

In this section we construct an algebra $A(Z)$. It will be a linking chain between YM theory and ambitwistors.

We construct a manifold Z as a direct product $\Pi F \times \Sigma$. Intuitively speaking, the algebra $\Omega^{0\bullet}(\widetilde{\Pi F})$ carries all the information about YM theory reduced to a point, whereas the algebra $C^\infty(\Sigma)$ contains similar information about the space Σ. The idea is that a tensor product $\Omega^{0\bullet}(\widetilde{\Pi F}) \otimes C^\infty(\Sigma)$ with a suitably twisted differential will contain all information about 4-D YM theory. Later we will interpret the same complex as tangential CR complex on the manifold Z.

The linear space

$$\Sigma^{\mathbb{C}} = V = W_l \otimes W_r \tag{35}$$

has coordinates $x^{ij}, 1 \le i, j \le 2$. The vector space V has an $SO(4, \mathbb{C})$ action compatible with the decomposition (36). It is induced from the $SO(4, \mathbb{R})$ action on Σ.

Define a differentiation of $\Omega^{0\bullet}(\widetilde{\Pi F}) \otimes C^\infty(\Sigma)$ as follows.

There is an $SO(4, \mathbb{C})$ action compatible with decomposition (36). It is induced from the $SO(4, \mathbb{R})$ action on Σ.

Define a differentiation of $\Omega^{0\bullet}(\widetilde{\Pi F}) \otimes C^\infty(\Sigma)$ as follows.

There is an $O(4)$ equivariant isomorphism $V \cong H^0(\mathcal{O}(1,1))$. The line bundle $\mathcal{O}(1,1)$ is generated by its global sections. We have a short exact sequence:

$$0 \to M \to V \xrightarrow{m} \Theta = \mathcal{O}(1,1) \to 0 \tag{36}$$

where V is considered as a trivial vector bundle with fiber V. The differentiation δ of $\Omega^{0\bullet}(\widetilde{\Pi F}) \otimes C^\infty(\Sigma)$ is equal $\delta = m(x^{ij})\frac{\partial}{\partial x^{ij}}$. We interpret $\frac{\partial}{\partial x^{ij}}$ as global sections of $T^{\mathbb{C}}\Sigma$.

Recall that the differential D in $\Omega^{0\bullet}(\widetilde{\Pi F})$ is equal to $\bar{\partial} + \alpha$. It becomes clear from explicit computation of cohomology that the coefficients β^{ij} in $\{\alpha, \delta\} = \beta^{ij}\frac{\partial}{\partial x^{ij}}$ are $\bar{\partial}$-exact.

Choose γ^{ij} such that $\bar{\partial}\gamma^{ij} = -\beta^{ij}$. We define a differentiation γ of $\Omega^{0\bullet}(\widetilde{\Pi F}) \otimes C^\infty(\Sigma)$ as $\gamma^{ij}\frac{\partial}{\partial x^{ij}}$ and zero on the rest of the generators.

It follows from our construction that $D_{ext} = D + \delta + \gamma$ satisfies $D_{ext}^2 = 0$. Set $D' = \delta + \gamma$.

By definition, the integral form on the manifold $\widetilde{\Pi F} \times \Sigma$ is

$$\mathrm{Vol} = \mu \otimes \bigotimes_{i,j=1}^{2} dx^{ij}. \tag{37}$$

Proposition 22. *Vol is invariant with respect to D, D' and therefore with respect to D_{ext}.*

Proof. Direct inspection. $\qquad\square$

4.2 Proof of the equivalence

Proposition 23. *Define $Z = \Pi F \times \Sigma$. We equip the algebra $A(Z) = \Omega^{0\bullet}(\widetilde{\Pi F}) \otimes C^\infty(\Sigma) \otimes \mathrm{Mat}_n$ with a trace*

$$\int a = \int_Z \mathrm{tr}(a)\mathrm{Vol} \tag{38}$$

The CS theory constructed by the triple $(A(Z), D_{ext}, \int)$ is classically equivalent to $N = 3$ euclidean YM with a gauge group $U(n)$.

The equivalence should hold also on the quantum level.

Proof. We shall only outline the basic ideas.

A precise mathematical statement is about quasi-isomorphism of certain A_∞ algebras. One of them is $A(Z)$. The reader should consult [MSch06] for information about A_∞ algebras with a trace corresponding to $N = 3$ YM theory.

Suppose $\psi : A \to B$ is a quasi-isomorphism of two A_∞ algebras. Let m be an associative algebra. Then we have a quasi-isomorphism of tensor products $\psi : A \otimes m \to B \otimes m$. Let a be a solution of the MC equation in $A \otimes m$ (the reader may consult [MSch05] for the definition). The map ψ transports a solution a to a solution $\psi(a)$ of the MC equation in $B \otimes m$.

In a formal interpretation of structure maps of the A_∞ algebra A as the Taylor coefficients of the noncommutative vector field on noncommutative space $\mathbb{A}$, a solution a of the MC equation corresponds to a zero of the vector field. We can expand the vector field into series at a and get some new A_∞ algebra. This construction is particularly transparent in the case of a dga. A solution of the MC equation defines a new differential $\tilde{d}x = dx + [a,x]$. It corresponds to a shift of a vacuum in physics jargon. Denote by A_a an A_∞ algebra constructed by the element a.

It is easy to see that the map ψ defines (under some mild assumptions in a) a quasi-isomorphism

$$\psi : A \otimes m_a \to B \otimes m_{\psi(a)}.$$

Denote by $\mathrm{Diff}(\Sigma)$ an algebra of differential operators on Σ. We would like to apply the construction from the previous paragraph to the tensor products $\Omega^{0\bullet}(\widetilde{\Pi F}) \otimes \mathrm{Diff}(\Sigma)$ and $C \otimes \mathrm{Diff}(\Sigma)$. We can interpret $\delta + \gamma$ as a solution of the MC equation for the algebra $\Omega^{0\bullet}(\widetilde{\Pi F})$ with coefficients in $\mathrm{Diff}(\Sigma)$. The quasi-isomorphism ψ maps $\delta + \gamma$ into an element $y^{ij} \frac{\partial}{\partial x^{ij}}$, where y^{ij} is a basis of $H^0(X, \Lambda^0(H^*) \otimes \Theta) \subset C$.

The rest is a matter of formal manipulations. It is straightforward to see that $C \otimes \mathrm{Diff}(\Sigma)_{\psi(a)}$ is an A_∞ mathematics counterpart of the YM equation where the gauge potential, spinors and matter fields have their coefficients not in functions on Σ but in $\mathrm{Diff}(\Sigma)$. This is not precisely what we had hoped to obtain. We shall address this issue presently.

Suppose an associative algebra m contains a subalgebra m'. The previous construction has a refinement. The noncommutative vector field on a space $\mathbb{A}_m$ corresponding to the A_∞ algebra $A \otimes m$ is tangential to the noncommutative subspace $\mathbb{A}_{m'}$ (because m' is closed under multiplication). We say that a solution of the MC $a \in A \otimes m$ is compatible with m' if the vector field defined by the algebra $A \otimes m$ is tangential to the space $a + \mathbb{A}_{m'} \subset \mathbb{A}_m$. This is merely another way to say that a linear space $A \otimes m'$ is a subalgebra of $(A \otimes m)_a$. If a is compatible with m', the map ψ (under some mild assumptions on a) induces a quasi-isomorphism $\psi : A \otimes m' \to B \otimes m'_{\psi(a)}$.

We apply this construction to the subalgebra $C^\infty(\Sigma) \subset \mathrm{Diff}(\Sigma)$ for which the above mentioned condition on $\delta + \gamma$ is met.

Suppose in addition that the algebras A, B, m' have a traces and a morphism $\psi : A \to B$ is compatible with the traces. Assume, moreover, that the induced A_∞ structure $A \otimes m'_a$ is compatible with the trace. Then the induced morphism $\psi : A \otimes m'_a \to B \otimes m'_{\psi(a)}$ is compatible with traces.

In our case, the operator D' preserves the integral form Vol, and the above conditions are met.

The CS theory associated with the algebra $C \otimes C^\infty(\Sigma)$ has the following even part of the Lagrangian:

$$\langle F_{ij}, F_{ij} \rangle + \langle \nabla_i \phi^\alpha, \nabla_i \phi_\alpha \rangle + \langle [\phi^\alpha, \phi_\beta], K_\alpha^\beta \rangle + \langle K_\alpha^\beta, K_\beta^\alpha \rangle.$$

We interpret F_{ij} as coefficients of the curvature of a connection ∇_i in a principal $U(n)$ bundle, ϕ^α, ϕ_β as the components of a matter field $\phi \in \mathrm{Ad} \otimes T$, $\phi^* \in \mathrm{Ad} \otimes T^*$, K_α^β as the components of an auxiliary field $K \in Ad \otimes T \otimes T^*$. This theory is equivalent to $N = 3$ YM as the odd parts of the Lagrangians coincide.

$\square$

4.3 Relation between a CR structure on Z and an algebra $A(Z)$

In this section we give a geometric interpretation of the algebra $A = \Omega^{0\bullet}(\widetilde{\Pi F}) \otimes C^\infty(\Sigma)$.

The fibers of the projection

$$p : X \times \Sigma \to \Sigma \tag{39}$$

have a holomorphic structure. Denote by T_{vert} a bundle of p-vertical vector fields. We define a distribution $\mathcal{G}$ as $T_{\text{vert}}^{1,0} \subset T_{\text{vert}}^{\mathbb{C}}$.

Choose a linear basis $e_1, \ldots, e_4 \in \Sigma$. Define $\frac{\partial}{\partial x^s}$ as differentiation in the direction of e_s. Restrict the map m from the short exact sequence (37) to $\Sigma \subset V$, then $m(e_s)$ is a set of holomorphic sections of Θ.

For any point $x \in X$ we have a subspace $\mathcal{H}_x \subset T_\Sigma^{\mathbb{C}}$ spanned by

$$\sum_{i=1}^{4} m(e_s)_x \frac{\partial}{\partial x^d}. \tag{40}$$

A union of such subspaces defines a complex distribution $\mathcal{H}$ on $X \times \Sigma$.

Define an integrable distribution $\mathcal{F} = \mathcal{G} + \mathcal{H} \subset T_{X \times \Sigma}^{\mathbb{C}}$.

The reader can see that the CR structure on the space $X \times \Sigma$ literally coincides with the CR structure on the ambitwistor space $A(\Sigma)$ defined in Section 2.2.

In light of this identification, an element θ (a local coordinate on Θ^*) can be interpreted as a local CR form with nonzero values on $\overline{\mathcal{H}}$.

The restriction of the vector bundle H^* (used in the construction of the supermanifold F in equation (21)) to $X \times \Sigma$ is holomorphic along $\mathcal{F}$. Additionally, we can interpret the component α in the differential D on $\Omega^{0\bullet}(\widetilde{\Pi F})$ as a contribution from a superextension of the CR structure defined in (19).

From this we deduce that the algebra $A(Z) = (\Omega^{0\bullet}(\widetilde{\Pi F}) \otimes C^\infty(\Sigma), D_{ext})$ we have constructed coincides with the CR tangential complex on an open subset of the manifold $\mathcal{R}_{\text{GL}(4|3)}$.

The integral form Vol is the only CR holomorphic form invariant with respect to $\text{SU}(2) \times \text{SU}(2) \ltimes \mathbb{R}^4$.

From this we deduce that it coincides (up to a multiplicative constant) with the integral form defined by vol.

It is possible to reconstruct an action of the super Poincaré group SP on A. The reader will find explanations of why the measure is not invariant with respect to the full superconformal group in Section 5.3.

5 Appendix

5.1 On the definition of a graded real superspace

A tensor category of complex superspaces $\mathcal{C}_{\mathbb{C}}$ (see [DMln] for an introduction to tensor categories) has two real forms. The first is a category of real superspaces $\mathcal{C}_{\mathbb{R}}$. It is more convenient to think about objects of this category as of complex superspaces, equipped with an antiholomorphic involution σ.

Another tensor category related to $\mathcal{C}_{\mathbb{C}}$ is formed by complex superspaces equipped with an antilinear map ρ.

Definition 24. *Suppose V is a $\mathbb{Z}_2$-graded vector space over the complex numbers. An antilinear map $\rho : v \to \bar{v}$ is a graded real structure if*

$$\rho^2 = \text{sid},$$
$$\text{sid}(v) = (-1)^{|v|} v. \tag{41}$$

We denote by $|v|$ the parity of v.

An element v is real iff $\bar{v} = v$. Only even elements can be real with respect to a graded real structure. A graded real superspace is a pair (V, ρ). Graded real superspaces form a tensor category $\mathcal{C}_{\mathbb{H}}$.

We shall be mostly interested in categories $\mathcal{C}_{\mathbb{C}}$ and $\mathcal{C}_{\mathbb{H}}$.

The categories $\mathcal{C}_{\mathbb{C}}$, $\mathcal{C}_{\mathbb{H}}$ are related by tensor functors.

The first functor is the complexification $\mathcal{C}_{\mathbb{H}} \Rightarrow \mathcal{C}_{\mathbb{C}}$, $V \Rightarrow V^{\mathbb{C}}$. It forgets about the map ρ.

The second functor is $\mathcal{C}_{\mathbb{C}} \Rightarrow \mathcal{C}_{\mathbb{H}}$, $V \Rightarrow V^{\mathbb{H}}$. The object $V^{\mathbb{H}}$ is the direct sum $V + \overline{V}$. There is an antilinear isomorphism $\sigma : V \to \overline{V}$. For $v = a + b + c + d \in V^0 + V^1 + \overline{V}^0 + \overline{V}^1$ define

$$\rho(a + b + c + d) = \sigma^{-1}(c) - \sigma^{-1}(d) + \sigma(a) + \sigma(b). \tag{42}$$

By construction, $\rho^2 = \text{sid}$.

The language of tensor categories can be used as a foundation for developing commutative algebra and algebraic geometry. If we start off in this direction with a category $\mathcal{C}_{\mathbb{C}}$, the result will turn to be algebraic super-geometry.

The category $\mathcal{C}_{\mathbb{H}}$ provides us with a real form of this geometry.

First of all, a $\mathcal{C}_{\mathbb{H}}$ or a real graded manifold is an algebraic supermanifold M defined over $\mathbb{C}$. The manifold M carries some additional structure. The manifold M_{red} is equipped with an antiholomorphic involution ρ_{red}. There is also an antilinear isomorphism of sheaves of algebras

$$\rho : \rho_{\text{red}}^* \mathcal{O} \to \mathcal{O} \tag{43}$$

such that $\rho^2 = \text{sid}$.

There is a C^∞ version of a $\mathcal{C}_{\mathbb{H}}$ manifold. It basically mimics the structure of the C^∞ completion of the algebraic $\mathcal{C}_{\mathbb{H}}$ manifold at the locus of ρ_{red} fixed points. Any C^∞ manifold admits a noncanonical splitting: $M \cong \Pi E$, where E is some complex vector bundle over M_{red}. A $\mathcal{C}_{\mathbb{H}}$-structure manifests itself in an antilinear automorphism ρ of E that satisfies $\rho^2 = -\text{id}$. Observe that ρ, together with multiplication on i, defines a quaternionic structure on E.

For any complex algebraic supermanifold M there is the underlying $\mathcal{C}_{\mathbb{H}}$ manifold. As an algebraic supermanifold it is equal to $M \times \overline{M}$. There is a canonical anti-involution on $(M \times \overline{M})_{\text{red}}$. The morphism of sheaves (44) in local charts is defined by formulas similar to (43).

This construction manifests itself in the C^∞-setting as follows. Any complex supermanifold M defines a C^∞ supermanifold $\tilde{M}$ that is holomorphic in odd directions. It is a C^∞-completion of $M \times \overline{M}_{red}$ near diagonal of $M_{red} \times \overline{M}_{red}$. Suppose ΠE is a splitting of $\tilde{M}$, where E is a complex vector bundle on $\tilde{M}_{red}$. The vector bundle $E + \overline{E}$ has a natural quaternionic structure and defines a $C^\infty \; \mathcal{C}_{\mathbb{H}}$-manifold $\Pi(E + \overline{E})$. This manifold is isomorphic to completion of $M \times \overline{M}$. Sometimes it is more convenient to work with the manifold $\tilde{M}$.

5.2 On homogeneous CR-structures

Suppose we are given an ordinary real Lie group and a closed subgroup $A \subset G$ with Lie algebras $\mathfrak{a} \subset \mathfrak{g}$. Additionally we have a complex subgroup $P \subset G^{\mathbb{C}}$ in complexification of G, with Lie algebras $\mathfrak{p} \subset \mathfrak{g}^{\mathbb{C}}$. We assume that the map

$$p : G/A \to G^{\mathbb{C}}/P \qquad (44)$$

is a local embedding. By construction $G^{\mathbb{C}}/P$ is a holomorphic homogeneous space. It tangent space $T_x, x \in G/A$ contain a subspace $H_x = T_x \cap JT_x$. The operator J is an operator of complex structure on $G^{\mathbb{C}}/P$. Due to G-homogeneity spaces H_x have constant rank and form a subbundle $H \subset T$. We can decompose $H \otimes \mathbb{C} = \mathcal{F} + \overline{\mathcal{F}}$. It follows from the fact that $\mathfrak{p}$ is a subalgebra that the constructed distribution $\mathcal{F}$ is integrable and defines a CR-structure.

A condition that the map p (44) is a local embedding is equivalent to $\mathfrak{g} \cap \mathfrak{p} = \mathfrak{a}$. In other words

$$\mathfrak{p} \cap \bar{\mathfrak{p}} = \mathfrak{a}^{\mathbb{C}} \qquad (45)$$

It is easy to see that a fiber $\mathcal{F}_x$ at a point x is isomorphic to $\mathfrak{p}/\mathfrak{a}^{\mathbb{C}}$

This construction of CR-structure can be extended to a supercase. A consistent way to derive such extension is to use a functorial language of ref. [DM].

However since this exercise, which we leave to the interested reader, is quite straightforward we provide only the upshot.

We start off with description of a data that defines a homogeneous space of a supergroup.

A complex homogeneous space X of a complex supergroup G with Lie algebra $\mathfrak{g}$ is encoded by:

1. Global data: An isotropy subgroup $H \subset G_{red}$ (closed, analytic, possibly nonconnected). This data defines an ordinary homogeneous space $X_{red} = G_{red}/H$;

2. Local data: a pair of complex super Lie algebras $\mathfrak{p} \subset \mathfrak{g}$ such that $\mathfrak{p}_0 = Lie(H), Lie(G_{red}) = \mathfrak{g}_0$.

In the cases when we specify only Lie algebra of isotropy subgroup is clear from the context.

A real graded structure on a homogeneous space $X = G/A$ is encoded by an antiholomorphic involution on G_{red} that leaves subgroup A_{red} invariant; a graded real structure ρ on $\mathfrak{g}$, such that $\rho(\mathfrak{a}) \subset \mathfrak{a}$.

If we are given a real subalgebra $(\mathfrak{a}, \rho) \subset (\mathfrak{g}, \rho)$ and a complex subalgebra $\mathfrak{p} \subset \mathfrak{g}$ such that $a = \mathfrak{p} \cap \rho(\mathfrak{p})$, we claim that a supermanifold G/A carries a (G, ρ)-homogeneous CR structure.

5.3 General facts about CR structures on supermanifolds

In this section we will discuss mostly general facts about CR structures specific to supergeometry. Suppose M is a supermanifold equipped with an integrable CR distribution $\mathcal{F}$. We present some basic examples of $\mathcal{F}$-holomorphic vector bundles on M.

Example 25. Sections of the vector bundle $T^{\mathbb{C}}/\overline{\mathcal{F}}$ form a module over the Lie algebra of the sections of $\overline{\mathcal{F}}$. Thus the gluing cocycle of this bundle is CR holomorphic. It implies that the bundle $\mathrm{Ber}((T^{\mathbb{C}}/\overline{\mathcal{F}})^*)$ is also CR holomorphic.

Suppose we have a trivial CR structure on $\mathbb{R}^{n_1|n_2} \times \mathbb{C}^{m_1|m_2}$. We assume that the space is equipped with global coordinates $x_i, \eta_j, z_k, \theta_l$. The algebra of the tangential CR complex $\Omega_{\mathcal{F}}^{\bullet}(\mathbb{R}^{n_1|n_2} \times \mathbb{C}^{m_1|m_2})$ has topological generators $x_i, \eta_j, z_k, \theta_l, \bar{z}_k, \bar{\theta}_l, d\bar{z}_k, d\bar{\theta}_l$. Denote by A the subalgebra generated by $x_i, \eta_j, z_k, \theta_l, \bar{z}_k, , d\bar{z}_k$ and by K the subalgebra generated by $\bar{\theta}_l, d\bar{\theta}_l$. We have $\Omega^{0\bullet} = A \otimes K$. The algebra K has trivial cohomology. As a result the projection $\Omega^{0\bullet} \to A$ is a quasi-isomorphism.

It turns out that this construction exists in the more general context of an arbitrary CR manifold. Informally, we may say that a super-CR manifold is affine in holomorphic odd directions. It parallels with the complex case.

The construction requires a choice of C^{∞}-splitting of the CR manifold M.

Suppose Y is an ordinary manifold, E a vector bundle. Let ΠE denote the supermanifold whose sheaf of functions coincides with the sheaf of sections of the Grassmann algebra of the bundle E^*. Such a supermanifold is said to be split. By construction, it admits a projection $p : \Pi E \to Y$. Any C^{∞} manifold is split, but the splitting is not unique. A space of global functions on ΠE is isomorphic to a space of sections of the Grassmann algebra ΛE^* of the vector bundle E.

To make a connection with our considerations we identify $Y = M_{\mathrm{red}}$.

Define $\overline{\mathcal{F}}_{\mathrm{red}} = \overline{\mathcal{F}}_{\mathrm{red}}^0 + \overline{\mathcal{F}}_{\mathrm{red}}^1$, the restriction of $\overline{\mathcal{F}}$ to M_{red}. In terms of the splitting operator, $\bar{\partial}$ can be encoded by a pair of operators of the first order $\bar{\partial}_{\mathrm{ev}} : C^{\infty}(M_{\mathrm{red}}) \to \Lambda E^* \otimes \overline{\mathcal{F}}_{\mathrm{red}}, \bar{\partial}_{\mathrm{odd}} : E^* \to \Lambda E^* \otimes \overline{\mathcal{F}}_{\mathrm{red}}$.

The lowest degree component in powers $\Lambda^i E^*$ of the operator $\bar{\partial}_{\mathrm{odd}}$ is $\bar{\partial}_{\mathrm{odd}}^0 E^* \to \overline{\mathcal{F}}_{\mathrm{red}}^1$. It is a $C^{\infty}(Y)$ linear map of constant value. The image S_0 of a splitting $\overline{\mathcal{F}}_{\mathrm{red}}^1 \to E^*$ can be used to generate a differential ideal (S) of $\Omega_{\mathcal{F}}$. Denote the quotient $\Omega_{\mathcal{F}}/S$ by $\Omega_{s\mathcal{F}}$.

The complex $\Omega_{s\mathcal{F}}$ is a differential graded algebra. We can interpret it as a space of functions on some superspace L. A possibility to split $\bar{\partial}_{\mathrm{odd}}^0$ implies smoothness of L.

Proposition 26. *The map* $\Omega_{\mathcal{F}} \to \Omega_{s\mathcal{F}}$ *is a quasi-isomorphism.*

Proof. Follows from consideration of a spectral sequence associated with filtration $F^i \Omega_{\mathcal{F}}^p = (S_0)^{i-p} \Omega_{\mathcal{F}}^p$ (we denote by $(S_0)^k$ the k-th power of the ideal generated by S_0). □

A CR manifold M is locally embeddable to $\mathbb{C}^{m|n}$ if, in a neighborhood of a point, there is a collection of $z_1, \ldots, z_m$ even and $\theta_1, \ldots, \theta_n$ odd function that are annihilated by $\bar{\partial}$ and whose Jacobian is nondegenerate.

Definition 27. *Let us assume that* $\mathcal{F}^1 \Big| M_{\mathrm{red}} + \overline{\mathcal{F}}^1 \Big|_{M_{\mathrm{red}}} = T^{\mathbb{C}1}$ *(the superscript 1 denotes the odd part), i.e., the dimension of the odd part of the CR distribution is the greatest possible.*

We can locally generate the ideal S_0 by elements $\bar{\theta}_1, \ldots, \bar{\theta}_n$ and take S as the $\bar{\partial}$ closure of S_0. It is not hard to check that under such assumptions, Proposition 26 holds. Denote by sM a submanifold specified by S_0.

Remark 28. The Lie algebra of infinitesimal automorphisms of the CR structure is equal to $\mathrm{Aut}_{\mathcal{F}} = \{a \in T^{\mathbb{C}}|[a, b] \in \mathcal{F}, \text{ for all } b \in \overline{\mathcal{F}}\}\}$ with $\mathrm{Out}_{\mathcal{F}} = \mathrm{Aut}_{\mathcal{F}}/\mathcal{F}$. In the purely complex case the quotient construction can be replaced by $\mathrm{Out}_{\mathrm{complex}} = \{a \in \mathcal{F}|[a, b] \in \overline{\mathcal{F}}, \text{ for all } b \in \overline{\mathcal{F}}\}\}$ and the extension

$$0 \to \mathrm{Inn}_{\mathcal{F}} \to \mathrm{Aut}_{\mathcal{F}} \to \mathrm{Out}_{\mathcal{F}} \to 0 \qquad (46)$$

has a splitting. By construction, the elements $\bar{\theta}_1, \ldots, \bar{\theta}_n$ are invariant along vector fields from the distribution $\mathcal{F}$. We can guarantee that the differential ideal generated by $\bar{\theta}_i$ is invariant with respect to the elements of $\mathrm{Out}_{\mathrm{complex}}$. As a result we can push the action of $\mathrm{Out}_{\mathrm{complex}}$ to $\Omega_{s\mathcal{F}}^{\bullet}$, this is a familiar fact from super complex geometry. This contrasts with absence of the action of $\mathrm{Aut}_{\mathcal{F}}$ or $\mathrm{Out}_{\mathcal{F}}$ on the ideal S and on $\Omega_{s\mathcal{F}}^{\bullet}$ for a general CR structure. One can prove, however, that $\Omega_{s\mathcal{F}}^{\bullet}$ admits an A_{∞} action of $\mathrm{Aut}_{\mathcal{F}}$. A partial remedy is to consider the subalgebra $\widetilde{\mathrm{Out}}_{\mathcal{F}} = \{a \in \mathcal{F}|[a, b] \in \overline{\mathcal{F}}, \text{ for all } b \in \overline{\mathcal{F}}\}\} \subset \mathrm{Out}_{\mathcal{F}}$. This subalgebra acts on $\Omega_{s\mathcal{F}}^{\bullet}$. However, this algebra is trivial if the Levi form of $\mathcal{F}$ is not degenerate.

Supergeometry provides us with a complex of CR integral forms. Let $\Lambda \overline{\mathcal{F}}$ be the super Grassmann algebra of $\overline{\mathcal{F}}$. Let Ber be the Berezinian line bundle of a real manifold M. Denote $^{\mathrm{int}}\Omega_{\mathcal{F}}^{-p}$ the tensor product $\mathrm{Ber} \otimes \Lambda \overline{\mathcal{F}}$. There is a pairing $(\omega, \nu) = \int_M \langle \omega, \nu \rangle$ between sections $\omega \in \Omega_{\mathcal{F}}^p$ and $\nu \in {}^{\mathrm{int}}\Omega_{\mathcal{F}}^{-p}$. The symbol $\langle ., . \rangle$ denotes contraction of a differential form with a polyvector field. The operation $\langle ., . \rangle$ takes values in sections of Ber. The value $\langle \omega, \nu \rangle$ can be used as an integrand for integration over M.

The orthogonal complement to the ideal $S \subset \Omega_{\mathcal{F}}^p$ is a subcomplex $^{\mathrm{int}}\Omega_{s\mathcal{F}}^{-p} \subset {}^{\mathrm{int}}\Omega_{\mathcal{F}}^{-p}$. It is a differential graded module over $\Omega_{s\mathcal{F}}^p$.

Proposition 29. *Let M be a supermanifold with a CR distribution $\mathcal{F}$ of dimension $(n|k)$. There is an isomorphism $i : {}^{\mathrm{int}}\Omega_{s\mathcal{F}}^{p-n} \to \Omega_{s\mathcal{F}}^p \mathrm{Ber}((T^{\mathbb{C}}/\overline{\mathcal{F}})^*)$ compatible with a stricture of $\Omega_{s\mathcal{F}}^{\bullet}$-module. The isomorphism is unique.*

380 Mikhail V. Movshev

Proof. Using C^∞ splitting we can identify $\Omega^\bullet_{s\mathcal{F}}\mathrm{Ber}((T^{\mathbb{C}}/\overline{\mathcal{F}})^*)$ and ${}^{\mathrm{int}}\Omega^{p-n}_{s\mathcal{F}}$ with sections of some vector bundles A_p and B_p over M_{red}. It is fairly straightforward to establish an isomorphism of A_p and B_p with the help of the splitting. In particular there is a C^∞ isomorphism $\mathrm{Ber}((T^{\mathbb{C}}/\overline{\mathcal{F}})^*) = {}^{\mathrm{int}}\Omega^{-n}_{s\mathcal{F}}$.

One can think about $\Omega^0_{s\mathcal{F}}$ as a space of functions on a supermanifold sM. Then the space of sections $\mathrm{Ber}(sM)$ coincide with $\Omega^n_{s\mathcal{F}}\mathrm{Ber}((T^{\mathbb{C}}/\overline{\mathcal{F}})^*))$. Its elements can be integrated over sM. The integral defines a pairing $(.,.)_s$:
$\Omega^\bullet_{s\mathcal{F}} \otimes \Omega^\bullet_{s\mathcal{F}}\mathrm{Ber}((T^{\mathbb{C}}/\overline{\mathcal{F}})^*)) \xrightarrow{\int (...)_s} \mathbb{C}$, which is nondegenerate.

An element $a \in {}^{\mathrm{int}}\Omega^0_{s\mathcal{F}}$ defines a functional $f \to \int_M af$ ($f \in C^\infty(sM)$). We may think about it as an integral $\int_{sM} fi(a)$, where $i(a) \in \Omega^n_{s\mathcal{F}}\mathrm{Ber}((T^{\mathbb{C}}/\overline{\mathcal{F}})^*)$. Such an interpretation of the integral uniquely specifies the map i. Since $\Omega^n_{s\mathcal{F}}$ is invertible, the induced isomorphism $i : \mathrm{Ber}((T^{\mathbb{C}}/\overline{\mathcal{F}})^*) = {}^{\mathrm{int}}\Omega^{-n}_{s\mathcal{F}}$ is compatible with $\bar{\partial}$ (use pairings $(.,.), (.,.)_s$ to check this). $\qquad\square$

Suppose that a super CR manifold M is CR embedded into a complex super manifold N. Denote by J an operator of complex structure in the tangent bundle TN. We assume that $TM + JTM = TN|_M$. Denote by $\mathrm{Ber}_{\mathbb{C}}(N)$ the complex Berezinian of N. An easy local computation shows that $\mathrm{Ber}_{\mathbb{C}}(N)|_M$ is isomorphic to $\mathrm{Ber}((T^{\mathbb{C}}/\overline{\mathcal{F}})^*)$. Suppose N is a Calabi–Yau manifold, i.e., it admits a global nonvanishing section vol of $\mathrm{Ber}_{\mathbb{C}}(N)$. The restriction of vol to M defined a global CR holomorphic section of $\mathrm{Ber}((T^{\mathbb{C}}/\overline{\mathcal{F}})^*)$. An isomorphism of Proposition 29 provides a $\bar{\partial}$ closed section of ${}^{\mathrm{int}}\Omega^{-n}_{s\mathcal{F}} \subset {}^{\mathrm{int}}\Omega^{-n}_{\mathcal{F}}$.

Remark 30. The proof Proposition 29 parallels the proof of Serre duality in the super case given in [HW]. Haske and Wells used a sheaf-theoretic description of a complex supermanifold, which significantly simplifies the argument. The main simplification comes from the local Poincaré lemma, which is absent in the CR case.

It is worthwhile to mention that there is no canonical ($\mathrm{Aut}_{\mathcal{F}}$ or $\mathrm{Out}_{\mathcal{F}}$ equivariant) map $\Omega^\bullet_{\mathcal{F}}\mathrm{Ber}((T^{\mathbb{C}}/\overline{\mathcal{F}})^*) \to {}^{\mathrm{int}}\Omega^\bullet_{\mathcal{F}}[-n]$. This seems to be one of fundamental distinctions of the purely even and super cases.

We think the reason is that the only known construction of this map is through the intermediate complex $\Omega^\bullet_{s\mathcal{F}}\mathrm{Ber}((T^{\mathbb{C}}/\overline{\mathcal{F}})^*)$. As have have already mentioned in Remark 28, the complex $\Omega^\bullet_{s\mathcal{F}}$ carries only the A_∞ action of $\mathrm{Aut}_{\mathcal{F}}$.

A real line bundle Ber on a complex super manifold is a tensor product $\mathrm{Ber}_{\mathbb{C}} \otimes \overline{\mathrm{Ber}}_{\mathbb{C}}$, where $\mathrm{Ber}_{\mathbb{C}}$ is a holomorphic Berezinian.

A decomposition $T^{\mathbb{C}} = \mathcal{T} = \overline{\mathcal{T}}$ implies that

$$
{}^{\mathrm{int}}\Omega^{-k} = \mathrm{Ber} \otimes \Lambda^k(\mathcal{T}) =
$$
$$
= \bigoplus_{i+j=k} \mathrm{Ber}_{\mathbb{C}} \otimes \Lambda^i(\mathcal{T}_{\mathbb{C}}) \otimes \overline{\mathrm{Ber}}_{\mathbb{C}} \otimes \Lambda^j(\overline{\mathcal{T}}_{\mathbb{C}}) = \bigoplus_{i+j=k} {}^{\mathrm{int}}\Omega^{-i,-j}. \qquad (47)
$$

Remark 31. Suppose M is a complex n-dimensional manifold, E is a k-dimensional vector bundle. On the total space ΠE there is a canonical section of $c_{\Pi E} \in \mathrm{Ber}_{\mathbb{C}} \otimes \Lambda^n(\mathcal{T})$. In local odd fiberwise coordinates θ_i, it is equal to

$$c_{\Pi V} = \theta_1 \dots \theta_k d\theta_1 \wedge \cdots \wedge d\theta_k. \tag{48}$$

Proposition 32. *The forms $c_{\Pi V}$ and $\bar{c}_{\Pi V}$ are $\bar{\partial}$-closed.*

Proof. Since the formula (49) does not depend on the choice of coordinates on M, one can do a local computation, which is trivial. $\square$

6 Acknowledgments

The author would like to thank P. Candelas, P. Deligne, M. Kontsevich, L. Maison, A.S. Schwarz for useful comments and discussions. This paper was written while the author was staying at Mittag-Leffler Institut, Max Planck Institute for Mathematics, and the Institute for Advanced Study. The author is grateful to these institutions for their hospitality and support. After the preprint version of this paper [Mov] had been published, the author learned about the work by L.J. Mason and D. Skinner [MS], where they treat a similar problem.

References

[AHDM] M. F. ATIYAH, N. J. HITCHIN, V. G. DRINFEL'D, YU. I. MANIN, *Construction of instantons* Phys. Lett. A 65 (1978), no. 3, 185–187.

[Berk] N. BERKOVITS, *Covariant Quantization of the Superparticle Using Pure Spinors* JHEP 0109 (2001) 016, hep-th/0105050.

[DMiln] P. DELIGNE, J. S. MILNE, *Tannakian categories, in Hodge cycles, motives and Shimura varieties*, LNM 900, 101–228.

[DM] P. DELIGNE, J. W. MORGAN, *Notes on supersymmetry (following Joseph Bernstein). Quantum fields and strings: a course for mathematicians*, Vol. 1, 2 (Princeton, NJ, 1996/1997), 4197, Amer. Math. Soc., Providence, RI, 1999.

[FP02] J. FRAUENDIENER, R. PENROSE, *Twistors and general relativity.* Mathematics unlimited—2001 and beyond, 479–505, Springer, Berlin, 2001.

[GH] Ph. GRIFFITHS, J. HARRIS, *Principles of algebraic geometry.* Pure and Applied Mathematics. Wiley-Interscience, John Wiley & Sons, New York, 1978. xii+813 pp.

[HW] C. HASKE, R. O. WELLS, JR., *Serre duality on complex supermanifolds* Duke Math. J. 54, no. 2 (1987), 493–500.

[IKKT] N. ISHIBASHI, H. KAWAI, I. KITAZAWA, A. TSUCHIYA, *A large-N reduced model as superstring* Nucl. Phys. B492 (1997).

[Kad] T. KADEISHVILI, *The algebraic structure in the homology of an A_∞-algebra* Soobshch. Akad. Nauk Gruzin. SSR, 1982.

[Man] YU. I. MANIN, *Gauge field theory and complex geometry* Grundlehren der Mathematischen Wissenschaften, 289, Springer-Verlag, Berlin, 1997.

[Markl] M. MARKL, *Transferring A_∞ (strongly homotopy associative) structures* math.AT/0401007.

[MS] L. J. MASON, D. SKINNER, *An ambitwistor Yang-Mills Lagrangian* Phys. Lett. B636 (2006) 60–67, hep-th/0510262.

[Mov] M. MOVSHEV, *Yang–Mills theory and a superquadric* hep-th/0411111.

[MSch05] M. MOVSHEV, A. SCHWARZ, *On maximally supersymmetric Yang–Mills theories* Nuclear Physics B, 681, no. 3, 324–350.

[MSch06] M. MOVSHEV, A. SCHWARZ, *Algebraic structure of Yang–Mills theory*, The Unity of Mathematics, 473–523, Progr. Math., 244, Birkhäuser Boston, Boston, MA, 2006, hep-th/0404183.

[Sch92] A. SCHWARZ, *Semiclassical approximation in Batalin–Vilkovisky formalism* Commun. Math. Phys. 158 (1993) 373–396, hep-th/9210115.

[Sch00] A. SCHWARZ, *Topological quantum field theories* hep-th/0011260.

[Serg] V. V. SERGANOVA, *Classification of simple real Lie superalgebras and symmetric superspaces* (Russian) Funktsional. Anal. i Prilozhen. 17 (1983) no. 3, 46–54.

[Pel] F. PELLEGRINI, *On real forms of complex Lie superalgebras and complex algebraic supergroups* math.RA/0311240.

[P90] R. PENROSE, *Twistor theory after 25 years—its physical status and prospects. Twistors in mathematics and physics* London Math. Soc. Lecture Note Ser., 156, Cambridge Univ. Press, Cambridge, 1990.

[W78] E. WITTEN, *An interpretation of classical Yang–Mills theory* preprint HUTP-78/A009.

[W03] E. WITTEN, *Gauge theory as a string theory in twistor space*, hep-th/0312171.

A Generalization of the Capelli Identity

E. Mukhin,[1] V. Tarasov[2] and A. Varchenko[3]

[1] Department of Mathematical Sciences, Indiana University – Purdue University Indianapolis, 402 North Blackford St, Indianapolis, IN 46202-3216, USA, mukhin@math.iupui.edu

[2] Department of Mathematical Sciences, Indiana University – Purdue University Indianapolis, 402 North Blackford St, Indianapolis, IN 46202-3216, USA vtarasov@math.iupui.edu, and
St. Petersburg Branch of Steklov Mathematical Institute, Fontanka 27, St. Petersburg, 191023, Russia, vt@pdmi.ras.ru

[3] Department of Mathematics, University of North Carolina at Chapel Hill, Chapel Hill, NC 27599-3250, USA, anv@email.unc.edu

To Yuri Ivanovich Manin on the occasion of his 70th birthday, with admiration

Summary. We prove a generalization of the Capelli identity. As an application we obtain an isomorphism of the Bethe subalgebras actions under the $(\mathfrak{gl}_N, \mathfrak{gl}_M)$ duality.

Key words: Capelli identity, Gaudin model, $(\mathfrak{gl}_N, \mathfrak{gl}_M)$ duality

2000 Mathematics Subject Classifications: 17B80, 22E45

1 Introduction

Let $\mathcal{A}$ be an associative algebra over the complex numbers. Let $A = (a_{ij})_{i,j=1}^{n}$ be an $n \times n$ matrix with entries in $\mathcal{A}$. The *row determinant of A* is defined by the formula

$$\operatorname{rdet}(A) := \sum_{\sigma \in S_n} \operatorname{sgn}(\sigma) a_{1\sigma_1} \cdots a_{n\sigma_n}.$$

Let x_{ij}, $i,j = 1, \ldots, M$, be commuting variables. Let $\partial_{ij} = \partial/\partial x_{ij}$,

$$E_{ij} = \sum_{a=1}^{M} x_{ia} \partial_{ja}. \tag{1}$$

Let $X = (x_{ij})_{i,j=1}^{M}$ and $D = (\partial_{ij})_{i,j=1}^{M}$ be $M \times M$ matrices.

Y. Tschinkel and Y. Zarhin (eds.), *Algebra, Arithmetic, and Geometry,*
Progress in Mathematics 270, DOI 10.1007/978-0-8176-4747-6_12,
© Springer Science+Business Media, LLC 2009

The classical Capelli identity [C1] asserts the following equality of differential operators:

$$\mathrm{rdet}\left(E_{ji} + (M - i)\delta_{ij}\right)^{M}_{i,j=1} = \det(X)\det(D).$$ (2)

This identity is a "quantization" of the identity

$$\det(AB) = \det(A)\det(B)$$

for any matrices A, B with commuting entries.

The Capelli identity has the following meaning in representation theory. Let $\mathbb{C}[X]$ be the algebra of complex polynomials in the variables x_{ij}. There are two natural actions of the Lie algebra $\mathfrak{gl}_M$ on $\mathbb{C}[X]$. The first action is given by operators from (1), and the second action is given by operators $\widetilde{E}_{ij} = \sum_{a=1}^{M} x_{ai}\partial_{aj}$. The two actions commute and the corresponding $\mathfrak{gl}_M \oplus \mathfrak{gl}_M$ action is multiplicity-free.

It is not difficult to see that the right-hand side of (2), considered as a differential operator on $\mathbb{C}[X]$, commutes with both actions of $\mathfrak{gl}_M$ and therefore lies in the image of the center of the universal enveloping algebra $U\mathfrak{gl}_M$ with respect to the first action. Then the left-hand side of the Capelli identity expresses the corresponding central element in terms of $U\mathfrak{gl}_M$ generators.

Many generalizations of the Capelli identity are known. One group of generalizations considers other elements of the center of $U\mathfrak{gl}_M$, called quantum immanants, and then expresses them in terms of $\mathfrak{gl}_M$ generators; see [C2], [N1], [O]. Another group of generalizations considers other pairs of Lie algebras in place of $(\mathfrak{gl}_M, \mathfrak{gl}_M)$, e.g., $(\mathfrak{gl}_M, \mathfrak{gl}_N)$, $(\mathfrak{sp}_{2M}, \mathfrak{gl}_2)$, $(\mathfrak{sp}_{2M}, \mathfrak{so}_N)$; see [MN], [HU]. The third group of generalizations produces identities corresponding not to pairs of Lie algebras, but to pairs of quantum groups [NUW] or superalgebras [N2].

In this paper we prove a generalization of the Capelli identity that seemingly does not fit the above classification.

Let $z = (z_1, \ldots, z_N)$, $\lambda = (\lambda_1, \ldots, \lambda_M)$ be sequences of complex numbers. Let $Z = (z_i\delta_{ij})^{N}_{ij=1}$, $\Lambda = (\lambda_i\delta_{ij})^{M}_{ij=1}$ be the corresponding diagonal matrices. Let X and D be the $M \times N$ matrices with entries x_{ij} and ∂_{ij}, $i = 1, \ldots, M$, $j = 1, \ldots, N$, respectively. Let $\mathbb{C}[X]$ be the algebra of complex polynomials in variables x_{ij}, $i = 1, \ldots, M$, $j = 1, \ldots, N$. Let $E_{ij}^{(a)} = x_{ia}\partial_{ja}$, where $i, j = 1, \ldots, M$, $a = 1, \ldots, N$.

In this paper we prove that

$$\prod_{a=1}^{N}(u - z_a)\,\mathrm{rdet}\left((\partial_u - \lambda_i)\delta_{ij} - \sum_{a=1}^{N}\frac{E_{ji}^{(a)}}{u - z_a}\right)^{M}_{i,j=1} = \mathrm{rdet}\begin{pmatrix} u - Z & X^t \\ D & \partial_u - \Lambda \end{pmatrix}.$$ (3)

The left-hand side of (3) is an $M \times M$ matrix, while the right-hand side is an $(M + N) \times (M + N)$ matrix.

Identity (3) is a "quantization" of the identity

$$\det \begin{pmatrix} A & B \\ C & D \end{pmatrix} = \det(A) \; \det(D - CA^{-1}B),$$

which holds for any matrices A, B, C, D with commuting entries, for the case when A and D are diagonal matrices.

By setting all z_i, λ_j, and u to zero, and $N = M$ in (3), we obtain the classical Capelli identity (2); see Section 2.4.

Our proof of (3) is combinatorial and reduces to the case of 2×2 matrices. In particular, it gives a proof of the classical Capelli identity, which may be new.

We invented identity (3) to prove Theorem 6 below, and Theorem 6 in turn was motivated by results of [MTV2]. In Theorem 6 we compare actions of two Bethe subalgebras.

Namely, consider $\mathbb{C}[X]$ as a tensor product of evaluation modules over the current Lie algebras $\mathfrak{gl}_M[t]$ and $\mathfrak{gl}_N[t]$ with evaluation parameters z and λ, respectively. The action of the algebra $\mathfrak{gl}_M[t]$ on $\mathbb{C}[X]$ is given by the formula

$$E_{ij} \otimes t^n = \sum_{a=1}^{N} x_{ia} \partial_{ja} z_a^n,$$

and the action of the algebra $\mathfrak{gl}_N[t]$ on $\mathbb{C}[X]$ is given by the formula

$$E_{ij} \otimes t^n = \sum_{a=1}^{M} x_{ai} \partial_{aj} \lambda_a^n.$$

In contrast to the previous situation, these two actions do not commute.

The algebra $U\mathfrak{gl}_M[t]$ has a family of commutative subalgebras $\mathcal{G}(M, \lambda)$ depending on parameters λ and called the Bethe subalgebras. For a given λ, the Bethe subalgebra $\mathcal{G}(M, \lambda)$ is generated by the coefficients of the expansion of the expression

$$\text{rdet} \left((\partial_u - \lambda_i)\delta_{ij} - \sum_{a=1}^{N} \sum_{s=1}^{\infty} \left(E_{ji}^{(a)} \otimes t^s \right) u^{-s-1} \right)_{i,j=1}^{M} \tag{4}$$

with respect to powers of u^{-1} and ∂_u; cf. Section 3. For different versions of definitions of Bethe subalgebras and relations between them, see [FFR], [T], [R], [MTV1].

Similarly, there is a family of Bethe subalgebras $\mathcal{G}(N, z)$ in $U\mathfrak{gl}_N[t]$ depending on parameters z.

For fixed λ and z, consider the action of the Bethe subalgebras $\mathcal{G}(M, \lambda)$ and $\mathcal{G}(N, z)$ on $\mathbb{C}[X]$ as defined above. In Theorem 6 we show that the actions of the Bethe subalgebras on $\mathbb{C}[X]$ induce the same subalgebras of endomorphisms of $\mathbb{C}[X]$.

The paper is organized as follows. In Section 2 we describe and prove formal Capelli-type identities, and in Section 3 we discuss the relations of the identities to the Bethe subalgebras.

Acknowledgment. We thank the referee for bringing to our attention the papers [KS] and [GR], which relate the Capelli identity to Jordan algebras and "quasideterminants," respectively.

Research of E.M. is supported in part by NSF grant DMS-0601005. Research of A.V. is supported in part by NSF grant DMS-0555327.

2 Identities

2.1 The main identity

We work over the field of complex numbers. However, all results of this paper hold over any field of characteristic zero.

Let $\mathcal{A}$ be an associative algebra. Let $A = (a_{ij})_{i,j=1}^n$ be an $n \times n$ matrix with entries in $\mathcal{A}$. Define the *row determinant of A* by the formula:

$$\mathrm{rdet}(A) := \sum_{\sigma \in S_n} \mathrm{sgn}(\sigma) a_{1\sigma_1} \cdots a_{n\sigma_n},$$

where S_n is the symmetric group on n elements.

Fix two natural numbers M and N and a complex number $h \in \mathbb{C}$. Consider noncommuting variables u, p_u, x_{ij}, p_{ij}, where $i = 1, \ldots, M$, $j = 1, \ldots, N$, such that the commutator of two variables equals zero except

$$[p_u, u] = h, \qquad [p_{ij}, x_{ij}] = h,$$

$i = 1, \ldots, M$, $j = 1, \ldots, N$.

Let X, P be two $M \times N$ matrices given by

$$X := (x_{ij})_{i=1,\ldots,M}^{j=1,\ldots,N}, \qquad P := (p_{ij})_{i=1,\ldots,M}^{j=1,\ldots,N}.$$

Let $\mathcal{A}_h^{(MN)}$ be the associative algebra whose elements are polynomials in p_u, x_{ij}, p_{ij}, $i = 1, \ldots, M$, $j = 1, \ldots, N$, with coefficients that are rational functions in u.

Let $\mathcal{A}^{(MN)}$ be the associative algebra of linear differential operators in u, x_{ij}, $i = 1, \ldots, M$, $j = 1, \ldots, N$, with coefficients in $\mathbb{C}(u) \otimes \mathbb{C}[X]$.

We often drop the dependence on M, N and write $\mathcal{A}_h$, $\mathcal{A}$ for $\mathcal{A}_h^{(MN)}$ and $\mathcal{A}^{(MN)}$, respectively.

For $h \neq 0$, we have the isomorphism of algebras

$$\iota_h : \mathcal{A}_h \to \mathcal{A}, \qquad (5)$$

$$u, x_{ij} \mapsto u, x_{ij},$$

$$p_u, p_{ij} \mapsto h\frac{\partial}{\partial u}, h\frac{\partial}{\partial x_{ij}}.$$

Fix two sequences of complex numbers $z = (z_1, \dots, z_N)$ and $\lambda = (\lambda_1, \dots, \lambda_M)$.

Define the $M \times M$ matrix $G_h = G_h(M, N, u, p_u, z, \lambda, X, P)$ by the formula

$$
G_h := \left((p_u - \lambda_i)\delta_{ij} - \sum_{a=1}^{N} \frac{x_{ja} p_{ia}}{u - z_a} \right)_{i,j=1}^{M}. \tag{6}
$$

Theorem 1. *We have*

$$
\prod_{a=1}^{N} (u - z_a)\ \mathrm{rdet}(G_h) =
$$
$$
\sum_{A,B,|A|=|B|} (-1)^{|A|} \prod_{a \notin B}(u - z_a) \prod_{b \notin A}(p_u - \lambda_b) \det(x_{ab})_{a \in A}^{b \in B}\ \det(p_{ab})_{a \in A}^{b \in B},
$$

where the sum is over all pairs of subsets $A \subset \{1, \dots, M\}$, $B \subset \{1, \dots, N\}$ such that A and B have the same cardinality, $|A| = |B|$. Here the sets A, B inherit the natural ordering from the sets $\{1, \dots, M\}$, $\{1, \dots, N\}$. This ordering define the determinants in the formula.

Theorem 1 is proved in Section 2.5.

2.2 A presentation as a row determinant of size $M + N$

Theorem 1 implies that the row determinant of G can be written as the row determinant of a matrix of size $M + N$.

Namely, let Z be the diagonal $N \times N$ matrix with diagonal entries $z_1, \dots, z_N$. Let Λ be the diagonal $M \times M$ matrix with diagonal entries $\lambda_1, \dots, \lambda_M$:

$$
Z := (z_i \delta_{ij})_{i,j=1}^{N}, \qquad \Lambda := (\lambda_i \delta_{ij})_{i,j=1}^{M}.
$$

Corollary 2. *We have*

$$
\prod_{a=1}^{N} (u - z_a)\ \mathrm{rdet}\ G = \mathrm{rdet} \begin{pmatrix} u - Z & X^t \\ P & p_u - \Lambda \end{pmatrix},
$$

where X^t denotes the transpose of the matrix X.

Proof. Define

$$
W := \begin{pmatrix} u - Z & X^t \\ P & p_u - \Lambda \end{pmatrix},
$$

The entries of the first N rows of W commute. The entries of the last M rows of W also commute. Write the Laplace expansion of $\mathrm{rdet}(W)$ with respect to the first N rows. Each term in this expansion corresponds to a choice of N

columns in the $N \times (N + M)$ matrix $(u - Z, X^T)$. We label such a choice by a pair of subsets $A \subset \{1, \ldots, M\}$ and $B \subset \{1, \ldots, N\}$ of the same cardinality. Namely, the elements of A correspond to the chosen columns in X^T, and the elements of the complement to B correspond to the chosen columns in $u - Z$. Then the term in the Laplace expansion corresponding to A and B is exactly the term labeled by A and B in the right-hand side of the formula in Theorem 1. Therefore, the corollary follows from Theorem 1. □

Let A, B, C, D be any matrices with commuting entries of sizes $N \times N, N \times M, M \times N$, and $M \times M$, respectively. Let A be invertible. Then we have the equality of matrices of sizes $(M + N) \times (M + N)$:

$$\begin{pmatrix} A & B \\ C & D \end{pmatrix} = \begin{pmatrix} A & 0 \\ C & D - CA^{-1}B \end{pmatrix} \begin{pmatrix} 1 & A^{-1}B \\ 0 & 1 \end{pmatrix}$$

and therefore

$$\det \begin{pmatrix} A & B \\ C & D \end{pmatrix} = \det(A) \det(D - CA^{-1}B). \tag{7}$$

The identity of Corollary 2 for $h = 0$ turns into identity (7) with diagonal matrices A and D. Therefore, the identity of Corollary 2 may be thought of as a "quantization" of identity (7) with diagonal A and D.

2.3 A Relation Between Determinants of Sizes M and N

Introduce new variables v, p_v such that $[p_v, v] = h$.

Let $\bar{A}_h$ be the associative algebra whose elements are polynomials in p_u, p_v, x_{ij}, p_{ij}, $i = 1, \ldots, M$, $j = 1, \ldots, N$, with coefficients in $\mathbb{C}(u) \otimes \mathbb{C}(v)$.

Let $e : \bar{A}_h \to \bar{A}_h$ be the unique linear map that is the identity map on the subalgebra of $\bar{A}_h$ generated by all monomials that do not contain p_u and p_v and that satisfies

$$e(ap_u) = e(a)v, \qquad e(ap_v) = e(a)u,$$

for any $a \in \bar{A}_h$.

Let $\bar{A}$ be the associative algebra of linear differential operators in u, v, x_{ij}, $i = 1, \ldots, M$, $j = 1, \ldots, N$, with coefficients in $\mathbb{C}(u) \otimes \mathbb{C}(v) \otimes \mathbb{C}[x_{ij}]$. Then for $h \neq 0$, we have an isomorphism of algebras extending the isomorphism (5):

$$\bar{\iota}_h : \bar{A}_h \to \bar{A},$$
$$u, v, x_{ij} \mapsto u, v, x_{ij},$$
$$p_u, p_v, p_{ij} \mapsto h \frac{\partial}{\partial u}, h \frac{\partial}{\partial v}, h \frac{\partial}{\partial x_{ij}}.$$

For $a \in \bar{A}$ and a function $f(u, v)$ let $a \cdot f(u, v)$ denote the function obtained by the action of a considered as a differential operator in u and v on the function $f(u, v)$.

We have
$$\bar{\iota}_h(e(a)) = \exp(-uv/h)\bar{\iota}_h(a) \cdot \exp(uv/h)$$
for any $a \in \bar{A}_h$ such that a does not depend on either p_u or p_v.

Define the $N \times N$ matrix $H_h = H_h(M, N, v, p_v, z, \lambda, X, P)$ by

$$H_h := \left((p_v - z_i)\delta_{ij} - \sum_{b=1}^{M} \frac{x_{bj} p_{bi}}{v - \lambda_b} \right)_{i,j=1}^{N} ; \tag{8}$$

cf. formula (6).

Corollary 3. *We have*

$$e\left(\prod_{a=1}^{N}(u - z_a)\,\mathrm{rdet}(G_h) \right) = e\left(\prod_{b=1}^{M}(v - \lambda_b)\,\mathrm{rdet}(H_h) \right).$$

Proof. Write the dependence on the parameters of the matrix G: $G_h = G_h(M, N, u, p_u, z, \lambda, X, P)$. Then

$$H_h = G_h(N, M, v, p_v, \lambda, z, X^T, P^T).$$

The corollary now follows from Theorem 1. □

2.4 A relation to the Capelli identity

In this section we show how to deduce the Capelli identity from Theorem 1.

Let s be a complex number. Let $\alpha_s : A_h \to A_h$ be the unique linear map that is the identity map on the subalgebra of A_h generated by all monomials that do not contain p_u and that satisfies

$$\alpha_s(aup_u) = s\alpha_s(a)$$

for any $a \in \bar{A}_h$.

We have
$$\bar{\iota}_h(\alpha_s(a)) = u^{-s/h}\bar{\iota}_h(a) \cdot u^{s/h}$$
for any $a \in \bar{A}_h$.

Consider the case $z_1 = \cdots = z_N = 0$ and $\lambda_1 = \cdots = \lambda_M = 0$ in Theorem 1.

Then it is easy to see that the row determinant $\mathrm{rdet}(G)$ can be rewritten in the following form:

$$u^M\,\mathrm{rdet}(G_h) = \mathrm{rdet}\left(h(up_u - M + i)\delta_{ij} - \sum_{a=1}^{N} x_{ja} p_{ia} \right)_{i,j=1}^{M}.$$

Applying the map α_s, we get

$$\alpha_s(u^M\,\mathrm{rdet}(G_h)) = \mathrm{rdet}\left(h(s - M + i)\delta_{ij} - \sum_{a=1}^{N} x_{ja} p_{ia} \right)_{i,j=1}^{M}.$$

Therefore, applying Theorem 1, we obtain the identity

$$\mathrm{rdet}\left(h(s-M+i)\delta_{ij} - \sum_{a=1}^{N} x_{ja}p_{ia}\right)_{i,j=1}^{M}$$
$$= \sum_{A,B,|A|=|B|} (-1)^{|A|} \prod_{b=0}^{M-|A|-1} (s-bh)\,\det(x_{ab})_{a\in A}^{b\in B}\,\det(p_{ab})_{a\in A}^{b\in B}.$$

In particular, if $M = N$ and $s = 0$, we obtain the famous Capelli identity:

$$\mathrm{rdet}\left(\sum_{a=1}^{M} x_{ja}p_{ia} + h(M-i)\delta_{ij}\right)_{i,j=1}^{M} = \det X \, \det P.$$

If $h = 0$ then all entries of X and P commute, and the Capelli identity reads $\det(XP) = \det(X)\det(P)$. Therefore, the Capelli identity can be thought of as a "quantization" of the identity $\det(AB) = \det(A)\det(B)$ for square matrices A, B with commuting entries.

2.5 Proof of Theorem 1

We define

$$E_{ij,a} := x_{ja}p_{ia}/(u - z_a).$$

We obviously have

$$[E_{ij,a}, E_{kl,b}] = \delta_{ab}(\delta_{kj}(E_{il,a})' - \delta_{il}(E_{kj,a})'),$$

where the prime denotes formal differentiation with respect to u.

Define also $F_{jk,a}^1 = -E_{jk,a}$ and $F_{jj,0}^0 = (p_u - \lambda_j)$.

Expand $\mathrm{rdet}(G)$. We get an alternating sum of terms,

$$\mathrm{rdet}(G_h) = \sum_{\sigma,a,c} (-1)^{\mathrm{sgn}(\sigma)} F_{1\sigma(1),a(1)}^{c(1)} F_{2\sigma(2),a(2)}^{c(2)} \cdots F_{M\sigma(M),a(M)}^{c(M)}, \qquad (9)$$

where the summation is over all triples σ, a, c such that σ is a permutation of $\{1, \ldots, M\}$ and a, c are maps $a : \{1, \ldots, M\} \to \{0, 1, \ldots, N\}, c : \{1, \ldots, M\} \to \{0, 1\}$ satisfying $c(i) = 1$ if $\sigma(i) \neq i$; $a(i) = 0$ if and only if $c(i) = 0$.

Let m be a product whose factors are of the form $f(u), p_u, p_{ij}, x_{ij}$, where $f(u)$ are some rational functions in u. Then the product m will be called *normally ordered* if all factors of the form p_u, p_{ij} are to the right of all factors of the form $f(u), x_{ij}$. For example, $(u-1)^{-2}x_{11}p_up_{11}$ is normally ordered and $p_u(u-1)^{-2}x_{11}p_{11}$ is not.

Given a product m as above, define a new normally ordered product $: m :$ as the product of all factors of m in which all factors of the form p_u, p_{ij} are placed to the right of all factors of the form $f(u), x_{ij}$. For example, $: p_u(u-1)^{-2}x_{11}p_{11} := (u-1)^{-2}x_{11}p_up_{11}$.

If all variables p_u, p_{ij} are moved to the right in the expansion of $\mathrm{rdet}(G)$, then we get terms obtained by normal ordering from the terms in (9) plus new terms created by the nontrivial commutators. We show that in fact all new terms cancel in pairs.

Lemma 4. *For $i = 1, \ldots, M$, we have*

$$\mathrm{rdet}(G_h) = \sum_{\sigma, a, c} (-1)^{\mathrm{sgn}(\sigma)} F^{c(1)}_{1\sigma(1), a(1)} \cdots F^{c(i-1)}_{(i-1)\sigma(i-1), a(i-1)}$$
$$\times \left(: F^{c(i)}_{i\sigma(i), a(i)} \cdots F^{c(M)}_{M\sigma(M), a(M)} : \right), \tag{10}$$

where the sum is over the same triples σ, a, c as in (9).

Proof. We prove the lemma by induction on i. For $i = M$ the lemma is a tautology. Assume that it is proved for $i = M, M - 1, \ldots, j$, and let us prove it for $i = j - 1$.

We have

$$F^1_{(j-1)r, a} : F^{c(j)}_{j\sigma(j), a(j)} \cdots F^{c(M)}_{M, \sigma(M), a(M)}$$
$$:= F^1_{(j-1)r, a} F^{c(j)}_{j\sigma(j), a(j)} \cdots F^{c(M)}_{M\sigma(M), a(M)} :$$
$$+ \sum_k : F^{c(j)}_{j\sigma(j), a(j)} \cdots (-E_{kr, a})' \cdots F^{c(M)}_{M\sigma(M), a(M)} :, \tag{11}$$

where the sum is over $k \in \{j, \ldots, M\}$ such that $a(k) = a$, $\sigma(k) = j - 1$, and $c(k) = 1$.

We also have

$$F^0_{(j-1)(j-1), 0} : F^{c(j)}_{j\sigma(j), a(j)} \cdots F^{c(M)}_{M\sigma(M), a(M)}$$
$$:= F^0_{(j-1)(j-1), 0} F^{c(j)}_{j\sigma(j), a(j)} \cdots F^{c(M)}_{M\sigma(M), a(M)} :$$
$$+ \sum_k : F^{c(j)}_{j\sigma(j), a(j)} \cdots (-E_{k\sigma(k), a(k)})' \cdots F^{c(M)}_{M\sigma(M), a(M)} :, \tag{12}$$

where the sum is over $k \in \{j, \ldots, M\}$ such that $c(k) = 1$.

Using (11), (12), rewrite each term in (10) with $i = j$. Then the k-th term obtained by using (11) applied to the term labeled by σ, c, a with $c(j-1) = 0$ cancels with the k-th term obtained by using (12) applied to the term labeled by $\tilde{\sigma}, \tilde{c}, \tilde{a}$ defined by the following rules:

$$\tilde{\sigma}(i) = \sigma(i) \ (i \neq j - 1, k), \qquad \tilde{\sigma}(j - 1) = j - 1, \qquad \tilde{\sigma}(k) = \sigma(j - 1),$$

$$\tilde{c}(i) = c(i) \ (i \neq j - 1), \qquad \tilde{c}(j - 1) = 0,$$
$$\tilde{a}(i) = a(i) \ (i \neq j - 1), \qquad \tilde{a}(j - 1) = 0.$$

After this cancellation we obtain the statement of the lemma for $i = j - 1$. $\square$

Remark 5. The proof of Lemma 4 implies that if the matrix σG_h is obtained from G_h by permuting the rows of G_h by a permutation σ, then $\mathrm{rdet}(\sigma G_h) = (-1)^{\mathrm{sgn}(\sigma)} \mathrm{rdet}(G_h)$.

Consider the isomorphism of vector spaces $\phi_h : A_h \to A_0$ that sends any normally ordered monomial m in A_h to the same monomial m in A_0.

By (10) with $i = 1$, the image $\phi_h(\mathrm{rdet}(G_h))$ does not depend on h and therefore can be computed at $h = 0$. Therefore Theorem 1 for all h follows from Theorem 1 for $h = 0$. Theorem 1 for $h = 0$ follows from formula (7).

3 The $(\mathfrak{gl}_M, \mathfrak{gl}_N)$ Duality and the Bethe Subalgebras

3.1 Bethe subalgebra

Let E_{ij}, $i, j = 1, \ldots, M$, be the standard generators of $\mathfrak{gl}_M$. Let $\mathfrak{h}$ be the Cartan subalgebra of $\mathfrak{gl}_M$,

$$\mathfrak{h} = \oplus_{i=1}^{M} \mathbb{C} \cdot E_{ii}.$$

We denote by $U\mathfrak{gl}_M$ the universal enveloping algebra of $\mathfrak{gl}_M$.

For $\mu \in \mathfrak{h}^*$, and a $\mathfrak{gl}_M$ module L denote by $L[\mu]$ the vector subspace of L of vectors of weight μ,

$$L[\mu] = \{v \in L \mid hv = \langle \mu, h \rangle v \text{ for any } h \in \mathfrak{h}\}.$$

We always assume that $L = \oplus_\mu L[\mu]$.

For any integral dominant weight $\Lambda \in \mathfrak{h}^*$, denote by L_Λ the finite-dimensional irreducible $\mathfrak{gl}_M$-module with highest weight Λ.

Recall that we fixed sequences of complex numbers $\boldsymbol{z} = (z_1, \ldots, z_N)$ and $\boldsymbol{\lambda} = (\lambda_1, \ldots, \lambda_M)$. From now on we will assume that $z_i \neq z_j$ and $\lambda_i \neq \lambda_j$ if $i \neq j$.

For $i, j = 1, \ldots, M,$, $a = 1, \ldots, N$, let $E_{ji}^{(a)} = 1^{\otimes(a-1)} \otimes E_{ji} \otimes 1^{\otimes(N-a)} \in (U\mathfrak{gl}_M)^{\otimes N}$.

Define the $M \times M$ matrix $\widetilde{G} = \widetilde{G}(M, N, \boldsymbol{z}, \boldsymbol{\lambda}, u)$ by

$$\widetilde{G}(M, N, \boldsymbol{z}, \boldsymbol{\lambda}, u) := \left(\left(\frac{\partial}{\partial u} - \lambda_i \right) \delta_{ij} - \sum_{a=1}^{N} \frac{E_{ji}^{(a)}}{u - z_a} \right)_{i,j=1}^{M}.$$

The entries of $\widetilde{G}$ are differential operators in u whose coefficients are rational functions in u with values in $(U\mathfrak{gl}_M)^{\otimes N}$.

Write

$$\mathrm{rdet}(\widetilde{G}(M, N, \boldsymbol{z}, \boldsymbol{\lambda}, u)) = \frac{\partial^M}{\partial u^M} + \widetilde{G}_1(M, N, \boldsymbol{z}, \boldsymbol{\lambda}, u) \frac{\partial^{M-1}}{\partial u^{M-1}} + \cdots$$
$$+ \widetilde{G}_M(M, N, \boldsymbol{z}, \boldsymbol{\lambda}, u).$$

The coefficients $\widetilde{G}_i(M, N, z, \lambda, u)$, $i = 1, \ldots, M$, are called the *transfer matrices of the Gaudin model*. The transfer matrices are rational functions in u with values in $(U\mathfrak{gl}_M)^{\otimes N}$.

The transfer matrices commute:

$$[\widetilde{G}_i(M, N, z, \lambda, u), \widetilde{G}_j(M, N, z, \lambda, v)] = 0,$$

for all i, j, u, v; see [T] and Proposition 7.2 in [MTV1].

The transfer matrices clearly commute with the diagonal action of $\mathfrak{h}$ on $(U\mathfrak{gl}_M)^{\otimes N}$.

For $i = 1, \ldots, M$, it is clear that $\widetilde{G}_i(M, N, z, \lambda, u) \prod_{a=1}^{N}(u - z_a)^i$ is a polynomial in u whose coefficients are pairwise commuting elements of $(U\mathfrak{gl}_M)^{\otimes N}$. Let $\mathcal{G}(M, N, z, \lambda) \subset (U\mathfrak{gl}_M)^{\otimes N}$ be the commutative subalgebra generated by the coefficients of polynomials $\widetilde{G}_i(M, N, z, \lambda, u) \prod_{a=1}^{N}(u - z_a)^i$, $i = 1, \ldots, M$. We call the subalgebra $\mathcal{G}(M, N, z, \lambda)$ the *Bethe subalgebra*.

Let $\mathcal{G}(M, \lambda) \subset U\mathfrak{gl}_M[t]$ be the subalgebra considered in the introduction. Let $U\mathfrak{gl}_M[t] \to (U\mathfrak{gl}_M)^{\otimes N}$ be the algebra homomorphism defined by $E_{ij} \otimes t^n \mapsto \sum_{a=1}^{N} E_{ij}^{(a)} z_a^n$. Then the subalgebra $\mathcal{G}(M, N, z, \lambda)$ is the image of the subalgebra $\mathcal{G}(M, \lambda)$ under that homomorphism.

The Bethe subalgebra clearly acts on any N-fold tensor product of $\mathfrak{gl}_M$ representations.

Define the *Gaudin Hamiltonians* $H_a(M, N, z, \lambda) \subset (U\mathfrak{gl}_M)^{\otimes N}$, $a = 1, \ldots, N$, by the formula

$$H_a(M, N, z, \lambda) = \sum_{b=1, b\neq a}^{N} \frac{\Omega^{(ab)}}{z_a - z_b} + \sum_{b=1}^{M} \lambda_b E_{bb}^{(a)},$$

where $\Omega^{(ab)} := \sum_{i,j=1}^{M} E_{ij}^{(a)} E_{ji}^{(b)}$.

Define the *dynamical Hamiltonians* $H_a^\vee(M, N, z, \lambda) \subset (U\mathfrak{gl}_M)^{\otimes N}$, $a = 1, \ldots, M$, by the formula

$$H_a^\vee(M, N, z, \lambda) = \sum_{b=1, \, b\neq a}^{M} \frac{(\sum_{i=1}^{N} E_{ab}^{(i)})(\sum_{i=1}^{N} E_{ba}^{(i)}) - \sum_{i=1}^{N} E_{aa}^{(i)}}{\lambda_a - \lambda_b} + \sum_{b=1}^{N} z_b E_{aa}^{(b)} .$$

It is known that the Gaudin Hamiltonians and the dynamical Hamiltonians are in the Bethe subalgebra; see, e.g., Appendix B in [MTV1]:

$$H_a(M, N, z, \lambda) \in \mathcal{G}(M, N, z, \lambda), \qquad H_b^\vee(M, N, z, \lambda) \in \mathcal{G}(M, N, z, \lambda),$$

$a = 1, \ldots, N$, $b = 1, \ldots, M$.

3.2 The $(\mathfrak{gl}_M, \mathfrak{gl}_N)$ Duality

Let $L_\bullet^{(M)} = \mathbb{C}[x_1, \ldots, x_M]$ be the space of polynomials in M variables. We define the $\mathfrak{gl}_M$-action on $L_\bullet^{(M)}$ by the formula

$$E_{ij} \mapsto x_i \frac{\partial}{\partial x_j}.$$

Then we have an isomorphism of $\mathfrak{gl}_M$ modules

$$L_\bullet^{(M)} = \bigoplus_{m=0}^{\infty} L_m^{(M)},$$

the submodule $L_m^{(M)}$ being spanned by homogeneous polynomials of degree m. The submodule $L_m^{(M)}$ is the irreducible $\mathfrak{gl}_M$ module with highest weight $(m, 0, \ldots, 0)$ and highest-weight vector x_1^m.

Let $L_\bullet^{(M,N)} = \mathbb{C}[x_{11}, \ldots, x_{1N}, \ldots, x_{M1}, \ldots, x_{MN}]$ be the space of polynomials of MN commuting variables.

Let $\pi^{(M)} : (U\mathfrak{gl}_M)^{\otimes N} \to \operatorname{End}(L_\bullet^{(M,N)})$ be the algebra homomorphism defined by

$$E_{ij}^{(a)} \mapsto x_{ia} \frac{\partial}{\partial x_{ja}}.$$

In particular, we define the $\mathfrak{gl}_M$ action on $L_\bullet^{(M,N)}$ by the formula

$$E_{ij} \mapsto \sum_{a=1}^{N} x_{ia} \frac{\partial}{\partial x_{ja}}.$$

Let $\pi^{(N)} : (U\mathfrak{gl}_N)^{\otimes M} \to \operatorname{End}(L_\bullet^{(M,N)})$ be the algebra homomorphism defined by

$$E_{ij}^{(a)} \mapsto x_{ai} \frac{\partial}{\partial x_{aj}}.$$

In particular, we define the $\mathfrak{gl}_N$ action on $L_\bullet^{(M,N)}$ by the formula

$$E_{ij} \mapsto \sum_{a=1}^{M} x_{ai} \frac{\partial}{\partial x_{aj}}.$$

We have isomorphisms of algebras

$$\left(\mathbb{C}[x_1, \ldots, x_M]\right)^{\otimes N} \to L_\bullet^{(M,N)}, \quad 1^{\otimes(j-1)} \otimes x_i \otimes 1^{\otimes(N-j)} \mapsto x_{ij},$$

$$\left(\mathbb{C}[x_1, \ldots, x_N]\right)^{\otimes M} \to L_\bullet^{(M,N)}, \quad 1^{\otimes(i-1)} \otimes x_j \otimes 1^{\otimes(M-i)} \mapsto x_{ij}. \tag{13}$$

Under these isomorphisms the space $L_\bullet^{(M,N)}$ is isomorphic to $\left(L_\bullet^{(M)}\right)^{\otimes N}$ as a $\mathfrak{gl}_M$ module and to $\left(L_\bullet^{(N)}\right)^{\otimes M}$ as a $\mathfrak{gl}_N$ module.

Fix $\boldsymbol{n} = (n_1, \dots, n_N) \in \mathbb{Z}_{\geq 0}^N$ and $\boldsymbol{m} = (m_1, \dots, m_M) \in \mathbb{Z}_{\geq 0}^M$ with $\sum_{i=1}^{N} n_i = \sum_{a=1}^{M} m_a$. The sequences $\boldsymbol{n}$ and $\boldsymbol{m}$ naturally correspond to integral $\mathfrak{gl}_N$ and $\mathfrak{gl}_M$ weights, respectively.

Let $\boldsymbol{L_m}$ and $\boldsymbol{L_n}$ be $\mathfrak{gl}_N$ and $\mathfrak{gl}_M$ modules, respectively, defined by the formulas

$$\boldsymbol{L_m} = \otimes_{a=1}^{M} L_{m_a}^{(N)}, \qquad \boldsymbol{L_n} = \otimes_{b=1}^{N} L_{n_b}^{(M)}.$$

The isomorphisms (13) induce an isomorphism of the weight subspaces,

$$\boldsymbol{L_n}[\boldsymbol{m}] \simeq \boldsymbol{L_m}[\boldsymbol{n}]. \tag{14}$$

Under the isomorphism (14) the Gaudin and dynamical Hamiltonians interchange,

$$\pi^{(M)} H_a(M, N, z, \lambda) = \pi^{(N)} H_a^\vee(N, M, \lambda, z),$$
$$\pi^{(M)} H_b^\vee(M, N, z, \lambda) = \pi^{(N)} H_b(N, M, \lambda, z),$$

for $a = 1, \dots, N$, $b = 1, \dots, M$; see [TV].

We prove the stronger statement that the images of $\mathfrak{gl}_M$ and $\mathfrak{gl}_N$ Bethe subalgebras in $\mathrm{End}\left(L_\bullet^{(M,N)}\right)$ are the same.

Theorem 6. *We have*

$$\pi^{(M)}(\mathcal{G}(M, N, z, \lambda)) = \pi^{(N)}(\mathcal{G}(N, M, \lambda, z)).$$

Moreover, we have

$$\prod_{a=1}^{N} (u - z_a) \pi^{(M)} \, \mathrm{rdet}(\widetilde{G}(M, N, z, \lambda, u)) = \sum_{a=1}^{N} \sum_{b=1}^{M} A_{ab}^{(M)} \, u^a \, \frac{\partial^b}{\partial u^b},$$

$$\prod_{b=1}^{M} (v - \lambda_b) \pi^{(N)} \, \mathrm{rdet}(\widetilde{G}(N, M, \lambda, z, v)) = \sum_{a=1}^{N} \sum_{b=1}^{M} A_{ab}^{(N)} \, v^b \, \frac{\partial^a}{\partial v^a},$$

where $A_{ab}^{(M)}$, $A_{ab}^{(N)}$ are linear operators independent of $u, v, \partial/\partial u$, and $\partial/\partial v$, and furthermore,

$$A_{ab}^{(M)} = A_{ab}^{(N)}.$$

Proof. We obviously have

$$\pi^{(M)}(\widetilde{G}(M, N, z, \lambda, u)) = \bar{\imath}_{h=1}(G_{h=1}),$$
$$\pi^{(N)}(\widetilde{G}(N, M, \lambda, z, v)) = \bar{\imath}_{h=1}(H_{h=1}),$$

where $G_{h=1}$ and $H_{h=1}$ are matrices defined in (6) and (8). Then the coefficients of the differential operators $\prod_{a=1}^{N} (u - z_a) \pi^{(M)} \, \mathrm{rdet}(\widetilde{G}(M, N, z, \lambda, u))$ and $\prod_{b=1}^{M} (v - \lambda_b) \pi^{(N)} \, \mathrm{rdet}(\widetilde{G}(N, M, \lambda, z, v))$ are polynomials in u and v of degrees N and M, respectively, by Theorem 1. The rest of the theorem follows directly from Corollary 3. $\qquad \square$

3.3 Scalar Differential Operators

Let $w \in \boldsymbol{L_n[m]}$ be a common eigenvector of the Bethe subalgebra $\mathcal{G}(M, N, \boldsymbol{z}, \boldsymbol{\lambda})$. Then the operator $\mathrm{rdet}(\widetilde{G}(M, N, \boldsymbol{z}, \boldsymbol{\lambda}, u))$ acting on w defines a monic scalar differential operator of order M with rational coefficients in the variable u. Namely, let $D_w(M, N, \boldsymbol{\lambda}, \boldsymbol{z})$ be the differential operator given by

$$D_w(M, N, \boldsymbol{z}, \boldsymbol{\lambda}, u) = \frac{\partial^M}{\partial u^M} + \widetilde{G}_1^w(M, N, \boldsymbol{z}, \boldsymbol{\lambda}, u)\frac{\partial^{M-1}}{\partial u^{M-1}} + \cdots + \widetilde{G}_M^w(M, N, \boldsymbol{z}, \boldsymbol{\lambda}, u),$$

where $\widetilde{G}_i^w(M, N, \boldsymbol{z}, \boldsymbol{\lambda}, u)$ is the eigenvalue of the ith transfer matrix acting on the vector w:

$$\widetilde{G}_i(M, N, \boldsymbol{z}, \boldsymbol{\lambda}, u)w = \widetilde{G}_i^w(M, N, \boldsymbol{z}, \boldsymbol{\lambda}, u)w.$$

Using isomorphism (14), consider w as a vector in $\boldsymbol{L_m[n]}$. Then by Theorem 6, w is also a common eigenvector for the algebra $\mathcal{G}(N, M, \boldsymbol{\lambda}, \boldsymbol{z})$. Thus, similarly, the operator $\mathrm{rdet}(\widetilde{G}(N, M, \boldsymbol{\lambda}, \boldsymbol{z}, v))$ acting on w defines a monic scalar differential operator of order N, $D_w(N, M, \boldsymbol{\lambda}, \boldsymbol{z}, v)$.

Corollary 7. *We have*

$$\prod_{a=1}^{N}(u - z_a)D_w(M, N, \boldsymbol{z}, \boldsymbol{\lambda}, u) = \sum_{a=1}^{N}\sum_{b=1}^{M} A_{ab,w}^{(M)}\, u^a \frac{\partial^b}{\partial u^b},$$

$$\prod_{b=1}^{M}(v - \lambda_b)D_w(N, M, \boldsymbol{\lambda}, \boldsymbol{z}, v) = \sum_{a=1}^{N}\sum_{b=1}^{M} A_{ab,w}^{(N)}\, v^b \frac{\partial^a}{\partial v^a},$$

where $A_{ab,w}^{(M)}$, $A_{ab,w}^{(N)}$ are numbers independent of $u, v, \partial/\partial u, \partial/\partial v$. Moreover,

$$A_{ab,w}^{(M)} = A_{ab,w}^{(N)}.$$

Proof. The corollary follows directly from Theorem 6. □

Corollary 7 was essentially conjectured in Conjecture 5.1 in [MTV2].

Remark 8. The operators $D_w(M, N, \boldsymbol{z}, \boldsymbol{\lambda})$ are useful objects; see [MV1], [MTV2], [MTV3]. They have the following three properties:

(i) The kernel of $D_w(M, N, \boldsymbol{z}, \boldsymbol{\lambda})$ is spanned by the functions $p_i^w(u)e^{\lambda_i u}$, $i = 1, \ldots, M$, where $p_i^w(u)$ is a polynomial in u of degree m_i.

(ii) All finite singular points of $D_w(M, N, \boldsymbol{z}, \boldsymbol{\lambda})$ are $z_1, \ldots, z_N$.

(iii) Each singular point z_i is regular and the exponents of $D_w(M, N, \boldsymbol{z}, \boldsymbol{\lambda})$ at z_i are $0, n_i + 1, n_i + 2, \ldots, n_i + M - 1$.

A converse statement is also true. Namely, if a linear differential operator of order M has properties (i)–(iii), then the operator has the form $D_w(M, N, \boldsymbol{z}, \boldsymbol{\lambda})$ for a suitable eigenvector w of the Bethe subalgebra. This statement may be deduced from Proposition 9 below.

We discuss the properties of such differential operators in [MTV4]; cf. also [MTV2] and Appendix A in [MTV3].

3.4 The Simple Joint Spectrum of the Bethe Subalgebra

It is proved in [R] that for any tensor product of irreducible $\mathfrak{gl}_M$ modules and for generic z, λ the Bethe subalgebra has a simple joint spectrum. We give here a proof of this fact in the special case of the tensor product L_n.

Proposition 9. *For generic values of λ, the joint spectrum of the Bethe subalgebra $\mathcal{G}(M, N, z, \lambda)$ acting in $L_n[m]$ is simple.*

Proof. We claim that for generic values of λ, the joint spectrum of the Gaudin Hamiltonians $H_a(M, N, z, \lambda)$, $a = 1, \ldots, N$, acting in $L_n[m]$ is simple. Indeed, fix z and consider λ such that $\lambda_1 \gg \lambda_2 \gg \cdots \gg \lambda_M \gg 0$. Then the eigenvectors of the Gaudin Hamiltonians in $L_n[m]$ will have the form $v_1 \otimes \cdots \otimes v_N + o(1)$, where $v_i \in L_{n_i}[m^{(i)}]$ and $m = \sum_{i=1}^{N} m^{(i)}$. The corresponding eigenvalue of $H_a(M, N, z, \lambda)$ will be $\sum_{j=1}^{M} \lambda_j m_j^{(a)} + O(1)$.

The weight spaces $L_{n_i}^{(M)}[m_i]$ all have dimension at most 1, and therefore the joint spectrum is simple in this asymptotic zone of parameters. Therefore it is simple for generic values of λ. $\square$

References

[C1] A. Capelli, *Über die Zurückführung der Cayley'schen Operation Ω auf gewöhnliche Polar-Operationen* (German) Math. Ann. **29** (1887), no. 3, 331–338.

[C2] A. Capelli, *Sur les opérations dans la théorie des formes algébriques* (French) Math. Ann. **37** (1890), no. 1, 1–37.

[FFR] B. Feigin, E. Frenkel, N. Reshetikhin, *Gaudin model, Bethe ansatz and critical level*, Comm. Math. Phys. **166** (1994), no. 1, 27–62.

[GR] I.M. Gelfand, V. Retakh, *Theory of noncommutative determinants, and characteristic functions of graphs* (Russian) Funktsional. Anal. i Prilozhen. **26** (1992), no. 4, 1–20, 96; translation in Funct. Anal. Appl. **26** (1992), no. 4, 231–246 (1993).

[HU] R. Howe, T. Umeda, *The Capelli identity, the double commutant theorem, and multiplicity-free actions*, Math. Ann. **290** (1991), no. 3, 565–619.

[KS] B. Kostant, S. Sahi, *Jordan algebras and Capelli identities*, Invent. Math. **112** (1993), no. 3, 657–664.

[MN] A. Molev, M. Nazarov, *Capelli identities for classical Lie algebras*, Math. Ann. *313* (1999), no. 2, 315–357.

[N1] M. Nazarov, *Capelli identities for Lie superalgebras*, Ann. Sci. École Norm. Sup. (4) **30** (1997), no. 6, 847–872.

[N2] M. Nazarov, *Yangians and Capelli identities*, Kirillov's seminar on representation theory, 139–163, Amer. Math. Soc. Transl. Ser. 2, **181**, Amer. Math. Soc., Providence, RI, 1998.

[NUW] M. Noumi, T. Umeda, M. Wakayama, *A quantum analogue of the Capelli identity and an elementary differential calculus on $\mathrm{GL}_q(n)$*, Duke Math. J. **76** (1994), no. 2, 567–594.

[MV1] E. Mukhin and A. Varchenko, *Spaces of quasi-polynomials and the Bethe Ansatz*, math.QA/0604048, 1–29.

[MTV1] E. Mukhin, V. Tarasov and A. Varchenko, *Bethe eigenvectors of higher transfer matrices*, J. Stat. Mech. (2006) P08002.

[MTV2] E. Mukhin, V. Tarasov and A. Varchenko, *Bispectral and $(\mathfrak{gl}_N, \mathfrak{gl}_M)$ Dualities*, Func. Anal. and Other Math., **1** (1), (2006), 55–80.

[MTV3] E. Mukhin, V. Tarasov and A. Varchenko, *The B. and M. Shapiro conjecture in real algebraic geometry and the Bethe ansatz*, math.AG/0512299, 1–18.

[MTV4] E. Mukhin, V. Tarasov and A. Varchenko, *Generating operator of XXX or Gaudin transfer matrices has quasi-exponential kernel*, math. QA/0703893, 1–36.

[O] A. Okounkov, *Quantum immanants and higher Capelli identities*, Transform. Groups *1* (1996), no. 1–2, 99–126.

[R] L. Rybnikov, *Argument Shift Method and Gaudin Model*, math.RT/ 0606380, 1–15.

[T] D. V. Talalaev, *The quantum Gaudin system* (Russian) Funktsional. Anal. i Prilozhen. 40 (2006), no. 1, 86–91; translation in Funct. Anal. Appl. 40 (2006), no. 1, 73–77.

[TV] V. Tarasov, A. Varchenko, *Duality for Knizhnik-Zamolodchikov and dynamical equations*, The 2000 Twente Conference on Lie Groups (Enschede). Acta Appl. Math. 73 (2002), no. 1–2, 141–154.

Hidden Symmetries in the Theory of Complex Multiplication

Jan Nekovář

Université Pierre et Marie Curie (Paris 6), Institut de Mathématiques de Jussieu,
Théorie des Nombres, Case 247, 4 place Jussieu, F-75252 Paris cedex 05, FRANCE
nekovar@math.jussieu.fr

To Yuri Manin on the occasion of his 70th birthday, with admiration

Summary. It is (almost) known that the Galois action on etale cohomology of a Hilbert modular variety extends to an action of a bigger group. We show that this bigger group acts on the set of CM points.

Key words: Tensor induction, complex multiplication, Taniyama group

2000 Mathematics Subject Classifications: 11G15, 11F41

0 Introduction

0.1

Let F be a totally real number field of degree d. It is well known that one can associate to any cuspidal Hilbert eigenform f over F of parallel weight 2 a compatible system of two-dimensional l-adic Galois representations $V_l(f)$ of $\Gamma_F = \mathrm{Gal}(\overline{\mathbf{Q}}/F)$ over $\overline{\mathbf{Q}}_l$ (having fixed embeddings $\overline{\mathbf{Q}} \hookrightarrow \mathbf{C}$ and $\overline{\mathbf{Q}} \hookrightarrow \overline{\mathbf{Q}}_l$).

0.2

On the other hand, the Shimura variety X associated to $R_{F/\mathbf{Q}}\mathrm{GL}(2)_F$ has reflex field $\mathbf{Q}$, which means that its étale cohomology groups give rise to l-adic representations of $\Gamma_{\mathbf{Q}} = \mathrm{Gal}(\overline{\mathbf{Q}}/\mathbf{Q})$. The action of $\Gamma_{\mathbf{Q}}$ on the intersection cohomology of the Baily–Borel compactification X^* of X was determined, up to semi-simplification, by Brylinski and Labesse [BL84]: non-primitive cohomology (into which we include IH^0) occurs in even degrees and decomposes as

$$IH_{et}^{2j}(X^* \otimes_{\mathbf{Q}} \overline{\mathbf{Q}}, \overline{\mathbf{Q}}_l)_{\text{non-prim}} \xrightarrow{\sim} \bigoplus_{\chi} \chi(-j),$$

Y. Tschinkel and Y. Zarhin (eds.), *Algebra, Arithmetic, and Geometry*,
Progress in Mathematics 270, DOI 10.1007/978-0-8176-4747-6_13,
© Springer Science+Business Media, LLC 2009

where each χ is a finite-order character of $\Gamma_{\mathbf{Q}}$. Primitive cohomology occurs only in degree d; it decomposes as

$$IH_{et}^d(X^* \otimes_{\mathbf{Q}} \overline{\mathbf{Q}}, \overline{\mathbf{Q}}_l)_{\mathrm{prim}} \xrightarrow{\sim} \bigoplus_f \pi(f) \otimes W_l(f),$$

where f is as above, $\pi(f)$ is the (non-archimedean part of the) automorphic representation of $\mathrm{GL}(2, \mathbf{A}_F)$ associated to f, and $W_l(f)$ is a 2^d-dimensional l-adic representation of $\Gamma_{\mathbf{Q}}$ whose semi-simplification $W_l(f)^{\mathrm{ss}}$ is isomorphic to the tensor induction of $V_l(f)$,

$$\bigotimes \mathrm{Ind}_{F/\mathbf{Q}} V_l(f),$$

which is defined as follows. A choice of coset representatives

$$\Gamma_{\mathbf{Q}} = \coprod_{i=1}^d g_i \Gamma_F \qquad (0.2.1)$$

defines an injective group homomorphism (see Section 1.1 below)

$$\Gamma_{\mathbf{Q}} \hookrightarrow S_d \ltimes \Gamma_F^d, \qquad g \mapsto (\sigma, (h_1, \ldots, h_d)), \qquad gg_i = g_{\sigma(i)} h_i, \qquad (0.2.2)$$

and $\bigotimes \mathrm{Ind}_{F/\mathbf{Q}} V_l(f)$ is obtained from the $(S_d \ltimes \Gamma_F^d)$-module $V_l(f)^{\otimes d}$ by pullback via the map (0.2.2).

0.3

In particular, the action of $\Gamma_{\mathbf{Q}}$ on $IH_{et}^d(X^* \otimes_{\mathbf{Q}} \overline{\mathbf{Q}}, \overline{\mathbf{Q}}_l)_{\mathrm{prim}}^{\mathrm{ss}}$ extends to an action of $S_d \ltimes \Gamma_F^d$. The same should be true for the action on $IH_{et}^d(X^* \otimes_{\mathbf{Q}} \overline{\mathbf{Q}}, \overline{\mathbf{Q}}_l)_{\mathrm{prim}}$, since general conjectures predict that $\Gamma_{\mathbf{Q}}$ should act semi-simply on $IH_{et}^*(Y \otimes_{\mathbf{Q}} \overline{\mathbf{Q}}, \overline{\mathbf{Q}}_l)$, for any proper scheme Y over $\mathrm{Spec}(\mathbf{Q})$.

The representations $\chi(-j)$ of $\Gamma_{\mathbf{Q}}$ occurring in the non-primitive cohomology of X^* do not extend to representations of $S_d \ltimes \Gamma_F^d$, but they extend to representations of the group $(S_d \ltimes \Gamma_F^d)_0$, which is defined as the fibre product

$$\begin{array}{ccc}
(S_d \ltimes \Gamma_F^d)_0 & \longrightarrow & S_d \ltimes \Gamma_F^d \\
\downarrow & & \downarrow \\
\Gamma_{\mathbf{Q}}^{ab} & \xrightarrow{V_{F/\mathbf{Q}}} & \Gamma_F^{ab},
\end{array} \qquad (0.3.1)$$

where the right vertical arrow is trivial on S_d and is given by the product map on Γ_F^d. Since the field F is totally real, the transfer map $V_{F/\mathbf{Q}}$ is injective (see 1.2.5 below), which means that we can (and will) consider $(S_d \ltimes \Gamma_F^d)_0$ as a subgroup of $S_d \ltimes \Gamma_F^d$. The inclusion (0.2.2) factors through an inclusion $\Gamma_{\mathbf{Q}} \hookrightarrow (S_d \ltimes \Gamma_F^d)_0$.

Question 0.4 *To sum up: the results of [BL84] combined with the semi-simplicity conjecture imply that the action of $\Gamma_{\mathbf{Q}}$ on $IH_{et}^*(X^* \otimes_{\mathbf{Q}} \overline{\mathbf{Q}}, \overline{\mathbf{Q}}_l)$ should extend to an action of $(S_d \ltimes \Gamma_F^d)_0$. Is there a geometric explanation of this hidden symmetry of $IH_{et}^*(X^* \otimes_{\mathbf{Q}} \overline{\mathbf{Q}}, \overline{\mathbf{Q}}_l)$?*

0.5

This question admits a more invariant formulation. Recall that the inclusion (0.2.2) depends on the choice of coset representatives (0.2.1). The same choice defines an isomorphism of F-algebras

$$F \otimes_{\mathbf{Q}} \overline{\mathbf{Q}} \xrightarrow{\sim} \overline{F}^d, \qquad a \otimes b \mapsto (a \otimes g_i^{-1}(b))_i,$$

hence a group isomorphism

$$S_d \ltimes \Gamma_F^d \xrightarrow{\sim} \mathrm{Aut}_{F-alg}(F \otimes_{\mathbf{Q}} \overline{\mathbf{Q}}), \tag{0.5.1}$$

the composition of which with (0.2.2) coincides with the canonical injective map

$$\Gamma_{\mathbf{Q}} = \mathrm{Aut}_{\mathbf{Q}-alg}(\overline{\mathbf{Q}}) \hookrightarrow \mathrm{Aut}_{F-alg}(F \otimes_{\mathbf{Q}} \overline{\mathbf{Q}}), \quad g \mapsto \mathrm{id}_F \otimes g. \tag{0.5.2}$$

The subgroup $\mathrm{Aut}_{F-alg}(F \otimes_{\mathbf{Q}} \overline{\mathbf{Q}})_0$ of $\mathrm{Aut}_{F-alg}(F \otimes_{\mathbf{Q}} \overline{\mathbf{Q}})$ corresponding to $(S_d \ltimes \Gamma_F^d)_0$ under the isomorphism (0.5.1) is independent of any choices, which means that we should restate Question 0.4 as follows.

Question 0.6 *Is there a geometric explanation of the fact that the action of $\Gamma_{\mathbf{Q}}$ on $IH_{et}^*(X^* \otimes_{\mathbf{Q}} \overline{\mathbf{Q}}, \overline{\mathbf{Q}}_l)$ extends to an action of $\mathrm{Aut}_{F-alg}(F \otimes_{\mathbf{Q}} \overline{\mathbf{Q}})_0$? For example, does $X^* \otimes_{\mathbf{Q}} \overline{\mathbf{Q}}$ (or a related space) admit an action of $\mathrm{Aut}_{F\text{-alg}}(F \otimes_{\mathbf{Q}} \overline{\mathbf{Q}})_0$?*

0.7 Idle speculation

The recipe (0.2.2) defines an inclusion

$$G \hookrightarrow S_d \ltimes H^d \tag{0.7.1}$$

(depending on chosen coset representatives of H in G) whenever H is a subgroup of index d of a group G.

If $p : Y \longrightarrow Z$ is an unramified covering of degree d between "nice" connected topological spaces and $H = \pi_1(Y, y)$, $G = \pi_1(Z, p(y))$, then there are at least two geometric incarnations of (0.7.1).

Firstly, if $\tilde{Z}$ is the universal covering of Z, then

$$G \xrightarrow{\sim} \mathrm{Aut}(\tilde{Z}/Z), \qquad S_d \ltimes H^d \xrightarrow{\sim} \mathrm{Aut}(Y \times_Z \tilde{Z}/Y),$$

and (0.7.1) comes from the canonical map

$$\mathrm{Aut}(\tilde{Z}/Z) \longrightarrow \mathrm{Aut}(Y \times_Z \tilde{Z}/Y), \qquad g \mapsto \mathrm{id}_Y \times g. \tag{0.7.2}$$

In our situation, the rôle of p (resp., by $\tilde{Z}$) is played by the structure map $\mathrm{Spec}(F) \longrightarrow \mathrm{Spec}(\mathbf{Q})$ (resp., by $\mathrm{Spec}(\overline{\mathbf{Q}})$), and (0.7.2) is nothing but (0.5.2).

Secondly, $S_d \ltimes H^d$ is closely related to $\pi_1(Y^d/S_d, p^{-1}(p(y)))$, and there is a canonical map

$$Z \longrightarrow Y^d/S_d, \qquad z \mapsto p^{-1}(z). \qquad (0.7.3)$$

In other words, the map induced by (0.7.3),

$$\pi_1(Z, z) \longrightarrow \pi_1(Y^d/S_d, p^{-1}(z)),$$

is an approximate version of (0.7.1).

In our situation, in which the role of Y (resp., of Z) is played by $\operatorname{Spec}(F)$ (resp., by $\operatorname{Spec}(\mathbf{Q})$), we are confronted with the fact that the analogue of Y^d/S_d should be the dth symmetric power of $\operatorname{Spec}(F)$ over the elusive absolute point $\operatorname{Spec}(\mathbf{F}_1)$. A Grothendieckean approach to Question 0.6 would then involve

- making sense of the dth symmetric power $\operatorname{Sym}^d(F/\mathbf{F}_1)$ of $\operatorname{Spec}(F)$ over $\operatorname{Spec}(\mathbf{F}_1)$;
- extending X^* to an object $\widetilde{X}^*$ defined over (a desingularisation of) $\operatorname{Sym}^d(F/\mathbf{F}_1)$;
- relating l-adic intersection cohomology groups of X^* and $\widetilde{X}^*$.[1]

At present, this seems beyond reach, but as A. Genestier pointed out to us, everything makes sense for Drinfeld modular varieties over global fields of positive characteristic.

0.8

Leaving speculations aside, in the present article we test Question 0.6 by studying the action of $\Gamma_{\mathbf{Q}}$ on the set of CM points. It is convenient to replace $R_{F/\mathbf{Q}}\operatorname{GL}(2)_F$ by the group G defined as the fibre product

$$
\begin{array}{ccc}
G & \longrightarrow & R_{F/\mathbf{Q}}(\operatorname{GL}(2)_F) \\
\downarrow & & \downarrow{\scriptstyle\det} \\
\mathbf{G}_{m,\mathbf{Q}} & \longrightarrow & R_{F/\mathbf{Q}}(\mathbf{G}_{m,F}),
\end{array}
$$

since the corresponding Shimura variety is a moduli space for polarised Hilbert–Blumenthal abelian varieties (HBAV) equipped with adelic level structures.

The first main result of the present article (see 2.2.5 below) is the following.

Theorem 0.9 *The group* $\operatorname{Aut}_{F-\mathrm{alg}}(F \otimes_{\mathbf{Q}} \overline{\mathbf{Q}})_0$ *acts naturally on the set of CM points of the Shimura variety* $\operatorname{Sh}(G, \mathscr{X})$ *associated to G. This action extends the natural action of $\Gamma_{\mathbf{Q}}$ and commutes with the action of $G(\mathbf{A}_f) = G(\widehat{\mathbf{Q}})$ on* $\operatorname{Sh}(G, \mathscr{X})$.

---

[1]Establishing a relation between de Rham cohomology of X^* and $\widetilde{X}^*$ would also be of interest, in view of potential applications to period relations for Hilbert modular forms.

The key point in the proof is to show that the (reverse) 1-cocycle f_Φ: $\Gamma_{\mathbf{Q}} \longrightarrow \widehat{K}^*/K^*$ ("the Taniyama element"), which describes the Galois action on the set of CM points by K, naturally extends to a 1-cocycle $\widetilde{f}_\Phi$: $\mathrm{Aut}_{F\text{-alg}}(F \otimes_{\mathbf{Q}} \overline{\mathbf{Q}})_0 \longrightarrow \widehat{K}^*/K^*$ (above, K is a totally imaginary quadratic extension of F, $\widehat{K}$ is the ring of finite adèles of K and Φ is a CM type of K). In fact, f_Φ extends to a 1-cocycle $\widetilde{f}_\Phi$ defined on a slightly bigger subgroup $\mathrm{Aut}_{F\text{-alg}}(F \otimes_{\mathbf{Q}} \overline{\mathbf{Q}})_1$ of $\mathrm{Aut}_{F\text{-alg}}(F \otimes_{\mathbf{Q}} \overline{\mathbf{Q}})$, which corresponds to the fibre product

$$
\begin{array}{ccc}
(S_d \ltimes \Gamma_F^d)_1 & \longrightarrow & S_d \ltimes \Gamma_F^d \\
\downarrow & & \downarrow {\scriptstyle (1,\mathrm{prod})} \\
\Gamma_{\mathbf{Q}}^{ab}/\langle c \rangle & \stackrel{\overline{V}_{F/\mathbf{Q}}}{\hookrightarrow} & \Gamma_F^{ab}/\langle c_1, \ldots, c_d \rangle,
\end{array}
$$

where $c \in \Gamma_{\mathbf{Q}}^{ab}$ (resp., $c_1, \ldots, c_d \in \Gamma_F^{ab}$) is the complex conjugation (resp., are the complex conjugations at the infinite primes of F). We have

$$
\mathrm{Aut}_{F\text{-alg}}(F \otimes_{\mathbf{Q}} \overline{\mathbf{Q}})_1/\mathrm{Aut}_{F\text{-alg}}(F \otimes_{\mathbf{Q}} \overline{\mathbf{Q}})_0 \xrightarrow{\sim} (\mathbf{Z}/2\mathbf{Z})^{d-1},
$$

but only the elements of $\mathrm{Aut}_{F\text{-alg}}(F \otimes_{\mathbf{Q}} \overline{\mathbf{Q}})_0$ preserve the positivity of the polariations.

0.10

A more abstract formulation of this result (Section 2.4) involves a generalisation of the Taniyama group $\mathscr{T}$ and its finite-level quotients $_K\mathscr{T}$. More precisely, in the special case that K is a Galois extension of $\mathbf{Q}$, the maps $\widetilde{f}_\Phi$ factor through $\mathrm{Aut}_{F\text{-alg}}(F \otimes_{\mathbf{Q}} K^{ab})_1 = \mathrm{Im}\left(\mathrm{Aut}_{F\text{-alg}}(F \otimes_{\mathbf{Q}} \overline{\mathbf{Q}})_1 \longrightarrow \mathrm{Aut}_{F\text{-alg}}(F \otimes_{\mathbf{Q}} K^{ab})\right)$ and can be put together, yielding a 1-cocycle

$$
\widetilde{f} : \mathrm{Aut}_{F\text{-alg}}(F \otimes_{\mathbf{Q}} K^{ab})_1 \longrightarrow {}_K\mathscr{S}(\widehat{K})/{}_K\mathscr{S}(K), \tag{0.10.1}
$$

where $_K\mathscr{S}$ is the Serre torus associated to K (see Section 1.5).

Our second main result (see 2.4.2–3 below) states that the coboundary of $\widetilde{f}$ gives rise to an exact sequence of affine group schemes over $\mathbf{Q}$,

$$
1 \longrightarrow {}_K\mathscr{S} \xrightarrow{\widetilde{i}} {}_K\mathscr{T} \xrightarrow{\widetilde{\pi}} \mathrm{Aut}_{F-\mathrm{alg}}(F \otimes_{\mathbf{Q}} K^{ab})_1' \longrightarrow 1, \tag{0.10.2}
$$

where $\mathrm{Aut}_{F\text{-alg}}(F \otimes_{\mathbf{Q}} K^{ab})_1'$ is a certain $F/\mathbf{Q}$-form of the constant group scheme $\mathrm{Aut}_{F\text{-alg}}(F \otimes_{\mathbf{Q}} K^{ab})_1$. Moreover, there is a group homomorphism $\widetilde{sp}$: $\mathrm{Aut}_{F\text{-alg}}(F \otimes_{\mathbf{Q}} K^{ab})_1 \longrightarrow {}_K\widetilde{\mathscr{T}}(\widehat{F})$ satisfying $\widetilde{\pi} \circ \widetilde{sp} = \mathrm{id}$. The pull-back of (0.10.2) to $\mathrm{Aut}_{\mathbf{Q}-\mathrm{alg}}(K^{ab}) = \mathrm{Gal}(K^{ab}/\mathbf{Q})$ is the level-K Taniyama extension

$$
1 \longrightarrow {}_K\mathscr{S} \xrightarrow{i} {}_K\mathscr{T} \xrightarrow{\pi} \mathrm{Gal}(K^{ab}/\mathbf{Q}) \longrightarrow 1.
$$

For varying K, the 1-cocycles $\widetilde{f}$ are compatible. When put together, they give rise to an exact sequence of affine group schemes over $\mathbf{Q}$,

$$1 \longrightarrow \mathscr{S} \longrightarrow \widetilde{\mathscr{T}} \longrightarrow \varinjlim_{F} \operatorname{Aut}_{F\text{-alg}}(F \otimes_{\mathbf{Q}} \overline{\mathbf{Q}})_1' \longrightarrow 1 \qquad (0.10.3)$$

(where $\mathscr{S}$ is the inverse limit of the tori $_K\mathscr{S}$ with respect to the norm maps, and the direct limit is taken with respect to the transition maps $\operatorname{id}_{F'} \otimes_F -$, for $F \subseteq F'$), whose pull-back to $\Gamma_{\mathbf{Q}}$ coincides with the Taniyama extension

$$1 \longrightarrow \mathscr{S} \longrightarrow \mathscr{T} \longrightarrow \Gamma_{\mathbf{Q}} \longrightarrow 1.$$

Question 0.11 *As shown in [Del82], the Taniyama group $\mathscr{T}$ has a natural Tannakian interpretation. Does $\widetilde{\mathscr{T}}$, or its subgroup scheme $\widetilde{\mathscr{T}_0} \subset \widetilde{\mathscr{T}}$ sitting in the exact sequence*

$$1 \longrightarrow \mathscr{S} \longrightarrow \widetilde{\mathscr{T}_0} \longrightarrow \varinjlim_{F} \operatorname{Aut}_{F-\text{alg}}(F \otimes \overline{\mathbf{Q}})_0' \longrightarrow 1,$$

have a similar interpretation?

0.12

If A is a polarised HBAV over $\overline{\mathbf{Q}}$, then $H_{dR}^1(A/\overline{\mathbf{Q}})$ is a free $F \otimes_{\mathbf{Q}} \overline{\mathbf{Q}}$-module of rank 2, and for each prime p the $F \otimes_{\mathbf{Q}} \overline{\mathbf{Q}} \otimes_{\mathbf{Q}} \mathbf{Q}_p$-module $H_{dR}^1(A/\overline{\mathbf{Q}}) \otimes_{\mathbf{Q}} \mathbf{Q}_p$ has an additional crystalline structure. The comparison theorems between étale and crystalline cohomology together with Faltings's isogeny theorem imply that the F-linear isogeny class of A is determined by $H_{dR}^1(A/\overline{\mathbf{Q}})$ with this additional structure. It is very likely (even though we have not checked this) that the action (0.9) of $\operatorname{Aut}_{F\text{-alg}}(F \otimes_{\mathbf{Q}} \overline{\mathbf{Q}})_0$ on the set of CM points of $\operatorname{Sh}(G, \mathscr{X})$ is compatible, via the functor $A \mapsto H_{dR}^1(A/\overline{\mathbf{Q}})$, with the natural action of $\operatorname{Aut}_{F\text{-alg}}(F \otimes_{\mathbf{Q}} \overline{\mathbf{Q}})$ on the category of $F \otimes_{\mathbf{Q}} \overline{\mathbf{Q}}$-modules.

Question 0.13 *What happens for non-CM points? In other words, for what $g \in \operatorname{Aut}_{F\text{-alg}}(F \otimes_{\mathbf{Q}} \overline{\mathbf{Q}})_0$ is there a polarised HBAV A' over $\overline{\mathbf{Q}}$ such that*

$$H_{dR}^1(A'/\overline{\mathbf{Q}}) = g^* H_{dR}^1(A/\overline{\mathbf{Q}}),$$

with all the additional structure?

1 Background material

In Sections 1.4–1.7 of this chapter we recall the main results of the theory of complex multiplication. In Section 1.1–1.3 we collect some elementary background material.

Notation and conventions: An action of a group on a set always means a left action. We write $A \otimes B$ instead of $A \otimes_{\mathbf{Z}} B$ and denote by $\overline{\mathbf{Q}}$ the algebraic closure of $\mathbf{Q}$ in $\mathbf{C}$. By a number field we always understand a subfield of $\overline{\mathbf{Q}}$ of finite degree over $\mathbf{Q}$. The ring of integers of a number field k will be denoted by O_k. For each subfield L of $\overline{\mathbf{Q}}$ we put $\Gamma_L = \mathrm{Gal}(\overline{\mathbf{Q}}/L)$ and $X(L) = \mathrm{Hom}_{\mathbf{Q}-\mathrm{alg}}(L, \overline{\mathbf{Q}})$. The restriction map $g \mapsto g|_L$ defines an isomorphism of left $\Gamma_{\mathbf{Q}}$-sets $\Gamma_{\mathbf{Q}}/\Gamma_L \xrightarrow{\sim} X(L)$. Denote by $c \in \Gamma_{\mathbf{Q}}$ the complex conjugation. For any abelian group A, put $\widehat{A} = A \otimes \widehat{\mathbf{Z}}$. If A is a ring, so is $\widehat{A}$ (if k is a number field, then $\widehat{k}$ is the ring of finite adèles of k).

1.1 Wreath products and Galois theory

1.1.1 Notation

If X and Y are sets, we denote by $Y^X = \{f : X \longrightarrow Y\}$ the set of maps from X to Y. If Y is a group, so is Y^X. The group of permutations of the set X, denoted by $S_X = \{\text{bijective maps } \sigma : X \longrightarrow X\}$, acts on Y^X by $^\sigma f = f \circ \sigma^{-1}$. For any group H, the semidirect product of H^X and S_X (with respect to this action of S_X on H^X) is equal to

$$S_X \ltimes H^X = \{(\sigma, h) \mid \sigma \in S_X, h : X \longrightarrow H\}, \quad (\sigma, h)(\sigma', h') = (\sigma\sigma', (h \circ \sigma')h').$$

If Y is a left H-set, then Y^X is a left $(S_X \ltimes H^X)$-set via the action

$$(\sigma, h)(y) = (hy) \circ \sigma^{-1}, \qquad h \in H^X, \qquad y \in Y^X, \qquad (hy)(x) = (h(x))(y(x)). \tag{1.1.1.1}$$

1.1.2 Basic construction

Let H be a subgroup of a group G. Fix a section $s : X = G/H \longrightarrow G$ of the natural projection $G \longrightarrow G/H$. Left multiplication by $g \in G$ defines a permutation $\sigma = (x \mapsto gx) \in S_X$. For each $x \in X$,

$$gs(x) = s(gx)h(x), \qquad h(x) \in H,$$

and the map

$$g \mapsto (\sigma, h) = ((x \mapsto gx), (x \mapsto s(gx)^{-1}gs(x))) \in S_X \ltimes H^X$$

is an injective group homomorphism

$$\rho_s : G \hookrightarrow S_X \ltimes H^X \qquad\qquad (X = G/H). \tag{1.1.2.1}$$

If $s' : X = G/H \longrightarrow G$ is another section, then $s' = st$, $t \in H^X$, and

$$\forall g \in G, \qquad \rho_{s'}(g) = (1, t)^{-1}\rho_s(g)(1, t). \tag{1.1.2.2}$$

If $(G : H) < \infty$, then the diagram

$$
\begin{array}{ccc}
G & \xrightarrow{\ \rho_s\ } & S_X \ltimes H^X \\
\downarrow & & \downarrow {\scriptstyle (1,\mathrm{prod})} \\
G^{ab} & \xrightarrow{\ V\ } & H^{ab}
\end{array}
\qquad (1.1.2.3)
$$

is commutative, where prod is the product map $h \mapsto \prod_{x \in X} h(x) \pmod{[H, H]}$ and V is the transfer. The map ρ_s factors through an injective group homomorphism

$$
G \hookrightarrow (S_X \ltimes H^X)_0,
$$

where $(S_X \ltimes H^X)_0$ is the group defined as the fibre product

$$
\begin{array}{ccc}
(S_X \ltimes H^X)_0 & \longrightarrow & S_X \ltimes H^X \\
\downarrow & & \downarrow {\scriptstyle (1,\mathrm{prod})} \\
G^{ab} & \xrightarrow{\ V\ } & H^{ab}.
\end{array}
\qquad (1.1.2.4)
$$

If V is injective, we can (and will) identify $(S_X \ltimes H^X)_0$ with its image in $S_X \ltimes H^X$.

Proposition 1.1.3 *Let k'/k be a Galois extension (not necessarily finite) and X a finite set. The action of $\Gamma_{k'/k} = \mathrm{Gal}(k'/k) = \mathrm{Aut}_{k-\mathrm{alg}}(k')$ on k' gives rise, as in (1.1.1.1), to an action of $S_X \ltimes \Gamma_{k'/k}^X$ on $(k')^X$ by k-algebra automorphisms, and each k-algebra automorphism of $(k')^X$ arises in this way:*

$$
S_X \ltimes \Gamma_{k'/k}^X = \mathrm{Aut}_{k-\mathrm{alg}}((k')^X), \qquad (\sigma, h) \mapsto (a \mapsto (ha) \circ \sigma^{-1}).
$$

Proof. Any k-algebra automorphism f of $(k')^X$ must permute the set of irreducible idempotents $\{1_x \mid x \in X\}$ of $(k')^X$: $f(1_x) = 1_{\sigma(x)}$, $\sigma \in S_X$. This implies that $(\sigma, 1) \circ f$ preserves the decomposition $(k')^X = \prod_{x \in X} k' \cdot 1_x$; hence $(\sigma, 1) \circ f \in \mathrm{Aut}_{k-\mathrm{alg}}(k')^X = \Gamma_{k'/k}^X$, which implies that $f \in S_X \ltimes \Gamma_{k'/k}^X$.

Proposition 1.1.4 *Let k'/k be as in Proposition 1.1.3. Let F/k be a finite subextension of k'/k; put $X = \mathrm{Hom}_{k-\mathrm{alg}}(F, k')$. Fix a section $s : X \longrightarrow \Gamma_{k'/k}$ of the restriction map $\Gamma_{k'/k} \longrightarrow \Gamma_{k'/k}/\Gamma_{k'/F} = X$, $g \mapsto g|_F$. The chosen section induces an isomorphism of k-algebras*

$$
s : (k')^X \longrightarrow (k')^X, \qquad u \mapsto (x \mapsto s(x)(u(x))).
$$

(i)   *The map*

$$
\alpha : F \otimes_k k' \longrightarrow (k')^X, \qquad a \otimes b \mapsto (x \mapsto x(a)b)
$$

is an isomorphism of k-algebras.

(ii)   *The map*

$$
\beta_s : F \otimes_k k' \xrightarrow{\ \alpha\ } (k')^X \xleftarrow{\ s\ } (k')^X, \qquad a \otimes b \mapsto (x \mapsto as(x)^{-1}(b))
$$

is an isomorphism of F-algebras.

(iii) *The map β_s satisfies*

$$\forall g \in \mathrm{Aut}_{k-\mathrm{alg}}(k') = \Gamma_{k'/k} \qquad \beta_s \circ (\mathrm{id}_F \otimes g) = \rho_s(g)\beta_s,$$

hence induces a group isomorphism

$$\beta_{s*} : \mathrm{Aut}_{F\text{-}\mathrm{alg}}(F \otimes_k k') \xrightarrow{\sim} \mathrm{Aut}_{F\text{-}\mathrm{alg}}((k')^X) = S_X \ltimes \Gamma^X_{k'/F}, \quad f \mapsto \beta_s \circ f \circ \beta_s^{-1}$$

satisfying $\beta_{s*}(\mathrm{id}_F \otimes g) = \rho_s(g)$, *for all* $g \in \Gamma_{k'/k}$.
(iv) *If* $s' = st : X \longrightarrow \Gamma_{k'/k}$ *is another section of the restriction map* $g \mapsto g|_F$
($t : X \longrightarrow \Gamma_{k'/F}$*), then*

$$\forall g \in \mathrm{Aut}_{F\text{-}\mathrm{alg}}(F \otimes_k k') \qquad \beta_{s'*}(g) = (1,t)^{-1}\beta_{s*}(g)\,(1,t),$$

i.e., $\beta_{s*} = \mathrm{Ad}(1,t) \circ \beta_{s'*}$.

Proof. (i) This is a well-known fact from Galois theory.

(ii) The map β_s is an isomorphism of k-algebras, by (i). For each $a \in F$, we have $\beta_s(a) : x \mapsto a$, which means that β_s is a morphism of F-algebras.
(iii) Let $a \in F$, $b \in k'$, $g \in \Gamma_{k'/k} = G$, $H = \Gamma_{k'/F}$; put $\rho_s(g) = (\sigma, h)$. For each $x \in X$ we have $\sigma(x) = gx$ and

$$h(x) = s(gx)^{-1}gs(x) = s(\sigma(x))^{-1}gs(x) \in H, \quad \beta_s(a \otimes b)(x) = as(x)^{-1}(b),$$

hence

$$\beta_s \circ (\mathrm{id}_F \otimes g)(a \otimes b) = \beta_s(a \otimes g(b)) : x \mapsto as(x)^{-1}(g(b)).$$

On the other hand,

$$(\sigma, h) \circ \beta_s(a \otimes b) : x \mapsto h(\sigma^{-1}(x))\left(a\,s(\sigma^{-1}(x))^{-1}(b)\right) = a\left(s(x)^{-1}g\right)(b),$$

which proves that $\beta_s \circ (\mathrm{id}_F \otimes g) = \rho_s(g) \circ \beta_s$, as claimed.

(iv) We have $\beta_{s'} = t^{-1}\beta_s$, since

$$\forall x \in X \qquad \beta_{s'}(a \otimes b)(x) = at(x)^{-1} \circ s(x)^{-1}(x) = t(x)^{-1}\left(as(x)^{-1}(b)\right)$$
$$= t(x)^{-1}\left(\beta_s(a \otimes b)(x)\right),$$

in the notation of the proof of (iii). It follows that

$$\beta_{s'*}(g) = \beta_{s'} \circ g \circ \beta_{s'}^{-1} = t^{-1}\beta_s \circ g \circ \beta_s^{-1}t = t^{-1}\beta_{s*}(g)t,$$

as claimed.

1.1.5

To sum up the discussion from 1.1.3–4, the natural map

$$(\mathrm{id}_F \otimes -) : \Gamma_{k'/k} = \mathrm{Aut}_{k-\mathrm{alg}}(k') \longrightarrow \mathrm{Aut}_{F-\mathrm{alg}}(F \otimes_k k'), \qquad g \mapsto \mathrm{id}_F \otimes g$$

is a canonical incarnation of the morphism $\rho_s : \Gamma_{k'/k} \hookrightarrow S_X \ltimes \Gamma^X_{k'/F}$, since $\beta_{s*} \circ (\mathrm{id}_F \otimes -) = \rho_s$.

Proposition 1.1.6 *Let* $k \subset F \subset k'$ *and* $s : X \longrightarrow \Gamma_{k'/k}$ *be as in Proposition 1.1.4. Given* $\widetilde{u} \in \Gamma_{k'/k}$, *put* $u = \widetilde{u}|_F$, $F' = u(F)$, *and* $X' = \mathrm{Hom}_{k-\mathrm{alg}}(F', k')$. *The bijection* $X \xrightarrow{\sim} X'$ $(x \mapsto x' = xu^{-1})$ *gives rise to a section* $s' : X' \longrightarrow \Gamma_{k'/k}$ *of the restriction map* $g \mapsto g|_{F'}$, *given by* $s'(x') = s(x)\widetilde{u}^{-1}$.
(i) *The map*

$$\widetilde{u}_* : S_X \ltimes \Gamma^X_{k'/F} \longrightarrow S_{X'} \ltimes \Gamma^{X'}_{k'/F'}, \qquad (\sigma, h) \mapsto (\sigma', h'),$$
$$\sigma'(x') = \sigma(x)' \quad (\iff \sigma'(xu^{-1}) = \sigma(x)u^{-1}),$$
$$h'(x') = \widetilde{u}h(x)\widetilde{u}^{-1} \quad (\iff h'(xu^{-1}) = \widetilde{u}h(x)\widetilde{u}^{-1})$$

is a group isomorphism satisfying $\widetilde{u}_* \circ \rho_s = \rho_{s'}$.
(ii) $\forall \widetilde{u}, \widetilde{u}' \in \Gamma_{k'/k}, \qquad \widetilde{u}'_*\widetilde{u}_* = (\widetilde{u}'\widetilde{u})_*$.

Proof. Easy calculation.

Proposition 1.1.7 *In the situation of Proposition 1.1.6,*
(i) *the map*

$$[u] : \mathrm{Aut}_{F-\mathrm{alg}}(F \otimes_k k') \longrightarrow \mathrm{Aut}_{F'-\mathrm{alg}}(F' \otimes_k k')$$
$$g \mapsto (u \otimes \mathrm{id}_{k'}) \circ g \circ (u^{-1} \otimes \mathrm{id}_{k'})$$

is a group isomorphism satisfying $[u'u] = [u'] \circ [u]$ *and*

$$\forall g \in \Gamma_{k'/k} \qquad [u](\mathrm{id}_F \otimes g) = \mathrm{id}_{F'} \otimes g.$$

(ii) *The following diagram is commutative:*

$$
\begin{array}{ccc}
\mathrm{Aut}_{F-\mathrm{alg}}(F \otimes_k k') & \xrightarrow{\beta_{s*}} & S_X \ltimes \Gamma^X_{k'/F} \\
\downarrow{[u]} & & \downarrow{\widetilde{u}_*} \\
\mathrm{Aut}_{F'-\mathrm{alg}}(F' \otimes_k k') & \xrightarrow{\beta_{s'*}} & S_{X'} \ltimes \Gamma^{X'}_{k'/F'}
\end{array}
$$

(iii) *If* $F' = F$, *then the group automorphism*

$$\beta_{s*} \circ [u] \circ \beta_{s*}^{-1} : S_X \ltimes \Gamma^X_{k'/F} \longrightarrow S_X \ltimes \Gamma^X_{k'/F}$$

is given by the formula $(\sigma, h) \mapsto (\sigma_u, h_u)$, *where for each* $x \in X$,

$$\sigma_u(x) = \sigma(xu)u^{-1}, \qquad h_u(x) = s\left(\sigma_u(x)\right)^{-1} s\left(\sigma_u(x)u\right) h(xu)s(xu)^{-1}s(x).$$

(iv) *If* F *is a Galois extension of* k, *then the maps* $[u]$ *define an action of* $\Gamma_{F/k}$ *on* $\mathrm{Aut}_{F-\mathrm{alg}}(F \otimes_k k')$, *whose set of fixed points is equal to* $\mathrm{id}_F \otimes \Gamma_{k'/k}$.

Proof. (i) Straightforward. (ii) Let $g \in \mathrm{Aut}_{F\text{-alg}}(F \otimes_k k')$; put $(\sigma, h) = \beta_{s*}(g)$ and $(\sigma', h') = \tilde{u}_*(\sigma, h)$. For $a \otimes b \in F \otimes_k k'$, write $g(1 \otimes b) = \sum a_i \otimes b_i$; then $g(a \otimes b) = \sum a a_i \otimes b_i$. Since $\beta_s(a \otimes b)(x) = a s(x)^{-1}(b)$, the equalities

$$\beta_s(g(a \otimes b))(x) = ((\sigma, h)\beta_s(a \otimes b))(x) \qquad (x \in X)$$

read as

$$\sum a a_i s(x)^{-1}(b_i) = a h(\sigma^{-1}(x)) s(\sigma^{-1}(x))^{-1}(b) \qquad (x \in X). \quad (1.1.7.1)$$

Since $([u](g))(1 \otimes b) = \sum u(a_i) \otimes b_i$, the statement to be proved, namely

$$\forall x' \in X' \ \forall a' \in F' \ \forall b \in k', \quad \beta_{s'}(([u](g))(a' \otimes b))(x') \stackrel{?}{=} ((\sigma', h')\beta_{s'}(a' \otimes b))(x'),$$

reads as

$$\sum a' u(a_i) s'(x')^{-1}(b_i) \stackrel{?}{=} a' h'(\sigma'^{-1}(x')) s'(\sigma'^{-1}(x'))^{-1}(b),$$

which is obtained from (1.1.7.1) (with $x = x'u$) by applying u, since

$$s'(x')^{-1} = \tilde{u} s(x)^{-1}, \qquad s'(\sigma'^{-1}(x'))^{-1} = \tilde{u} s(\sigma^{-1}(x))^{-1},$$
$$h'(\sigma'^{-1}(x')) = \tilde{u} h(\sigma^{-1}(x)) \tilde{u}^{-1}.$$

(iii) The assumption $F' = F$ implies that $s' = st$, where $t : X \longrightarrow \Gamma_{k'/F}$ is given by $t(x) = s(x)^{-1} s(xu)\tilde{u}^{-1}$. It follows from (ii) and Proposition 1.1.4 (iv) that

$$\beta_{s*} \circ [u] \circ \beta_{s*}^{-1} = \beta_{s*} \circ \beta_{s'*}^{-1} \circ \tilde{u}_* = \mathrm{Ad}(1, t) \circ \tilde{u}_*;$$

hence

$$(\sigma_u, h_u) = (1, t)(\sigma', h')(1, t)^{-1} = (\sigma', (t\circ\sigma')h't^{-1}), \quad \sigma_u(x) = \sigma'(x) = \sigma(xu)u^{-1},$$
$$h_u(x) = t(\sigma_u(x)) h'(x) t(x)^{-1} = s\left(\sigma_u(x)\right)^{-1} s\left(\sigma_u(x)u\right) h(xu) s(xu)^{-1} s(x).$$

Proposition 1.1.8 *In the situation of Proposition 1.1.4, let F'/F be a subextension of k'/F; put $X' = \mathrm{Hom}_{k-\mathrm{alg}}(F', k')$ and fix a section $s' : X' \longrightarrow \Gamma_{k'/k}$ of the restriction map $g \mapsto g|_{F'}$. For each $x' \in X'$, define $t(x') \in \Gamma_{k'/F}$ by the relation $s'(x') = s(x'|_F)t(x')$.*
(i) *The map*

$$\rho_{s,s'} : S_X \ltimes \Gamma_{k'/F}^X \longrightarrow S_{X'} \ltimes \Gamma_{k'/F'}^{X'}, \quad (\sigma, h) \mapsto (\sigma', h'),$$
$$\sigma'(x') = s(\sigma(x))h(x)s(x)^{-1}x', \quad h'(x') = t(\sigma'(x'))^{-1}h(x)t(x'), \qquad x = x'|_F$$

is a group homomorphism satisfying

$$\sigma'(x')|_F = \sigma(x), \quad s'(\sigma'(x'))h'(x')s'(x')^{-1} = s(\sigma(x))h(x)s(x)^{-1}.$$

(ii) *The following diagram is commutative:*

$$
\begin{array}{ccc}
\mathrm{Aut}_{F\text{-alg}}(F \otimes_k k') & \xrightarrow{\;\beta_{s*}\;} & S_X \ltimes \Gamma_{k'/F}^X \\
{\scriptstyle (\mathrm{id}_{F'} \otimes_F -)} \downarrow & & \downarrow {\scriptstyle \rho_{s,s'}} \\
\mathrm{Aut}_{F'\text{-alg}}(F' \otimes_k k') & \xrightarrow{\;\beta_{s'*}\;} & S_{X'} \ltimes \Gamma_{k'/F'}^{X'}
\end{array}
$$

Proof. (i) Easy calculation. (ii) As in the proof of Proposition 1.1.7, fix $a \otimes b \in F \otimes_k k'$, $g \in \mathrm{Aut}_{F-\mathrm{alg}}(F \otimes_k k')$ and put $(\sigma, h) = \beta_{s*}(g)$. Writing $g(1 \otimes b) = \sum a_i \otimes b_i$, then (1.1.7.1) (for $\sigma(x)$ instead of x) reads as

$$\sum a_i s(\sigma(x))^{-1}(b_i) = h(x)s(x)^{-1}(b) \qquad (x \in X). \qquad (1.1.8.1)$$

Define $(\sigma', h') := \rho_{s,s'}(\sigma, h)$; we must show that, for all $x' \in X', a' \in F', b \in k'$,

$$\beta_{s'}\left((\mathrm{id}_{F'} \otimes_F g)(a' \otimes b)\right)(x') \overset{?}{=} ((\sigma', h')\beta_{s'}(a' \otimes b))(x'),$$

which can be rewritten (again using (1.1.7.1) and replacing x' by $\sigma'(x')$) as follows:

$$\sum a' a_i s'(\sigma'(x'))^{-1}(b_i) \overset{?}{=} a' h'(x')s'(x')^{-1}(b) \qquad (x' \in X'). \qquad (1.1.8.2)$$

Since $\sigma'(x')|_F = \sigma(x'|_F)$, the equality (1.1.8.2) is obtained by multiplying (1.1.8.1) (for $x = x'|_F$) by $t(\sigma'(x'))^{-1}$ on the left.

1.2 Class Field Theory

1.2.1

Let k be a number field. Denote by

$$k_+^* = \mathrm{Ker}\left(k^* \longrightarrow \pi_0((k \otimes \mathbf{R})^*)\right), \qquad O_{k,+}^* = O_k^* \cap k_+^*,$$

the set of totally positive elements and the set of totally positive units of k, respectively. Let $\mathbf{A}_k$ be the adèle ring of k and $C_k = \mathbf{A}_k^*/k^*$ the idèle class group of k. The reciprocity map

$$\mathrm{rec}_k : C_k \longrightarrow \Gamma_k^{ab}$$

will be normalised by letting local uniformisers correspond to **geometric** Frobenius elements. Since rec_k induces an isomorphism $\pi_0(C_k) \overset{\sim}{\longrightarrow} \Gamma_k^{ab}$, its restriction to the group of finite idèles gives rise to a surjective continuous morphism

$$r_k : \widehat{k}^*/k_+^* \longrightarrow \Gamma_k^{ab}.$$

1.2.2

It follows from the structure of the connected component of C_k [AT68, ch. 9, Thm. 3] that the kernel of r_k is isomorphic, as an $\mathrm{Aut}(k/\mathbf{Q})$-module, to $O_{k,+}^* \otimes (\widehat{\mathbf{Z}}/\mathbf{Z}) = O_{k,+}^* \otimes (\widehat{\mathbf{Q}}/\mathbf{Q})$.

1.2.3

For $k = \mathbf{Q}$, the map $r_\mathbf{Q}$ is an isomorphism, and its composition with the canonical isomorphism $\widehat{\mathbf{Z}}^* \xrightarrow{\sim} \widehat{\mathbf{Q}}^*/\mathbf{Q}_+^*$ (induced by the inclusion of $\widehat{\mathbf{Z}}$ into $\widehat{\mathbf{Q}}$) is inverse to the cyclotomic character

$$\chi : \Gamma_\mathbf{Q}^{ab} \xrightarrow{\sim} \widehat{\mathbf{Z}}^*, \qquad g(\zeta) = \zeta^{\chi(g)} \qquad (\forall \zeta \text{ a root of unity in } \overline{\mathbf{Q}}).$$

1.2.4

If k'/k is a finite extension of number fields, then the inclusion $k \hookrightarrow k'$ and the norm $N_{k'/k} : k'^* \longrightarrow k^*$ induce commutative diagrams

$$
\begin{array}{ccc}
\widehat{k}^*/k_+^* & \xrightarrow{i_{k'/k}} & \widehat{k'}^*/k'^*_+ \\
\downarrow r_k & & \downarrow r'_k \\
\Gamma_k^{ab} & \xrightarrow{V_{k'/k}} & \Gamma_{k'}^{ab}
\end{array}
\qquad
\begin{array}{ccc}
\widehat{k'}^*/k'^*_+ & \xrightarrow{N_{k'/k}} & \widehat{k}^*/k_+^* \\
\downarrow r'_k & & \downarrow r_k \\
\Gamma_{k'}^{ab} & \xrightarrow{j_{k'/k}} & \Gamma_k^{ab},
\end{array}
\qquad (1.2.4.1)
$$

where $V_{k'/k}$ is the transfer map and $j_{k'/k}$ is given by the restriction map $g \mapsto g|_{k^{ab}}$.

Proposition 1.2.5 *For any number field L,*

$$\mathrm{Ker}\left(V_{L/\mathbf{Q}} : \Gamma_\mathbf{Q}^{ab} \longrightarrow \Gamma_L^{ab}\right) = \begin{cases} \{1, c\}, & \text{if } L \text{ is totally complex,} \\ \{1\}, & \text{otherwise.} \end{cases}$$

Proof. Let L' be the Galois closure of L over $\mathbf{Q}$. As

$$\mathrm{Im}\left(i_{L/\mathbf{Q}}\right) \cap \mathrm{Ker}(r_L) \subseteq \left(O^*_{L',+} \otimes \widehat{\mathbf{Q}}/\mathbf{Q}\right)^{\mathrm{Gal}(L'/\mathbf{Q})} = \mathbf{Z}^*_+ \otimes \widehat{\mathbf{Q}}/\mathbf{Q} = \{1\},$$

the first commutative diagram (1.2.4.1) for $L/\mathbf{Q}$ implies that $i_{L/\mathbf{Q}}^{-1}\left(\mathrm{Ker}(r_L)\right) = \mathrm{Ker}(r_L \circ i_{L/\mathbf{Q}})$ is equal to

$$\mathrm{Ker}\left(i_{L/\mathbf{Q}}\right) = \left(\mathbf{Q}^* \cap L_+^*\right)/\mathbf{Q}_+^* = \begin{cases} \mathbf{Q}^*/\mathbf{Q}_+^* = \{\pm 1\}, & L \text{ totally complex,} \\ \{1\}, & \text{otherwise.} \end{cases}$$

As $r_\mathbf{Q}$ is an isomorphism and $r_\mathbf{Q}(-1) = c$, the statement follows.

1.3 CM fields

Let K be a CM number field; let F be its maximal totally real subfield (in other words, $c(K) = K$, $\tau c = c\tau \neq \tau$ for all $\tau \in X(K)$, and $F = K^{c=1}$). Put $X = X(F)$.

1.3.1 Complex conjugations

Fix a section $s : X \longrightarrow \Gamma_{\mathbf{Q}}$ of the restriction map $g \mapsto g|_F$. For each $x \in X$, the image of the element $s(x)^{-1}cs(x) \in \Gamma_F$ in Γ_F^{ab} is independent of the chosen section; denote it by $c_x \in \Gamma_F^{ab}$ (this is the complex conjugation defined by the real place x of F). Denote by $\langle c_X \rangle$ the subgroup of Γ_F^{ab} generated by all c_x $(x \in X)$. The signs at the real places induce an isomorphism

$$(\operatorname{sgn} \circ x)_{x \in X} : F^*/F_+^* \xrightarrow{\sim} \{\pm 1\}^X.$$

Compatibility of the local and global reciprocity maps implies that

$$\forall a \in F^* \qquad r_F(aF_+^*) = \prod_{x \in X} c_x^{a_x}, \qquad (-1)^{a_x} = \operatorname{sgn}(x(a)).$$

As $\operatorname{Ker}(r_F)$ is a $\mathbf{Q}$-vector space, we have $\operatorname{Ker}(r_F) \cap F^*/F_+^* = \{1\}$, which means that r_F induces an isomorphism $F^*/F_+^* \xrightarrow{\sim} \langle c_X \rangle$.

1.3.2 Transfer maps

If we denote by

$$R : \Gamma_F \longrightarrow \Gamma_K, \qquad g, cg \mapsto g \qquad (g \in \Gamma_K)$$

the "retraction map" from Γ_F to Γ_K, then

$$\forall h \in \Gamma_F \qquad V_{K/F}\left(h|_{F^{ab}}\right) = V_{K/F}\left(ch|_{F^{ab}}\right) = hchc|_{K^{ab}} = {}^{1+c}\left(R(h)|_{K^{ab}}\right). \tag{1.3.2.1}$$

As noted in 1.2.5,

$$\operatorname{Ker}\left(V_{K/\mathbf{Q}} : \Gamma_{\mathbf{Q}}^{ab} \longrightarrow \Gamma_K^{ab}\right) = r_{\mathbf{Q}}\left(\operatorname{Ker}(i_{K/\mathbf{Q}})\right) = r_{\mathbf{Q}}\left(\mathbf{Q}^*/\mathbf{Q}_+^*\right) = \{1, c\} = \langle c \rangle. \tag{1.3.2.2}$$

The equality $\operatorname{Ker}(r_F) = O_{F,+}^* \otimes \widehat{\mathbf{Q}}/\mathbf{Q} = O_K^* \otimes \widehat{\mathbf{Q}}/\mathbf{Q} = \operatorname{Ker}(r_K)$ implies, thanks to (1.2.4.1), that

$$\operatorname{Ker}\left(V_{K/F} : \Gamma_F^{ab} \longrightarrow \Gamma_K^{ab}\right) = r_F\left(\operatorname{Ker}(i_{K/F})\right) = r_F\left(F^*/F_+^*\right) = \langle c_X \rangle. \tag{1.3.2.3}$$

As a result, the map

$$\overline{V}_{F/\mathbf{Q}} : \Gamma_{\mathbf{Q}}^{ab}/\langle c \rangle \hookrightarrow \Gamma_F^{ab}/\langle c_X \rangle \tag{1.3.2.4}$$

induced by $V_{F/\mathbf{Q}}$ is injective and

$$\{h \in \Gamma_F^{ab} \mid V_{K/F}(h) \in V_{K/\mathbf{Q}}(\Gamma_{\mathbf{Q}}^{ab})\} = \langle c_X \rangle V_{F/\mathbf{Q}}(\Gamma_{\mathbf{Q}}^{ab}). \tag{1.3.2.5}$$

It also follows that

$$V_{F/\mathbf{Q}}(\Gamma_{\mathbf{Q}}^{ab}) \cap \langle c_X \rangle = \langle V_{F/\mathbf{Q}}(c) \rangle \tag{1.3.2.6}$$

is the cyclic group of order 2 generated by $V_{F/\mathbf{Q}}(c) = \prod_{x \in X} c_x$.

1.3.3

As observed in [Tat81, Lemma 1], the finiteness of $O_K^*/O_{F,+}^*$ implies that c (resp., $1 + c$) acts trivially (resp., invertibly) on the $\mathbf{Q}$-vector space $\mathrm{Ker}(r_K)$.

Proposition 1.3.4 (i) *The continuous homomorphism (induced by r_K)*

$$\{a \in \widehat{K}^* \mid {}^{1+c}a \in \widehat{\mathbf{Z}}^* K^*\}/K^* \longrightarrow \{g \in \Gamma_K^{ab} \mid g|_{F^{ab}} \in \langle c_X \rangle V_{F/\mathbf{Q}}(\Gamma_{\mathbf{Q}}^{ab})\}$$

is bijective. Denote by ℓ_K its inverse; then ${}^{1+c}\ell_K(g) = \chi(u(g))K^*$, where $u(g) \in \Gamma_{\mathbf{Q}}^{ab}/\langle c \rangle$ is the (unique) element satisfying $\overline{V}_{F/\mathbf{Q}}(u(g)) = \langle c_X \rangle g|_{F^{ab}}$ (equivalently, $V_{K/\mathbf{Q}}(u(g)) = {}^{1+c}g$).
(ii) *More precisely, if $g \in \Gamma_K^{ab}$ satisfies*

$$g|_{F^{ab}} = V_{F/\mathbf{Q}}(u(g)) \prod_{x \in X} c_x^{a_x} \qquad (u(g) \in \Gamma_{\mathbf{Q}}^{ab}, \, a_x \in \mathbf{Z}/2\mathbf{Z}),$$

then $N_{K/F}(\ell_K(g)) = \chi(u(g))\alpha F_+^* \in \widehat{F}^*/F_+^*$, where $\alpha \in F^*$ and

$$\forall x \in X \qquad \mathrm{sgn}(x(\alpha)) = (-1)^{a_x}.$$

(iii) *The canonical morphism (induced by the inclusion $\widehat{O}_K \hookrightarrow \widehat{K}$)*

$$\{x \in \widehat{O}_K^* \mid {}^{1+c}x \in \widehat{\mathbf{Z}}^*\} \longrightarrow \{a \in \widehat{K}^* \mid {}^{1+c}a \in \widehat{\mathbf{Z}}^* K^*\}/K^*$$

has finite kernel and cokernel.
(iv) *The morphism ℓ_K defined in (i) admits a lift*

$$\widetilde{\ell}_K : \{g \in \Gamma_K^{ab} \mid g|_{F^{ab}} \in \langle c_X \rangle V_{F/\mathbf{Q}}(\Gamma_{\mathbf{Q}}^{ab})\} \longrightarrow \{a \in \widehat{K}^* \mid {}^{1+c}a \in \widehat{\mathbf{Z}}^* K^*\},$$

which is a homomorphism when restricted to a suitable open subgroup.

Proof. (i) In the following commutative diagram the right column is exact and $r_{\mathbf{Q}}$ is an isomorphism:

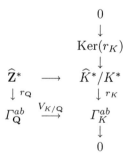

As $1 + c$ acts invertibly on $\mathrm{Ker}(r_K)$, the snake lemma implies that r_K induces an isomorphism between $\mathrm{Ker}\left(1 + c : \widehat{K}^*/K^* \longrightarrow \widehat{K}^*/K^*\widehat{\mathbf{Z}}^*\right)$ and

$$\mathrm{Ker}\left(1 + c = V_{K/F} \circ j_{K/F} : \Gamma_K^{ab} \longrightarrow \Gamma_K^{ab}/V_{K/\mathbf{Q}}(\Gamma_{\mathbf{Q}}^{ab})\right);$$

by (1.3.2.5), the latter group is equal to $\{g \in \Gamma_K^{ab} \mid g|_{F^{ab}} \in \langle c_X \rangle V_{F/\mathbf{Q}}(\Gamma_{\mathbf{Q}}^{ab})\}$. The remaining statement follows from the fact that

$$r_K\left({}^{1+c}\ell_K(g)\right) = {}^{1+c}g = V_{K/F} \circ j_{K/F}(g) = V_{K/F}\left(g|_{F^{ab}}\right)$$
$$= V_{K/F} \circ V_{F/\mathbf{Q}}\left(u(g)\right) = V_{K/\mathbf{Q}}\left(u(g)\right)$$
$$= r_K \circ i_{K/\mathbf{Q}} \circ r_{\mathbf{Q}}^{-1}\left(u(g)\right) = r_K\left(\chi(u(g))\right).$$

(ii) Let $a \in \widehat{K}^*$ be a lift of $\ell_K(g)$ such that ${}^{1+c}a = b\alpha'$, where $b \in \widehat{\mathbf{Z}}^*$, $\alpha' \in K^*$; then $\alpha' \in (K^*)^{c=1} = F^*$. As

$$g|_{F^{ab}} = r_F(N_{K/F}(a)) = r_F(b)r_F(\alpha') = V_{F/\mathbf{Q}}(r_{\mathbf{Q}}(b)) \prod_{x \in X} c_x^{a'_x},$$

where $(-1)^{a'_x} = \mathrm{sgn}(x(\alpha'))$, it follows from (1.3.2.6) that there is $t \in \mathbf{Z}/2\mathbf{Z}$ such that

$$u(g) = r_{\mathbf{Q}}(b)c^t, \qquad \forall x \in X \quad a'_x = a_x + t.$$

This implies that $\chi(u(g)) = b(-1)^t$ and

$$N_{K/F}\left(\ell_K(g)\right) = {}^{1+c}aF_+^* = \chi(u(g))\alpha F_+^*$$

with $\alpha = \alpha'(-1)^t$; hence

$$\forall x \in X \quad \mathrm{sgn}(x(\alpha)) = \mathrm{sgn}(x(\alpha'))(-1)^t = (-1)^{a'_x + t} = (-1)^{a_x}.$$

(iii) This follows from the finiteness of the groups $\mathrm{Ker}, \mathrm{Coker}(1 + c : O_K^* \longrightarrow O_K^*)$ and $Cl_K = \widehat{K}^*/\widehat{O}_K^* K^*$, combined with the snake lemma applied to the diagrams

$$
\begin{array}{ccccccccc}
0 & \longrightarrow & O_K^* & \longrightarrow & \widehat{O}_K^* & \longrightarrow & \widehat{O}_K^*/O_K^* & \longrightarrow & 0 \\
& & \downarrow{\scriptstyle 1+c} & & \downarrow{\scriptstyle 1+c} & & \downarrow{\scriptstyle 1+c} & & \\
0 & \longrightarrow & O_K^*/\mathbf{Z}^* & \longrightarrow & \widehat{O}_K^*/\widehat{\mathbf{Z}}^* & \longrightarrow & \widehat{O}_K^*/\widehat{\mathbf{Z}}^* O_K^* & \longrightarrow & 0
\end{array}
$$

and

$$
\begin{array}{ccccccccc}
0 & \longrightarrow & \widehat{O}_K^*/O_K^* & \longrightarrow & \widehat{K}^*/K^* & \longrightarrow & Cl_K & \longrightarrow & 0 \\
& & \downarrow{\scriptstyle 1+c} & & \downarrow{\scriptstyle 1+c} & & \downarrow{\scriptstyle 1+c} & & \\
0 & \longrightarrow & \widehat{O}_K^*/\widehat{\mathbf{Z}}^* O_K^* & \longrightarrow & \widehat{K}^*/\widehat{\mathbf{Z}}^* K^* & \longrightarrow & Cl_K & \longrightarrow & 0.
\end{array}
$$

Above, $\widehat{O}_K^*$ is a shorthand for $(\widehat{O_K})^*$. Note also that $\widehat{\mathbf{Z}}^* \cap O_K^* = \mathbf{Z}^*$ inside $\widehat{O}_K^*$.

(iv) By (i) and (iii), r_K induces a continuous homomorphism of profinite abelian groups

$$f \colon A = \{x \in \widehat{O}_K^* \mid {}^{1+c}x \in \widehat{\mathbf{Z}}^*\} \longrightarrow B = \{g \in \Gamma_K^{ab} \mid g|_{F^{ab}} \in \langle c_X \rangle V_{F/\mathbf{Q}}(\Gamma_{\mathbf{Q}}^{ab})\}$$

with finite kernel and cokernel. This implies that there exists an open subgroup (= a compact subgroup of finite index) $A' \subset A$ such that $A' \cap \mathrm{Ker}(f) = \{1\}$. Then $B' = f(A')$ is a compact subgroup of finite index (= an open subgroup) of B, and f induces a topological isomorphism $f' : A' \xrightarrow{\sim} B'$. Fix coset representatives $B = \bigcup_i b_i B'$ (disjoint union) and lifts $\widetilde{a}_i \in \widehat{K}^*$ of $\ell_K(b_i) \in \widehat{K}^*/K^*$ such that $b_{i_0} = 1$ and $\widetilde{a}_{i_0} = 1$; the map

$$\widetilde{\ell}_K : B \longrightarrow \widehat{K}^*, \qquad b_i f'(a') \mapsto \widetilde{a}_i a' \qquad (a' \in A')$$

has the required properties.

1.4 Tate's construction

Let Φ be a CM type of K, i.e., a subset $\Phi \subset X(K)$ such that $X(K) = \Phi \cup c\Phi$ (disjoint union).

1.4.1 Tate's half-transfer

Tate's half-transfer [Tat81] is the continuous map $F_\Phi : \Gamma_{\mathbf{Q}} \longrightarrow \Gamma_K^{ab}$ defined by the formula

$$F_\Phi(g) = \prod_{\varphi \in \Phi} w(g\varphi)^{-1} g w(\varphi) \pmod{\Gamma_{K^{ab}}}, \qquad (1.4.1.1)$$

where $w : X(K) \longrightarrow X(\overline{\mathbf{Q}}) = \Gamma_{\mathbf{Q}}$ is any section of the restriction map $g \mapsto g|_K$ satisfying $w(cy) = cw(y)$, for all $y \in X(K)$.

The restriction map $g \mapsto g|_F$ defines a bijection $\Phi \xrightarrow{\sim} X(F)$. Composing its inverse with w, we obtain a section $t : X(F) \longrightarrow X(\overline{\mathbf{Q}}) = \Gamma_{\mathbf{Q}}$ of the restriction map to F, which implies that

$$F_\Phi(g)|_{F^{ab}} = \prod_{x \in X(F)} t(gx)^{-1} c^{a(g,x)} g t(x) \pmod{\Gamma_{F^{ab}}} \in \langle c_X \rangle V_{F/\mathbf{Q}}(g) \quad (1.4.1.2)$$

(for some $a(g,x) \in \mathbf{Z}/2\mathbf{Z}$). The maps F_Φ satisfy

$$F_\Phi(gg') = F_{g'\Phi}(g) F_\Phi(g') \qquad (g, g' \in \Gamma_{\mathbf{Q}}) \qquad (1.4.1.3)$$

and

$$u \circ F_\Phi(g) \circ u^{-1} = F_{\Phi u^{-1}}(g) \qquad (g \in \Gamma_{\mathbf{Q}}), \qquad (1.4.1.4)$$

for any isomorphism of CM number fields $u : K \xrightarrow{\sim} K'$. In addition, if K' is a CM number field containing K and $\Phi' = \{y \in X(K') \mid y|_K \in \Phi\}$ is the CM type of K' induced from Φ, then

$$F_{\Phi'}(g) = V_{K'/K}(F_\Phi(g)) \qquad (g \in \Gamma_{\mathbf{Q}}). \qquad (1.4.1.5)$$

1.4.2 The Taniyama element

The Taniyama element is the map $f_\Phi : \Gamma_{\mathbf{Q}} \longrightarrow \widehat{K}^*/K^*$ defined as

$$f_\Phi(g) = \ell_K\left(F_\Phi(g)\right), \tag{1.4.2.1}$$

where

$$\ell_K : \{g \in \Gamma_K^{ab} \mid g|_{F^{ab}} \in \langle c_X \rangle V_{F/\mathbf{Q}}(\Gamma_{\mathbf{Q}}^{ab})\} \xrightarrow{\sim} \{a \in \widehat{K}^* \mid {}^{1+c}a \in \widehat{\mathbf{Z}}^* K^*\}/K^*$$

is the morphism from 1.3.4(i). It follows that

$${}^{1+c}F_\Phi(g) = V_{K/F}\left(F_\Phi(g)|_{F^{ab}}\right) = V_{K/F} \circ V_{F/\mathbf{Q}}(g) = V_{K/\mathbf{Q}}(g)$$

$$= r_K \circ i_{K/\mathbf{Q}} \circ r_{\mathbf{Q}}^{-1}\left(g|_{\mathbf{Q}^{ab}}\right) = r_K(\chi(g)).$$

As in the proof of 1.3.4(i), this implies that

$${}^{1+c}f_\Phi(g) = \chi(g)K^*, \qquad r_K\left(f_\Phi(g)\right) = F_\Phi(g). \tag{1.4.2.2}$$

In Tate's original definition, the properties (1.4.2.2) were used to characterise $f_\Phi(g)$.

The identities (1.4.1.3)–(1.4.1.5) imply that

$$f_\Phi(gg') = f_{g'\Phi}(g)f_\Phi(g') \qquad (g, g' \in \Gamma_{\mathbf{Q}}), \tag{1.4.2.3}$$

$${}^u f_\Phi(g) = f_{\Phi u^{-1}}(g) \qquad (g \in \Gamma_{\mathbf{Q}},\ u : K \xrightarrow{\sim} K') \tag{1.4.2.4}$$

and

$$f_{\Phi'}(g) = i_{K'/K}\left(f_\Phi(g)\right) \qquad (K \subset K',\ \Phi' \text{ induced from } \Phi). \tag{1.4.2.5}$$

1.4.3

Tate [Tat81] conjectured that the idèle class $f_\Phi(g)$ determines the action of $g \in \Gamma_{\mathbf{Q}}$ on abelian varieties with complex multiplication and on their torsion points (this was, essentially, the zero-dimensional case of an earlier conjecture of Langlands [Lan79] about conjugation of Shimura varieties). Building on earlier results of Shimura and Taniyama, Tate proved his conjecture up to an element of $\widehat{F}^*$ of square 1. The full conjecture was subsequently proved by Deligne [Lan83, ch. 7, Section 4].

More precisely, if A is a CM abelian variety of type (K, Φ, a, t) in the sense of [Lan83, ch. 7, Section 3] (see 2.2.5 below), then ${}^g A$ is of type $(K, g\Phi, af, t\chi(g)/^{1+c}f)$, where $f \in \widehat{K}^*$ is any lift of $f_\Phi(g)$. Furthermore, for each complex uniformisation

$$\theta : \mathbf{C}^\Phi/a \xrightarrow{\sim} A(\mathbf{C}),$$

there is a unique uniformisation

$$\theta' : \mathbf{C}^{g\Phi}/af \xrightarrow{\sim} {}^g A(\mathbf{C}),$$

such that the action of g on $A(\overline{\mathbf{Q}})_{\mathrm{tors}} = A(\mathbf{C})_{\mathrm{tors}}$ is given by

$$g : A(\overline{\mathbf{Q}})_{\mathrm{tors}} \xrightarrow{\theta^{-1}} K/a \xrightarrow{[\times f]} K/af \xrightarrow{\theta'} {}^g A(\overline{\mathbf{Q}})_{\mathrm{tors}}.$$

This implies that, for each full level structure $\eta : (F/O_F)^2 \xrightarrow{\sim} A(\overline{\mathbf{Q}})_{\mathrm{tors}}$, the level structure ${}^g \eta$ is equal to

$${}^g \eta : (F/O_F)^2 \xrightarrow{\eta} A(\overline{\mathbf{Q}})_{\mathrm{tors}} \xrightarrow{\theta^{-1}} K/a \xrightarrow{[\times f]} K/af \xrightarrow{\theta'} {}^g A(\overline{\mathbf{Q}})_{\mathrm{tors}}. \qquad (1.4.3.1)$$

1.5 The Serre torus

Let K be as in 1.3.

1.5.1

The torus ${}_K T = R_{K/\mathbf{Q}}(\mathbf{G}_m)$ represents the functor $A \mapsto {}_K T(A) = (K \otimes_{\mathbf{Q}} A)^*$ on $\mathbf{Q}$-algebras A. The $\Gamma_{\mathbf{Q}}$-equivariant bijections

$$
\begin{array}{ccccc}
(K \otimes_{\mathbf{Q}} \overline{\mathbf{Q}})^* & \xrightarrow{\sim} & \mathrm{Hom}_{Sets}(X(K), \overline{\mathbf{Q}}^*) & \xrightarrow{\sim} & \mathrm{Hom}_{\mathbf{Z}}(\mathbf{Z}[X(K)], \overline{\mathbf{Q}}^*) \\
a \otimes b & \mapsto & (y \mapsto y(a)b) & & (y \in X(K))
\end{array}
$$

imply that the character group of ${}_K T$ is equal to

$$X^*({}_K T) = \mathbf{Z}[X(K)] = \left\{ \sum_{y \in X(K)} n_y [y] \mid n_y \in \mathbf{Z} \right\},$$

with $g \in \Gamma_{\mathbf{Q}}$ acting on $X^*({}_K T)$ by

$$\lambda = \sum n_y [y] \mapsto {}^g \lambda = \sum n_y [gy] = \sum n_{g^{-1}y} [y]. \qquad (1.5.1.1)$$

1.5.2

The **Serre torus** of K is the quotient ${}_K \mathscr{S}$ of ${}_K T$ (defined over $\mathbf{Q}$) whose character group is equal to

$$X^*({}_K \mathscr{S}) = \{\lambda \in X^*({}_K T) \mid {}^{1+c}\lambda \in \mathbf{Z} \cdot N_{K/\mathbf{Q}}\} \qquad \left(N_{K/\mathbf{Q}} = \sum_{y \in X(K)} [y] \right).$$

Each CM type Φ of K defines a character $\lambda_\Phi \in X^*({}_K \mathscr{S})$: $\lambda_\Phi(y) = 1$ (resp., $= 0$) if $y \in \Phi$ (resp., if $y \in c\Phi$). Moreover, the abelian group $X^*({}_K \mathscr{S})$ is generated by the characters λ_Φ [Sch94, 1.3.2], and

$$\forall g \in \Gamma_{\mathbf{Q}}, \qquad {}^g \lambda_\Phi = \lambda_{g\Phi}.$$

1.5.3

Tate's half-transfer satisfies the following identity: if n is a function

$$n : \{\text{CM types of } K\} \longrightarrow \mathbf{Z}, \qquad \Phi \mapsto n_\Phi,$$

such that $\sum_\Phi n_\Phi \lambda_\Phi = 0$, then

$$\forall g \in \Gamma_{\mathbf{Q}} \qquad \prod_\Phi F_\Phi(g)^{n_\Phi} = 1 \in \Gamma_K^{ab}. \tag{1.5.3.1}$$

Applying ℓ_K, we deduce from (1.5.3.1) that

$$\forall g \in \Gamma_{\mathbf{Q}} \qquad \prod_\Phi f_\Phi(g)^{n_\Phi} = 1 \in \widehat{K}^*/K^*. \tag{1.5.3.2}$$

1.5.4

In the special case when K is a Galois extension of $\mathbf{Q}$, the action (1.5.1.1) of $\Gamma_{\mathbf{Q}}$ factors through $\mathrm{Gal}(K/\mathbf{Q})$, which implies that the tori $_K T$ and $_K \mathscr{S}$ are split over K.

In addition, the action of $\mathrm{Gal}(K/\mathbf{Q})$ on K induces an action of $\mathrm{Gal}(K/\mathbf{Q})$ on the $\mathbf{Q}$-group scheme $_K T$, which will be denoted by $t \mapsto g * t$ ($g \in \mathrm{Gal}(K/\mathbf{Q})$). The corresponding action on the character group

$$(h * \lambda)(t) = \lambda(h^{-1} * t) \qquad (\lambda \in X^*(_K T)) \tag{1.5.4.1}$$

is given by

$$\lambda = \sum n_y[y] \mapsto h * \lambda = \sum n_y[yh^{-1}] = \sum n_{yh}[y].$$

The two actions are related by

$$\iota(^h\lambda) = h * \iota(\lambda) \qquad (h \in \mathrm{Gal}(K/\mathbf{Q}), \lambda \in X^*(_K T)), \tag{1.5.4.2}$$

where

$$\iota : X^*(_K T) \longrightarrow X^*(_K T), \qquad \sum n_y[y] \mapsto \sum n_y[y^{-1}] = \sum n_{y^{-1}}[y] \tag{1.5.4.3}$$

is the involution induced by the inverse map $g \mapsto g^{-1}$ on $\mathrm{Gal}(K/\mathbf{Q}) = X(K)$. As $\iota(\lambda_\Phi) = \lambda_{\Phi^{-1}}$, the involution ι and the action (1.5.4.1) preserve $X^*(_K \mathscr{S})$, and we have

$$h * \lambda_\Phi = \lambda_{\Phi h^{-1}}. \tag{1.5.4.4}$$

We denote by

$$\iota : {}_K\mathscr{S}_K = {}_K\mathscr{S} \otimes_{\mathbf{Q}} K \longrightarrow {}_K\mathscr{S}_K$$

the morphism corresponding to ι.

1.6 Universal Taniyama elements [Mil90], [Sch94]

In this section we assume that K is a CM number field which is a Galois extension of $\mathbf{Q}$.

1.6.1

The two actions of $\mathrm{Gal}(K/\mathbf{Q})$ on $X^*(_K\mathscr{S})$ correspond to two actions of $\mathrm{Gal}(K/\mathbf{Q})$ on $_K\mathscr{S}(\widehat{K})$:
the Galois action $t \mapsto {}^g t$ and the algebraic action $t \mapsto h * t$, which commute with each other and satisfy

$$({}^g\lambda)({}^g t) = {}^g(\lambda(t)), \qquad (h * \lambda)(h * t) = \lambda(t) \qquad (\lambda \in X^*(_K\mathscr{S}),\, t \in {}_K\mathscr{S}(\widehat{K})),$$

respectively.

Proposition 1.6.2 (i) *There is a unique map* $f' : \Gamma_{\mathbf{Q}} \longrightarrow {}_K\mathscr{S}(\widehat{K})/{}_K\mathscr{S}(K)$ *such that* $\lambda_\Phi \circ f' = f_\Phi$, *for all CM types* Φ *of* K. *The map* f' *factors through* $\mathrm{Gal}(K^{ab}/\mathbf{Q})$.
(ii) *For each* $\lambda \in X^*(_K\mathscr{S})$, *put* $f'_\lambda = \lambda \circ f' : \Gamma_{\mathbf{Q}} \longrightarrow \widehat{K}^*/K^*$; *then* $f'_{\lambda+\mu}(g) = f'_\lambda(g)f'_\mu(g)$.
(iii) $\forall \lambda \in X^*(_K\mathscr{S})\ \forall g, g' \in \Gamma_{\mathbf{Q}}, \qquad f'_\lambda(gg') = f'_{g'\lambda}(g)f'_\lambda(g')$.
(iv) $\forall h \in \mathrm{Gal}(K/\mathbf{Q}), \qquad {}^h(f'_\lambda(g)) = f'_{h*\lambda}(g)$.

Proof. (i) As the torus $_K\mathscr{S}$ is split over K and $X^*(_K\mathscr{S})$ is a free abelian group generated by the CM characters λ_Φ, we have

$$\begin{aligned}
{}_K\mathscr{S}(\widehat{K})/{}_K\mathscr{S}(K) &= \mathrm{Hom}_{\mathbf{Z}}(X^*(_K\mathscr{S}),\widehat{K}^*)/\mathrm{Hom}_{\mathbf{Z}}(X^*(_K\mathscr{S}),K^*) \\
&= \mathrm{Hom}_{\mathbf{Z}}(X^*(_K\mathscr{S}),\widehat{K}^*/K^*) \\
&= \{\alpha : \{\text{CM types of } K\} \longrightarrow \widehat{K}^*/K^* \mid \textstyle\prod \alpha(\Phi)^{n_\Phi} \\
&= 1 \text{ whenever } \textstyle\sum n_\Phi \lambda_\Phi = 0\}.
\end{aligned}$$

The existence and uniqueness of f' then follow from (1.5.3.2). As K is a Galois extension of $\mathbf{Q}$, the maps F_Φ (hence f_Φ, too) factor through $\mathrm{Gal}(K^{ab}/\mathbf{Q})$.
(ii) This is a consequence of (the proof of) (i).
(iii), (iv) If $\lambda = \lambda_\Phi$, the statement of (iii) (resp., of (iv)) is just (1.4.2.3) (resp., (1.4.2.4)). The general case then follows from (ii).

Proposition 1.6.3 (i) *Define the map* $f : \Gamma_{\mathbf{Q}} \longrightarrow {}_K\mathscr{S}(\widehat{K})/{}_K\mathscr{S}(K)$ *by the formula* $f(g) = (\iota(f'(g)))^{-1}$. *The map* f *factors through* $\mathrm{Gal}(K^{ab}/\mathbf{Q})$ *and has the following properties:*
(ii) *The maps* $f_\lambda = \lambda \circ f : \Gamma_{\mathbf{Q}} \longrightarrow \widehat{K}^*/K^*$ $(\lambda \in X^*(_K\mathscr{S}))$ *satisfy*

$$f_{\lambda+\mu}(g) = f_\lambda(g)f_\mu(g), \qquad f_\lambda(g) = f'_{\iota(\lambda)}(g)^{-1}, \qquad f_\lambda(gg') = f_{g'*\lambda}(g)f_\lambda(g').$$

(iii) $\forall h \in \mathrm{Gal}(K/\mathbf{Q}),\ \forall g \in \Gamma_{\mathbf{Q}}, {}^h(f_\lambda(g)) = f_{h\lambda}(g),\ {}^h(f(g)) = f(g)$.
(iv) $\forall g, g' \in \Gamma_{\mathbf{Q}}, \quad f(gg') = \left(g'^{-1} * f(g)\right) f(g')$.

Proof. The statements of (i), (ii), and the first part of (iii) are immediate consequences of 1.6.2, thanks to (1.5.4.2). The second part of (iii) follows from

$$\left({}^h\lambda\right)\left({}^h(f(g))\right) = {}^h(\lambda\left(f(g)\right)) \overset{\text{(iii)}}{=} \left({}^h\lambda\right)(f(g)) \qquad (\lambda \in X^*({}_K\mathscr{S})),$$

while (iv) is a consequence of the last formula from (ii) and

$$\lambda\left(g'^{-1} * f(g)\right) = (g' * \lambda)\left(f(g)\right).$$

1.6.4

For each CM type Φ of K, the map f_{λ_Φ} is given by

$$f_{\lambda_\Phi}(g) = f_{\Phi^{-1}}(g)^{-1},$$

which implies that

$$r_K \circ f_{\lambda_\Phi}(g) = F_{\Phi^{-1}}(g)^{-1}.$$

In the notation of [Sch94, 4.2], we have $f_\lambda(g) = f_K(g, \lambda)$. The map f is the "universal Taniyama element" of [Mil90, I.5.7].

Proposition 1.6.5 *If K' is a CM number field, which is a Galois extension of $\mathbf{Q}$ and contains K, then the universal Taniyama elements $f_K : \Gamma_\mathbf{Q} \longrightarrow {}_K\mathscr{S}(\widehat{K})/{}_K\mathscr{S}(K)$ and $f_{K'} : \Gamma_\mathbf{Q} \longrightarrow {}_{K'}\mathscr{S}(\widehat{K'})/{}_{K'}\mathscr{S}(K')$ over K and K', respectively, satisfy $f_K = N_{K'/K} \circ f_{K'}$.*

Proof. As the map $i_{K'/K} : \widehat{K}^*/K^* \longrightarrow \widehat{K'}^*/K'^*$ is injective, it is enough to check that, for any CM type Φ of K and $g \in \Gamma_\mathbf{Q}$,

$$i_{K'/K} \circ \lambda_\Phi \circ f_K(g) \overset{?}{=} i_{K'/K} \circ \lambda_\Phi \circ N_{K'/K} \circ f_{K'}(g) \in \widehat{K'}^*/K'^*,$$

which follows from (1.4.2.5), since

$$i_{K'/K} \circ \lambda_\Phi \circ f_K(g) = i_{K'/K}\left(f_{\Phi^{-1}}(g)^{-1}\right) \overset{(1.4.2.5)}{=} f_{\Phi'^{-1}}(g)^{-1} = \lambda_{\Phi'} \circ f_{K'}(g)$$
$$= i_{K'/K} \circ \lambda_\Phi \circ N_{K'/K} \circ f_{K'}(g),$$

where Φ' is the CM type of K' induced from Φ.

1.7 The Taniyama group [Mil90], [MS82], [Sch94]

Let K be as in Section 1.6.

1.7.1

The **Taniyama group** of level K sits in an exact sequence of affine group schemes over $\mathbf{Q}$,

$$1 \longrightarrow {}_K\mathscr{S} \overset{i}{\longrightarrow} {}_K\mathscr{T} \overset{\pi}{\longrightarrow} \mathrm{Gal}(K^{ab}/\mathbf{Q}) \longrightarrow 1,$$

such that the action of (the constant group scheme) $\mathrm{Gal}(K^{ab}/\mathbf{Q})$ on ${}_K\mathscr{S}$ defined by this exact sequence is given by the algebraic action $(g, t) \mapsto g * t$. In addition, there exists a continuous group homomorphism

$$sp : \mathrm{Gal}(K^{ab}/\mathbf{Q}) \longrightarrow {}_K\mathscr{T}(\widehat{\mathbf{Q}})$$

satisfying $\pi \circ sp = \mathrm{id}$.

1.7.2

Choose a section

$$\alpha : \mathrm{Gal}(K^{ab}/\mathbf{Q}) \longrightarrow {}_K\mathscr{T}(K)$$

of the map ${}_K\mathscr{T}(K) \longrightarrow \mathrm{Gal}(K^{ab}/\mathbf{Q})$ (which is surjective, since the torus ${}_K\mathscr{S}$ is split over K and $H^1(K, \mathbf{G}_m) = 0$); the map

$$b : \mathrm{Gal}(K^{ab}/\mathbf{Q}) \longrightarrow {}_K\mathscr{S}(\widehat{K}), \qquad b(g) = sp(g)\alpha(g)^{-1},$$

has the following properties:

(1.7.2.1) The induced map $\overline{b} : \mathrm{Gal}(K^{ab}/\mathbf{Q}) \longrightarrow {}_K\mathscr{S}(\widehat{K})/{}_K\mathscr{S}(K)$ does not depend on the choice of α.
(1.7.2.2) $\forall g, g' \in \mathrm{Gal}(K^{ab}/\mathbf{Q}), \qquad \overline{b}(gg') = \big(g'^{-1} * \overline{b}(g)\big)\,\overline{b}(g').$
(1.7.2.3) $\forall h \in \mathrm{Gal}(K/\mathbf{Q})\ \forall g \in \mathrm{Gal}(K^{ab}/\mathbf{Q}), \qquad {}^h\big(\overline{b}(g)\big) = \overline{b}(g).$
(1.7.2.4) The "coboundary" $d_{g,g'} = \big(g'^{-1} * b(g)\big)\,b(g')\,b(gg')^{-1}$ is a locally constant function on $\mathrm{Gal}(K^{ab}/\mathbf{Q})^2$.

1.7.3

Conversely, any map b satisfying (1.7.2.1)–(1.7.2.4) gives rise to an object from 1.7.1 [MS82, Prop. 2.7]: firstly, the reverse 2-cocycle $d_{g,g'}$ with values in ${}_K\mathscr{S}(K)$ defines an exact sequence of affine group schemes over K,

$$1 \longrightarrow {}_K\mathscr{S}_K \overset{i}{\longrightarrow} G' \overset{\pi}{\longrightarrow} \mathrm{Gal}(K^{ab}/\mathbf{Q}) \longrightarrow 1, \tag{1.7.3.1}$$

equipped with a section $\alpha : \mathrm{Gal}(K^{ab}/\mathbf{Q}) \longrightarrow G'(K)$ such that

$$\forall g, g' \in \mathrm{Gal}(K^{ab}/\mathbf{Q}), \qquad \alpha(gg') = \alpha(g)\alpha(g')d_{g,g'}.$$

Secondly, the map

$$sp : \mathrm{Gal}(K^{ab}/\mathbf{Q}) \longrightarrow G'(\widehat{K}), \qquad sp(g) = b(g)\alpha(g),$$

is a group homomorphism satisfying $\pi \circ sp = \mathrm{id}$. Thirdly, each element $h \in \Gamma_K$ acts on $G'(\overline{\mathbf{Q}})$ by

$$^{h}(s\,\alpha(g)) = {}^{h}s\,\alpha(g) \qquad (s \in {}_K\mathscr{S}(\overline{\mathbf{Q}})). \qquad (1.7.3.2)$$

In order to descend the sequence (1.7.3.1) to an exact sequence of group schemes over $\mathbf{Q}$,

$$1 \longrightarrow {}_K\mathscr{S} \xrightarrow{\ i\ } G \xrightarrow{\ \pi\ } \mathrm{Gal}(K^{ab}/\mathbf{Q}) \longrightarrow 1,$$

it is enough to extend the action of Γ_K from (1.7.3.2) to an action of $\Gamma_{\mathbf{Q}}$ compatible with i and π. This is done by putting, for $h \in \Gamma_{\mathbf{Q}}$ and $g \in \mathrm{Gal}(K^{ab}/\mathbf{Q})$,

$$^{h}(s\,\alpha(g)) = c_h(g)\,{}^{h}s\,\alpha(g), \qquad c_h(g) = b(g)\,{}^{h}(b(g))^{-1} \in {}_K\mathscr{S}(K).$$

As $^{h}(sp(g)) = sp(g)$ for all $h \in \Gamma_{\mathbf{Q}}$ and $g \in \mathrm{Gal}(K^{ab}/\mathbf{Q})$, the map sp has values in $G(\widehat{\mathbf{Q}})$. Up to isomorphism, the quadruple (G, i, π, sp) obtained by this method depends only on $\overline{b}$, not on its lift b.

1.7.4

The Taniyama group $_K\mathscr{T}$ of level K is defined by applying the construction from 1.7.3 to the universal Taniyama element f, which satisfies (1.7.2.2)–(1.7.2.3), by Proposition 1.6.3. The existence of a lift b of f satisfying (1.7.2.4) is established in the following proposition.

Proposition 1.7.5 *There exists a lift* $b : \mathrm{Gal}(K^{ab}/\mathbf{Q}) \longrightarrow {}_K\mathscr{S}(\widehat{K})$ *of* f *whose "coboundary"* $d_{g,g'} = (g'^{-1} * b(g))\, b(g')\, b(gg')^{-1}$ *is a locally constant function on* $\mathrm{Gal}(K^{ab}/\mathbf{Q})^2$.

Proof. Let $\widetilde{\ell}_K$ be as in 1.3.4(iv). As the maps F_{Φ} (which factor through $\mathrm{Gal}(K^{ab}/\mathbf{Q})$) are continuous, there exists an open subgroup $U \subset \Gamma_K^{ab}$ such that $\widetilde{\ell}_K$, when restricted to $\bigcup_{\Phi} F_{\Phi}(U)$, is a homomorphism. If $n_{\Phi} \in \mathbf{Z}$ satisfy $\sum_{\Phi} n_{\Phi}\lambda_{\Phi} = 0$, then the relation (1.5.3.1) implies that

$$\forall u \in U \qquad \prod_{\Phi} \widetilde{\ell}_K\left(F_{\Phi}(u)\right)^{n_{\Phi}} = 1 \in \widehat{K}^*.$$

As in the proof of 1.6.2(i), we conclude that, for each $u \in U$, there exists a unique element $b'(u) \in {}_K\mathscr{S}(\widehat{K})$ satisfying $\lambda_{\Phi}(b'(u)) = \widetilde{\ell}_K\left(F_{\Phi}(u)\right)$. Fix coset representatives $\mathrm{Gal}(K^{ab}/\mathbf{Q}) = \bigcup_j g_j U$ (disjoint union) and lifts $\widetilde{s}_j \in {}_K\mathscr{S}(\widehat{K})$ of $f'(g_j) \in {}_K\mathscr{S}(\widehat{K})/{}_K\mathscr{S}(K)$ such that $g_{j_0} = 1$ and $\widetilde{s}_{j_0} = 1$; define a map $b' : \mathrm{Gal}(K^{ab}/\mathbf{Q}) \longrightarrow {}_K\mathscr{S}(\widehat{K})$ by

$$b'(g_j u) = \widetilde{s}_j b'(u) \qquad (u \in U).$$

The map $b(g) := (\iota(b'(g)))^{-1}$ then has the required property.

1.7.6

Proposition 1.6.5 implies that the pull-backs of the group schemes $_K\mathcal{T}$ via $\Gamma_{\mathbf{Q}} \longrightarrow \mathrm{Gal}(K^{ab}/\mathbf{Q})$ form, for varying K, a projective system compatible with the norm maps $N_{K'/K} : {}_{K'}\mathcal{S} \longrightarrow {}_K\mathcal{S}$. In the limit, they give rise to an exact sequence

$$1 \longrightarrow \mathcal{S} \overset{i}{\longrightarrow} \mathcal{T} \overset{\pi}{\longrightarrow} \Gamma_{\mathbf{Q}} \longrightarrow 1 \qquad\qquad (1.7.6.1)$$

equipped with a splitting $sp : \Gamma_{\mathbf{Q}} \longrightarrow \mathcal{T}(\widehat{\mathbf{Q}})$. The main result of [Del82] states that the affine group scheme $\mathcal{T}$ (= the Taniyama group) is the Tannaka dual of the category $CM_{\mathbf{Q}}$ of CM motives (for absolute Hodge cycles) defined over $\mathbf{Q}$. The group scheme $_K\mathcal{T}$ corresponds to the full Tannakian subcategory of $CM_{\mathbf{Q}}$ consisting of objects with reflex field in K.

2 Hidden symmetries in the CM theory

Throughout this chapter, K and F are as in 1.3. Put $X = X(F)$. In Section 2.1 (resp., Section 2.2) we extend Tate's half-transfer F_Φ (resp., the Taniyama element f_Φ) from $\Gamma_{\mathbf{Q}}$ to $\mathrm{Aut}_{F\text{-alg}}(F \otimes \overline{\mathbf{Q}})$ (resp., to $\mathrm{Aut}_{F\text{-alg}}(F \otimes \overline{\mathbf{Q}})_1$). In Section 2.3–2.4 we use our generalisation of the Taniyama element to construct a generalised Taniyama group.

2.1 Generalised half-transfer

2.1.1

Fix a section $s : X \longrightarrow \Gamma_{\mathbf{Q}}$ of the restriction map $g \mapsto g|_F$. As in 1.1.2–1.1.4, the choice of s determines the following objects:

(2.1.1.1) An injection $\rho_s : \Gamma_{\mathbf{Q}} \hookrightarrow S_X \ltimes \Gamma_F^X$.

(2.1.1.2) An isomorphism $\beta_{s*} : \mathrm{Aut}_{F\text{-alg}}(F \otimes \overline{\mathbf{Q}}) \overset{\sim}{\longrightarrow} \mathrm{Aut}_{F\text{-alg}}(\overline{\mathbf{Q}}^X) = S_X \ltimes \Gamma_F^X$ satisfying $\beta_{s*}(\mathrm{id}_F \otimes g) = \rho_s(g)$.
In addition, we obtain

(2.1.1.3) A bijection between $(\mathbf{Z}/2\mathbf{Z})^X$ and the set of CM types of K: a function $\alpha : X \longrightarrow \mathbf{Z}/2\mathbf{Z}$ corresponds to the CM type $\{c^{\alpha(x)}s(x)|_K = s(x)c^{\alpha(x)}|_K\}_{x \in X}$.

(2.1.1.4) A section $w_s : X(K) \longrightarrow \Gamma_{\mathbf{Q}}$ of the restriction map $g \mapsto g|_K$ satisfying $w_s(cy) = cw_s(y)$, namely $w_s(c^a s(x)|_K) = c^a s(x)$ ($x \in X$, $a \in \mathbf{Z}/2\mathbf{Z}$).

For $h \in \Gamma_F^X$, we denote by $\overline{h} : X \longrightarrow \mathbf{Z}/2\mathbf{Z}$ the image of h in $\mathrm{Gal}(K/F)^X \overset{\sim}{\longrightarrow} (\mathbf{Z}/2\mathbf{Z})^X$. In other words,

$$\forall x \in X, \qquad h(x)|_K = c^{\overline{h}(x)}, \qquad R(h(x)) = c^{\overline{h}(x)}h(x),$$

where $R : \Gamma_F \longrightarrow \Gamma_K$ is the retraction map from 1.3.2. We let $S_X \ltimes \Gamma_F^X$ act on $(\mathbf{Z}/2\mathbf{Z})^X$ via (1.1.1.1) and the natural projection $(\sigma, h) \mapsto (\sigma, \overline{h})$:

$$(\sigma, h)\alpha = (\alpha + \overline{h}) \circ \sigma^{-1}. \tag{2.1.1.5}$$

2.1.2 Rewriting Tate's Half-Transfer in Terms of ρ_s

Let Φ be a CM type of K. If $g \in \Gamma_\mathbf{Q}$, then $\rho_s(g) = (\sigma, h) \in S_X \ltimes \Gamma_F^X$, where

$$\forall x \in X, \quad \sigma(x) = gx, \quad h(x) = s(gx)^{-1}gs(x) = s(\sigma(x))^{-1}gs(x) \in \Gamma_F.$$

Let $\alpha \in (\mathbf{Z}/2\mathbf{Z})^X$ correspond to Φ, as in (2.1.1.3). For each $x \in X$, the element

$$\varphi_x = c^{\alpha(x)}s(x)|_K = s(x)c^{\alpha(x)}|_K \in \Phi$$

satisfies $w_s(\varphi_x) = c^{\alpha(x)}s(x)$ and

$$g\varphi_x = gs(x)c^{\alpha(x)}|_K = s(\sigma(x))h(x)c^{\alpha(x)}|_K = c^{\alpha(x)+\overline{h}(x)}s(\sigma(x))|_K,$$

which implies that $w_s(g\varphi_x) = c^{\alpha(x)+\overline{h}(x)}s(\sigma(x))$ and

$$\begin{aligned}
w_s(g\varphi_x)^{-1}gw_s(\varphi_x) &= s(\sigma(x))^{-1}c^{\alpha(x)+\overline{h}(x)}gc^{\alpha(x)}s(x) \\
&= s(\sigma(x))^{-1}c^{\alpha(x)+\overline{h}(x)}s(\sigma(x))h(x)s(x)^{-1}c^{\alpha(x)}s(x) \\
&= \left[s(\sigma(x))^{-1}c^{\alpha(x)+\overline{h}(x)}s(\sigma(x))c^{\alpha(x)+\overline{h}(x)}\right] \\
&\quad \times \left[c^{\alpha(x)+\overline{h}(x)}h(x)c^{\alpha(x)}\right] \cdot \left[c^{\alpha(x)}s(x)^{-1}c^{\alpha(x)}s(x)\right].
\end{aligned}$$

Denote by $\gamma_{x,s}$ the image of $s(x)^{-1}cs(x)c \in \Gamma_K$ in Γ_K^{ab}. Since each of the three elements in square brackets lies in Γ_K, we have

$$\begin{aligned}
F_\Phi(g) &= \prod_{x \in X} w_s(g\varphi_x)^{-1}gw_s(\varphi_x)|_{K^{ab}} \\
&= \prod_{x \in |(\sigma,h)\alpha|} \gamma_{x,s} \prod_{x \in |\alpha|} \gamma_{x,s}^{-1} \prod_{x \in X} c^{\alpha(x)}R(h(x))c^{\alpha(x)}|_{K^{ab}},
\end{aligned}$$

where we have denoted by $|\alpha| = \{x \in X \mid \alpha(x) \neq 0\}$ the support of α. This calculation justifies the following:

Proposition–Definition 2.1.3 *For each $\alpha \in (\mathbf{Z}/2\mathbf{Z})^X$, the formula*

$$\begin{aligned}
{}_s\widetilde{F}_\alpha(\sigma, h) &= \prod_{x \in X} s(\sigma(x))^{-1}c^{\alpha(x)+\overline{h}(x)}s(\sigma(x))h(x)s(x)^{-1}c^{\alpha(x)}s(x)|_{K^{ab}} \\
&= \prod_{x \in |(\sigma,h)\alpha|} \gamma_{x,s} \prod_{x \in |\alpha|} \gamma_{x,s}^{-1} \prod_{x \in X} c^{\alpha(x)}R(h(x))c^{\alpha(x)}|_{K^{ab}}
\end{aligned}$$

defines a map

$$_s\widetilde{F}_\alpha : S_X \ltimes \Gamma_F^X \longrightarrow \Gamma_K^{ab}$$

(depending on s and α) satisfying ${}_s\widetilde{F}_\alpha \circ \rho_s = F_\Phi$, where Φ is the CM type corresponding to α, as in (2.1.1.3).

Proposition 2.1.4 *The maps $_s\widetilde{F}_\alpha$ have the following properties:*

(i) $\forall g, g' \in S_X \ltimes \Gamma_F^X$, $_s\widetilde{F}_\alpha(gg') = {_s\widetilde{F}_{g'\alpha}(g)} \, {_s\widetilde{F}_\alpha(g')}$.

(ii)   *For each $(\sigma, h) \in S_X \ltimes \Gamma_F^X$,*

$$_s\widetilde{F}_\alpha(\sigma, h)|_{F^{ab}} = \prod_{x \in |(\sigma,h)\alpha|} c_x \prod_{x \in |\alpha|} c_x \prod_{x \in X} h(x)|_{F^{ab}},$$

$$^{1+c}\left({_s\widetilde{F}_\alpha(\sigma, h)}\right) = \widetilde{V}_{K/F}(\sigma, h) = \prod_{x \in X} {^{1+c}R(h(x))}|_{K^{ab}},$$

where $\widetilde{V}_{K/F}(\sigma, h) = \prod_{x \in X} V_{K/F}(h(x)|_{F^{ab}})$.

(iii)   *Each map $_s\widetilde{F}_\alpha$ factors through $S_X \ltimes \mathrm{Gal}(K^{ab}/F)^X$.*

(iv)   *If $g = (\sigma, h) \in S_X \ltimes \Gamma_F^X$ satisfies $g\alpha = \alpha$, then*

$$_s\widetilde{F}_\alpha(g) = \prod_{x \in X} c^{\alpha(x)} R(h(x)) c^{\alpha(x)}|_{K^{ab}}.$$

(v) $\forall (\sigma, h) \in S_X \ltimes \Gamma_K^X$, $_s\widetilde{F}_0(\sigma, h) = \prod_{x \in X} h(x)|_{K^{ab}}$.

(vi) $\forall \alpha \in (\mathbf{Z}/2\mathbf{Z})^X$, $_s\widetilde{F}_0(1, c^\alpha) = \prod_{x \in |\alpha|} \gamma_{x,s}$.

Proof. (i)   If $g = (\sigma, h)$ and $g' = (\sigma', h')$, then $gg' = (\sigma\sigma', (h \circ \sigma')h')$ and $\alpha' := g'\alpha = (\alpha + \overline{h'}) \circ \sigma'^{-1}$, which implies that $_s\widetilde{F}_\alpha(gg') \, {_s\widetilde{F}_\alpha(g')^{-1}} \, {_s\widetilde{F}_{g'\alpha}(g)^{-1}}$ is equal to

$$\prod_{x \in X} \left(c^{\alpha(x) + \overline{h}(\sigma'(x)) + \overline{h'}(x)} h(\sigma'(x)) h'(x) c^{\alpha(x)} \right)$$

$$\times \left(c^{\alpha(x) + \overline{h'}(x)} h'(x) c^{\alpha(x)} \right)^{-1} \cdot \left(c^{\alpha'(x) + \overline{h}(x)} h(x) c^{\alpha'(x)} \right)^{-1}$$

$$= \prod_{x \in X} \left(c^{\alpha'(\sigma'(x)) + \overline{h}(\sigma'(x))} h(\sigma'(x)) c^{\alpha'(\sigma'(x))} \right) \left(c^{\alpha'(x) + \overline{h}(x)} h(x) c^{\alpha'(x)} \right)^{-1} = 1.$$

(ii)   The first formula is a consequence of the fact that

$$\forall x \in X, \qquad \gamma_{x,s}|_{F^{ab}} = c_x c, \qquad c^{\alpha(x)} R(h(x)) c^{\alpha(x)}|_{F^{ab}} = c^{\overline{h}(x)} h(x)|_{F^{ab}};$$

applying (1.3.2.1), we obtain the second formula.

The statements (iii)–(vi) follow directly from the definitions.

2.1.5 Change of s

Let $s, s' \longrightarrow \Gamma_{\mathbf{Q}}$ be two sections of the restriction map $g \mapsto g|_F$. We have $s' = st$, where $t : X \longrightarrow \Gamma_F$. As in 2.1.1, we write, for each $x \in X$, $t(x)|_K = c^{\overline{t}(x)}$ ($\overline{t}(x) \in \mathbf{Z}/2\mathbf{Z}$); then $R(t(x)) = c^{\overline{t}(x)} t(x) \in \Gamma_K$. The recipe (2.1.1.3), applied to s and s', respectively, associates to each CM type Φ of K two functions $\alpha = \alpha_{\Phi,s}, \alpha' = \alpha_{\Phi,s'} : X \longrightarrow \mathbf{Z}/2\mathbf{Z}$ such that

$$\Phi = \{c^{\alpha(x)} s(x)|_K\} = \{c^{\alpha'(x)} s'(x)|_K\} \qquad (\Longrightarrow \alpha' = \alpha + \overline{t}).$$

426   Jan Nekovář

According to Proposition 1.1.4, the following diagram is commutative:

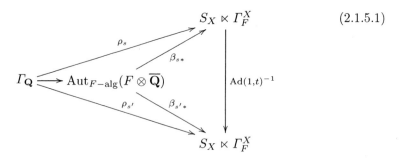

$$(2.1.5.1)$$

For $(\sigma, h) \in S_X \ltimes \Gamma_F^X$, put

$$(\sigma', h') := \operatorname{Ad}(1,t)^{-1}(\sigma, h) = (1,t)^{-1}(\sigma, h)(1,t) = (\sigma, (t \circ \sigma)^{-1} ht) \in S_X \ltimes \Gamma_F^X.$$
$$(2.1.5.2)$$

The map $\widetilde{V}_{K/F}$ from Proposition 2.1.4(ii) satisfies $\widetilde{V}_{K/F}(\sigma, h) = \widetilde{V}_{K/F}(\sigma', h')$, which means that the map

$$\widetilde{V}_{K/F} \circ \beta_{s*} : \operatorname{Aut}_{F-\mathrm{alg}}(F \otimes \overline{\mathbf{Q}}) \longrightarrow \Gamma_K^{ab} \qquad (2.1.5.3)$$

does not depend on s; we denote it again by $\widetilde{V}_{K/F}$. The equalities

$$(\sigma, h)\alpha = (\alpha + \overline{h}) \circ \sigma^{-1}, \quad (\sigma', h')\alpha' = (\alpha' + \overline{h}') \circ \sigma^{-1} = (\alpha + \overline{h}) \circ \sigma^{-1} + \overline{t} \in (\mathbf{Z}/2\mathbf{Z})^X$$

imply that the action of $S_X \ltimes \Gamma_F^X$ on $(\mathbf{Z}/2\mathbf{Z})^X$ defined in (2.1.1.5) gives rise to an action of the group $\operatorname{Aut}_{F-\mathrm{alg}}(F \otimes \overline{\mathbf{Q}})$ on the set of CM types of K, which is characterised by

$$\forall g \in \operatorname{Aut}_{F-\mathrm{alg}}(F \otimes \overline{\mathbf{Q}}) \qquad \alpha_{g\Phi, s} = \beta_{s*}(g)\alpha_{\Phi, s}, \qquad (2.1.5.4)$$

but which does not depend on s.

Proposition 2.1.6 $_s\widetilde{F}_\alpha(\sigma, h) = {}_{s'}\widetilde{F}_{\alpha'}(\sigma', h') \in \Gamma_K^{ab}$, *in the notation of* (2.1.5.2).

Proof. The relations $s' = st$, $(\sigma', h') = (\sigma, (t \circ \sigma)^{-1} ht)$, $\overline{h}' = \overline{h} + \overline{t} + \overline{t} \circ \sigma$, $\alpha' = \alpha + \overline{t}$, $(\sigma, h)\alpha = (\alpha + \overline{h}) \circ \sigma^{-1}$, and $(\sigma', h')\alpha' = (\sigma, h)\alpha + \overline{t}$ imply that

$$
\begin{aligned}
&{}_{s'}\widetilde{F}_{\alpha'}(\sigma', h') \\
&= \prod_{x \in X} t(\sigma(x))^{-1} s(\sigma(x))^{-1} c^{(\alpha + \overline{h})(x) + \overline{t}(\sigma(x))} s(\sigma(x)) h(x) s(x)^{-1} c^{(\alpha + \overline{t})(x)} s(x) t(x)|_{K^{ab}} \\
&= \prod_{x \in X} A'((\sigma, h)\alpha, x)^{-1} B(\alpha, x) A'(\alpha, x),
\end{aligned}
$$

where

$$A'(\alpha, x) = c^{\alpha(x)} s(x)^{-1} c^{(\alpha+\bar{t})(x)} s(x) t(x)|_{K^{ab}}, \quad B(\alpha, x) = c^{\alpha(x)} R(h(x)) c^{\alpha(x)}|_{K^{ab}}.$$

As

$$_s\widetilde{F}_\alpha(\sigma, h) = \prod_{x \in X} A((\sigma, h)\alpha, x)^{-1} B(\alpha, x) A(\alpha, x),$$

where

$$A(\alpha, x) = c^{\alpha(x)} s(x)^{-1} c^{\alpha(x)} s(x)|_{K^{ab}},$$

the equality $_s\widetilde{F}_\alpha(\sigma, h) = {}_{s'}\widetilde{F}_{\alpha'}(\sigma', h')$ follows from the fact that

$$\forall x \in X, \qquad A(\alpha, x)^{-1} A'(\alpha, x) = s(x)^{-1} c^{\bar{t}(x)} s(x) t(x)|_{K^{ab}}$$

does not depend on α.

Proposition–Definition 2.1.7 *In the notation of 2.1.5, the map*

$$\widetilde{F}_\Phi = {}_s\widetilde{F}_\alpha(\sigma, h) \circ \beta_{s*} : \mathrm{Aut}_{F-\mathrm{alg}}(F \otimes \overline{\mathbf{Q}}) \longrightarrow \Gamma_K^{ab}$$

depends on Φ, but not on s; it has the following properties:
(i) $\forall g \in \Gamma_{\mathbf{Q}}, \qquad \widetilde{F}_\Phi(\mathrm{id}_F \otimes g) = F_\Phi(g)$.
(ii) $\forall g, g' \in \mathrm{Aut}_{F-\mathrm{alg}}(F \otimes \overline{\mathbf{Q}}), \qquad \widetilde{F}_\Phi(gg') = \widetilde{F}_{g'\Phi}(g)\widetilde{F}_\Phi(g')$.
(iii) $\forall g \in \mathrm{Aut}_{F-\mathrm{alg}}(F \otimes \overline{\mathbf{Q}}), \qquad {}^{1+c}\widetilde{F}_\Phi(g) = \widetilde{V}_{K/F}(g)$ *(in the notation of (2.1.5.3)).*

Proof. The independence of $\widetilde{F}_\Phi$ of the choice of s follows from Proposition 2.1.6 and the commutative diagram (2.1.5.1). The remaining statements are consequences of Proposition 2.1.4.

2.1.8 Galois functoriality of $\widetilde{F}_\Phi$

Given an element $\widetilde{u} \in \Gamma_{\mathbf{Q}}$, define $u := \widetilde{u}|_K$, $u_F := u|_F$, $K' := u(K)$, $F' = u_F(F)$ and $X' = X(F')$. As in Proposition 1.1.6 (for $k = \mathbf{Q}$ and $k' = \overline{\mathbf{Q}}$), a fixed section $s : X \longrightarrow \Gamma_{\mathbf{Q}}$ of the restriction map $g \mapsto g|_F$ defines a section $s' : X' \longrightarrow \Gamma_{\mathbf{Q}}$ of the restriction map $g \mapsto g|_{F'}$, given by

$$s'(x') = s'(xu_F^{-1}) = s(x) \circ \widetilde{u}^{-1} \qquad (x \in X).$$

Proposition 2.1.9 *For each $\alpha : X \longrightarrow \mathbf{Z}/2\mathbf{Z}$, the diagram*

$$
\begin{array}{ccc}
S_X \ltimes \Gamma_F^X & \xrightarrow{\;{}_s\widetilde{F}_\alpha\;} & \Gamma_K^{ab} \\
\downarrow{\scriptstyle \widetilde{u}_*} & & \downarrow{\scriptstyle u} \\
S_{X'} \ltimes \Gamma_{F'}^{X'} & \xrightarrow{\;{}_{s'}\widetilde{F}_{\alpha'}\;} & \Gamma_{K'}^{ab}
\end{array}
$$

is commutative, where $\widetilde{u}_*$ is the map defined in Proposition 1.1.6, $\alpha' : X' \longrightarrow \mathbf{Z}/2\mathbf{Z}$ is given by $\alpha'(x') = \alpha(x)$ $(x = x'u_F)$, and the right vertical map (which depends only on u) is given by $g \mapsto \widetilde{u}g\widetilde{u}^{-1}$.

Proof. For $(\sigma, h) \in S_X \ltimes \Gamma_F^X$, we have $\tilde{u}_*(\sigma, h) = (\sigma', h')$, where $\sigma'(x') = \sigma(x) u_F^{-1}$, $h'(x') = \tilde{u} h(x) \tilde{u}^{-1}$ $(x' = x u_F^{-1})$. The relations $\alpha'(x') = \alpha(x)$, $s'(\sigma'(x')) = s(\sigma(x)) \tilde{u}^{-1}$, $s'(x') = s(x) \tilde{u}^{-1}$, and $\overline{h}'(x') = \overline{h}(x)$ imply that $_{s'}\tilde{F}_{\alpha'}(\sigma', h')$ is equal to

$$\prod_{x' \in X'} s'(\sigma'(x'))^{-1} c^{\alpha'(x') + \overline{h}'(x')} s'(\sigma'(x')) h'(x') s'(x')^{-1} c^{\alpha'(x')} s'(x')|_{K'^{ab}}$$

$$= \tilde{u} \prod_{x \in X} s(\sigma(x))^{-1} c^{\alpha(x) + \overline{h}(x)} s(\sigma(x)) h(x) s(x)^{-1} c^{\alpha(x)} s(x)|_{K^{ab}} \tilde{u}^{-1}$$

$$= u \left(_s\tilde{F}_\alpha(\sigma, h) \right).$$

Corollary 2.1.10 *For each CM type Φ of K, the diagram*

$$\begin{array}{ccc} \mathrm{Aut}_{F\text{-alg}}(F \otimes \overline{\mathbf{Q}}) & \xrightarrow{\tilde{F}_\Phi} & \Gamma_K^{ab} \\ \downarrow{[u_F]} & & \downarrow{u} \\ \mathrm{Aut}_{F'-\mathrm{alg}}(F' \otimes \overline{\mathbf{Q}}) & \xrightarrow{\tilde{F}_{\Phi u^{-1}}} & \Gamma_{K'}^{ab} \end{array}$$

is commutative, where $[u_F]$ is the map defined in Proposition 1.1.7(i).

Proof. This follows from Proposition 2.1.9 combined with Proposition 1.1.7 (ii) (for $k = \mathbf{Q}$ and $k' = \overline{\mathbf{Q}}$), if we take into account the fact that

$$\{c^{\alpha'(x')} s'(x')|_{K'}\}_{x' \in X'} = \{c^{\alpha(x)} s(x)|_K u^{-1}\}_{x \in X}.$$

2.2 Generalised Taniyama elements

2.2.1

Let $(S_X \ltimes \Gamma_F^X)_1$ be the group defined as the fibre product

$$\begin{array}{ccc} (S_X \ltimes \Gamma_F^X)_1 & \longrightarrow & S_X \ltimes \Gamma_F^X \\ \downarrow & & \downarrow{(1, \mathrm{prod})} \\ \Gamma_{\mathbf{Q}}^{ab}/\langle c \rangle & \xrightarrow{\overline{V}_{F/\mathbf{Q}}} & \Gamma_F^{ab}/\langle c_X \rangle. \end{array}$$

As the morphism $\overline{V}_{F/\mathbf{Q}}$ is injective (1.3.2.4), we can (and will) identify $(S_X \ltimes \Gamma_F^X)_1$ with its image in $S_X \ltimes \Gamma_F^X$. The group $(S_X \ltimes \Gamma_F^X)_0$, defined in (1.1.2.4), sits in an exact sequence

$$1 \longrightarrow (S_X \ltimes \Gamma_F^X)_0 \longrightarrow (S_X \ltimes \Gamma_F^X)_1 \longrightarrow \langle c_X \rangle / V_{F/\mathbf{Q}}(\langle c \rangle) \longrightarrow 1.$$

For $i = 0, 1$, the subgroups $\beta_{s*}^{-1}\left((S_X \ltimes \Gamma_F^X)_i\right)$ of $\mathrm{Aut}_{F-\mathrm{alg}}(F \otimes \overline{\mathbf{Q}})$ are independent of the choice of a section $s : X \longrightarrow \Gamma_{\mathbf{Q}}$; we denote them by

$$\mathrm{Aut}_{F\text{-alg}}(F \otimes \overline{\mathbf{Q}})_0 \subset \mathrm{Aut}_{F\text{-alg}}(F \otimes \overline{\mathbf{Q}})_1 \subset \mathrm{Aut}_{F\text{-alg}}(F \otimes \overline{\mathbf{Q}}).$$

Definition 2.2.2 *For each CM type Φ of K, define a map*

$$\widetilde{f}_\Phi : \operatorname{Aut}_{F-\mathrm{alg}}(F \otimes \overline{\mathbf{Q}})_1 \longrightarrow \widehat{K}^*/K^*$$

by

$$\widetilde{f}_\Phi(g) = \ell_K\left(\widetilde{F}_\Phi(g)\right),$$

where ℓ_K is the morphism from Proposition 1.3.4(i). [This definition makes sense, by Proposition 2.1.4(ii).]

Proposition 2.2.3 *The maps $\widetilde{f}_\Phi$ have the following properties:*
(i) $r_K \circ \widetilde{f}_\Phi = \widetilde{F}_\Phi$.
(ii) $\forall g \in \Gamma_{\mathbf{Q}}, \qquad \widetilde{f}_\Phi(\mathrm{id}_F \otimes g) = f_\Phi(g)$.
(iii) *Each map $\widetilde{f}_\Phi$ factors through*

$$\operatorname{Aut}_{F-\mathrm{alg}}(F \otimes K^{ab})_1 := \operatorname{Im}\left(\operatorname{Aut}_{F-\mathrm{alg}}(F \otimes \overline{\mathbf{Q}})_1 \longrightarrow \operatorname{Aut}_{F-\mathrm{alg}}(F \otimes K^{ab})\right).$$

(iv) $\forall g, g' \in \operatorname{Aut}_{F-\mathrm{alg}}(F \otimes \overline{\mathbf{Q}})_1, \qquad \widetilde{f}_\Phi(gg') = \widetilde{f}_{g'\Phi}(g)\widetilde{f}_\Phi(g')$.
(v) *If $u : K \xrightarrow{\sim} K'$ is an isomorphism of CM number fields, then*

$$\widetilde{f}_{\Phi u^{-1}} \circ [u|_F] = u \circ \widetilde{f}_\Phi.$$

(vi) *For $g \in \operatorname{Aut}_{F-\mathrm{alg}}(F \otimes \overline{\mathbf{Q}})_1$, denote by $u(g) \in \Gamma_{\mathbf{Q}}^{ab}/\langle c \rangle$ the unique element satisfying $V_{K/\mathbf{Q}}(u(g)) = {}^{1+c}\widetilde{F}_\Phi(g)$; then ${}^{1+c}\widetilde{f}_\Phi(g) = \chi(u(g))K^*$.*
(vii) *For $g \in \operatorname{Aut}_{F-\mathrm{alg}}(F \otimes \overline{\mathbf{Q}})_0$, denote by $u(g) \in \Gamma_{\mathbf{Q}}^{ab}$ the unique element satisfying $V_{F/\mathbf{Q}}(u(g)) = \widetilde{F}_\Phi(g)|_{F^{ab}}$; then $N_{K/F}(\widetilde{f}_\Phi(g)) = \chi(u(g))\alpha F_+^* \in \widehat{F}^*/F_+^*$, where $\alpha \in F^*$ satisfies*

$$\forall x \in X \quad \operatorname{sgn}(x(a)) = \begin{cases} 1, & \text{if } \Phi \text{ and } g\Phi \text{ agree at } x, \\ -1, & \text{if } \Phi \text{ and } g\Phi \text{ do not agree at } x \end{cases}$$

(we say that two CM types Φ and Φ' of K agree at $x \in X$ if the unique element of Φ whose restriction to F is x is equal to the unique element of Φ' whose restriction to F is x).

Proof. Statement (i) holds by definition, while (ii)–(v) follow from the corresponding assertions for $\widetilde{F}_\Phi$, proved in Proposition 2.1.7 and Corollary 2.1.10. Property (vi) (resp., (vii)) is a consequence of Proposition 1.3.4(i) (resp., 1.3.4(ii)) combined with the second (resp., the first) formula in Proposition 2.1.4(ii).

Proposition 2.2.4 *Let K' be a CM number field containing K; put $X' = X(F')$, where F' is the maximal totally real subfield of K'. If Φ is a CM type of K and Φ' is the induced CM type of K', then:*
(i) $\forall g \in \operatorname{Aut}_{F-\mathrm{alg}}(F \otimes \overline{\mathbf{Q}}), \qquad \widetilde{F}_{\Phi'}(\mathrm{id}_{F'} \otimes_F g) = V_{K'/K}\left(\widetilde{F}_\Phi(g)\right) \in \Gamma_{K'}^{ab}$.
(ii) $\forall i = 0, 1 \ \forall g \in \operatorname{Aut}_{F-\mathrm{alg}}(F \otimes \overline{\mathbf{Q}})_i, \qquad \mathrm{id}_{F'} \otimes_F g \in \operatorname{Aut}_{F'-\mathrm{alg}}(F' \otimes \overline{\mathbf{Q}})_i$.
(iii) $\forall g \in \operatorname{Aut}_{F-\mathrm{alg}}(F \otimes \overline{\mathbf{Q}})_1, \qquad \widetilde{f}_{\Phi'}(\mathrm{id}_{F'} \otimes_F g) = i_{K'/K}\left(\widetilde{f}_\Phi(g)\right) \in \widehat{K}'^*/K'^*$.

Proof. (i) Fix a section $s : X \longrightarrow \Gamma_{\mathbf{Q}}$; let $\alpha : X \longrightarrow \mathbf{Z}/2\mathbf{Z}$ correspond to Φ, as in (2.1.1.3). The sets $\Gamma_K/\Gamma_{K'}$ and $\Gamma_F/\Gamma_{F'}$ are canonically identified. Fix a section $u : \Gamma_K/\Gamma_{K'} = \mathrm{Hom}_{K-\mathrm{alg}}(K', \overline{\mathbf{Q}}) \longrightarrow \Gamma_K$ of the restriction map $g \mapsto g|_K$ and define a section $s' : X' \longrightarrow \Gamma_{\mathbf{Q}}$ by

$$s'(s(x)y|_{F'}) = s(x)u(y) \qquad (x \in X,\, y \in \Gamma_F/\Gamma_{F'});$$

then Φ' corresponds to $\alpha' = \alpha \circ p : X' \longrightarrow \mathbf{Z}/2\mathbf{Z}$, where we have denoted by $p : X' \longrightarrow X$ the restriction map $g \mapsto g|_F$. Proposition 1.1.8 implies that the elements

$$(\sigma, h) = \beta_{s*}(g) \in S_X \ltimes \Gamma_F^X, \qquad (\sigma', h') = \beta_{s'*}(\mathrm{id}_F \otimes_F g) \in S_{X'} \ltimes \Gamma_{F'}^{X'}$$

are related by

$$\sigma'(s(x)y|_{F'}) = s(\sigma(x))h(x)y|_{F'}, \qquad s'(\sigma'(s(x)y|_{F'})) = s(\sigma(x))u(h(x)y),$$
$$h'(s(x)y|_{F'}) = u(h(x)y)^{-1}h(x)u(y) \qquad (x \in X,\, y \in \Gamma_F/\Gamma_{F'})$$

hence $\overline{h}' = \overline{h} \circ p$. For $x \in X$ and $x' \in X'$, put

$$k(x) = s(\sigma(x))^{-1}c^{\alpha(x)+\overline{h}(x)}s(\sigma(x))h(x)s(x)^{-1}c^{\alpha(x)}s(x) \in \Gamma_K,$$
$$k'(x') = s'(\sigma'(x'))^{-1}c^{\alpha'(x')+\overline{h}'(x')}s'(\sigma'(x'))h'(x')s'(x')^{-1}c^{\alpha'(x')}s'(x') \in \Gamma_{K'}.$$

By definition,

$$\widetilde{F}_\Phi(g) = {}_s\widetilde{F}_\alpha(\sigma, h) = \prod_{x \in X} k(x)|_{K^{ab}} \in \Gamma_K^{ab},$$

$$\widetilde{F}_{\Phi'}(\mathrm{id}_F \otimes_F g) = {}_{s'}\widetilde{F}_{\alpha'}(\sigma', h') = \prod_{x' \in X'} k'(x')|_{K'^{ab}} \in \Gamma_{K'}^{ab}.$$

For each $x \in X$ and $y \in \Gamma_F/\Gamma_{F'}$,

$$k'(s(x)y|_{F'}) = u(h(x)y)^{-1}k(x)u(y) \in \Gamma_{K'},$$

which implies that $k(x)y = k(x)u(y)|_{K'} = u(h(x)y)|_{K'} = h(x)y$; hence $u(h(x)y) = u(k(x)y)$ and

$$\prod_{x' \in p^{-1}(x)} k'(x')|_{K'^{ab}} = \prod_{y \in \Gamma_K/\Gamma_{K'}} u(k(x)y)^{-1}k(x)u(y)|_{K'^{ab}} = V_{K'/K}\left(k(x)|_{K^{ab}}\right).$$

Taking the product over all $x \in X$ yields (i). The statement (ii) follows from the fact that, in the notation used in the proof of (i),

$$\prod_{x' \in p^{-1}(x)} h'(x')|_{F'^{ab}} = \prod_{y \in \Gamma_F/\Gamma_{F'}} u(h(x)y)^{-1}h(x)u(y)|_{F'^{ab}} = V_{F'/F}\left(h(x)|_{F^{ab}}\right).$$

Finally, (iii) follows by applying $\ell_{K'}$ to the statement of (i) (which makes sense, by (ii) for $i = 1$).

2.2.5 Action of $\mathrm{Aut}_{F\text{-alg}}(F \otimes \overline{\mathbf{Q}})_0$ on CM points of Hilbert modular varieties

In this section we prove Theorem 0.9. Given a polarised HBAV (Hilbert–Blumenthal abelian variety) A relative to F with CM, then A is defined over $\overline{\mathbf{Q}}$, and there exist

- a CM field K of degree 2 over F;
- a CM type Φ of K (which defines an embedding $K \hookrightarrow \mathbf{C}^\Phi$, $\alpha \mapsto (\varphi \mapsto \varphi(\alpha))_{\varphi \in \Phi}$);
- a fractional ideal a of K;
- an element $t \in K^*$ such that $t \notin F^*$, $t^2 \in F^*$, and $\forall \varphi \in \Phi$, $\quad \mathrm{Im}(\varphi(t)) < 0$;
- an O_K-linear isomorphism $\theta : \mathbf{C}^\Phi/a \xrightarrow{\sim} A(\mathbf{C})$ such that the Riemann form of the pull-back of the polarisation of A by θ is induced by the form $E_t(x, y) = \mathrm{Tr}_{K/\mathbf{Q}}(tx\,{}^c y)$ on K.

One says that A is a CM abelian variety of type (K, Φ, a, t) (via θ). The type is determined up to transformations $(K, \Phi, a, t) \mapsto (K, \Phi, a\alpha, t/{}^{1+c}\alpha)$ $(\alpha \in K^*)$, and it determines A with its polarisation up to isomorphism.

Given $g \in \mathrm{Aut}_{F\text{-alg}}(F \otimes \overline{\mathbf{Q}})_0$, let $u(g) \in \Gamma_{\mathbf{Q}}^{ab}$ be as in Proposition 2.2.3(vii). Fix a lift $\widetilde{f} \in \widehat{K}^*$ of $\widetilde{f}_\Phi(g) \in \widehat{K}^*/K^*$ and define $A' = \mathbf{C}^{g\Phi}/a\widetilde{f}$, with polarisation given by $E_{t'}$, where

$$ t' = t\,\chi(u(g))/{}^{1+c}\widetilde{f} \in K^* $$

(t' satisfies $t' \notin F^*$, $t'^2 \in F^*$ and $\forall \varphi' \in g\Phi$, $\quad \mathrm{Im}(\varphi'(t')) < 0$, the last condition by Proposition 2.2.3(vii)).

Given, in addition, a full level structure $\eta : (F/O_F)^2 \xrightarrow{\sim} A(\overline{\mathbf{Q}})_{\mathrm{tors}}$ of A under which the Weil pairing associated to the given polarisation is a $\widehat{\mathbf{Q}}^*$-multiple of the standard form $\mathrm{Tr}_{\widehat{F}/\widehat{\mathbf{Q}}} \circ \det_{\widehat{F}}$ on $\widehat{F}^2$, let η' be the following level structure of A':

$$ \eta' : (F/O_F)^2 \xrightarrow{\eta} A(\mathbf{C})_{\mathrm{tors}} \xrightarrow{\theta^{-1}} K/a \xrightarrow{[\times \widetilde{f}]} K/a\widetilde{f} = A'(\mathbf{C})_{\mathrm{tors}}. $$

The isomorphism class of the triple $(A', E_{t'}, \eta')$ depends only on g and on the isomorphism class $[(A, E_t, \eta)]$ of (A, E_t, η). Proposition 2.2.3 implies that the assignment

$$ {}^g[(A, E_t, \eta)] = [(A', E_{t'}, \eta')] $$

defines an action of $\mathrm{Aut}_{F\text{-alg}}(F \otimes \overline{\mathbf{Q}})_0$ on the isomorphism classes of polarised HBAV (relative to F) with CM, equipped with a full level structure. Moreover, this action commutes with the action of $G(\widehat{F})$ on η (by $\gamma : \eta \mapsto \eta \circ \gamma$), where G is the fibre product

$$ \begin{array}{ccc} G & \longrightarrow & R_{F/\mathbf{Q}}(\mathrm{GL}(2)_F) \\ \downarrow & & \downarrow {\scriptstyle \det} \\ \mathbf{G}_{m,\mathbf{Q}} & \longrightarrow & R_{F/\mathbf{Q}}(\mathbf{G}_{m,F}). \end{array} $$

In view of the results of Tate and Deligne that were recalled in 1.4.3, it follows from Proposition 2.2.3 that the action of $\mathrm{Aut}_{F-\mathrm{alg}}(F \otimes \overline{\mathbf{Q}})_0$ we have just defined extends the usual Galois action of $\Gamma_{\mathbf{Q}}$.

Recall that $\widetilde{f}_{\Phi}(g)$ is defined even for $g \in \mathrm{Aut}_{F-\mathrm{alg}}(F \otimes \overline{\mathbf{Q}})_1$. However, the positivity of polarisations implies that the above recipe makes sense only if the conlusion of Proposition 2.2.3(vii) is satisfied, namely if $g \in \mathrm{Aut}_{F-\mathrm{alg}}(F \otimes \overline{\mathbf{Q}})_0$.

Proposition–Definition 2.2.6 *Fix* $s : X \longrightarrow \Gamma_{\mathbf{Q}}$ *as in 2.1.1; then* $X(K) = \{s(x)c^a|_K \mid x \in X, a \in \mathbf{Z}/2\mathbf{Z}\}$.
(i) *Let* $g \in \mathrm{Aut}_{F\text{-}\mathrm{alg}}(F \otimes \overline{\mathbf{Q}})$; *put* $(\sigma, h) = \beta_{s*}(g) \in S_X \rtimes \Gamma_F^X$. *The formula*

$$^g\left(s(x)c^a|_K\right) := s(\sigma(x))c^{\overline{h}(x)+a}|_K = s(\sigma(x))h(x)c^a|_K$$

defines an action of $\mathrm{Aut}_{F\text{-}\mathrm{alg}}(F \otimes \overline{\mathbf{Q}})$ *on* $X(K)$. *The action of* g *on* $X(K)$ *depends only on the image of* (σ, h) *in* $S_X \rtimes \mathrm{Gal}(K/F)^X$.
(ii) *This action does not depend on the choice of* s.
(iii) *For each CM type* $\Phi \subset X(K)$ *of* K, *the set* $^g\Phi = \{^gy \mid y \in \Phi\}$ *coincides with* $g\Phi$, *defined in* (2.1.5.4).
(iv) *If* $g = \mathrm{id}_F \otimes u$, $u \in \Gamma_{\mathbf{Q}}$, *then* $^gy = u \circ y = uy$, *for each* $y \in X(K)$.

Proof. Easy calculation.

Corollary–Definition 2.2.7 (i) *The induced action of* $\mathrm{Aut}_{F\text{-}\mathrm{alg}}(F \otimes \overline{\mathbf{Q}})$ *on* $X^*(_KT) = \mathbf{Z}[X(K)]$,

$$\lambda = \sum n_y[y] \mapsto {}^g\lambda = \sum n_y[{}^gy] \qquad (g \in \mathrm{Aut}_{F-\mathrm{alg}}(F \otimes \overline{\mathbf{Q}})),$$

extends the action (1.5.1.1) *of* $\Gamma_{\mathbf{Q}}$ *and leaves stable the subgroup* $X^*(_K\mathscr{S})$ *of* $X^*(_KT)$ *spanned by the CM characters* λ_{Φ}.
(ii) *In the special case that* K *is a Galois extension of* $\mathbf{Q}$, *the involution* ι *from* (1.5.4.3) *gives rise to another action of* $\mathrm{Aut}_{F-\mathrm{alg}}(F \otimes \overline{\mathbf{Q}})$ *on* $X^*(_KT)$, *namely*

$$g * \iota(\lambda) = \iota(^g\lambda) \qquad (\lambda \in X^*(_KT)).$$

This action extends the action (1.5.4.1) *of* $\Gamma_{\mathbf{Q}}$ *and leaves stable* $X^*(_K\mathscr{S})$.

Proposition 2.2.8 *Let* $n : \{\text{CM types of } K\} \longrightarrow \mathbf{Z}$ *be a function satisfying*

$$\sum_{\Phi} n_{\Phi}\lambda_{\Phi} = w \cdot N_{K/\mathbf{Q}} = w \sum_{y \in X(K)} [y] \in X^*(_K\mathscr{S}) \qquad (w \in \mathbf{Z}).$$

Then: (i) $\forall g \in \mathrm{Aut}_{F-\mathrm{alg}}(F \otimes \overline{\mathbf{Q}})$, $\prod_{\Phi} \widetilde{F}_{\Phi}(g)^{n_{\Phi}} = \widetilde{V}_{K/F}(g)^w$.
(ii) *If* $w = 0$, *then* $\forall g \in \mathrm{Aut}_{F-\mathrm{alg}}(F \otimes \overline{\mathbf{Q}})_1$, $\prod_{\Phi} \widetilde{f}_{\Phi}(g)^{n_{\Phi}} = 1 \in \widehat{K}^*/K^*$.

Proof. (i) Fix $s : X \longrightarrow \Gamma_{\mathbf{Q}}$ as in 2.1.1, and parameterize the CM types by functions $\alpha : X \longrightarrow \mathbf{Z}/2\mathbf{Z}$, as in (2.1.1.3): we write $\Phi_\alpha = \{s(x)c^{\alpha(x)}|_K\}_{x \in X}$, $n_\alpha = n_{\Phi_\alpha}$ and $\lambda_\alpha = \lambda_{\Phi_\alpha}$. The condition $\sum_\Phi n_\Phi \lambda_\Phi = w \cdot N_{K/\mathbf{Q}}$ is equivalent to

$$\forall x \in X \qquad \sum_\alpha n_\alpha \lambda_\alpha(x) = \sum_\alpha n_\alpha(1 - \lambda_\alpha(x)) = w.$$

Statement (i) follows from the fact that, for each $g = (\sigma, h) \in S_X \ltimes \Gamma_F^X$,

$$\prod_\alpha {}_s\widetilde{F}_\alpha(g)^{n_\alpha} = \prod_{x \in X} \gamma_{x,s}^{\sum_\alpha (n_\alpha \lambda_{g\alpha}(x) - n_\alpha \lambda_\alpha(x))} R(h(x))^{\sum_\alpha n_\alpha(1 - \lambda_\alpha(x))} .$$

$$\times \prod_{x \in X} ({}^c R(h(x)))^{\sum_\alpha n_\alpha \lambda_\alpha(x)}$$

$$= \prod_{x \in X} {}^{1+c}R(h(c))^w = \widetilde{V}_{K/F}(g)^w.$$

If $w = 0$, statement (ii) follows by applying ℓ_K to (i).

2.3 Generalised universal Taniyama elements

As in Section 1.6, we assume that K is a CM number field which is a Galois extension of $\mathbf{Q}$.

Proposition 2.3.1 (i) *There exists a unique map* $\widetilde{f}' : \mathrm{Aut}_{F\text{-alg}}(F \otimes \overline{\mathbf{Q}})_1 \longrightarrow {}_K\mathscr{S}(\widehat{K})/{}_K\mathscr{S}(K)$ *such that* $\lambda_\Phi \circ \widetilde{f}' = \widetilde{f}_\Phi$, *for all CM types Φ of K. The map* $\widetilde{f}'$ *factors through* $\mathrm{Aut}_{F\text{-alg}}(F \otimes K^{ab})_1$.
(ii) *For each* $\lambda \in X^*({}_K\mathscr{S})$, *put* $\widetilde{f}'_\lambda = \lambda \circ \widetilde{f}' : \mathrm{Aut}_{F\text{-alg}}(F \otimes \overline{\mathbf{Q}})_1 \longrightarrow \widehat{K}^*/K^*$; *then* $\widetilde{f}'_{\lambda+\mu}(g) = \widetilde{f}'_\lambda(g)\widetilde{f}'_\mu(g)$.
(iii) $\forall \lambda \in X^*({}_K\mathscr{S})$ $\forall g, g' \in \mathrm{Aut}_{F\text{-alg}}(F \otimes \overline{\mathbf{Q}})_1,$ $\widetilde{f}'_\lambda(gg') = \widetilde{f}'_{g'\lambda}(g)\widetilde{f}'_\lambda(g').$
(iv) $\forall u \in \mathrm{Gal}(K/\mathbf{Q}),$ ${}^u(\widetilde{f}'_\lambda(g)) = \widetilde{f}'_{u*\lambda}([u|_F]g).$
(v) $\forall g \in \Gamma_{\mathbf{Q}},$ $\widetilde{f}'(\mathrm{id}_F \otimes g) = f'(g).$

Proof. Statements (i) and (ii) follow from Proposition 2.2.8(ii) by the same argument as in the proof of Proposition 1.6.2. If $\lambda = \lambda_\Phi$, then (iii) (resp., (iv)) is just the statement of Proposition 2.2.3 (iv) (resp., (v)); the general case follows from (ii). Finally, (v) is a consequence of the uniqueness of f', since

$$\forall \Phi \qquad \lambda_\Phi(\widetilde{f}'(\mathrm{id}_F \otimes g)) = \widetilde{f}_\Phi(\mathrm{id}_F \otimes g) = f_\Phi(g) = \lambda_\Phi(f'(g)),$$

by Proposition 2.2.3(ii).

Proposition 2.3.2 (i) *Let* $\widetilde{f} : \mathrm{Aut}_{F\text{-alg}}(F \otimes \overline{\mathbf{Q}})_1 \longrightarrow {}_K\mathscr{S}(\widehat{K})/{}_K\mathscr{S}(K)$ *be the map defined by the formula* $\widetilde{f}(g) = (\iota(\widetilde{f}'(g)))^{-1}$. *This map factors through* $\mathrm{Aut}_{F\text{-alg}}(F \otimes K^{ab})_1$ *and has the following properties:*
(ii) *The maps* $\widetilde{f}_\lambda = \lambda \circ \widetilde{f} : \mathrm{Aut}_{F\text{-alg}}(F \otimes \overline{\mathbf{Q}})_1 \longrightarrow \widehat{K}^*/K^*$ $(\lambda \in X^*({}_K\mathscr{S}))$ *satisfy*

$$\widetilde{f}_{\lambda+\mu}(g) = \widetilde{f}_\lambda(g)\widetilde{f}_\mu(g), \qquad \widetilde{f}_\lambda(g) = \widetilde{f}'_{\iota(\lambda)}(g)^{-1}, \qquad \widetilde{f}_\lambda(gg') = \widetilde{f}_{g'*\lambda}(g)\widetilde{f}_\lambda(g').$$

(iii) $\forall u \in \mathrm{Gal}(K/\mathbf{Q})$ $\forall g \in \mathrm{Aut}_{F\text{-alg}}(F \otimes \overline{\mathbf{Q}})_1,$ $^u(\widetilde{f}_\lambda(g)) = \widetilde{f}_{u\lambda}([u|_F]g),$
$^u(\widetilde{f}(g)) = \widetilde{f}([u|_F]g).$
(iv) $\forall g, g' \in \mathrm{Aut}_{F\text{-alg}}(F \otimes \overline{\mathbf{Q}})_1,$ $\widetilde{f}(gg') = (g'^{-1} * \widetilde{f}(g))\,\widetilde{f}(g').$
(v) $\forall g \in \Gamma_{\mathbf{Q}},$ $\widetilde{f}(\mathrm{id}_F \otimes g) = f(g).$

Proof. As in the proof of Proposition 1.6.3, everything follows from Proposition 2.3.1.

Proposition 2.3.3 *There exists a lift* $\widetilde{b} : \mathrm{Aut}_{F\text{-alg}}(F \otimes K^{ab})_1 \longrightarrow {}_K\mathscr{S}(\widehat{K})$ *of* $\widetilde{f}$ *whose "coboundary"* $\widetilde{d}_{g,g'} = (g'^{-1} * \widetilde{b}(g))\,\widetilde{b}(g')\,\widetilde{b}(gg')^{-1}$ *is a locally constant function on* $(\mathrm{Aut}_{F\text{-alg}}(F \otimes K^{ab})_1)^2.$

Proof. The argument from the proof of Proposition 1.7.5 applies.

Proposition 2.3.4 *If K' is a CM number field, which is a Galois extension of $\mathbf{Q}$ and contains K, then the generalised universal Taniyama elements $\widetilde{f}_K$:* $\mathrm{Aut}_{F\text{-alg}}(F \otimes \overline{\mathbf{Q}})_1 \longrightarrow {}_K\mathscr{S}(\widehat{K})/{}_K\mathscr{S}(K)$ *and* $\widetilde{f}_{K'} : \mathrm{Aut}_{F'\text{-alg}}(F' \otimes \overline{\mathbf{Q}})_1 \longrightarrow {}_{K'}\mathscr{S}(\widehat{K'})/{}_{K'}\mathscr{S}(K')$ *over K and K', respectively, satisfy*

$$\forall g \in \mathrm{Aut}_{F\text{-alg}}(F \otimes \overline{\mathbf{Q}})_1 \qquad \widetilde{f}_K(g) = N_{K'/K}\left(\widetilde{f}_{K'}(\mathrm{id}_{F'} \otimes_F g)\right).$$

Proof. This follows from Proposition 2.2.4(iii), as in the proof of Proposition 1.6.5.

2.4 Generalised Taniyama group

Let K be as in Section 2.3.

2.4.1

Let us try to apply the method of [MS82, Prop. 2.7] (see 1.7.3 above) to the generalised universal Taniyama element $\widetilde{f}$ and its lift $\widetilde{b}$. The reverse 2-cocycle $\widetilde{d}_{g,g'}$ with values in ${}_K\mathscr{S}(K)$ gives rise to an exact sequence of affine group schemes over K,

$$1 \longrightarrow {}_K\mathscr{S}_K \xrightarrow{\widetilde{i}} \widetilde{G}' \xrightarrow{\widetilde{\pi}} \mathrm{Aut}_{F\text{-alg}}(F \otimes K^{ab})_1 \longrightarrow 1 \qquad (2.4.1.1)$$

(where the term on the right is considered as a constant group scheme), equipped with a section $\widetilde{\alpha} : \mathrm{Aut}_{F\text{-alg}}(F \otimes K^{ab})_1 \longrightarrow \widetilde{G}'(K)$ such that

$$\forall g, g' \in \mathrm{Aut}_{F\text{-alg}}(F \otimes K^{ab})_1, \qquad \widetilde{\alpha}(gg') = \widetilde{\alpha}(g)\widetilde{\alpha}(g')\widetilde{d}_{g,g'}.$$

The map

$$\widetilde{sp} : \mathrm{Aut}_{F\text{-alg}}(F \otimes K^{ab})_1 \longrightarrow \widetilde{G}'(\widehat{K}), \qquad \widetilde{sp}(g) = \widetilde{b}(g)\widetilde{\alpha}(g),$$

is a group homomorphism satisfying $\widetilde{\pi} \circ \widetilde{sp} = \mathrm{id}.$

2.4.2

Each element $u \in \Gamma_K$ acts on $\widetilde{G}'(\overline{\mathbf{Q}})$ by

$$^u(s\,\widetilde{\alpha}(g)) = {}^u s\,\widetilde{\alpha}(g) \qquad (s \in {}_K\mathscr{S}(\overline{\mathbf{Q}})). \qquad (2.4.2.2)$$

We extend this action to an action of $\Gamma_{\mathbf{Q}}$: for $u \in \Gamma_{\mathbf{Q}}$ and $g \in \mathrm{Aut}_{F\text{-alg}}(F \otimes K^{ab})_1$, put

$$\widetilde{c}_u(g) = \widetilde{b}([u|_F]g)\, {}^u(\widetilde{b}(g))^{-1} \in {}_K\mathscr{S}(K).$$

As

$$\forall u, u' \in \Gamma_{\mathbf{Q}} \ \forall g \in \mathrm{Aut}_{F\text{-alg}}(F \otimes K^{ab})_1, \qquad \widetilde{c}_{uu'}(g) = \widetilde{c}_u([u'|_F]g)\, {}^u(\widetilde{c}_{u'}(g)),$$

the formula

$$^u(s\,\widetilde{\alpha}(g)) = \widetilde{c}_u(g)\, {}^u s\,\widetilde{\alpha}(g) \qquad (s \in {}_K\mathscr{S}(\overline{\mathbf{Q}}),\ g \in \mathrm{Aut}_{F-\mathrm{alg}}(F \otimes K^{ab})_1) \qquad (2.4.2.3)$$

defines an action of $\Gamma_{\mathbf{Q}}$ on $\widetilde{G}'(\overline{\mathbf{Q}})$ which extends the action (2.4.2.2) of Γ_K.

We define ${}_K\widetilde{\mathscr{T}}$ to be the affine group scheme over $\mathbf{Q}$ such that ${}_K\widetilde{\mathscr{T}}(\overline{\mathbf{Q}}) = \widetilde{G}'(\overline{\mathbf{Q}})$, with the $\Gamma_{\mathbf{Q}}$-action given by (2.4.2.3). The exact sequence (2.4.1.1) descends to an exact sequence

$$1 \longrightarrow {}_K\mathscr{S} \stackrel{\widetilde{i}}{\longrightarrow} {}_K\widetilde{\mathscr{T}} \stackrel{\widetilde{\pi}}{\longrightarrow} \mathrm{Aut}_{F\text{-alg}}(F \otimes K^{ab})'_1 \longrightarrow 1, \qquad (2.4.2.4)$$

where we have denoted by $\mathrm{Aut}_{F\text{-alg}}(F \otimes K^{ab})'_1$ a twisted form of the constant group scheme $\mathrm{Aut}_{F\text{-alg}}(F \otimes K^{ab})_1$, for which $u \in \Gamma_{\mathbf{Q}}$ acts on

$$\mathrm{Aut}_{F\text{-alg}}(F \otimes K^{ab})'_1(\overline{\mathbf{Q}}) = \mathrm{Aut}_{F\text{-alg}}(F \otimes K^{ab})_1$$

by $[u|_F]$. Note that

$$\mathrm{Aut}_{F\text{-alg}}(F \otimes K^{ab})'_1(\mathbf{Q}) = \mathrm{id}_F \otimes \mathrm{Gal}(K^{ab}/\mathbf{Q}), \qquad (2.4.2.5)$$

by Proposition 1.1.6(iv).

2.4.3

As $\widetilde{f}$ extends f (and the restriction of $\widetilde{b}$ to $\mathrm{Gal}(K^{ab}/\mathbf{Q})^2$ satisfies 1.7.5), there is a commutative diagram of affine group schemes over $\mathbf{Q}$ with exact rows:

$$
\begin{array}{ccccccccc}
1 & \longrightarrow & {}_K\mathscr{S} & \stackrel{i}{\longrightarrow} & {}_K\mathscr{T} & \stackrel{\pi}{\longrightarrow} & \mathrm{Gal}(K^{ab}/\mathbf{Q}) & \longrightarrow & 1 \\
& & \| & & \Big\downarrow & & \Big\downarrow {\scriptstyle (\mathrm{id}_F \otimes -)} & & \\
1 & \longrightarrow & {}_K\mathscr{S} & \stackrel{\widetilde{i}}{\longrightarrow} & {}_K\widetilde{\mathscr{T}} & \stackrel{\widetilde{\pi}}{\longrightarrow} & \mathrm{Aut}_{F-\mathrm{alg}}(F \otimes K^{ab})'_1 & \longrightarrow & 1.
\end{array}
$$

Moreover, there is a commutative diagram of groups

$$
\begin{array}{ccc}
{}_K\mathscr{T}(\widehat{\mathbf{Q}}) & \xleftarrow{\ sp\ } & \mathrm{Gal}(K^{ab}/\mathbf{Q}) \\
\downarrow & & \downarrow{\scriptstyle(\mathrm{id}_F\otimes-)} \\
{}_K\widetilde{\mathscr{T}}(\widehat{K}) & \xleftarrow{\ \widetilde{sp}\ } & \mathrm{Aut}_{F-\mathrm{alg}}(F\otimes K^{ab})_1
\end{array}
$$

such that $\pi\circ sp=\mathrm{id}$, $\widetilde{\pi}\circ\widetilde{sp}=\mathrm{id}$. As

$$
{}^u\widetilde{sp}(g)={}^u(\widetilde{b}(g)\widetilde{\alpha}(g))=\widetilde{c}_u(g)\,{}^u\widetilde{b}(g)\,\widetilde{\alpha}([u|_F]g)=\widetilde{b}([u|_F]g)\,\widetilde{\alpha}([u|_F]g)=\widetilde{sp}([u|_F]g)
$$

for all $u\in\Gamma_{\mathbf{Q}}$ and $g\in\mathrm{Aut}_{F-\mathrm{alg}}(F\otimes K^{ab})_1$, the map $\widetilde{sp}$ is $\Gamma_{\mathbf{Q}}$-equivariant. As $[u|_F]$ depends only on the image of u in $\mathrm{Gal}(F/\mathbf{Q})$, it follows that the image of $\widetilde{sp}$ is contained in ${}_K\widetilde{\mathscr{T}}(\widehat{F})$, and that $\widetilde{sp}$ is $\mathrm{Gal}(F/\mathbf{Q})$-equivariant.

2.4.4

Proposition 2.3.4 implies that the pull-backs of ${}_{K'}\widetilde{\mathscr{T}}$ to $\mathrm{Aut}_{F-\mathrm{alg}}(F\otimes\overline{\mathbf{Q}})'_1$ (for varying $K'\supset K$) give rise to an extension of $\mathrm{Aut}_{F-\mathrm{alg}}(F\otimes\overline{\mathbf{Q}})'_1$ by $\mathscr{S}$. These extensions for varying F are again compatible; they give rise to an extension of affine group schemes over $\mathbf{Q}$,

$$
1\longrightarrow\mathscr{S}\longrightarrow\widetilde{\mathscr{T}}\longrightarrow\varinjlim_F\mathrm{Aut}_{F-\mathrm{alg}}(F\otimes\overline{\mathbf{Q}})'_1\longrightarrow 1,
$$

whose pull-back to $\Gamma_{\mathbf{Q}}$ coincides with (1.7.6.1). The direct limit is taken with respect to the transition maps $\mathrm{id}_{F'}\otimes_F-$ (for $F\subseteq F'$).

2.4.5

It would be of interest to give an "abstract" definition of $\widetilde{\mathscr{T}}$ along the lines of [Del82]. As observed in 2.2.5, it is the group $\mathrm{Aut}_{F-\mathrm{alg}}(F\otimes\overline{\mathbf{Q}})_0$ rather than $\mathrm{Aut}_{F-\mathrm{alg}}(F\otimes\overline{\mathbf{Q}})_1$ which has a geometric significance, which means that one should rather consider the subgroup scheme $\widetilde{\mathscr{T}_0}\subset\widetilde{\mathscr{T}}$ sitting in the exact sequence

$$
1\longrightarrow\mathscr{S}\longrightarrow\widetilde{\mathscr{T}_0}\longrightarrow\varinjlim_F\mathrm{Aut}_{F-\mathrm{alg}}(F\otimes\overline{\mathbf{Q}})'_0\longrightarrow 1.
$$

References

[AT68] E. Artin and J. Tate, *Class Field Theory*. Benjamin, New York–Amsterdam, 1968.

[BL84] J.-L. Brylinski and J.-P. Labesse, Cohomologie d'intersection et fonctions L de certaines variétés de Shimura. *Ann. Sci. de l'E.N.S.*, 17:361–412, 1984.

[Del82] P. DELIGNE, Motifs et groupe de Taniyama. In P. Deligne, J.S. Milne, A. Ogus, and K. Shih, editors, *Hodge Cycles, Motives and Shimura Varieties*, volume 900 of *Lect. Notes in Math.*, pages 261–279, Berlin-New York, 1982. Springer.

[Lan79] R. LANGLANDS, Automorphic Representations, Shimura Varieties, and Motives. Ein Märchen. In A. Borel and W. Casselman, editors, *Automorphic forms, representations and L-functions (Corvallis 1977)*, volume 33/2 of *Proc. Symp. Pure Math.*, pages 205–246, Providence, 1979. A.M.S.

[Lan83] S. LANG, *Complex Multiplication*, volume 255 of *Grund. math. Wiss.* Springer, New York, 1983.

[Mil90] J.S. MILNE, Canonical models of (mixed) Shimura varieties and automorphic vector bundles. In L. Clozel and J.S. Milne, editors, *Automorphic forms, Shimura varieties and L-functions, Vol. I*, volume 10 of *Perspect. in Math.*, pages 283–414, Boston, 1990. Academic Press.

[MS82] J.S. MILNE AND K. SHIH, Langlands's construction of the Taniyama group. In P. Deligne, J.S. Milne, A. Ogus, and K. Shih, editors, *Hodge Cycles, Motives and Shimura Varieties*, volume 900 of *Lect. Notes in Math.*, pages 229–260, Berlin-New York, 1982. Springer.

[Sch94] N. SCHAPPACHER, CM motives and the Taniyama Group. In U. Jannsen, S. Kleiman, and J.-P. Serre, editors, *Motives (Seattle, 1991)*, volume 55/1 of *Proc. Symp. Pure Math.*, pages 485–508, Providence, 1994. A.M.S.

[Tat81] J. TATE, On conjugation of abelian varieties of CM type. Unpublished manuscript, 1981.

Self-Correspondences of K3 Surfaces via Moduli of Sheaves

Viacheslav V. Nikulin

Department of Pure Mathematics, The University of Liverpool, Liverpool,
L69 3BX, UK; Steklov Mathematical Institute, ul. Gubkina 8,
Moscow 117966, GSP-1, Russia
vnikulin@liv.ac.uk, vvnikulin@list.ru

To Yuri Ivanovich Manin for his 70th birthday

Summary. Let X be an algebraic K3 surface with Picard lattice $N(X)$, and $M_X(v)$ the moduli space of sheaves on X with given primitive isotropic Mukai vector $v = (r, H, s)$. In [14] and [3], we described all the divisors in moduli of polarized K3 surfaces (X, H) (that is, all pairs $H \in N(X)$ with rank $N(X) = 2$) for which $M_X(v) \cong X$. These provide certain Mukai self-correspondences of X.

Applying these results, we show that there exists a Mukai vector v and a codimension-2 subspace in moduli of (X, H) (that is, a pair $H \in N(X)$ with rank $N(X) = 3$) for which $M_X(v) \cong X$, but such that this subspace does not extend to a divisor in moduli having the same property. There are many similar examples.

Aiming to generalize the results of [14] and [3], we discuss the general problem of describing all subspaces of moduli of K3 surfaces with this property, and the Mukai self-correspondences defined by these and their composites, in an attempt to outline a possible general theory.

Key words: K3 surfaces, moduli of sheaves, moduli of vector bundles, algebraic cycles, correspondences, integral quadratic forms

2000 Mathematics Subject Classifications: 14D20, 14J28, 14J60, 14J10, 14C25, 11E12

1 Introduction

We consider algebraic K3 surfaces X over $\mathbb{C}$; recall that a nonsingular projective algebraic (or compact Kähler) surface X is a K3 surface if its canonical class K_X is zero and its irregularity $q = \dim \Omega^1[X]$ is 0. We write $N(X)$ for the Picard lattice of X, $\rho(X) = \operatorname{rank} N(X)$ for its rank, and $T(X)$ for the transcendental lattice.

Y. Tschinkel and Y. Zarhin (eds.), *Algebra, Arithmetic, and Geometry*,
Progress in Mathematics 270, DOI 10.1007/978-0-8176-4747-6_14,
© Springer Science+Business Media, LLC 2009

Consider a primitive isotropic Mukai vector on X,

$$v = (r, l, s), \quad \text{with } r \in \mathbb{N}, \ s \in \mathbb{Z} \text{ and } l \in N(X) \text{ such that } l^2 = 2rs, \qquad (1)$$

and denote by $Y = M_X(v) = M_X(r, l, s)$ the K3 surface obtained as the minimal resolution of singularities of the moduli space of sheaves on X with Mukai vector v. For details, see Mukai [4]–[7] and Yoshioka [19]. Under these assumptions, by results of Mukai [5], the quasi-universal sheaf on $X \times Y$ and its Chern class defines a 2-dimensional algebraic cycle on $X \times Y$ and a correspondence between X and Y with nice geometric properties. For more details, see Section 5.

If $Y \cong X$, this provides an important 2-dimensional algebraic cycle on $X \times X$, and a correspondence from X to itself; the question of when $Y \cong X$ is thus very interesting. The answer when $\rho(X) = 1$, probably already known to specialists, is given in Section 2.

Tensoring by any $D \in N(X)$ gives a natural isomorphism

$$T_D \colon M_X(r, l, s) \cong M_X(r, l + rD, s + \tfrac{1}{2}rD^2 + D \cdot l)$$
$$\text{defined by } \mathcal{E} \mapsto \mathcal{E} \otimes \mathcal{O}(D).$$

For $r, s > 0$, we have an isomorphism called *reflection*,

$$\delta \colon M_X(r, l, s) \cong M_X(s, l, r);$$

see for example [5] and [16], [17], [20]. For integers $d_1, d_2 > 0$ with $(d_1, d_2) = (d_1, s) = (r, d_2) = 1$, we have an isomorphism

$$\nu(d_1, d_2) \colon M_X(r, l, s) \cong M_X(d_1^2 r, d_1 d_2 l, d_2^2 s)$$

and its inverse $\nu(d_1, d_2)^{-1}$; see [5], [6], [14], [3].

In Theorem 2.1 and Corollary 2.2, we show that if $\rho(X) = 1$ and X is general, then for two primitive isotropic Mukai vectors v_1 and v_2, the moduli spaces $M_X(v_1)$ and $M_X(v_2)$ are isomorphic if and only if there exists an isomorphism between them obtained by composing the above three natural isomorphisms. They give *universal isomorphisms* between moduli of sheaves on X.

For $l \in N(X)$ with $\pm l^2 > 0$, it is known that we have a *Tyurin isomorphism* (see for example Tyurin [17])

$$\text{Tyu} = \text{Tyu}(\pm l) \colon M_X(\pm l^2/2, l, \pm 1) \cong X. \qquad (2)$$

Corollary 2.6 shows that if $\rho(X) = 1$, then $M_X(r, H, s)$ and X are isomorphic if and only if there exists an isomorphism between them that is a composite of the above three universal isomorphisms between moduli of sheaves, and a Tyurin isomorphism (see also Remark 3.4). Compare [14] for a similar result.

We showed in [14] and interpreted geometrically in [3] (together with Carlo Madonna) that analogous results hold if $\rho(X) = 2$ and X is general with its Picard lattice, i.e., the automorphism group of the transcendental periods is trivial: $\text{Aut}\big(T(X), H^{2,0}(X)\big) = \pm 1$. (See also [1], [2], [13] about important particular cases of these results.) We review these results in Section 3;

see Theorems 3.1, 3.2, and 3.3 for precise statements. These results show that in this case (i.e., when $\rho(X) = 2$ and X is general with its Picard lattice), $M_X(r, H, s) \cong X$ if and only if there exists an isomorphism between $M_X(r, H, s)$ and X that is a composite of the universal isomorphisms T_D, δ, and $\nu(d_1, d_2)$ between moduli of sheaves on X and a Tyurin isomorphism between moduli spaces of sheaves on X and X itself. The above results for $\rho(X) = 1$ clarify the appearance of the natural isomorphisms T_D, δ, $\nu(d_1, d_2)$, Tyu in these results for Picard number 2.

The importance of the results for $\rho(X) = 2$ and general X is that they describe all the divisorial conditions on moduli of algebraic polarized K3 surfaces (X, H) that imply $M_X(r, H, s) \cong X$. More exactly, the results for $\rho(X) = 2$ describe all abstract polarized Picard lattices $H \in N$ with rank $N = 2$ such that $H \in N \subset N(X)$ and $N \subset N(X)$ is primitive implies $M_X(r, H, s) \cong X$. Recall that such X have codimension 1 in the 19-dimensional moduli of polarized K3 surfaces. Applying these results, we give in Theorems 3.6 and 3.8 a necessary condition on a Mukai vector (r, H, s) and a polarized K3 surface X in order for the isomorphism $M_X(r, H, s) \cong X$ to follow from a divisorial condition on the moduli of X. In Example 3.7, we give an exact numerical example when this necessary condition is not satisfied. Thus for the K3 surfaces X in this example, the isomorphism $M_X(r, H, s) \cong X$ is not a consequence of any divisorial condition on moduli of polarized K3 surfaces. In other words, $M_X(r, H, s) \cong X$, but this isomorphism cannot be deduced from any divisorial condition on K3 surfaces X' implying $M_{X'}(r, H, s) \cong X'$.

Applying these results, in Section 4, Theorem 4.1, we give an exact example of a type of primitive isotropic Mukai vector (r, H, s) and a pair $H \in N$ of an (abstract) polarized K3 Picard lattice with rank $N = 3$ such that for any polarized K3 surface (X, H) with $H \in N \subset N(X)$ and primitive $N \subset N(X)$ one has $M_X(r, H, s) \cong X$, but this isomorphism does not follow from any divisorial condition (i.e., from Picard number 2) on the moduli of polarized K3 surfaces. Thus these polarized K3 surfaces have codimension 2 in moduli, and they cannot be extended to a divisor in moduli of polarized K3 surfaces preserving the isomorphism $M_X(r, H, s) \cong X$. *This is the main result of this paper.* Section 4 gives many similar examples for Picard numbers $\rho(X) \geq 3$.

These results give important corollaries for higher Picard numbers $\rho(X) \geq 3$ of the above results for Picard numbers 1 and 2; they also show that the case $\rho(X) \geq 3$ is very nontrivial. These are the main subjects of this paper. Another important aim is to formulate some general concepts, and predict the general structure of possible results for higher Picard numbers $\rho(X) \geq 3$.

At the end of Section 4, for a type (r, H, s) of primitive isotropic Mukai vector, we introduce a notion of *critical polarized K3 Picard lattice $H \in N$* (critical for the problem of K3 self-correspondences). Roughly speaking, it means that $M_X(r, H, s) \cong X$ for any polarized K3 surface X with $H \in N \subset N(X)$ where $N \subset N(X)$ is primitive, but the same does not hold for any primitive strict sublattice $H \in N_1 \subset N$. Thus the corresponding moduli space of K3 surfaces has dimension $20 - \operatorname{rank} N$, and is not a specialization of higher-dimensional moduli spaces of K3 surfaces.

The classification of critical polarized K3 Picard lattices is the main problem of self-correspondences of a K3 surface via moduli of sheaves. Our results for $\rho = 1$ and $\rho = 2$ can be interpreted as a classification of all critical polarized K3 Picard lattices of ranks one and two. The example of Theorem 4.1 mentioned above gives an example of a rank-3 critical polarized K3 Picard lattice N. In Theorem 4.10 we prove that a critical polarized K3 Picard lattice N has $\operatorname{rank} N \leq 12$. In Problem 4.11, we raise the problem of the exact bound for the rank of a critical polarized K3 Picard lattice for a fixed type of primitive isotropic Mukai vector. This problem is now solved only for very special types: we know all primitive isotropic Mukai vectors when the exact bound is one.

In Section 5, we interpret the above results in terms of isometric actions of correspondences and their composites on $H^2(X, \mathbb{Q})$. For example, the Tyurin isomorphisms of (2) give reflections in elements $l \in N(X)$, and generate the full automorphism group $O(N(X) \otimes \mathbb{Q})$. Every isotropic primitive Mukai vector (r, H, s) on X with $M_X(r, H, s) \cong X$ then generates some class of isometries in $O(N(X) \otimes \mathbb{Q})$. See Section 5 for exact statements. Thus the main problem of self-correspondences of X via moduli of sheaves is to find all these generators and the relations between them. In this connection, we state problems (1–4) at the end of Section 5; these show that in principle, the general results for any $\rho(X)$ should look similar to the now known results for $\rho(X) = 1, 2$.

Our general idea should be clear: for a K3 surface X that is general for its Picard lattice, the very complicated structure of self-correspondences of X via moduli of sheaves is hidden inside the abstract lattice $N(X)$; we try to recover this structure. This should lead to some nontrivial constructions involving the abstract Picard lattice $N(X)$, and should relate it more closely to the geometry of the K3 surface.

Acknowledgments. I am grateful to D.O. Orlov for useful discussions. I also would like to thank the referees for many helpful suggestions. This work was supported by EPSRC grant EP/D061997/1.

1.1 Preliminary Notation for Lattices

We use the notation and terminology of [10] for lattices, and their discriminant groups and forms. A *lattice* L is a nondegenerate integral symmetric bilinear form. That is, L is a free $\mathbb{Z}$-module of finite rank with a symmetric pairing $x \cdot y \in \mathbb{Z}$ for $x, y \in L$, assumed to be nondegenerate. We write $x^2 = x \cdot x$. The *signature* of L is the signature of the corresponding real form $L \otimes \mathbb{R}$. The lattice L is called *even* if x^2 is even for any $x \in L$. Otherwise, L is called *odd*. The *determinant* of L is defined to be $\det L = \det(e_i \cdot e_j)$, where $\{e_i\}$ is some basis of L. The lattice L is *unimodular* if $\det L = \pm 1$. The *dual* *lattice* of L is $L^* = \operatorname{Hom}(L, \mathbb{Z}) \subset L \otimes \mathbb{Q}$. The *discriminant group* of L is $A_L = L^*/L$; it has order $|\det L|$, and is equipped with a *discriminant bilinear form* $b_L \colon A_L \times A_L \to \mathbb{Q}/\mathbb{Z}$ and, if L is even, with a *discriminant quadratic form* $q_L \colon A_L \to \mathbb{Q}/2\mathbb{Z}$. To define these, we extend the form on L to a form on the dual lattice L^* with values in $\mathbb{Q}$.

An embedding $M \subset L$ of lattices is called *primitive* if L/M has no torsion. Similarly, a nonzero element $x \in L$ is called *primitive* if $\mathbb{Z}x \subset L$ is a primitive sublattice.

2 Isomorphisms Between $M_X(v)$ and X for a General K3 Surface X and a Primitive Isotropic Mukai Vector v

We consider algebraic K3 surfaces X over $\mathbb{C}$. Further, $N(X)$ denotes the Picard lattice of X, and $T(X)$ its transcendental lattice. We consider primitive isotropic Mukai vectors (1) on X. We denote by $Y = M_X(v) = M_X(r, l, s)$ the K3 surface obtained as the minimal resolution of singularities of the moduli space of sheaves on X with Mukai vector v. Compare Mukai [4]–[7] and Yoshioka [19].

In this section, we say that an algebraic K3 surface is *general* if its Picard number $\rho(X) = \operatorname{rank} N(X)$ is 1 and the automorphism group of the transcendental periods of X is trivial over $\mathbb{Q}$: $\operatorname{Aut}\big(T(X) \otimes \mathbb{Q}, \ H^{2,0}(X)\big) = \pm 1$.

We now consider the following question: for a general algebraic K3 surface X and two primitive isotropic Mukai vectors $v_1 = (r_1, l_1, s_1)$ and $v_2 = (r_2, l_2, s_2)$, when are the moduli spaces $M_X(v_1)$ and $M_X(v_2)$ isomorphic?

We have the following three *universal isomorphisms* between moduli spaces of sheaves over a K3 surface. (Here universal means that they are valid for all algebraic K3 surfaces.)

Let $D \in N(X)$. Then one has the natural isomorphism given by the tensor product

$$T_D \colon M_X(r, l, s) \cong M_X(r, l + rD, s + r(D^2/2) + D \cdot l), \quad \mathcal{E} \mapsto \mathcal{E} \otimes \mathcal{O}(D).$$

Moreover, here the Mukai vectors

$$v = (r, l, s) \quad \text{and} \quad T_D(v) = (r, l + rD, s + r(D^2/2) + D \cdot l)$$

have the same general common divisor and the same square under the Mukai pairing. In particular, one is primitive and isotropic if and only if the other is.

Taking $D = kH$ for H a hyperplane section and $k > 0$, using the isomorphisms T_D, we can always replace $M_X(r, l, s)$ by an isomorphic $M_X(r, l', s')$ where l' is ample, so that $l'^2 > 0$. Thus in our problem, we can also assume that $v = (r, l, s)$, where $r > 0$ and l is ample. Then $l^2 = 2rs > 0$ and $r, s > 0$.

For r, $s > 0$, one has an isomorphism called *reflection*:

$$\delta \colon M_X(r,l,s) \cong M_X(s,l,r).$$

See, for example, [5] and [16], [17], [20]. Thus using reflection, we can also assume that $0 < r \le s$.

For integers $d_1 > 0$, $d_2 > 0$ such that $(d_1, d_2) = (d_1, s) = (r, d_2) = 1$, one has an isomorphism

$$\nu(d_1, d_2) \colon M_X(r,l,s) \cong M_X(d_1^2 r, d_1 d_2 l, d_2^2 s)$$

and its inverse $\nu(d_1, d_2)^{-1}$; see [5], [6], [14], [3]. Using the isomorphisms $\nu(d_1, d_2)$, $\nu(d_1, d_2)^{-1}$ and reflection δ, we can always assume that the primitive isotropic Mukai vector $v = (r, l, s)$ satisfies

$$0 < r \le s, \quad l^2 = 2rs, \quad \text{and} \quad l \in N(X) \text{ is primitive and ample.} \tag{3}$$

We call such a primitive isotropic Mukai vector a *reduced primitive isotropic Mukai vector* (for $\rho(X) = 1$).

We have the following result.

Theorem 2.1. *Let X be a general algebraic K3 surface, i.e., $N(X) = \mathbb{Z}H$, where H is a primitive polarization of X and $\mathrm{Aut}\big(T(X) \otimes \mathbb{Q}, H^{2,0}(X)\big) = \pm 1$. Let $v = (r, H, s)$ and $v' = (r', H, s')$ be two reduced primitive isotropic Mukai vectors on X (see (3)), i.e., $0 < r \le s$ and $0 < r' \le s'$.*

Then $M_X(v) \cong M_X(v')$ if and only if $v = v'$, i.e., $r' = r$, $s' = s$.

It follows that the above universal isomorphisms T_D, δ, and $\nu(d_1, d_2)$ are sufficient to find all the isomorphic moduli spaces of sheaves with primitive isotropic Mukai vectors for a general K3 surface.

Corollary 2.2. *Let X be a general algebraic K3 surface and v, v' primitive isotropic Mukai vectors on X. Then $M_X(v) \cong M_X(v')$ if and only if there exists an isomorphism between $M_X(v)$ and $M_X(v')$ that is a composite of the universal isomorphisms T_D, δ and $\nu(d_1, d_2)$.*

Proof. The following considerations are similar to the more general and difficult calculations of [14], Section 2.3. We have

$$N(X) = \mathbb{Z}H = \big\{ x \in H^2(X, \mathbb{Z}) \mid x \cdot H^{2,0}(X) = 0 \big\},$$

and the transcendental lattice of X is

$$T(X) = N(X)^{\perp}_{H^2(X,\mathbb{Z})}.$$

The lattices $N(X)$ and $T(X)$ are orthogonal complements to one another in the unimodular lattice $H^2(X, \mathbb{Z})$, and $N(X) \oplus T(X) \subset H^2(X, \mathbb{Z})$ is a sublattice of finite index; here and in what follows $\oplus$ denotes the orthogonal sum. Since $H^2(X, \mathbb{Z})$ is unimodular and $N(X) = \mathbb{Z}H$ a primitive sublattice, there exists $u \in H^2(X, \mathbb{Z})$ such that $u \cdot H = 1$.

We denote the dual lattices by $N(X)^* = \mathbb{Z} \cdot \frac{1}{2rs}H \subset N(X) \otimes \mathbb{Q}$ and $T(X)^* \subset T(X) \otimes \mathbb{Q}$. Then $H^2(X, \mathbb{Z}) \subset N(X)^* \oplus T(X)^*$, and

$$u = \frac{1}{2rs}H + t^*(H) \quad \text{with} \quad t^*(H) \in T(X)^*.$$

The element

$$t^*(H) \mod T(X) \in T(X)^*/T(X) \cong \mathbb{Z}/2rs\mathbb{Z}$$

is canonically defined by the primitive element $H \in H^2(X, \mathbb{Z})$. Obviously,

$$H^2(X, \mathbb{Z}) = \left[N(X), T(X), u = \tfrac{1}{2rs}H + t^*(H)\right],$$

where $[\cdot]$ means "generated by." The element $t^*(H) \mod T(X)$ distinguishes the different polarized K3 surfaces with Picard number one and the same transcendental periods; more precisely, for another polarized K3 surface (X', H') with transcendental periods $\left(T(X'), H^{2,0}(X')\right)$, the periods of X and X' are isomorphic (and then $X \cong X'$ by the global Torelli theorem [15]) if and only if there exists an isomorphism of transcendental lattices $\phi \colon T(X) \cong T(X')$ such that $(\phi \otimes \mathbb{C})(H^{2,0}(X)) = H^{2,0}(X')$ and

$$(\phi \otimes \mathbb{Q})(t^*(H)) \mod T(X) = t^*(H') \mod T(X').$$

Thus the calculation of the periods of X in terms its transcendental periods is contained in the following statement.

Proposition 2.3. *Let (X, H) be a polarized K3 surface with a primitive polarization H such that $H^2 = 2rs$. Assume that $N(X) = \mathbb{Z}H$ (i.e., $\rho(X) = 1$). Then*

$$H^2(X, \mathbb{Z}) = \left[N(X) = \mathbb{Z}H, T(X), \tfrac{1}{2rs}H + t^*(H)\right],$$

where $t^*(H) \in T(X)^*$. The element $t^*(H) \mod T(X)$ is uniquely defined.

Moreover, $H^{2,0}(X) \subset T(X) \otimes \mathbb{C}$. (More generally, for $\rho(X) \geq 1$, one should replace $T(X)$ by $H^{\perp}_{H^2(X, \mathbb{Z})}$.)

Let $Y = M_X(r, H, s)$. We calculate the periods of Y. The Mukai lattice of X is defined by

$$\widetilde{H}(X, \mathbb{Z}) = H^0(X, \mathbb{Z}) + H^2(X, \mathbb{Z}) + H^4(X, \mathbb{Z}) = U \oplus H^2(X, \mathbb{Z}),$$

where $+$ is direct sum, and $\oplus$ the orthogonal direct sum of lattices. Here $H^2(X, \mathbb{Z})$ is the cohomology lattice of X with its intersection pairing and $U = \mathbb{Z}e_1 + \mathbb{Z}e_2$ is the hyperbolic plane, where canonically $\mathbb{Z}e_1 = H^0(X, \mathbb{Z})$ and $\mathbb{Z}e_2 = H^4(X, \mathbb{Z})$ with the Mukai pairing $e_1^2 = e_2^2 = 0$ and $e_1 \cdot e_2 = -1$.

We have

$$v = re_1 + se_2 + H. \tag{4}$$

By Mukai [5], we have

$$H^2(Y, \mathbb{Z}) = v^\perp / \mathbb{Z}v, \tag{5}$$

and $H^{2,0}(Y) = H^{2,0}(X)$ by the canonical projection. This determines the periods of the K3 surface Y and its isomorphism class (by the global Torelli theorem [15]). We calculate the periods of Y as in Proposition 2.3.

Any element f of $\widetilde{H}(X, \mathbb{Z})$ can be written in a unique way as

$$f = xe_1 + ye_2 + \alpha \tfrac{1}{2rs} H + \beta t^*, \quad \text{with } x, y, \alpha \in \mathbb{Z} \text{ and } t^* \in T(X)^*.$$

We have $f \cdot v = -sx - ry + \alpha$, so $f \in v^\perp$ if and only if $-sx - ry + \alpha = 0$, and then

$$f = xe_1 + ye_2 + (sx + ry) \tfrac{1}{2rs} H + \beta t^*.$$

By Proposition 2.3, $f \in \widetilde{H}(X, \mathbb{Z})$ if and only if $t^* = (sx + ry) t^*(H) \bmod T(X)$. Since $T(X) \subset v^\perp$, we can write

$$f = xe_1 + ye_2 + (sx + ry) \left(\tfrac{1}{2rs} H + t^*(H) \right) \bmod T(X) \quad \text{with } x, y \in \mathbb{Z}.$$

Set

$$c = (r, s), \quad a = r/c, \quad b = s/c.$$

Then $(a, b) = 1$. We have $h = -ae_1 + be_2 \in v^\perp$ and $h^2 = 2ab = 2rs/c^2$. Moreover, $h \perp T(X)$ and then $h \perp H^{2,0}(X)$. Thus

$$h \bmod \mathbb{Z}v = -ae_1 + be_2 \bmod \mathbb{Z}v \tag{6}$$

gives an element of the Picard lattice $N(Y)$. We have

$$e_1 = \frac{v - ch - H}{2r}, \quad e_2 = \frac{v + ch - H}{2s}.$$

It follows that

$$f = \frac{sx + ry}{2rs} v + \frac{c(-sx + ry)}{2rs} h + (sx + ry) t^*(H) \bmod T(X), \tag{7}$$

for some $x, y \in \mathbb{Z}$. Here $f \bmod \mathbb{Z}v$ gives all the elements of $H^2(Y, \mathbb{Z})$, and $H^{2,0}(Y) = H^{2,0}(X) \subset T(X) \otimes \mathbb{C}$.

It follows that $f \bmod \mathbb{Z}v \in T(Y)$ (where $\mathbb{Z}v$ gives the kernel of $v^\perp$ and $H^2(Y, \mathbb{Z}) = v^\perp / \mathbb{Z}v$) if and only if $-sx + ry = 0$. Equivalently, $-bx + ay = 0$, or (since $(a, b) = 1$) $x = az$, $y = bz$, where $z \in \mathbb{Z}$, and then

$$(sx + ry) t^*(H) = z(sa + rb) t^*(H) = z \, 2abc \, t^*(H) \quad \text{for some } z \in \mathbb{Z}.$$

It follows that

$$T(Y) = [T(X), 2abc \, t^*(H)]. \tag{8}$$

Since $t^*(H) \bmod T(X)$ has order $2rs = 2abc^2$ in $T(X)^* / T(X) \cong \mathbb{Z}/2rs\mathbb{Z}$, it follows that $[T(Y): T(X)] = c$ (this is a result of Mukai [5]).

By (7) and (8), we have $f \perp H^{2,0}(Y) = H^{2,0}(X)$, that is, $f \mod \mathbb{Z}v \in N(Y)$, if and only if

$$f = \frac{sx + ry}{2rs}v + \frac{c(-sx + ry)}{2rs}h,$$

where $sx + ry \equiv 0 \mod 2abc$. Thus $acx + bcy \equiv 0 \mod 2abc$ and $ax + by \equiv 0 \mod 2ab$. Since $(a, b) = 1$, it follows that $x = b\tilde{x}$, $y = a\tilde{y}$, where $\tilde{x}$, $\tilde{y} \in \mathbb{Z}$, and $\tilde{x} + \tilde{y} \equiv 0 \mod 2$. Thus $\tilde{y} = -\tilde{x} + 2k$, where $k \in \mathbb{Z}$. It follows that

$$f = \frac{k}{c}v + (-\tilde{x} + k)h, \quad \text{for some } \tilde{x}, k \in \mathbb{Z}.$$

Thus $h \mod \mathbb{Z}v$ generates the Picard lattice $N(Y)$, and we can consider $h \mod \mathbb{Z}v$ as the polarization of Y (or $-h \mod \mathbb{Z}v$, which makes no difference from the point of view of periods and isomorphism class of Y).

Let us calculate $t^*(h) \in T(Y)^*$. Then in (7) we should take an element f with $c(-sx + ry)/(2rs) = 1/(2ab)$. Thus $-sx + ry = c$ or $-bx + ay = 1$. Then

$$t^*(h) = (sx + ry)t^*(H) \mod T(Y).$$

By (8), $T(Y)^* = [T(X), ct^*(H)]$ and $T(Y)^*/T(Y) \cong \mathbb{Z}/2ab\mathbb{Z}$.

Thus $t^*(h) = (bx + ay)(ct^*(H) \mod [T(X), 2ab(ct^*(H))])$ is defined by $m \equiv bx + ay \mod 2ab$. Since $-bx + ay = 1$, we have $m \equiv 2ay - 1 \equiv -1 \mod 2a$ and $m \equiv 2bx + 1 \equiv 1 \mod 2b$. This defines $m \mod 2ab$ uniquely. We call such $m \mod 2ab$ a Mukai element (compare with [6]). Thus $m(a, b) \mod 2ab$ is called a *Mukai element* if

$$m(a, b) \equiv -1 \mod 2a \quad \text{and} \quad m(a, b) \equiv 1 \mod 2b. \tag{9}$$

Thus $t^*(h) = m(a, b) ct^*(H) \mod [T(X), 2abc t^*(H)]$.

Thus we have finally completed the calculation of the periods of Y in terms of those of X (see Proposition 2.3).

Proposition 2.4. *Let (X, H) be a polarized K3 surface with a primitive polarization H such that $H^2 = 2rs$ with $r, s > 0$. Assume that $N(X) = \mathbb{Z}H$ (i.e., $\rho(X) = 1$). Let $Y = M_X(r, H, s)$ and set $c = (r, s)$ and $a = r/c$, $b = s/c$. Then $N(Y) = \mathbb{Z}h$, where $h^2 = 2ab$,*

$$T(Y) = [T(X), 2abc t^*(H)], \quad T(Y)^* = [T(X), ct^*(H)],$$

and $t^*(h) \mod T(Y) = m(a, b)ct^*(H) \mod T(Y)$, where $m(a, b) \mod 2ab$ is the Mukai element: $m(a, b) \equiv -1 \mod 2a$, $m(a, b) \equiv 1 \mod 2b$. Thus

$$H^2(Y, \mathbb{Z}) = \left[N(Y), T(Y), \tfrac{1}{2ab}h + t^*(h)\right]$$

$$= \left[\mathbb{Z}h, [T(X), 2abc t^*(H)], \tfrac{1}{2ab}h + m(a, b)ct^*(H)\right].$$

(More generally, when $\rho(X) \geq 1$, one should replace $T(X)$ by $H^\perp_{H^2(X, \mathbb{Z})}$ and $T(Y)$ by $h^\perp_{H^2(Y, \mathbb{Z})}$.)

Now let us prove Theorem 2.1. We need to recover r and s from the periods of Y. By Proposition 2.4, we have $N(Y) = \mathbb{Z}h$, where $h^2 = 2ab$. Thus we recover ab. Since $c^2 = 2rs/2ab$, we recover c.

We have $(T(X) \otimes \mathbb{Q}, H^{2,0}(X)) \cong (T(Y) \otimes \mathbb{Q}, H^{2,0}(Y))$. Since X is general, there exists only one such isomorphism up to multiplication by ± 1. It follows that (up to multiplication by ± 1) there exists only one embedding $T(X) \subset T(Y)$ of lattices that identifies $H^{2,0}(X)$ and $H^{2,0}(Y)$. By Proposition 2.4, then $t^*(h) \mod T(Y) = \tilde{m}(a,b)ct^*(H) \mod T(Y)$, where $\tilde{m}(a,b) \equiv \pm m(a,b)$ $\mod 2ab$ and $m(a,b)$ is the Mukai element. Assume $p^\alpha \mid ab$ and $p^{\alpha+1}$ does not divide ab, where p is prime and $\alpha > 0$. Then $\tilde{m}(a,b) \equiv \pm 1 \mod 2p^\alpha$. Clearly, only one sign ± 1 is possible here; we denote by a the product of all the p^α having $\tilde{m}(a,b) \equiv -1 \mod 2p^\alpha$, and by b the product of all the other p^α having $\tilde{m}(a,b) \equiv 1 \mod 2p^\alpha$. If $a > b$, we must exchange a and b. Thus we recover a and b and the reduced primitive Mukai vector $(r, H, s) = (ac, H, bc)$ such that the periods of $M_X(r, H, s)$ are isomorphic to the periods of Y.

This completes the proof.   □

Remark 2.5. Propositions 2.3 and 2.4 and their proofs remain valid for any algebraic K3 surface X and a primitive element $H \in N(X)$ with $H^2 = 2rs \neq 0$, provided we replace $T(X)$ by the orthogonal complement $H^{\perp}_{H^2(X,\mathbb{Z})}$.

As an example of an application of Theorem 2.1, let us consider the case that $M_X(r, l, s) \cong X$. It is known (see for example [17]) that for $l \in N(X)$ and $\pm l^2 > 0$, one has the Tyurin isomorphism

$$\text{Tyu} = \text{Tyu}(\pm l) : M_X(\pm l^2/2, l, \pm 1) \cong X . \tag{10}$$

The existence of such an isomorphism follows at once from the global Torelli theorem for K3 surfaces [15] using Propositions 2.3, 2.4 and Remark 2.5.

Thus for a general K3 surface X and a primitive isotropic Mukai vector $v = (r, H, 1)$, where $r = H^2/2$, we have $M_X(r, H, 1) \cong X$. By Theorem 2.1, we then obtain the following result, where we also use the well-known fact that $\text{Aut}(T(X), H^{2,0}(X)) = \pm 1$ if $\rho(X) = 1$ (see (33) below); it is sufficient to consider the automorphism group over $\mathbb{Z}$ for this result.

Corollary 2.6. *Let X be an algebraic K3 surface with $\rho(X) = 1$, i.e., $N(X) = \mathbb{Z}H$, where H is a primitive polarization of X. Let $v = (r, H, s)$ be a reduced primitive isotropic Mukai vector on X (see (3)), i.e., $0 < r \le s$. Then $M_X(v) \cong X$ if and only if $v = (1, H, H^2/2)$, i.e., $r = 1$, $s = H^2/2$.*

3 Isomorphisms Between $M_X(v)$ and X for X a General K3 Surface with $\rho(X) = 2$

We now consider general K3 surfaces X with $\rho(X) = \text{rank} N(X) = 2$; here a *K3 surface X is called general with its Picard lattice* if the transcendental periods have trivial automorphism group, $\text{Aut}(T(X), H^{2,0}(X)) = \pm 1$.

For $\rho(X) \geq 2$, we do not know when $M_X(v_1) \cong M_X(v_2)$ for primitive isotropic Mukai vectors v_1 and v_2 on X. But we still have the universal isomorphisms T_D, $D \in N(X)$, the reflection δ, the isomorphism $\nu(d_1, d_2)$, and the Tyurin isomorphism Tyu considered in Section 2. They are *universal isomorphisms*, i.e., they are defined for all K3 surfaces.

We start by reviewing the results of [14] and [3], where we found all the primitive isotropic Mukai vectors v with $M_X(v) \cong X$ for general K3 surfaces X with $\rho(X) = 2$. In particular, we know when $M_X(v_1) \cong M_X(v_2)$ in the case that both moduli spaces are isomorphic to X. The result is that $M_X(v) \cong X$ if and only if there exists such an isomorphism that is a composite of the universal isomorphisms δ, T_D, and $\nu(d_1, d_2)$ between moduli of sheaves over X and the Tyurin isomorphism Tyu between moduli of sheaves over X and X itself. More exactly, the results are as follows.

Using the universal isomorphisms T_D, we can assume that the primitive isotropic Mukai vector is

$$v = (r, H, s), \quad \text{with } r > 0,\ s > 0 \text{ and } H^2 = 2rs.$$

(We can even assume that H is ample.) We are interested in the case that $Y = M_X(r, H, s) \cong X$.

We set $c = (r, s)$ and $a = r/c$, $b = s/c$. Then $(a, b) = 1$. Suppose that H is divisible by $d \in \mathbb{N}$, where $\tilde{H} = H/d$ is primitive in $N(X)$. The primitivity of $v = (r, H, s)$ means that $(r, d, s) = (c, d) = 1$. Since $\tilde{H}^2 = 2abc^2/d^2$ is even, we have $d^2 \mid abc^2$. Since $(a, b) = (c, d) = 1$, it follows that $d = d_a d_b$, where $d_a = (d, a)$, $d_b = (d, b)$, and we can introduce integers

$$a_1 = \frac{a}{d_a^2} \quad \text{and} \quad b_1 = \frac{b}{d_b^2},$$

obtaining $\tilde{H}^2 = 2a_1b_1c^2$. Define $\gamma = \gamma(\tilde{H})$ by $\tilde{H} \cdot N(X) = \gamma\mathbb{Z}$, in other words, $H \cdot N(X) = \gamma d\mathbb{Z}$. Clearly, $\gamma \mid \tilde{H}^2 = 2a_1b_1c^2$. We write

$$n(v) = (r, s, d\gamma) = (r, s, \gamma). \tag{11}$$

By Mukai [5], we have $T(X) \subset T(Y)$, and

$$n(v) = [T(Y) : T(X)], \tag{12}$$

where $T(X)$ and $T(Y)$ are the transcendental lattices of X and Y. Thus

$$Y \cong X \implies n(v) = (r, s, d\gamma) = (c, d\gamma) = (c, \gamma) = 1. \tag{13}$$

Assuming that $Y \cong X$ and then $n(v) = 1$, we have $\gamma \mid 2a_1b_1$, and we can introduce

$$\gamma_a = (\gamma, a_1), \quad \gamma_b = (\gamma, b_1), \quad \text{and} \quad \gamma_2 = \frac{\gamma}{\gamma_a\gamma_b}. \tag{14}$$

Clearly, $\gamma_2 \mid 2$.

In [14], Theorem 4.4, we obtained the following general theorem (see important particular cases of it in [1], [2], and [13]). In the theorem, we use the notation c, a, b, d, d_a, d_b, a_1, b_1 introduced above. The same notation γ, γ_a, γ_b, and γ_2 as above is used when we replace $N(X)$ by a 2-dimensional primitive sublattice $N \subset N(X)$, e.g., $\widetilde{H} \cdot N = \gamma\mathbb{Z}$ with $\gamma > 0$. We write $\det N = -\gamma\delta$ and $\mathbb{Z}f(\widetilde{H})$ for the orthogonal complement to $\widetilde{H}$ in N.

Theorem 3.1. *Let X be a K3 surface and H a polarization of X such that $H^2 = 2rs$, where $r, s \in \mathbb{N}$. Assume that the Mukai vector (r, H, s) is primitive. Let $Y = M_X(r, H, s)$ be the K3 surface that is the moduli of sheaves over X with isotropic Mukai vector $v = (r, H, s)$. Let $\widetilde{H} = H/d$ for $d \in \mathbb{N}$ be the corresponding primitive polarization.*

We have $Y \cong X$ if there exists $\widetilde{h}_1 \in N(X)$ such that $\widetilde{H}$ and $\widetilde{h}_1$ belong to a 2-dimensional primitive sublattice $N \subset N(X)$ such that $\widetilde{H} \cdot N = \gamma\mathbb{Z}$, $\gamma > 0$, $(c, d\gamma) = 1$, and the element $\widetilde{h}_1$ belongs to the a-series or the b-series described below:

$\widetilde{h}_1$ *belongs to the a-series if*

$$\widetilde{h}_1^2 = \pm 2b_1c, \quad \widetilde{H} \cdot \widetilde{h}_1 \equiv 0 \bmod \gamma(b_1/\gamma_b)c, \quad f(\widetilde{H}) \cdot \widetilde{h}_1 \equiv 0 \bmod \delta b_1 c \quad (15)$$

(where $\gamma_b = (\gamma, b_1)$);
$\widetilde{h}_1$ *belongs to the b-series if*

$$\widetilde{h}_1^2 = \pm 2a_1c, \quad \widetilde{H} \cdot \widetilde{h}_1 \equiv 0 \bmod \gamma(a_1/\gamma_a)c, \quad f(\widetilde{H}) \cdot \widetilde{h}_1 \equiv 0 \bmod \delta a_1 c \quad (16)$$

(where $\gamma_a = (\gamma, a_1)$).

These conditions are necessary to have $Y \cong X$ if $\rho(X) \leq 2$ and X is a general K3 surface with its Picard lattice.

In [3], we interpreted Theorem 3.1 geometrically as follows.

Theorem 3.2. *Let X be a K3 surface and H a polarization of X such that $H^2 = 2rs$, where $r, s \in \mathbb{N}$. Assume that the Mukai vector (r, H, s) is primitive. Let $Y = M_X(r, H, s)$ be the K3 surface that is the moduli of sheaves over X with isotropic Mukai vector $v = (r, H, s)$. Let $\widetilde{H} = H/d$ with $d \in \mathbb{N}$ be the corresponding primitive polarization.*

Assume that there exists $\widetilde{h}_1 \in N(X)$ such that $\widetilde{H}$ and $\widetilde{h}_1$ belong to a 2-dimensional primitive sublattice $N \subset N(X)$ such that $\widetilde{H} \cdot N = \gamma\mathbb{Z}$, $\gamma > 0$, $(c, d\gamma) = 1$, and the element $\widetilde{h}_1$ belongs to the a-series or to the b-series described in (15) and (16) above.

If $\widetilde{h}_1$ belongs to the a-series, then

$$\widetilde{h}_1 = d_2\widetilde{H} + b_1cD \quad \text{for some} \quad d_2 \in \mathbb{N}, \, D \in N, \quad (17)$$

which defines an isomorphism

$$\mathrm{Tyu}(\pm \widetilde{h}_1) \cdot T_D \cdot \nu(1, d_2) \cdot \delta \cdot \nu(d_a, d_b)^{-1} \colon Y = M_X(r, H, s) \cong X. \qquad (18)$$

If $\widetilde{h}_1$ belongs to the b-series, then

$$\widetilde{h}_1 = d_2 \widetilde{H} + a_1 c D \quad \textit{for some} \quad d_2 \in \mathbb{N}, \ D \in N, \qquad (19)$$

which defines an isomorphism

$$\mathrm{Tyu}(\pm \widetilde{h}_1) \cdot T_D \cdot \nu(1, d_2) \cdot \nu(d_a, d_b)^{-1} \colon Y = M_X(r, H, s) \cong X. \qquad (20)$$

Since the conditions of Theorems 3.1, 3.2 are necessary for general K3 surfaces with $\rho(X) \le 2$, we obtain the following result.

Theorem 3.3. *Let X be a K3 surface with a polarization H such that $H^2 = 2rs$, $r, s \ge 1$ and the Mukai vector (r, H, s) is primitive. Let $Y = M_X(r, H, s)$ be the moduli space of sheaves over X with isotropic Mukai vector (r, H, s). Assume that $\rho(X) \le 2$ and X is general with its Picard lattice. Let $\widetilde{H} = H/d$, $d \in \mathbb{N}$, be the corresponding primitive polarization.*

Then $Y = M_X(r, H, s)$ is isomorphic to X if and only if there exist $d_2 \in \mathbb{N}$ and $D \in N = N(X)$ such that either

$$\widetilde{h}_1 = d_2 \widetilde{H} + b_1 c D \quad \textit{has} \quad \widetilde{h}_1^2 = \pm 2 b_1 c, \qquad (21)$$

defining an isomorphism

$$\mathrm{Tyu}(\pm \widetilde{h}_1) \cdot T_D \cdot \nu(1, d_2) \cdot \delta \cdot \nu(d_a, d_b)^{-1} \colon Y = M_X(r, H, s) \cong X, \qquad (22)$$

or

$$\widetilde{h}_1 = d_2 \widetilde{H} + a_1 c D \quad \textit{has} \quad \widetilde{h}_1^2 = \pm 2 a_1 c, \qquad (23)$$

defining an isomorphism

$$\mathrm{Tyu}(\pm \widetilde{h}_1) \cdot T_D \cdot \nu(1, d_2) \cdot \nu(d_a, d_b)^{-1} \colon Y = M_X(r, H, s) \cong X. \qquad (24)$$

Theorem 2.1 clarifies the appearance of the isomorphisms T_D, δ, $\nu(d_1, d_2)$, and Tyu in these results for Picard number 2. These are universal and exist for all K3 surfaces; moreover, they are all the isomorphisms that are necessary to obtain all isomorphisms from moduli spaces $M_X(v)$ to X for isotropic Mukai vectors v on a general K3 surface X (i.e., with $\rho(X) = 1$). Thus the appearance of the isomorphisms T_D, δ, $\nu(d_1, d_2)$, and Tyu is very natural in the above results.

Remark 3.4. For Picard number $\rho(X) = 1$, Theorems 3.1, 3.2, and 3.3 are formally equivalent to Corollary 2.6. In fact, for $\rho(X) = 1$ we have $\gamma = 2 a_1 b_1 c^2$. Thus $(\gamma, c) = 1$ implies that $c = 1$. Then $\gamma = 2 a_1 b_1$ and $\gamma_2 = 2$, $\gamma_a = a_1$, $\gamma_b = b_1$. The conditions of Theorem 3.1 can be satisfied only for $\widetilde{h}_1 = \widetilde{H}$, which implies that $a_1 = 1$ for the a-series and $b_1 = 1$ for the b-series (we can formally put $f(\widetilde{H}) = 0$).

Thus for $\rho(X) = 1$ we have $Y \cong X$ if and only if $c = 1$ and either $a_1 = 1$ or $b_1 = 1$. This is equivalent to Corollary 2.6.

Under the conditions of Theorem 3.1, assume that for a primitive rank-2 sublattice $N \subset N(X)$ an element $\widetilde{h}_1 \in N$ with $\widetilde{h}_1^2 = \pm 2b_1 c$ belongs to the a-series. This is equivalent to the condition (17) of Theorem 3.2. Replacing $\widetilde{h}_1$ by $-\widetilde{h}_1$ if necessary, we see that (17) is equivalent to

$$\widetilde{h}_1 = d_2 \widetilde{H} + b_1 c \widetilde{D}, \quad d_2 \in \mathbb{Z}, \ \widetilde{D} \in N. \tag{25}$$

Since $\widetilde{H}$ is primitive, the lattice N has a basis $\widetilde{H}$, $D \in N$, i.e., $N = [\widetilde{H}, D]$. Since $\widetilde{H} \cdot N = \gamma \mathbb{Z}$ where $(\gamma, c) = 1$, the matrix of N in this basis is

$$\begin{pmatrix} \widetilde{H}^2 & \widetilde{H} \cdot D \\ \widetilde{H} \cdot D & D^2 \end{pmatrix} = \begin{pmatrix} 2a_1 b_1 c^2 & \gamma k \\ \gamma k & 2t \end{pmatrix}, \tag{26}$$

where $k, t \in \mathbb{Z}$ and $\gamma \mid 2a_1 b_1$, $(\gamma, c) = 1$ and $(2a_1 b_1 c^2 / \gamma, k) = 1$.

The condition of a-series (25) is then equivalent to the existence of $\widetilde{h}_1 \in [\widetilde{H}, b_1 cN] = [\widetilde{H}, b_1 cD]$ with $\widetilde{h}_1^2 = \pm 2b_1 c$. Thus the lattice $N_1 = [\widetilde{H}, b_1 cD]$ with the matrix

$$\begin{pmatrix} 2a_1 b_1 c^2 & b_1 c \gamma k \\ b_1 c \gamma k & b_1^2 c^2 2t \end{pmatrix} \tag{27}$$

must have $\widetilde{h}_1$ with $\widetilde{h}_1^2 = \pm 2b_1 c$. Writing $\widetilde{h}_1$ as $\widetilde{h}_1 = x\widetilde{H} + yb_1 cD$, we obtain that the quadratic equation $a_1 cx^2 + \gamma kxy + b_1 cty^2 = \pm 1$ must have an integral solution. Similarly, for b-series we obtain the equation $b_1 cx^2 + \gamma kxy + a_1 cty^2 = \pm 1$. Thus we finally obtain a very elementary reformulation of the above results.

Lemma 3.5. *For the matrix (26) of the lattice N in Theorems 3.1, 3.2, and 3.3, the conditions of a-series are equivalent to the existence of an integral solution of the equation*

$$a_1 cx^2 + \gamma kxy + b_1 cty^2 = \pm 1, \tag{28}$$

and for b-series of the equation

$$b_1 cx^2 + \gamma kxy + a_1 cty^2 = \pm 1. \tag{29}$$

This calculation has a very important corollary. Assume that $p \mid \gamma_b = (\gamma, b_1)$ for a prime p. Then (28) gives a congruence $a_1 cx^2 \equiv \pm 1 \bmod p$. Thus $\pm a_1 c$ is a quadratic residue mod p. Similarly, for the equation (29), we obtain that $\pm b_1 c$ is a quadratic residue mod p for a prime $p \mid \gamma_a = (\gamma, a_1)$.

We thus obtain an important necessary condition for $Y = M_X(v) \cong X$ when $\rho(X) = 2$.

Theorem 3.6. *Let X be a K3 surface with a polarization H such that $H^2 = 2rs$, $r, s \geq 1$, and the Mukai vector (r, H, s) is primitive. Let $Y = M_X(r, H, s)$ be the moduli of sheaves over X with isotropic Mukai vector (r, H, s). Assume* *that $\rho(X) \leq 2$ and X is general with its Picard lattice. Let $\widetilde{H} = H/d$, $d \in \mathbb{N}$, be the corresponding primitive polarization, $\widetilde{H} \cdot N(X) = \gamma\mathbb{Z}$ and $(\gamma, c) = 1$. Then $Y = M_X(r, H, s) \cong X$ implies that for one of $\pm$ either*

$$\forall\, p \mid \gamma_b \implies \left(\frac{\pm a_1 c}{p}\right) = 1 \tag{30}$$

or

$$\forall p \mid \gamma_a \implies \left(\frac{\pm b_1 c}{p}\right) = 1. \tag{31}$$

Here p means any prime, and $\left(\frac{x}{2}\right) = 1$ means that $x \equiv 1 \bmod 8$.
  Thus if for either choice of ± 1,

$$\exists p \mid \gamma_b \text{ such that } \left(\frac{\pm a_1 c}{p}\right) = -1 \quad \text{and} \quad \exists p \mid \gamma_a \text{ such that } \left(\frac{\pm b_1 c}{p}\right) = -1,$$
$$\tag{32}$$

then $Y = M_X(r, H, s)$ is not isomorphic to X for X a K3 surface with $\rho(X) \leq 2$ that is general with its Picard lattice.

Example 3.7. Assume that $a_1 = 5$, $b_1 = 13$, $c = 1$, and $\gamma = 5 \cdot 13$ (or $\gamma = 2 \cdot 5 \cdot 13$). Then (32) obviously holds. Thus for

$$v = (5, H, 13), \quad H^2 = 2 \cdot 5 \cdot 13, \text{ and } \gamma = 5 \cdot 13 \text{ or } 2 \cdot 5 \cdot 13$$

(then H is always primitive), for any general K3 surface X with $\rho(X) = 2$ and any $H \in N(X)$ with $H^2 = 2 \cdot 5 \cdot 13$ and $H \cdot N(X) = \gamma\mathbb{Z}$, the moduli space $Y = M_X(v)$ is not isomorphic to X.
  There are many such Picard lattices given by (26).

  In [13], we showed that any primitive isotropic Mukai vector $v = (r, H, s)$ with $H^2 = 2rs$ and $\gamma = 1$ is realized by a general K3 surface with Picard number 2 and $Y = M_X(v) \cong X$. Theorem 3.6 may possibly give all the necessary conditions for a similar result to hold for any γ; we hope to return to this problem later.
  The importance of these results for general K3 surfaces X with $\rho(X) = 2$ is that they describe *all divisorial conditions on moduli of polarized K3 surfaces that imply $Y = M_X(r, H, s) \cong X$*. Let us consider the corresponding simple general arguments.
  It is well known (see [9] and [11] where, it seems, it was first observed) that $\mathrm{Aut}\big(T(X), H^{2,0}(X)\big) \cong C_m$ is a finite cyclic group of order $m > 1$, and its representation in $T(X) \otimes \mathbb{Q}$ is the sum of irreducible representations of dimension $\phi(m)$ (where ϕ is the Euler function). $H^{2,0}(X)$ is a line in one of the eigenspaces of C_m. In particular, $\phi(m) \mid \mathrm{rank}\, T(X)$, and if $m > 2$ the dimension of moduli of these X is equal to

$$\dim \mathrm{Mod}(X) = \mathrm{rank}\, T(X)/\phi(m) - 1. \tag{33}$$

If $m = 2$, then $\dim \mathrm{Mod}(X) = \mathrm{rank}\, T(X) - 2$.

Consider polarized K3 surfaces (X, H) with $H^2 = 2rs$ and a primitive Mukai vector (r, H, s) with $r, s > 0$. Assume $Y = M_X(r, H, s) \cong X$.

If $\rho(X) = 1$, then $\operatorname{rank} T(X) = 21$ and $\phi(m) \mid 21$. Since 21 is odd, it follows that $m = 2$. Thus $\operatorname{Aut}(T(X), H^{2,0}(X)) = \pm 1$, and then $c = 1$ and either $a_1 = 1$ or $b_1 = 1$ by Corollary 2.6 (or Remark 3.4). By the specialization principle (see [14], Lemma 2.1.1), then $Y \cong M_X(r, H, s)$ for all K3 surfaces X and a Mukai vector with these invariants:

$$c = 1 \quad \text{and} \quad \text{either } a_1 = 1 \text{ or } b_1 = 1. \tag{34}$$

Now assume that (r, H, s) does not satisfy (34), but $Y = M_X(r, H, s) \cong X$; then $\rho(X) \neq 1$ by Corollary 2.6. Hence $\rho(X) \geq 2$ and

$$\dim \operatorname{Mod}(X) \leq 20 - \rho(X) \leq 18.$$

Thus a divisorial condition on moduli or polarized K3 surfaces (X, H) to have $Y = M_X(r, H, s) \cong X$ means that $\rho(X) = 2$ for a general K3 surface satisfying this condition. All these conditions are described by the isomorphism classes of $H \in N(X)$ where $\operatorname{rank} N(X) = 2$ and $H \in N(X)$ satisfies Theorems 3.1, 3.2, or 3.3 (which in this case are all equivalent). If $H \in N \subset N(X)$ is a primitive sublattice of rank two and $H \in N$ satisfies the equivalent Theorems 3.1 and 3.2, then $Y = M_X(r, H, s) \cong X$ by the specialization principle. This means that X belongs to the closure of the divisor defined by the moduli of polarized K3 surfaces (X', H) with Picard lattice $N(X') = N$ of rank two. Thus $Y' = M_X(r, H, s) \cong X'$ because X' satisfies the divisorial condition $H \in N$, where $H \in N \subset N(X')$.

By Theorem 3.6 we obtain the following result.

Theorem 3.8. *For $r, s \geq 1$ let*

$$v = (r, H, s), \quad d, \quad H^2 = 2rs, \quad (c, d) = 1, \quad d^2 \mid ab$$

be a type of primitive isotropic Mukai vector on K3, and $\gamma \mid 2a_1 b_1$ and $(\gamma, c) = 1$.

Then if (32) holds, there does not exist any divisorial condition on moduli of polarized K3 surfaces (X, H) that implies $Y = M_X(r, H, s) \cong X$ and $H \cdot N(X) = \gamma \mathbb{Z}$. Thus these K3 surfaces have codimension ≥ 2 in the 19-dimensional moduli space of polarized K3 surfaces (X, H).

For example, this holds for $r = 5$, $s = 13$ (then H is primitive and $d = 1$), and $\gamma = 5 \cdot 13$ (or $\gamma = 2 \cdot 5 \cdot 13$).

In the next section, we will show that the numerical example of Theorem 3.8 can be satisfied by K3 surfaces X with $\rho(X) = 3$ and $Y = M_X(r, H, s) \cong X$. Thus these K3 surfaces define a 17-dimensional submanifold in the moduli of polarized K3 surfaces that does not extend to a divisor in moduli preserving the condition $Y = M_X(r, H, s) \cong X$.

4 Isomorphisms Between $M_X(v)$ and X for a General K3 Surface X with $\rho(X) \geq 3$

Here we show that it is interesting and nontrivial to generalize the results of the previous section to $\rho(X) \geq 3$.

Let $K = [e_1, e_2, (e_1 + e_2)/2]$ be a negative definite 2-dimensional lattice with $e_1^2 = -6$, $e_2^2 = -34$, and $e_1 \cdot e_2 = 0$. Then $((e_1 + e_2)/2)^2 = (-6 - 34)/4 = -10$ is even, and the lattice K is even. Since $6x^2 + 34y^2 = 8$ has no integral solutions, it follows that K has no elements $\delta \in K$ with $\delta^2 = -2$. Consider the lattice

$$S = \mathbb{Z}H \oplus K,$$

which is the orthogonal sum of $\mathbb{Z}H$ with $H^2 = 2 \cdot 5 \cdot 13$ and the lattice K. By standard results about K3 surfaces, there exists a polarized K3 surface (X, H) with the Picard lattice S and the polarization $H \in S$. (E.g., see [15] and [9].) We then have $H \cdot S = 2 \cdot 5 \cdot 13\,\mathbb{Z}$. Thus $\gamma = 2 \cdot 5 \cdot 13$.

Let $Y = M_X(5, H, 13)$. We have the following result, perhaps the main result of the paper.

Theorem 4.1. *For any polarized K3 surface (X, H) with $N(X) = S$, where S is the hyperbolic lattice of rank 3 defined above, one has $Y = M_X(5, H, 13) \cong X$, which gives a 17-dimensional moduli space M_S of polarized K3 surfaces (X, H) with $Y = M_X(5, H, 13) \cong X$.*

On the other hand, M_S is not contained in any 18-dimensional moduli space M_N of polarized K3 surfaces (X', H) where $H \in N(X') = N \subset S$ is a primitive sublattice of rank $N = 2$ and $M_{X'}(5, H, 13) \cong X'$. Thus M_S is not defined by any divisorial condition on moduli of polarized K3 surfaces (X, H), implying $M_X(5, H, 13) \cong X$. (That is, M_S is not a specialization of a divisor with this condition.)

Proof. For this case, $c = (5, 13) = 1$ and $(\gamma, c) = 1$. By Mukai's results (5) and (12), the transcendental periods $(T(X), H^{2,0}(X))$ and $(T(Y), H^{2,0}(Y))$ are then isomorphic. The discriminant group $A_S = S^*/S$ of the lattice $S = T(X)^\perp$ is a cyclic group $\mathbb{Z}/(2 \cdot 5 \cdot 13 \cdot 3 \cdot 17)$. Thus the minimal number $l(A_S)$ of generators of A_S is one. Thus $l(A_S) \leq \operatorname{rank} S - 2$. By [10], Theorem 1.14.4, a primitive embedding of $T(X)$ into the cohomology lattice of K3 (which is an even unimodular lattice of signature $(3, 19)$) is then unique, up to isomorphisms. It follows that the isomorphism between transcendental periods of X and Y can be extended to an isomorphism of periods of X and Y. By the global Torelli theorem for K3 surfaces [15], the K3 surfaces X and Y are isomorphic. (These considerations are now standard.)

Let $H \in N \subset S$ be a primitive sublattice with rank $N = 2$. Since $H \cdot S = H \cdot H\mathbb{Z} = 2 \cdot 5 \cdot 13\mathbb{Z}$, it follows that $H \cdot N = 2 \cdot 5 \cdot 13\mathbb{Z}$, and the invariant $\gamma = 2 \cdot 5 \cdot 13$ is the same for any sublattice $N \subset S$ containing H. By Theorem 3.8, $M_{X'}(r, H, s)$ is not isomorphic to X' for any general K3 surface (X', H) with $N(X') = N$.

This completes the proof.   □

Similar arguments can be used to prove the following general statement for $\rho(X) \geq 12$. This shows that there are many cases in which $Y = M_X(r, H, s) \cong X$, that do not follow from divisorial conditions on moduli. Its first statement is well known (see for example [1], Proposition 2.2.1).

Theorem 4.2. *Let (X, H) be a polarized K3 surface with $\rho(X) \geq 12$, and for $r, s \geq 1$, let (r, H, s) be a primitive isotropic Mukai vector on X, i.e., $H^2 = 2rs$ and $(c, d) = 1$. Assume that $H \cdot N(X) = \gamma \mathbb{Z}$.*
  Then $Y = M_X(r, H, s) \cong X$ if $(\gamma, c) = 1$ (Mukai's necessary condition).
  On the other hand, if (32) holds, the isomorphism $Y = M_X(r, H, s) \cong X$ does not follow from any divisorial condition on moduli of polarized K3 surfaces. That is, for any primitive 2-dimensional sublattice $H \in N \subset N(X)$, there exists a polarized K3 surface (X', H) with $N(X') = N$ such that $Y' = M_{X'}(r, H, s)$ is not isomorphic to X'.

Proof. Since $\rho(X) \geq 12$,

$$\operatorname{rank} T(X) \leq 22 - 12 = 10 \quad \text{and} \quad l(A_{T(X)}) \leq \operatorname{rank} T(X) = 10.$$

Since $N(X)$ and $T(X)$ are orthogonal complements to one another in the unimodular lattice $H^2(X, \mathbb{Z})$, it follows that $A_{N(X)} \cong A_{T(X)}$ and $l(A_{N(X)}) \leq 10 \leq \operatorname{rank} N(X) - 2$. By [10], Theorem 1.14.4, a primitive embedding of $T(X)$ into the cohomology lattice of K3 is then unique up to isomorphisms. As in the proof of Theorem 4.1, it follows that $Y \cong X$.
  We prove the second statement. Since $H \cdot N(X) = \gamma \mathbb{Z}$ and $H \in N \subset N(X)$, it follows that $H \cdot N = \gamma(N) \mathbb{Z}$, where $\gamma \mid \gamma(N)$. Let X' be a general K3 surface with $N(X') = N$. If $(c, \gamma(N)) > 1$, then $Y' = M_{X'}(r, H, s)$ is not isomorphic to X' because $[T(Y') : T(X')] = (c, \gamma(N)) > 1$ by Mukai's result (12). Assume $(c, \gamma(N)) = 1$. Obviously, (32) for γ implies (32) for $\gamma(N)$. By Theorem 3.6, $Y' = M_{X'}(r, H, s)$ is not isomorphic to X'.
  This completes the proof. $\square$

Theorems 4.1 and 4.2 can be unified in the answer to the following question, which is the most general known: when does $Y = M_X(r, H, s) \cong X$ hold for any primitive isotropic Mukai vector on X satisfying Mukai's necessary condition? We read that two lattices have the same genus if they are isomorphic over $\mathbb{R}$ and rings $\mathbb{Z}_p$ of p-adic integers for all prime p.

Theorem 4.3. *Let X be a K3 surface. Assume that the Picard lattice $N(X)$ is unique in its genus, and the natural homomorphism*

$$O(N(X)) \rightarrow O(q_{N(X)})$$

is surjective, where $q_{N(X)}$ is the discriminant quadratic form of $N(X)$. Equivalently, any isomorphism of the transcendental periods of X and another K3 surface extends to an isomorphism of the periods of X and the other K3 surface.

Then for any primitive isotropic Mukai vector $v = (r, H, s)$ on X such that $(c, \gamma) = 1$ (Mukai's necessary condition), one has $Y = M_X(r, H, s) \cong X$.

On the other hand, if X is general with its Picard lattice and (32) holds, then the isomorphism $Y = M_X(r, H, s) \cong X$ does not follow from any divisorial condition on moduli of polarized K3 surfaces (X, H). That is, for any primitive 2-dimensional sublattice $H \in N \subset N(X)$, there exists a polarized K3 surface (X', H) with $N(X') = N$ such that $Y' = M_{X'}(r, H, s)$ is not isomorphic to X'.

These results and those of Section 3 suggest the following general notions. Let $r \in \mathbb{N}$ and $s \in \mathbb{Z}$. We formally put $H^2 = 2rs$ and introduce $c = (r, s)$ and $a = r/c$, $b = s/c$. Let $d \in \mathbb{N}$, $(d, c) = 1$ and $d^2 \mid ab$. We call

$$(r, H, s), \quad H^2 = 2rs, \quad d, \tag{35}$$

a *type of primitive isotropic Mukai vector for a K3*. Clearly, a Mukai vector of type (35) on a K3 surface X is just an element $H \in N(X)$ such that $H^2 = 2rs$ and $\widetilde{H} = H/d$ is primitive. As above, we introduce $d_a = (d, a)$, $d_b = (d, b)$ and put $a_1 = a/d_a^2$, $b_1 = b/d_b^2$. Then $\widetilde{H}^2 = 2a_1 b_1 c^2$.

Let N be a lattice that embeds primitively into the Picard lattice of some algebraic K3 surface (equivalently, there exists a Kähler K3 surface with this Picard lattice). It is equivalent to say that N is either negative definite, or negative semi-definite with 1-dimensional kernel, or hyperbolic (i.e., N has signature $(1, \rho - 1)$), and has a primitive embedding into an even unimodular lattice of signature $(3, 19)$. Moreover, we say that N is an *abstract K3 Picard lattice* (or just a K3 Picard lattice). Let $H \in N$; we say that $H \in N$ is a *polarized (abstract) K3 Picard lattice* (despite the fact that H^2 can be nonpositive). We consider such pairs up to natural isomorphism. Another polarized K3 Picard lattice $H' \in N'$ is *isomorphic* to $H \in N$ if there exists an isomorphism $f : N \cong N'$ of lattices with $f(H) = H'$.

Definition 4.4. Fix a type (35) of primitive isotropic Mukai vector of K3. A polarized K3 Picard lattice $H \in N$ is *critical for self-correspondences of a K3 surface via moduli of sheaves for the type* (35) *of Mukai vector* if $H^2 = 2rs$ and $\widetilde{H} = H/d \in N$ is primitive and $H \in N$ satisfies the following two conditions:

(a) For any K3 surface X such that $H \in N \subset N(X)$ is a primitive sublattice, one has $Y = M_X(r, H, s) \cong X$.
(b) The above condition (a) does not hold if one replaces $H \in N$ by $H \in N_1$ for any primitive sublattice $H \in N_1 \subset N$ of N of strictly smaller rank rank $N_1 <$ rank N.

In what follows we abbreviate this, saying that $H \in N$ is a *critical polarized K3 Picard lattice for the type* (35).

On the one hand [14], Theorem 2.3.3 gives a criterion for a polarized K3 Picard lattice $H \in N$, for a general (and then any) K3 surface with

$H \in N = N(X)$ to have $Y = M_X(r, H, s) \cong X$. On the other hand, by the specialization principle (Lemma 2.1.1 in [14]), if this criterion is satisfied, then $Y = M_X(r, H, s) \cong X$ for any K3 surface X such that $H \in N \subset N(X)$ is a primitive sublattice. Thus for the problem of describing, in terms of Picard lattices, all K3 surfaces X such that $Y = M_X(r, H, s) \cong X$, the main problem is as follows.

Problem 4.5. For a given type of primitive isotropic Mukai vector (35) for a K3, describe all *critical polarized K3 Picard lattices* $H \in N$ (for the problem of self-correspondences of a K3 surface via moduli of sheaves).

Now we have the following examples of solutions of this problem.

By (10), or Corollary 2.6, or Remark 3.4, we have classified the critical polarized K3 Picard lattices of rank one.

Example 4.6. For the type (r, H, s), $H^2 = 2rs$, d where $c = 1$ and either $a_1 = 1$ or $b_1 = \pm 1$, we obtain that $N = \mathbb{Z}\widetilde{H}$ where $\widetilde{H}^2 = 2a_1 b_1$ gives all critical polarized K3 Picard lattices $H = d\widetilde{H} \in N$ of rank one.

Example 4.7. For the type of Mukai vector that is different from Example 4.6, classification of the critical polarized K3 Picard lattices of rank 2 is given by equivalent Theorems 3.1, 3.2, and 3.3.

Example 4.8. For the Mukai vector of type $(5, H, 13)$ with $H^2 = 2 \cdot 5 \cdot 13$ and $d = 1$, the polarized Picard lattice $H \in S$ of Theorem 4.1 is critical with rank $S = 3$, by Theorem 4.1. Obviously, there are plenty of similar examples. It would be very interesting and nontrivial to find all critical polarized K3 Picard lattices $H \in S$ of rank 3.

Example 4.9. By Theorem 4.2, we should expect that there exist critical polarized K3 Picard lattices of the rank more than 3. On the other hand, the same Theorem 4.2 gives that the rank of a critical polarized K3 Picard lattice is ≤ 12.

Theorem 4.10. *For any type* (r, H, s), $H^2 = 2rs$, *and* d *of a primitive isotropic Mukai vector of K3, the rank of a critical polarized K3 Picard lattice* $H \in N$ *is at most 12: we have* rank $N \leq 12$.

Proof. Let $H \in N$ be a critical polarized K3 Picard lattice of this type and rank $N \geq 13$. Let us take any primitive sublattice $H \in N' \subset N$ of rank $N' = 12$ such that $\widetilde{H} \cdot N' = \widetilde{H} \cdot N$. Obviously, it does exist. Let X be an algebraic K3 surface such that $H \in N' \subset N(X)$. Then rank $N(X) \geq 12$ and $Y = M_X(r, H, s) \cong X$ by Theorem 4.2.

Then the condition (b) of Definition 4.4 is not satisfied, and we get a contradiction. Thus rank $N \leq 12$.

This completes the proof. $\qquad\qquad\qquad\qquad\qquad\qquad\qquad\qquad\square$

It would be very interesting to give an exact estimate for the rank of critical polarized K3 Picard lattices.

Problem 4.11. For a given primitive isotropic Mukai vector of K3 of type (35), give the exact estimate of rank H? of a critical polarized K3 Picard lattice $H \in N$ of this type (for the problem of self-correspondences of K3 surfaces).

We don't know the answer to this problem for any type (35) different from Example 4.6.

5 Composing Self-Correspondences of a K3 Surface via Moduli of Sheaves and the General Classification Problem

We want to interpret the above results in terms of the action of correspondences on the 2-dimensional cohomology lattice of a K3 surface. Moreover, we attempt to formulate the general problem of classification of self-correspondences of a K3 surface via moduli of sheaves.

Let $v = (r, H, s)$ be a primitive isotropic Mukai vector on a K3 surface X and $Y = M_X(r, H, s)$. Write π_X, π_Y for the projections of $X \times Y$ to X and Y. By Mukai [5], Theorem 1.5, the algebraic cycle

$$Z_{\mathcal{E}} = (\pi_X^* \sqrt{\mathrm{td}_X}) \cdot \mathrm{ch}(\mathcal{E}) \cdot (\pi_Y^* \sqrt{\mathrm{td}_Y})/\sigma(\mathcal{E}) \tag{36}$$

arising from the quasi-universal sheaf $\mathcal{E}$ on $X \times Y$ defines an isomorphism of the full cohomology groups

$$f_{Z_{\mathcal{E}}} : H^*(X, \mathbb{Q}) \to H^*(Y, \mathbb{Q}), \quad t \mapsto \pi_{Y*}(Z_{\mathcal{E}} \cdot \pi_X^* t) \tag{37}$$

with their Hodge structures (see [5], Theorem 1.5, for details). Moreover, according to Mukai, it defines an isomorphism of lattices (an isometry)

$$f_{Z_{\mathcal{E}}} : v^{\perp} \to H^4(Y, \mathbb{Z}) \oplus H^2(Y, \mathbb{Z}),$$

where $f_{Z_{\mathcal{E}}}(v) = w \in H^4(Y, \mathbb{Z})$ is the fundamental cocycle, and the orthogonal complement $v^{\perp}$ is taken in the Mukai lattice $\widetilde{H}(X, \mathbb{Z})$. This gives the relation (5) already used in Section 2.

In particular, composing $f_{Z_{\mathcal{E}}}$ with the projection $\pi : H^4(Y, \mathbb{Z}) \oplus H^2(Y, \mathbb{Z}) \to H^2(Y, \mathbb{Z})$ gives an embedding of lattices

$$\pi \cdot f_{Z_{\mathcal{E}}} : H^{\perp}_{H^2(X, \mathbb{Z})} \to H^2(Y, \mathbb{Z})$$

that extends to an isometry

$$\widetilde{f}_{Z_{\mathcal{E}}} : H^2(X, \mathbb{Q}) \to H^2(Y, \mathbb{Q}) \tag{38}$$

of quadratic forms over $\mathbb{Q}$ by Witt's theorem. If $H^2 = 0$, this extension is unique.

If $H^2 \neq 0$, there are two such extensions, differing by ± 1 on $\mathbb{Z}H$. We agree to take

$$\widetilde{f}_{Z\varepsilon}(\widetilde{H}) = ch, \tag{39}$$

where h is defined in (6), and we use Proposition 2.4 to relate the periods of X and Y.

The Hodge isometry (38) can be viewed as a minor modification of Mukai's algebraic cycle (36) to give an isometry in H^2. It is also clearly defined by some algebraic cycle, because it changes the Mukai isomorphism (37) in only the algebraic part.

By Proposition 2.4, the isomorphism $\widetilde{f}_{Z\varepsilon}$ is given by embeddings

$$\widetilde{H}^\perp \subset h^\perp = \left[\widetilde{H}^\perp, 2a_1 b_1 ct^*(\widetilde{H})\right], \quad \mathbb{Z}\widetilde{H} \subset \mathbb{Z}h, \quad \widetilde{H} = ch,$$
$$\text{and} \quad H^{2,0}(X) = H^{2,0}(Y). \tag{40}$$

This identifies the quadratic forms $H^2(X, \mathbb{Q}) = H^2(Y, \mathbb{Q})$ over $\mathbb{Q}$, and the lattices $H^2(X, \mathbb{Z})$, $H^2(Y, \mathbb{Z})$ as two sublattices of this. Let

$$O(H^2(X, \mathbb{Q}))_0 = \left\{ f \in O(H^2(X, \mathbb{Q})) \mid f|T(X) = \pm 1 \right\}$$
$$\cong O(N(X) \otimes \mathbb{Q}) \times \left\{ \pm 1_{T(X)} \right\},$$

and

$$O(H^2(X, \mathbb{Z}))_0 = O(H^2(X, \mathbb{Z})) \cap O(H^2(X, \mathbb{Q}))_0.$$

By the global Torelli theorem for K3 surfaces of [15], we obtain at once the following:

Proposition 5.1. *If a K3 surface X is general with its Picard lattice, then $Y = M_X(r, H, s) \cong X$ if and only if there exists an automorphism $\phi(r, H, s) \in O(H^2(X, \mathbb{Q}))_0$ such that $\phi(H^2(X, \mathbb{Z})) = H^2(Y, \mathbb{Z})$.*

If $Y \cong X$ we can give the following definition.

Definition 5.2. If $Y = M_X(r, H, s) \cong X$ and X is general with its Picard lattice, then the isomorphism of Proposition 5.1,

$$\phi(r, H, s) \mod O(H^2(X, \mathbb{Z}))_0 \in O(H^2(X, \mathbb{Q}))_0 / O(H^2(X, \mathbb{Z}))_0,$$

is called the *action* on $H^2(X, \mathbb{Q})$ of the self-correspondence of a general K3 surface X (general with its Picard lattice) via moduli of sheaves $Y = M_X(r, H, s)$ with primitive isotropic Mukai vector $v = (r, H, s)$.

By the global Torelli theorem for K3 surfaces [15], the group $O(H^2(X, \mathbb{Z}))_0$ mod ± 1 can be considered as generated by correspondences defined by graphs of automorphisms of X and by the reflections in elements $\delta \in N(X)$ with $\delta^2 = -2$ given by $s_\delta : x \mapsto x + (x \cdot \delta)\delta$ for $x \in H^2(X, \mathbb{Z})$. By the Riemann–Roch theorem for K3 surfaces, $\pm\delta$ contains an effective curve E. If $\Delta \subset X \times X$ is the diagonal, the effective 2-dimensional algebraic cycle $\Delta + E \times E \subset X \times X$ acts as the reflection s_δ in $H^2(X,\mathbb{Z})$ (I learned this from Mukai [8]). Thus considering actions of correspondences modulo $O(H^2(X,\mathbb{Z}))$ mod ± 1 is very natural.

Consider the Tyurin isomorphism (10) defined by the Mukai vector $v = (\pm H^2/2, H, \pm 1)$, where $H \in N(X)$ has $\pm H^2 > 0$. Then $M_X(\pm H^2/2, H, \pm 1) \cong M_X(\pm\widetilde{H}^2/2, \widetilde{H}, \pm 1)$, where $\widetilde{H} = H/d$ is primitive.

Then $c = 1$, $a_1 = \pm\widetilde{H}^2/2$ and $b_1 = \pm 1$, $m(a_1, b_1) \equiv -1 \bmod 2a_1 b_1$, $h = \widetilde{H}$, and we have

$$H^2(X,\mathbb{Z}) = \left[\mathbb{Z}\widetilde{H}, \widetilde{H}^\perp, \widetilde{H} + t^*(\widetilde{H})\right] \quad \text{and} \quad H^2(Y,\mathbb{Z}) = \left[\mathbb{Z}\widetilde{H}, \widetilde{H}^\perp, \widetilde{H} - t^*(\widetilde{H})\right].$$

Then the reflection $s_{\widetilde{H}}$ in $\widetilde{H}$,

$$s_{\widetilde{H}}(x) = x - \frac{2(x \cdot \widetilde{H})\widetilde{H}}{\widetilde{H}^2} \quad \text{for } x \in H^2(X,\mathbb{Q}),$$

belongs to $O(H^2(X,\mathbb{Q}))_0$, and $s_{\widetilde{H}}(H^2(X,\mathbb{Z})) = H^2(Y,\mathbb{Z})$. Moreover, the reflections s_H and $s_{\widetilde{H}}$ coincide.

This gives the following result.

Proposition 5.3. *For a K3 surface X and $H \in N(X)$ with $\pm H^2 > 0$, the Tyurin isomorphism*

$$M_X(\pm H^2, H, \pm 1) \cong X$$

defines a self-correspondence of X with the action

$$s_H \bmod O(H^2(X,\mathbb{Z})_0),$$

where s_H is the reflection in H.

By classical and well-known results, their composites generate the full group $O(H^2(X,\mathbb{Q}))_0$ mod ± 1.

5.1 General Problem of Classifying Self-Correspondences of a K3 Surface via Moduli of Sheaves

We need some notation. For a primitive sublattice $N \subset N(X)$, we introduce

$$O(N \otimes \mathbb{Q})_0 = \left\{ f \in O(H^2(X,\mathbb{Q})) \mid f|N^\perp_{H^2(X,\mathbb{Z})} = \pm 1 \right\}$$

and

$$O(N)_0 = O(H^2(X,\mathbb{Z})) \cap O(N \otimes \mathbb{Q})_0.$$

We denote by $[\,\cdot\,]_{pr} \subset L$ the primitive sublattice of L generated by.

Let X be a K3 surface that is general with its Picard lattice $N(X)$. From our current point of view, the problem of classifying self-correspondences of X via moduli of sheaves consists of the following problems:

(1) Find all primitive isotropic Mukai vectors (r, H, s) on X such that $Y = M_X(r, H, s) \cong X$.
(2) For a primitive isotropic Mukai vector (r, H, s) as in (1), find all primitive critical polarized Picard sublattices $H \in N(r, H, s) \subset N(X)$.

For either of these problems, the action $\phi(r, H, s)$ of Definition 5.2 can be taken to be in $O(N(r, H, s) \otimes \mathbb{Q})_0$. We denote it by $\phi_{N(r,H,s)}$, and it looks like a reflection with respect to $N(r, H, s)$. For two critical polarized Picard sublattices $H \in N(r, H, s)$ and $H' \in N'(r, H, s)$ as in (2), the automorphisms $\phi_{N(r,H,s)}$ and $\phi_{N'(r,H,s)}$ differ by an automorphism in $O(H^2(X, \mathbb{Z}))_0$.

(3) *The structures (1) and (2) are important for the following reason:* given any two primitive isotropic Mukai vectors (r, H, s) and (r', H', s') as in (1) and two critical polarized Picard sublattices $H \in N(r, H, s)$ and $H' \in N(r', H', s')$ for them as in (2), the isomorphism

$$\phi(r', H', s') \circ \phi(r, H, s)^{-1} \colon M_X(r, H, s) \to M_X(r', H', s')$$

comes from K3 surfaces with the Picard sublattice

$$[N(r, H, s) + N(r', H', s')]_{pr} \subset N(X),$$

and it can be viewed as a natural isomorphism between these moduli.
(4) All these generators $\phi_{N(r,H,s)}$ mod $O(N(r, H, s))_0$ can be considered as natural generators for self-correspondences of X via moduli of sheaves, together with automorphisms of X and reflections s_δ, $\delta \in N(X)$, and $\delta^2 = -2$. They and their relations are the natural subject to study.

Problems (1)–(4) are solved for $\rho(X) = 1$ and 2 in Sections 2 and 3. The results of Section 4 show that these problems are very nontrivial for $\rho(X) \geq 3$.

As an example, take a general K3 surface X with the rank-3 Picard lattice $N(X) = S$ of Theorem 4.1 (or any other Picard lattice of rank 3 satisfying Theorem 4.3). Let $v = (r, H, s)$ be a primitive isotropic Mukai vector on X. Then $Y = M_X(r, H, s) \cong X$ if and only if $(\gamma, c) = 1$, where $\widetilde{H} \cdot S = \gamma \mathbb{Z}$. Moreover, we have three cases:

(a) If $c = 1$ and either $a_1 = 1$ or $b_1 = \pm 1$ (Tyurin's case), then the critical sublattice is $N(v) = \mathbb{Z}\widetilde{H}$; it has rank one and is unique. The corresponding $\phi_{N(v)}$ is equal to s_H mod $O(H^2(X, \mathbb{Z}))_0$.
(b) If $v = (r, H, s)$ is different from (a), but the critical sublattice $N(v)$ has rank two (the divisorial case), then all critical sublattices $N(v)$ are primitively generated by $\widetilde{H}$ and $\widetilde{h}_1 \in [\widetilde{H}, a_1 c N(X)]$ with $\widetilde{h}_1^2 = \pm 2 a_1 c$ or $\widetilde{h}_1 \in [\widetilde{H}, b_1 c N(X)]$ with $\widetilde{h}_1^2 = \pm 2 b_1 c$ (see the theorems of Section 3). All these $N(v)$ give automorphisms $\phi_{N(v)}$ that differ by elements of $O(H^2(X, \mathbb{Z}))_0$.

(c) If $v = (r, H, s)$ is different from (a) and (b), then the critical sublattice $N(v) = N(X)$ has rank 3. These cases really happen by Theorem 4.1. We get $\phi_{N(v)} \bmod O(H^2(X, \mathbb{Z}))_0$.

Any two v_1, v_2 satisfying one of these conditions (a–c), together with any two critical sublattices $N(v_1)$, $N(v_2)$ for them, generate natural isomorphisms $\phi_{N(v_2)} \circ \phi_{N(v_1)}^{-1}$ between the corresponding moduli spaces of sheaves over X (all of which are isomorphic to X), which are specializations of isomorphisms from the Picard sublattice $[N(v_1) + N(v_2)]_{pr} \subset N(X)$.

References

1. C. MADONNA AND V.V. NIKULIN, *On a classical correspondence between K3 surfaces*, Proc. Steklov Inst. of Math. **241** (2003), 120–153; (see also math.AG/0206158).

2. C. MADONNA AND V.V. NIKULIN, *On a classical correspondence between K3 surfaces II*, in: M. Douglas, J. Gauntlett, M. Gross (eds.) Clay Mathematics Proceedings, Vol. 3 (Strings and Geometry), 2004, pp. 285–300; (see also math.AG/0304415).

3. C. MADONNA AND V.V. NIKULIN, *Explicit correspondences of a K3 surface with itself*, Izv. Math. (2008), Vol. 72 Number 3, pp. 497–509.

4. S. MUKAI, *Symplectic structure of the moduli space of sheaves on an Abelian or K3 surface*, Inv. Math. **77** (1984), 101–116.

5. S. MUKAI, *On the moduli space of bundles on K3 surfaces I*, in: Vector bundles on algebraic varieties, Tata Inst. Fund. Res. Studies in Math. no. **11** (1987), 341–413.

6. S. MUKAI, *Duality of polarized K3 surfaces*, in: K. Hulek (ed.) New trends in algebraic geometry. Selected papers presented at the Euro conference, Warwick, UK, July 1996, Cambridge University Press. London Math. Soc. Lect. Notes Ser. **264**, Cambridge, 1999, pp. 311–326.

7. S. MUKAI, *Vector bundles on a K3 surface*, Proc. ICM 2002 in Beijing, Vol. 3, pp. 495–502.

8. S. MUKAI, *Cycles on product of two K3 surfaces*, Lecture in the University of Liverpool, February 2002.

9. V.V. NIKULIN, *Finite automorphism groups of Kählerian surfaces of type K3*, Trans. Moscow Math. Soc. **38** (1980), 71–135.

10. V.V. NIKULIN, *Integral symmetric bilinear forms and some of their geometric applications*, Math. USSR Izv. **14** (1980), no. 1, 103–167.

11. V.V. NIKULIN, *On the quotient groups of the automorphism groups of hyperbolic forms by the subgroups generated by 2-reflections. Algebraic-geometric applications*, J. Soviet Math. **22** (1983), 1401–1476.

12. V.V. NIKULIN, *On correspondences between K3 surfaces*, Math. USSR Izv. **30** (1988), no.2, 375–383.

13. V.V. NIKULIN, *On correspondences of a K3 surface with itself. I*, Proc. Steklov Inst. Math. **246** (2004), 204–226 (see also math.AG/0307355).

14. V.V. NIKULIN, *On Correspondences of a K3 surfaces with itself. II*, in: JH. Keum and Sh. Kondo (eds.) Algebraic Geometry. Korea-Japan Conf. in Honor of Dolgachev, 2004, Contemporary mathematics **442**. AMS, 2007, pp. 121– 172 (see also math.AG/0309348).

15. I.I. PIATETSKI-SHAPIRO AND SHAFAREVICH, *A Torelli theorem for algebraic surfaces of type K3*, Math. USSR Izv. **5** (1971), no. 3, 547–588.

16. A.N. TYURIN, *Cycles, curves and vector bundles on algebraic surfaces*, Duke Math. J. **54** (1987), no. 1, 1–26.

17. A.N. TYURIN, *Special 0-cycles on a polarized K3 surface*, Math. USSR Izv. **30** (1988), no. 1, 123–143.

18. A.N. TYURIN, *Symplectic structures on the varieties of moduli of vector bundles on algebraic surfaces with $p_g > 0$*, Math. USSR Izv. **33** (1989), no. 1, 139–177.

19. K. YOSHIOKA, *Irreducibility of moduli spaces of vector bundles on K3 surfaces*, Preprint math.AG/9907001, 21 pages.

20. K. YOSHIOKA, *Some examples of Mukai's reflections on K3 surfaces*, J. reine angew. Math. **515** (1999), 97–123 (see also math.AG/9902105).

Foliations in Moduli Spaces of Abelian Varieties and Dimension of Leaves

Frans Oort

Department of Mathematics, Utrecht University, P.O. Box. 80.010, NL - 3508 TA
Utrecht, The Netherlands
f.oort@uu.nl

Dedicated to Yuri Manin on his seventieth birthday

Summary. In moduli spaces of abelian varieties and of p-divisible groups in characteristic p we have various foliations and statifications. In this paper we compute the dimensions of central leaves. We give three different proofs of these results, where every proof presents a different flavour of this beautiful topic. Components of Newton polygon strata for one fixed Newton polygon may have various different dimensions, according to properties of the polarizations considered; we show which dimensions do appear for a given Newton polygon. Hence dimensions of isogeny leaves can be computed this way.

Key words: moduli of abelian varieties, p-divisible groups, stratifications and foliations of moduli spaces, truncated Barsotti-Tate groups

2000 Mathematics Subject Classifications: 11G15, 14L05, 14L15

Introduction

In the theory of foliations in moduli spaces of abelian varieties, as developed in [32], we study *central leaves*. Consider a p-divisible group X_0 over a field K, and let $\mathrm{Def}(X_0) = \mathrm{Spf}(\Gamma)$ and $D(X_0) = \mathrm{Spec}(\Gamma)$. Consider $g \in \mathbb{Z}_{>0}$ and consider $\mathcal{A}_g \otimes \mathbb{F}_p$, the moduli space of polarized abelian varieties (in this paper to be denoted by $\mathcal{A}_g$); choose $[(A, \lambda)] = x \in \mathcal{A}$ and $(A, \lambda)[p^\infty] = (X, \lambda)$. Here is the central question of this paper: determine

$$\text{unpolarized case: what is} \quad \dim(\mathcal{C}_{X_0}(D(X_0)));$$

$$\text{polarized case: what is} \quad \dim(\mathcal{C}_{(X,\lambda)}(\mathcal{A}))?$$

For the notation $\mathcal{C}_-(-)$ see 1.7. We give a combinatorial description of certain numbers associated with a Newton polygon, such as "dim$(-)$," "sdim$(-)$," "cdu$(-)$," "cdp$(-)$". We show that these give the dimension of a stratum or

Y. Tschinkel and Y. Zarhin (eds.), *Algebra, Arithmetic, and Geometry*,
Progress in Mathematics 270, DOI 10.1007/978-0-8176-4747-6_15,
© Springer Science+Business Media, LLC 2009

a leaf, in the unpolarized and in the principally polarized cases. We give three different proofs that these formulas for the dimension of a central leaf are correct:

$$\dim(\mathcal{C}_Y(D(X))) = \mathrm{cdu}(\beta), \quad \beta := \mathcal{N}(Y), \quad \text{see Theorem 4.5, and}$$

$$\dim\left(\mathcal{C}_{(X,\lambda)}(\mathcal{A}_g)\right) = \mathrm{cdp}(\xi), \quad \xi := \mathcal{N}(X), \quad \text{see Theorem 5.4.}$$

One proof is based on the theory of minimal p-divisible groups, as developed in [36], together with a result by T. Wedhorn, see [42], [43]; this was the proof I first had in mind, written up in the summer of 2002.

The second proof is based on the theory of Chai about Serre-Tate coordinates, a generalization from the ordinary case to central leaves in an arbitrary Newton polygon stratum, see [2]. This generalization was partly stimulated by the first proof, and the question to "explain" the dimension formula that came out of my computations.

A third proof, in the unpolarized case and in the polarized case ($p > 2$), is based on recent work by E. Viehmann, see [40], [41], where the dimension of Rapoport-Zink spaces, and hence the dimension of isogeny leaves, is computed in the (un)polarized case; the almost product structure of an open Newton polygon stratum by central and isogeny leaves, as in [32], see 7.17, finishes a proof of the results.

These results enable us to answer a question, settle a conjecture, about bounds of the dimension of components of a Newton polygon stratum; see Section 6.

These results find their natural place in joint work with Ching-Li Chai, which we expect finally to appear in [5]. I thank Chai for the beautiful things I learned from him, in particular for his elegant generalization of Serre-Tate canonical coordinates used in the present paper.

The results of this paper were already announced earlier, e.g., see [32] 3.17, [1] 7.10, 7.12.

Historical remarks. Moduli for polarized abelian varieites in positive characteristic were studied in fundamental work by Yuri Manin, see [21]. That paper was and is a great source of inspiration.

In summer 2000 I gave a talk in Oberwolfach on foliations in moduli spaces of abelian varieties. After my talk, in the evening of Friday 4-VIII-2000, Bjorn Poonen asked me several questions, especially related to the problem I raised to determine the dimensions of central leaves. Our discussion resulted in Problem 21 in [8]. His expectations coincided with computations I had made of these dimensions for small values of g. Then I jumped to the conclusion what those dimensions for an arbitrary Newton polygon could be; that is what was proved later, and reported on here, see 4.5, 5.4. I thank Bjorn Poonen for his interesting questions; our discussion was valuable for me.

A suggestion to the reader. The results of this paper are in sections 4, 5, and 6; we refer to the introductions of those sections. The reader could start reading those sections and refer to other sections whenever definitions or results are needed. In Section 1 we explain some of the concepts used in this paper. In sections 2 and 3 we describe preliminary results used in the proofs. In Section 7 we list some of the well-known methods and results we need for our proofs.

Various strata NP - EO - Fol. Here is a short survey of strata and foliations, to be defined, explained, and studied below. For an abelian variety A with a polarization (sometimes supposed to be principal) we can study the following objects:

NP $A \;\mapsto\; A[p^\infty] \;\mapsto\; A[p^\infty]/\sim_k$ over an algebraically closed field: the isogeny class of its p-divisible group; by the Dieudonné - Manin theorem, see 7.2, we can identify this isogeny class of p-divisible groups with the Newton polygon of A. We obtain the Newton polygon strata, see 1.4 and 7.8.

EO $(A,\lambda) \;\mapsto\; (A,\lambda)[p] \;\mapsto\; (A,\lambda)[p]/\cong_k$ over an algebraically closed field: we obtain EO-strata; see [30] and 1.6. Important feature (Kraft, Oort): the number of geometric isomorphism classes of group schemes of a given rank annihilated by p is *finite*.

Fol $(A,\lambda) \;\mapsto\; (A,\lambda)[p^\infty] \mapsto (A,\lambda)[p^\infty]/\cong_k$ over an algebraically closed field: we obtain a foliation of an open Newton polygon stratum; see [32] and 1.7. Note that for $f < g-1$ the number of (central) leaves is *infinite*.

Note: $X \cong Y \;\Rightarrow\; \mathcal{N}(X) = \mathcal{N}(Y)$; conclusion: every central leaf in **Fol** is contained in exactly one Newton polygon stratum in **NP**.
Note: $X \cong Y \;\Rightarrow\; X[p] = Y[p]$; conclusion: every central leaf in **Fol** is contained in exactly EO-stratum in **EO**.

However, a NP-stratum may contain many EO-strata, and an EO-stratum may intersect several NP-strata, see 8.6. Whether an EO-stratum equals a central leaf is studied and answered in the theory of minimal p-divisible groups, see 1.5 and 7.5.

Isogeny correspondences are finite-to-finite above central leaves, but may blow up and down subsets of isogeny leaves; see 7.23 and Section 6.

1 Notations

We fix a prime number p. All base schemes and base fields will be in characteristic p. We write K for a field, and we write k and Ω for algebraically closed fields of characteristic p.

We study the (coarse) moduli scheme $\mathcal{A}_g$ of polarized abelian varieties of dimension g in characteristic p; this notation is used instead of $\mathcal{A}_g \otimes \mathbb{F}_p$. We write $\mathcal{A}_{g,1}$ for the moduli scheme of principally polarized abelian varieties of dimension g in characteristic p. We will use letters like A, B to denote abelian varieties.

For the notion of a p-divisible group we refer to the literature, e.g., [13]; see also [3], 1.18. Instead of the term p-divisible group, the equivalent notion "Barsotti-Tate group" is used. We will use letters like X, Y to denote a p-divisible group. For an abelian variety A, or an abelian scheme, and a prime number p we write $A[p^\infty] = \cup_i A[p^i] = X$ for its p-divisible group.

For finite group schemes and for p-divisible groups over a perfect field in characteristic p we use the theory of *covariant* Dieudonné modules. In [21] the contravariant theory was developed. However, it turned out that the covariant theory was easier to handle in deformation theory; see [30], 15.3 for references.

A warning and a remark on notation. Under the *covariant* Dieudonné module theory the Frobenius morphism on a group scheme is transformed into the Verschiebung homomorphism on its Dieudonné module; this homomorphism is denoted by $\mathcal{V}$; the analogous statement for V being transformed into $\mathcal{F}$; in shorthand notation $\mathbb{D}(F) = \mathcal{V}$ and $\mathbb{D}(V) = \mathcal{F}$, see [30], 15.3. In order not to confuse F on group schemes and the Frobenius on modules we have chosen the notation $\mathcal{F}$ and $\mathcal{V}$. An example: for an abelian variety A over a perfect field, writing $\mathbb{D}(A[p^\infty]) = M$ we have $\mathbb{D}(A[F]) = M/\mathcal{V}M$.

1.1. Newton polygons. Suppose we are given integers $h, d \in \mathbb{Z}_{\geq 0}$; here $h =$ "height," $d =$ "dimension." In case of abelian varieties we will choose $h = 2g$, and $d = g$. A Newton polygon γ (related to h and d) is a polygon $\gamma \subset \mathbb{Q} \times \mathbb{Q}$ (or, if you wish, in $\mathbb{R} \times \mathbb{R}$), such that

- γ starts at $(0,0)$ and ends at (h,d);
- γ is lower convex;
- any slope β of γ has the property $0 \leq \beta \leq 1$;
- the breakpoints of γ are in $\mathbb{Z} \times \mathbb{Z}$; hence $\beta \in \mathbb{Q}$.

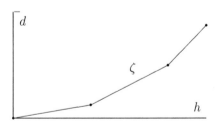

Note that a Newton polygon determines (and is determined by)

$$\beta_1, \cdots, \beta_h \in \mathbb{Q} \text{ with } 0 \leq \beta_1 \leq \cdots \leq \beta_h \leq 1 \quad \leftrightarrow \quad \zeta.$$

Sometimes we will give a Newton polygon by data $\sum_i (m_i, n_i)$; here $m_i, n_i \in \mathbb{Z}_{\geq 0}$, with $\gcd(m_i, n_i) = 1$, and $m_i/(m_i + n_i) \leq m_j/(m_j + n_j)$ for $i \leq j$, and $h = \sum_i (m_i + n_i)$, $d = \sum_i m_i$. From these data we construct the related Newton polygon by choosing the slopes $m_i/(m_i+n_i)$ with multiplicities $h_i = m_i + n_i$. Conversely clearly any Newton polygon can be encoded in a unique way in such a form.

Let ζ be a Newton polygon. Suppose that the slopes of ζ are $1 \geq \beta_1 \geq \cdots \geq \beta_h \geq 0$; this polygon has slopes $\beta_h, \cdots, \beta_1$ (nondecreasing order), and it is *lower convex*. We write ζ^* for the polygon starting at $(0,0)$ constructed using the slopes $\beta_1, \cdots, \beta_h$ (nonincreasing order); note that ζ^* is *upper convex*, and that the beginning and end point of ζ and of ζ^* coincide. Note that $\zeta = \zeta^*$ iff ζ is isoclinic (i.e., there is only one slope).

We say that ζ is *symmetric* if $h = 2g$ is even and the slopes $1 \geq \beta_1 \geq \cdots \geq \beta_h \geq 0$ satisfy $\beta_i = 1 - \beta_{h-i+1}$ for $1 \leq i \leq h$. We say that ζ is supersingular, and we write $\zeta = \sigma$, if all slopes are equal to $1/2$. A symmetric Newton polygon is isoclinic this is the case iff the Newton polygon is supersingular.

1.2. We will associate to a p-divisible group X over a field K its Newton polygon $\mathcal{N}(X)$. This will be the "Newton polygon of the characteristic polynomial of Frobenius on X"; this terminology is incorrect in case K is not the prime field $\mathbb{F}_p$. Here is a precise definition.

Let $m, n \in \mathbb{Z}_{\geq 0}$; we are going to define a p-divisble group $G_{m,n}$. We write $G_{1,0} = \mathbb{G}_m[p^\infty]$ and $G_{0,1} = \mathbb{Q}_p/\mathbb{Z}_p$. For positive, coprime values of m and n we choose a perfect field K, we write $M_{m,n} = R_K/R_K(\mathcal{V}^n - \mathcal{F}^m)$, where R_K is the Dieudonné ring. We define $G_{m,n}$ by $\mathbb{D}(G_{m,n}) = M_{m,n}$. Note that this works over any perfect field. This p-divisible group is defined over $\mathbb{F}_p$ and we will use the same notation over any field K, instead of writing $(G_{m,n})_K = (G_{m,n})_{\mathbb{F}_p} \otimes K$. Note that $M_{m,n}/\mathcal{V} \cdot M_{m,n}$ is a K-vector space of dimension m. Hence the dimension of $G_{m,n}$ is m. We see that the height of $G_{m,n}$ is $h = m + n$. We can show that under Serre-duality we have $G_{m,n}^t = G_{n,m}$.

We define $\mathcal{N}(G_{m,n})$ as the polygon that has slope $m/(m + n)$ with multiplicity $h = m + n$. Note this is the F-slope on $G_{m,n}$, and it is the $\mathcal{V}$-slope on $M_{m,n}$. Indeed, over $\mathbb{F}_p$ the Frobenius $F : G_{m,n} \to G_{m,n}$ has the property $F^{m+n} = F^m \mathcal{V}^m = p^m$.

Let X be a p-divisble group over a field K. Choose an algebraic closure $K \subset k$. Choose an isogeny $X_k \sim \quad \Pi_i (G_{m_i, n_i})$; see 7.1 and 7.2. We define $\mathcal{N}(X)$ as the "union" of these $\mathcal{N}(G_{m_i, n_i})$, i.e., take the slopes of these isogeny factors, and order all slopes in nondecreasing order. By the Dieudonné-Manin theorem we know that *over an algebraically closed field there is a bijective correspondence between isogeny classes of p-divisible groups on the one hand, and Newton polygons on the other hand;* see 7.2. For an abelian variety A we write $\mathcal{N}(A)$ instead of $\mathcal{N}(A[p^\infty])$.

For a commutative group scheme G over a field K we define the number $f = f(G)$ by $\mathrm{Hom}(\mu_p, G_k) \cong (\mathbb{Z}/p)^f$, where k is an algebraically closed field. For a p-divisble group X, respectively an abelian variety A, the number $f(X)$, respectively $f(A)$, is called the p-rank. Note that in these cases this number is the multiplicity of the slope equal to one in the Newton polygon.

For an abelian variety A, its Newton polygon ξ is *symmetric*; by definition this means that the multiplicity of the slope β in ξ is the same as the multiplicity of the slope $1 - \beta$. This was proved by Manin over finite fields. The general case follows from the duality theorem [28] 19.1; we see that $A^t[p^\infty] = A[p^\infty]^t$; using, moreover, $(G_{m,n})^t = G_{n,m}$ and the definition of the Newton polygon of a p-divisible group, we conclude that $\mathcal{N}(A)$ is symmetric.

1.3. The graph of Newton polygons. For Newton polygons we introduce a partial ordering.

We write $\zeta_1 \succ \zeta_2$ if ζ_1 is "below" ζ_2,
    i.e., if no point of ζ_1 is strictly above ζ_2.

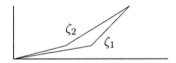

1.4. Newton polygon strata. If S is a base scheme and $\mathcal{X} \to S$ is a p-divisible group over S, we write

$$\mathcal{W}_\zeta(S) = \{s \in S \mid \mathcal{N}(\mathcal{X}_s) \prec \zeta\} \subset S$$

and

$$\mathcal{W}_\zeta^0(S) = \{s \in S \mid \mathcal{N}(\mathcal{X}_s) = \zeta\} \subset S.$$

Grothendieck showed in his Montreal notes [11] that "Newton polygons go up under specialization." The proof was worked out by Katz, see 7.8.

1.5. Minimal p-divisible groups. See [36] and [37]. In the isogeny class of $G_{m,n}$ we single out one p-divisible group $H_{m,n}$ specifically; for a description see [15], 5.3 - 5.7; the p-divisible group $H_{m,n}$ is defined over $\mathbb{F}_p$, it is isogenous with $G_{m,n}$, and

> *the endomorphism ring* $\mathrm{End}(H_{m,n} \otimes k)$ *is the maximal order*
> *in the endomorphism algebra* $\mathrm{End}(H_{m,n} \otimes k) \otimes \mathbb{Q}$;

these conditions determine $H_{m,n} \otimes \overline{\mathbb{F}_p}$ up to isomorphism. This p-divisible group $H_{m,n}$ is called *minimal*.

One can define $H_{m,n}$ over $\mathbb{F}_p$ by defining its (covariant) Dieudonné module by $\mathbb{D}(H_{(m,n),\mathbb{F}_p}) = M_{(m,n),\mathbb{F}_p}$; this module has a basis as a free module over $W = W_\infty(\mathbb{F}_p)$ given by $\{e_0, \cdots, e_{h-1}\}$, where $h = m + n$, write $p \cdot e_i = e_{i+h}$ inductively for all $i \geq 0$, there is an endomorphism $\pi \in \mathrm{End}(H_{(m,n),\mathbb{F}_p})$ with $\pi(e_i) = e_{i+1}$, and $\pi^n = \mathcal{F} \in \mathrm{End}(H_{(m,n),\mathbb{F}_p})$ and $\pi^m = \mathcal{V} \in \mathrm{End}(H_{(m,n),\mathbb{F}_p})$, hence $\pi^h = p \in \mathrm{End}(H_{(m,n),\mathbb{F}_p})$.

If $\zeta = \sum_i (m_i, n_i)$ we write $H(\zeta) := \sum_i H_{m_i,n_i}$, the minimal p-divisible group with Newton polygon equal to ζ. We write $G(\zeta) = H(\zeta)[p]$, the minimal BT_1 group scheme attached to ζ.

In case $\mu \in \mathbb{Z}_{>0}$ we write

$$H_{d,c} = (H_{m,n})^\mu, \quad \text{where} \quad d := \mu m, \quad c := \mu n, \quad \gcd(m,n) = 1.$$

For further information see 7.3.

1.6. EO-strata. Basic reference: [30]. We say that G is a BT_1 group scheme or a p-divisible group truncated at level one if is is annihilated by p, and the image of V and the kernel of F are equal; for more information see [13], 1.1. Let $X \to S$ be a p-divisble group over a base S (in characteristic p). We write

$$\mathcal{S}_G(S) := \{s \in S \mid \exists \Omega \ \ X_s[p] \otimes \Omega \cong G \otimes \Omega\}.$$

This is called the Ekedahl-Oort stratum defined by X/S. This is a locally closed subset in S. Polarizations can be considered, but are not taken into account in the definition of $\mathcal{S}_G(-)$. See [30], Section 9, for the case of principal polarizations.

Let G be a BT_1 group scheme over an algebraically closed field that is symmetric in the sense of [30], 5.1, i.e., there is an isomorphism $G \cong G^D$. To G we attached in [30], 5.6, an elementary sequence, denoted by $\mathrm{ES}(G)$. An important point is the fact (not easy in case $p = 2$) that a "principally polarized" BT_1 group scheme over an algebraically closed field is uniquely determined by this sequence; this was proved in [30], Section 9; in case $p > 2$ the proof is much easier, and the fact holds in a much more general situation, see [26], Section 5, in particular Corollary 5.4.

1.7. Foliations. Basic reference: [32]. Let X be a p-divisible group over a field K and let $\mathcal{Y} \to S$ be a p-divisible group over a base scheme S. We write

$$\mathcal{C}_X(S) = \{s \in S \mid \exists \Omega, \exists \mathcal{Y}_s \otimes \Omega \cong X \otimes \Omega\};$$

here Ω is an algebraically closed field containing $\kappa(s)$ and K. Consider a quasi-polarized p-divisible group (X, λ) over a field. Let $(\mathcal{Y}, \mu) \to S$ be a quasi-polarized p-divisible group over a base scheme S. We write

$$\mathcal{C}_{(X,\lambda)}(S) = \{s \in S \mid \exists \Omega, \exists (\mathcal{Y}, \mu)_s \otimes \Omega \cong (X, \lambda) \otimes \Omega\}.$$

In [4] we find a more precise definition, which also takes care of the behavior of the polarization at places prime to p. See 7.12 for the fact that *any central leaf is closed in an open Newton polygon stratum.*

We write $\mathcal{I}_X(S)$ and $\mathcal{I}_{(X,\lambda)}(S)$ for the notion of isogeny leaves introduced in [32], Section 4, see 4.10 and 4.11. We recall the definition in the polarized case $S = \mathcal{A}_g \otimes \mathbb{F}_p$. Let $x = [(X, \lambda)]$ be given over a perfect field. Write $\mathcal{H}_\alpha(x)$ for the set of points in $\mathcal{A}_g \otimes \mathbb{F}_p$ connected to x by iterated α_p-isogenies (over extension fields). In general this is not a closed subset of $\mathcal{A}_g \otimes \mathbb{F}_p$. However, the union of all irreducible components of $\mathcal{H}_\alpha(x)$ containing x is a closed subset; this subset with the induced reduced scheme structure is denoted by $\mathcal{I}_{(X,\lambda)}(\mathcal{A}_g \otimes \mathbb{F}_p)$; for the definition in the general (un)polarized case, and for existence theorems, see [32], Section 4. Note that formal completion of $\mathcal{I}_{(X,\lambda)}(\mathcal{A}_g \otimes \mathbb{F}_p)$ at the point x is the reduced, reduction mod p of the related Rapoport-Zink space; an analogous statement holds for the unpolarized case; for the definition of these spaces see [39], Section 2 for the unpolarized case and Chapter 3 for the polarized case.

1.8. Isogeny correspondences. Suppose $X \to S$ and $Y \to T$ are p-divisible groups. Consider triples $(f : U \to S,\ g : U \to T,\ \psi : X_f \to Y_g)$, where $f : U \to X$ and $g : U \to T$ are morphisms, and where $\psi : X_f = X \times_U S \to Y_g = Y \times_U S$ is an isogeny. An object representing such triples in the category of schemes over $S \times T$ is called an isogeny correspondence.

Consider polarized abelian schemes $(A, \mu) \to S$ and $(B, \nu) \to T$. Triples $(f : U \to X,\ g : U \to T,\ \psi : A_f \to B_g)$ such that $f^*(\mu) = g^*(\nu)$ define isogeny correspondences between families of polarized abelian varieites. These are also called Hecke correspondences. See [9], VII.3, for a slightly more general notion. See [3] for a discussion.

One important feature in our discussion is the fact that isogeny correspondence are finite-to-finite above central leaves. But note that isogeny correspondences in general blow up and down as correspondences in $(\mathcal{A}_g \otimes \mathbb{F}_p) \times (\mathcal{A}_g \otimes \mathbb{F}_p)$.

1.9. Local deformation spaces. Let X_0 be a p-divisible group over a field K. We write $\mathrm{Def}(X_0)$ for the local deformation space in characteristic p of X_0. By this we mean the following. Consider all local Artin rings R with a residue class homomorphism $R \to K$ such that $p{\cdot}1 = 0$ in R. Consider all p-divisble groups X over $\mathrm{Spec}(R)$ plus an identification $X \otimes_R K = X_0$. This functor on the category of such algebras is prorepresentable. The prorepresenting formal scheme is denoted by $\mathrm{Def}(X_0)$.

The prorepresenting formal p-divisible group can be written as $\mathcal{X} \to \mathrm{Def}(X_0) = \mathrm{Spf}(\Gamma)$. This affine formal scheme comes from a p-divisible group over $\mathrm{Spec}(\Gamma)$, e.g., see [14], 2.4.4. This object will be denoted by $X \to \mathrm{Spec}(\Gamma) =: D(X_0)$.

An analogous definition can be given for the local deformation space $\mathrm{Def}(X_0, \mu_0) = \mathrm{Spf}(\Gamma)$ of a quasi-polarized p-divisible group. In this case we will write $D(X_0, \mu_0) = \mathrm{Spec}((\Gamma))$.

Consider the local deformation space $\mathrm{Def}(A_0, \mu_0)$ of a polarized abelian variety (A_0, μ_0). By the Chow-Grothendieck algebraization theory, see [10], III[1].5.4, we know that there exists a polarized abelian scheme $(A, \mu) \rightarrow D(A_0, \mu_0) := \mathrm{Spec}(\Gamma)$ of which the corresponding formal scheme is the prorepresenting object of this deformation functor.

2 Computation of the dimension of automorphism schemes

Consider minimal p-divisible groups as in 1.5, and their BT_1 group schemes $H_{d,c}[p]$. Consider homomorphism group schemes between such automorphism group schemes and their dimensions. These automorphism group schemes are as defined in [42], 5.7, and the analogous definition for homomorphism group schemes. *In this section we compute the dimensions of* Hom-*schemes and of* Aut-*schemes.* In order to compute these dimensions it suffices to compute the dimension of such schemes of homomorphisms and automorphisms between Dieudonné modules, as explained in [42], 5.7. These computations use methods of proof, as in [23], Sections 4 and 5, [25], [37], 2.4. We carry out the proof of the first proposition, and leave the proof of the second, which is also a direct verification, to a future publication.

2.1. Proposition. *Suppose* $a, b, d, c \in \mathbb{Z}_{\geq 0}$; *assume that* $a/(a+b) \geq d/(d+c)$. *Then:*

$$\dim\left(\underline{\mathrm{Hom}}(H_{a,b}[p], H_{d,c}[p])\right) = bd = \dim\left(\underline{\mathrm{Hom}}(H_{d,c}[p], H_{a,b}[p])\right);$$

$$\dim\left(\underline{\mathrm{Aut}}(H_{d,c}[p])\right) = dc.$$

In fact, much more is true in case of minimal p-divisible groups. For $I, J \in \mathbb{Z}_{>0}$ we have

$$\dim(\underline{\mathrm{Hom}}(H_{a,b}[p^I], H_{d,c}[p^J])) \quad = \quad \dim(\underline{\mathrm{Hom}}(H_{a,b}[p], H_{d,c}[p])).$$

Proof. If $a' = \mu \cdot a$ and $b' = \mu \cdot b$, we have $H_{a',b'} \cong (H_{a,b})^\mu$. Hence it suffices to compute these dimensions in case $\gcd(a, b) = 1 = \gcd(d, c)$. From now on we suppose we are in this case. We distinguish three possibilities:
  (1) $1/2 \geq a/(a+b)$;
  (2) $a/(a+b) \geq 1/2 d/(d+c)$:
  (3) $a/(a+b) \geq d/(d+c) \geq 1/2$.
We will see that a proof of (2) is easy. Note that once (1) is proved, (3) follows by duality; indeed, $(H_{a,b})D = H_{b,a}$. Most of the work will be devoted to proving the case (1).

We remind the reader of some notation introduced in [37]. Finite words with letters $\mathcal{F}$ and $\mathcal{V}$ are considered. They are treated in a cyclic way, finite cyclic words repeat itself infinitely often. For such a word w a finite BT_1 group scheme G_w over a perfect field K is constructed by taking a basis for $\mathbb{D}(G_w) = \sum_{a \leq i \leq h} K.z_i$ of the same cardinality as the number h of letters in w. For $w = L_1 \cdots L_h$ we define

$$L_i = \mathcal{F} \quad \Rightarrow \quad \mathcal{F}z_i = z_{i+1}, \quad \mathcal{V}z_{i+1} = 0;$$

$$L_i = \mathcal{V} \quad \Rightarrow \quad \mathcal{V}z_{i+1} = z_i, \quad \mathcal{F}z_i = 0;$$

i.e., the $L_i = \mathcal{F}$ acting clockwise in the circular set $\{z_i, \cdots, z_h\}$ and $\mathcal{V}$ acting counterclockwise; see [37], page 282. A circular word w defines in this way a (finite) BT_1 group scheme. Moreover, over k a word w is indecomposable iff G_w is indecomposable, see [37], 1.5. By a theorem of Kraft, see [37], 1.5, this classifies all BT_1 group schemes over an algebraically closed field.

We define a *finite string* $\sigma : w' \to w$ between words as a pair $((\mathcal{V}s\mathcal{F}), (\mathcal{F}s\mathcal{V}))$ (see [37] page 283), where s is a finite noncyclic word, $(\mathcal{V}s\mathcal{F})$ is contained in w' and $(\mathcal{F}s\mathcal{V})$ is contained in w; note that "contained in w" means that it is a subword of $\cdots www \cdots$. In [37], 2.4, we see that for indecomposable words w', w a k-basis for $\mathrm{Hom}(G_{w'}, G_w)$ can be given by the set of strings from w' to w. From this we conclude that

$$\dim \left(\underline{\mathrm{Hom}}(G_{w'}, G_w)\right) \quad \text{equals the number of strings from } w' \text{ to } w.$$

For $G_{w'} = H_{a,b}$ we write $\mathbb{D}(H_{a,b}) = W \cdot e_0 \oplus \cdots \oplus W \cdot e_{a+b-1}$, with $\mathcal{F}e_i = e_{i+b}$ and $\mathcal{V}e_i = e_{i+a}$. For $G_{w'=} = H_{d,c}$ we write

$$\mathbb{D}(H_{a,b}) = W \cdot f_0 \oplus \cdots \oplus W \cdot f_{d+c-1},$$

$$\mathcal{F}e_i = e_{i+c}, \mathcal{V}e_i = e_{i+d}.$$

The number of symbols $\mathcal{V}$ in w' equals b; we choose some numbering $\{\mathcal{V} \mid \mathcal{V} \text{ in } w'\} = \{\nu_1, \cdots, \nu_b\}$. Also we choose $\{\mathcal{F} \mid \mathcal{F} \text{ in } w\} = \{\varphi_1, \cdots, \varphi_d\}$.

Claim. *For indices $1 \leq i \leq b$ and $1 \leq j \leq d$ there exists a unique noncyclic finite word s such that $((\nu_i \, s \, \mathcal{F}), (\varphi_j \, s \, \mathcal{V}))$ is a string from w' to w. This gives a bijective map*

$$\{\nu_1, \cdots, \nu_b\} \times \{\varphi_1, \cdots, \varphi_d\} \quad \longrightarrow \quad \{\text{string} \quad w' \to w\}.$$

Note that the claim proves the first equality in 2.1.

Proof of the Claim, case (2). In this case $b \geq a$ and $d \geq c$. We see that every $\mathcal{F}$ in w' is between letters $\mathcal{V}$, and every $\mathcal{V}$ in w is between letters $\mathcal{F}$. This shows that a string $((\mathcal{V}s\mathcal{F}), (\mathcal{F}s\mathcal{V}))$ can appear in this case only with the empty word s, and that any $(\nu_i \, \mathcal{F})$ and any j gives rise to a unique string $((\nu_i \, \mathcal{F}), (\varphi_j \, \mathcal{V}))$. Hence the claim follows in this case.

Proof of the Claim, case (1). First we note that for a finite word t of length at least the greatest common divisor C of $a + b$ and $d + c$ there is no string

$((\mathcal{V}t\mathcal{F}), (\mathcal{F}t\mathcal{V}))$ from w' to w. Indeed, after applying the first letter, and then C letters in t we should obtain the *same* action on the starting base elements of the string in $\mathbb{D}(G_{w'})$ and in $\mathbb{D}(G_w)$, a contradiction with $\mathcal{V} \neq \mathcal{F}$.

We start with some $\mathcal{V}$ in w' and some $\mathcal{F}$ in w and inductively consider words t such that $(\mathcal{V}t)$ is a subword of w'. We check whether $(\mathcal{F}t)$ is a subword of w. We know that this process stops. Let s be the last word for which $\mathcal{F}s$ is a subword of w. We are going to show that under these conditions $((\mathcal{V}s\mathcal{F}), (\mathcal{F}s\mathcal{V}))$ is a string from w' to w. Indeed, we make the following claim

(1a) *If $(\mathcal{V}t\mathcal{V})$ is contained in w' and $(\mathcal{F}t)$ is contained in w then $(\mathcal{F}t\mathcal{V})$ is contained in w.*

Note that this fact implies the claim; indeed, the first time the inductive process stops it is at $(\mathcal{V}s\mathcal{F})$ in w' and $(\mathcal{F}s\mathcal{V})$ in w.

Suppose that in (1a) the letter $\mathcal{F}$ appears γ times in t and $\mathcal{V}$ appears δ times in t. We see that

$$\mathcal{V}(e_x)t\mathcal{V} = e_N \quad \Longrightarrow \quad N = x - 2a + \gamma b - \delta a \geq 0.$$

Let us write

$$\mathcal{F}(f_y)t = f_M; \quad \text{hence} \quad M = y + c + \gamma c - \delta d.$$

We show that

$$N \geq 0 \quad \& \quad \frac{d}{c} \geq \frac{a}{b} \quad \Longrightarrow \quad M > d.$$

Indeed, since $x \leq a + b - 1$, we conclude that

$$N \geq 0 \quad \Rightarrow \quad a + b - 1 - 2a + \gamma b - \delta a \geq 0 \quad \Rightarrow$$

$$\Rightarrow \quad (\gamma + 1)b \geq (\delta + 1)a \quad \Rightarrow \quad \frac{d}{c} \leq \frac{a}{b} < \frac{\gamma + 1}{\delta + 1}.$$

Hence

$$M = y + (\gamma + 1)c - \delta a \geq (\gamma + 1)c - \delta a > d.$$

We see that $\mathcal{F}(f_M)$ is not defined; since $(\mathcal{F}t)$ is contained in w, say $\mathcal{F}(f_z)t = f_y$, we see that $\mathcal{F}(f_z)t\mathcal{V}$ is defined, i.e., $(\mathcal{F}t\mathcal{V})$ is contained in w. We see that claim (1a) follows. This ends the proof of the first equality in all cases.

For the proof of the second equality we number the symbols $\mathcal{F}$ in w', number the symbols $\mathcal{V}$ in w, and perform a proof analogous to the proof of the first equality. This shows the second equality.

For the third equality we observe that dim $(\underline{\mathrm{Aut}}(H_{d,c}[p]))$ equals the number of finite strings involved, and the result follows. This ends the proof of the proposition. $\qquad\square$

2.2. Proposition *Suppose $d, c \in \mathbb{Z}_{\geq 0}$ with $d > c$. Let λ be a principal quasi-polarization on $H_{d,c} \times H_{c,d}$. Then*

$$\dim(\underline{\mathrm{Aut}}((H_{d,c} \times H_{c,d}, \lambda)[p])) = c(c+1) + dc.$$

Moreover

$$\dim(\underline{\mathrm{Aut}}(((H_{1,1})^r, \lambda)[p])) = \frac{1}{2} \cdot r(r+1)$$

for a principal quasi-polarization λ.

The proof is a direct verification, with methods as in [23], Sections 4 and 5, [25], [37], 2.4.   □

3 Serre-Tate coordinates

See [2] and see [1], §7. For moduli of ordinary abelian varieties there exist canonical Serre-Tate parameters. Ching-Li Chai showed how to generalize that concept from the ordinary case to Serre-Tate parameters on a central leaf in $\mathcal{A}_{g,1}$. Results in this section are due to Chai.

3.1. The Serre-Tate theorem. Let A_0 be an abelian variety, and $X_0 = A_0[p^\infty]$. We obtain a natural morphism

$$\mathrm{Def}(A_0) \xrightarrow{\sim} \mathrm{Def}(X_0), \quad A \mapsto A[p^\infty];$$

a basic theorem of Serre and Tate says that this is an *isomorphism*. An analogous statement holds for (polarized abelian variety) $\mapsto$ (quasi-polarized p-divisible group). See [20], 6.ii; a proof first appeared in print in [22]; also see [7], [16]. See [3], Section 2.

3.2. Let (A, λ) be an *ordinary* principally polarized abelian variety; write $(X, \lambda) = (A, \lambda)[p^\infty]$. Deformations of (A, λ) are described by extensions of $(X, \lambda)_{\mathrm{et}}$ by $(X, \lambda)_{\mathrm{loc}}$. This shows that $\mathrm{Def}(X, \lambda)$ has the structure of a formal group. Let $n \in \mathbb{Z}_{\geq 3}$ be not divisible by p and let $[(A, \lambda, \gamma)] = a \in \mathcal{A}_{g,1,n} \otimes \mathbb{F}_p$. Write $(\mathcal{A}_{g,1,n} \otimes \mathbb{F}_p)^{/a}$ for the formal completion at a. Using the Serre-Tate theorem, see 3.1, we see that we have an isomorphism

$$(\mathcal{A}_{g,1,n} \otimes \mathbb{F}_p)^{/a} \quad \cong \quad (\mathbb{G}_m[p^\infty])^{g(g+1)/2},$$

canonically up to $\mathbb{Z}_p$-linear transformations: *the Serre-Tate canonical coordinates*; see [18]; see [24], Introduction.

Discussion. One can try to formulate an analogous result around a nonordinary point. Generalizations of Serre-Tate coordinates run into several difficulties. In an arbitrary deformation there is no reason that the slope filtration on the p-divisible group should remain constant (as it does in the ordinary case).

Even supposing that the slope filtration remains constant or supposing that the slope subfactors remain constant does not give the desired generalization. However, it turns out that if we suppose that *under deformation the geometric isomorphism type of the p-divisible group remains geometrically constant,* the slope filtration exists and is constant. Describing extensions, Chai arrives at a satisfactory generalization of Serre-Tate coordinates. Note that for the ordinary case and for $f = g - 1$ the leaf is the whole open Newton polygon stratum; however, for p-rank $= f < g - 1$, the inclusion $C(x) \subset W_\xi$ is proper; this can be seen by observing that in these cases isogeny leaves are positive-dimensional, or by using the computation of dimensions we carry out in this paper.

The input for this generalization is precisely the tool provided by the theory of *central leaves* as in [32]. We follow ideas basically due to Ching-Li Chai: we extract from [2], and from [1], §7, the information we need here.

Let Z be a p-divisible group, $\mathrm{Def}(Z) = \mathrm{Spf}(\Gamma)$ and $D(Z) = \mathrm{Spec}(\Gamma)$. Suppose that $Z = X_1 \times \cdots \times X_u$, where the summands are isoclinic of slopes $\nu_1 > \cdots > \nu_u$. Write $Z_{i,j} = X_i \times X_j$.

3.3. Proposition.

$$\dim\left(\mathcal{C}_Z(D(Z))\right) = \sum_{1 \leq i < j \leq u} \dim\left(\mathcal{C}_{Z_{i,j}}(D(Z_{(i,j)}))\right).$$

Note that the "group-like structure" on the formal completion at a point of the leaf $\mathcal{C}_Z(D)$ can be described using the notion of "cascades" as in [24], 0.4.

Let (Z, λ) be a principally quasi-polarized p-divisible group, and consider $D = D(Z, \lambda)$. Suppose that $Z = X_1 \times \cdots \times X_u$, where the summands are isoclinic of slopes $\nu_1 > \cdots > \nu_u$. Then the heights of X_i and X_{u+1-i} are equal and $\nu_i = 1 - \nu_{u+1-i}$. We have the following pairs of summands:
$X_i + X_j$, with $1 \leq i < j < u + 1 - i$ and $X_{u+1-j} + X_{u+1-i}$, and
$X_i + X_{u+1-i}$ for $1 \leq i \leq t/2$.
In this way all pairs are described. Note that
$Z_{i,j} := X_i + X_j + X_{u+1-j} + X_{u+1-i}$ for $1 \leq i < j < u + 1 - i$, and
$S_i := X_i + X_{u+1-i}$ for $1 \leq i \leq u/2$, and $S_{(u+1)/2}$ if u is odd
are principally quasi-polarized p-divisible groups (write the induced polarization again by λ on each of them).

3.4. Proposition.
$$\dim\left(\mathcal{C}_{(Z,\lambda)}(D(Z,\lambda))\right) =$$

$$= \sum_{1 \leq i < j < u+1-i} \dim\left(\mathcal{C}_{Z_{(i,j)}}(D(Z_{(i,j)}))\right) + \sum_{1 \leq i \leq u/2} \dim\left(\mathcal{C}_{(S_i,\lambda)}(D(S_i,\lambda))\right).$$

Note that

$$\{(i,j) \mid 1 \le i < j < u+1-i\} \xrightarrow{\sim} \{(I,J) \mid 1 \le I < J \quad \text{and} \quad u+1-I < J \le u\}$$

is a bijection under the map $(i,j) \mapsto (I = u + 1 - j, J = u + 1 - i)$. Indeed $i < j$ implies $I < J$ and $j < u + 1 - i$ gives $j = u + 1 - I < J = u + 1 - i$. In this case λ gives an isomorphism $X_i \times X_j \xrightarrow{\sim} X_J \times X_I$.

An example. The group structure on a leaf can be easily understood in the case of *two slopes*. This was the starting point for Chai to describe the relevant generalization of Serre-Tate coordinates from the ordinary case to the arbitrary case.

3.5. Theorem (Chai). *Let X be isoclinic of slope ν_X, height h_X and Y isoclinic of slope ν_Y and height h_Y. Suppose $\nu_Y > \nu_X$. Write $Z = Y \times X$. At every point of the central leaf $C = \mathcal{C}_Z(D(Z))$ the formal completion has the structure of a p-divisible group, isoclinic of slope $\nu_Y - \nu_X$, of height $h_X \cdot h_Y$, and*

$$\dim\left(\mathcal{C}_Z(D(Z))\right) \quad = \quad (\nu_Y - \nu_X) \cdot h_X \cdot h_Y.$$

Suppose, moreover, that there exists a principal quasi-polarization λ on Z; this implies $h_X = h_Y$ and $\nu_X = 1 - \nu_Y$. The central leaf $\mathcal{C}_{(Z,\lambda)}(\mathrm{Def}(Z,\lambda))$ has the structure of a p-divisible group, isoclinic of slope $\nu_Y - \nu_X$, of height $h_X \cdot (h_X + 1)/2$, and

$$\dim\left(\mathcal{C}_{(Z,\lambda)}(D(Z,\lambda))\right) \quad = \quad \frac{1}{2}(\nu_Y - \nu_X) \cdot h_X \cdot (h_X + 1).$$

See [1], 7.5.2.

3.6. Let Z be an *isoclinic* p-divisible group. Then $\dim\left(\mathcal{C}_Z(D(Z))\right) = 0$. This can also be seen from a generalization of the previous theorem: take $\nu_Y = \nu_X$. This fact was already known as the isogeny theorem, see [15], 2.17.

4 The dimension of central leaves, the unpolarized case

In this section we compute the dimension of a central leaf in the local deformation space of an (unpolarized) p-divisible group.

4.1. Notation. Let ζ be a Newton polygon, and $(x,y) \in \mathbb{Q} \times \mathbb{Q}$. We write
$(x,y) \prec \zeta$ if (x,y) is on or above ζ,
$(x,y) \not\succeq \zeta$ if (x,y) is strictly above ζ,
$(x,y) \succ \zeta$ if (x,y) is on or below ζ,
$(x,y) \not\preceq \zeta$ if (x,y) is strictly below ζ.

4.2. Notation. We fix integers $h \geq d \geq 0$, and we write $c := h - d$. We consider Newton polygons ending at (h, d). For such a Newton polygon ζ we write

$$\Diamond(\zeta) = \{(x, y) \in \mathbb{Z} \times \mathbb{Z} \mid y < d, \ y < x, \ (x, y) \prec \zeta\},$$

and we write

$$\boxed{\dim(\zeta) := \#(\Diamond(\zeta)).}$$

See 7.10 for an explanation of why we did choose this terminology.

Example:

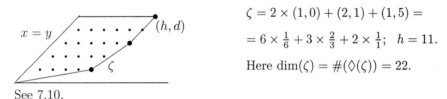

$$\zeta = 2 \times (1, 0) + (2, 1) + (1, 5) =$$

$$= 6 \times \tfrac{1}{6} + 3 \times \tfrac{2}{3} + 2 \times \tfrac{1}{1}; \ \ h = 11.$$

Here $\dim(\zeta) = \#(\Diamond(\zeta)) = 22$.

See 7.10.

4.3. Notation. We write

$$\Diamond(\zeta; \zeta^*) := \{(x, y) \in \mathbb{Z} \times \mathbb{Z} \mid (x, y) \prec \zeta, \ \ (x, y) \not\succeq \zeta^*\},$$

$$\boxed{\text{cdu}(\zeta) := \#(\Diamond(\zeta; \zeta^*))};$$

"cdu" = $\underline{\text{d}}$imension of $\underline{\text{c}}$entral leaf, $\underline{\text{u}}$npolarized case; see 4.5 for an explanation.

We suppose $\zeta = \sum_{1 \leq i \leq u} \mu_i \cdot (m_i, n_i)$, written in such a way that $\gcd(m_i, n_i) = 1$ for all i, and $\mu_i \in \mathbb{Z}_{>0}$, and $i < j \Rightarrow (m_i/(m_i + n_i)) > (m_j/(m_j + n_j))$. Write $d_i = \mu_i \cdot m_i$ and $c_i = \mu_i \cdot n_i$ and $h_i = \mu_i \cdot (m_i + n_i)$; write $\nu_i = m_i/(m_i + n_i) = d_i/(d_i + c_i)$ for $1 \leq i \leq u$. Note that the slope ν_i equals $\text{slope}(G_{m_i, n_i}) = m_i/(m_i + n_i) = d_i/h_i$: this Newton polygon is the "Frobenius-slopes" Newton polygon of $\sum (G_{m_i, n_i})^{\mu_i}$. Note that the slope ν_i appears h_i times; these slopes with these multiplicities give the set $\{\beta_j \mid 1 \leq j \leq h := h_1 + \cdots + h_u\}$ of all slopes of ζ.

4.4. Combinatorial Lemma, the unpolarized case. *The following numbers are equal*

$$\#(\Diamond(\zeta; \zeta^*)) =: \text{cdu}(\zeta) \quad = \quad \sum_{i=1}^{i=h} (\zeta^*(i) - \zeta(i)) =$$

$$= \sum_{1 \leq i < j \leq u} (d_i c_j - d_j c_i) = \sum_{1 \leq i < j \leq u} (d_i h_j - d_j h_i) = \sum_{1 \leq i < j \leq u} h_j \cdot h_i \cdot (\nu_i - \nu_j).$$

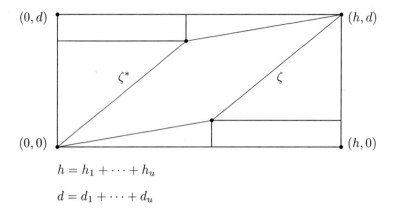

$$h = h_1 + \cdots + h_u$$

$$d = d_1 + \cdots + d_u$$

A **proof** for this lemma is not difficult. The equality $\mathrm{cdu}(\zeta) = \sum (\zeta^*(i) - \zeta(i))$ can be seen as follows. From every break point of ζ^* draw a vertical line up, and a horizontal line to the left; from every break point of ζ draw a vertical line down and a horizontal line to the right. This divides the remaining space of the $h \times d$ rectangle into triangles and rectangles. Pair opposite triangles to a rectangle. In each of these take lattice points, in the interior and in the lower or right hand sides; in this way all lattice points in the large rectangle belong to precisely one of the subspaces; for each of the subspaces we have the formula that the number of such lattice points is the total length of vertical lines. This proves the desired equality for $\mathrm{cdu}(\zeta)$. The other equalities follow by a straightforward computation.   □

4.5. Theorem. (Dimension formula, the unpolarized case.) *Let X_0 be a p-divisible group, $D = D(X_0)$; let $y \in D$, let Y be the p-divisible group given by y with $\beta = \mathcal{N}(Y) \succ \mathcal{N}(X_0)$;*

$$\dim(\mathcal{C}_Y(D)) \quad = \quad \mathrm{cdu}(\beta).$$

Example:

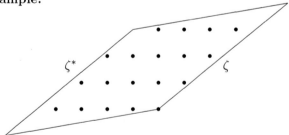

$$\dim(\mathcal{C}_X(D)) = \#\left((\lozenge(\zeta; \zeta^*))\right); \quad \left(\tfrac{4}{5} - \tfrac{1}{6}\right) \cdot 5 \cdot 6 = 19,$$

$$d_1 h_2 - d_2 h_1 = 4 \cdot 6 - 1 \cdot 5 = 19; \quad d_1 c_2 - d_2 c_1 = 4 \cdot 5 - 1 \cdot 1 = 19.$$

First proof. It suffices to prove this theorem in case $Y = X_0$. Write $\mathcal{N}(Y) = \zeta$. By 7.20 it suffices to prove this theorem in case $Y = H(\zeta)$. By 7.4, see 7.5, we know that in this case $\mathcal{C}_Y(D) = S_{Y[p]}(D)$. Let $\beta = \sum \mu_i \cdot (m_i, n_i)$; we suppose that $i < j \Rightarrow m_i/(m_i + n_i) > m_j/(m_j + n_j)$; write $d_i = \mu_i \cdot m_i$ and $d = \sum d_i$; write $c_i = \mu_i \cdot n_i$ and $c = \sum c_i$. We know that $\dim(\mathrm{Def}(Y)) = \dim Y \cdot \dim Y^t = dc$. By 7.28, using 2.1 and 4.4, we conclude that

$$\dim(\mathcal{C}_Y(D)) = \dim(S_{H(\beta)[p]}(D)) = \dim(\mathrm{Def}(Y)) - \dim(\underline{\mathrm{Aut}}(H(\beta)[p])) =$$

$$= (\sum_i d_i)(\sum c_i) - (\sum d_i \cdot c_i) - 2 \cdot \sum_{i<j}(c_i \cdot d_j) = \sum_{i<j}(d_i h_j - d_j h_i) = \mathrm{cdu}(\beta).$$

$$\square 4.5$$

Second proof. Assume, as above, that $Y = X_0 = H_\beta$. Write $Z_{i,j} = H_{d_i,c_i} \times H_{d_j,c_j}$. A **proof** of 4.5 follows from 3.5 using 3.3 and 7.20:

$$\dim(\mathcal{C}_Y(D)) = \sum_{i<j}\dim(\mathcal{C}_{Z_{i,j}}(D(Z_{i,j}))) = \sum_{i<j}h_j \cdot h_i \cdot (\nu_i - \nu_j),$$

where $\nu_i = d_i/(d_i + c_i) = m_i/(m_i + n_i)$ and $h_i = d_i + c_i$. Conclude by using 3.3. $\square 4.5$

4.6. Remark. A variant of the first proof can be given as follows. First prove 4.5 in the case of two slopes, as was done above. Then conclude using 3.3.

4.7. Remark: a third proof. We use a recent result by Eva Viehmann; see [40]. Write

$$\zeta = \sum_j (m_j, n_j), \quad \gcd(m_i, n_j) = 1, \quad h_j = m_j + n_j,$$

$$\lambda_j = m_j/h_j, \quad d = \sum m_j, \quad c = \sum n_j, \quad j < s \Rightarrow \lambda_j \geq \lambda_s.$$

We write $\mathrm{idu}(\zeta)$ for the dimension of the isogeny leaf, as in [32], of $Y = X_0$ in $D = D(X_0)$. By the theory of Rapoport-Zink spaces, as in [39], we see that the reduction modulo p completed at a point gives an isogeny leaf completed at that point. Hence $\mathrm{idu}(\zeta)$ is also the dimension of that Rapoport-Zink space modulo p defined by X_0. This dimension is computed in [40], Theorem B:

$$\mathrm{idu}(\zeta) = \sum_i (m_i - 1)(n_i - 1)/2 + \sum_{i>j} m_i n_j.$$

Let ρ be the ordinary Newton polygon, equal to $d(1,0) + c(0,1)$ in the case studied here. Note that

$$\{(x,y) \mid \rho^* \not\succeq (x,y) \prec \zeta^*\} \cup \{(x,y) \mid \zeta^* \not\succeq (x,y) \prec \zeta\} =$$

$$= \{(x,y) \mid \rho^* \not\succeq (x,y) \prec \zeta\}.$$

We know that $\dim(\zeta) = \mathrm{cdu}(\zeta) + \mathrm{idu}(\zeta)$ by the "almost product structure" on Newton polygon strata, see 7.17. By the computation of Viehmann we see that

$$\mathrm{idu}(\zeta) = \#\left(\{(x,y) \mid \rho^* \not\succeq (x,y) \prec \zeta^*\}\right).$$

Hence the dimension of the central leaf in this case equals

$$\#\left(\{(x,y) \mid \zeta^* \not\succeq (x,y) \prec \zeta\}\right).$$

This proves Theorem 4.5.   □

5 The dimension of central leaves, the polarized case

In this section we compute the dimension of a central leaf in the local deformation space of a polarized p-divisible group, and in the moduli space of polarized abelian varieties.

5.1. Notation. We fix an integer g. For every *symmetric* Newton polygon ξ of height $2g$ we define

$$\Delta(\xi) = \{(x,y) \in \mathbb{Z} \times \mathbb{Z} \mid y < x \le g, \ (x,y) \text{ on or above } \xi\},$$

and we write

$$\boxed{\mathrm{sdim}(\xi) := \#(\Delta(\xi)).}$$

See 7.11 for explanation of notation.

Example:

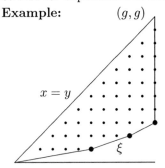

$\dim(\mathcal{W}_\xi(\mathcal{A}_{g,1} \otimes \mathbb{F}_p)) = \#(\Delta(\xi))$,
$\xi = (5,1) + (2,1) + 2 \cdot (1,1) + (1,2) + (1,5)$, $g = 11$;
slopes: $\{6 \times \frac{5}{6}, 3 \times \frac{2}{3}, 4 \times \frac{1}{2}, 3 \times \frac{1}{3}, 6 \times \frac{1}{6}\}$.
This case: $\dim(\mathcal{W}_\xi(\mathcal{A}_{g,1} \otimes \mathbb{F}_p)) = \mathrm{sdim}(\xi) = 48$.

5.2. Let ξ be a symmetric Newton polygon. For convenience we adapt notation to the symmetric situation:

$$\xi = \mu_1 \cdot (m_1, n_1) + \cdots + \mu_s \cdot (m_s, n_s) + r \cdot (1,1) + \mu_s \cdot (n_s, m_s) + \cdots + \mu_1 \cdot (n_1, m_1)$$

with
$$m_i > n_i \text{ and } \gcd(m_i, n_i) = 1 \text{ for all } i,$$
$$1 \leq i < j \leq s \Rightarrow (m_i/(m_i + n_i)) > (m_j/(m_j + n_j)),$$
$$r \geq 0 \text{ and } s \geq 0.$$
We write $d_i = \mu_i \cdot m_i$, and $c_i = \mu_i \cdot n_i$, and $h_i = d_i + c_i$. Write $g :=$
$\left(\sum_{1 \leq i \leq s}(d_i + c_i)\right) + r$ and write $u = 2s + 1$.

We write
$$\triangle(\xi; \xi^*) := \{(x, y) \in \mathbb{Z} \times \mathbb{Z} \mid (x, y) \prec \xi, \quad (x, y) \npreceq \xi^*, \quad x \leq g\},$$

$$\boxed{\mathrm{cdp}(\xi) := \#\left(\triangle(\xi; \xi^*)\right);}$$

"cdp" = dimension of central leaf, polarized case.

Write $\xi = \sum_{1 \leq i \leq u} \mu_i \cdot (m_i, n_i)$, i.e., $(m_j, n_j) = (n_{u+1-j}, m_{u+1-j})$ for $s < j \leq u$ and $r(1, 1) = \mu_{s+1}(m_{s+1}, n_{s+1})$. Write $\nu_i = m_i/(m_i + n_i)$ for $1 \leq i \leq u$; hence $\nu_i = 1 - \nu_{u+1-i}$ for all i.

5.3. Combinatorial Lemma, the polarized case. *The following numbers are equal*

$$\#\left(\triangle(\xi; \xi^*)\right) =: \mathrm{cdp}(\xi) = \frac{1}{2}\mathrm{cdu}(\xi) + \frac{1}{2}(\xi^*(g) - \xi(g)) = \sum_{1 \leq j \leq g} (\xi^*(j) - \xi(j)) =$$

$$= \sum_{1 \leq i \leq s} \left(\frac{1}{2} \cdot d_i(d_i + 1) - \frac{1}{2} \cdot c_i(c_i + 1)\right) + \sum_{\substack{1 \leq i < j \\ j \leq s}} (d_i - c_i)h_j + \left(\sum_{i=1}^{i=s}(d_i - c_i)\right) \cdot r =$$

$$= \frac{1}{2} \sum_{1 \leq i \leq s} (2\nu_i - 1)h_i(h_i + 1) + \frac{1}{2} \sum_{1 \leq i < j \neq u+1-i} (\nu_i - \nu_j)h_i h_j.$$

Example:

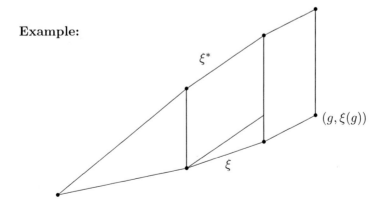

A **proof** of this lemma is not difficult. The first equalities follow from the unpolarized lemma, and from the definitions of cdu(-) and cdp(-). For a proof of the penultimate equality draw vertical lines connecting breakpoints, and then draw lines from the breakpoints of ξ with slopes and lengths as in ξ^*; this divides $\triangle(\xi; \xi^*)$ into subspaces, where lattice points are considered in the interior, and on the lower and righthand sides of the triangles and parallelograms created. Counting points in each of these gives all summands of the right hand side of the last equality.

For the last equality we see that

$$\frac{1}{2} \cdot d_i (d_i + 1) - \frac{1}{2} \cdot c_i (c_i + 1) = \frac{1}{2}(d_i - c_i)(d_i + c_i + 1) = \frac{1}{2}(2\nu_i - 1)h_i(h_i + 1);$$

for $1 \leq i \leq s$ we have

$$2 \cdot (d_i - c_i) \cdot r = \left((\nu_i - \frac{1}{2}) + (\frac{1}{2} - \nu_{u+1-i}) \right) \cdot h_i \cdot 2r =$$

$$= (\nu_i - \nu_{s+1})h_i h_{s+1} + (\nu_{s+1} - \nu_{u+i-1})h_{s+1}h_i;$$

for $1 \leq i < j \leq s$ we have

$$2(d_i - c_i)h_j = 2((d_i c_j - c_i d_j) + (d_i d_j - c_i c_j)) =$$

$$= (\nu_i - \nu_j)h_i h_j + (\nu_i - \nu_{u+1-j})h_i h_j + (\nu_j - \nu_{u+1-i})h_i h_j + (\nu_{u+1-j} - \nu_{u+1-i})h_i h_j;$$

this shows that

$$\sum_{1 \leq i < j}^{j \leq s}(d_i - c_i)h_j = \frac{1}{2}\sum_{1 \leq i < j \neq u+1-i, \ i \neq s+1, \ j \neq s+1} (\nu_i - \nu_j)h_i h_j.$$

Hence the last equality is proved. □

5.4. Theorem (Dimension formula, the polarized case). *Let (A, λ) be a polarized abelian variety. Let $(X, \lambda) = (A, \lambda)[p^\infty]$; write $\xi = \mathcal{N}(A)$; then*

$$\dim \left(\mathcal{C}_{(X,\lambda)}(A \otimes \mathbb{F}_p) \right) \quad = \quad \mathrm{cdp}(\xi).$$

Example:

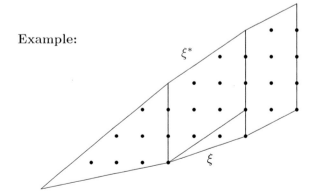

$$\dim(\mathcal{C}_{(A,\lambda)}(\mathcal{A}_g \otimes \mathbb{F}_p)) = \sum_{0 < i \leq g} (\xi^*(i) - \xi(i)),$$

slopes $1/5, 4/5, h = 5$: $\frac{1}{2}4{\cdot}5 - \frac{1}{2}1{\cdot}2 = 9,$

slopes $1/3, 2/3, h = 3$: $\frac{1}{2}2{\cdot}3 - \frac{1}{2}1{\cdot}2 = 2,$

$(d_1 - c_1)h_2 = 3{\cdot}3 = 9,$

$(d_1 + d_2 - c_1 - c_2)r = 4{\cdot}2 = 8,$

$\dim(\mathcal{C}_{(A,\lambda)}(\mathcal{A}_g \otimes \mathbb{F}_p)) = \#(\triangle(\zeta; \zeta^*)) = 28.$

5.5. Notation used in the proof of 5.4. Using 7.7 and 7.22 we need only prove Theorem 5.4 in case λ is a principal polarization on

$$A[p^\infty] = H(\xi) =: Z = Z_1 \times \cdots \times Z_s \times Z_{s+1} \times Z_{s+2} \times \cdots \times Z_u,$$

$$Y_i := Z_i = H_{d_i, c_i}, \quad Z_{s+1} = S_{s+1} = (H_{1,1})^r,$$

$$X_i := Z_{u+1-i} = H_{c_i, d_i} \quad 1 \leq i \leq s+1.$$

Write $S_i = H_{d_i, c_i} \times H_{c_i, d_i}$ for $i \leq s$, and we write λ for the induced quasi-polarization on S_i for $1 \leq i \leq s + 1$; note that $r \geq 0$. We have

$$Z = Y_1 \times \cdots \times Y_s \times Z_{s+1} \times X_s \times \cdots \times X_1.$$

First proof. We have

$$\dim(\underline{\mathrm{Aut}}((Z, \lambda)[p])) =$$

$$\sum_{i \leq s+1} \dim(\underline{\mathrm{Aut}}((S_i, \lambda)[p])) + \tfrac{1}{2} \cdot \sum_{i \neq j \neq u+1-i} \dim(\underline{\mathrm{Hom}}(Z_i, Z_j)) =$$

$$= \sum_{i \leq s+1} \dim(\underline{\mathrm{Aut}}((S_i, \lambda)[p])) \;+\; \sum_{1 \leq i < j \neq u+1-i} \dim(\underline{\mathrm{Hom}}(Z_i, Z_j)).$$

Using 2.1 and 2.2 and using the notation introduced, a computation shows:

$$\mathrm{cdp}(\xi) + \dim(\underline{\mathrm{Aut}}((Z, \lambda)[p])) = \frac{1}{2} \cdot g(g + 1).$$

Indeed, write

$$I = \sum_{1 \leq i \leq s} \left(\frac{1}{2} \cdot d_i(d_i + 1) - \frac{1}{2} \cdot c_i(c_i + 1) \right),$$

$$II = \sum_{\substack{1 \leq i < j \\ j \leq s}} (d_i - c_i)h_j, \quad III = \left(\sum_{i=1}^{i=s}(d_i - c_i) \right) \cdot r.$$

Note that

$$1 \leq i < j \leq s: \quad \dim(\underline{\mathrm{Hom}}(Y_i, Y_j)) = c_i \cdot d_j,$$

$$1 \leq i < j = s+1: \quad \dim(\underline{\mathrm{Hom}}(Y_i, Z_{s+1})) = c_i \cdot r,$$

$$1 \leq i < s < j: \quad \dim(\underline{\mathrm{Hom}}(Y_i, Z_j)) = c_i \cdot c_{u+1-j}, \quad Z_j = X_{u+1-j},$$

$$i = s+1 < j: \quad \dim(\underline{\mathrm{Hom}}(Z_{s+1}, Z_j)) = r \cdot c_{u+1-j},$$

$$s < i < j: \quad \dim(\underline{\mathrm{Hom}}(Z_i, Z_j)) = d_{u+1-i} \cdot c_{u+1-j}.$$

Direct verification gives

$$I + II + III + \sum_{i \leq s}(d_i c_i + c_i(c_i + 1)) \;+\; \frac{1}{2} \cdot r(r+1) \;+$$

$$+ \frac{1}{2} \cdot \sum_{i \neq j \neq u+1-i} \dim(\underline{\mathrm{Hom}}(Z_i, Z_j)) =$$

$$= (d_1 + \cdots + d_s + r + c_s + \cdots + c_1)(d_1 + \cdots + d_s + r + c_s + \cdots + c_1 + 1)/2.$$

First we suppose $p > 2$, and prove the theorem in this case. Indeed, using 7.6 and 7.29 we see that

$$\dim\left(\mathcal{C}_{(X,\lambda)}(\mathcal{A} \otimes \mathbb{F}_p)\right) = \dim(\mathcal{A} \otimes \mathbb{F}_p) - \dim(\underline{\mathrm{Aut}}((Z,\lambda)[p])) = \mathrm{cdp}(\xi).$$

Hence Theorem 5.4 is proved in case $p > 2$.

Let p and q be prime numbers, let ξ be a symmetric Newton polygon, and let $H^{(p)}(\xi)$ respectively $H^{(q)}(\xi)$ be this minimal p-divisible group in characteristic p, respectively in characteristic q, both with a principal quasi-polarization λ. Their elementary sequences as defined in [30] are equal:
Claim.

$$\varphi((H^{(p)}(\xi),\lambda)[p]) = \varphi((H^{(q)}(\xi),\lambda)[q]).$$

The proof is a direct verification: in the process of constructing the canonical filtration, the characteristic plays no role.   □Claim

In order to conclude the proof of Theorem 5.4 in case $p = 2$ we can follow two different roads. One is by using 7.6 and 7.29 we see that

$$\dim\left(\mathcal{C}_{(X,\lambda)}(\mathcal{A} \otimes \mathbb{F}_p)\right) =$$

$$= \dim(\mathcal{A} \otimes \mathbb{F}_p) - \dim(\underline{\mathrm{Aut}}((Z,\lambda)[p])) = \mathrm{cdp}(\xi).$$

This argument in the proof of 5.4 works in all characteristics by the generalization of Wedhorn's 7.29, see 7.30, see [43]; QED for 5.4.

One can also show that once 5.4 is proved in one characteristic, it follows in every characteristic. Here is the argument.
Next we assume $p > 2$ and $q = 2$, and we prove the theorem in characteristic $q = 2$. We have seen that the theorem holds in the case $p > 2$. In that case we know, using 7.6 and [30], Theorem 1.2, that

$$\mathrm{cdp}(\xi) = \dim\left(\mathcal{C}_{(X,\lambda)}(\mathcal{A} \otimes \mathbb{F}_p)\right) = \dim\left(\mathcal{S}_{(X,\lambda)[p]}(\mathcal{A} \otimes \mathbb{F}_p)\right);$$

we see that

$$\mathrm{cdp}(\xi) =| \mathrm{ES}((H^{(p)}(\xi),\lambda)[p]) | \,.$$

Hence
$$\dim\left(\mathcal{C}_{(X,\lambda)}(\mathcal{A}\otimes\mathbb{F}_q)\right)=\varphi((H^{(q)}(\xi),\lambda)[q])=\varphi((H^{(p)}(\xi),\lambda)[p])=\mathrm{cdp}(\xi).$$

This ends the first proof of Theorem 5.4.

5.6. (A proof of 5.4 in the case of two slopes). *Suppose* $\xi=(d,c)+(c,d)$ *with* $d>c$, *i.e.*, $s=1$ *and* $r=0$ *in the notation used above, i.e., the case of a symmetric Newton polygon with only two different slopes. Write* $g=d+c$. *In this case,*

$$\mathrm{cdp}(\xi)=\frac{1}{2}d(d+1)-\frac{1}{2}c(c+1)=$$

$$=\frac{1}{2}\cdot(d-c)(d+c+1)=(1+\cdots+g)\left(\frac{d}{d+c}-\frac{c}{d+c}\right).$$

We choose $X=H_{d,c}\times H_{c,d}$, and $G=X[p]$; let λ be the principal quasi-polarization on X over k. Note that this is unique up to isomorphism, see [32], Proposition 3.7. Let $\varphi(G)=\mathrm{ES}(G)$ be the elementary sequence of G, in the notation and terminology as in [30]. Then

$$\varphi=\{0,\cdots,\varphi(c)=0,1,2,\cdots,\varphi(d)=d-c,d-c,\cdots,d-c\}.$$

Hence in this case,

$$\dim\left(\mathcal{C}_{(X,\lambda)}(\mathcal{A}\otimes\mathbb{F}_p)\right)=c\cdot 0+(1+\cdots+d-c)+c\cdot(d-c)=\mathrm{cdp}(\xi).$$

Proof. In order to write down a final sequence for $(H_{m,n}\times H_{n,m})^\mu$ it suffice to know a canonical filtration for $Z=(H_{m,n}\times H_{n,m})$. Write $\mathbb{D}(H_{m,n})=M_{m,n}$, the covariant Dieudonné module; there is a W-basis $M_{m,n}=W\cdot e_0\oplus\cdots\oplus W\cdot e_{m+n-1}$, and $\mathcal{F}(e_i)=e_{i+n}$, and $\mathcal{V}(e_i)=e_{i+m}$, with the convention $e_{j+m+n}=pe_j$. Also we have $\mathbb{D}(H_{n,m})=M_{n,m}=W\cdot f_0\oplus\cdots\oplus W\cdot f_{m+n-1}$, and $\mathcal{F}(f_j)=f_{j+n}$ and $\mathcal{V}(f_j)=f_{i+m}$; the quasi-polarization can be given by $\langle e_i,f_j\rangle=\delta_{i,m+n-1-j}$. Consider the k-basis for $\mathcal{V}(\mathbb{D}(Z[p]))$ given by

$$\{x_1=e_{m+n-1},\cdots,x_n=e_m,x_{n+1}=e_{m-1},\cdots,x_m=e_n,$$

$$x_{m+1}=f_{m+n-1},\cdots,x_{m+n}=f_m\};$$

this can be completed to a symplectic basis for $\mathbb{D}(Z[p])$; write $N_j=k\cdot x_1\oplus\cdots\oplus k\cdot x_j$ for $1\le j\le m+n$. Direct verification shows that $0\subset N_1\subset\cdots\subset N_{m+n}$ plus the symplectic dual filtration is a final filtration of $\mathbb{D}(Z[p])$. From this we compute φ as indicated, and the result for $H_{m,n}\times H_{n,m}$ follows. This proves the lemma. $\qquad\square$

Remark. It seems attractive to prove 5.4 in the general case along these lines by computing $|\varphi|$. There is an algorithm for determining the canonical filtration in general, but I do not know a closed formula in ξ for computing $|\varphi|$, with $\varphi=\mathrm{ES}(H(\xi))$. Therefore, in the previous proof of 5.4 we made a detour via 7.28.

5.7. Lemma. *Let* $(Z = Y \times X, \lambda)$ *be a principally quasi-polarized p-divisible group, where X is isoclinic of slope ν_X, height h_X, and Y isoclinic of slope ν_Y and height h_Y. Suppose $1 \geq \nu_Y > \frac{1}{2} > \nu_X = 1 - \nu_Y \geq 0$. Write $d_x = h_X \cdot \nu_X$, and $\nu_X = d_X/c_X$; analogous notation for d_Y and c_Y; write $d = d_Y = c_X$, and $c = c_Y = d_X$ and $g = d + c$. Then*

$$\dim\left(\mathcal{C}_{(Z,\lambda)}(D(Z,\lambda))\right) = \frac{1}{2}(\nu_Y - \nu_X)\cdot h_X\cdot(h_X + 1) = \frac{1}{2}d(d+1) - \frac{1}{2}c(c+1).$$

First proof. By 7.22 it suffices to prove this lemma in case $X = H_{c,d}$ and $Y = H_{d,c}$. By 5.6 the result follows.
Second proof. The result follows from 3.5.   □5.7

5.8. Second proof. This **proof** of 5.4 follows from 3.4 using Lemma 4.5 and Lemma 5.7.

                                                            □5.4

5.9. Remark. Third proof in the polarized case; $p > 2$. In [41] the dimension of Rapoport-Zink spaces in the polarized case is computed. Here $p > 2$. Using our computation of cdp($-$), analogous to 4.7, a proof of 5.4 follows from this result by Viehmann.

6 The dimension of Newton polygon strata

The dimension of a Newton polygon stratum in $\mathcal{A}_{g,1}$ is known, see 7.27. However, it was unclear what the possible dimensions of Newton polygon strata in the non-principally polarized case could be. In this section we settle this question, partly solving an earlier conjecture.

6.1. We know that $\dim(\mathcal{W}_\xi(\mathcal{A}_{g,1})) = \mathrm{sdim}(\xi)$, see 5.1 and 7.11. We like to know what the dimension could be of an irreducible component of $\mathcal{W}_\xi^0(\mathcal{A}_g)$. Note that isogeny correspondences blow up and down in general, hence various dimensions a priori can appear.

Write $\mathcal{V}_f(\mathcal{A}_g)$ for the moduli space of polarized abelian varieties having p-rank at most f; this is a closed subset, and we give it the induced reduced scheme structure. By [27], Theorem 4.1 we know that every irreducible component of this space has dimension exactly equal to $(g(g+1)/2)-(g-f) = ((g-1)g/2)+f$ (it seems a miracle that under blowing up and down this locus after all has only components of exactly this dimension).

Let ξ be a symmetric Newton polygon. Let its p-rank be $f = f(\xi)$. This is the multiplicity of the slope 1 in ξ; for a symmetric Newton polygon it is also the multiplicity of the slope 0. Clearly

$$\mathcal{W}_\xi^0(\mathcal{A}_g) \subset \mathcal{V}_{f(\xi)}(\mathcal{A}_g).$$

Hence for every irreducible component

$$T \subset \mathcal{W}_\xi^0(\mathcal{A}_g) \quad \text{we have} \quad \dim(T) \le \frac{1}{2}(g-1)g + f.$$

In [31], 5.8, we find the conjecture that

for any ξ we expect there would be an irreducible component
T of $\mathcal{W}_\xi^0(\mathcal{A}_g)$ with $\dim(T) = ((g-1)g/2) + f(\xi)$.

In this section we settle this question completely by showing that this is true for many Newton polygons, but not true for all. The result is that a component can have the maximal possible (expected) dimension: for many symmetric Newton polygons the conjecture is correct (for those with $\delta(\xi) = 0$, for notation see below), but for every $g > 4$ there exists a ξ for which the conjecture fails (those with $\delta(\xi) > 0$); see 6.3 for the exact statement.

6.2. Notation. Consider $\mathcal{W}_\xi^0(\mathcal{A}_g)$ and consider every irreducible component of this locus; let $\mathrm{minsd}(\xi)$ be the minimum of $\dim(T)$, where T ranges through the set of such irreducible components of $\mathcal{W}_\xi^0(\mathcal{A}_g)$, and let $\mathrm{maxsd}(\xi)$ be the maximum. Write

$$\delta = \delta(\xi) := \lceil (\xi(g)) \rceil - \#(\{(x,y) \in \mathbb{Z} \times \mathbb{Z} \mid f(\xi) < x < g, \ (x,y) \in \xi\}) - 1,$$

where $\lceil b \rceil$ is the smallest integer not smaller than b. Note that $\xi(g) \in \mathbb{Z}$ iff the multiplicity of $(1,1)$ in ξ is even. Here δ stands for "discrepancy." We will see that $\delta \ge 0$. We will see that $\delta = 0$ and $\delta > 0$ are possible.

6.3. Theorem.

$$\mathrm{sdim}(\xi) = \mathrm{minsd}(\xi),$$

and

$$\mathrm{maxsd}(\xi) = \mathrm{cdp}(\xi) + \mathrm{idu}(\xi) = \frac{1}{2}(g-1)g + f(\xi) - \delta(\xi).$$

6.4. Corollary/Examples. *Suppose $\xi = \sum(m_i, n_i)$ with $\gcd(m_i, n_i) = 1$ for all i. Then*

$$\delta(\xi) = 0 \quad \Longleftrightarrow \quad \min(m_i, n_i) = 1, \ \forall i.$$

We see that $0 \le \delta(\xi) \le \lceil g/2 \rceil - 2$. We see that

$$\mathrm{maxsd}(\xi) = \frac{1}{2}(g-1)g/2 + f \quad \Longleftrightarrow \quad \delta(\xi) = 0.$$

We see that $\delta(\xi) > 0$ for example in the following cases:
 $g = 5$ and $\delta((3,2) + (2,3)) = 1$, $\quad g = 8$ and $\delta((4,3) + (1,1) + (3,4)) = 2$,
 more generally, $g = 2k + 1$, and $\delta((k+1,k) + (k, k+1)) = k - 1$,
 $g = 2k + 2$, and $\delta((k+1,k) + (1,1) + (k, k+1)) = k - 1$.

Knowing this theorem, one can construct many examples of pairs of symmetric Newton polygons $\zeta \prec \xi$ such that

$$\mathcal{W}_\zeta^0(\mathcal{A}_g) \quad \not\subset \quad \left(\mathcal{W}_\xi^0(\mathcal{A}_g)\right)^{\mathrm{Zar}}.$$

6.5. Proof of 6.3. Let T be an irreducible component of $\mathcal{W}^0_\xi(\mathcal{A}_g) \otimes k$. Let $\eta \in T$ be the generic point. There exist a finite extension $[L_1 : k(\eta)] < \infty$ and (B, μ) over L_1 such that $[(B, \mu)] = \eta$. There exist a finite extension $[L : L_1] < \infty$ and an isogeny $\varphi : (B_L, \mu_L) \to (C, \lambda)$, where (C, λ) is a principally polarized abelian variety over L. Let T' be the normalization of T in $k(\eta) \subset L$. Let $N = \mathrm{Ker}(\varphi)$. By flat extension there exists a dense open subscheme $T^0 \subset T'$, and a flat extension $\mathcal{N} \subset \mathcal{B}^0 \to T^0$ of $(N \subset B_L)/L$. Hence we arrive at a morphism $(\mathcal{B}^0, \mu) \to (\mathcal{C}, \lambda)$, with $\mathcal{C} := \mathcal{B}^0/\mathcal{N}$, of polarized abelian schemes over T^0. This gives the moduli morphism $\psi : T^0 \to \mathcal{W}^0_\xi(\mathcal{A}_{g,1}) \otimes k$.

We study Isog_g as in [9], VII.4. The morphism $\psi : T^0 \to \mathcal{W}^0_\xi(\mathcal{A}_{g,1}) \otimes k$ extends to an isogeny correspondence. This is proper in both its projections by [9], VII.4.3. Since T is an irreducible component of $\mathcal{W}^0_\xi(\mathcal{A}_g) \otimes k$ this implies that the image of ψ is dense in a component T'' of $\mathcal{W}^0_\xi(\mathcal{A}_{g,1}) \otimes k$. Hence $\dim(T) \geq \dim(T'')$. By 7.11 we have $\dim(T'') = \mathrm{sdim}(\xi)$. This proves the first claim of the theorem.

Let $(A_0, \mu_0) \in \mathcal{W}^0_\xi(\mathcal{A}_g) \otimes k$, and define $(X_0, \mu_0) = (A_0, \mu_0)[p^\infty]$. We obtain

$$\mathrm{Def}(A_0, \mu_0) = \mathrm{Def}(X_0, \mu_0) \subset \mathrm{Def}(X_0),$$

the first equality by the Serre-Tate theorem, and the inclusion is a closed immersion. Moreover, $\mathcal{I}_{(A_0,\mu_0)}(D(A_0, \mu_0)) \subset \mathcal{I}_{X_0}(D(X_0))$. This shows that

$$\mathrm{maxsd}(\xi) \leq \mathrm{cdp}(\xi) + \mathrm{idu}(\xi).$$

We show that in certain cases, for certain degrees of polarization, equality holds.

We choose A_0 such that $X_0 = A_0[p^\infty]$ is minimal. Let J be an irreducible component of $\mathcal{I}_{X_0}(D(X_0))$. Let $\varphi : (Y_0 \times J) \to X$ be the universal family over J defining this isogeny leaf. Let $q = p^n$ be the degree of φ. Define $r = p^{2gn}$. We are going to prove that in $\mathcal{I}_{X_0}\left(\mathcal{W}^0_\xi((\mathcal{A}_{g,r})_k)\right)$ there exists a component I with $I = J$. Hence inside $\mathcal{W}^0_\xi((\mathcal{A}_{g,r})_k)$ there is a component of dimension $\mathrm{cdp}(\xi) + \mathrm{idu}(\xi)$. Choose $[(A_0, \mu_0)] \in \mathcal{W}^0_\xi((\mathcal{A}_{g,r})_k)$ such that $\mathrm{Ker}(\mu_0) = A_0[p^n]$; since X_0 is minimal, this is possible by [32], 3.7.

Claim. In this case,

$$\mathcal{I}_{(A_0,\mu_0)}(D(A_0, \mu_0)) \supset I = J \subset \mathcal{I}_{X_0}(D(X_0)).$$

Let τ be the quasi-polarization on Y_0 obtained by pulling back μ_0 via $Y_0 \to X_0$. Note that the kernel of φ is totally isotropic under the form given by $\tau = \varphi^*(\mu)$. Hence the conditions imposed by the polarization do not give any restrictions and we have proved the claim. This finishes the proof of $\mathrm{maxsd}(\xi) = \mathrm{cdp}(\xi) + \mathrm{idu}(\xi)$.

By 4.5 and 7.17 and by 5.4 we see that $\mathrm{maxsd}(\xi) = \mathrm{cdp}(\xi) + \mathrm{idu}(\xi)$ is the cardinality of the set of (integral points) in the following regions:

$$\triangle(\xi) \cup \{(x,y) \mid (x,y) \not\succeq \xi^*, g < x,\ y < g\} \cup \{(x,y) \mid (x,y) \in \xi^*, g < x,\ y < g\}.$$

Note that

$$\{(x,y) \mid (x,y) \not\succeq \xi^*,\ g < x,\ y < g\} \quad \cong \quad \{(x,y) \mid (x,y) \not\succeq \xi, x < g,\ y > 0\},$$

and

$$\{(x,y) \mid (x,y) \in \xi^*,\ g < x,\ y < g\} \quad \cong \quad \{(x,y) \mid (x,y) \in \xi,\ f(\xi) < x < g\}.$$

Hence

$$\mathrm{cdp}(\xi) + \mathrm{idu}(\xi) \quad = \quad \frac{1}{2}(g-1)g + f - \delta(\xi).$$

$$\square 6.3$$

Remark. Let $q = p^n$ be as above. Actually, we can already construct inside $\mathcal{W}_\xi^0(\mathcal{A}_{g,q}) \otimes k$ a component of dimension equal to $\mathrm{maxsd}(\xi)$; in this way the relevant part of the proof above can be given.

6.6. Explanation. We see the curious fact that on a Newton polygon stratum *the dimension of a central leaf is independent of the degree of the polarization* (which supports the "feeling" that these leaves look like moduli spaces in characteristic zero), while the dimension of an isogeny leaf in general *depends on the degree of the polarization. As we know, Hecke correspondences are finite-to-finite above central leaves, and may blow up and down subsets of isogeny leaves.*

7 Some results used in the proofs

7.1. A basic theorem tells us that the isogeny class of a p-divisible group over an algebraically closed field $k \supset \mathbb{F}_p$ is "the same" as its Newton polygon, see below. Let X be a simple p-divisible group of dimension m and height h over k. In that case we define $\mathcal{N}(X)$ as the isoclinic polygon (all slopes are equal) of slope equal to m/h with multiplicity h. Such a simple p-divisible group exists, see the construction of $G_{m,n}$, [21], page 50; see 1.2; in the *covariant* theory of Dieudonné modules this group can be given (over any perfect field) by the module generated by one element e over the Dieudonné ring, with relation $(\mathcal{V}^n - \mathcal{F}^m)e$. Any p-divisible group X over an algebraically field closed k is isogenous with a product

$$X \quad \sim_k \quad \Pi_i\, (G_{m_i,n_i}),$$

where $m_i \geq 0$, $n_i \geq 0$, and $\gcd(m_i, n_i) = 1$ for every i. In this case the Newton polygon $\mathcal{N}(X)$ of X is defined by all slopes $m_i/(m_i + n_i)$ with multiplicity $h_i := m_i + n_i$.

7.2. Theorem (Dieudonné and Manin), see [21], "Classification theorem" on page 35.

$$\{X\}/\sim_k \quad \xrightarrow{\;\sim\;} \quad \{\text{Newton polygon}\}, \qquad X \mapsto \mathcal{N}(X).$$

This means that for every p-divisible group X over a field we define its Newton polygon $\mathcal{N}(X)$; over an algebraically closed field, every Newton polygon comes from a p-divisible group and

$$X \quad \sim_k \quad Y \quad \Longleftrightarrow \quad \mathcal{N}(X) = \mathcal{N}(Y).$$

7.3. Minimal p-divisible groups. In [36] and [37] we study the following question:

Starting from a p-divisible group X we obtain a BT_1 group scheme

$$[p] : \{X \mid p\text{-divisible group}\}/\cong_k \longrightarrow \{G \mid \text{a } \mathrm{BT}_1\}/\cong_k; X \mapsto G := X[p].$$

This map is known to be surjective. Does $G = X[p]$ determine the isomorphism class of X? This seems a strange question, and in general the answer is "no". It is the main theorem of [36] that the fiber of this map over G up to $\cong_k$ is precisely one p-divisible group X if G is minimal:

7.4. Theorem. *If $G = G(\zeta)$ is minimal over k, and X and Y are p-divisible groups with $X[p] \cong G \cong Y[p]$, then $X \cong Y$; hence $X \cong H(\zeta) \cong Y$.* $\square$

For the notation $H(\zeta)$ see 1.5.

However, things are different if G is not minimal; it is one of the main results of [37] that for a non-minimal BT_1 group scheme G there are infinitely many isomorphism classes X with $X[p] \cong G$.

Note the following important corollaries.

7.5. *Suppose X is a p-divisible group and $G = X[p]$; let $D = D(X)$. Study the inclusion $\mathcal{C}_X(D) \subset \mathcal{S}_G(D)$. Then*

$$X \quad \text{is minimal} \quad \Rightarrow \quad \mathcal{C}_X(D) = \mathcal{S}_G(D).$$

$\square$

7.6. Corollary. *Let (A_0, μ) be a polarized abelian variety. If $A_0[p]$ is minimal, then every irreducible component of $\mathcal{C}_{(A_0,\mu)[p^\infty]}(\mathcal{A}_g)$ is an irreducible component of $\mathcal{S}_{A[p]}(\mathcal{A}_g)$.* $\square$

7.7. Remark. *Let (X, λ') be a quasi-polarized p-divisible group over k, with $\mathcal{N}(X) = \xi$. There exists an isogeny between (X, ξ') and $(H(\xi), \lambda)$, where λ is a principal quasi-polarization.*

See [32], 3.7.

7.8. Newton polygon strata. A theorem by Grothendieck and Katz, see [17], Th. 2.3.1 on page 143, says that for any $\mathcal{X} \to S$ and for any Newton polygon ζ

$$\mathcal{W}_\zeta(S) \subset S \text{ is a } closed \text{ set.}$$

Hence

$$\mathcal{W}_\zeta^0(S) \subset S \text{ is a } locally\ closed \text{ set.}$$

Notation. We do not know a natural way of defining a scheme structure on these sets. These set can be considerd as schemes by introducing the *reduced scheme structure* on these sets.

Sometimes we will write $W_\xi = \mathcal{W}_\xi(\mathcal{A}_{g,1})$ and $W_\xi^0 = \mathcal{W}_\xi^0(\mathcal{A}_{g,1})$ for a symmetric Newton polygon ξ and the moduli space of *principally* polarized abelian varieties.

7.9. Remark. For $\xi = \sigma$, the *supersingular* Newton polygon, the locus W_σ has many components (for $p \gg 0$), see [19], 4.9. However in [4] we find that for a *non-supersingular* Newton polygon the locus $W_\xi = \mathcal{W}_\xi(\mathcal{A}_{g,1})$ is geometrically irreducible.

7.10. Theorem (the dimension of Newton polygon strata in the unpolarized case), see [29], Theorem 3.2 and [31], Theorem 2.10. *Let X_0 be a p-divisible group over a field K; let $\zeta \succ \mathcal{N}(X_0)$. Then:*

$$\dim(\mathcal{W}_\zeta(D(X_0))) \quad = \quad \dim(\zeta).$$

See 4.2 for the definition of $\dim(\zeta)$. □

7.11. Theorem (the dimension of Newton polygon strata in the principally polarized case), see [29], Theorem 3.4 and [31], Theorem 4.1. *Let ξ be a symmetric Newton polygon. Then*

$$\dim\left(\mathcal{W}_\xi(\mathcal{A}_{g,1} \otimes \mathbb{F}_p)\right) = \mathrm{sdim}(\xi).$$

□

See 5.1 for the definition of $\mathrm{sdim}(\xi)$. See Section 6 for what happens for *non-principally* polarized abelian varieties and Newton polygon strata in their moduli spaces.

7.12. Theorem. see [32], Theorem 2.3.

$$\mathcal{C}_X(S) \subset \mathcal{W}_{\mathcal{N}(X)}^0(S)$$

is a closed set.

A proof can be given using 7.13, 7.14, and 7.15. □

7.13. Definition. *Let S be a scheme, and let $X \to S$ be a p-divisible group. We say that X/S is* geometrically fiberwise constant, *abbreviated* gfc *if there exist a field K, a p-divisible group X_0 over K, a morphism $S \to \mathrm{Spec}(K)$, and for every $s \in S$ an algebraically closed field $k \supset \kappa(s) \supset K$ containing the residue class field of s and an isomorphism $X_0 \otimes k \cong_k X_s \otimes k$.*

The analogous terminology will be used for quasi-polarized p-divisible groups and for (polarized) abelian schemes.

See [32], 1.1.

7.14. Theorem (T. Zink & F. Oort). *Let S be an integral, normal Noetherian scheme. Let $\mathcal{X} \to S$ be a p-divisible group with constant Newton polygon. Then there exist a p-divisible $\mathcal{Y} \to S$ and an S-isogeny $\varphi : \mathcal{Y} \to \mathcal{X}$ such that $\mathcal{Y}/S$ is gfc.*

See [44], [38], 2.1, and [32], 1.8. □

7.15. Theorem. *Let S be a scheme that is integral and such that the normalization $S' \to S$ gives a noetherian scheme. Let $\mathcal{X} \to S$ be a p-divisible group; let $n \in \mathbb{Z}_{\geq 0}$. Suppose that $\mathcal{X} \to S$ is gfc. Then there exists a finite surjective morphism $T_n = T \to S$ such that $\mathcal{X}[p^n] \times_S T$ is constant over T.*

See [32], 1.3. □

Note that we gave a "point-wise" definition of $\mathcal{C}_X(S)$; we can consider $\mathcal{C}_X(S) \subset S$ as a closed set, or as a subscheme with induced reduced structure; however is this last definition "invariant under base change"? It would be much better to have a "functorial definition" and a naturally given scheme structure on $\mathcal{C}_X(S)$.

Note that the proof of Theorem 7.12 is quite involved. One of the ingredients is the notion of "completely slope divisible p-divisible groups" introduced by T. Zink, and theorems on p-divisible groups over a normal base, see [44] and [38].

Considering the situation in the moduli space with enough level structure in order to obtain a fine moduli scheme, we see that $C(x) = \mathcal{C}_{(A,\lambda)[p^\infty]}(\mathcal{A}_{g,*,n} \otimes \mathbb{F}_p)$ is regular (as a stack, or regular as a scheme in case sufficiently high level structure is taken into account).

We write $\mathcal{C}_x$ for the irreducible component of $\mathcal{C}_{(A,\lambda)[p^\infty]}\mathcal{A}$ passing through $[(A,\lambda)] = x$.

Remark/Theorem (C.-L. Chai and F. Oort). In fact, for $\mathcal{N}(A) \neq \sigma$, i.e., A is not supersingular, it is known that $\mathcal{C}_{(A,\lambda)[p^\infty]}(\mathcal{A})$ is *geometrically irreducible in every irreducible component of $\mathcal{A}_g$*; see [4].

7.16. We study central and isogeny leaves in a deformation space. We give additional results on deformation spaces of p-divisible groups analogously to the results in the polarized case in [32]. We choose a p-divisible group over a perfect field K. We write $D = D(X)$.

7.17. Proposition. *The central leaf $\mathcal{C}_X(D) \subset D$ is closed. There exists an isogeny leaf (a maximal H_α-subscheme as in [32], §4), $\mathcal{I}_X(D) = I(X) \subset D$. The intersection $\mathcal{C}_X(D) \cap I(X) \subset D$ equals the closed point $[X] = 0 \in D$. There is a natural, finite epimorphism $\mathcal{C}_X(D) \times I(X) \to D$. Hence*

$$\mathrm{cdu}(\zeta) + \mathrm{idu}_X(\zeta) = \dim(D).$$

Here $\mathrm{idu}_X(\zeta)$ is the dimension of $I(X) \subset D$.
The proof of this proposition follows as in [32], §4, (5.1), (5.3).   □

7.18. Corollary. *Let ζ be a Newton polygon. There exists a number $\mathrm{idu}(\zeta)$ such that for every X with $\mathcal{N}(X) = \zeta$ the isogeny leaf in $D = D(X)$ has pure dimension equal to $\mathrm{idu}(\zeta)$.*

This follows because $\dim(\mathcal{C}_X(D))$ and $\dim(D)$ depend only on ζ, see 4.5 and 7.22.   □

7.19. Theorem. Isogeny correspondences, unpolarized case. *Let $\psi : X \to Y$ be an isogeny between p-divisible groups. Then the isogeny correspondence contains an integral scheme T with two finite surjective morphisms*

$$\mathcal{C}_X(D(X)) \quad \twoheadleftarrow \quad T \quad \twoheadrightarrow \quad \mathcal{C}_Y(D(Y))$$

such that T contains a point corresponding with ψ.

7.20. *The dimension of $\mathcal{C}_X(D(X))$ depends only on the isogeny class of X.*   □

7.21. Isogeny correspondences, polarized case. *Let $\psi : A \to B$ be an isogeny, and let λ respectively μ be a polarization on A, respectively on B, and suppose there exists an integer $n \in \mathbb{Z}_{>0}$ such that $\psi^*(\mu) = n \cdot \lambda$. Then there exist finite surjective morphisms*

$$\mathcal{C}_{(A,\lambda)[p^\infty]}(\mathcal{A}_g \otimes \mathbb{F}_p) \quad \twoheadleftarrow \quad T \quad \twoheadrightarrow \quad \mathcal{C}_{(B,\mu)[p^\infty]}(\mathcal{A}_g \otimes \mathbb{F}_p).$$

See [32], 3.16.

7.22. *The dimension of $\mathcal{C}_{(X,\lambda)}(\mathcal{A}_g \otimes \mathbb{F}_p)$ only depends on the isogeny class of (X, λ).*   □

Remark/Notation. In fact, this dimension depends only on the isogeny class of X. We write

$$c(\xi) := \dim\left(\mathcal{C}_{(X,\lambda)}(\mathcal{A}_g \otimes \mathbb{F}_p)\right), \quad X = A[p^\infty], \quad \xi := \mathcal{N}(X);$$

this is well defined: all irreducible components have the same dimension.

A proof of all previous results on isogeny correspondences and the independence of the dimension of the leaf in an isogeny class can be given using 7.13, 7.14, and 7.15; see [32], 2.7 and 3.13.

7.23. Remark. Isogeny correspondences in characteristic p in general blow up and down in a rather wild pattern. The dimensions of Newton polygon strata and of EO-strata in general depends very much on the degree of the polarization under consideration. However, for p-rank strata the dimension in the whole of $\mathcal{A}_g \otimes \mathbb{F}_p$ depends solely on the p-rank; see [27], Theorem 4.1. It seems a miracle that the dimension on central leaves does not depend on the degree of a polarization. See 6.6.

In [32] we also find the definition of *isogeny leaves*, and we see that any irreducible component of $\mathcal{W}_\xi(\mathcal{A}_g \otimes \mathbb{F}_p)$ up to a finite morphism is isomorphic to the product of a central leaf and an isogeny leaf; see [32], 5.3. Note that all central leaves with the same Newton polygon have the same dimension; see 7.20 and 7.22. However, for Newton polygon strata, and hence also for isogeny leaves, the dimension in general depends very much on the degree of the polarization; for more information see Section 6.

7.24. Cayley-Hamilton. See [29]. We would like to compute the dimension of a Newton polygon stratum. In 7.10 and 7.11 we have seen "easy" formulas to compute these dimensions (in the unpolarized or in the principally polarized case). However, up to now there seems to be no really easy proof that these are indeed the correct formulas. In [19] the dimension of the supersingular locus $W_\sigma \subset \mathcal{A}_{g,1} \otimes \mathbb{F}_p$ is computed: every irreducible component of W_σ has dimension equal to $[g^2/4]$. Using purity, see [15], and if we knew we would have a proof that Newton polygon strata in $\mathcal{A}_{g,1} \otimes \mathbb{F}_p$ are nested as predicted by the Newton polygon graph, we would have a proof of 7.11. However, *proofs work the other way around.*

7.25. For a group scheme G over a perfect field K we write $a(G) := \dim_K(\mathrm{Hom}(\alpha_p, G))$. For a local-local p-divisible group X the fact $a(X) = 1$ implies that this Dieudonné module $\mathbb{D}(X)$ is generated by one element over the Dieudonné ring. It turns out that Newton polygon strata are smooth around points where $a = 1$ (in the local deformation space in the unpolarized case, and in the principally polarized case). In this case the local dimension of the deformation space is computed in [29]. More precisely:

7.26. Theorem (CH - unpolarized). *Let X_0 be a p-divisible group over a perfect field K. Suppose $a(X_0) = 1$. Let $\gamma = \mathcal{N}(X_0)$, and let $\gamma \prec \beta$. Let ρ be the ordinary Newton polygon of the same dimension and height as X_0. Define R_β by*

$$\mathcal{W}_\beta(\mathrm{Def}(X_0)) = \mathrm{Spf}(R_\beta);$$

then

$$R_\beta \;\cong\; \frac{K[[z_{x,y} \mid (x,y) \in \Diamond(\rho)]]}{(z_{x,y} \mid (x,y) \notin \Diamond(\beta))} \;\cong\; K[[z_{x,y} \mid (x,y) \in \Diamond(\beta)]].$$

$\square$

7.27. Theorem (CH - polarized). *Let (A_0, λ) be a principally polarized abelian variety over a perfect field K. Suppose $a(A_0) = 1$. Let $\zeta = \mathcal{N}(A_0)$, and let $\zeta \prec \xi$. Let ρ be the ordinary Newton polygon. Define R_ξ by*

$$\mathcal{W}_\xi(\mathrm{Def}(A_0, \lambda_0)) = \mathrm{Spf}(R_\xi);$$

then

$$R_\xi \quad \cong \quad \frac{K[[z_{x,y} \mid (x,y) \in \triangle(\rho)]]}{(z_{x,y} \mid (x,y) \notin \triangle(\beta))} \quad \cong \quad K[[z_{x,y} \mid (x,y) \in \triangle(\beta)]].$$

□

A theorem by Torsten Wedhorn on the dimension of EO-strata, see [42]. *Let X be a p-divisible group, and $G := X[p]$. Consider the EO-stratum $S_G(D(X))$.*

7.28. Theorem (Wedhorn).

$$\dim(S_G(D(X))) = \dim(\mathrm{Def}(X)) - \dim\left(\underline{\mathrm{Aut}}(G)\right).$$

See [42], 6.10.

□

Let (X, λ) be a principally quasi-polarized p-divisible group over a field of characteristic $\mathrm{char}(k) = p > 2$. Write $(G, \lambda) = (X, \lambda)[p]$.

7.29. Theorem (Wedhorn) $\boxed{p > 2}$.

$$\dim(S_{(G,\lambda)}(D(X, \lambda))) = \dim(\mathrm{Def}(X, \lambda)) - \dim\left(\underline{\mathrm{Aut}}((G, \lambda))\right).$$

See [42], 2.8 and 6.10.

□

7.30. In [43] we find a theorem that shows that the previous result also holds in case the characteristic of the base field equals 2.

8 Some questions and some remarks

8.1. In general, the number of lattice points in a region need not be equal to its volume. For example, in the case $\rho = g(1,0) + g(0,1)$ and $\triangle(\rho)$. The same remark holds for $\Diamond(\beta)$ and for $\triangle(\xi; \xi^*)$. However, we make the following observation.

Remark (I thank Cathy O'Neil for this observation). The number $\#(\Diamond(\zeta; \zeta^*)) = \mathrm{cdu}(\zeta)$ as defined and computed in Section 2 is equal to the volume of the region between ζ^* and ζ for every ζ.

8.2. Remark. Using the result 7.8 by Grothendieck and Katz we have defined open and closed Newton polygon strata. Suppose we have symmetric Newton polygons $\zeta \prec \xi$. Then by the definitions we see that

$$\mathcal{W}_\zeta^0(\mathcal{A}_g) \quad \subset \quad \mathcal{W}_\xi(\mathcal{A}_g) \quad \supset \quad \overline{\mathcal{W}_\xi^0(\mathcal{A}_g)}.$$

In general, the last inclusion is not an equality. For example, for $\zeta = \sigma$, the supersingular Newton polygon, and for $\xi = (2,1) + (1,2)$ we can see that $\mathcal{W}_\sigma^0(\mathcal{A}_{3,p^3}) = \mathcal{W}_\sigma(\mathcal{A}_{3,p^3})$ is not contained in the closure of $\mathcal{W}_\xi^0(\mathcal{A}_{3,p^3})$. Using the results of Section 6 we see that many more such examples do exist.

However, for every symmetric ξ in the *principally polarized* case we have

$$ \mathcal{W}_\xi(\mathcal{A}_g) \quad = \quad \overline{\mathcal{W}_\xi^0(\mathcal{A}_g)}. $$

We consider the central leaf $\mathcal{C}_Y(D(X_0) \subset D(X_0))$. Can we describe this locus in the coordinates $z_{x,y}$ given as in 7.26? I.e., does the inclusion $\triangle(\beta; \beta^*) \subset \triangle(\beta)$ induce the inclusion $\mathcal{C}_Y(D) \subset \mathcal{W}_\beta(D)$?

8.3. Question. *Under the identification given in 7.26 can the formal completion C of the central leaf $\mathcal{C}_Y(D(X_0))$ be described by*

$$ C \quad \overset{?}{=} \quad \mathrm{Spf}(K[[z_{x,y} \mid (x,y) \in \Diamond(\beta; \beta^*)]]) \subset \mathcal{W}_\beta(D)? $$

Here $\mathcal{W}_\beta(D) = \mathrm{Spf}(K[[z_{x,y} \mid (x,y) \in \Diamond(\beta)]])$.

8.4. Question. *Under the identification given in 7.27 is it true that the formal completion C of the central leaf $\mathcal{C}_{(B,\mu)}(D(A_0, \lambda))$ can be described by*

$$ C \quad \overset{?}{=} \quad \mathrm{Spf}(K[[z_{x,y} \mid (x,y) \in \triangle(\xi, \xi^*)]]) \subset \mathcal{W}_\xi(D)? $$

Here $\mathcal{W}_\xi(D) = \mathrm{Spf}(K[[z_{x,y} \mid (x,y) \in \triangle(\xi)]])$.

8.5. It seems desirable to have an explicit formula for the elementary sequence of a principally quasi-polarized minimal p-divisible group. If there are only two slopes this is easy. For every explicitly given Newton polygon this can be computed. In case there are more than two slopes, I do not know a general formula. However Harashita has proven, see [12], that for symmetric Newton polygons $\zeta \prec \xi$ and their minimal p divisible groups we have $\mathrm{ES}(H(\zeta)) \subset \mathrm{ES}(H(\xi))$.

8.6. Problem. *Give a simple criterion, in terms of φ and ξ, which decides when an elementary sequence φ appears on an open Newton polygon stratum, i.e., when $S_\varphi \cap W_\xi^0 \neq \emptyset$.*

Added in proof. It seems that this is settled now completely (S. Harashita, T. Wedhorn).

8.7. Conjecture. *Let $\psi_\xi = \mathrm{ES}(H(\xi))$. I expect:*

$$ S_\varphi \cap W_\xi^0 \neq \emptyset \quad \Rightarrow \quad \psi_\xi \subset \varphi. $$

The notation $\psi_\xi \subset \varphi$ stands for $S_{\psi_\xi} \subset \overline{S_\varphi}$, see [30], 14.3.
Added in proof. S. Harashita has proved this conjecture to be true.

References

1. C.-L. Chai – *Hecke orbits on Siegel modular varieties.* Progress in Math. **235**, Birkhäuser, 2004, pp. 71–107.
2. C.-L. Chai – *Canonical coordinates on leaves of p-divisible groups: the two-slope case.* Manuscript 10-I-2005. [Submitted for publication.]
3. C.-L. Chai & F. Oort – *Moduli of abelian varieties and p-divisble groups.* Lectures at the conference on Arithmetic Geometry, Göttingen, July-August 2006. [to appear]
   See arXiv   math.AG/0701479
4. C.-L. Chai & F. Oort – *Monodromy and irreducibility.* [To appear.]
5. C.-L. Chai & F. Oort – *Hecke orbits.* [In preparation]
6. M. Demazure – *Lectures on p-divisible groups.* Lect. Notes Math. 302, Springer–Verlag, Berlin 1972.
7. V. G. Drinfeld – *Coverings of p-adic symmetric domains.* [Funkcional. Anal. i Priložen. 10 (1976), no. 2, 29–40.] Functional Analysis and its Applications, 10 (1976), 107–115.
8. S. J. Edixhoven, B. J. J. Moonen, & F. Oort (Editors) – *Open problems in algebraic geometry.* Bull. Sci. Math. **125** (2001), 1–22.
9. G. Faltings & C.-L. Chai – *Degeneration of abelian varieties.* Ergebnisse Bd 22, Springer–Verlag, 1990.
10. A. Grothendieck & J. Dieudonné - *Eléments de géométrie algébrique. Ch. III^1: Etude cohomologique des faisceaux cohérents.* Publ. Math. 11, IHES 1961.
11. A. Grothendieck – *Groupes de Barsotti-Tate et cristaux de Dieudonné.* Sém. Math. Sup. **45**, Presses de l'Univ. de Montreal, 1970.
12. S. Harashita – *Configuration of the central streams in the moduli of abelian varieties.* Manuscript 33 pp.
    See: http://www.math.sci.hokudai.ac.jp/~harasita/
13. L. Illusie – *Déformations de groupes de Barsotti-Tate.* Exp.VI in: Séminaire sur les pinceaux arithmétiques: la conjecture de Mordell (L. Szpiro), Astérisque 127, Soc. Math. France 1985.
14. A. J. de Jong – *Crystalline Dieudonné module theory via formal rigid geometry.* Publ. Math. IHES **82** (1995), 5–96.
15. A. J. de Jong & F. Oort – *Purity of the stratification by Newton polygons.* Journ. Amer. Math. Soc. **13** (2000), 209–241. See: http://www.ams.org/jams
16. N. M. Katz – *Appendix to Expose V.* In: Surfaces algébriques (Ed. J. Giraud, L. Illusie, M. Raynaud). Lect. Notes Math. 868, Springer–Verlag, Berlin 1981; pp. 127 – 137.
17. N. M. Katz – *Slope filtration of F-crystals.* Journ. Géom. Alg. Rennes, Vol. I, Astérisque **63** (1979), Soc. Math. France, 113 - 164.
18. N. M. Katz – *Serre-Tate local moduli.* In: Surfaces algébriques (Ed. J. Giraud, L. Illusie, M. Raynaud). Lect. Notes Math. 868, Springer–Verlag, Berlin 1981; pp. 138–202.
19. K.-Z. Li & F. Oort - *Moduli of supersingular abelian varieties.* Lect. Notes Math. 1680, Springer–Verlag 1998.
20. J. Lubin, J-P. Serre, & J. Tate – *Elliptic curves and formal groups.* In: Lecture notes prepared in connection with the seminars held at the Summer Institute on Algebraic Geometry, Whitney Estate, Woods Hole, Massachusetts, July 6-July 31, 1964. Mimeographed notes.
    See http://www.ma.utexas.edu/users/voloch/lst.html

Frans Oort

21. Yu. I. Manin – *The theory of commutative formal groups over fields of finite characteristic*. Usp. Math. **18** (1963), 3–90; Russ. Math. Surveys **18** (1963), 1–80.

22. W. Messing – *The crystals associated to Barsotti-Tate groups: with applications to abelian schemes*. Lect. Notes Math. 264, Springer–Verlag 1972.

23. B. Moonen – *Group schemes with additional structures and Weyl group cosets*. In: *Moduli of abelian varieties*. (Ed. C. Faber, G. van der Geer, F. Oort). Progress Math. 195, Birkhäuser Verlag 2001; pp. 255–298 .

24. B. Moonen – *Serre-Tate theory for moduli spaces of* PEL *type*. Ann. Scient. Ec. Norm. Sup. 4e Série **37** (2004), 223–269.

25. B. Moonen – *A dimension formula for Ekedahl - Oort strata*. Ann. Inst. Fourier **54** (2004), 666–698.

26. B. Moonen & T. Wedhorn – *Discrete invariants of varieties in positive characteristic*. International Math. Research Notices **72** (2004), 3855–3903.

27. P. Norman & F. Oort – *Moduli of abelian varieties*. Ann. Math. **112** (1980), 413–439.

28. F. Oort – *Commutative group schemes*. Lect. Notes Math. 15, Springer–Verlag 1966.

29. F. Oort – *Newton polygons and formal groups: conjectures by Manin and Grothendieck*. Ann. Math. **152** (2000), 183–206.

30. F. Oort – *A stratification of a moduli space of polarized abelian varieties*. In: *Moduli of abelian varieties*. (Ed. C. Faber, G. van der Geer, F. Oort). Progress in Math. 195, Birkhäuser Verlag 2001; pp. 345–416.

31. F. Oort – *Newton polygon strata in the moduli space of abelian varieties*. In: *Moduli of abelian varieties*. (Ed. C. Faber, G. van der Geer, F. Oort). Progress in Math. 195, Birkhäuser Verlag 2001; pp. 417–440.

32. F. Oort – *Foliations in moduli spaces of abelian varieties*. Journ. Amer. Math. Soc. **17** (2004), 267–296.

33. F. Oort – *Monodromy, Hecke orbits and Newton polygon strata*. Talk Bonn, 24-II-2003. See: http://www.math.uu.nl/people/oort/

34. F. Oort – *Hecke orbits and stratifications in moduli spaces of abelian varieties*. Talk at the Orsay / SAGA, 14-X-2003. See: http://www.math.uu.nl/people/oort/

35. F. Oort – *Hecke orbits in moduli spaces of abelian varieties and foliations*. Talk at the ETH in Zürich, 2-IV-2004.
    See: http://www.math.uu.nl/people/oort/

36. F. Oort – *Minimal p-divisible groups*. Annals of Math. **161** (2005), 1–16.

37. F. Oort – *Simple p-kernels of p-divisible groups*. Advances in Mathematics **198** (2005), 275–310.

38. F. Oort & T. Zink – *Families of p-divisible groups with constant Newton polygon*. Documenta Mathematica **7** (2002), 183–201.
    See: http://www.mathematik.uni-bielefeld.de/documenta/vol-07/09.html

39. M. Rapoport & Th. Zink – *Period spaces for p-divisible groups*. Ann. Math. Studies **141**, Princeton University Press, 1996.

40. E. Viehmann – *Moduli spaces of p-divisible groups*. Journal of Algebraic Geometry, **17** (2008), 341–374.

41. E. Viehmann – *The global structure of moduli spaces of polarized p-divisible groups*. Document. Math. **13** (2008), 825–852.

42. T. Wedhorn – *The dimension of Oort strata of Shimura varieties of* PEL-*type. Moduli of abelian varieties.* (Ed. C. Faber, G. van der Geer, F. Oort). Progress in Math. 195, Birkhäuser Verlag 2001; pp. 441–471.

43. T. Wedhorn – *Specializations of F-Zips.* Manuscript 22 pp., 20-VI-2005.

44. T. Zink – *On the slope filtration.* Duke Math. Journ. **109** (2001), 79–95.

Derived Categories of Coherent Sheaves and Triangulated Categories of Singularities

Dmitri Orlov

Algebra Section, Steklov Mathematical Institute RAN, Gubkin str. 8, Moscow
119991, RUSSIA
orlov@mi.ras.ru

Dedicated to Yuri Ivanovich Manin on the occasion of his 70th birthday

Summary. In this paper we establish an equivalence between the category of
graded D-branes of type B in Landau–Ginzburg models with homogeneous superpotential W and the triangulated category of singularities of the fiber of W over
zero. The main result is a theorem that shows that the graded triangulated category
of singularities of the cone over a projective variety is connected via a fully faithful
functor to the bounded derived category of coherent sheaves on the base of the cone.
This implies that the category of graded D-branes of type B in Landau–Ginzburg
models with homogeneous superpotential W is connected via a fully faithful functor
to the derived category of coherent sheaves on the projective variety defined by the
equation $W = 0$.

Key words: Triangulated categories of singularities, derived categories of
coherent sheaves, branes, Landau–Ginzburg models

2000 Mathematics Subject Classifications: 18E30, 81T30, 14B05

Introduction[1]

With any algebraic variety X one can naturally associate two triangulated
categories: the bounded derived category $\mathbf{D}^b(\mathrm{coh}(X))$ of coherent sheaves and
the triangulated subcategory $\mathfrak{Perf}(X) \subset \mathbf{D}^b(\mathrm{coh}(X))$ of perfect complexes
on X. If the variety X is smooth, then these two categories coincide. For
singular varieties this is no longer true. In [22] we introduced a new invariant of
a variety X the triangulated category $\mathbf{D}_{\mathrm{Sg}}(X)$ of the singularities of X as the
quotient of $\mathbf{D}^b(\mathrm{coh}(X))$ by the full subcategory of perfect complexes $\mathfrak{Perf}(X)$.
The category $\mathbf{D}_{\mathrm{Sg}}(X)$ captures many properties of the singularities of X.

---

[1]This work was done with partial financial support from the Weyl Fund, from
grant RFFI 05-01-01034, from grant CRDF Award No RUM1-2661-MO-05.

Y. Tschinkel and Y. Zarhin (eds.), *Algebra, Arithmetic, and Geometry*,
Progress in Mathematics 270, DOI 10.1007/978-0-8176-4747-6_16,
© Springer Science+Business Media, LLC 2009

Similarly we can define a triangulated category of singularities $\mathbf{D}_{\mathrm{Sg}}(A)$ for any Noetherian algebra A. We set $\mathbf{D}_{\mathrm{Sg}}(A) = \mathbf{D}^b(\mathrm{mod}\text{-}A)/\mathfrak{Perf}(A)$, where $\mathbf{D}^b(\mathrm{mod}\text{-}A)$ is the bounded derived category of finitely generated right A-modules and $\mathfrak{Perf}(A)$ is its triangulated subcategory consisting of objects that are quasi-isomorphic to bounded complexes of projectives. We will again call $\mathfrak{Perf}(A)$ the subcategory of perfect complexes, but usually we will write $\mathbf{D}^b(\mathrm{proj}\text{-}A)$ instead of $\mathfrak{Perf}(A)$, since this category can also be identified with the derived category of the exact category proj-A of finitely generated right projective A-modules (see, e.g., [19]).

The investigation of triangulated categories of singularities not only is connected with a study of singularities but is mainly inspired by the homological mirror symmetry conjecture [20]. More precisely, the objects of these categories are directly related to D-branes of type B (B-branes) in Landau–Ginzburg models. Such models arise as a mirrors to Fano varieties [15]. For Fano varieties one has the derived categories of coherent sheaves (B-branes), and given a symplectic form, one can propose a suitable Fukaya category (A-branes). Mirror symmetry should interchange these two classes of D-branes. Thus, to extend the homological mirror symmetry conjecture to the Fano case, one should describe D-branes in Landau–Ginzburg models.

To specify a Landau–Ginzburg model in general one needs to choose a target space X, and a holomorphic function W on X called a superpotential. The B-branes in the Landau–Ginzburg model are defined as W-twisted $\mathbb{Z}_2$-periodic complexes of coherent sheaves on X. These are chains $\{\cdots \xrightarrow{d} P_0 \xrightarrow{d} P_1 \xrightarrow{d} P_0 \xrightarrow{d} P_1 \xrightarrow{d} P_0 \xrightarrow{t} \cdots\}$ of coherent sheaves in which the composition of differentials is no longer zero, but is equal to multiplication by W (see, e.g., [17, 22, 23]). In the paper [22] we analyzed the relationship between the categories of B-branes in Landau–Ginzburg models and triangulated categories of singularities. Specifically, we showed that for an affine X the product of the triangulated categories of singularities of the singular fibers of W is equivalent to the category of B-branes of (X, W).

In this paper we consider the graded case. Let $A = \bigoplus_i A_i$ be a graded Noetherian algebra over a field $\mathbf{k}$. We can define the triangulated category of singularities $\mathbf{D}_{\mathrm{Sg}}^{\mathrm{gr}}(A)$ of A as the quotient $\mathbf{D}^b(\mathrm{gr}\text{-}A)/\mathbf{D}^b(\mathrm{grproj}\text{-}A)$, where $\mathbf{D}^b(\mathrm{gr}\text{-}A)$ is the bounded derived category of finitely generated graded right A-modules and $\mathbf{D}^b(\mathrm{grproj}\text{-}A)$ is its triangulated subcategory consisting of objects that are isomorphic to bounded complexes of projectives.

The graded version of the triangulated category of singularities is closely related to the category of B-branes in Landau–Ginzburg models (X, W) equipped with an action of the multiplicative group $\mathbf{k}^*$ for which W is semi-invariant. The notion of grading on D-branes of type B was defined in the papers [16, 28]. In the presence of a $\mathbf{k}^*$-action one can construct a category of graded B-branes in the Landau–Ginzburg model (X, W) (Definition 30 and Section 3.3). Now our Theorem 39 gives an equivalence between the category of graded B-branes and the triangulated category of singularities $\mathbf{D}_{\mathrm{Sg}}^{\mathrm{gr}}(A)$, where A is such that $\mathbf{Spec}(A)$ is the fiber of W over 0.

This equivalence allows us to compare the category of graded B-branes and the derived category of coherent sheaves on the projective variety that is defined by the superpotential W. For example, suppose X is the affine space $\mathbb{A}^N$ and W is a homogeneous polynomial of degree d. Denote by $Y \subset \mathbb{P}^{N-1}$ the projective hypersurface of degree d that is given by the equation $W = 0$. If $d = N$, then the triangulated category of graded B-branes $\mathrm{DGrB}(W)$ is equivalent to the derived category of coherent sheaves on the Calabi–Yau variety Y. Furthermore, if $d < N$ (i.e., Y is a Fano variety), we construct a fully faithful functor from $\mathrm{DGrB}(W)$ to $\mathbf{D}^b(\mathrm{coh}(Y))$, and if $d > N$ (i.e., Y is a variety of general type), we construct a fully faithful functor from $\mathbf{D}^b(\mathrm{coh}(Y))$ to $\mathrm{DGrB}(W)$ (see Theorem 40).

This result follows from a more general statement for graded Gorenstein algebras (Theorem 16). It gives a relation between the triangulated category of singularities $\mathbf{D}_{\mathrm{Sg}}^{\mathrm{gr}}(A)$ and the bounded derived category $\mathbf{D}^b(\mathrm{qgr}\,A)$, where qgr A is the quotient of the abelian category of graded finitely generated A-modules by the subcategory of torsion modules. More precisely, for Gorenstein algebras we obtain a fully faithful functor between $\mathbf{D}_{\mathrm{Sg}}^{\mathrm{gr}}(A)$ and $\mathbf{D}^b(\mathrm{qgr}\,A)$, and the direction of this functor depends on the Gorenstein parameter a of A. In particular, when the Gorenstein parameter a is equal to zero, we obtain an equivalence between these categories. Finally, the famous theorem of Serre that identifies $\mathbf{D}^b(\mathrm{qgr}\,A)$ with $\mathbf{D}^b(\mathrm{coh}(\mathbf{Proj}\,(A)))$ when A is generated by its first component allows us to apply this result to geometry.

I am grateful to Alexei Bondal, Anton Kapustin, Ludmil Katzarkov, Alexander Kuznetsov, Tony Pantev, and Johannes Walcher for very useful discussions.

1 Triangulated Categories of Singularities for Graded Algebras

1.1 Localization in Triangulated Categories and Semiorthogonal Decomposition

Recall that a triangulated category $\mathcal{D}$ is an additive category equipped with the following additional data:

(a) an additive autoequivalence $[1] : \mathcal{D} \longrightarrow \mathcal{D}$, which is called a translation functor,

(b) a class of exact (or distinguished) triangles

$$X \xrightarrow{u} Y \xrightarrow{v} Z \xrightarrow{w} X[1],$$

which must satisfy a certain set of axioms (see [27], also [12, 19, 21]).

A functor $F : \mathcal{D} \longrightarrow \mathcal{D}'$ between two triangulated categories is called exact if it commutes with the translation functors, i.e., $F \circ [1] \cong [1] \circ F$, and transforms exact triangles into exact triangles.

With any pair $\mathcal{N} \subset \mathcal{D}$, where $\mathcal{N}$ is a full triangulated subcategory in a triangulated category $\mathcal{D}$, we can associate the quotient category $\mathcal{D}/\mathcal{N}$. To construct it let us denote by $\Sigma(\mathcal{N})$ a class of morphisms s in $\mathcal{D}$ fitting into an exact triangle

$$X \xrightarrow{s} Y \longrightarrow N \longrightarrow X[1]$$

with $N \in \mathcal{N}$. It can be checked that $\Sigma(\mathcal{N})$ is a multiplicative system. Define the quotient $\mathcal{D}/\mathcal{N}$ as the localization $\mathcal{D}[\Sigma(\mathcal{N})^{-1}]$ (see [10, 12, 27]). It is a triangulated category. The translation functor on $\mathcal{D}/\mathcal{N}$ is induced from the translation functor in the category $\mathcal{D}$, and the exact triangles in $\mathcal{D}/\mathcal{N}$ are the triangles isomorphic to the images of exact triangles in $\mathcal{D}$. The quotient functor $Q : \mathcal{D} \longrightarrow \mathcal{D}/\mathcal{N}$ annihilates $\mathcal{N}$. Moreover, any exact functor $F : \mathcal{D} \longrightarrow \mathcal{D}'$ between triangulated categories for which $F(X) \simeq 0$ when $X \in \mathcal{N}$ factors uniquely through Q. The following lemma is obvious.

Lemma 1. *Let $\mathcal{N}$ and $\mathcal{N}'$ be full triangulated subcategories of triangulated categories $\mathcal{D}$ and $\mathcal{D}'$ respectively. Let $F : \mathcal{D} \to \mathcal{D}'$ and $G : \mathcal{D}' \to \mathcal{D}$ be an adjoint pair of exact functors such that $F(\mathcal{N}) \subset \mathcal{N}'$ and $G(\mathcal{N}') \subset \mathcal{N}$. Then they induce functors*

$$\overline{F} : \mathcal{D}/\mathcal{N} \longrightarrow \mathcal{D}'/\mathcal{N}', \qquad \overline{G} : \mathcal{D}'/\mathcal{N}' \longrightarrow \mathcal{D}/\mathcal{N},$$

which are adjoint as well. Moreover, if the functor $F : \mathcal{D} \to \mathcal{D}'$ is fully faithful, then the functor $\overline{F} : \mathcal{D}/\mathcal{N} \longrightarrow \mathcal{D}'/\mathcal{N}'$ is also fully faithful.

Now recall some definitions and facts concerning admissible subcategories and semiorthogonal decompositions (see [7,8]). Let $\mathcal{N} \subset \mathcal{D}$ be a full triangulated subcategory. The right orthogonal to $\mathcal{N}$ is the full subcategory $\mathcal{N}^{\perp} \subset \mathcal{D}$ consisting of all objects M such that $\operatorname{Hom}(N, M) = 0$ for any $N \in \mathcal{N}$. The left orthogonal $^{\perp}\mathcal{N}$ is defined analogously. The orthogonals are also triangulated subcategories.

Definition 2. *Let $I : \mathcal{N} \hookrightarrow \mathcal{D}$ be an embedding of a full triangulated subcategory $\mathcal{N}$ in a triangulated category $\mathcal{D}$. We say that $\mathcal{N}$ is right admissible (respectively left admissible) if there is a right (respectively left) adjoint functor $Q : \mathcal{D} \to \mathcal{N}$. The subcategory $\mathcal{N}$ will be called admissible if it is right and left admissible.*

Remark 3. For the subcategory $\mathcal{N}$ the property of being right admissible is equivalent to requiring that for each $X \in \mathcal{D}$ there be an exact triangle $N \to X \to M$, with $N \in \mathcal{N}, M \in \mathcal{N}^{\perp}$.

Lemma 4. *Let $\mathcal{N}$ be a full triangulated subcategory in a triangulated category $\mathcal{D}$. If $\mathcal{N}$ is right (respectively left) admissible, then the quotient category $\mathcal{D}/\mathcal{N}$ is equivalent to $\mathcal{N}^{\perp}$ (respectively $^{\perp}\mathcal{N}$). Conversely, if the quotient functor $Q : \mathcal{D} \longrightarrow \mathcal{D}/\mathcal{N}$ has a left (respectively right) adjoint, then $\mathcal{D}/\mathcal{N}$ is equivalent to $\mathcal{N}^{\perp}$ (respectively $^{\perp}\mathcal{N}$).*

If $\mathcal{N} \subset \mathcal{D}$ is a right admissible subcategory, then we say that the category $\mathcal{D}$ has a weak semiorthogonal decomposition $\langle \mathcal{N}^\perp, \mathcal{N} \rangle$. Similarly, if $\mathcal{N} \subset \mathcal{D}$ is a left admissible subcategory, we say that $\mathcal{D}$ has a weak semiorthogonal decomposition $\langle \mathcal{N}, {}^\perp \mathcal{N} \rangle$.

Definition 5. *A sequence of full triangulated subcategories $(\mathcal{N}_1, \ldots, \mathcal{N}_n)$ in a triangulated category $\mathcal{D}$ will be called a* weak semiorthogonal decomposition *of $\mathcal{D}$ if there is a sequence of left admissible subcategories $\mathcal{D}_1 = \mathcal{N}_1 \subset \mathcal{D}_2 \subset \cdots \subset \mathcal{D}_n = \mathcal{D}$ such that $\mathcal{N}_p$ is left orthogonal to $\mathcal{D}_{p-1}$ in $\mathcal{D}_p$. We will write $\mathcal{D} = \langle \mathcal{N}_1, \ldots, \mathcal{N}_n \rangle$. If all N_p are admissible in $\mathcal{D}$ then the decomposition $\mathcal{D} = \langle \mathcal{N}_1, \ldots, \mathcal{N}_n \rangle$ is called* semiorthogonal.

The existence of a semiorthogonal decomposition on a triangulated category $\mathcal{D}$ clarifies the structure of $\mathcal{D}$. In the best scenario, one can hope that $\mathcal{D}$ has a semiorthogonal decomposition $\mathcal{D} = \langle \mathcal{N}_1, \ldots, \mathcal{N}_n \rangle$ in which each elementary constituent $\mathcal{N}_p$ is as simple as possible, i.e., is equivalent to the bounded derived category of finite-dimensional vector spaces.

Definition 6. *An object E of a $\mathbf{k}$-linear triangulated category $\mathcal{T}$ is called* exceptional *if $\mathrm{Hom}(E, E[p]) = 0$ when $p \neq 0$, and $\mathrm{Hom}(E, E) = \mathbf{k}$. An exceptional collection in $\mathcal{T}$ is a sequence of exceptional objects $(E_0, \ldots, E_n)$ satisfying the semiorthogonality condition $\mathrm{Hom}(E_i, E_j[p]) = 0$ for all p when $i > j$.*

If a triangulated category $\mathcal{D}$ has an exceptional collection $(E_0, \ldots, E_n)$ that generates the whole of $\mathcal{D}$ then we say that the collection is full. In this case $\mathcal{D}$ has a semiorthogonal decomposition with $\mathcal{N}_p = \langle E_p \rangle$. Since E_p is exceptional, each of these categories is equivalent to the bounded derived category of finite-dimensional vector spaces. In this case we write $\mathcal{D} = \langle E_0, \ldots, E_n \rangle$.

Definition 7. *An exceptional collection $(E_0, \ldots, E_n)$ is called* strong *if in addition, $\mathrm{Hom}(E_i, E_j[p]) = 0$ for all i and j when $p \neq 0$.*

1.2 Triangulated Categories of Singularities for Algebras

Let $A = \bigoplus_{i \geq 0} A_i$ be a Noetherian graded algebra over a field $\mathbf{k}$. Denote by mod-A and gr-A the category of finitely generated right modules and the category of finitely generated graded right modules respectively. Note that morphisms in gr-A are homomorphisms of degree zero. These categories are abelian. We will also use the notation Mod-A and Gr-A for the abelian categories of all right modules, and all graded right modules and we will often omit the prefix "right." Left A-modules are will be viewed as right A°-modules and A-B bimodules as right A°-B-modules, where A° is the opposite algebra.

The twist functor (p) on the category gr-A is defined as follows: it takes a graded module $M = \oplus_i M_i$ to the module $M(p)$ for which $M(p)_i = M_{p+i}$ and takes a morphism $f : M \longrightarrow N$ to the same morphism viewed as a morphism between the twisted modules $f(p) : M(p) \longrightarrow N(p)$.

Consider the bounded derived categories $\mathbf{D}^b(\text{gr-}A)$ and $\mathbf{D}^b(\text{mod-}A)$. They can be endowed with natural structures of triangulated categories. The categories $\mathbf{D}^b(\text{gr-}A)$ and $\mathbf{D}^b(\text{mod-}A)$ have full triangulated subcategories consisting of objects that are isomorphic to bounded complexes of projectives. These subcategories can also be considered as the derived categories of the exact categories of projective modules $\mathbf{D}^b(\text{grproj-}A)$ and $\mathbf{D}^b(\text{proj-}A)$ respectively (see, e.g., [19]). They will be called the subcategories of perfect complexes. Observe also that the category $\mathbf{D}^b(\text{gr-}A)$ (respectively $\mathbf{D}^b(\text{mod-}A)$) is equivalent to the category $\mathbf{D}^b_{\text{gr-}A}(\text{Gr-}A)$ (respectively $\mathbf{D}^b_{\text{mod-}A}(\text{Mod-}A)$) of complexes of arbitrary modules with finitely generated cohomologies (see [5]). We will tacitly use this equivalence throughout our considerations.

Definition 8. *We define triangulated categories of singularities* $\mathbf{D}^{\text{gr}}_{\text{Sg}}(A)$ *and* $\mathbf{D}_{\text{Sg}}(A)$ *as the quotient* $\mathbf{D}^b(\text{gr-}A)/\mathbf{D}^b(\text{grproj-}A)$ *and* $\mathbf{D}^b(\text{mod-}A)/\mathbf{D}^b(\text{proj-}A)$ *respectively.*

Remark 9. As in the commutative case [22, 23], the triangulated categories of singularities $\mathbf{D}^{\text{gr}}_{\text{Sg}}(A)$ and $\mathbf{D}_{\text{Sg}}(A)$ will be trivial if A has finite homological dimension. Indeed, in this case any A-module has a finite projective resolution, i.e., the subcategories of perfect complexes coincide with the full bounded derived categories of finitely generated modules.

Homomorphisms of (graded) algebras $f : A \to B$ induce functors between the associated derived categories of singularities. Furthermore, if B has a finite Tor-dimension as an A-module, then we get the functor $\overset{\mathbf{L}}{\otimes}_A B$ between the bounded derived categories of finitely generated modules that maps perfect complexes to perfect complexes. Therefore, we get functors between triangulated categories of singularities

$$\overset{\mathbf{L}}{\otimes}_A B : \mathbf{D}^{\text{gr}}_{\text{Sg}}(A) \longrightarrow \mathbf{D}^{\text{gr}}_{\text{Sg}}(B) \quad \text{and} \quad \overset{\mathbf{L}}{\otimes}_A B : \mathbf{D}_{\text{Sg}}(A) \longrightarrow \mathbf{D}_{\text{Sg}}(B).$$

If, in addition, B is finitely generated as an A-module, then these functors have right adjoints induced from the functor that sends a complex of B-modules to itself considered as a complex of A-modules.

More generally, suppose $_A\underline{M}^{\bullet}_B$ is a complex of graded A-B bimodules that as a complex of graded B-modules is quasi-isomorphic to a perfect complex. Suppose that $_A\underline{M}^{\bullet}$ has a finite Tor-amplitude as a left A-module. Then we can define the derived tensor product functor $\overset{\mathbf{L}}{\otimes}_A \underline{M}^{\bullet}_B : \mathbf{D}^b(\text{gr}-A) \longrightarrow \mathbf{D}^b(\text{gr}-B)$. Moreover, since $\underline{M}^{\bullet}_B$ is perfect over B, this functor sends perfect complexes to perfect complexes. Therefore, we get an exact functor

$$\overset{\mathbf{L}}{\otimes}_A \underline{M}^{\bullet}_B : \mathbf{D}^{\text{gr}}_{\text{Sg}}(A) \longrightarrow \mathbf{D}^{\text{gr}}_{\text{Sg}}(B).$$

In the ungraded case we also get the functor $\overset{\mathbf{L}}{\otimes}_A \underline{M}^{\bullet}_B : \mathbf{D}_{\text{Sg}}(A) \longrightarrow \mathbf{D}_{\text{Sg}}(B).$

1.3 Morphisms in Categories of Singularities

In general, it is not easy to calculate spaces of morphisms between objects in a quotient category. The following lemma and proposition provide some information about the morphism spaces in triangulated categories of singularities.

Lemma 10. *For any object $T \in \mathbf{D}^{\mathrm{gr}}_{\mathrm{Sg}}(A)$ (respectively $T \in \mathbf{D}_{\mathrm{Sg}}(A)$) and for any sufficiently large k, there is a module $M \in \mathrm{gr}\text{-}A$ (respectively $M \in \mathrm{mod}\text{-}A$) depending on T and k and such that T is isomorphic to the image of $M[k]$ in the triangulated category of singularities. If, in addition, the algebra A has finite injective dimension, then for any sufficiently large k the corresponding module M satisfies $\mathrm{Ext}^i_A(M, A) = 0$ for all $i > 0$.*

Proof. The object T is represented by a bounded complex of modules $\underline{T}^{\bullet}$. Choose a bounded-above projective resolution $\underline{P}^{\bullet} \xrightarrow{\sim} \underline{T}^{\bullet}$ and a sufficiently large $k \gg 0$. Consider the stupid truncation $\sigma^{\geq -k+1}\underline{P}^{\bullet}$ of $\underline{P}^{\bullet}$. Denote by M the cohomology module $H^{-k+1}(\sigma^{\geq -k+1}\underline{P}^{\bullet})$. Clearly $T \cong M[k]$ in $\mathbf{D}^{\mathrm{gr}}_{\mathrm{Sg}}(A)$ (respectively $\mathbf{D}_{\mathrm{Sg}}(A)$).

If now A has finite injective dimension, then morphism spaces $\mathrm{Hom}(\underline{T}^{\bullet}, A[i])$ in $\mathbf{D}^b(\mathrm{gr}\text{-}A)$ (respectively $\mathbf{D}^b(\mathrm{mod}\text{-}A)$) are trivial for all but finitely many $i \in \mathbb{Z}$. So if M corresponds to T and a sufficiently large k, then we will have $\mathrm{Ext}^i_A(M, A) = 0$ for all $i > 0$. $\qquad\square$

Proposition 11. *Let M be an A-module such that $\mathrm{Ext}^i_A(M, A) = 0$ for all $i > 0$. Then for any A-module N we have*

$$\mathrm{Hom}_{\mathbf{D}_{\mathrm{Sg}}(A)}(M, N) \cong \mathrm{Hom}_A(M, N)/\mathcal{R},$$

where $\mathcal{R}$ is the subspace of elements factoring through a projective module i.e., $e \in \mathcal{R}$ iff $e = \beta\alpha$ with $\alpha : M \to P$ and $\beta : P \to N$, where P is projective. If M is a graded module, then for any graded A-module N,

$$\mathrm{Hom}_{\mathbf{D}^{\mathrm{gr}}_{\mathrm{Sg}}(A)}(M, N) \cong \mathrm{Hom}_{\mathrm{gr}\text{-}A}(M, N)/\mathcal{R}.$$

Proof. We will discuss only the graded case. By the definition of localization any morphism from M to N in $\mathbf{D}^{\mathrm{gr}}_{\mathrm{Sg}}(A)$ can be represented by a pair

$$M \xrightarrow{a} \underline{T}^{\bullet} \xleftarrow{s} N \tag{1}$$

of morphisms in $\mathbf{D}^b(\mathrm{gr}\text{-}A)$ such that the cone $\underline{C}^{\bullet}(s)$ is a perfect complex. Consider a bounded-above projective resolution $\underline{Q}^{\bullet} \to N$ and its stupid truncation $\sigma^{\geq -k}\underline{Q}^{\bullet}$ for sufficiently large k. There is an exact triangle

$$E[k] \longrightarrow \sigma^{\geq -k}\underline{Q}^{\bullet} \longrightarrow N \xrightarrow{s'} E[k+1],$$

where E denotes the module $H^{-k}(\sigma^{\geq -k}Q^{\bullet})$. Choosing k to be sufficiently large, we can guarantee that $\mathrm{Hom}(\underline{C}^{\bullet}(s), E[i]) = 0$ for all $i > k$. Using the triangle

$$\underline{C}^{\bullet}(s)[-1] \longrightarrow N \xrightarrow{s} \underline{T}^{\bullet} \longrightarrow \underline{C}^{\bullet}(s),$$

we find that the map $s' : N \to E[k+1]$ can be lifted to a map $\underline{T}^{\bullet} \to E[k+1]$. The map $\underline{T}^{\bullet} \to E[k+1]$ induces a pair of the form

$$M \xrightarrow{a'} E[k+1] \xleftarrow{s'} N, \tag{2}$$

and this pair gives the same morphism in $\mathbf{D}_{\mathrm{Sg}}(A)$ as the pair (1). Since $\mathrm{Ext}^{i}(M, P) = 0$ for all $i > 0$ and any projective module P, we obtain

$$\mathrm{Hom}(M, (\sigma^{\geq -k}\underline{Q}^{\bullet})[1]) = 0.$$

Hence, the map $a' : M \to E[k+1]$ can be lifted to a map f that completes the diagram

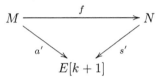

Thus, the map f is equivalent to the map (2) and, as a consequence, to the map (1). Hence, any morphism from M to N in $\mathbf{D}_{\mathrm{Sg}}^{\mathrm{gr}}(A)$ is represented by a morphism from M to N in the category $\mathbf{D}^{b}(\mathrm{gr}\text{-}A)$.

Now if f is the 0-morphism in $\mathbf{D}_{\mathrm{Sg}}^{\mathrm{gr}}(A)$, then without loss of generality we can assume that the map a is the zero map. In this case we will have $a' = 0$ as well. This implies that f factors through a morphism $M \to \sigma^{\geq -k}Q^{\bullet}$. By the assumption on M, any such morphism can be lifted to a morphism $M \to Q^{0}$. Hence, if f is the 0-morphism in $\mathbf{D}_{\mathrm{Sg}}^{\mathrm{gr}}(A)$, then it factors through Q^{0}. The same proof works in the ungraded case (see [22]). $\qquad\square$

Next we describe a useful construction utilizing the previous statements. Let $\underline{M}^{\bullet}$ and $\underline{N}^{\bullet}$ be two bounded complexes of (graded) A-modules. Assume that $\mathrm{Hom}(\underline{M}^{\bullet}, A[i])$ in the bounded derived categories of A-modules are trivial except for a finite number of $i \in \mathbb{Z}$. By Lemma 10, for sufficiently large k there are modules $M, N \in \mathrm{gr}\text{-}A$ (resp. $M, N \in \mathrm{mod}\text{-}A$) such that $\underline{M}^{\bullet}$ and $\underline{N}^{\bullet}$ are isomorphic to the images of $M[k]$ and $N[k]$ in the triangulated category of singularities. Moreover, it follows immediately from the assumption on $\underline{M}^{\bullet}$ and the construction of M that for any sufficiently large k we have $\mathrm{Ext}_{A}^{i}(M, A) = 0$ whenever $i > 0$. Hence, by Proposition 11, we get

$$\mathrm{Hom}_{\mathbf{D}_{\mathrm{Sg}}^{\mathrm{gr}}(A)}(\underline{M}^{\bullet}, \underline{N}^{\bullet}) \cong \mathrm{Hom}_{\mathbf{D}_{\mathrm{Sg}}^{\mathrm{gr}}(A)}(M, N) \cong \mathrm{Hom}_{A}(M, N)/\mathcal{R},$$

where $\mathcal{R}$ is the subspace of elements factoring through a projective module. This procedure works in the ungraded situation as well.

2 Categories of Coherent Sheaves and Categories of Singularities

2.1 Quotient Categories of Graded Modules

Let $A = \bigoplus_{i \geq 0} A_i$ be a Noetherian graded algebra. We suppose that A is connected, i.e., $A_0 = \mathbf{k}$. Denote by tors-A the full subcategory of gr-A, which consists of all graded A-modules that are finite-dimensional over $\mathbf{k}$.

An important role will be played by the quotient abelian category qgr $A = $ gr-$A/$ tors-A. It has the following explicit description. The objects of qgr A are the objects of gr-A (we denote by πM the object in qgr A that corresponds to a module M). The morphisms in qgr A are given by

$$\mathrm{Hom}_{\mathrm{qgr}}(\pi M, \pi N) := \varinjlim_{M'} \mathrm{Hom}_{\mathrm{gr}}(M', N), \tag{3}$$

where M' runs over submodules of M such that M/M' is finite-dimensional.

Given a graded A-module M and an integer p, the graded A-submodule $\bigoplus_{i \geq p} M_i$ of M is denoted by $M_{\geq p}$ and is called the pth tail of M. In the same way, we can define the pth tail $\underline{M}^{\bullet}_{\geq p}$ of any complex of modules $\underline{M}^{\bullet}$. Since A is Noetherian, we have

$$\mathrm{Hom}_{\mathrm{qgr}}(\pi M, \pi N) = \varinjlim_{p \to \infty} \mathrm{Hom}_{\mathrm{gr}}(M_{\geq p}, N).$$

We will also identify M_p with the quotient $M_{\geq p}/M_{\geq p+1}$.

Similarly, we can consider the subcategory Tors-$A \subset$ Gr-A of torsion modules. Recall that a module M is called torsion if for any element $x \in M$ one has $xA_{\geq p} = 0$ for some p. Denote by QGr A the quotient category Gr-$A/$ Tors-A. The category QGr A contains qgr A as a full subcategory. Sometimes it is convenient to work in QGr A instead of qgr A.

Denote by Π and π the canonical projections of Gr-A to QGr A and of gr-A to qgr A respectively. The functor Π has a right adjoint Ω, and moreover, for any $N \in$ Gr-A,

$$\Omega\Pi N \cong \bigoplus_{n=-\infty}^{\infty} \mathrm{Hom}_{\mathrm{QGr}}(\Pi A, \Pi N(n)). \tag{4}$$

For any $i \in \mathbb{Z}$ we can consider the full abelian subcategories Gr$-A_{\geq i} \subset$ GrA and gr-$A_{\geq i} \subset$ gr$A$, which consist of all modules M such that $M_p = 0$ when $p < i$. The natural projection functor $\Pi_i : $ Gr-$A_{\geq i} \longrightarrow$ QGr-A has a right adjoint Ω_i satisfying

$$\Omega_i \Pi_i N \cong \bigoplus_{n=i}^{\infty} \mathrm{Hom}_{\mathrm{QGr}}(\Pi A, \Pi_i N(n)).$$

Since the category QGr A is an abelian category with enough injectives, there is a right derived functor

$$\mathbf{R}\Omega_i : \mathbf{D}^{+}(\mathrm{QGr}\, A) \longrightarrow \mathbf{D}^{+}(\mathrm{Gr}\text{-}A_{\geq i})$$

defined as

$$R\Omega_i M \cong \bigoplus_{k=i}^{\infty} R\operatorname{Hom}_{\mathrm{QGr}}(\Pi A, M(k)). \tag{5}$$

Assume now that the algebra A satisfies condition "χ" from [1, Sec. 3]. We recall that by definition, a connected Noetherian graded algebra A satisfies condition "χ" if for every $M \in \mathrm{gr}\text{-}A$ the grading on the space $\mathrm{Ext}^i_A(\mathbf{k}, M)$ is right bounded for all i. In this case it was proved in [1, Prop. 3.14] that the restrictions of the functors Ω_i to the subcategory $\mathrm{qgr}\,A$ give functors $\omega_i : \mathrm{qgr}\,A \longrightarrow \mathrm{gr}\text{-}A_{\geq i}$ that are right adjoint to π_i. Moreover, it follows from [1, Th. 7.4] that the functor ω_i has a right derived

$$\mathbf{R}\omega_i : \mathbf{D}^+(\mathrm{qgr}\,A) \longrightarrow \mathbf{D}^+(\mathrm{gr}\text{-}A_{\geq i})$$

and all $\mathbf{R}^j\omega_i \in \mathrm{tors}\text{-}A$ for $j > 0$.

If, in addition, the algebra A is Gorenstein (i.e., if it has a finite injective dimension n and $D(\mathbf{k}) = \mathbf{R}\operatorname{Hom}_A(\mathbf{k}, A)$ is isomorphic to $\mathbf{k}(a)[-n]$), we obtain the right derived functor

$$\mathbf{R}\omega_i : \mathbf{D}^b(\mathrm{qgr}\,A) \longrightarrow \mathbf{D}^b(\mathrm{gr}\text{-}A_{\geq i})$$

between bounded derived categories (see [30, Cor. 4.3]). It is important to note that the functor $\mathbf{R}\omega_i$ is fully faithful because $\pi_i\mathbf{R}\omega_i$ is isomorphic to the identity functor ([1, Prop. 7.2]).

2.2 Triangulated Categories of Singularities for Gorenstein Algebras

The main goal of this section is to establish a connection between the triangulated category of singularities $\mathbf{D}^{\mathrm{gr}}_{\mathrm{Sg}}(A)$ and the derived category $\mathbf{D}^b(\mathrm{qgr}\,A)$, in the case of a Gorenstein algebra A.

When the algebra A has finite injective dimension as 60th a right and left module over itself (i.e., A is a dualizing complex for itself) we get two functors

$$D := \mathbf{R}\operatorname{Hom}_A(-, A) : \mathbf{D}^b(\mathrm{gr}\text{-}A\ )^\circ \longrightarrow \mathbf{D}^b(\mathrm{gr}\text{-}A^\circ), \tag{6}$$

$$D^\circ := \mathbf{R}\operatorname{Hom}_{A^\circ}(-, A) : \mathbf{D}^b(\mathrm{gr}\text{-}A^\circ)^\circ \longrightarrow \mathbf{D}^b(\mathrm{gr}\text{-}A\ ), \tag{7}$$

which are quasi-inverse triangulated equivalences (see [29, Prop. 3.5]).

Definition 12. *We say that a connected graded Noetherian algebra A is Gorenstein if it has a finite injective dimension n and $D(\mathbf{k}) = \mathbf{R}\operatorname{Hom}_A(\mathbf{k}, A)$ is isomorphic to $\mathbf{k}(a)[-n]$ for some integer a, which is called the Gorenstein parameter of A. (Such an algebra is also called AS-Gorenstein, where "AS" stands for "Artin–Schelter.")*

Remark 13. It is known (see [30, Cor. 4.3]) that any Gorenstein algebra satisfies condition "χ" and for any Gorenstein algebra A and for any $i \in \mathbb{Z}$ we have derived functors

$$\mathbf{R}\omega_i : \mathbf{D}^b(\text{qgr } A) \longrightarrow \mathbf{D}^b(\text{gr-}A_{\geq i})$$

that are fully faithful.

Now we describe the images of the functors $\mathbf{R}\omega_i$. Denote by $\mathcal{D}_i$ the subcategories of $\mathbf{D}^b(\text{gr-}A)$ that are the images of the composition of $\mathbf{R}\omega_i$ and the natural inclusion of $\mathbf{D}^b(\text{gr-}A_{\geq i})$ to $\mathbf{D}^b(\text{gr-}A)$. All $\mathcal{D}_i$ are equivalent to $\mathbf{D}^b(\text{qgr-}A)$. Further, for any integer i denote by $\mathcal{S}_{<i}(A)$ (or simply $\mathcal{S}_{<i}$) the full triangulated subcategory of $\mathbf{D}^b(\text{gr-}A)$ generated by the modules $\mathbf{k}(e)$ with $e > -i$. In other words, the objects of $\mathcal{S}_{<i}$ are complexes $\underline{M}^{\bullet}$ for which the tail $\underline{M}^{\bullet}_{\geq i}$ is isomorphic to zero. Analogously, we define $\mathcal{S}_{\geq i}$ as the triangulated subcategory that is generated by the objects $\mathbf{k}(e)$ with $e \leq -i$. In other words, the objects of $\mathcal{S}_{\geq i}$ are complexes of torsion modules from gr-$A_{\geq i}$. It is clear that $\mathcal{S}_{<i} \cong \mathcal{S}_{<0}(-i)$ and $\mathcal{S}_{\geq i} \cong \mathcal{S}_{\geq 0}(-i)$.

Furthermore, denote by $\mathcal{P}_{<i}$ the full triangulated subcategory of $\mathbf{D}^b(\text{gr-}A)$ generated by the free modules $A(e)$ with $e > -i$ and denote by $\mathcal{P}_{\geq i}$ the triangulated subcategory that is generated by the free modules $A(e)$ with $e \leq -i$. As above, we have $\mathcal{P}_{<i} \cong \mathcal{P}_{<0}(-i)$ and $\mathcal{P}_{\geq i} \cong \mathcal{P}_{\geq 0}(-i)$.

Lemma 14. *Let $A = \bigoplus_{i \geq 0} A_i$ be a connected graded Noetherian algebra. Then the subcategories $\mathcal{S}_{<i}$ and $\mathcal{P}_{<i}$ are left and respectively right admissible for any $i \in \mathbb{Z}$. Moreover, there are weak semiorthogonal decompositions*

$$\mathbf{D}^b(\text{gr-}A) = \langle \mathcal{S}_{<i}, \mathbf{D}^b(\text{gr-}A_{\geq i}) \rangle, \qquad \mathbf{D}^b(\text{tors-}A) = \langle \mathcal{S}_{<i}, \mathcal{S}_{\geq i} \rangle, \qquad (8)$$

$$\mathbf{D}^b(\text{gr-}A) = \langle \mathbf{D}^b(\text{gr-}A_{\geq i}), \mathcal{P}_{<i} \rangle, \qquad \mathbf{D}^b(\text{grproj-}A) = \langle \mathcal{P}_{\geq i}, \mathcal{P}_{<i} \rangle. \qquad (9)$$

Proof. For any complex $\underline{M}^{\bullet} \in \mathbf{D}^b(\text{mod-}A)$ there is an exact triangle of the form

$$\underline{M}^{\bullet}_{\geq i} \longrightarrow \underline{M}^{\bullet} \longrightarrow \underline{M}^{\bullet}/\underline{M}^{\bullet}_{\geq i}.$$

By definition, the object $\underline{M}^{\bullet}/\underline{M}^{\bullet}_{\geq i}$ belongs to $\mathcal{S}_{<i}$, and the object $\underline{M}^{\bullet}_{\geq i}$ is in the left orthogonal $^{\perp}\mathcal{S}_{<i}$. Hence, by Remark 3, $\mathcal{S}_{<i}$ is left admissible. Moreover, $\underline{M}^{\bullet}_{\geq i}$ also belongs to $\mathbf{D}^b(\text{gr-}A_{\geq i})$, i.e., $\mathbf{D}^b(\text{gr-}A_{\geq i}) \cong {}^{\perp}\mathcal{S}_{<i}$ in the category $\mathbf{D}^b(\text{gr-}A)$. If $\underline{M}^{\bullet}$ is a complex of torsion modules, then $\underline{M}^{\bullet}_{\geq i}$ belongs to $\mathcal{S}_{\geq i}$. Thus, we obtain both decompositions of (8).

To prove the existence of the decompositions (9) we first note that due to the connectedness of A, any finitely generated graded projective A-module is free. Second, any finitely generated free module P has a canonical split decomposition of the form

$$0 \longrightarrow P_{<i} \longrightarrow P \longrightarrow P_{\geq i} \longrightarrow 0,$$

where $P_{<i} \in \mathcal{P}_{<i}$ and $P_{\geq i} \in \mathcal{P}_{\geq i}$. Third, any bounded complex of finitely generated A-modules $\underline{M}^{\bullet}$ has a bounded-above free resolution $\underline{P}^{\bullet} \to \underline{M}^{\bullet}$ such that $P^{-k} \in \mathcal{P}_{\geq i}$ for all $k \gg 0$. This implies that the object $\underline{P}^{\bullet}_{<i} \in \mathcal{P}_{<i}$ from the exact sequence of complexes

$$0 \longrightarrow \underline{P}^{\bullet}_{<i} \longrightarrow \underline{P}^{\bullet} \longrightarrow \underline{P}^{\bullet}_{\geq i} \longrightarrow 0$$

is a bounded complex. Since $\underline{P}^{\bullet}$ is quasi-isomorphic to a bounded complex, the complex $\underline{P}^{\bullet}_{\geq i}$ is also quasi-isomorphic to some bounded complex $\underline{K}^{\bullet}$ from $\mathbf{D}^b(\text{gr-}A_{\geq i})$. Thus, any object $\underline{M}^{\bullet}$ has a decomposition

$$\underline{P}^{\bullet}_{<i} \longrightarrow \underline{M}^{\bullet} \longrightarrow \underline{K}^{\bullet},$$

where $\underline{P}^{\bullet}_{<i} \in \mathcal{P}_{<i}$ and $\underline{K}^{\bullet} \in \mathbf{D}^b(\text{gr-}A_{\geq i})$. This proves the decompositions (9). $\qquad \square$

Lemma 15. *Let $A = \bigoplus_{i \geq 0} A_i$ be a connected graded Noetherian algebra that is Gorenstein. Then the subcategories $\mathcal{S}_{\geq i}$ and $\mathcal{P}_{\geq i}$ are right and respectively left admissible. Moreover, for any $i \in \mathbb{Z}$ there are weak semiorthogonal decompositions*

$$\mathbf{D}^b(\text{gr-}A_{\geq i}) = \langle \mathcal{D}_i, \mathcal{S}_{\geq i} \rangle, \quad \mathbf{D}^b(\text{gr-}A_{\geq i}) = \langle \mathcal{P}_{\geq i}, \mathcal{T}_i \rangle, \qquad (10)$$

where the subcategory $\mathcal{D}_i$ is equivalent to $\mathbf{D}^b(\text{qgr } A)$ under the functor $\mathbf{R}\omega_i$, and $\mathcal{T}_i$ is equivalent to $\mathbf{D}^{\text{gr}}_{\text{Sg}}(A)$.

Proof. The functor $\mathbf{R}\omega_i$ is fully faithful and has the left adjoint π_i. Thus, we obtain a semiorthogonal decomposition

$$\mathbf{D}^b(\text{gr-}A_{\geq i}) = \langle \mathcal{D}_i, {}^\perp\mathcal{D}_i \rangle,$$

where $\mathcal{D}_i \cong \mathbf{D}^b(\text{qgr } A)$. Furthermore, the orthogonal ${}^\perp\mathcal{D}_i$ consists of all objects $\underline{M}^{\bullet}$ satisfying $\pi_i(\underline{M}^{\bullet}) = 0$. Thus, ${}^\perp\mathcal{D}_i$ coincides with $\mathcal{S}_{\geq i}$. Hence, $\mathcal{S}_{\geq i}$ is right admissible in $\mathbf{D}^b(\text{gr-}A_{\geq i})$, which is right admissible in all of $\mathbf{D}^b(\text{gr-}A)$. This implies that $\mathcal{S}_{\geq i}$ is right admissible in $\mathbf{D}^b(\text{gr-}A)$ as well.

The functor D from (6) establishes an equivalence of the subcategory $\mathcal{P}_{\geq i}(A)^\circ$ with the subcategory $\mathcal{P}_{<-i+1}(A^\circ)$, which is right admissible by Lemma 14. Hence, $\mathcal{P}_{\geq i}(A)$ is left admissible and there is a decomposition of the form

$$\mathbf{D}^b(\text{gr-}A_{\geq i}) = \langle \mathcal{P}_{\geq i}, \mathcal{T}_i \rangle$$

with some $\mathcal{T}_i$.

Now applying Lemma 1 to the full embedding of $\mathbf{D}^b(\text{gr-}A_{\geq i})$ to $\mathbf{D}^b(\text{gr-}A)$ and using Lemma 4, we get a fully faithful functor from $\mathcal{T}_i \cong \mathbf{D}^b(\text{gr-}A_{\geq i})/\mathcal{P}_{\geq i}$ to $\mathbf{D}^{\text{gr}}_{\text{Sg}}(A) = \mathbf{D}^b(\text{gr-}A)/\mathbf{D}^b(\text{grproj-}A)$. Finally, since this functor is essentially surjective on objects, it is actually an equivalence. $\qquad \square$

Theorem 16. *Let* $A = \bigoplus_{i \geq 0} A_i$ *be a connected graded Noetherian algebra that is Gorenstein with Gorenstein parameter a. Then the triangulated categories* $\mathbf{D}_{\mathrm{Sg}}^{\mathrm{gr}}(A)$ *and* $\mathbf{D}^b(\mathrm{qgr}\, A)$ *are related as follows:*

(i) *if* $a > 0$, *there are fully faithful functors* $\Phi_i : \mathbf{D}_{\mathrm{Sg}}^{\mathrm{gr}}(A) \longrightarrow \mathbf{D}^b(\mathrm{qgr}\, A)$ *and semiorthogonal decompositions*

$$\mathbf{D}^b(\mathrm{qgr}\, A) = \langle \pi A(-i - a + 1), \ldots, \pi A(-i), \Phi_i \mathbf{D}_{\mathrm{Sg}}^{\mathrm{gr}}(A) \rangle,$$

where $\pi : \mathbf{D}^b(\mathrm{gr}{-}A) \longrightarrow \mathbf{D}^b(\mathrm{qgr}\, A)$ *is the natural projection;*

(ii) *if* $a < 0$, *there are fully faithful functors* $\Psi_i : \mathbf{D}^b(\mathrm{qgr}\, A) \longrightarrow \mathbf{D}_{\mathrm{Sg}}^{\mathrm{gr}}(A)$ *and semiorthogonal decompositions*

$$\mathbf{D}_{\mathrm{Sg}}^{\mathrm{gr}}(A) = \langle q\mathbf{k}(-i), \ldots, q\mathbf{k}(-i + a + 1), \Psi_i \mathbf{D}^b(\mathrm{qgr}\, A) \rangle,$$

where $q : \mathbf{D}^b(\mathrm{gr}{-}A) \longrightarrow \mathbf{D}_{\mathrm{Sg}}^{\mathrm{gr}}(A)$ *is the natural projection;*

(iii) *if* $a = 0$, *there is an equivalence* $\mathbf{D}_{\mathrm{Sg}}^{\mathrm{gr}}(A) \xrightarrow{\sim} \mathbf{D}^b(\mathrm{qgr}\, A)$.

Proof. Lemmas 14 and 15 gives us that the subcategory $\mathcal{T}_i$ is admissible in $\mathbf{D}^b(\mathrm{gr}{-}A)$ and the right orthogonal $\mathcal{T}_i^\perp$ has a weak semiorthogonal decomposition of the form

$$\mathcal{T}_i^\perp = \langle \mathcal{S}_{<i}, \mathcal{P}_{\geq i} \rangle. \tag{11}$$

Now let us describe the right orthogonal to the subcategory $\mathcal{D}_i$. First, since A is Gorenstein, the functor D takes the subcategory $\mathcal{S}_{\geq i}(A)$ to the subcategory $\mathcal{S}_{<-i-a+1}(A^\circ)$. Hence, D sends the right orthogonal $\mathcal{S}_{\geq i}^\perp(A)$ to the left orthogonal ${}^\perp \mathcal{S}_{<-i-a+1}(A^\circ)$, which coincides with the right orthogonal $\mathcal{P}_{<-i-a+1}^\perp(A^\circ)$ by Lemma 14. Therefore, the subcategory $\mathcal{S}_{\geq i}^\perp$ coincides with ${}^\perp \mathcal{P}_{\geq i+a}$. On the other hand, by Lemmas 14 and 15 we have that

$$ {}^\perp \mathcal{P}_{\geq i+a} = \mathcal{S}_{\geq i}^\perp \cong \langle \mathcal{S}_{<i}, \mathcal{D}_i \rangle. $$

This implies that the right orthogonal $\mathcal{D}_i^\perp$ has the following decomposition:

$$\mathcal{D}_i^\perp = \langle \mathcal{P}_{\geq i+a}, \mathcal{S}_{<i} \rangle. \tag{12}$$

Assume that $a \geq 0$. In this case, the decomposition (12) is not only semiorthogonal, but is in fact mutually orthogonal, because $\mathcal{P}_{\geq i+a} \subset \mathbf{D}^b(\mathrm{gr}{-}A_{\geq i})$. Hence, we can interchange $\mathcal{P}_{\geq i+a}$ and $\mathcal{S}_{<i}$, i.e.,

$$\mathcal{D}_i^\perp = \langle \mathcal{S}_{<i}, \mathcal{P}_{\geq i+a} \rangle.$$

Thus, we obtain that $\mathcal{D}_i^\perp \subset \mathcal{T}_i^\perp$ and, consequently, $\mathcal{T}_i$ is a full subcategory of $\mathcal{D}_i$. Moreover, we can describe the right orthogonal to $\mathcal{T}_i$ in $\mathcal{D}_i$. In fact, there is a decomposition

$$\mathcal{P}_{\geq i} = \langle \mathcal{P}_{\geq i+a}, \mathcal{P}_i^a \rangle,$$

where $\mathcal{P}_i^a$ is the subcategory generated by the modules $A(-i-a+1), \ldots, A(-i)$. Moreover, these modules form an exceptional collection. Thus, the category $\mathcal{D}_i$ has the semiorthogonal decomposition

$$\mathcal{D}_i = \langle A(-i-a+1), \ldots, A(-i), \mathcal{T}_i \rangle.$$

Since $\mathcal{D}_i \cong \mathbf{D}^b(\mathrm{qgr}\, A)$ and $\mathcal{T}_i \cong \mathbf{D}_{\mathrm{Sg}}^{\mathrm{gr}}(A)$, we obtain the decomposition

$$\mathbf{D}^b(\mathrm{qgr}\, A) \cong \langle \pi A(-i-a+1), \ldots, \pi A(-i), \Phi_i \mathbf{D}_{\mathrm{Sg}}^{\mathrm{gr}}(A) \rangle,$$

where the fully faithful functor Φ_i is the composition $\mathbf{D}_{\mathrm{Sg}}^{\mathrm{gr}}(A) \xrightarrow{\sim} \mathcal{T}_i \hookrightarrow \mathbf{D}^b(\mathrm{gr}-A) \xrightarrow{\pi} \mathbf{D}^b(\mathrm{qgr}\, A)$.

Assume now that $a \leq 0$. In this case, the decomposition (11) is not only semiorthogonal but is in fact mutually orthogonal, because the algebra A is Gorenstein and $\mathbf{R}\,\mathrm{Hom}_A(\mathbf{k}, A) = \mathbf{k}(a)[-n]$ with $a \leq 0$. Hence, we can interchange $\mathcal{P}_{\geq i}$ and $\mathcal{S}_{<i}$, i.e.,

$$\mathcal{T}_i^\perp = \langle \mathcal{P}_{\geq i}, \mathcal{S}_{<i} \rangle.$$

Now we see that $\mathcal{T}_i^\perp \subset \mathcal{D}_{i-a}^\perp$, and consequently, $\mathcal{D}_{i-a}$ is the full subcategory of $\mathcal{T}_i$. Moreover, we can describe the right orthogonal to $\mathcal{D}_{i-a}$ in $\mathcal{T}_i$. In fact, there is a decomposition of the form

$$\mathcal{S}_{<i-a} = \langle \mathcal{S}_{<i}, \mathbf{k}(-i), \ldots, \mathbf{k}(-i+a+1) \rangle.$$

Therefore, the category $\mathcal{T}_i \cong \mathbf{D}_{\mathrm{Sg}}^{\mathrm{gr}}(A)$ has a semiorthogonal decomposition of the form

$$\mathcal{T}_i = \langle \mathbf{k}(-i), \ldots, \mathbf{k}(-i+a+1), \mathcal{D}_{i-a} \rangle. \tag{13}$$

Since $\mathcal{D}_{i-a} \cong \mathbf{D}^b(\mathrm{qgr}\, A)$ and $\mathcal{T}_i \cong \mathbf{D}_{\mathrm{Sg}}^{\mathrm{gr}}(A)$, we obtain the decomposition

$$\mathbf{D}_{\mathrm{Sg}}^{\mathrm{gr}}(A) \cong \langle q\mathbf{k}(-i), \ldots, q\mathbf{k}(-i+a+1), \Psi_i \mathbf{D}^b(\mathrm{qgr}\, A) \rangle,$$

where the fully faithful functor Ψ_i can be defined as the composition $\mathbf{D}^b(\mathrm{qgr}\, A) \xrightarrow{\sim} \mathcal{D}_{i-a} \hookrightarrow \mathbf{D}^b(\mathrm{gr}-A) \xrightarrow{q} \mathbf{D}_{\mathrm{Sg}}^{\mathrm{gr}}(A)$. If $a = 0$, then we get equivalence.   □

Remark 17. It follows from the construction that the functor Ψ_{i+a} from the bounded derived category $\mathbf{D}^b(\mathrm{qgr}\, A)$ to $\mathbf{D}_{\mathrm{Sg}}^{\mathrm{gr}}(A)$ is the composition of the functor $\mathbf{R}\omega_i : \mathbf{D}^b(\mathrm{qgr}\, A) \longrightarrow \mathbf{D}^b(\mathrm{gr}-A_{\geq i})$, which is given by formula (5), the natural embedding $\mathbf{D}^b(\mathrm{gr}-A_{\geq i}) \hookrightarrow \mathbf{D}^b(\mathrm{gr}-A)$, and the projection $\mathbf{D}^b(\mathrm{gr}-A) \xrightarrow{q} \mathbf{D}_{\mathrm{Sg}}^{\mathrm{gr}}(A)$.

Let us consider two limiting cases. The first case is that the algebra A has finite homological dimension. In this case the triangulated category of singularities $\mathbf{D}_{\mathrm{Sg}}^{\mathrm{gr}}(A)$ is trivial, and hence the Gorenstein parameter a is non negative and the derived category $\mathbf{D}^b(\mathrm{qgr}\, A)$ has a full exceptional collection $\sigma = (\pi A(0), \ldots, \pi A(a-1))$. More precisely, we have the following:

Corollary 18. Let $A = \bigoplus_{i \geq 0} A_i$ be a connected graded Noetherian algebra that is Gorenstein with Gorenstein parameter a. Suppose that A has finite homological dimension. Then, $a \geq 0$ and the derived category $\mathbf{D}^b(\mathrm{qgr}\, A)$ has a full strong exceptional collection $\sigma = (\pi A(0), \ldots, \pi A(a-1))$. Moreover, the category $\mathbf{D}^b(\mathrm{qgr}\, A)$ is equivalent to the derived category $\mathbf{D}^b(\mathrm{mod-}\, \mathrm{Q}(A))$ of finite (right) modules over the algebra $\mathrm{Q}(A) := \mathrm{End}_{\mathrm{gr-}A}\left(\bigoplus_{i=0}^{a-1} A(i)\right)$ of homomorphisms of σ.

Proof. Since A has finite homological dimension, the category $\mathbf{D}^{\mathrm{gr}}_{\mathrm{Sg}}(A)$ is trivial. By Theorem 16 we get that $a \geq 0$ and that $\mathbf{D}^b(\mathrm{qgr}\, A)$ has a full exceptional collection $\sigma = (\pi A(0), \ldots, \pi A(a-1))$. Consider the object $P_\sigma = \bigoplus_{i=0}^{a-1} \pi A(i)$ and the functor

$$\mathrm{Hom}(P_\sigma, -) : \mathrm{qgr}\, A \longrightarrow \mathrm{mod-}\, \mathrm{Q}(A),$$

where

$$Q(A) = \mathrm{End}_{\mathrm{qgr}-A}\left(\bigoplus_{i=0}^{a-1} \pi A(i)\right) = \mathrm{End}_{\mathrm{gr}-A}\left(\bigoplus_{i=0}^{a-1} A(i)\right)$$

is the algebra of homomorphisms of the exceptional collection σ. It is easy to see that this functor has a right derived functor

$$\mathbf{R}\,\mathrm{Hom}(P_\sigma, -) : \mathbf{D}^b(\mathrm{qgr}\, A) \longrightarrow \mathbf{D}^b(\mathrm{mod-}\, Q(A))$$

(e.g., as a composition $\mathbf{R}\omega_0$ and $\mathrm{Hom}\left(\bigoplus_{i=0}^{a-1} A(i), -\right)$). The standard reasoning (see, e.g., [6] or [7]) now shows that the functor $\mathbf{R}\,\mathrm{Hom}(P_\sigma, -)$ is an equivalence. $\square$

Example 19. As an application we obtain a well-known result (see [4]) asserting the existence of a full exceptional collection in the bounded derived category of coherent sheaves on the projective space $\mathbb{P}^n$. This result follows immediately if we take $A = \mathbf{k}[x_0, \ldots, x_n]$ with its standard grading. More generally, if we take A to be the polynomial algebra $\mathbf{k}[x_0, \ldots, x_n]$ graded by $\deg x_i = a_i$, then we get a full exceptional collection $(\mathcal{O}, \ldots, \mathcal{O}(\sum_{i=0}^n a_i - 1))$ in the bounded derived category of coherent sheaves on the weighted projective space $\mathbb{P}(a_0, \ldots, a_n)$ considered as a smooth orbifold (see [2, 3]). It is also true for noncommutative (weighted) projective spaces [2].

Another limiting case is that the algebra A is finite-dimensional over the base field (i.e., A is a Frobenius algebra). In this case the category $\mathrm{qgr}\, A$ is trivial, and hence the triangulated category of singularities $\mathbf{D}^{\mathrm{gr}}_{\mathrm{Sg}}(A)$ has a full exceptional collection (compare with [13, 10.10]). More precisely, we get the following:

Corollary 20. Let $A = \bigoplus_{i \geq 0} A_i$ be a connected graded Noetherian algebra that is Gorenstein with Gorenstein parameter a. Suppose that A is finite-dimensional over the field $\mathbf{k}$. Then $a \leq 0$, and the triangulated category of singularities $\mathbf{D}^{\mathrm{gr}}_{\mathrm{Sg}}(A)$ has a full exceptional collection $(q\mathbf{k}(0), \ldots, q\mathbf{k}(a+1))$, where $q : \mathbf{D}^b(\text{gr-}A) \longrightarrow \mathbf{D}^{\text{gr}}_{\text{Sg}}(A)$ *is the natural projection. Moreover, the triangulated category* $\mathbf{D}^{\text{gr}}_{\text{Sg}}(A)$ *is equivalent to the derived category* $\mathbf{D}^b(\text{mod-}Q(A))$ *of finite (right) modules over the algebra* $Q(A) = \text{End}_{\text{gr-}A}\left(\bigoplus_{i=a+1}^0 A(i)\right)$.

Proof. Since A is finite-dimensional, the derived category $\mathbf{D}^b(\text{qgr }A)$ is trivial. By Theorem 16 we get that $a \le 0$ and $\mathbf{D}^{\text{gr}}_{\text{Sg}}(A)$ has a full exceptional collection $(q\mathbf{k}(0), \dots, q\mathbf{k}(a+1))$. Unfortunately, this collection is not strong. However, we can replace it by the dual exceptional collection, which is already strong (see Definition 7). By Lemma 15 there is a weak semiorthogonal decomposition $\mathbf{D}^b(\text{gr-}A_{\ge 0}) = \langle \mathcal{P}_{\ge 0}, \mathcal{T}_0 \rangle$, where $\mathcal{T}_0$ is equivalent to $\mathbf{D}^{\text{gr}}_{\text{Sg}}(A)$. Moreover, by formula (13) we have the following semiorthogonal decomposition for $\mathcal{T}_0$:

$$\mathcal{T}_0 = \langle \mathbf{k}(0), \dots, \mathbf{k}(a+1) \rangle.$$

Denote by E_i, where $i = 0, \dots, -a-1$, the modules $A(i+a+1)/A(i+a+1)_{\ge a}$. These modules belong to $\mathcal{T}_0$ and form a full exceptional collection

$$\mathcal{T}_0 = \langle E_0, \dots, E_{-(a+1)} \rangle.$$

Furthermore, this collection is strong, and the algebra of homomorphisms of this collection coincides with the algebra $Q(A) = \text{End}_{\text{gr}-A}\left(\bigoplus_{i=a+1}^0 A(i)\right)$.

As in the previous proposition, consider the object $E = \bigoplus_{i=0}^{-(a+1)} E_i$ and the functor

$$\mathbf{R}\,\text{Hom}(E, -) : \mathcal{T}_0 = \mathbf{D}^{\text{gr}}_{\text{Sg}}(A) \longrightarrow \mathbf{D}^b(\text{mod-}Q(A)).$$

Again the standard reasoning from [6,7] shows that the functor $\mathbf{R}\,\text{Hom}(E, -)$ is an equivalence of triangulated categories.   □

Example 21. The simplest example here is $A = \mathbf{k}[x]/x^{n+1}$. In this case the triangulated category of singularities $\mathbf{D}^{\text{gr}}_{\text{Sg}}(A)$ has a full exceptional collection and is equivalent to the bounded derived category of finite-dimensional representations of the Dynkin quiver of type $A_n : \underbrace{\bullet - \bullet - \cdots - \bullet}_{n}$, because in this case the algebra $Q(A)$ is isomorphic to the path algebra of this Dynkin quiver. This example is considered in detail in the paper [26].

Remark 22. There are other cases in which the triangulated category of singularities $\mathbf{D}^{\text{gr}}_{\text{Sg}}(A)$ has a full exceptional collection. It follows from Theorem 16 that if $a \le 0$ and the derived category $\mathbf{D}^b(\text{qgr }A)$ has a full exceptional collection, then $\mathbf{D}^{\text{gr}}_{\text{Sg}}(A)$ has a full exceptional collection as well. It happens, for example, in the case that the algebra A is related to a weighted projective line, an orbifold over $\mathbb{P}^1$ (see, e.g., [11]).

2.3 Categories of Coherent Sheaves for Gorenstein Schemes

Let X be a connected projective Gorenstein scheme of dimension n and let $\mathcal{L}$ be a very ample line bundle. Denote by A the graded coordinate algebra $\bigoplus_{i\ge 0} H^0(X, \mathcal{L}^i)$. The famous Serre theorem [25] asserts that the abelian category of coherent sheaves $\text{coh}(X)$ is equivalent to the quotient category qgr A.

Assume that the dualizing sheaf ω_X is isomorphic to $\mathcal{L}^{-r}$ for some $r \in \mathbb{Z}$ and assume also that $H^j(X, \mathcal{L}^k) = 0$ for all $k \in \mathbb{Z}$ when $j \neq 0, n$. (For example, if X is a complete intersection in $\mathbb{P}^N$ then it satisfies these conditions.) In this case, Theorem 16 allows us to compare the triangulated category of singularities $\mathbf{D}_{\mathrm{Sg}}^{\mathrm{gr}}(A)$ with the bounded derived category of coherent sheaves $\mathbf{D}^b(\mathrm{coh}(X))$. To apply that theorem we need the following lemma.

Lemma 23. *Let X be a connected projective Gorenstein scheme of dimension n. Let $\mathcal{L}$ be a very ample line bundle such that $\omega_X \cong \mathcal{L}^{-r}$ for some $r \in \mathbb{Z}$ and $H^j(X, \mathcal{L}^k) = 0$ for all $k \in \mathbb{Z}$ when $j \neq 0, n$. Then the algebra $A = \bigoplus_{i \geq 0} H^0(X, \mathcal{L}^i)$ is Gorenstein with Gorenstein parameter $a = r$.*

Proof. Consider the projection functor $\Pi : \mathrm{Gr}{-}A \to \mathrm{QGr}\, A$ and its right adjoint $\Omega : \mathrm{QGr}\, A \to \mathrm{Gr}{-}A$ which is given by the formula (4)

$$\Omega \Pi N \cong \bigoplus_{n=-\infty}^{\infty} \mathrm{Hom}_{\mathrm{QGr}}(\Pi A, \Pi N(n)).$$

The functor Ω has a right derived $\mathbf{R}\Omega$ that is given by the formula

$$\mathbf{R}^j \Omega(\Pi N) \cong \bigoplus_{n=-\infty}^{\infty} \mathrm{Ext}_{\mathrm{QGr}}^j(\Pi A, \Pi N(n))$$

(see, e.g. [1, Prop. 7.2]). The assumptions on X and $\mathcal{L}$ imply that $\mathbf{R}^j \Omega(\Pi A) \cong 0$ for all $j \neq 0, n$. Moreover, since X is Gorenstein and $\omega_X \cong \mathcal{L}^{-r}$, Serre duality for X yields that

$$\mathbf{R}^0 \Omega(\Pi A) \cong \bigoplus_{i=-\infty}^{\infty} H^0(X, \mathcal{L}^i) \cong A \quad \text{and} \quad \mathbf{R}^n \Omega(\Pi A) \cong \bigoplus_{i=-\infty}^{\infty} H^n(X, \mathcal{L}^i) \cong A^*(r),$$

where $A^* = \mathrm{Hom}_{\mathbf{k}}(A, \mathbf{k})$. As X is irreducible, the algebra A is connected. Since Π and $\mathbf{R}\Omega$ are adjoint functors we have

$$\mathbf{R}\,\mathrm{Hom}_{\mathrm{Gr}}(\mathbf{k}(s), \mathbf{R}\Omega(\Pi A)) \cong \mathbf{R}\,\mathrm{Hom}_{\mathrm{QGr}}(\Pi \mathbf{k}(s), \Pi A) = 0$$

for all s. Furthermore, we know that $\mathbf{R}\,\mathrm{Hom}_A(\mathbf{k}, A^*) \cong \mathbf{R}\,\mathrm{Hom}_A(A, \mathbf{k}) \cong \mathbf{k}$. This implies that $\mathbf{R}\,\mathrm{Hom}_A(\mathbf{k}, A) \cong \mathbf{k}(r)[-n-1]$. This isomorphism implies that the affine cone $\mathbf{Spec}A$ is Gorenstein at the vertex and the assumption on X now implies that $\mathbf{Spec}A$ is Gorenstein scheme ([14, V, §9,10]). Since $\mathbf{Spec}A$ has a finite Krull dimension, the algebra A is a dualizing complex for itself, i.e. it has a finite injective dimension. Thus, the algebra A is Gorenstein with parameter r. □

Theorem 24. *Let X be an connected projective Gorenstein scheme of dimension n. Let $\mathcal{L}$ be a very ample line bundle such that $\omega_X \cong \mathcal{L}^{-r}$ for some $r \in \mathbb{Z}$. Suppose $H^j(X, \mathcal{L}^k) = 0$ for all $k \in \mathbb{Z}$ when $j \neq 0, n$. Set $A := \bigoplus_{i \geq 0} H^0(X, \mathcal{L}^i)$. Then the derived category of coherent sheaves*

$\mathbf{D}^b(\mathrm{coh}(X))$ *and the triangulated category of singularities* $\mathbf{D}^{\mathrm{gr}}_{\mathrm{Sg}}(A)$ *are related as follows:*

(i) *if* $r > 0$, *i.e., if* X *is a Fano variety, then there is a semiorthogonal decomposition*

$$\mathbf{D}^b(\mathrm{coh}(X)) = \langle \mathcal{L}^{-r+1}, \ldots, \mathcal{O}_X, \mathbf{D}^{\mathrm{gr}}_{\mathrm{Sg}}(A) \rangle;$$

(ii) *if* $r < 0$, *i.e., if* X *is a variety of general type, then there is a semiorthogonal decomposition*

$$\mathbf{D}^{\mathrm{gr}}_{\mathrm{Sg}}(A) = \langle q\mathbf{k}(r+1), \ldots, q\mathbf{k}, \mathbf{D}^b(\mathrm{coh}(X)) \rangle,$$

where $q : \mathbf{D}^b(\mathrm{gr} - A) \longrightarrow \mathbf{D}^{\mathrm{gr}}_{\mathrm{Sg}}(A)$ *is the natural projection;*
(iii) *if* $r = 0$, *i.e., if* X *is a Calabi–Yau variety, then there is an equivalence*

$$\mathbf{D}^{\mathrm{gr}}_{\mathrm{Sg}}(A) \xrightarrow{\sim} \mathbf{D}^b(\mathrm{coh}(X)).$$

Proof. Since $\mathcal{L}$ is very ample, Serre's theorem implies that the bounded derived category $\mathbf{D}^b(\mathrm{coh}(X))$ is equivalent to the category $\mathbf{D}^b(\mathrm{qgr}\, A)$, where $A = \bigoplus_{i \geq 0} H^0(X, \mathcal{L}^i)$. Since $H^j(X, \mathcal{L}^k) = 0$ for $j \neq 0, n$ and all $k \in \mathbb{Z}$, Lemma 23 implies that A is Gorenstein. Now the theorem immediately follows from Theorem 16. □

Corollary 25. *Let* X *be an irreducible projective Gorenstein Fano variety of dimension* n *with at most rational singularities. Let* $\mathcal{L}$ *be a very ample line bundle such that* $\omega_X^{-1} \cong \mathcal{L}^r$ *for some* $r \in \mathbb{N}$. *Set* $A = \bigoplus_{i \geq 0} H^0(X, \mathcal{L}^i)$. *Then the category* $\mathbf{D}^b(\mathrm{coh}(X))$ *admits a semiorthogonal decomposition of the form*

$$\mathbf{D}^b(\mathrm{coh}(X)) = \langle \mathcal{L}^{-r+1}, \ldots, \mathcal{O}_X, \mathbf{D}^{\mathrm{gr}}_{\mathrm{Sg}}(A) \rangle.$$

Proof. The Kawamata–Viehweg vanishing theorem (see, e.g., [18, Th. 1.2.5]) yields $H^j(X, \mathcal{L}^k) = 0$ for $j \neq 0, n$ and all k. Hence, we can apply Theorem 24(i). □

Corollary 26. *Let* X *be a Calabi–Yau variety. That is,* X *is an irreducible projective variety with at most rational singularities, with trivial canonical sheaf* $\omega_X \cong \mathcal{O}_X$ *and such that* $H^j(X, \mathcal{O}_X) = 0$ *for* $j \neq 0, n$. *Let* $\mathcal{L}$ *be some very ample line bundle on* X. *Set* $A = \bigoplus_{i \geq 0} H^0(X, \mathcal{L}^i)$. *Then there is an equivalence*

$$\mathbf{D}^b(\mathrm{coh}(X)) \cong \mathbf{D}^{\mathrm{gr}}_{\mathrm{Sg}}(A).$$

Proof. The variety X has rational singularities hence it is Cohen–Macaulay. Moreover, X is Gorenstein, because $\omega_X \cong \mathcal{O}_X$. The Kawamata–Viehweg vanishing theorem ([18, Th. 1.2.5]) yields $H^j(X, \mathcal{L}^k) = 0$ for $j \neq 0, n$ and all $k \neq 0$. Since by assumption $H^j(X, \mathcal{O}_X) = 0$ for $j \neq 0, n$, we can apply Theorem 24 (iii). □

Proposition 27. *Let* $X \subset \mathbb{P}^N$ *be a complete intersection of* m *hypersurfaces* $D_1, \ldots, D_m$ *of degrees* $d_1, \ldots, d_m$ *respectively. Then* X *and* $\mathcal{L} = \mathcal{O}_X(1)$ *satisfy the conditions of Theorem 24 with Gorenstein parameter* $r = N + 1 - \sum_{i=1}^{m} d_i$.

Proof. Since the variety X is a complete intersection, it is Gorenstein. The canonical class ω_X is isomorphic to $\mathcal{O}(\sum d_i - N - 1)$. It can be easily proved by induction on m that $H^j(X, \mathcal{O}_X(k)) = 0$ for all k and $j \neq 0, n$, where $n = N - m$ is the dimension of X. Indeed, the base of the induction is clear. For the induction step, assume that for $Y = D_1 \cap \cdots \cap D_{m-1}$ these conditions hold. Then, consider the short exact sequence

$$0 \longrightarrow \mathcal{O}_Y(k - d_m) \longrightarrow \mathcal{O}_Y(k) \longrightarrow \mathcal{O}_X(k) \longrightarrow 0.$$

Since the cohomologies $H^j(Y, \mathcal{O}_Y(k))$ ane 0 for all k and $j \neq 0, n+1$, we obtain that $H^j(X, \mathcal{O}_X(k)) = 0$ for all k and $j \neq 0, n$. □

Theorem 24 can be extended to the case of quotient stacks. To do this we will need an appropriate generalization of Serre's theorem [25]. The usual Serre theorem says that if a commutative connected graded algebra $A = \bigoplus_{i \geq 0} A_i$ is generated by its first component, then the category qgr A is equivalent to the category of coherent sheaves $\mathrm{coh}(X)$ on the projective variety $X = \mathbf{Proj}\, A$. (Such equivalence holds for the categories of quasicoherent sheaves $\mathrm{Qcoh}(X)$ and $\mathrm{QGr}\, A$ too.)

Consider now a commutative connected graded $\mathbf{k}$-algebra $A = \bigoplus_{i \geq 0} A_i$ that is not necessarily generated by its first component. The grading on A induces an action of the group $\mathbf{k}^*$ on the affine scheme $\mathbf{Spec}A$. Let $\mathbf{0}$ be the closed point of $\mathbf{Spec}A$ that corresponds to the ideal $A_+ = A_{\geq 1} \subset A$. This point is invariant under the action.

Denote by $\mathbb{P}\mathrm{roj}\, A$ the quotient stack $[(\mathbf{Spec}A \backslash \mathbf{0})/\mathbf{k}^*]$. (Note that there is a natural map $\mathbb{P}\mathrm{roj}\, A \to \mathbf{Proj}\, A$, which is an isomorphism if the algebra A is generated by A_1.)

Proposition 28. (see also [2]) *Let $A = \oplus_{i \geq 0} A_i$ be a connected graded finitely generated algebra. Then the category of (quasi)coherent sheaves on the quotient stack $\mathbb{P}\mathrm{roj}\,(A)$ is equivalent to the category qgr A (respectively $\mathrm{QGr}\, A$).*

Proof. Let $\mathbf{0}$ be the closed point on the affine scheme $\mathbf{Spec}A$ that corresponds to the maximal ideal $A_+ \subset A$. Denote by U the complement $\mathbf{Spec}A \backslash \mathbf{0}$. We know that the category of (quasi)coherent sheaves on the stack $\mathbb{P}\mathrm{roj}\, A$ is equivalent to the category of $\mathbf{k}^*$-equivariant (quasi)coherent sheaves on U. The category of (quasi)coherent sheaves on U is equivalent to the quotient of the category of (quasi)coherent sheaves on $\mathbf{Spec}A$ by the subcategory of (quasi)coherent sheaves with support on $\mathbf{0}$ (see [9]). This is also true for the categories of $\mathbf{k}^*$-equivariant sheaves. But the category of (quasi)coherent $\mathbf{k}^*$-equivariant sheaves on $\mathbf{Spec}A$ is just the category gr-A (resp. Gr-A) of graded modules over A, and the subcategory of (quasi)coherent sheaves with support on $\mathbf{0}$ coincides with the subcategory tors-A (resp. Tors-A). Thus, we obtain that $\mathrm{coh}(\mathbb{P}\mathrm{roj}\, A)$ is equivalent to the quotient category qgr $A = $ gr-$A/$ tors-A (and $\mathrm{Qcoh}(\mathbb{P}\mathrm{roj}\, A)$ is equivalent to $\mathrm{QGr}\, A = $ Gr-$A/$ Tors-A). □

Corollary 29. *Assume that the Noetherian Gorenstein connected graded algebra A from Theorem 16 is finitely generated and commutative. Then in place
of the bounded derived category* $\mathbf{D}^b(\mathrm{qgr}\, A)$ *in Theorem 16 we can substitute the
category* $\mathbf{D}^b(\mathrm{coh}(\mathbb{P}\mathrm{roj}\, A))$, *where* $\mathbb{P}\mathrm{roj}\, A$ *the quotient stack* $\left[(\mathbf{Spec}\, A\backslash 0)/\mathbf{k}^*\right]$.

3 Categories of Graded D-branes of Type B in Landau–Ginzburg Models

3.1 Categories of Graded Pairs

Let $B = \bigoplus_{i \geq 0} B_i$ be a finitely generated connected graded algebra over a field
k. Let $W \in B_n$ be a central element of degree n that is not a zero-divisor,
i.e., $Wb = bW$ for any $b \in B$ and $bW = 0$ only for $b = 0$. Denote by J the
two-sided ideal $J := WB = BW$ and denote by A the quotient graded algebra
B/J.

With any such element $W \in B_n$ we can associate two categories: an exact
category $\mathrm{GrPair}(W)$ and a triangulated category $\mathrm{DGrB}(W)$.[2] Objects of these
categories are ordered pairs

$$\overline{P} := \left(P_1 \underset{p_0}{\overset{p_1}{\rightleftarrows}} P_0 \right),$$

where $P_0, P_1 \in \mathrm{gr}{-}B$ are finitely generated free graded right B-modules, p_1
is a map of degree 0, and p_0 is a map of degree n (i.e., a map from P_0 to
$P_1(n)$) such that the compositions $p_0 p_1$ and $p_1(n) p_0$ are the left multiplications
by the element W. A morphism $f : \overline{P} \to \overline{Q}$ in the category $\mathrm{GrPair}(W)$ is
a pair of morphisms $f_1 : P_1 \to Q_1$ and $f_0 : P_0 \to Q_0$ of degree 0 such
that $f_1(n) p_0 = q_0 f_0$ and $q_1 f_1 = f_0 p_1$. The morphism $f = (f_1, f_0)$ is null-
homotopic if there are two morphisms $s : P_0 \to Q_1$ and $t : P_1 \to Q_0(-n)$
such that $f_1 = q_0(n) t + s p_1$ and $f_0 = t(n) p_0 + q_1 s$. Morphisms in the category
$\mathrm{DGrB}(W)$ are the classes of morphisms in $\mathrm{GrPair}(W)$ modulo null-homotopic
morphisms.

In other words, objects of both categories are quasi–periodic infinite sequences

$$\underline{K}^{\bullet} := \{ \cdots \longrightarrow K^i \xrightarrow{k^i} K^{i+1} \xrightarrow{k^{i+1}} K^{i+2} \longrightarrow \cdots \}$$

of morphisms in $\mathrm{gr}{-}B$ of *free* graded right B-modules such that the composition
$k^{i+1} k^i$ of any two consecutive morphisms is equal to multiplication by W. The
quasi–periodicity property here means that $\underline{K}^{\bullet}[2] = \underline{K}^{\bullet}(n)$. In particular,

$$K^{2i-1} \cong P_1(i \cdot n), \; K^{2i} \cong P_0(i \cdot n), \; k^{2i-1} = p_1(i \cdot n), \; k^{2i} = p_0(i \cdot n).$$

---

[2] One can also construct a differential graded category whose homotopy category
is equivalent to DGrB .

A morphism $f : \underline{K}^\bullet \longrightarrow \underline{L}^\bullet$ in the category GrPair(W) is a family of morphisms $f^i : K^i \longrightarrow L^i$ in gr-B that is quasiperiodic, i.e., $f^{i+2} = f^i(n)$, and that commutes with k^i and l^i, i.e., $f^{i+1}k^i = l^i f^i$.

Morphisms in the category DGrB(W) are morphisms in GrPair(W) modulo null-homotopic morphisms, and we consider only quasiperiodic homotopies, i.e., families $s^i : K^i \longrightarrow L^{i-1}$ such that $s^{i+2} = s^i(n)$.

Definition 30. *The category* DGrB(W) *constructed above will be called* the category of graded D-branes of type B *for the pair* ($B = \bigoplus_{i \geq 0} B_i, W$).

Remark 31. If B is commutative, then we can consider the affine scheme **Spec**B. The grading on B corresponds to an action of the algebraic group $\mathbf{k}^*$ on **Spec**B. The element W can be viewed as a regular function on **Spec**B that is semi-invariant with respect to this action. This way, we get a singular Landau–Ginzburg model (**Spec**B, W) with an action of the torus $\mathbf{k}^*$. Thus, Definition 30 is a definition of the category of *graded* D-branes of type B for this model (see also [16, 28]).

It is clear that the category GrPair(W) is an exact category (see [24] for the definition) with monomorphisms and epimorphisms being the componentwise monomorphisms and epimorphisms. The category DGrB(W) can be endowed with a natural structure of a triangulated category. To exhibit this structure we have to define a translation functor [1] and a class of exact triangles.

The translation functor is usually defined as a functor that takes an object $\underline{K}^\bullet$ to the object $\underline{K}^\bullet[1]$, where $K[1]^i = K^{i+1}$ and $d[1]^i = -d^{i+1}$, and takes a morphism f to the morphism $f[1]$, which coincides with f componentwise.

For any morphism $f : \underline{K}^\bullet \to \underline{L}^\bullet$ from the category GrPair(W) we define a mapping cone $\underline{C}^\bullet(f)$ as an object

$$\underline{C}^\bullet(f) = \{\cdots \longrightarrow L^i \oplus K^{i+1} \xrightarrow{c^i} L^{i+1} \oplus K^{i+2} \xrightarrow{c^{i+1}} L^{i+2} \oplus K^{i+3} \longrightarrow \cdots\}$$

such that

$$c^i = \begin{pmatrix} l^i & f^{i+1} \\ 0 & -k^{i+1} \end{pmatrix}.$$

There are maps $g : \underline{L}^\bullet \to \underline{C}^\bullet(f)$, $g = (\text{id}, 0)$ and $h : \underline{C}^\bullet(f) \to \underline{K}^\bullet[1]$, $h = (0, -\text{id})$.

Now we define a standard triangle in the category DGrB(W) as a triangle of the form

$$\underline{K}^\bullet \xrightarrow{f} \underline{L}^\bullet \xrightarrow{g} \underline{C}^\bullet(f) \xrightarrow{h} \underline{K}^\bullet[1]$$

for some $f \in$ GrPair(W).

Definition 32. *A triangle* $\underline{K}^\bullet \to \underline{L}^\bullet \to \underline{M}^\bullet \to \underline{K}^\bullet[1]$ *in* DGrB(W) *will be called an exact (distinguished) triangle if it is isomorphic to a standard triangle.*

Proposition 33. *The category* DGrB(W) *endowed with the translation functor* [1] *and the above class of exact triangles becomes a triangulated category.*

We omit the proof of this proposition, which is more or less the same as the proof of the analogous result for a usual homotopic category (see, e.g., [12]).

3.2 Categories of Graded Pairs and Categories of Singularities

With any object $\underline{K}^{\bullet}$ as above, one associates a short exact sequence

$$0 \longrightarrow K^{-1} \overset{k^{-1}}{\longrightarrow} K^0 \longrightarrow \operatorname{Coker} k^{-1} \longrightarrow 0. \tag{14}$$

We can attach to an object $\underline{K}^{\bullet}$ the right B-module $\operatorname{Coker} k^{-1}$. It can be easily checked that the multiplication by W annihilates it. Hence, the module $\operatorname{Coker} k^{-1}$ is naturally a right A-module, where $A = B/J$ with $J = WB = BW$. Any morphism $f : \underline{K}^{\bullet} \to \underline{L}^{\bullet}$ in GrPair(W) induces a morphism between cokernels. This construction defines a functor Cok : GrPair(W) $\longrightarrow$ gr-A. Using the functor Cok we can construct an exact functor between triangulated categories DGrB(W) and $\mathbf{D}_{\mathrm{Sg}}^{\mathrm{gr}}(A)$.

Proposition 34. *There is a functor F that completes the following commutative diagram:*

$$
\begin{array}{ccc}
\mathrm{GrPair}(W) & \overset{\mathrm{Cok}}{\longrightarrow} & \mathrm{gr}\text{-}A \\
\downarrow & & \downarrow \\
\mathrm{DGrB}(W) & \overset{F}{\longrightarrow} & \mathbf{D}_{\mathrm{Sg}}^{\mathrm{gr}}(A).
\end{array}
$$

Moreover, the functor F is an exact functor between triangulated categories.
Proof. We have the functor GrPair(W) $\longrightarrow \mathbf{D}_{\mathrm{Sg}}^{\mathrm{gr}}(A)$, which is the composition of Cok and the natural functor from gr-A to $\mathbf{D}_{\mathrm{Sg}}^{\mathrm{gr}}(A)$. To prove the existence of a functor F we need to show that any morphism $f : \underline{K}^{\bullet} \to \underline{L}^{\bullet}$ that is null-homotopic goes to the 0-morphism in $\mathbf{D}_{\mathrm{Sg}}^{\mathrm{gr}}(A)$. Fix a homotopy $s = (s^i)$ with $s^i : K^i \to L^{i-1}$. Consider the following decomposition of f:

$$
\begin{array}{ccccc}
K^{-1} & \overset{k^{-1}}{\longrightarrow} & K^0 & \longrightarrow & \operatorname{Coker} k^{-1} \\
{\scriptstyle (s^{-1}, f^{-1})}\big\downarrow & & {\scriptstyle (s^0, f^0)}\big\downarrow & & \big\downarrow \\
L^{-2} \oplus L^{-1} & \overset{u^{-1}}{\longrightarrow} & L^{-1} \oplus L^0 & \longrightarrow & L^0 \otimes_B A \\
{\scriptstyle pr}\big\downarrow & & {\scriptstyle pr}\big\downarrow & & \big\downarrow \\
L^{-1} & \overset{l^{-1}}{\longrightarrow} & L^0 & \longrightarrow & \operatorname{Coker} l^{-1}
\end{array}
$$

$$\text{where} \quad u^{-1} = \begin{pmatrix} -l^{-2} & \mathrm{id} \\ 0 & l^{-1} \end{pmatrix},$$

This yields a decomposition of $F(f)$ through a locally free object $L^0 \otimes_B A$. Hence, $F(f) = 0$ in the category $\mathbf{D}_{\mathrm{Sg}}^{\mathrm{gr}}(A)$. By Lemma 36, which is proved below, the tensor product $\underline{K}^{\bullet} \otimes_B A$ is an acyclic complex. Hence, there is an exact sequence $0 \to \operatorname{Coker} k^{-1} \to K^1 \otimes_B A \to \operatorname{Coker} k^0 \to 0$. Since $K^1 \otimes_B A$ is free, we have $\operatorname{Coker} k^0 \cong \operatorname{Coker} k^{-1}[1]$ in $\mathbf{D}_{\mathrm{Sg}}^{\mathrm{gr}}(A)$. But $\operatorname{Coker} k^0 = F(\underline{K}^{\bullet}[1])$. Hence, the functor F commutes with translation functors. It is easy to see that F takes a standard triangle in DGrB(W) to an exact triangle in $\mathbf{D}_{\mathrm{Sg}}^{\mathrm{gr}}(A)$. Thus, F is exact. □

Lemma 35. *The functor* Cok *is full.*

Proof. Any map $g :$ Coker $k^{-1} \to$ Coker l^{-1} between A-modules can be considered as the map of B-modules and can be extended to a map of short exact sequences

$$
\begin{array}{ccccccccc}
0 & \longrightarrow & K^{-1} & \xrightarrow{k^{-1}} & K^0 & \longrightarrow & \text{Coker } k^{-1} & \longrightarrow & 0 \\
& & \downarrow{\scriptstyle f^{-1}} & & \downarrow{\scriptstyle f^0} & & \downarrow{\scriptstyle g} & & \\
0 & \longrightarrow & L^{-1} & \xrightarrow{l^{-1}} & L^0 & \longrightarrow & \text{Coker } l^{-1} & \longrightarrow & 0,
\end{array}
$$

because K^0 is free. This gives us a sequence of morphisms $f = (f^i), i \in \mathbb{Z}$, where $f^{2i} = f^0(in)$ and $f^{2i-1} = f^{-1}(in)$. To prove the lemma it is sufficient to show that the family f is a map from $\underline{K}^\bullet$ to $\underline{L}^\bullet$, i.e., $f^1 k^0 = l^0 f^0$. Consider the sequence of equalities

$$l^1(f^1 k^0 - l^0 f^0) = f^2 k^1 k^0 - W f^0 = f^2 W - W f^0 = f^0(2)W - W f^0 = 0.$$

Since l^1 is an embedding, we obtain that $f^1 k^0 = l^0 f^0$. □

Lemma 36. *For any sequence $\underline{K}^\bullet \in \mathrm{GrPair}(W)$ the tensor product $\underline{K}^\bullet \otimes_B A$ is an acyclic complex of A-modules and the A-module* Coker k^{-1} *satisfies the condition*
$$\mathrm{Ext}_A^i(\text{Coker } k^{-1}, A) = 0 \qquad \text{for all} \quad i > 0.$$

Proof. It is clear that $\underline{K}^\bullet \otimes_B A$ is a complex. Applying the snake lemma to the commutative diagram

$$
\begin{array}{ccccccccc}
0 & \longrightarrow & K^{i-2} & \xrightarrow{k^{i-2}} & K^{i-1} & \longrightarrow & \text{Coker } k^{i-2} & \longrightarrow & 0 \\
& & \downarrow{\scriptstyle W} & & \downarrow{\scriptstyle W} & & \downarrow{\scriptstyle 0} & & \\
0 & \longrightarrow & K^i & \xrightarrow{k^i} & K^{i+1} & \longrightarrow & \text{Coker } k^i & \longrightarrow & 0,
\end{array}
$$

we obtain an exact sequence

$$0 \to \text{Coker } k^{i-2} \longrightarrow K^i \otimes_B A \xrightarrow{k_i | w} K^{i+1} \otimes_B A \longrightarrow \text{Coker } k^i \to 0.$$

This implies that $\underline{K}^\bullet \otimes_B A$ is an acyclic complex.

Further, consider the dual sequence of left B-modules $\underline{K}^{\bullet\vee}$, where $\underline{K}^{\bullet\vee} \cong \mathrm{Hom}_B(\underline{K}^\bullet, B)$. For the same reasons as above, $A \otimes_B \underline{K}^{\bullet\vee}$ is an acyclic complex. On the other hand, the cohomologies of the complex $\{(K^0)^\vee \longrightarrow (K^{-1})^\vee \longrightarrow (K^{-2})^\vee \longrightarrow \cdots\}$ are isomorphic to $\mathrm{Ext}_A^i(\text{Coker } k^{-1}, A)$. And so, by the acyclicity of $A \otimes_B \underline{K}^{\bullet\vee}$, they are equal to 0 for all $i > 0$. □

Lemma 37. *If $F\underline{K}^\bullet \cong 0$, then $\underline{K}^\bullet \cong 0$ in* $\mathrm{DGrB}(W)$.

Proof. If $F\underline{K}^{\bullet} \cong 0$, then the A-module Coker k^{-1} is perfect as a complex of A-modules. Let us show that Coker k^{-1} is projective in this case. Indeed, there is a natural number m such that $\mathrm{Ext}^i_A(\mathrm{Coker}\ k^{-1}, N) = 0$ for any A-module N and any $i \geq m$. Considering the exact sequence

$$0 \longrightarrow \mathrm{Coker}\ k^{-2m-1} \longrightarrow K^{-2m} \otimes_B A \longrightarrow \cdots \longrightarrow K^{-1} \otimes_B A \longrightarrow K^0 \otimes_B A$$
$$\longrightarrow \mathrm{Coker}\ k^{-1} \longrightarrow 0$$

and taking into account that all A-modules $K^i \otimes_B A$ are free, we find that for all modules N, $\mathrm{Ext}^i_A(\mathrm{Coker}\ k^{-2m-1}, N) = 0$ when $i > 0$. Hence, Coker k^{-2m-1} is a projective A-module. This implies that Coker k^{-1} is also projective, because it is isomorphic to Coker $k^{-2m-1}(-mn)$.

Since Coker k^{-1} is projective, there is a map $f : \mathrm{Coker}\ k^{-1} \to K^0 \otimes_B A$ that splits the epimorphism pr $: K^0 \otimes_B A \to \mathrm{Coker}\ k^{-1}$. It can be lifted to a map from the complex $\{K^{-1} \xrightarrow{k^{-1}} K^0\}$ to the complex $\{K^{-2} \xrightarrow{W} K^0\}$. Denote the lift by (s^{-1}, u). Consider the following diagram:

$$
\begin{array}{ccccc}
K^{-1} & \xrightarrow{k^{-1}} & K^0 & \longrightarrow & \mathrm{Coker}\ k^{-1} \\
\downarrow{\scriptstyle s^{-1}} & & \downarrow{\scriptstyle u} & & \downarrow{\scriptstyle f} \\
K^{-2} & \xrightarrow{W} & K^0 & \longrightarrow & K^0 \otimes_B A \\
\downarrow{\scriptstyle k^{-2}} & & \downarrow{\scriptstyle \mathrm{id}} & & \downarrow{\scriptstyle \mathrm{pr}} \\
K^{-1} & \xrightarrow{k^{-1}} & K^0 & \longrightarrow & \mathrm{Coker}\ k^{-1}.
\end{array}
$$

Since the composition pr f is identical, the map $(k^{-2}s^{-1}, u)$ from the pair $\{K^{-1} \xrightarrow{k^{-1}} K^0\}$ to itself is homotopic to the identity map. Hence, there is a map $s^0 : K^0 \to K^{-1}$ such that

$$\mathrm{id}_{K^{-1}} - k^{-2}s^{-1} = s^0 k^{-1} \qquad \text{and} \qquad k^{-1}s^0 = \mathrm{id}_{K^0} - u.$$

Moreover, we have the following equalities:

$$0 = (uk^{-1} - Ws^{-1}) = (uk^{-1} - s^{-1}(n)W) = (u - s^{-1}(n)k^0)k^{-1}.$$

This gives us that $u = s^{-1}(n)k^0$, because there are no maps from Coker k^{-1} to K^0. Finally, we get the sequence of morphisms $s^i : K^i \longrightarrow K^{i-1}$, where $s^{2i-1} = s^{-1}(in), s^{2i} = s^0(in)$, such that $k^{i-1}s^i + k^i s^{i+1} = \mathrm{id}$. Thus the identity morphism of the object $\underline{K}^{\bullet}$ is null-homotopic. Hence, the object $\underline{K}^{\bullet}$ is isomorphic to the zero object in the category $\mathrm{DGrB}_0(W)$. $\square$

Theorem 38. *The exact functor* $F : \mathrm{DGrB}(W) \longrightarrow \mathrm{D}^{\mathrm{gr}}_{\mathrm{Sg}}(A)$ *is fully faithful.*

Proof. By Lemma 36 we have $\mathrm{Ext}^i_A(\mathrm{Coker}\ k^{-1}, A) = 0$ for $i > 0$. Now, Proposition 11 gives an isomorphism

$$\mathrm{Hom}_{\mathrm{D}^{\mathrm{gr}}_{\mathrm{Sg}}(A)}(\mathrm{Coker}\ k^{-1}, \mathrm{Coker}\ l^{-1}) \cong \mathrm{Hom}_{\mathrm{gr}-A}(\mathrm{Coker}\ k^{-1}, \mathrm{Coker}\ l^{-1})/\mathcal{R},$$

where $\mathcal{R}$ is the subspace of morphisms factoring through projective modules. Since the functor Cok is full, we get that the functor F is also full.

Next we show that F is faithful. The reasoning is standard. Let $f : \underline{K}^{\bullet} \to \underline{L}^{\bullet}$ be a morphism for which $F(f) = 0$. Include f in an exact triangle $\underline{K}^{\bullet} \xrightarrow{f} \underline{L}^{\bullet} \xrightarrow{g} \underline{M}^{\bullet}$. Then the identity map of $F\underline{L}^{\bullet}$ factors through the map $F\underline{L}^{\bullet} \xrightarrow{Fg} F\underline{M}^{\bullet}$. Since F is full, there is a map $h : \underline{L}^{\bullet} \to \underline{L}^{\bullet}$ factoring through $g : \underline{L}^{\bullet} \to \underline{M}^{\bullet}$ such that $Fh = \mathrm{id}$. Hence, the cone $\underline{C}^{\bullet}(h)$ of the map h goes to zero under the functor F. By Lemma 37 the object $\underline{C}^{\bullet}(h)$ is the zero object as well, so h is an isomorphism. Thus $g : \underline{L}^{\bullet} \to \underline{M}^{\bullet}$ is a split monomorphism and $f = 0$. □

Theorem 39. *Suppose that the algebra B has a finite homological dimension. Then the functor $F : \mathrm{DGrB}(W) \longrightarrow \mathbf{D}^{\mathrm{gr}}_{\mathrm{Sg}}(A)$ is an equivalence.*
Proof. We know that F is fully faithful. To prove the theorem we need to show that each object $T \in \mathbf{D}^{\mathrm{gr}}_{\mathrm{Sg}}(A)$ is isomorphic to $F\underline{K}^{\bullet}$ for some $\underline{K}^{\bullet} \in \mathrm{DGrB}(W)$.

The algebra B has a finite homological dimension and as a consequence, it has a finite injective dimension. This implies that $A = B/J$ has a finite injective dimension too. By Lemma 10 any object $T \in \mathbf{D}^{\mathrm{gr}}_{\mathrm{Sg}}(A)$ is isomorphic to the image of an A-module M such that $\mathrm{Ext}^i_A(M, A) = 0$ for all $i > 0$. This means that the object $D(M) = \mathbf{R}\,\mathrm{Hom}_A(M, A)$ is a left A-module. We can consider a projective resolution $\underline{Q}^{\bullet} \to D(M)$. The dual of $\underline{Q}^{\bullet}$ is a right projective A-resolution

$$0 \longrightarrow M \longrightarrow \{P^0 \longrightarrow P^1 \longrightarrow \cdots\}.$$

Consider M as B-module and chose an epimorphism $K^0 \twoheadrightarrow M$ from the free B-module K^0. Denote by $k^{-1} : K^{-1} \to K^0$ the kernel of this map.

The short exact sequence $0 \to B \xrightarrow{W} B \to A \to 0$ implies that for a projective A-module P and any B-module N we have equalities $\mathrm{Ext}^i_B(P, N) = 0$ when $i > 1$. This also yields that $\mathrm{Ext}^i_B(M, N) = 0$ for $i > 1$ and any B-module N, because M has a right projective A-resolution and the algebra B has finite homological dimension. Therefore, $\mathrm{Ext}^i_B(K^{-1}, N) = 0$ for $i > 0$ and any B-module N, i.e., B-module K^{-1} is projective. Since A is connected and finitely generated, any graded projective module is free. Hence, K^{-1} is free.

Since the multiplication on W gives the zero map on M, there is a map $k^0 : K^0 \to K^{-1}(n)$ such that $k^0 k^{-1} = W$ and $k^{-1}(n)k^0 = W$. This way, we get a sequence $\underline{K}^{\bullet}$ with

$$K^{2i} \cong K^0(i \cdot n), \quad K^{2i-1} = K^{-1}(i \cdot n), \quad k^{2i} = k^0(i \cdot n), \quad k^{2i-1} = k^{-1}(i \cdot n),$$

and this sequence is an object of $\mathrm{DGrB}(W)$ for which $F\underline{K}^{\bullet} \cong T$.   □

3.3 Graded D-branes of Type B and Coherent Sheaves

By a Landau–Ginzburg model we mean the following data: a smooth variety X equipped with a symplectic Kähler form ω, a closed real 2-form $\mathcal{B}$, which is called a B-field, and a regular nonconstant function W on X. The function W is called the superpotential of the Landau–Ginzburg model. Since for the definition of D-branes of type B a symplectic form and B-field are not needed, we do not fix them.

With any point $\lambda \in \mathbb{A}^1$ we can associate a triangulated category $DB_\lambda(W)$. We give a construction of these categories under the condition that $X = \mathbf{Spec}(B)$ is affine (see [17, 22]). The category of coherent sheaves on an affine scheme $X = \mathbf{Spec}(B)$ is the same as the category of finitely generated B-modules. The objects of the category $DB_\lambda(W)$ are ordered pairs $\overline{P} := (\, P_1 \underset{p_0}{\overset{p_1}{\rightleftarrows}} P_0 \,)$, where P_0, P_1 are finitely generated projective B-modules and the compositions $p_0 p_1$ and $p_1 p_0$ are the multiplications by the element $(W - \lambda) \in B$. The morphisms in the category $DB(W)$ are the classes of morphisms between pairs modulo null-homotopic morphisms, where a morphism $f : \overline{P} \to \overline{Q}$ between pairs is a pair of morphisms $f_1 : P_1 \to Q_1$ and $f_0 : P_0 \to Q_0$ such that $f_1 p_0 = q_0 f_0$ and $q_1 f_1 = f_0 p_1$. The morphism f is null-homotopic if there are two morphisms $s : P_0 \to Q_1$ and $t : P_1 \to Q_0$ such that $f_1 = q_0 t + s p_1$ and $f_0 = t p_0 + q_1 s$.

We define a category of D-branes of type B (B-branes) on $X = \mathbf{Spec}(B)$ with the superpotential W as the product $DB(W) = \prod_{\lambda \in \mathbb{A}^1} DB_\lambda(W)$.

It was proved in the paper [22, Cor. 3.10] that the category $DB_\lambda(W)$ for smooth affine X is equivalent to the triangulated category of singularities $\mathbf{D}_{\mathrm{Sg}}(X_\lambda)$, where X_λ is the fiber over $\lambda \in \mathbb{A}^1$. Therefore, the category of B-branes $DB(W)$ is equivalent to the product $\prod_{\lambda \in \mathbb{A}^1} \mathbf{D}_{\mathrm{Sg}}(X_\lambda)$. For nonaffine X the category $\prod_{\lambda \in \mathbb{A}^1} \mathbf{D}_{\mathrm{Sg}}(X_\lambda)$ can be considered as a definition of the category of D-branes of type B. Note that in the affine case, X_λ is $\mathbf{Spec}(A_\lambda)$, where $A_\lambda = B/(W - \lambda)B$, and hence the triangulated categories of singularities $\mathbf{D}_{\mathrm{Sg}}(X_\lambda)$ is the same as the category $\mathbf{D}_{\mathrm{Sg}}(A_\lambda)$.

Assume now that there is an action of the group $\mathbf{k}^*$ on the Landau–Ginzburg model (X, W) such that the superpotential W is semi-invariant of weight d. Thus, $X = \mathbf{Spec}(B)$ and $B = \bigoplus_i B_i$ is a graded algebra. The superpotential W is an element of B_d. Let us assume that B is positively graded and connected. In this case, we can consider the triangulated category of graded B-branes $DGrB(W)$, which was constructed in Section 3.1 (see Definition 30).

Denote by A the quotient graded algebra B/WB. We see that the affine variety $\mathbf{Spec}(A)$ is the fiber X_0 of W over the point 0. Denote by Y the quotient stack $[(\mathbf{Spec}(A) \setminus \mathbf{0})/\mathbf{k}^*]$, where $\mathbf{0}$ is the point on $\mathbf{Spec}(A)$ corresponding to the ideal A_+. Theorems 16, 39 and Proposition 28 allow us to establish a relation between the triangulated category of graded B-branes $DGrB(W)$ and the bounded derived category of coherent sheaves on the stack Y.

First, Theorem 39 gives us the equivalence F between the triangulated category of graded B-branes $DGrB(W)$ and the triangulated category of singularities $\mathbf{D}_{\mathrm{Sg}}^{\mathrm{gr}}(A)$. Second, Theorem 16 describes the relationship between the category $\mathbf{D}_{\mathrm{Sg}}^{\mathrm{gr}}(A)$ and the bounded derived category $\mathbf{D}^b(\mathrm{qgr}\,A)$. Third, the category $\mathbf{D}^b(\mathrm{qgr}\,A)$ is equivalent to the derived category $\mathbf{D}^b(\mathrm{coh}(Y))$ by Proposition 28. In the particular case, that X is the affine space $\mathbb{A}^N$ with the standard action of the group $\mathbf{k}^*$, we get the following result.

Theorem 40. *Let X be the affine space $\mathbb{A}^N$ and let W be a homogeneous polynomial of degree d. Let $Y \subset \mathbb{P}^{N-1}$ be the hypersurface of degree d that is given by the equation $W = 0$. Then, there is the following relation between the triangulated category of graded B-branes $\mathrm{DGrB}(W)$ and the derived category of coherent sheaves $\mathbf{D}^b(\mathrm{coh}(Y))$:*

(i) *if $d < N$, i.e. if Y is a Fano variety, there is a semiorthogonal decomposition*

$$\mathbf{D}^b(\mathrm{coh}(Y)) = \langle \mathcal{O}_Y(d - N + 1), \ldots, \mathcal{O}_Y, \mathrm{DGrB}(W) \rangle;$$

(ii) *if $d > N$, i.e., if X is a variety of general type, there is a semiorthogonal decomposition*

$$\mathrm{DGrB}(W) = \langle F^{-1}q(\mathbf{k}(r + 1)), \ldots, F^{-1}q(\mathbf{k}), \mathbf{D}^b(\mathrm{coh}(Y)) \rangle,$$

where $q : \mathbf{D}^b(\mathrm{gr}-A) \longrightarrow \mathbf{D}_{\mathrm{Sg}}^{\mathrm{gr}}(A)$ is the natural projection, and $F : \mathrm{DGrB} \xrightarrow{\sim} \mathbf{D}_{\mathrm{Sg}}^{\mathrm{gr}}(A)$ is the equivalence constructed in Proposition 34.

(iii) *if $d = N$, i.e., if Y is a Calabi–Yau variety, there is an equivalence*

$$\mathrm{DGrB}(W) \xrightarrow{\sim} \mathbf{D}^b(\mathrm{coh}(Y)).$$

Remark 41. We can also consider a weighted action of the torus $\mathbf{k}^*$ on the affine space $\mathbb{A}^N$ with positive weights $(a_1, \ldots, a_N)$, $a_i > 0$ for all i. If the superpotential W is quasi–homogeneous, then we have the category of graded B-branes $\mathrm{DGrB}(W)$. The polynomial W defines an orbifold (quotient stack) $Y \subset \mathbb{P}^{N-1}(a_1, \ldots, a_N)$. The orbifold Y is the quotient of $\mathbf{Spec}(A)\backslash \mathbf{0}$ by the action of $\mathbf{k}^*$, where $A = \mathbf{k}[x_1, \ldots, x_N]/W$. Proposition 28 gives the equivalence between $\mathbf{D}^b(\mathrm{coh}(Y))$ and $\mathbf{D}^b(\mathrm{qgr}\,A)$. And Theorem 16 shows that we get an analogue of Theorem 40 for the weighted case as well.

References

1. M. ARTIN, J. ZHANG, *Noncommutative projective schemes*, Adv. Math. **109** (1994), no. 2, 248–287.
2. D. AUROUX, L. KATZARKOV, D. ORLOV, *Mirror symmetry for weighted projective planes and their noncommutative deformations*, Ann. of Math. (2)**167** (2008), no.3, 867–943.
3. D. BAER, *Tilting sheaves in representation theory of algebras*, Manuscripta Math. **60** (1988), no. 3, 323–247.
4. A. BEILINSON, *Coherent sheaves on $\mathbb{P}^n$ and problems in linear algebra*, Funct. Anal. Appl. **12** (1978), no. 3, 68–69.

5. P. BERTHELOT, A. GROTHENDIECK, L. ILLUSIE, Théorie des intersections et théoreme de Riemann-Roch, Lectere Notes in Mathematics, vol. 225, Springer, 1971.

6. A. BONDAL, *Representation of associative algebras and coherent sheaves*, Izv. Akad. Nauk SSSR **53** (1989), no. 1, 25–44.

7. A. BONDAL, M. KAPRANOV, *Enhanced triangulated categories*, Matem. Sb. **181** (1990), no. 5, 669–683.

8. A. BONDAL, D. ORLOV, *Semiorthogonal decomposition for algebraic varieties*, preprint MPIM 95/15, 1995, `arXiv:math.AG/9506012`.

9. P. GABRIEL, *Des Catégories Abéliennes*, Bull. Soc. Math. Fr. **90** (1962), 323–448.

10. P. GABRIEL, M. ZISMAN, Calculus of fractions and homotopy theory, Ergebnisse der Mathematik und ihrer Grenzgebiete, Band 35, Springer-Verlag New York, New York, 1967.

11. W. GEIGLE, H. LENZING, *A class of weighted projective curves arising in representation theorey of finite dimensional algebras*, Singularities, representation of algebras, and vector bundles (Proc. Symp., Lambrecht/Pfalz/FRG 1985), Lect. Notes Math., vol. 1273, 1987, 265–297.

12. S. GELFAND, Y. MANIN, Homological algebra, algebra v, Encyclopaedia Math. Sci., vol. 38, Springer-Verlag, 1994.

13. D. HAPPEL, *On the derived categories of a finite-dimensional algebra*, Comment. Math. Helv **62** (1987), 339–389.

14. R. HARTSHORNE, Residues and Duality, Lecture Notes in Mathematics, vol. 20, Springer, 1966.

15. K. HORI, C. VAFA, *Mirror Symmetry*, `arXiv:hep-th/0002222`.

16. K. HORI, J. WALCHER, *F-term equation near Gepner points*, `arXiv: hep-th/0404196`.

17. A. KAPUSTIN, Y. LI, *D-branes in Landau-Ginzburg models and algebraic geometry*, J. High Energy Physics, JHEP **12** (2003), no. 005, `arXiv:hep-th/0210296`.

18. Y. KAWAMATA, K. MATSUDA, K. MATSUKI, *Introduction to the minimal model program*, Adv. Stud. in Pure Math. **10** (1987), 283–360.

19. B. KELLER, *Derived categories and their uses*, Handbook of Algebra, vol. 1, 671–701, North-Holland, Amsterdam, 1996.

20. M. KONTSEVICH, *Homological algebra of mirror symmetry*, Proceedings of ICM, Zurich 1994 (Basel), Birkhauser, 1995, 120–139.

21. A. NEEMAN, Triangulated categories, Ann. of Math. Studies, vol. 148, Princeton University Press, 2001.

22. D. ORLOV, *Triangulated categories of singularities and D-branes in Landau-Ginzburg models*, Trudy Steklov Math. Institute **246** (2004), 240–262.

23. D. ORLOV, *Triangulated categories of singularities and equivalences between Landau-Ginzburg models*, Matem. Sbornik, (2006), 197(12):p.1827.

24. D. QUILLEN, Higher Algebraic K-theory I, Springer Lecture Notes in Math., vol. 341, Springer-Verlag, 1973.

25. J. P. SERRE, *Modules projectifs et espaces fibrés à fibre vectorielle*, Séminaire Dubreil–Pisot, vol. 23, Paris, 1958.

26. A. TAKAHASHI, *Matrix factorizations and representations of quivers 1*, arXiv:math.AG/0506347.

27. J. L. VERDIER, *Categories derivées*, SGA 4 1/2, Lecture Notes in Math., vol. 569, Springer-Verlag, 1977.

28. J. WALCHER, *Stability of Landau-Ginzburg branes*, arXiv:hep-th/0412274.

29. A. YEKUTIELI, *Dualizing complexes over noncommutative graded algebras*, J. Algebra **153** (1992), 41–84.

30. A. YEKUTIELI, J. J. ZHANG, *Serre duality for noncommutative projective schemes*, Proc. Amer. Math. Soc. **125** (1997), 697–707.

Rankin's Lemma of Higher Genus and Explicit Formulas for Hecke Operators

Alexei Panchishkin[1] and Kirill Vankov[2]

[1] Institut Fourier, Université de Grenoble I, BP 74, 38402 Saint-Martin d'Hères, France
`Alexei.Pantchichkine@ujf-grenoble.fr`
[2] Institut Fourier, Université de Grenoble I, BP 74, 38402 Saint-Martin d'Hères, France
`kvankov@fourier.ujf-grenoble.fr`

To dear Yuri Ivanovich Manin for his seventieth birthday with admiration

Summary. We develop explicit formulas for Hecke operators of higher genus in terms of spherical coordinates. Applications are given to summation of various generating series with coefficients in local Hecke algebras and in a tensor product of such algebras. In particular, we formulate and prove Rankin's lemma in genus two. An application to a holomorphic lifting from $\mathrm{GSp}_2 \times \mathrm{GSp}_2$ to GSp_4 is given using Ikeda–Miyawaki constructions.

Key words: Hecke operators, Siegel modular forms, Eisenstein series, Spinor zeta functions, motives

2000 Mathematics Subject Classifications: 11F46, 11F66

1 Introduction: Generating Series for the Hecke Operators

Let p be a prime. The Satake isomorphism [Sa63] relates p-local Hecke algebras of reductive groups over $\mathbb{Q}$ to certain polynomial rings. Then one can use a computer in order to find interesting identities between Hecke operators, between their eigenvalues, and relations to Fourier coefficients of modular forms of higher degree.

The purpose of the present paper is to extend Rankin's lemma to the summation of Hecke series of higher genus using symbolic computation. We refer to [Ma-Pa77], where Rankin's lemma was used in the elliptic modular case for multiplicative and additive convolutions of Dirichlet series. That work was further developped in [Pa87], [Pa02]; see also [Ma-Pa05].

Y. Tschinkel and Y. Zarhin (eds.), *Algebra, Arithmetic, and Geometry*,
Progress in Mathematics 270, DOI 10.1007/978-0-8176-4747-6_17,
© Springer Science+Business Media, LLC 2009

Recall that a classical method to produce L-functions for an algebraic group G over $\mathbb{Q}$ uses the generating series

$$\sum_{n=1}^{\infty} \lambda_f(n) n^{-s} = \prod_{p \text{ primes}} \sum_{\delta=0}^{\infty} \lambda_f(p^\delta) p^{-\delta s}$$

of the eigenvalues of Hecke operators on an automorphic form f on G. We study the generating series of Hecke operators $\mathbf{T}(n)$ for the symplectic group Sp_g, and $\lambda_f(n) = \lambda_f(\mathbf{T}(n))$.

Let $\Gamma = Sp_g(\mathbb{Z}) \subset SL_{2g}(\mathbb{Z})$ be the Siegel modular group of genus g, and let $[\mathbf{p}]_g = p\mathbf{I}_{2g} = \mathbf{T}(\underbrace{p, \dots, p}_{2g})$ be the scalar Hecke operator for Sp_g. According to Hecke and Shimura,

$$
D_p(X) = \sum_{\delta=0}^{\infty} \mathbf{T}(p^\delta) X^\delta
$$

$$
= \begin{cases}
\dfrac{1}{1 - \mathbf{T}(p)X + p[\mathbf{p}]_1 X^2}, & \text{if } g = 1 \\[2mm]
\qquad\qquad \text{(see [Hecke], and [Shi71], Theorem 3.21),} \\[4mm]
\dfrac{1 - p^2[\mathbf{p}]_2 X^2}{1 - \mathbf{T}(p)X + \{p\mathbf{T}_1(p^2) + p(p^2+1)[\mathbf{p}]_2\}X^2 - p^3[\mathbf{p}]_2\mathbf{T}(p)X^3 + p^6[\mathbf{p}]_2^2 X^4} \\[2mm]
\qquad\qquad \text{if } g = 2 \text{ (see [Shi63], Theorem 2),}
\end{cases}
$$

where $\mathbf{T}(p), \mathbf{T}_i(p^2)$ $(i = 1, \dots, g)$ are the $g+1$ generators of the corresponding Hecke ring over $\mathbb{Z}$ for the symplectic group Sp_g, in particular, $\mathbf{T}_g(p^2) = [\mathbf{p}]_g$.

It was established by A. N. Andrianov that there exist polynomials

$$E(X), F(X) \in \mathbb{Q}[\mathbf{T}(p), \mathbf{T}_1(p^2), \dots, \mathbf{T}_g(p^2), X]$$

such that

$$D(X) = \sum_{\delta=0}^{\infty} \mathbf{T}(p^\delta) X^\delta = \frac{E(X)}{F(X)},$$

with a polynomial $F(X) = \sum_{j=0}^{2^g} \mathbf{q}_j X^j$ of degree 2^g, and such that $E(X) = \sum_{j=0}^{2^g-2} \mathbf{u}_j X^j$ is a polynomial of degree $2^g - 2$ with leading term

$$(-1)^{g-1} p^{g(g+1)2^{g-2} - g^2} [\mathbf{p}]^{2^{g-1}-1} X^{2^g-2}$$

(as stated in Theorem 6 on p. 451 of [An70] and on p. 61 of §1.3, [An74]).

In the present paper we study explicit formulas for Hecke operators in higher genus in terms of spherical coordinates. Applications are given to summation of various generating series with coefficients in local Hecke algebras.

The question of computing such series explicitly was raised by Prof. S. Friedberg during the first author's talk at the conference "Zeta Functions" (The Independent Moscow University, September 18–22, 2006).

In particular, we formulate and prove Rankin's lemma in genus two for generating series with coefficients in a tensor product of local Hecke algebras. We prove in Theorem 4 that for $g = 2$,

$$\sum_{\delta=0}^{\infty} \mathbf{T}(p^{\delta}) \otimes \mathbf{T}(p^{\delta})\, X^{\delta} = \frac{(1 - p^6[\mathbf{p}] \otimes [\mathbf{p}]X^2) \cdot R(X)}{S(X)},$$

for certain two polynomials

$$R(X) = 1 + r_1 X + \cdots + r_{11}X^{11} + r_{12}X^{12} \text{ with } r_1 = r_{11} = 0,$$
$$S(X) = 1 + s_1 X + \cdots + s_{16}X^{16},$$

with coefficients explicitly expressed through Hecke operators given in the appendix. A motivic interpretation of the polynomial S is given in terms of the tensor product of motives.

For the group $\mathrm{Gl}(n)$, a simpler result could be obtained using the Tamagawa generating series (see [Tam]) and its expansion into simple fractions:

$$\frac{1}{(1 - x_1 X)\cdots(1 - x_n X)} = \sum_{i=1}^{n} \frac{\alpha_i}{1 - x_i X}, \text{ where } \alpha_i = \frac{x_i^{n-1}}{\prod_{j \neq i}^{n}(x_j - x_i)},$$

see Remark 6.

Also, it is helpful for the reader to remember cases in which the Hecke series do have simple numerators: Andrianov (GSp(2), see [An74]), Tamagawa (GL(n), see [Tam]), Böcherer (for standard-L-function, see [BHam]).

2 Results

2.1 Preparation: A Formula for the Total Hecke Operator $\mathbf{T}(p^{\delta})$ of Genus 2

We establish first the following useful formula (in spherical variables x_0, x_1, x_2):

$$\Omega_x^{(2)}(\mathbf{T}(p^{\delta})) \tag{1}$$

$$= p^{-1} x_0^{\delta}(p\, x_1^{(3+\delta)}\, x_2 - p\, x_1^{(2+\delta)} - p\, x_1^{(3+\delta)}\, x_2^{(2+\delta)} + p\, x_1^{(2+\delta)}\, x_2^{(3+\delta)}$$
$$- p\, x_1\, x_2^{(3+\delta)} + p\, x_2^{(2+\delta)} + p\, x_1 - p\, x_2 - x_1^{(2+\delta)}\, x_2^2 + x_1^{(1+\delta)}\, x_2$$
$$+ x_1^{(2+\delta)}\, x_2^{(1+\delta)} - x_1^{(1+\delta)}\, x_2^{(2+\delta)} + x_1^2\, x_2^{(2+\delta)} - x_1\, x_2^{(1+\delta)} - x_1^2\, x_2 + x_1\, x_2^2)/$$
$$((1 - x_1)\,(1 - x_2)\,(1 - x_1\, x_2)\,(x_1 - x_2))$$
$$= -p^{-1} x_0^{\delta}((1 - x_1\, x_2)\,(p\, x_1 - x_2)\, x_1^{(\delta+1)} + (1 - x_1\, x_2)\,(x_1 - p\, x_2)\, x_2^{(\delta+1)}$$
$$- (1 - p\, x_1\, x_2)\,(x_1 - x_2)\,(x_1\, x_2)^{(\delta+1)} - (p - x_1\, x_2)(x_1 - x_2))/$$
$$((1 - x_1)\,(1 - x_2)\,(1 - x_1\, x_2)\,(x_1 - x_2)).$$

Andrianov's Generating Series

The expression (1) comes from Andrianov's generating series

$$\sum_{\delta=0}^{\infty} \Omega_x^{(2)}(\mathbf{T}(p^\delta))\, X^\delta = \frac{1 - \frac{x_0^2\, x_1\, x_2}{p}\, X^2}{(1 - x_0\, X)\,(1 - x_0\, x_1\, X)\,(1 - x_0\, x_2\, X)\,(1 - x_0\, x_1\, x_2\, X)}$$

after developing and a simplification using change of summation.

Note that the formula (1) makes it possible to treat higher generating series of the following type:

$$D_{p,m}(X) = \sum_{\delta=0}^{\infty} \Omega_x^{(2)}(\mathbf{T}(p^{m\delta}))\, X^\delta \quad (m = 2, 3, \dots)$$

(in spherical variables x_0, x_1, x_2).

2.2 Rankin's Generating Series in Genus 2

Let us use the spherical variables x_0, x_1, x_2 and y_0, y_1, y_2 for the Hecke operators.

Note that there are two types of convolutions: the first one is defined through the Fourier coefficients (it was used by [An-Ka] for the analytic continuation of the standard L- function), and the second one is defined through the eigenvalues of Hecke operators, and it is more suitable for treating the L-functions attached to tensor products of representations of the Langlands group. However, a link between the two types is known only for $g = 1$.

In order to state a multiplicative analogue of Rankin's lemma in genus two we need to write the corrseponding formula for the Hecke operator $\mathbf{T}(p^\delta)$ (in spherical variables y_0, y_1, y_2):

$$\begin{aligned}
\Omega_y^{(2)}(\mathbf{T}(p^\delta)) = p^{-1}\, y_0^\delta (p\, y_1^{(3+\delta)}\, y_2 &- p\, y_1^{(2+\delta)} - p\, y_1^{(3+\delta)}\, y_2^{(2+\delta)} + p\, y_1^{(2+\delta)}\, y_2^{(3+\delta)} \\
&- p\, y_1\, y_2^{(3+\delta)} + p\, y_2^{(2+\delta)} + p\, y_1 - p\, y_2 - y_1^{(2+\delta)}\, y_2^2 + y_1^{(1+\delta)}\, y_2 \\
&+ y_1^{(2+\delta)}\, y_2^{(1+\delta)} - y_1^{(1+\delta)}\, y_2^{(2+\delta)} + y_1^2\, y_2^{(2+\delta)} - y_1\, y_2^{(1+\delta)} - y_1^2\, y_2 \\
&+ y_1\, y_2^2) / ((1 - y_1)\,(1 - y_2)\,(1 - y_1\, y_2)\,(y_1 - y_2))
\end{aligned}$$

Then we have that the product of the above polynomials is given by

$$\begin{aligned}
\Omega_x^{(2)}&(\mathbf{T}(p^\delta)) \cdot \Omega_y^{(2)}(\mathbf{T}(p^\delta)) \\
&= p^{-2}\, x_0^\delta\, y_0^\delta (p\, x_1^{(3+\delta)}\, x_2 - p\, x_1^{(2+\delta)} - p\, x_1^{(3+\delta)}\, x_2^{(2+\delta)} + p\, x_1^{(2+\delta)}\, x_2^{(3+\delta)} \\
&\quad - p\, x_1\, x_2^{(3+\delta)} + p\, x_2^{(2+\delta)} + p\, x_1 - p\, x_2 - x_1^{(2+\delta)}\, x_2^2 + x_1^{(1+\delta)}\, x_2 \\
&\quad + x_1^{(2+\delta)}\, x_2^{(1+\delta)} - x_1^{(1+\delta)}\, x_2^{(2+\delta)} + x_1^2\, x_2^{(2+\delta)} - x_1\, x_2^{(1+\delta)} - x_1^2\, x_2 + x_1\, x_2^2) \\
&\quad \cdot (p\, y_1^{(3+\delta)}\, y_2 - p\, y_1^{(2+\delta)} - p\, y_1^{(3+\delta)}\, y_2^{(2+\delta)} + p\, y_1^{(2+\delta)}\, y_2^{(3+\delta)}
\end{aligned}$$

$$-p\, y_1\, y_2^{(3+\delta)} + p\, y_2^{(2+\delta)} + p\, y_1 - p\, y_2 - y_1^{(2+\delta)}\, y_2^2 + y_1^{(1+\delta)}\, y_2$$
$$+y_1^{(2+\delta)}\, y_2^{(1+\delta)} - y_1^{(1+\delta)}\, y_2^{(2+\delta)} + y_1^2\, y_2^{(2+\delta)} - y_1\, y_2^{(1+\delta)} - y_1^2\, y_2 + y_1\, y_2^2)/$$
$$((1 - x_1)(1 - x_2)(1 - x_1 x_2)(x_1 - x_2)(1 - y_1)(1 - y_2)(1 - y_1 y_2)(y_1 - y_2)).$$

We wish to compute the generating series

$$\sum_{\delta=0}^{\infty} \Omega_x^{(2)}(\mathbf{T}(p^\delta)) \cdot \Omega_y^{(2)}(\mathbf{T}(p^\delta))\, X^\delta \in \mathbb{Q}[x_0, x_1, x_2, y_0, y_1, y_2][\![X]\!].$$

The answer is given by the following multiplicative analogue of Rankin's lemma in genus two:

Theorem 1. *The following equality holds:*

$$\sum_{\delta=0}^{\infty} \Omega_x^{(2)}(\mathbf{T}(p^\delta)) \cdot \Omega_y^{(2)}(\mathbf{T}(p^\delta))\, X^\delta = \tag{2}$$

$$-\frac{(p\, x_1 - x_2)\,(1 - p\, y_1\, y_2)\, x_1\, y_1\, y_2}{p^2\,(1 - x_1)\,(1 - x_2)\,(x_1 - x_2)\,(1 - y_1)\,(1 - y_2)\,(1 - y_1\, y_2)\,(1 - x_0\, x_1\, y_0\, y_1\, y_2\, X)}$$

$$+\frac{x_2\, y_1\,(x_1 - p\, x_2)\,(p\, y_1 - y_2)}{p^2\,(1 - x_1)\,(1 - x_2)\,(x_1 - x_2)\,(1 - y_1)\,(1 - y_2)\,(y_1 - y_2)\,(1 - x_0\, x_2\, y_0\, y_1\, X)}$$

$$+\frac{x_2\, y_2\,(x_1 - p\, x_2)(y_1 - p\, y_2)}{p^2\,(1 - x_1)\,(1 - x_2)\,(x_1 - x_1)\,(1 - y_1)\,(1 - y_2)\,(y1 - y_2)\,(1 - x_0\, y_0\, x_2\, y_2\, X)}$$

$$-\frac{x_2\, y_1\, y_2\,(x_1 - p\, x_2)\,(1 - p\, y_1\, y_2)}{p^2\,(1 - x_1)\,(1 - x_2)\,(x_1 - x_2)\,(1 - y_1)\,(1 - y_2)\,(1 - y_1\, y_2)\,(1 - x_0\, x_2\, y_0\, y_1\, y_2\, X)}$$

$$-\frac{x_1\,(p\, x_1 - x_2)\,(p - y_1\, y_2)}{p^2\,(1 - x_1)\,(1 - x_2)\,(x_1 - x_2)\,(1 - y_1)\,(1 - y_2)\,(1 - y_1\, y_2)\,(1 - x_0\, x_1\, y_0\, X)}$$

$$-\frac{x_1\, x_2\, y_1\,(1 - p\, x_1\, x_2)\,(p\, y_1 - y_2)}{p^2\,(1 - x_1)\,(1 - x_2)\,(1 - x_1\, x_2)\,(1 - y_1)\,(1 - y_2)\,(y_1 - y_2)\,(1 - x_0\, x_1\, x_2\, y_0\, y_1\, X)}$$

$$-\frac{x_1\, x_2\, y_2\,(1 - p\, x_1\, x_2)\,(y_1 - p\, y_2)}{p^2\,(1 - x_1)\,(1 - x_2)\,(1 - x_1\, x_2)\,(1 - y_1)\,(1 - y_2)\,(y_1 - y_2)\,(1 - x_0\, x_1\, x_2\, y_0\, y_2\, X)}$$

$$+\frac{y_1\, y_2\,(p - x_1\, x_2)\,(1 - p\, y_1\, y_2)}{p^2\,(1 - x_1)\,(1 - x_2)\,(1 - x_1\, x_2)\,(1 - y_1)\,(1 - y_2)\,(1 - y_1\, y_2)\,(1 - x_0\, y_0\, y_1\, y_2\, X)}$$

$$+\frac{x_1\, x_2\,(1 - p\, x_1\, x_2)\,(p - y_1\, y_2)}{p^2\,(1 - x_1)\,(1 - x_2)\,(1 - x_1\, x_2)\,(1 - y_1)\,(1 - y_2)\,(1 - y_1\, y_2)\,(1 - x_0\, x_1\, x_2\, y_0\, X)}$$

$$-\frac{x_1\, y_1\,(p\, x_1 - x_2)\,(p\, y_1 - y_2)}{p^2\,(1 - x_1)\,(1 - x_2)\,(x_1 - x_2)\,(1 - y_1)\,(1 - y_2)\,(y_1 - y_2)\,(1 - x_0\, x_1\, y_0\, y_1\, X)}$$

$$+\frac{x_1\, y_2\,(p\, x_1 - x_2)\,(y_1 - p\, y_2)}{p^2\,(1 - x_1)\,(1 - x_2)\,(x_1 - x_2)\,(1 - y_1)\,(1 - y_2)\,(y_1 - y_2)\,(1 - x_0\, x_1\, y_0\, y_2\, X)}$$

$$-\frac{x_2\,(x_1 - p\, x_2)\,(p - y_1\, y_2)}{p^2\,(1 - x_1)\,(1 - x_2)\,(x_1 - x_2)\,(1 - y_1)\,(1 - y_2)\,(1 - y_1\, y_2)\,(1 - x_0\, x_2\, y_0\, X)}$$

$$+\frac{x_1\, x_2\, y_1\, y_2\,(1 - p\, x_1\, x_2)\,(1 - p\, y_1\, y_2)}{p^2\,(1 - x_1)\,(1 - x_2)\,(1 - x_1\, x_2)\,(1 - y_1)\,(1 - y_2)\,(1 - y_1\, y_2)\,(1 - x_0\, x_1\, x_2\, y_0\, y_1\, y_2\, X)}$$

$$+\frac{(p - x_1\, x_2)\,(p - y_1\, y_2)}{p^2\,(1 - x_1)\,(1 - x_2)\,(1 - x_1\, x_2)\,(1 - y_1)\,(1 - y_2)\,(1 - y_1\, y_2)\,(1 - x_0\, y_0\, X)}$$

$$-\frac{y_1\,(p - x_1\,x_2)\,(p\,y_1 - y_2)}{p^2\,(1 - x_1)\,(1 - x_2)\,(1 - x_1\,x_2)\,(1 - y_1)\,(1 - y_2)\,(y_1 - y_2)\,(1 - x_0\,y_0\,y_1\,X)}$$

$$-\frac{y_2\,(p - x_1\,x_2)\,(y_1 - p\,y_2)}{p^2\,(1 - x_1)\,(1 - x_2)\,(1 - x_1\,x_2)\,(1 - y_1)\,(1 - y_2)\,(y_1 - y_2)\,(1 - x_0\,y_0\,y_2\,X)}.$$

Remark 2 (On the denominator of series (2)). One finds using a computer that the polynomials not depending on X in the denominators of (2) cancel after the simplification in the ring $\mathbb{Q}[x_0, x_1, x_2, y_0, y_1, y_2][\![X]\!]$, so that the common denominator becomes

$$(1 - x_0\,y_0 X)(1 - x_0\,y_0\,x_1\,X)(1 - x_0\,y_0\,y_1\,X)(1 - x_0\,y_0\,x_2\,X)(1 - x_0\,y_0\,y_2\,X)$$
$$(1 - x_0\,y_0\,x_1\,y_1\,X)(1 - x_0\,y_0\,x_1\,x_2\,X)(1 - x_0\,y_0\,x_1\,y_2\,X)(1 - x_0\,y_0\,y_1\,x_2\,X)$$
$$(1 - x_0\,y_0\,y_1\,y_2\,X)(1 - x_0\,y_0\,x_2\,y_2\,X)(1 - x_0\,y_0\,x_1\,y_1\,x_2\,X)(1 - x_0\,y_0\,x_1\,y_1\,y_2\,X)$$
$$(1 - x_0\,y_0\,x_1\,x_2\,y_2\,X)(1 - x_0\,y_0\,y_1\,x_2\,y_2\,X)(1 - x_0\,y_0\,x_1\,y_1\,x_2\,y_2\,X).$$

Remark 3 (Comparison with $g = 1$). It turns out by direct computation that the numerator of the rational fraction (2) is a product of the factor $1 - x_0^2 y_0^2\,x_1 y_1 x_2 y_2 X^2$ by a polynomial of degree 12 in X with coefficients in $\mathbb{Q}[x_0, x_1, x_2, y_0, y_1, y_2]$ with the constant term equal to 1 and leading term

$$\frac{x_0^{12} y_0^{12} x_1^6 x_2^6 y_1^6 y_2^6}{p^2} X^{12}.$$

Moreover, the factor of degree 12 does not contain terms of degree 1 and 11 in X. The factor of degree 2 in X is very similar to one in the case $g = 1$ (this series was studied and used in [Ma-Pa77]):

$$\sum_{\delta=0}^{\infty} \Omega_x^{(1)}(\mathbf{T}(p^\delta)) \cdot \Omega_y^{(1)}(\mathbf{T}(p^\delta))\,X^\delta = \sum_{\delta=0}^{\infty} \frac{x_0^\delta\,(1 - x_1^{(1+\delta)})}{1 - x_1} \cdot \frac{y_0^\delta\,(1 - y_1^{(1+\delta)})}{1 - y_1} X^\delta$$

$$= \frac{1}{(1 - x_1)\,(1 - y_1)\,(1 - x_0\,y_0\,X)} - \frac{y_1}{(1 - x_1)\,(1 - y_1)\,(1 - x_0\,y_0\,y_1\,X)}$$

$$- \frac{x_1}{(1 - x_1)\,(1 - y_1)\,(1 - x_0\,y_0\,x_1\,X)} + \frac{x_1\,y_1}{(1 - x_1)\,(1 - y_1)\,(1 - x_0\,y_0\,x_1\,y_1\,X)}$$

$$= \frac{1 - x_0^2\,y_0^2\,x_1\,y_1\,X^2}{(1 - x_0\,y_0\,x_1\,y_1\,X)\,(1 - x_0\,y_0\,x_1\,X)\,(1 - x_0\,y_0\,y_1\,X)\,(1 - x_0\,y_0\,X)}.$$

2.3 Symmetric Square Generating Series in Genus 2

Using the same method, one can evaluate the symmetric square generating series and the cubic generating series of higher genus. Note that this series, written here in spherical variables x_0, x_1, x_2, is different from the one studied by Andrianov–Kalinin, and has the form

$$\sum_{\delta=0}^{\infty} \Omega_x^{(2)}(\mathbf{T}(p^{2\delta})) X^\delta = \left(1 - \frac{x_0^2 x_1 x_2}{p} X\right)$$

$$\times \frac{(1 + x_0^2 x_1 X + x_0^2 x_2 X + 2x_0^2 x_1 x_2 X + x_0^2 x_1 x_2^2 X + x_0^2 x_1^2 x_2 X + x_0^4 x_1^2 x_2^2 X^2)}{(1 - x_0^2 x_1^2 x_2^2 X)(1 - x_0^2 x_1^2 X)(1 - x_0^2 x_2^2 X)(1 - x_0^2 X)}.$$

2.4 Cubic Generating Series in Genus 2

The cubic generating series of higher genus, written here in spherical variables x_0, x_1, x_2, has the form

$$\sum_{\delta=0}^{\infty} \Omega_x^{(2)}(\mathbf{T}(p^{3\,\delta})) \, X^\delta = p^{-1}(-p + x_0^6 \, x_1^4 \, x_2^2 \, X^2 + x_0^6 \, x_1^2 \, x_2^4 \, X^2 + 2\, x_0^6 \, x_1^2 \, x_2^3 \, X^2$$

$$- p \, x_0^6 \, x_1^4 \, x_2^4 \, X^2 - p \, x_0^6 \, x_1^2 \, x_2^4 \, X^2 - 2 \, p \, x_0^3 \, x_1^2 \, x_2 \, X + x_0^6 \, x_1 \, x_2^3 \, X^2$$

$$+ x_0^6 \, x_1^3 \, x_2 \, X^2 + x_0^6 \, x_1^3 \, x_2^5 \, X^2 + x_0^6 \, x_1^5 \, x_2^3 \, X^2 + 3 \, x_0^6 \, x_1^3 \, x_2^3 \, X^2$$

$$+ x_0^6 \, x_1^2 \, x_2^2 \, X^2 + 2 \, x_0^6 \, x_1^3 \, x_2^2 \, X^2 - p \, x_0^3 \, x_1^2 \, X - p \, x_0^3 \, x_2^2 \, X - p \, x_0^6 \, x_1^4 \, x_2^2 \, X^2$$

$$- 2 \, p \, x_0^3 \, x_1 \, x_2^2 \, X - p \, x_0^6 \, x_1^2 \, x_2^2 \, X^2 + x_0^3 \, x_1^2 \, x_2^2 \, X + x_0^3 \, x_1 \, x_2 \, X$$

$$- p \, x_0^6 \, x_1^2 \, x_2^3 \, X^2 - p \, x_0^6 \, x_1^3 \, x_2^2 \, X^2 - p \, x_0^3 \, x_1^2 \, x_2^3 \, X - p \, x_0^3 \, x_1^3 \, x_2^2 \, X$$

$$- 2 \, p \, x_0^3 \, x_1^2 \, x_2^2 \, X - p \, x_0^3 \, x_1^3 \, x_2 \, X + x_0^3 \, x_1^2 \, x_2 \, X + x_0^9 \, x_1^4 \, x_2^4 \, X^3$$

$$- 2 \, p \, x_0^6 \, x_1^3 \, x_2^3 \, X^2 - 2 \, p \, x_0^3 \, x_1 \, x_2 \, X + x_0^9 \, x_1^4 \, x_2^5 \, X^3 + x_0^9 \, x_1^5 \, x_2^4 \, X^3$$

$$- p \, x_0^6 \, x_1^3 \, x_2^4 \, X^2 + x_0^3 \, x_1 \, x_2^2 \, X + x_0^9 \, x_1^5 \, x_2^5 \, X^3 + x_0^6 \, x_1^4 \, x_2^4 \, X^2$$

$$- p \, x_0^6 \, x_1^4 \, x_2^3 \, X^2 - p \, x_0^3 \, x_1 \, x_2^3 \, X - p \, x_0^3 \, x_2 \, X - p \, x_0^3 \, x_1 \, X + 2 \, x_0^6 \, x_1^4 \, x_2^3 \, X^2$$

$$+ 2 \, x_0^6 \, x_1^3 \, x_2^4 \, X^2)/((1 - x_0^3 \, X)(1 - x_0^3 x_1^3 \, X)(1 - x_0^3 x_2^3 \, X)(1 - x_0^3 x_1^3 x_2^3 X)).$$

3 Proofs: Formulas for the Hecke Operators of Sp_g

3.1 Satake's Spherical Map Ω

Our result is based on the use of the Satake spherical map Ω, by applying the spherical map Ω to elements $\mathbf{T}(p^\delta) \in \mathcal{L}_{\mathbb{Z}}$ of the Hecke ring $\mathcal{L}_{\mathbb{Z}} = \mathbb{Z}[\mathbf{T}(p), \mathbf{T}_1(p^2), \dots, \mathbf{T}_n(p^2)]$ for the symplectic group; see [An87], Chapter 3.

- Case $\mathbf{T}_1(p^2)$: In genus 2 (in spherical variables x_0, x_1, x_2), we obtain using Andrianov's formulas

$$\Omega(\mathbf{T}_1(p^2)) = \frac{x_0^2 \left((x_1^2 \, x_2 + x_1 \, x_2^2) \, p^2 + x_1 \, x_2 \, p^2 - x_1 \, x_2 + (x_1 + x_2) \, p^2\right)}{p^3}.$$

- Cases $\mathbf{T}_2(p^2) = [\mathbf{p}]_2$ and $\mathbf{T}(p)$:

$$\Omega(\mathbf{T}_2(p^2)) = \frac{x_0^2 \, x_1 \, x_2}{p^3}, \quad \Omega(\mathbf{T}(p)) = x_0(1 + x_1)(1 + x_2).$$

3.2 Use of Andrianov's Generating Series in Genus 2

We refer to [An87], p. 164, (3.3.75), for the following celebrated summation formula:

$$\sum_{\delta=0}^{\infty} \Omega^{(2)}(\mathbf{T}(p^{\delta})) \, X^{\delta} = \frac{1 - \frac{x_0^2 \, x_1 \, x_2}{p} X^2}{(1 - x_0 \, X)\,(1 - x_0 \, x_1 \, X)(1 - x_0 \, x_2 \, X)\,(1 - x_0 \, x_1 \, x_2 \, X)} \tag{3}$$

gives after development and simplification the formula

$$\Omega^{(2)}(\mathbf{T}(p^{\delta})) = p^{-1} \, x_0^{\delta} \, (p \, x_1^{(3+\delta)} \, x_2 - p \, x_1^{(2+\delta)} - p \, x_1^{(3+\delta)} \, x_2^{(2+\delta)} + p \, x_1^{(2+\delta)} \, x_2^{(3+\delta)}$$
$$- p \, x_1 \, x_2^{(3+\delta)} + p \, x_2^{(2+\delta)} + p \, x_1 - p \, x_2 - x_1^{(2+\delta)} \, x_2^2 + x_1^{(1+\delta)} \, x_2$$
$$+ x_1^{(2+\delta)} \, x_2^{(1+\delta)} - x_1^{(1+\delta)} \, x_2^{(2+\delta)} + x_1^2 \, x_2^{(2+\delta)} - x_1 \, x_2^{(1+\delta)} - x_1^2 \, x_2 + x_1 \, x_2^2)/$$
$$((1 - x_1)\,(1 - x_2)\,(1 - x_1 \, x_2)\,(x_1 - x_2)).$$

Then we use two groups of variables $x_0, \ldots, x_n$ and $y_0, \ldots, y_n$ in two copies Ω_x, Ω_y of the sperical map in order to treat the tensor product of two local Hecke algebras.

Next, in order to carry out the summation of the series

$$\sum_{\delta=0}^{\infty} \Omega_x^{(2)}(\mathbf{T}(p^{\delta})) \cdot \Omega_y^{(2)}(\mathbf{T}(p^{\delta})) \, X^{\delta}$$

on a computer, we used a subdivision of each summand (over δ) into smaller parts. These parts correspond to symbolic monomials in x_1^{δ}, y_1^{δ}, x_2^{δ}, y_2^{δ}, $(x_1 x_2)^{\delta}$, $(y_1 y_2)^{\delta}$.

3.3 Rankin's Lemma of Genus 2 (Compare with [Jia96])

Let us compute the series

$$D_p(X) = \sum_{\delta=0}^{\infty} \mathbf{T}(p^{\delta}) \otimes \mathbf{T}(p^{\delta}) \, X^{\delta} \in \mathcal{L}_{2,\mathbb{Z}} \otimes \mathcal{L}_{2,\mathbb{Z}}[\![X]\!]$$

in terms of the generators of the Hecke algebra $\mathcal{L}_{2,\mathbb{Z}} \otimes \mathcal{L}_{2,\mathbb{Z}}$ given by the following operators:

$$\mathbf{T}(p) \otimes 1, \mathbf{T}_1(p^2) \otimes 1, [\mathbf{p}] \otimes 1, 1 \otimes \mathbf{T}(p), 1 \otimes \mathbf{T}_1(p^2), 1 \otimes [\mathbf{p}] \in \mathcal{L}_{2,\mathbb{Z}} \otimes \mathcal{L}_{2,\mathbb{Z}}[\![X]\!].$$

Theorem 4. *For $g = 2$, we have the following explicit representation:*

$$D_p(X) = \sum_{\delta=0}^{\infty} \mathbf{T}(p^{\delta}) \otimes \mathbf{T}(p^{\delta}) \, X^{\delta} = \frac{(1 - p^6[\mathbf{p}] \otimes [\mathbf{p}]X^2) \cdot R(X)}{S(X)},$$

where

$$R(X), S(X) \in \mathcal{L}_{2,\mathbb{Z}} \otimes \mathcal{L}_{2,\mathbb{Z}}[X]$$

are given by the equalities (4) and (5):

$$R(X) = 1 + r_2 X^2 + \cdots + r_{10} X^{10} + r_{12} X^{12} \in \mathcal{L}_{2,\mathbb{Z}} \otimes \mathcal{L}_{2,\mathbb{Z}}[X] \qquad (4)$$

with $r_1 = r_{11} = 0$,

$$S(X) = 1 + s_1 X + \cdots + s_{16} X^{16} \qquad (5)$$
$$= 1 - (\mathbf{T}(p) \otimes \mathbf{T}(p))X + \cdots + (p^6[\mathbf{p}] \otimes [\mathbf{p}])^8 X^{16} \in \mathcal{L}_{2,\mathbb{Z}} \otimes \mathcal{L}_{2,\mathbb{Z}}[X],$$

with r_i *and* s_i *given in the Appendix. Moreover, there is an easy functional equation (similar to [An87], p. 164, (3.3.79)):*

$$s_{16-i} = (p^6[\mathbf{p}] \otimes [\mathbf{p}])^{8-i} s_i \quad (i = 0, \ldots, 8).$$

Remark 5 (Comparison with the case $g = 1$). . The corresponding result in the case $g = 1$, written in terms of Hecke operators, looks as follows:

$$\sum_{\delta=0}^{\infty} \mathbf{T}(p^\delta) \otimes \mathbf{T}(p^\delta) X^\delta = (1 - p^2[\mathbf{p}] \otimes [\mathbf{p}]X^2)/$$
$$(1 - \mathbf{T}(p) \otimes \mathbf{T}(p)X + (p(\mathbf{T}(p)^2 \otimes [\mathbf{p}] + [\mathbf{p}] \otimes \mathbf{T}(p)^2) - 2p^2[\mathbf{p}] \otimes [\mathbf{p}])X^2$$
$$- p^2 \mathbf{T}(p)[\mathbf{p}] \otimes \mathbf{T}(p)[\mathbf{p}]X^3 + p^4[\mathbf{p}]^2 \otimes [\mathbf{p}]^2 X^4).$$

Indeed, this follows directly from Remark 3.

Remark 6. For the group $GL(n)$, a simpler result could be obtained using the Tamagawa generating series (see [Tam]) and its expansion into simple fractions:

$$\frac{1}{(1 - x_1 X) \ldots (1 - x_n X)} = \sum_{i=1}^{n} \frac{\alpha_i(x)}{1 - x_i X} = \sum_{i=1}^{n} \sum_{r=0}^{\infty} \alpha_i(x)(x_i X)^r,$$

$$\frac{1}{(1 - y_1 X) \ldots (1 - y_n X)} = \sum_{j=1}^{n} \frac{\alpha_j(y)}{1 - y_j X} = \sum_{j=1}^{n} \sum_{r=0}^{\infty} \alpha_j(y)(y_j X)^r,$$

where

$$\alpha_i(x) = \frac{x_i^{n-1}}{\prod_{k \neq i}^{n}(x_k - x_i)}, \quad \alpha_j(y) = \frac{y_j^{n-1}}{\prod_{l \neq j}^{n}(y_l - y_j)},$$

hence

$$\sum_{i,j=1}^{n} \sum_{r=0}^{\infty} \alpha_i(x)\alpha_j(y)(x_i y_j X)^r = \sum_{i,j=1}^{n} \frac{\alpha_i(x)\alpha_j(y)}{1 - x_i y_j X}.$$

If $n = 3$, this gives after simplification the following fraction:

$$(y_2^2 x_2^2 y_1^2 x_3^2 y_3^2 x_1^2 X^6 - y_2 x_2 y_1 y_3^2 x_1 x_3^2 X^4 - y_2 x_2^2 y_1 y_3^2 x_1 x_3 X^4$$

$$- y_2\, x_2\, y_1\, x_1^2\, x_3\, y_3{}^2\, X^4 - y_2^2\, x_2\, y_1\, y_3\, x_1\, x_3{}^2\, X^4 - y_2^2\, x_2^2\, y_1\, y_3\, x_1\, x_3\, X^4$$
$$- y_2^2\, x_2\, y_1\, x_1^2\, x_3\, y_3\, X^4 - y_2\, x_2^2\, y_1^2\, x_3\, y_3\, x_1\, X^4 - y_2\, x_2\, y_1^2\, x_3^2\, y_3\, x_1\, X^4$$
$$- y_2\, x_2\, y_1^2\, x_3\, x_1^2\, y_3\, X^4 + y_2\, x_3\, y_1\, x_1^2\, y_3\, X^3 + y_2\, x_2\, y_3^2\, x_1\, x_3\, X^3$$
$$+ y_2^2\, x_2\, y_3\, x_1\, x_3\, X^3 + y_3\, x_2\, y_1^2\, x_3\, x_1\, X^3 + y_2\, x_2\, y_1\, x_1^2\, y_3\, X^3$$
$$+ y_2\, x_2\, y_1\, y_3\, x_3^2\, X^3 + y_2\, x_2^2\, y_1\, y_3\, x_3\, X^3 + y_2^2\, x_2\, y_1\, x_1\, x_3\, X^3$$
$$+ 4\, y_2\, x_2\, y_1\, y_3\, x_1\, x_3\, X^3 + y_2\, x_2^2\, y_1\, y_3\, x_1\, X^3 + y_2\, x_3^2\, y_1\, y_3\, x_1\, X^3$$
$$+ y_2\, x_2\, y_1^2\, x_3\, x_1\, X^3 + x_2\, y_1\, y_3^2\, x_3\, x_1\, X^3 - y_2\, x_2\, y_1\, x_3\, X^2 - y_3\, x_2\, y_1\, x_1\, X^2$$
$$- y_2\, x_2\, y_3\, x_1\, X^2 - y_2\, x_3\, y_3\, x_1\, X^2 - y_2\, x_3\, y_1\, x_1\, X^2 - y_3\, x_3\, y_1\, x_1\, X^2$$
$$- y_3\, x_2\, y_1\, x_3\, X^2 - y_2\, x_2\, y_1\, x_1\, X^2 - y_2\, x_2\, y_3\, x_3\, X^2 + 1 \Big) \Big/ \big((1 - x_3\, y_3\, X)$$
$$(1 - x_2\, y_3\, X)\, (1 - x_1\, y_3\, X)(1 - x_3\, y_2\, X)\, (1 - x_2\, y_2\, X)\, (1 - x_1\, y_2\, X)$$
$$(1 - x_3\, y_1\, X)\, (1 - x_2\, y_1\, X)\, (1 - x_1\, y_1\, X)\big).$$

4 Relations with L-Functions and Motives for Sp_n

The modest purpose of this section is to recall the motivic origin of the L-function produced by the denominator of our Hecke series.

L-functions, Functional Equation, and Motives for Sp_n (see [Pa94], [Yosh01])

One defines

- $$Q_{f,p}(X) = (1 - \alpha_0 X) \prod_{r=1}^{n} \prod_{1 \le i_1 < \cdots < i_r \le n} (1 - \alpha_0 \alpha_{i_1} \cdots \alpha_{i_r} X),$$

- $$R_{f,p}(X) = (1 - X) \prod_{i=1}^{n} (1 - \alpha_i^{-1} X)(1 - \alpha_i X) \in \mathbb{Q}[\alpha_0^{\pm 1}, \ldots, \alpha_n^{\pm 1}][X].$$

Then the spinor L-function $L(Sp(f), s)$ and the standard L-function $L(St(f), s, \chi)$ of f (for $s \in \mathbb{C}$, and for all Dirichlet characters χ) are defined as the Euler products

- $$L(Sp(f), s, \chi) = \prod_p Q_{f,p}(\chi(p)p^{-s})^{-1},$$

- $$L(St(f), s, \chi) = \prod_p R_{f,p}(\chi(p)p^{-s})^{-1}.$$

Motivic L-Functions

Following [Pa94] and [Yosh01], these functions are conjectured to be motivic for all $k > n$: $L(Sp(f), s, \chi) = L(M(Sp(f))(\chi), s), L(St(f), s) =$

$L(M(St(f))(\chi), s)$, where the motives $M(Sp(f))$ and $M(St(g))$ are conjectured to be *pure* if f is a genuine cusp form (not coming from a lifting of a smaller genus):

- $M(Sp(f))$ is a motive over $\mathbb{Q}$ with coefficients in $\mathbb{Q}(\lambda_f(n))_{n\in\mathbb{N}}$ of rank 2^n, of weight $w = kn - n(n+1)/2$, and of Hodge type $\oplus_{p,q} H^{p,q}$, with

$$p = (k - i_1) + (k - i_2) + \cdots + (k - i_r), \tag{6}$$
$$q = (k - j_1) + (k - j_2) + \cdots + (k - j_s), \text{ where } r + s = n,$$
$$1 \le i_1 < i_2 < \cdots < i_r \le n, 1 \le j_1 < j_2 < \cdots < j_s \le n,$$
$$\{i_1, \ldots, i_r\} \cup \{j_1, \ldots, j_s\} = \{1, 2, \ldots, n\};$$

- $M(St(g))$ is a motive over $\mathbb{Q}$ with coefficients in $\mathbb{Q}(\lambda_f(n))_{n\in\mathbb{N}}$ of rank $2n + 1$, of weight $w = 0$, and of Hodge type $H^{0,0} \oplus_{i=1}^{n} (H^{-k+i,k-i} \oplus H^{k-i,-k+i})$.

A Functional Equation

Following Deligne's general conjecture [De79] on the motivic L-functions, the L-functions satisfy a functional equation determined by the Hodge structure of a motive:

$$\Lambda(Sp(f), kn - n(n+1)/2 + 1 - s) = \varepsilon(f)\Lambda(Sp(f), s),$$

where

$$\Lambda(Sp(f), s) = \Gamma_{n,k}(s)L(Sp(f), s), \varepsilon(f) = (-1)^{k2^{n-2}},$$
$$\Gamma_{1,k}(s) = \Gamma_{\mathbb{C}}(s) = 2(2\pi)^{-s}\Gamma(s), \Gamma_{2,k}(s) = \Gamma_{\mathbb{C}}(s)\Gamma_{\mathbb{C}}(s - k + 2),$$

and $\Gamma_{n,k}(s) = \prod_{p<q} \Gamma_{\mathbb{C}}(s-p)\Gamma_{\mathbb{R}}^{a_+}(s - (w/2))\Gamma_{\mathbb{R}}(s+1-(w/2))^{a_-}$ for some nonnegative integers a_+ and a_-, with $a_+ + a_- = w/2$ and $\Gamma_{\mathbb{R}}(s) = \pi^{-s/2}\Gamma(s/2)$.

Motive of the Rankin Product of Genus $g = 2$

Let f and g be two Siegel cusp eigenforms of weights k and l, $k > l$, and let $M(Sp(f))$ and $M(Sp(g))$ be the spinor motives of f and g. Then $M(Sp(f))$ is a motive over $\mathbb{Q}$ with coefficients in $\mathbb{Q}(\lambda_f(n))_{n\in\mathbb{N}}$ of rank 4, of weight $w = 2k - 3$, and of Hodge type $H^{0,2k-3} \oplus H^{k-2,k-1} \oplus H^{k-1,k-2} \oplus H^{2k-3,0}$, and $M(Sp(g))$ is a motive over $\mathbb{Q}$ with coefficients in $\mathbb{Q}(\lambda_g(n))_{n\in\mathbb{N}}$ of rank 4, of weight $w = 2l - 3$, and of Hodge type $H^{0,2l-3} \oplus H^{l-2,l-1} \oplus H^{l-1,l-2} \oplus H^{2l-3,0}$.

The tensor product $M(Sp(f)) \otimes M(Sp(g))$ is a motive over $\mathbb{Q}$ with coefficients in $\mathbb{Q}(\lambda_f(n), \lambda_g(n))_{n\in\mathbb{N}}$ of rank 16, of weight $w = 2k + 2l - 6$, and of Hodge type

$$H^{0,2k+2l-6} \oplus H^{l-2,2k+l-4} \oplus H^{l-1,2k+l-5} \oplus H^{2l-3,2k-3}$$
$$\oplus H^{k-2,k+2l-4} \oplus H^{k+l-4,k+l-2} \oplus H^{k+l-3,k+l-3}_+ \oplus H^{k+2l-5,k-1}$$
$$\oplus H^{k-1,k+2l-5} \oplus H^{k+l-3,k+l-3}_- \oplus H^{k+l-2,k+l-4} \oplus H^{k+2l-4,k-2}$$
$$\oplus H^{2k-3,2l-3} \oplus H^{2k+l-5,l-1} \oplus H^{2k+l-4,l-2} \oplus H^{2k+2l-6,0}.$$

Motivic L-Functions: Analytic Properties

Following Deligne's conjecture [De79] on motivic L-functions, applied to a Siegel cusp eigenform F for the Siegel modular group $\mathrm{Sp}_4(\mathbb{Z})$ of genus $n = 4$ and of weight $k > 5$, one has $\Lambda(Sp(F), s) = \Lambda(Sp(F), 4k - 9 - s)$, where

$$\Lambda(Sp(F), s) = \Gamma_{\mathbb{C}}(s)\Gamma_{\mathbb{C}}(s - k + 4)\Gamma_{\mathbb{C}}(s - k + 3)\Gamma_{\mathbb{C}}(s - k + 2)\Gamma_{\mathbb{C}}(s - k + 1)$$
$$\times \Gamma_{\mathbb{C}}(s - 2k + 7)\Gamma_{\mathbb{C}}(s - 2k + 6)\Gamma_{\mathbb{C}}(s - 2k + 5)L(Sp(F), s)$$

(compare this functional equation with that given in [An74], p. 115).

On the other hand, for $n = 2$ and for two cusp eigenforms f and g for $\mathrm{Sp}_2(\mathbb{Z})$ of weights k, l, $k > l + 1$, $\Lambda(Sp(f) \otimes Sp(g), s) = \varepsilon(f, g)\Lambda(Sp(f) \otimes Sp(g), 2k + 2l - 5 - s)$, $|\varepsilon(f, g)| = 1$, where

$$\Lambda(Sp(f) \otimes Sp(g), s) = \Gamma_{\mathbb{C}}(s)\Gamma_{\mathbb{C}}(s - l + 2)\Gamma_{\mathbb{C}}(s - l + 1)\Gamma_{\mathbb{C}}(s - k + 2)$$
$$\times \Gamma_{\mathbb{C}}(s - k + 1)\Gamma_{\mathbb{C}}(s - 2l + 3)\Gamma_{\mathbb{C}}(s - k - l + 2)$$
$$\times \Gamma_{\mathbb{C}}(s - k - l + 3)L(Sp(f) \otimes Sp(g), s).$$

We used here the Gauss duplication formula $\Gamma_{\mathbb{C}}(s) = \Gamma_{\mathbb{R}}(s)\Gamma_{\mathbb{R}}(s + 1)$. Notice that $a_+ = a_- = 1$ in this case, and the conjectural motive $M(Sp(f)) \otimes M(Sp(g))$ does not admit critical values.

5 A Holomorphic Lifting from $\mathbf{GSp_2} \times \mathbf{GSp_2}$ to $\mathbf{GSp_4}$: A Conjecture

(compare with constructions in [BFG06], [BFG92], [Jia96] for generic automorphic forms.)

Our computation makes it possible to compare the spinor Hecke series of genus 4 given in [VaSp4] (in variables u_0, u_1, u_2, u_3, u_4) with the Rankin product of two Hecke series of genus 2 (in variables $x_0, x_1, x_2, y_0, y_1, y_2$). It follows from our computation that if we make the substitution $u_0 = x_0 y_0$, $u_1 = x_1$, $u_2 = x_2$, $u_3 = y_1$, $u_4 = y_2$, then the denominator of the series

$$\sum_{\delta=0}^{\infty} \Omega_u^{(4)}(\mathbf{T}(p^\delta))X^\delta$$

coincides with the denominator of the Rankin product

$$\sum_{\delta=0}^{\infty} \Omega_x^{(2)}(\mathbf{T}(p^\delta)) \cdot \Omega_y^{(2)}(\mathbf{T}(p^\delta)) X^\delta \in \mathbb{Q}[x_0, x_1, x_2, y_0, y_1, y_2][\![X]\!].$$

On the basis of this equality we would like to push forward the following conjecture.

Conjecture 7 (on a lifting from $\mathrm{GSp}_2 \times \mathrm{GSp}_2$ to GSp_4). *Let f and g be two Siegel modular forms of genus 2 and of weights $k > 4$ and $l = k - 2$. Then there exists a Siegel modular form F of genus 4 and of weight k with the Satake parameters*

$$\gamma_0 = \alpha_0 \beta_0, \quad \gamma_1 = \alpha_1, \quad \gamma_2 = \alpha_2, \quad \gamma_3 = \beta_1, \quad \gamma_4 = \beta_2,$$

for a suitable choice of Satake's parameters $\alpha_0, \alpha_1, \alpha_2$ and $\beta_0, \beta_1, \beta_2$ of f and g.

Remark 8. Evidence for the conjecture comes from Ikeda–Miyawaki constructions ([Ike01], [Mur02], [Ike06]): let k be an even positive integer, $h \in S_{2k}(\Gamma_1)$ a normalized Hecke eigenform of weight $2k$, $F_2(h) \in S_{k+1}(\Gamma_2) = \mathrm{Maass}(h)$ the Maass lift of h, and $F_{2n} \in S_{k+n}(\Gamma_{2n})$ the Ikeda lift of h (we assume $k \equiv n \bmod 2$, $n \in \mathbb{N}$).

Next let $f \in S_{k+n+r}(\Gamma_r)$ be an arbitrary Siegel cusp eigenform of genus r and weight $k + n + r$, with $n, r \geq 1$. Then according to Ikeda–Miyawaki (see [Ike06]) there exists a Siegel eigenform $\mathcal{F}_{h,f} \in S_{k+n+r}(\Gamma_{2n+r})$ such that

$$L(s, \mathcal{F}_{h,f}, St) = L(s, f, St) \prod_{j=1}^{2n} L(s + k + n - j, h) \tag{7}$$

(under a nonvanishing condition; see Theorem 2.3 on p. 63 in [Mur02]). The form $\mathcal{F}_{h,f}$ is given by the integral

$$\mathcal{F}_{h,f}(Z) = \langle F_{2n+2r}(\mathrm{diag}(Z, Z')), f(Z') \rangle_{Z'}.$$

If we take $n = 1, r = 2, k := k + 1$, then an example of the validity of the conjecture is given by $g = F_2(h)$,

$$(f, g) = (f, F_2(h)) \mapsto \mathcal{F}_{f,h} \in S_{k+3}(\Gamma_4),$$
$$(f, g) = (f, F_2(h)) \in S_{k+3}(\Gamma_2) \times S_{k+1}(\Gamma_2).$$

Remark 9. Notice that the Satake parameters of the Ikeda lift $F = F_{2m}(h)$ of h can be taken in the form $\beta_0, \beta_1, \ldots, \beta_{2m}$, where

$$\beta_0 = p^{mk - m(m+1)/2}, \quad \beta_i = \alpha p^{i-1/2}, \quad \beta_{m+i} = \alpha^{-1} p^{i-1/2} \ (i = 1, \cdots, m),$$

and

$$(1 - \alpha p^{k-1/2} X)(1 - \alpha^{-1} p^{k-1/2} X) = 1 - a(p)X + p^{2k-1} X^2, h = \sum_{n=1}^{\infty} a(n) q^n;$$

see [Mur02].

The L-function of degree 16 in Conjecture 7 is related to the tensor product L-function as in [Jia96]. In the example of Remark 8 it coincides with the product of two shifted L-functions of degree 8 of Boecherer–Heim [BoeH06].

Conjecture 10 (on a lifting from $GSp_{2m} \times GSp_{2m}$ to GSp_{4m}).
Here is a version of Conjectire 7 for any even genus $r = 2m$. Let f and g be two Siegel modular forms of genus $2m$ and of weights $k > 2m$ and $l = k - 2m$. Then there exists a Siegel modular form F of genus $4m$ and of weight k with the Satake parameters $\gamma_0 = \alpha_0 \beta_0, \gamma_1 = \alpha_1, \gamma_2 = \alpha_2, \ldots, \gamma_{2m} = \alpha_{2m}, \gamma_{2m+1} = \beta_1, \ldots, \gamma_{4m} = \beta_{2m}$ for suitable choices $\alpha_0, \alpha_1, \ldots, \alpha_{2m}$ and $\beta_0, \beta_1, \ldots, \beta_{2m}$ of Satake's parameters of f and g.

One readily checks that the Hodge types of $M(Sp(f)) \otimes M(Sp(g))$ and $M(Sp(F))$ are again the same (of rank 2^{4m}) (it follows from the above description (6), and from Künneth-type formulas).

Evidence for this version of the conjecture comes again from Ikeda–Miyawaki constructions ([Ike01], [Mur02], [Ike06]): let k be an even positive integer, $h \in S_{2k}(\Gamma_1)$ a normalized Hecke eigenform of weight $2k$, $F_{2n} \in S_{k+n}(\Gamma_{2n})$ the Ikeda lift of h of genus $2n$ (we assume $k \equiv n \bmod 2$, $n \in \mathbb{N}$).

Next let $f \in S_{k+n+r}(\Gamma_r)$ be an arbitrary Siegel cusp eigenform of genus r and weight $k + n + r$, with $n, r \geq 1$. If we take in (7) $n = m, r = 2m$, $k := k + m$, $k + n + r := k + 3m$, then an example of the validity of this version of the conjecture is given by

$$
\begin{aligned}
(f, g) &= (f, F_{2m}(h)) \mapsto \mathcal{F}_{h,f} \in S_{k+3m}(\Gamma_{4m}), \\
(f, g) &= (f, F_{2m}) \in S_{k+3m}(\Gamma_{2m}) \times S_{k+m}(\Gamma_{2m}).
\end{aligned}
$$

Further evidence comes from Siegel–Eiseinstein series

$$
f = E_k^{2m} \text{ and } g = E_{k-2m}^{2m}
$$

of even genus $2m$ and weights k and $k - 2m$: we have then

$$
\begin{aligned}
\alpha_0 &= 1, \ \alpha_1 = p^{k-2m}, \ \ldots, \ \alpha_{2m} = p^{k-1}, \\
\beta_0 &= 1, \ \beta_1 = p^{k-4m}, \ \ldots, \ \beta_{2m} = p^{k-2m-1},
\end{aligned}
$$

and then we have that

$$
\gamma_0 = 1, \ \gamma_1 = p^{k-4m}, \ \ldots, \ \gamma_{2m} = p^{k-1},
$$

are the Satake parameters of the Siegel–Eisenstein series $F = E_k^{4m}$.

Remark 11. If we compare the L-function of the conjecture (given by the Satake parameters $\gamma_0 = \alpha_0 \beta_0, \gamma_1 = \alpha_1, \gamma_2 = \alpha_2, \ldots, \gamma_{2m} = \alpha_{2m}, \gamma_{2m+1} = \beta_1, \ldots, \gamma_{4m} = \beta_{2m}$ for suitable choices $\alpha_0, \alpha_1, \ldots, \alpha_{2m}$ and $\beta_0, \beta_1, \ldots, \beta_{2m}$ of Satake's parameters of f and g), we see that it corresponds to the tensor product of spinor L-functions and is *not of the same type* as that of the Yoshida's lifting [Yosh81], which is a certain product of Hecke's L-functions.

We would like to mention in this context Langlands's functoriality: The denominators of our L-series belong to local Langlands L-factors (attached to representations of L-groups). If we consider the homomorphisms

$$^{L}\mathrm{GSp}_{2m} = \mathrm{GSpin}(4m+1) \to \mathrm{GL}_{2^{2m}}, \quad ^{L}\mathrm{GSp}_{4m} = \mathrm{GSpin}(8m+1) \to \mathrm{GL}_{2^{4m}},$$

we see that our conjecture is compatible with the homomorphism of L-groups

$$\mathrm{GL}_{2^{2m}} \times \mathrm{GL}_{2^{2m}} \to \mathrm{GL}_{2^{4m}}, \ (g_1, g_2) \mapsto g_1 \otimes g_2, \mathrm{GL}_n(\mathbb{C}) = {}^{L}\mathrm{GL}_n.$$

However, it is unclear to us whether Langlands's functoriality predicts a holomorphic Siegel modular form as a lift.

Appendix: Coefficients of the Polynomials $R(X)$ and $S(X)$

We give here explicit expressions for the coefficients of the polynomials $R(X)$ and $S(X)$ from Theorem 4. From these formulas one can observe some nice divisibility properties (by certain powers of p and the elements $[\mathbf{p}] \otimes [\mathbf{p}] \in \mathcal{L}_{2,\mathbb{Z}} \otimes \mathcal{L}_{2,\mathbb{Z}}$):

$$R(X) = 1 + r_2 X^2 + \cdots + r_{10} X^{10} + r_{12} X^{12} \in \mathcal{L}_{2,\mathbb{Z}} \otimes \mathcal{L}_{2,\mathbb{Z}}[X]$$
$$\text{with } r_1 = r_{11} = 0,$$

$$S(X) = 1 + s_1 X + \cdots + s_{16} X^{16}$$
$$= 1 - (\mathbf{T}(p) \otimes \mathbf{T}(p))X + \cdots + (p^6[\mathbf{p}] \otimes [\mathbf{p}])^8 X^{16} \in \mathcal{L}_{2,\mathbb{Z}} \otimes \mathcal{L}_{2,\mathbb{Z}}[X],$$

with r_i and s_i given as follows:

$$r_2 = p^2((2p-1)(p^2+1)[\mathbf{p}] \otimes [\mathbf{p}] - (p^2-p+1)(\mathbf{T}_1(p^2) \otimes [\mathbf{p}] + [\mathbf{p}] \otimes \mathbf{T}_1(p^2))$$
$$- (\mathbf{T}_1(p^2) \otimes \mathbf{T}_1(p^2) + \mathbf{T}(p)^2 \otimes [\mathbf{p}] + [\mathbf{p}] \otimes \mathbf{T}(p)^2),$$

$$r_3 = p^3(p+1)(2[\mathbf{p}] \otimes [\mathbf{p}] + \mathbf{T}_1(p^2) \otimes [\mathbf{p}] + [\mathbf{p}] \otimes \mathbf{T}_1(p^2))\mathbf{T}(p) \otimes \mathbf{T}(p),$$

$$r_4 = -p^5((p^7 + 2p^6 - 2p^5 + 6p^4 + p^3 + 6p^2 + p + 2)[\mathbf{p}]^2 \otimes [\mathbf{p}]^2$$
$$- (p^2+1)(p^3 - 3p^2 - p - 3)(\mathbf{T}_1(p^2) \otimes [\mathbf{p}] + [\mathbf{p}] \otimes \mathbf{T}_1(p^2))[\mathbf{p}] \otimes [\mathbf{p}]$$
$$+ (p+4)(p^2+1)\mathbf{T}_1(p^2)[\mathbf{p}] \otimes \mathbf{T}_1(p^2)[\mathbf{p}] - (p^3 - p^2 - 1)(\mathbf{T}_1(p^2)^2 \otimes [\mathbf{p}]^2$$
$$+ [\mathbf{p}]^2 \otimes \mathbf{T}_1(p^2)^2) + (\mathbf{T}_1(p^2) \otimes [\mathbf{p}] + [\mathbf{p}] \otimes \mathbf{T}_1(p^2))\mathbf{T}_1(p^2) \otimes \mathbf{T}_1(p^2)$$
$$- p(p^3 + 2p^2 - p + 2)(\mathbf{T}(p)^2 \otimes [\mathbf{p}] + [\mathbf{p}] \otimes \mathbf{T}(p)^2)[\mathbf{p}] \otimes [\mathbf{p}] - 2p(\mathbf{T}(p)^2 \otimes \mathbf{T}_1(p^2)$$
$$+ \mathbf{T}_1(p^2) \otimes \mathbf{T}(p)^2)[\mathbf{p}] \otimes [\mathbf{p}] + p^2(\mathbf{T}(p)^2\mathbf{T}_1(p^2) \otimes [\mathbf{p}]^2 + [\mathbf{p}]^2 \otimes \mathbf{T}(p)^2\mathbf{T}_1(p^2))$$
$$+ (p+2)\mathbf{T}(p)^2[\mathbf{p}] \otimes \mathbf{T}(p)^2[\mathbf{p}]),$$

$$r_5 = -p^7(2(p+1)(2p^4 - p^3 + p^2 - 1)[\mathbf{p}] \otimes [\mathbf{p}] + (p+1)(p-2)(\mathbf{T}_1(p^2) \otimes [\mathbf{p}]$$
$$+ [\mathbf{p}] \otimes \mathbf{T}_1(p^2)) - 2\mathbf{T}_1(p^2) \otimes \mathbf{T}_1(p^2) - p(p+1)(\mathbf{T}(p)^2 \otimes [\mathbf{p}]$$
$$+ [\mathbf{p}] \otimes \mathbf{T}(p)^2))\mathbf{T}(p)[\mathbf{p}] \otimes \mathbf{T}(p)[\mathbf{p}],$$

$$r_6 = -p^{10} \left(p \left(p^2 + 1 \right) (p^5 - 2p^3 - 8p^2 - p - 4)[\mathbf{p}]^3 \otimes [\mathbf{p}]^3 \right.$$
$$- p \left(p^5 + 4p^4 + 2p^3 + 12p^2 + p + 6 \right) (\mathbf{T}_1(p^2) \otimes [\mathbf{p}] + [\mathbf{p}] \otimes \mathbf{T}_1(p^2))[\mathbf{p}]^2 \otimes [\mathbf{p}]^2$$
$$+ p \left(p - 4 \right) (p^2 + 1) \mathbf{T}_1(p^2)[\mathbf{p}]^2 \otimes \mathbf{T}_1(p^2)[\mathbf{p}]^2$$
$$- p \left(p + 4 \right) (p^2 + 1) (\mathbf{T}_1(p^2)^2 \otimes [\mathbf{p}]^2 + [\mathbf{p}]^2 \otimes \mathbf{T}_1(p^2)^2)[\mathbf{p}] \otimes [\mathbf{p}]$$
$$- p \left(\mathbf{T}_1(p^2) \otimes [\mathbf{p}] + [\mathbf{p}] \otimes \mathbf{T}_1(p^2) \right) \mathbf{T}_1(p^2)[\mathbf{p}] \otimes \mathbf{T}_1(p^2)[\mathbf{p}]$$
$$- p \left(\mathbf{T}_1(p^2)^3 \otimes [\mathbf{p}]^3 + [\mathbf{p}]^3 \otimes \mathbf{T}_1(p^2)^3 \right)$$
$$- (p^5 - 4p^2 - p - 2)(\mathbf{T}(p)^2 \otimes [\mathbf{p}] + [\mathbf{p}] \otimes \mathbf{T}(p)^2)[\mathbf{p}]^2 \otimes [\mathbf{p}]^2$$
$$+ (p^2 + 3)(\mathbf{T}(p)^2 \otimes \mathbf{T}_1(p^2) + \mathbf{T}_1(p^2) \otimes \mathbf{T}(p)^2)[\mathbf{p}]^2 \otimes [\mathbf{p}]^2$$
$$+ (\mathbf{T}(p)^2[\mathbf{p}] \otimes \mathbf{T}_1(p^2)^2 + \mathbf{T}_1(p^2)^2 \otimes \mathbf{T}(p)^2[\mathbf{p}])[\mathbf{p}] \otimes [\mathbf{p}]$$
$$+ (p^3 + 3p^2 + p + 1)(\mathbf{T}(p)^2\mathbf{T}_1(p^2) \otimes [\mathbf{p}]^2 + [\mathbf{p}]^2 \otimes \mathbf{T}(p)^2\mathbf{T}_1(p^2))[\mathbf{p}] \otimes [\mathbf{p}]$$
$$+ (\mathbf{T}(p)^2 \otimes [\mathbf{p}] + [\mathbf{p}] \otimes \mathbf{T}(p)^2)\mathbf{T}_1(p^2)[\mathbf{p}] \otimes \mathbf{T}_1(p^2)[\mathbf{p}]$$
$$+ (p^2 + 1)\mathbf{T}(p)^2[\mathbf{p}]^2 \otimes \mathbf{T}(p)^2[\mathbf{p}]^2 \Big),$$

$$r_7 = -p^{13} \left(2(p+1)(p^3 + p - 1)[\mathbf{p}] \otimes [\mathbf{p}] - (p+1)(p^2 - 2p + 2)(\mathbf{T}_1(p^2) \otimes [\mathbf{p}] \right.$$
$$+ [\mathbf{p}] \otimes \mathbf{T}_1(p^2)) - 2\mathbf{T}_1(p^2) \otimes \mathbf{T}_1(p^2) - (p+1)(\mathbf{T}(p)^2 \otimes [\mathbf{p}]$$
$$+ [\mathbf{p}] \otimes \mathbf{T}(p)^2))\mathbf{T}(p)[\mathbf{p}]^2 \otimes \mathbf{T}(p)[\mathbf{p}]^2 ,$$

$$r_8 = -p^{16}(p\,(2p^6 + 3p^5 + 6p^4 - p^3 + 6p^2 - p + 2)[\mathbf{p}]^2 \otimes [\mathbf{p}]^2$$
$$+ p\,(p^2 + 1)(p^3 + 3p^2 - p + 3)(\mathbf{T}_1(p^2) \otimes [\mathbf{p}] + [\mathbf{p}] \otimes \mathbf{T}_1(p^2))[\mathbf{p}] \otimes [\mathbf{p}]$$
$$+ p\,(p + 4)(p^2 + 1)\mathbf{T}_1(p^2)[\mathbf{p}] \otimes \mathbf{T}_1(p^2)[\mathbf{p}]$$
$$+ p\,(p^2 - p + 1)(\mathbf{T}_1(p^2)^2 \otimes [\mathbf{p}]^2 + [\mathbf{p}]^2 \otimes \mathbf{T}_1(p^2)^2)$$
$$+ p\,(\mathbf{T}_1(p^2) \otimes [\mathbf{p}] + [\mathbf{p}] \otimes \mathbf{T}_1(p^2))\mathbf{T}_1(p^2) \otimes \mathbf{T}_1(p^2)$$
$$- p\,(2p^3 + p^2 + 2p - 1)(\mathbf{T}(p)^2 \otimes [\mathbf{p}] + [\mathbf{p}] \otimes \mathbf{T}(p)^2)[\mathbf{p}] \otimes [\mathbf{p}]$$
$$- 2p^2(\mathbf{T}(p)^2 \otimes \mathbf{T}_1(p^2) + \mathbf{T}_1(p^2) \otimes \mathbf{T}(p)^2)[\mathbf{p}] \otimes [\mathbf{p}]$$
$$+ p\,(\mathbf{T}(p)^2\mathbf{T}_1(p^2) \otimes [\mathbf{p}]^2 + [\mathbf{p}]^2 \otimes \mathbf{T}(p)^2\mathbf{T}_1(p^2))$$
$$+ (2p + 1)\mathbf{T}(p)^2[\mathbf{p}] \otimes \mathbf{T}(p)^2[\mathbf{p}])[\mathbf{p}]^2 \otimes [\mathbf{p}]^2 ,$$

$$r_9 = p^{20}(p + 1)(2[\mathbf{p}] \otimes [\mathbf{p}] + \mathbf{T}_1(p^2) \otimes [\mathbf{p}] + [\mathbf{p}] \otimes \mathbf{T}_1(p^2))\mathbf{T}(p)[\mathbf{p}]^3 \otimes \mathbf{T}(p)[\mathbf{p}]^3$$

$$r_{10} = p^{24}((p^2 + 1)(p^4 + 2p^3 - p^2 - 1)[\mathbf{p}] \otimes [\mathbf{p}] + (p^3 - p^2 - 1)(\mathbf{T}_1(p^2) \otimes [\mathbf{p}]$$
$$+ [\mathbf{p}] \otimes \mathbf{T}_1(p^2)) - \mathbf{T}_1(p^2) \otimes \mathbf{T}_1(p^2) - p^2(\mathbf{T}(p)^2 \otimes [\mathbf{p}] + [\mathbf{p}] \otimes \mathbf{T}(p)^2))[\mathbf{p}]^4 \otimes [\mathbf{p}]^4,$$

$$r_{11} = 0,$$

$$r_{12} = p^{34}[\mathbf{p}]^6 \otimes [\mathbf{p}]^6.$$

As for the coefficients of $S(X)$, one has

$$S(X) = 1 - (\mathbf{T}(p) \otimes \mathbf{T}(p))X + \cdots + (p^6[\mathbf{p}] \otimes [\mathbf{p}])^8 X^{16} \in \mathcal{L}_{2,\mathbb{Z}} \otimes \mathcal{L}_{2,\mathbb{Z}}[X],$$

where

$$s_1 = -\mathbf{T}(p) \otimes \mathbf{T}(p),$$
$$s_2 = -p(2\,p\,(p^2 + 1)^2\,[\mathbf{p}] \otimes [\mathbf{p}] + 2\,p\,(p^2 + 1)\,(\mathbf{T}_1(p^2) \otimes [\mathbf{p}] + [\mathbf{p}] \otimes \mathbf{T}_1(p^2))$$

$$+ 2\,p\,\mathbf{T}_1(p^2) \otimes \mathbf{T}_1(p^2) - (p^2+1)\,(\mathbf{T}(p)^2 \otimes [\mathbf{p}] + [\mathbf{p}] \otimes \mathbf{T}(p)^2)$$

$$- (\mathbf{T}(p)^2 \otimes \mathbf{T}_1(p^2) + \mathbf{T}_1(p^2) \otimes \mathbf{T}(p)^2)),$$

$$s_3 = p^2((2\,p^4 + 4\,p^2 - 1)\,[\mathbf{p}] \otimes [\mathbf{p}] + (2\,p^2 - 1)\,(\mathbf{T}_1(p^2) \otimes [\mathbf{p}] + [\mathbf{p}] \otimes \mathbf{T}_1(p^2))$$

$$- \mathbf{T}_1(p^2) \otimes \mathbf{T}_1(p^2) - p\,(\mathbf{T}(p)^2 \otimes [\mathbf{p}] + [\mathbf{p}] \otimes \mathbf{T}(p)^2),)\mathbf{T}(p) \otimes \mathbf{T}(p)),$$

$$s_4 = p^4((p^8 + 12\,p^6 + 10\,p^4 + 4\,p^2 + 1)\,[\mathbf{p}]^2 \otimes [\mathbf{p}]^2$$

$$+ 2\,(3\,p^6 + 5\,p^4 + 3\,p^2 + 1)\,(\mathbf{T}_1(p^2) \otimes [\mathbf{p}] + [\mathbf{p}] \otimes \mathbf{T}_1(p^2))[\mathbf{p}] \otimes [\mathbf{p}]$$

$$+ 4\,(p^2+1)^2\,\mathbf{T}_1(p^2)[\mathbf{p}] \otimes \mathbf{T}_1(p^2)[\mathbf{p}]$$

$$+ (3\,p^4 + 2\,p^2 + 1)\,(\mathbf{T}_1(p^2)^2 \otimes [\mathbf{p}]^2 + [\mathbf{p}]^2 \otimes \mathbf{T}_1(p^2)^2)$$

$$+ 2\,(p^2+1)\,(\mathbf{T}_1(p^2) \otimes [\mathbf{p}] + [\mathbf{p}] \otimes \mathbf{T}_1(p^2))\mathbf{T}_1(p^2) \otimes \mathbf{T}_1(p^2)$$

$$+ \mathbf{T}_1(p^2)^2 \otimes \mathbf{T}_1(p^2)^2$$

$$- 2\,p\,(p^4 + 4\,p^2 + 1)\,(\mathbf{T}(p)^2 \otimes [\mathbf{p}] + [\mathbf{p}] \otimes \mathbf{T}(p)^2)\,[\mathbf{p}] \otimes [\mathbf{p}]$$

$$- 4\,p\,(p^2+1)\,(\mathbf{T}(p)^2 \otimes \mathbf{T}_1(p^2) + \mathbf{T}_1(p^2) \otimes \mathbf{T}(p)^2)[\mathbf{p}] \otimes [\mathbf{p}]$$

$$- 2\,p\,(\mathbf{T}(p)^2[\mathbf{p}] \otimes \mathbf{T}_1(p^2)^2 + \mathbf{T}_1(p^2)^2 \otimes \mathbf{T}(p)^2[\mathbf{p}])$$

$$- 4\,p^3\,(\mathbf{T}(p)^2\mathbf{T}_1(p^2) \otimes [\mathbf{p}]^2 + [\mathbf{p}]^2 \otimes \mathbf{T}(p)^2\mathbf{T}_1(p^2))$$

$$+ (p^2 + 2)\,\mathbf{T}(p)^2[\mathbf{p}] \otimes \mathbf{T}(p)^2[\mathbf{p}]$$

$$+ (\mathbf{T}_1(p^2) \otimes [\mathbf{p}] + [\mathbf{p}] \otimes \mathbf{T}_1(p^2))\mathbf{T}(p)^2 \otimes \mathbf{T}(p)^2$$

$$+ p^2\,(\mathbf{T}(p)^4 \otimes [\mathbf{p}]^2 + [\mathbf{p}]^2 \otimes \mathbf{T}(p)^4)),$$

$$s_5 = -p^6((6\,p^6 + 2\,p^4 - p^2 + 2)\,[\mathbf{p}]^2 \otimes [\mathbf{p}]^2$$

$$+ (p^4 - p^2 + 3)\,(\mathbf{T}_1(p^2) \otimes [\mathbf{p}] + [\mathbf{p}] \otimes \mathbf{T}_1(p^2))[\mathbf{p}] \otimes [\mathbf{p}]$$

$$+ (3\,p^2 + 4)\,\mathbf{T}_1(p^2)[\mathbf{p}] \otimes \mathbf{T}_1(p^2)[\mathbf{p}]$$

$$- (2\,p^2 - 1)\,(\mathbf{T}_1(p^2)^2 \otimes [\mathbf{p}]^2 + [\mathbf{p}]^2 \otimes \mathbf{T}_1(p^2)^2)$$

$$+ (\mathbf{T}_1(p^2) \otimes [\mathbf{p}] + [\mathbf{p}] \otimes \mathbf{T}_1(p^2))\,\mathbf{T}_1(p^2) \otimes \mathbf{T}_1(p^2)$$

$$- p\,(2\,p^2 + 1)\,(\mathbf{T}(p)^2 \otimes [\mathbf{p}] + [\mathbf{p}] \otimes \mathbf{T}(p)^2)[\mathbf{p}] \otimes [\mathbf{p}]$$

$$- 2\,p\,(\mathbf{T}(p)^2 \otimes \mathbf{T}_1(p^2) + \mathbf{T}_1(p^2) \otimes \mathbf{T}(p)^2)[\mathbf{p}] \otimes [\mathbf{p}]$$

$$+ p\,(\mathbf{T}(p)^2\mathbf{T}_1(p^2) \otimes [\mathbf{p}]^2 + [\mathbf{p}]^2 \otimes \mathbf{T}(p)^2\mathbf{T}_1(p^2))$$

$$+ \mathbf{T}(p)^2[\mathbf{p}] \otimes \mathbf{T}(p)^2[\mathbf{p}])\mathbf{T}(p) \otimes \mathbf{T}(p)),$$

$$s_6 = -p^8(2\,p^2\,(p^8 + 6\,p^6 + 11\,p^4 + 8\,p^2 + 2)\,[\mathbf{p}]^3 \otimes [\mathbf{p}]^3$$

$$+ 2\,p^2\,(5\,p^4 + 12\,p^2 + 6)\,\mathbf{T}_1(p^2)[\mathbf{p}]^2 \otimes \mathbf{T}_1(p^2)[\mathbf{p}]^2$$

$$+ (3\,p^4 + 10\,p^2 - 1)\,\mathbf{T}(p)^2[\mathbf{p}]^2 \otimes \mathbf{T}(p)^2[\mathbf{p}]^2 - \mathbf{T}(p)^2\mathbf{T}_1(p^2)[\mathbf{p}] \otimes \mathbf{T}(p)^2\mathbf{T}_1(p^2)[\mathbf{p}]$$

$$+ 2\,p^2\,(3\,p^6 + 11\,p^4 + 12\,p^2 + 4)\,(\mathbf{T}_1(p^2) \otimes [\mathbf{p}] + [\mathbf{p}] \otimes \mathbf{T}_1(p^2))\,[\mathbf{p}]^2 \otimes [\mathbf{p}]^2$$

$$+ 6\,p^2\,(p^2 + 1)^2\,(\mathbf{T}_1(p^2)^2 \otimes [\mathbf{p}]^2 + [\mathbf{p}]^2 \otimes \mathbf{T}_1(p^2)^2)[\mathbf{p}] \otimes [\mathbf{p}]$$

$$+ 6\,p^2\,(p^2 + 1)\,(\mathbf{T}_1(p^2) \otimes [\mathbf{p}] + [\mathbf{p}] \otimes \mathbf{T}_1(p^2))\,\mathbf{T}_1(p^2)[\mathbf{p}] \otimes \mathbf{T}_1(p^2)[\mathbf{p}]$$

$$+ 2\,p^2\,(p^2 + 1)\,(\mathbf{T}_1(p^2)^3 \otimes [\mathbf{p}]^3 + [\mathbf{p}]^3 \otimes \mathbf{T}_1(p^2)^3)$$

$$+ 2\,p^2\,(\mathbf{T}_1(p^2)^2 \otimes [\mathbf{p}]^2 + [\mathbf{p}]^2 \otimes \mathbf{T}_1(p^2)^2)\mathbf{T}_1(p^2) \otimes \mathbf{T}_1(p^2)$$

$$- p \, (5 \, p^6 + 13 \, p^4 + 10 \, p^2 + 2) \, (\mathbf{T}(p)^2 \otimes [\mathbf{p}] + [\mathbf{p}] \otimes \mathbf{T}(p)^2)[\mathbf{p}]^2 \otimes [\mathbf{p}]^2$$

$$- p \, (7 \, p^4 + 12 \, p^2 + 4) \, (\mathbf{T}(p)^2 \otimes \mathbf{T}_1(p^2) + \mathbf{T}_1(p^2) \otimes \mathbf{T}(p)^2)[\mathbf{p}]^2 \otimes [\mathbf{p}]^2$$

$$- 3p \, (p^2 + 1) \, (\mathbf{T}(p)^2[\mathbf{p}] \otimes \mathbf{T}_1(p^2)^2 + \mathbf{T}_1(p^2)^2 \otimes \mathbf{T}(p)^2[\mathbf{p}])[\mathbf{p}] \otimes [\mathbf{p}]$$

$$- p \, (\mathbf{T}(p)^2[\mathbf{p}]^2 \otimes \mathbf{T}_1(p^2)^3 + \mathbf{T}_1(p^2)^3 \otimes \mathbf{T}(p)^2[\mathbf{p}]^2)$$

$$- 2 \, p \, (3 \, p^4 + 4 \, p^2 + 1) \, (\mathbf{T}(p)^2 \mathbf{T}_1(p^2) \otimes [\mathbf{p}]^2 + [\mathbf{p}]^2 \otimes \mathbf{T}(p)^2 \mathbf{T}_1(p^2))[\mathbf{p}] \otimes [\mathbf{p}]$$

$$- 2 \, p \, (3 \, p^2 + 1) \, (\mathbf{T}(p)^2 \otimes [\mathbf{p}] + [\mathbf{p}] \otimes \mathbf{T}(p)^2)\mathbf{T}_1(p^2)[\mathbf{p}] \otimes \mathbf{T}_1(p^2)[\mathbf{p}]$$

$$- p \, (p^2 + 1) \, (\mathbf{T}(p)^2 \mathbf{T}_1(p^2)^2 \otimes [\mathbf{p}]^3 + [\mathbf{p}]^3 \otimes \mathbf{T}(p)^2 \mathbf{T}_1(p^2)^2)$$

$$- p \, (\mathbf{T}(p)^2 \mathbf{T}_1(p^2) \otimes [\mathbf{p}]^2 + [\mathbf{p}]^2 \otimes \mathbf{T}(p)^2 \mathbf{T}_1(p^2))\mathbf{T}_1(p^2) \otimes \mathbf{T}_1(p^2)$$

$$+ (5 \, p^2 - 1) \, (\mathbf{T}_1(p^2) \otimes [\mathbf{p}] + [\mathbf{p}] \otimes \mathbf{T}_1(p^2)) \, \mathbf{T}(p)^2[\mathbf{p}] \otimes \mathbf{T}(p)^2[\mathbf{p}]$$

$$+ 2 \, p^2 \, (p^2 + 1) \, (\mathbf{T}(p)^4 \otimes [\mathbf{p}]^2 + [\mathbf{p}]^2 \otimes \mathbf{T}(p)^4)[\mathbf{p}] \otimes [\mathbf{p}]$$

$$+ 2 \, p^2 \, (\mathbf{T}(p)^4 \otimes \mathbf{T}_1(p^2)[\mathbf{p}] + \mathbf{T}_1(p^2)[\mathbf{p}] \otimes \mathbf{T}(p)^4) \, [\mathbf{p}] \otimes [\mathbf{p}]$$

$$- p \, (\mathbf{T}(p)^4 \otimes \mathbf{T}(p)^4[\mathbf{p}] + \mathbf{T}(p)^4[\mathbf{p}] \otimes \mathbf{T}(p)^4) \, [\mathbf{p}] \otimes [\mathbf{p}]),$$

$$s_7 = p^{11}(p \, (5 \, p^6 - 2 \, p^4 + 2) \, \mathbf{T}(p)[\mathbf{p}]^3 \otimes \mathbf{T}(p)[\mathbf{p}]^3$$

$$+ 8 \, p \, \mathbf{T}(p)\mathbf{T}_1(p^2)[\mathbf{p}]^2 \otimes \mathbf{T}(p)\mathbf{T}_1(p^2)[\mathbf{p}]^2$$

$$+ p \, \mathbf{T}(p)^3[\mathbf{p}]^2 \otimes \mathbf{T}(p)^3[\mathbf{p}]^2$$

$$- p \, (p^4 - 3) \, (\mathbf{T}_1(p^2) \otimes [\mathbf{p}] + [\mathbf{p}] \otimes \mathbf{T}_1(p^2)) \, \mathbf{T}(p)[\mathbf{p}]^2 \otimes \mathbf{T}(p)[\mathbf{p}]^2$$

$$- p \, (\mathbf{T}_1(p^2)^2 \otimes [\mathbf{p}]^2 + [\mathbf{p}]^2 \otimes \mathbf{T}_1(p^2)^2)\mathbf{T}(p)[\mathbf{p}] \otimes \mathbf{T}(p)[\mathbf{p}]$$

$$+ 2 \, p \, (\mathbf{T}_1(p^2) \otimes [\mathbf{p}] + [\mathbf{p}] \otimes \mathbf{T}_1(p^2)) \, \mathbf{T}(p)\mathbf{T}_1(p^2)[\mathbf{p}] \otimes \mathbf{T}(p)\mathbf{T}_1(p^2)[\mathbf{p}]$$

$$- p \, (\mathbf{T}_1(p^2)^3 \otimes [\mathbf{p}]^3 + [\mathbf{p}]^3 \otimes \mathbf{T}_1(p^2)^3)\mathbf{T}(p) \otimes \mathbf{T}(p)$$

$$- (3 \, p^4 - 3 \, p^2 + 2) \, (\mathbf{T}(p)^2 \otimes [\mathbf{p}] + [\mathbf{p}] \otimes \mathbf{T}(p)^2)\mathbf{T}(p)[\mathbf{p}]^2 \otimes \mathbf{T}(p)[\mathbf{p}]^2$$

$$+ (p^2 - 3) \, (\mathbf{T}(p)^2 \otimes \mathbf{T}_1(p^2) + \mathbf{T}_1(p^2) \otimes \mathbf{T}(p)^2)\mathbf{T}(p)[\mathbf{p}]^2 \otimes \mathbf{T}(p)[\mathbf{p}]^2$$

$$- (\mathbf{T}(p)^2[\mathbf{p}] \otimes \mathbf{T}_1(p^2)^2 + \mathbf{T}_1(p^2)^2 \otimes \mathbf{T}(p)^2[\mathbf{p}])\mathbf{T}(p)[\mathbf{p}] \otimes \mathbf{T}(p)[\mathbf{p}]$$

$$+ (2 \, p^2 - 1) \, (\mathbf{T}(p)^2 \mathbf{T}_1(p^2) \otimes [\mathbf{p}]^2 + [\mathbf{p}]^2 \otimes \mathbf{T}(p)^2 \mathbf{T}_1(p^2))\mathbf{T}(p)[\mathbf{p}] \otimes \mathbf{T}(p)[\mathbf{p}]$$

$$- (\mathbf{T}(p)^2 \otimes [\mathbf{p}] + [\mathbf{p}] \otimes \mathbf{T}(p)^2)\mathbf{T}(p)\mathbf{T}_1(p^2)[\mathbf{p}] \otimes \mathbf{T}(p)\mathbf{T}_1(p^2)[\mathbf{p}]),$$

$$s_8 = p^{14}(2 \, p^2 \, (2 \, p^8 + 4 \, p^6 + 14 \, p^4 + 12 \, p^2 + 3) \, [\mathbf{p}]^4 \otimes [\mathbf{p}]^4$$

$$+ 4 \, p^2 \, (p^6 + 7 \, p^4 + 9 \, p^2 + 3) \, (\mathbf{T}_1(p^2) \otimes [\mathbf{p}] + [\mathbf{p}] \otimes \mathbf{T}_1(p^2)) \, [\mathbf{p}]^3 \otimes [\mathbf{p}]^3$$

$$+ 16 \, p^2 \, (p^2 + 1)^2 \, \mathbf{T}_1(p^2)[\mathbf{p}]^3 \otimes \mathbf{T}_1(p^2)[\mathbf{p}]^3$$

$$+ 2 \, p^2 \, (3 \, p^4 + 10 \, p^2 + 5) \, (\mathbf{T}_1(p^2)^2 \otimes [\mathbf{p}]^2 + [\mathbf{p}]^2 \otimes \mathbf{T}_1(p^2)^2) \, [\mathbf{p}]^2 \otimes [\mathbf{p}]^2$$

$$+ 8 \, p^2 \, (p^2 + 1) \, (\mathbf{T}_1(p^2) \otimes [\mathbf{p}] + [\mathbf{p}] \otimes \mathbf{T}_1(p^2)) \, \mathbf{T}_1(p^2)[\mathbf{p}]^2 \otimes \mathbf{T}_1(p^2)[\mathbf{p}]^2$$

$$+ 4 \, p^2 \, \mathbf{T}_1(p^2)^2[\mathbf{p}]^2 \otimes \mathbf{T}_1(p^2)^2[\mathbf{p}]^2$$

$$+ 4 \, p^2 \, (p^2 + 1) \, (\mathbf{T}_1(p^2)^3 \otimes [\mathbf{p}]^3 + [\mathbf{p}]^3 \otimes \mathbf{T}_1(p^2)^3) \, [\mathbf{p}] \otimes [\mathbf{p}]$$

$$+ p^2 \, (\mathbf{T}_1(p^2)^4 \otimes [\mathbf{p}]^4 + [\mathbf{p}]^4 \otimes \mathbf{T}_1(p^2)^4)$$

$$- 4 \, p \, (2 \, p^6 + 3 \, p^4 + 4 \, p^2 + 1) \, (\mathbf{T}(p)^2 \otimes [\mathbf{p}] + [\mathbf{p}] \otimes \mathbf{T}(p)^2)[\mathbf{p}]^3 \otimes [\mathbf{p}]^3$$

$$- 8 \, p \, (p^2 + 1)^2 \, (\mathbf{T}(p)^2 \otimes \mathbf{T}_1(p^2) + \mathbf{T}_1(p^2) \otimes \mathbf{T}(p)^2)[\mathbf{p}]^3 \otimes [\mathbf{p}]^3$$

$$- 4 \, p \, (p^2 + 1) \, (\mathbf{T}(p)^2[\mathbf{p}] \otimes \mathbf{T}_1(p^2)^2 + \mathbf{T}_1(p^2)^2 \otimes \mathbf{T}(p)^2[\mathbf{p}])[\mathbf{p}]^2 \otimes [\mathbf{p}]^2$$

$$- 4\,p\,(p^4 + 4\,p^2 + 1)\,(\mathbf{T}(p)^2\mathbf{T}_1(p^2) \otimes [\mathbf{p}]^2 + [\mathbf{p}]^2 \otimes \mathbf{T}(p)^2\mathbf{T}_1(p^2))[\mathbf{p}]^2 \otimes [\mathbf{p}]^2$$
$$- 8\,p\,(p^2 + 1)\,(\mathbf{T}(p)^2 \otimes [\mathbf{p}] + [\mathbf{p}] \otimes \mathbf{T}(p)^2)\mathbf{T}_1(p^2)[\mathbf{p}]^2 \otimes \mathbf{T}_1(p^2)[\mathbf{p}]^2$$
$$- 4\,p\,(\mathbf{T}(p)^2 \otimes \mathbf{T}_1(p^2) + \mathbf{T}_1(p^2) \otimes \mathbf{T}(p)^2)\mathbf{T}_1(p^2)[\mathbf{p}]^2 \otimes \mathbf{T}_1(p^2)[\mathbf{p}]^2$$
$$- 4\,p^3\,(\mathbf{T}(p)^2\mathbf{T}_1(p^2)^2 \otimes [\mathbf{p}]^3 + [\mathbf{p}]^3 \otimes \mathbf{T}(p)^2\mathbf{T}_1(p^2)^2) + [\mathbf{p}] \otimes [\mathbf{p}]$$
$$+ 2\,(5\,p^4 + 2\,p^2 + 2)\,\mathbf{T}(p)^2[\mathbf{p}]^3 \otimes \mathbf{T}(p)^2[\mathbf{p}]^3$$
$$+ 2\,(p^2 + 2)\,(\mathbf{T}_1(p^2) \otimes [\mathbf{p}] + [\mathbf{p}] \otimes \mathbf{T}_1(p^2))\,\mathbf{T}(p)^2[\mathbf{p}]^2 \otimes \mathbf{T}(p)^2[\mathbf{p}]^2$$
$$+ 2\,\mathbf{T}(p)^2\mathbf{T}_1(p^2)[\mathbf{p}]^2 \otimes \mathbf{T}(p)^2\mathbf{T}_1(p^2)[\mathbf{p}]^2$$
$$+ (\mathbf{T}_1(p^2)^2 \otimes [\mathbf{p}]^2 + [\mathbf{p}]^2 \otimes \mathbf{T}_1(p^2)^2)\mathbf{T}(p)^2[\mathbf{p}] \otimes \mathbf{T}(p)^2[\mathbf{p}]$$
$$+ (3\,p^4 + 2\,p^2 + 1)\,(\mathbf{T}(p)^4 \otimes [\mathbf{p}]^2 + [\mathbf{p}]^2 \otimes \mathbf{T}(p)^4)[\mathbf{p}]^2 \otimes [\mathbf{p}]^2$$
$$+ 2\,(p^2 + 1)\,(\mathbf{T}(p)^4 \otimes \mathbf{T}_1(p^2)[\mathbf{p}] + \mathbf{T}_1(p^2)[\mathbf{p}] \otimes \mathbf{T}(p)^4)\,[\mathbf{p}]^2 \otimes [\mathbf{p}]^2$$
$$+ (\mathbf{T}(p)^4 \otimes \mathbf{T}_1(p^2)^2 + \mathbf{T}_1(p^2)^2 \otimes \mathbf{T}(p)^4)\,[\mathbf{p}]^2 \otimes [\mathbf{p}]^2$$
$$- 2\,p\,(\mathbf{T}(p)^2 \otimes [\mathbf{p}] + [\mathbf{p}] \otimes \mathbf{T}(p)^2)\mathbf{T}(p)^2[\mathbf{p}]^2 \otimes \mathbf{T}(p)^2[\mathbf{p}]^2).$$

Then we find the remaining coefficients $s_9, \ldots, s_{16}$, using an easy functional equation (similar to [An87], p. 164, (3.3.79)):

$$s_{16-i} = (p^6[\mathbf{p}] \otimes [\mathbf{p}])^{8-i}s_i \quad (i = 0, \ldots, 8).$$

To conclude, we give the Newton polygons of $R(X)$ and $S(X)$ with respect to powers of p and X (see Figure 1). It follows from our comutation that all slopes are *integral*. We hope that these polygons might help to find some geometric objects attached to the polynomials $R(X)$ and $S(X)$, in the spirit of a recent work of C. Faber and G. Van Der Geer [FVdG].

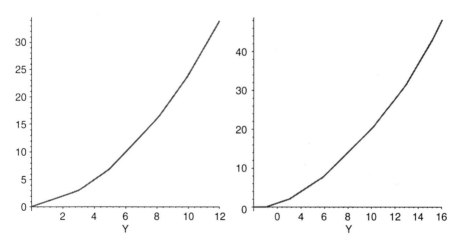

Fig. 1. Newton polygons of $R(X)$ and $S(X)$ with respect to powers of p and X, of heights 34 and 48, resp.

Acknowledgments

We are very grateful to the Institute Fourier (UJF, Grenoble-1) for the always excellent working environment.

It is a great pleasure for us to thank Siegfried Boecherer, Stephen Gelbart, Solomon Friedberg, and Mikhail Tsfasman for valuable discussions and observations.

Our special thanks go to Yuri Ivanovich Manin, for providing us with advice and encouragement.

References

[An67] ANDRIANOV, A.N., *Shimura's conjecture for Siegel's modular group of genus 3*, Dokl. Akad. Nauk SSSR 177 (1967), 755–758 = Soviet Math. Dokl. 8 (1967), 1474–1478.

[An68] ANDRIANOV, A.N., *Rationality of multiple Hecke series of a complete linear group and Shimura's hypothesis on Hecke's series of a symplectic group* Dokl. Akad. Nauk SSSR 183 (1968), 9–11 = Soviet Math. Dokl. 9 (1968), 1295–1297.

[An69] ANDRIANOV, A.N., *Rationality theorems for Hecke series and zeta functions of the groups GL_n and SP_n over local fields*, Izv. Akad. Nauk SSSR, Ser. Mat., Tom 33 (1969), No. 3 (Math. USSR – Izvestija, Vol. 3 (1969), No. 3, pp. 439–476).

[An70] ANDRIANOV, A.N., *Spherical functions for GL_n over local fields and summation of Hecke series*, Mat. Sbornik, Tom 83 (125) (1970), No 3 (Math. USSR Sbornik, Vol. 12 (1970), No. 3, pp. 429–452).

[An74] ANDRIANOV, A.N., *Euler products corresponding to Siegel modular forms of genus 2*, Russian Math. Surveys, 29:3 (1974), pp. 45–116 (Uspekhi Mat. Nauk 29:3 (1974) pp. 43–110).

[An87] ANDRIANOV, A.N., *Quadratic Forms and Hecke Operators*, Springer-Verlag, Berlin, Heidelberg, New York, London, Paris, Tokyo, 1987.

[An-Ka] ANDRIANOV, A.N., KALININ, V.L., *On Analytic Properties of Standard Zeta Functions of Siegel Modular Forms*, Mat. Sbornik 106 (1978), pp 323–339 (in Russian).

[AnZh95] ANDRIANOV, A.N., ZHURAVLEV, V.G., *Modular Forms and Hecke Operators*, Translations of Mathematical Monographs, Vol. 145, AMS, Providence, Rhode Island, 1995.

[AsSch] ASGARI, M., SCHMIDT, R., *Siegel modular forms and representations*. Manuscripta Math. 104 (2001), 173–200.

[BHam] BÖCHERER, S., *Ein Rationalitätssatz für formale Heckereihen zur Siegelschen Modulgruppe*. Abh. Math. Sem. Univ. Hamburg 56 (1986), 35–47.

[BoeH06] BÖCHERER, S. and HEIM B.E. *Critical values of L-functions on $GSp_2 \times GL_2$*. Math.Z, 254, 485–503 (2006).

[BFG06] BUMP, D.; FRIEDBERG, S.; GINZBURG, D. *Lifting automorphic representations on the double covers of orthogonal groups*. Duke Math. J. 131 (2006), no. 2, 363–396.

[BFG92] BUMP, D.; FRIEDBERG, S.; GINZBURG, D. *Whittaker-orthogonal models, functoriality, and the Rankin-Selberg method.* Invent. Math. 109 (1992), no. 1, 55–96.

[CourPa] COURTIEU, M., PANCHISHKIN, A.A., *Non-Archimedean L-Functions and Arithmetical Siegel Modular Forms,* Lecture Notes in Mathematics 1471, Springer-Verlag, 2004 (2nd augmented ed.).

[De79] DELIGNE P., *Valeurs de fonctions L et périodes d'intégrales,* Proc. Sympos. Pure Math. vol. 55. Amer. Math. Soc., Providence, RI, 1979, 313–346

[Evd] EVDOKIMOV, S.A., *Dirichlet series, multiple Andrianov zeta-functions in the theory of Euler modular forms of genus 3* (Russian), Dokl. Akad. Nauk SSSR 277 (1984), no. 1, 25–29.

[FVdG] FABER, C., VAN DER GEER, G., Sur la cohomologie des systÃ¨mes locaux sur les espaces de modules des courbes de genre 2 et des surfaces abÃ©liennes. I, II C. R. Math. Acad. Sci. Paris 338, (2004) No.5, pp. 381–384 and No.6, 467–470.

[Hecke] HECKE, E., *Über Modulfunktionen und die Dirichletschen Reihen mit Eulerscher Produktenwickelung, I, II.* Math. Annalen 114 (1937), 1–28, 316–351 (Mathematische Werke. Göttingen: Vandenhoeck und Ruprecht, 1959, 644–707).

[Ike01] IKEDA, T., *On the lifting of elliptic cusp forms to Siegel cusp forms of degree 2n,* Ann. of Math. (2) 154 (2001), 641–681.

[Ike06] IKEDA, T., *Pullback of the lifting of elliptic cusp forms and Miyawaki's Conjecture* Duke Mathematical Journal, **131**, 469–497 (2006).

[Jia96] JIANG, D., *Degree 16 standard L-function of* GSp(2) × GSp(2). Mem. Amer. Math. Soc. 123 (1996), no. 588.

[Ku88] KUROKAWA, N., *Analyticity of Dirichlet series over prime powers.* Analytic number theory (Tokyo, 1988), 168–177, Lecture Notes in Math., 1434, Springer, Berlin, 1990.

[La79] R. LANGLANDS, *Automorphic forms, Shimura varieties, and L-functions. Ein Märchen.* Proc. Symp. Pure Math. 33, part 2 (1979), 205–246.

[Maa76] MAASS, H., *Indefinite Quadratische Formen und Eulerprodukte.* Comm. on Pure and Appl. Math, 19, 689–699 (1976).

[Ma-Pa77] MANIN, YU.I. and PANCHISHKIN, A.A., *Convolutions of Hecke series and their values at integral points,* Mat. Sbornik, 104 (1977) 617–651 (in Russian); Math. USSR, Sb. 33, 539–571 (1977) (English translation). In: Selected Papers of Yu. I .Manin, World Scientific, 1996, pp. 325–357.

[Ma-Pa05] MANIN, YU.I. and PANCHISHKIN, A.A., *Introduction to Modern Number Theory,* Encyclopaedia of Mathematical Sciences, vol. 49 (2nd ed.), Springer-Verlag, 2005.

[Mur02] MUROKAWA, K., *Relations between symmetric power L-functions and spinor L-functions attached to Ikeda lifts,* Kodai Math. J. 25, 61–71 (2002).

[Pa87] PANCHISHKIN, A., *A functional equation of the non-Archimedean Rankin convolution,* Duke Math. J. 54 (special volume in Honor of Yu.I. Manin on his 50th birthday) (1987) 77–89.

[Pa94] PANCHISHKIN, A., *Admissible Non-Archimedean standard zeta functions of Siegel modular forms*, Proceedings of the Joint AMS Summer Conference on Motives, Seattle, July 20–August 2 1991, Seattle, Providence, R.I., 1994, vol.2, 251–292.

[Pa02] PANCHISHKIN, A., *A new method of constructing p-adic L-functions associated with modular forms*, Moscow Mathematical Journal, 2 (2002), Number 2 (special issue in Honor of Yu.I. Manin on his 65th birthday), 1–16.

[PaGRFA] PANCHISHKIN, A., *Produits d'Euler attachés aux formes modulaires de Siegel.* Exposé au Séminaire Groupes Réductifs et Formes Automorphes à l'Institut de Mathématiques de Jussieu, June 22, 2006.

[PaHakuba5] PANCHISHKIN, A.A., *Triple products of Coleman's families and their periods (a joint work with S. Boecherer)* Proceedings of the 8th Hakuba conference "Periods and related topics from automorphic forms," September 25–October 1, 2005.

[PaSerre6] PANCHISHKIN, A.A., *p-adic Banach modules of arithmetical modular forms and triple products of Coleman's families*, (for a special volume of Quarterly Journal of Pure and Applied Mathematics dedicated to Jean-Pierre Serre), 2006.

[PaVa] PANCHISHKIN, A., VANKOV, K. *Explicit Shimura's conjecture for Sp3 on a computer.* Math. Res. Lett. 14, No. 2, (2007), 173–187.

[SchR03] SCHMIDT, R., *On the spin L-function of Ikeda's lifts.* Comment. Math. Univ. St. Pauli 52 (2003), 1–46.

[Sa63] SATAKE, I., *Theory of spherical functions on reductive groups over p-adic fields*, Publ. mathématiques de l'IHES 18 (1963), 1–59.

[Shi63] SHIMURA, G., *On modular correspondences for Sp(n, Z) and their congruence relations*, Proc. Nat. Acad. Sci. U.S.A. 49 (1963), 824–828.

[Shi71] SHIMURA, G., *Introduction to the Arithmetic Theory of Automorphic Functions*, Princeton Univ. Press, 1971.

[Tam] TAMAGAWA, T., *On the ζ-function of a division algebra*, Ann. of Math. 77 (1963), 387–405.

[VaSp4] VANKOV, K., *Explicit formula for the symplectic Hecke series of genus four*, Arxiv, math.NT/0606492 (2006).

[Yosh81] YOSHIDA, H., *Siegel's Modular Forms and the Arithmetic of Quadratic Forms*, Inventiones math. 60, 193–248 (1980).

[Yosh01] YOSHIDA, H., *Motives and Siegel modular forms*, American Journal of Mathematics, 123 (2001), 1171–1197.

Rank-2 Vector Bundles on ind-Grassmannians

Ivan Penkov[1] and Alexander S. Tikhomirov[2]

[1] Jacobs University Bremen, Germany
   i.penkov@jacobs-university.de
[2] State Pedagogical University Yaroslavl, Russia
   astikhomirov@mail.ru

To Yuri Ivanovich Manin on the occasion of his 70th birthday

Summary. An ind-Grassmannian $\mathbf{X} = \varinjlim G(k_i; V^{n_i})$ is *linear* if almost all defining embeddings $\varphi_m : G(k_m; V^{n_m}) \longrightarrow G(k_{m+1}; V^{n_{m+1}})$ are of degree 1, and is *twisted* if infinitely many defining embeddings φ_m are of degree higher than 1. In this paper we give a complete description of finite-rank vector bundles on any linear ind-Grassmannian, and prove that any vector bundle of rank 2 on a twisted ind-Grassmannian is trivial. Our work extends work by W. Barth, J. Donin, I. Penkov, E. Sato, A. Tyurin, and A. Van de Ven.

Key words: ind-Grassmannian, vector bundle

2000 Mathematics Subject Classifications: 14M15, 14F05, 14M17, 32L10

1 Introduction

The simplest example of an ind-Grassmannian is the infinite projective space $\mathbf{P}^\infty$. The Barth–Van de Ven–Tyurin (BVT) theorem, proved more than 30 years ago [BV], [T], [Sa1] (see also a recent proof by A. Coandă and G. Trautmann, [CT]), claims that any vector bundle of finite-rank on $\mathbf{P}^\infty$ is isomorphic to a direct sum of line bundles. In the last decade, natural examples of infinite flag varieties (or flag ind-varieties) have arisen as homogeneous spaces of locally linear ind-groups, [DPW], [DiP]. In the present note we concentrate our attention on the special case of ind-Grassmannians, i.e., inductive limits of Grassmannians of growing dimension. If $V = \bigcup_{n>k} V^n$ is a countable-dimensional vector space, then the ind-variety $\mathbf{G}(k; V) = \varinjlim G(k; V^n)$ (or simply $\mathbf{G}(k; \infty)$) of k-dimensional subspaces of V is of course an ind-Grassmannian: this is the simplest example beyond $\mathbf{P}^\infty = \mathbf{G}(1; \infty)$.

Y. Tschinkel and Y. Zarhin (eds.), *Algebra, Arithmetic, and Geometry*,
Progress in Mathematics 270, DOI 10.1007/978-0-8176-4747-6_18,
© Springer Science+Business Media, LLC 2009

A significant difference between $\mathbf{G}(k; V)$ and a general ind-Grassmannian $\mathbf{X} = \varinjlim G(k_i; V^{n_i})$ defined via a sequence of embeddings

$$G(k_1; V^{n_1}) \xrightarrow{\varphi_1} G(k_2; V^{n_2}) \xrightarrow{\varphi_2} \cdots \xrightarrow{\varphi_{m-1}} G(k_m; V^{n_m}) \xrightarrow{\varphi_m} \cdots \quad (1)$$

is that in general, the morphisms φ_m can have arbitrary degrees. We say that the ind-Grassmannian $\mathbf{X}$ is *twisted* if $\deg \varphi_m > 1$ for infinitely many m, and that $\mathbf{X}$ is *linear* if $\deg \varphi_m = 1$ for almost all m.

Conjecture 1.1. *Let the ground field be $\mathbb{C}$, and let $\mathbf{E}$ be a vector bundle of rank $r \in \mathbf{Z}_{>0}$ on an ind-Grassmannian $\mathbf{X} = \varinjlim G(k_m; V^{n_m})$, i.e., $\mathbf{E} = \varprojlim E_m$, where $\{E_m\}$ is an inverse system of vector bundles of (fixed) rank r on $G(k_m; V^{n_m})$. Then*

(i) $\mathbf{E}$ is semisimple: it is isomorphic to a direct sum of simple vector bundles on $\mathbf{X}$, i.e., vector bundles on $\mathbf{X}$ with no non trivial subbundles;

(ii) for $m \gg 0$ the restriction of each simple bundle $\mathbf{E}$ to $G(k_m, V^{n_m})$ is a homogeneous vector bundle;

(iii) each simple bundle $\mathbf{E}'$ has rank 1 unless $\mathbf{X}$ is isomorphic to $\mathbf{G}(k; \infty)$ for some k: in the latter case, $\mathbf{E}'$, twisted by a suitable line bundle, is isomorphic to a simple subbundle of the tensor algebra $T^{\cdot}(\mathbf{S})$, $\mathbf{S}$ being the tautological bundle of rank k on $\mathbf{G}(k; \infty)$;

(iv) each simple bundle $\mathbf{E}$ (and thus each vector bundle of finite-rank on $\mathbf{X}$) is trivial whenever $\mathbf{X}$ is a twisted ind-Grassmannian.

The BVT theorem and Sato's theorem about finite-rank bundles on $\mathbf{G}(k; \infty)$, [Sa1], [Sa2], as well as the results in [DP], are particular cases of the above conjecture. The purpose of this note is to prove Conjecture 1.1 for vector bundles of rank 2, and also for vector bundles of arbitrary rank r on linear ind-Grassmannians $\mathbf{X}$.

In the 1970s and 1980s, Yuri Ivanovich Manin taught us mathematics in (and beyond) his seminar, and the theory of vector bundles was a recurring topic (among many others). In 1980, he asked one of us (I.P.) to report on A. Tyurin's paper [T], and most importantly to try to understand this paper. This current note is a very preliminary progress report.

Acknowledgments. We acknowledge the support and hospitality of the Max Planck Institute for Mathematics in Bonn, where the present note was conceived. A. S. T. also acknowledges partial support from Jacobs University Bremen. Finally, we thank the referee for a number of sharp comments.

2 Notation and Conventions

The ground field is $\mathbb{C}$. Our notation is mostly standard: if X is an algebraic variety (over $\mathbb{C}$), $\mathcal{O}_X$ denotes its structure sheaf, Ω_X^1 (respectively T_X) denotes the cotangent (resp. tangent) sheaf on X under the assumption that X is smooth etc. If F is a sheaf on X, its cohomologies are denoted by $H^i(F)$, $h^i(F) := \dim H^i(F)$, and $\chi(F)$ stands for the Euler characteristic of F. The Chern classes of F are denoted by $c_i(F)$. If $f : X \to Y$ is a morphism, f^* and f_* denote respectively the inverse and direct image functors of $\mathcal{O}$-modules. All vector bundles are assumed to have finite-rank. We denote the dual of a sheaf of $\mathcal{O}_X$-modules F (or that of a vector space) by the superscript $^\vee$. Furthermore, in what follows, for any ind-Grassmannian $\mathbf{X}$ defined by (1), no embedding φ_i is an isomorphism.

We fix a finite-dimensional space V and denote by X the Grassmannian $G(k; V)$ for $k < \dim V$. In the sequel we sometimes write $G(k; n)$ to indicate simply the dimension of V. Below, we will often consider (parts of) the following diagram of flag varieties:

$$
\begin{array}{ccc}
Z := Fl(k-1, k, k+1; V) & \xrightarrow{\ \pi_2\ } & X := G(k; V) \\[1ex]
\ \ \downarrow{\scriptstyle \pi_1} & & \\[1ex]
Y := Fl(k-1, k+1; V) & \xrightarrow{\ p_2\ } & Y^2 := G(k+1; V) \\[1ex]
\ \ \downarrow{\scriptstyle p_1} & & \\[1ex]
Y^1 := G(k-1; V) & &
\end{array}
\tag{2}
$$

under the assumption that $k + 1 < \dim V$. Moreover, we reserve the letters X, Y, Z for the varieties in the above diagram. By S_k, S_{k-1}, S_{k+1} we denote the tautological bundles on X, Y, and Z, whenever they are defined (S_k is defined on X and Z, S_{k-1} is defined on Y^1, Y, and Z, etc.). By $\mathcal{O}_X(i)$, $i \in \mathbf{Z}$, we denote the isomorphism class (in the Picard group $\mathrm{Pic}\, X$) of the line bundle $(\Lambda^k(S_k^\vee))^{\otimes i}$, where Λ^k stands for the kth exterior power (in this case the maximal exterior power as $\mathrm{rk} S_k^\vee = k$). The Picard group of Y is isomorphic to the direct product of the Picard groups of Y^1 and Y^2, and by $\mathcal{O}_Y(i, j)$ we denote the isomorphism class of the line bundle $p_1^*(\Lambda^{k-1}(S_{k-1}^\vee))^{\otimes i} \otimes_{\mathcal{O}_Y} p_2^*(\Lambda^{k+1}(S_{k+1}^\vee))^{\otimes j}$.

If $\varphi : X = G(k; V) \to X' := G(k; V')$ is an embedding, then $\varphi^* \mathcal{O}_{X'}(1) \simeq \mathcal{O}_X(d)$ for some $d \in \mathbf{Z}_{\geq 0}$: by definition, d is the *degree* $\deg \varphi$ of φ. We say that φ is linear if $\deg \varphi = 1$. By a *projective subspace* (in particular a *line*, i.e., a 1-dimensional projective subspace) of X we mean a linearly embedded projective space into X. It is well known that all such are Schubert varieties of the form $\{V^k \in X | V^{k-1} \subset V^k \subset V^t\}$ or $\{V^k \in X | V^i \subset V^k \subset V^{k+1}\}$, where V^k is a variable k-dimensional subspace of V, and V^{k-1}, V^{k+1}, V^t, V^i are fixed subspaces of V of respective dimensions $k-1$, $k+1$, t, i. (Here and in what follows V^t always denotes a vector space of dimension t). In other words, all projective subspaces of X are of the form $G(1; V^t/V^{k-1})$ or $G(k-i, V^{k+1}/V^i)$. Note also that $Y = Fl(k-1, k+1; V)$ is the variety of lines in $X = G(k; V)$.

3 The Linear Case

We consider the cases of linear and twisted ind-Grassmannians separately. In the case of a linear ind-Grassmannian, we show that Conjecture 1.1 is a straightforward corollary of existing results combined with the following proposition. We recall [DP] that a *standard extension* of Grassmannians is an embedding of the form

$$G(k; V) \to G(k + a; V \oplus \hat{W}), \quad \{V^k \subset \mathbb{C}^n\} \mapsto \{V^k \oplus W \subset V \oplus \hat{W}\}, \quad (3)$$

where W is a fixed a-dimensional subspace of a finite-dimensional vector space $\hat{W}$.

Proposition 3.1. *Let* $\varphi : X = G(k; V) \to X' := G(k'; V')$ *be an embedding of degree 1. Then* φ *is a standard extension, or* φ *factors through a standard extension* $\mathbb{P}^r \to G(k'; V')$ *for some* r.

Proof. We assume that $k \leq n - k$, $k \leq n' - k'$, where $n = \dim V$ and $n' = \dim V'$, and use induction on k. For $k = 1$ the statement is obvious, since the image of φ is a projective subspace of $G(k'; V')$ and hence φ is a standard extension. Assume that the statement is true for $k - 1$. Since $\deg \varphi = 1$, φ induces an embedding $\varphi_Y : Y \to Y'$, where $Y = Fl(k - 1, k + 1; V)$ is the variety of lines in X and $Y' = Fl(k' - 1, k' + 1; V')$ is the variety of lines in X'. Moreover, clearly we have a commutative diagram of natural projections and embeddings

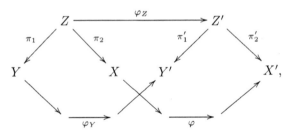

where $Z := Fl(k - 1, k, k + 1; V)$ and $Z' := Fl(k' - 1, k', k' + 1; V')$.

We claim that there is an isomorphism

$$\varphi_Y^* \mathcal{O}_{Y'}(1, 1) \simeq \mathcal{O}_Y(1, 1). \quad (4)$$

Indeed, $\varphi_Y^* \mathcal{O}_{Y'}(1, 1)$ is determined up to isomorphism by its restriction to the fibers of p_1 and p_2 (see diagram (2)), and therefore it is enough to check that

$$\varphi_Y^* \mathcal{O}_{Y'}(1, 1)_{|p_1^{-1}(V^{k-1})} \simeq \mathcal{O}_{p_1^{-1}(V^{k-1})}(1), \quad (5)$$

$$\varphi_Y^* \mathcal{O}_{Y'}(1, 1)_{|p_2^{-1}(V^{k+1})} \simeq \mathcal{O}_{p_2^{-1}(V^{k+1})}(1) \quad (6)$$

for some fixed subspaces $V^{k-1} \subset V$, $V^{k+1} \subset V$. Note that the restriction of φ to the projective subspace $G(1; V/V^{k-1}) \subset X$ is simply an isomorphism of

$G(1; V/V^{k-1})$ with a projective subspace of X'; hence the map induced by φ on the variety $G(2; V/V^{k-1})$ of projective lines in $G(1; V/V^{k-1})$ is an isomorphism with the Grassmannian of 2-dimensional subspaces of an appropriate subquotient of V'. Note furthermore that $p_1^{-1}(V^{k-1})$ is nothing but the variety of lines $G(2; V/V^{k-1})$ in $G(1; V/V^{k-1})$, and that the image of $G(2; V/V^{k-1})$ under φ is nothing but $\varphi_Y(p_1^{-1}(V^{k-1}))$. This shows that the restriction of $\varphi_Y^* \mathcal{O}_{Y'}(1,1)$ to $G(2; V/V^{k-1})$ is isomorphic to the restriction of $\mathcal{O}_Y(1,1)$ to $G(2; V/V^{k-1})$, and we obtain (5). The isomorphism (6) follows from a very similar argument.

The isomorphism (4) leaves us with two alternatives:

$$\varphi_Y^* \mathcal{O}_{Y'}(1,0) \simeq \mathcal{O}_Y \text{ or } \varphi_Y^* \mathcal{O}_{Y'}(0,1) \simeq \mathcal{O}_Y \tag{7}$$

or

$$\varphi_Y^* \mathcal{O}_{Y'}(1,0) \simeq \mathcal{O}_Y(1,0) \text{ or } \varphi_Y^* \mathcal{O}_{Y'}(1,0) \simeq \mathcal{O}_Y(0,1). \tag{8}$$

Let (7) hold; more precisely, let $\varphi_Y^* \mathcal{O}_{Y'}(1,0) \simeq \mathcal{O}_Y$. Then φ_Y maps each fiber of p_2 into a single point in Y' (depending on the image in Y^2 of this fiber), say $((V')^{k'-1} \subset (V')^{k'+1})$, and moreover, the space $(V')^{k'-1}$ is constant. Thus φ maps X into the projective subspace $G(1; V'/(V')^{k'-1})$ of X'. If $\varphi_Y^* \mathcal{O}_{Y'}(0,1) \simeq \mathcal{O}_Y$, then φ maps X into the projective subspace $G(1; (V')^{k'+1})$ of X'. Therefore, the proposition is proved in the case that (7) holds.

We assume now that (8) holds. It is easy to see that (8) implies that φ induces a linear embedding φ_{Y^1} of $Y^1 := G(k-1; V)$ into $G(k'-1; V')$ or $G(k'+1; V')$. Assume that $\varphi_{Y^1} : Y^1 \to (Y')^1 := G(k'-1; V')$ (the other case is completely similar). Then, by the induction assumption, φ_{Y^1} is a standard extension or factors through a standard extension $\mathbb{P}^r \to (Y')^1$. If φ_{Y^1} is a standard extension corresponding to a fixed subspace $W \subset \hat{W}$, then $\varphi_{Y^1}^* S_{k'-1} \simeq S_{k-1} \oplus (W \otimes_{\mathbb{C}} \mathcal{O}_{Y^1})$ and we have a vector bundle monomorphism

$$0 \to \pi_1^* p_1^* \varphi_{Y^1}^* S_{k'-1} \to \pi_2^* \varphi^* S_{k'}. \tag{9}$$

By restricting (9) to the fibers of π_1 we see that the quotient line bundle $\pi_2^* \varphi^* S_{k'} / \pi_1^* p_1^* \varphi_{Y^1}^* S_{k'-1}$ is isomorphic to $S_k/S_{k-1} \otimes \pi_1^* p_1^* \mathcal{L}$, where $\mathcal{L}$ is a line bundle on Y^1. Applying π_{2*}, we obtain

$$0 \to W \otimes_{\mathbb{C}} \mathcal{O}_X \to \pi_{2*}(\pi_2^* \varphi^* S_{k'}) = \varphi^* S_{k'} \to \pi_{2*}((S_k/S_{k-1}) \otimes \pi_1^* p_1^* \mathcal{L}) \to 0. \tag{10}$$

Since $\mathrm{rk}\varphi^* S_{k'} = k'$ and $\dim W = k'-k$, we have $\mathrm{rk}\pi_{2*}((S_k/S_{k-1}) \otimes \pi_1^* p_1^* \mathcal{L}) = k$, which implies immediately that $\mathcal{L}$ is trivial. Hence (10) reduces to $0 \to W \otimes_{\mathbb{C}} \mathcal{O}_X \to \varphi^* S_{k'} \to S_k \to 0$, and thus

$$\varphi^* S_{k'} \simeq S_k \oplus (W \otimes_{\mathbb{C}} \mathcal{O}_X), \tag{11}$$

since there are no nontrivial extensions of S_k by a trivial bundle. Now (11) implies that φ is a standard extension.

It remains to consider the case that φ_{Y^1} maps Y^1 into a projective subspace $\mathbb{P}^s$ of $(Y')^1$. Then $\mathbb{P}^s$ is of the form $G(1; V'/(V')^{k'-2})$ for some $(V')^{k'-2} \subset V'$, or of the form $G(k'-1; (V')^{k'})$ for some $(V')^{k'} \subset V'$. The second case is clearly impossible because it would imply that φ maps X into the single point $(V')^{k'}$. Hence $\mathbb{P}^s = G(1; V'/(V')^{k'-2})$ and φ maps X into the Grassmannian $G(2; V'/(V')^{k'-2})$ in $G(k'; V')$. Let S_2' be the rank-2 tautological bundle on $G(2; V'/(V')^{k'-2})$. Then its restriction $S'' := \varphi^* S_2'$ to any line l in X is isomorphic to $\mathcal{O}_l \oplus \mathcal{O}_l(-1)$, and we claim that this implies one of two alternatives:

$$S'' \simeq \mathcal{O}_X \oplus \mathcal{O}_X(-1) \tag{12}$$

or

$$S'' \simeq S_2 \text{ and } k = 2, \text{ or } S'' \simeq (V \otimes_{\mathbb{C}} \mathcal{O}_X)/S_2 \text{ and } k = n - k = 2. \tag{13}$$

Let $k \geq 2$. The evaluation map $\pi_1^* \pi_{1*} \pi_2^* S'' \to \pi_2^* S''$ is a monomorphism of the line bundle $\pi_1^* \mathcal{L} := \pi_1^* \pi_{1*} \pi_2^* S''$ into $\pi_2^* S''$ (here $\mathcal{L} := \pi_{1*} \pi_2^* S''$). Restricting this monomorphism to the fibers of π_2, we see immediately that $\pi_1^* \mathcal{L}$ is trivial when restricted to those fibers and is hence trivial. Therefore $\mathcal{L}$ is trivial, i.e., $\pi_1^* \mathcal{L} = \mathcal{O}_Z$. Pushdown to X yields

$$0 \to \mathcal{O}_X \to S'' \to \mathcal{O}_X(-1) \to 0, \tag{14}$$

and hence (14) splits as $\mathrm{Ext}^1(\mathcal{O}_X(-1), \mathcal{O}_X) = 0$. Therefore (12) holds. For $k = 2$, there is an additional possibility for the above monomorphisms to be of the form $\pi_1^* \mathcal{O}_Y(-1, 0) \to \pi_2^* S$ (or of the form $\pi_1^* \mathcal{O}_Y(0, -1) \to \pi_2^* S$ if $n - k = 2$), which yields the option (13).

If (12) holds, φ maps X into an appropriate projective subspace of $G(2; V'/(V')^{k'-2})$, which is then a projective subspace of X', and if (13) holds, φ is a standard extension corresponding to a zero-dimensional space W. The proof is now complete. $\qquad \square$

We are ready now to prove the following theorem.

Theorem 3.2. *Conjecture 1.1 holds for any linear ind-Grassmannian* $\mathbf{X}$.

Proof. Assume that $\deg \varphi_m = 1$ for all m, and apply Proposition 3.1. If infinitely many φ_m's factor through respective projective subspaces, then $\mathbf{X}$ is isomorphic to $\mathbf{P}^\infty$ and the BVT theorem implies Conjecture 1.1. Otherwise, all φ_m's are standard extensions of the form (3). There are two alternatives: $\lim_{m \to \infty} k_m = \lim_{m \to \infty} (n_m - k_m) = \infty$, or one of the limits $\lim_{m \to \infty} k_m$ or $\lim_{m \to \infty} (n_m - k_m)$ equals l for some $l \in \mathbf{N}$. In the first case, the claim of Conjecture 1.1 is proved in [DP]: Theorem 4.2. In the second case $\mathbf{X}$ is isomorphic to $\mathbf{G}(l; \infty)$, and therefore Conjecture 1.1 is proved in this case by E. Sato in [Sa2]. $\qquad \square$

4 Auxiliary Results

In order to prove Conjecture 1.1 for rank-2 bundles $\mathbf{E}$ on a twisted ind-Grassmannian $\mathbf{X} = \varinjlim G(k_m; V^{n_m})$, we need to prove that the vector bundle $\mathbf{E} = \varprojlim E_m$ of rank 2 on $\mathbf{X}$ is trivial, i.e., that E_m is a trivial bundle on $G(k_m; V^{n_m})$ for each m. From this point on, we assume that none of the Grassmannians $G(k_m; V^{n_m})$ is a projective space, since for a twisted projective ind-space, Conjecture 1.1 is proved in [DP] for bundles of arbitrary rank r.

The following known proposition gives a useful triviality criterion for vector bundles of arbitrary rank on Grassmannians.

Proposition 4.1. *A vector bundle E on $X = G(k; n)$ is trivial iff its restriction $E_{|l}$ is trivial for every line l in $G(k; n)$, $l \in Y = Fl(k-1, k+1; n)$.*

Proof. We recall the proof given in [P]. It uses the well known fact that the Proposition holds for any projective space, [OSS, Theorem 3.2.1]. Let first $k = 2$, $n = 4$, i.e. $X = G(2; 4)$. Since E is linearly trivial, $\pi_2^* E$ is trivial along the fibers of π_1 (we refer here to diagram (2)). Moreover, $\pi_{1*}\pi_2^* E$ is trivial along the images of the fibers of π_2 in Y. These images are of the form $\mathbb{P}_1^1 \times \mathbb{P}_2^1$, where $\mathbb{P}_1^1$ (respectively $\mathbb{P}_2^1$) are lines in $Y^1 := G(1; 4)$ and $Y^2 := G(3; 4)$. The fiber of p_1 is filled by lines of the form $\mathbb{P}_2^1$, and thus $\pi_{1*}\pi_2^* E$ is linearly trivial and hence trivial along the fibers of p_1. Finally the lines of the form $\mathbb{P}_1^1$ fill Y^1, hence $p_{1*}\pi_{1*}\pi_2^* E$ is also a trivial bundle. This implies that $E = \pi_{2*}\pi_1^* p_1^*(p_{1*}\pi_{1*}\pi_2^* E)$ is also trivial.

The next case is the case when $k = 2$ and n is arbitrary, $n \geq 5$. Then the above argument goes through by induction on n since the fiber of p_1 is isomorphic to $G(2; n-1)$. The proof is completed by induction on k for $k \geq 3$: the base of p_1 is $G(k-1; n)$ and the fiber of p_1 is $G(2; n-1)$. $\square$

If $C \subset N$ is a smooth rational curve in an algebraic variety N and E is a vector bundle on N, then by a classical theorem of Grothendieck, $E_{|C}$ is isomorphic to $\bigoplus_i \mathcal{O}_C(d_i)$ for some $d_1 \geq d_2 \geq \cdots \geq d_{rkE}$. We call the ordered rkE-tuple $(d_1, \ldots, d_{rkE})$ *the splitting type* of $E_{|C}$ and denote it by $\mathbf{d}_E(C)$. If $N = X = G(k; n)$, then the lines on N are parametrized by points $l \in Y$, and we obtain a map

$$Y \to \mathbf{Z}^{rkE} : l \mapsto \mathbf{d}_E(l).$$

By semicontinuity (cf. [OSS, Ch.I, Lemma 3.2.2]), there is a dense open set $U_E \subset Y$ of lines with minimal splitting type with respect to the lexicographical ordering on $\mathbf{Z}^{rkE}$. Denote this minimal splitting type by $\mathbf{d}_E$. By definition, $U_E = \{l \in Y \mid \mathbf{d}_E(l) = \mathbf{d}_E\}$ is the set of *non-jumping* lines of E, and its complement $Y \setminus U_E$ is the proper closed set of *jumping* lines.

A coherent sheaf F over a smooth irreducible variety N is called *normal* if for every open set $U \subset N$ and every closed algebraic subset $A \subset U$ of codimension at least 2 the restriction map $F(U) \to F(U \setminus A)$ is surjective.

It is well known that, since N is smooth, hence normal, a normal torsion-free sheaf F on N is reflexive, i.e. $F^{\vee\vee} = F$. If in addition F has finite rank, then F is necessarily a line bundle (see [OSS, Ch.II, 1.1.12 and 1.1.15]).

Theorem 4.2. *Let E be a rank r vector bundle of splitting type $\mathbf{d}_E = (d_1, ..., d_r)$, $d_1 \geq ... \geq d_r$, on $X = G(k; n)$. If $d_s - d_{s+1} \geq 2$ for some $s < r$, then there is a normal subsheaf $F \subset E$ of rank s with the following properties: over the open set $\pi_2(\pi_1^{-1}(U_E)) \subset X$ the sheaf F is a subbundle of E, and for any $l \in U_E$*

$$F_{|l} \simeq \bigoplus_{i=1}^{s} \mathcal{O}_l(d_i).$$

Proof. It is similar to the proof of Theorem 2.1.4 in [OSS, Ch. II]. Consider the vector bundle $E' = E \otimes \mathcal{O}_X(-d_s)$ and the evaluation map $\Phi : \pi_1^* \pi_{1*} \pi_2^* E' \to \pi_2^* E'$. The definition of U_E implies that $\Phi_{|\pi_1^{-1}(U_E)}$ is a morphism of constant rank s and that its image $\mathrm{im}\Phi \subset \pi_2^* E'$ is a subbundle of rank s over $\pi_1^{-1}(U_E)$. Let $M := \pi_2^* E'/\mathrm{im}\Phi$, let $T(M)$ be the torsion subsheaf of M, and $F' := \ker(\pi_2^* E' \to M' := M/T(M))$. Consider the singular set $\mathrm{Sing}\, F'$ of the sheaf F' and set $A := Z \setminus \mathrm{Sing}\, F'$. By the above, A is an open subset of Z containing $\pi_1^{-1}(U_E)$, and $f = \pi_{2|A} : A \to B := \pi_2(A)$ is a submersion with connected fibers.

Next, take any point $l \in Y$ and put $L := \pi_1^{-1}(l)$. By definition, $L \simeq \mathbb{P}^1$, and we have

$$T_{Z/X}{}_{|L} \simeq \mathcal{O}_L(-1)^{\oplus(n-2)}, \tag{15}$$

where $T_{Z/X}$ is the relative tangent bundle of Z over X. The construction of the sheaves F' and M implies that for any $l \in U_E$: $F'^{\vee}_{|L} = \oplus_{i=1}^{s} \mathcal{O}_L(-d_i + d_s)$, $M'_{|L} = \oplus_{i=s+1}^{r} \mathcal{O}_L(d_i - d_s)$. This, together with (15) and the condition $d_s - d_{s+1} \geq 2$, immediately implies that $H^0(\Omega^1_{A/B} \otimes F'^{\vee} \otimes M'_{|L}) = 0$. Hence $H^0(\Omega^1_{A/B} \otimes F'^{\vee} \otimes M'_{|\pi_1^{-1}(U_E)}) = 0$, and thus, since $\pi_1^{-1}(U_E)$ is dense open in Z, $\mathrm{Hom}(T_{A/B}, \mathcal{H}om(F', M'_{|A})) = H^0(\Omega^1_{A/B} \otimes F'^{\vee} \otimes M'_{|A}) = 0$. Now we apply the descent lemma (see [OSS, Ch. II, Lemma 2.1.3]) to the data $(f_{|\pi_1^{-1}(U_E)} : \pi_1^{-1}(U_E) \to V_E, \; F'_{|\pi_1^{-1}(U_E)} \subset E'_{|\pi_1^{-1}(U_E)})$. Then $F := (\pi_{2*}F') \otimes \mathcal{O}_X(-d_s)$ is the desired sheaf. $\qquad\square$

5 The Case rkE = 2

In the following, considering a twisted ind-Grassmannian $\mathbf{X} = \varinjlim G(k_m; V^{n_m})$, we set $G(k_m; V^{n_m}) = X_m$. Theorem 4.2 now yields the following corollary.

Corollary 5.1. *Let $\mathbf{E} = \varprojlim E_m$ be a rank-2 vector bundle on a twisted ind-Grassmannian $\mathbf{X} = \varinjlim X_m$. Then there exists $m_0 \geq 1$ such that $\mathbf{d}_{E_m} = (0,0)$ for any $m \geq m_0$.*

Proof. Note first that the fact that $\mathbf{X}$ is twisted implies

$$c_1(E_m) = 0, \; m \geq 1. \tag{16}$$

Indeed, $c_1(E_m)$ is nothing but the integer corresponding to the line bundle $\Lambda^2(E_m)$ in the identification of $\operatorname{Pic} X_m$ with $\mathbf{Z}$. Since $\mathbf{X}$ is twisted, $c_1(E_m) = \deg \varphi_m \deg \varphi_{m+1} \cdots \deg \varphi_{m+k} c_1(E_{m+k+1})$ for any $k \geq 1$; in other words, $c_1(E_m)$ is divisible by larger and larger integers and hence $c_1(E_m) = 0$ (cf. [DP, Lemma 3.2]). Suppose that for any $m_0 \geq 1$ there exists $m \geq m_0$ such that $\mathbf{d}_{E_m} = (a_m, -a_m)$ with $a_m > 0$. Then Theorem 4.2 applies to E_m with $s = 1$, and hence E_m has a normal rank-1 subsheaf F_m such that

$$F_{m|l} \simeq \mathcal{O}_l(a_m) \tag{17}$$

for a certain line l in X_m. Since F_m is a torsion-free normal subsheaf of the vector bundle E_m, the sheaf F_m is a line bundle, i.e., $F_m \simeq \mathcal{O}_{X_m}(a_m)$. Therefore we have a monomorphism

$$0 \to \mathcal{O}_{X_m}(a_m) \to E_m, \quad a_m \geq 1. \tag{18}$$

This is clearly impossible. In fact, this monomorphism implies in view of (16) that any rational curve $C \subset X_m$ of degree $\delta_m := \deg \varphi_1 \ldots \deg \varphi_{m-1}$ has splitting type $\mathbf{d}_{E_m}(C) = (a'_m, -a'_m)$, where $a'_m \geq a_m \delta_m \geq \delta_m$. Hence, by semicontinuity, any line $l \in X_1$ has splitting type $\mathbf{d}_{E_1}(l) = (b, -b)$, $b \geq \delta_m$. Since $\delta_m \to \infty$ as $m_0 \to \infty$, this is a contradiction. $\qquad\square$

We now recall some standard facts about the Chow rings of $X_m = G(k_m; V^{n_m})$ (see, e.g., [F, 14.7]):

(i) $A^1(X_m) = \operatorname{Pic}(X_m) = \mathbb{Z}[\mathbb{V}_m]$, $A^2(X_m) = \mathbb{Z}[\mathbb{W}_{1,m}] \oplus \mathbb{Z}[\mathbb{W}_{2,m}]$, where $\mathbb{V}_m, \mathbb{W}_{1,m}, \mathbb{W}_{2,m}$ are the following Schubert varieties: $\mathbb{V}_m := \{V^{k_m} \in X_m | \dim(V^{k_m} \cap V_0^{n_m-k_m}) \geq 1$ for a fixed subspace $V_0^{n_m-k_m-1}$ of $V^{n_m}\}$, $\mathbb{W}_{1,m} := \{V^{k_m} \in X_m | \dim(V^{k_m} \cap V_0^{n_m-k_m-1}) \geq 1$ for a fixed subspace $V_0^{n_m-k_m-1}$ in $V^{n_m}\}$, $\mathbb{W}_{2,m} := \{V^{k_m} \in X_m | \dim(V^{k_m} \cap V_0^{n_m-k_m+1}) \geq 2$ for a fixed subspace $V_0^{n_m-k_m+1}$ of $V^{n_m}\}$;

(ii) $[\mathbb{V}_m]^2 = [\mathbb{W}_{1,m}] + [\mathbb{W}_{2,m}]$ in $A^2(X_m)$;

(iii) $A_2(X_m) = \mathbb{Z}[\mathbb{P}_{1,m}^2] \oplus \mathbb{Z}[\mathbb{P}_{2,m}^2]$, where the projective planes $\mathbb{P}_{1,m}^2$ (called α-*planes*) and $\mathbb{P}_{2,m}^2$ (called β-*planes*) are respectively the Schubert varieties $\mathbb{P}_{1,m}^2 := \{V^{k_m} \in X_m | V_0^{k_m-1} \subset V^{k_m} \subset V_0^{k_m+2}$ for a fixed flag $V_0^{k_m-1} \subset V_0^{k_m+2}$ in $V^{n_m}\}$, $\mathbb{P}_{2,m}^2 := \{V^{k_m} \in X_m | V_0^{k_m-2} \subset V^{k_m} \subset V_0^{k_m+1}$ for a fixed flag $V_0^{k_m-2} \subset V_0^{k_m+1}$ in $V^{n_m}\}$;

(iv) the bases $[\mathbb{W}_{i,m}]$ and $[\mathbb{P}_{j,m}^2]$ are dual in the standard sense that $[\mathbb{W}_{i,m}] \cdot [\mathbb{P}_{j,m}^2] = \delta_{i,j}$.

Lemma 5.2. *There exists* $m_1 \in \mathbf{Z}_{>0}$ *such that for any* $m \geq m_1$ *one of the following holds:*

(1) $c_2(E_{m|\mathbb{P}^2_{1,m}}) > 0,\ c_2(E_{m|\mathbb{P}^2_{2,m}}) \leq 0,$
(2) $c_2(E_{m|\mathbb{P}^2_{2,m}}) > 0,\ c_2(E_{m|\mathbb{P}^2_{1,m}}) \leq 0,$
(3) $c_2(E_{m|\mathbb{P}^2_{1,m}}) = 0,\ c_2(E_{m|\mathbb{P}^2_{2,m}}) = 0.$

Proof. According to (i), for any $m \geq 1$ there exist $\lambda_{1m}, \lambda_{2m} \in \mathbf{Z}$ such that

$$c_2(E_m) = \lambda_{1m}[\mathbb{W}_{1,m}] + \lambda_{2m}[\mathbb{W}_{2,m}]. \tag{19}$$

Moreover, (iv) implies

$$\lambda_{jm} = c_2(E_{m|\mathbb{P}^2_{j,m}}), \quad j = 1, 2. \tag{20}$$

Next, (i) yields

$$\varphi_m^*[\mathbb{W}_{1,m+1}] = a_{11}(m)[\mathbb{W}_{1,m}] + a_{21}(m)[\mathbb{W}_{2,m}], \tag{21}$$
$$\varphi_m^*[\mathbb{W}_{2,m+1}] = a_{12}(m)[\mathbb{W}_{1,m}] + a_{22}(m)[\mathbb{W}_{2,m}], \tag{22}$$

where $a_{ij}(m) \in \mathbb{Z}$. Consider the 2×2 matrix $A(m) = (a_{ij}(m))$ and the column vector $\Lambda_m = (\lambda_{1m}, \lambda_{2m})^t$. Then, in view of (iv), the relation (21) gives $\Lambda_m = A(m)\Lambda_{m+1}$. Iterating this equation and denoting by $A(m, i)$ the 2×2 matrix $A(m) \cdot A(m+1) \cdots A(m+i)$, $i \geq 1$, we obtain

$$\Lambda_m = A(m, i)\Lambda_{m+i+1}. \tag{23}$$

The twisting condition $\varphi_m^*[\mathbb{V}_{m+1}] = \deg \varphi_m[\mathbb{V}_m]$ together with (ii) implies $\varphi_m^*([\mathbb{W}_{1,m+1}] + [\mathbb{W}_{2,m+1}]) = (\deg \varphi_m)^2([\mathbb{W}_{1,m}] + [\mathbb{W}_{2,m}])$. Substituting (21) into the last equality, we have $a_{11}(m) + a_{12}(m) = a_{21}(m) + a_{22}(m) = (\deg \varphi_m)^2, m \geq 1$. This means that the column vector $v = (1,1)^t$ is an eigenvector of $A(m)$ with eigenvalue $(\deg \varphi_m)^2$. Hence, it is an eigenvector of $A(m, i)$ with the eigenvalue $d_{m,i} = (\deg \varphi_m)^2(\deg \varphi_{m+1})^2 \cdots (\deg \varphi_{m+i})^2$:

$$A(m, i)v = d_{m,i}v. \tag{24}$$

Notice that the entries of $A(m)$, $m \geq 1$, are nonnegative integers (in fact, from the definition of the Schubert varieties $\mathbb{W}_{j,m+1}$ it immediately follows that $\varphi_m^*[\mathbb{W}_{j,m+1}]$ is an effective cycle on X_m, so that (21) and (iv) give $0 \leq \varphi_m^*[\mathbb{W}_{i,m+1}] \cdot [\mathbb{P}^2_{j,m}] = a_{ij}(m)$); hence also the entries of $A(m, i)$, $m, i \geq 1$, are nonnegative integers). Besides, clearly $d_{m,i} \to \infty$ as $i \to \infty$ for any $m \geq 1$. This, together with (23) and (24), implies that for $m \gg 1$, λ_{1m} and λ_{2m} cannot both be nonzero and have the same sign. This together with (20) is equivalent to the statement of the lemma. $\qquad \square$

In what follows we denote the α-planes and the β-planes on $X = G(2; 4)$ respectively by $\mathbb{P}^2_\alpha$ and $\mathbb{P}^2_\beta$.

Proposition 5.3. *There exists no rank-2 vector bundle E on the Grassmannian $X = G(2; 4)$ such that*

(a) $c_2(E) = a[\mathbb{P}^2_\alpha], \quad a > 0,$
(b) $E_{|\mathbb{P}^2_\beta}$ *is trivial for a generic β-plane $\mathbb{P}^2_\beta$ on X.*

Proof. Now assume that there exists a vector bundle E on X satisfying conditions (a) and (b) of the proposition. Fix a β-plane $P \subset X$ such that

$$E_{|P} \simeq \mathcal{O}_P^{\oplus 2}. \tag{25}$$

Since X is the Grassmannian of lines in $\mathbb{P}^3$, the plane P is the dual plane of a certain plane $\tilde{P}$ in $\mathbb{P}^3$. Next, fix a point $x_0 \in \mathbb{P}^3 \smallsetminus \tilde{P}$ and denote by S the variety of lines in $\mathbb{P}^3$ that contain x_0. Consider the variety $Q = \{(x, l) \in \mathbb{P}^3 \times X \mid x \in l \cap \tilde{P}\}$ with natural projections $p : Q \to S : (x, l) \mapsto \mathrm{Span}(x, x_0)$ and $\sigma : Q \to X : (x, l) \mapsto l$. Clearly, σ is the blowing up of X at the plane P, and the exceptional divisor $D_P = \sigma^{-1}(P)$ is isomorphic to the incidence subvariety of $P \times \tilde{P}$. Moreover, one easily checks that $Q \simeq \mathbb{P}(\mathcal{O}_S(1) \oplus T_S(-1))$, so that the projection $p : Q \to S$ coincides with the structure morphism $\mathbb{P}(\mathcal{O}_S(1) \oplus T_S(-1)) \to S$. Let $\mathcal{O}_Q(1)$ be the Grothendieck line bundle on Q such that $p_*\mathcal{O}_Q(1) = \mathcal{O}_S(1) \oplus T_S(-1)$. Using the Euler exact triple on Q,

$$0 \to \Omega^1_{Q/S} \to p^*(\mathcal{O}_S(1) \oplus T_S(-1)) \otimes \mathcal{O}_Q(-1) \to \mathcal{O}_Q \to 0, \tag{26}$$

we find the p-relative dualizing sheaf $\omega_{Q/S} := \det(\Omega^1_{Q/S})$:

$$\omega_{Q/S} \simeq \mathcal{O}_Q(-3) \otimes p^*\mathcal{O}_S(2). \tag{27}$$

Set $\mathcal{E} := \sigma^* E$. By construction, for each $y \in S$ the fiber $Q_y = p^{-1}(y)$ is a plane such that $l_y = Q_y \cap D_P$ is a line, and, by (25),

$$\mathcal{E}_{|l_y} \simeq \mathcal{O}_{l_y}^{\oplus 2}. \tag{28}$$

Furthermore, $\sigma(Q_y)$ is an α-plane in X, and from (28) it follows clearly that $h^0(\mathcal{E}_{|Q_y}(-1)) = \mathcal{E}^\vee_{|Q_y}(-1)) = 0$. Hence, in view of condition (a) of the proposition, the sheaf $\mathcal{E}_{|Q_y}$ is the cohomology sheaf of a monad

$$0 \to \mathcal{O}_{Q_y}(-1)^{\oplus a} \to \mathcal{O}_{Q_y}^{\oplus(2a+2)} \to \mathcal{O}_{Q_y}(1)^{\oplus a} \to 0 \tag{29}$$

(see [OSS, Ch. II, Ex. 3.2.3]). This monad immediately implies the equalities

$$h^1(\mathcal{E}_{|Q_y}(-1)) = h^1(\mathcal{E}_{|Q_y}(-2)) = a, \quad h^1(\mathcal{E}_{|Q_y} \otimes \Omega^1_{Q_y}) = 2a + 2, \tag{30}$$

$$h^i(\mathcal{E}_{|Q_y}(-1)) = h^i(\mathcal{E}_{|Q_y}(-2)) = h^i(\mathcal{E}_{|Q_y} \otimes \Omega^1_{Q_y}) = 0, \quad i \neq 1.$$

Consider the sheaves of $\mathcal{O}_S$-modules

$$E_{-1} := R^1 p_*(\mathcal{E} \otimes \mathcal{O}_Q(-2) \otimes p^* \mathcal{O}_S(2)), \tag{31}$$

$$E_0 := R^1 p_*(\mathcal{E} \otimes \Omega^1_{Q/S}), \tag{32}$$

$$E_1 := R^1 p_*(\mathcal{E} \otimes \mathcal{O}_Q(-1)). \tag{33}$$

The equalities (30) imply via base change [H] that E_{-1}, E_1, and E_0 are locally free $\mathcal{O}_S$-modules, and $\mathrm{rk}(E_{-1}) = \mathrm{rk}(E_1) = a$, and $\mathrm{rk}(E_0) = 2a+2$. Moreover,

$$R^i p_*(\mathcal{E} \otimes \mathcal{O}_Q(-2)) = R^i p_*(\mathcal{E} \otimes \Omega^1_{Q/S}) = R^i p_*(\mathcal{E} \otimes \mathcal{O}_Q(-1)) = 0 \tag{34}$$

for $i \neq 1$. Note that $\mathcal{E}^\vee \simeq \mathcal{E}$ as $c_1(\mathcal{E}) = 0$ and $\mathrm{rk}\mathcal{E} = 2$. Furthermore, (27) implies that the nondegenerate pairing (p-relative Serre duality) $R^1 p_*(\mathcal{E} \otimes \mathcal{O}_Q(-1)) \otimes R^1 p_*(\mathcal{E}^\vee \otimes \mathcal{O}_Q(1) \otimes \omega_{Q/S}) \to R^2 p_* \omega_{Q/S} = \mathcal{O}_S$ can be rewritten as $E_1 \otimes E_{-1} \to \mathcal{O}_S$, thus giving an isomorphism

$$E_{-1} \simeq E_1^\vee. \tag{35}$$

Similarly, since $\mathcal{E}^\vee \simeq \mathcal{E}$ and $\Omega^1_{Q/S} \simeq T_{Q/S} \otimes \omega_{Q/S}$, p-relative Serre duality yields a nondegenerate pairing $E_0 \otimes E_0 = R^1 p_*(\mathcal{E} \otimes \Omega^1_{Q/S}) \otimes R^1 p_*(\mathcal{E} \otimes \Omega^1_{Q/S}) = R^1 p_*(\mathcal{E} \otimes \Omega^1_{Q/S}) \otimes R^1 p_*(\mathcal{E}^\vee \otimes T_{Q/S} \otimes \omega_{Q/S}) \to R^2 p_* \omega_{Q/S} = \mathcal{O}_S$. Therefore E_0 is self-dual, i.e., $E_0 \simeq E_0^\vee$, and in particular, $c_1(E_0) = 0$.

Now let J denote the fiber product $Q \times_S Q$ with projections $Q \xleftarrow{pr_1} J \xrightarrow{pr_2} Q$ such that $p \circ pr_1 = p \circ pr_2$. Put $F_1 \boxtimes F_2 := pr_1^* F_1 \otimes pr_2^* F_2$ for sheaves F_1 and F_2 on Q, and consider the standard $\mathcal{O}_J$-resolution of the structure sheaf $\mathcal{O}_\Delta$ of the diagonal $\Delta \hookrightarrow J$,

$$0 \to \mathcal{O}_Q(-1) \otimes p^* \mathcal{O}_S(2) \boxtimes \mathcal{O}_Q(-2) \to \Omega^1_{Q/S}(1) \boxtimes \mathcal{O}_Q(-1) \to \mathcal{O}_J \to \mathcal{O}_\Delta \to 0. \tag{36}$$

Twist this sequence by the sheaf $(\mathcal{E} \otimes \mathcal{O}_Q(-1)) \boxtimes \mathcal{O}_Q(1)$ and apply the functor $R^i pr_{2*}$ to the resulting sequence. In view of (31) and (34) we obtain the following monad for $\mathcal{E}$:

$$0 \to p^* E_{-1} \otimes \mathcal{O}_Q(-1) \xrightarrow{\lambda} p^* E_0 \xrightarrow{\mu} p^* E_1 \otimes \mathcal{O}_Q(1) \to 0, \qquad \ker(\mu)/\mathrm{im}(\lambda) = \mathcal{E}. \tag{37}$$

Put $R := p^* h$, where h is the class of a line in S. Furthermore, set $H := \sigma^* H_X$, $[\mathbb{P}_\alpha] := \sigma^*[\mathbb{P}^2_\alpha]$, $[\mathbb{P}_\beta] := \sigma^*[\mathbb{P}^2_\beta]$, where H_X is the class of a hyperplane section of X (via the Plücker embedding), and respectively, $[\mathbb{P}^2_\alpha]$ and $[\mathbb{P}^2_\beta]$ are the classes of an α- and β-plane. Note that clearly, $\mathcal{O}_Q(H) \simeq \mathcal{O}_Q(1)$. Thus, taking into account the duality (35), we rewrite the monad (37) as

$$0 \to p^* E_1^\vee \otimes \mathcal{O}_Q(-H) \xrightarrow{\lambda} p^* E_0 \xrightarrow{\mu} p^* E_1 \otimes \mathcal{O}_Q(H) \to 0, \quad \ker(\mu)/\mathrm{im}(\lambda) \simeq \mathcal{E}. \tag{38}$$

In particular, it becomes clear that (37) is a relative version of the monad (29).

As a next step, we are going to express all Chern classes of the sheaves in (38) in terms of a. We start by writing down the Chern polynomials of the bundles $p^*E_1 \otimes \mathcal{O}_Q(H)$ and $p^*E_1^\vee \otimes \mathcal{O}_Q(-H)$ in the form

$$c_t(p^*E_1 \otimes \mathcal{O}_Q(H)) = \prod_{i=1}^a (1 + (\delta_i + H)t), \quad c_t(p^*E_1^\vee \otimes \mathcal{O}_Q(-H)) = \prod_{i=1}^a (1 - (\delta_i + H)t),$$

$$(39)$$

where δ_i are the Chern roots of the bundle p^*E_1. Thus

$$cR^2 = \sum_{i=1}^a \delta_i^2, \quad dR = \sum_{i=1}^a \delta_i \tag{40}$$

for some $c, d \in \mathbb{Z}$. Next we invoke the following easily verified relations in $A^\cdot(Q)$:

$$H^4 = RH^3 = 2[pt], \tag{41}$$

$$R^2H^2 = R^2[\mathbb{P}_\alpha] = RH[\mathbb{P}_\alpha] = H^2[\mathbb{P}_\alpha] = RH[\mathbb{P}_\beta] = H^2[\mathbb{P}_\beta] = [pt], \tag{42}$$

$$[\mathbb{P}_\alpha][\mathbb{P}_\beta] = R^2[\mathbb{P}_\beta] = R^4 = R^3H = 0,$$

where $[pt]$ is the class of a point. This, together with (40), gives

$$\sum_{1 \le i < j \le a} \delta_i^2 \delta_j^2 = \sum_{1 \le i < j \le a} (\delta_i^2 \delta_j + \delta_i \delta_j^2)H = 0, \tag{43}$$

$$\sum_{1 \le i < j \le a} \delta_i \delta_j H^2 = \frac{1}{2}(d^2 - c)[pt], \tag{44}$$

$$\sum_{1 \le i \le a} (\delta_i + \delta_j)H^3 = 2(a-1)d[pt]. \tag{45}$$

Note that since $c_1(E_0) = 0$,

$$c_t(p^*E_0) = 1 + bR^2t^2 \tag{46}$$

for some $b \in \mathbb{Z}$. Furthermore,

$$c_t(\mathcal{E}) = 1 + a[\mathbb{P}_\alpha]t^2 \tag{47}$$

by the hypotheses of the proposition. Substituting (46) and (47) into the polynomial $f(t) := c_t(\mathcal{E})c_t(p^*E_1 \otimes \mathcal{O}_Q(H))c_t(p^*E_1^\vee \otimes \mathcal{O}_Q(-H))$, we have $f(t) = (1 + a[\mathbb{P}_\alpha]t^2)\prod_{i=1}^a (1 - (\delta_i + H)^2t^2)$. Expanding $f(t)$ in t and using (40)–(43), we obtain

$$f(t) = 1 + (a[\mathbb{P}_\alpha] - cR^2 - 2dRH - aH^2)t^2 + e[pt]t^4, \tag{48}$$

where

$$e = -3c - a(2d + a) + (a-1)(a + 4d) + 2d^2. \tag{49}$$

Next, the monad (38) implies $f(t) = c_t(p^*E_0)$. A comparison of (48) with (46) yields

$$c_2(\mathcal{E}) = a[\mathbb{P}_\alpha] = (b+c)R^2 + 2dRH + aH^2, \tag{50}$$

$$e = c_4(p^*E_0) = 0. \tag{51}$$

The relation (51) is the crucial relation that enables us to express the Chern classes of all sheaves in (38) just in terms of a. More precisely, (50) and (41) give $0 = c_2(\mathcal{E})[\mathbb{P}_\beta] = 2d + a$, hence $a = -2d$. Substituting these latter equalities into (49), we get $e = -a(a-2)/2 - 3c$. Hence $c = -a(a-2)/6$ by (51). Since $a = -2d$, (40) and the equality $c = -a(a-2)/6$ give $c_1(E_1) = -a/2$, $c_2(E_1) = (d^2 - c)/2 = a(5a-4)/24$. Substituting this into the standard formulas $e_k := c_k(p^*E_1 \otimes \mathcal{O}_Q(H)) = \sum_{i=0}^{2} \binom{a-i}{k-i} R^i H^{k-i} c_i(E_1)$, $1 \le k \le 4$, we obtain

$$e_1 = -aR/2 + aH, \quad e_2 = (5a^2/24 - a/6)R^2 + (a^2 - a)(-RH + H^2)/2, \tag{52}$$

$$e_3 = (5a^3/24 - 7a^2/12 + a/3)R^2H + (-a^3/4 + 3a^2/4 - a/2)RH^2$$
$$+ (a^3/6 - a^2/2 + a/3)H^3,$$

$$e_4 = (-7a^4/144 + 43a^3/144 - 41a^2/72 + a/3)[pt].$$

It remains to write down explicitly $c_2(p^*E_0)$: (41), (50), and the relations $a = -2d$, $c = -a(a-2)/6$ give $a = c_2(\mathcal{E})[\mathbb{P}_\alpha] = b + c$, hence

$$c_2(E_0) = b = (a^2 + 4a)/6 \tag{53}$$

by (46).

Our next and final step will be to obtain a contradiction by computing the Euler characteristic of the sheaf $\mathcal{E}$ in two different ways. We first compute the Todd class $td(T_Q)$ of the bundle T_Q. From the exact triple dual to (26) we obtain $c_t(T_{Q/S}) = 1 + (-2R + 3H)t + (2R^2 - 4RH + 3H^2)t^2$. Next, $c_t(T_Q) = c_t(T_{Q/S})c_t(p^*T_S)$. Hence $c_1(T_Q) = R + 3H$, $c_2(T_Q) = -R^2 + 5RH + 3H^2$, $c_3(T_Q) = -3R^2H + 9H^2R$, $c_4(T_Q) = 9[pt]$. Substituting into the formula for the Todd class of T_Q, $td(T_Q) = 1 + \frac{1}{2}c_1 + \frac{1}{12}(c_1^2 + c_2) + \frac{1}{24}c_1c_2 - \frac{1}{720}(c_1^4 - 4c_1^2c_2 - 3c_2^2 - c_1c_3 + c_4)$, where $c_i := c_i(T_Q)$ (see, e.g., [H, p. 432]), we get

$$td(T_Q) = 1 + \frac{1}{2}R + \frac{3}{2}H + \frac{11}{12}RH + H^2 + \frac{1}{12}HR^2 + \frac{3}{4}H^2R + \frac{3}{8}H^3 + [pt]. \tag{54}$$

Next, by the hypotheses of the proposition, $c_1(\mathcal{E}) = 0$, $c_2(\mathcal{E}) = a[\mathbb{P}_\alpha]$, $c_3(\mathcal{E}) = c_4(\mathcal{E}) = 0$. Substituting this into the general formula for the Chern character of a vector bundle F yields

$$ch(F) = rk(F) + c_1 + (c_1^2 - 2c_2)/2 + (c_1^3 - 3c_1c_2 - 3c_3)/6$$
$$+ (c_1^4 - 4c_1^2c_2 + 4c_1c_3 + 2c_2^2 - 4c_4)/24,$$

$c_i := c_i(F)$ (see, e.g., [H, p. 432]), and using (54), we obtain by the Riemann–Roch theorem for $F = \mathcal{E}$,

$$\chi(\mathcal{E}) = \frac{1}{12}a^2 - \frac{23}{12}a + 2. \tag{55}$$

In a similar way, using (52), we obtain

$$\chi(p^* E_1 \otimes \mathcal{O}_Q(H)) + \chi(p^* E_1^\vee \otimes \mathcal{O}_Q(-H)) = \frac{5}{216}a^4 - \frac{29}{216}a^3 - \frac{1}{54}a^2 + \frac{113}{36}a. \tag{56}$$

Next, in view of (53) and the equality $c_1(E_0) = 0$, the Riemann–Roch theorem for E_0 easily gives

$$\chi(p^* E_0) = \chi(E_0) = -\frac{1}{6}a^2 + \frac{4}{3}a + 2. \tag{57}$$

Together with (55) and (56) this yields

$$\Phi(a) := \chi(p^* E_0) - (\chi(\mathcal{E}) + \chi(p^* E_1 \otimes \mathcal{O}_Q(H)) + \chi(p^* E_1^\vee \otimes \mathcal{O}_Q(-H)))$$
$$= -\frac{5}{216}a(a-2)(a-3)\left(a - \frac{4}{5}\right).$$

The monad (38) now implies $\Phi(a) = 0$. The only positive integer roots of the polynomial $\Phi(a)$ are $a = 2$ and $a = 3$. However, (55) implies $\chi(\mathcal{E}) = -\frac{3}{2}$ for $a = 2$, and (57) implies $\chi(p^* E_0) = \frac{9}{2}$ for $a = 3$. This is a contradiction, since the values of $\chi(\mathcal{E})$ and $\chi(p^* E_0)$ are integers by definition.   □

We need a last piece of notation. Consider the flag variety $Fl(k_m - 2, k_m + 2; V^{n_m})$. Any point $u = (V^{k_m-2}, V^{k_m+2}) \in Fl(k_m-2, k_m+2; V^{n_m})$ determines a standard extension

$$i_u : X = G(2;4) \hookrightarrow X_m, \tag{58}$$

$$W^2 \mapsto V^{k_m-2} \oplus W^2 \subset V^{k_m+2} \subset V^{n_m} = V^{k_m-2} \oplus W^4 \subset V^{n_m}, \tag{59}$$

where $W^2 \in X = G(2; W^4)$ and an isomorphism $V^{k_m-2} \oplus W^4 \simeq V^{k_m+2}$ is fixed (clearly i_u does not depend on the choice of this isomorphism modulo $\mathrm{Aut}(X_m)$). We clearly have isomorphisms of Chow groups

$$i_u^* : A^2(X_m) \xrightarrow{\sim} A^2(X), \quad i_{u*} : A_2(X) \xrightarrow{\sim} A_2(X_m), \tag{60}$$

and the flag variety $Y_m := Fl(k_m - 1, k_m + 1; V^{n_m})$ (respectively, $Y := Fl(1,3;4)$) is the set of lines in X_m (respectively, in X).

Theorem 5.4. *Let* $\mathbf{X} = \varinjlim X_m$ *be a twisted ind-Grassmannian. Then any vector bundle* $\mathbf{E} = \varprojlim E_m$ *on* $\mathbf{X}$ *of rank 2 is trivial, and hence Conjecture 1.1(iv) holds for vector bundles of rank 2.*

Proof. Fix $m \geq \max\{m_0, m_1\}$, where m_0 and m_1 are as in Corollary 5.1 and Lemma 5.2. For $j = 1, 2$, let $E^{(j)}$ denote the restriction of E_m to a projective plane of type $\mathbb{P}^2_{j,m}$, let $T^j \simeq Fl(k_m - j, k_m + 3 - j, V^{n_m})$ be the variety of planes of the form $\mathbb{P}^2_{j,m}$ in X_m, and suppose that $\varPi^j := \left\{ \mathbb{P}^2_{j,m} \in T^j \mid E_{m|\mathbb{P}^2_{j,m}} \right.$ is properly unstable (i.e., not semistable)$\left.\right\}$. Since semistability is an open condition, $\varPi^j$ is a closed subset of T^j.

(i) Assume that $c_2(E^{(1)}) > 0$. Then, since $m \geq m_1$, Lemma 5.2 implies $c_2(E^{(2)}) \leq 0$.

(i.1) Suppose that $c_2(E^{(2)}) = 0$. If $\varPi^2 \neq T^2$, then for any $\mathbb{P}^2_{2,m} \in T^2 \smallsetminus \varPi^2$ the corresponding bundle $E^{(2)}$ is semistable; hence $E^{(2)}$ is trivial, since $c_2(E^{(2)}) = 0$; see [DL, Prop. 2.3,(4)]. Thus, for a generic point $u \in Fl(k_m - 2, k_m + 2; V^{n_m})$, the bundle $E = i_u^* E_m$ on $X = G(2; 4)$ satisfies the conditions of Proposition 5.3, which is a contradiction.

We therefore assume $\varPi^2 = T^2$. Then for any $\mathbb{P}^2_{2,m} \in T^2$ the corresponding bundle $E^{(2)}$ has a maximal destabilizing subsheaf $0 \to \mathcal{O}_{\mathbb{P}^2_{2,m}}(a) \to E^{(2)}$. Moreover, $a > 0$. In fact, otherwise the condition $c_2(E^{(2)}) = 0$ would imply that $a = 0$ and $E^{(2)}/\mathcal{O}_{\mathbb{P}^2_{2,m}} = \mathcal{O}_{\mathbb{P}^2_{2,m}}$, i.e., $E^{(2)}$ would be trivial, in particular semistable. Hence

$$\mathbf{d}_{E^{(2)}} = (a, -a). \tag{61}$$

Since any line in X_m is contained in a plane $\mathbb{P}^2_{2,m} \in T^2$, (61) implies $\mathbf{d}_{E_m} = (a, -a)$ with $a > 0$ for $m > m_0$, contrary to Corollary 5.1.

(i.2) Assume $c_2(E^{(2)}) < 0$. Since $E^{(2)}$ is not stable for any $\mathbb{P}^2_{2,m} \in T^2$, its maximal destabilizing subsheaf $0 \to \mathcal{O}_{\mathbb{P}^2_{2,m}}(a) \to E^{(2)}$ clearly satisfies the condition $a > 0$, i.e., $E^{(2)}$ is properly unstable; hence $\varPi^2 = T^2$. Then we again obtain a contradiction as above.

(ii) Now we assume that $c_2(E^{(2)}) > 0$. Then, replacing $E^{(2)}$ by $E^{(1)}$ and vice versa, we arrive at a contradiction by the same argument as in case (i).

(iii) We must therefore assume $c_2(E^{(1)}) = c_2(E^{(2)}) = 0$. Set

$$D(E_m) := \{l \in Y_m \mid \mathbf{d}_{E_m}(l) \neq (0,0)\}$$

and

$$D(E) := \{l \in Y \mid \mathbf{d}_E(l) \neq (0,0)\}.$$

By Corollary 5.1, $\mathbf{d}_{E_m} = (0,0)$; hence $\mathbf{d}_E = (0,0)$ for a generic embedding $i_u : X \hookrightarrow X_m$. Then by deformation theory [B], $D(E_m)$ (resp., $D(E)$) is an effective divisor on Y_m (resp., on Y). Thus, $\mathcal{O}_Y(D(E)) = p_1^* \mathcal{O}_{Y^1}(a) \otimes p_2^* \mathcal{O}_{Y^2}(b)$ for some $a, b \geq 0$, where p_1, p_2 are as in diagram (2). Note that each fiber of p_1 (respectively, of p_2) is a plane $\tilde{\mathbb{P}}^2$ dual to some α-plane $\mathbb{P}^2_\alpha$ (respectively, a plane $\tilde{\mathbb{P}}^2_\beta$ dual to some β-plane $\mathbb{P}^2_\beta$). Thus, setting

$$D(E_{|\mathbb{P}^2}) := \{l \in \tilde{\mathbb{P}}^2_\alpha \mid \mathbf{d}_E(l) \neq (0,0)\},$$

$$D(E_{|\mathbb{P}^2}) := \{l \in \tilde{\mathbb{P}}^2_\beta \mid \mathbf{d}_E(l) \neq (0,0)\},$$

we obtain

$$\mathcal{O}_{\check{\mathbb{P}}_\alpha^2}(D(E_{|\mathbb{P}_\alpha^2})) = \mathcal{O}_Y(D(E))_{|\check{\mathbb{P}}_\alpha^2} = \mathcal{O}_{\check{\mathbb{P}}_\alpha^2}(b),$$

$$\mathcal{O}_{\check{\mathbb{P}}_\beta^2}(D(E_{|\mathbb{P}_\beta^2})) = \mathcal{O}_Y(D(E))_{|\check{\mathbb{P}}_\beta^2} = \mathcal{O}_{\check{\mathbb{P}}_\beta^2}(a).$$

Now if $E_{|\mathbb{P}_\alpha^2}$ is semistable, a theorem of Barth [OSS, Ch. II, Theorem 2.2.3] implies that $D(E_{|\mathbb{P}_\alpha^2})$ is a divisor of degree $c_2(E_{|\mathbb{P}_\alpha^2}) = a$ on $\mathbb{P}_\alpha^2$. Hence $a = c_2(E^{(1)}) = 0$ for a semistable $E_{|\mathbb{P}_\alpha^2}$. If $E_{|\mathbb{P}_\alpha^2}$ is not semistable, it is unstable, and the equality $\mathbf{d}_E(l) = (0,0)$ yields $\mathbf{d}_{E_{|\mathbb{P}_\alpha^2}} = (0,0)$. Then the maximal destabilizing subsheaf of $E_{|\mathbb{P}_\alpha^2}$ is isomorphic to $\mathcal{O}_{\mathbb{P}_\alpha^2}$, and since $c_2(E_{|\mathbb{P}_\alpha^2}) = 0$, we obtain an exact triple $0 \to \mathcal{O}_{\mathbb{P}_\alpha^2} \to E_{|\mathbb{P}_\alpha^2} \to \mathcal{O}_{\mathbb{P}_\alpha^2} \to 0$, so that $E_{|\mathbb{P}_\alpha^2} \simeq \mathcal{O}_{\mathbb{P}_\alpha^2}^{\oplus 2}$ is semistable, a contradiction. This shows that $a = 0$ whenever $c_2(E^{(1)}) = c_2(E^{(2)}) = 0$. Similarly, $b = 0$. Therefore $D(E_m) = \emptyset$, and Proposition 4.1 implies that E_m is trivial. Therefore $\mathbf{E}$ is trivial as well.   □

In [DP], Conjecture 1.1 (iv) was proved not only when $\mathbf{X}$ is a twisted projective ind-space, but also for finite-rank bundles on special twisted ind-Grassmannians defined through certain homogeneous embeddings φ_m. These include embeddings of the form

$$G(k; n) \to G(ka; nb),$$

$$V^k \subset V \mapsto V^k \otimes W^a \subset V \otimes W^b,$$

where $W^a \subset W^b$ is a fixed pair of finite-dimensional spaces with $a < b$, or of the form

$$G(k; n) \to G\left(\frac{k(k+1)}{2}; n^2\right),$$

$$V^k \subset V \mapsto S^2(V^k) \subset V \otimes V.$$

More precisely, Conjecture 1.1 (iv) was proved in [DP] for twisted ind-Grassmannians whose defining embeddings are homogeneous embeddings satisfying some specific numerical conditions relating the degrees $\deg \varphi_m$ to the pairs of integers (k_m, n_m). There are many twisted ind-Grassmannians for which those conditions are not satisfied. For instance, this applies to the ind-Grassmannians defined by iterating each of the following embeddings:

$$G(k; n) \to G\left(\frac{k(k+1)}{2}; \frac{n(n+1)}{2}\right),$$

$$V^k \subset V \mapsto S^2(V^k) \subset S^2(V),$$

$$G(k; n) \to G\left(\frac{k(k-1)}{2}; \frac{n(n-1)}{2}\right),$$

$$V^k \subset V \mapsto \Lambda^2(V^k) \subset \Lambda^2(V).$$

Therefore the resulting ind-Grassmannians $\mathbf{G}(k, n, S^2)$ and $\mathbf{G}(k, n, \Lambda^2)$ are examples of twisted ind-Grassmannians for which Theorem 5.4 is new.

References

[BV] W. BARTH, A. VAN DE VEN, *On the geometry in codimension 2 in Grassmann manifolds*, Lecture Notes in Math. **412** (1974), 1–35.

[B]   E. BRIESKORN, *Über holomorphe $\mathbb{P}_n$-bündel über $\mathbb{P}_1$*, Math. Ann. **157** (1967), 351–382.

[CT] I. COANDĂ, G. TRAUTMANN, *The splitting criterion of Kempf and the Babylonian tower theorem*, Comm. Algebra **34** (2006), 2485–2488.

[DiP] I. DIMITROV, I. PENKOV, *Ind-varieties of generalized flags as homogeneous spaces for classical ind-groups*, IMRN 2004, no 55, 2935–2953.

[DL] J. M. DREZET, J. LE POTIER, *Fibrés stables et fibrés exceptionnels sur $\mathbb{P}_2$*, Ann. Sci. Ec. Norm. Supér. IV. Sér. **18** (1985), 193–243.

[DP] J. DONIN, I. PENKOV, *Finite rank vector bundles on inductive limits of Grassmannians*, IMRN 2003, no 34, 1871–1887.

[DPW] I. DIMITROV, I. PENKOV, J. A. WOLF, *A Bott-Borel-Weil theorem for direct limits of algebraic groups.* Amer. J. of Math. **124** (2002), 955–998.

[F]   W. FULTON, *Intersection Theory,* Springer, Berlin, 1998.

[H]   R. HARTSHORNE, *Algebraic Geometry,* Springer, New York, 1977.

[OSS] C. OKONEK, M. SCHNEIDER, H. SPINDLER, *Vector Bundles on Complex Projective Spaces*, Birkhäuser, 1980.

[P]   I. PENKOV, *The Penrose transform on general Grassmannians*, C. R. Acad. Bulg. des Sci. **22** (1980), 1439–1442.

[Sa1] E. SATO, *On the decomposability of infinitely extendable vector bundles on projective spaces and Grassmann varieties,* J. Math. Kyoto Univ. **17** (1977), 127–150.

[Sa2] E. SATO, *On infinitely extendable vector bundles on G/P*, J. Math. Kyoto Univ. **19** (1979), 171–189.

[T]   A. N. TYURIN, *Vector bundles of finite-rank over infinite varieties*, Math. USSR Izvestija **10** (1976), 1187–1204.

Massey Products on Cycles of Projective Lines and Trigonometric Solutions of the Yang–Baxter Equations

A. Polishchuk

Department of Mathematics, University of Oregon, Eugene, OR 97403, USA
`apolish@uoregon.edu`

To Yuri Ivanovich Manin on his 70th birthday

Summary. We show that a nondegenerate unitary solution $r(u, v)$ of the associative Yang–Baxter equation (AYBE) for $\mathrm{Mat}(N, \mathbb{C})$ (see [7]) with the Laurent series at $u = 0$ of the form $r(u, v) = \frac{1 \otimes 1}{u} + r_0(v) + \cdots$ satisfies the quantum Yang–Baxter equation, provided the projection of $r_0(v)$ to $\mathrm{sl}_N \otimes \mathrm{sl}_N$ has a period. We classify all such solutions of the AYBE, extending the work of Schedler [8]. We also characterize solutions coming from triple Massey products in the derived category of coherent sheaves on cycles of projective lines.

Key words: Associative Yang–Baxter equation, Quantum Yang–Baxter equation, Belavin–Drinfeld triple, simple vector bundle

2000 Mathematics Subject Classifications: 16W30, 14F05, 18E30

Introduction

This paper is concerned with solutions of the associative Yang–Baxter equation (AYBE)

$$r^{12}(-u', v)r^{13}(u+u', v+v') - r^{23}(u+u', v')r^{12}(u, v) + r^{13}(u, v+v')r^{23}(u', v') = 0,$$
(1)

where $r(u, v)$ is a meromorphic function of two complex variables (u, v) in a neighborhood of $(0, 0)$ taking values in $A \otimes A$, where $A = \mathrm{Mat}(N, \mathbb{C})$ is the matrix algebra. Here we use the notation $r^{12} = r \otimes 1 \in A \otimes A \otimes A$, etc. We will refer to a solution of (1) as an *associative r-matrix*. This equation was introduced in the above form in [7] in connection with triple Massey products for simple vector bundles on elliptic curves and their degenerations. It is usually coupled with the *unitarity* condition

$$r^{21}(-u, -v) = -r(u, v).$$
(2)

Y. Tschinkel and Y. Zarhin (eds.), *Algebra, Arithmetic, and Geometry*,
Progress in Mathematics 270, DOI 10.1007/978-0-8176-4747-6_19,
© Springer Science+Business Media, LLC 2009

Note that the constant version of (1) was independently introduced in [1] in connection with the notion of infinitesimal bialgebra (where A can be any associative algebra). The AYBE is closely related to the classical Yang–Baxter equation (CYBE) with spectral parameter

$$[r^{12}(v), r^{13}(v + v')] - [r^{23}(v'), r^{12}(v)] + [r^{13}(v + v'), r^{23}(v')] = 0 \quad (3)$$

for the Lie algebra sl_N (so $r(v)$ takes values in $\mathrm{sl}_N \otimes \mathrm{sl}_N$) and also with the quantum Yang–Baxter equation (QYBE) with spectral parameter

$$R^{12}(v)R^{13}(v + v')R^{23}(v') = R^{23}(v')R^{13}(v + v')R^{12}(v), \quad (4)$$

where $R(v)$ takes values in $A \otimes A$. In the seminal work [3] Belavin and Drinfeld made a thorough study of the CYBE for simple Lie algebras. In particular, they showed that all nondegenerate solutions are equivalent to either elliptic, trigonometric, or rational solutions, and gave a complete classification in the elliptic and trigonometric cases. In the present paper we extend some of their results and techniques to the AYBE. In addition, we show that often solutions of the AYBE are automatically solutions of the QYBE (for fixed u).

We will be mostly studying unitary solutions of the AYBE (i.e., solutions of (1) and (2)) that have the Laurent expansion at $u = 0$ of the form

$$r(u, v) = \frac{1 \otimes 1}{u} + r_0(v) + u r_1(v) + \cdots . \quad (5)$$

It is easy to see that in this case, $r_0(v)$ is a solution of the CYBE. Hence, denoting by $\mathrm{pr} : \mathrm{Mat}(N, \mathbb{C}) \to \mathrm{sl}_N$ the projection along $\mathbb{C} \cdot 1$, we obtain that $\bar{r}_0(v) = (\mathrm{pr} \otimes \mathrm{pr}) r_0(v)$ is a solution of the CYBE for sl_N. We prove that if $r(u, v)$ is nondegenerate (i.e., the tensor $r(u, v) \in A \otimes A$ is nondegenerate for generic (u, v)) then so is $\bar{r}_0$. Thus, $\bar{r}_0$ falls within Belavin–Drinfeld classification. Furthermore, we show that if $\bar{r}_0$ is either elliptic or trigonometric, then $r(u, v)$ is uniquely determined by $\bar{r}_0$ up to certain natural transformations. The natural question raised in [7] is which solutions of the CYBE for sl_N extend to unitary solutions of the AYBE of the form (5). In [7] we showed that this is the case for all elliptic solutions and gave some examples with trigonometric solutions. In [8] Schedler studied further this question for trigonometric solutions of the CYBE of the form $r_0(v) = \frac{r + e^v r^{21}}{1 - e^v}$, where r is a constant solution of the CYBE. He discovered that not all trigonometric solutions of the CYBE can be extended to solutions of the AYBE, and found a nice combinatorial structure that governs the situation (called *associative BD triples*). In this paper we complete the picture by giving the answer to the above question for arbitrary trigonometric solutions of the CYBE (see Theorem 0.2 below). We will also prove that every nondegenerate unitary solution $r(u, v)$ of the AYBE with the Laurent expansion at $u = 0$ of the form (5) satisfies the QYBE with spectral parameter for fixed u, provided $\bar{r}_0(v)$ either has a period (i.e., it is either elliptic or trigonometric) or has no infinitesimal symmetries (see Theorem 1.5). Thus, our work on extending trigonometric classical r-matrices (with spectral parameter) to solutions of the AYBE leads to explicit formulas for the corresponding quantum r-matrices. The connection with the QYBE was noticed before for elliptic solutions constructed in [7] (because they are given essentially by Belavin's elliptic R-matrix) and also for those trigonometric solutions that are constructed in [8].

An important input for our study of trigonometric solutions of the AYBE is the geometric picture with Massey products developed in [7] that involves considering simple vector bundles on elliptic curves and their rational degenerations. In that article we constructed all elliptic solutions in this way and some trigonometric solutions coming from simple vector bundles on the union of two projective lines glued at two points. In this paper we consider the case of bundles on a cycle of projective lines of arbitrary length. We compute explicitly corresponding solutions of the AYBE. Then we notice that a similar formula makes sense in a more general context and prove this by a direct calculation. The completeness of the obtained list of trigonometric solutions is then checked by combining the arguments of [8] with those of [3] (modified appropriately for the case of the AYBE). It is interesting that contrary to the initial expectation expressed in [7], not all trigonometric solutions of the AYBE can be obtained from the triple Massey products on cycles of projective lines (see Theorem 5.5). This makes us wonder whether there is some generalization of our geometric setup.

Another question that seems to be worth pursuing is the connection between the combinatorics of simple vector bundles on a cycle of projective lines X and the Belavin–Drinfeld combinatorics. Namely, the discrete type of a vector bundle on X is described by the splitting type on each component of X. As was observed in [4], Theorem 5.3, simplicity of a vector bundle corresponds to a certain combinatorial condition on these splitting types (see also Lemma 3.1). In this paper we show that this condition allows us to associate with such a splitting type a Belavin–Drinfeld triple (or rather enhanced combinatorial data described below). It seems that this connection might provide additional insight into the problem of classifying discrete types of simple vector bundles on X.

In [6] Mudrov constructs solutions of the QYBE from certain algebraic data that should be viewed as associative analogues of Manin triples. Elsewhere we will show how solutions of the AYBE give rise to such data and will study the corresponding associative algebras that are related to both the classical and quantum sides of the story.

Now let us present the combinatorial data on which our trigonometric solutions of the AYBE depend (generalizing Belavin–Drinfeld triples with associative structure considered in [8]). Let S be a finite set. To equip S with a cyclic order is the same as to fix a transitive cyclic permutation $C_0 : S \to S$. We denote by $\Gamma_{C_0} := \{(s, C_0(s)) \mid s \in S\}$ the graph of C_0.

Definition 0.1. An *associative BD-structure* on a finite set S is given by a pair of transitive cyclic permutations $C_0, C : S \to S$ and a pair of proper subsets $\Gamma_1, \Gamma_2 \subset \Gamma_{C_0}$ such that $(C \times C)(\Gamma_1) = \Gamma_2$, where $(C \times C)(i, i') = (C(i), C(i'))$.

We can identify Γ_{C_0} with the set of vertices Γ of the affine Dynkin diagram $\widetilde{A}_{N-1}$, where $N = |S|$ (preserving the cyclic order). Then we get from the above structure a Belavin–Drinfeld triple $(\Gamma_1, \Gamma_2, \tau)$ for $\widetilde{A}_{N-1}$, where the bijection $\tau : \Gamma_1 \to \Gamma_2$ is induced by $C \times C$. It is clear that τ preserves the inner product. The nilpotency condition on τ is satisfied automatically. Indeed, choose $(s_1, C_0(s_1)) \in \Gamma_S \setminus \Gamma_1$. Then for every $(s, C_0(s)) \in \Gamma_1$ there exists $k \geq 1$ with $C^k(s) = s_1$, so that $(C \times C)^k(s, C_0(s)) \notin \Gamma_1$.

We extend the bijection τ to a bijection $\tau : P_1 \to P_2$ induced by $C \times C$, where

$$P_\iota = \{(s, C_0^k(s)) \mid (s, C_0(s)) \in \Gamma_\iota, (C_0(s), C_0^2(s)) \in \Gamma_\iota, \ldots,$$
$$(C_0^{k-1}(s), C_0^k(s)) \in \Gamma_\iota\}, \ \iota = 1, 2.$$

For a finite set S let us denote by A_S the algebra of endomorphisms of the $\mathbb{C}$-vector space with the basis $(\mathbf{e}_i)_{i \in S}$, so that $A_S \simeq \mathrm{Mat}(N, \mathbb{C})$, where $N = |S|$. We denote by $e_{ij} \in A_S$ the endomorphism defined by $e_{ij}(\mathbf{e}_k) = \delta_{jk}\mathbf{e}_i$. We denote by $\mathfrak{h} \subset A_S$ the subalgebra of diagonal matrices (i.e., the span of $(e_{ii})_{i \in S}$). Now we can formulate our result about trigonometric solutions of the AYBE.

Theorem 0.2. *(i) Let $(C_0, C, \Gamma_1, \Gamma_2)$ be an associative BD-structure on a finite set S. Consider the $A_S \otimes A_S$-valued function*

$$r(u, v) = \frac{1}{1 - \exp(-v)} \sum_i e_{ii} \otimes e_{ii}$$

$$+ \frac{1}{\exp(u) - 1} \sum_{0 \leq k < N, i} \exp\left(\frac{ku}{N}\right) e_{C^k(i), C^k(i)} \otimes e_{ii}$$

$$+ \frac{1}{\exp(v) - 1} \sum_{0 < m < N, j = C_0^m(i)} \exp\left(\frac{mv}{N}\right) e_{ij} \otimes e_{ji}$$

$$+ \sum_{0 < m < N, k \geq 1; j = C_0^m(i), \tau^k(i,j) = (i', j')} \left[\exp\left(-\frac{ku + mv}{N}\right) e_{ji} \otimes e_{i'j'} \right.$$
$$\left. - \exp\left(\frac{ku + mv}{N}\right) e_{i'j'} \otimes e_{ji} \right],$$

where i, i', j, j' denote elements of S, and the summation in the last sum is taken only over those (i, j) for which τ^k is defined on (i, j). Then $r(u, v)$ satisfies (1) and (2). Furthermore, let us set

$$R(u, v) = \left(\left[\exp\left(\frac{u}{2}\right) - \exp\left(-\frac{u}{2}\right) \right]^{-1} + \left[\exp\left(\frac{v}{2}\right) - \exp\left(-\frac{v}{2}\right) \right]^{-1} \right)^{-1} \cdot r(u, v).$$

$$(6)$$

Then $R(u,v)$ satisfies the QYBE with spectral parameter (4) (for fixed u) and the unitarity condition

$$R(u,v)R^{21}(u,-v) = 1 \otimes 1. \tag{7}$$

(ii) Assume that $N > 1$. Then every nondegenerate unitary solution of the AYBE for $A = \mathrm{Mat}(N, \mathbb{C})$ with the Laurent expansion at $u = 0$ of the form (5), where $\bar{r}_0(v)$ is a trigonometric solution of the CYBE for sl_N, is equal to

$$c \exp(\lambda uv) \exp[u(1 \otimes a) + v(b \otimes 1)] r(cu, c'v) \exp[-u(a \otimes 1) - v(b \otimes 1)],$$

where $r(u,v)$ is obtained from one of the solutions from (i) by applying an algebra isomorphism $A_S \simeq A$, λ, c, and c' are constants ($c \neq 0$, $c' \neq 0$), and $a, b \in \mathfrak{h}$ are infinitesimal symmetries of $r(u,v)$, i.e.,

$$[a \otimes 1 + 1 \otimes a, r(u,v)] = [b \otimes 1 + 1 \otimes b, r(u,v)] = 0.$$

Note that the complete list of scalar unitary solutions of the AYBE was obtained in Theorem 5 of [7]. The solution obtained from Theorem 0.2(i) in the case $N = 1$ coincides with the basic trigonometric solution from that list (up to changing v to $-v$).

We will also deduce the following result about solutions of the AYBE not depending on the variable u.

Theorem 0.3. *Assume that $N > 1$. Let $r(v)$ be a nondegenerate unitary solution of the AYBE for $A = \mathrm{Mat}(N, \mathbb{C})$ not depending on the variable u. Then*

$$r(v) = \bar{r}(v) + b \otimes 1 + 1 \otimes b + \frac{c \cdot 1 \otimes 1}{Nv},$$

where $\bar{r}(v)$ is equivalent to a rational nondegenerate solution of the CYBE for sl_N, $b \in \mathrm{sl}_N$ is an infinitesimal symmetry of $\bar{r}(v)$, $c \in \mathbb{C}^*$. Also,

$$R(u,v) = \left(1 + \frac{cu}{v}\right)^{-1} \cdot (1 + ur(v))$$

is a unitary solution of the QYBE with spectral parameter for fixed u (hence, the same is true for $\frac{v}{c} r(v) = \lim_{u \to \infty} R(u,v)$).

The case of nondegenerate unitary solutions of the AYBE not depending on v turns out to be much easier — in this case, we get a complete list of solutions (see Proposition 1.1). Note that there are no constant nondegenerate solutions of the AYBE for $A = \mathrm{Mat}(N, \mathbb{C})$ (unitary or not), as follows from Proposition 2.9 of [2].

The paper is organized as follows. In Section 1 we discuss nondegeneracy conditions for solutions of the AYBE and show how to deduce the QYBE in Theorem 1.5. After recalling in Section 2 the geometric setup leading to solutions of the AYBE, we calculate these solutions associated with simple vector bundles on cycles of projective lines in Sections 3 and 4 (the result is given by formulas (32), (33)). Then in Section 5 we consider associative BD-structures on completely ordered sets and classify such structures coming from simple vector bundles on cycles of projective lines (see Theorem 5.5). In Section 6 we prove the first part of Theorem 0.2. In Section 7 we establish a meromorphic continuation in v for a class of solutions of the AYBE and derive some additional information about these solutions. Finally, in Section 8 we prove the second part of Theorem 0.2 and Theorem 0.3.

Acknowledgments. I am grateful to Pavel Etingof for crucial help with organizing my initial computations into a nice combinatorial pattern. I also thank him and Travis Schedler for useful comments on the first draft of the paper and the subsequent helpful discussions. Parts of this work were done while the author enjoyed the hospitality of the Max-Planck-Institute für Mathematik in Bonn and of the SISSA in Trieste. This work was partially supported by the NSF grant DMS-0601034.

1 The AYBE and the QYBE

Recall that we denote $A = \mathrm{Mat}(N, \mathbb{C})$. Let $r(u, v)$ be a meromorphic $A \otimes A$-function in a neighborhood of $(0, 0)$. We say that $r(u, v)$ is *nondegenerate* if the tensor $r(u, v)$ is nondegenerate for generic (u, v).

We start by collecting some facts about nondegenerate unitary solutions of the AYBE. First, let us consider the case when $r(u, v)$ does not depend on v. Then the AYBE reduces to

$$r^{12}(-u')r^{13}(u + u') - r^{23}(u + u')r^{12}(u) + r^{13}(u)r^{23}(u') = 0, \qquad (8)$$

and the unitarity condition becomes $r^{21}(-u) = -r(u)$.

Let us set $P = \sum_{i,j} e_{ij} \otimes e_{ji}$.

Proposition 1.1. *All nondegenerate unitary solutions of* (8) *have form*

$$r(u) = (\phi_a(cu) \otimes \mathrm{id})(P),$$

where $c \in \mathbb{C}^*$, $a \in \mathrm{sl}_N$, $\phi_a(u) \in \mathrm{End}(A)$ *is the linear operator on* A *defined from the equation*

$$u\phi_a(u)(X) + [a, \phi_a(u)(X)] = X.$$

Proof. Let us write $r(u, v)$ in the form $r(u) = (\phi(u) \otimes \mathrm{id})(e)$, where $e \in A^* \otimes A$ is the canonical element, $\phi(u) : A^* \to A$ is an operator, nondegenerate for generic u. Now set $B(u)(X, Y) = (X, \phi(u)^{-1}(Y))$ for $X, Y \in A$. It is easy to see that the equation (8) together with the unitarity condition are equivalent to the following equations on $B(u)$:

$$B(-u)(XY, Z) + B(-u')(YZ, X) + B(u + u')(ZX, Y) = 0, \qquad (9)$$

$$B(u)(X, Y) + B(-u)(Y, X) = 0. \qquad (10)$$

Substituting $Z = 1$ in the first equation we find

$$B(-u)(XY, 1) + (B(u + u') - B(u'))(X, Y) = 0, \text{ i.e.,}$$

$$B(u + u')(X, Y) = \xi(u)(XY) + B(u')(X, Y),$$

where $\xi(u)(X) = -B(-u)(X, 1)$. Exchanging u and u' we get that $C(X, Y) = B(u)(X, Y) - \xi(u)(XY)$ does not depend on u. Substituting $B(u)(X, Y) = \xi(u)(XY) + C(X, Y)$ into the previous equation we get

$$\xi(u + u') = \xi(u) + \xi(u');$$

hence $\xi(u) = u \cdot \xi$ for some $\xi \in A^*$. Now substituting $B(u)(X, Y) = u \cdot \xi(XY) + C(X, Y)$ into (10), we derive that $\xi(XY) = \xi(YX)$ and C is skew-symmetric. Therefore, $\xi = c \cdot \text{tr}$. Finally, equation (9) reduces to the equation

$$C(XY, Z) + C(YZ, X) + C(ZX, Y) = 0.$$

Together with the skew-symmetry of C this implies that $C(X, 1) = C(1, X) = 0$ and the restriction of C to $\text{sl}_N \times \text{sl}_N$ is a 2-cocycle. Hence, $C(X, Y) = l(XY - YX)$ for some linear functional l on sl_N. Conversely, for C of this form the above equation is satisfied. Thus, all solutions of (9) and (10) are given by

$$B(u)(X, Y) = cu \, \text{tr}(X, Y) + l(XY - YX),$$

where $c \in \mathbb{C}^*$ and l is a linear functional on sl_N. Let us identify A with A^* using the metric $\text{tr}(XY)$. Then we can view $\phi(u)$ as an operator from A to A such that $B(u)(X, Y) = \text{tr}(X\phi(u)^{-1}(Y))$. Representing the functional l in the form $l(X) = -\text{tr}(Xa)$, we obtain the formula

$$\phi(u)^{-1}(Y) = cuY + [a, Y].$$

$\square$

Remark 1.2. It is easy to see that $\phi_a(u)$ (and hence the corresponding associative r-matrix) always has a pole at $u = 0$ with order equal to the maximal k such that there exists $X \in A$ with $\text{ad}^k(a)(X) = 0$ and $\text{ad}^{k-1}(a)(X) \neq 0$. Indeed, $\phi_a(u)$ cannot be regular at $u = 0$, since this would give $[a, \phi_a(0)(1)] = 1$. Let

$$\phi_a(u) = \frac{\psi_{-k}}{u^k} + \frac{\psi_{-k+1}}{u^{k-1}} + \cdots$$

be the Laurent expansion of $\phi_a(u)$. Then we have

$$\psi_{i-1} + \text{ad}(a) \circ \psi_i = 0$$

for $i \neq 0$ and

$$\psi_{-1} + \text{ad}(a) \circ \psi_0 = \text{id}.$$

Decomposing $\mathrm{End}(A)$ into generalized eigenspaces of the operator $\psi \mapsto \mathrm{ad}(a) \circ \psi$, we see that ψ_{-1} is the component of $\mathrm{id} \in \mathrm{End}(A)$ corresponding to the zero eigenvalue. This immediately implies our claim. For example, if a is semisimple then $\phi_a(u)$ has a simple pole at $u = 0$. More precisely, taking the diagonal matrix $a = \sum_i a_i e_{ii}$, we get the associative r-matrix

$$r(u) = \sum_{ij} \frac{1}{u + a_i - a_j} e_{ij} \otimes e_{ji}.$$

The proofs of the next two results are parallel to those of Propositions 2.2 and 2.1 in [3], respectively.

Lemma 1.3. *Let $r(u, v)$ be a nondegenerate unitary solution of the AYBE. Assume that $r(u, v)$ does not have a pole at $v = 0$. Then $r(u, 0)$ is still nondegenerate, and hence has the form described in Proposition 1.1.*

Proof. Let us fix v_0 such that $r(u, v)$ does not have a pole at $v = v_0$ and $r(u, v_0)$ is nondegenerate for generic u. Then we can define a meromorphic function $\phi(u, v)$ with values in $\mathrm{End}_{\mathbb{C}}(A)$ by the condition

$$(\phi(u, v) \otimes \mathrm{id})(r(u, v_0)) = r(u, v).$$

We claim that this function satisfies the identity

$$\phi(u + u', v)(XY) = \phi(u, v)(X)\phi(u', v)(Y), \tag{11}$$

where $X, Y \in A$. Indeed, since $r(u, v)$ does not have a pole at $v = 0$, substituting $v' = 0$ in (1), we get

$$r^{12}(-u', v)r^{13}(u + u', v) = r^{23}(u + u', 0)r^{12}(u, v) - r^{13}(u, v)r^{23}(u', 0).$$

Note that the right-hand side is obtained by applying $\phi(u, v) \otimes \mathrm{id} \otimes \mathrm{id}$ to the right-hand side for $v = v_0$. Applying the above equation for $v = v_0$, we deduce that it is equal to

$$(\phi(u, v) \otimes \mathrm{id} \otimes \mathrm{id})(r^{12}(-u', v_0)r^{13}(u + u', v_0)).$$

On the other hand, the left-hand side can be rewritten as

$$[(\phi(-u', v) \otimes \mathrm{id})r(-u', v_0)]^{12}[(\phi(u + u', v) \otimes \mathrm{id})r(u + u', v)]^{13}.$$

Thus, if we write $r(u, v_0) = \sum K^\alpha(u) \otimes \mathbf{e}_\alpha$, where $\mathbf{e}_\alpha$ is a basis of A, then we derive

$$\phi(-u', v)(K^\alpha(-u'))\phi(u + u', v)(K^\beta(u + u')) = \phi(u, v)(K^\alpha(-u')K^\beta(u + u')).$$

By nondegeneracy of $r(u, v_0)$ this implies (11). Taking $Y = 1$ in this equation, we obtain

$$\phi(u + u', v)(X) = \phi(u, v)(X)\phi(u', v)(1). \tag{12}$$

Similarly, we deduce that

$$\phi(u + u', v)(Y) = \phi(u', v)(1)\phi(u, v)(Y).$$

Comparing these equations, we see that $\phi(u', v)(1)$ commutes with $\phi(u, v)(X)$ for any $X \in A$. Using nondegeneracy of $r(u, v)$ we derive that $\phi(u, v)(1) = f(u, v) \cdot 1$ for some scalar meromorphic function $f(u, v)$. Furthermore, we should have

$$f(u + u', v) = f(u, v)f(u', v),$$

which implies that $f(u, v) = \exp(g(v)u)$ for some function $g(v)$ holomorphic near $v = 0$. Next, from (12) we obtain that $\exp(-g(v)u)\phi(u, v)$ does not depend on u. Thus, all solutions of (11) have the form $\phi(u, v) = \exp(g(v)u)\psi(v)$, where for every v, $\psi(v)$ is an *algebra automorphism* of A or zero. By our assumption, $\phi(u, v)$ does not have a pole at $v = 0$. Therefore, $\psi(v)$ is holomorphic near $v = 0$. Now we use the fact that every algebra automorphism of A is inner, and hence has determinant equal to 1 (it is enough to check this for the conjugation with a diagonalizable matrix). Since, $\psi(v_0) = \text{id}$, this implies that $\det \psi(v) = 1$ identically. Therefore, $\det \psi(0) = 1$ and $\phi(u, 0)$ is invertible. $\square$

Lemma 1.4. *Let $r(u, v)$ be a nondegenerate unitary solution of the* AYBE. *Assume that $r(u, v)$ has a pole at $v = 0$. Then this pole is simple and $\lim_{v \to 0} vr(u, v) = cP$ for some nonzero constant c.*

Proof. Let $r(u, v) = \frac{\theta(u)}{v^k} + \frac{\eta(u)}{v^{k-1}} + \cdots$ be the Laurent expansion of $r(u, v)$ near $v = 0$. Considering the polar parts as $v' \to 0$ (resp., $v \to 0$) in (1), we get

$$- \theta^{23}(u + u')r^{12}(u, v) + r^{13}(u, v)\theta^{23}(u') = 0, \tag{13}$$
$$\theta^{12}(-u')r^{13}(u + u', v') - r^{23}(u + u', v')\theta^{12}(u) = 0. \tag{14}$$

Let $V \subset A$ be the minimal subspace such that $\theta(u) \in V \otimes A$ (for all u where $\theta(u)$ is defined). Then we have $r^{13}(u, v)\theta^{23}(u') \in A \otimes V \otimes A$. Hence, from (13) we get $\theta^{23}(u + u')r^{12}(u, v) \in A \otimes V \otimes A$. This implies that $r^{12}(u, v) \in A \otimes A_1$, where

$$A_1 = \{a \in A \ : \ (a \otimes 1)\theta(u) \in V \otimes A \text{ for all } u\}.$$

By nondegeneracy we get $A_1 = A$, hence $AV \subset V$. Similarly, using (14) we derive that $VA \subset V$. Thus, V is a nonzero two-sided ideal in A, so we have $V = A$. Now let us prove that the order k of a pole cannot be greater that 1. Indeed, assuming that $k > 1$ and considering the coefficient of v^{1-k} in the expansion of (1) near $v = 0$, we get

$$\eta^{12}(-u')r^{13}(u + u', v') - r^{23}(u + u', v')\eta^{12}(u) + \theta^{12}(-u')\frac{\partial r^{13}}{\partial v}(u + u', v') = 0.$$

Now looking at polar parts at $v' = 0$ we get $\theta^{12}(-u')\theta^{13}(u + u') = 0$, which contradicts the equality $V = A$ established above. Therefore, $k = 1$. Now let us look at (13) again. Let us fix u and consider the subspace

$$A(u) = \{x \in A \ : \ \theta(u + u')(x \otimes 1) = (1 \otimes x)\theta(u') \text{ for all } u'\}.$$

Then from (13) we get that $r(u, v) \in A \otimes A(u)$. By nondegeneracy this implies that $A(u) = A$ for generic u, so we get the identity

$$\theta(u + u')(x \otimes 1) = (1 \otimes x)\theta(u')$$

for all $x \in A$. Taking $x = 1$, we see that $\theta(u) = \theta$ is constant. Finally, any tensor $\theta \in A \otimes A$ with the property $\theta(x \otimes 1) = (1 \otimes x)\theta$ is proportional to P. $\square$

Recall that if $r(u, v)$ is a solution of the AYBE with the Laurent expansion at $u = 0$ of the form (5), then $r_0(v)$ is a unitary solution of the CYBE (see proof of Lemma 1.2 in [7], or Lemma 2.9 of [8]). The same is true for $\bar{r}_0(v) = (\mathrm{pr} \otimes \mathrm{pr})(r_0(v)) \in \mathrm{sl}_N \otimes \mathrm{sl}_N$. We will show below that the nondegeneracy of $r(u, v)$ implies that $\bar{r}_0(v)$ is also nondegenerate; hence it is either elliptic, trigonometric, or rational. The first two cases are distinguished from the third by the condition that $\bar{r}_0(v)$ is periodic with respect to $v \mapsto v + p$ for some $p \in \mathbb{C}^*$.

Recall that by an infinitesimal symmetry of an $A \otimes A$-valued function $f(x)$ we mean an element $a \in A$ such that $[a \otimes 1 + 1 \otimes a, f(x)] = 0$ for all x.

Theorem 1.5. *Let $r(u, v)$ be a nondegenerate unitary solution of the AYBE with the Laurent expansion at $u = 0$ of the form (5), and let $\bar{r}_0(v) = (\mathrm{pr} \otimes \mathrm{pr})(r_0(v))$. Then*

(i) $\bar{r}_0(v)$ is a nondegenerate unitary solution of the CYBE.
(ii) The following conditions are equivalent:
  (a) $r(u, v)$ satisfies the QYBE (4) in v (for fixed u);
  (b) the product $r(u, v)r(-u, v)$ is a scalar multiple of $1 \otimes 1$;
  (c) $\frac{d}{dv}(r_0(v) - \bar{r}_0(v))$ is a scalar multiple of $1 \otimes 1$.
  (d) $(\mathrm{pr} \otimes \mathrm{pr} \otimes \mathrm{pr})[\bar{r}_0^{12}(v)\bar{r}_0^{13}(v+v') - \bar{r}_0^{23}(v')\bar{r}_0^{12}(v) + \bar{r}_0^{13}(v+v')\bar{r}_0^{23}(v')] = 0.$
(iii) The equivalent conditions in (ii) hold when $\bar{r}_0(v)$ either admits a period or has no infinitesimal symmetries in sl_N.

Remarks 1.6. 1. In fact, our proof shows that equivalent conditions in (ii) hold under the weaker assumption that the system

$$[\bar{r}_0(v), a^1 + a^2] = [\bar{r}_0(v), b^1 + b^2 + va^1] = [b, a] = 0$$

on $a, b \in \mathrm{sl}_N$ implies that $a = 0$.
2. Note that the implication (b) $\Longrightarrow$ (a) in part (ii) of the theorem holds for any unitary solution of the AYBE (as follows easily from Lemma 1.8 below). It is plausible that one can check condition (b) in other situations than those

considered in the above theorem. For example, we have nondegenerate unitary solutions of the AYBE of the form

$$r(u, v) = \frac{\omega}{u^n} + \frac{P}{v},$$

where $n \geq 1$, and $\omega \in A \otimes A$ satisfies $\omega^{12}\omega^{13} = 0$ and $\omega^{21} = (-1)^{n-1}\omega$. It is easy to see that these solutions satisfy $r(u, v)r(-u, v) = 1 \otimes 1/v^2$, so they also satisfy the QYBE. On the other hand, the solutions of the AYBE constructed in Proposition 1.1 do not satisfy the QYBE in general.

Lemma 1.7. *Assume that $N > 1$. Let $r(u, v)$ be a nondegenerate unitary solution of the AYBE with the Laurent expansion at $u = 0$ of the form (5). Then $r(u, v)$ has a simple pole at $v = 0$ with the residue $c \cdot P$, where $c \in \mathbb{C}^*$.*

Proof. By Lemma 1.4 we only have to rule out the possibility that $r(u, v)$ has no pole at $v = 0$. Assume this is the case. Then $r(u, 0)$ is the solution of (8) that has a simple pole at $u = 0$ with the residue $1 \otimes 1$. Let $\phi(u) : A \to A$ be the linear operator such that $r(u, 0) = (\phi(u) \otimes \mathrm{id})(P)$. Then $\phi(u)$ has a Laurent expansion at $u = 0$ of the form

$$\phi(u)(X) = \frac{\mathrm{tr}(X) \cdot 1}{u} + \psi(X) + \cdots$$

for some operator $\psi : A \to A$. By Lemma 1.3 and Proposition 1.1 we have

$$cu\phi(u)(X) + [a, \phi(u)(X)] = X$$

for some $c \in \mathbb{C}^*$ and $a \in \mathrm{sl}_N$. Considering the constant terms of the expansions at $u = 0$ we get

$$c\,\mathrm{tr}(X) \cdot 1 + [a, \psi(X)] = X.$$

It follows that $[a, \psi(X)] = X$ for all $X \in \mathrm{sl}_N$. Hence, the operator $\mathrm{pr}\,\psi|_{\mathrm{sl}_N} : \mathrm{sl}_N \to \mathrm{sl}_N$ is invertible. Taking in the above equality $X \in \mathrm{sl}_N$ such that $\mathrm{pr}\,\psi(X) = a$, we derive that $a = 0$, which leads to a contradiction. $\square$

The next two lemmas constitute the core of the proof of Theorem 1.5.

Lemma 1.8. *For a triple of variables u_1, u_2, u_3 (resp., v_1, v_2, v_3) set $u_{ij} = u_i - u_j$ (resp., $v_{ij} = v_i - v_j$). Then for every unitary solution of the AYBE one has*

$$r^{12}(u_{12}, v_{12})r^{13}(u_{23}, v_{13})r^{23}(u_{12}, v_{23}) - r^{23}(u_{23}, v_{23})r^{13}(u_{12}, v_{13})r^{12}(u_{23}, v_{12})$$
$$= s^{23}(u_{23}, v_{23})r^{13}(u_{13}, v_{13}) - r^{13}(u_{13}, v_{13})s^{23}(u_{21}, v_{23})$$
$$= r^{13}(u_{13}, v_{13})s^{12}(u_{32}, v_{12}) - s^{12}(u_{12}, v_{12})r^{13}(u_{13}, v_{13}),$$

where $s(u, v) = r(u, v)r(-u, v)$.

Proof. In the following proof we will use the shorthand notation $r^{ij}(u)$ for $r^{ij}(u, v_{ij})$. The AYBE can be rewritten as

$$r^{12}(u_{12})r^{13}(u_{23}) - r^{23}(u_{23})r^{12}(u_{13}) + r^{13}(u_{13})r^{23}(u_{21}) = 0. \qquad (15)$$

On the other hand, switching indices 1 and 2 and using the unitarity condition, we obtain

$$r^{23}(u_{23})r^{13}(u_{12}) - r^{12}(u_{12})r^{23}(u_{13}) + r^{13}(u_{13})r^{12}(u_{32}) = 0. \qquad (16)$$

Multiplying (16) by $r^{12}(u_{23})$ on the right, we get

$$r^{23}(u_{23})r^{13}(u_{12})r^{12}(u_{23}) - r^{12}(u_{12})r^{23}(u_{13})r^{12}(u_{23}) + r^{13}(u_{13})s^{12}(u_{32}) = 0.$$

On the other hand, switching u_1 and u_2 in (15) and multiplying the obtained equation by $r^{12}(u_{12})$ on the left, we obtain

$$s^{12}(u_{12})r^{13}(u_{13}) - r^{12}(u_{12})r^{23}(u_{13})r^{12}(u_{23}) + r^{12}(u_{12})r^{13}(u_{23})r^{23}(u_{12}) = 0.$$

Taking the difference between these cubic equations gives

$$r^{23}(u_{23})r^{13}(u_{12})r^{12}(u_{23}) - r^{12}(u_{12})r^{13}(u_{23})r^{23}(u_{12})$$
$$= s^{12}(u_{12})r^{13}(u_{13}) - r^{13}(u_{13})s^{12}(u_{32}).$$

The other half of the required equation is obtained by switching the indices 1 and 3 and using the unitarity condition.   □

Lemma 1.9. *Let $r(u, v)$ be a unitary solution of the AYBE with the Laurent expansion (5) at $u = 0$. Assume also that $r(u, v)$ has a simple pole at $v = 0$ with residue cP. Then one has*

$$s(u, v) = r(u, v)r(-u, v) = a \otimes 1 + 1 \otimes a + (f(u) + g(v)) \cdot 1 \otimes 1$$

with

$$g(v) = -\frac{c}{N}(\text{tr} \otimes \text{tr})\left(\frac{dr_0(v)}{dv}\right),$$

$$f(u) = \frac{1}{N} \text{tr}\, \mu\left(\frac{\partial r(u, 0)}{\partial u}\right),$$

$$a = \text{pr}\, \mu\left(\frac{\partial r(u, 0)}{\partial u}\right),$$

where $\mu : A \otimes A \to A$ denotes the product. Furthermore, $a \in \text{sl}_N$ is an infinitesimal symmetry of $r(u, v)$, and if we write

$$r_0(v) = \overline{r}_0(v) + \alpha(v) \otimes 1 - 1 \otimes \alpha(-v) + h(v) \cdot 1 \otimes 1,$$

where $\overline{r}_0(v) \in \text{sl}_N \otimes \text{sl}_N$ and $\alpha(v) \in \text{sl}_N$, then

$$\alpha(v) = \alpha(0) - \frac{v}{cN}a.$$

Proof. Let us write $r(u,v) = \frac{cP}{v} + \tilde{r}(u,v)$, where $\tilde{r}(u,v)$ does not have a pole at $v = 0$. Then we can rewrite the AYBE as follows (where $v_{ij} = v_i - v_j$):

$$r^{13}(u, v_{13})r^{23}(-u + h, v_{23})$$
$$= r^{23}(h, v_{23})r^{12}(u, v_{12}) - r^{12}(u - h, v_{12})r^{13}(h, v_{13})$$
$$= r^{23}(h, v_{23})r^{12}(u, v_{12}) - r^{12}(u, v_{12})r^{13}(h, v_{13})$$
$$+ [r^{12}(u, v_{12}) - r^{12}(u - h, v_{12})]r^{13}(h, v_{13})$$
$$= \frac{r^{23}(h, v_{23}) - r^{23}(h, v_{13})}{v_{12}}cP^{12} + [r^{23}(h, v_{23})\tilde{r}^{12}(u, v_{12})$$
$$- \tilde{r}^{12}(u, v_{12})r^{13}(h, v_{13})] + [\tilde{r}^{12}(u, v_{12}) - \tilde{r}^{12}(u - h, v_{12})]r^{13}(h, v_{13}).$$

Passing to the limit $v_2 \to v_1$ we derive

$$r^{13}(u, v)r^{23}(-u + h, v)$$
$$= -\frac{\partial r^{23}}{\partial v}(h, v)cP^{12} + [r^{23}(h, v)\tilde{r}^{12}(u, 0) - \tilde{r}^{12}(u, 0)r^{13}(h, v)]$$
$$+ [\tilde{r}^{12}(u, 0) - \tilde{r}^{12}(u - h, 0)]r^{13}(h, v).$$

Next, we are going to apply the operator $\mu \otimes \mathrm{id} : A \otimes A \otimes A \to A \otimes A$, where μ is the product on A. We use the following easy observations:

$$(\mu \otimes \mathrm{id})(x^{13}y^{23}) = xy,$$
$$(\mu \otimes \mathrm{id})(x^{23}y^{12} - y^{12}x^{13}) = 0,$$
$$(\mu \otimes \mathrm{id})(x^{23}P^{12}) = 1 \otimes \mathrm{tr}_1(x),$$

where $x, y \in A \otimes A$, $\mathrm{tr}_1 = \mathrm{tr} \otimes \mathrm{id} : A \otimes A \to A$ (the last property follows from the identity $\sum_{ij} e_{ij}ae_{ji} = \mathrm{tr}(a) \cdot 1$ for $a \in A$). Thus, applying $\mu \otimes \mathrm{id}$ to the above equation, we get

$$r(u, v)r(-u + h, v) = -c \cdot 1 \otimes \mathrm{tr}_1\left(\frac{\partial r}{\partial v}(h, v)\right)$$
$$+ (\mu \otimes \mathrm{id})\left([\tilde{r}^{12}(u, 0) - \tilde{r}^{12}(u - h, 0)]r^{13}(h, v)\right).$$

Finally, taking the limit $h \to 0$, we derive

$$s(u, v) = -c \cdot 1 \otimes \mathrm{tr}_1\left(\frac{dr_0(v)}{dv}\right) + \mu\left(\frac{\partial r(u, 0)}{\partial u}\right) \otimes 1, \qquad (17)$$

where we used the equalities $\frac{\partial r(0,v)}{\partial v} = \frac{dr_0(v)}{dv}$ and $\frac{\partial \tilde{r}(u,v)}{\partial u} = \frac{\partial r(u,v)}{\partial u}$. Hence, we can write $s(u, v)$ in the form

$$s(u, v) = a(u) \otimes 1 + 1 \otimes b(v) + (f(u) + g(v))1 \otimes 1,$$

where $a(u)$ and $b(v)$ take values in sl_N, and

$$b(v) = -c\,\mathrm{pr}\,\mathrm{tr}_1\left(\frac{dr_0(v)}{dv}\right).$$

The unitarity condition on $r(u, v)$ implies that $s^{21}(-u, -v) = s(u, v)$. This immediately gives the required form of $s(u, v)$ with some $a \in sl_N$, as well as the formulas for $g(v)$, $f(u)$, a, and $\alpha(v)$. The fact that a is an infinitesimal symmetry of $r(u, v)$ follows from the second equality in the identity of Lemma 1.8.   □

Lemma 1.10. *Let $r_0(v) \in A \otimes A$ be a unitary solution of the CYBE of the form*

$$r_0(v) = \bar{r}_0(v) + \alpha(v) \otimes 1 - 1 \otimes \alpha(-v) + h(v) \cdot 1 \otimes 1,$$

where $\bar{r}_0(v) \in sl_N \otimes sl_N$ and $\alpha(v) \in sl_N$. Then

$$[\bar{r}_0(v - v'), \alpha(v) \otimes 1 + 1 \otimes \alpha(v')] = [\alpha(v), \alpha(v')] = 0.$$

In particular, if $\alpha(v)$ depends linearly on v, i.e., $\alpha(v) = b + v \cdot a$, then

$$[\bar{r}_0(v), a \otimes 1 + 1 \otimes a] = [\bar{r}_0(v), b \otimes 1 + 1 \otimes b + va \otimes 1] = [b, a] = 0.$$

Proof. Applying $\text{pr} \otimes \text{pr} \otimes \text{pr}$ to both sides of the CYBE, we see that $\bar{r}_0$ itself satisfies the CYBE. Taking this into account, the equation can be rewritten as

$$[\bar{r}_0^{12}(v_{12}), \alpha^1(v_{13}) + \alpha^2(v_{23})] + [\alpha^1(v_{12}), \alpha^1(v_{13})] + c.p.(1, 2, 3) = 0,$$

where $v_{ij} = v_i - v_j$ (the omitted terms are obtained by cyclically permuting $1, 2, 3$). Applying the operator $\text{pr} \otimes \text{pr} \otimes \text{id}$ gives

$$[\bar{r}_0^{12}(v_{12}), \alpha^1(v_{13}) + \alpha^2(v_{23})] = 0.$$

Now returning to the above equality and applying $\text{pr} \otimes \text{id} \otimes \text{id}$, we derive that $[\alpha(v), \alpha(v')] = 0$.   □

Proof of Theorem 1.5. (i) In the case $N = 1$ the statement is vacuous, so we can assume that $N > 1$. By Lemma 1.7, $r_0(v)$ has a simple pole at $v = 0$ with residue cP, where $c \in \mathbb{C}^*$. Projecting to sl_N, we deduce that $\bar{r}_0(v)$ is nondegenerate.

(ii) By Lemma 1.8, $r(u, v)$ satisfies the QYBE iff

$$s^{23}(u, v_{23})r^{13}(2u, v_{13}) = r^{13}(2u, v_{13})s^{23}(-u, v_{23}).$$

Using the formula for $s(u, v)$ from Lemma 1.9, we see that this is equivalent to the equality

$$[r(u, v), 1 \otimes a] = 0,$$

which is equivalent to $a = 0$ by the nondegeneracy of $r(u, v)$. Note that by Lemma 1.9, both conditions (b) and (c) are also equivalent to the equality

$a = 0$. It remains to show the equivalence of (d) with this equality. To this end we use the identity

$$r_0^{12}(v)r_0^{13}(v+v') - r_0^{23}(v')r_0^{12}(v) + r_0^{13}(v+v')r_0^{23}(v') = r_1^{12}(v) + r_1^{13}(v+v') + r_1^{23}(v'),$$
(18)

deduced by substituting the Laurent expansions in the first variable into (1). Let us denote the expression in the left-hand-side of (18) by $\text{AYBE}[r_0](v, v')$. Using the relation between $r_0(v)$ and $\bar{r}_0(v)$ from Lemma 1.9, we obtain

$$-cN \cdot (\text{pr} \otimes \text{pr} \otimes \text{pr})\,(\text{AYBE}[r_0](v, v') - \text{AYBE}[\bar{r}_0](v, v'))$$
$$= v\bar{r}_0^{12}(v)a^3 + v'\bar{r}_0^{23}(v')a^1 + (v+v')\bar{r}_0^{13}(v+v')a^2,$$

where $a^1 = a \otimes 1 \otimes 1$, etc. Note that (18) implies that $(\text{pr} \otimes \text{pr} \otimes \text{pr})\,\text{AYBE}[r_0]$ $(v, v') = 0$. Therefore, it suffices to prove that the equation

$$v\bar{r}_0^{12}(v)a^3 + v'\bar{r}_0^{23}(v')a^1 + (v+v')\bar{r}_0^{13}(v+v')a^2 = 0$$

on $a \in \text{sl}_N$ implies that $a = 0$. Passing to the limit as $v \to 0$ and $v' \to 0$, we deduce from the above equality that

$$(\text{pr} \otimes \text{pr} \otimes \text{pr})[P^{12}a^3 + P^{23}a^1 + P^{13}a^2] = 0.$$

Let $a = \sum a_{ij}e_{ij}$. Looking at the coefficient with $e_{ij} \otimes e_{ji} \otimes e_{ij}$, we deduce that $a_{ij} = 0$ for $i \neq j$. Finally, looking at the projection to $e_{12} \otimes e_{21} \otimes \text{sl}_N$, we deduce that a_{ii} does not depend on i; hence $a = 0$.   □

(iii) It suffices to prove that under our assumptions the infinitesimal symmetry $a \in \text{sl}_N$ appearing in Lemma 1.9 is equal to zero. We only have to consider the case in which $\bar{r}_0$ has a period, i.e., $\bar{r}_0(v + p) = \bar{r}_0(v)$ for some $p \in \mathbb{C}^*$. By Lemma 1.10, it remains to check that the equation

$$[\bar{r}_0(v), b \otimes 1 + 1 \otimes b + va \otimes 1] = 0$$

on $a, b \in \text{sl}_N$ implies that $a = 0$. From the periodicity of $\bar{r}_0$ we derive that

$$[\bar{r}_0(v), a \otimes 1] = 0.$$

By the nondegeneracy of $\bar{r}_0$, it follows that $a = 0$.   □

2 Solutions of the AYBE Associated with Simple Vector Bundles on Degenerations of Elliptic Curves

Now let us review how solutions of the AYBE arise from geometric structures on elliptic curves and their degenerations. Let X be a nodal projective curve over $\mathbb{C}$ of arithmetic genus 1 such that the dualizing sheaf on X is isomorphic to $\mathcal{O}_X$. Let us fix such an isomorphism. Recall that a vector bundle V on X is called *simple* if $\text{End}(V) = \mathbb{C}$. The following result follows from Theorems 1 and 4 of [7].

Theorem 2.1. *Let V_1, V_2 be a pair of simple vector bundles on X such that $\mathrm{Hom}^0(V_1, V_2) = \mathrm{Ext}^1(V_1, V_2) = 0$. Let y_1, y_2 be a pair of distinct smooth points of X. Consider the tensor*

$$r_{y_1,y_2}^{V_1,V_2} \in \mathrm{Hom}(V_{1,y_1}^*, V_{2,y_1}^*) \otimes \mathrm{Hom}(V_{2,y_2}^*, V_{1,y_2}^*)$$

corresponding to the following composition:

$$\mathrm{Hom}(V_{1,y_1}, V_{2,y_1}) \xrightarrow{\mathrm{Res}_{y_1}^{-1}} \mathrm{Hom}(V_1, V_2(y_1)) \xrightarrow{\mathrm{ev}_{y_2}} \mathrm{Hom}(V_{1,y_2}, V_{2,y_2}),$$

where $V_{i,y}$ denotes the fiber of V_i at a point $y \in X$, the map

$$\mathrm{Res}_y : \mathrm{Hom}(V_1, V_2(y)) \widetilde{\to} \mathrm{Hom}(V_{1,y}, V_{2,y})$$

is obtained by taking the residue at y, and the map ev_y is the evaluation at y. Then for a triple of simple bundles (V_1, V_2, V_3) such that each pair satisfies the above assumptions and for a triple of distinct points (y_1, y_2, y_3) one has

$$\left(r_{y_1 y_2}^{V_3 V_2}\right)^{12} \left(r_{y_1 y_3}^{V_1 V_3}\right)^{13} - \left(r_{y_2 y_3}^{V_1 V_3}\right)^{23} \left(r_{y_1 y_2}^{V_1 V_2}\right)^{12} + \left(r_{y_1 y_3}^{V_1 V_2}\right)^{13} \left(r_{y_2 y_3}^{V_2 V_3}\right)^{23} = 0 \quad (19)$$

in $\mathrm{Hom}(V_{1,y_1}^*, V_{2,y_1}^*) \otimes \mathrm{Hom}(V_{2,y_2}^*, V_{3,y_2}^*) \otimes \mathrm{Hom}(V_{3,y_3}^*, V_{1,y_3}^*)$. In addition the following unitarity condition holds:

$$\left(r_{y_1 y_2}^{V_1 V_2}\right)^{21} = -r_{y_2 y_1}^{V_2 V_1}. \quad (20)$$

Remark 2.2. The tensor $r_{y_1,y_2}^{V_1,V_2}$ in the above theorem is a certain triple Massey product in the derived category of X, and the equation (19) follows from the appropriate A_∞-axiom (see [7]).

We are going to apply the above theorem for bundles V_i of the form $V_i = V \otimes L_i$, where V is a fixed simple vector bundle of rank N on X and L_i are line bundles in $\mathrm{Pic}^0(X)$, the neutral component of $\mathrm{Pic}(X)$. Also, we let points y_i vary in a connected component X_0 of X. Uniformizations of $X_0 \cap X^{reg}$ and of $\mathrm{Pic}^0(X)$ allow to describe V_i's and y_i's by complex parameters. Thus, using trivializations of the bundles V_{i,y_j}^* we can view the tensor $r_{y_1,y_2}^{V_1,V_2}$ in the above theorem as a function of complex variables $r(u_1, u_2; v_1, v_2) \in A \otimes A$, where $A = \mathrm{Mat}(N, \mathbb{C})$, u_i describes V_i, v_j describes y_j. Note that equation (19) reduces to the AYBE in the case when r depends only on the differences of variables, i.e., $r(u_1, u_2; v_1, v_2) = r(u_1 - u_2, v_1 - v_2)$.

A different choice of trivializations of V_{i,y_i}^* would lead to the tensor $\widetilde{r}(u_1, u_2, v_1, v_2)$ given by

$$\widetilde{r}(u_1, u_2; v_1, v_2) = (\varphi(u_2, v_1) \otimes \varphi(u_1, v_2)) r(u_1, u_2; v_1, v_2) (\varphi(u_1, v_1) \otimes \varphi(u_2, v_2))^{-1}$$

where $\varphi(u, v)$ is a function with values in $\mathrm{GL}_N(\mathbb{C})$. We say that tensor functions $\widetilde{r}$ and r related in this way are *equivalent*. Note that the condition for functions to depend only on the differences $u_1 - u_2$ and $v_1 - v_2$ is not preserved under these equivalences in general. However, if (a, b) is a pair of commuting infinitesimal symmetries of $r(u_1 - u_2, v_1 - v_2)$ then taking $\varphi(u, v) = \exp(ua + vb)$ we do get a tensor function $\widetilde{r}$ that depends only on the differences, namely,

$$\widetilde{r}(u, v) = \exp[u(1 \otimes a) + v(b \otimes 1)]r(u, v)\exp[-u(a \otimes 1) - v(b \otimes 1)]$$

(this kind of equivalence shows up in Theorem 0.2(ii)).

Since we are interested in trigonometric solutions, we will be using the multiplicative variables $x_i = \exp(u_i)$, $y_i = \exp(v_i)$. The solutions of (19) that we are going to construct in the next section will be equivalent to those depending only on the differences $u_1 - u_2$, $v_1 - v_2$. It will be convenient for us also to work with the intermediate form of the AYBE

$$r^{12}((x')^{-1}; y_1, y_2)r^{13}(xx'; y_1, y_3) - r^{23}(xx'; y_2, y_3)r^{12}(x; y_1, y_2) + r^{13}(x; y_1, y_3)r^{23}(x'; y_2, y_3) = 0 \qquad (21)$$

for the tensor $r(x; y_1, y_2) \in A \otimes A$, obtained from (19) in the case when $r(x_1, x_2; y_1, y_2) = r(x_1/x_2; y_1, y_2)$. The corresponding unitarity condition has form

$$r^{21}(x; y_1, y_2) = -r(x^{-1}; y_2, y_1). \qquad (22)$$

3 Simple Vector Bundles on Cycles of Projective Lines

Let $X = X_0 \cup X_1 \cup \cdots \cup X_{n-1}$ be the union of n copies of $\mathbb{P}^1$'s glued (transversally) in a configuration of type $\widetilde{A}_{n-1}$, so that the point ∞ on X_j is identified with the point 0 on X_{j+1} for $j = 0, \ldots, n - 1$ (where we identify indices with elements of $\mathbb{Z}/n\mathbb{Z}$). A vector bundle V of rank N on X is given by a collection of vector bundles V_j of rank N on X_j along with isomorphisms $(V_j)_\infty \simeq (V_{j+1})_0$. Since every vector bundle on $\mathbb{P}^1$ splits into a direct sum of line bundles, we can assume that

$$V_j = \mathcal{O}_{\mathbb{P}^1}(m_1^j) \oplus \cdots \oplus \mathcal{O}_{\mathbb{P}^1}(m_N^j)$$

for every $j = 0, \ldots, n-1$. Thus, the splitting types are described by the $N \times n$ matrix of integers (m_i^j).

Let $(z_0 : z_1)$ denote the homogeneous coordinates on $\mathbb{P}^1$. We will use the standard trivialization of the fiber of $\mathcal{O}_{\mathbb{P}^1}(1)$ at $0 = (1 : 0) \in \mathbb{P}^1$ (resp., at ∞) given by the generating section z_0 (resp., z_1). Note that a section $s \in \mathcal{O}_{\mathbb{P}^1}(1)$ is uniquely determined by its values $s(0)$ and $s(\infty)$ (namely, $s = s(0)z_0 + s(\infty)z_1$).

Let us fix a splitting type matrix $\mathbf{m} = (m_i^j)$. For every $\lambda \in \mathbb{C}^*$ we define the rank-N bundle $V^\lambda = V^\lambda(\mathbf{m})$ on X by using standard trivializations of $V_j = \oplus_{i=1}^n \mathcal{O}(m_i^j)$ at 0 and ∞ and setting the transition isomorphisms $(V_j)_\infty \simeq (V_{j+1})_0$ to be identical for $j = 0, \ldots, n - 2$, and the last transition map to be

$$\lambda C^{-1} : (V_0)_0 \to (V_{n-1})_\infty,$$

where C is the cyclic permutation matrix $Ce_i = e_{i-1}$, where we identify the set of indices with $\mathbb{Z}/N\mathbb{Z}$. Note that in this definition only the cyclic order on the indices $\{1, \ldots, N\}$ is used. In particular, if we cyclically permute the rows of the matrix (m_i^j) (by replacing m_i^j with m_{i+1}^j), then we get the same vector bundle.

Lemma 3.1 below provides a criterion for simplicity of $V^\lambda(\mathbf{m})$. This result is well known (see [4], Theorem 5.3). For completeness we include the proof. It is also known that every simple vector bundle on X is isomorphic to some $V^\lambda(\mathbf{m})$ (see [4]). It will be convenient to extend the $N \times n$ matrix (m_i^j) to the matrix with columns numbered by $j \in \mathbb{Z}$ using the rule $m_i^{j+n} = m_{i-1}^j$.

Lemma 3.1. *The vector bundle $V^\lambda(\mathbf{m})$ is simple iff the following two conditions are satisfied:*
(a) the differences $m_i^j - m_{i'}^j$ for $i, i' \in \mathbb{Z}/N\mathbb{Z}$ take values only $\{-1, 0, 1\}$;
(b) for every $i, i' \in \mathbb{Z}/N\mathbb{Z}$, $i \neq i'$, the nN-periodic infinite sequence

$$(m_i^j - m_{i'}^j), \quad j \in \mathbb{Z},$$

is not identically 0, and the occurrences of 1 and -1 in it alternate.
Furthermore, if (a) and (b) hold then $V^{\lambda_1}(\mathbf{m}) \simeq V^{\lambda_2}(\mathbf{m})$ iff $(\lambda_1/\lambda_2)^N = 1$.

Proof. First, we observe that if $m_i^j - m_{i'}^j = 2$ then there exists a nonzero morphism $\mathcal{O}_{\mathbb{P}^1}(m_{i'}^j) \to \mathcal{O}_{\mathbb{P}^1}(m_i^j)$ vanishing at 0 and ∞. Viewing it as an endomorphism of V_j, we obtain a nonscalar endomorphism of V^λ. Hence, the condition (a) is necessary. From now on let us assume that (a) is satisfied.

A morphism $V^{\lambda_1} \to V^{\lambda_2}$ is given by a collection of morphisms $A_j : V_j \to V_j$, $j = 0, \ldots, n-1$, such that $A_j(\infty) = A_{j+1}(0)$ for $j = 0, \ldots, n-2$ and

$$A_0(0) = \frac{\lambda_1}{\lambda_2} C A_{n-1}(\infty) C^{-1}.$$

We can write these maps as matrices $A_j = (a_{ii'}^j)_{1 \le i, i' \le N}$, where $a_{ii'}^j \in H^0(\mathbb{P}^1, \mathcal{O}(m_i^j - m_{i'}^j))$. Let us allow the index j to take all integer values by using the rule $a_{ii'}^{j+n} = a_{i-1, i'-1}^j$. Note that we still have $a_{ii'}^j \in H^0(\mathbb{P}^1, \mathcal{O}(m_i^j - m_{i'}^j))$ because of our convention on m_i^j for $j \in \mathbb{Z}$. Then the equations on (A_j) can be rewritten as

$$a_{ii'}^j(0) = x^{\delta(j)} a_{ii'}^{j-1}(\infty) \tag{23}$$

for all $i, i' \in \mathbb{Z}/N\mathbb{Z}$ and $j \in \mathbb{Z}$, where $x = \lambda_1/\lambda_2$, and $\delta(j) = 1$ for $j \equiv 0(n)$, $\delta(j) = 0$ otherwise. Due to condition (a) we have the following possibilities for each $a_{ii'}^j$:

(i) if $m_i^j < m_{i'}^j$ then $a_{ii'}^j = 0$;
(ii) if $m_i^j = m_{i'}^j$ then $a_{ii'}^j$ is a constant, so $a_{ii'}^j(0) = a_{ii'}^j(\infty)$;
(iii) if $m_i^j > m_{i'}^j$ then $a_{ii'}^j$ is a section of $\mathcal{O}(1)$, so it is uniquely determined by its values at 0 and ∞, and these values can be arbitrary.

From this we can immediately derive that (b) is necessary for V^λ to be simple. Indeed, if for some $i \neq i'$ we have $m_i^j = m_{i'}^j$ for all $j \in \mathbb{Z}$, then we can get a solution of (23) with $x = 1$ by setting $a_{i+k,i'+k}^j = 1$ for all $j, k \in \mathbb{Z}$ and letting the remaining entries be zero. This would give a nonscalar endomorphism of V^λ. Similarly, if for some $i \neq i'$ and some segment $[j, k] \subset \mathbb{Z}$ we have

$$(m_i^j - m_{i'}^j, m_i^{j+1} - m_{i'}^{j+1}, \ldots, m_i^k - m_{i'}^k) = (1, 0, \ldots, 0, 1),$$

then we get a solution of (23) with $x = 1$ by setting

$$a_{ii'}^j = z_1, \ a_{ii'}^{j+1} = 1, \ldots, a_{ii'}^{k-1} = 1, \ a_{ii'}^k = z_0$$

and letting the remaining entries be zero.

Conversely, assume (a) and (b) hold. Then one can easily derive that V^λ is simple by analyzing the system (23) (with $x = 1$). Indeed, let us show first that $a_{ii'}^j = 0$ for $i \neq i'$. It follows from (b) that in the case $m_i^j = m_{i'}^j$ we can either find a segment $[j_1, j] \subset \mathbb{Z}$ such that $m_i^k = m_{i'}^k$ for $j_1 < k < j$ and $m_{j_1}^k < m_{j_1}^k$, or a segment $[j, j_2] \subset \mathbb{Z}$ such that $m_i^k = m_{i'}^k$ for $j < k < j_2$ and $m_{j_2}^k < m_{j_2}^k$. In either case, applying iteratively (23), we derive that $a_{ii'}^j = 0$ (recall that in this case $a_{ii'}^j$ is a constant). In the case $m_i^j > m_{i'}^j$ we can find both segments $[j_1, j]$ and $[j, j_2]$ as above, so that (23) implies that $a_{ii'}^j(0) = a_{ii'}^j(\infty) = 0$. Hence, $a_{ii'}^j = 0$. The remaining part of the system (23) shows that all a_{ii}^j are equal to the same constant, i.e., V^λ has no nonscalar endomorphisms.

The above argument also shows that a morphism $(a_{ii'}^j) : V^{\lambda_1}(\mathbf{m}) \to V^{\lambda_2}(\mathbf{m})$ has $a_{ii'}^j = 0$ (assuming conditions (a) and (b) hold), while the remaining components $a_{ii}^j \in \mathbb{C}$ satisfy the equations

$$a_{ii}^j = x^{\delta(j)} a_{ii}^{j-1}, \ a_{ii}^{j+n} = a_{i-1,i-1}^j,$$

where $x = \lambda_1/\lambda_2$. This system has a nonzero solution iff $x^N = 1$, in which case the solution gives an isomorphism $V^{\lambda_1}(\mathbf{m}) \simeq V^{\lambda_2}(\mathbf{m})$. $\qquad \square$

4 Computation of the Associative r-Matrix Arising as a Massey Product

Henceforward, we always assume that the matrix (m_i^j) satisfies the conditions of Lemma 3.1. Given a pair of parameters $\lambda_1, \lambda_2 \in \mathbb{C}^*$ and a pair of points $y, y' \in X_0 \setminus \{0, \infty\}$, we want to describe explicitly the maps

$$\mathrm{Res}_y : \mathrm{Hom}(V^{\lambda_1}, V^{\lambda_2}(y)) \to \mathrm{Hom}(V_y^{\lambda_1}, V_y^{\lambda_2}),$$

$$\mathrm{ev}_{y'} : \mathrm{Hom}(V^{\lambda_1}, V^{\lambda_2}(y)) \to \mathrm{Hom}(V_{y'}^{\lambda_1}, V_{y'}^{\lambda_2}),$$

and especially the composition $ev_{y'} \circ Res_y^{-1}$ (for generic λ_1, λ_2). We will identify the target spaces of both maps with $N \times N$ matrices using trivializations of the relevant line bundles over y induced by the appropriate power of $z_0 \in H^0(\mathbb{P}^1, \mathcal{O}(1))$. We also use the global 1-form trivializing ω_X that restricts to dz/z on each $\mathbb{P}^1 \setminus \{0, \infty\}$ (where $z = z_1/z_0$).

A morphism $V^{\lambda_1} \to V^{\lambda_2}(y)$ is given by a collection of morphisms

$$A_0 : V_0 \to V_0(y), \; A_1 : V_1 \to V_1, \dots, A_{n-1} : V_{n-1} \to V_{n-1}$$

with same equations as before. Writing these maps as matrices, we can view $\mathrm{Hom}(V^{\lambda_1}, V^{\lambda_2}(y))$ as the space of solutions of (23), where $a_{ii'}^j \in H^0(\mathbb{P}^1, \mathcal{O}(m_i^j - m_{i'}^j))$ for $j \not\equiv 0(n)$ and $a_{ii'}^j \in H^0(\mathbb{P}^1, \mathcal{O}(m_i^j - m_{i'}^j)(y))$ for $j \equiv 0(n)$.

Since the component X_0 plays a special role, we will use a shorthand notation $m_i := m_i^0$, $a_{ii'} := a_{ii'}^0$. Let us also set $b_{ii'} = Res_y(a_{ii'})$. Recall that for every pair $i, i' \in \mathbb{Z}/N\mathbb{Z}$ we have the following three possibilities:

(i) If $m_i < m_{i'}$ then we have $a_{ii'} = \frac{yb_{ii'}}{z_1 - yz_0}$, so that

$$a_{ii'}(0) = -b_{ii'}, a_{ii'}(\infty) = yb_{ii'}. \tag{24}$$

(ii) If $m_i = m_{i'}$ then $a_{ii'} = \frac{a_{ii'}(\infty)z - a_{ii'}(0)y}{z-y}$ (where $z = z_1/z_0$), so we get the relation

$$a_{ii'}(\infty) - a_{ii'}(0) = b_{ii'}. \tag{25}$$

(iii) If $m_i > m_{i'}$ then $a_{ii'}$ is uniquely determined by $a_{ii'}(0)$, $a_{ii'}(\infty)$, and $b_{ii'}$. Namely, one can easily check that

$$a_{ii'} = \frac{z(b_{ii'} + a_{ii'}(0) - ya_{ii'}(\infty)) - ya_{ii'}(0)}{z - y} \cdot z_0 + \frac{za_{ii'}(\infty)}{z - y} \cdot z_1.$$

Note that in the above three cases we also have the following expressions for $a_{ii'}(y')$:

$$a_{ii'}(y') = \begin{cases} \frac{yb_{ii'}}{y'-y}, & m_i < m_{i'}, \\ \frac{y'b_{ii'}}{y'-y} + a_{ii'}(0) = \frac{yb_{ii'}}{y'-y} + a_{ii'}(\infty), & m_i = m_{i'}, \\ \frac{y'b_{ii'}}{y'-y} + a_{ii'}(0) + y'a_{ii'}(\infty), & m_i > m_{i'}. \end{cases} \tag{26}$$

To compute $ev_{y'} \circ Res_y^{-1}$ means to express all the entries $a_{ii'}(y')$ in terms of $(b_{ii'})$. The above formula gives such an expression in the case $m_i < m_{i'}$; in the case $m_i = m_{i'}$ we need to know either $a_{ii'}(0)$ or $a_{ii'}(\infty)$; and in the case $m_i > m_{i'}$ we need to know both. Of course, in the last two cases one has to use equations (23). Then condition (b) of Lemma 3.1 will guarantee that we get a closed formula for $a_{ii'}(y')$ in terms of all the entries $b_{ii'}$. To organize the computation it is convenient to use the complete order on the set of indices $\{1, \dots, N\}$ given by

($\star$) $i \prec i'$ if either $m_i < m_{i'}$ or $m_i = m_{i'}$ and the first nonzero term in the sequence $(m_i^j - m_{i'}^j)$, $j = 0, 1, \dots$, is negative.

The fact that this is a complete order follows immediately from condition (b) of Lemma 3.1. We will write $(ii') > 0$ if $i \prec i'$ and $(ii') < 0$ if $i \succ i'$. We will also use the notation $-(i, i') = (i', i)$.

Let us define a partially defined operation on pairs of distinct indices in $\mathbb{Z}/N\mathbb{Z}$ by setting

$$\tau(ii') = (i - 1, i' - 1) \text{ if } (i - 1) \prec (i' - 1) \text{ and } m_i^j = m_{i'}^j, \text{ for } 0 < j < n.$$

Note that τ is one-to-one. We denote by τ^{-1} the (partially defined) inverse and by τ^k the iterated maps. Condition (b) of Lemma 3.1 implies that for every pair of distinct indices (ii') there exists $k > 0$ such that τ^k is not defined on (ii').

Case 1. Assume that $i \prec i'$, i.e., $(ii') > 0$. Then either $m_i < m_{i'}$, or there exists $j > 0$ such that $m_i^{j'} = m_{i'}^{j'}$ for $0 \leq j' < j$ and $m_i^j < m_{i'}^j$. In the first case we can use formula (26). In the second case we have $a_{ii'}^{j'} = \text{const}$ for $0 < j' < j$, $j \not\equiv 0(n)$, while $a_{ii'}^j = 0$. Therefore, using (23) and (25) iteratively we get the following expression for $a_{ii'}(\infty)$:

$$-a_{ii'}(\infty) = \sum_{k \geq 1} x^{-k} b_{\tau^k(ii')}, \qquad (27)$$

where the summation is only over a finite number of k's for which $\tau^k(ii')$ is defined. This gives the formula

$$a_{ii'}(y') = \frac{yb_{ii'}}{y' - y} - \sum_{k \geq 1} x^{-k} b_{\tau^k(ii')} \qquad \text{if } (ii') > 0, \qquad (28)$$

which works also for the case $m_i < m_{i'}$ (since in this case τ is not defined on (ii')).

Case 2. Assume that $i \succ i'$, i.e., $(ii') < 0$. Then either $m_i > m_{i'}$, or there exists $j < 0$ such that $m_i^j = m_{i'}^j$ for $j < j' \leq 0$ and $m_i^j < m_{i'}^j$. Assume first that $m_i > m_{i'}$. Note that in this case there still exists $j < 0$ with the above property and in addition there is $k > 0$ such that $m_i^{j'} = m_{i'}^{j'}$ for $0 \leq j' < k$ and $m_i^k < m_{i'}^k$ (by condition (b) of Lemma 3.1). Using equations (23) and (25) we derive that (27) still holds and also we have

$$a_{ii'}(0) = \sum_{k \geq 1} y^{\epsilon(\sigma\tau^{-k}\sigma(ii'))} x^k b_{\sigma\tau^{-k}\sigma(ii')}, \qquad (29)$$

where σ is the transposition $\sigma(i, i') = (i', i)$, the summation is only over those k for which $\tau^{-k}\sigma(ii')$ is defined, $\epsilon(ii') = 1$ for $(ii') > 0$, and $\epsilon(ii') = 0$ otherwise. This gives

$$a_{ii'}(y') = \frac{y'b_{ii'}}{y' - y} + \sum_{k \geq 1} y^{\epsilon(\sigma\tau^{-k}\sigma(ii'))} x^k b_{\sigma\tau^{-k}\sigma(ii')} - y' \sum_{k \geq 1} x^{-k} b_{\tau^k(ii')} \quad \text{if } (ii') < 0. \qquad (30)$$

We observe that this formula still works in the case $m_i = m_{i'}$ (the second summation becomes empty in this case).

Case 3. Assume that $i = i'$. In this case we have relations

$$a_{ii}(0) = xa_{i+1,i+1}(0) + xb_{i+1,i+1}$$

for all $i \in \mathbb{Z}/N\mathbb{Z}$. Solving this linear system for $a_{ii}(0)$, we get

$$a_{ii}(0) = (1 - x^N)^{-1} \sum_{k=1}^{N} x^k b_{i+k,i+k}.$$

Finally, we derive

$$a_{ii}(y') = \frac{y}{y' - y} b_{ii} + (1 - x^N)^{-1} \sum_{k=0}^{N-1} x^k b_{i+k,i+k}. \tag{31}$$

Formulas (28), (30), and (31) completely determine the map $\mathrm{ev}_{y'} \circ \mathrm{Res}_y^{-1}$, so we can compute the associative r-matrix corresponding to the family of simple vector bundles V^λ on X:

$$r(x; y, y') = r_{\mathrm{const}}(x, y/y')$$

$$+ \sum_{\alpha > 0, k \geq 1} [-x^{-k} e_{-\tau^k(\alpha)} \otimes e_\alpha + y^{\epsilon(-\tau^{-k}(\alpha))} x^k e_{\tau^{-k}(\alpha)} \otimes e_{-\alpha}$$

$$- y' x^{-k} e_{-\tau^k(-\alpha)} \otimes e_{-\alpha}],$$

where

$$r_{\mathrm{const}}(x, z) = \frac{z}{1 - z} \sum_{\alpha > 0} e_{-\alpha} \otimes e_\alpha + \frac{1}{1 - z} \sum_{\alpha > 0} e_\alpha \otimes e_{-\alpha}$$

$$+ \frac{z}{1 - z} \sum_i e_{ii} \otimes e_{ii} + (1 - x^N)^{-1} \sum_i \sum_{k=0}^{N-1} x^k e_{i+k,i+k} \otimes e_{ii}. \tag{32}$$

In these formulas i is an element of $\mathbb{Z}/N\mathbb{Z}$, and α denotes a pair of distinct indices in $\mathbb{Z}/N\mathbb{Z}$. By a simple rearrangement of terms we can rewrite $r(x; y, y')$ in the following way:

$$r(x; y, y') = r_{\mathrm{const}}(x, y/y')$$

$$+ \sum_{\alpha > 0, k \geq 1} [x^k e_\alpha \otimes e_{-\tau^k(\alpha)} - x^{-k} e_{-\tau^k(\alpha)} \otimes e_\alpha$$

$$+ y x^k e_{-\alpha} \otimes e_{-\tau^k(-\alpha)} - y' x^{-k} e_{-\tau^k(-\alpha)} \otimes e_{-\alpha}]. \tag{33}$$

Recall that this is a solution of (21) with the unitarity condition (22).

Example 4.1. Assume that $n > N$ and the only nonzero entries of (m_i^j) are $m_1^N = m_2^{N-1} = \cdots = m_{N-1}^1 = 1$. Then the domain of definition of τ is empty, so in this case we have $r(x; y, y') = r_{\mathrm{const}}(x, y/y')$. Hence, $r_{\mathrm{const}}(\exp(u), \exp(v))$ is a solution of the AYBE.

Later we will show that $r(\exp(u); \exp(v_1), \exp(v_2))$ is equivalent to an r-matrix depending only on the difference $v_1 - v_2$ (see Lemma 6.1), so that it gives a solution of the AYBE.

5 Associative Belavin–Drinfeld Triples Associated with Simple Vector Bundles

The right-hand side of (33) depends only on the parameters x, y, y' and on a certain combinatorial structure on the set $S = \{1, \ldots, N\}$. We are going to show that this structure consists of an *associative BD-structure* as defined in the introduction together with a compatible complete order (see below). Later we will show that one can get rid of the dependence on a complete order by passing to an equivalent r-matrix (see Lemma 6.1). However, for purposes of studying splitting types of simple vector bundles on cycles of projective lines the full combinatorial structure described below may be useful.

Definition 5.1. We say that a *complete order on a set S is compatible with the cyclic order* given by a cyclic permutation C_0 (or simply *compatible with* C_0) if C_0 takes every non-maximal element to the next element in this order. In other words, if we identify S with the segment of integers $[1, N]$ preserving the complete order, then $C_0(i) = i + 1$ (where the indices are identified with $\mathbb{Z}/N\mathbb{Z}$). In this case we set $\alpha_0 = (s_{\max}, s_{\min}) \in \Gamma_{C_0}$, where $s_{\min}$ (resp., $s_{\max}$) is the minimal (resp., maximal) element of S. A choice of a complete order on S compatible with C_0 is equivalent to a choice of an element $\alpha_0 \in \Gamma_{C_0}$. By an *associative BD-structure on a completely ordered set S* we mean an associative BD-structure $(C_0, C, \Gamma_1, \Gamma_2)$ on S such that the complete order is compatible with C_0.

Note that a choice of an associative BD-structure on the completely ordered set $[1, N]$ such that $\alpha_0 \notin \Gamma_1$ and $\alpha_0 \notin \Gamma_2$, is equivalent to a choice of a Belavin–Drinfeld triple in A_{N-1} equipped with an *associative structure* as defined in [8].

We will need the following characterization of associative BD-structures on completely ordered sets such that $\alpha_0 \notin \Gamma_2$.

Lemma 5.2. Let $(S, <)$ be a completely ordered finite set equipped with a transitive cyclic permutation $C : S \to S$. Then to give an associative BD-structure on S with $\alpha_0 \notin \Gamma_2$ is equivalent to giving a pair of subsets P_1 and P_2 in the set of pairs of distinct elements of S such that $(C \times C)(P_1) = P_2$ and the following properties are satisfied:

(a) For every $(s, s') \in P_2$ one has $s < s'$.
(b) Assume that $s < s' < s''$. If $(s, s'') \in P_1$ then $(s, s'), (s', s'') \in P_1$. The same property holds for P_2. Also, if $(s', s) \in P_1$ then $(s', s''), (s'', s) \in P_1$ (resp., if $(s'', s') \in P_1$ then $(s'', s), (s, s') \in P_1$).

The proof is left for the reader. Let us observe only that property (b) ensures that P_ι is determined by $\Gamma_\iota = P_\iota \cap \Gamma_{C_0}$, where $\iota = 1, 2$.

Now let us check that in the setting of Section 4 we do get a completely ordered set with an associative BD-structure.

Lemma 5.3. *Let* (m_i^j) *be a* $N \times n$ *matrix satisfying the conditions of Lemma 3.1. Equip the set* $S = \{1, \ldots, N\}$ *with the complete order* $\prec$ *given by* $(\star)$ *and the cyclic permutation* $C(i) = i - 1$. *Also, let*

$$P_1 = \{(ii') \mid m_i^j = m_{i'}^j \quad for \ 0 < j < n \ and \ C(i) \prec C(i')\}.$$

Then these data define an associative BD-structure with $\alpha_0 \notin \Gamma_2$.

Proof. We use Lemma 5.2. The only question is why property (b) holds. Let $i \prec i' \prec i''$.

Assume first that $(i, i'') \in P_2$. Then $m_{i+1}^j = m_{i''+1}^j$ for $j \in [1, n-1]$. Suppose there exists $j \in [1, n-1]$ such that $m_{i+1}^j \neq m_{i'+1}^j$. Consider the maximal such j. We have either $m_{i+1}^j < m_{i'+1}^j$ or $m_{i'+1}^j < m_{i''+1}^j$. By condition (b) of Lemma 3.1, the former assumption contradicts $i \prec i'$, while the latter contradicts $i' \prec i''$. Hence, $m_{i+1}^j = m_{i'+1}^j = m_{i''+1}^j$ for all $j \in [1, n-1]$, so that $(i, i'), (i', i'') \in P_2$.

Assume that $(i, i'') \in P_1$. Then $m_i^j = m_{i''}^j$ for $j \in [1, n-1]$. Furthermore, since $i \prec i''$ and $i - 1 \prec i'' - 1$, we should have $m_i^0 = m_{i''}^0$ (by condition (b) of Lemma 3.1). Suppose there exists $j \in [0, n-1]$ such that $m_i^j \neq m_{i'}^j$. Consider the minimal such j. We have either $m_i^j > m_{i'}^j$ or $m_{i'}^j > m_{i''}^j$. But the former contradicts $i \prec i'$, and the latter contradicts $i' \prec i''$. Therefore, $m_i^j = m_{i'}^j = m_{i''}^j$ for all $j \in [0, n-1]$, so that $(i, i'), (i', i'') \in P_1$.

Finally, assume that $(i', i) \in P_1$ (resp., $(i'', i') \in P_1$). Then $m_i^j = m_{i'}^j$ (resp., $m_{i'}^j = m_{i''}^j$) for $j \in [1, n-1]$. Also, since $i' \succ i$ and $i' - 1 \prec i - 1$ (resp., $i'' \succ i'$ and $i'' - 1 \prec i' - 1$), we necessarily have $m_i^0 < m_{i'}^0$ (resp., $m_{i'}^0 < m_{i''}^0$). Hence, $m_{i'}^0 = m_{i''}^0$ (resp., $m_i^0 = m_{i'}^0$). Suppose there exists $j \in [1, n-1]$ such that $m_{i'}^j \neq m_{i''}^j$ (resp., $m_i^j \neq m_{i'}^j$). Consider the minimal such j. Since $i' \prec i''$ (resp., $i \prec i'$), we have $m_i^j = m_{i'}^j < m_{i''}^j$ (resp., $m_i^j < m_{i'}^j = m_{i''}^j$). But this contradicts condition (b) of Lemma 3.1 (applied to i and i''). Therefore, $m_i^j = m_{i'}^j = m_{i''}^j$ for all $j \in [1, n-1]$. Since $m_{i'}^0 = m_{i''}^0$ (resp., $m_i^0 = m_{i'}^0$), we have $i' - 1 \prec i'' - 1$ (resp., $i - 1 \prec i' - 1$), and hence $(i', i'') \in P_1$ (resp., $(i, i') \in P_1$). Also, $i'' - 1 \prec i - 1$ (by condition (b) of Lemma 3.1), so that $(i'', i) \in P_1$. □

We will need below the following two operations on associative BD-structures.

Definition 5.4. For an associative BD-structure $(C_0, C, \Gamma_1, \Gamma_2)$ on a finite set S we define
(i) the *opposite associative BD-structure* to be $(C_0^{-1}, C, \sigma(\Gamma_1), \sigma(\Gamma_2))$, where σ is the permutation of factors in $S \times S$ (note that $\sigma(\Gamma_{C_0}) = \Gamma_{C_0^{-1}}$);
(ii) the *inverse associative BD-structure* to be $(C_0, C^{-1}, \Gamma_2, \Gamma_1)$.

Note that under passing to the opposite associative BD-structure each set P_ι, $\iota = 1, 2$, gets replaced with $\sigma(P_\iota)$.

Theorem 5.5. *An associative BD-structure on a completely ordered finite set S is obtained by the construction of Lemma 5.3 from some matrix (m_i^j) (satisfying the conditions of Lemma 3.1) iff $\alpha_0 \notin \Gamma_2$ and $C = C_0^k$ for some $k \in \mathbb{Z}$ (relatively prime to $N = |S|$).*

Proof. **"Only if."** Let us denote by $t_i = \sum_{j=0}^{n-1} m_i^j$, $i = 1, \ldots, N$, the sums of entries in the rows of the matrix (m_i^j). Then we claim that for $i \prec i'$ one has

$$t_i - t_{i'} = \begin{cases} -1, & \text{if } i - 1 \succ i' - 1, \\ 0, & \text{otherwise.} \end{cases} \tag{34}$$

Indeed, assume first that $m_i^j = m_{i'}^j$ for all $j \in [0, n-1]$. Then $i-1 \prec i'-1$ and $t_i = t_{i'}$, so the above equation holds. Next, assume that $m_i^j \neq m_{i'}^j$ for some $j \in [0, n-1]$. Then the first nonzero term in the sequence $(m_i^j - m_{i'}^j)_{j \in [0, n-1]}$ is -1. Since -1's and 1's in this sequence alternate, we have $t_i - t_{i'} = 0$ (resp., $t_i - t_{i'} = -1$) iff the last nonzero term in the sequence $(m_i^j - m_{i'}^j)_{j \in [0, n-1]}$ is 1 (resp., -1). But this happens precisely when the first nonzero term in $(m_{i-1}^j - m_{i'-1}^j)_{j \geq 0}$ is -1 (resp., 1), so (34) follows.

Now assume that $i \prec i' \prec i''$. Then it follows from (34) that either $C(i) \prec C(i') \prec C(i'')$ or $C(i') \prec C(i'') \prec C(i)$ or $C(i'') \prec C(i) \prec C(i')$. Since this holds for every triple (i, i', i''), it is easy to deduce that $C = C_0^k$ for some $k \in \mathbb{Z}$.

"If." First, note that the construction of the associative BD-structure on a completely ordered set S given in Lemma 5.3 can be rewritten as follows. Assume we are given a transitive cyclic permutation C of S and a matrix (m_s^j), where $j \in [0, n]$, $s \in S$. Then we can extend the range of the index j to $\mathbb{Z}$ using the rule $m_s^{j+n} = m_{C(s)}^j$. Assuming that condition (b) of Lemma 3.1 holds for this extended matrix, we can proceed to define the complete order by $(\star)$ and the set P_1 as in Lemma 5.3. Of course, we can always identify S with $\{1, \ldots, N\}$ in such a way that $C(i) = i - 1$, so that we get to the setup of Lemma 5.3. The advantage of the new point of view is that we can also consider the set $S = \{1, \ldots, N\}$ with the cyclic permutation $C(i) = i - k$, where $k \in \mathbb{Z}/N\mathbb{Z}$ is relatively prime to N. Then as was noted above, we have to modify the definition of the extended matrix by using the rule $m_i^{j+n} = m_{i-k}^j$.

Note that changing (m_i^j) to $(-m_i^j)$ changes the associative BD-structure on S to the opposite BD-structure, and the complete order on S gets reversed. Let us denote by $w_0 : S \to S$ the permutation that reverses the order. Assume that we have $C = C_0^k$. Then conjugating by w_0 the BD-structure associated with $(-m_i^j)$ we get a BD-structure that is obtained from the original one by leaving the complete order the same, changing $C = C_0^k$ to C_0^{-k}, and replacing P_1 with $(w_0 \times w_0)\sigma(P_1)$. Therefore, it is enough to show that Lemma 5.3 produces all associative BD-structures with $C = C_0^{-k}$, where $N/2 \leq k < N$.

Next, we describe a construction of a class of matrices (m_i^j) satisfying the conditions of Lemma 3.1. Fix k, relatively prime to N, such that $N/2 \le k < N$. Start with a sequence $(a_1, \ldots, a_N)$ such that $a_1 = 1$, $a_N = n - 1$ (where $n > 1$), and for every $i \in [1, N-1]$ one has either $a_{i+1} = a_i$ or $a_{i+1} = a_i + 1$. Then set $m_i^0 = 1$ for $i \in [k+1, N]$, $m_{k+1-i}^{a_i} = 1$ for $i = 1, \ldots, N$, and let the remaining entries be zero. We are going to check that this matrix satisfies the conditions of Lemma 3.1 (with the modified definition of the extended matrix).

It is convenient to extend the range of the index i to $\mathbb{Z}$ by the rule $m_i^j = m_{i+N}^j$, so that we get a matrix (m_i^j) with rows and columns numbered by $\mathbb{Z}$. Let us consider the subset $\Lambda \subset [k+1-N, N] \times [0, n-1]$ defined by

$$\Lambda = ([k+1, N] \times \{0\}) \cup \{(k+1-i, a_i) \mid i = 1, \ldots, N\}.$$

Then we have

$$\{(i,j) \in \mathbb{Z} \times \mathbb{Z} \mid m_i^j \ne 0\} = \cup_{a \in \mathbb{Z}} \Lambda_a, \text{ where}$$
$$\Lambda_0 = \cup_{b \in \mathbb{Z}} (\Lambda + b(2N-k, n)), \quad \Lambda_a = \Lambda_0 + a(N, 0).$$

Note that each Λ_a intersects each row once, and if we denote by $(i, j_a(i))$ the intersection point of Λ_a with the ith row, then either $j_a(i-1) = j_a(i)$ or $j_a(i-1) = j_a(i) + 1$. In other words, as we go down one row, the point of intersection either stays in the same column or moves one step to the right. It follows that the intersection of Λ_a with each column is a line segment. Moreover, it is easy to see that the number of elements in this intersection is at most N. Indeed, for columns corresponding to $j \equiv 0(n)$ the intersection segment has $N - k$ elements. On the other hand, for $j \not\equiv 0(n)$ this number is equal to the number of $i \in [1, N]$ such that $j \equiv a_i(n)$, so it is at most N. This implies that Λ_a and $\Lambda_{a'}$ are disjoint for $a \ne a'$. Hence, $j_a(i) < j_{a+1}(i)$ for all $a \in \mathbb{Z}$ and $i \in \mathbb{Z}$.

Let us set $E_i = \{j_a(i) \mid a \in \mathbb{Z}\}$ for every $i \in \mathbb{Z}$. We have to check that for every pair of rows, the ith and the i'th, where $i < i' < i + N$, one has $E_i \ne E_{i'}$, and the subsets $E_i \setminus E_{i'}$ and $E_{i'} \setminus E_i$ in $\mathbb{Z}$ alternate.

To prove that $E_i \ne E_{i'}$ we recall that by the construction, for every $b \in \mathbb{Z}$ the intersection of Λ_0 with the bnth column is the segment $[k+1+b(2N-k), N+b(2N-k)]$. The intersections of other sets Λ_a with the same column are obtained from the above segment by shifts in $N\mathbb{Z}$. Since $2N - k$ is relatively prime to N, it follows that for appropriate $b \in \mathbb{Z}$ the intersection of $\cup_a \Lambda_a$ with the bnth column contains exactly one of the numbers i and i'. Hence, bn belongs to exactly one of the sets E_i and $E_{i'}$.

Finally, we have to prove that subsets $E_i \setminus E_{i'}$ and $E_{i'} \setminus E_i$ alternate. Note that for all a we have $j_a(i') \le j_a(i)$. Hence, our assertion will follow once we check that for every $a \in \mathbb{Z}$ one has $j_a(i) \le j_{a+1}(i')$. Suppose we have $j_{a+1}(i') < j_a(i)$. Then the intersection of Λ_{a+1} with the $j_a(i)$th column is a segment $[i_1, i_2]$, where $i < i_1 \le i_2 < i'$. Since $\Lambda_{a+1} = \Lambda_a + (N, 0)$, the intersection of Λ_a with the $j_a(i)$th column is $[i_1 - N, i_2 - N]$. Hence, $i \le i_2 - N < i' - N$, which contradicts our assumptions on i and i'.

Now given a BD-structure on a set $S = \{1, \ldots, N\}$ with the complete order $1 < 2 < \cdots < N$ and the cyclic permutation $C = C_0^{-k}$ (where $N/2 \le k < N$) we define the sequence $(a_1, \ldots, a_N)$ as follows. Set $a_1 = 1$, and for $i = 1, \ldots, N - 1$ set

$$a_{i+1} = \begin{cases} a_i & \text{if } \alpha_{k-i} \in \Gamma_1, \\ a_i + 1 & \text{otherwise} \end{cases}$$

where $\alpha_j = (j, j+1)$ (this uniquely defines n). It is easy to check that the corresponding matrix (m_i^j) realizes our BD-structure. $\square$

6 Solutions of the AYBE and Associative BD-Structures

Let $(S, <, C, \Gamma_1, \Gamma_2)$ be a completely ordered finite set with an associative BD-structure such that $\alpha_0 \notin \Gamma_2$. As in the introduction, for an element $\alpha = (i, j) \in S \times S$ we set $e_\alpha = e_{ij} \in A_S \simeq \mathrm{Mat}_N(\mathbb{C})$ (where $N = |S|$, the rows and columns are numbered by S). We write $(i, j) > 0$ (resp., $(i, j) < 0$) if $i < j$ (resp., $i > j$). Also, for $\alpha = (i, j)$ we set $-\alpha = (j, i)$. Mimicking formulas (32) and (33) we define

$$r_{\mathrm{const}}(x, z) = \frac{z}{1 - z} \sum_{\alpha > 0} e_{-\alpha} \otimes e_\alpha + \frac{1}{1 - z} \sum_{\alpha > 0} e_\alpha \otimes e_{-\alpha}$$
$$+ \frac{z}{1 - z} \sum_{i \in S} e_{ii} \otimes e_{ii}$$
$$+ (1 - x^N)^{-1} \sum_{i \in S} \sum_{k=0}^{N-1} x^k e_{i,i} \otimes e_{C^k(i), C^k(i)}, \quad (35)$$

$$r(x; y, y') = r_{\mathrm{const}}(x, y/y')$$
$$+ \sum_{\alpha > 0, k \ge 1} [x^k e_\alpha \otimes e_{-\tau^k(\alpha)} - x^{-k} e_{-\tau^k(\alpha)} \otimes e_\alpha]$$
$$+ \sum_{\alpha < 0, k \ge 1} [y x^k e_\alpha \otimes e_{-\tau^k(\alpha)} - y' x^{-k} e_{-\tau^k(\alpha)} \otimes e_\alpha]. \quad (36)$$

In the last formula we use the operation τ defined on $P_1 \subset S \times S$; the summation is extended only over those (k, α) for which $\tau^k(\alpha)$ is defined. Below we will show that $r(x; y, y')$ is a solution of (21) (see Theorem 6.2). To deduce from this Theorem 0.2(i) we will use the following simple observation.

Lemma 6.1. *In the above situation the $A_S \otimes A_S$-valued function*

$$-r\left(\exp\left(\frac{u_1 - u_2}{N}\right); \exp(v_1), \exp(v_2)\right)$$

is equivalent to the one given in Theorem 0.2(i) for the inverse associative BD-structure $(C_0, C^{-1}, \Gamma_2, \Gamma_1)$, where $u = u_1 - u_2$ and $v = v_1 - v_2$.

Proof. We can assume that $S = [1, N]$ (the segment of natural numbers) with the standard order. Let us set

$$\varphi(v)e_j = \exp\left(-\frac{jv}{N}\right)e_j.$$

Then the corresponding equivalent matrix $\tilde{r}(u_1, u_2; v_1, v_2)$ is obtained from $r(\exp(\frac{u_1 - u_2}{N}); \exp(v_1), \exp(v_2))$ by multiplying each term $e_{ij} \otimes e_{j'i'}$ by $\exp(\frac{(j-i)v_1 - (j'-i')v_2}{N})$. Now we observe that $r_{\text{const}}(x, y/y')$ is a linear combination of $e_{ij} \otimes e_{j'i'}$, where $j - i = j' - i'$. Such a term gets multiplied by $\exp(\frac{(j-i)(v_1-v_2)}{N})$. The same is true about the terms in $r(x; y, y')$ not containing y or y'. Indeed, if $i < j$ and τ^k is defined on (i, j) then $C^k(j) - C^k(i) = j - i$. On the other hand, the terms involving $y = \exp(v_1)$ and $y' = \exp(v_2)$ are linear combinations of $e_{i,j} \otimes e_{j',i'}$, where $j' - i' = j - i + N$. Indeed, this follows from the fact that if $i > j$ and τ^k is defined on (i, j) then $C^k(j) - C^k(i) = j - i + N$ (the proof reduces to the case $(i, j) = (N, 1)$). The only other observation we use to rewrite $-\tilde{r}$ in the form given in Theorem 0.2(i) (with C replaced by C^{-1} and Γ_1 and Γ_2 exchanged) is that for $0 < m < N$ and for $i, j \in [1, N]$ we have $j - i \equiv m(N)$ iff either $i < j$ and $j = i + m$, or $i > j$ and $j = i + m - N$.   □

Since for every associative BD-structure on a finite set S we can choose a compatible complete order in such a way that $\alpha_0 \notin \Gamma_2$, Theorem 0.2(i) will follow easily from the above lemma and the next result.

Theorem 6.2. *Let $(S, <, C, \Gamma_1, \Gamma_2)$ be a completely ordered finite set with an associative BD-structure such that $\alpha_0 \notin \Gamma_2$. Then the function $r(x; y, y')$ given by (36) is a solution of (21) satisfying the unitarity condition (22).*

Remark 6.3. By Theorem 5.5 we already know the statement to be true if $C = C_0^k$. Also, the work [8] deals with the case in which in addition, $\alpha_0 \notin \Gamma_1$ (this fact will be used below).

The rest of this section will be occupied with the proof of Theorem 6.2 (in the end we will also explain how to deduce Theorem 0.2(i)).

Let us denote by $P = \sum_{i,j} e_{ij} \otimes e_{ji}$ the permutation tensor. Then we can rewrite our r-matrix in the form

$$r(x; y, y') = a(x) + yb(x) - y'c(x) + \frac{y}{y' - y}P,$$

where

$$a(x) = (1 - x^N)^{-1} \sum_{i \in S} \sum_{k=0}^{N-1} x^k e_{i,i} \otimes e_{C^k(i), C^k(i)} + \sum_{\alpha > 0} e_\alpha \otimes e_{-\alpha}$$

$$+ \sum_{\alpha > 0, k \geq 1} [x^k e_\alpha \otimes e_{-\tau^k(\alpha)} - x^{-k} e_{-\tau^k(\alpha)} \otimes e_\alpha],$$

$$b(x) = \sum_{\alpha<0, k\geq 1} x^k e_\alpha \otimes e_{-\tau^k(\alpha)},$$

$$c(x) = b^{21}(x^{-1}) = \sum_{\alpha<0, k\geq 1} x^{-k} e_{-\tau^k(\alpha)} \otimes e_\alpha.$$

Lemma 6.4. *Assume that $\alpha_0 \notin \Gamma_2$. Let us set $\Gamma_1' = \Gamma_1 \setminus \{\alpha_0\}$, $\Gamma_2' = \tau(\Gamma_1')$. Then*

$$a(x) + \frac{y}{y'-y}P$$

is exactly the r-matrix corresponding to the associative BD-structure $(S, <, C, \Gamma_1')$.

Proof. It is easy to see that $P_1' = \{\alpha \in P_1 \mid \alpha > 0\}$. Thus, the terms $b(x)$ and $c(x)$ in the r-matrix associated with the new associative BD-structure vanish. We claim that the term $a(x)$ for the new associative BD-structure is the same as for the old one. Indeed, it is enough to check that $\alpha \in P_1'$ is in the domain of definition of τ^k iff it is in the domain of $(\tau')^k$, where $\tau' : P_1' \to P_2'$ is the bijection induced by τ. But this follows immediately from the fact that P_2 consists only of $\alpha > 0$ (due to the assumption that $\alpha_0 \notin \Gamma_2$).   □

Let us denote by $\mathrm{AYBE}[r](x, x'; y_1, y_2, y_3)$ the left-hand side of (21).

Lemma 6.5. *Consider the r-matrix of the form*

$$r(x; y_1, y_2) = a(x) + y_1 b(x) - y_2 c(x) + \frac{y_1}{y_2 - y_1} P, \tag{37}$$

where $a^{21}(x^{-1}) + a(x) = P$ and $b^{21}(x^{-1}) = c(x)$. Then r satisfies the unitarity condition (22). Also, $\mathrm{AYBE}[r] = 0$ iff the following equations are satisfied:

(i) $\mathrm{AYBE}[a] = 0$;
(ii) $b^{12}(x)b^{13}(x') = 0$;
(iii) $b^{13}(x)b^{23}(x') = b^{21}(x')b^{13}(xx') + b^{23}(xx')b^{12}(x)$;
(iv) $c^{13}(x)a^{23}(x') + a^{12}((x')^{-1})c^{13}(xx') = c^{23}(xx')a^{12}(x) - a^{13}(x)c^{23}(x')$.

Proof. The unitarity condition follows immediately from our assumptions on $a(x)$, $b(x)$, and $c(x)$. It is easy to check that

$$\mathrm{AYBE}[a(x) + y_1 b(x) - y_2 c(x) + \frac{y_1}{y_2 - y_1} P](x, x'; y_1, y_2, y_3)$$
$$= \mathrm{AYBE}[a(x) + y_1 b(x) - y_2 c(x)](x, x'; y_1, y_2, y_3)$$
$$- y_1 c^{21}(x') P^{13} - y_2 c^{13}(x) P^{23} - y_1 b^{23}(xx') P^{12}.$$

Now the conditions (i)–(iv) are obtained by equating to zero coefficients with various monomials in y_1, y_2, and y_3 (of degree ≤ 2). Namely, (i) is obtained by looking at the constant term (i.e., by substituting $y_i = 0$). Conditions (ii), (iii), and (iv) are obtained by looking at the coefficients with y_1^2, $y_1 y_2$, and y_3, respectively. To see that these conditions imply AYBE[r] = 0 we can use the identity

$$\text{AYBE}[r](x, x'; y_2, y_3, y_1)^{231} = \text{AYBE}[r]((xx')^{-1}, x; y_1, y_2, y_3),$$

which holds for any r satisfying the unitarity condition (22).   □

Let us introduce the following notation. For every $k \geq 1$ we denote by $P(k) \subset P_1$ the domain of definition of τ^k and by $P(k)^+ \subset P(k)$ (resp., $P(k)^- \subset P(k)$) the set of all $\alpha > 0$ (resp., $\alpha < 0$) contained in $P(k)$. Note that $P(1) = P_1$. The assumption $\alpha_0 \notin \Gamma_2$ implies that $\tau(P(k)) \subset P(k-1)^+$. Using this notation we can rewrite our formulas for $a(x)$, $b(x)$, and $c(x)$ as follows:

$$a(x) = (1 - x^r)^{-1} \sum_{0 \leq k < r, i} x^k e_{i,i} \otimes e_{C^k(i), C^k(i)} + \sum_{i < j} e_{i,j} \otimes e_{j,i}$$

$$+ \sum_{(i,j) \in P(k)^+} [x^k e_{i,j} \otimes e_{C^k(j), C^k(i)} - x^{-k} e_{C^k(j), C^k(i)} \otimes e_{i,j}],$$

$$b(x) = \sum_{k \geq 1, (i,j) \in P(k)^-} x^k e_{i,j} \otimes e_{C^k(j), C^k(i)},$$

$$c(x) = \sum_{k \geq 1, (i,j) \in P(k)^-} x^{-k} e_{C^k(j), C^k(i)} \otimes e_{i,j}.$$

The following two combinatorial observations are also going to be useful in the proof.

Lemma 6.6. *Let (i_1, i_2, i_3) be a triple of elements of S and let $k \geq 1$. Then the following two conditions are equivalent:*

(a) $(i_1, i_3) \in P(k)^-$ and $i_1 < i_2$ (resp., $i_2 < i_3$);
(b) $(i_1, i_2) \in P(k)^+$ and $(i_2, i_3) \in P(k)^-$ (resp., $(i_1, i_2) \in P(k)^-$ and $(i_2, i_3) \in P(k)^+$).

The proof is straightforward and is left to the reader.

Lemma 6.7. *Let $k \geq 1$. Then for every $(i_1, i_2) \in P(k)^-$ one has a decomposition $S = S_1 \sqcup S_2$, where*

$$S_1 = \{i \mid i < i_1, C^k(i) > C^k(i_1)\}, \quad S_2 = \{i \mid i > i_2, C^k(i) < C^k(i_2)\}.$$

Proof. We can assume that $S = [1, N]$ with the standard order. Note that the map C^k restricts to a bijection

$$[i_1, N] \sqcup [1, i_2] \widetilde{\rightarrow} [C^k(i_1), C^k(i_2)].$$

Passing to the complements, we derive that the open segment (i_2, i_1) is the disjoint union of its intersections with S_1 and S_2. Next, if $i \leq i_2$ then $(i_1, i) \in$

$P(k)^-$ (by Lemma 6.6), so that $C^k(i_1) < C^k(i)$. Hence, $[1, i_2] \subset S_1 \setminus S_2$. Similarly, $[i_1, N] \subset S_2 \setminus S_1$.   □

Proof of Theorem 6.2. Let us check that equations (i)–(iv) of Lemma 6.5 hold in our case. Equation (i) follows from Lemma 6.4 and Theorem 3.4 of [8]. More precisely, one can easily check that in the case when $\alpha_0 \notin \Gamma_1$ and $\alpha_0 \notin \Gamma_2$, our r-matrix coincides with the associative r-matrix constructed in [8] for the opposite associative BD-structure on S. Equation (ii) follows from the fact that for any $(i, j), (i', j') \in P(1)^-$ one has $i' > j$ and $i > j'$ (otherwise we would have $\Gamma_1 = \Gamma_S$). To check equation (iii) we write

$$b^{13}(x)b^{23}(x') = \sum_{\substack{k \geq 1, m \geq 1; (i,j) \in P(k)^-, \\ (i',j') \in P(m)^-; C^k(i) = C^m(j')}} x^k(x')^m e_{i,j} \otimes e_{i',j'} \otimes e_{C^k(j),C^m(i')}.$$

Note that we cannot have $k = m$, since this would imply that $i = j'$, contradicting the assumption that $(i, j) \in P(k)^- \subset P(1)^-$ and $(i', j') \in P(m)^- \subset P(1)^-$. Hence, we can split the summation into two parts: one with $k > m$ and one with $k < m$. Denoting $k - m$ (resp., $m - k$) by l in the first (resp., second) case, we can rewrite these sums as

$$\Sigma_1 = \sum_{\substack{l \geq 1, m \geq 1; (i,j) \in P(m+l)^-, \\ (i',C^l(i)) \in P(m)^-}} x^l(xx')^m e_{i,j} \otimes e_{i',C^l(i)} \otimes e_{C^{m+l}(j),C^m(i')},$$

$$\Sigma_2 = \sum_{\substack{l \geq 1, m \geq 1; (i',j') \in P(m+l)^-, \\ (C^l(j'),j) \in P(m)^-}} (xx')^m(x')^l e_{C^l(j'),j} \otimes e_{i',j'} \otimes e_{C^m(j),C^{m+l}(i')}.$$

On the other hand, we have

$$b^{23}(xx')b^{12}(x) = \sum_{\substack{l \geq 1, m \geq 1; (i,j) \in P(l)^-, \\ (i',C^l(j)) \in P(m)^-}} x^l(xx')^m e_{i,j} \otimes e_{i',C^l(i)} \otimes e_{C^{m+l}(j),C^m(i')}.$$

We claim that this is equal to Σ_1. Indeed, the condition $(i, j) \in P(m+l)^-$ is equivalent to the conjuction of $(i, j) \in P(l)^-$ and $(C^l(i), C^l(j)) \in P(m)^+$. Now our claim follows from Lemma 6.6 applied to the triple $(i', C^l(i), C^l(j))$ (recall that $C^l(i) < C^l(j)$, since $(i, j) \in P(l)$). Similarly, we check that $b^{21}(x')b^{13}(xx') = \Sigma_2$, which finishes the proof of equation (iii).

Finally, let us verify equation (iv). We can split both terms in the left-hand side of this equation into four sums according to the four pieces that constitute $a(x)$:

$$c^{13}(x)a^{23}(x') = L_1 + L_2 + L_3 - L_4, \quad a^{12}((x')^{-1})c^{13}(xx') = -L_5 + L_6 + L_7 - L_8,$$

where

$$L_1 = (1 - (x')^N)^{-1}$$
$$\times \sum_{0 \leq m < N, k \geq 1; (i,j) \in P(k)^-} x^{-k}(x')^m e_{C^k(j),C^k(i)} \otimes e_{C^{N-m}(j),C^{N-m}(j)} \otimes e_{i,j},$$

$$L_2 = \sum_{m \geq 1; i < j, (i',j) \in P(m)^-} x^{-m} e_{C^m(j),C^m(i')} \otimes e_{i,j} \otimes e_{i',i},$$

$$L_3 = \sum_{\substack{k\geq 1, m\geq 1;(i,j)\in P(k)+, \\ (i',C^k(j))\in P(m)-}} x^{-m}(x')^k e_{C^{k+m}(j),C^m(i')} \otimes e_{i,j} \otimes e_{i',C^k(i)},$$

$$L_4 = \sum_{\substack{k\geq 1, m\geq 1;(i,i')\in P(m)-, \\ (i',j)\in P(k)+}} x^{-m}(x')^{-k} e_{C^m(i'),C^m(i)} \otimes e_{C^k(j),C^k(i')} \otimes e_{i,j},$$

$$L_5 = (1-(x')^N)^{-1}$$
$$\times \sum_{\substack{0\leq m<N,k\geq 1; \\ (i,j)\in P(k)-}} (x')^{N-m}(xx')^{-k} e_{C^k(j),C^k(i)} \otimes e_{C^{k+m}(j),C^{k+m}(j)} \otimes e_{i,j},$$

$$L_6 = \sum_{k\geq 1,(i,j)\in P(k)-,i'<C^k(j)} (xx')^{-k} e_{i',C^k(i)} \otimes e_{C^k(j),i'} \otimes e_{i,j},$$

$$L_7 = \sum_{\substack{k\geq 1, m\geq 1;(i,j)\in P(k)-, \\ (i',C^k(j))\in P(m)+}} (x')^{-m}(xx')^{-k} e_{i',C^k(i)} \otimes e_{C^{k+m}(j),C^m(i')} \otimes e_{i,j},$$

$$L_8 = \sum_{\substack{k\geq 1, m\geq 1;(i,j)\in P(k)-, \\ (i',j')\in P(m)+,C^m(i')=C^k(j)}} (x')^m(xx')^{-k} e_{C^m(j'),C^k(i)} \otimes e_{i',j'} \otimes e_{i,j}.$$

We split each of the sums L_4 and L_8 into three parts according to the ranges of summation $k = m$, $k > m$, and $k < m$ (in the last two cases we make substitutions $k \mapsto k+m$ and $m \mapsto k+m$, respectively):

$$L_4 = L_{4,1} + L_{4,2} + L_{4,3}, \quad L_8 = L_{8,1} + L_{8,2} + L_{8,3},$$

where

$$L_{4,1} = \sum_{k\geq 1,(i,j)\in P(k)-,i'<j} (xx')^{-k} e_{C^k(i'),C^k(i)} \otimes e_{C^k(j),C^k(i')} \otimes e_{i,j},$$

$$L_{4,2} = \sum_{\substack{k\geq 1, m\geq 1;(i,i')\in P(m)-, \\ (i',j)\in P(k+m)+}} x^{-m}(x')^{-k-m} e_{C^m(i'),C^m(i)} \otimes e_{C^{k+m}(j),C^{k+m}(i')} \otimes e_{i,j},$$

$$L_{4,3} = \sum_{\substack{k\geq 1, m\geq 1;(i,i')\in P(k+m)-, \\ (i',j)\in P(k)+}} x^{-k-m}(x')^{-k} e_{C^{k+m}(i'),C^{k+m}(i)} \otimes e_{C^k(j),C^k(i')} \otimes e_{i,j},$$

$$L_{8,1} = \sum_{k\geq 1,(i,j')\in P(k)-,j<j'} x^{-k} e_{C^k(j'),C^k(i)} \otimes e_{j,j'} \otimes e_{i,j},$$

$$L_{8,2} = \sum_{\substack{k\geq 1, m\geq 1;(i,j)\in P(k+m)-, \\ (C^k(j),j')\in P(m)+}} x^{-k-m}(x')^{-k} e_{C^m(j'),C^{k+m}(i)} \otimes e_{C^k(j),j'} \otimes e_{i,j},$$

$$L_{8,3} = \sum_{\substack{k\geq 1, m\geq 1;(i,C^m(i'))\in P(k)-, \\ (i',j')\in P(k+m)+}} x^{-k}(x')^m e_{C^{k+m}(j'),C^k(i)} \otimes e_{i',j'} \otimes e_{i,C^m(i')}.$$

Making appropriate substitutions of the summation variables and using Lemma 6.6, one can easily check that

$$L_2 = L_{8,1}, \quad L_3 = L_{8,3}.$$

It follows that the left-hand side of (iv) is equal to

$$(L_1 - L_5) + (L_6 - L_{4,1}) + (L_7 - L_{4,2}) - (L_{4,3} + L_{8,2}).$$

Next, making the substitution $m \mapsto N - k - m$ in the sum L_5, we obtain

$$L_1 - L_5 = - \sum_{0<m<k,(i,j)\in P(k)^-} x^{-k}(x')^{-m} e_{C^k(j),C^k(i)} \otimes e_{C^m(j),C^m(j)} \otimes e_{i,j}.$$

Also, substituting i' by $C^k(i')$ in L_6, switching k and m in $L_{4,2}$, and using Lemma 6.6, we find that

$$L_6 - L_{4,1} = \sum_{\substack{k\geq 1,(i,j)\in P(k)^-,\\ i'>j,C^k(i')<C^k(j)}} (xx')^{-k} e_{C^k(i'),C^k(i)} \otimes e_{C^k(j),C^k(i')} \otimes e_{i,j},$$

$$L_7 - L_{4,2} =$$

$$\sum_{\substack{k\geq 1,m\geq 1,(i,j)\in P(k)^-,\\ (C^k(i'),C^k(j))\in P(m)^+,i'>j}} x^{-k}(x')^{-k-m} e_{C^k(i'),C^k(i)} \otimes e_{C^{k+m}(j),C^{k+m}(i')} \otimes e_{i,j}.$$

Finally, we can rewrite the sum of the other remaining terms as follows:

$$L_1 - L_5 - L_{4,3} - L_{8,2} =$$

$$-\sum_{k\geq 1,m\geq 1,(i,i',j)\in \Pi(k,m)} x^{-k-m}(x')^{-k} e_{C^{k+m}(i'),C^{k+m}(i)} \otimes e_{C^k(j),C^k(i')} \otimes e_{i,j},$$

where $\Pi(k,m)$ is the subset of $\{(i,i',j) \mid (i,j) \in P(k)^-, (C^k(i),C^k(i')) \in P(m)^+\}$ consisting of (i,i',j) such that either $i' \leq j$ or $C^k(j) < C^k(i')$. It follows from Lemma 6.7 that

$$\Pi(k,m) = \{(i,i',j) \mid (i,j) \in P(k)^-, (C^k(i),C^k(i')) \in P(m)^+, i' < i\}.$$

We deal similarly with the right-hand side of equation (iv). Namely, we write

$$c^{23}(xx')a^{12}(x) = R_1 + R_2 + R_3 - R_4, \quad a^{13}(x)c^{23}(x') = R_5 + R_6 + R_7 - R_8,$$

where the parts correspond to the summands in $a(x)$. We also have a decomposition $R_3 = R_{3,1} + R_{3,2} + R_{3,3}$ (resp., $R_8 = R_{8,1} + R_{8,2} + R_{8,3}$) obtained by collecting terms with $x^k(xx')^{-m}$ (resp., $x^{-k}(x')^{-m}$) with $k = m$, $k > m$ and $k < m$. Now one can easily check that

$$R_6 = R_{3,1}, \quad R_7 = R_{3,2}.$$

Also, we have

$$R_1 - R_5 = \sum_{\substack{m\geq 1,0<k\leq m;\\ (i,j)\in P(m)^-}} x^{-k}(x')^{-m} e_{C^k(i),C^k(i)} \otimes e_{C^m(j),C^m(i)} \otimes e_{i,j}.$$

We denote by $(R_1 - R_5)_{k=m}$ and by $(R_1 - R_5)_{k<m}$ parts of this sum corresponding to the ranges $k = m$ and $k < m$. Then we have

$$(R_1 - R_5)_{k=m} + R_2 + R_{8,1}$$
$$= \sum_{\substack{k \geq 1, (i,j) \in P(k)^-, \\ i \leq i' \text{ or } C^k(i') < C^k(i)}} (xx')^{-k} e_{C^k(i'), C^k(i)} \otimes e_{C^k(j), C^k(i')} \otimes e_{i,j}.$$

Using Lemma 6.7 it is easy to see that the condition on (i, j, i') in this summation can be replaced by the conjunction of $(i, j) \in P(k)^-$, $j < i'$, and $C^k(i') < C^k(j)$ (same as in the formula for $L_6 - L_{4,1}$). Finally, we have

$$R_{8,2} - R_4 =$$
$$- \sum_{\substack{k \geq 1, m \geq 1, (i,j) \in P(m)^-, \\ j' < i, (C^m(i), C^m(j')) \in P(k)^+}} x^{-k-m} (x')^{-m} e_{C^{k+m}(j'), C^{k+m}(i)} \otimes e_{C^m(j), C^m(j')} \otimes e_{i,j},$$

$$(R_1 - R_5)_{k<m} + R_{3,3} + R_{8,3} =$$
$$\sum_{k \geq 1, m \geq 1, (i,j',j) \in \Pi'(k,m)} x^{-k} (x')^{-k-m} e_{C^k(j'), C^k(i)} \otimes e_{C^{k+m}(j), C^{k+m}(j')} \otimes e_{i,j},$$

where $\Pi'(k, m)$ is the subset of $\{(i, j', j) \mid (i, j) \in P(k)^-, (C^k(j'), C^k(j)) \in P(m)^+\}$ consisting of (i, j', j) such that either $i \leq j'$ or $C^k(j') < C^k(i)$. By Lemma 6.7, we get

$$\Pi'(k, m) = \{(i, j', j) \mid (i, j) \in P(k)^-, (C^k(j'), C^k(j)) \in P(m)^+, j < j'\}.$$

Now it is easy to see that parts of the left-hand side and the right-hand side of equation (iv) match as follows:

$$L_6 - L_{4,1} = (R_1 - R_5)_{k=m} + R_2 + R_{8,1},$$
$$L_7 - L_{4,2} = (R_1 - R_5)_{k<m} + R_{3,3} + R_{8,3},$$
$$L_1 - L_5 - L_{4,3} - L_{8,2} = R_{8,2} - R_4.$$

$\square$

Proof of Theorem 0.2(i). As was already observed, the fact that $r(u, v)$ is a unitary solution of the AYBE follows from Lemma 6.1 and Theorem 6.2. It follows from Theorem 1.5 that $r(u, v)$ also satisfies the QYBE for fixed u. It remains to check the unitarity condition for the quantum R-matrix given by (6). In view of the unitarity of $r(u, v)$, this boils down to proving the identity

$$s(u, v) = \left(\left[\exp \left(\frac{v}{2} \right) - \exp \left(-\frac{v}{2} \right) \right]^{-2} - \left[\exp \left(\frac{u}{2} \right) - \exp \left(-\frac{u}{2} \right) \right]^{-2} \right) \cdot 1 \otimes 1. \tag{38}$$

To this end we observe that from Theorem 1.5 and Lemma 1.9 we know that

$$s(u,v) = (f(u) + g(v)) \cdot 1 \otimes 1,$$

where $f(u) = \frac{1}{N} \operatorname{tr} \mu(\frac{\partial r(u,0)}{\partial u})$ and $g(v) = -\frac{1}{N}(\operatorname{tr} \otimes \operatorname{tr})(\frac{dr_0(v)}{dv})$. Now (38) follows immediately from the equalities

$$f(u) = \frac{d}{du}\left(\frac{1}{\exp(u) - 1}\right) = -\left[\exp\left(\frac{u}{2}\right) - \exp\left(-\frac{u}{2}\right)\right]^{-2},$$

$$g(v) = \frac{d}{dv}\left(\frac{1}{\exp(-v) - 1}\right) = \left[\exp\left(\frac{v}{2}\right) - \exp\left(-\frac{v}{2}\right)\right]^{-2}.$$

$\square$

Remark 6.8. The following interesting observation is due to T. Schedler. Assume that Γ_1 does not contain two consecutive elements of Γ_{C_0}, say, $(C_0^{-1}(i_0), i_0)$ and $(i_0, C_0(i_0))$. Then the function $r(u,v)$ given by Theorem 0.2(i) is equivalent to a function of the form $\frac{1 \otimes 1}{\exp(u)-1} + r(v)$. Indeed, let us denote by $O(i_0, i)$ the minimal $k \geq 0$ such that $C^k(i_0) = i$. Then one can easily check that

$$a = \sum_i \frac{O(i_0, i)}{N} e_{ii}$$

is an infinitesimal symmetry of $r(u,v)$ and

$$\exp[u(1 \otimes a)]r(u,v)\exp[-u(a \otimes 1)] = \frac{1 \otimes 1}{\exp(u) - 1} + r(v),$$

where $r(v)$ depends only on v. Note that the fact that $r(u,v)$ is a unitary solution of the AYBE is equivalent to the following equations on $r(v)$:

$$\text{AYBE}[\mathbf{r}](v, v') = \mathbf{r}^{13}(v + v'), \quad \mathbf{r}^{21}(-v) + \mathbf{r}(v) = 1 \otimes 1.$$

7 Meromorphic Continuation

As was shown in the proof of Theorem 6 of [7] (see also Lemma 4.14 of [8]), a unitary solution of the AYBE with the Laurent expansion (5) at $u = 0$ is uniquely determined by $r_0(v)$. Therefore, it is not surprising that some of the results from [3] about solutions of the CYBE (such as meromorphic continuation) can be extended to solutions of the AYBE.

First, we apply the above uniqueness principle to infinitesimal symmetries.

Lemma 7.1. *Let $r(u,v)$ be a nondegenerate unitary solution of the AYBE with the Laurent expansion (5) at $u = 0$. Then the algebras of infinitesimal symmetries of $r(u,v)$ and of $r_0(v)$ are the same (and are contained in the algebra of infinitesimal symmetries of $\bar{r}_0$). If in addition $\bar{r}_0$ has a period, then these coincide with the commutative algebra of infinitesimal symmetries of $\bar{r}_0$.*

Proof. Let $a \in A$ be an infinitesimal symmetry of $r_0(v)$. Then for any $t \in \mathbb{C}$ the function

$$\exp[t(a \otimes 1 + 1 \otimes a)]r(u,v)\exp[-t(a \otimes 1 + 1 \otimes a)]$$

is a solution of the AYBE with the same r_0-term in the Laurent expansion at $u = 0$. By the uniqueness mentioned above this implies that $\exp[t(a \otimes 1 + 1 \otimes a)]$ commutes with $r(u,v)$, so a is an infinitesimal symmetry of $r(u,v)$. Recall that by Theorem 1.5(i), $\overline{r}_0(v)$ is nondegenerate. It is easy to see that if $\overline{r}_0$ is either elliptic or trigonometric then the algebra of infinitesimal symmetries of $\overline{r}_0$ is commutative. Indeed, in the elliptic case this algebra is trivial (see Lemma 5.1 of [7]). In the trigonometric case this follows from the fact proven in [3] that there exists a pole γ of $\overline{r}_0$ such that

$$\overline{r}_0(v + \gamma) = (\phi \otimes \mathrm{id})(\overline{r}_0(v)),$$

where ϕ is a Coxeter automorphism of sl_N. Thus, any infinitesimal symmetry is contained in the commutative algebra of ϕ-invariant elements.   □

Proposition 7.2. *Assume $N > 1$. Let $r(u,v)$ be a nondegenerate unitary solution of the AYBE with the Laurent expansion (5) at $u = 0$ such that the equivalent conditions of Theorem 1.5(ii) hold. Then $r(u,v)$ admits a meromorphic continuation to $D \times \mathbb{C}$, where D is a neighborhood of 0 in $\mathbb{C}$. If $r(u,v)$ has a pole at $v = \gamma$ then this pole is simple and $\overline{r}_0(v)$ also has a pole at $v = \gamma$.*

Proof. Note that $\overline{r}_0(v)$ has a meromorphic continuation to $\mathbb{C}$ with at most simple poles by Theorem 1.1 of [3]. First, we want to deduce a meromorphic continuation for $r_0(v)$. From the condition (c) in Theorem 1.5(ii) we know that

$$r_0(v) = \overline{r}_0(v) + b \otimes 1 - 1 \otimes b + h(v) \cdot 1 \otimes 1, \tag{39}$$

where b is an infinitesimal symmetry of $\overline{r}_0(v)$ (by Lemma 1.10). Note that b is also an infinitesimal symmetry of $r_0(v)$. Hence, by Lemma 7.1, b is an infinitesimal symmetry of $r(u,v)$. Applying the equivalence transformation

$$r(u,v) \mapsto \exp[u(1 \otimes b)]r(u,v)\exp[-u(b \otimes 1)],$$

we can assume that $b = 0$. In this case we have

$$\mathrm{AYBE}[r_0](v_{12}, v_{23}) - \mathrm{AYBE}[\overline{r}_0](v_{12}, v_{23})$$
$$\equiv [h(v_{13}) - h(v_{23})]\overline{r}_0^{12}(v_{12}) + c.p.(1,2,3) \mod (\mathbb{C} \cdot 1 \otimes 1 \otimes 1),$$

where we use the notation from the proof of Theorem 1.5 (the omitted terms are obtained by cyclically permuting $(1,2,3)$; we set $v_{ij} = v_i - v_j$). Applying $\mathrm{pr} \otimes \mathrm{pr} \otimes \mathrm{id}$ and using (18), we obtain

$$[h(v + v') - h(v')]\bar{r}_0^{12}(v) = [(\mathrm{pr} \otimes \mathrm{pr})r_1(v)]^{12} - (\mathrm{pr} \otimes \mathrm{pr} \otimes \mathrm{id})\mathrm{AYBE}[\bar{r}_0](v, v').$$
$$(40)$$

Note that $\mathrm{AYBE}[\bar{r}_0](v, v')$ is meromorphic on all of $\mathbb{C} \times \mathbb{C}$ and has at most simple poles at $v = \gamma$, $v' = \gamma$, and $v + v' = \gamma$, where γ is a pole of $\bar{r}_0(v)$. Also, by Lemma 1.7, $r_1(v)$ is holomorphic near $v = 0$. Choose a small disk D around zero such that $r_1(v)$ is holomorphic in D and $\bar{r}_0(v)$ has no poles or zeros in $D \setminus \{0\}$. Assume that we already have a meromorphic continuation of $h(z)$ to some open subset $U \subset \mathbb{C}$ containing zero. Then the above formula gives a meromorphic continuation of $h(z)$ to $U + D$. Iterating this process, we continue $h(z)$ meromorphically to the entire complex plane. Furthermore, it is clear from (40) that $h(v)$ has only simple poles and is holomorphic outside the set of poles of $\bar{r}_0(v)$. Therefore, the same is true for $r_0(v)$.

Next, considering the constant terms of the Laurent expansions of the AYBE in u' (keeping u fixed), we get

$$r_0^{12}(v_{12})r^{13}(u, v_{13}) + r^{13}(u, v_{13})r_0^{23}(v_{23}) - r^{23}(u, v_{23})r^{12}(u, v_{12}) = \frac{\partial r^{13}}{\partial u}(u, v_{13}).$$
$$(41)$$

Since we already know that $r_0(v)$ is meromorphic on all of $\mathbb{C}$, we can use this equation to get a meromorphic continuation of $r(u, v)$. Indeed, assume that $r(u, v)$ is meromorphic in $D \times D$ for some open disk $D \subset \mathbb{C}$ around zero. For fixed $v_{13} \in D$ the above equation gives a meromorphic extension of

$$r^{23}(u, v_{21} + v_{13})r^{21}(-u, v_{21}) = -r^{23}(u, v_{23})r^{12}(u, v_{12})$$

to $D \times \mathbb{C}$. By the nondegeneracy of $r(u, v)$ this allows us to extend $r(u, v)$ meromorphically from $D \times U$ to $D \times (U + D)$. Iterating this process, we get the required meromorphic extension. The assertion about poles follows easily from (41) by fixing v_{13} such that $r(u, v)$ has no pole at $v = v_{13}$ and $r(u, v_{13} - \gamma)$ is nondegenerate, and considering the polar parts at $v_{12} = \gamma$. ☐

The argument in the following lemma is parallel to that in Proposition 4.3 of [3].

Lemma 7.3. *With the same assumptions as in Proposition 7.2, for every pole γ of $\bar{r}_0(v)$ there exists an algebra automorphism ϕ_γ of A and a constant $\lambda \in \mathbb{C}$ such that*

$$r(u, v + \gamma) = \exp(\lambda u)(\phi_\gamma \otimes \mathrm{id})(r(u, v)).$$

Proof. From Proposition 7.2 we know that the pole of $r(u, v)$ at $v = \gamma$ is simple. Set $\tau(u) = \lim_{v \to \gamma}(v - \gamma)r(u, v)$. Recall that $\lim_{v \to 0} vr(u, v) = cP$ for $c \in \mathbb{C}^*$. Let us define an operator $\phi(u) \in \mathrm{End}(A)$ by the equality

$$\tau(u) = (\phi(u) \otimes \mathrm{id})(cP).$$

Considering polar parts near $v = \gamma$ in (1), we get

$$\tau^{12}(-u')r^{13}(u + u', v' + \gamma) = r^{23}(u + u', v')\tau^{12}(u).$$

The right-hand side can be rewritten as follows:

$$r^{23}(u + u', v')\tau^{12}(u)$$
$$= c(\phi(u) \otimes \text{id} \otimes \text{id})(r^{23}(u + u', v')P^{12})$$
$$= c(\phi(u) \otimes \text{id} \otimes \text{id})(P^{12}r^{13}(u + u', v')).$$

Hence, we have

$$\tau^{12}(-u')r^{13}(u + u', v' + \gamma) = c(\phi(u) \otimes \text{id} \otimes \text{id})(P^{12}r^{13}(u + u', v')). \quad (42)$$

Taking the residues at $v' = 0$, we obtain

$$\tau^{12}(-u')\tau^{13}(u + u') = c^2(\phi(u) \otimes \text{id} \otimes \text{id})(P^{12}P^{13}).$$

This means that $\phi(u)$ satisfies the identity

$$\phi(u_1 + u_2)(XY) = \phi(u_1)(X)\phi(u_2)(Y),$$

where $X, Y \in A$. Let D be a small disk around zero in $\mathbb{C}$ such that $\phi(u)$ is holomorphic on $D \setminus \{0\}$. For every $u \in D \setminus \{0\}$ we denote by $I(u) \subset A$ the kernel of $\phi(u)$. Then from the above identity we derive that $I(u)A \subset I(u+u')$ and $AI(u) \subset I(u+u')$ whenever $u, u', u+u' \in D \setminus \{0\}$. In particular, we deduce that $I(u) \subset I(u + u')$, so $I(u) = I \subset A$ does not depend on $u \in D \setminus \{0\}$. It follows that I is a two-sided ideal in A. Since $\phi(u)$ is not identically zero, we derive that $I = 0$. Therefore, $\phi(u)$ is invertible for every $u \in D \setminus \{0\}$. Now as in the proof of Lemma 1.3 we derive that

$$\phi(u) = \exp(\lambda u)\phi_\gamma$$

for some $\lambda \in \mathbb{C}$, where ϕ_γ is an algebra automorphism of A. Applying $\phi_\gamma^{-1} \otimes \text{id} \otimes \text{id}$ to (42), we derive

$$\exp(-\lambda u')P^{12}(\phi_\gamma^{-1} \otimes \text{id} \otimes \text{id})r^{13}(u + u', v' + \gamma) = \exp(\lambda u)P^{12}r^{13}(u + u', v').$$

This implies the required identity. $\qquad \square$

Lemma 7.4. *Keep the same assumptions as in Proposition 7.2. Assume that $\bar{r}_0(v+p) = \bar{r}_0(v)$ for some $p \in \mathbb{C}^*$. Then $r(u, v+p) = \exp(\lambda u)r(u, v)$ for some constant $\lambda \in \mathbb{C}$.*

Proof. Consider the decomposition (39) again. The identity (40) implies that $h(v + v') - h(v')$ is periodic in v' with the period p. Hence, $h(v + p) = h(v) + \lambda$ for some $\lambda \in \mathbb{C}$. It follows that $r_0(v + p) = r_0(v) + \lambda \cdot 1 \otimes 1$. Applying the rescaling $r(u, v) \mapsto \exp(-\lambda uv)r(u, v)$, we can assume that $r_0(v + p) = r_0(v)$. Now Lemma 7.3 implies that $r(u, v + p) = (\phi_p \otimes \text{id})r(u, v)$, where ϕ_p is an automorphism of A. Since $r_0(v)$ is nondegenerate (as follows from Lemma 1.7), we derive that $\phi_p = \text{id}$. $\qquad \square$

We will use the following result in the proof of Theorem 0.3.

Proposition 7.5. *Assume $N > 1$. Let $r(u,v)$ be a nondegenerate unitary solution of the AYBE with the Laurent expansion at $u = 0$ of the form (5) such that the equivalent conditions of Theorem 1.5(ii) hold. Then one has*

$$r_0(v) = \bar{r}_0(v) + b \otimes 1 + 1 \otimes b + h(v) \cdot 1 \otimes 1,$$
$$h(v) = \lambda v + ch_0(c'v),$$

where $b \in sl_N$ is an infinitesimal symmetry of $\bar{r}_0(v)$, $\lambda \in \mathbb{C}$, $c, c' \in \mathbb{C}^*$, and $h_0(v)$ is one of the following three functions: Weierstrass zeta function $\zeta(v)$ associated with a lattice in $\mathbb{C}$; $\frac{1}{2}\coth(\frac{v}{2})$; $\frac{1}{v}$. Furthermore, if $\bar{r}_0(v)$ is equivalent to a rational solution of the CYBE then $h_0(v) = \frac{1}{v}$.

Proof. Equation (18) implies that

$$[r_0^{12}(v_{12}) + r_0^{23}(v_{23}) + r_0^{31}(v_{31})]^2 = x^{12}(v_{12}) + x^{23}(v_{23}) + x^{31}(v_{31}), \quad (43)$$

where $x(v) = r_0(v)^2 - 2r_1(v)$ (and $v_{ij} = v_i - v_j$). On the other hand, it is easy to see that $x(v)$ is the constant term of the Laurent expansion of $s(u,v) = r(u,v)r(-u,v)$ at $u = 0$. Rescaling $r(u,v)$, we can assume that its residue at $v = 0$ is equal to P (see Lemma 1.7). Then we have

$$s(u,v) = [f(u) + g(v)] \cdot 1 \otimes 1,$$

where

$$f(u) = \frac{1}{N} \operatorname{tr} \mu \left(\frac{\partial r(u,0)}{\partial u} \right), \quad g(v) = -\frac{1}{N}(\operatorname{tr} \otimes \operatorname{tr}) \left(\frac{dr_0(v)}{dv} \right)$$

(see Lemma 1.9). If we change $r(u,v)$ to $\exp(\lambda uv)r(u,v)$ for some $\lambda \in \mathbb{C}$, then $f(u)$ changes to $f(u) + N\lambda$ (this operation also changes $r_0(v)$ to $r_0(v) + \lambda v \cdot 1 \otimes 1$). Therefore, we can assume that $f(u)$ has no constant term in the Laurent expansion at $u = 0$. In this case we obtain $x(v) = g(v) \cdot 1 \otimes 1$. Hence, setting

$$T(v_1, v_2, v_3) = r_0^{12}(v_{12}) + r_0^{23}(v_{23}) + r_0^{31}(v_{31}),$$

we can rewrite (43) as follows:

$$T(v_1, v_2, v_3)^2 = [g(v_{12}) + g(v_{23}) + g(v_{31})] \cdot 1 \otimes 1. \quad (44)$$

Viewing $T(v_1, v_2, v_3) \in A \otimes A \otimes A$ as an endomorphism of $V \otimes V \otimes V$, where $A = \operatorname{End}(V)$, we obtain

$$T(v_1, v_2, v_3) = T_0(v_1, v_2, v_3) + [h(v_{12}) + h(v_{23}) + h(v_{31})] \cdot \operatorname{id}_{V \otimes V \otimes V}, \quad (45)$$

where T_0 is a traceless endomorphism and $h(v)$ is defined from the decomposition (39). Note also that for fixed (generic) v_2 and v_3 we have

$$\lim_{v_1 \to v_2} (v_1 - v_2)T(v_1, v_2, v_3) = P^{12}.$$

The latter operator has $S^2V \otimes V$ and $\wedge^2 V \otimes V$ as eigenspaces. Therefore, for v_1 close to v_2 we have a decomposition

$$V \otimes V \otimes V = W_1 \oplus W_2,$$

where $\dim W_1 = N^2(N+1)/2$, $\dim W_2 = N^2(N-1)/2$, and

$$(T(v_1, v_2, v_3) - \lambda \operatorname{id})(W_1) = 0, \quad (T(v_1, v_2, v_3) + \lambda \operatorname{id})(W_2) = 0,$$

where

$$\lambda^2 = g(v_{12}) + g(v_{23}) + g(v_{31}).$$

Comparing the traces of both sides of (45), we derive

$$\lambda = N[h(v_{12}) + h(v_{23}) + h(v_{31})].$$

Since $g(v) = -Nh'(v)$, we obtain

$$N[h(v_{12}) + h(v_{23}) + h(v_{31})]^2 + h'(v_{12}) + h'(v_{23}) + h'(v_{31}) = 0.$$

Replacing $h(v)$ by $h(Nv)$ we get

$$[h(v_{12}) + h(v_{23}) + h(v_{31})]^2 + h'(v_{12}) + h'(v_{23}) + h'(v_{31}) = 0. \tag{46}$$

We are interested in solutions of this equation for an odd meromorphic function $h(v)$ in a neighborhood of zero having a simple pole at $v = 0$. It is easy to see that the Laurent expansion of $h(v)$ at $v = 0$ should have the form $h(v) = 1/v + c_3v^3 + \cdots$. As shown in the proof of Theorem 5 of [7], all such solutions of (46) have the form $c \cdot h_0(cv)$, where h_0 is one of the three functions described in the formulation.[1]

Finally, if $\overline{r}_0(v)$ is rational, then its only pole is $v = 0$ (see [3]). Therefore, by Proposition 7.2, $r_0(v)$ also cannot have poles outside zero, which implies that $h_0(v) = \frac{1}{v}$. $\qquad \square$

Remark 7.6. In the case when $\overline{r}_0(v)$ is either elliptic or trigonometric, the assertion of the above proposition can also be deduced from the explicit formulas for $r(u, v)$ (the elliptic case is discussed in [7], Section 2; the trigonometric case is considered in Theorem 0.2).

8 Classification of Trigonometric Solutions of the AYBE

Recall (see [3]) that to every nondegenerate trigonometric solution $\overline{r}_0(v)$ of the CYBE for sl_N with poles exactly at $2\pi i \mathbb{Z}$ Belavin and Drinfeld associate an automorphism of the Dynkin diagram A_{N-1} by considering the class of the automorphism ϕ of sl_N defined by

$$\overline{r}_0(v + 2\pi i) = (\phi \otimes \operatorname{id})(\overline{r}_0(v)). \tag{47}$$

---

[1]Solutions of (46) were first described by L. Carlitz in [5].

They also show that ϕ is a Coxeter automorphism. The next lemma shows that in the case of trigonometric solutions coming from a solution of the AYBE the automorphism of the Dynkin diagram is always trivial.

Lemma 8.1. *Let $r(u,v)$ be a nondegenerate unitary solution of the AYBE with the Laurent expansion (5) at $u = 0$. Assume that $\bar{r}_0(v)$ is a trigonometric solution of the CYBE with poles exactly at $2\pi i\mathbb{Z}$. Then the automorphism ϕ in (47) is inner.*

Proof. This follows immediately from Lemma 7.3, since every algebra automorphism of A is inner.   □

Now let us recall the Belavin–Drinfeld classification of trigonometric solutions of the CYBE for sl_N corresponding to the trivial automorphism of A_{N-1}. Let us denote by $\mathfrak{h}_0 \subset \mathrm{sl}_N$ the subalgebra of traceless diagonal matrices. For every Belavin–Drinfeld triple $(\Gamma_1, \Gamma_2, \tau)$ for $\widetilde{A}_{N-1}$ we have the corresponding series of solutions

$$\bar{r}_0(v) = t + \frac{1}{\exp(v) - 1}(\mathrm{pr} \otimes \mathrm{pr}) \sum_{0 \leq m < N, j-i \equiv m(N)} \exp\left(\frac{mv}{N}\right) e_{ij} \otimes e_{ji}$$

$$+ \sum_{0 < m < N, k \geq 1; j-i \equiv m(N), \tau^k(i,j)=(i',j')} [\exp\left(-\frac{mv}{N}\right) e_{ji} \otimes e_{i'j'}$$

$$- \exp\left(\frac{mv}{N}\right) e_{i'j'} \otimes e_{ji}], \tag{48}$$

where $t \in \mathfrak{h}_0 \otimes \mathfrak{h}_0$ satisfies

$$t^{12} + t^{21} = (\mathrm{pr} \otimes \mathrm{pr})P^0, \tag{49}$$

$$[\tau(\alpha) \otimes \mathrm{id} + \mathrm{id} \otimes \alpha]t = 0, \quad \alpha \in \Gamma_1, \tag{50}$$

where $P^0 = \sum_i e_{ii} \otimes e_{ii}$. The result of Belavin and Drinfeld in [3] is that every nondegenerate unitary trigonometric solution of the CYBE for sl_N that has poles exactly at $2\pi i\mathbb{Z}$ and the residue $(\mathrm{pr} \otimes \mathrm{pr})P$ at 0 is conjugate to

$$\exp[v(b \otimes 1)]\bar{r}_0(v)\exp[-v(b \otimes 1)],$$

where $\bar{r}_0(v)$ is one of the solutions of the form (48) and $b \in \mathrm{sl}_N$ is an infinitesimal symmetry of $\bar{r}_0$.

It is easy to see that the solution of the CYBE for sl_N obtained from the associative r-matrix in Theorem 0.2(i) for $S = [1, N]$ and $C_0(i) = i + 1$ is given by the above formula with

$$t = \frac{1}{2}(\mathrm{pr} \otimes \mathrm{pr})P^0 + s_C, \tag{51}$$

where

$$s_C = \sum_{0 < k < N, i} \left(\frac{1}{2} - \frac{k}{N}\right) e_{ii} \otimes e_{C^k(i), C^k(i)} \in \mathfrak{h}_0 \wedge \mathfrak{h}_0.$$

The proof of the next result is almost identical to that of Lemmas 4.19 and 4.20 in [8]. Let us denote by $e_i : \mathfrak{h} \to \mathbb{C}$ the functional on diagonal matrices given by $e_i(e_{jj}) = \delta_{ij}$.

Lemma 8.2. *Let $r(u, v)$ be a nondegenerate unitary solution of the AYBE with the Laurent expansion (5), such that $\bar{r}_0(v)$ is given by (48). Then there exists a unique transitive cyclic permutation C of $[1, N]$ such that (51) holds. Furthermore, for any $(i, i + 1) \in \Gamma_1$ with $\tau(i, i + 1) = (i', i' + 1)$ one has $C(i) = i'$ and $C(i + 1) = i' + 1$ (i.e., τ is induced by $C \times C$).*

Proof. We will make use of the identity

$$(\mathrm{pr} \otimes \mathrm{pr} \otimes \mathrm{pr})(\mathrm{AYBE}[\bar{r}_0]) = 0 \tag{52}$$

that follows from Theorem 1.5. First, considering the projection of $\mathrm{AYBE}[\bar{r}_0]$ to $\mathfrak{h} \otimes \mathfrak{h} \otimes \mathfrak{h}$ we get

$$(\mathrm{pr} \otimes \mathrm{pr} \otimes \mathrm{pr})(\mathrm{AYBE}[\frac{1}{\exp(v) - 1}P^0 + t]) = 0.$$

Using the fact that $t^{12} + t^{21} \equiv P^0 \mod (\mathbb{C} \cdot 1 \otimes 1)$ this can be rewritten as

$$(\mathrm{pr} \otimes \mathrm{pr} \otimes \mathrm{pr})(\mathrm{AYBE}[t]) = 0.$$

Therefore, we have

$$[(e_i - e_1) \otimes (e_j - e_1) \otimes (e_k - e_1)](\mathrm{AYBE}[t]) = 0 \tag{53}$$

for all i, j, k. Set $t = \sum_{i,j} t_{ij} e_{ii} \otimes e_{jj}$. Note that $t_{ij} + t_{ji} = 0$ for $i \neq j$ and $t_{ii} = \frac{1}{2}$ for all i. Let us set $t'_{ij} = t_{ij} - t_{1j} - t_{i1}$. Then substituting $t_{ij} = t'_{ij} + t_{1j} - t_{1i}$ into t and then into (53) we deduce that

$$t'_{ij} t'_{ik} - t'_{jk} t'_{ij} + t'_{ik} t'_{jk} = \frac{1}{4}, \quad 1 < i, j, k \leq N. \tag{54}$$

As shown in the proof of Lemma 4.20 in [8], the above equation implies that $t'_{ij} = \pm \frac{1}{2}$ for $1 < i, j \leq N$, $i \neq j$, and there is a unique complete order $\prec$ on $[2, N]$ such that $t'_{ij} = \frac{1}{2}$ iff $i \prec j$ (for $i, j \in [2, N]$, $i \neq j$). We define the cyclic permutation C of $[1, N]$ by the condition that it sends each element of $[2, N]$ to the next element with respect to this complete order. As in the proof of Lemma 4.20 in [8] this easily implies that $t - \frac{1}{2}P^0 \equiv s_C \mod (\mathbb{C} \cdot 1 \otimes 1)$.

Next, we want to check that τ is induced by $C \times C$. Assume that $\tau(i, i+1) = (j, j+1)$ and consider the coefficient A_{ijk} with $e_{i+1,i} \otimes e_{j,j+1} \otimes e_{kk}$ in $\mathrm{AYBE}[\bar{r}_0]$. Let us denote by $\langle e_{lm} \otimes e_{np}, r(v) \rangle$ the coefficient with $e_{lm} \otimes e_{np}$ in $r(v)$. Then we have

$$A_{ijk} = \langle e_{i+1,i} \otimes e_{j,j+1}, r(v_{12}) \rangle \langle e_{ii} \otimes e_{kk}, r(v_{13}) \rangle$$
$$- \langle e_{jj} \otimes e_{kk}, r(v_{23}) \rangle \langle e_{i+1,i} \otimes e_{j,j+1}, r(v_{12}) \rangle$$

$$+\langle e_{i+1,i} \otimes e_{k,k+1}, r(v_{13})\rangle\langle e_{j,j+1} \otimes e_{k+1,k}, r(v_{23})\rangle. \tag{55}$$

Note that we cannot have $\tau^n(j+1,j) = (i+1,i)$, since this would imply that Γ_1 (resp., Γ_2) is the complement to $(j, j+1)$ (resp., $(i, i+1)$), N is even, $j - i \equiv N/2(N)$, and $\tau(l, l+1) = (l + N/2, l+1+N/2)$, in which case the nilpotency condition is not satisfied. Therefore,

$$\langle e_{i+1,i} \otimes e_{j,j+1}, r(v)\rangle = \exp\left(-\frac{v}{N}\right),$$

$$\langle e_{j,j+1} \otimes e_{i+1,i}, r(v)\rangle = -\exp\left(\frac{v}{N}\right).$$

Next, we claim that the third summand in the right-hand side of (55) is zero unless $k = i$ or $k = j$. Indeed, τ (resp., τ^{-1}) cannot be defined on both $(k, k+1)$ and $(k+1, k)$. This leaves only two possibilities with $k \neq i$ and $k \neq j$: either $\tau^{n_1}(i, i+1) = (k, k+1)$ and $\tau^{n_2}(k, k+1) = (j, j+1)$, or $\tau^{n_1}(j+1, j) = (k+1, k)$ and $\tau^{n_2}(k+1, k) = (i, i+1)$ (where $n_1, n_2 > 0$). The latter case is impossible since $j \neq k$. In the former case we derive that $\tau^{n_1+n_2}(i, i+1) = (j, j+1)$, which contradicts our assumption that $\tau(i, i+1) = (j, j+1)$ (since $n_1 + n_2 \geq 2$). Thus, recalling that

$$\langle e_{i+1,i} \otimes e_{i,i+1}, r(v)\rangle = \frac{\exp(\frac{(N-1)v}{N})}{\exp(v) - 1},$$

$$\langle e_{j,j+1} \otimes e_{j+1,j}, r(v)\rangle = \frac{\exp(\frac{v}{N})}{\exp(v) - 1},$$

we can rewrite (55) as follows:

$$A_{ijk} = \exp\left(-\frac{v_{12}}{N}\right)\left[t_{ik} - t_{jk} + \frac{\delta_{ik}}{\exp(v_{13}) - 1} - \frac{\delta_{jk}}{\exp(v_{23}) - 1}\right]$$

$$-\delta_{ik} \exp\left(\frac{v_{23}}{N}\right)\frac{\exp(\frac{(N-1)v_{13}}{N})}{\exp(v_{13}) - 1} + \delta_{jk} \exp(-\frac{v_{13}}{N})\frac{\exp\left(\frac{v_{23}}{N}\right)}{\exp(v_{23}) - 1}.$$

Hence,

$$\exp\left(\frac{v_{12}}{N}\right) A_{ijk} = t_{ik} - t_{jk} - \delta_{ik}.$$

Since $\mathrm{pr} \otimes \mathrm{pr} \otimes \mathrm{pr}(\mathrm{AYBE}[\bar{r}_0]) = 0$, it follows that A_{ijk} does not depend on k. Therefore,

$$t_{ik} - t_{jk} - \delta_{ik} = [(e_i - e_j) \otimes e_k]s_C - \frac{1}{2}(e_i + e_j, e_k)$$

does not depend on k (note that $(e_i, e_j) = \delta_{ij}$), i.e.,

$$[(e_i - e_j) \otimes \alpha]s_C = \frac{1}{2}(e_i + e_j, \alpha) \tag{56}$$

for all roots $\alpha \in \Gamma$. Repeating the above argument for the coefficient with $e_{j,j+1} \otimes e_{i+1,i} \otimes e_{kk}$ in $\mathrm{AYBE}[\bar{r}_0]$, we derive that

$$[(e_{i+1} - e_{j+1}) \otimes \alpha]s_C = \frac{1}{2}(e_{i+1} + e_{j+1}, \alpha) \tag{57}$$

for all $\alpha \in \Gamma$. As shown in the proof of Lemma 4.20 in [8], (56) and (57) imply that $C(i) = j$ and $C(i + 1) = j + 1$.   □

Lemma 8.3. *Assume that $N > 1$. Then a nondegenerate unitary solution $r(u, v)$ of the* AYBE *with Laurent expansion at $u = 0$ of the form (5) such that $r_0(v) \equiv \bar{r}_0(v) \mod (\mathbb{C} \cdot 1 \otimes 1)$ is uniquely determined by $\bar{r}_0$, up to rescaling $r(u, v) \mapsto \exp(\lambda u v) r(u, v)$.*

Proof. This follows from the proof of Theorem 6 in [7]: one has only to observe that $\bar{r}_0(v)$ is nondegenerate by Theorem 1.5(i), so it has rank > 2 generically.

□

Proof of Theorem 0.2(ii). Let $r(u, v)$ be a nondegenerate unitary solution of the AYBE with the Laurent expansion at $u = 0$ of the form (5) such that $\bar{r}_0(v)$ is trigonometric. Changing $r(u, v)$ to $cr(cu, c'v)$, we can assume that $\bar{r}_0(v)$ has poles exactly at $2\pi i \mathbb{Z}$ and $\lim_{v \to 0} v \bar{r}_0(v) = (\mathrm{pr} \otimes \mathrm{pr})P$. Recall that we are allowed to change $r(u, v)$ to an equivalent solution

$$\tilde{r}(u, v) = \exp[u(1 \otimes a) + v(b \otimes 1)] r(u, v) \exp[-u(a \otimes 1) - v(b \otimes 1)],$$

where a and b are infinitesimal symmetries of $r(u, v)$ (note that a and b always commute by Lemma 7.1). This operation changes $\bar{r}_0(v)$ to an equivalent solution in the sense of Belavin–Drinfeld [3] and also changes $r_0(v)$ to $r_0(v) - a \otimes 1 + 1 \otimes a$. Therefore, in view of Lemma 8.1 and of (39), changing $r(u, v)$ to an equivalent solution, we can achieve that $r_0(v) \equiv \bar{r}_0(v) \mod (\mathbb{C} \cdot 1 \otimes 1)$ and $\bar{r}_0(v)$ has the form (48). Note that in this case any infinitesimal symmetry of $\bar{r}_0(v)$ is diagonal (since it has to commute with the corresponding Coxeter automorphism ϕ from (47)). It remains to use Lemmas 8.2 and 8.3.   □

Proof of Theorem 0.3. Let $r(v)$ be a nondegenerate unitary solution of the AYBE, not depending on u. Then one can easily check that

$$r(u, v) = \frac{1 \otimes 1}{u} + r(v)$$

is also a nondegenerate unitary solution of the AYBE. By Lemma 1.4, $r(u, v)$ (and hence, $r(v)$) has a simple pole at $v = 0$ with residue cP, where $c \neq 0$. Now applying Lemma 1.9 we obtain

$$s(u, v) = -\frac{1 \otimes 1}{u^2} + g(v) \cdot 1 \otimes 1,$$

where $g(v) = -\frac{c}{N}(\mathrm{tr} \otimes \mathrm{tr})(\frac{dr(v)}{dv})$. Hence, by Theorem 1.5, $r(u, v)$ is a solution of the QYBE and $\bar{r}(v)$ is a nondegenerate solution of the CYBE. It is easy to see that $\bar{r}(v)$ cannot be equivalent to an elliptic or a trigonometric solution. Indeed, if this were the case then $r(u,v)$ would have a pole of the form $u = u_0$ with $u_0 \neq 0$ (in the elliptic case this follows from the explicit formulas for elliptic solutions in [7], Section 2; in the trigonometric case this follows from Theorem 0.2(ii)). Now Proposition 7.5 gives the required decomposition of $r(v)$. Therefore, $g(v) = -c^2/v^2$, which shows that $R(u,v) = (1/u + c/v)^{-1} r(u,v)$ satisfies unitarity condition (7). $\square$

Remark 8.4. The function of the form $r(v) = \frac{p}{v} + r$, where $r \in A \otimes A$ does not depend on v, is a unitary solution of the AYBE iff r is a skew-symmetric constant solution of the AYBE for $A = \mathrm{Mat}(N, \mathbb{C})$. Some information about such solutions can be found in [2], Section 2 (including the classification for $N = 2$, see Ex. 2.8 of [2]).

References

1. M. AGUIAR, *Infinitesimal Hopf algebras*, New trends in Hopf algebra theory (La Falda, 1999), 1–29. Contemp. Math. 267, AMS, 2000.
2. M. AGUIAR, *On the associative analog of Lie bialgebras*, J. Algebra 244 (2001) 492–532.
3. A. A. BELAVIN, V. G. DRINFELD, *Solutions of the classical Yang-Baxter equation for simple Lie algebras*, Funct. Anal. and its Appl. 16 (1982), 1–29.
4. I. BURBAN, YU. DROZD, G.-M. GREUEL, *Vector bundles on singular projective curves*, in *Applications of Algebraic Geometry to Coding Theory, Physics and Computation (Eilat, Israel, 2001)*, Kluwer, 2001.
5. L. CARLITZ, *A functional equation for the Weierstrass ζ-function*, Math. Student 21 (1953), 43–45.
6. A. MUDROV, *Associative triples and Yang-Baxter equation*, Israel J. Math. 139 (2004), 11–28.
7. A. POLISHCHUK, *Classical Yang-Baxter equation and the A_∞-constraint*, Advances in Math. 168 (2002), 56–95.
8. T. SCHEDLER, *Trigonometric solutions of the associative Yang-Baxter equation*, Math. Res. Lett. 10 (2003), 301–321.

On Linnik and Selberg's Conjecture About Sums of Kloosterman Sums

Peter Sarnak[1,2] and Jacob Tsimerman[1]

[1] Department of Mathematics, Princeton University, Princeton, NJ, USA
[2] Institute for Advanced Study, Princeton, NJ, USA
sarnak@math.princeton.edu

Dedicated to Y. Manin on the Occasion of his 70th Birthday

Summary. We examine the Linnik and Selberg Conjectures concerning sums of Kloosterman sums, in all its aspects (x, m and n). We correct the precise form of the Conjecture and establish an analogue of Kuznetzov's 1/6 exponent in the mn aspect. This, perhaps somewhat surprisingly, is connected with the transional ranges associated with asymptotics of Bessel Functions of large order.

Key words: Kloosterman sums, Ramanujan-Selberg conjectures, Kuznetzov Formula

2000 Mathematics Subject Classifications: 11Fxx, 11Lxx

1 Statements

Linnik [Li] and later Selberg [Se] put forth far-reaching conjectures concerning cancellations due to signs of Kloosterman sums. Both point to the connection between this and the theory of modular forms. For example, the generalization of their conjectures to sums over arithmetic progressions imples the general Ramanujan conjectures for $GL_2/\mathbb{Q}$. Selberg exploited this connection to give a nontrivial bound to his well-known "$\lambda_1 \geq \frac{1}{4}$" conjecture. Later, Kuznetzov [Ku] used his summation formula, which gives an explicit relation between these sums and the spectrum of $GL_2/\mathbb{Q}$ modular forms, to prove a partial result to Linnik's conjecture. Given that we now have rather strong bounds to the Ramanaujan conjectures for $GL_2/\mathbb{Q}$ [L-R-S], [Ki-Sh], [Ki-Sa], it seems timely to revisit the Linnik and Selberg conjectures.

---
[1]The first author was partially supported by NSF grant DMS-0758299.

Y. Tschinkel and Y. Zarhin (eds.), *Algebra, Arithmetic, and Geometry*,
Progress in Mathematics 270, DOI 10.1007/978-0-8176-4747-6_20,
© Springer Science+Business Media, LLC 2009

The Kloosterman sum $S(m, n; c)$, for $c \geq 1$, $m \geq 1$, and $n \neq 0$, is defined by

$$S(m, n; c) = \sum_{\substack{x \bmod c \\ x\bar{x} \equiv 1(c)}} e\left(\frac{mx + n\bar{x}}{c}\right) \tag{1}$$

(here $e(z) = e^{2\pi i z}$).

It follows from Weil's bound [Wei] for $S(m, n; p)$, p prime, and elementary considerations that

$$|S(m, n; c)| \leq \tau(c)(m, n, c)^{1/2}\sqrt{c}, \tag{2}$$

where $\tau(c)$ is the number of divisors of c. Linnik's conjecture asserts that for $\epsilon > 0$ and $x \geq \sqrt{|n|}$,

$$\sum_{c \leq x} \frac{S(1, n; c)}{c} \ll_\epsilon x^\epsilon. \tag{3}$$

Note that from (2) we have that

$$\sum_{c \leq x} \frac{|S(m, n; c)|}{c} \ll_\epsilon \sqrt{x}(x(m, n))^\epsilon. \tag{4}$$

On the other hand, for m, n fixed, Michel [Mi] shows that

$$\sum_{c \leq x} \frac{|S(m, n; c)|}{c} \gg_\epsilon x^{\frac{1}{2} - \epsilon}, \tag{5}$$

and hence if (3) is true it represents full "square-root" cancellation due to the signs of the Kloosterman sums.

Selberg puts forth a much stronger conjecture, which has been replicated in many places and which reads

For $x \geq (m, n)^{1/2}$,

$$\sum_{c \leq x} \frac{S(m, n; c)}{c} \ll_\epsilon x^\epsilon. \tag{6}$$

As stated, this is false (see Section 2). It needs to be modified to incorporate an "ϵ-safety valve" in all parameters and not only in x and (m, n). For example, the following modification seems okay:

$$\sum_{c \leq x} \frac{S(m, n; c)}{c} \ll_\epsilon (|mn|x)^\epsilon. \tag{7}$$

We call the range $x \geq \sqrt{|mn|}$ the Linnik range and $x \leq \sqrt{|mn|}$ the Selberg range. Obtaining nontrivial bounds (i.e., a power saving beyond (4)) for the sums (3) and (7) in the Selberg range is quite a bit more difficult.

By a smooth dyadic sum of the type (7) we mean

$$\sum_c \frac{S(m,n;c)}{c} F\left(\frac{c}{x}\right) \tag{8}$$

where $F \in C_0^\infty(\mathbb{R}^+)$ is of compact support and where the estimates for (8) as x, m, n vary are allowed to depend on F. Summation by parts shows that an estimate for the left-hand side of (7) will give a similar one for (8), but not conversely. In particular, estimates for (7) imply bounds for nondyadic sums $\sum_{x \leq c \leq x+h} \frac{S(m,n;c)}{c}$ with $x^\beta \leq h \leq x$ and $\beta < 1$. We note that for many applications, understanding smooth dyadic sums is good enough, and in the Linnik range these smooth dyadic sums are directly connected to the Ramanujan conjectures; see Deshouilliers and Iwaniec [D-I] and Iwaniec and Kowalski [I-K pp. 415–418].

Let $\pi \cong \otimes_v \pi_v$ be an automorphic cuspidal representation of $GL_2(\mathbb{Q})\backslash GL_2(\mathbb{A})$ with a unitary central character. Here v runs over all primes p and ∞. Let $\mu_1(\pi_v)$, $\mu_2(\pi_v)$ denote the Satake parameters of π_v at an unramified place v of π. We normalize as in [Sa] and use the results described there. For $0 \leq \theta < \frac{1}{2}$ we denote by H_θ the following hypothesis: for any π as above,

$$|\Re(\mu_j(\pi_v))| \leq \theta, \quad j = 1, 2. \tag{9}$$

Thus H_0 is the Ramanujan–Selberg conjecture and H_θ is known for $\theta = \frac{7}{64}$ [Ki-Sa].

Concerning (7), the main result is still that of Kuznetzov [Ku], who using his formula together with the elementary fact that $\lambda_1(SL_2(\mathbb{Z})\backslash\mathbb{H}) \geq \frac{1}{4}$ proved the following theorem (the barrier $\frac{1}{6}$ is discussed at the end of this paper):

Theorem 1 (Kuznetzov) *Fix m and n. Then*

$$\sum_{c \leq x} \frac{S(m,n;c)}{c} \underset{m,n}{\ll} x^{1/6}(\log x)^{1/3}.$$

One can ask for similar bounds when c varies over an arithmetic progression. For this one applies the Kuznetzov formula for $\Gamma = \Gamma_0(q)$ in place of $SL_2(\mathbb{Z})$ (see [D-I]) or one can use the softer method in [G-S]. What is needed is the $v = \infty$ version of H_θ with $\theta \leq \frac{1}{6}$. This was first established in [Ki-Sh] and leads to the following result:

Theorem 1'. *Fix m, n, q. Then*

$$\sum_{\substack{c \equiv 0(q) \\ c \leq x}} \frac{S(m,n;c)}{c} \underset{m,n,q}{\ll} x^{\frac{1}{6}}(\log x)^{1/3}.$$

Consider now what happens if m and n are allowed to be large, as asked by Linnik and Selberg. We examine the case $q = 1$; the general case with q

also varying deserves a similar investigation. We also assume that $mn > 0$; the $mn < 0$ case is similar except that some results such as Theorem 4 are slightly weaker, since when $mn > 0$ we can exploit H_0, which is known for the Fourier coefficients of holomorphic cusp forms [De]. The analogue of Theorem 1 that we seek is

$$\sum_{c \leq x} \frac{S(m,n;c)}{c} \ll_\epsilon \left(x^{\frac{1}{6}} + (mn)^{\frac{1}{6}} \right) (mnx)^\epsilon . \tag{10}$$

As with Theorem 1, we show below that the exponent $\frac{1}{6}$ in the mn aspect constitutes a natural barrier. We will establish (10) assuming the general Ramanujan conjectures for $GL_2/\mathbb{Q}$ (i.e., assuming H_0), and unconditionally we come close to proving it.

Theorem 2 *Assuming H_θ we have*

$$\sum_{c \leq x} \frac{S(m,n;c)}{c} \ll_\epsilon \left(x^{\frac{1}{6}} + (mn)^{\frac{1}{6}} + (m+n)^{1/8}(mn)^{\theta/2} \right) (xmn)^\epsilon .$$

Corollary 3 *Assuming $H_{\frac{1}{12}}$ (and a fortiori H_0), (10) is true. Unconditionally, Theorem 2 is true with $\theta = 7/64$, and in particular, (10) is true if m and n are close in the sense that $n^{5/43} \leq m \leq n$ or $m^{5/43} \leq n \leq m$.*

The proof of Theorem 2 uses Kuznetzov's formula to study the dyadic sums

$$\sum_{x \leq c \leq 2x} \frac{S(m,n;c)}{c} ;$$

see Section 2. The dyadic pieces with $x = (mn)^{1/3}$ or smaller are estimated trivially using (2) and give the $(mn)^{1/6}$ term in (10). For $x \leq (mn)^{1/3}$ we don't have any nontrivial bound for the sum even in smooth dyadic form. Indeed, applying the Kuznetzov formula to such smooth sums in the range $x \leq (mn)^{1/2-\delta}$ for some $\delta > 0$, one finds that the main terms on the spectral side localize in the transitional range for the Bessel functions of large order and argument. The analysis becomes a delicate one with the Airy function. In this $mn > 0$ case the main terms involve only Fourier coefficients of holomorphic modular forms. This indicates that one should be able to obtain the result backward using the Petersson formula [I-K] together with the smooth k-averaging technique [I-L-S, p. 102]. Indeed, this is so, and we give the details of both methods in Section 2. Another bonus that comes from this holomorphic localization is that we can appeal to Deligne's theorem, that is, H_0, for these forms.

Theorem 4 *Fix $F \in C_0^\infty(0, \infty)$ and $\delta > 0$. For $mn > 0$ and $x \leq (mn)^{\frac{1}{2}-\delta}$,*

$$\sum_c \frac{S(m,n;c)}{c} F \left(\frac{c}{x} \right) \ll_{F,\delta} \frac{\sqrt{mn}}{x} .$$

The bound in Theorem 4 is nontrivial only in the range $x \geq (mn)^{1/3}$. To extend this to smaller x requires establishing cancellations in short spectral sums for Fourier coefficients of modular forms (see equation (34)), which is an interesting challenge. In any case, this analysis of the smooth dyadic sums explains the $(mn)^{1/6}$ barrier in (10).

2 Proofs

As we remarked at the beginning, the "randomness" in the signs of the Kloosterman sums in the form (7) implies the general Ramanujan conjectures for $GL_2/\mathbb{Q}$. Since there seems to be no complete proof of this in the literature, we give one here. For the archimedian q-aspect, that is, the $\lambda_1 \geq 1/4$ conjecture, this follows from Selberg [Se], since his "zeta function" $Z(m, n, s)$ has poles with nonzero residues in $\Re(s) > \frac{1}{2}$ if there are exceptional eigenvalues (i.e., $\lambda < 1/4$). For the case of Fourier coefficients of holomorphic modular forms the implication is derived in [Mu, pp. 240–242]. So we focus on the Fourier coefficients of Maass forms and restrict to level $q = 1$ (the case of general q is similar). We apply Kuznetzov's formula with $m = n$ for $\Gamma = SL_2(\mathbb{Z})$, see [I-K, pp. 409] for the notation:

$$\sum_{j=1}^{\infty} |\rho_j(n)|^2 \, \frac{h(t_j)}{\cosh(\pi t_j)} + \frac{1}{4\pi} \int_{-\infty}^{\infty} |\tau(n, t)|^2 \, \frac{h(t)}{\cosh(\pi t)} \, dt$$

$$= g_0 + \sum_{c=1}^{\infty} \frac{S(m, n; c)}{c} \, g\left(\frac{4\pi n}{c}\right). \tag{11}$$

Here h is entire and decays faster than $(1 + |t|)^{-2-\delta}$, with $\delta > 0$ as $t \to \pm\infty$, and

$$g_0 = \frac{1}{\pi^2} \int_{-\infty}^{\infty} r \, h(r) \, \tan h(\pi r) \, dr$$

and

$$g(x) = \frac{2i}{\pi} \int_{-\infty}^{\infty} J_{2ir}(x) \, \frac{r \, h(r)}{\cosh(\pi r)} \, dr \, .$$

We want to prove that for a fixed $j = j_0$, we have

$$\rho_{j_0}(n) = O_\epsilon\left(|n|^\epsilon\right) \text{ as } |n| \to \infty. \tag{12}$$

It is well known how to deduce the various forms of H_0 for ϕ_{j_0} from (12). We choose h with $h(t) \geq 0$ for $t \in \mathbb{R}$ and $h(t_{j_0}) \geq 1$ and such that $g \in C^\infty(0, \infty)$ is rapidly decreasing at ∞ and g vanishes at $x = 0$ to order 1. Now let $|n| \to \infty$ in (11). Since $|\tau(n, t)| = O_\epsilon(|n|^\epsilon)$, we have

$$|\rho_{j_0}(n)|^2 \ll |n|^\epsilon + \left| \sum_{c=1}^{\infty} \frac{S(n, n; c)}{c} \, g\left(\frac{4\pi n}{c}\right) \right| \, .$$

If we assume (7) then we can estimate the sum on c by summing by parts, and we obtain the desired bound (12).

We turn next to (6) and show that as stated it is false. Let x be a large integer. For each prime p in $(x, 2x)$ there is an integer a_p such that

$$\frac{S(1, a_p; p)}{\sqrt{p}} \geq \frac{1}{10}. \tag{13}$$

This follows from the easily verified identities

$$\frac{1}{p} \sum_{a(p)} \frac{S(1, a; p)}{\sqrt{p}} = 0,$$
$$\frac{1}{p} \sum_{a(p)} \left(\frac{S(1, a; p)}{\sqrt{p}} \right)^2 = 1 - \frac{1}{p}. \tag{14}$$

We note in passing that the asymptotics of any moment $\frac{1}{p} \sum_a \left(\frac{S(1, a; p)}{\sqrt{p}} \right)^m$, in the limit as $p \to \infty$, have been determined by Katz [Ka], the mth moment being that of the Sato–Tate measure. Now let n be an integer satisfying $n \equiv 0 \pmod{(x!)^2}$ and $n \equiv a_p \pmod{p}$ for $x < p < 2x$. Such an n can be chosen, since $(x!)^2$ and the p's are all relatively prime to each other. For $0 < c \leq x$, $n \equiv 0 \pmod{c}$, and hence $S(1, n; c) = \mu(c)$, the Möbius function. For $x < c < 2x$ and c not a prime, we have $S(1, n; c) = S(1, n; c_1 c_2)$ with $1 < c_1, c_2 < x$. Hence $n \equiv 0 \pmod{c}$ and $S(1, n; c) = \mu(c)$. It follows that

$$\sum_{c \leq 2x} \frac{S(1, n; c)}{c} = \sum_{c \leq x} \frac{\mu(c)}{c} + \sum_{\substack{x < c \leq 2x \\ c \text{ not prime}}} \frac{\mu(c)}{c} + \sum_{x < p < 2x} \frac{S(1, a_p; c)}{c}$$

$$\geq \frac{1}{10} \sum_{x < p < 2x} \frac{1}{\sqrt{p}} + O(\log x)$$

$$\geq \frac{x^{1/2}}{\log x} + O(\log x). \tag{15}$$

Hence (6) is false.

Of course the n constructed above is very large, and the right-hand side in (12) is certainly $O_\epsilon(n^\epsilon)$ for any $\epsilon > 0$. Thus with the n^ϵ in (7) this is no longer a counterexample.

We turn next to Theorem 4, that is, the analysis of the smooth dyadic sums in the Selberg range $x \leq (mn)^{1/3}$. We need Kuznetzov's formula [I-K]. Let $f \in C_0^\infty(\mathbb{R}^+)$ and

$$M_f(t) = \frac{\pi i}{\sinh(2\pi t)} \int_0^\infty (J_{2it}(x) - J_{-2it}(x)) f(x) \frac{dx}{x} \tag{16}$$

and

$$N_f(k) = \frac{4(k-1)!}{(4\pi i)^k} \int\limits_0^\infty J_{k-1}(x)\, f(x)\, \frac{dx}{x}.$$ (17)

Let $f_{j,k}(z)$, $j = 1, \ldots, \dim S_k(\Gamma)$, be an orthonormal basis of Hecke eigenforms for the space of holomorphic cusp forms of weight k for $\Gamma = \mathrm{SL}_2(\mathbb{Z})$. Let $\psi_{j,k}(n)$ denote the corresponding Fourier coefficients normalized by

$$f_{j,k}(z) = \sum_{k=1}^\infty \psi_{j,k}(m)\, m^{\frac{k-1}{2}}\, e(mz).$$ (18)

Let $\tau(m,t)$ be the nth Fourier coefficient of the unitary Eisenstein series $E(z, \tfrac{1}{2} + it)$. Then for $mn > 0$,

$$\sum_c \frac{S(m,n;c)}{c}\, f\left(\frac{4\pi\sqrt{mn}}{c}\right)$$

$$= \sum_j M_f(t_j)\, \overline{\rho_j(m)}\, \rho_j(n) + \frac{1}{4\pi} \int\limits_{-\infty}^\infty M_f(t)\, \overline{\tau}(m,t)\, \tau(n,t)\, dt$$

$$+ \sum_{k \equiv 0(2)} N_f(k) \sum_{1 \le j \le \dim S_k(\Gamma)} \overline{\psi_{j,k}(m)}\, \psi_{j,k}(n).$$ (19)

For $(mn)^\delta \le Y \le (mn)^{1/2}$ we apply (19) with

$$f_Y(x) = f_0\left(\frac{x}{Y}\right)$$ (20)

and $f_0 \in C_0^\infty (\mathbb{R} > 0)$ fixed. The left-hand side of the formula is

$$\sum_c \frac{S(m,n;c)}{c}\, f_0\left(\frac{4\pi\sqrt{mn}}{cY}\right) = \sum_c \frac{S(m,n;c)}{c}\, F_0\left(\frac{c}{X}\right),$$

where $F_0(\xi) = f_0(1/\xi)$ and $X = \frac{4\pi\sqrt{mn}}{Y}$, which is what we are interested in when Y is large. One can analyze $N_{f_Y}(t)$ and $M_{f_Y}(t)$ as $Y \longrightarrow \infty$ using the known asymptotics of the Bessel functions $J_{it}(x)$ and $J_{k-1}(x)$ for x and t large. Using repeated integrations by parts one can show that the main term comes from the holomorphic forms only, with their contribution coming from the transition range:

$$J_k(Yx), \text{ with } |Yx - k| \ll k^{1/3}.$$ (21)

Using the approximations [Wa, p. 249] by the Airy function,

$$J_k(x) \sim \frac{1}{\pi}\left(\frac{2(k-x)}{3x}\right)^{1/2} K_{\frac{1}{3}}\left(\frac{2^{3/2}(k-x)^{3/2}}{3x^{1/2}}\right) \text{ for } k - k^{1/3} \ll x < k$$ (22)

and

$$J_k(x) \sim \frac{1}{3} \left(\frac{2(x-k)}{x} \right)^{1/2} \left(J_{1/3} + J_{-1/3} \right) \left(\frac{2^{3/2}(x-k)^{3/2}}{3x^{1/2}} \right)$$

$$\text{for } k < x \ll k + k^{1/3}, \quad (23)$$

and keeping only the leading terms of all asymptotics, one finds that

$$\sum_c \frac{S(m,n;c)}{c} f_0 \left(\frac{4\pi \sqrt{mn}}{cY} \right)$$

$$\sim \frac{1}{\pi} \left(\int_{-\infty}^{\infty} \mathrm{Ai}(\xi)\, d\xi \right) \sum_k \frac{1}{k} f_0 \left(\frac{k}{Y} \right) \frac{4(k-1)!}{(4\pi i)^k} \sum_{1 \leq j \leq \dim S_\Gamma(\Gamma)} \overline{\psi_{j,k}(m)}\, \psi_{j,k}(n).$$

$$(24)$$

Here $\mathrm{Ai}(\xi) = \displaystyle\int_{-\infty}^{\infty} \cos\left(t^3 + \xi t\right) dt$, is the Airy function.

The above derivation is tedious and complicated, but once we see the form of the answer and especially that it involves only holomorphic modular forms, another approach suggests itself, and fortunately it is simpler. We start with the Peterson formula [I-K]

$$\frac{\Gamma(k-1)}{(4\pi)^{k-1}} \sum_{j=1,\dots,\dim S_k(\Gamma)} \overline{\psi_j(m)}\, \psi_j(n)$$

$$= \delta(m,n) + 2\pi i^k \sum_{c=1}^{\infty} \frac{S(m,n;c)}{c} J_{k-1} \left(\frac{4\pi \sqrt{mn}}{c} \right).$$

Setting

$$\rho_f(n) = \left(\frac{\Gamma(k-1)}{(4\pi)^{k-1}} \right)^{1/2} \psi_f(n), \quad (25)$$

we have

$$\sum_{f \in H_k(\Gamma)} \overline{\rho_f(m)}\, \rho_f(n) = \delta(m,n) = 2\pi i^k \sum_{c=1}^{\infty} \frac{S(m,n;c)}{c} J_{k-1} \left(\frac{4\pi \sqrt{mn}}{c} \right),$$

$$(26)$$

where $H_k(\Gamma)$ denotes any orthonormal basis of $S_k(\Gamma)$ and in particular the Hecke basis that we choose.

In this case,

$$\rho_f(m) = \rho_f(1) \lambda_f(m), \quad (27)$$

where $|\lambda_f(m)| \leq \tau(m)$, by Deligne's proof of the holomorphic Ramanujan conjectures [De]. With this normalization, (24) reads

$$\sum_c \frac{S(m,n;c)}{c} f_0\left(\frac{4\pi\sqrt{mn}}{cY}\right) \sim (\text{const}) \sum_k f_0\left(\frac{k}{Y}\right) \sum_{f\in H_k(\Gamma)} \overline{\rho_f(m)}\,\rho_f(n).$$

$$(28)$$

From (26) with $m = n = 1$ we have that for k large,

$$\sum_{f\in H_k(\Gamma)} |\rho_f(1)|^2 = 1 + \text{ small}.$$

$$(29)$$

We use the k-averaging formula in [I-L-S, p.102], which gives for $K \geq 1$, $x > 0$, and $h \in C_0^\infty(\mathbb{R} > 0)$ that

$$h_k(x) := \sum_{k\equiv 0(2)} h\left(\frac{k-1}{K}\right) J_{k-1}(x) = -iK \int_{-\infty}^{\infty} \hat{h}(tK)\sin\left(x\sin 2\pi t\right) dt.$$

$$(30)$$

From this it follows easily that if $x \geq K^{1+\epsilon_0}$, then for any fixed $N \geq 1$,

$$\sum_{k\equiv 0(2)} h\left(\frac{k-1}{K}\right) J_{k-1}(x) \ll_N K^{-N},$$

$$(31)$$

while for $0 \leq x \leq K^{1+\epsilon_0}$ and for any fixed N,

$$h_k(x) = h_0\left(\frac{x}{K}\right) + \frac{1}{K^2} h_1\left(\frac{x}{K}\right) + \cdots + \frac{1}{K^{2N}} h_N\left(\frac{x}{K}\right) + O_N\left(K^{-2N-1}\right),$$

$$(32)$$

where $h_0(\xi) = h(\xi)$ and $h_1, \dots, h_N$ involve derivatives of $h(\xi)$ and are also in $C_0^\infty(\mathbb{R} > 0)$. From (26) we have that

$$\sum_{k\equiv 0(2)} i^k h\left(\frac{k-1}{K}\right) \sum_{f\in H_k(\Gamma)} \overline{\rho_f(m)}\,\rho_f(n)$$

$$= \sum_{k\equiv 0(2)} (i)^k h\left(\frac{k-1}{K}\right) \delta(m,n) - 2\pi \sum_{c=1}^{\infty} \frac{S(m,n;c)}{c} h_K\left(\frac{4\pi\sqrt{mn}}{c}\right). \quad (33)$$

We assume that $(mn)^\delta \leq K \leq \sqrt{mn}$ with $\delta > 0$ and fixed. Then for N large enough (depending on δ) we have from (27), (28), and (29) that

$$\sum_{j=0}^{N} K^{-2j} \sum_c \frac{S(m,n;c)}{c} h_j\left(\frac{4\pi\sqrt{mn}}{cK}\right)$$

$$= \sum_{k\equiv 0(2)} i^k h\left(\frac{k-1}{K}\right) \sum_{f\in H_k(\Gamma)} \overline{\rho_f(m)}\,\rho_f(n) + O(1).$$

$$(34)$$

Thus the lead term $j = 0$ in (34) recovers the asymptotics (24) and in a much more precise form. We can estimate the right-hand side of (34) as being at most

$$RHS \leq \sum_{k} |h| \left(\frac{k-1}{K}\right) \sum_{f \in H_k(\Gamma)} |\lambda_f(m)\lambda_f(n)| \, |\rho_f(1)|^2 \ll K \tau(m) \tau(n),$$

$$(35)$$

by (27) and (29). It follows that

$$\sum_{j=0}^{N} K^{-2j} \sum_{c} \frac{S(m,n;c)}{c} h_j \left(\frac{4\pi \sqrt{mn}}{cK}\right) \ll K \tau(m) \tau(n). \qquad (36)$$

From this it follows by estimating the $j \geq 1$ sums trivially that

$$\sum_{c} \frac{S(m,n;c)}{c} h \left(\frac{4\pi \sqrt{mn}}{cK}\right) \ll K \tau(m) \tau(n) + \frac{(mn)^{1/4}}{K^{5/2}}. \qquad (37)$$

Using this bound, which is now valid for $h_1, \ldots, h_N$, we can feed it back into (36) to get

$$\sum_{c} \frac{S(m,n;c)}{c} h \left(\frac{4\pi \sqrt{mn}}{Kc}\right)$$

$$\ll K \tau(m) \tau(n) + K^{-2} \left(K \tau(m) \tau(n) + \frac{(mn)^{1/4}}{K^{5/2}}\right)$$

$$\ll K \tau(m) \tau(n) + \frac{(mn)^{1/4}}{K^{5/2}}. \qquad (38)$$

Replicating this iteration a finite number of times yields that for $(mn)^\delta \leq K \leq \sqrt{mn}$,

$$\sum_{c} \frac{S(m,n;c)}{c} h \left(\frac{4\pi \sqrt{mn}}{Kc}\right) \ll K \tau(m) \tau(n), \qquad (39)$$

or what is the same thing,

$$\sum_{c} \frac{S(m,n;c)}{c} F \left(\frac{c}{x}\right) \ll_\epsilon \frac{(mn)^{1/2} \tau(m) \tau(n)}{x}, \quad \text{for } x \leq (mn)^{1/2-\delta}. \qquad (40)$$

This completes the proof of Theorem 4. □

Finally we turn to the proof of Theorem 2. The dependence on mn in Kuznetzov's argument was examined briefly in Huxley [Hu]. In order for us to bring in H_θ effectively, we first examine the dyadic sums $\sum_{x \leq c \leq 2x} \frac{S(m,n;c)}{c}$ for x in various ranges.

Proposition 5 *Assume H_θ. Then for $\epsilon > 0$,*

$$\sum_{x \leq c \leq 2x} \frac{S(m,n;c)}{c} \ll_\epsilon (mnx)^\epsilon \left(x^{1/6} + \frac{\sqrt{mn}}{x} + (m+n)^{1/8}(mn)^{\theta/2}\right).$$

Theorem 2 follows from this proposition by breaking the sum $1 \leq c \leq x$ into at most $\log x$ dyadic pieces $y \leq c \leq 2y$ with $(mn)^{1/3} \leq y \leq x$ and estimating the initial segment $1 \leq c \leq \min((mn)^{1/3}, x)$ using (4). We conclude by proving Proposition 5. In formula (19) we choose the test function f to be $\phi(t)$ depending on mn, x, and a parameter $x^{1/3} \leq T \leq x^{2/3}$ to be chosen later. The function ϕ is smooth on $(0, \infty)$, taking values in $[0, 1]$ and satisfying

(i) $\phi(t) = 1$ for $\frac{a}{2x} \leq t \leq \frac{a}{x}$ where $a = 4\pi\sqrt{mn}$.
(ii) $\phi(t) = 0$ for $t \leq \frac{a}{2x+2T}$ and $t \geq \frac{a}{x-T}$.
(iii) $\phi'(t) \ll \left(\frac{a}{x-T} - \frac{a}{x} \right)^{-1}$.
(iv) ϕ and ϕ' are piecewise monotone on a fixed number of intervals.

For ϕ chosen this way we have that

$$
\left| \sum_{c=1}^{\infty} \frac{S(m, n; c)}{c} \phi\left(\frac{4\pi\sqrt{mn}}{c} \right) - \sum_{x \leq c \leq 2x} \frac{S(m, n; , c)}{c} \right|
$$

$$
\leq \sum_{\substack{x-T \leq c \leq x \\ 2x \leq c \leq 2x+T}} \left| \frac{S(m, n; c)}{c} \right|
$$

$$
\ll_{\epsilon} \frac{(mn)^{\epsilon}}{\sqrt{x}} \sum_{\substack{x-T \leq c \leq x \\ 2x \leq c \leq 2x+T}} \tau(c) \ll \frac{(mn)^{\epsilon} T \log x}{\sqrt{x}}, \tag{41}
$$

where we have used (2) and the mean value bound for the divisor function. We estimate $\sum_{c=1}^{\infty} \frac{S(m,n;c)}{c} \phi\left(\frac{4\pi\sqrt{mn}}{c} \right)$ using (19), and to this end, according to (16), we first estimate $\hat{\phi}(r) = \cosh(\pi r) M_{\phi}(r)$. We follow Kuznetzov, keeping track of the dependence on mn.

The following asymptotic expansion of the Bessel function is uniform; see [Du]:

$$
J_{ir}(y) = \frac{c\, e^{irw(y/r) + \frac{\pi r}{2}}}{\sqrt{y^2 + r^2}} \left(1 + \frac{a_1}{\sqrt{r^2 + y^2}} + \cdots \right),
$$

where c, a_1 are constants and

$$
w(s) = \sqrt{1 + s^2} + \log\left(\frac{1}{s} - \sqrt{\frac{1}{s^2} + 1} \right). \tag{42}
$$

We analyze the leading term; the lower-order terms are treated similarly and their contribution is smaller. For $|r| \leq 1$ we have

$$
\hat{\phi}(r) \ll |r|^{-2}, \tag{43}
$$

as is clear from the Taylor expansion for J_{ir} when $\frac{a}{x} \leq 1$ and from (42) otherwise. So assume that $r \geq 1$ (or ≤ -1), and making the substitution $y = rs$, we are reduced to bounding

$$r^{-1/2} \int_0^\infty \frac{e^{irw(s)}}{(s^2+1)^{1/4}} \phi(rs) \frac{ds}{s} \tag{44}$$

$$= r^{-3/2} \int_0^\infty \left(e^{irw(s)} rw'(s) \right) \frac{\phi(rs)}{w'(s) s(s^2+1)^{1/4}}. \tag{45}$$

Now $w'(s)$ is bounded away from zero uniformly on $(0,\infty)$ and approaches 2 as $s \to \infty$ and behaves like $\frac{1}{s}$ as $s \to 0$. We may then apply the following easily proven mean value estimate: If F and G are defined on $[A,B]$ with G monotonic and taking values in $[0,1]$, then

$$\left| \int_A^B F(x) G(x) \, dx \right| \le 2 \sup_{A \le C \le B} \left| \int_A^C F(x) dx \right|. \tag{46}$$

This yields that the quantity in (45) is at most $O(r^{-3/2})$ and hence that

$$\hat{\phi}(r) \ll r^{-3/2} \text{ for } |r| \ge 1. \tag{47}$$

For r large we seek a better bound, which one gets by integration by parts in (45). This gives

$$-r^{-3/2} \int_0^\infty e^{irw(s)} \frac{d}{ds} \left(\frac{\phi(rs)}{w'(s)s} \right) ds$$

$$= O(r^{-5/2}) + r^{-3/2} \int_0^\infty \frac{e^{irw(s)} \theta'(s)}{(s^2+1)^{1/4} sw'(s)} ds,$$

where the first term follows as in (47) and $\theta(s) = \phi(rs)$. Applying the mean value estimate to the last integral and using property (iii) of ϕ yields

$$\hat{\phi}(r) \ll \frac{x}{T} r^{-5/2}. \tag{48}$$

The bounds (47) and (48) are the same as those obtained by Kuznetzov, only now they are uniform in x and nm.

For the term involving $\tau(m,t)$ in (19) and our choice of test function we have

$$\int_{-\infty}^\infty \frac{\hat{\phi}(r)}{|\rho(1+2ir)|^2} \, dr \ll 1, \tag{49}$$

which follows immediately from (43).

The term in (19) involving the k-sum over Fourier coefficients of holomorphic forms is handled using the Ramanujan conjecture for these. We find that this sum is bounded by

$$(mn)^\epsilon \sum_{\substack{k \equiv 0(2) \\ k > 2}} 4(k-1) \left| N_\phi(k) \right|.$$ (50)

Now for $x \geq \sqrt{mn}$ it is immediate from the decay of the Bessel function at small argument that $N_\phi(k) \ll 1/k!$. Hence the sum in (50) is $O_\epsilon((mn)^\epsilon)$, which is a lot smaller than the upper bounds that we derive for the right-hand side of (19).

For $x \leq \sqrt{mn}$ we need to investigate the transition ranges for the Bessel functions $J_k(y)$, that is, the ranges $y \leq k - k^{1/3}$, $k - k^{1/3} \leq y \leq k + k^{1/3}$ and $y \geq k + k^{1/3}$. We invoke the formula for the leading-term asymptotic behavior of these in each region, (see [Ol]) for uniform asymptotics which allows one to connect the ranges. For our ϕ we break the integral (16) defining $N_\phi(k)$ into the corresponding ranges. In the range $(0, k - k^{1/3})$ the Bessel function is exponentially small, and so the contribution is negligible. In the transitional range we use (22) and bound the integrand in absolute value. The contribution from this part to $(k-1) N_\phi(k)$ is $O(1)$, and since there are $O\left(\frac{\sqrt{mn}}{x}\right)$ values of k for which the transitional range is present, we conclude that the contribution to the sum (50) from the transitional range is

$$O\left(\frac{\sqrt{mn}}{x}(mn)^\epsilon\right).$$ (51)

We are left with the contribution to (50) from the range $y \geq k + k^{1/3}$ in the integral defining $N_\phi(k)$. In this range we have

$$J_k(ks) \sim \frac{e^{ikW(s)}}{\sqrt{k}(s^2 - 1)^{1/4}},$$ (52)

where

$$W((s) = \sqrt{s^2 - 1} - \arctan\sqrt{s^2 - 1}.$$ (53)

In particular,

$$W'(s) = \frac{\sqrt{s^2 - 1}}{s}.$$ (54)

Changing variables for the integral in the range in question leads one to consider

$$\int_{1+k^{-2/3}}^{\infty} \frac{\phi(ks) e^{ikW(s)}}{\sqrt{k} \, s(s^2 - 1)^{1/4}} \, ds.$$ (55)

We argue as with $\hat{\phi}(r)$ and the elementary mean value estimate. It is enough to bound

$$\int_{1+k^{-2/3}}^{c} \frac{e^{ikW(s)}}{s\sqrt{k}(k^2 - 1)^{1/4}} \, ds$$ (56)

independent of c.

Multiply and divide by $kW'(s)$ and integrate by parts. The boundary terms are $e^{ikW(s)} / \left(k^{3/2}(s^2 - 1^{3/4}) \right)$ evaluated at $1 + k^{2/3}$ and c. Hence they are $O(1/k)$. The resulting new integral is

$$- k^{-3/2} \int\limits_{1+k^{-2/3}}^{c} \frac{3s\, e^{ikW(s)}}{2(s^2 - 1)^{7/4}}\, ds. \tag{57}$$

Now bound this trivially by estimating the integrand in absolute value. This also gives a contribution of $O(1/k)$. The number of k's for which this range intersects the support of ϕ is again $O\left(\frac{\sqrt{mn}}{x} \right)$. It follows that the contribution to the k sum from this range is

$$O\left((mn)^\epsilon \frac{\sqrt{mn}}{x} \right). \tag{58}$$

That is, we have shown that

$$\sum_{k \equiv 0(2)} N_\phi(k) \sum_{1 \le j \le \dim S_k(\Gamma)} \overline{\psi_{j,k}(m)}\, \psi_{j,k(n)} \ll (mn)^\epsilon \left(1 + \frac{\sqrt{mn}}{x} \right). \tag{59}$$

Relations (49) and (59) give us the desired bounds for the last two terms in (19). In order to complete the analysis we must estimate the first term on the right-hand side of (19). It is here that we invoke H_θ. Consider the dyadic sums

$$\sum_{A \le t_j \le 2A} \frac{\rho_j(n)\, \overline{\rho_j(m)}}{\cosh \pi t_j}\, \hat{\phi}(t_j). \tag{60}$$

One can treat these in two ways. Firstly we can use H_θ directly, from which it follows (for a Hecke basis of Maass forms) that

$$|\rho_j(n)| \le \tau(n)\, n^\theta |\rho_j(1)|. \tag{61}$$

Hence

$$\sum_{A \le t_j \le 2A} \left| \frac{\rho_j(n)\, \overline{\rho_j(m)}}{\cosh \pi t_j} \right| \le \tau(n)\, \tau(m)(nm)^\theta \sum_{A \le t_j \le 2A} \frac{|\rho_j(1)|^2}{\cosh \pi t_j}. \tag{62}$$

We recall Kuznetzov's mean value estimate

$$\sum_{t_j \le y} \frac{|\rho_j(n)|^2}{\cosh \pi t_j} = \frac{y^2}{\pi} + O_\epsilon \left(y \log y + y n^\epsilon + n^{\frac{1}{2}+\epsilon} \right). \tag{63}$$

Applying (63) with $n = 1$ in (62) yields

$$\left| \sum_{A \le t_j \le 2A} \frac{\rho_j(n)\, \overline{\rho_j(m)}}{\cosh \pi t_j} \right| \ll_\epsilon (nm)^{\theta+\epsilon}\, A^2. \tag{64}$$

Alternatively, we can estimate (60) directly using (63) via Cauchy–Schwarz and obtain

$$\sum_{A \le t_j \le 2A} \left| \frac{\rho_j(n)\,\overline{\rho_j(m)}}{\cosh(\pi t_j)} \right| \le \left(\sum_A^{2A} \frac{|\rho_j(n)|^2}{\cosh \pi t_j} \right)^{1/2} \left(\sum_A^{2A} \frac{|\rho_j(m)|^2}{\cosh \pi t_j} \right)^{1/2}$$

$$\ll_{\epsilon} (A + m^{1/4+\epsilon})(A + n^{1/4+\epsilon}). \tag{65}$$

With these we have

$$\left| \sum_j \hat{\phi}(t_j) \frac{\rho_j(m)\,\overline{\rho_j(n)}}{\cosh \pi t_j} \right| \le \sum_j \frac{|\hat{\phi}(t_j)\,\rho_j(n)\,\rho_j(m)|}{\cosh \pi t_j}, \tag{66}$$

and breaking this into dyadic pieces applying (47), (48), (64), or (65), one obtains

$$\sum_A^{2A} \left| \hat{\phi}(t_j) \frac{\rho_j(n)\,\overline{\rho_j(m)}}{\cosh \pi t_j} \right| \ll (mn)^{\epsilon} \min\left(1, \frac{x}{TA}\right)$$

$$\times \min\left(\sqrt{A}(nm)^{\theta}, \ \sqrt{A} + \left(n^{\frac{1}{4}} + m^{1/4}\right) A^{-1/2} \right.$$

$$\left. + (mn)^{\frac{1}{4}} A^{-3/2} \right) \tag{67}$$

$$\ll (mn)^{\epsilon} \min\left(1, \frac{x}{TA}\right) \left(\sqrt{A} + \left(m^{1/8} + n^{1/8}\right)(mn)^{\theta/2} \right). \tag{68}$$

Hence,

$$\sum_A^{2A} \left| \hat{\phi}(t_j) \frac{\rho_j(n)\overline{\rho_j(m)}}{\cosh \pi t_j} \right| \ll (mn)^{\epsilon} \left(\left(m^{1/8} + n^{1/8}\right)(mn)^{\theta/2} \right.$$

$$\left. + \min\left(\sqrt{A}, \frac{x}{T\sqrt{A}} \right) \right). \tag{69}$$

Combining the dyadic pieces yields

$$\sum_j \hat{\phi}(t_j) \frac{\rho_j(n)\,\overline{\rho_j(m)}}{\cosh \pi t_j} \ll (mn)^{\epsilon} \left(\left(m^{1/8} + n^{1/8}\right)(mn)^{\theta/2} + \frac{\sqrt{x}}{T} \right). \tag{70}$$

Putting this together with (41) and (51) in (19) yields

$$\sum_{x \le c \le 2x} \frac{S(m,n;c)}{c} \ll (xmn)^{\epsilon} \left(\frac{T}{\sqrt{x}} + \frac{\sqrt{mn}}{x} + \left(m^{1/8} + n^{1/8}\right)(nm)^{\theta/2} + \sqrt{\frac{x}{T}} \right).$$

Finally, choosing $T = x^{2/3}$ yields

$$\sum_{x \le c \le 2x} \frac{S(m,n;c)}{c} \ll (xmn)^{\epsilon} \left(x^{1/6} + \frac{\sqrt{mn}}{x} + \left(m^{1/8} + n^{1/8}\right)(mn)^{\theta/2} \right). \tag{71}$$

This completes the proof of Proposition 5.

634   Peter Sarnak and Jacob Tsimerman

In Theorem 4 we explained the $(mn)^{1/6}$ in Theorem 2. The $x^{1/6}$ barrier is similar, that is, in the proof of Proposition 5, if we want to go beyond the exponent $1/6$ (ignoring the mn dependence) we would need to capture cancellations in sums of the type

$$\sum_{t_j \sim x^{1/3}} \frac{|\rho_j(1)|^2}{\cosh \pi t_j} x^{it_j} . \tag{72}$$

This appears to be quite difficult. A similar feature appears with the exponent of $1/3$ in the remainder term in the hyperbolic circle problem, which has resisted improvements (see [L-P] and [Iw]).

Acknowledgments

We thank D. Hejhal for his comments on an earlier draft of this paper.

References

[D-I]   J. Deshouillers and H. Iwaniec, "Kloosterman sums and Fourier coefficients of cusp forms," *Invent. Math.*, **70** (1982/83), no. 2, 219–288.

[De]   P. Deligne, La conjecture de Weil. I (French), *Inst. Hautes Études Sci. Publ. Math*, **43** (1974), 273–307.

[Du]   T.M. Dunster, "Bessel functions of purely imaginary order with applications to second order linear differential equations," *Siam Jnl. Math. Anal.*, **21**, No. 4 (1990).

[G-S]   D. Goldfeld and P. Sarnak, "Sums of Kloosterman sums," *Invent. Math.*, **71** (1983), no. 2, 243–250.

[Hu]   M.N. Huxley, "Introduction to Kloosertmania" in Banach Center Publications, Vol. 17 (1985), Polish Scientific Publishers.

[Iw]   H. Iwaniec, "Introduction to the spectral theory of automorphic forms," *AMS* (2002).

[I-K]   H. Iwaniec and F. Kowalski, "Analytic Number Theory," *AMS*, Coll. Publ., Vol. 53 (2004).

[I-L-S]   H. Iwaniec, W. Luo and P. Sarnak, "Low lying zeros of families of *L*-functions," *Inst. Hautes Études Sci. Publ. Math.*, **91**, (2000), 55–131 (2001).

[Ka]   N. Katz, "Gauss Sums, Kloosterman Sums and Monodromy," P.U. Press (1988).

[Ki-Sh]   H. Kim and F. Shahidi, "Functional products for $GL_2 \times GL_3$ and the symmetric cube for GL_2," *Annals of Math.* (2), **155**, no. 3 (2002), 837–893.

[Ki-Sa]   H. Kim and P. Sarnak, Appendix to Kim's paper, "Functoriality for the exterior square of GL_4 and the symmetric fourth of GL_2, *J. Amer. Math. Soc.*, **16** (2003), no. 1, 139–183.

[Ku]   N. Kuznetzov, "The Petersson conjecture for cusp forms of weight zero and the Linnik conjecture. Sums of Kloosterman sums," *Mat. Sb. (N.S)*, **111** (1980), 334–383, *Math. USSR-Sb.*, **39** (1981), 299–342.

[L-P]   P. Lax and R. Phillips, "The asymptotic distribution of lattice points in Euclidean and non-Euclidean spaces," *Jnl Funct. Anal.*, **46**, (1982), no. 3, 280–350.

[Li]   Y. Linnik, "Additive problems and eigenvalues of the modular operators," (1963), *Proc. Internat. Congr. Mathematicians (Stockholm, 1962)*, 270–284.

[L-R-S]   W. Luo, Z. Rudnick, and P. Sarnak, "On Selberg's eigenvalue conjecture," *Geom. Funct. Anal.*, **5** (1995), no. 2, 387–401.

[Mi]   P. Michel,"Autour de la conjecture de Sato-Tate pour les sommes de Kloosterman. I" (French) ["On the Sato-Tate conjecture for Kloosterman sums. I"], *Invent. Math.*, **121** (1995), no. 1, 61–78.

[Mu]   R. Murty, in "Lectures on Automorphic *L*-functions," Fields Institute Monographs, vol. 20 (2004).

[Ol]   F.W.J. Olver, "The asymptotic expansion of Bessel functions of large order,"*Philos. Trans. Royal Soc. London. Ser A*, **247** (1954), 328–368.

[Sa]   P. Sarnak, "Notes on the Generalized Ramanujan Conjectures," *Clay Math. Proc.*, Vol. 4, (2005), 659–685.

[Se]   A. Selberg, "On the estimation of Fourier coefficients of modular forms," *Proc. Sympos. Pure Math.*, **8**, (1965), 1–15.

[Wa]   G. Watson, "A treatise on the Theory of Bessel Functions," Cambridge Press (1966).

[We]   A. Weil, "On some exponential sums," *Proc. Nat. Acad. Sci.*, U.S.A., **34**, (1948), 204–207.

Une Algèbre Quadratique Liée à la Suite de Sturm

Oleg Ogievetsky[1] and Vadim Schechtman[2]

[1] Centre de Physique Théorique (Unité Mixte de Recherche 6207 du CNRS et des Universités Aix–Marseille I, Aix–Marseille II et du Sud Toulon – Var; laboratoire affilié à la FRUMAM, FR 2291), Luminy, 13288 Marseille, France
and
Institut de Physique P.N. Lebedev, Leninsky prospekt 53, 119991, Moscou, Russie; oleg@cpt.univ-mrs.fr
[2] Laboratoire Emile Picard, UFR MIG, Université Paul Sabatier, 31062 Toulouse, France; schechtman@math.ups-tlse.fr

A Yuri Ivanovich Manin, à l'occasion de son 70-ème anniversaire

Summary. An algebra given by quadratic relations in a polynomial algebra on infinite set of generators is introduced. Using it, we prove some explicit formulas for the coefficients of the Sturm sequence of a polynomial. In the second part we discuss a numerical example of polynomials studied by Euler. There, the Hilbert matrices and the Cauchy determinants appear in the asymptotics of the Sturm sequence.

Key words: Sturm theorem, Quadratic algebras, Hankel matrices, Cauchy determinants

2000 Mathematics Subject Classifications: 12D10, 16S37, 47B35

PREMIÈRE PARTIE

FORMULES

§ 1 Introduction

1.1 Cet article est une variation sur un thème de [Jacobi].
Soit

$$f(x) = a_n x^n + a_{n-1} x^{n-1} + \ldots + a_0$$

Y. Tschinkel and Y. Zarhin (eds.), *Algebra, Arithmetic, and Geometry*,
Progress in Mathematics 270, DOI 10.1007/978-0-8176-4747-6_21,
© Springer Science+Business Media, LLC 2009

un polynôme de degré $n > 0$ à coefficients dans un corps de base $\mathfrak{k}$ de caractéristique 0. Rappelons que *la suite de Sturm* de f,

$$\mathfrak{f} = (f_0, f_1, f_2, \ldots) \,,$$

est définie par récurrence : on pose $f_0(x) = f(x)$, $f_1(x) = f'(x)$ et pour $j \geq 1$ f_{j+1} est le reste de la division euclidienne de f_{j-1} par f_j, avec le signe opposé :

$$f_{j-1}(x) = q_{j-1}(x)f_j(x) - f_{j+1}(x), \tag{1.1.1}$$

$\deg f_{j+1}(x) < \deg f_j(x)$, cf. le célèbre mémoire [Sturm].

Dans cette note on propose des formules explicites pour les coefficients des polynômes f_j en termes des coefficients de f. Plus généralement, on donnera des formules analogues pour les membres de l'algorithme d'Euclide correspondant à deux polynômes quelconques f_1, f_2 de degrés $n-1, n-2$.

Notre point de départ est une algèbre $\mathfrak{B}$, quotient de l'anneau de polynômes en variables $b(i)_j$ ($i \geq 1$, $j \geq 2i$) par certains rélations quadratiques, cf. (1.7.1) ci-dessous. Nos formules sont des conséquences des identités dans $\mathfrak{B}$, analogues des rélations de Plücker.

1.2 Pour énoncer le résultat, introduisons les quantités quadratiques

$$b(j)_i = n \sum_{p=0}^{j-1} (i - 2p)a_{n-p}a_{n-i+p} - j(n-i+j)a_{n-j}a_{n+j-i},$$

$j \geq 1$, $i \geq 2j$. Ici on pose $a_i = 0$ pour $i < 0$. Par exemple,

$$b(1)_i = nia_n a_{n-i} - (n-i+1)a_{n-1}a_{n-i+1}.$$

1.3 Ensuite on introduit, pour $m \geq 2$, les matrices $(m-1) \times (m-1)$ symétriques

$$C(m) = \begin{pmatrix} b(1)_2 & b(1)_3 & b(1)_4 & b(1)_5 & \ldots & b(1)_m \\ b(1)_3 & b(2)_4 & b(2)_5 & b(2)_6 & \ldots & b(2)_{m+1} \\ b(1)_4 & b(2)_5 & b(3)_6 & b(3)_7 & \ldots & b(3)_{m+2} \\ b(1)_5 & b(2)_6 & b(3)_7 & b(4)_8 & \ldots & b(4)_{m+3} \\ & & \cdot & \cdot & \cdot & \\ b(1)_m & b(2)_{m+1} & b(3)_{m+2} & b(4)_{m+3} & \ldots & b(m-1)_{2m-2} \end{pmatrix}.$$

De plus, pour $i \geq 0$ on définit une matrice "décalée" $C(m)_i$: elle est obtenue en remplaçant dans $C(m)$ la dernière ligne par

$$\begin{pmatrix} b(1)_{m+i} & b(2)_{m+i+1} & b(3)_{m+i+2} & b(4)_{m+i+3} & \ldots & b(m-1)_{2m+i-2} \end{pmatrix}.$$

Donc $C(m)_0 = C(m)$. On pose

$$c(m)_i := \det C(m)_i, \quad c(m) := c(m)_0.$$

En particulier,

$$c(2)_i = b(1)_{i+2}$$

Il est commode de poser

$$c(1)_i := \frac{(n-i)a_{n-i}}{na_n},$$

$i \geq 0$, $c(1) := c(1)_0 = 1$.

1.4 Puis on définit les nombres γ_j, $j \geq 1$ par récurrence :

$$\gamma_1 = na_n, \quad \gamma_2 = -\frac{1}{n^2 a_n}, \quad \gamma_{j+1} = \gamma_{j-1} \cdot \frac{c(j-1)^2}{c(j)^2},$$

$j \geq 2$. Autrement dit,

$$\gamma_j = (-1)^{j+1} \epsilon_j \cdot \prod_{i=1}^{j-2} c(j-i)^{2(-1)^i},$$

où $\epsilon_j = na_n$ si j est impair et $1/(n^2 a_n)$ sinon.

Les nombres $\gamma_1, \ldots, \gamma_j$ sont donc bien définis si tous les nombres $c(2), c(3), \ldots, c(j-1)$ sont différents de zéro.

1.5 *Théorème.* Supposons que $\deg f_j = n - j$, donc $\deg f_i = n - i$ pour $i \leq j$.

Alors pour tous $i \leq j$, on a $c(i) \neq 0$ et

$$f_i(x) = \gamma_i \cdot \sum_{p=0}^{n-i} c(i)_p x^{n-i-p}.$$

En particulier, le coefficient dominant de $f_i(x)$ est égal à $\gamma_i c(i)$.

1.6 On vérifie aussitôt que

$$b(k)_i - b(k-1)_i = c(1)_{k-1} b(1)_{i-k+1} - c(1)_{i-k} b(1)_k \qquad (1.6.1)$$

pour tous $k \geq 2$, $i \geq 2k - 2$. Par exemple,

$$b(2)_i - b(1)_i = c(1)_1 b(1)_{i-1} - c(1)_{i-2} b(1)_2,$$
$$b(3)_i - b(2)_i = c(1)_2 b(1)_{i-2} - c(1)_{i-3} b(1)_3,$$

etc. Il s'en suit que tous les $b(j)_i$, $j \geq 2$, sont expressibles en termes de $c(1)_p$ et $c(2)_p = b(1)_{p+2}$, $p \geq 0$.

1.7 Les formules (1.6.1) impliquent que les nombres $b(i)_j$ satisfont aux relations quadratiques suivantes :

$$\big(b(k)_i - b(k-1)_i\big) \cdot b(1)_j$$

$$= \big(b(j)_{i-k+j} - b(j-1)_{i-k+j}\big) \cdot b(1)_k - \big(b(j)_{k+j-1} - b(j-1)_{k+j-1}\big) \cdot b(1)_{i-k+1}$$

$$(1.7.1)$$

On verra que la preuve de 1.5 ne dépend que des relations (1.7.1).

On formalise la situation en introduisant une algèbre quadratique correspondante, cf. §2 ci-dessous.

1.8 Maintenant soient

$$f_1(x) = \alpha_0 x^{n-1} + \alpha_1 x^{n-2} + \dots$$

et

$$f_2(x) = \beta_0 x^{n-2} + \beta_1 x^{n-3} + \dots$$

deux polynômes arbitraires de degrés $n-1, n-2$. On définit f_j, $j \geq 3$ à partir de f_1, f_2 par les formules de l'algorithme d'Euclide (1.1.1).

Posons

$$c(1)_i := \frac{\alpha_i}{\alpha_0}, \ b(1)_{i+2} := \beta_i, \ i \geq 0.$$

Définissons les nombres $b(k)_i$, $k \geq 2$ par récurrence sur k, à partir des formules (1.6.1).

Définissons les nombres $c(m)_i$, $m \geq 2$, par les formules 1.3.

Enfin, on pose :

$$\tilde{\gamma}_1 = \alpha_0 \ , \ \tilde{\gamma}_2 = 1 \ , \ \tilde{\gamma}_{j+1} = \tilde{\gamma}_{j-1} \frac{c(j-1)^2}{c(j)^2}$$

Alors on a

1.9 *Théorème.* Supposons que $\deg f_j = n - j$, d'où $\deg f_i = n - i$ pour $i \leq j$.

Alors pour tous $i \leq j$, on a $c(i) \neq 0$ et

$$f_i(x) = \tilde{\gamma}_i \cdot \sum_{p=0}^{n-i} c(i)_p x^{n-i-p} \ .$$

En particulier, le coefficient dominant de $f_i(x)$ est égal à $\tilde{\gamma}_i c(i)$.

Cf. [Jacobi], section 15.

1.10 Dans la Deuxième Partie on présente un exemple numérique. Là, les déterminants de Cauchy apparaissent dans les asymptotiques des coefficients dominants de la suite de Sturm pour les polynômes d'Euler.

§ 2 Algèbre $\mathfrak{B}$

2.1 On peut réécrire les relations (1.7.1) sous la forme suivante :

$$
\det \begin{pmatrix} b(1)_j & b(1)_k \\ b(j-1)_{i+j-k} & b(k-1)_i \end{pmatrix} - \det \begin{pmatrix} b(1)_j & b(j-1)_{j+k-1} \\ b(1)_{i-k+1} & b(k)_i \end{pmatrix}
$$

$$
+ \det \begin{pmatrix} b(1)_k & b(j)_{j+k-1} \\ b(1)_{i-k+1} & b(j)_{i+j-k} \end{pmatrix} = \Delta(k,j)_i - \Delta'(k,j)_i + \Delta''(k,j)_i = 0.
$$

$$(2.1.1)$$

2.2 On définit une algèbre quadratique $\mathfrak{B}$ comme une $\mathfrak{k}$-algèbre commutative engendrée par les lettres $b(i)_j$, $i,j \in \mathbb{Z}$, modulo les relations (2.1.1), où $i,j,k \in \mathbb{Z}$.

(D'ailleurs, dans tout le paragraphe qui suit on peut remplacer le corps de base $\mathfrak{k}$ par un anneau commutatif quelconque.)

2.3 Le but de ce paragraphe est d'écrire certaines relations entre les déterminants $n \times n$ dans $\mathfrak{B}$ qui généralisent (2.1.1).

On fixe un nombre entier $n \geq 2$. Soient $m_1, \ldots, m_n, i$ des entiers.

On définit $2n+2$ vecteurs $v_j, w_j \in \mathfrak{k}^n$, $j = 1, \ldots, n+1$:

$$
w_1 = (b(1)_{m_1}, b(1)_{m_2}, \ldots, b(1)_{m_n}),
$$

$$
w_{j+1} = (b(1)_{m_1}, \ldots, \hat{b}(1)_{m_{n+1-j}}, \ldots, b(1)_{m_n}, b(1)_{i-m_n+1}),
$$

$1 \leq j \leq n$ (suivant l'usage, $\hat{x}$ signifie que l'on omet la composante x).

Puis

$$
v_1 = (b(m_1-1)_{i+m_1-m_n}, b(m_2-1)_{i+m_2-m_n}, \ldots, b(m_{n-1}-1)_{i+m_{n-1}-m_n}, b(m_n-1)_i),
$$

$$
v_2 = (b(m_1-1)_{m_1+m_n-1}, b(m_2-1)_{m_2+m_n-1}, \ldots, b(m_{n-1}-1)_{m_{n-1}+m_n-1}, b(m_n)_i),
$$

$$
v_3 = (b(m_1-1)_{m_1+m_{n-1}-1}, b(m_2-1)_{m_2+m_{n-1}-1}, \ldots, b(m_{n-2}-1)_{m_{n-2}+m_{n-1}-1},
$$
$$
b(m_{n-1})_{m_{n-1}+m_n-1}, b(m_{n-1})_{i+m_{n-1}-m_n}),
$$

$$
v_4 = (b(m_1-1)_{m_1+m_{n-2}-1}, b(m_2-1)_{m_2+m_{n-2}-1}, \ldots, b(m_{n-3}-1)_{m_{n-3}+m_{n-2}-1},
$$
$$
b(m_{n-2})_{m_{n-2}+m_{n-1}-1}, b(m_{n-2})_{m_{n-2}+m_n-1}, b(m_{n-2})_{i+m_{n-2}-m_n}),
$$

$$
\cdots
$$

$$
v_n = (b(m_1-1)_{m_1+m_2-1}, b(m_2)_{m_2+m_3-1}, b(m_2)_{m_2+m_4-1}, \ldots, b(m_2)_{m_2+m_n-1},
$$
$$
b(m_2)_{i+m_2-m_n})
$$
$$
v_{n+1} = (b(m_1)_{m_1+m_2-1}, b(m_1)_{m_1+m_3-1}, \ldots, b(m_1)_{m_1+m_n-1}, b(m_1)_{i+m_1-m_n}).
$$

2.4 Soit

$$
M = \begin{pmatrix} x_{11} & \cdots & x_{1,n+1} \\ \cdot & \cdots & \cdot \\ x_{n-2,1} & \cdots & x_{n-2,n+1} \end{pmatrix}
$$

une matrice $(n-2) \times (n+1)$ sur $\mathfrak{B}$; soit M_i, $i = 1, \ldots, n+1$, ses sous-matrices $(n-2) \times n$. Pour écrire M_i, on enlève donc la i-ième colonne de M.

Maintenant on va définir $n+1$ matrices $n \times n$

$$D_j = D_j(m_1, \ldots, m_n; M_{n+2-j})_i,$$

$j = 1, \ldots, n+1$. On pose :

$$D_1 = \begin{pmatrix} w_1 \\ M_{n+1} \\ v_1 \end{pmatrix}, D_j = \begin{pmatrix} w_j^t & M_{n+2-j}^t & v_j^t \end{pmatrix},$$

$j = 2, \ldots, n+1$. Ici $(.)^t$ désigne la matrice transposée.

Enfin, on pose

$$\Delta_j = \Delta_j(m_1, \ldots, m_n; M_{n+2-j})_i = \det D_j(m_1, \ldots, m_n; M_{n+2-j})_i,$$

$j = 1, \ldots, n+1$.

Considérons la somme alternée

$$R(n; m_1, \ldots, m_n; M)_i = \sum_{j=1}^{n+1} (-1)^{j+1} \Delta_j(m_1, \ldots, m_n; M_{n+2-j})_i.$$

2.5 *Exemple.* $n = 2$. Dans ce cas il n'y a pas de matrice M ; trois nombres entiers sont donnés : m_1, m_2 et i. On aura 6 vecteurs :

$$w_1 = (b(1)_{m_1}, b(1)_{m_2}), \ w_2 = (b(1)_{m_1}, b(1)_{i-m_2+1}), \ w_3 = (b(1)_{m_2}, b(1)_{i-m_2+1})$$

et

$$v_1 = (b(m_1-1)_{i+m_1-m_2}, b(m_2-1)_i), \ v_2 = (b(m_1-1)_{m_1+m_2-1}, b(m_2)_i),$$

$$v_2 = (b(m_1)_{m_1+m_2-1}, b(m_1)_{i+m_1-m_2}).$$

Il s'ensuit :

$$R(2; m_1, m_2)_i = \det \begin{pmatrix} b(1)_{m_1} & b(1)_{m_2} \\ b(m_1-1)_{i+m_1-m_2} & b(m_2-1)_i \end{pmatrix}$$

$$-\det \begin{pmatrix} b(1)_{m_1} & b(m_1-1)_{m_1+m_2-1} \\ b(1)_{i-m_2+1} & b(m_2)_i \end{pmatrix} + \det \begin{pmatrix} b(1)_{m_2} & b(m_1)_{m_1+m_2-1} \\ b(1)_{i-m_2+1} & b(m_2)_{i+m_1-m_2} \end{pmatrix}$$

On reconnaît là la partie gauche de (2.1.1) pour $(j, k) = (m_1, m_2)$. Il en découle que $R(2; m_1, m_2)_i = 0$.

2.6 *Exemple.* $n = 3$. Dans ce cas la matrice M se réduit à 4 éléments :

$$M = \begin{pmatrix} x_1 & x_2 & x_3 & x_4 \end{pmatrix}.$$

L'expression $R(3; m_1, m_2, m_3; M)_i$ prend la forme

$$
R(3; m_1, m_2, m_3; M)_i = \det \begin{pmatrix} b(1)_{m_1} & b(1)_{m_2} & b(1)_{m_3} \\ x_1 & x_2 & x_3 \\ b(m_1-1)_{i+m_1-m_3} & b(m_2-1)_{i+m_2-m_3} & b(m_3-1)_i \end{pmatrix}
$$

$$
- \det \begin{pmatrix} b(1)_{m_1} & x_1 & b(m_1-1)_{m_1+m_3-1} \\ b(1)_{m_2} & x_2 & b(m_2-1)_{m_2+m_3-1} \\ b(1)_{i-m_3+1} & x_4 & b(m_3)_i \end{pmatrix}
$$

$$
+ \det \begin{pmatrix} b(1)_{m_1} & x_1 & b(m_1-1)_{m_1+m_2-1} \\ b(1)_{m_3} & x_3 & b(m_2)_{m_2+m_3-1} \\ b(1)_{i-m_3+1} & x_4 & b(m_2)_{i+m_2-m_3} \end{pmatrix}
$$

$$
- \det \begin{pmatrix} b(1)_{m_2} & x_2 & b(m_1)_{m_1+m_2-1} \\ b(1)_{m_3} & x_3 & b(m_1)_{m_1+m_3-1} \\ b(1)_{i-m_3+1} & x_4 & b(m_1)_{i+m_1-m_3} \end{pmatrix}.
$$

Calculons cette expression.

On développe le premier déterminant suivant la deuxième ligne et les autres suivant les deuxièmes colonnes :

$$
\Delta_1(3; m_1, m_2, m_3; M_4)_i = -x_1 \Delta_1(2; m_2, m_3)_i + x_2 \Delta_1(2; m_1, m_3)_i
$$
$$
- x_3 \Delta_1(2; m_1, m_2)_{i+m_2-m_3},
$$
$$
\Delta_2(3; m_1, m_2, m_3; M_3)_i = -x_1 \Delta_2(2; m_2, m_3)_i + x_2 \Delta_2(2; m_1, m_3)_i
$$
$$
- x_4 \Delta_1(2; m_1, m_3)_{m_2+m_3-1}.
$$

Puis

$$
\Delta_3(3; m_1, m_2, m_3; M_2)_i = -x_1 \Delta_3(2; m_2, m_3)_i + x_3 \Delta_2(2; m_1, m_2)_{i+m_2-m_3}
$$
$$
- x_4 \Delta_2(2; m_1, m_3)_{m_2+m_3-1}
$$

et

$$
\Delta_4(3; m_1, m_2, m_3; M_1)_i = -x_2 \Delta_3(2; m_1, m_3)_i + x_3 \Delta_3(2; m_1, m_2)_{i+m_2-m_3}
$$
$$
- x_4 \Delta_3(2; m_1, m_3)_{m_2+m_3-1}.
$$

Pour abréger les notations on introduit des vecteurs entiers :

$$
(i_1, i_2, i_3, i_4) := (i, i, i + m_2 - m_3, m_2 + m_3 - 1),
$$

$$
\mu = (m_1, m_2, m_3),
$$

$$
\mu_1 = (m_2, m_3), \quad \mu_2 = (m_1, m_3), \quad \mu_3 = (m_1, m_2).
$$

On peut réécire les formules ci-desssus sous une forme matricielle :

$$
\begin{pmatrix} \Delta_1(3; \mu; M_4)_i \\ -\Delta_2(3; \mu; M_3)_i \\ \Delta_3(3; \mu; M_2)_i \\ -\Delta_4(3; \mu; M_1)_i \end{pmatrix} = \begin{pmatrix} -\Delta_1(2; \mu_1)_{i_1} & \Delta_1(2; \mu_2)_{i_2} & -\Delta_1(2; \mu_3)_{i_3} & 0 \\ \Delta_2(2; \mu_1)_{i_1} & -\Delta_2(2; \mu_2)_{i_2} & 0 & \Delta_1(2; \mu_2)_{i_4} \\ -\Delta_3(2; \mu_1)_{i_1} & 0 & \Delta_2(2; \mu_3)_{i_3} & -\Delta_2(2; \mu_2)_{i_4} \\ 0 & \Delta_3(2; \mu_2)_{i_2} & -\Delta_3(2; \mu_3)_{i_3} & \Delta_3(2; \mu_2)_{i_4} \end{pmatrix} \cdot \begin{pmatrix} x_1 \\ x_2 \\ x_3 \\ x_4 \end{pmatrix}.
$$

En rajoutant :

$$R(3; m_1, m_2, m_3; M)_i = -x_1 \cdot \left\{ \Delta_1(2; \mu_1)_{i_1} - \Delta_2(2; \mu_1)_{i_1} + \Delta_3(2; \mu_1)_{i_1} \right\}$$

$$+x_2 \cdot \left\{ \Delta_1(2; \mu_2)_{i_2} - \Delta_2(2; \mu_2)_{i_2} + \Delta_3(2; \mu_2)_{i_2} \right\}$$

$$-x_3 \cdot \left\{ \Delta_1(2; \mu_3)_{i_3} - \Delta_2(2; \mu_3)_{i_3} + \Delta_3(2; \mu_3)_{i_3} \right\}$$

$$+x_4 \cdot \left\{ \Delta_1(2; \mu_2)_{i_4} - \Delta_2(2; \mu_2)_{i_4} + \Delta_3(2; \mu_2)_{i_4} \right\} = 0.$$

Le théorème ci-dessous généralise ces exemples.

2.7 *Théorème.* On a

$$R(n; m_1, \ldots, m_n; M)_i = 0$$

pour tous $n, m_1, \ldots, m_n, M$ et i.

Démonstration : elle se fait par récurrence sur n. Le cas $n = 2$ est l'exemple 2.5.

Le passage de $n - 1$ à n suit l'exemple 2.6.

Posons pour abréger

$$\mu = (m_1, \ldots, m_n).$$

À partir de cela, on introduit $n + 1$ vecteurs $\mu_j \in \mathbb{Z}^{n-1}$:

$$\mu_j := (m_1, \ldots, \hat{m}_j, \ldots, m_n),$$

$j = 1, \ldots, n$, et

$$\mu_{n+1} := (m_1, \ldots, \hat{m}_{n-1}, m_n) = \mu_{n-1}.$$

On définit le vecteur

$$(i_1, i_2, \ldots, i_{n+1}) := (\underbrace{i, i, \ldots, i}_{n-1 \text{ fois}}, i + m_{n-1} - m_n, m_{n-1} + m_n - 1) \in \mathbb{Z}^{n+1}.$$

En développant les déterminants $\Delta_j(n; \mu, M_{n+2-j})_i$, $2 \le j \le n + 1$ suivant la deuxième colonne et le déterminant $\Delta_1(n; \mu, M_{n+1})_i$ suivant la deuxième ligne, on obtient :

$$R(n; \mu; M)_i = \sum_{j=1}^{n+1} (-1)^j x_j R(n - 1; \mu_j; M_{1j})_{i_j}.$$

Ici M_{1j} est la matrice obtenue en enlevant la première ligne et la j-ième colonne de la matrice M.

Notre assertion en découle immédiatement par récurrence sur n.

2.8 On aura besoin d'un cas particulier de ces relations. Prenons
$$\mu = (m_1, m_2, \ldots, m_n) = (2, 3, \ldots, n+1).$$
Pour la matrice M, prenons
$$M = \begin{pmatrix} b(1)_3 & b(2)_4 & b(2)_5 & \cdots & b(2)_{n+2} & b(2)_{i+n+1} \\ b(1)_4 & b(2)_5 & b(3)_5 & \cdots & b(3)_{n+3} & b(3)_{i+n+2} \\ \cdot & \cdot & \cdot & \cdot & \cdot & \cdot \\ b(1)_n & b(2)_{n+1} & b(3)_{n+2} & \cdots & b(n-1)_{2n-1} & b(n-1)_{i+2n-2} \end{pmatrix}.$$
Alors le premier déterminant
$$\Delta_1(n; \mu; M_{n+1})_{i+2n} = c(n+1)_i.$$
On pose par définition :
$$c(n+1)'_i := \Delta_2(n; \mu; M_n)_{i+2n},$$
$$c(n+1)''_i := \Delta_3(n; \mu; M_{n-1})_{i+2n}.$$
Par contre, si $j \geq 4$ on voit que dans le déterminant $\Delta_j(n; \mu; M_{n+2-j})_{i+2n}$ la dernière colonne est égale à la $(n-j+3)$-ième colonne, d'où
$$\Delta_4(n; \mu; M_{n-2})_{i+2n} = \Delta_5(n; \mu; M_{n-3})_{i+2n} = \ldots = \Delta_{n+1}(n; \mu; M_1)_{i+2n} = 0.$$
Donc 2.7 entraîne

2.9 *Corollaire.* Pour tous $n \geq 3$
$$c(n)_i - c(n)'_i + c(n)''_i = 0.$$

§ 3 Début de la démonstration du théorème 1.5

3.1 On a
$$f(x) = a_n x^n + a_{n-1} x^{n-1} + \ldots + a_0.$$
La dérivée :
$$f_1(x) = f'(x) = n a_n x^{n-1} + (n-1) a_{n-1} x^{n-2} + \ldots + a_1$$
$$= n a_n \left\{ x^{n-1} + \frac{(n-1)a_{n-1}}{n a_n} x^{n-2} + \ldots + \frac{a_1}{n a_n} \right\}$$
$$= \gamma_1 (c(1)_0 x^{n-1} + c(1)_1 x^{n-2} + \ldots + c(1)_{n-1}).$$

3.2 Le quotient de la division euclidienne de deux polynômes $f(x)$ et $g(x) = a'_{n-1} x^{n-1} + a'_{n-2} x^{n-2} + \ldots$ est égal à
$$\frac{a_n}{a'_{n-1}} x + \frac{a'_{n-1} a_{n-1} - a'_{n-2} a_n}{(a'_{n-1})^2}.$$

On fait la division euclidienne :

$$f - (x/n + a_{n-1}/n^2 a_n)f' = \frac{2n a_n a_{n-2} - (n-1)a_{n-1}^2}{n^2 a_n}x^{n-2}$$
$$+ \frac{3n a_n a_{n-3} - (n-2)a_{n-1}a_{n-2}}{n^2 a_n}x^{n-3} + \dots$$

$$= \frac{1}{n^2 a_n} \cdot (b(1)_2 x^{n-2} + b(1)_3 x^{n-3} + \dots + b(1)_n)$$

$$= -\gamma_2 \cdot (c(2)_0 x^{n-2} + c(2)_1 x^{n-3} + \dots c(2)_{n-2}).$$

Donc

$$f_2(x) = \gamma_2 \cdot \sum_{i=0}^{n-2} c(2)_i x^{n-2-i}.$$

Cela démontre l'assertion 1.5 pour $j = 1, 2$, et l'on procède par récurrence par j.

3.3 On suppose que l'on a déjà trouvé :

$$f_{j-1}(x) = \gamma_{j-1} \cdot [c(j-1)x^{n-j+1} + c(j-1)_1 x^{n-j} + \dots + c(j-1)_i x^{n-j+1-i} + \dots]$$

et

$$f_j(x) = \gamma_j \cdot [c(j)x^{n-j} + c(j)_1 x^{n-j-1} + \dots + c(j)_i x^{n-j-i} + \dots].$$

On fait la division euclidienne :

$$f_{j-1}(x) - \left(\frac{\gamma_{j-1}c(j-1)}{\gamma_j c(j)}x + \gamma_{j-1}\left[c(j-1)_1 - \frac{c(j-1)}{c(j)} \cdot c(j)_1\right] \cdot \frac{1}{\gamma_j c(j)}\right)f_j(x) =$$

$$= \sum_{i=2}^{n-j+1} \gamma_{j-1}\left\{c(j-1)_i - \frac{c(j-1)}{c(j)} \cdot c(j)_i - \left[c(j-1)_1 - \frac{c(j-1)}{c(j)} \cdot c(j)_1\right] \cdot \frac{c(j)_{i-1}}{c(j)}\right\}$$
$$\cdot x^{n-j+1-i} =$$

$$= \frac{\gamma_{j-1}}{c(j)^2} \sum_{i=2}^{n-j+1} \left\{c(j-1)_i c(j)^2 - c(j-1)c(j)c(j)_i - c(j-1)_1 c(j)_{i-1}c(j)\right.$$

$$\left. + c(j-1)c(j)_1 c(j)_{i-1}\right\} \cdot x^{n-j+1-i}.$$

On pose :

$$Q(j)_i := c(j-1)_i c(j)^2 - c(j-1)c(j)c(j)_i - c(j-1)_1 c(j)_{i-1}c(j)$$
$$+ c(j-1)c(j)_1 c(j)_{i-1} \qquad (3.3.1)$$

Alors on a :

$$f_{j+1}(x) = -\frac{\gamma_{j-1}}{c(j)^2} \sum_{i=2}^{n-j+1} Q(j)_i x^{n-j+1-i}$$

Il faut montrer que

$$f_{j+1}(x) = \gamma_{j+1} \sum_{i=0}^{n-j-1} c(j+1)_i x^{n-j-1-i} = \gamma_{j+1} \sum_{i=0}^{n-j+1} c(j+1)_{i-2} x^{n-j+1-i}$$

où

$$\gamma_{j+1} = \gamma_{j-1} \cdot \frac{c(j-1)^2}{c(j)^2}.$$

Donc notre théorème est équivalent à l'identité suivante :

$$Q(j)_i = -c(j)^2 c(j+1)_i. \tag{3.3.2}$$

§ 4 Formule (A)

4.1 Revenons à notre algèbre $\mathfrak{B}$.

On considère la matrice $n \times n$

$$C(n+1)_{i-2} = \begin{pmatrix} b(1)_2 & b(1)_3 & \dots & b(1)_n & b(1)_{n+1} \\ b(1)_3 & b(2)_4 & \dots & b(2)_{n+1} & b(2)_{n+2} \\ \cdot & \cdot & \dots & \cdot & \cdot \\ b(1)_n & b(2)_{n+1} & \dots & b(n-1)_{2n-2} & b(n-1)_{2n-1} \\ b(1)_{n+i-1} & b(2)_{n+i} & \dots & b(n-1)_{2n+i-3} & b(n)_{2n+i-2} \end{pmatrix}.$$

Donc $c(n+1)_{i-2} = \det C(n+1)_{i-2}$.

Si l'on désigne par $C(n+1)_{i-2;\hat{p},\hat{q}}$ la matrice $C(n+1)_{i-2}$ avec la p-ième ligne et la q-ième colonne enlevée, on aura :

$$c(n) = \det C(n+1)_{i-2;\hat{n},\hat{n}},$$

$$c(n)_{i-1} = \det C(n+1)_{i-2;\widehat{n-1},\hat{n}}.$$

En plus, on a :

$$c(n)_i'' = \det C(n+1)_{i-2;\widehat{n-2},\hat{n}}$$

où $c(n)_i''$ a été introduit dans 2.9.

4.2 *Théorème.* Pour tous $n, i \in \mathbb{Z}$, $n \geq 3$, on a la relation suivante dans $\mathfrak{B}$

$$c(n-1)_i c(n)^2 - c(n-1)c(n)c(n)_i - c(n-1)_1 c(n)_{i-1}c(n) + c(n-1)c(n)_1 c(n)_{i-1}$$
$$= -c(n-1)^2 c(n+1)_{i-2} \tag{F}$$

On a vu que notre théorème principal 1.5 est une conséquence de (F) : en effet (F) coïncide avec la formule (3.3.2) (avec j remplacé par n).

À son tour, (F) est une conséquence immédiate de deux formules :

$$c(n-1)_i c(n) - c(n-1)_1 c(n)_{i-1} = -c(n-1)c(n)_i'' \tag{A}$$

ou bien

$$c(n-1)_i c(n) - c(n-1)_1 c(n)_{i-1} + c(n-1)c(n)_i'' = 0 \qquad (A')$$

et

$$\{c(n)_i + c(n)_i''\} \cdot c(n) - c(n)_1 c(n)_{i-1} = c(n-1)c(n+1)_{i-2}. \qquad (B)$$

La démonstration de (B) utilise les relations quadratiques entre les lettres $b(i)_j$. Par contre, (A) est "élémentaire", en ce sens que cette identité n'utilise pas de relations entre les lettres $b(i)_j$.

Pour démontrer (A), on applique le lemme suivant (une variante des relations de Plücker) :

4.3 *Lemme.* (A_n) Considérons n vecteurs de dimension $n-1$, $w_i = (w_{i1}, \ldots, w_{i,n-1})$, $i = 1, \ldots, n$. À partir d'eux, on définit n vecteurs de dimension $n-2 : v_i = (w_{i1}, \ldots, w_{i,n-2})$. On pose :

$$W_i = \det(w_1, \ldots, \hat{w}_i, \ldots, w_n)^t,$$

$$V_{ij} := \det(v_1, \ldots, \hat{v}_i, \ldots, \hat{v}_j, \ldots, v_n)^t.$$

Alors

$$V_{n-2,n-1} \cdot W_n - V_{n-2,n} \cdot W_{n-1} + V_{n-1,n} \cdot W_{n-2} = 0.$$

(B_n) Considérons n vecteurs de dimension $n-2$, $v_i = (v_{i1}, \ldots, v_{i,n-2})$, $i = 1, \ldots, n$. Considérons les mineurs

$$V_{ij} := \det(v_1, \ldots, \hat{v}_i, \ldots, \hat{v}_j, \ldots, v_n)^t.$$

Alors pour chaque $i < n-2$,

$$V_{n-2,n-1} \cdot V_{i,n} - V_{n-2,n} \cdot V_{i,n-1} + V_{n-1,n} \cdot V_{i,n-2} = 0.$$

En effet, en développant W_i par rapport à la dernière colonne, on obtient : $(B_n) \Rightarrow (A_n)$.

Par contre, pour vérifier (B_n), considérons la matrice $(n-1) \times (n-2)$, $W^\sim = V_i$. Alors on aura $V_{ij} = W_j^\sim$, $j = n, n-1, n-2$. D'un autre côté, en développant les mineurs dans (B_n) : V_{pq}, $n-2 \leq p < q \leq n$ par rapport à la i-ième ligne, on obtient les mineurs $V_{pq}^\sim$, où $V^\sim$ est obtenue de $W^\sim$ en enlevant la dernière colonne. On vérifie que (B_n) se réduit à (A_{n-1}) correspondant à $W^\sim$.

Il s'ensuit que $(A_{n-1}) \Rightarrow (B_n)$ et on conclut par récurrence.

4.4. Le lemme étant vérifié, l'assertion 4.2 (A) est 4.3 (A_n) pour la matrice W égale à $c(n+1)_{i-2}$ avec la dernière colonne enlevée.

§ 5 Formule (B)

5.1. Maintenant on s'occupe de la formule

$$P := \{c(n)_i + c(n)_i''\} \cdot c(n) - c(n)_1 c(n)_{i-1} = c(n-1)c(n+1)_{i-2}. \qquad (B)$$

On introduit n vecteurs de dimension $n-1$, $w_1, \ldots, w_n$ qui sont les lignes de la matrice $c(n+1)_{i-2}$ sans la dernière colonne :

$$
W = \begin{pmatrix}
b(1)_2 & b(1)_3 & \cdots & b(1)_{n-1} & b(1)_n \\
b(1)_3 & b(2)_4 & \cdots & b(2)_n & b(2)_{n+1} \\
\cdot & \cdot & \cdots & \cdot & \cdot \\
b(1)_{n-1} & b(2)_n & \cdots & b(n-2)_{2n-4} & b(n-2)_{2n-3} \\
b(1)_n & b(2)_{n+1} & \cdots & b(n-2)_{2n-3} & b(n-1)_{2n-2} \\
b(1)_{n+i-1} & b(2)_{n+i} & \cdots & b(n-2)_{2n+i-4} & b(n-1)_{2n+i-3}
\end{pmatrix}
$$

et n mineurs

$$
W_i = \det(w_1, \ldots, \hat{w}_i, \ldots, w_n)^t, \quad i = 1, \ldots, n.
$$

Par exemple, $W_n = c(n)$, $W_{n-1} = c(n)_{i-1}$, $W_{n-2} = c(n)''_i$. Donc,

$$
c(n+1)_{i-2} = b(n)_{2n+i-2}W_n - b(n-1)_{2n-1}W_{n-1}
$$

$$
+ b(n-2)_{2n-2}W_{n-2} - \ldots + (-1)^{n-1}b(1)_{n+1}W_1
$$

$$
= b(n)_{2n+i-2}W_n - b(n-1)_{2n-1}W_{n-1} + R \tag{5.1.1}
$$

où

$$
R = b(n-2)_{2n-2}W_{n-2} - b(n-3)_{2n-3}W_{n-3} + \ldots + (-1)^{n-1}b(1)_{n+1}W_1. \tag{5.1.2}
$$

5.2. On a $n-1$ relations linéaires entre les W_i : la i-ième est obtenue en ajoutant à W sa i-ième colonne et en développant le déterminant $= 0$ par rapport à la dernière colonne.

Explicitement :

$$
b(n-1)_{2n+i-3}W_n - b(n-1)_{2n-2}W_{n-1} + b(n-2)_{2n-3}W_{n-2} - \ldots + (-1)^{n-1}b(1)_nW_1 = 0,
$$
$$
b(n-2)_{2n+i-4}W_n - b(n-2)_{2n-3}W_{n-1} + b(n-2)_{2n-4}W_{n-2} - \ldots + (-1)^{n-1}b(1)_{n-1}W_1 = 0,
$$
$$
\cdots
$$
$$
b(2)_{n+i}W_n - b(2)_{n+1}W_{n-1} + b(2)_nW_{n-2} - \ldots + (-1)^{n-2}b(2)_4W_2 + (-1)^{n-1}b(1)_3W_1 = 0,
$$
$$
b(1)_{n+i-1}W_n - b(1)_nW_{n-1} + b(1)_{n-1}W_{n-2} - \ldots + (-1)^{n-2}b(1)_3W_2 + (-1)^{n-1}b(1)_2W_1 = 0.
$$

5.3. D'autre part, rappelons la matrice $c(n)_1$:

$$
c(n)_1 = \det \begin{pmatrix}
b(1)_2 & b(1)_3 & \cdots & b(1)_{n-1} & b(1)_n \\
b(1)_3 & b(2)_4 & \cdots & b(2)_n & b(2)_{n+1} \\
\cdot & \cdot & \cdots & \cdot & \cdot \\
b(1)_{n-1} & b(2)_n & \cdots & b(n-2)_{2n-4} & b(n-2)_{2n-3} \\
b(1)_{n+1} & b(2)_{n+2} & \cdots & b(n-2)_{2n-2} & b(n-1)_{2n-1}
\end{pmatrix}.
$$

On développe cette quantité par rapport à la dernière colonne :

$$c(n)_1 = b(n-1)_{2n-1}c(n-1) - b(n-2)_{2n-3}M_{n-2} + \ldots + (-1)^{n-1}b(2)_{n+1}M_2$$
$$+(-1)^n b(1)_n M_1.$$

Après la multiplication par $-c(n)_{i-1} = -W_{n-1}$ on obtient :

$$-c(n)_1 c(n)_{i-1} = -b(n-1)_{2n-1}c(n-1)W_{n-1} \ (*) + R'$$

où

$$R' = b(n-2)_{2n-3}W_{n-1}M_{n-2} - b(n-3)_{2n-4}W_{n-1}M_{n-3} + \ldots$$
$$+(-1)^n b(2)_{n+1}W_{n-1}M_2 + (-1)^{n-1}b(1)_n W_{n-1}M_1.$$

5.4. Maintenant remplaçons dans R' les termes $(-1)^i b(n-i)_{2n-i-1}W_{n-1}$ en utilisant les relations 5.2 :

$$b(n-2)_{2n-3}W_{n-1} = b(n-2)_{2n+i-4}W_n + b(n-2)_{2n-4}W_{n-2} - \ldots$$
$$+(-1)^{n-1}b(1)_{n-1}W_1,$$
$$\ldots$$
$$b(2)_{n+1}W_{n-1} = b(2)_{n+i}W_n + b(2)_n W_{n-2} - \ldots + (-1)^{n-2}b(2)_4 W_2$$
$$+(-1)^{n-1}b(1)_3 W_1,$$
$$b(1)_n W_{n-1} = b(1)_{n+i-1}W_n + b(1)_{n-1}W_{n-2} - \ldots + (-1)^{n-2}b(1)_3 W_2$$
$$+(-1)^{n-1}b(1)_2 W_1.$$

Alors on obtient :

$$-c(n)_1 c(n)_{i-1} = -b(n-1)_{2n-1}c(n-1)W_{n-1} \ (*) + \left\{ b(n-2)_{2n+i-4}M_{n-2} - \ldots \right.$$
$$\left. +(-1)^n b(2)_{n+i}M_2 + (-1)^{n+1}b(1)_{n+i-1}M_1 \right\} \cdot c(n) + R'',$$

où :

$$R'' = \left\{ b(n-2)_{2n-4}W_{n-2} - \ldots + (-1)^{n-1}b(1)_{n-1}W_1 \right\} \cdot M_{n-2} - \ldots$$
$$+(-1)^n \cdot \left\{ b(2)_n W_{n-2} - \ldots + (-1)^{n-2}b(2)_4 W_2 + (-1)^{n-1}b(1)_3 W_1 \right\} \cdot M_2$$
$$+(-1)^{n-1} \cdot \left\{ b(1)_{n-1}W_{n-2} - \ldots + (-1)^{n-2}b(1)_3 W_2 + (-1)^{n-1}b(1)_2 W_1 \right\} \cdot M_1.$$

5.5. Lemme. $R'' = c(n-1)R.$

Démonstration. On introduit les vecteurs de dimension $n-2$:

$$\mathcal{W} = \left((-1)^{n+1}W_1, (-1)^{n+2}W_2, \ldots, W_{n-2} \right),$$
$$\mathcal{M} = \left((-1)^{n+1}M_1, (-1)^{n+2}M_2, \ldots, M_{n-2} \right)$$

et

$$b = \big(b(1)_{n+1}, b(1)_{n+2}, \ldots, b(1)_{2n-2}\big).$$

Alors la définition de R'' se récrit :

$$R'' = \mathcal{M} \cdot C(n-1) \cdot \mathcal{W}^t \tag{5.5.1}$$

(où $c(n-1) = \det C(n-1)$, la matrice $C(n-2)$ étant symétrique) ; de plus,

$$R = b \cdot \mathcal{W}^t.$$

Maintenant développons les quantités M_i par rapport à la dernière ligne :

$$M_i = b(1)_{2n-2}M_{i,n-2} - b(1)_{2n-3}M_{i,n-3} + \ldots + (-1)^{n+2}b(1)_{n+2}M_{i2}$$
$$+(-1)^{n+1}b(1)_{n+1}M_{i1}$$
$$= \sum_{j=1}^{n-2} (-1)^{n+j}b(1)_{n+j}M_{ij}, \; i = 1, \ldots, n-2.$$

On remarque que les quantités M_{ij} sont les mineurs de la matrice $(n-2) \times (n-2)$ $C(n-1)$. Il vient :

$$\mathcal{M} = b \cdot \hat{C}(n-1)$$

où

$$\hat{C}(n-1) = \big((-1)^{i+j}M_{ij}\big),$$

donc $\hat{C}(n-1) \cdot C(n-1) = c(n-1)$. En substituant dans (5.5.1) :

$$R'' = b \cdot \hat{C}(n-1) \cdot C(n-1) \cdot \mathcal{W}^t = c(n-1) \cdot b \cdot \mathcal{W}^t = c(n-1)R,$$

cqfd.

5.6. Il s'ensuit que pour vérifier l'identité (B) il reste à démontrer que

$$c(n)_i + c(n)_i'' + b(n-2)_{2n+i-4}M_{n-2} - \ldots$$
$$+(-1)^n b(2)_{n+i}M_2 + (-1)^{n+1}b(1)_{n+i-1}M_1 = b(n)_{2n+i-2}c(n-1). \tag{5.6.1}$$

Par contre, la quantité

$$b(n)_{2n+i-2}c(n-1) - b(n-2)_{2n+i-4}M_{n-2} + \ldots + (-1)^{n+1}b(2)_{n+i}M_2$$
$$+(-1)^n b(1)_{n+i-1}M_1$$

n'est autre que le développement de $c(n)_i'$ suivant la dernière colonne, donc (5.6.1) est équivalent à

$$c(n)_i + c(n)_i'' = c(n)_i' \tag{5.6.2}$$

qui a été déjà prouvée, cf. Corollaire 2.9.

Ceci achève la démonstration du théorème 4.2, et donc du 1.5.

5.7. *Démonstration du théorème 1.9.* En fait, nous l'avons déjà montré : la démonstration de la récurrence principale (3.3.2) n'utilise que les relations dans l'algèbre $\mathfrak{B}$.

Ces relations sont vérifiées si l'on définit les variables $b(i)_j$ à partir de coefficients de polynômes $f_1(x)$ et $f_2(x)$ comme dans 1.8, d'où l'assertion.

DEUXIÈME PARTIE

POLYNÔMES D'EULER ET DÉTERMINANT DE CAUCHY

§ 1 Nombres $\beta(j)_i$

1.1. Rappelons que pour un polynôme

$$f(x) = a_n x^n + a_{n-1} x^{n-1} + \ldots + a_0$$

les nombres $b(j)_i$ sont définis par

$$b(j)_i = n \sum_{p=0}^{j-1} (i - 2p) a_{n-p} a_{n-i+p} - j(n - i + j) a_{n-j} a_{n+j-i}.$$

On introduit les quantités :

$$q_i := \frac{a_{i-1}}{a_i},$$

$$r_i := \frac{q_{i-1}}{q_i} = \frac{a_i a_{i-2}}{a_{i-1}^2},$$

puis

$$\beta(j)_i := \frac{b(j)_i}{(n-i+j) a_{n-j} a_{n+j-i}} = \sum_{p=0}^{j-1} \frac{n(i - 2p)}{n - i + j} \cdot \frac{a_{n-p} a_{n+p-i}}{a_{n-j} a_{n+j-i}} - j.$$

1.2. Par exemple :

$$\beta(1)_2 = \frac{2n}{n - 1} \cdot \frac{a_n a_{n-2}}{a_{n-1}^2} - 1 = \frac{2n}{n - 1} \cdot r_n - 1,$$

$$\beta(1)_i = \frac{ni}{n - i + 1} \cdot \frac{a_n a_{n-i}}{a_{n-1} a_{n-i+1}} - 1.$$

On remarque que

$$\frac{a_n a_{n-i}}{a_{n-1} a_{n-i+1}} = \frac{q_{n-i+1}}{q_n} = r_{n-i+2} r_{n-i+1} \cdots r_n.$$

On définit les quantités

$$\psi(i,j) := \prod_{p=i}^{j} r_p$$

(donc $\psi(i,j) = 1$ si $i > j$). Il s'ensuit :

$$\beta(1)_i = \frac{ni}{n-i+1} \cdot \psi(n-i+2, n) - 1.$$

1.3. De même :

$$\frac{a_n a_{n-i}}{a_{n-2} a_{n-i+2}} = \frac{a_n a_{n-i}}{a_{n-1} a_{n-i+1}} \cdot \frac{a_{n-1} a_{n-i+1}}{a_{n-2} a_{n-i+2}} = \psi(n-i+2, n)\psi(n-i+3, n-1).$$

Par exemple :

$$\frac{a_n a_{n-4}}{a_{n-2}^2} = \psi(n-2, n)\psi(n-1, n-1) = r_{n-2} r_{n-1}^2 r_n.$$

Il en découle :

$$\beta(2)_4 = \frac{4n}{n-2} \frac{a_n a_{n-4}}{a_{n-2}^2} + \frac{2n}{n-2} \frac{a_{n-1} a_{n-3}}{a_{n-2}^2} - 2 = \frac{4n}{n-2} r_{n-2} r_{n-1}^2 r_n$$

$$+ \frac{2n}{n-2} r_{n-1} - 2,$$

$$\beta(2)_i = \frac{ni}{n-i+2} \frac{a_n a_{n-i}}{a_{n-2} a_{n-i+2}} + \frac{n(i-2)}{n-i+2} \frac{a_{n-1} a_{n-i+1}}{a_{n-2} a_{n-i+2}} - 2$$

$$= \frac{ni}{n-i+2} \psi(n-i+2, n)\psi(n-i+3, n-1)$$

$$+ \frac{n(i-2)}{n-i+2} \psi(n-i+3, n-1) - 2.$$

1.4. Un autre exemple :

$$\frac{a_n a_{n-6}}{a_{n-3}^2} = \psi(n-4, n)\psi(n-3, n-1)\psi(n-2, n-2) = r_{n-4} r_{n-3}^2 r_{n-2}^3 r_{n-1}^2 r_n.$$

1.5. En général on pose :

$$\phi(n, j, i) := \frac{a_n a_{n-i}}{a_{n-j} a_{n-i+j}} = \prod_{q=0}^{j-1} \psi(n-i+j+q, n-q)$$

et l'on aura :

$$\beta(j)_i = \sum_{p=0}^{j-1} \frac{n(i-2p)}{n-i+j} \cdot \phi(n-p, j-p, i-p) - j.$$

1.6. Passons maintenant aux déterminants $c(n)$. On commence par un exemple :

$$c(4) = \det \begin{pmatrix} b(1)_2 & b(1)_3 & b(1)_4 \\ b(1)_3 & b(2)_4 & b(2)_5 \\ b(1)_4 & b(2)_5 & b(3)_6 \end{pmatrix}$$

$$= \det \begin{pmatrix} (n-1)a_{n-1}^2\beta(1)_2 & (n-2)a_{n-1}a_{n-2}\beta(1)_3 & (n-3)a_{n-1}a_{n-3}\beta(1)_4 \\ (n-2)a_{n-1}a_{n-2}\beta(1)_3 & (n-2)a_{n-2}^2\beta(2)_4 & (n-3)a_{n-2}a_{n-3}\beta(2)_5 \\ (n-3)a_{n-1}a_{n-3}\beta(1)_4 & (n-3)a_{n-2}a_{n-3}\beta(2)_5 & (n-3)a_{n-3}^2\beta(3)_6 \end{pmatrix}$$

$$= (a_{n-1}a_{n-2}a_{n-3})^2 \cdot \det \begin{pmatrix} (n-1)\beta(1)_2 & (n-2)\beta(1)_3 & (n-3)\beta(1)_4 \\ (n-2)\beta(1)_3 & (n-2)\beta(2)_4 & (n-3)\beta(2)_5 \\ (n-3)\beta(1)_4 & (n-3)\beta(2)_5 & (n-3)\beta(3)_6 \end{pmatrix}.$$

1.7. En général

$$c(m+1) = \left(\prod_{i=1}^{m} a_{n-i}\right)^2 \times$$

$$\times \det \begin{pmatrix} (n-1)\beta(1)_2 & (n-2)\beta(1)_3 & \ldots & (n-m)\beta(1)_{m+1} \\ (n-2)\beta(1)_3 & (n-2)\beta(2)_4 & \ldots & (n-m)\beta(2)_{m+2} \\ \cdot & \cdot & \ldots & \cdot \\ (n-m)\beta(1)_{m+1} & (n-m)\beta(2)_{m+2} & \ldots & (n-m)\beta(m)_{2m} \end{pmatrix}.$$

§ 2 Polynômes d'Euler et fonction hypergéométrique

2.1. Suivant [Euler], on définit les polynômes

$$E_n(x) = \frac{1}{2}\left\{(1 + ix/2n)^{2n} + (1 - ix/2n)^{2n}\right\}. \tag{2.1.1}$$

Donc, $E_n(x)$ est un polynôme de degré $2n$, avec le terme constant 1, ne contenant que des puissances paires de x. Plus précisément,

$$E_n(x) = \sum_{k=0}^{n} (-1)^k \binom{2n}{2k} \frac{x^{2k}}{(2n)^{2k}}. \tag{2.1.2}$$

Par exemple :

$$E_1(x) = 1 - \frac{1}{4}x^2,$$

$$E_2(x) = 1 - \frac{3}{8}x^2 + \frac{1}{256}x^4,$$

$$E_3(x) = 1 - \frac{5}{12}x^2 + \frac{5}{432}x^4 - \frac{1}{46656}x^6,$$

$$E_4(x) = 1 - \frac{7}{16}x^2 + \frac{35}{2048}x^4 - \frac{7}{65536}x^6 + \frac{1}{16777216}x^8.$$

2.2. Rappelons que la fonction hypergéométrique de Gauß est définie par

$$F(\alpha, \beta, \gamma, x) = 1 + \frac{\alpha\beta}{1 \cdot \gamma}x + \frac{\alpha(\alpha+1)\beta(\beta+1)}{1 \cdot 2 \cdot \gamma(\gamma+1)}x^2$$
$$+ \frac{\alpha(\alpha+1)(\alpha+2)\beta(\beta+1)(\beta+2)}{1 \cdot 2 \cdot 3 \cdot \gamma(\gamma+1)(\gamma+2)}x^3 + \dots$$
$$= \sum_{i=0}^{\infty} c_i(\alpha, \beta, \gamma)x^i,$$

où

$$c_i(\alpha, \beta, \gamma) = \frac{\alpha(\alpha+1)\dots(\alpha+i-1) \cdot \beta(\beta+1)\dots(\beta+i-1)}{i! \cdot \gamma(\gamma+1)\dots(\gamma+i-1)},$$

cf. [Gauß]. Il s'ensuit :

$$c_i(-n/2, -n/2+1/2, 1/2)$$
$$= \frac{(-n/2)(-n/2+1)\dots(-n/2+i-1)\cdot(-n/2+1/2)(-n/2+3/2)\dots(-n/2+i-1/2)}{i! \cdot (1/2)(1/2+1)\dots(1/2+i-1)}$$
$$= \frac{(-1)^i 2^{-i} n(n-2)\dots(n-2i+2)\cdot(-1)^i 2^{-i}(n-1)(n-3)\dots(n-2i+1)}{i! \cdot 2^{-i} \cdot 1 \cdot 3 \cdot 5 \dots (2i-1)}$$
$$= \frac{2^{-i} \cdot n(n-1)(n-2)\dots(n-2i+1)}{2^{-i} \cdot 2 \cdot 4 \dots 2i \cdot 1 \cdot 3 \cdot 5 \dots (2i-1)} = \binom{n}{2i}.$$

Donc

$$F(-n/2, -n/2+1/2, 1/2, x^2) = \sum_{i=0}^{[n/2]} \binom{n}{2i} x^{2i} = \frac{1}{2}\{(1+x)^n+(1-x)^n\}. \quad (2.2.1)$$

Il en découle :

$$t^n F(-n/2, -n/2+1/2, 1/2, u^2/t^2) = \frac{1}{2}\{(t+u)^n + (t-u)^n\}, \quad (2.2.2)$$

cf. [Gauß], no. 5, formula II.

2.3. La formule (2.2.1) implique :

$$E_n(x) = F(-n, -n+1/2, 1/2, -x^2/4n^2). \quad (2.3.1)$$

2.4. Si l'on écrit

$$E_n(x) = \sum_{k=0}^{n} e_{nk}t^{2k}, \quad e_{nk} := (-1)^k \binom{2n}{2k} \frac{1}{(2n)^{2k}}$$

alors

$$e_{nk} = (-1)^k \frac{2n(2n-1)\dots(2n-2k+1)}{(2k)!(2n)^{2k}}$$
$$= \frac{(-1)^k}{(2k)!} \cdot 1 \cdot \left(1-\frac{1}{2n}\right)\left(1-\frac{2}{2n}\right)\dots\left(1-\frac{2k-1}{2n}\right),$$

d'où

$$\lim_{n\to\infty} e_{nk} = \frac{(-1)^k}{(2k)!},$$

i.e.

$$\lim_{n\to\infty} E_n(x) = \sum_{k=0}^{\infty} \frac{(-1)^k}{(2k)!} x^{2k} = \cos x,$$

comme il faut. En d'autres termes,

$$\lim_{n\to\infty} F(-n, -n+1/2, 1/2, -x^2/4n^2) = \cos x,$$

ou, comme aurait pu écrire Gauß,

$$F(-k, k+1/2, 1/2, -x^2/4k^2) = \cos x,$$

k étant "un nombre infiniment grand" (*denotante k numerum infinite magnum*). En fait, Gauß écrivit

$$F(k, k', 1/2, -x^2/4kk') = \cos x,$$

denotante k, k' numeros infinite magnos, cf. [Gauß], no. 5, formula XII.

§ 3 Asymptotiques

3.1. On pose :

$$f_n(x) = \sum_{k=0}^{n} (-1)^k \binom{2n}{2k} \frac{x^k}{(2n)^{2k}} = \sum_{k=0}^{n} a_k^{(n)} x^k. \qquad (3.1.1)$$

Donc

$$E_n(x) = f_n(x^2).$$

On désigne les quantités $b(j)_i, r_i$, etc. qui correspondent au polynôme f_n en ajoutant l'indice (n) en haut : $b(j)_i^{(n)}, r_i^{(n)}$, etc.

Donc on aura :

$$c(m+1)^{(n)} = \left(\prod_{i=1}^{m} a_{n-i}^{(n)} \right)^2 \times$$

$$\times \det \begin{pmatrix} (n-1)\beta(1)_2^{(n)} & (n-2)\beta(1)_3^{(n)} & \dots & (n-m)\beta(1)_{m+1}^{(n)} \\ (n-2)\beta(1)_3^{(n)} & (n-2)\beta(2)_4^{(n)} & \dots & (n-m)\beta(2)_{m+2}^{(n)} \\ \cdot & \cdot & \dots & \cdot \\ (n-m)\beta(1)_{m+1}^{(n)} & (n-m)\beta(2)_{m+2}^{(n)} & \dots & (n-m)\beta(m)_{2m}^{(n)} \end{pmatrix}.$$

3.2. On a :

$$a_i^{(n)} = (-1)^i \binom{2n}{2i},$$

d'où

$$r_i^{(n)} = \frac{a_i^{(n)} a_{i-2}^{(n)}}{a_{i-1}^{(n)2}} = \frac{[(2i-2)!]^2 [(2n-2i+2)!]^2}{(2i)!(2n-2i)!(2i-4)!(2n-2i+4)!}$$

$$= \frac{(2i-2)(2i-3)}{2i(2i-1)} \cdot \frac{(2n-2i+1)(2n-2i+2)}{(2n-2i+3)(2n-2i+4)}.$$

En remplaçant i par $n-i$,

$$r_{n-i}^{(n)} = \frac{(2i+1)(2i+2)}{(2i+3)(2i+4)} \cdot \frac{(2n-2i-2)(2n-2i-3)}{(2n-2i)(2n-2i-1)}.$$

On s'interesse aux valeurs limites :

$$r_{\infty-i}^{(\infty)} := \lim_{n\to\infty} r_{n-i}^{(n)} = \frac{(2i+1)(2i+2)}{(2i+3)(2i+4)}.$$

Il s'ensuit :

$$\psi(\infty-i+2,\infty) := \lim_{n\to\infty} \psi(n-i+2,n) = \frac{1\cdot 2}{(2i-1)2i},$$

$$\psi(\infty-i+3,\infty-1) = \frac{3\cdot 4}{(2i-2)(2i-3)},$$

$$\psi(\infty-i+4,\infty-2) = \frac{5\cdot 6}{(2i-4)(2i-5)},$$

etc.

3.3. Maintenant on veut calculer

$$\beta(j)_i^{(\infty)} := \lim_{n\to\infty} \beta(j)_i^{(n)}.$$

Il est commode de poser :

$$B(j)_i^\infty := \beta(j)_i^{(\infty)} + j.$$

On a :

$$B(1)_i^{(\infty)} = i \cdot \psi(\infty-i+2,\infty) = \frac{1}{2i-1}$$

d'où

$$\beta(1)_i^{(\infty)} = -\frac{2(i-1)}{2i-1}.$$

Ensuite,

$$B(2)_i^{(\infty)} = i \cdot \psi(\infty-i+2,\infty)\psi(\infty-i+3,\infty-1)$$
$$+(i-2)\cdot\psi(\infty-i+3,\infty-1)$$
$$= \psi(\infty-i+3,\infty-1)\cdot\left\{B(1)_i^{(\infty)}+i-2\right\}$$
$$= \frac{3\cdot 4}{(2i-2)(2i-3)}\cdot\left\{\frac{1}{2i-1}+i-2\right\} = \frac{3\cdot 2}{2i-1},$$

d'où

$$\beta(2)_i^{(\infty)} = -\frac{4(i-2)}{2i-1}.$$

De même,

$$B(3)_i^{(\infty)} = \psi(\infty - i + 4, \infty - 2) \cdot \left\{ B(2)_i^{(\infty)} + i - 4 \right\}$$

$$= \frac{5 \cdot 6}{(2i-4)(2i-5)} \cdot \left\{ \frac{3 \cdot 2}{2i-1} + i - 4 \right\} = \frac{5 \cdot 3}{2i-1},$$

d'où

$$\beta(3)_i^{(\infty)} = -\frac{6(i-3)}{2i-1}.$$

3.4. En général, la récurrence évidente fournit

$$B(j)_i^{(\infty)} = \frac{(2j-1) \cdot j}{2i-1}$$

et

$$\beta(j)_i^{(\infty)} = -\frac{2j(i-j)}{2i-1}.$$

3.5. On définit les nombres

$$\mathfrak{c}(m+1)^\infty := \det \begin{pmatrix} \beta(1)_2^{(\infty)} & \beta(1)_3^{(\infty)} & \cdots & \beta(1)_{m+1}^{(\infty)} \\ \beta(1)_3^{(\infty)} & \beta(2)_4^{(\infty)} & \cdots & \beta(2)_{m+2}^{(\infty)} \\ \cdot & \cdot & \cdots & \cdot \\ \beta(1)_{m+1}^{(\infty)} & \beta(2)_{m+2}^{(\infty)} & \cdots & \beta(m)_{2m}^{(\infty)} \end{pmatrix}.$$

Donc on aura :

$$\left(\prod_{i=1}^m a_{n-i}^{(n)} \right)^{-2} \cdot c(m+1)^{(n)} = \mathfrak{c}(m+1)^\infty \cdot n^m + O(n^{m-1}).$$

Les calculs précédents fournissent par exemple :

$$\mathfrak{c}(4)^\infty = \det \begin{pmatrix} -\frac{2}{3} & -\frac{4}{5} & -\frac{6}{7} \\ -\frac{4}{5} & -\frac{8}{7} & -\frac{12}{9} \\ -\frac{6}{7} & -\frac{12}{9} & -\frac{18}{11} \end{pmatrix} = (-1)^3 \cdot 2 \cdot 4 \cdot 6 \cdot \det \begin{pmatrix} \frac{1}{3} & \frac{2}{5} & \frac{3}{7} \\ \frac{1}{5} & \frac{2}{7} & \frac{3}{9} \\ \frac{1}{7} & \frac{2}{9} & \frac{3}{11} \end{pmatrix}$$

$$= (-1)^3 \cdot 2^3 \cdot (3!)^2 \cdot \det \begin{pmatrix} \frac{1}{3} & \frac{1}{5} & \frac{1}{7} \\ \frac{1}{5} & \frac{1}{7} & \frac{1}{9} \\ \frac{1}{7} & \frac{1}{9} & \frac{1}{11} \end{pmatrix}.$$

En général on obtient

$$\mathfrak{c}(m+1)^{\infty} = (-1)^m \cdot 2^m \cdot (m!)^2 \cdot \det \begin{pmatrix} \frac{1}{3} & \frac{1}{5} & \cdots & \frac{1}{2m+1} \\ \frac{1}{5} & \frac{1}{7} & \cdots & \frac{1}{2m+3} \\ \cdot & \cdot & \cdots & \cdot \\ \frac{1}{2m+1} & \frac{1}{2m+3} & \cdots & \frac{1}{4m-1} \end{pmatrix}.$$

On remarque que la dernière matrice (une variante de la matrice de Hilbert) est du type Hankel.

3.6. Le déterminant

$$\mathfrak{C}(m+1) := \det \begin{pmatrix} \frac{1}{3} & \frac{1}{5} & \cdots & \frac{1}{2m+1} \\ \frac{1}{5} & \frac{1}{7} & \cdots & \frac{1}{2m+3} \\ \cdot & \cdot & \cdots & \cdot \\ \frac{1}{2m+1} & \frac{1}{2m+3} & \cdots & \frac{1}{4m-1} \end{pmatrix}$$

est un cas particulier du déterminant calculé par Cauchy (d'où le caractère $\mathfrak{C}$), cf. son *Mémoire sur les fonctions alternées et sur les sommes alternées*, pp. 173–182 dans [Cauchy].

Rappelons que, étant données deux suites $x_1, \ldots, x_m$ et $y_1, \ldots, y_m$, le théorème de Cauchy dit que

$$\det\big((x_i + y_j)^{-1}\big)_{i,j=1}^m = \frac{\prod_{1 \le i < j \le m} (x_j - x_i)(y_j - y_i)}{\prod_{i,j=1}^m (x_i + y_j)}$$

d'où, en posant $x_i = 2i - 2$, $y_i = 2i + 1$,

$$\mathfrak{C}(m+1) = \frac{\prod_{1 \le i < j \le m} (2j - 2i)^2}{\prod_{i,j=1}^m (2i + 2j - 1)}.$$

Bibliography

[Cauchy] Exercices d'Analyse et de Physique Mathématique, par le Baron Augustin Cauchy, Tome Deuxième, Paris, Bachelier, 1841; *Oeuvres complètes*, II-e série, tome **XII**, Gauthier-Villars, MCMXVI.

[Euler] L.Euler, De summis serierum reciprocarum ex potestatibus numerorum naturalium ortarum dissertatio altera in qua eaedem summationes ex fonte maxime diverso derivantur, *Miscellanea Berolinensia* **7**, 1743, pp. 172–192.

[Gauß] C.F.Gauß, Circa seriem infinitam $1 + \frac{\alpha\beta}{1.\gamma}x + \frac{\alpha(\alpha+1)\beta(\beta+1)}{1.2.\gamma(\gamma+1)}xx + \frac{\alpha(\alpha+1)(\alpha+2)\beta(\beta+1)(\beta+2)}{1.2.3.\gamma(\gamma+1)(\gamma+2)}x^3 +$ etc. Pars Prior, *Commentationes societatis regiae scientarum Gottingensis recentiores*, Vol. **II**, Gottingae MDCCCXIII.

[Jacobi] C.G.J. Jacobi, De eliminatione variabilis e duabus aequationibus algebraitis, *Crelle J. für reine und angewandte Mathematik*, **15**, 1836, ss. 101–124.

[Sturm] C.-F. Sturm, Mémoire sur la résolution des équations numériques, *Mémoires présentés par divers savants à l'Académie Royale des Sciences*, Sciences mathématiques et physiques, tome **VI**, 1835, pp. 271–318.

Fields of u-Invariant $2^r + 1$

Alexander Vishik

School of Mathematical Sciences, University of Nottingham, University Park,
Nottingham, NG7 2RD, United Kingdom
alexander.vishik@nottingham.ac.uk

To my teacher Yuri Ivanovich Manin with gratitude

Summary. In this article we provide a uniform construction of fields with all known u-invariants. We also obtain the new values for the u-invariant: $2^r + 1$, for $r > 3$. The main tools here are the new discrete invariant of quadrics (so-called *elementary discrete invariant*), and the methods of [14] (which permit one to reduce the questions of *rationality* of elements of the Chow ring over the base field to that over bigger fields, the generic point of a quadric).

Key words: Quadratic forms, Grassmannians, Chow groups, Algebraic cobordisms, Landweber–Novikov operations, Steenrod operations

2000 Mathematics Subject Classifications: 11E04, 14C15, 14C35, 55S05, 55N22

1 Introduction

The u-invariant of a field is defined as the maximal dimension of anisotropic quadratic forms over it. The problem of describing values of this invariant is one of the major open problems in the theory of quadratic forms. Using elementary methods it is easy to establish that the u-invariant cannot take the values 3, 5, and 7. The conjecture of Kaplansky (1953) suggested that the only possible values are the powers of 2 (by that time, examples of fields with u-invariant being any power of 2 were known). This conjecture was disproved by A. Merkurjev in 1991, who constructed fields with all even u-invariants. The next challenge was to find out whether fields with odd u-invariant > 1 are possible at all. The breakthrough here was made by O. Izhboldin, who in 1999 constructed a field of u-invariant 9; see [4]. Still the question of other possible values remained open. This paper suggests a new uniform method of

Y. Tschinkel and Y. Zarhin (eds.), *Algebra, Arithmetic, and Geometry,*
Progress in Mathematics 270, DOI 10.1007/978-0-8176-4747-6_22,
© Springer Science+Business Media, LLC 2009

constructing fields with various u-invariants. In particular, we get fields with any even u-invariant without using the *index reduction formula* of Merkurjev. We also construct fields with u-invariant $2^r + 1$, for all $r \geqslant 3$. It should be mentioned that O. Izhboldin conjectured the existence of fields with such u-invariant, and suggested ideas on how to prove the conjecture. However, this paper employs very different new ideas. One can see the difference in the example of the u-invariant 9. I would say that our method uses substantially coarser invariants (such as the *generic discrete invariant of quadrics*), while the original construction used very subtle ones (such as the *cokernel on the unramified cohomology*). Thus, this paper amply demonstrates that u-invariant questions can be solved just with the help of "coarse" invariants. The method is based on the new so-called *elementary discrete invariant* of quadrics (introduced in this paper). This invariant contains important pieces of information about the particular quadric, and at the same time, is quite easy to work with. The field with the given u-invariant is constructed using the standard field tower of A. Merkurjev. And the central problem is to control the behavior of the elementary discrete invariant while passing from the base field to a generic point of a (sufficiently large) quadric. This is done using the general statement from [14] concerning the question of rationality of small-codimensional classes in the Chow ring of an arbitrary smooth variety under similar passage. The driving force behind all of this comes from the *symmetric operations* in algebraic cobordism [12], [15].

Acknowledgments. I am very grateful to V. Chernousov, I. Fesenko, J. Minac, and U. Rehmann for very useful discussions and helpful suggestions, and to M. Zhykhovich for finding a mistake in the original version. This text was partially written while I was visiting Indiana University, and I would like to express my gratitude to this institution for their support and excellent working conditions. The support of CRDF award RUM1-2661-MO-05, INTAS 05-1000008-8118, and RFBR grant 06-01-72550 is gratefully acknowledged. Finally, I want to thank the referees for their very useful suggestions and remarks.

2 Elementary Discrete Invariant

In this section we assume that the base field k has characteristic different from 2. We will fix an algebraic closure $\overline{k}$ of k.

For the nondegenerate quadratic form q we will denote by the capital letter Q the respective smooth projective quadric. The same applies to forms $p, p', q', \dots$. The dimension of a quadric Q will be denoted by N_Q, and if there is no ambiguity, simply by N. We also set $d_Q := [N_Q/2]$ (respectively, $d := [N/2]$). For the smooth variety X we will denote by $\mathrm{CH}^*(X)$ the Chow ring of algebraic cycles modulo rational equivalence on X, and by $\mathrm{Ch}^*(X)$ the Chow ring modulo 2 (see [3] for details).

To each smooth projective quadric Q/k of dimension N one can assign the so-called *generic discrete invariant* $\mathrm{GDI}(Q)$ see [13], which is defined as the collection of subrings

$$\mathrm{GDI}(Q, i) := \mathrm{image}(\mathrm{Ch}^*(F(Q, i)) \to \mathrm{Ch}^*(F(Q, i)|_{\overline{k}})),$$

for all $0 \leqslant i \leqslant d$, where $F(Q, i)$ is the Grassmannian of i-dimensional projective subspaces on Q, and the map is induced by the restriction of scalars $k \to \overline{k}$. Note that $F(Q, 0)$ is the quadric Q itself, and $F(Q, d)$ is the last Grassmannian.

For $J \subset I \subset \{0, \ldots, d\}$ let us denote the natural projection between partial flag varieties $F(Q, I) \to F(Q, J)$ by π with subindex I with J underlined inside it. In particular, we have projections

$$F(Q, i) \overset{\pi_{(0,i)}}{\leftarrow} F(Q, 0, i) \overset{\pi_{(\underline{0},i)}}{\to} Q.$$

The Chow ring of a split quadric is a free $\mathbb{Z}$-module with basis h^s, l_s, $0 \leqslant s \leqslant d$, where $l_s \in \mathrm{CH}_s(Q|_{\overline{k}})$ is the class of a projective subspace of dimension s, and $h^s \in \mathrm{CH}^s(Q|_{\overline{k}})$ is the class of a plane section of codimension s; see [11, Lemma 8].

In $\mathrm{CH}^*(F(Q, i)|_{\overline{k}})$ we have special classes: $Z_j^{\boxed{i-d}} \in \mathrm{CH}^j$, $N - d - i \leqslant j \leqslant N - i$, and $W_j^{\boxed{i-d}} \in \mathrm{CH}^j$, $0 \leqslant j \leqslant d - i$, defined by

$$Z_j^{\boxed{i-d}} := (\pi_{(0,\underline{i})})_*(\pi_{(\underline{0},i)})^*(l_{N-i-j}); \qquad W_j^{\boxed{i-d}} := (\pi_{(0,\underline{i})})_*(\pi_{(\underline{0},i)})^*(h^{i+j}).$$

Let us denote by $z_j^{\boxed{i-d}}$ and $w_j^{\boxed{i-d}}$ the same classes in Ch^*. We will call classes $z_j^{\boxed{i-d}}$ *elementary*. Notice, that the classes $w_j^{\boxed{i-d}}$ always belong to $\mathrm{GDI}(Q, i)$.

Let $\mathcal{T}_i$ be the tautological $(i + 1)$-dimensional vector bundle on $F(Q, i)$. The following proposition explains the meaning of our classes.

Proposition 2.1. *For any $0 \leqslant i \leqslant d$ and $N - d - i \leqslant j \leqslant N - i$,*

$$c_\bullet(-\mathcal{T}_i) = \sum_{j=0}^{d-i} W_j^{\boxed{i-d}} + 2 \sum_{d-i < j \leqslant N-i} Z_j^{\boxed{i-d}}.$$

Proof. Since $\mathcal{T}_0 = \mathcal{O}(-1)$ on Q, the statement is true for $i = 0$. Consider the projections

$$F(Q, i) \overset{\pi_{(0,\underline{i})}}{\leftarrow} F(Q, 0, i) \overset{\pi_{(\underline{0},i)}}{\to} Q.$$

Notice that $F(Q, 0, i)$ is naturally identified with the projective bundle $\mathbb{P}_{F(Q,i)}(\mathcal{T}_i)$, and the sheaf $\pi_{(\underline{0},i)}^*(\mathcal{T}_0)$ is naturally identified with $\mathcal{O}(-1)$. Thus,

$$(\pi_{(0,\underline{i})})_*(\pi_{(\underline{0},i)})^*(c_\bullet(-\mathcal{T}_0)) = (\pi_{(0,\underline{i})})_*(c_\bullet(-\mathcal{O}(-1))) = c_\bullet(-\mathcal{T}_i).$$

$\square$

Remark 2.2. In particular, for $i = d$ we get another proof of [13, Theorem 2.5(3)].

Definition 2.3. *Define the elementary discrete invariant* $\mathrm{EDI}(Q)$ *as the collection of subsets* $\mathrm{EDI}(Q, i)$ *consisting of those* j *such that* $z_j^{\boxed{i-d}} \in \mathrm{GDI}(Q, i)$.

One can visualize $\mathrm{EDI}(Q)$ as the coordinate $d \times d$ square, where some integral nodes are marked, each row corresponds to particular Grassmannian, and the codimension of a "node" is decreasing up and to the right. The lower row corresponds to the quadric itself, and the upper one to the last Grassmannian. The southwest corner is marked if and only if Q is isotropic.

Example 2.4. $\mathrm{EDI}(Q)$ for the 10-dimensional excellent form looks like this:

$$
\begin{array}{ccccc}
\bullet & \bullet & \bullet & \bullet & \circ \\
\bullet & \bullet & \circ & \bullet & \circ \\
\bullet & \bullet & \circ & \circ & \circ \\
\circ & \bullet & \circ & \circ & \circ \\
\circ & \circ & \circ & \circ & \circ
\end{array}
$$

The following statement puts serious constraints on possible markings.

Proposition 2.5. *Let* $0 \leqslant i < d$ *and* $j \in \mathrm{EDI}(Q, i)$. *Then* $j, j - 1 \in \mathrm{EDI}(Q, i + 1)$.

This can be visualized as follows:

Proof. The proposition easily follows from the next lemma. Let us temporarily denote $\pi_{(\underline{i}, i+1)}$ by α, and $\pi_{(i, \underline{i+1})}$ by β.

Lemma 2.6.

$$\alpha^*(Z_j^{\boxed{i-d}}) = \beta^*(Z_j^{\boxed{i+1-d}}) + c_1(\mathcal{O}(1)) \cdot \beta^*(Z_{j-1}^{\boxed{i+1-d}});$$

$$\alpha^*(W_j^{\boxed{i-d}}) = \beta^*(W_j^{\boxed{i+1-d}}) + c_1(\mathcal{O}(1)) \cdot \beta^*(W_{j-1}^{\boxed{i+1-d}}), 0 \leqslant j < d - i;$$

$$\alpha^*(W_{d-i}^{\boxed{i-d}}) = 2\beta^*(Z_{d-i}^{\boxed{i+1-d}}) + c_1(\mathcal{O}(1)) \cdot \beta^*(W_{d-i-1}^{\boxed{i+1-d}}),$$

where $\mathcal{O}(1)$ *is the standard sheaf on the projective bundle*

$$F(Q, i, i+1) = \mathbb{P}_{F(Q, i+1)}(\mathcal{T}_{i+1}^\vee),$$

for the vector bundle dual to the tautological one.

Proof. By definition, $Z_j^{\boxed{i-d}}, W_j^{\boxed{i-d}}$ have the form $(\pi_{(0,i)})_*(\pi_{(\underline{0},i)})^*(x)$, for certain $x \in \mathrm{CH}^*(Q|_{\overline{k}})$. Since the square

$$
\begin{array}{ccc}
F(Q, i, i+1) & \xleftarrow{\;\pi_{(0,i,i+1)}\;} & F(Q, 0, i, i+1) \\[2mm]
\Big\downarrow{\scriptstyle \pi_{(\underline{i},i+1)}} & & \Big\downarrow{\scriptstyle \pi_{(\underline{0},i,i+1)}} \\[2mm]
F(Q, i) & \xleftarrow[\;\pi_{(0,\underline{i})}\;]{} & F(Q, 0, i)
\end{array}
$$

is transversal Cartesian, $(\pi_{(\underline{i},i+1)})^*$ of such an element is equal to

$$(\pi_{(0,\underline{i},i+1)})_*(\pi_{(\underline{0},i,i+1)})^*(\pi_{(\underline{0},i)})^*(x) = (\pi_{(0,\underline{i},i+1)})_*(\pi_{(\underline{0},i,i+1)})^*(\pi_{(\underline{0},i+1)})^*(x).$$

Variety $F(Q, 0, i, i+1)$ is naturally a divisor D on the transversal product $F(Q, 0, i+1) \times_{F(Q,i+1)} F(Q, i, i+1)$ with $\mathcal{O}(D) = \pi^*_{(0,i,i+1)}(\mathcal{O}(h)) \otimes \pi^*_{(0,\underline{i},i+1)}(\mathcal{O}(1))$, where $\mathcal{O}(h)$ is the sheaf given by the hyperplane section on Q. Then

$$
\begin{aligned}
(\pi_{(0,\underline{i},i+1)})_*(\pi_{(\underline{0},i,i+1)})^*(\pi_{(\underline{0},i+1)})^*(x) = \\
c_1(\mathcal{O}(1)) \cdot (\pi_{(i,i+1)})^*(\pi_{(0,i+1)})_*(\pi_{(\underline{0},i+1)})^*(x) + \\
(\pi_{(i,i+1)})^*(\pi_{(0,i+1)})_*(\pi_{(\underline{0},i+1)})^*(h \cdot x).
\end{aligned}
$$

It remains to plug in the appropriate x. $\qquad \square$

Notice, that the projective bundle theorem

$$\mathrm{Ch}^*(\mathbb{P}_{F(Q,i+1)}(\mathcal{T}^{\vee}_{i+1})) = \oplus_{l=0}^{i} c_1(\mathcal{O}(1))^l \cdot \mathrm{Ch}^*(F(Q, i+1))$$

implies that the element of this group is defined over k if and only if all of it's coordinates are. Since the cycle $z_j^{\boxed{i-d}}$ is defined over k, by Lemma 2.6 the cycles $z_j^{\boxed{i+1-d}}$ and $z_{j-1}^{\boxed{i+1-d}}$ are defined too. $\qquad \square$

The following Proposition describes the EDI of the isotropic quadric.

Proposition 2.7. *Let $p' = p \perp \mathbb{H}$ be isotropic quadratic form (here $\mathbb{H}$ is a 2-dimensional hyperbolic form $\langle 1, -1 \rangle$). Then EDI of P and P' are related as follows: for any $N_P - d_P - i \leqslant j \leqslant N_P - i$*

$$z_j^{\boxed{i-d_P}}(P) \text{ is defined} \;\Leftrightarrow\; z_j^{\boxed{i+1-d_{P'}}}(P') \text{ is defined}.$$

In other words, $EDI(P)$ fits well into $EDI(P')$, if we glue their N-E corners.

Proof. The quadric P can be identified with the quadric of lines on P' passing through the given rational point x. Then we have natural regular embedding

$$e : F(P, i) \to F(P', i+1), \text{ with } e^*(z_j^{\boxed{i+1-d_{P'}}}) = z_j^{\boxed{i-d_P}} \text{ and } (\Leftarrow) \text{ follows}.$$

We have natural maps $F(P,i) \xleftarrow{f} (F(P',i+1)\backslash F(P,i)) \xrightarrow{g} F(P',i+1)$, where the map f sends $(i+1)$-dimensional plane π_{i+1} to $T_{x,P'} \cap (x + \pi_{i+1}))$ (expression in the projective space), and g is an open embedding. It is an exercise for the reader, to show that $f^*(z_j^{\boxed{i-d_{P}}}(P)) = g^*(z_j^{\boxed{i+1-d_{P'}}}(P'))$. It remains to observe that $F(P,i)$ has codimension $(N_P - i + 1) > j$ inside $F(P',i+1)$, and thus g^* is an isomorphism on Ch^j. This proves $(\Rightarrow)$. $\qquad\square$

If we have a codimension-1 subquadric P of a quadric Q, the EDI's of them are related by the following:

Proposition 2.8.

$$z_j^{\boxed{i-d_{Q}}}(Q) \text{ is defined} \Rightarrow z_j^{\boxed{i-d_{P}}}(P) \text{ is defined};$$

$$z_j^{\boxed{i-d_{P}}}(P) \text{ is defined} \Rightarrow z_j^{\boxed{i+1-d_{Q}}}(Q) \text{ is defined}.$$

Proof. Consider the natural embedding: $e : F(P,i) \to F(Q,i)$. Then, it follows from the definition that $e^*(z_j^{\boxed{i-d_{Q}}}) = z_j^{\boxed{i-d_{P}}}$. To prove the second statement just observe that Q is a codimension-1 subquadric in P', where $p' = p \perp \mathbb{H}$ (if $q = p \perp \langle a \rangle$, then $p' = q \perp \langle -a \rangle$), and apply Proposition 2.7. $\qquad\square$

Unfortunately, the Steenrod operations (see [1, 17]), in general, do not act on $\mathrm{EDI}(Q,i)$, since they do not preserve *elementary classes*. But they act in the lower and the upper rows: for the quadric itself, and for the last Grassmannian. Also, it follows from [13, Main Theorem 5.8] that $\mathrm{EDI}(Q,d)$ carries the same information as $\mathrm{GDI}(Q,d)$. The same is true about $\mathrm{EDI}(Q,0)$ and $\mathrm{GDI}(Q,0)$ for the obvious reasons.

The action of the Steenrod operations on the *elementary classes* can be described as follows.

Proposition 2.9. *Let* $0 \leqslant i \leqslant d$, *and* $N - d - i \leqslant j \leqslant N - i$. *Then*

$$S^m(z_j^{\boxed{i-d}}) = \sum_{k=0}^{d-i} \binom{j-k}{m-k} z_{j+m-k}^{\boxed{i-d}} \cdot w_k^{\boxed{i-d}},$$

where elementary classes of codimension more than $(N-i)$ *are assumed to be 0.*

Proof. We recall from [1] that on the Chow groups modulo 2 of a smooth variety X one has the action of Steenrod operations $S^\bullet$ and $S_\bullet$, where the former commute with the pullbacks for all morphisms, and the latter commute with the pushforwards for proper morphisms. The relation between the upper and the lower operations is given by

$$S^\bullet = S_\bullet \cdot c_\bullet(T_X).$$

From this (and the description of the tangent bundle for the quadric and the projective space) one gets that $S^\bullet(l_s) = (1+h)^{N-s+1} l_s$. Since $(\pi_{\underline{0},i})^*(\mathcal{O}(1))$ is the sheaf $\mathcal{O}(1)$ on $F(Q,0,i) = \mathbb{P}_{F(Q,i)}(\mathcal{T}_i)$,

$$S_\bullet(\pi_{(\underline{0},i)})^*(l_{N-i-j}) = c_\bullet(-T_{F(Q,0,i)}) \cdot (1+H)^{i+j+1} \cdot (\pi_{(\underline{0},i)})^*(l_{N-i-j}),$$

where $H = c_1(\mathcal{O}(1))$. Since $S_\bullet$ commutes with the pushforward morphisms,

$$S^\bullet(z_j^{\boxed{i-d}}) = S^\bullet(\pi_{(0,\underline{i})})_*(\pi_{(\underline{0},i)})^*(l_{N-i-j})$$
$$= (\pi_{(0,\underline{i})})_*(c_\bullet(-T_{\text{fiber}}) \cdot (1+H)^{i+j+1} \cdot (\pi_{(\underline{0},i)})^*(l_{N-i-j})).$$

Recall that if $\mathcal{V}$ is a virtual vector bundle of virtual dimension M, and $c_\bullet(\mathcal{V})(t) := \sum_{k \geqslant 0} c_k(\mathcal{V}) \cdot t^{M-k}$, then for the divisor H, $c_\bullet(\mathcal{V} \otimes \mathcal{O}(H))(t) = c_\bullet(\mathcal{V})(t+H)$, and $c_\bullet(\mathcal{V} \otimes \mathcal{O}(H)) = c_\bullet(\mathcal{V} \otimes \mathcal{O}(H))(1) = c_\bullet(\mathcal{V})(1+H)$.

Since $c_\bullet(-T_{\text{fiber}}) = c_\bullet(-\mathcal{T}_i \otimes \mathcal{O}(1))$, by Proposition 2.1, $(\bmod\, 2)$ this is equal to $\sum_{k=0}^{d-i} w_k^{\boxed{i-d}}(1+H)^{-i-1-k}$. Thus,

$$S^\bullet(z_j^{\boxed{i-d}}) = \left(\sum_{k=0}^{d-i} w_k^{\boxed{i-d}} \right)(\pi_{(0,\underline{i})})_*(\pi_{(\underline{0},i)})^*((1+h)^{j-k} l_{N-i-j})$$
$$= \sum_{r \geqslant 0} \sum_{k=0}^{d-i} \binom{j-k}{r-k} z_{j+r-k}^{\boxed{i-d}} w_k^{\boxed{i-d}}.$$

$\square$

Remark 2.10. In particular, for $i = d$ we get a new proof of [13, Theorem 4.1].

The following fact is well known (see, for example, [2]). We will give an independent proof below.

Proposition 2.11. *The ring* $\mathrm{CH}^*(F(Q,i)|_{\overline{k}})$ *is generated by the classes*

$$Z_j^{\boxed{i-d}}, \quad N - d - i \leqslant j \leqslant N - i, \quad \text{and} \quad W_j^{\boxed{i-d}}, \quad 0 \leqslant j \leqslant d - i.$$

Proof. For $0 \leqslant l \leqslant i$, let us denote the pullback of $Z_j^{\boxed{l-d}}$ to $F(Q,0,\ldots,i)$ by the same symbol. On this flag variety we have natural line bundles $\mathcal{L}_k := \mathcal{T}_k/\mathcal{T}_{k-1}$. Let us define $h_k := c_1(\mathcal{L}_k^{-1})$.

Lemma 2.12. *Let* E/k *be some field extension. Suppose that* $Q|_E$ *is split. Then the ring* $\mathrm{CH}^*(F(Q,0,\ldots,i)|_E)$ *is generated by* $W_j^{\boxed{l-d}}, 0 \leqslant l \leqslant i$, $1 \leqslant j \leqslant d - l$, *and* $Z_j^{\boxed{l-d}}, 0 \leqslant l \leqslant i, N - d - l \leqslant j \leqslant N - l$.

Proof. Induction on i. For $i = 0$ the statement is evident.

Statement 2.13. *Let $\pi : Y \to X$ be a smooth morphism to a smooth variety X. For $x \in X^{(r)}$, let Y_x be the fiber over the point x. Let ζ denote the generic point of X, and $s_x : \mathrm{CH}^*(Y_\zeta) \to \mathrm{CH}^*(Y_x)$ the specialization map. Let $B \subset \mathrm{CH}^*(Y)$ be a subgroup. Suppose:*

(a) the map $B \to \mathrm{CH}^*(Y_\zeta)$ is surjective;
(b) all the maps s_x are surjective.

Then $\mathrm{CH}^*(Y) = B \cdot \pi^*(\mathrm{CH}^*(X))$.

Proof. On $\mathrm{CH}^*(Y)$ we have decreasing filtration $F^\bullet$, where F^r consists of classes, having a representative with the image under π of codimension $\geqslant r$. This gives the surjection:

$$\oplus_r \oplus_{x \in X^{(r)}} \mathrm{CH}^*(Y_x) \to gr_{F^\bullet} \mathrm{CH}^{r+*}(Y).$$

Let $[x] \in \mathrm{CH}^r(X)$ be the class represented by the closure of x. Clearly, the image of $\pi^*([x]) \cdot B$ covers the image of $\mathrm{CH}^*(Y_x)$ in F^r/F^{r+1}. ☐

Consider the projection

$$\pi_{(\underline{0}, \ldots, i-1, i)} : F(Q, 0, \ldots, i-1, i) \to F(Q, 0, \ldots, i-1).$$

Let $Q_{\{i\},x}/E(x)$ be the fiber of this projection over the point x. It is a split quadric of dimension $N - 2i$. Thus, the condition (b) of the Statement 2.13 is satisfied. Since $[\mathcal{T}_i|_{Q_{\{i\},\zeta}}] = [\mathcal{L}_i] + i \cdot [\mathcal{O}] = [\mathcal{O}(-h_i)] + i \cdot [\mathcal{O}]$ in $K_0(Q_{\{i\},\zeta})$, it follows from Proposition 2.1 that

$$Z_j^{\boxed{i-d}}|_{Q_{\{i\},\zeta}} = l_{N-2i-j}, \qquad W_j^{\boxed{i-d}}|_{Q_{\{i\},\zeta}} = h_i^j.$$

We can take B additively generated by $Z_j^{\boxed{i-d}}, N - d - i \leqslant j \leqslant N - 2i$, and $W_j^{\boxed{i-d}}, 0 \leqslant j \leqslant d - i$. Then the condition (a) will be satisfied too. The induction step follows. ☐

Lemma 2.6 implies that the $Z_j^{\boxed{l-d}}, W_j^{\boxed{l-d}}$, for $l < i$ are expressible in terms of $Z_k^{\boxed{i-d}}, W_k^{\boxed{i-d}}$ and $h_m, 0 \leqslant m \leqslant i$. Let $A \subset \mathrm{CH}^*(F(Q, i))$ be the subring generated by $Z_j^{\boxed{i-d}}, W_j^{\boxed{i-d}}$. Since $F(Q, 0, \ldots, i)$ is a variety of complete flags of subspaces of the vector bundle $\mathcal{T}_i$ on $F(Q, i)$, the ring $\mathrm{CH}^*(F(Q, 0, \ldots, i))$ is isomorphic to

$$\mathrm{CH}^*(F(Q, i))[h_0, \ldots, h_i]/(\sigma_r(h) - c_r(\mathcal{T}_i^\vee), 1 \leqslant r \leqslant i + 1),$$

where $\sigma_r(h)$ are elementary symmetric functions on h_k. But $c_r(T_i^\vee) \in A$, by Proposition 2.1. Since A and $h_m, 0 \leqslant m \leqslant i$ generate $\mathrm{CH}^*(F(Q, 0, \ldots, i))$, A must coincide with $\mathrm{CH}^*(F(Q, i))$. $\qquad \square$

In particular, since the cycles $W_j^{\boxed{i-d}}$ are defined over k, we have:

Corollary 2.14. *The graded part of* $\mathrm{CH}^*(F(Q, i)|_{\overline{k}})$ *of degree less than* $(N - d - i)$ *consists of classes which are defined over* k.

Notice that for $i = d$, $\mathrm{Ch}^*(F(Q, d)|_{\overline{k}})$ is generated as a ring by $z_j^{\boxed{0}}$, and moreover, $\mathrm{GDI}(Q, d)$ is always generated as a ring by the subset of $z_j^{\boxed{0}}$ contained in it; see [13, Main Theorem 5.8].

We will need one more simple fact.

Statement 2.15. *The class of a rational point on* $F(Q, i)|_{\overline{k}}$ *is given by the product*

$$\prod_{j=i}^{2i} z_{N-j}^{\boxed{i-d}}.$$

Proof. Use induction on N. Let x be a fixed rational point on $Q|_{\overline{k}}$. Then we have a natural regular embedding $e : F(P, i - 1) \to F(Q, i)$, where P is the $(N - 2)$-dimensional quadric of lines on Q passing through x, with $e_*(1) = z_{N_Q - i}^{\boxed{i - d_Q}}(Q)$, and $e^*(z_{N_Q - j}^{\boxed{i - d_Q}}(Q)) = z_{N_P - j + 1}^{\boxed{i - 1 - d_P}}(P)$. Thus the induction step follows from the projection formula. The base of induction is trivial. $\quad \square$

3 Generic Points of Quadrics and Chow Groups

Everywhere below we will assume that the base field k has characteristic 0. Although many things work for odd characteristics as well, the use of the algebraic cobordism theory of M. Levine and F. Morel will require such an assumption.

In this section I would like to recall the principal result of [14]. Let Q be a smooth projective quadric, Y a smooth quasiprojective variety, and $\overline{y} \in \mathrm{Ch}^m(Y|_{\overline{k}})$. This will be our main tool in the construction of fields with various u-invariants.

Theorem 3.1. ([14, Corollary 3.5], [15, Theorem 4.3].)
Suppose $m < N_Q - d_Q$. *Then*

$$\overline{y}|_{\overline{k(Q)}} \text{ is defined over } k(Q) \quad \Leftrightarrow \quad \overline{y} \text{ is defined over } k.$$

Example 3.2. Let $\alpha = \{a_1, \ldots, a_n\} \in K_n^M(k)/2$ be a nonzero pure symbol, and let Q_α be the respective anisotropic Pfister quadric. Then in $\mathrm{EDI}(Q_\alpha)$ the marked nodes will be exactly those that live above the main (northwest to southeast) diagonal:

Indeed, over its own generic point $k(Q_\alpha)$, the quadric Q_α becomes hyperbolic, and so all the elementary cycles are defined there. Then the cycles above the main diagonal were defined already over the base field, since their codimension is smaller than $N_{Q_\alpha} - d_{Q_\alpha}$. On the other hand, the northwest corner could not be defined over k, since otherwise all the elementary cycles on the last Grassmannian of Q_α would be defined over k, but by Statement 2.15, the product of all these cycles is the class of a rational point on this Grassmannian. Since Q_α is not hyperbolic over k, this is impossible. The rest of the picture follows from Proposition 2.5.

The proof of Theorem 3.1 uses algebraic cobordisms of M. Levine and F. Morel. Let me say few words about the latter.

3.1 Algebraic Cobordisms

In [7] M. Levine and F. Morel have constructed a universal oriented generalized cohomology theory Ω^* on the category of smooth quasiprojective varieties over the field k of characteristic 0, called *algebraic cobordism*.

For any smooth quasiprojective variety X over k, the additive group $\Omega^*(X)$ is generated by the classes $[v : V \to X]$ of projective maps from smooth varieties V subject to certain relations, and the upper grading is the codimensional one. There is a natural morphism of theories $pr : \Omega^* \to \mathrm{CH}^*$. The main properties of Ω^* are:

(1) $\Omega^*(\mathrm{Spec}(k)) = \mathbb{L} = MU(pt)$, the Lazard ring, and the isomorphism is given by the topological realization functor;
(2) $\mathrm{CH}^*(X) = \Omega^*(X)/\mathbb{L}^{<0} \cdot \Omega^*(X)$.

On Ω^* there is the action of the Landweber–Novikov operations (see [7, Example 4.1.25]). Let $R(\sigma_1, \sigma_2, \ldots) \in \mathbb{L}[\sigma_1, \sigma_2, \ldots]$ be some polynomial, where we assume $\deg(\sigma_i) = i$. Then $S_{L-N}^R : \Omega^* \to \Omega^{*+\deg(R)}$ is given by

$$S_{L-N}^R([v : V \to X]) := v_*(R(c_1, c_2, \ldots) \cdot 1_V),$$

where $c_j = c_j(\mathcal{N}_v)$, and $\mathcal{N}_v := -T_V + v^* T_X$ is the *virtual normal bundle*.

If $R = \sigma_i$, we will denote the respective operation simply by S^i_{L-N}. The following statement follows from the definition of Steenrod and Landweber–Novikov operations; see P. Brosnan [1], A. Merkurjev [10], and M. Levine [6].

Proposition 3.3. *There is commutative square*

$$
\begin{array}{ccc}
\Omega^*(X) & \xrightarrow{\;S^i_{L-N}\;} & \Omega^{*+i}(X) \\
\downarrow & & \downarrow \\
\mathrm{Ch}^*(X) & \xrightarrow{\;S^i\;} & \mathrm{Ch}^{*+i}(X),
\end{array}
$$

where S^i is the Steenrod operation (mod 2) [1, 17].

In particular, using the results of P. Brosnan on S^i (see [1]), we get the following:

Corollary 3.4. (1) $pr \circ S^i_{L-N}(\Omega^m) \subset 2 \cdot \mathrm{CH}^{i+m}$, *if $i > m$;*
(2) $pr \circ (S^m_{L-N} - \Box)(\Omega^m) \in 2 \cdot \mathrm{CH}^{2m}$, *where $\Box$ is the square operation.*

This implies that (modulo 2-torsion) we have well-defined maps $\frac{pr \circ S^i_{L-N}}{2}$ and $\frac{pr \circ (S^m_{L-N} - \Box)}{2}$. In reality, these maps can be lifted to well-defined so-called *symmetric operations* $\Phi^{t^{i-m}} : \Omega^m \to \Omega^{m+i}$, see [15]. Since over an algebraically closed field all our varieties are cellular, and thus the Chow groups of them are torsion-free, we will not need such subtleties, but we will keep the notation from [15], and denote our maps by $\phi^{t^{i-m}}$.

3.2 Beyond Theorem 3.1

Below we will need to study the relation between the rationality of $\overline{y}$ and $\overline{y}|_{\overline{k(Q)}}$ for $\mathrm{codim}(\overline{y})$ slightly bigger than $N_Q - d_Q$. The methods involved are just the same as are employed for the proof of Theorem 3.1.

Let Y be a smooth quasiprojective variety, Q a smooth projective quadric. Let $v \in \mathrm{Ch}^*(Y \times Q)$ be some element, and $w \in \Omega^*(Y \times Q)$ its arbitrary lifting via pr. Over $\overline{k}$, the quadric Q becomes a cellular variety with a basis of Chow groups and cobordisms given by the set $\{l_i, h^i\}_{0 \leqslant i \leqslant d_Q}$ of projective subspaces and plane sections. This implies that

$$
\mathrm{CH}^*(Y \times Q|_{\overline{k}}) = \oplus_{i=0}^{d_Q}(\mathrm{CH}^*(Y|_{\overline{k}}) \cdot l_i \oplus \mathrm{CH}^*(Y|_{\overline{k}}) \cdot h^i)
$$

and

$$
\Omega^*(Y \times Q|_{\overline{k}}) = \oplus_{i=0}^{d_Q}(\Omega^*(Y|_{\overline{k}}) \cdot l_i \oplus \Omega^*(Y|_{\overline{k}}) \cdot h^i);
$$

see [16, Section 2]. In particular,

$$
\overline{v} = \sum_{i=0}^{d_Q}(\overline{v}^i \cdot h^i + \overline{v}_i \cdot l_i) \quad \text{and} \quad \overline{w} = \sum_{i=0}^{d_Q}(\overline{w}^i \cdot h^i + \overline{w}_i \cdot l_i).
$$

Denote by $\widetilde{\mathrm{Ch}}^*$ the ring $\mathrm{CH}^*/(2, 2\text{-torsion})$.

Proposition 3.5. *Let Q be a smooth projective quadric of dimension $\geqslant 4n-1$, Y a smooth quasiprojective variety, and $v \in \mathrm{Ch}^{2n+1}(Y \times Q)$ some element. Then the class*

$$\overline{v}^0 + S^1(\overline{v}^1) + \overline{v}^1 \cdot \overline{v}_{N_Q - 2n} + \overline{v}^0 \cdot \overline{v}_{N_Q - 2n - 1}$$

in $\widetilde{\mathrm{Ch}}^{2n+1}(Y|_{\overline{k}})$ is defined over k.

Corollary 3.6. *Let Q be a smooth projective quadric of dimension $\geqslant 4n - 1$, Y a smooth quasiprojective variety, and $\overline{y} \in \mathrm{Ch}^{2n+1}(Y|_{\overline{k}})$ defined over $k(Q)$. Then, either*

(a) $z_{2n+1}^{\boxed{-d_Q}}(Q|_{k(Y)})$ *is defined; or*

(b) *for certain $\overline{v}^1 \in \mathrm{Ch}^{2n}(Y|_{\overline{k}})$, and for a certain divisor $\overline{v}_{N_Q - 2n} \in \mathrm{Ch}^1(Y|_{\overline{k}})$, the element*

$$\overline{y} + S^1(\overline{v}^1) + \overline{v}^1 \cdot \overline{v}_{N_Q - 2n}$$

in $\widetilde{\mathrm{Ch}}^{2n+1}(Y|_{\overline{k}})$ is defined over k.

Proof. Since $\overline{y}$ is defined over $k(Q)$, there is $x \in \mathrm{Ch}^{2n+1}(Y|_{k(Q)})$ such that $\overline{x} = \overline{y}|_{\overline{k(Q)}}$. Using the surjection $\mathrm{CH}^*(Y \times Q) \twoheadrightarrow \mathrm{CH}^*(Y|_{k(Q)})$, lift the x to an element $v \in \mathrm{Ch}^{2n+1}(Y \times Q)$. Then $\overline{v} = \sum_{i=0}^{d_Q}(\overline{v}^i \cdot h^i + \overline{v}_i \cdot l_i)$, where $\overline{v}^0|_{\overline{k(Q)}} = \overline{y}|_{\overline{k(Q)}}$. But for any extension of fields $F/\overline{k}$ (with the smaller one algebraically closed), the restriction morphism on Chow groups (with any coefficients) is injective by the specialization arguments. Thus, $\overline{v}^0 = \overline{y}$. It remains to apply Proposition 3.5 and observe that if $\overline{v}_{N_Q - 2n - 1} \in \mathrm{Ch}^0(Y|_{\overline{k}}) = \mathbb{Z}/2 \cdot 1$ is nonzero, then the class $l_{N_Q - 2n - 1} = z_{2n+1}^{\boxed{-d_Q}}$ is defined over $k(Y)$. In fact, this class is just equal to $\overline{v}|_{\overline{k(Y)}}$. $\square$

Remark 3.7. One can get rid of factoring (2-*torsion*) in the statements above by using the genuine *symmetric operations* (see [15], cf. [14]) instead of the Landweber–Novikov operations. As was explained above, for our purposes it is irrelevant.

Before proving the proposition let us study a bit some special power series. Denote by $\gamma(t) \in \mathbb{Z}/2[[t]]$ the power series $1 + \sum_{i \geqslant 0} t^{2^i}$. Then $\gamma(t)$ satisfies the equation

$$\gamma^2 - \gamma = t$$

and generates quadratic extension of $\mathbb{Z}/2(t)$. In particular, for any $m \geqslant 0$, $\gamma^m = a_m \gamma + b_m$ for certain unique $a_m, b_m \in \mathbb{Z}/2(t)$. The following statement is clear.

Observation 3.8. (1) $a_{m+1} = a_m + b_m$, $\quad b_{m+1} = t a_m$;
(2) a_m and b_m are polynomials in t of degree $\leqslant [m - 1/2]$ and $[m/2]$, respectively.

For the power series $\beta(t)$ let us denote by $(\beta)_{\leqslant l}$ the polynomial $\sum_{j=0}^{l} \beta_j t^j$, and by $(\beta)_{>l}$ the remaining part $\beta - (\beta)_{\leqslant l}$.

Lemma 3.9.

$$a_m = (\gamma^m)_{\leqslant [m/2]} = (\gamma^m)_{\leqslant [m-1/2]}.$$

Proof. Let $m = 2^k + m_1$, where $0 \leqslant m_1 < 2^k$. Then $\gamma^m = \gamma^{2^k} \cdot \gamma^{m_1} = (a_{m_1}\gamma + b_{m_1}) + O(t^{2^k}) = (1 + \sum_{i=0}^{k-1} t^{2^i})a_{m_1} + b_{m_1} + O(t^{2^k})$. Observation 3.8 implies that $(\gamma^m)_{\leqslant [m/2]} = (1 + \sum_{i=0}^{k-1} t^{2^i})a_{m_1} + b_{m_1}$. On the other hand, $\gamma^{2^k} = \gamma + (\sum_{j=1}^{k-1} t^{2^j})$; thus γ^m is equal to

$$\left(\gamma + \left(\sum_{j=1}^{k-1} t^{2^j}\right)\right) (a_{m_1}\gamma + b_{m_1})$$

$$= a_{m_1}\gamma + a_{m_1}t + \left(\sum_{j=1}^{k-1} t^{2^j}\right) a_{m_1}\gamma + b_{m_1}\gamma + \left(\sum_{j=1}^{k-1} t^{2^j}\right) b_{m_1}$$

$$= \left(\left(1 + \sum_{i=0}^{k-1} t^{2^i}\right) a_{m_1} + b_{m_1}\right) \gamma + \left(ta_{m_1} + \left(\sum_{j=1}^{k-1} t^{2^j}\right) b_{m_1}\right).$$

Hence, $a_m = ((1 + \sum_{i=0}^{k-1} t^{2^i})a_{m_1} + b_{m_1}) = (\gamma^m)_{\leqslant [m/2]}$. The second equality follows from Observation 3.8(2).   □

Lemma 3.9 implies that

$$\gamma^m = (\gamma^m)_{\leqslant [m-1/2]} \cdot \gamma + t(\gamma^{m-1})_{\leqslant [m-2/2]}.$$

Lemma 3.10.

$$(\gamma^m)_{>[m/2]} = t^m(1 + mt) + O(t^{m+2})$$

Proof. Use induction on m and on the number of 1's in the binary presentation of m. For $m = 2^k$ the statement is clear. Let now $m = 2^k + m_1$, where $0 < m_1 < 2^k$. We have $(\gamma^m)_{>[m/2]} = ((\gamma^{m+1})_{\leqslant [m/2]})_{>[m/2]} + t((\gamma^m)_{\leqslant [m-1/2]} \cdot \gamma^{-1})_{>([m/2]-1)} = t(a_m \cdot \gamma^{-1})_{>[m/2]-1}$.

We also have $a_m = (\gamma^m)_{\leqslant [m/2]} = (\gamma^{2^k} \cdot \gamma^{m_1})_{\leqslant [m/2]} = (\gamma^{m_1})_{\leqslant [m/2]} = (a_{m_1}\gamma + b_{m_1})_{\leqslant [m/2]}$, and since the degrees of a_{m_1} and b_{m_1} are no more than $[m_1/2]$, this expression should be equal to $\gamma^{m_1} + a_{m_1}t^{2^k} + O(t^{2^{k+1}})$. Then

$$a_m \cdot \gamma^{-1} = \gamma^{m_1-1} + a_{m_1}\gamma^{-1}t^{2^k} + O(t^{2^{k+1}})$$
$$= (a_{m_1-1}\gamma + b_{m_1-1}) + a_{m_1}\gamma^{-1}t^{2^k} + O(t^{2^{k+1}})$$
$$= (a_{m_1-1}\left(1 + \sum_{i=0}^{k-1} t^{2^i}\right) + b_{m_1-1}) + t^{2^k}(a_{m_1-1} + a_{m_1}\gamma^{-1}) + O(t^{2^{k+1}}).$$

Since the degree of a_{m_1-1} is no more than $[m_1/2] - 1$, using Observation 3.8(1), we get

$$
\begin{aligned}
(a_m \cdot \gamma^{-1})_{>[m/2]-1} &= t^{2^k}(a_{m_1-1} + a_{m_1}\gamma^{-1}) + O(t^{2^{k+1}}) \\
&= t^{2^k}(\gamma^{m_1-2} + a_{m_1-1}\gamma^{-1}) + O(t^{2^{k+1}}) \\
&= t^{2^k}\gamma^{-1}(\gamma^{m_1-1} + (\gamma^{m_1-1})_{\leqslant[m_1-1/2]}) + O(t^{2^{k+1}}).
\end{aligned}
$$

Consequently, $(\gamma^m)_{>[m/2]} = t^{2^k+1}(\gamma^{m_1-1})_{>[m_1-1/2]} \cdot \gamma^{-1} + O(t^{2^{k+1}})$. And by the inductive hypothesis, this is equal to

$$
t^{2^k+1}(t^{m_1-1}(1 + (m_1 - 1)t))\gamma^{-1} + O(t^{2^k+m_1+2}) = t^m(1 + mt) + O(t^{m+2}).
$$

$\square$

Corollary 3.11.

$$
(a_m \cdot \gamma^{-1})_{>[m/2]-1} = t^{m-1}(1 + mt) + O(t^{m+1}).
$$

Observe now that $\gamma^{-1}(t) = \sum_{i\geqslant 0} t^{2^i-1}$. Denote by $\delta(t)$ the polynomial $a_{2n+1}(t)$. Then

$$
\delta(t)\gamma^{-1}(t) = \alpha(t) + t^{2n} + t^{2n+1} + O(t^{2n+2}), \tag{1}
$$

where $\delta(t)$ and $\alpha(t)$ are polynomials of degree $\leqslant n$. Observation 3.8(1) shows that $\delta = 1 + t + \cdots$. For us it will be important that $\delta(t)\gamma^{-1}(t)$ does not contain monomials of degrees from $(n + 1)$ to $(2n - 1)$, but contains t^{2n} and t^{2n+1}.

Proof (of Proposition 3.5). The idea of the proof is the following: having some element $v \in \mathrm{Ch}^{2n}(Y \times Q)$, we first lift it via pr to some $w \in \Omega^*(Y \times Q)$, then restrict w to $Y \times Q_s$ for various subquadrics Q_s of Q, and apply to these restrictions the combination of the *symmetric operations* ϕ^{t^i} and $(\pi_{Y,s})_*$ (see below) in a different order. The point is that by adding the results with the appropriate coefficients one can get the expression in question. In particular, all the choices made while lifting to Ω^* will be canceled out. And the needed coefficients are provided by the power series $\delta(t)$ above.

The case of $\dim(Q) \geqslant 4n - 1$ can be reduced to that of $\dim(Q) = 4n - 1$ by considering an arbitrary subquadric $Q' \subset Q$ of dimension $4n - 1$, and restricting v to $Y \times Q'$. So, we will assume that $\dim(Q) = 4n - 1$.

Let $Q_s \xrightarrow{e_s} Q$ be an arbitrary smooth subquadric of Q of dimension s. Denote by $w(s)$ the class $(id \times e_s)^*(w) \in \Omega^{2n+1}(Q_s \times Y)$. Then

$$
\overline{w(s)} = \sum_{0\leqslant i\leqslant \min(2n-1,s)} \overline{w}^i \cdot h^i + \sum_{4n-s-1\leqslant j\leqslant 2n-1} \overline{w}_j \cdot l_{j-4n+s+1}.
$$

Let $\pi_{Y,s} : Q_s \times Y \to Y$ be the natural projections.

Consider the element

$$u := (\pi_{Y,2n+1})_* \phi^{t^0}(w(2n+1)) + \sum_{p=n+1}^{2n+1} \delta_{2n+1-p} \phi^{t^{2p-(2n+1)}}(\pi_{Y,p})_* (w(p))$$

in $\mathrm{Ch}^{2n+1}(Y)$, where δ_j are the coefficients of the power series δ above. Let us compute $\bar{u}$. Since we are computing modulo 2-torsion, it is sufficient to compute $2\bar{u}$, which is equal to the Chow trace of

$$(\pi_{Y,2n+1})_* (S_{L-N}^{2n+1} - \square)(\overline{w}(2n+1)) + \sum_{p=n+1}^{2n+1} \delta_{2n+1-p} S_{L-N}^p (\pi_{Y,p})_* (\overline{w}(p)).$$

Using the multiplicative properties of the Landweber–Novikov operations,

$$S_{L-N}^a(x \cdot y) = \sum_{b+c=a} S_{L-N}^b(x) S_{L-N}^c(y),$$

and Proposition 3.3, we get (modulo 4)

$$\begin{aligned}
pr(\pi_{Y,2n+1})_* S_{L-N}^{2n+1}(\overline{w}(2n+1)) &= \sum_{j=0}^{2n-1} \binom{j}{2n-j+1} \cdot 2 \cdot S^j(\overline{v}^j) \\
&+ pr(\binom{2n+1}{1}) \cdot S_{L-N}^{2n}(\overline{w}_{2n-1}) \\
&+ \binom{2n+2}{0}) \cdot S_{L-N}^{2n+1}(\overline{w}_{2n-2})).
\end{aligned}$$

The codimension of $\overline{v}^j$ is $2n+1-j$; thus either $\binom{j}{2n-j+1}$ is zero, or $S^j(\overline{v}^j)$ is, and our expression is equal to $pr(S_{L-N}^{2n}(\overline{w}_{2n-1}) + S_{L-N}^{2n+1}(\overline{w}_{2n-2}))$. Also, (modulo 4),

$$pr(\pi_{Y,2n+1})_* \square(\overline{w}(2n+1)) = 2 \cdot pr(\overline{w}^0 \overline{w}_{2n-2} + \overline{w}^1 \overline{w}_{2n-1}).$$

In the same way (modulo 4),

$$\begin{aligned}
pr S_{L-N}^p(\pi_{Y,p})_* (\overline{w}(p)) &= \sum_{j=0}^{\min(2n-1,p)} \binom{-(p+2-j)}{p-j} \cdot 2 \cdot S^j(\overline{v}^j) \\
&+ pr(\sum_{i=0}^{p-2n} \binom{-(i+1)}{i}) S_{L-N}^{p-i}(\overline{w}_{i+4n-1-p})).
\end{aligned}$$

Observe that the second sum is empty for $p < 2n$, is equal (modulo 4) to $pr S_{L-N}^{2n}(\overline{w}_{2n-1})$ for $p = 2n$, and to $pr S_{L-N}^{2n+1}(\overline{w}_{2n-2})$ for $p = 2n+1$ (we used here Corollary 3.4).

Since the coefficient $\binom{-(l+2)}{l}$ is odd if and only if $l = 2^k - 1$, for some k, the first sum is equal to

$$2 \sum_{j=0}^{\min(2n-1,p)} (\gamma^{-1})_{p-j} \cdot S^j(\overline{v}^j).$$

Taking into account that $\delta(t) = 1 + t + \cdots$, we get

$$pr \sum_{p=n+1}^{2n+1} \delta_{2n+1-p} S_{L-N}^p (\pi_{Y,p})_* (\overline{w}(p))$$

$$= 2 \sum_{p=n+1}^{2n+1} \sum_{j=0}^{min(2n-1,p)} \delta_{2n+1-p} (\gamma^{-1})_{p-j} \cdot S^j(\overline{v}^j)$$

$$+ (pr S_{L-N}^{2n}(\overline{w}_{2n-1}) + pr S_{L-N}^{2n+1}(\overline{w}_{2n-2}))$$

$$= 2 \sum_{j=0}^{2n-1} (\delta \cdot \gamma^{-1})_{2n+1-j} S^j(\overline{v}^j) + (pr S_{L-N}^{2n}(\overline{w}_{2n-1}) + pr S_{L-N}^{2n+1}(\overline{w}_{2n-2}))$$

$$= 2(\overline{v}^0 + S^1(\overline{v}^1)) + (pr S_{L-N}^{2n}(\overline{w}_{2n-1}) + pr S_{L-N}^{2n+1}(\overline{w}_{2n-2})),$$

in light of formula (1) and Corollary 3.4.

Putting things together (and again using Corollary 3.4), we obtain

$$2\overline{u} = 2(\overline{v}^0 + S^1(\overline{v}^1)) + \overline{v}^1 \cdot \overline{v}_{2n-1} + \overline{v}^0 \cdot \overline{v}_{2n-2}).$$

Since u is defined over the base field k, the proposition is proven.   □

There is another result that extends a bit Theorem 3.1.

Proposition 3.12. ([14, Statement 3.8]) *Let Y be a smooth quasiprojective variety, Q a smooth projective quadric over k. Let $\overline{y} \in \mathrm{Ch}^m(Y|_{\overline{k}})$. Suppose $z_{N_Q - d_Q}^{\boxed{0}} (Q)$ is defined. Then for $m \leqslant N_Q - d_Q$,*

$$\overline{y}|_{\overline{k(Q)}} \text{ is defined over } k(Q) \Leftrightarrow \overline{y} \text{ is defined over } k.$$

Proposition 3.12 extends Theorem 3.1 in the direction of the following:

Conjecture 3.13. ([14, Conjecture 3.11]) *In the notation of Theorem 3.1, suppose $z_l^{\boxed{N_Q - d_Q - l}}(Q)$ is defined. Then for any $m \leqslant l$,*

$$\overline{y}|_{\overline{k(Q)}} \text{ is defined over } k(Q) \Leftrightarrow \overline{y} \text{ is defined over } k.$$

This conjecture is known for $l = N_Q - d_Q, N_Q - 1, N_Q$.

3.3 Some Auxiliary Facts

For our purposes it will be important to be able (under certain conditions) to get rid of the last term in the formula from Proposition 3.5. For this we will need the following facts.

Proposition 3.14. *Let $0 \leqslant i \leqslant d_R$, and let $F(R, i) \overset{\alpha}{\leftarrow} F(R, 0, i) \overset{\beta}{\rightarrow} R$ be the natural projections. Let $z_{N_R-i}^{\boxed{i-d_R}}$ be defined. Let $t \in \text{Ch}_{N_R-i}(F(R, i))$ be such that $\beta_* \alpha^*(t) = 1 \in \text{Ch}^0(R)$. Then $\beta_* \alpha^*(t \cdot z_{N_R-i}^{\boxed{i-d_R}}) = l_i \in \text{Ch}_i(R)$.*

Proof. By the definition, $z_{N_R-i}^{\boxed{i-d_R}} = \alpha_* \beta^*(l_0)$. By the projection formula,

$$\alpha_* \beta^*(t \cdot z_{N_R-i}^{\boxed{i-d_R}}) = \beta_* \alpha^*(t \cdot z_{N_R-i}^{\boxed{i-d_R}}) = \beta_* \alpha^* \alpha_* (\alpha^*(t) \cdot \beta^*(l_0)).$$

Again by the projection formula, $\beta_*(\alpha^*(t) \cdot \beta^*(l_0)) = l_0$. Thus, $\alpha^*(t) \cdot \beta^*(l_0)$ is a zero cycle of degree 1 on $F(R, 0, i)$, and $\alpha_*(\alpha^*(t) \cdot \beta^*(l_0))$ is a zero cycle of degree 1 on $F(R, i)$. The proposition follows. $\square$

Let $v \in \text{Ch}^m(Y \times Q)$ be some element. Then

$$\overline{v} = \sum_{i=0}^{d_Q} (\overline{v}^i \cdot h^i + \overline{v}_i \cdot l_i).$$

Statement 3.15. *Suppose $z_m^{\boxed{N_Q-m-d}}(Q)$ is defined. Then for any v as above, there exists $u \in \text{Ch}^m(Y \times Q)$ such that $\overline{u}^0 = \overline{v}^0$, and $\overline{u}_{N_Q-m} = 0$.*

Proof. If $\overline{v}_{N_Q-m} = 0$, there is nothing to prove. Otherwise, the class $l_{N_Q-m} \in \text{Ch}_{N_Q-m}(Q|_{k(Y)})$ is defined. Indeed, let

$$\rho_X : \text{Ch}^*(Y \times X) \twoheadrightarrow \text{Ch}^*(X|_{k(Y)})$$

be the natural restriction. Then $\rho_Q(\overline{v}) = l_{N_Q-m}$ plus $\lambda \cdot h^m$, if $2m = N_Q$ (notice that $\overline{v}_{N_Q-m} \in \text{Ch}^0$). This implies that the class $l_{N_Q-m} = z_{N_Q-m}^{\boxed{-d_Q}}$ is defined on $Q|_{k(Y)}$. Using Proposition 2.5 and Statement 2.15, we get that the class of a rational point is defined on $F(Q, N_Q - m)|_{k(Y)}$ (this proof is somewhat longer than the standard one, but it does not use the theorem of Springer (see [5])!). Let $x \in \text{Ch}_{\dim(Y)}(F(Q, N_Q - m) \times Y)$ be an arbitrary lifting of this class with respect to $\rho_{F(Q, N_Q-m)}$. Let

$$F(Q, N_Q - m) \overset{\alpha}{\leftarrow} F(Q, 0, N_Q - m) \overset{\beta}{\rightarrow} Q$$

be the natural projections. Consider $u' := (\beta \times id)_* (\alpha \times id)^*(x) \in \text{Ch}^m(Q \times Y)$. Proposition 3.14 implies that the (defined over k) cycle

$$u'' := \pi_Y^*(\pi_Y)_*((h^{N_Q-m} \times 1_Y) \cdot (\beta \times id)_*(\alpha \times id)^*(x \cdot z_m^{\boxed{N_Q-m-d}}(Q)))$$

satisfies: $\overline{u''}^0 = \overline{u'}^0$, and (evidently) $\overline{u''}_{N_Q-m} = 0$. Since $\overline{u'}_{N_Q-m} = 1 = \overline{v}_{N_Q-m}$, it remains to take: $u := v - u' + u''$. $\square$

4 Even u-invariants

The fields of any given even u-invariant were constructed by A. Merkurjev in [9] using his *index-reduction formula* for central simple algebras. The idea of such a construction is based on the so-called *Merkurjev tower of fields*, which was first used in [8]. In our case, this tower is constructed as follows: Let F be any field, and let S_F be the set of all (isomorphism classes of) quadrics over F of dimesion $> (M - 2)$. Let $F' := \lim_{I \subset S_F} F(\times_{i \in I} Q_i)$, where the limit is taken over all finite subsets of S_F via the natural restriction maps. Then starting from an arbitrary field k one constructs the tower of fields $k = k_0 \to k_1 \to \cdots \to k_r \to \cdots$, where $k_{r+1} := (k_r)'$. One gets the huge field $k_\infty := \lim_r k_r$ having the property that all forms of dimension $> M$ over it are isotropic, and thus $u(k_\infty) \leqslant M$. But to get a field whose u-invariant is exactly M, one has to start with some special field k, and since one wants to have some anisotropic M-dimensional form p over k_∞, it is better to have it already over k, and then check that p will not become isotropic while passing from k to k_∞. Of course, to be able to control this, we need to know something interesting about p. That is, we need to control some other property that implies ours. More precisely, for a given base field k, on the set of field extensions E/k we should define two properties A and B, where

$$A(E) \text{ is satisfied } \Leftrightarrow p|_E \text{ is anisotropic,}$$

so that the following conditions are satisfied:

(1) $B \Rightarrow A$.
(2) $B(E) \Rightarrow B(E(Q))$, for an arbitrary quadric Q/E of dimension $> \dim(P)$.
(3) Let $\{E_j\}_{j \in J}$ be the directed system of field extensions with the limit E_∞.
  Then $B(E_j)$ for all j implies $B(E_\infty)$.

Then $B(k) \Rightarrow A(k_\infty)$. So if one finds a quadratic form p of dimension M over k and some property B satisfying the above conditions (and such that $B(k)$ is satisfied), then k_∞ will have u-invariant M.

In the case that $M = 2n$ is even A. Merkurjev takes $p \in I^2$ and the following property B:

$$B(E) \text{ is satisfied } \Leftrightarrow C_0^+(p|_E) \text{ is a division algebra,}$$

where $C_0^+(p)$ is the "half" of the *even Clifford algebra* $C_0(p) = C_0^+(p) \times C_0^-(p)$ (both factors are isomorphic here). Of course, for $B(k)$ to be satisfied one has to start with some form p for which C_0^+ is a division algebra over the base field. The *generic* form from I^2 (that is, the form $\langle a_1, \ldots, a_{2n-1}, (-1)^n a_1 \cdots a_{2n-1} \rangle / k = k_0(a_1, \ldots, a_{2n-1})$) will do the job. The condition (1) is satisfied; since the isotropy of $p|_E$ gives the matrix factor in $C_0^+(p|_E)$ ([5]), and the condition (3) is clear. The only nontrivial fact here is the condition (2), which follows from the index reduction formula of Merkurjev, claiming that over the generic point of a quadric Q the index of a division algebra D can drop at most by the factor 2, and the latter happens if and only if $C_0(q)$ can be mapped to D. Indeed, if p is of dimension $2n$, then $C_0^+(p)$ is a central simple algebra of rank 2^{n-1}, and $C_0(q)$ is either a simple algebra, or a product of two simple algebras of large rank (this will not be true for odd-dimensional p!); thus there is no ring homomorphisms $C_0(q) \to C_0^+(p)$, and the index of $C_0^+(p|_E)$ is equal to that of $C_0^+(p|_{E(Q)})$.

Let me give another construction, which does not use the *index reduction formula*. Instead, I will use the northwest corner of the EDI and the property

$$B(E) \text{ is satisfied} \quad \Leftrightarrow \quad z_{N_P - d_P}^{\boxed{0}}(P|_E) \text{ is not defined.}$$

Statement 4.1. *Let $M = \dim(p)$ be even. Then our property B satisfies the above conditions $(1 - 3)$.*

Proof. Condition (1) follows from Proposition 2.5, since the property $A(E)$ is satisfied $\Leftrightarrow$ the southwest corner of the EDI is not defined for $P|_E$. Condition (3) is clear, since $\mathrm{CH}^*(X|_{E_\infty}) = \lim_j \mathrm{CH}^*(X|_{E_j})$. Finally, suppose $z_{n-1}^{\boxed{0}}(P|_E)$ is not defined. Then, by Theorem 3.1, for any form q of dimension $> M$, $z_{n-1}^{\boxed{0}}(P|_{E(Q)})$ is not defined as well (will not work for M odd). Thus, condition (2) is satisfied. $\qquad\square$

Corollary 4.2. (A. Merkurjev, [9]) *For each $M = 2n$ there is a field of the u-invariant M.*

Proof. Take any form p/k of dimension M such that $z_{n-1}^{\boxed{0}}(P)$ is not defined. One can use the generic form; see [14, Statement 3.6]. Then $B(k)$ is satisfied, and hence $A(k_\infty)$ is satisfied too. $\qquad\square$

5 Odd u-invariants

Let us analyze a bit the above construction. Instead of working with the cycle $z_{N_P}^{\boxed{-d_P}}$, the class of a rational point on a quadric P, we worked with the (smaller codimensional!) cycle $z_{N_P - d_P}^{\boxed{0}}$, and used the fact that rationality of the former implies rationality of the latter (Proposition 2.5).

Unfortunately, for odd-dimensional forms we cannot use the class $z_{N_P - d_P}^{\boxed{0}}$.

Indeed, if p is any such form, then for $q := p \perp \langle \det_\pm(p) \rangle$, $z_{N_P - d_P}^{\boxed{0}}(P|_{k(Q)})$ will be defined, since the rationality of this class is equivalent to the rationality of $z_{N_Q - d_Q}^{\boxed{0}}(Q|_{k(Q)})$, observe that

$$G(Q, d_Q) = G(P, d_P) \coprod G(P, d_P),$$

and the rationality of the latter follows from the rationality of the class $z_{N_Q}^{\boxed{-d_Q}}(Q|_{k(Q)})$ (isotropy of $Q|_{k(Q)}$). So, even if we start from the form where our class is not defined, over the generic point of some bigger-dimensional form it will become rational, and we cannot control anisotropy of P.

But the rationality of $z_{N_P}^{\boxed{-d_P}}$ implies rationality not just of $z_{N_P-d_P}^{\boxed{0}}$, but of all the west edge $z_{N_P-d_P+s}^{\boxed{-s}}$, $0 \leqslant s \leqslant d_P$. So, let us use these other cycles.

Let the form p have dimension $2^r + 1$. In this case, one can use the next-to-last Grassmannian and the class $z_{2^{r-1}+1}^{\boxed{-1}}$ on it.

Theorem 5.1. *Let* $\dim(p) = 2^r + 1$, $r \geqslant 3$, *and suppose* $\mathrm{EDI}(P)$ *looks like this:*

$$
\begin{array}{ccccc}
\text{⑦} & \circ & \cdots & \circ \\
\circ & \circ & \cdots & \circ \\
\cdots & \cdots & \cdots & \cdots \\
\circ & \circ & \cdots & \circ
\end{array}
$$

Let $\dim(q) > \dim(p)$. *Then* $\mathrm{EDI}(P|_{k(Q)})$ *has the same property.*

Corollary 5.2. *For any* $r \geqslant 3$ *there is a field of the u-invariant* $2^r + 1$.

Proof. Start with the *generic form* p over $k = k_0(a_1, \ldots, a_{2^r+1})$ and the property

$$B(E) \text{ is satisfied } \Leftrightarrow \mathrm{EDI}(P|_E) \text{ is as in Theorem 5.1,}$$

Then $\mathrm{EDI}(P)$ is empty. This follows from Proposition 2.5 and [14, Statement 3.6]. Thus, $B(k)$ is satisfied. Condition (1) is satisfied by the definition of A and B. Condition (3) is satisfied since $\mathrm{CH}^*(X|_{E_\infty}) = \lim_j \mathrm{CH}^*(X|_{E_j})$. And the condition (2) is equivalent to Theorem 5.1. Then, as we know, $A(k_\infty)$ is satisfied, and $u(k_\infty) = 2^r + 1$. □

Proof (of Theorem 5.1). Let $d := d_P = 2^{r-1} - 1$. It follows from Theorem 3.1 that the cycles $z_j^{\boxed{0}}(P|_{k(Q)})$, $1 \leqslant j \leqslant d$ are not defined. That is, we have o's to the right of ⑦. In light of Proposition 2.5, it remains only to treat the case of $z_{d+2}^{\boxed{-1}}(P|_{k(Q)})$ (that is, the node just below the ⑦).

This is done as follows. If this cycle is defined over $k(Q)$, we can lift it to a cycle v on $F(P, d-1) \times Q$. Over $\bar{k}$, quadric Q becomes cellular, and our cycle decomposes in a standard way, producing coordinates $\bar{v}^i, \bar{v}_i$. Using the fact that $\dim(q) = 2^r + 1, r \geqslant 3$, one can correct v in such a way that the "last" of these coordinates will be zero. This is done by some play with elementary classes using Theorem 3.1, Proposition 3.12, Proposition 2.9, and other major statements, and is the most delicate part of the proof (in particular, it is the only place where the high specific of the dimension is used; everything else works for $\dim(q) \equiv 1 \,(\mathrm{mod}\,4)$. After this is achieved, one can apply Proposition 3.5, and get a k-rational class on $F(P, d-1)$ given by the sum of Four terms, the last of which will be zero because of our choice of v. Now, from knowledge of the action of the Steenrod operations it is not difficult to prove that the obtained k-rational class is nonzero. Finally, we use the information about elementary classes on $F(P, d)$ and the main result of [13] to conclude that our nonzero k-rational class on $F(P, d-1)$ should be $z_{d+2}^{\boxed{-1}}$. This contradiction proves the theorem.

In more detail, suppose $z_{d+2}^{\boxed{-1}}(P|_{k(Q)})$ is defined. We clearly can assume that $\dim(q) = \dim(p) + 1 = 2^r + 2$. Let us denote $F(P, d-1)$ temporarily by Y. We have $y \in \mathrm{Ch}^{d+2}(Y|_{k(Q)})$ such that $\overline{y} = z_{d+2}^{\boxed{-1}} \in \mathrm{Ch}^{d+2}(Y|_{\overline{k(Q)}})$. Let us lift it to $v \in \mathrm{Ch}^{d+2}(Y \times Q)$ via the natural projection $\mathrm{Ch}^{d+2}(Y \times Q) \xrightarrow{\rho_Y} \mathrm{Ch}^{d+2}(Y|_{k(Q)})$.

Statement 5.3. *There exists* $v \in \mathrm{Ch}^{d+2}(Y \times Q)$ *such that* $\overline{v}^0 = \overline{y}$ *and* $\overline{v}_d = 0 \in \mathrm{Ch}^0(Y|_{\overline{k}})$.

Proof. Let v be an arbitrary lifting of y with respect to ρ_Y. If $\overline{v}_d = 0$, there is nothing to prove. Otherwise, let us show that $z_{2^{r-1}+1}^{\boxed{-1}}(Q)$ is defined over k.

Suppose $\overline{v}_d \neq 0$; then it is equal to $1 \in \mathrm{Ch}^0(Y|_{\overline{k}})$. Then $\overline{\rho_Q(v)} = l_d \in \mathrm{Ch}_d(Q|_{\overline{k(Y)}}) = \mathbb{Z}/2 \cdot l_d$. Thus, $z_{d+2}^{\boxed{-d_Q}}(Q|_{k(Y)})$ is defined. By Proposition 2.5, $z_3^{\boxed{-2}}(Q|_{k(Y)})$ is defined too (it lives in the same column above). Let us say that $\tilde{z}$ is defined if either $z_3^{\boxed{-2}}$, or $z_3^{\boxed{-2}} + z_2^{\boxed{-2}} w_1^{\boxed{-2}}$ is defined. In particular, we know that $\tilde{z}(Q|_{k(Y)})$ is defined. We want to show that $\tilde{z}(Q)$ is defined.

Consider the two towers of fibrations

$$\mathrm{Spec}(k) \leftarrow P \leftarrow \ldots \leftarrow F(P, 0, 1, \ldots, d-1),$$
$$\mathrm{Spec}(k) \leftarrow Q \leftarrow \ldots \leftarrow F(Q, 0, 1, \ldots, d-1),$$

with the generic fibers quadrics $P = P_1, \ldots, P_d$, $Q = Q_1, \ldots, Q_d$ of dimension $2d+1, 2d-1, \ldots, 3$, and $2d+2, 2d, \ldots, 4$, respectively. Let us define $k_{a,b} := k(F(P, 0, \ldots, a-1) \times F(Q, 0, \ldots, b-1))$. Then

$$k_{a+1,b} = k_{a,b}(P_a) \quad \text{and} \quad k_{a,b+1} = k_{a,b}(Q_b).$$

Since we have embeddings of fields

$$k \subset k(Y) = k(F(P, d-1)) \subset k(F(P, 0, \ldots, d-1)) = k_{d,0},$$

$z_3^{\boxed{-2}}(Q|_{k_{d,0}})$ is defined. Then by Proposition 2.5, $z_1^{\boxed{0}}(Q|_{k_{d,0}})$ is defined. By Theorem 3.1, $z_1^{\boxed{0}}(Q)$ and $z_2^{\boxed{0}}(Q) = (z_1^{\boxed{0}}(Q))^2 = S^1(z_1^{\boxed{0}}(Q))$ are defined.

It follows from Propositions 2.9 and 2.11 that for arbitrary elements $\overline{\alpha}, \overline{\beta} \in \mathrm{Ch}^*(F(Q, d-1)|_{\overline{k_{a-1,0}}})$ of codimension 2 and 1, respectively, the class $S^1(\overline{\alpha}) + \overline{\alpha} \cdot \overline{\beta}$ modulo classes defined over any field is either equal to 0, or to $z_2^{\boxed{-2}} w_1^{\boxed{-2}}$. It follows from Corollary 3.6 that

$$\tilde{z}(Q|_{k_{a,0}}) \text{ is defined} \Rightarrow \begin{cases} \text{either} & \tilde{z}(Q|_{k_{a-1,0}}) \text{ is defined;} \\ \text{or} & z_3^{\boxed{-d_{P_a}}}(P_a|_{k_{a-1,d}}) \text{ is defined.} \\ & \text{and } d_{P_a} \leqslant 3. \end{cases}$$

Let us show that the second case is impossible. By Proposition 2.5,

$$z_3^{\boxed{-d_{P_a}}}(P_a|_{k_{a-1,d}}) \text{ is defined} \Rightarrow z_{3-d_{P_a}}^{\boxed{0}}(P_a|_{k_{a-1,d}}) \text{ is defined}$$

Since $3 - d_{P_a} \leqslant 2$, $\dim(Q_b) \geqslant 4$, and $z_2^{\boxed{0}}(Q_d)$ is defined, by Proposition 3.12 and Theorem 3.1,

$$z_{3-d_{P_a}}^{\boxed{0}}(P_a|_{k_{a-1,b}}) \text{ is defined} \Rightarrow z_{3-d_{P_a}}^{\boxed{0}}(P_a|_{k_{a-1,b-1}}) \text{ is defined}$$

Then $z_{3-d_{P_a}}^{\boxed{0}}(P_a/k_{a-1,0})$ is defined, and $z_{3-d_{P_a}}^{\boxed{0}}(P)$ is defined (by Theorem 3.1). This contradicts the conditions of our theorem (here we are using the fact that $r \geqslant 3$). Thus,

$$\tilde{z}(Q|_{k_{a,0}}) \text{ is defined} \Rightarrow \tilde{z}(Q|_{k_{a-1,0}}) \text{ is defined,}$$

and consequently, $\tilde{z}(Q)$ is defined. Denoting $\pi_{(d-1,d)}$ as γ, and $\pi_{(d-1,\underline{d})}$ as δ, from Lemma 2.6, we have: $\gamma^*(\tilde{z})$ is equal either to

$$\delta^*(z_3^{\boxed{-1}}) + c_1(\mathcal{O}(1)) \cdot \delta^*(z_2^{\boxed{-1}}), \text{ or to}$$

$$\delta^*(z_3^{\boxed{-1}} + z_2^{\boxed{-1}} w_1^{\boxed{-1}}) + c_1(\mathcal{O}(1)) \cdot \delta^*(z_1^{\boxed{-1}} w_1^{\boxed{-1}}) + c_1(\mathcal{O}(1))^2 \cdot \delta^*(z_1^{\boxed{-1}})$$

Using the arguments from the proof of Proposition 2.5, we get, in the first case, that $z_3^{\boxed{-1}}(Q)$ is defined, and in the second, that $z_1^{\boxed{-1}}$ is defined, hence $z_2^{\boxed{-1}} = S^1(z_1^{\boxed{-1}}) + z_1^{\boxed{-1}} w_1^{\boxed{-1}}$ is defined, and, finally, $z_3^{\boxed{-1}} = (z_3^{\boxed{-1}} + z_2^{\boxed{-1}} w_1^{\boxed{-1}}) + z_2^{\boxed{-1}} w_1^{\boxed{-1}}$ is defined. In any case, $z_3^{\boxed{-1}}(Q)$ is defined. Proposition 2.9 implies that

$$z_{2^{r-1}+1}^{\boxed{-1}}(Q) = S^{2^{r-2}} S^{2^{r-3}} \cdots S^2(z_3^{\boxed{-1}}(Q))$$

is also defined over k. Since $z_{2^{r-1}+1}^{\boxed{-1}}(Q)$ is defined, everything follows from Statement 3.15. $\qquad\square$

Consider $v \in \mathrm{Ch}^{d+2}(Y \times Q)$ satisfying the conditions of Statement 5.3. As above, $\overline{v} = \sum_{i=0}^{2^r-1} (\overline{v}^i \cdot h^i + \overline{v}_i \cdot l_i)$. Then, by Proposition 3.5, the class $\overline{v}^0 + S^1(\overline{v}^1) + \overline{v}^1 \overline{v}_{2^r-1} + \overline{v}^0 \overline{v}_{2^r-1-1}$ is defined over k. But $\overline{v}_{2^r-1-1} = 0$. Thus on Y we have the class $\overline{v}^0 + S^1(\overline{v}^1) + \overline{v}^1 \overline{v}_{2^r-1}$ defined over k.

Now it is time to use the specifics of Y and v. Our Y is a Grassmannian $F(P, d-1)$. In particular, it is a geometrically cellular variety, and the map $\mathrm{Ch}^*(Y|_{\overline{k}}) \to \mathrm{Ch}^*(Y|_{\overline{k(Q)}})$ is an isomorphism. Thus, $\overline{v}^0 = z_{d+2}^{\boxed{-1}}$. On the other hand, $\overline{v}_{2^r-1} = \overline{v}_{d+1}$ belongs to $\mathrm{Ch}^1(Y|_{\overline{k}})$, and so is equal either to 0, or to $w_1^{\boxed{-1}}$. So, on $F(P, d-1)$ we have a class either of the form $z_{d+2}^{\boxed{-1}} + S^1(\overline{v}^1)$, or of the form $z_{d+2}^{\boxed{-1}} + S^1(\overline{v}^1) + \overline{v}^1 w_1^{\boxed{-1}}$ defined over k. The following statement shows that such a class should be nonzero.

Statement 5.4. *Let R be a smooth projective quadric of dimension $4n - 1$. Then $z_{N_R-d_R+1}^{\boxed{-1}}(R)$ belongs to the image of neither of the following two maps:*

$$S^1, (S^1 + w_1^{\boxed{-1}} \cdot \ ) : \mathrm{Ch}^{N_R-d_R}(F(R, d_R - 1)) \to \mathrm{Ch}^{N_R-d_R+1}(F(R, d_R - 1)).$$

Proof. We can assume that $k = \overline{k}$. Consider the natural projections

$$F(R, d_R - 1) \xleftarrow{\alpha} F(R, d_R - 1, d_R) \xrightarrow{\beta} F(R, d_R).$$

The map β provides $F(R, d_R-1, d_R)$ with the structure of the projective bundle $\mathbb{P}_{F(R, d_R)}(\mathcal{T}_{d_R}^{\vee})$, and the Chern classes of $\mathcal{T}_{d_R}$ are divisible by 2; see Proposition 2.1 (and [13]). Thus, $\mathrm{Ch}^*(F(R, d_R-1, d_R)) = \mathrm{Ch}^*(F(R, d_R))[h]/(h^{d_R+1})$, where $h = c_1(\mathcal{O}(1))$. By Lemma 2.6,

$$\alpha^*(z_{N_R-d_R+1}^{\boxed{-1}}) = \beta^*(z_{N_R-d_R}^{\boxed{0}}) \cdot h, \text{ and } h = \alpha^*(w_1^{\boxed{-1}}).$$

The first fact now is simple, since

$$S^1(\alpha^*(z_{N_R-d_R+1}^{\boxed{-1}})) = S^1(\beta^*(z_{N_R-d_R}^{\boxed{0}}) \cdot h) = \beta^*(z_{N_R-d_R}^{\boxed{0}}) \cdot h^2,$$

by Proposition 2.9, and the latter element is nonzero. Thus, even $\alpha^*(z_{N_R-d_R+1}^{\boxed{-1}})$ cannot be in the image of S^1, since $S^1 \circ S^1 = 0$.

To prove the second fact, observe that

$$\alpha^*(z_{N_R-d_R+1}^{\boxed{-1}}) = \beta^*(z_{N_R-d_R}^{\boxed{0}}) \cdot h = (S^1 + h \cdot \ )(\beta^*(z_{N_R-d_R}^{\boxed{0}})).$$

Let $u \in \mathrm{Ch}^{N_R-d_R}(F(R, d_R - 1))$ be such that $(S^1 + w_1^{\boxed{-1}} \cdot \ )(u) = z_{N_R-d_R+1}^{\boxed{-1}}$. Then $(S^1 + h \cdot \ )(\beta^*(z_{N_R-d_R}^{\boxed{0}}) - \alpha^*(u)) = 0$. Since d_R is odd, the differential $(S^1 + h \cdot \ )$ acts without cohomology on $\mathrm{Ch}^*(F(R, d_R))[h]/(h^{d_R+1})$.

Consequently, $(\beta^*(z_{N_R-d_R}^{\boxed{0}}) - \alpha^*(u)) = (S^1 + h \cdot \ )(w)$, for certain $w \in \mathrm{Ch}^{N_R-d_R-1}(F(R, d_R - 1, d_R))$. This implies

$$\alpha_*\beta^*(z_{N_R-d_R}^{\boxed{0}}) = \alpha_*(S^1 + h \cdot \ )(w),$$

since $\alpha_*\alpha^* = 0$. Notice that $\alpha : F(R, d_R - 1, d_R) \to F(R, d_R - 1)$ is a conic bundle with relative tangent sheaf $\alpha^*(\mathcal{O}(w_1^{\boxed{-1}})) = \mathcal{O}(h)$. Thus, $\alpha_*(S^1 + h \cdot \ )(w) = S^1(\alpha_*(w))$, and $\alpha^*\alpha_*(\beta^*(z_{N_R-d_R}^{\boxed{0}})) = S^1(\alpha^*\alpha_*(w))$. But $\alpha^*\alpha_*(\beta^*(z_{N_R-d_R}^{\boxed{0}})) = h^{N_R-d_R-1} = h^{d_R}$, and this element is not in the image of S^1, as one can easily see. The contradiction shows that u as above does not exist. □

It follows from Statement 5.4 that in $\mathrm{Ch}^{d+2}(F(P, d-1)|_{\overline{k}})$ we have nonzero class x defined over k. Then $\alpha^*(x) \in \mathrm{Ch}^{d+2}(F(P, d-1, d)|_{\overline{k}})$ will also be a nonzero class defined over k. But the subring of k-rational classes in $\mathrm{Ch}^*(F(P, d-1, d)|_{\overline{k}})$ is $\mathrm{GDI}(P, d)[h]/(h^{d+1})$, and by the main result of [13], $\mathrm{GDI}(P, d)$ as a ring is generated by the elementary classes $z_j^{\boxed{0}}$ contained in it. By the conditions of our theorem, among such classes only $z_{d+1}^{\boxed{0}}$ could be defined over k. Then the degree $= (d+2)$ component of the subring of k-rational classes in $\mathrm{Ch}^*(F(P, d-1, d)|_{\overline{k}})$ is contained in $\mathbb{Z}/2 \cdot (\beta^*(z_{d+1}^{\boxed{0}}) \cdot h) = \mathbb{Z}/2 \cdot \alpha^*(z_{d+2}^{\boxed{-1}})$. Thus, if x is nonzero, it must be $z_{d+2}^{\boxed{-1}}$. But this class is not defined over k by the condition of the theorem. And this contradiction shows that the class $z_{d+2}^{\boxed{-1}}$ is not defined over $k(Q)$ as well. Theorem 5.1 is proven. □

References

1. Brosnan, P.: Steenrod operations in Chow theory. Trans. Amer. Math. Soc. **355**, no.5, 1869–1903 (2003).
2. Boe, B., Hiller, H.: Pieri formula for $SO2n + 1 = Un$ and $SPn = Un$. Adv. Math., **62**, 49–67 (1986).
3. Fulton, W.: Intersection Theory. Springer-Verlag (1984).
4. Izhboldin, O.T.: Fields of u-invariant 9. Annals of Math., **154**, no.3, 529–587 (2001).
5. Lam, T.Y.: Algebraic theory of quadratic forms (2nd ed.). Addison-Wesley (1980).
6. Levine, M.: Steenrod operations, degree formulas and algebraic cobordism. Preprint, 1–11 (2005).
7. Levine, M., Morel, F.: Algebraic cobordism. Springer Monographs in Mathematics. Springer-Verlag (2007).

8. Merkurjev, A.: On the norm residue symbol of degree 2. Dokl. Akad. Nauk SSSR, **261**, no.3, 542–547 (1981).

9. Merkurjev, A.: Simple algebras and quadratic forms (in Russian). Izv. Akad. Nauk SSSR, **55**, 218–224 (1991); English translation: Math. USSR - Izv. **38**, 215–221 (1992).

10. Merkurjev, A.: Steenrod operations and degree formulas. J. Reine Angew. Math., **565**, 13–26 (2003).

11. Rost, M.: The motive of a Pfister form. Preprint. 1–13 (1998).

12. Vishik, A.: Symmetric operations (in Russian). Trudy Mat. Inst. Steklova **246**, Algebr. Geom. Metody, Svyazi i Prilozh., 92–105 (2004); English transl.: Proc. of the Steklov Institute of Math. **246**, 79–92 (2004).

13. Vishik, A.: On the Chow Groups of Quadratic Grassmannians. Documenta Math. **10**, 111–130 (2005).

14. Vishik, A.: Generic points of quadrics and Chow groups. Manuscr. Math. **122**, no.3, 365–374 (2007).

15. Vishik, A.: Symmetric operations in algebraic cobordism. Adv. Math. **213**, 489–552 (2007).

16. Vishik, A., Yagita, N.: Algebraic cobordisms of a Pfister quadric, J. London Math. Soc. **76**, 586–604 (2007).

17. Voevodsky, V.: Reduced power operations in motivic cohomology. Publ. Math. IHES **98**, 1–57 (2003).

Cubic Surfaces and Cubic Threefolds, Jacobians and Intermediate Jacobians

Yuri Zarhin

Department of Mathematics, Pennsylvania State University, University Park, PA 16802, USA and Steklov Mathematical Institute, Russian Academy of Sciences, Moscow, Russia
zarhin@math.psu.edu

To my teacher Yuri Ivanovich Manin with admiration and gratitude

Summary. In this paper we study principally polarized Abelian varieties that admit an automorphism of order 3. It turns out that certain natural conditions on the multiplicities of its action on the differentials of the first kind guarantee that those polarized varieties are not Jacobians of curves. As an application, we get another proof of the (already known) fact that intermediate Jacobians of certain cubic threefolds are not Jacobians of curves.

Key words: Cubic threefolds, Del Pezzo surfaces, jacobians, intermediate jacobians, Lüroth problem

2000 Mathematics Subject Classifications: 14J45, 14K30

1 Principally Polarized Abelian Varieties That Admit an Automorphism of Order 3

Let $\zeta_3 = \frac{-1+\sqrt{-3}}{2}$ be a primitive (complex) cube root of unity. It generates the multiplicative order-3 cyclic group μ_3 of cube roots of unity.

Let $g > 1$ be an integer and (X, λ) a principally polarized g-dimensional abelian variety over the field $\mathbf{C}$ of complex numbers, δ an automorphism of (X, λ) that satisfies the cyclotomic equation $\delta^2 + \delta + 1 = 0$ in $\mathrm{End}\,(X)$. In other words, δ is a periodic automorphism of order 3, whose set of fixed points is finite. This gives rise to the embeddings

$$\mathbf{Z}[\zeta_3] \hookrightarrow \mathrm{End}(X), 1 \mapsto 1_X, \ \zeta_3 \mapsto \delta,$$

$$\mathbf{Q}(\zeta_3) \hookrightarrow \mathrm{End}^0(X), 1 \mapsto 1_X, \ \zeta_3 \mapsto \delta.$$

Y. Tschinkel and Y. Zarhin (eds.), *Algebra, Arithmetic, and Geometry,*
Progress in Mathematics 270, DOI 10.1007/978-0-8176-4747-6_23,
© Springer Science+Business Media, LLC 2009

By functoriality, $\mathbf{Q}(\zeta_3)$ acts on the g-dimensional complex vector space $\Omega^1(X)$ of differentials of the first kind on X. This provides $\Omega^1(X)$ with a structure of a $\mathbf{Q}(\zeta_3) \otimes_\mathbf{Q} \mathbf{C}$-module. Clearly,

$$\mathbf{Q}(\zeta_3) \otimes_\mathbf{Q} \mathbf{C} = \mathbf{C} \oplus \mathbf{C},$$

where the summands correspond to the embeddings $\mathbf{Q}(\zeta_3) \to \mathbf{C}$ that send ζ_3 to ζ_3 and ζ_3^{-1} respectively. So, $\mathbf{Q}(\zeta_3)$ acts on $\Omega^1(X)$ with multiplicities a and b that correspond to the two embeddings of $\mathbf{Q}(\zeta_3)$ into $\mathbf{C}$. Clearly, a and b are nonnegative integers with $a + b = g$.

Theorem 1.1. *If $g+2 < 3 \mid a-b \mid$, then (X, λ) is not the Jacobian of a smooth projective irreducible genus g curve with canonical principal polarization.*

Proof. Suppose that $(X, \lambda) \cong (J(C), \Theta)$, where C is an irreducible smooth projective genus g curve, $J(C)$ its Jacobian with canonical principal polarization Θ. It follows from the Torelli theorem in Weil's form [10, 11] that there exists an automorphism $\phi : C \to C$ that induces (by functoriality) either δ or $-\delta$ on $J(C) = X$. Replacing ϕ by ϕ^4 and taking into account that δ^3 is the identity automorphism of $X = J(C)$, we may and will assume that ϕ induces δ. Clearly, ϕ^3 is the identity automorphism of C, because it induces the identity map on $J(C)$ and $g > 1$. The action of ϕ on C gives rise to the embedding

$$\mu_3 \hookrightarrow \mathrm{Aut}(C), \quad \zeta_3 \mapsto \phi.$$

Let $P \in C$ be a fixed point of ϕ. Then ϕ induces the automorphism of the corresponding (one-dimensional) tangent space $T_P(C)$, which is multiplication by a complex number c_P. Clearly, c_P is a cube root of unity.

Lemma 1.2. *Every fixed point P of ϕ is nondegenerate, i.e., $c_P \neq 1$.*

Proof (of Lemma 1.2). The result is well known. However, I failed to find a proper reference.

Suppose that $c_P = 1$. Let $\mathcal{O}_P$ be the local ring at P and $\mathfrak{m}_P$ its maximal ideal. We write ϕ_* for the automorphism of $\mathcal{O}_P$ induced by ϕ. Clearly, ϕ_*^3 is the identity map. Since ϕ is *not* the identity map, there are no ϕ_*-invariant local parameters at P. Clearly, $\phi_*(\mathfrak{m}_P) = \mathfrak{m}_P, \phi_*(\mathfrak{m}_P^2) = \mathfrak{m}_P^2$. Since $T_P(C)$ is the dual of $\mathfrak{m}_P/\mathfrak{m}_P^2$ and $c_p = 1$, we conclude that ϕ_* induces the identity map on $\mathfrak{m}_P/\mathfrak{m}_P^2$. This implies that if $t \in \mathfrak{m}_P$ is a local parameter at t (i.e., its image $\bar{t}$ in $\mathfrak{m}_P/\mathfrak{m}_P^2$ is *not* zero) then $t' := t + \phi_*(t) + \phi_*^2(t)$ is ϕ_*-invariant and its image in $\mathfrak{m}_P/\mathfrak{m}_P^2$ equals $3\bar{t} \neq 0$. This implies that t' is a ϕ_*-invariant local parameter at P. Contradiction.

Corollary 1.3. *$D := C/\mu_3$ is a smooth projective irreducible curve. The map $C \to D$ has degree 3, its ramification points are exactly the images of fixed points of ϕ, and all the ramification indices are 3.*

Lemma 1.4. *D is biregularly isomorphic to the projective line.*

Proof (of Lemma 1.4). The map $C \to D$ induces, by Albanese functoriality, the surjective homomorphism of the corresponding Jacobians $J(C) \to J(D)$ that kills all the divisor classes of the form $(Q) - (\phi(Q))$ $(Q \in C)$. This implies that it kills $(1 - \delta)J(C)$. On the other hand, $1 - \delta : J(C) \to J(C)$ is, obviously, an isogeny. This implies that the image of $J(C)$ in $J(D)$ is zero and the surjectivity implies that $J(D) = 0$. This means that the genus of D is 0.

Corollary 1.5. *The number h of fixed points of ϕ is $g + 2$.*

Proof (of Corollary 1.5). Applying Hurwitz's, formula to $C \to D$, we get

$$2g - 2 = 3 \cdot (-2) + 2 \cdot h.$$

Lemma 1.6. *Let $\phi^* : \Omega^1(C) \to \Omega^1(C)$ be the automorphism of $\Omega^1(C)$ induced by ϕ and τ its trace. Then*

$$\tau = a\zeta_3 + b\zeta_3^{-1}.$$

Proof (of Lemma 1.6). Pick a ϕ-invariant point P_0 and consider the regular map

$$\alpha : C \to J(C), Q \mapsto \mathrm{cl}((Q) - (P_0)).$$

It is well known that α induces an isomorphism of complex vector spaces

$$\alpha^* : \Omega^1(X) \cong \Omega^1(C).$$

Clearly,

$$\phi^* = \alpha^* \delta^* \alpha^{*-1},$$

where $\delta^* : \Omega^1(J(C)) = \Omega^1(J(C))$ is the automorphism induced by δ. This implies that the traces of ϕ^* and δ^* coincide. Now the very definition of a and b implies that the trace of ϕ^* equals $a\zeta_3 + b\zeta_3^{-1}$.

End of proof of Theorem 1.1. Let B be the set of fixed points of ϕ. We know that $\#(B) = g + 2$. By the holomorphic Lefschetz fixed-point formula [2, Th. 2], [6, Ch. 3, Sect. 4] (see also [9, Sect. 12.2 and 12.5]) applied to ϕ,

$$1 - \bar{\tau} = \sum_{P \in B} \frac{1}{1 - c_P},$$

where $\bar{\tau}$ is the complex conjugate of τ. Recall that every c_P is a (primitive) cube root of unity and therefore

$$|1 - c_P| = \sqrt{3}, \quad \left| \frac{1}{1 - c_P} \right| = \frac{1}{\sqrt{3}}$$

and

$$| 1 - \bar{\tau} | \le \frac{g+2}{\sqrt{3}}.$$

Now

$$| 1 - \bar{\tau} |^2 = \frac{(a+b+2)^2 + 3(a-b)^2}{4} = \frac{(g+2)^2 + 3(a-b)^2}{4}.$$

This implies that

$$\frac{(g+2)^2}{3} \ge \frac{(g+2)^2 + 3(a-b)^2}{4}.$$

It follows that $(g+2)^2 \ge 9(a-b)^2$ and we are done.

2 Cubic Threefolds

Let $S : F(x_0, x_1, x_2, x_3) = 0 \subset \mathbf{P}^3$ be a smooth projective cubic surface over $\mathbf{C}$ [7]. (In particular, F is an irreducible homogeneous cubic polynomial in x_0, x_1, x_2, x_3 with complex coefficients.) Then the equation

$$y^3 = F(x_0, x_1, x_2, x_3)$$

defines a smooth projective threefold $T \subset \mathbf{P}^4$ provided with the natural action of μ_3 that arises from multiplication of y by cube roots of unity [1] (see also [3, 8]). We have the μ_3-invariant Hodge decomposition

$$\mathrm{H}^3(T, \mathbf{C}) = \mathrm{H}^3(T, \mathbf{Z}) \otimes \mathbf{C} = \mathrm{H}^{1,2}(T) \oplus \mathrm{H}^{2,1}(T)$$

and the μ_3-invariant nondegenerate alternating intersection pairing

$$(,) : \mathrm{H}^3(T, \mathbf{C}) \times \mathrm{H}^3(T, \mathbf{C}) \to \mathbf{C}.$$

In addition, both $\mathrm{H}^{1,2}(T)$ and $\mathrm{H}^{2,1}(T)$ are 5-dimensional isotropic subspaces and μ_3 acts on $\mathrm{H}^{2,1}(T)$ with multiplicities $(4, 1)$, i.e., $\zeta_3 \in \mu_3$ acts as diagonalizable linear operator in $\mathrm{H}^{2,1}(T)$ with eigenvalue ζ_3 of multiplicity 4 and eigenvalue ζ_3^{-1} of multiplicity 1 ([3, Sect. 5], [1, Sect. 2.2 and Lemma 2.6]). (The proof is based on [5, Th. 8.3 on p. 488]; see also [4, pp. 338–339].)

Since both $\mathrm{H}^{1,2}(T)$ and $\mathrm{H}^{2,1}(T)$ are isotropic and the intersection pairing is nondegenerate, its restriction to $\mathrm{H}^{1,2}(T) \times \mathrm{H}^{2,1}(T)$ gives rise to the nondegenerate μ_3-invariant $\mathbf{C}$-bilinear pairing

$$(,) : \mathrm{H}^{1,2}(T) \times \mathrm{H}^{2,1}(T) \to \mathbf{C}. \tag{1}$$

It follows that μ_3 acts on $\mathrm{H}^{1,2}(T)$ with multiplicities $(1, 4)$. (This assertion also follows from the fact that $\mathrm{H}^{1,2}(T)$ is the complex the conjugate of $\mathrm{H}^{2,1}(T)$.) In particular, the action of μ_3 on $\mathrm{H}^{1,2}(T)$ extends to the embedding

$$\mathbf{Z}[\mu_3] \hookrightarrow \mathrm{End}_{\mathbf{C}}(\mathrm{H}^{1,2}(T)). \tag{2}$$

3 Intermediate Jacobians

Let $(J(T), \theta_T)$ be the *intermediate Jacobian* of the cubic threefold T [4, Sect. 3]; it is a principally polarized five-dimensional complex Abelian variety. By functoriality, μ_3 acts on $J(T)$ and respects the principal polarization θ_T. As a complex torus,

$$J(T) = \mathrm{H}^{1,2}(T)/p(\mathrm{H}^3(T, \mathbf{Z})), \tag{3}$$

where

$$p : \mathrm{H}^3(T, \mathbf{C}) = \mathrm{H}^3(T, \mathbf{Z}) \otimes \mathbf{C} = \mathrm{H}^{1,2}(T) \oplus \mathrm{H}^{2,1}(T) \to \mathrm{H}^{1,2}(T)$$

is the projection map that kills $\mathrm{H}^{2,1}(T)$. The imaginary part of the Riemann form of the polarization coincides with the intersection pairing on $\mathrm{H}^3(T, \mathbf{Z}) \cong p(\mathrm{H}^3(T, \mathbf{Z}))$.

It follows from (2) that the action of μ_3 on $J(T)$ extends to the embedding

$$\mathbf{Z}[\mu_3] \hookrightarrow \mathrm{End}(J(T)).$$

Combining (1) and (3), we conclude that the μ_3-modules $\Omega^1(J(T)) = \mathrm{Hom}_{\mathbf{C}}(\mathrm{H}^{1,2}(T), \mathbf{C})$ and $\mathrm{H}^{2,1}(T)$ are canonically isomorphic. Now the assertions of Section 2 about multiplicities imply that $\mathbf{Z}[\zeta_3]$ acts on $\Omega^1(J(T))$ with multiplicities $(4, 1)$.

Since $3 \times \mid 4 - 1 \mid > 5 + 2$, it follows from Theorem 1.1 that $(J(T), \theta_T)$ is not isomorphic to the canonically polarized Jacobian of a curve. Of course, this assertion was proven by completely different methods in [4] for arbitrary smooth projective cubic threefolds.

References

1. D. ALLCOCK, J.A. CARLSON, D. TOLEDO, *The complex hyperbolic geometry of the moduli space of cubic surfaces.* J. Algebraic Geometry **11** (2002), 659–724.
2. M. F. ATIYAH, R. BOTT, *A Lefschetz fixed point formula for elliptic differential operators.* Bull. Amer. Math. Soc. **72** (1966), 245–250.
3. J.A. CARLSON, D. TOLEDO, *Discriminant complements and kernels of monodromy representations.* Duke Math. J. **97** (1999), 621–648.
4. C.H. CLEMENS, PH. GRIFFITHS, *The intermediate Jacobian of the cubic threefold.* Ann. of Math. (2) **95** (1972), 281–356.
5. PH. GRIFFITHS, *On the periods of certain rational integrals: I and II.* Ann. of Math. (2) **90** (1969), 460–541.
6. PH. GRIFFITHS, J. HARRIS, *Principles of algebraic geometry.* John Wiley and Sons, New York, 1978.
7. YU. I. MANIN, *Cubic forms*, second edition. North Holland, 1986.
8. K. MATSUMOTO, T. TERASOMA, *Theta constants associated to cubic threefolds.* J. Algebraic Geometry **12** (2003), 741–775.
9. J. MILNOR, *Dynamics in one complex variable.* Vieweg, Braunschweig/Wiesbaden, 1999.
10. A. WEIL, *Zum Beweis des Torellischen Satzes.* Gött. Nachr. 1957, no. 2, pp. 33–53; Œuvres, vol. III, [1957a].
11. A. WEIL, *Sur le théorème de Torelli.* Séminaire Bourbaki **151** (Mai 1957).

De Jong-Oort Purity for p-Divisible Groups

Thomas Zink

Faculty of Mathematics, University of Bielefeld, POB 100131, D 33501 Bielefeld
zink@math.uni-bielefeld.de

Dedicated to Professor Yuri Manin

Summary. We present a short proof of de Jong-Oort purity.

Key words: p-divisible groups, displays, purity

2000 Mathematics Subject Classifications: 14L05, 14F30

1 Introduction

De Jong-Oort purity states that for a family of p-divisible groups $X \to S$ over a noetherian scheme S, the geometric fibers have all the same Newton polygon if this is true outside a set of codimension bigger than 2. A more general result was first proved in [JO] and an alternative proof is given in [V1]. We present here a short proof that is based on the fact that a formal p-divisible group may be defined by a display [Z1], [Me2]. There are two other ingredients of the proof that have been known for a long time. One is the boundedness principle for crystals over an algebraically closed field [O], [V1], [V2] and the other is the existence of a slope filtration for a p-divisible group over a non-perfect field [Z2]. The last fact was already mentioned in a letter of Grothendieck to Barsotti [G]. The boundedness property is also an important ingredient in the proof given by Vasiu in [V1].

We discuss in detail some elementary consequences of the display structure. The other two ingredients can be found in the literature above. Therefore we discuss them only briefly. I thank W. Messing for pointing out the correct formulation of Proposition 3 below. I also thank the referees of this paper for many helpful suggestions.

Y. Tschinkel and Y. Zarhin (eds.), *Algebra, Arithmetic, and Geometry*,
Progress in Mathematics 270, DOI 10.1007/978-0-8176-4747-6_24,
© Springer Science+Business Media, LLC 2009

2 Frobenius Modules

We fix a prime number p. Let R be a commutative ring such that p is nilpotent in R. The ring of Witt vectors with respect to p is denoted by $W(R)$. We write $I_R = VW(R)$ for the Witt vectors whose first component is 0. The Witt polynomials are denoted by $\mathbf{w}_n : W(R) \to R$. The truncated Witt vectors of length n are denoted by $W_n(R)$. If $pR = 0$ the Frobenius endomorphism F of the ring $W(R)$ induces an endomorphism $F : W_n(R) \to W_n(R)$.

Definition 1. *A Frobenius module over R is a pair (M, F), where M is a projective finitely generated $W(R)$-module of some fixed rank h and $F : M \to M$ is a Frobenius linear homomorphism such that $\det F = p^d \epsilon$ locally for the Zariski topology on R, where $\epsilon : \det M \to \det M$ is a Frobenius linear isomorphism and $d \geq 0$ is some integer. We call h the height of the Frobenius module and d the dimension.*

This definition implies that the factorization $\det F = p^d \epsilon$ exists even globally, but we will never use this. Since the kernel of $\mathbf{w}_0 : W(R) \to R$ is in the radical of $W(R)$, there is always a covering $\operatorname{Spec} R = \bigcup_i \operatorname{Spec} R_{f_i}$ such that $W(R_{f_i}) \otimes_{W(R)} M$ is a free $W(R_{f_i})$-module for each i. Therefore we will often consider the case where M is a free $W(R)$-module. If we choose a basis of M we may view $\det F$ as an element of $W(R)$. Then (M, F) is a Frobenius module iff $\det F = p^d \eta$ for some unit $\eta \in W(R)$. In a question that is local on $\operatorname{Spec} R$ we will consider $\det F$ as an element of $W(R)$ without futher notice.

In this article a display over R is a 3n-display in the sense of [Z1]. The displays of [Z1] are called nilpotent displays. If $\mathcal{P} = (P, Q, F, F_1)$ is a display over R then (P, F) is a Frobenius module over R.

Let X be a p-divisible group over R and assume that p is nilpotent in R. If we evaluate the Grothendieck-Messing crystal of X at $W(R)$ we obtain a finitely generated locally free $W(R)$-module M_X, which is endowed with a Frobenius linear map $F : M_X \to M_X$. If X is the formal p-divisible group associated to a nilpotent display $\mathcal{P}$, then $(M_X, F) = (P, F)$ is a Frobenius module. The pair (M_Y, F) is also a Frobenius module if Y is an extension of an étale p-divisible group by X.

If we assume, moreover, that R is a complete local noetherian ring (M_X, F) is a Frobenius module for an arbitrary p-divisible group X over R. Indeed, if the special fiber of X has no étale part, then (M_X, F) comes from a display and is therefore a Frobenius module. Since X is an extension of an étale p-divisible group by a p-divisible group with no étale part in the special fiber, we see that (M_X, F) is a Frobenius module in general.

By these remarks, any (M_X, F) appearing in this work are Frobenius modules.

We add that Lau [L] in a forthcoming paper will associate a display to any p-divisible group over a ring R, where p is nilpotent. Thereby he obtains a functor from p-divisible groups to Frobenius modules. If we could use this functor it would be more satisfying then the remark above.

The following lemma is mainly a motivation for the definitions we are going to make:

Lemma 2. *Let $\mathcal{P}$ and $\mathcal{P}'$ be displays over a ring R of the same height and dimension. Let $\alpha : \mathcal{P} \to \mathcal{P}'$ be a homomorphism.*
Locally on $\operatorname{Spec} R$ the element $\det \alpha \in W(R)$ satisfies an equation

$$^{F}\!\det \alpha = \varepsilon \cdot \det \alpha,$$

where $\varepsilon \in W(R)^{*}$ is a unit.

Proof: We choose normal decompositions

$$\begin{aligned}
P &= L \oplus T, & Q &= L \oplus I_R T \\
P' &= L' \oplus T', & Q' &= L' \oplus I_R T'.
\end{aligned}$$

Without loss of generality we may assume that L, L', T, T' are free $W(R)$-modules. We choose identifications

$$L \simeq W(R)^l \simeq L', \quad T \simeq W(R)^t \simeq T'.$$

Then operators F_1 and F_1' are given by invertible block matrices with coefficient in $W(R)$:

$$F_1 \left(v \frac{x}{y} \right) = \begin{pmatrix} X & Y \\ Z & W \end{pmatrix} \begin{pmatrix} {}^F x \\ y \end{pmatrix},$$

$$F_1' \left(v \frac{x}{y} \right) = \begin{pmatrix} X' & Y' \\ Z' & W' \end{pmatrix} \begin{pmatrix} {}^F x \\ y \end{pmatrix}.$$

The block matrices are invertible by the definition of a display. We also represent α by a block matrix

$$\alpha \left(v \frac{x}{y} \right) = \begin{pmatrix} A & B \\ {}^V C & D \end{pmatrix} \begin{pmatrix} x \\ v \frac{x}{y} \end{pmatrix}$$

Since α commutes with the operators F_1 and F_1', we obtain

$$\begin{pmatrix} X' & Y' \\ Z' & W' \end{pmatrix} \begin{pmatrix} {}^F A & p\,{}^F B \\ C & {}^F D \end{pmatrix} = \begin{pmatrix} A & B \\ {}^V C & D \end{pmatrix} \begin{pmatrix} X & Y \\ Z & W \end{pmatrix}. \tag{1}$$

We see that

$$^{F}\!\begin{pmatrix} A & B \\ {}^V C & D \end{pmatrix} = \begin{pmatrix} {}^F A & {}^F B \\ pC & {}^F D \end{pmatrix}$$

has the same determinant as

$$\begin{pmatrix} {}^F A & p\,{}^F B \\ C & {}^F D \end{pmatrix}.$$

But then taking determinants in (1) gives the result. $\qquad\square$

Proposition 3. *Let R be a noetherian ring such that $\operatorname{Spec} R$ is connected. We assume that $pR = 0$. Let $\alpha : \mathcal{P} \to \mathcal{P}'$ be a homomorphism of displays of the same height h and the same dimension d.*

If $\det \alpha \neq 0$, then there is a nonnegative integer u such that locally on $\operatorname{Spec} R$ the following equation holds:

$$\det \alpha = p^u \varepsilon, \quad \text{where} \quad \varepsilon \in W(R)^*, \; u \in \mathbb{Z}_{\geq 0}.$$

Proof: If the number u exists locally, it is clearly a locally constant function. Therefore the question is local. We may replace $\operatorname{Spec} R$ by a small affine connected neighborhood.

We set $\eta = \det \alpha$. By the last proposition we obtain

$$^F\eta = \zeta \cdot \eta \text{ for some } \zeta \in W(R)^*. \tag{2}$$

We write $\eta = {}^{V^t}\xi$, such that $\mathbf{w}_0(\xi) \neq 0$. We claim that (2) implies:

$$^F\xi = {}^{F^t}\zeta \cdot \xi. \tag{3}$$

To verify this we may assume that $t > 0$. We obtain

$$^{FV^t}\xi = \zeta \, {}^{V^t}\xi = {}^{V^t}(\, {}^{F^t}\zeta \xi).$$

Since $pR = 0$, the operators F and V acting on $W(R)$ commute. Therefore we deduce (3).

Let $\mathbf{w}_0(\xi) = x$ and $\mathbf{w}_0(\, {}^{F^t}\zeta) = e \in R^*$. We apply $\mathbf{w}_0$ to equation (3) and obtain

$$x^p = ex. \tag{4}$$

Since the product

$$x(x^{p-1} - e) = 0$$

has relatively prime factors, it follows that

$$D(x) \cup D(x^{p-1} - e) = \operatorname{Spec} R,$$
$$D(x) \cap D(x^{p-1} - e) = \emptyset.$$

Hence by connectedness either $D(x) = \operatorname{Spec} R$ or $D(x) = \emptyset$. In the first case x is nilpotent. But then we have $x = 0$, by iterating the equation (4). This is a contradiction to our choices. Therefore $D(x) = \operatorname{Spec} R$ and x is a unit. Then ξ is a unit too. We obtain

$$^{F^t}\eta = {}^{F^t V^t}\xi = p^t \xi.$$

But by (2), $^{F^t}\eta$ may be expressed as the product of η by a unit. This proves the result. $\qquad\square$

Definition 4. *A homomorphism as in the proposition is called an isogeny of displays.*

Let R be a ring such that $pR = 0$. Assume that the ideal of nilpotent elements of R is nilpotent. Let $\alpha : \mathcal{P} \to \mathcal{P}'$ be a homomorphism of nilpotent displays of the same height and dimension. By the functor from the category of nilpotent displays to the category of formal p-divisible groups ([Z1] 3.1) we obtain from α a morphism $\phi : X \to X'$ of p-divisible groups. It follows from Proposition 66 and Proposition 99 of [Z1] that α is an isogeny iff ϕ is an isogeny of p-divisible groups.

Since $pR = 0$ the Frobenius endomorphism on $W(R)$ induces a Frobenius endomorphism on the truncated Witt vectors $F : W_n(R) \to W_n(R)$. Therefore we may consider truncated Frobenius modules. We are going to prove a version of Proposition 3 for truncated Frobenius modules.

Definition 5. *Let R be a ring such that $pR = 0$. A truncated Frobenius module of level n, dimension d, and height h over R is a finitely generated projective $W_n(R)$-module M of rank h equipped with a Frobenius linear operator $F : M \to M$ such that locally on $\operatorname{Spec} R$ the determinant has the form*

$$\det F = p^d \varepsilon,$$

where $\varepsilon : \det M \to \det M$ is a Frobenius linear isomorphism.

A Frobenius module M over R induces a truncated Frobenius module, if we tensor it by $W_n(R)$.

Definition 6. *Let M and N be truncated Frobenius modules of level n and of the same dimension d and height h. A morphism of Frobenius modules $\alpha : M \to N$ is called an isogeny if there is a natural number $u < n$ such that the determinant of α has locally on $\operatorname{Spec} R$ the form*

$$^{F^d}\det \alpha = p^u \varepsilon, \quad \varepsilon \in W_n(R)^*.$$

The number u is called the height of the isogeny.

Proposition 7. *Let M and N be truncated Frobenius modules of level n and of the same dimension d and height h over a ring R such that $\operatorname{Spec} R$ is connected and $pR = 0$.*

Let $u \geq 0$ be an integer such that $n > u + d$. Let $\alpha : M \to N$ be a homomorphism of Frobenius modules such that

$$^{F^d}\det \alpha \notin V^{u+1} W_{n-u-1}(R). \tag{5}$$

Then α becomes an isogeny if we truncate it to level $n - d$:

$$\alpha[n - d] : M[n - d] \to N[n - d].$$

Proof: We may assume that M and N are free $W_n(R)$-modules. We choose isomorphisms

$$\det M \simeq W_n(R) \simeq \det N$$

and view $\theta := \det \alpha$ as an element of $W(R)$. Then we obtain a commutative diagram

$$
\begin{array}{ccc}
\det M & \xrightarrow{\;\theta\;} & \det N \\
{\scriptstyle p^d \tau_M F} \downarrow & & \downarrow {\scriptstyle p^d \tau_N F} \\
\det M & \xrightarrow{\;\theta\;} & \det N,
\end{array}
$$

where $\tau_M, \tau_N \in W_n(R)^*$ are units. We obtain

$$
p^d \tau_N \,{}^F \theta = \theta p^d \tau_M. \tag{6}
$$

Using $p^d = V^d F^d$ in $W_n(R)$, we can divide (6) by V^d. We then obtain an equality in $W_{n-d}(R)$

$$
{}^{F^{d+1}} \theta[n-d] = {}^{F^d} \theta[n-d]\rho. \tag{7}
$$

Here $\theta[n-d]$ denotes the image of θ by the natural restriction $W_n(R) \to W_{n-d}(R)$ and $\rho \in W_{n-d}(R)^*$ is a unit.

On the other hand we may write by assumption

$$
{}^{F^d} \theta = {}^{V^{u_1}} \sigma, \tag{8}
$$

where $u_1 \leq u$, and $\mathbf{w}_0(\sigma) = s_0 \neq 0$. Clearly we may assume $u = u_1$. Since $n - d > u$ we obtain from equation (7)

$$
s_0^p = s_0 e
$$

for some unit $e \in R^*$. As in the proof of Proposition 3 (see: (4)) we conclude that s_0 is a unit. Then σ is a unit too. From (8) we obtain

$$
{}^{F^{d+u}} \theta = p^u \sigma.
$$

We truncate this equation to $W_{n-d}(R)$ and use (7) to obtain

$$
{}^{F^d} \theta[n-d] = p^u \varepsilon
$$

for some unit $\varepsilon \in W_{n-d}(R)^*$.   □

Let $n > u$ be natural numbers. It is clear that a morphism of displays $\alpha : \mathcal{P} \to \mathcal{P}'$ is an isogeny of height u, iff the map of the truncated Frobenius modules $\alpha[n] : (P[n], F) \to (P'[n], F)$ is an isogeny of height u.

3 Proof of Purity

For the proof of the purity theorem of de Jong and Oort for p-divisible groups we need to recall a few facts on completely slope divisible p-divisible groups (abbreviated: c.s.d. groups) from [Z2] and [OZ] Definition 1.2. We will use truncated Frobenius modules of p-divisible groups over any scheme U. These are locally free $W_n(\mathcal{O}_U)$-modules.

Lemma 8. *Let Y be a c.s.d. group over a normal noetherian scheme U over $\bar{\mathbb{F}}_p$. Let n be a natural number. Then there is a finite morphism $U' \to U$, such that the truncated Frobenius module $M_Y[n]$ of Y over U' is obtained by base change from a truncated Frobenius module over $\bar{\mathbb{F}}_p$, i.e. we can find a Frobenius module N over $\bar{\mathbb{F}}_p$ such that there is an isomorphism of Frobenius modules*

$$W_n(\mathcal{O}_{U'}) \otimes_{W_n(\mathcal{O}_U)} M_Y[n] \simeq W_n(\mathcal{O}_{U'}) \otimes_{W(\bar{\mathbb{F}}_p)} N. \tag{9}$$

Proof: This is an immediate consequence of [OZ], Proposition 1.3, since it says that this is true if we take for U' the perfect hull of the universal pro-étale cover of U. Another proof is obtained by substituting in the proof of loc.cit. Frobenius modules. □

Proposition 9. *Let T be a regular connected 1-dimensional scheme over $\mathbb{F}_p$. Then any p-divisible group X with constant Newton polygon over T is isogenous to a c.s.d. group.*

Proof: This follows from the main result of [OZ], Thm. 2.1. for any normal noetherian scheme T. But under under the assumptions made the proof is much easier (compare [Z2], proof of Thm. 7). Indeed let $K = K(T)$ be the function field of T. Then we find over K an isogeny to a c.s.d. group:

$$X_K \to \overset{\circ}{Y}. \tag{10}$$

Let $\overset{\circ}{G}$ be the finite group scheme that is the kernel of (10) and let $G \subset X$ be its scheme theoretic closure. We set $Y = X/G$. Using the fact that X has constant Newton polygon one proves that Y is c.s.d. □

The third ingredient is the boundedness principle, which seems to have been known for a long time [M].

Proposition 10. *Let k be an algebraically closed field of characteristic p. Let h be a natural number. Then there is a constant $c \in \mathbb{N}$ with the following property:*
Let M_1 and M_2 be Frobenius modules of height $\leq h$ over k. Let $n \in \mathbb{N}$ be arbitrary and let $\bar{\alpha} : M_1/p^n M_1 \to M_2/p^n M_2$ be a morphism of truncated Frobenius modules that lifts to a morphism of truncated Frobenius modules $M_1/p^{n+c}M_1 \to M_2/p^{n+c}M_2$. Then $\bar{\alpha}$ lifts to a morphism of Frobenius modules $\alpha : M_1 \to M_2$.

A weaker version of this is contained in [O], where the existence of the constant c is asserted only for given modules M_1 and M_2. But one can show that for given modules N_1 resp. N_2 in the isogeny class of M_1 resp. M_2, there are always isogenies $N_1 \to M_1$ resp. $N_1 \to M_1$ whose degrees are bounded by a constant depending only on h. This is another well-known boundedness principle. As an alternative to this proof the reader may use the much stronger results discussed in the introduction of [V2].

Theorem 11. *(de Jong-Oort)   Let R be a noetherian local ring of Krull dimension ≥ 2 with $p \cdot R = 0$. Let $U = \operatorname{Spec} R \setminus \{\mathfrak{m}\}$, the complement of the closed point. A p-divisible group X over $\operatorname{Spec} R$ that has constant Newton polygon over U has constant Newton polygon over $\operatorname{Spec} R$.*

Proof: It is not difficult to reduce to the case that R is complete, normal of Krull dimension 2 with algebraically closed residue class field $k = R/\mathfrak{m}$ ([JO]). Then U is a 1-dimensional regular scheme. We obtain by Proposition 9 a c.s.d. group Y over U and an isogeny

$$\alpha : Y \to X_{|U}, \tag{11}$$

Let d be the dimension of X let u be the height of α and let c be the number from Proposition 10. We choose a natural number $n > c + u + d$. After a finite extension of R we may assume by Lemma 8 that the truncated Frobenius module of Y is constant

$$M_Y[n] \simeq W_n(\mathcal{O}_U) \otimes_{W(\bar{\mathbb{F}}_p)} N, \tag{12}$$

where N is a Frobenius module over $\bar{\mathbb{F}}_p$. In particular the Newton polygons of N and Y must be the same by the boundedness principle applied to the field $\bar{K}$, where K is the field of fractions of R.

Combining (11) and (12) gives an isogeny of height u of truncated Frobenius modules

$$W_n(\mathcal{O}_U) \otimes_{W(\bar{\mathbb{F}}_p)} N \to W_n(\mathcal{O}_U) \otimes_R M_X[n]. \tag{13}$$

By the normality of R we have $\Gamma(U, W_n(\mathcal{O}_U)) = W_n(R)$. Taking the global section of (13) over U we obtain a morphism of truncated Frobenius modules

$$W_n(R) \otimes_{W(\bar{\mathbb{F}}_p)} N \to M_X[n]. \tag{14}$$

We know that (14) is an isogeny over K of height u. Therefore Proposition 3 is applicable to the morphism (14). We obtain therefore an isogeny of height u of truncated Frobenius modules over R:

$$W_{n-d}(R) \otimes_{W(\bar{\mathbb{F}}_p)} N \to M_X[n-d],$$

It is clear that the base change of an isogeny of truncated Frobenius modules is again an isogeny. Making the base change $R \to k$ we obtain an isogeny:

$$W_{n-d}(k) \otimes_{W(\bar{\mathbb{F}}_p)} N \to W_{n-d}(k) \otimes_{W(R)} M_X[n-d] = M_{X_k}[n-d].$$

The boundedness principle shows that X_k and N have the same Newton polygon. $\square$

References

[JO] A.J. de Jong and F. Oort, *Purity of the stratification by Newton polygons*, J. Amer. Math. Soc. **13** (2000), 209–241.

[G] A. Grothendieck, *Groupes de Barsotti-Tate et cristaux de Dieudonné*, Sém. Math. Sup. **45**, Presses de l'Univ. de Montreal, 1970.

[L] E. Lau, in preparation.

[M] Yu. I. Manin, *The theory of commutative formal groups over fields of finite characteristic*, Usp. Math. **18** (1963), 3–90, Russ. Math. Surveys **18** (1963), 1–80.

[Me1] W. Messing, *The crystals associated to Barsotti-Tate groups*, LNM **264**, Springer 1972.

[Me2] W. Messing, *Travaux de Zink*, Sém. Bourbaki no. 964, 2005/2006.

[O] F. Oort, *Foliations of moduli spaces of abelian varieties*, J. Amer. Math. Soc. **17** (2004), 267–296.

[OZ] F. Oort, T. Zink, *Families of p-divisible groups with constant Newton polygon*, Doc. Math. **7** (2002), 183–201.

[V1] A. Vasiu, *Crystalline boundedness principle*, Ann. Sci. École Norm. Sup.(4) **39** (2006), 245–300.

[V2] A. Vasiu, *Reconstructing p-divisible groups from their truncations of small level*, arXiv:math/0607268.

[Z1] Th. Zink, *The display of a formal p-divisible group*, in: Cohomologies p-adiques et applications arithmétiques, I. Astérisque no. **278** (2002), 127–248.

[Z2] Th. Zink, *On the slope filtration*, Duke Math. J. **109** (2001), 79–95.

Breinigsville, PA USA
19 April 2010
236341BV00001B/1/P